Lecture Notes in Artificial Intelligence 7106

Subseries of Lecture Notes in Computer Science

Dianhui Wang Mark Reynolds (Eds.)

AI 2011: Advances in Artificial Intelligence

24th Australasian Joint Conference
Perth, Australia, December 5-8, 2011
Proceedings

Series Editors

Randy Goebel, University of Alberta, Edmonton, Canada
Jörg Siekmann, University of Saarland, Saarbrücken, Germany
Wolfgang Wahlster, DFKI and University of Saarland, Saarbrücken, Germany

Volume Editors

Dianhui Wang
La Trobe University
School of Engineering and Mathematical Sciences
Melbourne, VIC 3086, Australia
E-mail: dh.wang@latrobe.edu.au

Mark Reynolds
The University of Western Australia
School of Computer Science and Software Engineering
Perth, WA 6009, Australia
E-mail: mark@csse.uwa.edu.au

ISSN 0302-9743 e-ISSN 1611-3349
ISBN 978-3-642-25831-2 e-ISBN 978-3-642-25832-9
DOI 10.1007/978-3-642-25832-9
Springer Heidelberg Dordrecht London New York

Library of Congress Control Number: Applied for

CR Subject Classification (1998): I.2, H.3, H.4, F.1, H.2.8, I.4-5

LNCS Sublibrary: SL 7 – Artificial Intelligence

Typesetting: Camera-ready by author, data conversion by Scientific Publishing Services, Chennai, India

Printed on acid-free paper

Springer is part of Springer Science+Business Media (www.springer.com)

Preface

The Australasian Joint Conference on Artificial Intelligence (AI) is an annual conference that brings together researchers and practitioners in areas related to artificial intelligence for promoting research and scientific discussions. This volume contains 82 papers presented at the 24th AI conference held in Perth, Western Australia, December 5–8, 2011.

The AI 2011 call for papers solicited contributions across a wide range of areas of research in artificial intelligence. This year's conference received 193 submissions, including papers from authors in Australia, Japan, Malaysia, China, Iran, New Zealand, India, USA, Poland, Canada, Singapore and 19 other countries. The topics addressed by the submitted papers illustrated the broadness of the discipline.

The conference featured three distinguished keynote speakers, Witold Pedrycz (Department of Electrical and Computer Engineering, University of Alberta, Edmonton, Canada), Kit Po Wong (Department of Electrical Engineering, Hong Kong Polytechnic University) and Kay Chen Tan (Department of Electrical and Computer Engineering, National University of Singapore). Their talks were of great interest to the attendees.

This year, AI 2011 incorporated the 5th Australian Conference on Artificial Life (ACAL11). ACAL papers were submitted and reviewed along with the AI 2011 papers, and accepted papers appear here in the AI 2011 proceedings. AI 2011 also included two tutorial workshops, the Australasian Ontology Workshop (AOW 2011) and the First Australian Workshop on Artificial Intelligence in Health (AIH 2011). There is a growing community of researchers in Australia and New Zealand working on various aspects of ontologies. The primary aim of AOW 2011 was to bring together ontology researchers in the region. The AIH workshop was a first-of-its-kind national initiative to bring together scholars and practitioners in the field of artificial intelligence-driven health informatics to present and discuss their research, share their knowledge and experiences, define key research challenges and explore possible collaborations so as to advance e-Health development nationally and internationally. These tutorials and workshops together provided an excellent start to the event.

The contributed talks of AI 2011 were organized into 18 sessions including two special sessions, a special session on "Modern Machine Learning in Intelligent Image Processing" and a special session on "Information Processing, Inference and Learning". The latter also included a special session keynote presented by Martin Riedmiller (Albert Ludwigs University Freiburg).

The Program Committee consisted of 135 highly regarded academics from 20 countries including Australia, New Zealand, China, UK and Germany. All papers were peer reviewed by at least three Program Committee members, and, in some cases, external reviewers. Of the 193 papers submitted, 82 were selected for presentation at the conference. We would like to thank all authors who submitted papers and all conference participants for helping to make the conference a success. We also would like to thank the members of the Program Committee and the external referees for their expertise in carefully reviewing the papers. We were grateful to Springer for its assistance in the production of the proceedings.

Many thanks to the School of Information Technology at Murdoch University for providing all the local support for organizing the conference. We would like to express our appreciation to the General Conference Chairs, Kevin Wong, Lance Fung and Hussein Abbas, for their tireless work. Thanks too, to the Publicity Chairs, Yew Soon Ong, Yasufumi Takama, and Wanquan Liu. Last, but not least, we express gratitude to our hosts in Perth and in particular Hong Xie and Shri Rai, the local Organizing Chairs and their helpers.

October 2011

Dianhui Wang
Mark Reynolds

Organization

Program Committee

Hussein Abbass	UNSW@ADFA, Australia
Yun Bai	University of Western Sydney, Australia
James Bailey	University of Melbourne, Australia
Timothy Baldwin	University of Melbourne, Australia
Peter Baumgartner	National ICT Australia, Australia
Lubica Benuskova	University of Otago, New Zealand
Ghassan Beydoun	University of Wollongong, Australia
Richard Booth	University of Luxembourg, Luxembourg
Sebastian Brand	University of Melbourne, Australia
Thomas Braunl	University of Western Australia, Australia
Bob Brown	University of Wollongong, Australia
Jinhai Cai	University of South Australia, Australia
Lawrence Cavedon	NICTA and RMIT University, Australia
Stephan Chalup	The University of Newcastle, Australia
Chia-Yen Chen	The University of Auckland, New Zealand
Ling Chen	University of Technology, Sydney, Australia
Songcan Chen	Nanjing University of Aeronautics and Astronautics, China
Andrew Chiou	Central Queensland University, Australia
Dominique Chu	University of Kent, UK
Vic Ciesielski	RMIT University, Australia
David Cornforth	CSIRO Energy Technology, Australia
Stephen Cranefield	University of Otago, New Zealand
Michael Cree	University of Waikato, New Zealand
Corbett Daniel	DARPA, USA
Hepu Deng	RMIT University, Australia
Jeremiah D. Deng	University of Otago, New Zealand
Grant Dick	University of Otago, New Zealand
Xiangjun Dong	Shandong Institute of Light Industry, China
David Dowe	Monash University, Australia
Atilla Elci	Toros University, Turkey
Mark Ellison	University of Western Australia, Australia
Esra Erdem	Sabanci University, Turkey
Cesar Ferri	DSIC, Universitat Politecnica de València, Spain
Lance Fung	Murdoch University, Australia
Alfredo Gabaldon	New University of Lisbon, Portugal
Junbin Gao	Charles Sturt University, Australia

Yang Gao	Nanjing University, China
Manolis Gergatsoulis	Ionian University, Greece
Guido Governatori	NICTA, Australia
Charles Gretton	University of Birmingham, UK
Hans W. Guesgen	Massey University, New Zealand
Fikret Gurgen	Bogazici University, Turkey
Patrik Haslum	Australia National University, Australia
Tim Hendtlass	Swinburne University, Australia
Bernhard Hengst	UNSW, CAS, NICTA, Australia
Jose Hernandez-Orallo	Universitat Politecnica de Valencia, Spain
Geoffrey Holmes	University of Waikato, New Zealand
Wei-Chiang Hong	Oriental Institute of Technology, India
Guang-Bin Huang	Nanyang Technological University, Singapore
Xiaodi Huang	Charles Sturt University, Australia
Huidong Jin	CSIRO, Australia
Yaochu Jin	University of Surrey, UK
Zhi Jin	Peking University, China
Byeong Ho Kang	University of Tasmania, Australia
Paul Kennedy	University of Technology, Sydney, Australia
Philip Kilby	NICTA and ANU, Australia
Les Kitchen	University of Melbourne, Australia
Reinhard Klette	The University of Auckland, New Zealand
Alistair Knott	University of Otago, New Zealand
Mario Koeppen	Kyushu Institute of Technology, Japan
Rudolf Kruse	University of Magdeburg, Germany
Rex Kwok	The University of New South Wales, Australia
Willem Labuschagne	University of Otago, New Zealand
Gerhard Lakemeyer	RWTH Aachen University, Germany
Jérôme Lang	LAMSADE, France
Maria Lee	Shih Chien University, Taiwan
Nung Kion Lee	Universiti Malaysia Sarawak, Malaysia
Bin Li	University of Science and Technology of China, China
Jiuyong Li	University of South Australia, Australia
Wei Li	Central Queensland University, Australia
Xiaodong Li	RMIT University, Australia
Jing Liu	Xidian University, China
Wan Quan Liu	Curtin University of Technology, Australia
Ashesh Mahidadia	University of New South Wales, Australia
Zhihong Man	Swinburne University of Technology, Australia
Eric Martin	University of New South Wales, Australia
Rodrigo Martínez-Béjar	University of Murcia, Spain
Alexandre Mendes	University of Newcastle, Australia
Kathryn Merrick	University of New South Wales, Australian Defence Force Academy, Australia
Thomas Meyer	Meraka, South Africa

Zbigniew Michalewicz	University of Adelaide, Australia
Diego Molla	Macquarie University, Australia
John Morris	The University of Auckland, New Zealand
Saeid Nahavandi	Deakin University, Australia
Detlef Nauck	BT Plc, UK
Richi Nayak	QUT, Australia
Oliver Obst	CSIRO, Australia
Kouzou Ohara	Aoyama Gakuin University, Japan
Mehmet Orgun	Macquarie University, Australia
Maurice Pagnucco	The University of New South Wales, Australia
Bernard Pailthorpe	The University of Queensland, Australia
Francis Jeffry Pelletier	Simon Fraser University, Canada
Pavlos Peppas	University of Patras, Greece
Duc-Nghia Pham	NICTA/Griffith University, Australia
David Powders	Flinders University, Australia
Mikhail Prokopenko	CSIRO, Australia
Mark Reynolds	University of Western Australia, Australia
Panos Rondogiannis	University of Athens, Greece
Malcolm Ryan	Centre for Autonomous Systems
Rafal Rzepka	Hokkaido University, Japan
Sebastian Sardina	RMIT University, Australia
Abdul Sattar	Griffith Univeristy, Australia
Torsten Schaub	University of Potsdam, Germany
Steven Shapiro	University of Toronto, Canada
Maolin Tang	Queensalnd University of Technology, Australia
Dacheng Tao	UTS, Australia
Andrea Torsello	Università CA Foscari, Italy
Andrew Turpin	The University of Melbourne, Australia
Miroslav Velev	Aries Design Automation
Svetha Venkatesh	Curtin University of Technology, Australia
Toby Walsh	NICTA and UNSW, Australia
Dianhui Wang	La Trobe University, Australia
Kewen Wang	Griffith University, Australia
Xizhao Wang	Hebei University, China
Renata Wassermann	University of Sao Paulo, Brazil
Peter Whigham	University of Otago, New Zealand
Graham Williams	Togaware
Bill Wilson	University of New South Wales, Australia
Wayne Wobcke	University of New South Wales, Australia
Kevin Wong	Murdoch University, Australia
Brendon J. Woodford	University of Otago, New Zealand
Roland Yap	National University of Singapore, Singapore
Lean Yu	Chinese Academy of Sciences, China
Xinghuo Yu	RMIT University, Australia
Chengqi Zhang	UTS, Australia

Daoqiang Zhang	Nanjing University of Aeronautics and Astronautics, China
Dongmo Zhang	University of Western Sydney, Australia
Jun Zhang	Sun Yat-sen University, China
Mengjie Zhang	Victoria University of Wellington, New Zealand
Min-Ling Zhang	Southeast University, China
Shichao Zhang	Guangxi Normal University, China
Zili Zhang	Deakin University, Australia
Yanchang Zhao	Centrelink, Australia
Yi Zhou	University of Western Sydney, Australia
Xingquan Zhu	UTS, Australia

Additional Reviewers

Abello, Manuel
Aker, Erdi
Albayrak, Mehmet
Alhamdoosh, Monther
Belle, Vaishak
Bian, Wei
Braunl, Thomas
Buckley, Muneer
Budiono, Tri
Cai, Xiongcai
Celik, Duygu
Cerexhe, Timothy
Charalambidis, Angelos
Chen, Xiaoming
Dong, Zhao
Dybala, Pawel
Ferrein, Alexander
Finger, Marcelo
Ghandar, Adam
Guan, Naiyang
Handjopoulos, Konstantinos
Hawes, Nick
Heap, Bradford
Held, Pascal
Horne, Tertia
Hsu, Wynne
Jalalian, Arash
Jeatrakul, Piyasak
Kim, Yang Sok
Kountouriotis, Vassilios
Le, Thuc Duy
Lee, Nung Kion
Li, Billy
Li, Jun
Li, Wei
Li, Weitao
Li, Xi
Liu, Wei
Lopes, Fabrício M.
Luo, Chao
Martinez-Plumed, Fernando
Mccane, Brendan
Meng, Qinxue
Moewes, Christian
Mutluergil, Suha Orhun
Oztok, Umut
Paireekreng, Worapat
Ptaszynski, Michal
Rahnama, Behnam
Ramirez-Quintana, Maria Jose
Ramírez-Quintana, Maria Jose
Rens, Gavin
Robles-Kelly, Antonio
Ruß, Georg
Sanner, Scott
Sattar, A.H.M. Sarowar
Schiffer, Stefan
Schwitter, Rolf
Silva, Paulo-Js
Slaney, John
Song, Mingli
Songram, Panida
Tafavogh, Siamak
Tan, Kian Lee
Tapan, Sarwar
Varzinczak, Ivan
Wagner, Markus
Wang, Zhe
Wang, Zheng Qun
Wei, Jingxuan
Yu, Jun
Zender, Hendrik
Zhao, Qi
Zhou, Tianyi
Zhu, Junwu

Table of Contents

Session 1: Data Mining and Knowledge Discovery

Session 2: Machine Learning

Session 3: Evolutionary Computation and Optimization

Session 4: Intelligent Agent Systems

Session 5: Logic and Reasoning

Session 6: Vision and Graphics

Session 7: Image Processing

Session 8: Natural Language Processing

Session 9: Cognitive Modeling and Simulation Technology

Session 10: AI Applications

Guided Rule Discovery in XCS for High-Dimensional Classification Problems

Mani Abedini and Michael Kirley

Department of Computer Science and Software Engineering
The University of Melbourne, Australia
{mabedini,mkirley}@csse.unimelb.edu.au

Abstract. XCS is a learning classifier system that combines a reinforcement learning scheme with evolutionary algorithms to evolve a population of classifiers in the form of condition-action rules. In this paper, we investigate the effectiveness of XCS in high-dimensional classification problems where the number of features greatly exceeds the number of data instances – common characteristics of microarray gene expression classification tasks. We introduce a new guided rule discovery mechanisms for XCS, inspired by feature selection techniques commonly used in machine learning. The extracted feature quality information is used to bias the evolutionary operators. The performance of the proposed model is compared with the standard XCS model and a number of well-known machine learning algorithms using benchmark binary classification tasks and gene expression data sets. Experimental results suggests that the guided rule discovery mechanism is computationally efficient, and promotes the evolution of more accurate solutions. The proposed model performs significantly better than comparative algorithms when tackling high-dimensional classification problems.

1 Introduction

Learning classifier systems (LCS) combine machine learning with metaheuristics to build models that learn to solve a particular classification problem (see [8,18] for detailed reviews). The eXtended classifier system (XCS) is a well-known LCS that maintains a population of classifiers [21]. Each classifier consists of a *condition-action-prediction* rule with an associated fitness value, which represents the accuracy of the predicted reward. Through an iterative learning process, the population of classifiers evolves. A key step in this iterative process is the rule discovery component that creates new classifiers that are added to the bounded population pool.

One of the challenges when designing a LCS, is to build flexibility and robustness into the model such that it is capable of handling large scale data mining and classification problems. Consider a prototypical high-dimensional data set, such as a microarrray gene expression data set, that has several thousands genes (features) but only a small number of samples [24]. Standard XCS implementations, and many other machine learning algorithms for that matter, are typically

D. Wang and M. Reynolds (Eds.): AI 2011, LNAI 7106, pp. 1–10, 2011.

less effective in this high-dimensional space. It is difficult to effectively explore the solution space and build an appropriate classification model. In such circumstances, feature selection pre-processing can be used [14]. Such approaches can reduce the negative effect of the irrelevant features on the learning task, and speed up the learning process significantly.

In this paper, a new guided rule discovery mechanisms is proposed for XCS for high-dimensional classification problems. Our model (GRD-XCS), is inspired by feature selection techniques commonly used in machine learning. Typically, filtering techniques assess the relevance of features in the data set, with the low-scoring features subsequently being removed. A subset of the "more important" features is then presented as input to the classification algorithm. However, in our model the filtering process is used to build a probability distribution that biases the evolutionary operators encapsulated in the rule discovery component of XCS. This probability distribution can be thought of as a mask that biases the uniform crossover and mutation operators. This flexible approach is scalable, thus the enhanced XCS can be used to tackle high-dimensional classification tasks without reducing the dimensionality of the data set.

To test the efficacy of the new GRD-XCS, a systematic set of experiments were carried out using benchmark binary classification tasks and a suite of gene expression microarray data sets. The proposed model was compared to XCS and a range of well-known machine learning algorithms. The results show that the new guided rule discovery mechanisms leads to improved accuracy, particularly for high dimensional binary classification problems.

The remainder of this paper is organized as follows: In section 2 we present background material related to XCS and related work. In section 3 we describe the guided rule discovery mechanism in detail. The experiments and results appear in Section 4. We conclude the paper by summarizing the contributions and identifying the possible future directions.

2 Background

2.1 XCS Overview

In this section, we provide a brief overview of the functionality XCS. See [21,8,18] for detailed discussion of LCS in general.

XCS maintains a population of classifiers. Each classifier consists of a *condition-action-prediction* rule, which maps input features to the output signal (or class). A ternary representation of the form 0,1,# (where # is don't care) for the condition and 0,1 for the action can be used. In addition, real encoding can also be used [22].

At each time step, the classifier system receives a problem instance – input in the form of a vector of features – which requires a decision, that is an action to be performed next. A *match set* $[M]$ is created consisting of rules (classifiers) that can be "triggered" by the given data instance. A covering operator is used to create new matching classifiers when $[M]$ is empty. A prediction array $[PA]$ is calculated for $[M]$ that contains an estimation of the corresponding rewards

Algorithm 1. High level overview of XCS

Require: Input data:σ, Population:$[\Delta]$
repeat
$\sigma \leftarrow env$
$[M] \leftarrow GetMatchSet(\sigma,[\Delta])$
$[PA] \leftarrow CreatePredictionArray([M])$
$act \leftarrow SelectAnAction([PA])$
$[A] \leftarrow CreateActionSet([M], act)$
$R \leftarrow ExecutingActionOnENV(act)$
$[A] \leftarrow UpdateSet([A], R)$
$[\Delta] \leftarrow RuleDiscovery([A],[\Delta])$
until terminating conditions are not met

for each of the possible actions. Based on the values in the prediction array, an action, *act*, is selected. Those classifiers which support the predicted action make up the *Action Set* $[A]$. In response to *act*, the reinforcement mechanism is invoked and the prediction (p), prediction error, accuracy, and fitness of the $[A]$ classifiers are updated. The corresponding numerical reward is distributed to the rules accountable for it so as to improve the estimates of the action values (see algorithm 1).

The rule discovery module is a key component of XCS. During the evolutionary process, fitness-proportionate selection is applied to $[A]$. Standard evolutionary operators, *uniform crossover* and *mutation*, are then applied to the selected individuals. In addition, a second mutation-style operator – the *don't care* operator – is used to randomly modify a condition part of a classifier to the don't care value #. The newly created offspring (classifiers) are then added to the bounded population. A form of niching is then used to determine if the offspring survive in the population and/or which of the old members of the population are deleted to make room for the new classifiers (offspring). A subsumption mechanism combines similar classifiers and a randomized deletion mechanism removes classifiers with a low fitness from the population.

It is important to note that the XCS population consists of a set unique *macro-classifiers* – a set of classifiers that have same condition part and same action part. Every macro-classifier has an associated *numerosity* value, which records how many instances of that specific classifier actually exists in the population.

2.2 Related Work

It is well documented in the evolutionary computation literature that the implementation of the genetic operators can influence the trajectory of the evolving population. However, there has been a paucity of studies focussed specifically on the impact of selected evolutionary operator implementations in LCS. We briefly discuss some of the key studies related to XCS/LCS below.

In one of the first studies focussed on the rule discovery component specifically for XCS, Butz et al. [6] have shown that uniform crossover can ensure successful learning in many tasks. In subsequent work, Butz et al. [5] introduced an informed crossover operator, which extended the usual uniform operator such that exchanges of effective building blocks occurred. This approach helped to avoid the over-generalization phenomena inherent in XCS [13]. In other work, Bacardit et al. [3] customized the GAssist crossover operator to switch between the standard crossover or a new simple crossover, SX, randomly. SX is a heuristic selection approach to take a minimum number of rules from the parents (more than two), which can obtain maximum accuracy. Morales-Ortigosa et al. [16] have also proposed a new XCS crossover operator, BLX, which allowed for the creation of multiple offspring with a diversity parameter to control differences between offspring and parents. Finally, in a more comprehensive overview paper, Morales-Ortigosa et al. [17] present a systematic experimental analysis of the rule discovery component in LCS. Subsequently, they developed crossover operators to enhance the discovery component based on evolution strategies with significant performance improvements.

Other work focussed on biased evolutionary operators in LCS include the work of Luis et al. [12], who introduced a hybridized GA - Tabu Search (GA-TS) method that employed modified mutation and crossover operators. Here, the operator probabilities were tuned by analyzing all the fitness values of individuals during the evolution process. Wang et. al. [20] used *information gain* as the fitness function in a GA. They reported improved results when comparing their model to other machine learning algorithms. Recently, Huerta et al. [4] combined *linear discriminant analysis* with a GA to evaluate the fitness of individuals and associated discriminate coefficients for crossover and mutation operators. Moore et al. [15] argue that the biasing of the initial population, based on expert knowledge preprocessing, should lead to improved performance in LCS. In their approach, a statistical method, Tuned ReliefF, was used to determine the dependencies between features to seed the initial population. A modified fitness function and a new guided mutation operator based on features dependency was also introduced, leading to significantly improved performance.

3 Model

The motivation behind the design and development of the GRD-XCS was to improve classifier performance especially for high-dimensional classification problems. Our goal was to make the overall task computationally faster, without degrading accuracy. To meet this goal, GRD-XCS introduces a probabilistically guided rule discovery mechanism for XCS. Here, two distinct phases are used. In the pre-processing phase, each feature is examined independently of all others and assigned a rank. This rank is then used when generating the probability distribution used to bias the evolutionary operators, which are deployed during the second phase – the generation of classifiers in XCS.

The evolution process is regulated by a rule discovery probability vector, *RDP*, which controls the bitwise *crossover*, *mutation* and *don't care* operators. Each value in the vector is associated with the corresponding feature, and is allocated a value in the range [0, 1.0]. The *RDP* values are determined based on a ranked *Information Gain* (IG) measure [11]. The IG measure is defined as *entropy* reduction:

$$IG = H(C) - H(C|f_i) \tag{1}$$

where H represent entropy, $F = \{f_0, f_1, ...f_i, ..., f_n\}$ is the feature set, and C the classes in this context. Entropy is a measure to quantify the information content, it is calculated using the formula:

$$H(C) = \sum_{j \in C} p_j \log_2 p_j \tag{2}$$

where p_j is the probability of having j in C, and the conditional entropy is calculated as:

$$H(C|f_i) = \sum_{j \in C} p_j \log_2 H(C|f_i = j) \tag{3}$$

The actual values in the *RDP* vector are calculated based on the IG values as described below:

$$RDP_i = \begin{cases} \frac{1-\gamma}{\Lambda} \times (\Lambda - i) + \gamma & \text{if } i \leq \Lambda \\ \xi & \text{otherwise} \end{cases} \tag{4}$$

where i represents the rank index in ascending order for the selected features. The probability values associated with the other features are given a very low value (ξ). Thus, all features have a chance to participate in the rule discovery process. However, the Λ-top ranked features have a greater chance of being selected (see figure 1).

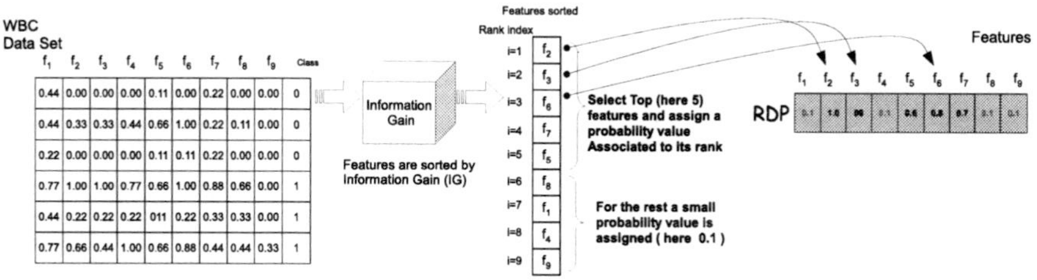

f_1	f_2	f_3	f_4	f_5	f_6	f_7	f_8	f_9	Class
0.44	0.00	0.00	0.00	0.11	0.00	0.22	0.00	0.00	0
0.44	0.33	0.33	0.44	0.66	1.00	0.22	0.11	0.00	0
0.22	0.00	0.00	0.00	0.11	0.11	0.22	0.00	0.00	0
0.77	1.00	1.00	0.77	0.66	1.00	0.88	0.66	0.00	1
0.44	0.22	0.22	0.22	011	0.22	0.33	0.33	0.00	1
0.77	0.66	0.44	1.00	0.66	0.88	0.44	0.44	0.33	1

Fig. 1. Information Gain is used to rank the features. The top Λ features (in this example Λ =5) are selected and allocated relatively large probability values $\in [\gamma, 1]$. The *RDP* vector maintains these values. The highest ranked feature value is set to 1.0. Other features receive smaller values relative to their rank (in this example γ =0.5). Features that are not selected based on information gain, are assigned very small probability values (in this example ξ = 0.1).

GRD-XCS uses the probability values recorded in the *RDP* vector in the pre-processing phase to bias the evolutionary operators used in the rule discovery phase of XCS. The modified algorithms describing the *crossover*, *mutation* and *don't care* operators in GRD-XCS are very similar to standard XCS operators:

GRD-XCS *crossover* operator: The crossover operator is a hybrid uniform /n-point function. Here, an additional check of each feature is carried out before exchange of genetic material. If $rand() < RDP[i]$ then feature i is swapped between the selected parents.

GRD-XCS *mutation* operator: Uses the *RDP* vector to determine if feature i undergoes a mutation, if the feature was randomly selected to be mutated.

GRD-XCS *don't care* operator: In this special mutation operator, the values in the *RDP* vector are used in the reverse order. That is, if the feature i has been selected to be mutated and $rand() < (1 - RDP[i])$, then feature i is changed to # (don't care).

The application of the *RDP* reduces the crossover and mutation probabilities for "uninformative" features. However, it increases the don't care operator probability for the same feature. Therefore, the more informative features (based on the Information Gain measure in this case) should appear in rules more often than the uninformative ones.

4 Experiments

A series of independent experiments were conducted to verify if the guided rule discovery mechanism for XCS was able to find accurate classifiers. In particular, we wished to establish if the proposed model had statistically significantly improved accuracy values when compared to the standard XCS across a suite of benchmark classification problems. All experiments have been conducted with N-fold cross validation over 100 trials. The average accuracy values for specific scenarios have been reported using the Area Under the ROC Curve (AUC) value. Paired t-tests are used for statistical comparisons.

4.1 Data Sets

Table 1 lists the data set characteristics used in the experiments. Two different types of data sets have been used in these experiments: data sets with either a small number of features with many samples (low-dimensional data set) obtained from the UCI [1] machine learning repository; and DNA Microarray Gene Expression data sets with a large number of features with few samples (high-dimensional data set). Gene expression profiles provide important insights into, and further our understanding of, biological processes. As such, they are key tools used in medical diagnosis, treatment and drug design [23].

Table 1. Data set details

Data Set	#Instances	#Features	Cross Validation	Reference
Low-dimensional data sets (UCI examples)				
Pima	768	8	10	[1]
WBC	699	9	10	[1]
Hepatits	155	19	10	[1]
Parkinson	197	23	10	[1]
High-dimensional data sets (Microarray DNA gene expression)				
Breast Cancer	22	3226	3	[10]
Colon Cancer	62	2000	10	[2]
Leukemia Cancer	72	7129	10	[9]
Prostate Cancer	136	12600	10	[19]

4.2 Parameters

Default parameter values as recommended in [7] have been used to configure the underlying XCS model in GRD-XCS. However, in the case of the high-dimensional data sets it was necessary to scale-up the population to 2000 individuals as compared with 1000 individuals for the low-dimensional data sets. The number of iterations was capped at 5000.

The guided rule discovery module relies on the ranking of feature. Here, we have limited the ranking to the top 64 features (Λ= 64) for the gene expression profiles classifications. For the low dimensional data sets, all features were used when building the probability models (see section 3). The limits used in probability values calculations in equation 4 were γ=0.5 and ξ=0.1.

4.3 Results

Tables 2 and 3 lists accuracy results for the low-dimensional data sets and high-dimensional gene expression data sets respectively. Results for GRD-XCS, the standard XCS and a range of machine learning algorithms (using default Weka implementations) are listed for each data set. The bold value in each column indicates the highest mean accuracy value over all trials. The †symbol indicates that the result for the classifier listed in the row was significantly better than the GRD-XCS result based on a paired t-test ($p < 0.05$).

For the low-dimensional data sets considered, the GRD-XCS results were better than the standard XCS, although this difference was not always statistically significant. When compared against the other machine learning algorithms, the GRD-XCS results were somewhat mixed. GRD-XCS performed best for one data set only – the Parkinson data set. In contrast, for the high-dimensional data sets the results for GRD-XCS were significantly better than the other machine learning algorithms based on paired t-tests. A direct comparison between GRD-XCS and the standard XCS clearly illustrates that the guided rule discovery mechanisms leads to improved performance.

Table 2. AUC results for low-dimensional data sets

Classifier	Pima	WBC	Hepatit	Parkinson
j48	0.75 ± 0.01 †	0.94 ± 0.01	0.60± 0.04	0.78 ± 0.03
SVM	0.71 ± 0.01 †	0.96± 0.01	0.75 ± 0.02	0.75 ± 0.01
Naive Bayes Classifier	**0.81 ± 0.01** †	**0.98 ± 0.01** †	**0.84± 0.01** †	0.85 ± 0.01
NBTree	0.80 ± 0.01 †	**0.98 ± 0.01** †	0.76± 0.03	0.88 ± 0.02
One Rule	0.65 ± 0.01	0.90 ± 0.01	0.56± 0.02	0.77 ± 0.01
Random Forest	0.79 ± 0.01 †	**0.98 ± 0.01** †	0.81± 0.02 †	**0.94 ± 0.01** †
XCS	0.70± 0.03	**0.98 ± 0.01** †	0.81 ± 0.11 †	0.93 ± 0.08
GRD-XCS	0.72± 0.03	**0.98 ± 0.01**	0.82 ± 0.13	**0.94 ± 0.07**

Table 3. AUC results for high-dimensional data sets

Classifier	Breast Cancer	Colon Cancer	Leukemia	Prostate Cancer
j48	0.43 ± 0.09	0.76 ± 0.04	0.79 ± 0.03	0.79 ± 0.02
SVM	0.63 ± 0.06	0.81 ± 0.03	0.97 ± 0.01	0.91 ± 0.01
Naive Bayes	0.55 ± 0.02	0.64 ± 0.02	0.98 ± 0.01	0.58 ± 0.01
NBTree	0.66 ± 0.03	0.75 ± 0.05	0.97 ± 0.01	0.90 ± 0.01
One Rule	0.42 ± 0.05	0.66 ± 0.04	0.82 ± 0.02	0.81 ± 0.02
Random Forest	0.67 ± 0.09	0.82 ± 0.03	0.92 ± 0.02	0.88 ± 0.01
XCS	0.66 ± 0.12	0.74 ± 0.18	0.93 ± 0.11	0.83 ± 0.09
GRD-XCS	**0.74 ± 0.19**	**0.86 ± 0.14**	**0.99 ± 0.01**	**0.93 ± 0.05**

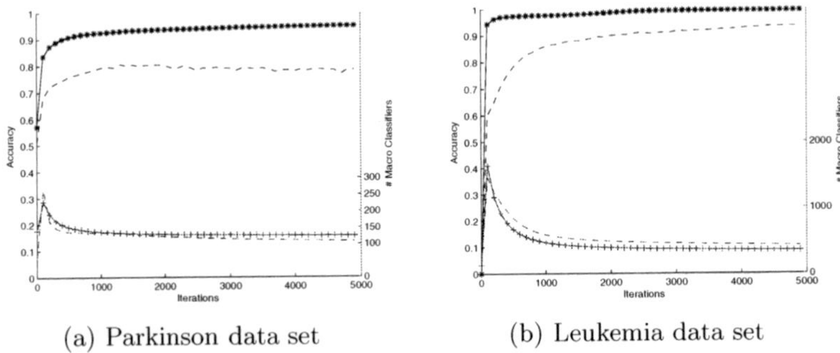

(a) Parkinson data set (b) Leukemia data set

Fig. 2. Accuracy and the number of macro-classifier versus the number of iteration comparisons for the base line XCS model and GRD-XCS for representative low-dimensional (dotted lines) and high-dimensional (solid lines) data sets

To further explore the efficacy of the proposed guided rule discovery enhancements, figure 2 plots time series values for overall accuracy and the number of macro-classifiers in the evolving population for both the GRD-XCS and the standard XCS for a representative low-dimensional dat set and a high-dimensional gene expression data set. Space constraints preclude the inclusion of plots for all data sets, however, the general trends for other data sets is qualitatively similar. There is a correlation between the accuracy of the model and the number of macro-classifiers in the population for high-dimensional classification problems examined. As expected, the number of unique classifiers (individuals) in the population for both XCS and GRD-XCS decreases over time. However, GRD-XCS typically maintains a smaller number of macro-classifiers.

5 Conclusions

In this paper, we have introduced a guided rule discovery component designed specifically for XCS when tackling high-dimensional classification problems. Here, a filtering or feature ranking process is used to build a probabilistic model of feature importance in a pre-processing phase. This probability distribution is then used to bias the evolutionary operators in the underlying XCS model. Comprehensive numerical simulations have shown that the guided rule discovery mechanism improves the performance of XCS in terms of accuracy and more generally in terms of classifier diversity in the population, particularly for high-dimensional classification problems.

We have limited the feature ranking process in this study to simple entropy analysis. In future work, we will explore the use of alternative metrics to rank the features. In the case of microarray data, there is scope to incorporate domain specific knowledge when building the probabilistic rule discovery mask. A second research direction that we will consider will focus on designing a distributed and parallel deployment of the scalable model.

References

1. UCI Machine Learning Repository, `http://archive.ics.uci.edu/ml/`
2. Alon, U., Barkai, N., Notterman, D.A., Gishdagger, K., Ybarradagger, S., Mackdagger, D., Levine, A.J.: Broad patterns of gene expression revealed by clustering analysis of tumor and normal colon tissues probed by oligonucleotide arrays. Proc. of the National Academy of Sciences of the USA 96, 6745–6750 (1999)
3. Bacardit, J., Krasnogor, N.: Smart crossover operator with multiple parents for a Pittsburgh learning classifier system. In: Proceedings of the 8th Conference on GECCO, pp. 1441–1448. ACM (2006)
4. Bonilla Huerta, E., Hernández Hernández, J.C., Hernández Montiel, L.A.: A New Combined Filter-Wrapper Framework for Gene Subset Selection with Specialized Genetic Operators. In: Martínez-Trinidad, J.F., Carrasco-Ochoa, J.A., Kittler, J. (eds.) MCPR 2010. LNCS, vol. 6256, pp. 250–259. Springer, Heidelberg (2010), `http://dx.doi.org/10.1007/978-3-642-15992-3_27`
5. Butz, M., Pelikan, M., Lloral, X., Goldberg, D.E.: Automated global structure extraction for effective local building block processing in XCS. Evolutionary Computation 14(3), 345–380 (2006)
6. Butz, M.V., Goldberg, D.E., Tharakunnel, K.: Analysis and improvement of fitness exploitation in XCS: bounding models, tournament selection, and bilateral accuracy. Evol. Comput. 11, 239–277 (2003)
7. Butz, M.V., Wilson, S.W.: An Algorithmic Description of XCS. In: Lanzi, P.L., Stolzmann, W., Wilson, S.W. (eds.) IWLCS 2000. LNCS (LNAI), vol. 1996, pp. 253–274. Springer, Heidelberg (2001)

8. Fernandndez, A., Garcianda, S., Luengo, J., Bernado-Mansilla, E., Herrera, F.: Genetics-based machine learning for rule induction: State of the art, taxonomy, and comparative study. IEEE Transactions on Evolutionary Computation 14(6), 913–941 (2010)
9. Golub, T.R., Slonim, D.K., Tamayo, P., Huard, C., Gaasenbeek, M., Mesirov, J.P., Coller, H., Loh, M.L., Downing, J.R., Caligiuri, M.A., Bloomfield, C.D.: Molecular classification of cancer: class discovery and class prediction by gene expression monitoring. Science 286, 531–537 (1999)
10. Hedenfalk, I., Duggan, D., Chen, Y., Radmacher, M., Bittner, M., Simon, R., Meltzer, P., Gusterson, B., Esteller, M., Kallioniemi, O.P., Wilfond, B., Borg, A., Trent, J.: Gene-Expression profiles in hereditary breast cancer. N. Engl. J. Med. 344(8), 539–548 (2001)
11. Isabelle Guyon, M.N., Gunn, S., Zadeh, L. (eds.): Feature Extraction, Foundations and Applications. Springer, Heidelberg (2006)
12. Jose-Revuelta, L.M.S.: A Hybrid GA-TS Technique with Dynamic Operators and its Application to Channel Equalization and Fiber Tracking. I-Tech Education and Publishing (2008)
13. Lanzi, P.L.: A Study of the Generalization Capabilities of XCS. In: Bäck, T. (ed.) Proceedings of the 7th International Conference on Genetic Algorithms, pp. 418–425. Morgan Kaufmann (1997)
14. Liu, H., Motoda, H.: Computational Methods of Feature Selection. Data Mining and Knowledge Discovery Series. Chapman & Hall/CRC (2007)
15. Moore, J.H., White, B.C.: Exploiting Expert Knowledge in Genetic Programming for Genome-Wide Genetic Analysis. In: Runarsson, T.P., Beyer, H.-G., Burke, E.K., Merelo-Guervós, J.J., Whitley, L.D., Yao, X. (eds.) PPSN 2006. LNCS, vol. 4193, pp. 969–977. Springer, Heidelberg (2006)
16. Morales-Ortigosa, S., Orriols-Puig, A., Bernadó-Mansilla, E.: New Crossover Operator for Evolutionary Rule Discovery in XCS. In: 8th International Conference on Hybrid Intelligent Systems, pp. 867–872. IEEE Computer Society (2008)
17. Morales-Ortigosa, S., Orriols-Puig, A., Bernadó-Mansilla, E.: Analysis and improvement of the genetic discovery component of XCS. In: International Joint Conference on Hybrid Intelligent Systems, vol. 6, pp. 81–95 (April 2009)
18. Orriols-Puig, A., Casillas, J., Bernadó-Mansilla, E.: Genetic-based machine learning systems are competitive for pattern recognition. Evolutionary Intelligence 1, 209–232 (2065), doi:10.1007/s12065-008-0013-9
19. Singh, D., Febbo, P.G., Ross, K., Jackson, D.G., Manola, J., Ladd, C., Tamayo, P., Renshaw, A.A.: Gene expression correlates of clinical prostate cancer behavior. Cancer Cell 1, 203–209 (2002)
20. Wang, P., Weise, T., Chiong, R.: Novel evolutionary algorithms for supervised classification problems: an experimental study. Evolutionary Intelligence 4(1), 3–16 (2011)
21. Wilson, S.W.: Classifier Fitness Based on Accuracy. Evolutionary Computation 3(2), 149–175 (1995), `http://prediction-dynamics.com/`
22. Wilson, S.W.: Get Real! XCS with Continuous-Valued Inputs. In: Lanzi, P.L., Stolzmann, W., Wilson, S.W. (eds.) IWLCS 1999. LNCS (LNAI), vol. 1813, pp. 209–222. Springer, Heidelberg (2000)
23. Wu, F.-X., Zhang, W., Kusalik, A.: On Determination of Minimum Sample Size for Discovery of Temporal Gene Expression Patterns. In: First International Multi-Symposiums on Computer and Computational Sciences, pp. 96–103 (2006)
24. Zhang, Y., Rajapakse, J.C.: Machine Learning in Bioinformatics, 1st edn. Wiley Series in Bioinformatics (2008)

Motif-Based Method for Initialization the *K*-Means Clustering for Time Series Data

Le Phu and Duong Tuan Anh

Faculty of Computer Science and Technology,
Ho Chi Minh City University of Technology, Vietnam
dtanh@cse.hcmut.edu.vn

Abstract. Time series clustering by *k*-Means algorithm still has to overcome the dilemma of choosing the initial cluster centers. In this paper, we present a new method for initializing the *k*-Means clustering algorithm of time series data. Our initialization method hinges on the use of time series motif information detected by a previous task in choosing *k* time series in the database to be the seeds. Experimental results show that our proposed clustering approach performs better than ordinary *k*-Means in terms of clustering quality, robustness and running time.

Keywords: time series, *k*-Means clustering, initialization, motif discovery.

1 Introduction

One of the crucial tasks in time series data mining which have received an increasing amount of attention lately is time series clustering. Given a set of unlabeled time series, it is often desirable to determine groups of similar time series in such a way that time series belonging to the same group are more "similar" to each other rather than time series from different groups. Although there have been several research works on clustering in general, most classic data mining algorithms do not work well for time series due to their unique characteristics. In particular, the high dimensionality not only slows the clustering process but also degrades it.

The *k*-Means is one of the popular algorithms for clustering time series data ([11], [8]). However, *k*-Means still suffers one shortcoming: the initial cluster centers are still chosen randomly and hence the clustering quality depends significantly on these initial centers. In this paper, we present a new method for initializing the *k*-Means clustering algorithm of time series data. Our initialization method hinges on the use of time series motifs discovered by a previous task in choosing *k* time series in the database to be the initial centroids. A motif of a time series is the most frequently occurring pattern in that time series ([7]). There exist a few efficient motif discovery algorithms for time series, exact as well as approximate algorithms ([2], [7], [9]). Although motif discovery algorithm could be used as subroutine in various other data mining tasks, so far, surprisingly, there have been so few applications of motifs in time series data mining tasks. Recently, only two research works are reported in literature: Jiang et al., 2009 in [3] proposed a method to apply motifs in financial time

D. Wang and M. Reynolds (Eds.): AI 2011, LNAI 7106, pp. 11–20, 2011.

series prediction and Buza and Thieme, 2010 in [1] proposed a method to apply motifs in time series classification using Bayesian networks and SVMs.

The use of motifs in initialization for *k*-Means clustering of time series can be seen as one of our attempts to utilize motif information in some time series data mining tasks. To the best of our knowledge, our initialization technique for *k*-Means time series clustering is completely novel. We experimented our proposed clustering scheme on synthetic and real world datasets. Experimental results show that our proposed clustering approach performs better than ordinary *k*-Means clustering for time series in terms of clustering quality, number of iterations and running time.

2 Background

2.1 Dimensionality Reduction

In order to adapt the various clustering algorithms to massive time series datasets, one can reduce the dimensionality of the time series. For time series, one tries to find a representation at a lower dimensionality that preserves the original information and describes the original shape of the time series data as closely as possible. Many methods have been proposed in the literature, including Discrete Fourier Transform (DFT), Discrete Wavelet Transform (DWT), Piecewise Aggregate Approximation (PAA), Adaptive Piecewise Constant Approximation (APCA). PAA ([4]) is chosen in our work due to its popularity and ease-to-implement property. In PAA, the time series is divided into equal sized segments and the vector consisting of the mean values of the segments becomes the reduced representation of the time series.

2.2 Clustering for Time Series Data

A fast method to perform clustering of time series data is *k*-Means and its variants ([11], [8]). The basic idea behind *k*-Means is the continuous reassignment of objects into different clusters, so that the within-cluster distance is minimized. Therefore, if $\boldsymbol{x}$ are the objects and $\boldsymbol{c}$ are the cluster centers, *k*-Means seeks to optimize the following objective function

$$F = \sum_{m=1}^{k} \sum_{i=1}^{N} \| x_i - c_m \|$$

Unfortunately, the *k*-Means algorithm is a local optimization strategy, therefore it is guaranteed to converge to a local but not necessarily global optimum. Besides, it is sensitive to the choices of the initial positions of the cluster centers. These initial center locations are often termed the *seeds* for the *k*-Means algorithm.

2.3 Time Series Motifs and the Brute-Force Algorithm for Finding Motifs

In this work, we use the definition of motif first introduced by Lin et al., 2002 ([7]). This nontrivial definition of motif requires some other terminology and definitions given as follows.

Definition 1. *Time series*: A time series $T = t_1\ t_2 \ldots,\ t_m$ is an ordered set of m real-valued points.

Definition 2. *Subsequence*: Given a time series T of length m, a subsequence C of T is a sampling of length $n < m$ of contiguous position from T, that is, $C = t_p \ldots t_{p+n-1}$ for $1 \leq p \leq m - n + 1$.

Definition 3. *Match*: Given a real number R and a time series T, containing a subsequence C beginning at position p and a subsequence M beginning at q, if $D(C,M) <= R$ then M is called a matching subsequence of C.

Definition 4. *Trivial match*: Given a time series T containing a subsequence C beginning at position p and a matching subsequence M beginning at q, we say that M is a trivial match to C if either $p = q$ or there does not exists a subsequence M' beginning at q' such that $D(C, M') > R$ and either $q < q' < p$ or $p < q' < q$.

Definition 5. *K-Motifs*: Given a time series T, a subsequence length n and a range R, the most significant motif in T (called thereafter 1-Motif) is a subsequence C_1 that has the highest count of non-trivial matches (ties are broken by choosing the motif whose matches have a lower variance). The Kth most significant motif in T (called thereafter K-Motif) is the subsequence C_k that has the highest count of non-trivial matches and satisfies $D(C_k, C_i) > 2R$, for all $1 < i < K$.

Lin et al. ([7]) also introduced the brute-force algorithm to find 1-motif (see Fig. 1). This brute-force algorithm works directly on raw time series and requires $O(m^2)$ calls to the distance function (m is the length of the time series).

```
Algorithm Find-1-Motif-Brute-Force(T, n, R)
best_motif_count_so_far = 0
best_motif_location_so_far = null;
for i = 1 to length(T) – n + 1
   count = 0;  pointers = null;
   for j = 1 to length(T) – n + 1
      if Non_Trivial_Match (C[i: i + n – 1], C[j: j + n – 1], R ) then
          count = count + 1;
          pointers = append (pointers, j);
      end
   end
   if count > best_motif_count_so_far then
      best_motif_count_so_far = count;
      best_motif_location_so_far = i;
      motif_matches = pointers;
   end
end
```

Fig. 1. The outline of brute-force algorithm for 1-motif discovery

3 The Proposed Clustering Method for Time Series Data

In this work, we propose a new method for initializing the *k*-Means clustering algorithm of time series data. Our initialization method is based on the use of time series motif information discovered by a previous step in choosing *k* time series in the database to be the initial centroids. Our proposed approach to perform clustering of time series data is outlined in Fig. 2.

Step 1: We find 1-motifs for all time series in the database using the brute-force 1-motif discovery algorithm.
Step 2: We apply *k*-Means clustering on the 1-motifs of all time series to obtain the clusters of motifs. From the centroids of the motif clusters, we derive the associated time series and choose these time series as initial centroids for the *k*-Means clustering in the step 4.
Step 3: PAA transform is computed for all time series data in the database to reduce their dimensionality.
Step 4: We perform *k*-Means algorithm on the reduced time series using the initial centroids obtained in Step 2.

Fig. 2. The algorithm for the proposed clustering approach

To have an efficient implementation of our proposed clustering approach, we try to speed up Step 1 and Step 2. To make the brute-force algorithm for finding 1-motif more efficiently in a clustering context, we apply some state-of-the-art techniques to improve it. Furthermore, we devise another technique to derive the initial centroids from the results of *k*-Means clustering on motifs.

3.1 How to Speed Up the Brute-Force Algorithm for Finding 1-Motifs

To speed up significantly the brute-force algorithm for finding 1-motif (Step 1 in our proposed method of time series clustering), we achieve four improvement techniques. Three among the four techniques are similar in spirit to those were used in the exact algorithm of time series motif discovery proposed by Mueen et al., 2009 ([9]) even though the definition of motif in our work is somewhat different from their simple definition of motif. Thank to the three techniques, the algorithm proposed by Mueen et al. (called MK [9]) is up to three orders of magnitude faster than brute-force algorithm.

Exploiting the Symmetry of Euclidean Distance

The brute-force algorithm requires approximately $O(m^2)$ calls of distance function (m is the length of time series). However, by exploiting the symmetry of Euclidean distance ([4]), that means $D(A, B) = D(B, A)$, we can prune off a half of the distance computations by storing $D(A, B)$ and reusing the value when it is necessary to find $D(B, A)$. Therefore, the algorithm only needs to compute and save $m(m-1)/2$ distance values.

Exploiting Triangular Inequality and Reference Point

In order to check whether two subsequences C_a and C_b are non-trivial matches, the brute-force algorithm has to check if $D(C_a, C_b)$ is greater than a given range R ($R > 0$) or not rather than to compute a real value for this distance. Basing on this observation, we can apply triangular inequality and the *reference point* technique proposed by Mueen et al. in [9] as described follows.

Given a *reference subsequence* Q we have to compute the distances from Q to all the subsequences in time series T_i. That means we have to compute $D(Q, t_i)$ for each subsequence t_i in the time series T_i.

Observe that given two subsequences C_a and C_b. By triangular inequality, we have $D(Q, C_a) \leq D(Q, C_b) + D(C_a, C_b)$. From that, we can derive: $D(C_{a,} C_b) \geq D(Q, C_a) - D(Q, C_b)$. Thus, if we want to check whether $D(C_a, C_b) \geq R$, we only need to look at $D(Q, C_a) - D(Q, C_b)$. If $D(Q, C_a) - D(Q, C_b) \geq R$, we can conclude that $D(C_{a,} C_b) \geq R$ since $D(C_{a,} C_b) \geq D(Q, C_a) - D(Q, C_b)$.

When applying reference point technique, we have to deal with the problem how to select an appropriate reference subsequence Q. To apply the triangular inequality for a tighter checking, we should select one subsequence Q to be reference subsequence such that the difference $D(Q, C_a) - D(Q, C_b)$ gets the large value. Hence, in this work we choose a subsequence Q which stays outside of all the other subsequences as reference subsequence. For more details about the reference point technique, interested readers can refer to [9].

Applying Early Abandoning

In the case the triangular inequality can not help to check if $D(C_{a,} C_b) \geq R$ or not since $D(Q, C_a) - D(Q, C_b) < R$, we have to compute the Euclidean distance $D(C_{a,} C_b)$ between them. In this case we can apply early abandoning technique. The idea of this technique is that we can abandon the Euclidean distance computation as soon as the cumulative sum during distance computation goes beyond the range R.

A Dynamic-Programming Technique to Improve *Non_Trivial_Match* Checking

Beside the three above-mentioned techniques, we devise a dynamic programming technique in order to improve the non-trivial match checking between two subsequences $C_{[i:\ i+n-1]}$ and $C_{[j:\ j+n-1]}$. This technique has the same spirit of dynamic programming but does not require any array to keep intermediate computed results.

To perform non-trivial match checking between two subsequences $C_{[i:\ i+n-1]}$ and $C_{[j:\ j+n-1]}$, we look at the non-trivial match check between two subsequences $C_{[i:\ i+n-1]}$ and $C_{[k:\ k+n-1]}$ for $k \in [i+1, j-1]$. At the previous step k we have already checked whether $D(C_{[i:\ i+n-1],} C_{[k:\ k+n-1]}) > R$. We can use a flag variable *Flag* initiated with *false* value. If there exists a value k for which we have $D(C_{[i:\ i+n-1],} C_{[k:\ k+n-1]}) > R$, then we set *Flag* = *true*. Later, when we need to perform non-trivial match checking between $C_{[i:\ i+n-1]}$ and $C_{[j:\ j+n-1]}$, we first check whether $D(C_{[i:\ i+n-1]}, C_{[j:\ j+n-1]}) > 0$ by using early abandoning technique. If $D(C_{[i:\ i+n-1],} C_{[j:\ j+n-1]}) > R$, we set *Flag* = true to serve the non-trivial match checking between two subsequences $C_{[i:\ i+n-1]}$ and $C_{[l:\ l+n-1]}$ for some

$l > j$ and conclude that $C_{[i:\, i+n-1]}$ is not a non-trivial match with $C_{[j:\, j+n-1]}$. Otherwise, if $D(C_{[i:\, i+n-1]}, C_{[k:\, k+n-1]}) \leq R$, we check the *Flag* variable.

- If *Flag* = true, then there exists some $k \in [i+1, j-1]$ such that $D(C_{[i:\, i+n-1]}, C_{[k:\, k+n-1]}) > R$. Therefore, we conclude that $C_{[i:\, i+n-1]}$ is a non-trivial match with $C_{[j:\, j+n-1]}$.
- Otherwise, we can not find any $k \in [i+1, j-1]$ such that $D(C_{[i:\, i+n-1]}, C_{[k:\, k+n-1]}) > R$, hence, we conclude that $C_{[i:\, i+n-1]}$ is not non-trivial match with $C_{[j:\, j+n-1]}$.

Now we can come to the final version of our brute-force algorithm for time series 1-motif discovery. But due to limited space, we could not describe the algorithm here. For more details about this algorithm, interested readers can refer to [6].

3.2 How to Derive Initial Centers from Results of *K*-Means Clustering on 1-Motifs

After applying *k*-Means clustering on the 1-motifs of all time series, we obtain the clusters of motifs and the centers of these motif clusters. Remember that the centers of these motif clusters are just computed subsequences, they have no connection to any time series in the database. Therefore for each such cluster, we find a particular motif which distance to the cluster center is smallest and then from this particular motif, we access to the corresponding time series and use this time series as one initial center for *k*-Means clustering in Step 4 of the clustering approach given in Figure 2.

To speed up the finding of the specific motif that is closest to its cluster center, we can apply the early abandoning technique as follows. Given a motif cluster G, we have to compute the distances from all motifs in G to its center. For a motif T_i in cluster G, if $D(\text{center}(G), T_i) = R$ and for some other motif T_j in G we want to check if $D(center(G), T_j)$ is greater than R or not. If the cumulative sum so far in the Euclidean distance computation for $D(center(G), T_j)$ goes beyond R, we can abandon the distance computation and conclude that $D(center(G), T_j) >= D(center(G), T_i)$.

4 Experimental Evaluation

In this experiment, we compare the performance of our proposed clustering approach with classical *k*-Means. We implemented the two algorithms with Microsoft Visual C# and conducted the experiments on a Intel(R) Pentium(R) Dual CPU T2370 & 1.73 GHz, 2GB RAM PC.

Clustering Quality Evaluation Criteria

We can use classified datasets and compare how good the clustered results fit with the data labels, which is the popular clustering evaluation method. Five objective clustering evaluation criteria were used in our experiments: Jaccard, Rand, FM, CSM and NMI. The definitions of these evaluation criteria can be found in [11].

Besides, since *k*-Means seeks to optimize the objective function by minimizing the sum of squared intra-cluster error, we can also evaluate the quality of clustering by using the objective function given in the subsection 2.2.

Data Description

We tested on two publicly available datasets : Heterogeneous and Stock. The Heterogeneous, which is obtained from the UCR Time Series Data Mining Archive [5], is the classified dataset, therefore we can use it to evaluate the accuracy of the clustering algorithms. The Stock dataset is the stock data of year 1998 from Historical Data for S&P 500 Stocks [10] and each stock time series consists of 252 points representing the open prices of a particular stock.

We conduct the experiments on the two datasets with cardinalities ranging from 1000 to 10000 for each dataset. In the Heterogeneous dataset, each time series consists of 1024 points. In the Stock dataset, the length of each time series has been set to 252.

The Heterogeneous dataset is generated from a mixture of 10 real time series data from the UCR Time Series Data Mining Archive. Using the 10 time-series as seeds, we produced variation of the original patterns by adding small time shifting (2-3 % of the series length), and interpolated Gaussian noise. Gaussian noisy peaks are interpolated using splines to create smooth random variations.

Experimental Results

For the Heterogeneous dataset we tested on 1000 time series. In this dataset, each time series consists of 1024 points. We used the brute-force 1-motif discovery algorithm with the parameter setting: the length of the motif $n = 16$ and the range $R = 0.008$. Here, we applied PAA as a feature extraction method with the length of each segment $l = 8$. We compare the performance of k-Means on time series reduced by PAA (called K-Means/PAA) and k-Means initialized with motif information on time series reduced by PAA (called K-Means/PAA+Motif). Since the Heterogeneous dataset is a classified dataset, we have the knowledge of correct clustering results in advance. In this case, we can compute the evaluation criteria such as Jaccard, Rand, FM, CSM, and NMI to assess the clustering quality of the competing algorithms.

Table 1. Evaluation criteria values obtained from two clustering algorithms with Heterogeneous dataset

Clustering Algorithm	Jaccard	Rand	FM	CSM	NMI
K-Means/ PAA	0.5623	0.9303	0.7351	0.8	0.8956
K-Means/ PAA+Motif	0.7899	0.9749	0.8848	0.9	0.9543

Table 1 shows the criteria values obtained from the experiments on the two algorithms: K-Means/PAA and K-Means/PAA + Motif. As we can see from Table 1, our new algorithm (K-Means/PAA+Motif) results in better criteria values than the classical K-Means (K-Means/PAA).

In Fig. 3 we show the objective function from the experiments with the two methods: K-Means/PAA and K-Means/PAA+Motif on the Heterogeneous dataset over different widths of PAA segment. Again, we can observe that K-Means/PAA+Motif produces a better objective function value than K-Means/PAA, especially when PAA segment is fixed with the appropriate width ($l = 8$). Besides, we can note that the high reduction ratio of the feature extraction stage may negatively impact on the quality of clustering when using motif information for initialization.

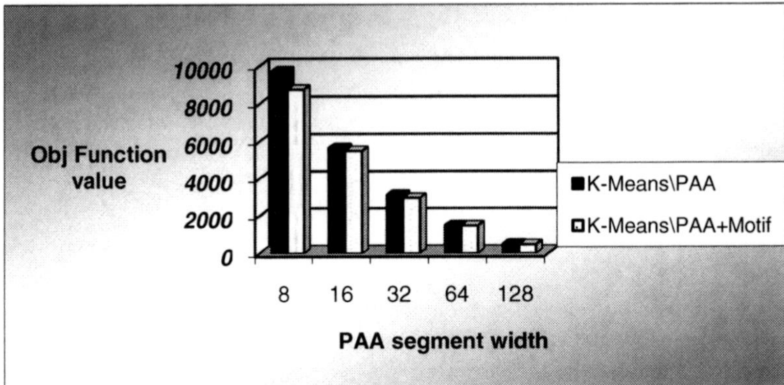

Fig. 3. K-Means/PAA vs. K-Means/PAA+Motif in terms of objective function (Heterogeneous dataset)

In Fig. 4 we show the number of iterations from the experiments with the two clustering methods: K-Means/PAA and K-Means/PAA+Motif on the Heterogeneous dataset, over different lengths of time series. Again, we can see that K-Means/PAA + Motif requires a smaller number of iterations than K-Means/PAA.

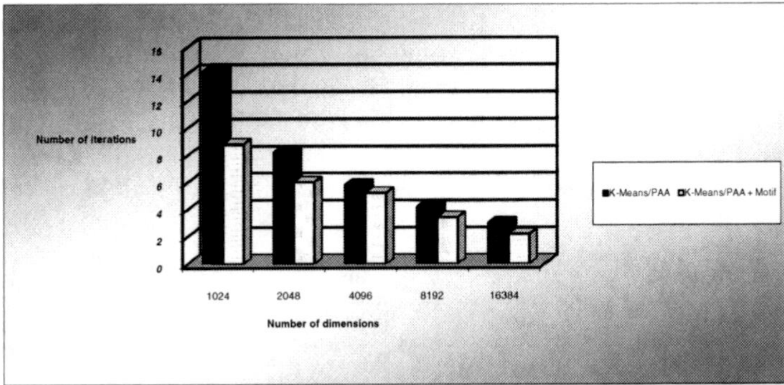

Fig 4. K-Means/PAA vs. K-Means/PAA+Motif in terms of the number of iterations over different lengths of time series (Heterogeneous dataset)

Since motif discovery is an important task in time series data mining, this task should be included in any time series data mining systems beside similarity search, classification, clustering, rule discovery, prediction and anomaly detection. In real world applications, motif discovery helps to provide the most representative pattern or a good summary about a time series. Therefore, it is a task that should be done before several other time series data mining tasks such as classification, clustering, rule discovery, and prediction. For that reason, when comparing the run time between K-Means/PAA+Motif and K-Means/PAA, we do not include the run time of time series motif discovery step in the clustering process.

Fig. 5.a and Fig. 5.b report the run times of two clustering methods on the Heterogeneous dataset with different numbers of time series in the database and different lengths of time series, respectively. Experiment results show that K-Means/PAA+Motif always runs much faster than K-Means/PAA.

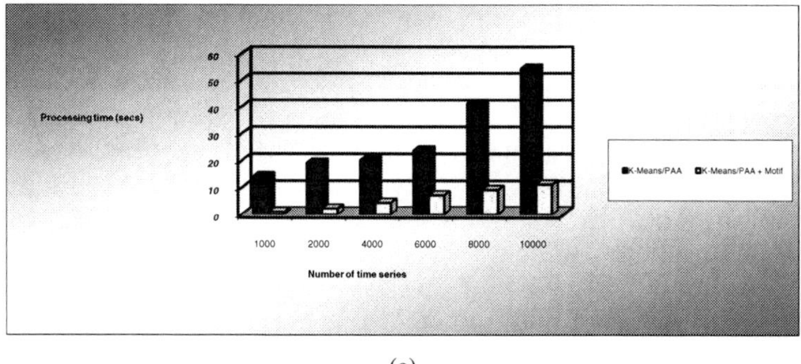

(a)

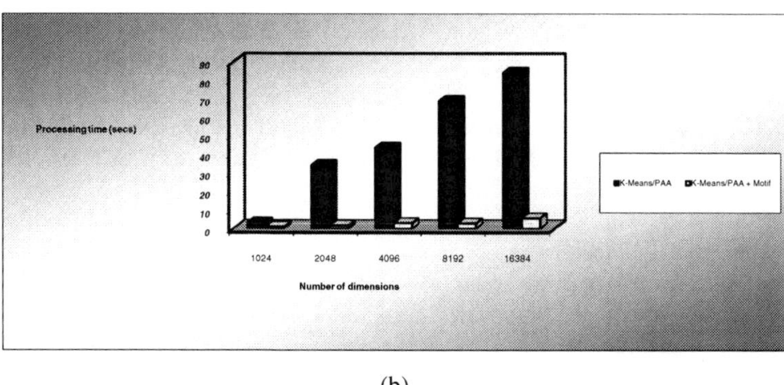

(b)

Fig 5. K-Means/PAA vs. K-Means/PAA+Motif in terms of run times over (a) different numbers of time series in the database and (b) different lengths of time series (Heterogeneous dataset)

Beside the Heterogeneous dataset, we repeated the similar experiments on the Stock dataset. Since the Stock dataset is unclassified dataset, we could not evaluate the quality of the two clustering methods using criteria values such as Rand, Jaccard, FM, CSM and NMI, but we could use objective function to evaluate. Except that, all the experimental results of two algorithms on the Stock dataset according to objective function, the number of iterations and run time are almost similar to those with the Heterogeneous dataset. Experimental results reveal that with the Stock dataset, K-Means/PAA+Motif also performs better than K-Means/PAA in terms of clustering quality, the number of iterations and running time.

5 Conclusions

Time series motif discovery is a crucial task in time series data mining since it brings out the most representative pattern of a time series. In this paper, we have presented a motif-based technique for initializing the k-Means algorithm for time series clustering. We use motif information to derive the initial centers for clustering rather than selecting them in a random manner. Examining the experimental results of the proposed clustering approach against classical k-Means algorithm, we see that our method of seed selection for k-Means clustering of time series data provides a fast and reasonably reliable way to initialization.

To make brute-force algorithm for 1-motif discovery applicable in clustering, we exploit some state-of-the-art techniques to speed up the algorithm such as applying the triangular inequality, the symmetry of Euclidean distance, early abandoning and using a reference point. We also devise a dynamic programming technique to improve the non-trivial match checking between two subsequences in the brute-force algorithm for 1-motif discovery.

References

1. Buza, K., Thieme, L.S.: Motif-based classification of time series with Bayesian networks and SVMs. In: Fink, A., et al. (eds.) Advances in Data Analysis, Data Handling and Business Intelligences, Studies in Classification, Data Analysis, Knowledge Organization, pp. 105–114. Springer, Heidelberg (2010)
2. Chiu, B., Keogh, E., Lonardi, S.: Probabilistic discovery of time series motifs. In: Proc. of 9th Int. Conf. on Knowledge Discovery and Data Mining (KDD 2003), pp. 493–498 (2003)
3. Jiang, Y., Li, C., Han, J.: Stock temporal prediction based on time series motifs. In: Proc. of 8th Int. Conf. on Machine Learning and Cybernetics, Baoding, July 12-15 (2009)
4. Keogh, E., Chakrabarti, K., Pazzani, M., Mehrotra, S.: Dimensionality reduction for fast similarity search in large time series database. Journal of Knowledge and Information Systems, 263–286 (2000)
5. Keogh, E., Folias, T.: The UCR time series data mining archive (2002), `http://www.cs.ucr.edu/~eamonn/TSDMA/index.html`
6. Le, P.: Motif-based method for k-Means clustering of time series data. Master Thesis, Faculty of Computer Science & Engineering, Ho Chi Minh City University of Technology, Vietnam (June 2011)
7. Lin, J., Keogh, E., Lonardi, S., Patel, P.: Finding motifs in time series. In: Proc. of 2nd Workshop on Temporal Data Mining, KDD 2002 (2002)
8. Lin, J., Vlachos, M., Keogh, E.J., Gunopulos, D.: Iterative Incremental Clustering of Time Series. In: Hwang, J., Christodoulakis, S., Plexousakis, D., Christophides, V., Koubarakis, M., Böhm, K. (eds.) EDBT 2004. LNCS, vol. 2992, pp. 106–122. Springer, Heidelberg (2004)
9. Mueen, A., Keogh, E., Zhu, Q., Cash, S., Westover, B.: Exact discovery of time series motifs. In: Proc. of SDM (2009)
10. Historical Data for S&P 500 Stocks, `http://kumo.swcp.com/stocks/`
11. Zhang, H., Ho, T.B., Zhang, Y., Lin, M.S.: Unsupervised feature extraction for time series clustering using orthogonal wavelet transform. Journal Informatica 30(3), 305–319 (2006)

Semi-Supervised Classification Using Tree-Based Self-Organizing Maps

César A. Astudillo[1,*] and B. John Oommen[2,**]

[1] Universidad de Talca, Km. 1 Camino a Los Niches, Curicó, Chile
castudillo@utalca.cl,
http://ing.utalca.cl/~castudillo/

[2] School of Computer Science, Carleton University, Ottawa, Canada, K1S 5B6
oommen@scs.carleton.ca
http://scs.carleton.ca/~oommen/

Abstract. This paper presents a classifier which uses a tree-based Neural Network (NN), and uses both, unlabeled and labeled instances. First, we learn the structure of the data distribution in an unsupervised manner. After convergence, and once labeled data become available, our strategy tags each of the clusters according to the evidence provided by the instances. Unlike other neighborhood-based schemes, our classifier uses only a small set of representatives whose cardinality can be much smaller than that of the input set. Our experiments show that, on average, the accuracy of such classifier is reasonably comparable to those obtained by some of the state-of-the-art classification schemes that only use labeled instances during the training phase. The experiments also show that improved levels of accuracy can be obtained by imposing trees with a larger number of nodes.

Keywords: Hierarchical SOM, Topology-Based Self-Organization, Pattern Recognition, Semi-Supervised Learning.

1 Introduction

The literature includes scores of algorithms which can achieve supervised Pattern Recognition (PR) [5]. Such schemes assume the full specification of the identity of each training instance. In the unsupervised model [10], the class labels of the instances are assumed to be unknown. Rather, the algorithm attempts to infer the distinct group of items, a process which might be time consuming, especially for large datasets. We propose using the Tree-based Topology Oriented SOM (TTOSOM) [1] for classification, attempting to bridge the two paradigms

* The work of this author was done while pursuing his Doctoral studies at the School of Computer Science, Carleton University, Ottawa, Canada : K1S 5B6.

** *Chancellor's Professor* ; *Fellow : IEEE* and *Fellow : IAPR*. This author is also an *Adjunct Professor* with the University of Agder in Grimstad, Norway. The work of this author was partially supported by NSERC, the Natural Sciences and Engineering Research Council of Canada.

D. Wang and M. Reynolds (Eds.): AI 2011, LNAI 7106, pp. 21–30, 2011.

by following the so-called semi-supervised approach, explained by Zhu [15]. Using an unsupervised approach, we will first train a TTOSOM in which the neural tree mimics the properties of the input set. Subsequently, we assign a class label to every single node in the NN[1] by using a voting scheme.

On receiving the testing data, one determines the closest neuron to the testing sample and assigns the sample to the corresponding class.

Once the TTOSOM has been computed, the complexity of the testing is linear, not in cardinality of the training set, but in the size of the neural tree.

Such a nearest-neighbor type testing is similar to the ones used by prototype reduction schemes [13]. However, in our case, the prototypes are not unrelated to each other. Rather, they are constrained by the tree structure.

We cannot expect an accuracy greater than that which a true nearest neighbor classifier yields, because we could be only using a small set (e.g., 25 prototypes) instead of the entire set, which could consist of thousands of points. Further, by starving the classifier of information of the class labels, one can expect the accuracy to be even less. What is astonishing, however, is the fact that our "semi-supervised" TTOSOM-based classifier achieves an accuracy which is only marginally less than state-of-the-art supervised classifiers reported.

The remainder of the paper is organized as follows: We first summarize the TTOSOM. Subsequently, we present the details of the design and implementation of our TTOSOM-based classifier. Thereafter, we provide the experimental results, and finally, we conclude the paper with a discussion of the results of our study.

2 The Tree-Based Topology Oriented SOM

The authors of [1] presented the Tree-based Topology Oriented SOM (TTOSOM), a technique by which the user can represent data points using prototypes, both with respect to the underlying distribution and an arbitrary tree-like topology. Since the topology can be fairly arbitrary, the TTOSOM defines a Bubble of Activity (BoA) different from the ones defined in the prior literature, both structurally and conceptually. As we can see, the map learned as a consequence of the training process is able to infer both the distribution and, simultaneously, the structured topology of the data. This was verified by extensive experiments. The strategy proposed reduces to the traditional 1-dimensional SOM when the tree is a linear sequence of nodes. In other words, the traditional SOM is a special case of the family of ANNs proposed in [1].

Acquiring information about a set of stimuli in an unsupervised manner usually demands the deduction of its structure. In general, the *topology* employed by any ANN possessing this ability has an important impact on the manner by which it will "absorb" and display the properties of the input set. Consider the following example: A user may want to devise an algorithm that is capable of learning a triangle-shaped distribution as the one depicted in Figure 1.

[1] This can be done in numerous ways, but we have chosen to do it using a simplistic Euclidean criterion.

The SOM tries to achieve this by defining an underlying grid-based topology and to fit the grid within the overall shape, as shown in Figure 1a (duplicated from [12]). However, according to the authors of [1], a grid-like topology does not naturally fit a triangular-shaped distribution, and thus, one experiences a deformation of the original lattice during the modeling phase. As opposed to this, Figure 1b, shows the result of applying the TTOSOM. As the reader can observe from Figure 1b, a 3-ary tree seems to be a far more superior choice for representing the particular shape in question.

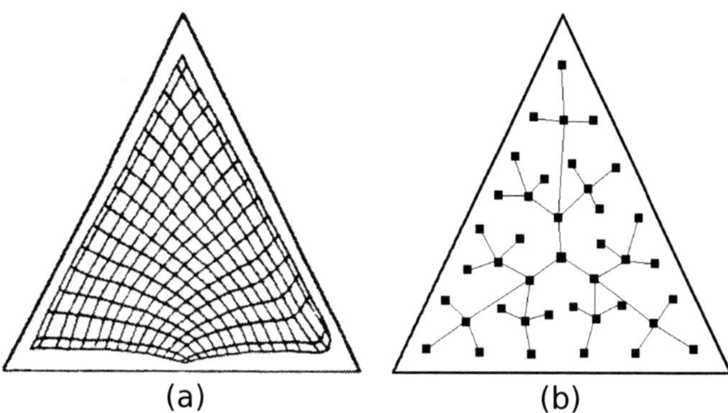

Fig. 1. How a triangle-shaped distribution is learned through unsupervised learning. (a) The grid learned by the SOM. (b) The tree learned by the TTOSOM.

On closer inspection, Figure 1b depicts how the complete tree fills in the triangle formed by the set of stimuli, and further, seems to do it *uniformly*. The final position of the nodes of the tree suggests that the underlying structure of the data distribution corresponds to the triangle. Additionally, the root of the tree is placed roughly in the center of mass of the triangle. It is also interesting to note that each of the three main branches of the tree, cover the areas directed towards a vertex of the triangle respectively, and their sub-branches fill in the surrounding space around them in a recursive manner, which the authors of [1] identify as being a holograph-like behavior. The results of [1] also showed how the TTOSOM can be used to obtain the skeleton structure of an image being examined, and its Pattern Recognition (PR) capabilities.

3 The TTOSOM-Based Classifier

Zhu, in [15], proposed the concept that clustering algorithms could be employed to perform pattern classification. As per his solution methodology, one alternative is to perform classification by applying the so-called Cluster-then-Label method. Prior research to the latter approach includes [3,4,9], among others.

Given a clustering algorithm $\mathcal{A}_C$, a set of labeled instances $\mathcal{X}_L$, a set of unlabeled instances $\mathcal{X}_U$, and a supervised learning algorithm $\mathcal{A}_S$, the Cluster-then-Label method works as follows: First, we identify the clusters of the input manifold using the clustering algorithm $\mathcal{A}_C$. Secondly, we determine which of the labeled samples fall in each cluster. For each cluster we determine a decision boundary based on the supervised algorithm $\mathcal{A}_S$, and the labeled samples assigned to that cluster, which, in turn, allows the prediction of the label of every cluster. Finally, each uncategorized item is labeled according to the predicted class of the cluster in which it is contained.

According to the author of [15], the performance of this approach is dependent on the capabilities of the clustering algorithm to mimic the properties of the original data distribution.

Our aim is to devise a classifier that works in 2 stages. First, we learn the stochastic properties of the data in an unsupervised manner. Secondly, we use some labeled items to tag the decision regions created previously. The resultant TTOSOM-based classifier is described in Algorithm 1.

In order to learn the decision boundaries, the TTOSOM algorithm is employed to train a tree structure so as to mimic the properties of the distribution of data points of *all* the classes, which is done without the necessity of providing the actual class labels of the items. This corresponds to line 1 of Algorithm 1. The output of this initial phase is a TTOSOM tree structure, where each of the neural nodes are optimally placed in the feature space so as to glean the properties of the data distribution. Our hypothesis is that these neurons represent regions of the hyper-space belonging to the same taxonomy, whose label is unknown. The problem then is to accurately guess the actual label of that taxonomy.

In the subsequent phase (see line 2 of Algorithm 1), our classifier determines which subset of the labeled instances are represented by each neuron. In an ideal scenario, where a neuron is the Best Matching Unit (BMU) of instances belonging to the *same* category, the decision of tagging the unlabeled instances falling into the region will be trivial. Unfortunately, as the authors of [3] point out, the latter does not occur necessarily. For this reason a general mechanism is required which permits the *a posterior* decision about the class to be assigned to each neuron. We thus maintain a statistical record of the number of instances belonging to each category that fall in a particular region where a neuron is the BMU.

The next phase (see line 3 of Algorithm 1), consists of a *supervised* phase in which class labels are assigned to each neuron in the tree. From a statistical perspective, when the functions that dictate the probability of finding an item in a certain region of the hyperspace are known, the problem of deciding the category of a particular sample in the area can be optimally determined by the function which maximizes its probability where the query item is positioned. However, as per our problem statement, these probability density functions are not known, and so if one employs an approach like the one described above, we must have an "approximation of sorts" of such functions. Fortunately, there is a simple way to have a rough estimation of the probability functions, i.e., by using the information provided by the labeled training set. Each neuron in the tree is thus assigned a label based on the k-Nearest Neighbor (k-NN) rule [5]. On closer inspection, the label of each neuron will be the one which occurs more frequently among the k nearest samples, where k is the number of data points for which the particular neuron is the BMU.

Algorithm 1. `TTOSOM-Build-Classifier`($\mathcal{X}_U$,$\mathcal{X}_L$)

Input:

i) $\mathcal{X}_U$, the set of unlabeled instances.

ii) $\mathcal{X}_L$, the set of labeled instances.

Output:

i) A set of labels $\mathcal{Y}_U$ of the unlabeled samples $\mathcal{X}_U$.

Method:

1: Train a TTOSOM tree using $\mathcal{X}_U \cup \mathcal{X}_L$.
2: Determine the subset $\mathcal{X}_L^i \in \mathcal{X}_L$ for which the neuron i is the BMU.
3: Label each neuron using $\mathcal{X}_L^i$ and the k-nearest neighbors rule, where $k = |\mathcal{X}_L^i|$.
4: Label each sample in $\mathcal{X}_U$ *as per* the label of its respective BMU.

End Algorithm

The final step (line 4 of Algorithm 1) consists in predicting the class label of each of the unlabeled instances. In our method, this is done by taking a particular instance referred to as the "query" instance and finding its, BMU, i.e., the closest neuron in the feature space, which is basically the notion of a Vector Quantization (VQ) query. The class label of the query instance will be same as the class label of the neuron which is "representing" it. Given the nature of the TTOSOM, some of the neurons act as a "joint" within the tree, reflecting the concentration of other smaller clusters in its vicinity. It is likely, that these joints may not represent any sample in particular, and therefore, one needs an additional assumption in order to define its class label. In our case, we have simply decided to exclude them from the competitive learning process. In that sense, the search for the BMU in the classifier is slightly different from the one utilized by the TTOSOM (and inherited from the SOM). In this case, the label of the neuron is examined, and when it is undefined, the respective neuron is excluded from the "competition" process, which is a phenomenon that we call *supervised BMU search.*

4 Experimental Setup

The Classifiers: The classifiers considered in this study are 5 supervised classifiers, namely, Bayesian Network (BN), Naïve Bayes (NB), C4.5, k-NN and LVQ1, and 2 "semi-supervised" classifiers, namely, the TTOSOM and the SOM. The reader may consult [5] for a general overview of these schemes.

Performance Metrics for Comparing Classifiers: In this study, we shall utilize the most simple and widely-used performance metric, i.e., the accuracy of the classifier [6]. Even though the accuracy measure is, in many contexts, inadequate, our experience is that the inferences gleaned from using it are identical to those obtained by using a more elaborate measure such as the Area Under the ROC Curve (AUC). For a comprehensive examination of metrics for quantifying the quality of a classifier, the interested reader is requested to consult [2,14].

Stochastic Sampling: In this study, we use the technique referred to as "Stratified 10-fold cross-validation". Here, the training samples are roughly divided into 10 equal partitions. Each fold is further used for testing the classifier, while the remainder 9 are employed for training. The process is then repeated for each of the folds. The term stratified, comes from the statistical concept known as "stratified sampling", which is a sampling method that draws items from the different categories so as to obtain relatively homogeneous subgroups.

The Datasets: To test the ability of the TTOSOM for classifying items belonging to the real world domain, we have 6 datasets from the UCI Machine Learning repository [8]. These datasets are Iris, Wisconsin Diagnostic Breast Cancer (WDBC), Wine, Yeast, Wine_Quality, and Glass.

The datasets used in these experiments have different numbers of output classes, ranging from 2 to 10. Additionally, their features pertain primarily to the continuous domain, whose dimensions varies from 4 up to 30. Table 1 describes the different aspects of each dataset, including its name, number of instances, number of attributes, number of output classes and problem type.

Table 1. Datasets selected for the comparison of the classifiers

Dataset	Instances	Attributes	Classes	Problem Type
Iris	150	4	3	classification
WDBC	569	30	2	classification
Wine	178	13	3	classification
Yeast	1,484	8	10	classification
Wine_Quality (red)	1,599	11	6	classification/regression
Glass	214	9	6	classification

The Parameters: The respective parameters for the algorithms were rendered to be the same across all the different datasets, and no algorithm possessed parameter values that were tuned for the datasets. In particular, the 3 strategies based on VQ, i.e., the TTOSOM, the SOM and the LVQ1 utilized the same number of iterations $(50,000)$. Additionally, they all used the same initial learning rate (0.5), and the radius of the BoA was chosen in such a way that initially, all the neurons were considered as part of the BoA, i.e., twice the depth of the tree in the case of the TTOSOM, and the width plus the height in the case of the SOM. Observe that LVQ1, as defined in [11], does not consider a BoA. As well, the three schemes utilized the same (linear) decaying schedule for its parameters.

5 Results

Comparison to Other Classifiers: The results of the performance of the different classifiers (columns) across all the dataset (rows) is summarized in Table 2. Specifically, we are interested in the performance of our classifier on problems across a diversity of domains in which labeled and unlabeled data is available[2]. For example, Table 2 shows that the TTOSOM classifier, using *only* 15 neurons, is able to accurately predict with an accuracy of 89.33% the correct label of the instances belonging the wine dataset. On the other hand, the SOM classifies correctly the same dataset with an accuracy of only 67.98%.

One possibility for quantifying the quality of our method is to consider the family of classifiers inheriting the VQ mechanism. One such strategy that belongs to the supervised family is the LVQ1, while the SOM and the TTOSOM primarily learn the distributions using the unsupervised learning paradigm. The three classifiers utilized the same parameters, which are described in Section 4. Besides, while the LVQ1 and the SOM utilized 128 neurons, the results shown for the TTOSOM include *only* 15 neurons. As per our results, the TTOSOM, using only a small percentage of the neurons used in the SOM and LVQ1 (almost 10%), outperforms their recognition capabilities in *all* six datasets.

Apart from the above, observe that the classification results offered by the TTOSOM are comparable to the ones obtained by the k-NN. However, both approaches present important differences in how they perform learning. First of all, the k-NN, being a supervised classifier, requires *all* the instances to be properly labeled. Secondly, due to its "laziness", the computations for the k-NN are left until a query is performed, which implies that the *whole* manifold is visited so as to create the ordering of the samples, as per their proximity to the query sampl6 years since the date of the TRes. On the other hand, the TTOSOM only requires a small subset of the tagged labels, and is able to learn from unlabeled samples. Also, the query is done by using the TTOSOM tree and the respective labels of the neurons, and only requires the comparison with the total number of neurons, which is usually significantly smaller than the entire dataset. Even though our method internally uses the k-NN to tag the neurons, we note that this is done only once, i.e., when the tree is being learned, and furthermore, the computations are performed only for each neuron instead of the whole dataset.

Another perspective by which we can compare the schemes is to consider the "most" competitive supervised classifiers. In this case, except for the LVQ1, they outperformed the accuracy produced by the unsupervised strategies. This is an expected behavior, because the supervised classifiers had access to the class labels of *all* the instances. However, in environments where only few tags are available, traditional supervised classifiers struggle to extract useful information from unlabeled instances. Indeed, experiments performed by Gabrys *et al.* [9], showed that when a sufficiently large number of labeled instances were utilized,

[2] Our hypothesis is that one should use as much labeled data as is available. Since the datasets mentioned above are all composed of labeled instances, we have opted to use *all* this information in the "supervised" phase of our algorithm.

the semi-supervised schemes included in their study achieved levels of accuracy that were comparable to the ones obtained by the supervised classifiers that incorporated a much higher number of labeled samples.

Table 2. General classification results of the methods investigated, reported in terms of accuracy (shown in percentages)

Dataset	TTOSOM15	BN	NB	C4.5	k-NN	LVQ1	SOM
iris	96.67	92.67	96.00	96.00	95.33	96.00	84.67
wdbc	93.32	95.08	93.15	93.15	96.66	92.09	90.51
glass	67.29	71.96	49.07	67.76	67.76	61.22	63.08
wine	89.33	98.88	97.19	93.82	94.94	74.16	67.98
yeast	54.18	56.74	57.61	55.86	54.78	24.33	46.16
wine_quality	51.91	57.72	55.03	62.91	57.79	44.15	49.59

Effect of the Number of Neurons: Another set of experiments were conducted so as to observe the effect of the *number* of neurons on the classification accuracy. To test this, we systematically increased the size of the TTOSOM tree. In order to retain the desired property that, initially, all the neurons are considered as part of the BoA, in each case we adjusted the radius to be twice the depth of the tree. Even though the size of the tree was increased, we decided to maintain the number of training iterations to be unchanged.

We identified an increase in the performance as the number of neurons is increased. For example, for the wine dataset, an accuracy of 64.61% was obtained when using 15 nodes, and increased to 76.40% when using 1023 nodes. Similarly, for the glass dataset, we obtained an accuracy of 69.16% when we used 15 nodes, which increased to 71.96% when the number of nodes was 127.

Additionally, we noted that a lesser number of neurons, which implies a lower computational requirement, outputs a fairly good approximation to the one offered by the reported supervised classifiers.

Changing the Distance Measure: In all the results presented so far, we assumed that the data was previously normalized. Specifically, the classifiers utilized the so-called Local Normalization [7], in which the range of every dimension was scaled to be between 0 and unity so as to have them equally weighted. We performed additional experiments so as to observe how the technique behave if we maintain all the parameters at their original values, and simultaneously not perform any type of normalization prior to the training process.

As a general remark we note that one observes differences with respect to the case when the data was normalized. For example, in the *glass* dataset it was possible to obtain an accuracy of 71.96% when using 127 neurons which is an index equivalent to the one provided by the best supervised classifier (BN) for this specific problem domain. It is even more interesting to see that when the number of neurons was increased to 1,023, the accuracy obtained was 74.30%, which is the *best* reported accuracy obtained for the glass dataset, when one includes *all* the supervised classifiers displayed in Table 2.

However, we have noticed as well, that in some problem sets, as in the case of the *wine* dataset, the classification accuracies are inferior to those obtained when an *a priori* normalization was invoked.

Our explanation for this phenomenon is that, when we do not normalize the feature vectors before processing them, the classifier weights those features with larger ranges for its values, more, and in certain cases it happens that these features are exactly the ones that help to advantageously discriminate between the different categories. This reasoning also explains the scenario when poorer results are obtained. This is apparently a consequence of weighting certain features (i.e., those which possess a high variance) more, i.e., those which offer inadequate discriminating aspects. Those features do not provide information that is too useful for effectively building the discrimination regions.

Using Trees Other Than Binary Trees: All the experiments presented previously in this section have employed a binary tree structure. To further investigate the power of the TTOSOM, we performed another set of experiments so as to test the effect of using trees with a higher branching factor, i.e., the number of children that a particular node had.

In particular we tested the algorithm using trees with a branching factor of 3. As far as we could observe, there were no noticeable changes in accuracy when the branching factor per node is increased from 2 to 3.

In [1], when we focused on the clustering properties on the TTOSOM, we showed how different branching factors led to a "better representation" of certain shapes. By better representation, in this case, we meant that the basic properties of some objects were preserved, so that the human eye could perceive the essential characteristics of the original object by merely looking at the learned structure. The above mentioned paper included examples, including a triangle and a rectangle, which were represented in a superior manner using specific branching factors (c.f., the representations in [12], which correspond to neural structures for the triangle using a grid and a line, respectively).

The clustering property mentioned above suggests that the symmetry presented in some data sets could be better exploited by a TTOSOM-based classifier using the adequate branching factor. However our preliminary evidence shows us that at least for the *real-world dataset* that we tested, the classifier is not noticeably affected by incrementing the number of branches in the tree. Instead, the number of neurons utilized, regardless of the branching factor of the tree, seems to be more pertinent when it concerns the resultant accuracy. This certainly is an avenue for further research.

6 Conclusions

The purpose of this paper was to design and present an experimental analysis of a novel PR scheme based on the TTOSOM. Our classifier combined the information provided by labeled and unlabeled instances simultaneously.

Our experimental results showed that the TTOSOM classifier possesses an improved classification accuracy in comparison to other VQ-based classifiers. Additionally, these accuracies are comparable to the one attained by the state-of-the-art schemes, even when the number of neurons utilized is only a small fraction of the cardinality of the dataset.

Moreover, increasingly superior recognition capabilities could be obtained when training trees with a larger number of neurons. In particular, our results suggest a "monotonic" improvement of the mean classifier performance as the size of the tree is increased. We believe that this occurs because of the desirable properties of the TTOSOM to mimic the underlying distribution of the points, and its capability to represent the stochastic and structural characteristics more accurately by utilizing a larger tree.

References

1. Astudillo, C.A., Oommen, B.J.: Imposing tree-based topologies onto self organizing maps. Information Sciences 181(18), 3798–3815 (2011)
2. Baeza-Yates, R.A., Ribeiro-Neto, B.: Modern Information Retrieval. Addison-Wesley Longman Publishing Co., Inc., Boston (1999)
3. Dara, R., Kremer, S., Stacey, D.: Clustering unlabeled data with SOMs improves classification of labeled real-world data. In: Proc. of the 2002 International Joint Conference on Neural Networks, IJCNN 2002, vol. 3, pp. 2237–2242 (2002)
4. Demiriz, A., Bennett, K., Embrechts, M.: Semi-supervised clustering using genetic algorithms. In: Artificial Neural Networks in Engineering (ANNIE 1999), pp. 809–814 (1999)
5. Duda, R., Hart, P.E., Stork, D.G.: Pattern Classification, 2nd edn. Wiley-Interscience (2000)
6. Fawcett, T.: An introduction to ROC analysis. Pattern Recogn. Lett. 27(8), 861–874 (2006)
7. Fayyad, U., Grinstein, G.G., Wierse, A.: Information Visualization in Data Mining and Knowledge Discovery. Morgan Kaufmann Publishers Inc., San Francisco (2001)
8. Frank, A., Asuncion, A.: UCI machine learning repository (2010), `http://archive.ics.uci.edu/ml`
9. Gabrys, B., Petrakieva, L.: Combining labelled and unlabelled data in the design of pattern classification systems. International Journal of Approximate Reasoning 35(3), 251–273 (2004), Integration of Methods and Hybrid Systems
10. Jain, A.K., Murty, M.N., Flynn, P.J.: Data clustering: a review. ACM Computing Surveys 31(3), 264–323 (1999), `citeseer.ist.psu.edu/jain99data.html`
11. Kohonen, T.: Improved versions of learning vector quantization. In: 1990 IJCNN International Joint Conference on Neural Networks, vol. 1, pp. 545–550 (June 1990)
12. Kohonen, T.: Self-Organizing Maps. Springer-Verlag New York, Inc., Secaucus (1995)
13. Lazebnik, S., Raginsky, M.: Supervised learning of quantizer codebooks by information loss minimization. IEEE Transactions on Pattern Analysis and Machine Intelligence 31(7), 1294–1309 (2009)
14. Manning, C.D., Raghavan, P., Schütze, H.: An Introduction to Information Retrieval. Cambridge University Press, Cambridge (2009), `http://nlp.stanford.edu/IR-book/information-retrieval-book.html`
15. Zhu, X., Goldberg, A.B.: Introduction to Semi-Supervised Learning. Morgan & Claypool Publishers (2009)

The Discovery and Use of Ordinal Information on Attribute Values in Classifier Learning

Alex Berry and Mike Cameron-Jones

School of Computing and Information Systems, University of Tasmania
{amberry,Michael.CameronJones}@utas.edu.au
http://www.cis.utas.edu.au

Abstract. Rule and tree based classifier learning systems can employ the idea of order on discrete attribute and class values to aid in classification. Much work has been done on using both orders on class values and monotonic relationships between class and attribute orders. In contrast to this, we examine the usefulness of order specifically on attribute values, and present and evaluate three new methods for recovering or discovering such orders, showing that under some circumstances they can significantly improve accuracy. In addition we introduce the use of classifier ensembles that use random value orders as a source of variation, and show that this can also lead to significant accuracy gains.

Keywords: machine learning, classifier learning, decision trees, ordinal attributes, ensemble classifiers, ordinal aggregation.

1 Introduction

Decision tree based classifier learners typically use two kinds of attributes, nominal and numeric. Nominal attributes are discrete and unordered. When creating a node involving a nominal attribute a tree classifier learner typically creates a branch for each different value. Numeric attributes are continuous and ordered. For numeric attributes two branches are typically created, one branch with lesser values and one with greater, with a split value determined by the classifier learner separating them. J4.8 [12], the tree classifier we have chosen to use as a reference, exhibits these behaviours.

There is also a third kind of attribute, ordinal. Ordinal attributes are both discrete and ordered. An example is T-shirt sizes (S, M, L, XL): they are logically discrete but also have an implied order, and if you are considering which one to buy then knowledge of that order is potentially useful.

A decision tree can use such order information by treating an ordinal attribute as a numeric attribute, and searching for binary splits of the values rather than splitting the tree one way for each different value [12]. This may be advantageous because by comparison with the one branch per value case, during the training process more test cases are available at each of the lower nodes to use in decision making, possibly increasing accuracy.

D. Wang and M. Reynolds (Eds.): AI 2011, LNAI 7106, pp. 31–40, 2011.

Several researchers have looked at the issue of ordinal class values in general [4,5]. Ordinal monotonicity, where class and attribute value orders have an enforced relationship, has also been a very active field of research [1,7]. It is therefore somewhat surprising that little general research has been done on the discovery and possible uses of orders in attribute values, although order discovery has been pursued in other contexts (e.g. [3]).

This paper investigates the usefulness of ordinal information for classifier learning, both on specific datasets and in general. We focus on decision tree learning, but the methods can also be applied in rule learning where similar threshold tests are used, and may be applicable in other classifier learning.

We present methods for predicting an order on a nominal attribute from a data set, and test these methods for effectiveness at improving classification accuracy for classifier learners.

Furthermore we present and investigate extensions to ensemble classifiers by applying random orders to nominal attributes as a novel form of creating variance through modifying the data presented to a classifier learner (which need not be able to handle weighted instances).

For the purpose of this paper we compared our method with Bagging [2], which seems the most directly comparable well known method, but there has been much other noteworthy work in the area of randomization, e.g [6].

2 Value of Ordinal Information

When there is a discrete ordinal attribute in a problem, it can be treated as either an ordinal attribute or a nominal attribute for the purposes of classification. Whether or not there is any benefit to treating ordinal attributes as ordinal rather than nominal is an important question which has received little general attention. There are two major factors which will affect how useful an ordinal attribute is - how relevant the order is to solving the problem, and how much of an advantage is gained by using binary splits. The difference between these two factors can be measured experimentally using the idea of a random ordinal attribute - this provides a binary split without the potentially beneficial order information.

2.1 Testing

To test the usefulness of ordinal attributes, a large number of datasets relying on primarily ordinal attributes is required. As there are not many such standard data sets, we chose to create ordinal data sets by discretizing the numeric attributes of primarily numeric standard data sets from the UCI collection [10]. This gives a large number of sets of data with orders implied from the original numeric nature of the attributes.

Experiments were run on these data sets with five distinct methods, shown below. We used the J4.8 version of C4.5 from WEKA 3-4 as the base tree classifier learner for all tests in this paper. We ran 10 separate 10 fold cross validations and used a two-tailed confidence test with 0.05 confidence to determine significant improvement or degradation in accuracy between two methods on a data set.

The methods used were with discretized nominal data (Nom), original undiscretized data (UD), discretized data marked as ordinal with the original order (S. Ord), discretized data marked as ordinal with randomized orders (R. Ord), and versions of S. Ord and R. Ord with both the (discretized) nominal and the modified ordinal version of the attribute available to the classifier learner.

Discretization was performed on numeric attributes using six bins and equal frequency binning. These attributes were then treated differently for each method tested. Nominal is the discretized data only, which is the baseline here for improvement or degradation.

2.2 Results

There was no general statistical difference across the datasets (in a win/loss sense) between the ordinal (S. Ord) and non-ordinal discretized (Nom) methods. However, on four of the datasets the ordinal method (S. Ord) significantly outperformed the unordered method (Nom), and on no datasets was it

Table 1. Discretized Numeric Order Discovery : Classifier Accuracy

Data Set	Nom	UD	S. Ord	R. Ord	S. O+N	R. O+N
anneal	98.92	98.57	98.78	98.62	98.76	98.86
autos	79.48	81.77	80.53	77.47	79.50	80.11
w-b-cancer	95.08	95.01	94.78	93.79	94.74	94.67
c-rating	87.22	85.57	86.26	84.78 •	86.59	86.61
g-credit	72.12	71.25	71.71	70.68	72.24	72.01
diabetes	75.22	74.49	74.18	71.46 •	74.13	73.18
glass	62.47	67.63	76.23 ○	64.34	72.79 ○	63.98
c-heart	78.92	76.94	76.70	76.81	76.73	77.53
h-heart	80.53	80.22	79.99	79.20	80.91	80.53
hepatitis	82.46	79.22	81.36	78.32	82.27	80.85
hypothyroid	94.17	99.54 ○	94.48	94.16	94.30	94.24
ionosphere	88.92	89.74	90.17	87.07	89.86	88.56
iris	93.40	94.73	92.80	93.47	92.67	93.40
letter	80.91	88.03 ○	87.20 ○	82.37 ○	86.50 ○	82.96 ○
segment	94.59	96.79 ○	95.93 ○	94.44	95.70 ○	94.75
sonar	68.20	73.61	74.39	67.03	75.98	68.98
vehicle	68.43	72.28	69.13	66.80	69.08	68.18
vowel	72.63	80.20 ○	79.91 ○	74.39	78.19 ○	73.22
Win/Draw/Loss vs Nominal						
	-	10/0/8	9/0/9	4/0/14	10/0/8	7/2/9
Win/Draw/Loss vs Random Ordinal						
	17/0/1	14/0/4	16/0/2	-	16/0/2	15/0/3

○ significant improvement against Nominal

• significant degradation against Nominal

significantly outperformed. This makes it clear that under some circumstances an ordinal style split is of great importance to classification.

The correct order method (S. Ord) outperformed the random method (R. Ord) 16 out of 18 times, across a wide variety of datasets; this is strong evidence that if an ordinal split is employed, correct ordinal information is vital to classification as opposed to just using random information. This is unsurprising but important.

These results add value to investigating if it is possible to recover or discover orders, since those orders can be of significant benefit in classification.

3 Discovering Orders

We developed three related methods for finding attribute orders on typical nominal attributes. To narrow the scope and focus on improving classification we made a fundamental assumption that the order of an attribute would be related to the class value, which is reasonable because if an attribute is present in a dataset it is likely there because of an initial suspicion that it is related to the solution. There are potential problems however, for example if we classify our T-shirts sizes by how well they fit a medium size person, we could end up with S and XXL on one end of an order and M on the other.

For the purpose of simplicity and generality we assume no known order on the class value.

3.1 Developing Methods

The basis for all our methods is a solution to problems with two class values only. In this case we use the Laplace equation, a probabilistic estimator, to measure the probability of each nominal value for an attribute being classified as the first class (this method has the disadvantage of making the assumption of an even spread of cases within the data set). This gives each value a positive probability between 0 and 1 which provides an obvious ranking for correlation to the class value.

This approach is simple but will not work for problems with multiple classes, since there is no obvious way to split the classes into a single ranking. Our first method solves this problem by using Principal Component Analysis (a linear algebra technique) [8]. For an n class problem each attribute value is represented by a point in $n-1$ dimensional space, with each dimension being the probability of classification as the mth class (the last class's probabilities would be completely correlated to the other $n-1$ probabilities and so are omitted). Principal Component Analysis gives a major axis of variance in this space, which in turn gives a ranking based on distance along that axis.

This method is mathematically clean but relies on every class having a mostly linear spread of points, which is an indication of a direct linear relationship between class and values. To offset this problem we also considered alternative approaches based on building up orders by combining orders from subproblems.

To combine subproblem orders we used a weighted voting system similar to the Borda Count [11]. Given a number of orders with a real number weighting, each order is converted to a ranking and the values are multiplied by the weighting for the order and then summed, and a new order is obtained from that sum (with arbitrary tiebreaking).

Given w_i is the weighting for subproblem i, n is the number of subproblem, $r_{j,i}$ is the ranking of element j in subproblem i, and $vote(a_j)$ is the final weighted score for element j:

$$vote(a_j) = \sum{}_{i=1}^{n}(w_i * r_{o,i}) \tag{1}$$

When merging orders generated from subproblems we also need to remember that the direction of the order is not well defined - inverted orders may need to be combined with direct orders, which will cause problems in our weighted sum method because they will cancel each other out. To solve this problem, each order is compared against the order with the most weight using a least squares difference. The smaller difference of the direct order and the reverse order is taken, and the corresponding order is used.

Our correspondence algorithm is as follows; given r, t are orders of the n values, $pos_k(i)$ is the position of element i in order k, and n is the number of attributes:

$$correspondence(r, t) = \sum{}_{i=1}^{n}(pos_r(i) - pos_t(i))^2 \tag{2}$$

Our first order combination method, multi-class discovery, uses the probability of an attribute classifying as each individual class as a subproblem. This gives a subproblem for each class value. Per class weightings for merging are the proportion of the class value in the test data.

The second order combination method, tree based discovery, creates a decision tree with large leaves using the J4.8 classifier learner and very heavy pruning with a minimum leaf size. Each leaf is treated as a separate subproblem, and the orders from leaf subproblems are merged, with the size of the leaf determining the weighting. When learning the tree to find the leaves, the attribute for which an order is being sought is not split on. For each leaf we take the probability of classification as the most common class in the leaf, a heuristic which assumes a much simplified decision structure, which is reasonable because the tree structure will reduce the effects of other attributes. The best leaf size is likely problem dependent, but we chose a flat minimum size of 50.

The main purpose of considering subproblems after a level of tree building is to handle probability distorting imbalances in the frequency of different areas of data, and to remove the skewing effect of highly distinguishing attributes.

3.2 Testing Order Discovery

The three new methods were tested in two ways, firstly on several datasets with known useful orders to determine whether correct orders could be recovered. These results are too long and database specific to present in full, but typically multi-class discovery performed best followed by PCA. Tree discovery was more variable due to highly branching attributes often leading to non useful division of the data into subproblems. When this was not a problem it performed equivalently to multi-class.

The second testing method was to apply the order learning methods to real datasets to determine if classification accuracy gains could be made by using these methods. The discretized numeric datasets seen earlier were tested first.

The six methods tested are all based on the discretized data. They are the discretized data seen previously without ordinal set (Nom), with ordinal set (S. Ord), the randomly ordered discretized data (R. Ord) based on a different seed to Table 1, the Principal Components Analysis method (PCA), the multi-class method (MC), and the tree method (Tree).

Table 2. Discretized Numeric Order Discovery: Classifier Accuracy

Data Set	Nom	S. Ord	R. Ord	PCA	MC	Tree
anneal	98.92	98.78	98.60	98.73	99.03	98.68
autos	79.48	80.53	78.46	79.31	77.45	79.45
w-b-cancer	95.08	94.78	94.25	95.01	95.01	95.14
c-rating	87.22	86.26	84.62 •	84.80 •	86.90	85.36
g-credit	72.12	71.71	70.86	71.99	71.99	71.28
diabetes	75.22	74.18	71.23 •	73.36	73.36	72.76
glass	62.47	76.23 ∘	63.89	70.24 ∘	70.51 ∘	68.91
c-heart	78.92	76.70	76.78	76.17	75.44	76.47
h-heart	80.53	79.99	79.47	79.06	79.79	80.02
hepatitis	82.46	81.36	80.37	81.61	81.78	81.78
hypothyroid	94.17	94.48	94.19	94.26	94.24	94.36
ionosphere	88.92	90.17	87.85	87.07	90.23	90.23
iris	93.40	92.80	93.87	92.87	93.20	93.80
letter	80.91	87.20 ∘	82.86 ∘	85.45 ∘	85.71 ∘	85.45 ∘
segment	94.59	95.93 ∘	94.48	95.21	95.26	95.39
sonar	68.20	74.39	68.90	71.57	74.92	74.92
vehicle	68.43	69.13	66.42	69.06	69.12	67.68
vowel	72.63	79.91 ∘	74.17	78.52 ∘	77.31 ∘	76.70 ∘
Win/Draw/Loss vs Standard Ordinal						
	9/0/9	-	2/0/16	4/0/14	8/0/10	6/0/12
Win/Draw/Loss vs Random Ordinal						
	12/0/6	16/0/2	-	14/0/4	15/0/3	16/0/2

∘ significant improvement against Nominal

• significant degradation against Nominal

The order methods showed strong accuracy ratios across the data sets against the random order method, evidence that regardless of correctness the orders recovered were almost always preferable to random orders. On the four datasets where correct ordinal splitting was previously shown to be of significant importance, that significance of improvement was retained on three datasets by multi-class and PCA, and on two by tree. This is evidence that significant improvements are possible using order recovery where an order is already known to be of value.

Multi-class was generally the strongest method, with a win/loss ratio competitive with the original known orders.

Finally, the order recovery methods were tested on 18 primarily nominal attribute datasets from the UCI collection where there was no general reason to suppose that an order existed.

In addition to the previously introduced methods we used WEKA's binary split setting (Bin. S), which tests single element subsets, to provide some comparison with an alternative form of binary split.

Table 3. Nominal Order Discovery : Classifier Accuracy

Data Set	Nom	Bin. S	R. Ord	PCA	MC	Tree
audiology	77.26	76.92	76.92	76.51	77.88	76.92
b-cancer	74.28	70.50	72.24	72.11	72.11	71.94
bridges	57.42	66.00	64.71	64.79	63.85	63.36
car-eval	92.22	96.63 ○	95.49 ○	96.74 ○	96.65 ○	96.96 ○
cmc	51.44	52.26	52.23	51.92	53.02	52.86
colic	85.16	85.34	85.18	83.93	83.93	84.69
c-rating	85.57	85.20	85.03	83.97	84.59	84.33
g-credit	71.25	70.64	70.75	70.98	70.98	71.36
dermatology	94.10	95.90	95.71	95.20	95.23	94.70
c-heart	76.94	78.17	78.22	78.74	78.05	78.18
h-heart	80.22	78.95	78.48	78.08	77.77	77.73
lymph	75.84	76.59	77.97	79.45	78.71	79.26
nursery	97.18	99.36 ○	98.92 ○	99.28 ○	99.46 ○	98.88 ○
postop	69.78	70.11	70.00	70.22	71.11	71.00
p-tumor	41.39	41.19	41.54	41.04	41.42	41.28
soybean	91.78	92.30	91.08	91.80	91.75	92.18
splice	94.03	94.48	93.03	93.45	93.81	93.21
tic-tac-toe	85.28	93.79 ○	92.51 ○	92.48 ○	93.78 ○	93.95 ○
Win/Draw/Loss vs Nominal						
	-	12/0/6	11/0/7	10/0/8	11/0/7	11/0/7
Win/Draw/Loss vs Binary Split						
	6/0/12	-	5/1/12	6/0/12	9/0/9	9/1/8
Win/Draw/Loss vs Random Ordinal						
	7/0/11	12/1/5	-	9/0/9	10/0/8	8/1/9

○ significant improvement against Nominal

• significant degradation against Nominal

The results show no general advantage to using order, with a mostly even ratio to the purely nominal method. Significant gains on three databases were observed over nominal, although these were shared by all binary split methods including the random order method. The general equivalence of the order discovery methods to the random ordinal method suggests that orders useful to classification were not generally found. This is perhaps not surprising as there is no reason to assume that these useful orders existed on these nominal data sets.

4 Random Orders for Ensemble Classifiers

To this point we have investigated the direct application of orders to single tree classification. We now look at using attribute orders in a different scenario, this time as a source of variance inside an ensemble classifier method.

The standard Bagging algorithm ([2]) induces variation between its base classifiers by randomly resampling the data presented to each classifier. Since elements are removed, this means that fewer of the original instances are available to each classifier, so the internal base classifiers are likely to be less accurate than a single classifier built on the original data. We propose an alternative, to mark all nominal attributes as ordinal and randomize the orders of the values between base classifier instances. This induces a novel form of randomization and differentiation without reducing the size of the test data. We call this method Ordinal Aggregation.

It is worth noting that this method will only be of direct use for data sets with largely nominal attributes, as only the nominal attributes will be randomly re-ordered. For other data sets it would be possible to discretize the numeric attributes first, but we did not attempt to test this.

We also considered a hybrid method which first applies the test data randomization of bagging, and then applies random ordinalization to all nominal attributes as per Ordinal Aggregation. This produces an ensemble classifier which combines the variance from both methods.

We compared the methods experimentally using the same primarily nominal attribute datasets as earlier. All ensemble classifiers had 10 internal J4.8 classifiers. We ran ten 10-fold cross validations for each method.

The methods tested were J4.8, Bagging (Bag), Bagging with WEKA binary splits (BagB), Ordinal Aggregation (OA), and the hybrid method (B/OA).

The results show that the ordinal method is competitive with standard bagging, with a fairly even 10 wins 8 losses and also, although not shown here, a statistically significant improvement on bagging on two datasets, with significant losses on none. The composite method on the other hand performed very well, beating bagging on 17 of the 18 datasets, and J4.8 on 14. Both these margins are statistically significant at a 0.05 confidence level using a one tailed sign test. This is a very promising sign. It could indicate that the methods were simply not introducing enough variance on their own. Since neither method would generally allow you to set the variance required arbitrarily it would be worth investigating

Table 4. Random Ensemble Orders : Classifier Accuracy

Data Set	J4.8	Bag	BagB	OA	B/OA
audiology	77.26	81.07 ∘	80.44 ∘	77.87	81.33 ∘
b-cancer	74.28	69.40 •	68.36 •	69.28 •	70.33
bridges	57.42	65.56 ∘	70.22 ∘	66.77	64.68
car-eval	92.22	93.98 ∘	97.54 ∘	97.23	96.90
cmc	51.44	51.20 ∘	52.03 ∘	51.88 ∘	52.08 ∘
colic	85.16	84.88	84.88	84.48	84.99
c-rating	85.57	84.93	85.12	84.03	86.07
g-credit	71.25	72.17	73.21	72.46	73.57
dermatology	94.10	96.67 ∘	96.29	97.05 ∘	97.62 ∘
c-heart	76.94	79.21	81.76 ∘	78.33	80.60
h-heart	80.22	78.93	78.68	78.18	78.99
lymph	75.84	78.53	80.28	80.81	80.28
nursery	97.18	98.57 ∘	99.66 ∘	99.68 ∘	99.64 ∘
postop	69.78	59.56 •	62.22 •	58.00 •	61.67 •
p-tumor	41.39	42.86	43.04	42.43	42.93
soybean	91.78	91.78	92.37	92.96	92.96
splice	94.03	94.16	95.25 ∘	95.62 ∘	95.36 ∘
tic-tac-toe	85.28	94.41 ∘	97.53 ∘	96.24 ∘	97.44 ∘
Win/Draw/Loss vs J4.8					
	-	11/1/6	13/0/5	13/0/5	14/0/4
Win/Draw/Loss vs Bagging					
	6/1/11	-	13/1/4	10/0/8	17/0/1
Win/Draw/Loss vs Bagging with Binary Splits					
	5/0/13	4/1/13	-	6/0/12	10/1/7

∘ significant improvement against J4.8
• significant degradation against J4.8

if there is a general value to this kind of composition. However, the comparison with Bagging with WEKA's binary splits shows that the binary splitting alone proves a useful enhancement to Bagging, so the extra value of the ordinal method may be more limited than the comparison with Bagging suggested.

Given the positive results with Bagging we also ran some tests on both standard and composite Multiboost classifiers [13]. The results suggested that adding ordinal randomization to boosting may give a slight general improvement to accuracy, but not on the order of that shown for Bagging.

5 Conclusions and Further Work

Ordinal attributes were found to be of significant benefit to accuracy in certain classification problems, but not in general across a wide range of problems. Where there was a reason to suggest an order existed, having the correct order appeared to be generally useful.

Three new methods for discovering orders on attribute values were presented and evaluated. The most effective method was the multi-class method, which correctly recovered known orders and gave classification accuracies almost equivalent to the original orders on problems where orders were known to be present.

Our results show that methods such as these could likely be applied to new problems to deduce or confirm useful orders on attribute values. If a new problem has unknown but relevant orders in its attributes then order discovery could be used to improve classification accuracy.

We also found that the idea of inducing variance inside ensemble classifiers using random orders could provide a benefit to classification accuracy.

This research throws up a lot of questions about how to best balance the use of ordinal attributes with nominal attributes (feature selection). The eventual goal of this line of investigation is to gain a better understanding of the internal organization and relationship between attributes, and to consider learning more descriptive and featureful attributes for the purpose of augmenting existing classifier learner methods, including methods other than decision tree learners.

References

1. Ben-David, A., Sterling, L., Tran, T.D.: Adding monotonicity to learning algorithms impair their accuracy. Expert Systems with Applications 36(3), 6627–6634 (2009)
2. Breiman, L.: Bagging predictors. Machine Learning 24(2), 123–140 (1996)
3. Cameron-Jones, R.M., Quinlan, J.R.: Avoiding pitfalls when learning recursive theories. In: 13th International Joint Conference on Artificial Intelligence, pp. 1050–1055. Morgan Kaufmann (1993)
4. Chu, W., Keerthi, S.S.: New approaches to support vector ordinal regression. In: International Conference on Machine Learning, pp. 07–11. ACM (2005)
5. Frank, E., Hall, M.: A Simple Approach to Ordinal Classification. In: Flach, P.A., De Raedt, L. (eds.) ECML 2001. LNCS (LNAI), vol. 2167, pp. 145–156. Springer, Heidelberg (2001)
6. Liu, T., Ting, K.M., Yu, Y., Zhou, Z.: Spectrum of Variable-Random Trees. Journal of Artificial Intelligence Research 32, 355–384 (2008)
7. Potharst, R.: Decision trees for ordinal classification. Intelligent Data Analysis 4(2), 97–111 (2000)
8. Jolliffe, I.: Principal component analysis. Wiley Online Library (2002)
9. Quinlan, J.R.: C4. 5: Programs for Machine Learning. Morgan Kaufmann, San Francisco (1993)
10. UCI machine learning repository, `http://www.ics.uci.edu/mlearn/MLRepository.html`
11. Van Erp, M., Schomaker, L.: Variants of the Borda count method for combining ranked classifier hypotheses. In: 7th International Workshop on Frontiers in Handwriting Recognition, pp. 443–452 (2000)
12. Witten, I., Frank, E.: Data Mining: Practical Machine Learning Tools and Techniques, 2nd edn. Morgan Kaufmann (2005)
13. Webb, G.I.: Multiboosting: A technique for combining boosting and wagging. Machine learning 40(2), 159–196 (2000)

Beyond Trees: Adopting MITI to Learn Rules and Ensemble Classifiers for Multi-Instance Data

Luke Bjerring and Eibe Frank

Department of Computer Science, University of Waikato
{lb54,eibe}@cs.waikato.ac.nz

Abstract. MITI is a simple and elegant decision tree learner designed for multi-instance classification problems, where examples for learning consist of bags of instances. MITI grows a tree in best-first manner by maintaining a priority queue containing the unexpanded nodes in the fringe of the tree. When the head node contains instances from positive examples only, it is made into a leaf, and any bag of data that is associated with this leaf is removed. In this paper we first revisit the basic algorithm and consider the effect of parameter settings on classification accuracy, using several benchmark datasets. We show that the chosen splitting criterion in particular can have a significant effect on accuracy. We identify a potential weakness of the algorithm—subtrees can contain structure that has been created using data that is subsequently removed—and show that a simple modification turns the algorithm into a rule learner that avoids this problem. This rule learner produces more compact classifiers with comparable accuracy on the benchmark datasets we consider. Finally, we present randomized algorithm variants that enable us to generate ensemble classifiers. We show that these can yield substantially improved classification accuracy.

1 Introduction

Multi-instance classification differs from standard propositional classification in that examples for learning consist of bags of instances. Potential application domains are drug activity prediction, where instances can be feature vectors describing different conformations of a molecule [5], and content-based image classification, where they are associated with different regions in an image [13]. In either case, a class label—indicating, e.g., whether a molecule is "active" or "inactive"—is available only for the entire example (i.e. bag), not the individual instances it contains, which renders this learning setting a challenging one.

In this paper we consider induction of decision trees and classification rules for multi-instance problems, and also consider ensemble learning. Decision tree induction is a popular learning method in standard propositional problems because of its computational efficiency and the interpretability of the output it generates. It can also yield highly competitive classification accuracy when used to learn ensembles. In this paper, we first revisit an existing decision tree induction method for multi-instance learning [2], called MITI, in Section 2, and

D. Wang and M. Reynolds (Eds.): AI 2011, LNAI 7106, pp. 41–50, 2011.

evaluate its performance on a collection of benchmark datasets based on different configurations of the algorithm. Then, in Section 3, we show how we can apply a simple modification to this algorithm to yield a rule learner, which we call MIRI, that yields a compact set of classification rules for multi-instance problems. Finally, in Section 4, we show how accurate ensembles can be learned using randomisation, and summarise our main findings in Section 5.

2 The MITI Algorithm

The standard assumption in multi-instance learning—based on classification problems with two classes, positive and negative—is that a bag is positive if and only if it contains a positive instance, and negative otherwise [5,12]. The key problem is that instance-level class labels are unknown for positive bags. All instances in negative bags must necessarily be true negative instances—otherwise the bag-level class label could not be negative. In contrast, it is possible that all but one of the instances in a positive bag are in fact false positives.

A common learning strategy under the standard assumption is to identify regions in instance space where positive bags overlap, i.e. regions of the instance space that contain positive instances from a non-trivial number of positive bags. This basic strategy was employed in the two oldest methods for multi-instance learning: in [5] a hyperrectangle is learned to describe the region where positive bags overlap and in [12] the maximum diverse density approach identifies parameters of a probabilistic model that is centered in such a region.

The multi-instance tree inducer (MITI) proposed by Blockeel *et al.* [2] is a learning algorithm based on the same standard assumption. It implements the top-down decision tree learning approach known from propositional tree inducers such as C4.5 [18], with two key modifications: (a) nodes are expanded in best-first order guided by a heuristic that aims to identify pure positive leaf nodes as quickly as possible, and (b) whenever a pure positive leaf node is created, all positive bags containing instances in this leaf node are deactivated.

A pure positive leaf node in this context is a node that only contains instances from positive bags. The assumption underlying this approach is that an instance's presence in a pure positive leaf is a strong indication that it is a true positive instance, and that all instances in the same bag that are not in the leaf should be eliminated from further consideration in the learning process because they are potentially false positives.

Pseudo code for MITI is shown in Algorithm 1. The algorithm is as originally presented in [2], with one small difference. On some of the datasets considered in our study, it can happen that the current node cannot be split any further because it contains identical instances from bags with different class labels. In that case, a leaf node is created based on the majority class.

Standard propositional tree induction normally proceeds in depth-first fashion, which can be implemented in a non-recursive fashion by storing unexpanded nodes in a last-in first-out (LIFO) queue. In MITI, this LIFO queue is replaced by a priority queue in which nodes are sorted in descending order according to

Algorithm 1. Pseudo code for MITI, based on [2]

```
let Q = root node
while Q is not empty do
  remove the first node N from Q
  if N is pure positive then
    make N a positive leaf and deactivate all bags with instances in N
  else if N is pure negative then
    make N a negative leaf
  else
    find the best split S for N
    if N cannot be split then
      make N a leaf with majority label and deactivate bags if necessary
    else
      split N according to S and add the child nodes of N to Q
    end if
  end if
  sort Q
end while
```

the proportion of positive instances they contain. When calculating this proportion, each instance is weighted by $1/|B|$, where $|B|$ is the size of the bag that contains the instance, to give each bag the same total weight, namely 1. Assuming w_p is the sum of weights of "positive" instances in the node concerned (note that this includes any potential false positives in the node), and w_n is the sum of weights of negative instances, the ratio $\frac{w_p}{(w_p+w_n+k)}$ is used to sort the nodes in the priority queue, where k is a parameter to the algorithm. This measure is called the *tozero(k)* estimate in [2].

Given numeric attributes, MITI applies binary splits to divide the data into two subsets at each internal node. Split selection is another important aspect of tree induction. [2] considers several measures to identify the best split at a particular node in the tree, but finds negligible differences for most of them. In the following, we consider two of the split selection criteria from [2]: *max-bepp*, which uses the maximum of the two estimated proportions of positives for the two subsets created by a split as the split quality score, and the standard *Gini index*, applied with the same estimate of the proportion of positives. Note that in the Prolog implementation of MITI kindly provided by the authors of [2], the Gini index is calculated without taking subset weights into account—the Gini-based impurity scores from each of the two subsets concerned are combined using a simple unweighted average. In this paper, we use branch weights in the standard fashion to combine the two subset scores for a split when calculating its Gini index.

2.1 Experimental Results

In [2], Blockeel *et al.* evaluate classification accuracy of MITI on synthetic data and two real-world multi-instance domains— the *musk* and *mutagenesis* problems—but splitting criteria are only compared on the synthetic data. In this section, we present a more extensive evaluation of MITI on benchmark data, including data from image classification problems, where we consider two splitting criteria (*max-bepp* and *Gini index*) and two values for the parameter k in the *tozero(k)* heuristic: 5, the default value from [2], and 0, which means that an unbiased estimate of the proportion of positives is applied. We also report tree size, which gives an indication of interpretability and is not considered in [2].

All experimental results presented in this paper are based on stratified 10-fold cross-validation, repeated 10 times, to yield 100 performance estimates for each dataset/algorithm combination. Tables show average accuracy as well as standard deviation across the 100 estimates. To test for statistical significance of individual differences, the corrected resampled paired t-test [16] is used, which is a conservative version of the standard paired t-test that is adjusted for dependency of estimates due to data reuse. This test is the standard test available in the Experimenter facility available in the WEKA workbench [10], which we used for the experiments. The significance level was left at the default value 0.05. Algorithms were implemented in Java and integrated into WEKA.

Table 1 shows estimated classification accuracy for the datasets included in our experiments. Table 2 shows tree size. The datasets used are those employed in [7].[1] These include mutagenicity prediction [19]—which was originally considered for multi-instance tree and rule learning in [21]—based on three different representations of molecules as bags of instances (*muta-atoms, muta-bonds, muta-chains*), the well-known trains problem from ILP [15] (*eastwest, westeast*), the two *musk* datasets [5], the *thioredoxin* protein identification task [20], and two groups of content-based image classification datasets (*elephant, fox, tiger* [1] and *bikes, cars, people* respectively, the latter group with Ohta-based features as in [14], derived from the GRAZ02 dataset [17]). We also included the synthetic *maron* problem [12,8]. In this problem, instances are uniformly distributed in a 2D space and bags are classified as positive if they contain at least one instance that is located in a small area in the center of this space.

Considering classification accuracy, we can see that adjusting the value of the parameter k is important when using the special-purpose *max-bepp* split selection heuristic from [2] in MITI. Using the raw estimated proportion of positives (k=0) yields significantly more accurate classifiers on the *thioredoxin* problem, but applying a biased estimate of proportion (k=5) produces significantly higher accuracy on all but one of the image classification problems. In contrast, the results obtained using the Gini index appear less sensitive to the choice of k, but $k = 0$ yields higher estimated accuracy for all datasets apart from *musk1*, with two significant differences (not shown in the table)—on *cars* and *thioredoxin* respectively. The results for the *Gini index* with $k = 0$ are the best ones overall: this method dominates the *max-bepp* baseline in 12 out of 15 cases and yields statistically significant improvements in two cases. The results provide evidence that (a) biasing the estimate of proportion is generally detrimental when using the standard *Gini index* in MITI, and (b) the special-purpose *max-bepp* heuristic is generally inferior to the *Gini index*.

The results in Table 2 reinforce this message: trees grown using the *Gini index* with $k = 0$ are often substantially smaller than those generated using the other three variants. This also has a strong impact on runtime (not shown here) because the smaller trees can be grown more quickly.

[1] Excluding the suramin data, which contains missing values.

Table 1. Classification accuracy for different parameter settings in MITI

Dataset	MITI *max-bepp* k=5	MITI *Gini index* k=5	MITI *max-bepp* k=0	MITI *Gini index* k=0
eastwest	55.5±32.5	60.5± 34.3	67.0±32.7	62.5± 35.1
westeast	30.5±31.7	52.5± 34.4	47.0±26.4	58.5± 35.6
musk1	70.4±16.5	83.3± 12.7 ∘	60.3±14.5	82.2± 12.7 ∘
musk2	71.0±15.3	73.2± 13.1	62.7±14.5	74.4± 14.1
muta-atoms	80.2± 8.2	82.0± 8.3	80.5± 8.4	84.3± 7.9
muta-bonds	80.6± 7.9	81.2± 8.3	85.7± 8.8	81.9± 8.4
muta-chains	83.4± 7.7	84.4± 7.2	83.5± 8.5	87.2± 8.5
maron	50.0±10.1	55.6± 17.4	48.6±22.4	56.2± 22.9
elephant	77.6± 9.4	77.4± 9.3	72.0±10.5	77.9± 9.5
fox	61.7± 8.7	60.8± 9.6	49.7±11.0 •	61.7± 10.4
tiger	74.7±10.0	70.3± 10.6	62.8±10.9 •	74.0± 9.8
bikes	76.5± 5.2	74.6± 5.0	68.4± 5.0 •	76.1± 5.1
cars	67.9± 4.3	63.7± 5.0	58.2± 5.8 •	69.6± 4.9
people	73.4± 5.6	73.0± 4.8	66.1± 5.7 •	74.8± 5.3
thioredoxin	35.7±11.0	62.7± 14.5 ∘	80.0± 8.1 ∘	82.1± 9.6 ∘

∘, •: statistically significant compared to 2nd column

Table 2. Tree size for different parameter settings in MITI

Dataset	MITI *max-bepp* k=5	MITI *Gini index* k=5	MITI *max-bepp* k=0	MITI *Gini index* k=0
eastwest	25.8± 6.5	10.6± 4.3 •	11.0± 1.9 •	10.0± 3.2 •
westeast	36.8± 5.3	23.4± 6.0 •	14.8± 3.0 •	12.8± 3.1 •
musk1	20.1± 2.2	20.6± 2.2	50.4± 2.2 ∘	22.8± 3.6 ∘
musk2	43.1± 16.0	41.3±13.2	44.7± 2.2	33.1± 3.7
muta-atoms	261.7± 16.9	163.1±10.0 •	62.7± 3.5 •	62.5± 4.3 •
muta-bonds	286.4± 30.5	157.8±11.4 •	67.9± 4.2 •	62.2± 6.0 •
muta-chains	419.7± 44.5	198.1±11.4 •	90.4± 7.5 •	55.6± 7.5 •
maron	902.5± 32.2	177.7±34.9 •	30.5± 2.4 •	17.5± 3.0 •
elephant	46.6± 22.0	218.7±39.0 ∘	102.0± 2.7 ∘	32.3± 3.5
fox	166.5± 50.7	167.4±18.5	107.8± 3.3 •	51.2± 4.8 •
tiger	55.8± 16.8	160.2±20.2 ∘	79.5± 5.0 ∘	32.4± 4.1 •
bikes	248.9± 83.1	634.7±96.3 ∘	288.3±10.3	84.4± 4.2 •
cars	219.4± 70.6	474.3±42.3 ∘	387.4±15.6 ∘	110.2± 5.5 •
people	165.9± 45.4	662.7±61.5 ∘	285.4±14.8 ∘	90.2± 5.3 •
thioredoxin	2202.2±207.5	250.4±43.7 •	42.2± 3.3 •	35.9± 5.7 •

∘, •: statistically significant compared to 2nd column

The results on the *maron* data are particularly noteworthy because here the standard multi-instance assumption is known to hold by construction. Note that the *max-bepp* split selection criterion requires only one of the two subsets created by a split to exhibit high purity for it to be rated highly. The other subset can be poor and may thus need to be expanded into a large subtree—unless positive data in this subset can be successfully deactivated before this happens. In contrast, the *Gini index* combines impurity scores from both subsets in a weighted fashion.

3 MIRI: Using MITI to Learn Rule Sets

Whenever a positive leaf node is created, the MITI algorithm disables all instances of all bags that are associated with this leaf: any positive bag that has at least one positive instance in the leaf is disabled. The corresponding data is removed from all unexpanded nodes waiting in the priority queue and will thus not influence subsequent tree growth. However, tree structure that has already been created is left untouched. Conceptually, this is a potential drawback of the algorithm because data is removed from partially grown subtrees elsewhere in the overall tree structure. Splitting and node selection decisions that generated those existing incomplete subtrees should be revised to accommodate the new data distribution. At the very least, one would expect this to produce a more compact classifier because the amount of relevant training data is reduced.

Implementing this idea yields an algorithm whose output can be more naturally represented as a set of classification rules: when a positive leaf is encountered in the basic MITI algorithm, all positive bags associated with the leaf are removed from the training data, the path from the root node to this leaf node is turned into an if-then rule, and the algorithm is restarted on the remaining data. The tree structure is discarded and grown from scratch on the reduced data. We call this algorithm MIRI, for multi-instance rule induction.

Clearly, this approach will not generate any output that is due to potentially suboptimal split and node selection decisions based on outdated data because the entire tree structure is discarded after a positive leaf node has been turned into an if-then rule. When no positive leaf node can be created, the algorithm stops and appends a final default rule to the rule set that predicts the negative class. This will normally only happen when all positive data has been exhausted because the priority queue used in the best-first expansion method is ordered based on the proportion of positive data in each node located in the queue. Consequently it is appropriate to create a "catch-all" rule that simply predicts the negative class when the first negative leaf node is encountered.

There are pathological scenarios where positive data remains that is not covered by any positive rule, namely when there are identical instances that are located in both positive and negative bags. In that case it can happen that a node has to be turned into a leaf node even if it contains both positive and negative data. If the sum of weights for the negative instances in this node is greater than the sum of weights for the positive instances, then the node is turned into a negative leaf and the algorithm stops. On the other hand, if the positive data outweighs the negative data, the node is turned into a positive leaf and the associated positive bags are deactivated in the standard manner. This heuristic does not appear to cause problems on the benchmark datasets we consider.

The algorithm just described implements the standard separate-and-conquer rule learning strategy, where a rule is generated, the data covered by this rule is removed (i.e. separated out), and the remaining data is used to generate further rules. In contrast to most separate-and-conquer rule learners, a partial tree structure is induced to find the next rule to add to the rule set. In the context of propositional rule learning, where each example for learning consists of a single instance, this strategy is used in the rule learner PART [9], which generates a partial decision tree using the C4.5 tree learner [18].

It appears wasteful to generate a partial tree just to subsequently discard it. In practice, on the datasets we consider, MIRI's runtimes are within an order of magnitude of MITI's ones (which never requires more than a few seconds

Table 3. Accuracy and classifier size for MITI and MIRI ($k = 0$, *Gini index*)

	Classification accuracy		Classifier size	
Dataset	MITI	MIRI	MITI	MIRI
eastwest	62.5±35.1	69.0±33.9	13.1± 4.7	7.9± 2.1 •
westeast	58.5±35.6	67.0±32.7	21.2± 6.0	11.6± 2.0 •
musk1	82.2±12.7	80.6±12.6	23.6± 3.8	21.8± 3.4 •
musk2	74.4±14.1	75.1±12.9	47.4± 7.9	36.4± 5.3 •
muta-atoms	84.3± 7.9	82.9± 8.3	241.7±27.9	95.2±11.2 •
muta-bonds	81.9± 8.4	81.6± 7.9	270.1±48.3	98.6±13.7 •
muta-chains	87.2± 8.5	83.3± 8.2	233.5±59.5	81.2±14.6 •
maron	56.2±22.9	59.6±25.1	54.3±16.7	18.7± 4.3 •
elephant	77.9± 9.5	78.7± 9.3	75.2±16.1	31.9± 3.5 •
fox	61.7±10.4	59.9±10.8	166.1±34.0	61.1± 6.1 •
tiger	74.0± 9.8	75.7± 9.9	66.4±17.3	31.4± 3.9 •
bikes	76.1± 5.1	76.5± 5.0	295.4±33.7	102.2± 7.0 •
cars	69.6± 4.9	67.9± 4.5	435.3±48.6	145.1±10.9 •
people	74.8± 5.3	73.5± 4.9	326.2±36.5	115.7± 8.2 •
thioredoxin	82.1± 9.6	82.9± 8.5	84.3±27.7	48.0±13.2 •

∘, •: statistically significant difference

to generate a tree when using $k = 0$ and the *Gini index*): MIRI is never more than five times slower. The best-first node expansion strategy is very effective in homing in on positive leaf nodes, which means that little additional tree structure is generated before a rule can be obtained. In the best case, only one path is created because only nodes leading to the relevant leaf node are expanded. Note also that many challenging multi-instance problems exhibit large bags of instances, which means that creation of a rule removes a substantial amount of instance-level data that will not need to be considered in subsequent iterations.

3.1 Experimental Results

Table 3 shows classification accuracy and classifier size for MITI and MIRI. Classifier size is measured by counting the number of tests in all positive rules included in the classifier. In MITI, positive rules correspond to leafs with a positive classification. In both cases, $k = 0$ was used (no bias in the estimated proportion of positives), and the *Gini index* was applied for split selection.

The results paint a clear picture: there is no statistically significant difference in classification accuracy between MITI and MIRI on the benchmark datasets we consider, but the classifiers learned by MIRI are significantly more compact in all cases. Hence, MIRI's ability to discard structure grown from outdated data does not have a significant impact on classification accuracy. Nevertheless, for data mining practitioners who are concerned with interpretability of the output, MIRI appears to provide a useful alternative to MITI.

4 Building Ensemble Classifiers

Although individual decision trees and rule sets can provide valuable insight into the structure underlying a dataset, and are thus an important tool for descriptive data mining, they are known to be inferior to ensemble classifiers in predictive tasks. A well-known strategy for generating an ensemble classifier is randomisation [6], in which the learning algorithm is randomised such that different classification models can be obtained from the same dataset, thereby yielding an ensemble. Predictions are then commonly obtained by voting.

In the propositional context, the random forest method [3] has proven particularly successful. Consequently we apply the basic strategy of this method to the multi-instance learning algorithms discussed above and evaluate whether similar gains in predictive accuracy can be obtained. In the random forest method, a decision tree learner is randomised by introducing non-determinism in the attribute selection step that is performed at each node. More specifically, rather than choosing the best split amongst all m available attributes, l attributes are selected at random first, where this randomly chosen subset can be different for each node, and then the best split amongst those l attributes is picked (where split quality is measured using a standard criterion such as the *Gini index*).

We can directly apply this method in MITI, and, consequently, also in MIRI. Large values of l decrease randomness and thus diversity, small values increase diversity but may yield ensemble members that are individually not very accurate. Both, accuracy of individual ensemble members and their diversity, will affect the accuracy of the final vote-based ensemble classifier.

4.1 Experimental Results

We generated empirical results using 100 ensemble members based on two values of l by applying WEKA's *RandomCommittee* method in conjunction with both MITI and MIRI as the base learner, yielding four configurations in total. Recent versions of WEKA allow parallel computation of ensemble members using *RandomCommittee* on multiple cores and this was exploited to obtain the results in a timely manner. The *Gini index* was used in MITI and MIRI and an unbiased estimate of positive proportion was applied ($k = 0$). The two values for l we consider are $l = 1$, which implies completely random attribute selection, and $l = \sqrt{m} + 1$, where m is the number of attributes in the dataset concerned, yielding a semi-random strategy. Results are provided in Table 4.

These results show that there is no noteworthy difference between MITI and MIRI ensembles in the case of semi-random attribute selection. However, when selecting attributes completely randomly, the MIRI-based ensemble performs worse. Thus selecting informative attributes appears more important when MIRI is used. Comparing semi-random selection with completely random selection, we can see that the latter strategy generally performs worse. The win/loss ratio is 10/4 in favour of semi-random selection in the case of MITI, although none of the differences are individually statistically significant. The semi-random selection method appears to have an edge on the datasets with a larger number of attributes (the image datasets and the *musk* problems) but on the datasets with a small number of attributes (*maron* and *mutagenesis*) there is no advantage. This makes sense intuitively: when there are many attributes, it is less likely that any one of them will be relevant to the classification.

Comparing these results to the ones in Tables 3 for individual trees and rule sets, we can see that the ensemble approach yields substantial improvements

Table 4. Classification accuracy for MITI and MIRI ensembles

Dataset	MITI Ensemble semi-r.	MITI Ensemble random	MIRI Ensemble semi-r.	MIRI Ensemble random
eastwest	72.5±31.3	72.5±27.9	75.5±32.2	72.5±29.6
westeast	37.5±32.9	31.5±33.1	48.5±34.4	30.5±33.3
musk1	86.5±11.5	80.7±12.6	85.0±11.7	78.6±12.3 •
musk2	79.1±10.9	74.2±12.5	77.5±11.7	72.4±13.1
muta-atoms	85.3± 7.9	86.9± 7.7	85.4± 7.7	87.5± 7.1
muta-bonds	85.4± 7.5	86.3± 7.6	85.0± 7.5	85.7± 7.4
muta-chains	87.7± 8.6	87.1± 8.4	85.6± 8.4	86.4± 7.6
maron	56.2±22.9	66.4±22.7	59.6±25.1	60.6±21.5
elephant	88.5± 7.1	87.7± 6.8	86.9± 7.7	82.8± 8.5
fox	68.3± 8.8	61.4±10.7	66.3± 8.7	55.8±10.4 •
tiger	82.8± 8.1	80.6± 8.4	83.2± 8.2	78.5± 9.7
bikes	84.1± 4.8	83.4± 4.6	84.9± 4.8	82.9± 4.8
cars	77.8± 4.1	76.3± 4.4	77.9± 4.1	74.6± 4.7 •
people	81.9± 4.1	82.2± 4.0	82.4± 4.0	81.3± 4.1
thioredoxin	90.4± 6.1	89.4± 5.5	87.9± 7.0	86.9± 4.7

○, •: statistically significant compared to 2nd column

most cases. Thus it is clear that the success of randomisation in the propositional case translates into the realm of multi-instance problems.

It is also interesting to compare these results to those that can be obtained with other high-performance multi-instance classifiers on the same datasets. As an indicative baseline, we can draw on the results for various variants of the well-known MILES method for multi-instance learning [4] that are presented in [7], and the best results for two simple propositionalisation methods that can also be found in [7]. The estimated accuracies for the *muta-atoms, muta-chains, thioredoxin, elephant, fox, bikes* and *cars* datasets obtained from the semi-random MITI ensembles are greater than the best ones in [7], which were generated under exactly the same experimental conditions. The only real-world dataset where accuracy is noticeably below the best result in [7] is *musk2.*

On the *elphant, fox, tiger* and *musk* datasets, we can also compare to the results in [11], which are for the so-called *MIForest* method (Table 1 in [11]). This method generates a random forest ensemble for multi-instance learning using optimisation based on deterministic annealing. The estimated accuracy for our semi-random MITI ensemble is greater for four of the five datasets, indicating that our method is indeed competitive.

5 Conclusions

In this paper we have (a) presented a comparison of multi-instance decision trees learned by different MITI configurations on a collection of benchmark datasets, (b) shown how a simple modification enables us to learn rule sets rather than trees—yielding the MIRI algorithm—and (c) considered the effect of randomisation for ensemble learning using both MITI and MIRI.

Our results provide evidence that the standard *Gini index* is an appropriate splitting criterion for MITI, in particular if an unbiased estimate is used for the proportion of positives ($k = 0$): trees are generally more accurate and compact than those learned using the special-purpose *max-bepp* criterion. We have also shown that MIRI generates even more compact classifiers than MITI while maintaining comparable accuracy. Finally, we obtained highly competitive classification accuracy by applying randomisation to generate MITI and MIRI-based ensembles.

References

1. Andrews, S., Tsochantaridis, I., Hofmann, T.: Support vector machines for multiple-instance learning. In: NIPS, pp. 561–568. MIT Press (2003)
2. Blockeel, H., Page, D., Srinivasan, A.: Multi-instance tree learning. In: ICML, pp. 57–64. ACM (2005)
3. Breiman, L.: Random forests. ML 45(1), 5–32 (2001)
4. Chen, Y., Bi, J., Wang, J.Z.: MILES: Multiple-instance learning via embedded instance selection. IEEE PAMI 28(12), 1931–1947 (2006)
5. Dietterich, T.G., Lathrop, R.H., Lozano-Perez, T.: Solving the multiple instance problem with axis-parallel rectangles. AI 89(1-2), 31–71 (1997)
6. Dietterich, T.G.: An experimental comparison of three methods for constructing ensembles of decision trees: Bagging, boosting, and randomization. ML 40(2), 139–157 (2000)
7. Foulds, J., Frank, E.: Revisiting multiple-instance learning via embedded instance selection. In: AUS-AI, pp. 300–310. Springer, Berlin (2008)
8. Foulds, J.R., Frank, E.: Speeding Up and Boosting Diverse Density Learning. In: Pfahringer, B., Holmes, G., Hoffmann, A. (eds.) DS 2010. LNCS, vol. 6332, pp. 102–116. Springer, Heidelberg (2010)
9. Frank, E., Witten, I.H.: Generating accurate rule sets without global optimization. In: ICML, pp. 144–151. Morgan Kaufmann (1998)
10. Hall, M., Frank, E., Holmes, G., Pfahringer, B., Reutemann, P., Witten, I.H.: The WEKA data mining software: an update. SIGKDD Explor. 11(1), 10–18 (2009)
11. Leistner, C., Saffari, A., Bischof, H.: MIForests: Multiple-Instance Learning with Randomized Trees. In: Daniilidis, K., Maragos, P., Paragios, N. (eds.) ECCV 2010. LNCS, vol. 6316, pp. 29–42. Springer, Heidelberg (2010)
12. Maron, O., Lozano-Pérez, T.: A framework for multiple-instance learning. In: NIPS, pp. 570–576. MIT Press (1998)
13. Maron, O., Ratan, A.L.: Multiple-instance learning for natural scene classification. In: ICML, pp. 341–349. Morgan Kaufmann (1998)
14. Mayo, M.: Effective classifiers for detecting objects. In: CIRAS (2007)
15. Michie, D., Muggleton, S., Page, D., Srinivasan, A.: To the international computing community: A new East-West challenge. Tech. rep., Oxford University (1994)
16. Nadeau, C., Bengio, Y.: Inference for the Generalization Error. ML 52(3), 239–281 (2003)
17. Opelt, A., Pinz, A., Fussenegger, M., Auer, P.: Generic object recognition with boosting. IEEE PAMI 28(3), 416–431 (2006)
18. Quinlan, J.R.: C4.5: Programs for Machine Learning. Morgan Kaufmann (1993)
19. Srinivasan, A., Muggleton, S., King, R., Sternberg, M.: Mutagenesis: ILP experiments in a non-determinate biological domain. In: ILP, pp. 217–232. GMD (1994)
20. Wang, C., Scott, S., Zhang, J., Tao, Q., Fomenko, D., Gladyshev, V.: A study in modeling low-conservation protein superfamilies. Tech. rep., Department of Comp. Sci., University of Nebraska-Lincoln (2004)
21. Zucker, J., Chevaleyre, Y.: Solving multiple-instance and multiple-part learning problems with decision trees and decision rules. Application to the mutagenesis problem. In: Proc. Conf. of the Canadian Society for Computational Studies of Intelligence, pp. 204–214 (2001)

Automatically Measuring the Quality of User Generated Content in Forums

Kevin Chai, Chen Wu, Vidyasagar Potdar, and Pedram Hayati

Digital Ecosystems and Business Institute, Curtin University
International Centre for Radio Astronomy Research,
University of Western Australia
{k.chai,p.hayati}@curtin.edu.au, chen.wu@uwa.edu.au,
vidysagar.potdar@cbs.curtin.edu.au

Abstract. The amount of user generated content on the Web is growing and identifying high quality content in a timely manner has become a problem. Many forums rely on its users to manually rate content quality but this often results in gathering insufficient rating. Automated quality assessment models have largely evaluated linguistic features but these techniques are less adaptive for the diverse writing styles and terminologies used by different forum communities. Therefore, we propose a novel model that evaluates content, usage, reputation, temporal and structural features of user generated content to address these limitations. We employed a rule learner, a fuzzy classifier and Support Vector Machines to validate our model on three operational forums. Our model outperformed the existing models in our experiments and we verified that our performance improvements were statistically significant.

Keywords: content quality assessment, user generated content, forums.

1 Introduction

Forums websites allows people to engage in online discussions. There are millions of forums on the Web and each forum can host large volumes of User Generated Content (UGC). However, forum users are being overwhelmed with excessive amounts of UGC and it is becoming more difficult to identify high quality content in a timely manner. Currently, many forums and Web 2.0 websites rely on its users to manually rate the quality of content to handle this problem [5]. However, there are a number of problems with relying solely on user ratings. Firstly, rating is voluntary so a large percentage of content often receives a lack of rating [22,19]. Secondly, users may not have sufficient knowledge and expertise to provide accurate ratings [17]. Lastly, reliance on manual user ratings becomes an ongoing problem if UGC is created at a faster speed than which it can be sufficiently rated [4]. Therefore, the objective of this paper is to propose a novel model that automatically measures the quality of UGC in forums. More specifically, the contributions of this paper are to:

D. Wang and M. Reynolds (Eds.): AI 2011, LNAI 7106, pp. 51–60, 2011.

- Present a model that evaluates content, usage, reputation, temporal and structural features for assessing forum post quality.
- Validate our model against three operational forums using supervised machine learning techniques and compare its performance against existing models in the literature.

2 Problem Definition

We formally define the problem of measuring the quality of forum posts as a multi-class classification problem. The forum dataset is described by a set of posts $P = \{p_1, p_2, ..., p_i, ..., p_{|P|}\}$ and a set of post quality classes $C = \{c_1 = low, c_2 = medium, c_3 = high\}$ where p_i is the i-th post in P. Furthermore, posts are represented as a set of content quality features $F = \{f_1, f_2, ..., f_j, ..., f_{|F|}\}$ in our model as defined for p_i in 1.

$$p_i = \{f_1^i, f_2^i, ..., f_j^i, ..., f_{|F|}^i\} \quad (1)$$

$\phi(p_i, c_k)$ is a Boolean function that is used to determine whether p_i belongs to c_k where $k = \{1, 2, 3\}$ as defined in 2.

$$\phi(p_i, c_k) : P \times C \rightarrow \{True, False\} \quad (2)$$

The task of performing automated post quality classification is to evaluate this function for all posts in a given forum dataset.

3 UGCQ Assessment Model

In recent work [4] we proposed a model that measures the quality of forum posts based upon its usage within a forum community. We extend this work by proposing a UGC Quality (UGCQ) model that evaluates content, usage, reputation, structural and temporal features for quality assessment.

Content features represent intrinsic information about the forum post such as features related to its textual content. Usage features represent the popularity of postings and usage data is obtained using the post usage tracking framework developed in our previous work [4]. Usage features evaluate view counts, dwell time as well as mouse and keyboard interactions between users and posts.

Reputation features evaluate the activeness, accountability and authority of post authors to gauge their overall reputation for quality assessment. Temporal features represent time-based characteristics of postings and evaluate the timeliness of when a forum post is created and edited. Structural features evaluate the position and visibility of postings within a forum thread.

As a result, we propose 46 post quality features based on these categories in the UGCQ model as presented in Table 1. An in-depth explanation of each feature and how it is measured is provided in [3].

Table 1. UGCQ Model Features

ID	Name
Content	
f_1	Word count
f_2	Unique word count
f_3	Ratio word count to average word count in thread
f_4	Quoted word count
f_5	Original word count
f_6	Ratio original word count to word count
f_7	Formatting tag count
f_8	Ratio formatting tag count to formatting tag count in thread
f_9	Hyperlink count
f_{10}	External hyperlink count
f_{11}	Internal hyperlink count
f_{12}	Ratio hyperlink count to hyperlink count in thread
f_{13}	Attachment count
f_{14}	Ratio attachment count to attachment count in the thread
f_{15}	Attachment download count
f_{16}	Ratio attachment download count to thread downloads
f_{17}	Post edit count
f_{18}	Post reported count
f_{19}	Is post created by thread author
Usage	
f_{20}	Post view count
f_{21}	Distinct user view count
f_{22}	Distinct users that revisit in different sessions count
f_{23}	Total dwell time
f_{24}	Average dwell time
f_{25}	Text selection count
f_{26}	Total number of characters selected
f_{27}	Average number of characters selected
f_{28}	Text copy count
f_{29}	Total number of characters copied
f_{30}	Average number of characters copied
Reputation	
f_{31}	First name, last name and location provided
f_{32}	E-mail displayed to public
f_{33}	Website URL provided
f_{34}	Membership group (member, moderator, administrator)
f_{35}	Number of posts created by user
f_{36}	Membership age
Temporal	
f_{37}	Age
f_{38}	Post edit time difference
f_{39}	Previous post time difference
f_{40}	Previous post time difference to thread average difference
f_{41}	Following post time difference
Structural	
f_{42}	Is first post
f_{43}	Is displayed on first thread page
f_{44}	Is last post
f_{45}	Is displayed on last thread page
f_{46}	Thread position to thread post count

4 Experiment

4.1 Datasets

We obtain three forum datasets for evaluating the performance of the UGCQ model. Firstly, data from `http://remnantsguild.com/` was collected from the July 21, 2009 to October 16 2009. Secondly, data from `http://nabble.com/` [22] and `http://slashdot.org/` [21] are obtained for experimentation. The Nabble dataset contains data from April 1, 2002 to July 24, 2006. The Slashdot dataset contains posts created from September 10, 2007 to September 24, 2007. Details of the datasets are displayed in Table 2.

Table 2. Forum Datasets

	Remnantsguild	Nabble	Slashdot
Users	54	1,832	3,893
Topics	114	2,956	191
Rated posts	531	4,291	7,847
Low quality posts	288 (54%)	2,037 (48%)	4,026 (51%)
Medium quality posts	166 (31%)	515 (12%)	2,693 (34%)
High quality posts	77 (15%)	1,739 (40%)	1,128 (15%)

Hsu *et al.* (2003) [12] showed that they could improve the performance of their Support Vector Machines (SVM) classifier by performing data normalisation. Therefore, we adopt a min-max data normalisation approach to scale feature values to a range of [0, 1] to avoid features in larger numeric ranges from dominating those in smaller ranges. Additionally, classifier performance can be improved when continuous features are discretised into ranked intervals [8]. Therefore, we use the Fayyad & Iranis Minimum Description Length method [9] for data discretisation. The datasets are split into complementary training and test sets using 10 fold cross-validation in our experiments.

4.2 Feature Selection

A number of features in the UGCQ model could not be evaluated for the Nabble and Slashdot datasets due to missing data. For example, usage data was not collected from Nabble and Slashdot because the datasets were provided to us. We had collected usage data from the Remnantsguild forum with our post usage tracking framework we proposed in [4]. As a result, the set of features evaluated for each dataset is (refer to Table 1 for feature names):

- **Remnantsguild:** 46 features $\{f_1\text{-}f_{46}\}$
- **Nabble:** 24 features $\{f_1\text{-}f_{12}, f_{19}, f_{35}, f_{37}\text{-}f_{46}\}$ with 22 features missing
- **Slashdot:** 23 features $\{f_1\text{-}f_{12}, f_{19}, f_{35}, f_{37}, f_{39}\text{-}f_{46}\}$ with 23 features missing

We perform feature selection using a sequential forward selection approach. The purpose of conducting feature selection is to identify the set of most important and relevant features for classifying the quality of forum posts. Waikato Environment for Knowledge Analysis (WEKA) [11] is a data mining tool that we use to perform feature selection and classification in our experiments. The selected feature sets generated for each forum dataset are:

- **Remnantsguild:** 8 features $\{f_1\text{-}f_3, f_5, f_8, f_9, f_{24}, f_{38}\}$
- **Nabble:** 4 features $\{f_5, f_{12}, f_{35}, f_{37}\}$
- **Slashdot:** 4 features $\{f_{35}, f_{39}, f_{45}, f_{46}\}$.

4.3 Performance Evaluation

We use the classification accuracy and Matthews Correlation Coefficient (MCC) to evaluate the performance of our model and existing models in the literature. MCC is considered one of the best for evaluating classifier performance on the imbalanced data [2] as in our experiment (See Table 2). This metric provides a correlation value between -1 to 1 where -1 represents perfect inverse prediction, 0 represents random prediction and 1 represents perfect prediction.

A MCC value is calculated from a classifiers confusion matrix for each quality class (i.e. low, medium and high). The MCC performance measure is defined in 3 where TP = true positives, TN = true negatives, FP = false positives and FN = false negatives.

$$MCC = \frac{TP \times TN - FP \times FN}{\sqrt{(TP+FP)(TP+FN)(TN+FP)(TN+FN)}} \tag{3}$$

4.4 Post Quality Classification

We developed the UGCQ model into a working prototype and implemented the models proposed by Weimer & Gurevych (2007) [22] and Wanas *et al.* (2008) [21]. We classify the quality of forum posts from each dataset using WEKA [11]. More specifically, we use WEKA's implementation of the Sequential Minimal Optimisation (SMO) algorithm [18] for SVM, the rule based learner JRIP which is based on Repeated Incremental Pruning to Produce Error Reduction (RIPPER) [6] as well as the fuzzy rule learner FURIA [13].

We perform a number of classification experiments so we introduce a naming scheme to label each experiment in the form of [model_name]_[classifier]_[encoding]_[selection] and the values of each of these fields is displayed in Table 3. For example, the experiment with the UGCQ framework, FURIA algorithm, normalisation and feature selection is labelled as UGCQ_FURIA_N_FS.

Table 3. Forum Post Classification Experiment Labels

Field	Values	Comments
model_name	UGCQ, Weimer, Wanas, Baseline	Baseline refers to the majority class classifier
classifier	JRIP, FURIA, SVM	Repeated Incremental Pruning to Produce Error Reduction (RIPPER) variant algorithm, Fuzzy Unordered Rule Induction Algorithm, Support Vector Machines
encoding	N, D	Normalisation, discrestiation
selection	FS	Feature selection

5 Results

The experimental results obtained from the Remnantsguild, Nabble and Slashdot datasets are presented in Table 4. The UGCQ model using the JRIP on the normalised Remnantsguild dataset achieved the best results with 68.55% accuracy and an average MCC value of 0.45. The Weimer model also achieved 68.55% accuracy but with a lower average MCC of 0.43 while the Wanas model achieved 63.84% with an average MCC value of 0.30. All CQA models outperformed the majority class baseline of 54.24% for Remnantsguild.

The UGCQ model using SVM on the discretised Nabble dataset achieved the best results with 69.98% accuracy and an average MCC value of 0.40. Additionally, the Weimer model achieved 65.07% accuracy with an average MCC of 0.28 while the Wanas model achieved 58.24% with an average MCC value of 0.19. All CQA models outperformed the majority class baseline of 47.47% for Nabble.

The UGCQ model using SVM on the normalised Slashdot dataset achieved the best accuracy of 53.94% accuracy but with an average MCC value of 0.12. The FURIA algorithm on the discretised dataset however achieved the highest average MCC value of 0.15 but with a lower accuracy of 51.20%. Additionally, the Weimer model achieved 51.29% accuracy with an average MCC of 0 while the Wanas model achieved 51.31% with an average MCC value of 0.01. The UGCQ model using SVM on the normalised dataset slightly outperformed the majority class baseline of 51.31% while the Weimer model under performed and the Wanas model achieved equivalent performance to the baseline for Slashdot.

5.1 Friedman Test

Demšar (2006) [7] surveyed papers published from the International Conference of Machine Learning in 1999 to 2003 and discovered that the majority of authors did not statistically verify whether their classifier(s) produced significant performance improvements.Therefore, a number of suitable statistical tests were

Table 4. Ranking Comparison of Classifiers over all Datasets

Classifier	Remnantsguild	Nabble	Slashdot	$Rank_{avg}$
UGCQ_SVM_D	66.29% (4)	69.98% (1)	52.66% (2.5)	2.5
UGCQ_JRIP_N	68.55% (1.5)	68.49% (3)	52.31% (4.5)	3
UGCQ_SVM_D_FS	65.16% (7)	68.10% (4)	52.66% (2.5)	4.5
UGCQ_JRIP_D	66.85% (3)	66.98% (7)	52.31% (4.5)	4.83
UGCQ_FURIA_N	65.35% (6)	67.78% (5)	52.03% (6)	5.67
UGCQ_JRIP_N_FS	63.47% (10)	68.93% (2)	51.56% (7)	6.33
UGCQ_SVM_N	63.65% (9)	64.18% (12)	53.94% (1)	7.33
Weimer	68.55% (1.5)	65.07% (11)	51.29% (13)	8.5
UGCQ_FURIA_D	65.72% (5)	66.25% (8)	51.20% (14)	9
UGCQ_JRIP_D_FS	61.39% (13)	66.16% (9)	51.31% (8)	10
UGCQ_FURIA _D_FS	62.90% (11)	66.05% (10)	51.31% (10.5)	10.5
Wanas	63.84% (8)	58.24% (14)	51.31% (10.5)	10.83
UGCQ_FURIA_N_FS	62.71% (12)	67.28% (6)	50.69% (15)	11
UGCQ_SVM_N_FS	59.89% (14)	63.34% (13)	51.31% (10.5)	12.5
Baseline	54.24% (15)	47.47% (15)	51.31% (10.5)	13.50

recommended based on the characteristics of a given experiment. We follow this recommendation by performing the Friedman test [10] for verifying if there is a significant statistical difference between the performance of multiple classifiers over multiple datasets.

Firstly, we rank classifiers within each dataset in terms of their classification accuracy. We use accuracy rather than the MCC average to include the baseline classifier for evaluation. The average rank for each classifier over all the datasets is presented in Table 4 in decreasing order of rank. Secondly, we evaluate the null hypothesis H_0 and alternate hypothesis H_a to determine if the average ranks of these classifiers over all datasets are significantly different:

- H_0: There is no difference in the average ranks for classifiers over the datasets.
- H_a: A difference exists in the average ranks for classifiers over the datasets.

We use statistical analysis tool, R [20] and conducted the Friedman test [10] to obtain a chi-squared χ^2 value of 24.84 with 14 degrees of freedom *df* and a p-value of 0.03618. The critical value of α based on the χ^2 value and *df* for the χ^2 distribution is 0.05. Therefore, we reject H_0 and accept H_a because 0.03618 (p-value) < 0.05 (α).

5.2 Nemenyi Test

We discovered from the Friedman test that some classifiers are significantly different to others but we do not know which specific classifiers are different. Therefore, we can use the Nemenyi test [16] to evaluate all pairs of classifiers ($\sum_{i=1}^{k-1} i$ permutations) to determine which classifiers are significantly different to each other. The critical distance q_α for the two-tailed Nemenyi test with $\alpha = 0.05$ (significance level) and $k = 15$ (number of classifiers) is 3.391.

We first calculate the distance between the average ranks between all pairs of classifiers. The distance between the average ranks of two classifiers must be ≥ 3.391 to be considered as significant with 95% probability. 56 out of 105 significant differences were identified from the pair-wise comparisons between the classifiers.

We compare our top UGCQ classifier (UGCQ_SVM_D) along with the existing models in the literature as shown in Table 5. The number shown in parenthesis depicts the rank of the classifier over all datasets identified from Table 4. These results show that the performance of the UGCQ classifier is significantly different from these models while the differences between the Weimer and Wanas classifier and, the Wanas and Baseline classifier are not significant.

Table 5. Comparisons between UGCQ and Existing CQA Models

Classifier A	Classifier B	Difference	Sig. (diff $\geq$ 3.391)
UGCQ_SVM_D (1)	Weimer (8)	6.00	Yes
UGCQ_SVM_D (1)	Wanas (12)	8.30	Yes
UGCQ_SVM_D (1)	Baseline (15)	11.00	Yes
Weimer (8)	Wanas (12)	2.33	No
Weimer (8)	Baseline (15)	5.00	Yes
Wanas (12)	Baseline (15)	2.67	No

6 Discussion

Seven out of twelve UGCQ classifiers outperformed the CQA models proposed by [22], [21] over the three datasets as shown in Table 4. Additionally, we statistically verified our highest ranking UGCQ classifier (UGCQ_SVM_D) significantly outperformed these models as highlighted in Table 5.

A large number of UGCQ features were not evaluated for the Nabble and Slashdot datasets due to missing data. For example, the average dwell time was identified as an important quality feature on the Remnantsguild dataset but could not be evaluated for the other datasets. The inclusion of our missing features could further improve the performance of the UGCQ classifiers.

We calculate the average MCC low, medium and high values excluding the baseline classifier for each dataset. The results indicate that CQA models performed better in classifying low and high quality posts than medium quality. This supports our intuition of how classifiers could misclassify low and high quality posts that neighbour closely with the medium quality class and vice versa.

7 Related Work

Chai *et al.* (2009) [5] conducted a comprehensive review of 19 content quality related assessment frameworks for forums, question & answering (Q&A) websites, blogs and wikis. Additionally, Zhu *et al.* 2009 [23] proposed and validated a multidimensional framework for assessing the quality of answers in Q&A websites.

Weimer & Gurevych (2007) [22] was first to propose a model for measuring the quality of forum posts and classified posts into two quality classes (high and low) by assessing surface, lexical, syntactic, similarity and forum specific post features. This work was extended by Wanas *et al.* (2008) [21] by classifying posts into 3 quality classes (low, medium and high) and evaluated features such as relevance, originality, post component, surface and forum-specific features. Lui & Baldwin (2009) [15] evaluated bag-of-words features and features proposed by [21] on the dataset collected by [22] for classifying good and bad posts.

Agichtein *et al.* (2008) [1] evaluated usage statistics of questions and answers in Yahoo! Answers to find high quality content. Additionally, the number of times an answer was copied by users was proposed as a feature by Jeon *et al.* (2006) [14] for measuring the quality of answers in Naver! (Korean Q&A website). Our previous work, Chai *et al.* (2010) [4] extended these ideas to track how users interact with forum posts to predict its quality.

We gained a number of insights from these related studies to propose our UGCQ model that measures the content, usage, reputation, temporal and structural features of UGC for quality assessment. We provide a detailed review of the related work in the area of content quality assessment in Chai (2011) [3].

8 Conclusion

We have proposed the UGCQ model that evaluates the content, usage, reptuation, temporal and structural features of forum UGC for quality assessment. We implemented our model into a prototype and validated its performance on the Remnantsguild, Nabble and Slashdot forums. Additionally, we implemented two existing models in the literature for performance comparison with the UGCQ model. We discovered that our model outperformed the existing models in the literature over all forum datasets and the performance increase was statistically significantly.

References

1. Agichtein, E., Castillo, C., Donato, D., Gionis, A., Mishne, G.: Finding High-Quality content in social media. In: Proceedings of the International Conference on Web Search and Web Data Mining (WSDM), pp. 183–194 (2008)
2. Baldi, P., Brunak, S., Chauvin, Y., Andersen, C.A.F., Nielsen, H.: Assessing the accuracy of prediction algorithms for classification: An overview. Bioinformatics 16(5), 412–424 (2000)
3. Chai, K.: A Machine Learning-based Approach for Automated Quality Assessment of User Generated Content in Web Forums. Ph.D. thesis, Curtin University (2011)
4. Chai, K., Hayati, P., Potdar, V., Wu, C., Talevski, A.: Assessing post usage for measuring the quality of forum posts. In: Proceedings of the 4th IEEE International Conference on Digital Ecosystems and Technologies, DEST (2010)

5. Chai, K., Potdar, V., Dillon, T.: Content Quality Assessment Related Frameworks for Social Media. In: Gervasi, O., Taniar, D., Murgante, B., Laganà, A., Mun, Y., Gavrilova, M.L. (eds.) ICCSA 2009. LNCS, vol. 5593, pp. 800–814. Springer, Heidelberg (2009)
6. Cohen, W.W.: Fast effective rule induction. In: Proceedings of the 12th International Conference on Machine Learning, p. 115 (1995)
7. Demšar, J.: Statistical comparisons of classifiers over multiple data sets. The Journal of Machine Learning Research 7, 1–30 (2006)
8. Dougherty, J., Kohavi, R., Sahami, M.: Supervised and unsupervised discretization of continuous features. In: Proceedings of the 12th International Conference on Machine Learning, pp. 194–202 (1995)
9. Fayyad, U., Irani, K.: Multi-interval discretization of continuous-valued attributes for classification learning. In: Proceedings of the International Joint Conference on Uncertainty in Artifical Intelligence, pp. 1022–1027 (1993)
10. Friedman, M.: The use of ranks to avoid the assumption of normality implicit in the analysis of variance. Journal of the American Statistical Association 32(200), 675–701 (1937)
11. Hall, M., Frank, E., Holmes, G., Pfahringer, B., Reutemann, P., Witten, I.H.: The WEKA data mining software: An update. Special Interest Group on Knowledge Discovery and Data Mining (SIGKDD) Explorations 11(1) (2009)
12. Hsu, C.W., Chang, C.C., Lin, C.J.: A practical guide to support vector classification. Tech. rep., National Taiwan University (2003), `http://www.csie.ntu.edu.tw/cjlin/papers/guide/guide.pdf`
13. Hühn, J., Hüllermeier, E.: FURIA: an algorithm for unordered fuzzy rule induction. Data Mining and Knowledge Discovery 19(3), 293–319 (2009)
14. Jeon, J., Croft, W.B., Lee, J.H., Park, S.: A framework to predict the quality of answers with Non-Textual features. In: Proceedings of the 29th Annual International ACM SIGIR Conference on Research and Development in Information Retrieval, pp. 228–235 (2006)
15. Lui, M., Baldwin, T.: You are what you post: User-level features in threaded discourse. In: Proceedings of the Fourteenth Australasian Document Computing Symposium (ADCS 2009), pp. 98–105 (2009)
16. Nemenyi, P.: Distribution-free multiple comparisons. Ph.D. thesis, Princeton University (1963)
17. Nussbaum, M.E., Hartley, K., Sinatra, G.M., Reynolds, R.E., Bendixe, L.D.: Enhancing the quality of On-Line discussions. In: Paper Presented at the Annual Meeting of the American Educational Research Association (2002)
18. Platt, J.: Sequential minimal optimization: A fast algorithm for training support vector machines. Advances in Kernel Methods Support Vector Learning 208(MSR-TR-98-14), 1–21 (1998)
19. Suryanto, M., Lim, E.P., Sun, A., Chiang, R.: Quality-Aware collaborative question answering: Methods and evaluation. In: Proceedings of the Second ACM International Conference on Web Search and Data Mining, pp. 142–151 (2009)
20. Team, R.D.C.: R: A Language and Environment for Statistical Computing. Vienna, Austria (2011), `http://www.R-project.org`
21. Wanas, N., El-Saban, M., Ashour, H., Ammar, W.: Automatic scoring of online discussion posts. In: Proceeding of the 2nd ACM Workshop on Information Credibility on the Web, pp. 19–26 (2008)
22. Weimer, M., Gurevych, I.: Predicting the perceived quality of web forum posts. In: Proceedings of the Conference on Recent Advances in Natural Language Processing (2007)
23. Zhu, Z., Bernhard, D., Gurevych, I.: A Multi-Dimensional model for assessing the quality of answers in social Q&A sites. Tech. rep., Ubiquitous Knowledge Processing Lab (2009)

Genetically Enhanced Feature Selection of Discriminative Planetary Crater Image Features

Joseph Paul Cohen, Siyi Liu, and Wei Ding

Department of Computer Science
The University of Massachusetts Boston
100 Morrissey Blvd. - Boston, MA 02125-3393
{joecohen,silu,ding}@cs.umb.edu

Abstract. Using gray-scale texture features has recently become a new trend in supervised machine learning crater detection algorithms. To provide better classification of craters in planetary images, feature subset selection is used to reduce irrelevant and redundant features. Feature selection is known to be NP-hard. To provide an efficient suboptimal solution, three genetic algorithms are proposed to use greedy selection, weighted random selection, and simulated annealing to distinguish discriminate features from indiscriminate features. A significant increase in the classification ability of a Bayesian classifier in crater detection using image texture features.

Keywords: machine learning, genetic algorithms, crater detection, bayesian classifier.

1 Introduction

Impact craters are structures formed by collisions of meteoroids with the planetary surface. The importance of impact craters stems from the wealth of information that detailed analysis of their distributions and morphology can bring forth. Crater counting is the only technique for establishing relative chronology of different planetary surfaces. However, crater detection from planetary images is a difficult problem because of the complex geological surface structure of remote planets. If an acceptable solution is found it will enable many studies including determining the geologically active regions of a planet, relatively dating sections of a planet, and determining both landing and exploration sites for interplanetary robots. The challenge currently is to achieve an acceptable level of accuracy as required by planetary domain scientists [11] [5] [10] [6] [12].

The state of the art method of crater detection involves utilizing the texture and contrast of the crater image [3]. This is achieved by extracting numerical features from an image, each representing a particular texture or contrast, and then applying machine learning to decide if potential crater images are in fact craters. Haar features, a gray-scale image texture features, are especially useful because of their ability to be calculated efficiently using a data structure called integral images [13].

D. Wang and M. Reynolds (Eds.): AI 2011, LNAI 7106, pp. 61–71, 2011.

The challenge in using Haar features is that the number of Haar features can easily be tens of thousands. Many Haar features are redundant or even irrelevant. The curse of dimensionality is inevitable if we do not select subsets of features that are useful to build a good classifier. All features generated from the image can be broken down into discriminate features and indiscriminate features. Discriminate features contain information that is useful during classification. Indiscriminate features provide no information to the classifier or misguiding information.

The goal of feature selection is to select the optimal subset of features for some classifier. Feature subset selection is known to be NP-hard. Exhaustive search is the only way to find the optimal subset of a set of features. To find, for certain, the optimal solution all permutations must be considered. The search space is 2^f where f is the number of features. For an example with only 58 features it would take 91,336,645.5 years to compute all classifiers if a classifier took 0.10 seconds to create and evaluate.

Three algorithms are presented in this paper, using approaches of highest fitness selection, weighted random selection, and simulated annealing, to select discriminate features that a classifier will use to classify images. These algorithms are given a set of features and return a suboptimal subset. The algorithms presented in this paper aim to reduce the search space automatically. They will automatically create a relevant subset of features utilizing a wrapped classifier fitness function. A significant increase in the classification ability of a Bayesian classifier in crater detection using image texture features.

2 Related Work

There have been many methods used to automatically detect craters. R. Honda [4] built a SOM using Hough transforms to extract geometric features. They then perform best parameter selection to reduce duplicate detections using a genetic algorithm based on the center location and radius of the detected crater. In this work we use genetic algorithms to choose the best features to build a classifier with which is different from R. Honda's method.

This paper uses standard classification methods to determine the probability that an image is in fact a crater as inspired by W. Ding [3] and L. Banderia [1]. Y. Cheng [2]. They used the concept of a confidence evaluation to detect craters and J. Kim [5] used a fitness check to determine if the candidate was a crater or not.

This work utilizes Haar features to perform candidate image classification. Haar features were used by W. Ding [3] L. Banderia [1], and S. Liu [7] and are the state of the art in crater detection because of their adaptive and discriminative ability. They are used in this work as crater features.

3 Genetically Enhanced Feature Selection

This section presents three genetic algorithms used for feature subset selection. Each algorithm builds upon the one before it in an attempt to achieve better results. The first algorithm is explained in detail and then only modifications are explained for the next two algorithms. First the main concepts of genetic algorithms, genetic representation, and fitness are discussed. For each algorithm there is an initial plan, explanation of steps, and a complexity analysis.

The three proposed methods of genetically enhanced feature selection are shown in Algorithms 3, 4, and 5. These vary in the way feature subsets are chosen to be crossed over. The goal is to pick the best feature subsets so that when they are combined will generate a feature subset with a higher fitness score than either of the original. The first algorithm attempts to choose the best two feature subsets and use them as parents while the later algorithms attempt to introduce controlled randomness. Controlled randomness is introduced by randomly selecting from feature subsets that are weighted based on their fitness score. Later simulated annealing is used to introduce more randomness at the beginning of the algorithm.

3.1 Genetic Representation

In these algorithms the genetic representation is a subset of features that are used in building a classifier. This is referred to in this work as a feature subset but is also called an individual or chromosome. The representation is treated as a subset and as a vector. This is achieved using the concept of a bitvector to set features as on or off. Each index of the bitvector represents one feature. The contents of a subset are the on features represented by the bitvector. The magnitude of the subset is the sum of all possible features.

3.2 Wrapped Classifier Fitness Function

The fitness function used in the following algorithms is modeled as an evaluation function in the F1 search space. It can be said that the fitness function is wrapped around a classifier. The F1 fitness metric, $fitness = F1 = \frac{2}{\frac{1}{recall}+\frac{1}{Precision}}$, takes into account two important attributes of a classifier, precision and recall. This allows us to compare two classifiers using one value. Using the F1 measure also allows the priority queue datastructure to be used. The calculation for precision, $Precision = \frac{truepositives}{truepositives+falsepositives}$, takes into account how generously the classifier predicted something was a positive example. This metric fails to describe the situation when the classifier has ignored many positive examples. Recall, $Recall = \frac{truepositives}{truepositives+falsenegatives}$, fixes this problem by describing the ability of the classifier to predict all the positive examples correctly. It can be thought of as describing the coverage of classifier. Recall by itself cannot describe the classifier because it does not take into account the case where the classifier marked everything as a positive instance.

3.3 Random Crossover

Instead of splitting the feature subset somewhere in the middle so the order is preserved at a loss for feature equality [9], we use random crossover to ensure

Algorithm 1. Preform Random Crossover $f1 \bigotimes f2$

Input: Feature Subset Vector $f1$
Feature Subset Vector $f2$
Output: Feature Subset Vector $f3$
for $|f1|$ **do**
  if $0.5 < Random(0, 1)$ **then**
    $f3_i = f1_i$
  else
    $f3_i = f2_i$

Algorithm 2. Preform Random Mutation $M(f1, \delta)$

Input: Feature subset $f1$
Percentage to mutate δ
Output: Feature subset vector $f2$
for δ *of* $|f3|$ **do**
  $r = Random(0, |f3|)$ // random index
  $f3_r = \neg f3_r$

that element of the feature subset has an equal chance of being preserved. This change is due to the absence of order that is involved with features.

The random crossover used, as shown in Algorithm 1, is the process of merging two parent subsets to make a new child subset. This resulting child subset is composed of the parent subsets. This process is used to simulate the mixing of chromosomes during natures genetic process. If a feature is enabled in both parents then it will be enabled in the child. If it varies in the parents then there is a 50% chance it will be preserved in the child. If a feature is turned off in both parents it will be turned off in the child. If this method was used exclusively in the algorithms then the children would converge to a local or global maximum fitness. This convergence is not desired because we want to avoid local maximums so the feature subsets are also mutated.

3.4 Mutation

Mutation is used to avoid the convergence of algorithms at a local maximum. As shown in Algorithm 2; mutation involves randomly flipping a percentage of bits in the feature subset to enable or disable features. Mutation as described here takes a percentage as an argument and mutates that percentage of the feature subset. This is used to simulate natures genetic mutations.

3.5 Highest Fitness (Greedy) Selection

Also called GHF (Genetic Highest Fitness), the initial plan for this algorithm was to greedily select the best two feature subsets. This would then concentrate the randomness to the crossover and mutation phases. The expectation is that combining good feature subsets will produce better feature subsets.

Algorithm 3. Highest Fitness (Greedy) Selection (GHF)

Input: Features Γ
Iterations I
Inital random subsets s
Percentage to mutate m
Maximum size of Υ $size$
Output: Feature subset Γ'

1 Add a full instance of Γ to Υ
2 Add s random subsets of Γ to Υ
3 **for** *each i in I* **do**
4 $\quad \{f1 | f \in \Upsilon, fitness(f1) \geq fitness(f)\}$
5 $\quad \{f2 | f \in (\Upsilon \cap \neg\{f1\}), fitness(f2) \geq fitness(f)\}$
6 $\quad f3 = f1 \bigotimes f2$ // crossover subset
7 $\quad f3' = M(f3, m)$ // mutate subset m percent
8 $\quad \{\Upsilon | f \in \Upsilon \cap \{f3'\}\}$
9 $\{\Gamma' | \Gamma' \in \Upsilon, f \in \Upsilon, fitness(\Gamma') \geq fitness(f)\}$

The GHF steps are explained now. Step 1: seed the algorithm with a subset that contains all features. Step 2: seed the algorithm with an initial set of feature subsets. Step 3: we loop some number of times to simulate many generations of evolution. Step 4: we select a parent from the set of feature subsets that has the best fitness score. Step 5: remove the already selected feature subset and then select the feature subset with the highest fitness score. Step 6: randomly crossover f1 and f2 to create f3. Step 7: mutate m percent of this new feature subset. Step 8: define upsilon to contain the mutated f3. Step 9: define Gamma prime to be a feature subset that has the highest fitness score in upsilon.

3.6 Weighted Random Selection

Also called GWR (Genetic Weighted Random), the initial plan for this algorithm was to increase the chance that a feature subset with better complementing features would be chosen. This is implemented using a weighting method. The feature subsets are selected at random but weighted based on their fitness values. To explain this sample data is shown in Figure 2. The sample data points are the values 0.95, 0.90, 0.80 and 0.70. To the right of the image those values have been scaled to 0.284, 0.269, 0.239, and 0.209. This allows a random number between 0 and 1 to select a feature. Figure 1 shows the result of the highest fitness score removed. The fitness values are normalized again to handle this change so that a new random number between 0 and 1 will select a feature subset.

Algorithm 4 has it's feature subset selection method modified. Steps 3 and 4 are changed now to call a GetWeightedSubset method. In GetWeightedSubset Step 10: defines a sub routine. Step 11: defines that f_n is a element of upsilon. Step 12: produces a normalized version of the feature subset and a fitness value to be used below only. Step 13: generates an r between 0 and 1 that we will subtract from to pick a feature subset. Step 14: loops for each feature subset.

Algorithm 4. Weighted Random Selection

Input: Features Γ
Iterations I
Inital random subsets s
Percentage to mutate m
Maximum size of Υ $size$
Output: Feature subset Γ'

1 Add a full instance of Γ to Υ
2 Add s random subsets of Γ to Υ
3 **for** *each i in I* **do**
4     $f1 = GetWeightedSubset(\Upsilon)$
5     $f2 = GetWeightedSubset(\Upsilon \cap \neg\{f1\})$
6     $f3 = f1 \bigotimes f2$ // crossover subset
7     $f3' = M(f3, m)$ // mutate subset m percent
8     $\{\Upsilon | f \in \Upsilon \cap \{f3'\}\}$
9 $\{\Gamma' | \Gamma' \in \Upsilon, f \in \Upsilon, fitness(\Gamma') \geq fitness(f)\}$

10 $GetWeightedSubset$:
Input: Current set of feature subsets Υ
Output: Feature subset $f2$
11 $fn \in \Upsilon$
12 $\{\hat{fn} | \sum_n fn = 1\}$
13 $r = Random(0, 1)$
14 **foreach** $\hat{fn}$ **do**
15     $r = r - \hat{fn}$
16   **if** $r < 0$ **then**
17         $returnf$

Step 15: subtracts the current feature subsets normalized fitness value from r. Step 16+17: if r has gone below 0 then we select that feature subset.

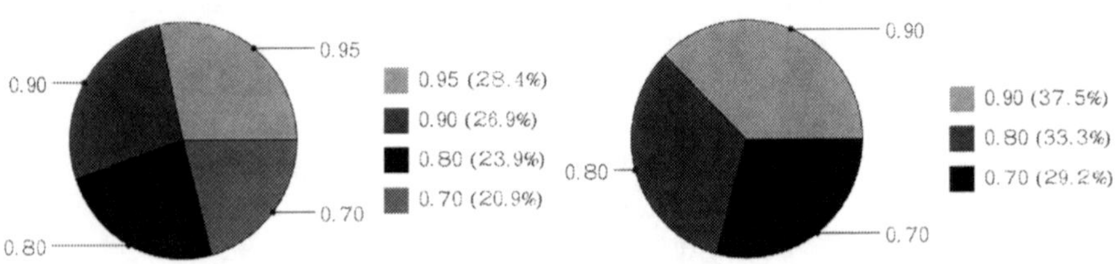

Fig. 1. Left: Sample values weighted based on fitness score, Right: Sample values weighted based on fitness score after one feature subset is removed

3.7 Weighted Random Selection with Simulated Annealing

Also called GWRSA (Genetic Weighted Random w/ Simulated Annealing), the idea for this algorithm is to initially weight all feature subsets the same during feature subset selection and then gradually weight them by their fitness score as the number of iterations increases.

As shown in Algorithm 5 the GetWeightedSubset method has been modified to take the current iteration as an argument. Steps 1 and 2 use this iteration value to drive the weighting to be close to equal at the beginning and properly weighted when iterations are at the end.

Algorithm 5. GetWeightedSubset w/ Simulated Annealing

Input: Current set of feature subsets Υ
Input: Current iteration i
Output: Feature subset $f2$

1 $annealing = I - i$
2 $\{\hat{fn} | \sum_n fn + annealing = 1\}$
3 $r = Random(0, 1)$
4 **foreach** $\hat{fn}$ **do**
5     $r = r - \hat{fn}$
6   **if** $r < 0$ **then**
7         $returnf$

3.8 Complexity

GHF is the most efficient algorithm out of the three. GWR and GWRSA have an added penalty due to the way they use randomness to avoid local maximums. The complexity of GHF is $O(ic\Gamma)$ where i is the number of iterations, c is the complexity of the classifier used to calculate the fitness score, and Γ is the number of features used. In GWR the change from selecting the feature subset with the highest fitness score to weighting and selecting causes the algorithm to increase in complexity. All the fitness values must now be added to create a normalization term. This increases the complexity to $O(ic\Gamma^2)$. A limit on the number of feature subsets in Γ would reduce the complexity but that method is not used in this algorithm. The complexity of GWRSA does not increase the complexity of GWR because the only change is an addition during the computation of the normalization term.

4 Experimental Results

This section analyses the advantages of using these feature selection algorithms. This is done by creating a super set of features that contain discriminate and

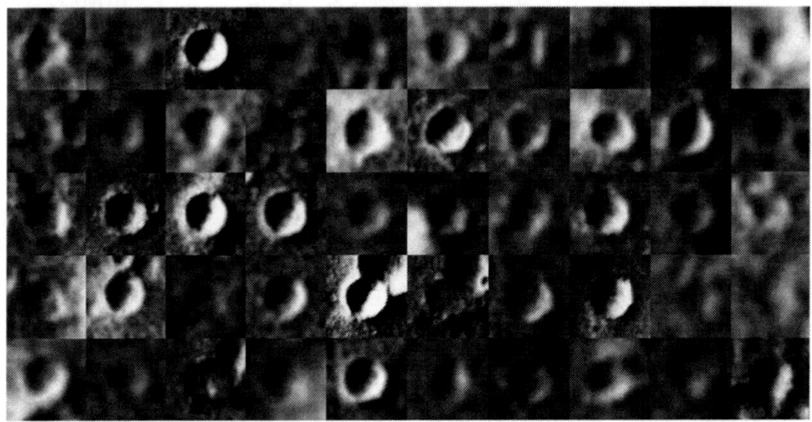

Fig. 2. Images used for positive examples

indiscriminate features. The goal of the feature selection algorithm will be to remove the indiscriminate features and return a feature subset with only discriminate ones.

A challenge of experimenting with these algorithms is finding their optimal parameters. This section will also analyze the mutation rate, number of iterations, maximum number of feature subsets accumulated, and the number of feature subsets generated for the initial pool. This section will also compare this algorithm to a variety of other algorithms applied to the same dataset.

In these experiments the **training set** consists of 166 positive examples of craters and 343 negative examples of ground without cratering are used from the HRSC h0905_0000 nadir panchromatic image. This training set used is selected to simulate crater detection and was inspired by the ones used by W. Ding [3] and L. Banderia [1]. Positive examples contain craters centered and cropped as shown in Figure 2. This training set provides the ability to analyze the algorithms without dealing with the size and complexity of applied crater detection.

To evaluate the proposed algorithms Haar **features** are used in combination with a Bayesian classifier. Haar features have a proven ability to detect craters [3]. A Bayesian classifier is used because of it's naive use of all features. Haar features were first proposed by Papageorgiou [8], then applied to face detection by Viola and Jones [13], and then applied to crater detection by W. Ding [3]. Haar features are described using feature masks that specify white and black regions. The masks are overlaid on the crater image and the sum of each region's pixel values are calculated and then the difference is taken. In Figure 3 the masks above A are basic Haar feature masks. The masks above B and C are horizontally and vertically scaled to capture contrast and texture that will not fit into a square. The features extracted depend on the image format for precision and size. Haar features can be optimized for speed using a technique discussed by Viola and Jones [13] that allows for O(1) calculations of Haar features from an image that has had a corresponding Integral Image computed.

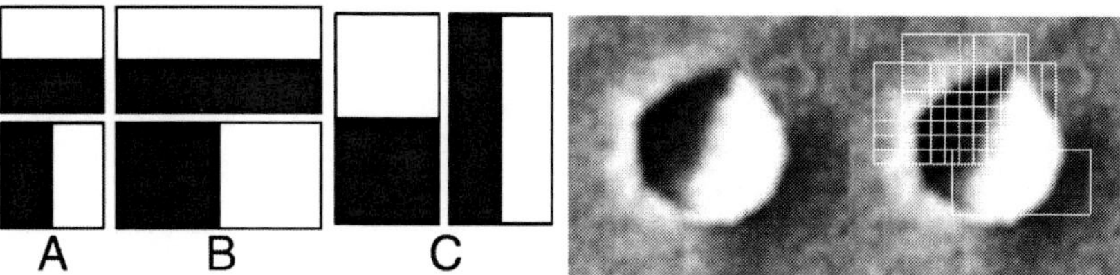

Fig. 3. Left: Haar Feature Masks Used, Right: Coverage of Haar Features on Crater Image

In the following experiments 58 Haar features are used. Figure 3 shows the outlines of these features to specify the coverage area. They were chosen to provide a challenge to the classifier while still providing discriminating features.

The **initial pool size** is the set of feature subsets that are given to the algorithm to start the process. There needs to be two or more feature subsets to start. Values from 10 to 700 are used over 1000 iterations at 5% mutation to determine the optimal value. There does not seem to be any advantage to varying this parameter.

The **maximum feature subsets accumulated** variable is the limit of feature subsets that will be maintained in memory during the program execution. This is only used for the Weighted Random selection and the Weighted Random Simulated annealing feature selections. Values are sampled from 3 to 1000 during 10000 iterations at 5% mutation. The scale starts at 3 because otherwise it is the Highest Fitness Score feature selection. The Highest Fitness Score feature selection method keeps only 2 feature subsets in memory so there is no collection of feature subsets to vary. Figure 4 shows that the optimal values appears to be around 10.

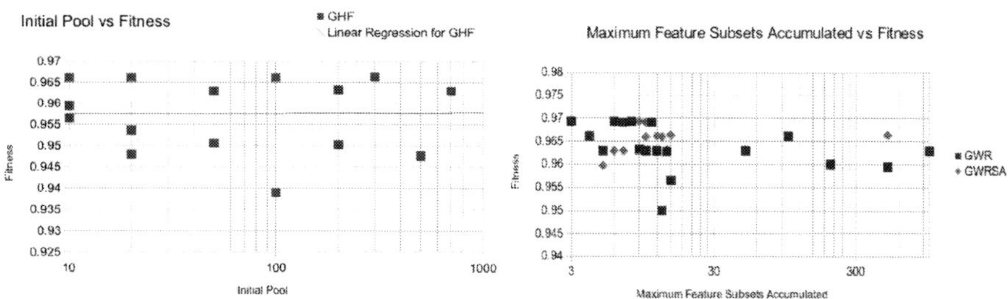

Fig. 4. Left: Initial Pool Sized vs Fitness, Right: Maximum Feature Subsets Accumulated vs Fitness

A **mutation** rate needs to be chosen for GHF, GWR, and GWRSA. The rate is the percentage of the feature subset that will be randomly turned on or off. A constant percentage is used for every iteration. The elements that are

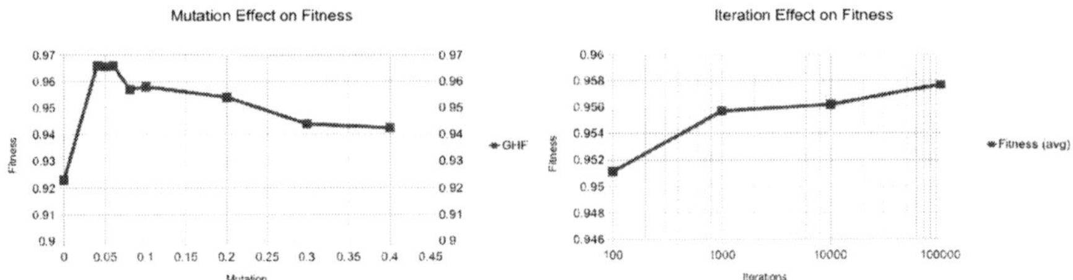

Fig. 5. Left: Average Percentage of Mutation Effectiveness, Right: Average Iteration Effect on Fitness

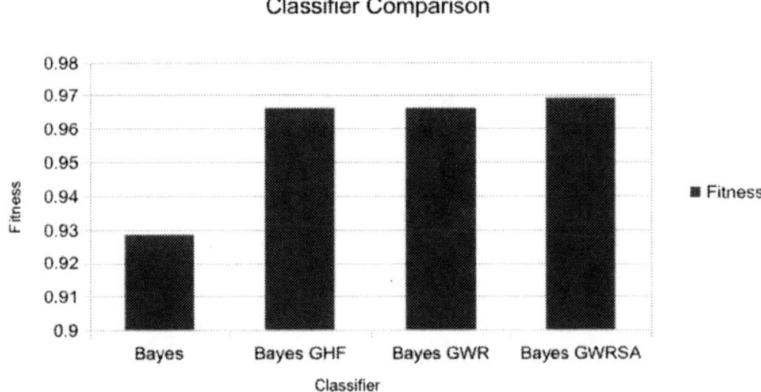

Fig. 6. Comparison of Classifiers

changed are randomly selected each iteration. An experiment was performed using 10,000 iterations, 10 randomly generated initial feature subsets, and a Naive Bayes Classifier. In Figure 5, 5% is shown to be the best mutation rate.

The **number of iterations** used would be a limiting factor in the application of these algorithms. Figure 5 shows all three proposed algorithm's fitness score grouped by iterations but varying in configurations.

A **classifier comparison** is shown in Figure 6. The best performance of these algorithms is compared to the standard Naive Bayes classifier result. The algorithms always start with the standard Naive Bayes result because they use all the features which ensures the result will never decline. The results show that genetically enhanced feature selection offers a significant increase in the classification ability to the standard Naive Bayes classifier.

5 Conclusion

This paper presented three feature selection algorithms that increase the classification ability of the Naive Bayes classifier. This is necessary because during applications of machine learning the classifier is presented with discriminate and indiscriminate features. This increase in classification ability is caused by training the classifier with a subset of features containing discriminate features. This algorithm is shown to boost the classification ability of a classifier that does not perform feature selection itself.

Acknowledgment. This research was supported in part by NASA Grant NNX09AK86G and NSF Grant 1062749.

References

1. Bandeira, L., Ding, W., Stepinski, T.F.: Automatic detection of sub-km craters using shape and texture information. In: Proceedings of the 41st Lunar and Planetary Science Conference, The Woodlands, Texas (March 2010)
2. Cheng, Y., Johnson, A.E., Matthies, L.H., Olson, C.F.: Optical landmark detection for spacecraft navigation. Advances in the Astronautical Sciences 114, 1785–1803 (2003)
3. Ding, W., Stepinski, T.F., Mu, Y., Bandeira, L., Ricardo, R., Wu, Y., Lu, Z., Cao, T., Wu, X.: Sub-kilometer crater discovery with boosting and transfer learning. ACM Transactions on Intelligent Systems and Technology 2(4) (July 2011)
4. Honda, R., Iijima, Y., Konishi, O.: Mining of Topographic Feature from Heterogeneous Imagery and its Application to Lunar Craters. In: Arikawa, S., Shinohara, A. (eds.) Progress in Discovery Science. LNCS (LNAI), vol. 2281, pp. 395–407. Springer, Heidelberg (2002)
5. Kim, J.R., Muller, J., van Gasselt, S., Morley, J.G., Neukum, G., et al.: Automated crater detection, a new tool for mars cartography and chronology. Photogrammetric Engineering and Remote Sensing 71(10), 1205 (2005)
6. Leroy, B., Medioni, G., Johnson, E., Matthies, L.: Crater detection for autonomous landing on asteroids. Image and Vision Computing 19(11), 787–792 (2001)
7. Liu, S., Ding, W., Cohen, J.P., Simovici, D., Stepinski, T.F.: Bernoulli trials based feature selection for crater detection. In: Proceedings of the 19th ACM SIGSPATIAL International Conference on Advances in Geographic Information Systems, Chicago, Illinois. ACM (November 2011)
8. Papageorgiou, C.P., Oren, M., Poggio, T.: A general framework for object detection. In: Center for Biological and Computational Learning Artificial Intelligence Laboratory (1998)
9. Russell, S.: Artificial intelligence, 3rd edn. Pearson Education, Upper Saddle River (2009)
10. Salamuniccar, G., Loncaric, S.: Open framework for objective evaluation of crater detection algorithms with first test-field subsystem based on MOLA data. Advances in Space Research 42(1), 6–19 (2008)
11. Urbach, E.R., Stepinski, T.F.: Automatic detection of sub-km craters in high resolution planetary images. Planetary and Space Science 57(7), 880–887 (2009)
12. Vinogradova, T., Burl, M., Mjolsness, E.: Training of a crater detection algorithm for mars crater imagery. In: 2002 IEEE Aerospace Conference (March 2002)
13. Viola, P., Jones, M.J.: Robust real-time face detection. International Journal of Computer Vision 57(2), 137–154 (2004)

Penalized Least Squares for Smoothing Financial Time Series

Adrian Letchford, Junbin Gao, and Lihong Zheng

School of Computing and Mathematics, Charles Sturt University, Australia

Abstract. Modeling of financial time series data by methods of artificial intelligence is difficult because of the extremely noisy nature of the data. A common and simple form of filter to reduce the noise originated in signal processing, the finite impulse response (FIR) filter. There are several of these noise reduction methods used throughout the financial instrument trading community. The major issue with these filters is the delay between the filtered data and the noisy data. This delay only increases as more noise reduction is desired. In the present marketplace, where investors are competing for quality and timely information, this delay can be a hindrance. This paper proposes a new FIR filter derived with the aim of maximizing the level of noise reduction and minimizing the delay. The model is modified from the old problem of time series graduation by penalized least squares. Comparison between five different methods has been done and experiment results have shown that our method is significantly superior to the alternatives in both delay and smoothness over short and middle range delay periods.

Keywords: Penalized least squares, Time series analysis, Financial analysis, Finite impulse response, Time series data mining.

1 Introduction

The presence of noise in time series data severely limits the ability to extract useful information [21]. Two types of noise have been identified, dynamical [8, 22] and measurement [8, 22, 24] noise. Dynamical noise is within the system and measurement noise is the result of less than perfect measurement instruments. Noise reduction is a broad term where the goal is to remove from a time series some unwanted component which itself is made up of noise [8].

Noise reduction of time series can be placed into four groups; graduation, prediction, filtering and smoothing. Graduation assumes that the signal has finished, thus allowing the use of all the data to be used to reduce the noise. This has been a very big area of research with models such as wavelets [2, 3], singular value decomposition [29, 34], empirical mode decomposition [1, 4], particle filters [10, 20], and singular spectrum analysis [12, 13]. Prediction involves estimating the future noise free values using old data. A very common and simple series of models for this purpose are the exponential smoothing models [9]. Filter models estimate the current noise free price using all available information. A famous

D. Wang and M. Reynolds (Eds.): AI 2011, LNAI 7106, pp. 72–81, 2011.

filter, the Kalman filter [17], has been used since it's derivation in 1960. Smoothing models are identical to filters with the exception of an added delay [26], they use some future data to reduce the noise such as in [32]. The models provide more accurate estimates at the cost of using some future data (the delay).

For noise reduction of financial data, it would appear that smoothing models are the most ideal. They are calculated in real time as the financial data stream is received, and they provide the best estimate in comparison to filters or predictors. The problem with these models, however, is the obvious lag. For example, the smoothed value at time t reflects the correct smoothed value for time $t-l$, where l is the lag. With the reduction of lag comes reduction of smoothness.

It has been shown that perfectly reducing the noise of the streaming time series increases the performance of data mining and forecasting methods [19, 31]. Investors use various combinations of these filters to produce trading rules, based on the reduced level of noise, to assist with buy and sell decisions. A comparison of two types of filters for this purpose was performed by [7] while [15] optimized the rules with a particle swarm algorithm. This paper will be concerned with a form of filter that is in wide spread use for security price analysis, finite impulse response (FIR) filters – or more commonly known in the financial industry as moving averages. The current methods will be presented and a new method with theoretical basis will be proposed to address the issue of lag within the limitation of the FIR filter.

The rest of this paper is outlined as follows, Sect. 2 will show the variations on the finite impulse response filter that are used. Section 3 proposes a new model with a theoretical basis. Section 4 describes the experiment performed to compare the various models and the results are presented in Sect. 5. Finally, Sect. 6 discusses the conclusions.

2 Current Methods

Finite impulse response (FIR) filters in the financial literature are more commonly known as moving averages, they can be generalized as:

$$\hat{y}_t = \sum_{i=1}^{n} \alpha_i y_{t-n+i} \tag{1}$$

Where $\boldsymbol{\alpha} = [\alpha_1, \alpha_2, \cdots, \alpha_n]$ is the set of model coefficients. The number of coefficients is denoted n, otherwise known as the FIR window size. There are only a handful of different methods of selecting these coefficients. The **simple moving average** (SMA) [5] sets $\alpha_1 = \alpha_2 = \ldots = \alpha_n = 1/n$. Analysts changed the coefficients to increase the weight on the most recent data with the aim of reducing the lag. The **weighted moving average** (WMA) [5] is one of these changes which sets the vector $\boldsymbol{\alpha} = [1, 2, \ldots, n] \cdot [n(n+1)2^{-1}]^{-1}$. The **hull moving average** (HMA) [16] is a modification of the WMA, which has less lag. Given that $\mathbf{WMA}(y, n)$ is the WMA of series $\mathbf{y}$ with n coefficients, the HMA is calculated as:

$$\hat{\mathbf{y}} = \mathbf{WMA}(2 \cdot \mathbf{WMA}(\mathbf{y}, \frac{n}{2}) - \mathbf{WMA}(\mathbf{y}, n), \sqrt{n}) \tag{2}$$

A Gaussian implementation, where $\boldsymbol{\alpha}$ is selected from a Gaussian kernel [11], is commonly known as the **Arnaud Legoux moving average** (ALMA) [23] which uses an offset O:

$$\hat{y}_t = \frac{\sum_{i=0}^{n-1} K_\sigma(i-O) y_{t-i}}{\sum_{i=0}^{n-1} K_\sigma(i-O)}, K_\sigma(x) = e^{-\frac{x^2}{2\sigma^2}} \tag{3}$$

There are other methods for selecting the coefficient vector $\boldsymbol{\alpha}$, however, they are unsuitable for financial data. For example, the least mean squares filter [30] is an adaptive moving average, the coefficients change with time. To calculate $\boldsymbol{\alpha}$, one must first know the smoothed series, quite impossible in finance and economics. The FIR wiener filter [18] also requires knowledge of the smoothed series.

Each of these FIR designs aims to maintain a smooth output while attempting to reduce lag. The following section shows a derivation of the coefficient vector $\boldsymbol{\alpha}$ which is optimized to give the smoothest curve on a training data set after specifying the FIR window size.

3 Our Proposed Method

Our proposed method for real time noise reduction is based on the **penalized least squares** (PLS) graduation method [6, 14, 33]. The PLS method balances two conflicting attributes of the final curve: (1) the accuracy of the curve to the original series and (2) the smoothness of the curve. The accuracy is expressed in matrix notation with the normal least squares method $||\mathbf{y}-\hat{\mathbf{y}}||^2$. The smoothness can be measured with differencing where $\nabla\hat{y}_x = \hat{y}_x - \hat{y}_{x-1}$ and $\nabla^2\hat{y}_x = \nabla(\nabla\hat{y}_x)$. The differencing can be expressed in matrix notation where D is a matrix such that $\mathbf{D}_d\hat{\mathbf{y}} = \nabla^d\hat{\mathbf{y}}$ where $d \in \mathbb{Z}$. For example, if the size of the $\mathbf{y}$ vector is 5 and $d = 1$ then:

$$\mathbf{D}_1 = \begin{bmatrix} -1 & 1 & 0 & 0 & 0 \\ 0 & -1 & 1 & 0 & 0 \\ 0 & 0 & -1 & 1 & 0 \\ 0 & 0 & 0 & -1 & 1 \end{bmatrix} \tag{4}$$

The problem is then expressed in least squares form as:

$$\mathbf{Q} = ||\mathbf{y} - \hat{\mathbf{y}}||^2 + \lambda||\mathbf{D}_d\hat{\mathbf{y}}||^2 \tag{5}$$

Where λ is a smoothing factor. Differentiating both sides with respect to $\hat{\mathbf{y}}$ and setting to zero leads to the following solution where $\hat{\mathbf{y}}$ is a graduation of $\mathbf{y}$:

$$\hat{\mathbf{y}} = (\mathbf{I} + \lambda\mathbf{D}_d^T\mathbf{D}_d)^{-1}\mathbf{y} \tag{6}$$

Penalized least squares moving average (PLSMA) is the proposed model which modifies the PLS method to calculate optimal moving average coefficients. To change the problem to a moving average model the underlying time series needs to be represented in a trajectory matrix $\bar{\mathbf{y}}$ and the corresponding time series vector $\mathbf{y}$ needs to be adjusted to match. The trajectory matrix is calculated as follows, considering the time series $\mathbf{y} = [y_1, y_2, \ldots, y_N]$ let n be the number of coefficients in the model, then:

$$\bar{\mathbf{y}} = \begin{bmatrix} y_1 & y_2 & \cdots & y_n \\ y_2 & y_3 & \cdots & y_{n+1} \\ \vdots & \vdots & \ddots & \vdots \\ y_{N-n+1} & y_{N-n+2} & \cdots & y_N \end{bmatrix} \quad (7)$$

While the corresponding time series vector $\mathbf{y}$ is the last column of $\bar{\mathbf{y}}$.

The model coefficients are represented in a column vector $\boldsymbol{\alpha}$, consistent with (1), and $\hat{\mathbf{y}}$ is then replaced by $\bar{\mathbf{y}}\boldsymbol{\alpha}$ in (5):

$$\mathbf{Q} = ||\mathbf{y} - \bar{\mathbf{y}}\boldsymbol{\alpha}||^{\mathbf{2}} + \lambda||\mathbf{D_d}\bar{\mathbf{y}}\boldsymbol{\alpha}||^{\mathbf{2}} \quad (8)$$

Differentiating both sides with respect to $\boldsymbol{\alpha}$ and setting to zero gives the solution:

$$\boldsymbol{\alpha} = [\bar{\mathbf{y}}^T\bar{\mathbf{y}} + \lambda(\mathbf{D_d}\bar{\mathbf{y}})^{\mathbf{T}}\mathbf{D_d}\bar{\mathbf{y}}]^{-\mathbf{1}}\bar{\mathbf{y}}^{\mathbf{T}}\mathbf{y} \quad (9)$$

Now, $\boldsymbol{\alpha}$ are FIR coefficients. While training data is needed to compute these coefficients, they can be used to smooth future data in an online fashion with increased smoothness (reduced lag) over the given data.

This raw method does come with some problems. (1) As λ increases the curve gets smoother until a point is reached where it cannot be any smoother and still remain on the same scale as $\mathbf{y}$. Then, $\hat{\mathbf{y}} \to 0$ as $\lambda \to \infty$. (2) As $\lambda \to \infty$ the matrix $\bar{\mathbf{y}}^T\bar{\mathbf{y}} + \lambda(\mathbf{D_d}\bar{\mathbf{y}})^{\mathbf{T}}\mathbf{D_d}\bar{\mathbf{y}}$ becomes singular – non-invertible. (3) most of the current filters have one or two inputs, this method has three inputs, FIR size, d, and λ.

The first problem is solved by normalizing $\boldsymbol{\alpha}$ by the sum of $\boldsymbol{\alpha}$. The second problem is rectified by noting that λ is used to change the proportion of the least squares equation by increasing the smoothness penalty. This ratio is maintained if the error part of the equation is multiplied by λ^{-1} and the smoothness penalty is left without a multiplier. Thus, (8) & (9) become:

$$\mathbf{Q} = \lambda^{-1}||\mathbf{y} - \bar{\mathbf{y}}\boldsymbol{\alpha}||^{\mathbf{2}} + ||\mathbf{D_d}\bar{\mathbf{y}}\boldsymbol{\alpha}||^{\mathbf{2}} \quad (10)$$

$$\boldsymbol{\alpha} = [\lambda^{-1}\bar{\mathbf{y}}^T\bar{\mathbf{y}} + (\mathbf{D_d}\bar{\mathbf{y}})^{\mathbf{T}}\mathbf{D_d}\bar{\mathbf{y}}]^{-\mathbf{1}}\lambda^{-\mathbf{1}}\bar{\mathbf{y}}^{\mathbf{T}}\mathbf{y} \quad (11)$$

Because of normalization, (11) can drop the second λ^{-1}:

$$\boldsymbol{\alpha} = [\lambda^{-1}\bar{\mathbf{y}}^T\bar{\mathbf{y}} + (\mathbf{D_d}\bar{\mathbf{y}})^{\mathbf{T}}\mathbf{D_d}\bar{\mathbf{y}}]^{-\mathbf{1}}\bar{\mathbf{y}}^{\mathbf{T}}\mathbf{y} \quad (12)$$

The third problem is overcome by noting that the goal is to achieve the greatest smoothing. Thus, λ ought to be maximized. Taking the limit:

$$\boldsymbol{\alpha} = \lim_{\lambda \to \infty} [\lambda^{-1} \bar{\mathbf{y}}^T \bar{\mathbf{y}} + (\mathbf{D_d} \bar{\mathbf{y}})^\mathbf{T} \mathbf{D_d} \bar{\mathbf{y}}]^{-1} \bar{\mathbf{y}}^\mathbf{T} \mathbf{y} \tag{13}$$

$$= [(\mathbf{D_d} \bar{\mathbf{y}})^\mathbf{T} \mathbf{D_d} \bar{\mathbf{y}}]^{-1} \bar{\mathbf{y}}^\mathbf{T} \mathbf{y} \tag{14}$$

4 Experiment Description

The models were compared over several data sets with a cross validation method. To calculate the performance of each model two measures were developed. One for measuring how smooth the new time series is and the other to calculate how much lag it has. The rest of this section presents the details of this experiment and these statistical measures.

4.1 Data

Six real world time series were used for these experiments; AUD/USD, EUR/USD, GOOG, INDU, NASDAQ, and XAU/USD, all daily prices each around 2000 samples. Table 1 shows the range and description of each series. In addition, two randomly generated series were also used. Both have 2000 random prices with returns generated from the standard normal distribution.

Table 1. Names and description of the time series used in the experiment

Series Name	Range	Description
AUD/USD	20/10/2003 - 14/06/2011	Australian Dollar to U.S.A. Dollar
EUR/USD	15/10/2003 - 31/05/2011	Euro to the U.S.A. Dollar
GOOG	25/10/2004 - 14/06/2011	Stock for Google
INDU	21/07/2003 - 14/06/2011	Index for Dow Jones Industrial Average
NASDAQ	10/04/2003 - 14/06/2011	NASDAQ market index
XAU/USD	22/10/2003 - 14/06/2011	Gold to U.S.A. Dollar
Random 1		2000 random prices
Random 2		2000 random prices

4.2 Smoothness (Noise) Function

Previously, to calculate the level of noise reduction, the signal to noise ratio (SNR) would be used [27]. However, it assumes that the clean signal is known, and assumes that $\hat{\mathbf{y}}$ has no delay. In previous research, measures have been used which do not hold these assumptions. For example, autocorrelation and power spectrum are used in [25]. Unfortunately, these methods output the result in a large dimension resulting in comparison issues when processing thousands of comparisons. The measure for smoothness used here builds upon $||\mathbf{D_d}\hat{\mathbf{y}}||^\mathbf{2}$ used in the PLS equation. Some considerations are made, if $d = 1$, then the error stems from using the previous value of $\hat{\mathbf{y}}$, similarly, $d = 2$ is using the previous rate of change (ROC) to forecast. However, the ROC may be smooth, where

$d = 3$ would result in a smaller error. Thus, the smoothness of $\hat{\mathbf{y}}$ is the minimum of the following function with respect to d normalized by the smoothness of $\mathbf{y}$:

$$S(\hat{\mathbf{y}}) = min\left\{||\mathbf{D_d}\hat{\mathbf{y}}||\right\}, d \in \mathbb{N} \tag{15}$$

$$S(\mathbf{y}, \hat{\mathbf{y}}) = 1 - \frac{S(\hat{\mathbf{y}})}{S(\mathbf{y})} \tag{16}$$

Which can be interpreted as the percentage of noise filtered from the original series $\mathbf{y}$ to produce the smooth curve $\hat{\mathbf{y}}$. Unlike the SNR, the S function does not assume that the clean signal is known, and does not make assumptions about the lag.

Usually, noise is measured as an error between values such as in prediction problems or when using the SNR. However, as this paper is not dealing with estimating exact unknown quantities, this is redundant. Instead, the aim is the online reduction of noise in known noisy data. Thus, a natural conclusion would be to reduce the variance between values. As this would result in producing a straight line, instead of following the time series, the smoothness function extends this to reducing the variance at the best derivative level. As a result, small values for $S(\mathbf{y})$ means that $\mathbf{y}$ is smooth and takes on the form of a curve.

4.3 Lag Function

Cross correlation is adapted to calculate the lag between $\mathbf{y}$ and a given $\hat{\mathbf{y}}$. After calculating the smoothed series $\hat{\mathbf{y}}$ of a testing data, the entire training-testing window ($\mathbf{y}$) and $\hat{\mathbf{y}}$ are lined up by their right side. This is lag 0 and the correlation is calculated between $\hat{\mathbf{y}}$ and the adjacent values in $\mathbf{y}$. Then $\hat{\mathbf{y}}$ is shifted left by 1, corresponding to lag 1, and the correlation is again calculated. This process is continued and the lag with the highest correlation is taken to be the lag of the smoothed series.

4.4 Cross Validation

The five models in Sect. 2 and 3 were compared by using a cross validation method. The best window size for the training data was 800 and the testing data was 400. Due to the large size of this combined window (1,200), it was shifted by 100 rather than 400 to maintain enough sample optimizations.

The aim of the experiment was to find out which model has greater smoothing for a given amount of lag. There is no direct input for lag, however, as FIR filters, the lag is related to the size of the filter. Thus, the size (n) was iterated between 2-150 and the remaining variables were optimized at each iteration.

As has been shown, the SMA, WMA, and HMA have a single input, the FIR window size (n). As a result, these three models do not need optimization. The smoothed series is simply calculated over each testing set, and the smoothness and lag are averaged for each value of n.

For each $n \in [2-150]$, the ALMA is optimized 5 times over each set of training-testing data using $1 - S(\mathbf{y}, \hat{\mathbf{y}})$. The best parameters out of the 5 are chosen for that data set. The variables σ and O are optimized over the ranges 1 to 50 and -50 to 50 respectively. The optimization algorithm is simulated annealing, see [28]. Put simply, simulated annealing takes an initial starting point and "jitters" it around the error surface with a tendency to move around local minima. The jittering gradually comes to a halt where the point is expected to be in a local minima. The standard MatLab algorithm with default parameters was used.

The PLSMA model optimization is performed differently. The only parameter to be optimized is d and this is an integer. After a few trials of different FIR window sizes up to 150 it seemed that the optimal d did not go over 10. d was evaluated over the range [1-10] and the best value according to the smoothness measure was selected.

Figure 1 is the pseudocode of the cross validation algorithm.

```
foreach model
    foreach time series
        for n = 2-150
            foreach CV window
                Optimize model on training data
                Apply model to testing data
                calculate smoothness and lag
            Calculate average smoothness and lag over the CV windows
        Calculate average smoothness for each lag
```

Fig. 1. Pseudocode for the cross-validation algorithm

Once the average smoothness for each lag had been obtained for each of the models on each of the time series, summary statistics were compiled. The percentage of superior lags in comparison to the other models on each time series is calculated. The model with the highest percentage of superior lags is considered to be the best model.

5 Results

A clear indication of each model's performance is shown in Tbl. 2. The %Lags column shows the percentage of lags for which that model is superior, and the Range column shows the range of those lags. These comparisons were different for each model, as each model spans a different range in relation to the others. For example, the HMA only goes as far as 17 delay periods on the AUD/USD data. Thus, these figures are for the comparable range of each model on each data set.

The SMA is the worst model with no suitable lag periods except for the XAU/USD series where it is superior for lag 1. The WMA falls next being only

Table 2. Percent improvement, lag range, and algorithm complexity

	SMA		WMA		HMA		ALMA		PLSMA	
	%Lags	Range	%Lags	Range	%Lags	Range	%Lags	Range	%Lags	Range
Random 1	0.0%	[]	3.0%	[1]	5.3%	[2]	31.8%	[31-44]	58.3%	[3-30]
Random 2	0.0%	[]	2.4%	[1]	11.1%	[2-3]	24.0%	[39-50]	57.4%	[4-38]
INDU	0.0%	[]	2.6%	[1]	11.8%	[2-3]	28.0%	[37-50]	49.3%	[5-36]
AUD/USD	0.0%	[]	2.4%	[1]	11.8%	[2-3]	30.0%	[36-50]	50.8%	[4-35]
EUR/USD	0.0%	[]	0.0%	[]	23.5%	[1-4]	28.0%	[37-50]	47.1%	[5-36]
GOOG	0.0%	[]	0.0%	[]	11.8%	[2-3]	24.0%	[39-50]	48.6%	[1, 4-38]
NASDAQ	0.0%	[]	2.4%	[1]	11.8%	[2-3]	26.0%	[38-50]	47.2%	[4-37]
XAU/USD	2.1%	[1]	0.0%	[]	11.1%	[2-3]	19.0%	[35-42]	66.0%	[4-34]
Average	**0.3%**		**1.6%**		**12.3%**		**26.4%**		**53.1%**	
Complexity	$O(n)$		$O(n)$		$O(n^2)$		$O(n)$		$O(n^3)$	

superior on average by lag 1. The HMA is approximately on the range 2-3. The ALMA takes a much wider range of about 37-48 lag periods. The PLSMA model (our proposed model) is shown to be the best smoother. Being the most smoothest model for 48.6%+ of the lag periods. It appears that the PLSMA is superior over short to middle term lag periods of about 4-36 while the ALMA smoother is best for longer term lag periods.

Once the FIR coefficients for each model has been calculated, applying the filter to the financial data stream is of $O(n)$ complexity. However, the models do have varying degrees of complexity for the calculation of the FIR coefficients. PLSMA excluded, the best model is the ALMA which is of complexity $O(n)$. The improvement that PLSMA brings comes at a complexity cost, with the model sitting at $O(n^3)$. However, this is not a setback in online applications as the FIR coefficients are calculated offline. The complexity for each model is shown in Tbl 2.

6 Conclusions

In this paper, we have shown some of the different FIR filters used by investors to smooth security prices. It is noted that the output of a FIR filter is delayed with respect to the underlying time series. In addition, there is a positive relationship between the smoothness of the resulting curve and the lag which is undesirable. A method was proposed to derive an impulse response which maximizes the smoothness and minimizes the delay. As there is no assurance of optimality over any future data the filter may be applied to, this model was compared against five common models with a cross validation process. It was discovered that the proposed model achieves greater overall smoothing, more specifically for the short to middle range lag periods. While the very short term (4 or less periods) and longer term (37+ periods) were dominated by other models.

Future research will expand the analysis in this paper to include noise reduction models that are not otherwise used for financial pre-processing. Further experiments will also be conducted to discover the level of improvement for data mining and forecasting algorithms as previous research implies.

References

1. Boudraa, A.O., Cexus, J.C.: EMD-based signal filtering. IEEE Transactions on Instrumentation and Measurement 56(6), 2196–2202 (2007)
2. Donoho, D.L., Johnstone, I.M.: Adapting to unknown smoothness via wavelet shrinkage. Journal of the American Statistical Association 90(432), 1200–1224 (1995)
3. Donoho, D.L., Johnstone, I.M.: Minimax estimation via wavelet shrinkage. The Annals of Statistics 26(3), 879–921 (1998)
4. Drakakis, K.: Empirical mode decomposition of financial data. International Mathematical Forum 3(25), 1191–1202 (2008)
5. Ehlers, J.F.: Rocket science for traders: Digital signal processing applications. John Wiley & Sons, Inc., New York (2001)
6. Eilers, P.H.C.: A perfect smoother. Analytical Chemistry 75(14), 3631–3636 (2003)
7. Ellis, C.A., Parbery, S.A.: Is smarter better? a comparison of adaptive, and simple moving average trading strategies. Research in International Business and Finance 19(3), 399–411 (2005)
8. Farmer, D.J., Sidorowich, J.J.: Optimal shadowing and noise reduction. Physica D: Nonlinear Phenomena 47(3), 373–392 (1991)
9. Gardner, J.E.S.: Exponential smoothing: The state of the art–part II. International Journal of Forecasting 22(4), 637–666 (2006)
10. Godsill, S.J., Doucet, A., West, M.: Monte Carlo smoothing for nonlinear time series. Journal of the American Statistical Association 99(465), 156–168 (2004)
11. Hale, D.: Recursive gaussian filters. Tech. rep., Center for Wave Phenomena (2006)
12. Hassani, H.: Singular spectrum analysis: Methodology and comparison. Journal of Data Sciences 5, 239–257 (2007), mPRA Paper
13. Hassani, H., Soofi, A.S., Zhigljavsky, A.A.: Predicting daily exchange rate with singular spectrum analysis. Nonlinear Analysis: Real World Applications 11(3), 2023–2034 (2010)
14. Hodrick, R.J., Prescott, E.C.: Postwar u.s. business cycles: An empirical investigation. Journal of Money, Credit and Banking 29(1), 1–16 (1997)
15. Hsu, L.-Y., Horng, S.-J., He, M., Fan, P., Kao, T.-W., Khan, M.K., Run, R.-S., Lai, J.-L., Chen, R.-J.: Mutual funds trading strategy based on particle swarm optimization. Expert Systems with Applications 38(6), 7582–7602 (2011)
16. Hull, A.: Hull moving average HMA (2011), `http://www.justdata.com.au/Journals/AlanHull/hull_ma.htm`
17. Kalman, R.E.: A new approach to linear filtering and prediction problems. Transactions of the ASME Journal of Basic Engineering 82(Series D), 35–45 (1960)
18. Kamen, E.W., Su, J.K.: Introduction to optimal estimation. Springer, London (1999)
19. Karunasingha, D.S.K., Liong, S.Y.: Enhancement of chaotic time series prediction with real-time noise reduction. In: International Conference on Small Hydropower - Hydro Sri Lanka (2007)
20. Klaas, M., Briers, M., Freitas, N.d., Doucet, A., Maskell, S., Lang, D.: Fast particle smoothing: if I had a million particles. In: Proceedings of the 23rd International Conference on Machine Learning, pp. 481–488 (2006)
21. Kostelich, E.J., Yorke, J.A.: Noise reduction in dynamical systems. Physical Review A 38(3), 1649 (1988)
22. Kostelich, E.J., Yorke, J.A.: Noise reduction: Finding the simplest dynamical system consistent with the data. Physica D: Nonlinear Phenomena 41(2), 183–196 (1990)

23. Legoux, A.: ALMA Arnaud Legous Moving Average (2009), http://www.arnaudlegoux.com/wp-content/uploads/2011/03/ALMA-Arnaud-Legoux-Moving-Average.pdf (2011)
24. Li, T., Li, Q., Zhu, S., Ogihara, M.: A survey on wavelet applications in data mining. SIGKDD Explor. Newsl. 4(2), 49–68 (2002), 772870
25. Liu, Y., Liao, X.: Adaptive chaotic noise reduction method based on dual-lifting wavelet. Expert Systems with Applications 38(3), 1346–1355 (2011)
26. Moore, J.B.: Discrete-time fixed-lag smoothing algorithms. Automatica 9(2), 163–173 (1973)
27. Nikpour, M., Nadernejad, E., Ashtiani, H., Hassanpour, H.: Using pde's for noise reduction in time series. International Journal of Computing and ICT Research 3(1), 42–48 (2009)
28. Russell, S.J., Norvig, P.: Artificial intelligence: A modern approach. Pretice Hall series in artificial intelligence. Prentice Hall, N.J (2010)
29. Sadasivan, P.K., Dutt, D.N.: SVD based technique for noise reduction in electroencephalographic signals. Signal Processing 55(2), 179–189 (1996)
30. Shynk, J.J.: Frequency-domain and multirate adaptive filtering. IEEE Signal Processing Magazine 9(1), 14–37 (1992)
31. Soofi, A.S., Cao, L.: Nonlinear forecasting of noisy financial data. In: Soofi, A.S., Cao, L. (eds.) Modeling and Forecasting Financial Data: Techniques of Nonlinear Dynamics, pp. 455–465. Kluwer Academic Publishers, Boston (2002)
32. Vandewalle, N., Ausloos, M., Boveroux, P.: The moving averages demystified. Physica A: Statistical Mechanics and its Applications 269(1), 170–176 (1999)
33. Whittaker, E.T.: On a new method of graduation. Proceedings of the Edinburgh Mathematical Society 41(-1), 63–75 (1923)
34. Zehtabian, A., Hassanpour, H.: A non-destructive approach for noise reduction in time domain. World Applied Sciences Journal 6(1), 53–63 (2009)

Logistic Regression with the Nonnegative Garrote

Enes Makalic and Daniel F. Schmidt

Centre for MEGA Epidemiology, The University of Melbourne
Carlton VIC 3053, Australia
{emakalic,dschmidt}@unimelb.edu.au

Abstract. Logistic regression is one of the most commonly applied statistical methods for binary classification problems. This paper considers the nonnegative garrote regularization penalty in logistic models and derives an optimization algorithm for minimizing the resultant penalty function. The search algorithm is computationally efficient and can be used even when the number of regressors is much larger than the number of samples. As the nonnegative garrote requires an initial estimate of the parameters, a number of possible estimators are compared and contrasted. Logistic regression with the nonnegative garrote is then compared with several popular regularization methods in a set of comprehensive numerical simulations. The proposed method attained excellent performance in terms of prediction rate and variable selection accuracy on both real and artificially generated data.

1 Introduction

Logistic regression is one of the most commonly applied statistical methods for binary classification problems. Here, one observes n data samples

$$\{(y_i, x_{i1}, \ldots, x_{ip}), i = 1, \ldots, n\}$$

comprising p predictor variables and a binary class indicator $y \in \{-1, +1\}$ which denotes the class membership of the observed predictors. The conditional probability that a vector of covariates $\mathbf{x} = (x_1, \ldots, x_p)'$ is assigned to class y is

$$p(y = \pm 1 | \mathbf{x}, \boldsymbol{\beta}) = \frac{1}{1 + \exp(-y\mathbf{x}'\boldsymbol{\beta})} \tag{1}$$

where $\boldsymbol{\beta} = (\beta_1, \ldots, \beta_p)' \in \mathbb{R}^p$ are the regression coefficients. The log-likelihood of a logistic regression is then given by

$$l(\boldsymbol{\beta}) = -\sum_{i=1}^{n} \log\left(1 + \exp(-y_i \mathbf{x}_i' \boldsymbol{\beta})\right) \tag{2}$$

which is a function of the regression parameters $\boldsymbol{\beta}$. The regression coefficients determine the probability of the target variable. That is, a positive regression coefficient for a predictor implies that the predictor is associated with an increased

D. Wang and M. Reynolds (Eds.): AI 2011, LNAI 7106, pp. 82–91, 2011.

probability of the response ($y = +1$), while a negative parameter coefficient reduces the response probability. A regression coefficient of zero has no effect on the conditional class probability and should ideally be excluded from the final model. The main task in inference of logistic models is to estimate the parameter coefficients $\boldsymbol{\beta}$ and select which of the p observed covariate vectors, if any, are useful in explaining the target variable.

This paper applies the nonnegative garrote to logistic regression models and examines the performance of the resulting procedure under various settings of sample size, number of predictors and regressor correlation. There are three main contributions in the paper: (1) an efficient algorithm for the implementation of NNG in logistic regression models, (2) empirical evaluation of several initial estimators for the NNG, and (3) extensive performance comparison in terms of prediction and variable selection of the NNG procedure with several popular regularization algorithms.

2 Nonnegative Garrote

Let $\boldsymbol{\beta}^* \in \mathbb{R}^p$ be an initial estimate of the logistic regression parameters, for example, the maximum likelihood estimate or a ridge regression estimate. Denote a shrunken estimate of $\boldsymbol{\beta}^*$ as $\tilde{\boldsymbol{\beta}}(\mathbf{c}) = (\mathbf{c} \odot \boldsymbol{\beta}^*)$ where $\mathbf{c} = (c_1, \ldots, c_p)'$ and the operator $\odot$ is the Hadamard (element-wise) product. The nonnegative garrote estimate [1] is defined as the solution to

$$\boldsymbol{\beta}_\lambda = \underset{\tilde{\boldsymbol{\beta}}(\mathbf{c})}{\arg\max} \left\{ l(\tilde{\boldsymbol{\beta}}) \right\} = \underset{\tilde{\boldsymbol{\beta}}(\mathbf{c})}{\arg\min} \left\{ \sum_{i=1}^{n} \log\left(1 + \exp(-y_i \mathbf{x}_i' \tilde{\boldsymbol{\beta}})\right) + \lambda \sum_{j=1}^{p} c_j \right\} \tag{3}$$

subject to the constraints

$$c_j \geq 0, \quad (j = 1, 2, \ldots, p) \tag{4}$$

and assuming that the initial parameter estimate $\boldsymbol{\beta}^*$ is kept fixed. The NNG shrinks the initial parameter estimates by varying the multiplier $\mathbf{c}$. The regularization parameter $\lambda > 0$ controls the amount of shrinkage that is applied to the initial parameter estimates. Increasing λ (tightening the garrote) results in more of the initial parameters being set to zero and greater shrinkage of the non-zero components of $\boldsymbol{\beta}^*$. In contrast, decreasing the regularization parameter induces less shrinkage leading to a final solution that is closer to the starting parameter estimates. In this way, the NNG allows for both parameter shrinkage and variable selection automatically. In practice, the regularization parameter may be selected using a model selection criterion such as the Bayesian information criterion (BIC) [2].

There is no clear consensus as to which initial estimator should be used with the NNG. Breiman [1] originally advocated the maximum likelihood estimator to be used as the initial estimate in linear models. There are three disadvantages of this approach in logistic regression: (1) the maximum likelihood estimator cannot

Algorithm 1. Cyclic coordinate descent for nonnegative garrote (nng)

input : data matrix $\mathbf{X} \in \mathbb{R}^{n \times p}$, target vector $\mathbf{y} \in \{-1, +1\}^n$, initial estimate $\boldsymbol{\beta}^* \in \mathbb{R}^p$, regularization parameter $\lambda > 0$
output: NNG estimate $\boldsymbol{\beta} \in \mathbb{R}^p$

initialize $\Delta_j \leftarrow 1$ for $j = 1, \ldots, p$, $\quad \Delta r_i \leftarrow 0$ for $i = 1, \ldots, n$
$r \leftarrow y \odot X\boldsymbol{\beta}^*$ `(`$\odot$ `denotes element-wise product)`
$\mathbf{x}_i \leftarrow \mathbf{x}_i \odot \boldsymbol{\beta}^* \quad (i = 1, \ldots, n)$ `(rescale data)`
$\beta \leftarrow (1, \ldots, 1)'$ `(start search from` $\boldsymbol{\beta}^*$`)`
for $t \leftarrow 1, 2, \ldots$ **to** convergence **do**
  for $j \leftarrow 1, 2, \ldots$ **to** p **do**
    $F_i \leftarrow \min(0.25, 1/(2\exp(-\Delta_j|x_{ij}|) + \exp(r_i - \Delta_j|x_{ij}|) + \exp(\Delta_j|x_{ij}| - r_i)) \quad (i = 1, \ldots, n)$
    $\Delta v_j \leftarrow \left(\sum_{i=1}^n x_{ij} y_i / (1 + \exp(r_i)) - \lambda\right) / \left(\sum_{i=1}^n x_{ij}^2 F_i\right)$ `(Newton--Raphson update)`
    if $\beta_j = 0$ **then**
      if $\Delta v_j \leq 0$ **then**
        $\Delta v_j = 0$
      end
    else
      if $\beta_j + \Delta v_j < 0$ **then**
        $\Delta v_j = -\beta_j$ `(if sign change, set` β_j `to zero)`
      end
    end
    $\Delta\beta_j \leftarrow \min(\max(\Delta v_j, -\Delta_j), \Delta_j)$ `(limit step size to trust region)`
    $\Delta r_i \leftarrow \Delta\beta_j X_{ij} y_i, \quad r_i \leftarrow r_i + \Delta r_i \quad (i = 1, \ldots, n)$
    $\beta_j \leftarrow \beta_j + \Delta\beta_j$
    $\Delta_j \leftarrow \max(2|\Delta\beta_j|, \Delta_j/2)$ `(update trust region size)`
  end
end
$\boldsymbol{\beta} \leftarrow \boldsymbol{\beta} \odot \boldsymbol{\beta}^*$ `(use original scale)`

be used if the number of predictors is greater than the number of samples ($p > n$ setting) or the covariates are highly correlated, (2) the maximum likelihood estimator performs poorly when the sample size is small, and (3) the maximum likelihood estimator does not exist if the data is quasicompletely or completely separable [3]. In this paper, following [4], we compare and contrast a number of alternative initial estimators.

A software implementation of NNG logistic regression requires some thought since the standard convex programming solution to (3)-(4) is not feasible when the number of covariates is large. The NNG solution was originally implemented using constrained least squares minimization in the linear regression setting. This approach is however not possible in logistic regression models and subsequently a number of alternative optimization routines have been proposed [5,6,7]. This paper employs a numerical optimization routine based on the cyclic coordinate descent method detailed in [5]. The cyclic coordinate descent method was chosen because of the low computational complexity and the fact that the algorithm can be used when the number of predictors is large, potentially much larger than the sample size. The pseudo-code for the proposed optimization routine is shown in in Algorithm 1.

Our modified algorithm, henceforth `NNG_OPT`, begins by transforming the data matrix $\mathbf{X}$ and the initial parameter vector $\boldsymbol{\beta}^*$ in such a way that the transformed regression parameters are restricted to be positive (lines 3–4). Contrary to the LASSO where a variable can change signs during the optimization, the NNG multiplier factor $\mathbf{c}$ is strictly positive. Subsequently, our algorithm checks for sign changes in variables (lines 9–17) and does not allow negative multiplier parameters. A Newton–Raphson update is then performed for each regression parameter, while all the remaining parameters are kept fixed (lines 19–20). During the optimization, the variable r is used to keep track of the product $y \odot X\boldsymbol{\beta}$ for speed purposes. The optimization steps are performed until convergence is reached; `NNG_OPT` uses the convergence criterion recommended in [5]. Note, when implementing `NNG_OPT`, special care needs to be taken for the constant regressor which should not be subject to shrinkage. A MATLAB™ implementation of `NNG_OPT` is available from the authors upon request.

3 Simulation

This section examines finite sample performance of the NNG estimator using artificially generated data (see Section 3.1) as well as real data (see Section 3.4). Since the performance of the NNG estimator depends on the initial estimate [4], four different initial estimates are considered: (1) stepwise forward selection (`fwd`), (2) ridge regression (`rr`) [8], (3) the least angle shrinkage and selection operators (`lasso`) [9], and (4) the elastic net (`enet`) [10]. For completeness, a method that uses several possible ridge regression estimates, denoted `nng`, is also considered. The resulting NNG estimates are compared against the standard stepwise forward selection (`fwd`), ridge regression (`rr`), LASSO (`lasso`) and elastic net (`enet`) estimates. Furthermore, we have also included the iterated LASSO estimate (`ilasso`) [11] in all our comparisons, though it is not used as an initial estimate for an NNG solution. The performance of each method is measured using a variety of metrics including classification accuracy, size of the final model, the mean number of false positive regressors and the mean number of false negative regressors. Note that the constant regressor was included in all subsequent simulation runs but was not used when tabulating results.

3.1 Simulated Data

3.2 Path Consistency

The first simulation examined how often the NNG estimator and the popular LASSO estimator select only the true regression coefficients from artificially generated data. This property is known as path consistency. The simulation closely followed the setup in ([4], Example 1) for linear regression models. Here, the regressor matrix $\mathbf{X}$ consists of four regressors ($p = 4$) generated from:

$$\mathbf{x}_i = (1, X_1, X_2, X_3)' \qquad X_1, X_2 \sim N(0,1), \quad X_3 \sim N(\alpha(X_1+X_2), 1-2\alpha^2) \quad (5)$$

where $N(\cdot,\cdot)$ denotes the univariate normal distribution, $\alpha \in \{0.35, 0.55\}$ and $i = (1, \ldots, n)$. The true regression coefficients were set to

$$\boldsymbol{\beta} = (0, 1, 1, 0)' \quad (6)$$

In all simulations, the class indicators $\mathbf{y} \in \{-1, +1\}^n$ were independently generated with probability given by (1). For each value of α, we generated training data with the following sample sizes $n = \{20, 50, 100, 200, 500\}$. The regularization parameter for both the LASSO and NNG algorithms was selected using the log-likelihood of an independently generated validation data set. The validation data set was of the same size as the corresponding training data set. The range of regularization parameters considered was chosen to comprise 1000 values of λ uniformly spaced between 10^{-5} and 10^2. The simulation comprised 100 training and validation data sets generated for each (n, α) pair. For each run, we recorded the number of times the LASSO and NNG correctly identified the true regression coefficients and excluded the noise variables. The NNG estimator was selected as follows: (1) train a number of initial models using ridge regression estimates, (2) obtain a NNG solution for each ridge regression model, and (3) use the validation data set to select the NNG model with the largest log-likelihood. Figure 1 depicts the frequency of true model identification by both LASSO and NNG.

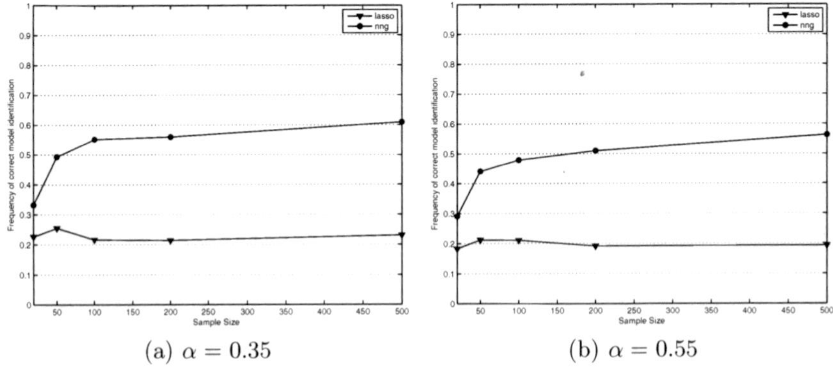

(a) $\alpha = 0.35$ (b) $\alpha = 0.55$

Fig. 1. Path consistency of the LASSO and NNG estimators

When $\alpha = 0.35$, the NNG and LASSO select the true model with frequencies of 30% and 20% respectively for smaller samples ($n < 50$). As the sample size is increased, the frequency of true model detection dramatically increases for the NNG and increases only slightly for the LASSO. At $n = 500$, for example, the NNG correctly identifies the true model 60% of the time, compared to about 25% for the LASSO. As noted in [4], increasing the value of α increases the difficulty of true model identification. At $\alpha = 0.55$, the LASSO is no longer path consistent and selects the true model approximately 20% of the time irrespective of the sample size. In contrast, the NNG remains path consistent and selects the true model with increasing frequency for larger sample sizes.

3.3 Initial Estimates for the NNG

The simulation involved generating 1000 training and validation data sets of $n \in \{20, 50, 100\}$ samples. The regularization parameters for each run were selected based on the log-likelihood of an independently generated validation set. The best subset for the stepwise forward selection method was selected using the same approach. In all simulations, the class indicators $\mathbf{y} \in \{-1, +1\}^n$ were generated independently with probability given by (1). Performance metrics recorded in each test run were: (1) negative log-likelihood, (2) model size, (3) the number of false positive regressors, and (4) the number of false negative regressors included in the best model. A regressor that was inferred to be zero is deemed to be a false positive if the data generating model has the corresponding coefficient set to a non-zero value, and similarly for false negative regressors. The following two simulation models were considered:

1. The true regression coefficients were set to $\boldsymbol{\beta} = (3, 1{\cdot}5, 0, 0, 2, 0, 0, 0)'$. The pairwise correlation between predictors i and j was $\text{corr}(i, j) = 0.5^{|i-j|}$ [9].
2. Same as Example (1), except $\beta_i = 0.85$ for all i [9].

The constant regressor was included in all simulations but was not used when tabulating results.

Example 1. The data generating model in this scenario is of medium sparsity with five out of eight regressors being noise. It is expected that ridge regression will not perform as well as the alternative methods given the level of sparsity. In contrast, the LASSO should do quite well as the data generating model does not contain highly correlated regressors. Simulation results for the test are shown in Table 1. The stepwise forward regression method (`fwd`) achieved the worst median negative log-likelihood from all the methods tested, with the performance being clearly inferior to other methods when the sample size was small ($n = 20$). From the four initial estimators tested, the `grr` method (NNG initialized with ridge regression) achieved the best median log-likelihood for the smallest sample size. However, as the sample size was increased, all four methods performed approximately equally well. In comparison to the starting estimates, the corresponding NNG method exhibited slightly more false positive regressors and significantly less false negative regressors. Additionally, the models selected by the NNG methods were generally smaller than any of the initial models.

Table 1. Simulation results for Example 1; median negative log-likelihood (NLL), mean model size (Size), mean number of false positive regressors (FP) and mean number of false negative regressors (FN) included in the selected model. Tests are based on 1000 iterations with standard errors included in parentheses.

Methods	$n = 20$				$n = 50$				$n = 100$			
	NLL	Size	FP	FN	NLL	Size	FP	FN	NLL	Size	FP	FN
fwd	30·70	1·24	1·96	0·20	7·12	2·79	0·63	0·42	2·92	3·46	0·11	0·57
	(0·80)	(0·04)	(0·03)	(0·02)	(0·07)	(0·04)	(0·03)	(0·03)	(0·01)	(0·04)	(0·01)	(0·04)
gfwd	26·64	1·18	1·97	0·15	6·74	2·75	0·64	0·38	2·84	3·36	0·12	0·47
	(0·41)	(0·04)	(0·03)	(0·02)	(0·05)	(0·04)	(0·02)	(0·03)	(0·01)	(0·04)	(0·01)	(0·03)
lasso	21·13	4·53	0·50	2·03	6·50	5·78	0·05	2·83	2·91	6·20	0·01	3·20
	(0·15)	(0·05)	(0·02)	(0·04)	(0·04)	(0·04)	(0·01)	(0·04)	(0·01)	(0·04)	(0·00)	(0·04)
glasso	22·40	2·89	0·93	0·82	6·44	3·97	0·23	1·20	2·85	4·33	0·02	1·35
	(0·18)	(0·04)	(0·03)	(0·03)	(0·05)	(0·04)	(0·01)	(0·04)	(0·01)	(0·04)	(0·00)	(0·04)
rr	21·13	8·00	0·00	5·00	6·76	8·00	0·00	5·00	3·00	8·00	0·00	5·00
	(0·14)	(0·00)	(0·00)	(0·00)	(0·03)	(0·00)	(0·00)	(0·00)	(0·01)	(0·00)	(0·00)	(0·00)
grr	21·86	3·37	0·77	1·14	6·45	4·32	0·17	1·49	2·86	4·51	0·02	1·53
	(0·21)	(0·05)	(0·02)	(0·03)	(0·03)	(0·04)	(0·01)	(0·04)	(0·01)	(0·04)	(0·00)	(0·04)
enet	20·64	6·10	0·21	3·31	6·50	6·40	0·02	3·42	2·92	6·50	0·00	3·51
	(0·12)	(0·05)	(0·01)	(0·05)	(0·03)	(0·04)	(0·00)	(0·04)	(0·01)	(0·04)	(0·00)	(0·04)
genet	22·20	3·07	0·85	0·92	6·42	4·06	0·20	1·26	2·85	4·34	0·02	1·36
	(0·22)	(0·04)	(0·02)	(0·03)	(0·04)	(0·04)	(0·01)	(0·04)	(0·01)	(0·04)	(0·00)	(0·04)
ilasso	22·41	2·90	0·93	0·83	6·42	3·98	0·22	1·20	2·85	4·33	0·02	1·35
	(0·19)	(0·04)	(0·02)	(0·03)	(0·05)	(0·04)	(0·01)	(0·04)	(0·01)	(0·04)	(0·00)	(0·04)
nng	21·34	3·50	0·66	1·16	6·34	4·35	0·11	1·46	2·84	4·51	0·01	1·52
	(0·25)	(0·04)	(0·02)	(0·03)	(0·03)	(0·04)	(0·01)	(0·04)	(0·01)	(0·04)	(0·00)	(0·04)

As the sample size was increased to $n = 100$, all methods performed equally well in terms of log-likelihood. Interestingly, the `nng` solution appeared to perform the same as using only the single best ridge regression estimate (`grr`) in this example.

Example 2. The true model is now dense and does not include any noise regressors which would make ridge regression the ideal solution for this type of problem. Table 2 depicts the corresponding simulation results. As in Example 1, all methods performed equally well given enough data. It is therefore of interest to examine regularization performance under small sample sizes ($n = 20$). Stepwise forward selection (`fwd`) attained the highest negative log-likelihood of all the regularization methods tested. The performance of `fwd` was especially poor when $n = 20$, obtaining the highest negative log-likelihood and largest number of false positives. Ridge regression achieved the smallest negative log-likelihood (`rr`) of all the methods tested for all sample sizes. This is not surprising give that the generating model is dense and ridge regression cannot zero out individual regressors. In contrast, using the NNG with LASSO, ridge regression and elastic net resulted in a somewhat worse log-likelihood and a significantly sparser solution, compared to the original model. This indicates that the NNG is producing models that are too sparse which agrees with the findings in [4]. Models superior to the corresponding initial estimates were obtained only when the NNG was

Table 2. Simulation results for Example 2; median negative log-likelihood (NLL), mean model size (Size), mean number of false positive regressors (FP) and mean number of false negative regressors (FN) included in the selected model. Tests are based on 1000 iterations with standard errors included in parentheses.

Methods	$n = 20$				$n = 50$				$n = 100$			
	NLL	Size	FP	FN	NLL	Size	FP	FN	NLL	Size	FP	FN
fwd	35·10	1·17	6·83	0·00	10·20	4·27	3·73	0·00	3·73	6·94	1·06	0·00
	(0·08)	(0·05)	(0·05)	(0·00)	(0·09)	(0·07)	(0·07)	(0·00)	(0·02)	(0·05)	(0·05)	(0·00)
gfwd	34·66	1·11	6·89	0·00	9·19	4·05	3·94	0·00	3·68	6·69	1·31	0·00
	(0·03)	(0·04)	(0·04)	(0·00)	(0·08)	(0·07)	(0·07)	(0·00)	(0·02)	(0·05)	(0·05)	(0·00)
lasso	24·83	4·95	3·06	0·00	7·76	6·94	1·06	0·00	3·50	7·76	0·23	0·00
	(0·19)	(0·05)	(0·05)	(0·00)	(0·04)	(0·03)	(0·03)	(0·00)	(0·01)	(0·01)	(0·01)	(0·00)
glasso	28·24	3·14	4·86	0·00	8·44	5·66	2·34	0·00	3·62	7·24	0·76	0·00
	(0·19)	(0·05)	(0·05)	(0·00)	(0·05)	(0·05)	(0·05)	(0·00)	(0·02)	(0·03)	(0·03)	(0·00)
rr	21·23	8·00	0·00	0·00	7·18	8·00	0·00	0·00	3·38	8·00	0·00	0·00
	(0·11)	(0·00)	(0·00)	(0·00)	(0·03)	(0·00)	(0·00)	(0·00)	(0·01)	(0·00)	(0·00)	(0·00)
grr	26·97	3·80	4·20	0·00	8·23	6·10	1·90	0·00	3·59	7·42	0·58	0·00
	(0·19)	(0·05)	(0·05)	(0·00)	(0·04)	(0·04)	(0·04)	(0·00)	(0·02)	(0·03)	(0·03)	(0·00)
enet	21·59	7·40	0·60	0·00	7·24	7·88	0·12	0·00	3·38	7·98	0·02	0·00
	(0·12)	(0·04)	(0·04)	(0·00)	(0·02)	(0·01)	(0·01)	(0·00)	(0·01)	(0·00)	(0·00)	(0·00)
genet	27·20	3·65	4·35	0·00	8·24	6·06	1·94	0·00	3·59	7·42	0·58	0·00
	(0·22)	(0·05)	(0·05)	(0·00)	(0·04)	(0·04)	(0·04)	(0·00)	(0·02)	(0·03)	(0·03)	(0·00)
ilasso	28·23	3·15	4·85	0·00	8·44	5·67	2·33	0·00	3·62	7·24	0·76	0·00
	(0·21)	(0·05)	(0·05)	(0·00)	(0·05)	(0·05)	(0·05)	(0·00)	(0·02)	(0·03)	(0·03)	(0·00)
nng	26·49	4·08	3·92	0·00	8·05	6·33	1·67	0·00	3·54	7·57	0·43	0·00
	(0·23)	(0·05)	(0·05)	(0·00)	(0·04)	(0·04)	(0·04)	(0·00)	(0·01)	(0·02)	(0·02)	(0·00)

used with stepwise forward selection. The `nng` approach attained the best negative log-likelihood and somewhat larger models in contrast to alternative NNG strategies.

3.4 Real Data

This section examines the performance of logistic regression regularization solutions using six real data sets from the UCI Machine Learning repository. The number of regressors, excluding the constant regressor, ranged from small ($p = 4$ in "transfusion") to moderate ($p = 60$ in "sonar"). All data sets were standardized to have $||\mathbf{x}_j|| = 1$ ($j = 1, \ldots, p$), where $|| \cdot ||$ denotes the Euclidean norm. For each data set, we randomly split the available data into a training, a validation and a test subset. The training data set was used to infer the parameter estimates, while the validation data set was used for selecting the regularization parameters. All tabulated results are based only on the test set. There were 100 simulation runs for each data set. For each iteration, two performance metrics were recorded: (1) classification accuracy, and (2) model size in terms of the number of regressors remaining. Stepwise forward selection was not included in this test due to its poor performance in previous experiments as well as the relatively high computational complexity of the method. The simulation results are shown in Table 3.

On the pima data set, the best classification accuracy of all the methods tested was attained by `nng`, closely followed by `grr` and `genet`. Although `rr` resulted

Table 3. Simulation results for real data. Median classification accuracy (in percent) is shown along with bootstrap estimates of standard error. Mean model size is included in parentheses. Tests are based on 100 iterations.

Methods	Datasets					
	pima	wdbc	spambase	ionosphere	transfusion	sonar
lasso	74.82 ± 0.30	95.53 ± 0.18	91.62 ± 0.09	79.68 ± 0.69	78.46 ± 0.29	71.02 ± 0.29
	(6.52)	(7.78)	(48.09)	(7.73)	(3.68)	(10.81)
glasso	74.82 ± 0.28	95.12 ± 0.20	91.79 ± 0.09	79.68 ± 0.58	78.35 ± 0.29	69.32 ± 0.29
	(4.61)	(4.64)	(35.22)	(3.67)	(3.42)	(4.57)
rr	74.30 ± 0.32	95.93 ± 0.16	91.45 ± 0.12	81.27 ± 0.53	78.57 ± 0.20	75.00 ± 0.20
	(8.00)	(30.00)	(57.00)	(32.00)	(4.00)	(60.00)
grr	75.18 ± 0.29	95.39 ± 0.15	91.75 ± 0.11	79.88 ± 0.32	78.79 ± 0.33	69.89 ± 0.33
	(4.77)	(6.21)	(39.04)	(5.53)	(3.53)	(7.54)
enet	74.65 ± 0.31	96.21 ± 0.16	91.48 ± 0.11	81.27 ± 0.40	78.57 ± 0.19	75.00 ± 0.19
	(7.05)	(23.03)	(51.61)	(22.27)	(3.92)	(53.06)
genet	75.00 ± 0.23	95.12 ± 0.16	91.77 ± 0.09	80.28 ± 0.41	78.79 ± 0.35	68.75 ± 0.35
	(4.70)	(5.82)	(37.47)	(5.10)	(3.52)	(7.20)
ilasso	74.82 ± 0.27	95.12 ± 0.20	91.77 ± 0.08	79.88 ± 0.55	78.35 ± 0.29	69.32 ± 0.29
	(4.61)	(4.82)	(35.28)	(3.79)	(3.44)	(4.75)
nng	75.35 ± 0.21	95.66 ± 0.19	91.77 ± 0.08	80.48 ± 0.36	78.35 ± 0.37	71.59 ± 0.37
	(4.80)	(6.63)	(40.49)	(6.53)	(3.49)	(7.94)

in the lowest classification accuracy on this data set, the difference between `rr` and `nng` was only about 1%. It is clear that applying the NNG to any initial estimate has again resulted in a more parsimonious model, which is still highly predictive. For example, `lasso` models have on average 6.5 regressors compared to 4.6 for the `glasso` for about the same classification accuracy. Although the iterated LASSO did not improve on the LASSO, it generally inferred sparser models. The elastic net obtained the best classification accuracy on the wdbc data set out of all the methods tested. However, the average model inferred by `enet` was approximately four times the size of the average `nng` model, while the classification accuracy of `nng` was only slightly smaller (95.6% for `nng` versus 96.2% for `enet`).

All methods performed equally well on the spambase dataset in terms of classification accuracy. In terms of model complexity, the LASSO and the elastic net resulted in the largest models, while the NNG based methods as well as the iterative LASSO inferred models with about 10 regressors less, on average. Similar findings can be noted for the ionosphere, transfusion and sonar data sets. We did observe an interesting anomaly on the sonar dataset. The elastic net and ridge regression obtained significantly higher prediction accuracy and larger average model size, in contrast to all other methods considered. For example, the `nng` inferred models were significantly simpler and resulted in reduced classification accuracy of about 5%. Given the size of the test data, an increase in accuracy of 5% equates to four extra samples being correctly classified by `enet` and `rr`, which is not a highly significant improvement.

3.5 Discussion and Recommendations

Simulations in Section 3.3 clearly show that stepwise forward selection commonly resulted in models which generalize poorly, especially given small to medium sample sizes. Predictive performance of stepwise methods remained poor irrespective of the sparsity of the data generating model. The nonnegative garrote, or the iterated LASSO, is recommended if the data generating model is expected to be (highly) sparse. While most of the considered regularization strategies showed promising performance in the sparse setting, the consistency of NNG and the iterated LASSO (see Section 3.2) make the techniques highly suitable. Models inferred by the NNG were consistently simpler and attained significantly smaller numbers of false negatives in contrast to most other methods considered. Interestingly, the iterated LASSO has outperformed the original LASSO in terms of prediction and model size and is thus recommended for logistic regression if a LASSO-type penalty is desired. Although the elastic net achieved similar classification performance to the NNG, the models inferred by the elastic net consisted of significantly more regressors.

A ridge regression estimate is recommended as the starting point for NNG over maximum likelihood or LASSO-type solutions. Ridge regression allows the NNG to be applied to collinear models which is otherwise not possible with the maximum likelihood approach. Unlike ridge regression, LASSO and the elastic net generate sparse models which implies that some coefficients will be set to zero prior to running the NNG. Due to the form of the NNG penalty, regression coefficients are not altered once set to zero. Thus, if a sparse solution, like the LASSO, is used for the initial estimates, coefficients that the NNG would normally retain may be rendered insignificant. Our recommendation of ridge regression as an initial solution to NNG is in agreement with the findings published in [4].

References

1. Breiman, L.: Better subset regression using the nonnegative garrote. Technometrics 37, 373–384 (1995)
2. Xiong, S.: Some notes on the nonnegative garrote. Technometrics 52(3), 349–361 (2010)
3. Albert, A., Anderson, J.A.: On the existence of maximum likelihood estimates in logistic regression models. Biometrika 71(1), 1–10 (1984)
4. Yuan, M., Lin, Y.: On the non-negative garrotte estimator. Journal of the Royal Statistical Society (Series B) 69(2), 143–161 (2007)
5. Genkin, A., Lewis, D.D., Madigan, D.: Large-scale Bayesian logistic regression for text categorization. Technometrics 49(3), 291–304 (2007)
6. Park, M.Y., Hastie, T.: L_1-regularization path algorithm for generalized linear models. Journal of the Royal Statistical Society (Series B) 69(4), 659–677 (2007)
7. Friedman, J., Hastie, T., Tibshirani, R.: Regularized paths for generalized linear models via coordinate descent. Journal of Statistical Software 33(1) (2010)
8. Cessie, S.L., Houwelingen, J.C.V.: Ridge estimators in logistic regression. Journal of the Royal Statistical Society (Series C) 41(1), 191–201 (1992)
9. Tibshirani, R.: Regression shrinkage and selection via the lasso. Journal of the Royal Statistical Society (Series B) 58(1), 267–288 (1996)
10. Zou, H., Hastie, T.: Regularization and variable selection via the elastic net. Journal of the Royal Statistical Society (Series B) 67(2), 301–320 (2005)
11. Huang, J., Ma, S., hui Zhang, C.: The iterated lasso for high-dimensional logistic regression. Technical Report 392, The University of Iowa (2008)

Identification of Breast Cancer Subtypes Using Multiple Gene Expression Microarray Datasets

Alexandre Mendes

Centre for Bioinformatics, Biomarker Discovery and Information-Based Medicine
School of Electrical Engineering and Computer Science
Faculty of Engineering and Built Environment
The University of Newcastle, Callaghan, NSW, 2308, Australia
Alexandre.Mendes@newcastle.edu.au

Abstract. This work is motivated by the need for consensus clustering methods using multiple datasets, applicable to microarray data. It introduces a new method for clustering samples with similar genetic profiles, in an unsupervised fashion, using information from two or more datasets. The method was tested using two breast cancer gene expression microarray datasets, with 295 and 249 samples; and 12,325 common genes. Four subtypes with similar genetic profiles were identified in both datasets. Clinical information was analysed for the subtypes found and they confirmed different levels of tumour aggressiveness, measured by the time of metastasis, thus indicating a connection between different genetic profiles and prognosis. Finally, the subtypes identified were compared to already established subtypes of breast cancer. That indicates that the new approach managed to detect similar gene expression profile patterns across the two datasets without any *a priori* knowledge. The two datasets used in this work, as well as all the figures, are available for download from the website http://www.cs.newcastle.edu.au/~mendes/BreastCancer.html.

Keywords: Bioinformatics, breast cancer, data mining, genetic algorithms.

1 Introduction

The introduction of the microarray technology imposed a series of new challenges in terms of producing relevant and statistically sound results. Current research indicates that with the amount of data publicly available, the use of a single dataset is no longer acceptable to justify new medical discoveries. Comparisons with previous, similar studies need to be carried out. A problem that arises in this situation is that microarray data is highly heterogeneous, noisy, and in general, different unsupervised techniques will find different configurations of clusters for the same dataset. In addition, clusters found using a specific dataset sometimes are not observed in other datasets. Consensus clustering techniques try to overcome these problems, with two main types being found in the literature.

D. Wang and M. Reynolds (Eds.): AI 2011, LNAI 7106, pp. 92–101, 2011.

The first deals with single datasets and proposes the concurrent use of several unsupervised clustering techniques, which will likely produce different partitions of the samples. A consensus clustering is then determined using information from all clusters found, usually based on some similarity measure among elements [15,10,4]. The second type of consensus clustering involves finding clusters which have similar profiles across multiple datasets. This is the goal of the method introduced in this paper, and two previous works should be cited. First, Filkov and Skiena (2003) [2] modeled the consensus clustering of multiple datasets as a median partition problem and use three types of heuristics (local search, greedy and simulated annealing) to address it. Then, in Hoshida et al. (2007) [5], the authors use a statistical test to find the consensus clusters. The literature on consensus clustering and microarrays is extensive and even though several methods are available, no single approach dominates the scientific literature.

This paper offers a new consensus clustering technique, which differs from the previous ones mainly because it optimizes three criteria at once. Those are the number of biomarkers that characterize the clusters; consistency of the clusters across datasets; and statistical relevance of the clusters, measured by a classification test.

The method introduced in this work extends the study in Mendes (2008) [8]. It uses a Genetic Algorithm as the search engine and was tested with two well-known datasets from previous breast cancer studies. The first contains 24,158 probes, 295 samples and was introduced in Vijver et al. (2002) [16]. The second dataset has 44,928 probes, 249 samples and was introduced in Miller et al. (2005) [9].

The results presented in this work show the clustering of breast cancer samples into four subtypes. These subtypes were then compared to subtypes of breast cancer already established in the medical literature, using well-known markers. Finally, the subtypes are justified from a clinical standpoint as well, by performing an analysis of the *time of metastasis* associated to the samples in each subtype. Even though such clinical information was not directly used in the determination of the subtypes by our method, the Kaplan-Meier curves of the time of metastasis are consistently distinct in both datasets. In other words, the subtypes found share similar genotypical and phenotypical profiles in the two datasets, even though the method only uses genotypical information.

2 The Consensus Clustering Problem

The consensus clustering problem addressed in this work can be described as follows. Given k input datasets (D_1, D_2,..., D_k), identify partitions of the samples in D_1, D_2,..., D_k into two clusters, which:

- Are supported by the same set of biomarkers; and preferably by a large number of them (higher statistical significance of genetic signatures);
- Have a high accuracy classification of the samples in each dataset (higher intra-cluster similarity and inter-cluster dissimilarity);

- Present a similar proportion of samples in both clusters for all datasets (both clusters should be observed in all datasets to indicate consistency).

The large number of biomarkers indicate that the clusters are not product of a statistical artifact. Although in practice biologists will use just a small number of biomarkers for classification purposes, or when designing a diagnostic kit, a relatively large number of biomarkers is generally recommended for the determination of subtypes. That follows the 'data-driven' approach to biomarker discovery, which is discussed in reference [17] (i.e. analyzing the entire genome rather that working from a hypothesis about one or few candidate genes).

The second characteristic is the classification accuracy obtained with a cross-validation procedure, associated to a classification model. High classification accuracy can be associated to high intra-cluster similarity and inter-cluster dissimilarity, and will reflect on the accuracy of future classifiers for prognosis.

Finally, the third characteristic is the proportion of the samples in each subtype and in each dataset, which reflects the consistency of these subtypes across datasets.

These three characteristics are combined into a single objective function used to assess the quality of putative partitions of the samples. Next, we formalize the objective function, but before doing so, consider the following notation:

- $D = \{D_1, D_2,...,D_k\}$: Set of k datasets;
- $C = \{c_1, c_2\}$: Set of classes. In every iteration, the samples are partitioned into two classes: c_1/c_2;
- S_{D_i}: Set of samples in D_i; $|S_{D_i}| = m_{D_i}$;
- $S_{D_i(c_j)}$: set of samples in D_i that belong to class c_j; $|S_{D_i(c_j)}| = m_{D_i(c_j)}$.

The identification of breast cancer subtypes is done iteratively. Initially, the samples are divided into two clusters. Then, those two clusters are further divided into four, and so on, resulting in a binary tree structure. The criterion to stop the division was based on the clinical analysis of the time of metastasis for the samples in each cluster. When no significant difference is observed between two new clusters, in terms of the time of metastasis, we consider that they actually represent the same subtype of the disease, and stop the division.

2.1 Objective Function

The objective function takes into account three characteristics that should be observed in high quality partitions.

- Partitions should be supported by a large number of biomarkers: In each division, the partitions should be supported by the same set of biomarkers in all datasets; and preferably be composed of a large number of them. The method implementation played an important role in this aspect. If we considered all k datasets separately and tried putative partitions for each of them, the search space would be prohibitively large and the sets of biomarkers would be considerably different for each partition in each dataset; i.e. there would be no consistency between biomarkers for any given subtype across datasets.

To overcome this, first we force all datasets D_1, D_2,...,D_k to contain the same genes; i.e. *any gene that is not present in all datasets is removed from the analysis*. Then, one of the datasets is selected as the *main dataset*. This main dataset will have its samples partitioned first, and this partition will induce the partitions in the other $k-1$ datasets.

Let the main dataset chosen be D_1. Given a putative partition for the samples in D_1 into classes c_1 and c_2, a t-Student statistical test is used to determine the $n_{markers}$ associated biomarkers ($p < 0.01$). The biomarkers for the partition in D_1 are then used to induce partitions in the other datasets D_2,...,D_k. A Nearest Neighbor classification model [19] is created with the biomarkers and samples in D_1 and then used to assign the samples in D_2,...,D_k either to class c_1 or c_2.

- High accuracy classification of samples in all datasets: The high accuracy classification of the samples in all datasets acts as a proxy for high intra-cluster similarity and inter-cluster dissimilarity. Given the nearest neighbor-based classification model from D_1 and the partitions of the samples in D_1, D_2,...,D_k, we perform a 10-fold cross-validation [19] in all datasets D_i, calculating the accuracy of each classification acc_{D_i}. The overall accuracy $\overline{acc}_D$ is:

$$\overline{acc}_D = \frac{1}{k}\sum_{i=1}^{k} acc_{D_i} \tag{1}$$

- Similar proportion of samples in clusters across all datasets: It is arguably recommended to have a similar proportion of samples in each cluster, across all datasets. First, this would indicate that subtypes of diseases identified are present in all datasets. Moreover, the proportion of the number of samples in each class indicates that a subtype of the disease, more/less common in a dataset, should be more/less common in all other datasets as well. This is a strong assumption, which only holds if different cohorts share similar sampling characteristics. The balance of the partition of the samples is denoted as B, and is calculated as follows. First, let:

$$\overline{m}_{c_j} = \frac{1}{k}\sum_{i=1}^{k} \frac{m_{D_i(c_j)}}{m_{D_i}} \tag{2}$$

be the average proportion of samples in class c_j in all datasets. The balance should be optimum when $m_{D_i(c_j)}/m_{D_i}$, i.e. the proportion of samples clustered in c_j is the same in every dataset D_i. The equation for the balance is:

$$B = \sum_{i=1}^{k} \left| \overline{m}_{c_1} - \frac{m_{D_i(c_1)}}{m_{D_i}} \right| + \sum_{i=1}^{k} \left| \overline{m}_{c_2} - \frac{m_{D_i(c_2)}}{m_{D_i}} \right| \tag{3}$$

Finally, the objective function used in this work is stated as:

$$obj = n_{markers} * \overline{acc}_D * \frac{1}{B + \epsilon} \tag{4}$$

The objective function aims at a trade-off between large number of biomarkers for D_1 ($n_{markers}$); high average accuracy of the classification across datasets ($\overline{acc}_D$); and good balance of classes across datasets ($B \approx 0$).

2.2 The Genetic Algorithm

The problem of finding the partition of the samples that maximizes Eq. 4 was addressed using a Genetic Algorithm (GA). GAs are population-based search methods [3] where a population of solutions evolves through the application of special operators (recombination and mutation), and selection pressure.

- **Representation:** The search space of the consensus clustering problem consists of all the possible partitions of the samples in the dataset D_1 into two classes. In terms of genetic algorithm implementation, a partition P is represented as a binary array $P = [p_1, p_2, ..., p_{m_{D_1}}]$, with $p_i \in \{0, 1\}$.

- **Population structure:** The GA employs a population structure that follows a complete ternary tree with three levels, i.e. 13 individuals. This structure was object of study in the past, and genetic/memetic algorithms using it performed better compared to non-structured approaches in several combinatorial optimization problems [1,11]. Also, the use of fewer individuals is critical because, in this problem, the objective function calculation is very time-consuming, as it involves several, complex steps.

- **Mutation:** The mutation operator implemented was the bit-swap. A sample is chosen uniformly at random and moves from a class to another, i.e. either $c_1 \rightarrow c_2$ or $c_2 \rightarrow c_1$. This 1-bit mutation is applied to 10% of the offspring created, also chosen uniformly at random.

- **Recombination and acceptance policy:** The recombination operator chosen was the uniform crossover (UX) [12]. In every generation, a number of individuals equal to the size of the population is created and evaluated. Offspring that are better than at least one of their parents survive to the next generation, directly replacing their worst parent. Even though this scheme creates a strong evolutionary pressure, premature convergence is controlled by checking population diversity and applying restart procedures.

- **Population diversity and restart:** The diversity check procedure verifies at every generation whether any offspring created was better than at least one of its parents. If none was better, a population restart follows, which keeps the current best solution within the population (*elitist restart*), and replaces all others by randomly-generated solutions. Indeed, if no solution created within a generation was better than one of its parents, that indicates that the current population has evolved enough generations to be consisted of high-quality individuals only, which are also likely to be very similar.

3 The Breast Cancer Datasets

Two breast cancer microarray datasets were used in this work. The first one (dataset D_1) is from a study with 295 patients diagnosed with primary breast carcinomas presented in Vijver et al. (2002) [16]. A 25,000-gene cDNA array consisting of 24,479 probes was used for each patient. The second dataset (D_2) comes from a study comprising 259 primary breast cancer patients presented in Miller et al. (2005) [9]. Each patient was sampled using an Affymetrix genechip with 38,061 probes.

A first pre-processing procedure removed duplicate genes from both datasets, resulting in D_1 keeping 14,547 unique genes; and D_2 keeping 18,342 unique genes. A second step involved forcing the two datasets to contain exactly the same genes (to enforce consistency of classifiers' attributes). Using the gene symbols as identifiers, there was a total of 12,325 common genes.

4 Results

After applying the clustering algorithm to the two datasets, a binary tree with the partition of the breast cancer samples was produced. It is shown in Figure 1 and depicts the types found, the biomarkers for the partitions found in both datasets, and the Kaplan-Meier (K-M) curves for the time of metastasis associated to the types identified.

Samples were first divided into two subtypes (Types 1 and 2) and then into four others (Types 3 to 6). Note that the biomarkers in each specific division are the same for the two datasets D_1 and D_2, and the types have a similar clinical profile in terms of prognosis. Type 2 is more aggressive than Type 1; and Types 3 and 5 are more aggressive as well, compared to Types 4 and 6.

Additional divisions of Types 3 to 6 into more subtypes were tested, but the clinical profiles obtained were not consistent across the two datasets. The classification shown in Figure 1 contains only those subtypes that present consistent clinical profiles.

4.1 Comparison with Existing Subtypes

There are five subtypes of breast cancer broadly accepted by the medical community: *normal breast-like*, *basal*, *luminal A*, *luminal B*, and *HER2+/ER-*. In order to compare the four subtypes identified in this work with them, we analyzed a number of genetic markers associated to breast cancer, collected from the following studies:

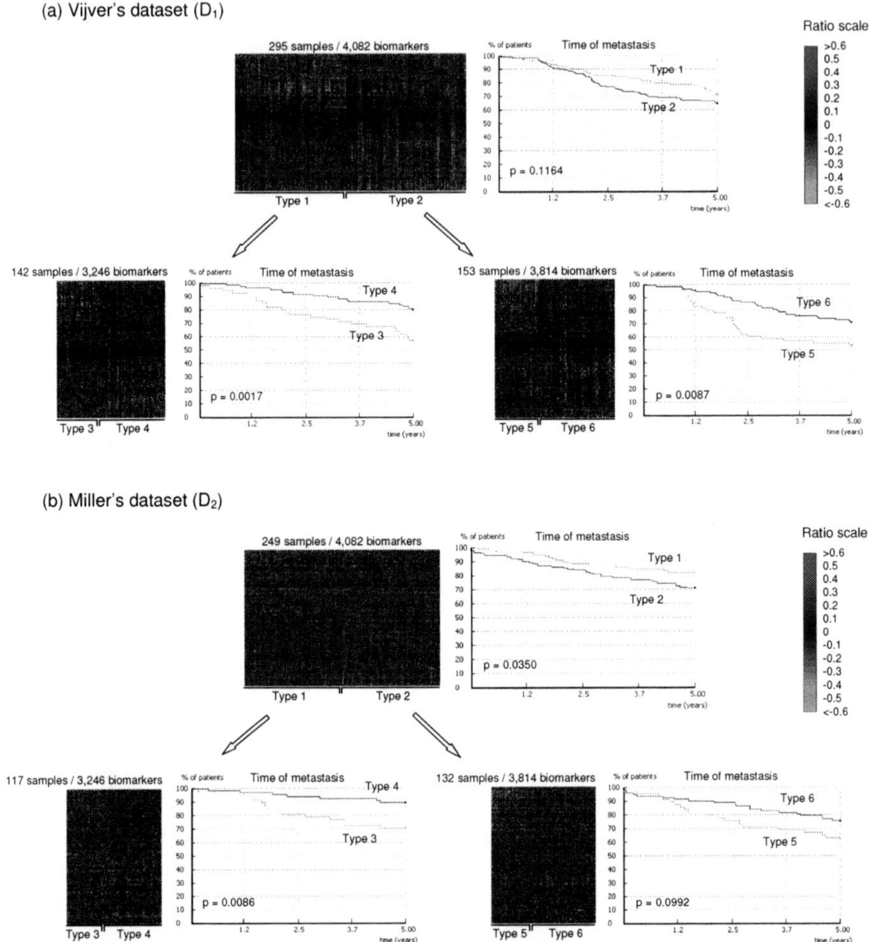

Fig. 1. (a) Classification of breast cancer subtypes for Vijver's dataset. Samples were initially divided into two subtypes – Types 1 and 2 – which were further divided into the final four subtypes – Types 3 to 6. For each division we present a genetic signature with the biomarkers obtained by a t-student statistical test ($p < 0.01$). Next to each signature we present the associated Kaplan-Meier curves for the time of metastasis. (b) Classification of breast cancer subtypes for Miller's dataset. The subtypes are analogous to the ones identified in (a).

- *Perreard et al. (2006)* [14]: 53 biomarkers for different subtypes of breast cancer – 37 so-called 'intrinsic' genes to classify the subtypes, plus PGR, EGFR and 14 proliferation-related genes.
- *Hu et al. (2009)* [6]: 9 oncogenes and tumor suppressor genes.
- *Paik et al. (2004)* [13] – *Oncogene DX*: a breast cancer prognosis kit based on 21 genes for ER+, lymph node-negative patients.

The expression profiles of the genetic markers mentioned in the three studies above are shown in Figure 2. They were divided according to the study and the dataset. Samples are ordered from Type 3 to Type 6, in all figures.

Basal tumors are characterized by being ESR1, PGR and ERBB2 negative, i.e. these three markers are under-expressed. This subtype is also referred to as triple receptor negative [7]. Type 5 is the cluster where those genes are the least expressed. This is an aggressive subtype and that behaviour agrees with K-M curves in Figure 1a-b. Therefore, we can associate Type 5 to the basal breast cancer subtype.

Two other types are also identifiable: *Luminal A* and *luminal B*. These types are molecularly similar, being characterized by the over-expression of ESR1, PGR, GATA3 and FOXA1. That occurs in both Types 4 and 6. The main

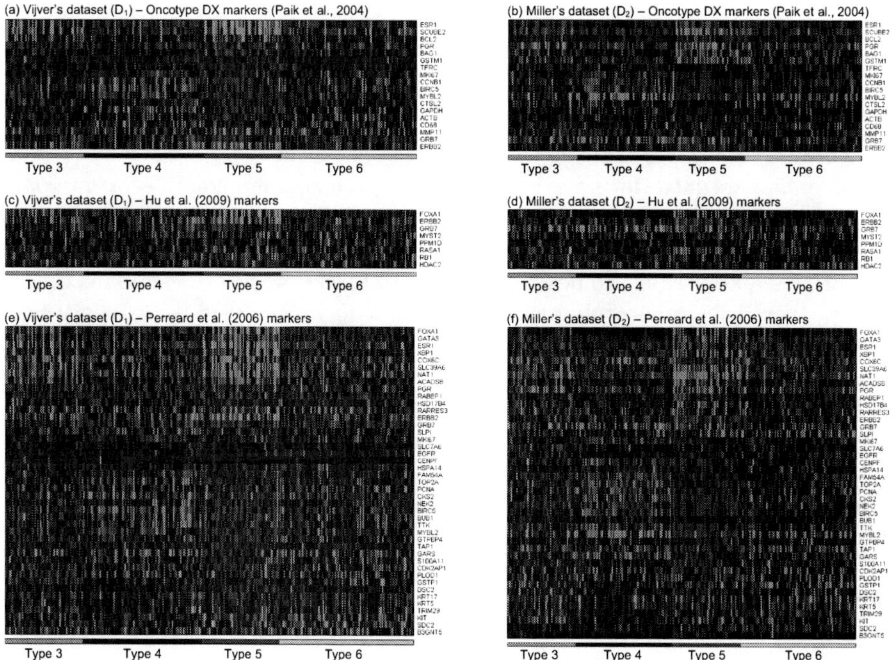

Fig. 2. Gene expression of breast cancer markers found in the literature, considering the four subtypes identified in Figure 1a-b. Three sets of genetic markers are compared: *(a-b)* Oncotype DX [13], *(c-d)* Hu et al. (2009) [6] and *(e-f)* Perreard et al. (2006) [14]. Based on the markers' expression, we can make the following mapping: Type 5 corresponds to *basal* samples. Basal is one of the most aggressive subtypes of breast cancer, which is in agreement with the Kaplan-Meier curves in Figure 1a-b. Types 4 and 6 correspond to *Luminal A* and *Luminal B* samples, respectively. They are similar with respect to the biomarkers, but proliferation-related genes are under-regulated in Type 4 and over-regulated in Type 3. Finally, Type 3 corresponds to HER2+/ER- tumors, which is also a very aggressive subtype – again showing agreement with the K-M curves in Figure 1a-b.

difference between luminal A and B is that *proliferation-related genes* are under-expressed and over-expressed in those subtypes, respectively [18]. The proliferation genes that we refer to are listed in Perreard et al. (2006) [14] (HSPA14, GTPBP4, PCNA, CKS2, NEK2, TOP2A, BUB1, TTK, FAM54A, MKI67, MYBL2, BIRC5 and CENPF). This difference indicates that Type 4 corresponds to *luminal A* and Type 6 to *luminal B.*

Finally, Type 3 appears to correspond to HER2+/ER- tumors. This type is characterized by the under-expression of ESR1 and PGR; and over-expression of ERBB2. In addition, proliferation-related genes are over-expressed. From the clinical standpoint, HER2+/ER- is, together with basal, one of the most aggressive breast cancer tumor subtype. That would be in agreement with the K-M curves in Figure 2. These findings illustrate how the method managed to identify, across two distinct datasets, four subtypes broadly accepted by the scientific community. Moreover, the clinical aspects have also shown consistency across datasets and agreed with the scientific literature for the subtypes.

5 Conclusion

In this paper we introduce a new method to perform classification of microarray samples using multiple datasets, and test the approach using two publicly available breast cancer datasets. Four subtypes were identified and presented similar gene expression profiles across both datasets, as well as similar clinical profiles (based on time of metastasis). A subsequent analysis comparing those four subtypes with the currently accepted subtypes of breast cancer in the scientific community provided a mapping between them. The types basal, luminal A, luminal B and HER2+/ER- were mapped into the four subtypes identified by our algorithm by analyzing the expression profile of several markers reported in the literature. That result was also corroborated by the analysis of the time of metastasis, which shows that the types mapped into basal and HER2+/ER- subtypes have a more aggressive behavior.

It is worth emphasizing that the method introduced in this study successfully discovered subtypes in an unsupervised, unbiased (data-driven) fashion, using data from a genetically heterogeneous disease. It has the potential to impact the discovery of subtypes of other heterogeneous diseases for which microarray data is available.

References

1. Buriol, L., Franca, P., Moscato, P.: A new memetic algorithm for the asymmetric traveling salesman problem. Journal of Heuristics 10, 483–506 (2004)
2. Filkov, V., Skiena, S.: Integrating microarray data by consensus clustering. In: Proceeding of the 15th IEEE International Conference on Tools with Artificial Intelligence, pp. 418–426. IEEE Computer Society (2003)
3. Glover, F., Kochenberger, G.: Handbook of Metaheuristics. Springer, USA (2003)

4. Grotkjaer, T., Winther, O., Regenberg, B., Nielsen, J., Hansen, L.: Robust multiscale clustering of large dna microarray datasets with the consensus algorithm. Bioinformatics 22, 58–67 (2006)
5. Hoshida, Y., Brunet, J., Tamayo, P., Golub, T., Mesirov, J.: Subclass mapping: Identifying common subtypes in independent disease data sets. PLoS ONE 2, e1195 (2007)
6. Hu, X., Stern, H.M., Ge, L., O'Brien, C., Haydu, L., Honchell, C.D., Haverty, P.M., Wu, B.P.T., Amler, L.C., Chant, J., Stokoe, D., Lackner, M.R., Cavet, G.: Genetic alterations and oncogenic pathways associated with breast cancer subtypes. Molecular Cancer Research 7, 511–522 (2009)
7. Irvin Jr., W., Carey, L.: What is triple-negative breast cancer? European Journal of Cancer 44, 2799–2805 (2008)
8. Mendes, A.: Consensus clustering of gene expression microarray data using genetic algorithms. In: Proceedings of PRIB 2008 - Third IAPR International Conference on Pattern Recognition in Bioinformatics (Supp. volume), pp. 181–192 (2008)
9. Miller, L.D., Smeds, J., George, J., Vega, V.B., Vergara, L., Ploner, A., Pawitan, Y., Hall, P., Klaar, S., Liu, E.T., Bergh, J.: An expression signature for p53 status in human breast cancer predicts mutation status, transcriptional effects, and patient survival. Proceedings of the National Academy of Sciences 102, 13550–13555 (2005)
10. Monti, S., Mesirov, P.T.J., Golub, T.: Consensus clustering: A resampling-based method for class discovery and visualization of gene expression microarray data. Machine Learning 52, 91–118 (2003)
11. Moscato, P., Mendes, A., Berretta, R.: Benchmarking a memetic algorithm for ordering microarray data. Biosystems 88, 56–75 (2007)
12. Olariu, S., Zomaya, A.: Handbook of Bioinspired Algorithms and Applications. Chapman & Hall/CRC, USA (2005)
13. Paik, S., Shak, S., Tang, G., Kim, C., Baker, J., Cronin, M., Baehner, F.L., Walker, M.G., Watson, D., Park, T., Hiller, W., Fisher, E.R., Wickerham, L., Bryant, J., Wolmark, N.: A multigene assay to predict recurrence of tamoxifen-treated, node-negative breast cancer. The New England Journal of Medicine 351, 2817–2826 (2004)
14. Perreard, L., Fan, C., Quackenbush, J., Mullins, M., Gauthier, N., Nelson, E., Mone, M., Hansen, H., Buys, S., Rasmussen, K., Orrico, A., Dreher, D., Walters, R., Parker, J., Hu, Z., He, X., Palazzo, J., Olopade, O., Szabo, A., Perou, C.M., Bernard, P.: Classification and risk stratification of invasive breast carcinomas using a real-time quantitative rt-pcr assay. Breast Cancer Research 8, R23 (2006)
15. Swift, S., Tucker, A., Vinciotti, V., Martin, N., Orengo, C., Liu, X., Kellam, P.: Consensus clustering and functional interpretation of gene-expression data. Genome Biology 5, R94 (2004)
16. van de Vijver, M., He, Y., van't Veer, L., Dai, H., Hart, A., Voskuil, D., Schreiber, G., Peterse, J., Roberts, C., Marton, M., Parrish, M., Atsma, D., Witteveen, A., Glas, A., Delahaye, L., van der Velde, T., Bartelink, H., Rodenhuis, S., Rutgers, E., Friend, S., Bernards, R.: A gene expression signature as a predictor of survival in breast cancer. The New England Journal of Medicine 347, 1999–2009 (2002)
17. van't Veer, L., Bernards, R.: Enabling personalized cancer medicine through analysis of gene-expression patterns. Nature 452, 564–570 (2008)
18. Weigelt, B., Baehner, F., Reis-Filho, J.: The contribution of gene expression profiling to breast cancer classification, prognostication and prediction: a retrospective of the last decade. Journal of Pathology 220, 263–280 (2010)
19. Witten, I., Frank, E.: Data Mining: Practical Machine Learning Tools and Techniques. Morgan Kaufmann, USA (2005)

Combining Instantaneous and Time-Delayed Interactions between Genes - A Two Phase Algorithm Based on Information Theory

Nizamul Morshed and Madhu Chetty

Monash University,
Australia
{nizamul.morshed,madhu.chetty}@monash.edu

Abstract. Understanding the way how genes interact is one of the fundamental questions in systems biology. The modeling of gene regulations currently assumes that genes interact either instantaneously or with a certain amount of time delay. In this paper, we propose an information theory based novel two-phase gene regulatory network (GRN) inference algorithm using the Bayesian network formalism that can model both instantaneous and single-step time-delayed interactions between genes simultaneously. We show the effectiveness of our approach by applying it to the analysis of synthetic data as well as the *Saccharomyces cerevisiae* gene expression data.

Keywords: Information theory, Bayesian network, Gene regulatory network.

1 Introduction

Development of high-throughput DNA microarray technologies has made the deciphering of regulatory interactions between genes feasible. Due to the fact that the system level view of gene functionalities provided by such gene regulatory networks (GRN) can aid in complex disease treatment, drug discovery, and also in designing environment friendly and efficient production of biofuels, this problem has received considerable interest in the recent years.

Bayesian networks (BN)[1, 2] have been used extensively for the inference of gene regulatory networks. Due to the firm statistical foundation, a BN can deal with the stochastic aspects of gene expression and the noisy measurements of microarray data in a natural way[2]. However, temporal dynamic aspects of gene regulation are not considered in BN-based models[1]. Dynamic Bayesian networks (DBN)[3–6], an extension of static Bayesian networks, can effectively deal with the temporal aspects of gene regulation. This, in effect, enables it to model feedback loops, which are an integral part of regulatory networks.

Although there is not much study on the type of interactions that occur among genes, it is natural that the interactions can be (almost) instantaneous or time-delayed. The biological intuition behind the first type of regulation is that, the

D. Wang and M. Reynolds (Eds.): AI 2011, LNAI 7106, pp. 102–111, 2011.

effect of a change in the expression level of a regulator gene can be carried on to the regulated gene (almost) instantaneously and in these scenarios, the effect will be reflected (almost) immediately in the regulated genes expression level. On the other hand, in cases where regulatory interactions are slower, the effect may be seen on the regulated gene after some time. To our knowledge, the currently existing BN-based techniques that use time series data assume either of the two, but not both.

To achieve the objective of capturing both types of interactions, we first describe a framework that can represent both types of interactions. A novel gene regulatory network reconstruction algorithm employing information theoretic quantities is then proposed. The approach is validated by carrying out experiments using both synthetic and real-life data. The comparison with other methods shows the effectiveness of our approach.

The rest of the paper is organized as follows. In Section 2, brief background information on Bayesian networks and information theoretic quantities is provided. Section 3 explains the proposed methodology and its formalization. Section 4 discusses the synthetic and real-life networks used for assessing our approach and also its comparison with other techniques. Section 5 concludes with some observations and remarks.

2 Background

In this section, we briefly discuss the formalizations behind the main concepts involved in this paper: Bayesian networks (BN), dynamic Bayesian Networks (DBN) and information theoretic quantities.

2.1 Bayesian Network (BN)

Formally, a Bayesian network for a set of random variables $\boldsymbol{X} = \{X_1, X_2, \cdots, X_n\}$, is represented by $B = \{G, \boldsymbol{\theta}\}$, where $G = \{\boldsymbol{V}; \boldsymbol{E}\}$ is a directed acyclic graph (DAG), having $\boldsymbol{V}$ as the vertex set and $\boldsymbol{E}$ as the edge set; $\boldsymbol{\theta} = \{\theta_1, \cdots, \theta_n\}$ corresponds to the parameter set storing the conditional joint probability distribution over $\boldsymbol{X}$ and $\theta_i = \theta(X_i|Pa(X_i))$ is the conditional probability distribution of X_i given all the parents of X_i (denoted by $P(X_i|Pa(X_i))$).

In a BN, the joint distribution can be decomposed in the product form:

$$P(X_1, \cdots, X_n) = \prod_{i=1}^{n} P(X_i|Pa(X_i)) \quad (1)$$

where $Pa(X_i)$ is the parent set of gene X_i in G.

2.2 Dynamic Bayesian Network (DBN)

Considering $\boldsymbol{X}$ to be a set of attributes changing in a temporal process of T time slices, a DBN represents the joint probability distribution over the variables

$X[0] \bigcup X[1] \bigcup \cdots \bigcup X[T-1]$, where random variable $X_i[t]$ denotes the value of node X_i at time slice t, and $X[t]$ denotes the set of variables $\{X_i[t] | 1 \leq i \leq n\}$, for $0 \leq t \leq T-1$ [3, 4].

In this paper we work with first-order Markov DBN, which is based on the following two assumptions:

1. *First Order Markov Property*:

$$P(X[t]|X[t-1], \cdots, X[0]) = P(X[t]|X[t-1]) \tag{2}$$

   the equation means that the value of a variable at a time point depends only on the previous time point.
2. *Stationarity*:
   $P(X[t]|X[t-1])$ is independent of t.

2.3 Information Theoretic Quantities

Decomposition Property of Mutual Information. In a BN, if $Pa(X_i)$ is the parent set of a node X_i ($X_{ik} \in Pa(X_i), k = 1, \ldots s_i$), and the cardinality of the set is s_i, the following identity holds[9]:

$$MI(X_i, Pa(X_i)) = MI(X_i, X_{i1}) + \sum_{j=2}^{s_i} MI\left(X_i, X_{ij} | \{X_{i1}, \cdots, X_{i(j-1))}\}\right) \tag{3}$$

Directionality Index (DI). The Directionality Index[10] between genes X and Y is defined as:

$$DI_{XY} = \frac{MI_{X \to Y} - MI_{Y \to X}}{MI_{X \to Y} + MI_{Y \to X}} \tag{4}$$

where the quantities $MI_{X \to Y}$ and $MI_{Y \to X}$ are defined by the following equations:

$$MI_{X \to Y} = \frac{1}{N} \sum_{\delta=1}^{N} MI^{\delta}_{X \to Y} \tag{5}$$

$$MI_{Y \to X} = \frac{1}{N} \sum_{\delta=1}^{N} MI^{\delta}_{Y \to X} \tag{6}$$

Here, the quantities in the left side of the equation (5) and (6) quantify the information that is gained from the gene X (or Y) about the gene Y (or X) at some later point in time and N is the maximal later point.

If we assume that the quantity X_δ (or Y_δ) is an observable derived from the state of the gene X (or Y) δ steps in the future, i.e. $X_\delta : x_{t+\delta} = x_t$ (or $Y : y_{t+\delta} = y_t$), $MI^{\delta}_{X \to Y}$ and $MI^{\delta}_{Y \to X}$ can be defined in terms of Conditional Mutual Information (CMI) by the following equations:

$$MI^{\delta}_{X \to Y} = MI(X, Y_\delta | Y) \tag{7}$$

$$MI^{\delta}_{Y \to X} = MI\left(Y, X_{\delta}|X\right) \tag{8}$$

The value of DI_{XY} ranges from -1 to +1. A positive value means that the direction of regulation between X and Y is from X to Y, whereas a negative value implies the inverse direction.

3 The Method

3.1 The Framework for Representation

We employ information theoretic quantities to the problem of building a Bayesian Network from data which can capture both instantaneous and time delayed interactions. Let us model a GRN containing n genes (denoted by X_1, X_2, X_n), with a corresponding microarray dataset having t_n time points. A DBN-based method would try to find associations between genes X_i and X_j by taking into consideration the data $x_{i1}, \ldots, x_{i(t_n-1)}$ and $x_{j2}, \ldots, x_{jt_n}$ or vice versa (small case letters mean data values in the microarray). This will effectively enable it to capture single-step time delayed interactions. On the other hand, a BN-based strategy would use the whole t_n time points and it will capture regulations that are effective instantaneously.

Now, let us double the number of nodes. The first n nodes of this new network model will correspond to the data $x_{k1}, \ldots, x_{k(t_n-1)}$ whereas the second half will contain $x_{k2}, \ldots, x_{kt_n}$, $k = 1, 2, \ldots, 2n$. So, from this data, if we use the BN formalism to construct a final network where we see, for example, edge $X_1 \to X_{n+2}$, we conclude that the inter-slice arc (or time-delayed interaction) between X_1 and X_2 is recovered. Similarly, if we find that $X_2 \to X_5$, we say that the intra-slice arc (or instantaneous interaction) between X_2 and X_5 is recovered. In this way, we can capture both types of interactions. However, the following two conditions must be satisfied in any resulting network: (i) The network must be a DAG, (ii) The inter-slice arcs must go in the correct direction (no backward arc). Finally, the *stationarity* assumption must also hold.

3.2 Finding the Appropriate Search Strategy

Let us consider the mutual information between a gene X_i and its parents, $Pa(X_i)$ in a DAG $\boldsymbol{G}$. According to the decomposition property of MI (equation (3)), the elements in the decomposition on the right side can be interpreted as follows:

We find the best parent for gene X_i (first term in the right side of equation 5) by calculating its MI with all the other potential parent candidates and select the gene X_j for which $MI(X_i, X_j)$ is maximum. This approach helps us in discarding potential indirect regulators (parents) of gene X_i. This is because of the Data Processing Inequality (DPI)[8], which states that if genes X_i and X_k are connected through an intermediate gene X_j,

$$MI(X_i, X_k) = \min\{MI(X_i, X_j), MI(X_j, X_k)\} \tag{9}$$

i.e., the lowest MI value corresponds to either the indirect relationship or another less strong regulatory relationship. Next, while adding subsequent parents, we calculate how much additional information we get about X_i by adding a candidate parent (X_j) as a parent of this gene, using $MI(X_i, X_{CP_k(X_i)}|Pa_c(X_i))$, where $X_{CP_k(X_i)}$ represents genes that are in the current candidate parent set of the gene X_i and $Pa_c(X_i)$ represents the current parent set of X_i. The candidate gene which can best explain the unexplained uncertainty of X_i relative to the current parent set of this gene is added as the parent of X_i.

However, merely getting a high MI value does not suffice to make it statistically significant. To assess whether the gain in information is statistically significant, we use a theorem of Kullbak[11]. According to the theorem, for a particular confidence level α, determining the value of $\chi(\alpha, df_{ik})$ such that

$$p(\chi^2(df_{ik}) \leq \chi(\alpha, df_{ik})) = \alpha \tag{10}$$

represents a statistical test of conditional independence[9]. Here df_{ik} is the degrees of freedom defined by the following equation:

$$df_{ik} = \begin{cases} (r_i - 1)(r_{ik} - 1) \prod_{m=1}^{k-1} r_{im}, & k \geq 2 \\ (r_i - 1)(r_{ik} - 1), & k = 1 \end{cases} \tag{11}$$

where r_{im} is defined by:

$$\begin{aligned} r_{im} &= config(X_m), \\ X_m &\in Pa_c(X_i) \end{aligned} \tag{12}$$

here $config(X_m)$ is the number of possible states/values that gene X_m can take. Based on the theorem, we can say that the test for statistical significance would assert that the genes are dependent, if in a data set containing N elements,

$$2N.MI(X_i, X_{CP_k(X_i)}|Pa_c(X_i)) \gg \chi(\alpha, df_{ik}) \tag{13}$$

Conversely, the genes are conditionally independent if

$$2N.MI(X_i, X_{CP_k(X_i)}|Pa_c(X_i)) < \chi(\alpha, df_{ik}) \tag{14}$$

Thus, if the maximum CMI value for the current candidate parent set fails this test, we stop adding parents to gene X_i.

3.3 Finding the Intra-slice Arc Directions

The inter-slice arcs in the network can be deduced uniquely since for this part, we are effectively calculating $MI(X_k[t], X_i[t+1]|Pa_c(X_i[t]))$. However, this is not the case with intra-slice arc additions. Since MI is symmetric, the directions of the intra-slice arcs cannot be uniquely determined. To determine the direction of the intra-slice arcs, we use the directionality index, DI_{XY}.

Although Directionality Indices can be used for deducing the direction of regulation, due to finite size of the data, it may be erroneous. As a result, while applying the direction suggested by the directionality index, if any of the conditions listed in part 1 of this Section is violated (e.g., the direction violates the DAG property), we reverse the direction suggested by the directionality index and if it does not violate the properties, we apply that direction to the corresponding edge .

The approaches described in the previous paragraphs are summarized in Table 1 as a 2-phase algorithm. In the first phase, the inter-slice portion of the network is built. The second phase builds the intra-slice portion and the directionality index is applied to each intra-slice edge to determine the direction of interactions. The two networks are then combined to give a final gene regulatory network.

Table 1. The Algorithm

Phase 1:

for each gene $X_i \in X_{n+1,\ldots,2n}$ **do**
  $CP(X_i) \leftarrow findParentCandidates(X_i, Pa_c(X_i))$
  find $X_k \in CP(X_i)$ for which $MI(X_k[t], X_i[t+1]|Pa_c(X_i[t]))$ is maximum
  if $((maximumMI \geq \chi(\alpha, df_{ik}))$ **and** $graphRemainValid(X_k, X_i))$ **then**
    $Pa_c(X_i) \leftarrow Pa_c(X_i) \cup X_k$
  end if
  continue inclusion **until** the above test fails
end for

Phase 2:

for each gene $X_i \in X_{1,\ldots,n}$ **do**
  $CP(X_i) \leftarrow findParentCandidates(X_i, Pa_c(X_i))$
  find gene $X_k \in CP(X_i)$ for which $MI(X_k, X_i|Pa_c(X_i))$ is maximum
  if (maximum $MI \geq \chi(\alpha, df_{ik}))$) **then**
    if ($DI_{X_k X_i} > 0$ **and** $graphRemainValid(X_k, X_i))$ **then**
      $Pa_c(X_i) \leftarrow Pa_c(X_i) \cup X_k$
    else if $(graphRemainValid(X_k, X_i))$ **then**
      $Pa_c(X_k) \leftarrow Pa_c(X_k) \cup X_i$
    end if
  end if
  continue inclusion **until** the above test fails
end for

combine the two networks and get final network, ***G***

4 Simulation and Results

We evaluate our proposed method by both synthetic network and real-life biological network of Saccharomyces cerevisiae (yeast). We applied four widely known performance measures, namely Sensitivity (Se), Specificity (Sp), Precision (Pr)

and *F-Score* (F) and compared our method with some recent methods as well as some traditional methods.

Our method uses discrete data for the statistical significance tests and continuous data for the Directionality Index calculations. We used the Persist[12] algorithm to discretize the data into 3 levels. The value of confidence level (α) used was 0.9. We used the Gaussian Kernel estimator to calculate MI[13, 14] from continuous data. The maximum value of the lag-parameter (δ) was set to 5. For all the experiments related to synthetic network, we used 3 different datasets for each experiment and combined these 3 datasets using the procedure described in[15].

4.1 Synthetic Network

As a first step towards evaluating the performance of our method, we consider the 5 gene target network given in Figure 1[15–17]. We use R-K integration method to obtain 3 sets of time series data, each having 30 time points. We use 5 such different 'combined' datasets in our simulations and calculate the above four performance measures using our technique and compare the performance with four other DBN-based techniques, namely, BITGRN[18], DBN(DP)[6], dynamic differential Bayesian network (DDBN)[15], and DBN(NPR)[5]. The results are shown in Table 2, where we observe that the values of *Se* and *F-Score* of our method are higher than the corresponding values of the other methods. The *Sp* and *Pr* values are also comparable to the other methods.

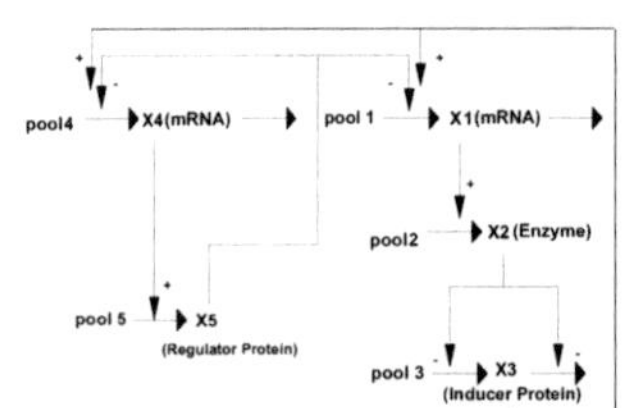

Fig. 1. 5-gene target network [17]

Table 2. Performance comparison of our method with, BITGRN, DBN (DP), DDBN and, DBN (NPR)

	Se	*Sp*	*Pr*	*F*
Our Method (Best)	0.83	0.85	0.83	0.83
Our Method (Average)	0.8	0.82	0.80	0.80
BITGRN	0.67	0.86	0.82	0.74
DBN (DP)	0.5	0.89	0.82	0.62
DDBN	0.75	0.85	0.82	0.78
DBN (NPR)	0.67	0.77	0.73	0.70

Effect of the Size of the Network, Number of Data Points and Noise. To study the effect of the size of the network, we use the network shown in Figure 2[19]. The network is composed of 20 nodes. We used the same parameters as described in[19] for data generation. The number of data points was varied to observe the effect of data points (20 and 30 data points for each dataset). To study the effect of noise, we added 6 different levels of noise (random Gaussian noise with zero mean and variance, $\sigma^2 = 0, 0.01, 0.02, 0.05, 0.1, 0.2$). Each experiment was done using 5 different datasets and the averages of these results are shown in Figure 3. Rectangles are used in the figure for the results from the 20-datapoints experiment whereas triangles represent results from the 30-datapoints experiment. Vertical lines denote standard deviation. From the figure, we observe that increasing the number of samples increases both the accuracy of the method and the noise performance. For higher levels of noise, the more the number of data points, the better is the performance. Moreover, for low values of noise, the performance measures are similar for both the datasets, indicating that the method is not very data hungry in these cases.

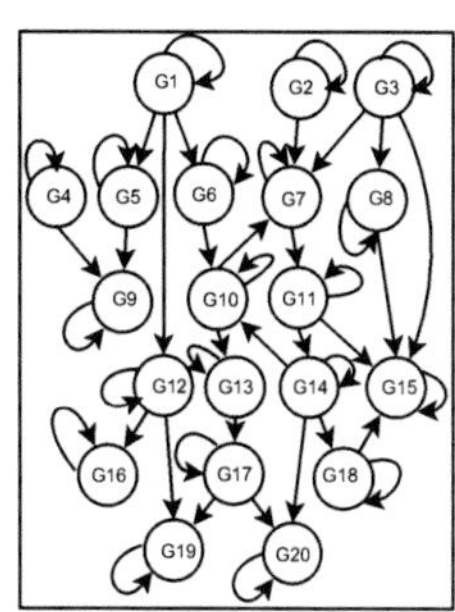

Fig. 2. 20-node target network

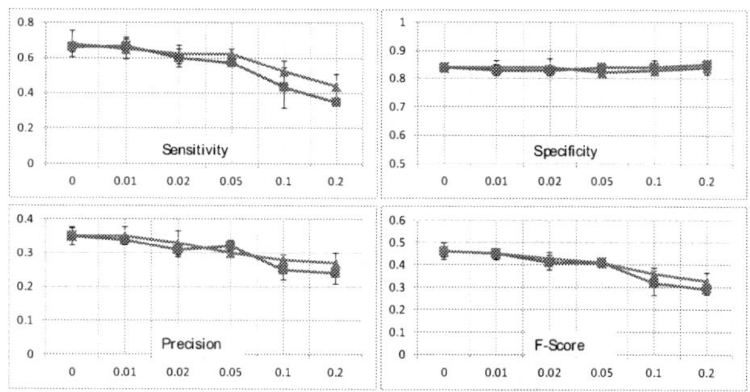

Fig. 3. Effect of noise and data points on the performance of the algorithm. X axes represent the variance values of the 6 noise levels used. Y axes represent the corresponding performance measure.

4.2 Real-Life Biological Data

To validate our method with a real-life biological gene regulatory network, we investigate a recent network reported in[20]. The network is composed of five genes regulating each other; it is also negligibly affected by endogenous genes. There are two sets of gene profiles called Switch ON and Switch OFF for this network, each containing 16 and 21 time series data points, respectively. A 'simplified' network, ignoring some protein level interactions, is also reported. We compare our reconstruction method with 5 other methods, namely, BITGRN[18], TDARACNE[13], NIR and TSNI[21], BANJO[2] and ARACNE[14]. These methods have been successfully used for reconstructing the networks under consideration previously.

IRMA ON Dataset. There are a total of 8 arcs in the original IRMA network. Using the ON dataset, our method correctly identified 6 arcs, corresponding to a Sensitivity, Precision and *F-Score* of 0.75. For the simplified network, the method correctly recovered 4 arcs. The performance comparison amongst various methods is shown in Table 3. From the table, we can clearly see that the overall performance of our method is quite satisfactory.

IRMA OFF Dataset. Due to the lack of 'stimulus', it is difficult to reconstruct the exact network from the OFF dataset[13]. The overall performances of all the

Table 3. Performance comparison based on IRMA ON dataset

	Original Network				Simplified Network			
	Se	*Sp*	*Pr*	*F*	*Se*	*Sp*	*Pr*	*F*
Our Method	0.75	0.88	0.75	0.75	0.67	0.89	0.67	0.67
BITGRN	0.63	0.94	0.83	0.71	0.67	1	1	0.80
TDARACNE	0.63	0.88	0.71	0.67	0.67	0.90	0.80	0.73
NIR & TSNI	0.50	0.94	0.80	0.62	0.67	1	1	0.80
BANJO	0.25	0.76	0.33	0.27	0.50	0.70	0.50	0.50
ARACNE	0.60	-	0.50	0.54	0.50	-	0.50	0.50

Table 4. Performance comparison based on IRMA OFF dataset

	Original Network				Simplified Network			
	Se	*Sp*	*Pr*	*F*	*Se*	*Sp*	*Pr*	*F*
Our Method	0.63	0.82	0.56	0.59	0.83	0.84	0.63	0.71
BITGRN	0.63	0.71	0.50	0.56	0.67	0.60	0.50	0.56
TDARACNE	0.60	-	0.37	0.46	0.75	-	0.50	0.60
NIR & TSNI	0.38	0.88	0.60	0.47	0.50	0.90	0.75	0.60
BANJO	0.38	0.88	0.60	0.46	0.33	0.90	0.67	0.44
ARACNE	0.33	-	0.25	0.28	0.60	-	0.50	0.54

algorithms suffer as a result. The comparison is shown in Table 4. As we can see, the four performance measures of our method are either higher or comparable to the other methods, thereby outperforming them.

5 Conclusion

Accurate reconstruction of gene regulatory networks is considered difficult due to various difficulties and challenges. In this paper we have proposed a novel mutual information based algorithm for reconstructing gene regulatory networks that can detect both instantaneous and time-delayed interactions between genes. The performance as measured by the four widely accepted performance measures show the effectiveness of our employed approach in discovering meaningful regulatory relationships. Due to the computational efficiency of the approach, we are focusing our current research on its application for inferring large networks.

References

1. Friedman, N., Linial, M., Nachman, I., Pe'er, D.: Using Bayesian networks to analyze expression data. Journal of Computational Biology 7(3-4), 601–620 (2000)
2. Yu, J., Smith, V., et al.: Advances to Bayesian network inference for generating causal networks from observational biological data. Bioinformatics 20(18), 3594 (2004)
3. Friedman, N., Murphy, K., Russell, S.: Learning the structure of dynamic probabilistic networks. In: Proc. UAI (UAI 1998), pp. 139–147. Citeseer (1998)

4. Xing, Z., Wu, D.: Modeling multiple time units delayed gene regulatory network using dynamic Bayesian network. In: Proc. ICDM - Workshops (ICDM 2006). pp. 190–195. IEEE (2006)
5. Kim, S., Imoto, S., Miyano, S.: Dynamic Bayesian network and nonparametric regression for nonlinear modeling of gene networks from time series gene expression data. Biosystems 75(1-3), 57–65 (2004)
6. Eaton, D., Murphy, K.: Bayesian structure learning using dynamic programming and MCMC. In: Proc. UAI, UAI 2007 (2007)
7. Chaitankar, V., Ghosh, P., et al.: A novel gene network inference algorithm using predictive minimum description length approach. BMC Systems Biology 4(suppl. 1), S7 (2010)
8. Cover, T., Thomas, J.: Elements of information theory, vol. 306. Wiley Online Library (1991)
9. de Campos, L.: A scoring function for learning Bayesian networks based on mutual information and conditional independence tests. The Journal of Machine Learning Research 7, 2149–2187 (2006)
10. Li, X., Ouyang, G.: Estimating coupling direction between neuronal populations with permutation conditional mutual information. NeuroImage 52(2), 497–507 (2010)
11. Kullback, S.: Information theory and statistics. Wiley (1968)
12. Morchen, F., Ultsch, A.: Optimizing time series discretization for knowledge discovery. In: Proc. ACM SIGKDD (SIGKDD 2005), pp. 660–665. ACM (2005)
13. Zoppoli, P., Morganella, S., Ceccarelli, M.: TimeDelay-ARACNE: Reverse engineering of gene networks from time-course data by an information theoretic approach. BMC Bioinformatics 11(1), 154 (2010)
14. Margolin, A., Nemenman, I., et al.: ARACNE: an algorithm for the reconstruction of gene regulatory networks in a mammalian cellular context. BMC Bioinformatics 7(suppl. 1), S7 (2006)
15. Sugimoto, N., Iba, H.: Inference of gene regulatory networks by means of dynamic differential bayesian networks and nonparametric regression. Genome Informatics Series 15(2), 121 (2004)
16. Sakamoto, E., Iba, H.: Inferring a system of differential equations for a gene regulatory network by using genetic programming. In: Proceedings of the 2001 Congress on Evolutionary Computation, vol. 1, pp. 720–726. IEEE (2001)
17. Savageau, M.: 20 Years of S-systems. In: Canonical Nonlinear Modeling. S-systems Approach to Understand Complexity, pp. 1–44 (1991)
18. Morshed, N., Chetty, M.: Information theoretic dynamic bayesian network approach for reconstructing genetic networks. In: Proc. AIA (AIA 2011), pp. 236–243 (2011)
19. Noman, N., Iba, H.: Inferring gene regulatory networks using differential evolution with local search heuristics. IEEE/ACM Transactions on Computational Biology and Bioinformatics, 634–647 (2007)
20. Cantone, I., Marucci, L., et al.: A yeast synthetic network for in vivo assessment of reverse-engineering and modeling approaches. Cell 137(1), 172–181 (2009)
21. Della Gatta, G., Bansal, M., et al.: Direct targets of the TRP63 transcription factor revealed by a combination of gene expression profiling and reverse engineering. Genome Research 18(6), 939 (2008)

A Sparse-Grid-Based Out-of-Sample Extension for Dimensionality Reduction and Clustering with Laplacian Eigenmaps

Benjamin Peherstorfer, Dirk Pflüger, and Hans-Joachim Bungartz

Technische Universität München, Department of Informatics
Boltzmannstr. 3, 85748 Garching, Germany

Abstract. Spectral graph theoretic methods such as Laplacian Eigenmaps are among the most popular algorithms for manifold learning and clustering. One drawback of these methods is, however, that they do not provide a natural out-of-sample extension. They only provide an embedding for the given training data. We propose to use sparse grid functions to approximate the eigenfunctions of the Laplace-Beltrami operator. We then have an explicit mapping between ambient and latent space. Thus, out-of-sample points can be mapped as well. We present results for synthetic and real-world examples to support the effectiveness of the sparse-grid-based explicit mapping.

Keywords: spectral methods, manifold learning, clustering, sparse grids.

1 Introduction

Spectral methods have emerged as promising techniques for dimensionality reduction and clustering. All of these methods use the eigenvector of some affinity matrix to derive a low-dimensional embedding or cluster assignment.

Today, many algorithms can be seen as spectral methods. In particular, a whole family of spectral clustering algorithms exists, see the survey [12]. In the following, we will concentrate on Laplacian Eigenmaps (LE) because this method provides a low-dimensional embedding of the training data which can be used not only for a clustering assignment, but also for dimensionality reduction [2].

One drawback of LE is that it learns only the embedding (or clustering) of the training data at hand. There is no natural way to treat out-of-sample (or test) points, i.e. assign new points, which are not available during the computation of the eigenvectors, to clusters. Besides the obvious situation where not all data is available from the beginning, out-of-sample extensions can also be helpful if there are *too many* points to create the graph, the matrices, and to finally solve the eigenproblem in feasible time. In such situations, it is convenient if one can partition the data points into a (small) training and a (large) test data set. The LE are then computed for the training set, and the out-of-sample extension is used to approximate the embedding for the test data.

The most common out-of-sample extension for LE and other spectral methods is based on the Nyström method. It has been proposed as an out-of-sample

D. Wang and M. Reynolds (Eds.): AI 2011, LNAI 7106, pp. 112–121, 2011.

extension for spectral methods in [3]. Since then it has been applied to various problems, see, e.g., [6,17]. This out-of-sample extension is based on the assumption that we can represent the similarity measure as a kernel function K. In [3], to each data point a kernel function is assigned. They are then utilized to compute the embedding of out-of-sample points.

Besides this kernel-based method, there have been efforts to create methods based on linear projections, see, e.g., [8,10]. However, since the underlying linearity assumption cannot always be assumed that easily, they already fail for simple problems (e.g. swiss roll, see Sec. 5) [10]. A more sophisticated approach based on polynomials has been presented in [15]. But as far as computational complexity is concerned, the method becomes infeasible very quickly.

In contrast, we propose to approximate the eigenfunctions of the Laplace-Beltrami operator by functions discretized on sparse grids. We then have an explicit mapping between ambient and latent space and can treat out-of-sample points in a natural way. Usually, grid-based approaches are not feasible in high dimensional settings because the number of grid points grows exponentially with the number of dimensions for straightforward discretizations. However, sparse grids allow us to cope with this so-called curse of dimensionality to some extent. They have been applied in various fields of application, see the survey [5].

2 Laplacian Eigenmaps and Spectral Clustering

For a given data set $\{x_1, \ldots, x_M\} \subset \mathbb{R}^d$ of M points we construct a weighted graph $G = (V, E)$ where the weights $w_{ij} \geq 0$ correspond to some similarity measure between the data points x_i and x_j. The (weighted) adjacency matrix of the graph is the matrix W with entries w_{ij}. The degree d_i of the i-th vertex of the graph is defined by $d_i = \sum_j w_{ij}$. We can then define the degree matrix D with $d_1, \ldots, d_M$ on the diagonal and zero elsewhere. If not otherwise stated, we will use the well-known Gaussian kernel with bandwidth σ as the similarity measure.

$$w_{ij} = e^{-\frac{\|x_i - x_j\|^2}{\sigma^2}} . \tag{1}$$

Note that we always assume that the graph is connected.

In the following we need the so-called (unnormalized) *graph Laplacian* $L = D - W$. An overview of many properties of L can be found in [12] and the references therein. We only want to recall that the smallest eigenvalue of L is 0 and the corresponding eigenvector is the constant vector $\mathbf{1}$.

The bipartitioning of a graph G means the division of the set of vertices V into two disjoint sets $A, B \subset V$ with $A \cup B = V$ based on a cut criterion. In our case, we want to minimize the sum of weights ("flow") of edges connecting the points between the two sets A and B, i.e. we want to minimize

$$\mathrm{cut}(A, B) = \sum_{i \in A, j \in B} w_{ij} .$$

In order to avoid a partitioning of the graph which just cuts off outliers, we consider the Normalized Cut between A and B

$$\mathrm{Ncut}(A,B) = \mathrm{cut}(A,B)\left(\frac{1}{\mathrm{vol}(A)} + \frac{1}{\mathrm{vol}(B)}\right),$$

where $\mathrm{vol}(A) = \sum_{i \in A} d_i$ is the so-called volume of A. The minimization of the Ncut is a well-studied problem. The result is a vector $y \in \{0,1\}^M$ which indicates if a vertex belongs to either A or B. Although the problem as stated is NP-complete [2], the relaxation to real values leads to a problem which can be solved in polynomial time [12]. Let $y \in \mathbb{R}^M$ and let L be the unnormalized graph Laplacian, then the solution can be found by minimizing the Rayleigh quotient

$$\frac{y^T L y}{y^T D y}, \tag{2}$$

under the constraint $y^T D\mathbf{1} = 0$. Thus, the minimum is achieved for the eigenvector corresponding to the second smallest eigenvalue of the generalized eigenvalue problem

$$Ly = \lambda D y, \tag{3}$$

see, e.g., [2,12]. In the following, we always assume an ascending order of the eigenvalues.

Since the components of the eigenvectors are real-valued they do not provide a clear assignment of the data points into clusters. However, due to the properties of the graph Laplacian, the change of representation from data points x_i to the components y_i of the eigenvectors enhances the cluster-properties of the data set [12]. That is why we can simply apply k-means to obtain the final cluster indicators.

The solution of (2) can also be considered as an optimal embedding of the data points $\{x_1, \ldots, x_M\}$ preserving local information in the following sense: If x_i and x_j are close with respect to the weight w_{ij}, their embedding y_i and y_j should be close as well [2]. From that point of view we compute a low-dimensional embedding of the high-dimensional data points $\{x_1, \ldots, x_M\}$. In this context of dimensionality reduction the above described approach is usually referred to as *Laplacian Eigenmaps* (LE).

As far as computational costs are concerned, the by far most expensive part is the solution of the generalized eigenproblem (3). Since this is usually in $\mathcal{O}(M^3)$, large amounts of data are impractical to process with this method.

3 Sparse Grids

As we have already mentioned at the beginning, very often the Nyström method is used for a *data-based* out-of-sample extension for LE. A kernel is associated to each data point. These are then combined to approximate the eigenfunctions corresponding to the eigenproblem (3). We propose a *grid-based* approach, i.e. our functions $f_N \in V_N$ can be represented as a linear combination with coefficients α_i

$$f_N(x) = \sum_{i=1}^{N} \alpha_i \phi_i(x)\,,$$

where the basis $\Phi = \{\phi_i\}_{i=1}^N$ comes from a grid and spans the function space V_N. Hence, the number of basis functions does *not* increase with the number of data points in contrast to classical approaches. Unfortunately, a straightforward conventional discretization with N grid points in each dimension suffers the curse of dimensionality: The number of grid points is of the order $\mathcal{O}(N^d)$, depending exponentially on the dimension d. In our case, the dimension d equals the dimension of the ambient space, i.e. the usually high-dimensional space where our data points come from. For sufficiently smooth functions, sparse grids enable us to reduce the number of grid points by orders of magnitude to only $\mathcal{O}(N \log(N)^{d-1})$ while keeping a similar accuracy as in the full grid case. Even though theory requires certain smoothness assumptions, sparse grids have also been successfully applied for not so smooth functions, see, e.g., [14,13,16]. In the following, we describe the very basics of sparse grids as briefly as possible, see [5,13] for more details.

The underlying principle of sparse grids is a one-dimensional hierarchical system of basis functions (see Fig. 1, left) which is then extended to the d-dimensional case by taking the product of the one-dimensional basis functions. These span subspaces $W_{\boldsymbol{l}}$ where the level $\boldsymbol{l} = (l_1, \ldots, l_d)$ determines the mesh size in each direction. We show in Fig. 1 (middle) the grids of the two-dimensional hierarchical increments $W_{\boldsymbol{l}}$ up to level 3 in each dimension.

Starting from a hierarchical scheme as in Fig. 1 we select only those subspaces that contribute most to the overall solution. The optimal choice is to cut off the tableau in Fig. 1 along the diagonal if the error is measured in the L^2- or maximum norm, see [5]. Thus, the sparse grid space of level n is

$$V_n^{(1)} := \bigoplus_{|\boldsymbol{l}|_1 \leq n+d-1} W_{\boldsymbol{l}}\,,$$

where $|\boldsymbol{l}|_1$ denotes the sum of the one-dimensional levels. In the example of Fig. 1 (middle) we can neglect the grayed out subspaces (many grid points with little contribution) and obtain the regular sparse grid in Fig. 1 (right top).

In order to reach higher dimensions further considerations are needed. First, we can use spatial (local) adaptivity to further reduce the number of unknowns needed to solve a problem up to some required accuracy. We start with a rather coarse sparse grid and use a suitable adaptivity criterion to add points in those regions of the domain that are most important, compare Fig. 1 (right bottom). A simple (though typically very effective) criterion for adaptive refinement, which we use in the following, is to select the refinement candidates with the highest absolute values of their hierarchical surpluses (coefficients α_i). Second, the ordinary basis functions at the boundary become infeasible in higher dimensions.

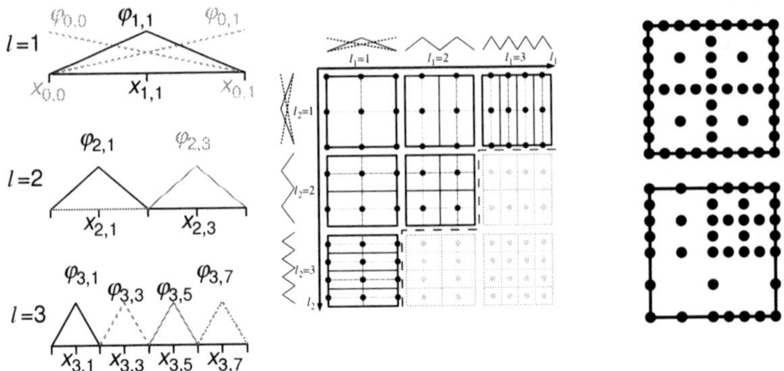

Fig. 1. One-dimensional hierarchical basis (*left*), the tableau of hierarchical increments W_l up to level 3 (*middle*) and the corresponding regular sparse grid (*right top*) and an adaptively refined sparse grid (*right bottom*)

Therefore, we omit the basis functions at the boundary and use modified basis functions adjacent to the boundary. These then extrapolate linearly towards the boundary, for details see [13].

4 Sparse-Grid-Based Out-of-Sample Extension

In this section, we present an out-of-sample extension which uses sparse grid functions $f \in V_n^{(1)}$ to approximate the eigenfunctions corresponding to the LE embedding.

Let f be a function of the sparse grid space $V_n^{(1)}$. We can represent such a function as a linear combination of the hierarchical basis $\Phi = \{\phi_i\}_{i=1}^N$ with coefficients $\alpha = (\alpha_1, \ldots, \alpha_N)$

$$f(x) = \sum_{i=1}^{N} \alpha_i \phi_i(x) \,. \tag{4}$$

Let $\tilde{f} = (f(x_1), \ldots, f(x_M))$ be the vector of the function values at the data points $X = \{x_1, \ldots, x_M\}$. We then look for coefficients α such that the Rayleigh quotient

$$\frac{\tilde{f}^T L \tilde{f}}{\tilde{f}^T D \tilde{f}} \,, \tag{5}$$

is minimized, where L and D are the graph Laplacian and degree matrix, respectively. Again we find the minimum by solving an eigenproblem. Instead of $Ly = \lambda Dy$ as in (3) we now have

$$B^T L B \alpha = \lambda B^T D B \alpha \,, \tag{6}$$

where B is a $M \times N$ matrix with $b_{ij} = \phi_j(x_i)$. Note that matrix B becomes singular if there exists a basis function which is not "hit" by any data point. More precisely, let ϕ_k be a basis function with support s_k. If $s_k \cap \{x_1, \ldots, x_M\} = \emptyset$ then column k of B has only zero entries. Hence, B is singular.

We show that even if B is singular, the eigenproblem (6) need not be singular and can thus be solved. For that purpose we rewrite the eigenproblem (6) as

$$\mu B^T LB\alpha = \nu B^T DB\alpha\,, \tag{7}$$

with $\lambda = \nu/\mu$, and we assume the graph is not empty ($V \neq \emptyset$) and connected, i.e. $y^T Ly = 0 \iff y = \mathbf{1}$. The eigenproblem is called singular if $\nu = \mu = 0$ [1]. Suppose we have an eigenvector $\alpha \in \mathbb{R}^N$ with $\nu = \mu = 0$. This means $B^T LB\alpha = 0$ and $B^T DB\alpha = 0$. Because our graph is connected we have $\alpha^T B^T LB\alpha = 0 \Rightarrow y^T Ly = 0 \Rightarrow y = \mathbf{1}$. But then we obtain $B^T Dy = B^T d$ with $d = (d_{11}, \ldots, d_{MM}) > 0$ component-wise. Since B^T has only non-negative entries and at least one entry > 0, $B^T DB\alpha$ has to have at least one entry > 0. Hence $\mu \neq 0$. So there cannot be an eigenvector $\alpha \in \mathbb{R}^N$ with $\mu = \nu = 0$.

Note that we simply ignore eigenvectors to eigenvalues $\lambda = \nu/\mu$ with $\mu = 0$. Note further that the graph Laplacian property $y^T Ly = 0 \iff y = \mathbf{1}$ still holds for $B^T LB$.

It is common in other settings to add a smoothness constraint to the minimization problem, e.g., the identity matrix, see, e.g., [13,7] for the case of classification with sparse grids. That is why we add a regularization term C and obtain the minimization problem

$$\frac{\alpha^T \left(\gamma C + B^T LB\right)\alpha}{\alpha^T B^T DB\alpha}\,, \tag{8}$$

with the corresponding eigenproblem

$$\left(\gamma C + B^T LB\right)\alpha = \lambda B^T DB\alpha\,,$$

under the constraint that we skip the (constant) eigenvector with eigenvalue 0. With the *regularization parameter* γ we can balance the demand for closeness to the minimum and the smoothness constraint. However, our experiments have shown that γ only slightly influences the result.

The algorithm for the proposed out-of-sample extension can be summarized as follows. Let X_{train} and X_{test} be the set of training and test points, respectively. We obtain the sparse grid function f by solving the minimization problem (8) for the training set X_{train}. The low-dimensional embedding for the training points is given by the function values. We can further compute a cluster assignment by using k-means on the function values of f at $x \in X_{train}$. However, we now can also evaluate f at test points $x \in X_{test}$ and thus find an embedding and cluster assignment for these out-of-sample points.

Finally, we want to discuss the computational complexity of our out-of-sample extension. Let us first consider the eigenproblem. With our grid-based approach, the dimension of the eigenproblem becomes data-independent, i.e. the solution of the eigenproblem can be obtained in $\mathcal{O}(N^3)$, where N is the number of grid points, rather than in $\mathcal{O}(M^3)$. For huge data sets where $M \gg N$ (e.g. image segmentation) this is a distinct improvement. The same holds for the out-of-sample extension. In our case, the computation of an embedding of an out-of-sample point just results in a function evaluation. For sparse grid functions in

$V_n^{(1)}$ with level n, this can be accomplished in $\mathcal{O}(n^d)$, thus, the complexity of a function evaluation is again data-independent. Furthermore, this can be done with very efficient parallel methods suitable for huge data sets [9].

5 Experiments

We start with the two moons data set, see Fig. 2 (right). We generate five training sets with 500 points each and compute the cluster assignment as discussed in the previous section. We then determine the ARI (adjusted rand index) [11] with the proposed sparse-grid-based out-of-sample extension on five test sets with 5,000 points each. The mean over all these ARI values is plotted in Fig. 2 against the σ^2 of the Gaussian kernel. We compare functions on sparse grids of level 4, 5 and 6.

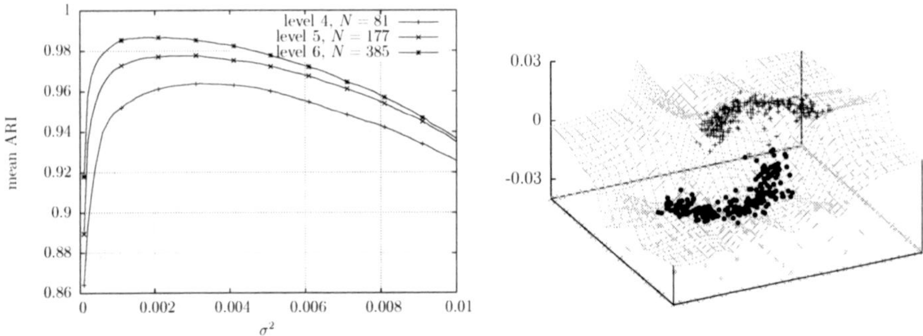

Fig. 2. The mean ARI for the two moons data set against the σ^2 of the Gaussian kernel for functions on sparse grids of level 4, 5 and 6 (*left*). The approximation of the second eigenfunction on a sparse grid of level 5 (*right*).

We clearly see that we already achieve a mean ARI of more than 0.95 with only 81 grid points. Hence, only very few grid points are required for a reasonable approximation of the eigenfunction. However, we also see that really high accuracies can be obtained if we increase the grid level n. This gives a good indication that our approximation really converges towards the eigenfunction. The approximation of the second eigenfunction on a sparse grid of level 5 is shown in Fig. 2 (right).

The next example is the swiss roll with 2000 training and 2000 test data points, see Fig. 3 (left). It is a two-dimensional submanifold of $\mathbb{R}^3$ and it has been shown that it can be "unrolled" with LE [2]. In order to compute the two-dimensional embedding we have to approximate the second and third eigenfunctions corresponding to the eigenproblem of LE. We then obtain a function $f_n : \mathbb{R}^3 \to \mathbb{R}^2$ with

$$f_n(x_1, x_2, x_3) = \begin{pmatrix} f_n^1(x_1, x_2, x_3) \\ f_n^2(x_1, x_2, x_3) \end{pmatrix},$$

where f_n^1 and f_n^2 are the sparse grid functions approximating the second and third eigenfunction, respectively. In Fig. 3 we show the two-dimensional embedding of the swiss roll. If we compare the proposed out-of-sample extension with the results shown in [2], we already obtain good results in the case of level 5 (705 grid points). Again we have distinctly less grid points than data points, thus, the dimension of the eigenproblem is reduced.

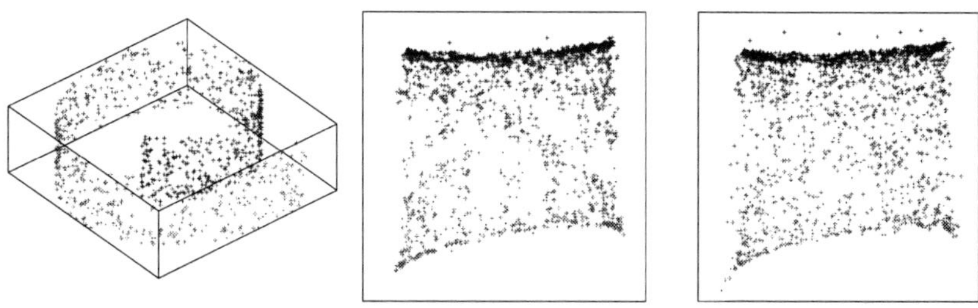

Fig. 3. Swiss roll (*left*) embedding of training (*middle*) and test (*right*) data with sparse grid functions on grid of level 5 (705 grid points). The parameter σ^2 of the Gaussian kernel is set to 6.0.

Next we consider the oil flow data set introduced in [4]. This is a 12-dimensional data set with three classes called "configurations". Because the data points corresponding to the "stratified configuration" are scattered into several smaller clusters, we consider the data points of the two remaining configurations. Furthermore, we add the validation data points to our training data. We then have a data set of 1318 training and 687 test data points representing two non-linearly separable clusters.

In order to cope with the 12-dimensional problem, we use the modified basis functions near the boundary. First, we consider the results for the regular grid of level 4 with $N = 3249$ grid points, see Fig. 4. We see that it performs better than the Nyström method for both training and test data. However, in this case we have more grid than training points. Therefore we employ adaptivity. We start with the (12-dimensional) sparse grid of level 2 and apply three refinement steps with the criterion from Sec. 3. In each refinement step we refine another 10 percent of grid points. For the reasonable kernel bandwidths (σ^2 between 0.08 to 0.1) we do not have more grid than training data points. With the adaptively refined grid we get about the same behavior as with the regular grid. Hence, the proposed method achieves, again, a higher ARI than the Nyström method. Furthermore, we do not have more grid than training data points, thus, we reduce the dimension of the eigenproblem as well.

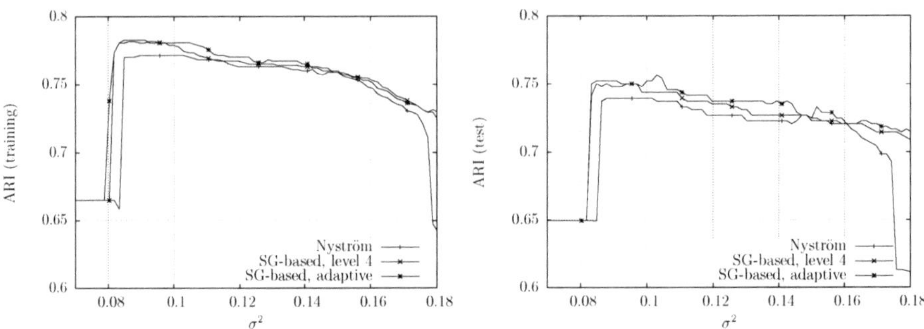

Fig. 4. Results for the oil flow data set. Comparison of the Nyström and proposed method with regular and adaptive sparse grids. Mean ARI for training (*left*) and test (*right*) data set.

6 Conclusion

We presented a novel out-of-sample extension for Laplacian Eigenmaps. It uses sparse grids to compute an explicit mapping between ambient and latent space. This grid-based approach is feasible because sparse grids allow us to cope with the curse of dimensionality to some extent. The advantages of the grid-based approach are obvious. Both the dimension of the eigenproblem as well as the computation of an embedding for an out-of-sample point are independent from the number of training data points.

We illustrated the sparse-grid-based out-of-sample extension by various examples. We have studied synthetic benchmark data sets for clustering as well as dimensionality reduction. The results have shown that the out-of-sample points are mapped reasonably from ambient to latent space. The same holds for the real-world example where we studied the 12-dimensional oil flow data set. The proposed grid-based method achieved better results than the usually used Nyström method.

Overall, the experiments have validated the effectiveness of the proposed out-of-sample extension based on sparse grids.

References

1. Anderson, E., Bai, Z., Bischof, C., Blackford, L.S., Demmel, J., Dongarra, J.J., Croz, J.D., Hammarling, S., Greenbaum, A., McKenney, A., Sorensen, D.: LAPACK Users' guide, 3rd edn. Society for Industrial and Applied Mathematics, Philadelphia (1999)
2. Belkin, M., Niyogi, P.: Laplacian eigenmaps for dimensionality reduction and data representation. Tech. Rep. TR-2002-01, The University of Chicago CS (January 2002)
3. Bengio, Y., Paiement, J., Vincent, P.: Out-of-Sample extensions for LLE, isomap, MDS, eigenmaps, and spectral clustering. In: Advances in Neural Information Processing Systems, pp. 177–184. MIT Press (2003)

4. Bishop, C.M., James, G.D.: Analysis of multiphase flows using dual-energy gamma densitometry and neural networks. Nuclear Instruments and Methods in Physics Research Section A: Accelerators, Spectrometers, Detectors and Associated Equipment 327(2-3), 580–593 (1993)
5. Bungartz, H.-J., Griebel, M.: Sparse grids. Acta Numerica 13 (2004)
6. Fowlkes, C., Belongie, S., Chung, F., Malik, J.: Spectral grouping using the nystrom method. IEEE Transactions on Pattern Analysis and Machine Intelligence 26(2), 214–225 (2004)
7. Garcke, J., Griebel, M., Thess, M.: Data mining with sparse grids. Computing 67(3), 225–253 (2001)
8. He, X., Yan, S., Hu, Y., Zhang, H.: Learning a locality preserving subspace for visual recognition. In: Ninth IEEE International Conference on Computer Vision, vol. 1, pp. 385–392 (2003)
9. Heinecke, A., Pflüger, D.: Multi- and many-core data mining with adaptive sparse grids. In: Proceedings of the 2011 ACM International Conference on Computing Frontiers (May 2011)
10. Huang, D., Zhang, X., Huang, G., Pang, Y., Zhang, L., Liu, Z., Yu, N., Li, H.: Neighborhood Preserving Projections (NPP): a Novel Linear Dimension Reduction Method. In: Huang, D.-S., Zhang, X.-P., Huang, G.-B. (eds.) ICIC 2005. LNCS, vol. 3644, pp. 117–125. Springer, Heidelberg (2005)
11. Hubert, L., Arabie, P.: Comparing partitions. Journal of Classification 2(1), 193–218 (1985)
12. von Luxburg, U.: A tutorial on spectral clustering. Statistics and Computing 17, 395–416 (2007)
13. Pflüger, D.: Spatially Adaptive Sparse Grids for High-Dimensional Problems. Dissertation, Institut für Informatik, Technische Universität München, München (February 2010)
14. Pflüger, D., Peherstorfer, B., Bungartz, H.J.: Spatially adaptive sparse grids for high-dimensional data-driven problems. Journal of Complexity 26(5), 508–522 (2010)
15. Qiao, H., Zhang, P., Wang, D., Zhang, B.: An explicit nonlinear mapping for manifold learning. CoRR abs/1001.2605 (2010)
16. Schraufstetter, S., Benk, J.: A general pricing technique based on theta-calculus and sparse grids. In: Proceedings of the ENUMATH 2009 Conference, Uppsala (December 2009)
17. Williams, C., Seeger, M.: Using the nyström method to speed up kernel machines. In: Advances in Neural Information Processing Systems, vol. 13, pp. 682–688. MIT Press (2001)

Distribution Based Data Filtering for Financial Time Series Forecasting

Goce Ristanoski and James Bailey

The University of Melbourne, Melbourne, Australia
g.ristanoski@pgrad.unimelb.edu.au, baileyj@unimelb.edu.au

Abstract. Changes in the distribution of financial time series, particularly stock market prices, can happen at a very high frequency. Such changes make the prediction of future behavior very challenging. Application of traditional regression algorithms in this scenario is based on the assumption that all data samples are equally important for model building. Our work examines the use of an alternative data pre-processing approach, whereby knowledge of distribution changes is used to pre-filter the training dataset. Experimental results indicate that this simple and efficient technique can produce effective results and obtain improvements in prediction accuracy when used in conjunction with a range of forecasting techniques.

Keywords: Time series classification, regression, distribution change.

1 Introduction

Prediction techniques for the behavior of financial time series have been intensively studied [1][2]. A prime example is the forecasting of stock prices, which aims to forecast the future values of the price of a stock, in order to obtain information about its trends and direction of movement and thus allow the development of buying/selling strategies to gain competitive advantage.

Classic and popular methods for stock price forecasting [3][4] for both univariate and multivariate time series data include linear regression, hidden markov models, neural networks [11] and support vector machines [7].

The underlying data for financial time series may span a frequency as small as hourly or as long as several years. The longer the time interval, the more likely it is that the data samples will not follow the same distribution [8]. The classic statistical [13] and data mining time series prediction methods [14], at least in their simple form, do not take into consideration that such changes in distribution over time may occur with financial time series data. This can lead to a loss in prediction accuracy, since the prediction model that is built places equal value on all samples, even those whose distribution is not close to the distribution of the samples in the most recent past.

In this paper, we address the challenge of forecasting the behavior of time series using distribution change. In particular, we propose a technique for filtering the samples in such time series, in order to project out those samples which appear least relevant and retain those samples which appear most relevant for prediction. Our

D. Wang and M. Reynolds (Eds.): AI 2011, LNAI 7106, pp. 122–131, 2011.

proposed **Distribution Based Samples Removing (DBSR)** algorithm operates by i) initially analyzing the time series to determine its different distributions, and then ii) reducing the time series by filtering out the samples whose distribution is furthest from the recent past. We develop two versions of the algorithm, one parametric and the other non-parametric. Our approach is designed to work for regression with univariate series that use a five day relative difference in percentage of price (RDP) format [16], but the approach can also be applied to original univariate time regressed on itself, as well as multivariate time series.

Our proposed data filtering method has a number of desirable properties: i) it is clean, simple and intuitive, ii) it is easy to implement and runs efficiently, since it is a data pre-processing step and thus iii) it can be used in conjunction with many existing time series prediction methods. Finally, we find that iv) it can help obtain improvements in prediction performance when used as a prior step to produce input for classic time series prediction algorithms.

2 Related Work

There is a large amount of literature dealing with classification and regression for financial time series. Descriptions of classic methods can be found in standard textbooks such as [1][2][3]. Instead we briefly review related work that can be used for dataset filtering or pre-processing, since this is an essential feature of our approach.

Selecting samples from a set can be performed by simple random sampling, cluster sampling, systematic sampling, or load shedding [5], but most of these methods do not consider the time element that is present when dealing with financial time series. Efforts have been by [21][22] to improve these methods and to include the time element, by using strategies based on sliding windows [22]. Nevertheless, sample selection in time series mostly consists of only selecting a continuous sample set, without the possibly of removing non contiguous ranges of samples from the set.

Investigating the changes in distribution that occur over time within the financial time series data and including them in the learning process is an ongoing research direction [9] [10] [12]. The benefits of the research in this area are not only algorithms that are adjusted to cope with the time element present in the data, but also algorithms that run online and can process data streams as well [15].

3 Distribution Based Samples Removing Algorithm

The notion of examining the nature of distribution change in a time series and using it to filter the data samples is inspired by the technique of load shedding [22] using sliding windows. In order to develop an algorithm that can filter based on distribution change, we will first need to decide on an appropriate statistic for measuring differences in distribution.

We choose to use the Wilcoxon rank sum method (WXN) [13], which is a non-parametric test that assesses whether two sets of data samples follow the same distribution. It is easy to implement, efficient and a well known statistical test. We adopt the WXN method and use the change points it detects. The WXN paradigm is as follows: we set a fixed window on *n* points, [1,n], and starting after it, a sliding window of *n* points as well, [n+1, 2n], as shown in Figure 1. We move the second window and compare if the samples in both windows follow same distribution: if that is the case, we continue moving the second window, until the distribution changes. The change point will be at the last sample of the second window (point 2n+k); we move the first window just after that point [2n+k+1, 3n+k], the second window comes after the first one [3n+k+1, 4n+k] and we repeat the process for the rest of the dataset.

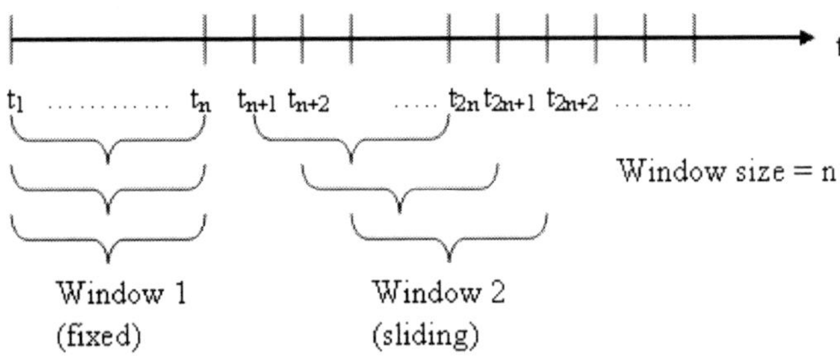

Fig. 1. The Wilcoxon method with fixed reference window

After the WXN method has detected all the distribution change points in the training set of the time series, the mean (average) value of each window is calculated and compared to the mean value of the last (most recent in time) window: the difference between the mean value for a given window with index *j*, and the mean value of the last window, called $\Delta avg[j,last]=mean_j-mean_{last}$ is calculated, all differences are then normalized into the range of [0,1], giving us the value for $d_j=\Delta avg[j,last]_{Normalized}$.

$$d_j = \Delta avg[j,last]_{Normalized} = abs\left(\frac{\Delta avg[j,last]}{\max_i(\Delta avg[i,last])}\right) \tag{1}$$

To gain an idea about likely behavior, we ran the WXN method on several real life time series (described in detail later in the paper) and the results showed the general pattern of Figure 2: some samples in the distant past were more similar to the most recent window than were some samples in the more recent past. We can see from Figure 2 moving left to right, there are windows in the most distant past with very similar distribution (windows 1 and 2) to the last window, and also windows in the not so distant past with quite different distribution (window 8) to the distribution of the last window. This confirmed our belief that many real time series are non-stationary, and that it is potentially promising to investigate methods for the filtering of samples based on similarity of distribution.

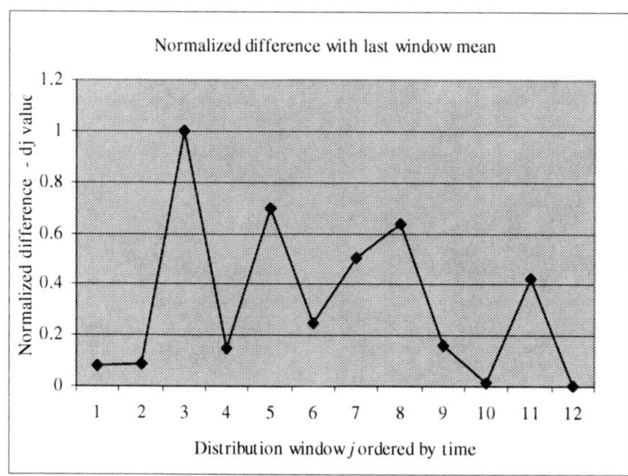

Fig. 2. Example of $d_j = \Delta \text{avg}[j, last]_{Normalized}$ value between the distribution windows

We develop two versions of a **Distribution Based Samples Removing (DBSR) Algorithm**, one parametric and the other non-parametric. They both use information about the distribution changes in the time series for making the decision about which samples of the dataset to remove.

3.1 Distance Value – Threshold Based Decision

The parametric based DBSR (P-DBSR) algorithm requires the user to analyze the distribution change data: the size of the windows and the value of the distance to the most recent window. It requires a threshold value, between 0 and 1, and removes the samples from the windows where the distance to the most recent window is above the threshold value. The structure of the P-DBSR algorithm is as follows:

Algorithm P-DBSR

Stage 1: change point detection

Input: time series dataset $X = \{x_i \mid i = 1..m\}$
Output: reduced time series $X = \{x_i \mid i = 1..k, k<m\}$
Initial: reference windows W_1 and W_2, window size n, threshold value p, $W_1=\{x_1, ..x_n\}$, $W_2=\{x_{n+1},..,X_{2n}\}$, number of change points cPoints=0, distance values $d_j = \Delta \text{avg}[j, last]_{Normalized}$.
While not the end of dataset
    Compare distribution for W_1 and W_2
If W_1 and W_2 from the same distribution
    Move W_2 one sample forward
Else
    Detect change point, cPoints += 1
    Set W_1 to start after W_2, then W_2 after W_1
EndIf
EndWhile

```
Stage 2: parameter based dataset reduction
    Calculate normalized distance values to the last window d_j, j=1.. cPoints+1
    For all windows
        If window distance d_j > p value
            Remove the current window
    EndIf
    EndFor
    Return reduced dataset X = {x_i | i = 1..k, k<m }
```

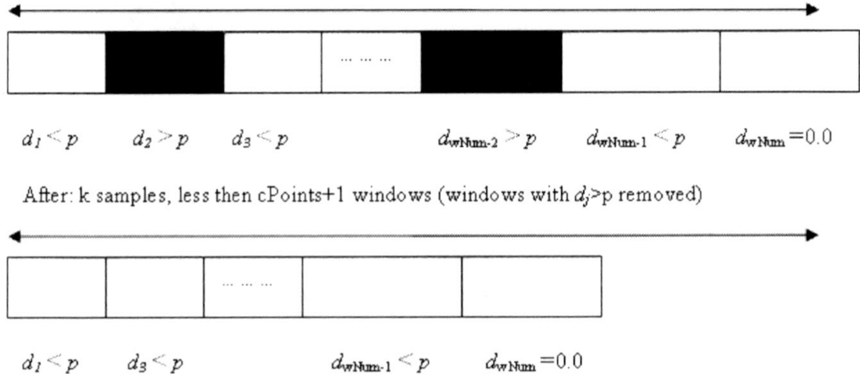

Fig. 3. The parametric DBSR datasets, before and after removing the windows

This version of the algorithm has several advantages: the user has access to the detailed information about the distribution, and can see how it changes over time, therefore getting insight into the volatility of the samples that will be used for forecasting; it will also indicate regions where the data may be noisy, and thus beneficial to remove.

We choose such value for p that would result in an amount is large enough for us to expect the final regression to be significantly different. Shown in Figure 3, the samples where the normalized distance was greater than the p value are in the black sections, and are removed at the end of the algorithm.

We assessed the algorithm over a range of values for the threshold - between 0.3 and 0.8. Some datasets had many windows with distributions similar to that of the last window, and in order to remove a significant amount of samples (around 30-35 %), those datasets required the threshold value set low. The datasets where there were windows with distribution quite different from the one of the last windows needed a threshold value set usually around 0.7 to remove the same percentage (30-35%) of samples. Even though the value for p was different for each dataset, the amount of samples removed was roughly the same for all datasets. We did so as we prefer to have same ratio of before and after dataset size, in order to test if removing such large amount of samples would be beneficial, regardless off the dataset.

3.2 Distance Value – Percentage Based Decision

Our non-parametric DBSR (NP-DBSR) algorithm again accesses information about the distribution change and distribution distance with respect to the most recent window. As the distance is normalized in the range of 0-1, the algorithm uses that value to determine the portion of the window to be removed – e.g. if the normalized distance value for a given window is 0.7, the algorithm will remove 70% of the samples of that window. In other words, the samples from each window are filtered in proportion to the amount of their dissimilarity to the last window. This gives windows with a moderate value (moderate dissimilarity) for the distance some chance to contribute samples. Since distances are normalized, it will result in the most distant window having all of its instances removed, and the most recent window having no instances removed. Shown in Figure 4, we can see we have the same windows with different distributions (marked with different patterns) before and after, with the windows after being smaller, as the have samples being removed from them.

The structure of the NP-DBSR algorithm is as follows:

Algorithm NP-DBSR

Stage 1: change point detection (lines 1 - 12)
Stage 2: parameter free dataset reduction

13: **Calculate** normalized distance values to the last window *d* $d_j=\Delta avg[j,last]_{Normalized \cdot j}$, j=1.. cPoints+1
14: **For** all windows
15: Remove d_j *100 percent of the samples of the current window
16: **EndFor**
17: **Return** reduced dataset X = $\{x_i \mid i = 1..k, k<m\}$

Before: m samples, cPoints+1 windows

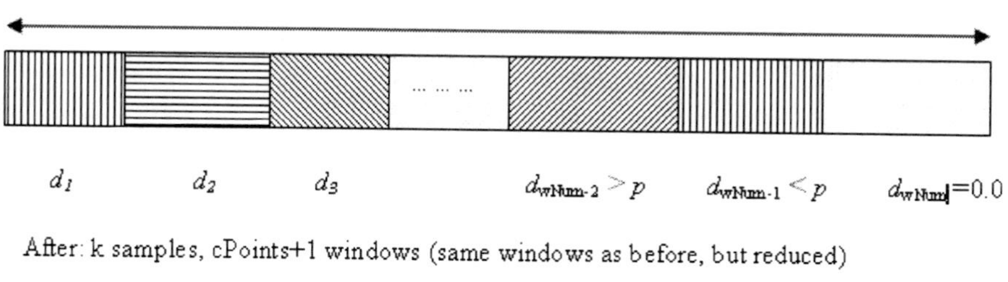

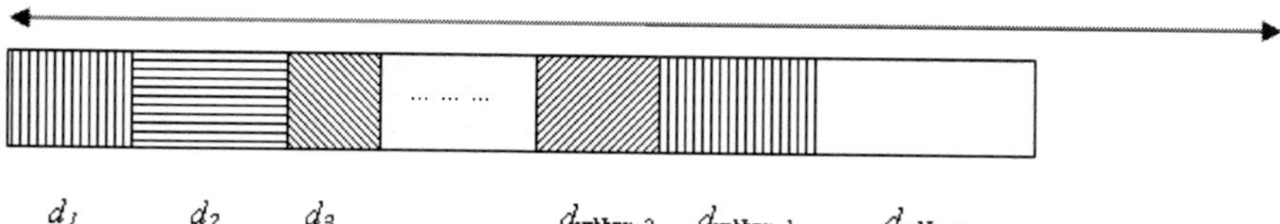

Fig. 4. Non parametric DBSR, before and after reducing the windows

4 Datasets

Our research was focused on forecasting stock market prices, as they are continuous series that can change very quickly, and are of great interest to both investors and researchers. We tested stock market prices of 12 random companies, with each dataset containing between 290 and 700 samples, recorded daily from a randomly chosen period between 1997 and 2010 [17][19]. We also tested a simulated dataset, where there did not exist many changes in the distribution, as well as the S&P quarterly index time series [18]. The stock market datasets were divided into a training and testing set, in the ratio of 9:1. We only focused on short term forecasting, so that the learning time of the machine learning models was small. The names of the companies, along with the number of samples and windows (changes) detected are listed in Table 1.

Since our technique focuses on data pre-processing, it can be used in conjunction with a large class of existing algorithms for time series prediction. We evaluated the use of our technique in conjunction with Linear Regression (LR), Pace Regression (PR), Support Vector machines (SVM) and Multilayer Perceptron (MLP). We did not evaluate the use of the popular ARIMA model, since that required an incompatible dataset format. We used the WEKA [19] software to run our experiments.

Table 1. Datasets used in the experiments

ID	Name	Samples/ windows	ID	Name	Samples/ windows (changes)
1	Amazon.com	422/13	8	Hewlett-Packard	612/27
2	Apple Computer	461/23	9	IBM	309/16
3	American Express	415/20	10	Island Pacific, Inc.	520/17
4	British Airways (ADS)	260/8	11	Johnson & Johnson	406/16
5	Colgate-Palmolive Co.	462/25	12	Simulated Dataset	475/22
6	eBay Inc.	520/22	13	S&P Quarterly Index	323/17
7	FedEx	423/17	14	Walt Disney Company	428/13

We used the five day relative difference in percentage of price (RDP) format [16]. The attributes by which the forecasted value was calculated were the 5, 10, 15 and 20 past days difference in percentage (RDP-5, RDP-10, RDP-15 and RDP-20), as well as a 15 day exponential moving average (EMA15). This type of transformation makes the data more symmetrical and closer to a normal distribution. The formulas that describe the RDP data format are listed in Table 2.

Table 2. RDP data format - attributes and forecast output

Input variables		Output variable	
EMA15	p(t)-EMA_{15}(t)	RDP+5	$(\overline{p(i+5)} - \overline{p(i)}) / \overline{p(i)} * 100$ $\overline{p(i)} = EMA_3(i)$
RDP-5	(p(t)-p(t-5))/p(t-5)*100		
RDP-10	(p(t)-p(t-10))/p(t-10) *100		
RDP-15	(p(t)-p(t-15))/p(t-15)*100		
RDP-20	(p(t)-p(t-20))/p(t-20)*100		

5 Experiments

The performances of the two versions of the algorithm were evaluated through the root mean square error (RMSE) metric. The results presented in Table 3 show the change in the RMSE value as captured in the formula:

(DBSR reduced dataset RMSE value) / (Full dataset RMSE value) * 100,

for both versions of the algorithm. i.e. The relative error using the filtered time series compared to using the full time series. In many cases for the machine learning methods, both versions of datasets filtered with our algorithms performed better than the machine learning methods trained on the full dataset, and in virtually all of them, employing at least one version of the algorithm resulted in a RMSE smaller than the methods trained on the full dataset.

The parametric method often yielded a smaller RMSE than the non-parametric method. The results in Table 3 also highlight some stability properties of the learning methods. As we can see from the RMSE reductions, the Linear Regression, Pace Regression and Support Vector Machines performed very similar when trained on the full datasets and on the reduced datasets as well, while Multilayer Perceptron performed poorly when trained on the full dataset, but had quite an improvement in performance when trained on some datasets filtered by the DBSR algorithm, but also had a large decrease in other cases.

Table 3. DBSMR algorithm change in RMSE values. Performances show percentage of RMSE error using our filtering approach compared to error without our filtering approach. Lower numbers indicate better performance for our filtering approach.

ID	P-DBSR p-val	LR P-DBSR	LR NP-DBSR	PR P-DBSR	PR NP-DBSR	SVM P-DBSR	SVM NP-DBSR	MLP P-DBSR	MLP NP-DBSR
1	0.5	98.30	95.72	98.07	93.82	100.2	96.56	61.32	64.04
2	0.3	92.53	95.11	91.86	95.49	86.98	96.87	114.6	103.6
3	0.4	92.06	83.91	97.35	85.34	103.6	94.82	120.4	106.6
4	0.7	94.95	99.93	97.28	99.55	97.19	96.89	84.51	102.3
5	0.8	98.72	101.1	98.12	98.89	98.52	99.37	105.3	98.56
6	0.6	99.32	104.7	100.1	103.4	100.9	107.1	99.00	126.9
7	0.6	98.77	98.51	97.77	99.65	95.71	99.86	181.3	105.8
8	0.5	95.77	97.38	95.72	96.37	99.14	100.2	60.59	91.52
9	0.7	95.22	97.80	98.44	97.07	94.31	99.93	100.6	86.12
10	0.5	100.9	99.27	102.9	99.69	96.90	95.73	78.31	83.74
11	0.7	91.66	97.82	94.37	100.6	95.05	103.3	95.54	95.68
12	0.5	96.81	94.57	96.78	94.53	98.53	95.22	84.92	107.1
13	0.5	93.40	93.98	91.75	90.96	90.75	125.8	158.8	139.6
14	0.5	98.70	104.2	98.97	102.4	97.23	102.8	81.08	109.6

6 Conclusion

Samples in financial time series datasets can be from different distributions and this creates challenges and opportunities for forecasting. We have developed data filtering algorithms that assess the importance of samples from a time series and retain those with most similarity to the recent past. Our experimental results show that the distribution of the data is indeed an important factor to consider, as we achieved reductions in forecasting error for time series with both few and many changes in the distribution. We believe our proposed DBSR algorithm is a simple and promising way to employ information about the distribution in the learning and prediction process.

In the future, we plan to investigate alternative methods to the Wilcoxon test for detecting distribution change and also investigate methods for stronger coupling of the distribution detection and prediction stages.

References

1. Tsay, R.S.: Analysis of Financial Time Series. Wiley-Interscience (2005)
2. Chatfield, C.: The Analysis of Time Series: an Introduction. Chapman & Hall/CRC (2004)
3. Witten, I.H., Frank, E.: Data mining: Practical Machine Learning Tools and Techniques. Morgan Kaufmann (2005)
4. Alpaydin, E.: Introduction to Machine Learning. The MIT Press (2004)
5. Gaber, M.M., Zaslavsky, A., Krishnaswamy, S.: Mining Data Streams: A Review. SIGMOD Record 24(2), 18–26 (2005)
6. Xindong, W., Yu, P.S., et al.: Data Mining: How Research Meets Practical Development? Knowledge and Information Systems 5(2), 248–261 (2003)
7. Yoo, P.D., Kim, M.H., et al.: Machine Learning Techniques and Use of Event Information for Stock Market Prediction: A Survey and Evaluation. CIMCA-IAWTIC (2005)
8. Hulten, G., Spencer, L., et al.: Mining time-changing data streams. In: Proceedings of the Seventh ACM SIGKDD International Conference on Knowledge Discovery and Data Mining, San Francisco, California, pp. 97–106 (2001)
9. Dong, G., Han, J., et al.: Online mining of changes from data streams: Research problems and preliminary results. In: Proceedings of the 2003 ACM SIGMOD Workshop on Management and Processing of Data Streams (2003)
10. Chen, J., Gupta, A.K.: Testing and locating variance changepoints with application to stock prices. Journal of the American Statistical Association 92(438), 739–747 (1997)
11. Adya, M., Collopy, F.: How effective are neural networks at forecasting and prediction? A review and evaluation. Journal of Forecasting 17(5-6), 481–495 (1998)
12. Kifer, D., Ben-David, S., et al.: Detecting change in data streams. In: Proceedings of the Thirtieth International Conference on Very Large Data Bases,Toronto, Canada, vol. 30, pp. 180–191. VLDB Endowment (2004)
13. Hollander, M., Wolfe, D.: Nonparametric Statistical Methods, 2nd edn. Wiley-Interscience (1999)
14. Kecman, V.: Learning and Soft Computing: support vector machines, neural networks, and fuzzy logic models. MIT Press (2001)
15. Liu, X., Zhang, R., et al.: Incremental Detection of Distribution Change in Stock Order Streams. In: 26th International Conference on Data Engineering Conference (ICDE), Long Beach, California, USA (2010)
16. Thomason, M.: The Practitioner Methods and Tools. Journal of Computational Intelligence in Finance 7(3), 36–45 (1999)
17. Web enabled scientific services and applications (2011), `http://www.wessa.net/stocksdata.wasp`
18. Hyndman, R.J.: S&P quarterly index online database (2008), `http://robjhyndman.com/tsdldata/data/9-17b.dat`
19. Tsay, R.S.: Analysis of Financial Time Series datasets (2002), `http://faculty.chicagobooth.edu/ruey.tsay/teaching/fts/d-ibmln.dat`
20. Waikato Environment for Knowledge Analysis, WEKA (2011), `http://www.cs.waikato.ac.nz/ml/weka/`
21. Ganti, V., Gehrke, J., Ramakrishnan, R.: DEMON: mining and monitoring evolving data. IEEE Transactions on Knowledge and Data Engineering 13(1) (2001)
22. Babcock, B., Datar, M., Motwani, R.: Load Shedding in Data Stream Systems. In: Proc. of the 2003 Workshop on Management and Processing of Data Streams, MPDS (2003)

Sequential Feature Selection for Classification

Thomas Rückstieß[1], Christian Osendorfer[1], and Patrick van der Smagt[2]

[1] Technische Universität München, 85748 Garching, Germany
{rueckst i,osendorf}@in.tum.de

[2] German Aerospace Center / DLR, 82230 Wessling, Germany
smagt@dlr.de

Abstract. In most real-world information processing problems, data is not a free resource; its acquisition is rather time-consuming and/or expensive. We investigate how these two factors can be included in supervised classification tasks by deriving classification as a sequential decision process and making it accessible to Reinforcement Learning. Our method performs a sequential feature selection that learns which features are most informative at each timestep, choosing the next feature depending on the already selected features and the internal belief of the classifier. Experiments on a handwritten digits classification task show significant reduction in required data for correct classification, while a medical diabetes prediction task illustrates variable feature cost minimization as a further property of our algorithm.

Keywords: reinforcement learning, feature selection, classification.

1 Introduction

In recent times, an enormous increase in data has been observed, without a corresponding growth of the information contained within them. In other words, the *redundancy* of data continuously increases. An example of such effects can be found in medical imaging. Diagnostic methods can be improved by increasing the amount of MRI, CT, EMG, and other imaging data yet the amount of underlying information does not increase. Even worse, the redundancy of such data seems to negatively impact the performance of associated classification methods. Indeed, common engineering practices employ data-driven methods (including dimensionality reduction, nonlinear PCA, etc.) to reduce data redundancy.

On the other hand, obtaining qualitatively good data gets increasingly expensive. Again, medical data serves as a good example: not only do the costs of the above-mentioned medical imaging techniques explode—MRT scans are performed at the end user price of several thousands of US dollars per hour—but also diagnostics tests are getting increasingly intricate and therefore costly, to the point that a selection of the right diagnostic methods while maintaining the level of diagnostic certainty is of high value.

Also, from a computer scientist's perspective, the amount of processable data grows faster than processor speed. According to various studies[1], recent years

[1] E.g., Gartner's survey at http://www.gartner.com/it/page.jsp?id=1460213.

D. Wang and M. Reynolds (Eds.): AI 2011, LNAI 7106, pp. 132–141, 2011.

showed an annual 40–60% increase of commercial storage needs and a 40+-fold increase is expected in the next decade. Though this may, just like the integration density of processors, follow Moore's law, the increase of computer speed is well below that.

In short, an improved approach *feature selection* (FS) is needed, which not only optimally spans the input space, but optimizes with respect to data consumption. All of these arguments clearly demonstrate the advantage of carefully selecting relevant portions of data. Going beyond traditional FS methods, in this paper we lay out and demonstrate an approach of selecting features in sequence, making the decision which feature to select next *dependent* on previously selected features and the current internal state of the supervised method that it interacts with. In particular, our sequential feature selection (SFS) will embed Reinforcement Learning (RL) into classification tasks, with the objective to reduce data consumption and associated costs of features during classification. The question we address in this paper is: "*Where do I have to look next, in order to keep data consumption and expenses low while maintaining high classification results?*"

Feature selection with RL has been addressed previously [5], yet the novelty of our approach lies in its sequential decision process. Our work is based on and inspired by existing research, combining aspects of online FS [17,11] and attentional control policy learning [1,14]. A similar concept, Online Streaming FS [17] has features streaming in one at a time, where the control mechanism can accept or reject the feature. While we adopt the idea of sequential feature selection, our scenario differs in that it allows access to all features with the subgoal of minimizing data consumption. A similar approach to ours is outlined in [10], where RL is used to create an ordered list of image segments based on their importance for a face recognition task. However, their decision process is not dependent on the internal state of the classifier, which brings their method closer to conventional FS.

Our framework is mapped out in Section 2. After introducing the general idea, we formally define sequential classifiers and rephrase the problem as a Partially Observable Markov Decision Process (POMDP). In addition, a novel action selection mechanism without replacement is introduced. Section 3 then demonstrates our approach, both on problems with redundant (handwritten digit classification) and costly (diabetes classification) data and discusses the results.

2 Framework

2.1 General Idea

In machine learning, solving a classification problem means to map an input x to one of a finite set of class labels $\mathcal{C}$. Classification algorithms are trained on labelled training samples $I = \{(x^1, c^1), \ldots, (x^n, c^n)\}$, while the quality of such a learned algorithm is determined by the generalization error on a separate test set. We regard features as disjunct portions (scalars or vectors) of the input pattern x, with feature labels $f_i \in F$ and feature values $f_i(x)$ for feature f_i. One key ingredient for good classification results is feature selection (also called *feature subset selection*): filtering out irrelevant, noisy, misleading or redundant features. FS is therefore a combinatorial optimization problem that tries to identify those features which will minimize the generalization error. In particular, FS tries to reduce the amount of useless or redundant data to process.

We want to take this concept even further and focus on minimizing *data consumption*, as outlined in the introduction. For this purpose, however, FS is not ideal. Firstly, the FS process on its own commonly assumes free access to the full dataset, which defeats the purpose of minimizing data access in most real-world scenarios. But more significantly, FS determines for *any* input the *same subset* of features that should be used for a subsequent classification. We argue that this limitation is not only unnecessary, but in fact disadvantageous in terms of minimizing data consumption.

We believe that by turning classification into a sequential decision process, we can further significantly reduce the amount of data to process, as FS and classification then become a closely intertwined process: deciding which feature to select next depends on the previously-selected features and the behaviour of the classifier on them. This will be achieved by using a fully trained classifier as an environment for an RL agent, that learns which feature to access next, receiving reward on successful classification of the partially uncovered input pattern.

2.2 Sequential Classification

A first step towards our goal is to re-formulate classification as a Partially Observable Markov Decision Process[2] (POMDP), making the problem sequential and thus accessible to Reinforcement Learning algorithms. We additionally require the following notation: ordered sequences are denoted by $(\cdot)$, unordered sets are denoted by $\{\cdot\}$, appending an element e to a sequence s is written as $s \circ e$. Related to power sets, we define a *power sequence* $\text{powerseq}(M)$ of a set M to be the set of all permutations of all elements of the power set of M, including the empty sequence $()$. As an example, for $M = \{1, 2\}$, the resulting $\text{powerseq}(M) = \{(), (1), (2), (1,2), (2,1)\}$. During an episode, the feature history $h_t \in \text{powerseq}(F)$ is the sequence of all previously selected features in an episode up to and including the current feature at time t. Costs associated with accessing a feature f are represented as negative scalars $r_f^- \in \mathbb{R}, r_f^- < 0$. We further introduce a non-negative global reward $r^+ \in \mathbb{R}, r^+ \geq 0$ for correctly classifying an input.

Classifiers in general are denoted with the symbol K. We define a *sequential* classifier $\widetilde{K}$ to be a functional mapping from the power sequence of feature values to a set of classes, i.e., $\widetilde{K} : \text{powerseq}\left(\{f(x)\}_{f \in F}\right) \rightarrow \mathcal{C}$. An additional requirement is to process the sequence one input a time in an online fashion, rather than classifying the whole sequence at once, and to output a class label after

[2] A *partially observable* MDP is a MDP with limited access to its states, i.e., the agent does not receive the full state information but only an incomplete observation based on the current state.

each input. Therefore, $\widetilde{K}$ requires some sort of memory. Recurrent Neural Networks (RNN) [7] are known to have implicit memory that can store information about inputs seen in the past. If the classifier does not possess such a memory, it can be provided explicitly: at timestep t, instead of presenting only the t-th feature value $f_t(x)$ to the classifier, the whole sequence $(f_1(x), \ldots, f_t(x))$ up to time t is presented instead.

As it turns out, the above approach of providing explicit memory is not limited to sequential classifiers. Any classifier, that can handle *missing values* [13] can be converted to a sequential classifier. For a given input x and a set F_1 of selected features, $F_1 \subseteq F$, the values of the features not chosen, i.e., $F \backslash F_1$, are defined as *missing*. Each episode starts with a vector of only missing values $(\phi, \phi, \ldots)$, where ϕ can be the mean over all values in the dataset, or simply consist of all zeros. At each timestep, the current feature gradually uncovers the original pattern x more. As an example, assuming scalar features f_1, f_4 and f_6 were selected from an input pattern $x \in \mathbb{R}^6$, the input to the classifier K would then be: $(f_1(x), \phi, \phi, f_4(x), \phi, f_6(x))$. This method allows us to use existing, pretrained non-sequential classifiers as well, that will remain unchanged and only act as an environment in which the SFS agent learns.

As we deal with a *partially observable* MDP, we need to extract an observation from the classifier, that summarizes the past into a stationary belief. Most classifiers base their class decision on some internal belief state. A Feed Forward Network (FFN) for example often uses a softmax output representation, returning a probability p_i in [0,1] for each of the classes with $\sum_{i=1}^{|C|} p_i = 1$. And if this is not the case (e.g., for purely discriminative functions like a Support Vector Machine), a straightforward belief representation of the current class is a k-dimensional vector with a 1-of-k coding.

To finally map the original problem of classification under the objective to minimize data consumption to a POMDP, we define each of the elements of the 6-tuple $(S, A, O, \mathcal{P}, \Omega, \mathcal{R})$, which describes a POMDP, as follows: the state $s \in S$ at timestep t comprises the current input x, the classifier $\widetilde{K}$, and the previous feature history h_{t-1}, so that $s_t = (x, \widetilde{K}, h_{t-1})$. This triple suffices to fully describe the decision process at any point in time. Actions $a_t \in A$ are chosen from the set of features $F \backslash h_{t-1}$, i.e., previously chosen features are not available. Section 2.3 describes, how this can be implemented practically. The observation is represented by the classifier's internal belief of the class after seeing the values of all features in h_{t-1}, written as $o_t = b(x, \widetilde{K}, h_{t-1}) = b(s_t)$. In the experiments section, we will demonstrate examples with FFN, RNN and Naive Bayes classifiers. Each of these architectures allows us to use the aforementioned softmax belief over the classes as belief state for the POMDP. The probabilities p_i for each class serve as an observation to the agent: $o_t = b(x, \widetilde{K}, h_{t-1}) = (p_1, p_2, \ldots, p_{|C|})$.

Assuming a fixed x and a deterministic, pretrained classifier $\widetilde{K}$, the state and observation transition probabilities $\mathcal{P}$ and Ω collapse and can be described by a deterministic transition function T, resulting in next state $s_{t+1} = T_x(s_t, a_t) = (x, \widetilde{K}, h_{t-1} \circ a_t)$ and next observation $o_{t+1} = b(s_{t+1})$. Lastly, the reward function

$\mathcal{R}^a_{ss'}$ returns the reward r_t at timestep t for transitioning from state s_t to s_{t+1} with action a_t. Given c as the correct class label, it is defined as:

$$r_t = \begin{cases} r^+ + r^-_{a_t} & \text{if } \widetilde{K}\left((h_\tau(x))_{0<\tau\leq t}\right) = c \\ r^-_{a_t} & \text{else} \end{cases} \tag{1}$$

2.3 Action Selection without Replacement

In this specific task we must ensure that an action (a feature) is only chosen at most once per episode, i.e., the set of available actions at each given decision step is dependent on the history h_t of all previously selected actions in an episode. Note that this does not violate the Markov assumption of the underlying MDP, because no information about available actions flows back into the state and therefore the decision does not depend on the feature history.

Value-based RL offers an elegant solution to this problem. By manually changing all action-values $Q(o, a_t)$ to $-\infty$ after choosing action a_t, we can guarantee that all actions not previously chosen in the current episode will have a larger value and be preferred over a_t. A compatible exploration strategy for this action selection without replacement is Boltzmann exploration. Here, the probability of choosing an action is proportional to its value under the given observation:

$$p(a_t|o_t) = \frac{e^{Q(o_t,a_t)/\tau}}{\sum_a e^{Q(o_t,a)/\tau}}, \tag{2}$$

where τ is a temperature parameter that is slowly reduced during learning for greedier selection towards the end. Thus, when selecting action a_{t+1}, all actions in h_t have a probability of $e^{-\infty} = 0$ of being chosen again. At the end of an episode, the original Q-values are restored.

2.4 Solving the POMDP

Having defined the original task of classification with minimal data consumption as a POMDP and solved the problem of action selection without replacement, we can revert to existing solutions for this class of problems. Since the transition function is unknown to the agent, it needs to learn from experience, and a second complication is the continuous observation space. For regular MDPs, a method well-suited to tackle both of these issues is Fitted Q-Iteration (FQI) [3]. The sequential classifier $\widetilde{K}$ then takes care of the PO part of the POMDP, yielding a static belief over the sequential input stream.

FQI uses a batch-trained function approximator (FA) as action-value function. Various types of non-linear function approximators have been successfully used with FQI, e.g., Neural Networks [12], Gaussian Processes [2], and others [9]. In this paper, we will use Locally Weighted Projection Regression (LWPR) [15] as the value function approximator of choice, as it is a fast robust online method that can handle large amounts of data.

Algorithm 1. Sequential Feature Selection (SFS)

Require: labelled inputs I, agent A, sequential classifier $\widetilde{K}$

1: **repeat**
2:     choose $(x,c) \in I$ randomly
3:     $h_0 \leftarrow (\phi)$
4:     $o_1 \leftarrow b(x, \widetilde{K}, h_0)$
5:     **for** $t = 1$ to $|F|$ **do**
6:         $a_t \leftarrow \mathrm{A}(o_t)$
7:         $h_t \leftarrow h_{t-1} \circ a_t$
8:         $o_{t+1} \leftarrow b(x, \widetilde{K}, h_t)$
9:         **if** $\widetilde{K}\left((h_\tau(x))_{0<\tau\leq t}\right) = c$ **then**
10:             $r_t \leftarrow (r^+ + r^-_{a_t})$
11:             **break**
12:         **else**
13:             $r_t \leftarrow r^-_{a_t}$
14:         **end if**
15:     **end for**
16:     train A with $(o_1, a_1, r_1, \ldots, r_t, o_{t+1})$
17: **until** convergence

The details of the algorithm are presented in Listing 1. The history is always initialized with the missing value ϕ (line 3). This gives the system the chance to pick the first feature before seeing any real data. The SFS agent is trained after every episode (line 16), which ends either with correct classification (line 9–11) or when the whole input pattern was uncovered (line 15), i.e., all features were accessed.

3 Experiments and Discussion

We evaluate the proposed method on two different datasets: the MNIST handwritten digits classification task, and a medical dataset for diabetes prediction. Each experiment was repeated 25 times, the plots below show single runs (gray) and the mean value over all runs (black).

3.1 Handwritten MNIST Digit Classification

In this experiment we looked at the well-known MNIST handwritten digit classification task [8], consisting of 60,000 training and 10,000 validation examples. Each pattern is an image of 28×28 pixels of gray values in [0,1], the task is to map each image to one of the digits 0–9. We split every image into 16 non-overlapping 7×7 patches, each patch representing a feature.

We present results for an FFN as a non-sequential classifier and an RNN with Long Short Term Memory (LSTM) cells [6] as a sequential classifier with implicit memory. The FFN was chosen because it is a well-understood simple method, widely used for classification. The RNN was chosen to investigate, how naturally

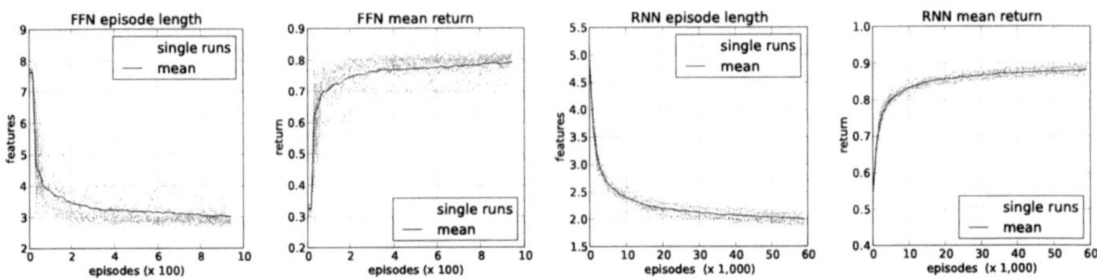

Fig. 1. Results of MNIST with FFN (left two plots) and RNN (right two plots). For each classifier, mean episode length and mean return over training episodes are shown.

sequential classifiers work with SFS. Throughout this experiment, rewards were set to $r^+ = 1.0$ and $r_k^- = -0.1 \; \forall k$.

The FFN has one hidden layer with sigmoid activation, the architecture is 784-300-10. The output layer uses softmax activation with a 1-of-n coding. Pretraining of the classifier was executed online with a learning rate $\alpha = 0.1$ on the full training dataset. After 30 epochs of presenting all 60,000 digits to the network, the error rate on the test dataset is 1.18%, slightly better than reported in [8]. However, this result is secondary, as the network acts merely as an environment for the SFS agent. During SFS training, each episode uses a random sample from the test dataset. Figure 1 (left two plots) shows the development of episode lengths and returns during training of the SFS agent. The average number of features required to correctly classify dropped from initially 7.65 (random order) to 3.06 (trained SFS). The rate of incorrectly classified images was 0.77%.

The architecture of the RNN classifier is 49-50-10 with LSTM cells in the hidden layer. The output activation function is softmax with a 1-of-n coding. The RNN was pretrained with Backpropagation Through Time (BPTT) (see, e.g., [16]), with a learning rate of $\alpha = 0.01$ and a random order of features. The results are illustrated in Figure 1 (right two plots). The average number of required features decreases from 4.91 features (random order) to 1.99 (trained SFS). The rate of incorrectly classified images was 1.71%.

3.2 Diabetes Dataset with Naive Bayes Classification

For the second experiment, we chose a more practical example from the medical field, the Pima Indians Diabetes data set [4]. We also decided on a Naive Bayes classification, to demonstrate the flexibility of the proposed method in terms of classifiers. The data set consists of 768 samples with 8 features (real-valued and integer) and two target classes (diabetes, no diabetes). Pretraining with a Naive Bayes classifier resulted in 73% correct prediction. There are two interesting aspects in this dataset. Firstly, it contains missing values, which should be handled well as we already use missing values to turn classification into a sequential process. Secondly, the features represent very different attributes of the (all female) patients. Some are simple questions (e.g., age, number of times pregnant), others are more complex medical tests (e.g., plasma glucose concentration after 2h in

Table 1. Assigned feature costs for diabetes dataset

Feature	# pregnant	2h glucose concentration	blood pressure	skin fold thickness	2h serum insulin	BMI	diabetes pedigree fct.	age
Cost	-1	-120	-5	-5	-120	-5	-60	-1

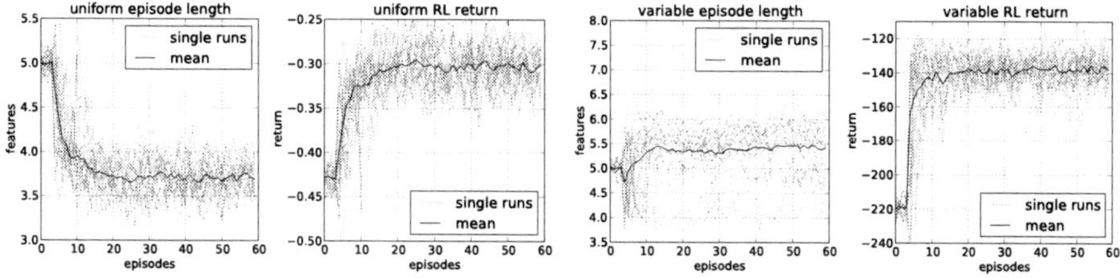

Fig. 2. Results of the PIMA diabetes dataset with Naive Bayes classification. Left two figures: episode lengths and mean returns for uniform feature costs. Right two figures: episode lengths and mean returns for feature costs according to Table 1.

an oral glucose tolerance test). While the MNIST experiment used uniform costs r_k^- for all features f_k, this experiment demonstrates another property of SFS: the feature costs can be weighted, representing cheaper and more expensive features. To investigate the difference between uniform and variable feature costs, two sets of experiments were conducted: The first uses uniform costs $r_k^- = -0.1\,\forall k$, with a final number of required features of 3.7 on average. The second variant uses variable, estimated costs[3] shown in Table 1. Number of features *increased* from 4.99 to 5.66 on average, while the average return increased from -218 to -141. Figure 2 shows the results of both variants graphically.

3.3 Discussion

The MNIST experiment with FFN classifier demonstrates a significant reduction of data consumption in two ways. Firstly, by making the decision process sequential, which enables the classifier to make decisions before all features have been looked at. This step alone reduces the average number of required features from all 16 features down to 7.65 (a reduction to 48%), and indicates that there is in fact a lot of redundancy in the MNIST images. Secondly, consumption is reduced further by learning the dependency of current belief and next feature, instead of accessing them in random order. After training the SFS agent, data consumption decreases to 3.06 on average, 19% of the full data.

[3] These costs represent a rough estimate of the time in minutes it takes to acquire the feature on a real patient. The estimates are based on oral communication with a local GP.

It is important to note that the stated error rates (1.18% for static and 0.77% for sequential classification) cannot be compared directly, because of the very different nature of the sequential approach. Sequential classification replaces the conventional error rates as performance measure based on the binary success of each sample (classified / not classified) with a scalar value (how many features until classified). In order to compare both classification methods, we would have to additionally learn when to stop the decision process, without using the class label. This could be achieved with a confidence threshold (e.g., if max(belief) reaches a certain value) or by explicitly learning when to stop with either supervised or RL methods. In this paper, we focussed on the RL feature selection process with existing classifiers rather than the performance of sequential classifiers. This issue will be addressed in a future publication.

Another aspect we investigated was the use of RNNs as naturally sequential classifiers. Where static classifiers still need to look at a full input (at least in terms of dimension, even though most of the pattern is filled with missing values), RNNs can make use of their intrinsic memory and achieve similar results with significantly fewer nodes in input and hidden layer and therefore even less data processing. They also converge with lower variance and reduce data consumption to a mere 12% on the MNIST task.

Finally, the Pima diabetes data set illustrates the use of variable feature costs, a variant that is naturally supported in our framework. The left two plots in Figure 2 show the development of episode length (i.e., number of selected features until correct classification) and mean return of the uniform cost experiment. As expected, episode lengths decrease with increasing returns, as the only objective for the agent is: *select those features first, that lead to correct classification.* However, if the reward scheme is changed (right two plots in Figure 2), we witness a *growth* of episode lengths in most of the 25 trials and on average. Still, all trials increase their returns (rightmost plot), which indicates that the agent does indeed learn and improve its performance. Comparing the final return average of -141 and the worst final return of -160 to the individual costs of Table 1, it becomes clear that in all runs, only one of the three most expensive features (number 2, 5 and 7) was selected. This behavior was caused by the different objective: *minimize the overall costs associated with the features.* In other words, it is okay to select many features, as long as they are cheap.

4 Conclusion

We have derived classification as a POMDP and thus made it accessible to RL methods. The application we focussed on was minimization of data consumption, by training an RL agent to pick features first that lead to quick classification. We presented results for different classifiers (both static and sequential) on vision and medical tasks. Our approach reduces the number of necessary features to access to a fraction of the full input, down to 12% with RNN classifiers. We also demonstrated that SFS is able to deal with weighted feature costs, a property that exists in plenty of real-world applications. A new action selection method was introduced that draws actions without replacement. It should prove useful in other ordering tasks as well, such as scheduling problems. Lastly, we would like to point out that our approach is not limited to classification but easily extends to regression or other supervised tasks.

References

1. Bazzani, L., de Freitas, N., Larochelle, H., Murino, V., Ting, J.A.: Learning attentional policies for tracking and recognition in video with deep networks. In: Proceedings of the 28th International Conference on Machine Learning (2011)
2. Deisenroth, M.P., Rasmussen, C.E., Peters, J.: Gaussian process dynamic programming. Neurocomputing 72(7-9), 1508–1524 (2009)
3. Ernst, D., Geurts, P., Wehenkel, L.: Tree-based batch mode reinforcement learning. Journal of Machine Learning Research 6(1), 503 (2005)
4. Frank, A., Asuncion, A.: UCI Machine Learning Repository. University of California, Irvine, CA (October 2011), `http://archive.ics.uci.edu/ml/`
5. Gaudel, R., Sebag, M.: Feature selection as a one-player game. In: Proceedings of the 2nd NIPS Workshop on Optimization for Machine Learning (2009)
6. Hochreiter, S., Schmidhuber, J.: Long short-term memory. Neural Computation 9(8), 1735–1780 (1997)
7. Hüsken, M., Stagge, P.: Recurrent neural networks for time series classification. Neurocomputing 50, 223–235 (2003)
8. LeCun, Y., Bottou, L., Bengio, Y., Haffner, P.: Gradient-based learning applied to document recognition. Proceedings of the IEEE 86(11), 2278–2324 (1998)
9. Neumann, G., Pfeiffer, M., Hauser, H.: Batch reinforcement learning methods for point to point movements. Technical report, Graz University of Technology (2006)
10. Norouzi, E., Nili Ahmadabadi, M., Nadjar Araabi, B.: Attention control with reinforcement learning for face recognition under partial occlusion. Machine Vision and Applications, 1–12 (2010)
11. Perkins, S., Theiler, J.: Online feature selection using grafting. In: Proceedings of the 20th International Conference on Machine Learning (2003)
12. Riedmiller, M.: Neural Fitted Q Iteration - First Experiences with a Data Efficient Neural Reinforcement Learning Method. In: Gama, J., Camacho, R., Brazdil, P.B., Jorge, A.M., Torgo, L. (eds.) ECML 2005. LNCS (LNAI), vol. 3720, pp. 317–328. Springer, Heidelberg (2005)
13. Saar-Tsechansky, M., Provost, F.: Handling missing values when applying classification models. Journal of Machine Learning Research 8(1625-1657), 9 (2007)
14. Schmidhuber, J., Huber, R.: Learning to generate artificial fovea trajectories for target detection. International Journal of Neural Systems 2(1), 135–141 (1991)
15. Vijayakumar, S., Schaal, S.: Locally weighted projection regression: An O(n) algorithm for incremental real time learning in high dimensional space. In: Proceedings of the Seventeenth International Conference on Machine Learning (2000)
16. Williams, R.J., Peng, J.: An efficient gradient-based algorithm for on-line training of recurrent network trajectories. Neural Computation 2(4), 490–501 (1990)
17. Wu, X., Yu, K., Wang, H., Ding, W.: Online streaming feature selection. In: Proceedings of the 27nd International Conference on Machine Learning (2010)

Long-Tail Recommendation Based on Reflective Indexing

Andrzej Szwabe, Michal Ciesielczyk, and Pawel Misiorek

Institute of Control and Information Engineering, Poznan University of Technology,
M. Sklodowskiej-Curie Square 5, 60-965 Poznan, Poland
{andrzej.szwabe,michal.ciesielczyk,pawel.misiorek}@put.poznan.pl

Abstract. We propose a collaborative filtering data processing method based on reflective vector-space retraining, referred to as Progressive Reflective Indexing (PRI). We evaluate the method's ability to provide recommendations of items from a long tail. In order to reflect 'real-world' demands, in particular those regarding non-triviality of recommendations, our evaluation is novelty-oriented. We compare PRI with a few widely-referenced collaborative filtering methods based on SVD and with Reflective Random Indexing (RRI) - a reflective data processing method established in the area of Information Retrieval. To demonstrate the superiority of PRI over other methods in long tail recommendation scenarios, we use the probabilistically interpretable AUROC measure. To show the relation between the structural properties of the user-item matrix and the optimal number of reflections we model the analyzed data sets as bipartite graphs.

Keywords: information retrieval, machine learning, e-commerce applications, collaborative filtering, long tail, RRI, SVD.

1 Introduction

Coping with heavily-tailed behavioral data is regarded as one of the main challenges of the research on collaborative filtering, and of the research on recommender systems in general [4], [8]. Heavily-tailed data distributions are known to be typical of popular Internet applications [2], [4]. Despite that, recommendation systems that are proposed as a means for long-tail recommendation, are based on Singular Value Decomposition (SVD) - the method that has been used for collaborative filtering for more than 10 years by many authors not mentioning the long-tail phenomenon [1], [10], [13]. Moreover, even researchers focusing on long-tail recommendation systems very rarely use specialized methods for analyzing heavily-tailed data distributions [10], [11].

On the other hand, in the last few years, we observe an increasing interest of the IR community in reflective matrix data processing. At least in some application scenarios, the most widely known reflective indexing method, RRI, is more useful than 'classical' methods based on SVD and dimensionality reduction [7]. Moreover, applications of reflective indexing have been shown as a very effective means for collaborative filtering [5], [6].

D. Wang and M. Reynolds (Eds.): AI 2011, LNAI 7106, pp. 142–151, 2011.

1.1 Novelty as an Important Value of Long-Tail Recommendations

The behavioral data collected by on-line retailers indicate that the majority of successful recommendations are based on the recommendations of items from the long tail [2], [8]. Trivial (popularity-based) recommendations are not of assistance for users who expect some novelty from recommendations [8], [10].

In this paper, we define a recommendation system as aimed at providing non-trivial recommendations of items from the long tail of the dataset. We believe that this kind of definition is most suitable to model the recommendation task in real-world e-commerce application scenarios [2], [4].

1.2 Methodological Assumptions

In the present paper, apart from the long-tail phenomenon, we also take into account the need to provide multiple contextual recommendations. Inspired by recommendation services for online retailers, we assume that the practical value of contextual recommendations strongly depends on the system's ability to identify as many attractive items as possible, rather than just top-n items [10]. To emulate a contextual recommendation scenario, while still being able to use a widely-referenced (non-contextual) data set, we follow the 'recommendation as classification' approach [3]. In doing so, we assume that the purpose of a recommender system is to identify, for each (equally important) user, all the items rated above the user's average rate.

The assumptions stated above have some important implications. Firstly, the original data set is pre-processed in order to make it more 'objective' and novelty-oriented, i.e., for each user/item rating a user average rating and an item average rating is subtracted. As a result, we are able to treat all users as equally important, and target the long tail scenario and the 'non-obviousness' of the recommendation [10]. As a consequence, we evaluate the system's ability to provide recommendations independently from the items' 'global' popularity. Secondly, we use recommendation quality measures that are classification-oriented, i.e., enable us to evaluate the systems' ability to distinguish between attractive and non-attractive items, rather than to predict absolute rating values [8], [10].

We use Area Underneath an ROC curve - referred to as AUROC (also known as Swets' A measure) [10] – because it is a probabilistically interpretable classification quality measure. In other words, AUROC indicates the probability that the recommendation system will choose a relevant item from a set of two items, one randomly chosen from a set of relevant items and the other randomly chosen from a set of irrelevant items [10]. Another important property of AUROC is its robustness to the test set sparsity: unknown ratings do not influence the measurement result. Although the results presented in many papers on recommendation systems are expressed using the F1 measure, we choose to use the AUROC measure. This decision is motivated by the fact that an F1 measurement requires the use of the $@n$ parameter, which makes the analysis of the results unclear, especially when the datasets have various sparsity and various heavy-tailness. Moreover, F1 (applied by the authors of [13]) does not allow to distinguish between the case of a negative rating and the case of no rating, which makes it impractical for using together with datasets evaluated in this paper.

In addition to the recommendation quality evaluation, we use the bipartite graph model to perform a structural analysis of data sets, in which nodes represent users and items, and edges encode the existence of interactions between them [9]. In particular, we investigate selected structural properties of the data, such as the distribution of node degrees and the distribution of lengths of the shortest paths between nodes.

1.3 Contribution of the Paper

To our knowledge, the solution presented in this paper is the first successful application of a reflective matrix data processing to long-tail collaborative filtering. The advantages of the method are demonstrated in scenarios of different density. We show that as a result of the unique features of PRI (including its 'sensitivity' to local dependencies in a dataset, and the applicability of information-theoretic optimization) our method outperforms a few widely-referenced methods, including those based on SVD and dimensionality reduction.

We present context vectors retraining as a process enabling the 'discovery' of indirect correspondences between objects. We focus on features of the reflective indexing, which, in contrast to SVD and RRI, provide a recommendation system with 'structure-sensitivity'.

2 Algebraic Model for PRI

We introduce system S representing user-item dependencies as a 'classical' user-item matrix. System S models the relations between objects from the set of users K and the set of items L. Let us assume that $|K| = m$ and $|L| = n$. Let us define the matrix $A_{i,j} = [a_{i,j}]_{m \times n}$ containing the user ratings on items. Let $R_1 \leq a_{i,j} \leq R_2$, where R_1 and R_2 are the minimum and the maximum rating value available in the system, respectively. Moreover, we define matrix $X = [x_{i,j}]_{(m+n)\times(m+n)}$ - the matrix of context vectors for objects described in S.

2.1 Modeling User-Item Dependencies as a Probability Space

We identify two features characterizing the model of S as particularly important. Firstly, each 'object' is modeled in the system in a way that ensures its initial independence. We treat all users and all items as equally important by subtracting average user ratings and average item ratings from the original data set.

Secondly, the probability space (Ω, F, P) provides a functional interpretation for states of system S. The space represents conditional dependencies between each user and each item under the condition that the matrix contains the whole information about the user-item system. The sample space $\Omega = \{E_{i,j}\}$ of the probability space is a set of events such as $E_{i,j}$ is an event of the dependency occurring between user i ($i \in K$) and item j ($j \in L$) (of probability representing the amount of information stored in the system). F is defined as a set of all possible subsets of Ω. Finally, the probability measure $P : F \rightarrow [0,1]$ is defined according to a distribution which can be presented as the following matrix $B_{i,j} = [b_{i,j}]_{m\times n}$, where $b_{i,j} = P[E_{i,j}]$ for $i = 1..m$, $j = 1..n$.

The values of matrix B are obtained by taking the squares of values from A and normalizing them in order to ensure that $\sum_{i=1}^{m}\sum_{j=1}^{n} b_{i,j} = 1$.

We follow a probabilistic interpretation of matrix A as a representation of the system state, and of matrix X as a representation of the context vectors. In particular, for matrix X, we normalize context vectors after each reflection step in order to model a conditional probability space associated with the object corresponding to a given context vector for which squares of coordinates describe the probability distribution. Each time we interpret a squared matrix entry value as some probability, we do so because we treat each matrix entry as representing an individual event that is independent of any other event, and because we follow the probabilistic interpretation of a vector space that has its roots in quantum mechanics [14]. According to this interpretation, the probability that a system is in a given state is equal to the squared length of the projection of the system state vector onto the given state vector.

2.2 Reflective Data Processing

Let $O = K \cup L$ denote the set of objects described in a system S. The goal of our method is to determine the correspondences between objects from O, based on the external information about relations between them.

In order to model the 'retraining' procedure, we construct a sequence of matrices X_i (where $i \geq 0$) containing context vectors for objects from O, defined in the r-dimensional vector space (where $r = m + n$). We assume that the first m rows of each matrix X_i describe users from set K, whereas the rows indexed from $m+1$ to r describe items from set L. For each matrix X_i, let us denote the top m rows by $m \times r$ matrix $\overline{X}_i$, and the bottom n rows by $n \times r$ matrix $\underline{X}_i$.

Since at the first step we assume that all the objects from set O are independent, we model the initial matrix X_0 as a diagonal matrix I_r. In the case when dimensionality reduction is needed for computational efficiency, matrix X_0 may be constructed using the random projection approach, as it is in the case of RI (Random Indexing) and RRI algorithms [7].

Matrices X_i (for $i \geq 1$) are obtained according to the PRI reflective retraining procedure. In contrast to RI/RRI methods, the PRI method uses the gain value g_i at each reflection step. In particular, for $i \geq 1$, the matrices of context vectors are defined as follows:

$$\overline{X}_i = \overline{X}_{i-1} \tag{1}$$

$$\underline{X}_i = (1 - g_i)\underline{X}_{i-1} + g_i A^T \overline{X}_{i-1}$$

for $i \geq 1, 3, 5, \ldots$, and

$$\overline{X}_i = (1 - g_i)\overline{X}_{i-1} + g_i A \underline{X}_{i-1} \tag{2}$$

$$\underline{X}_i = \underline{X}_{i-1}$$

for $i = 2, 4, 6, \ldots$.

As the first step of each reflection, rows of matrix X_i are normalized in order to ensure the equal importance of each context vector in the reflection process and their probabilistic interpretation: the squares of context vectors' entries constitute the probability distribution for the conditional probability space corresponding to a given object. Moreover, before each reflection step, we normalize the rows of matrices $A\underline{X}_{i-1}$ and $A^T\overline{X}_{i-1}$ in order to ensure the probabilistic interpretation of the retraining step.

3 The PRI Algorithm

The PRI algorithm is presented as the *reflect* procedure described in Algorithm 1 and is executed using the command *reflect(A, X_0, m, n, k)*, where k is the maximum length of the shortest path between nodes in the bipartite graph representing input matrix A. This number determines the number of reflection steps used in the PRI algorithm: PRI continues the procedure of reflecting data processing as long as there are some indirect connections in a dataset, which have not been explored.

Algorithm 1. reflect(A, X, m, n, k)

$norm2(X^T), norm2(X), norm2(A^T\overline{X}), k \Leftarrow k-1,$
for $g \in <0,1>$ **do**
  $\overline{X}(g) = \overline{X}$
  $\underline{X}(g) = A^T\overline{X} \times g + \underline{X} \times (1-g)$
  $norm2(\underline{X}(g))$
end for
$gain = argmax_{g\in<0,1>}\{findEntropy(X(g), m+n)\}$
$X \Leftarrow X(gain)$
if $k = 0$ **then return**
end if
$norm2(X^T), norm2(X), norm2(A\underline{X}), k \Leftarrow k-1,$
for $g \in <0,1>$ **do**
  $\underline{X}(g) = \underline{X}$
  $\overline{X}(g) = A\underline{X} \times g + \overline{X} \times (1-g)$
  $norm2(\overline{X}(g))$
end for
$gain = argmax_{g\in<0,1>}\{findEntropy(X(g), m+n)\}$
$X \Leftarrow X(gain)$
if $k = 0$ **then return**
end if
$reflect(A, X, m, n, k)$

In order to find the optimal values of g_i, the algorithm uses the information-theoretic criterion based on the calculation of the average entropy of context vectors. The *findEntropy(X,r)* procedure is defined in Algorithm 2. The detailed analysis of the criterion is beyond the scope of this paper. The normalization of rows of matrix $X = [x_{i,j}]$ (the *norm2* function in the *reflect* procedure) is conducted according to the formula $x'_{i,j} = x_{i,j}/(\sum_{j=1}^{r} x_{i,j}{}^2)^{0.5}$.

Algorithm 2. findEntropy(X, r)

$USV \Leftarrow SVD(X)$
v_i = i-th column of V
$D \Leftarrow \sum_{i=1}^{r} \left[v_i (v_i)^T (s_{i.i})^2 / \sum_{j=1}^{r} (s_{j,j})^2 \right]$
$D \Leftarrow XD$
d_i = i-th row of D
return $r^{-1} \sum_{i=1}^{r} entropy(d_i)$

The entropy of a given vector $V = [v_i]$ of length r is calculated according to the formula:

$$-\sum_{i=1}^{r} p_i \log_2(p_i), \tag{3}$$

where $p_i = {v_i}^2 / \sum_{j=1}^{r} {v_j}^2$ for $i = 1, \ldots, r$.

Similarly as for other collaborative filtering methods, PRI is based on the estimation of the missing entries of the input matrix. At the last step of the PRI algorithm, matrix A is reconstructed (using Equation 4). Finally, the ordered list of recommendations, for the purposes of the AUROC measure, is calculated based on the reconstructed matrix.

$$newA = \overline{X}\underline{X}^T \tag{4}$$

4 Evaluation

We have compared the accuracy of the PRI method to the accuracy of other methods presented in the relevant literature, such as the SVD-based kNN recommendation method [13] (in this paper referred to as 'SVD-kNN'), the popularity-based recommendation performed according to the number of ratings (referred to as 'Popularity') [8], the RRI technique [7], the PureSVD algorithm [8] and the RSVD-RRI method [5]. For each of these methods we have set all necessary parameters (e.g., k-cut of SVD) optimally in order to provide the best possible results. In contrast, the PRI algorithm does not require to set any parameter.

In order to perform the comparison, we have developed our implementations of the methods mentioned above and compared our results against the reported ones. As shown in [5], we were able to achieve results that are very close to the original ones. We have used the same implementations in the experiments reported in this paper.

As explained in Subsection 1.2, we have pre-processed the dataset by subtracting from each user/item rating the average of a user average rating and an item average rating. Only ratings that are positive after the pre-processing step have been regarded as hits for the recommendation system.

4.1 Data Sets

Due to a lack of freely available e-commerce data sets [12] and to make our results comparable with those presented in relevant papers, we have evaluated all the methods using the MovieLens ML100k set, one of the most widely referenced CF datasets

Table 1. Number of ratings in the train sets $ML(x, lt)$

	$lt = 1$	$lt = 0.945$	$lt = 9$	$lt = 0.8$	$lt = 0.7$
$x = 0.8$	80000	57282	45838	28215	17486
$x = 0.2$	20000	14320	11459	7053	4271

[5], [13]. The majority of the ML100k ratings are condensed in a small set of the most popular items (called the short head). However, the real-world datasets [2] are much more heavily-tailed. As observed in [8], recommending the most popular items is usually not beneficial for the users. Therefore, when preparing data sets for our experiments, we have followed a similar approach as the one presented in [8]: we have removed a specified number of the most popular items from the dataset, according to parameter lt, which determines the percentage of items remaining in the dataset after removing items from the short head. The data set obtained using a given value of parameter lt is denoted as $ML(lt)$. Unlike the authors of [8], we have removed the ratings from the whole dataset instead of just removing ratings from the test set. This way we have simulated more realistically the sparsity of a 'real-world' dataset (usually much sparser than the MovieLens dataset). Moreover, we have randomly partitioned each data set into two pairs of a train set and a test set (one pair for each value of the training ratio x) as proposed in [13]. A train set constructed for a given x and a given lt is denoted as $ML(x, lt)$. In our experiments we have used $x \in \{0.2, 0.8\}$ and $lt \in \{0.7, 0.8, 0.9, 0.945, 1\}$. In particular, we have examined $lt = 0.945$ since this case was also evaluated in [8]. Table 1 shows the number of ratings left from 100000 ratings of ML100k when preparing various $ML(x, lt)$ train sets.

Following the dataset representation as a bipartite graph, we have analyzed the distribution of the shortest path lengths between objects forming user-user, item-item as well

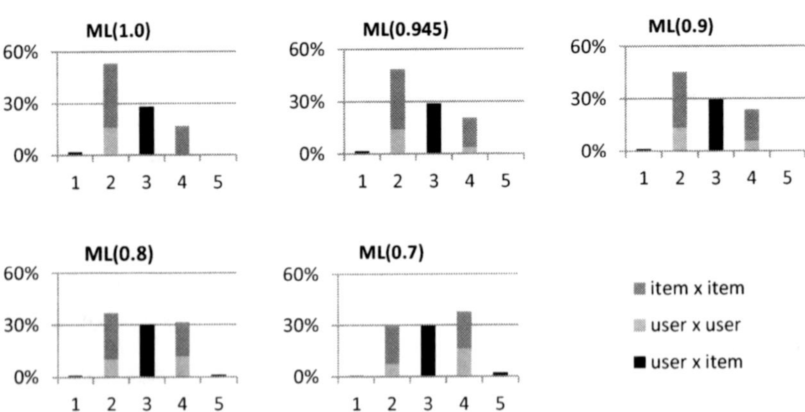

Fig. 1. The histogram of the shortest path lengths in the bipartite graphs representing various data sets $ML(lt)$

Table 2. The maximum shortest path length in $ML(x, lt)$

	$lt = 1$	$lt = 0.945$	$lt = 9$	$lt = 0.8$	$lt = 0.7$
$x = 0.8$	5	5	5	7	7
$x = 0.2$	7	7	7	9	11

as user-item pairs. We have observed that the structural properties closely correspond to the optimal configuration of recommendation algorithms based on reflective data processing. Figure 1 shows that in the case of the sparse long-tail data set - $ML(0.7)$ - cases of longer shortest paths are more frequent. As shown in the next subsection, the PRI algorithm applied to such a dataset estimates the bigger number of reflections as more optimal. Table 2 presents the values of the maximum lengths of the shortest paths between nodes of a bipartite graph representing various $ML(x, lt)$ train sets, i.e. the values of k determining the number of reflection steps used in PRI.

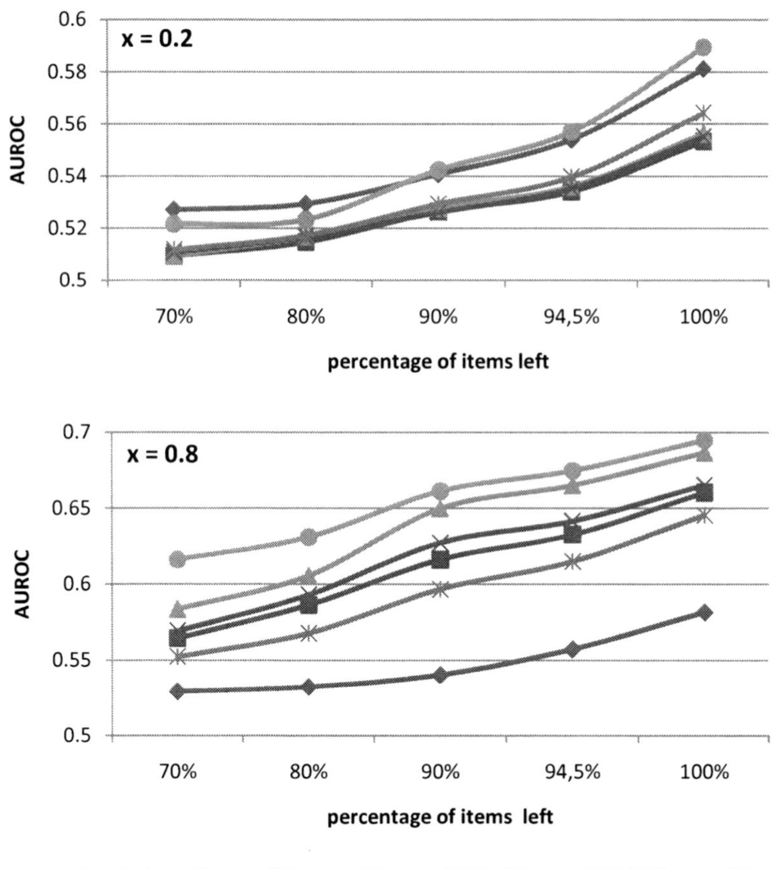

Fig. 2. The AUROC results

4.2 Recommendation Quality Evaluation

Figure 2 shows the comparison of the PRI algorithm with alternative approaches (*Popularity*, *RRI*, *RSVD-RRI*, *pureSVD*, *SVD-kNN*) performed using the AUROC measure and datasets of various sparsity. The results demonstrate the superiority of PRI over other methods. It can be observed that the advantage of PRI is especially visible in the case of both heavily-tailed and sparse datasets. As one may expect, the more long-tailed a dataset is, the worse the recommendation results are. On the other hand, the very low effective data set density (i.e., being the results of both setting lt and x) enables the popularity-based method to achieve a comparatively high accuracy. It may be observed that in the case of a sparser train set and a denser test set (i.e, for $x = 0.2$), the popularity-based algorithm provides comparatively good results, whereas in the opposite case (i.e, for $x = 0.8$), it operates poorly. One may realize that the popularity-based algorithm, instead of modeling users' preference profiles, simply reflects the ratio between positive ratings (hits) and negative ratings (misses) for the most popular items in a given dataset. Since we use a random procedure to divide the data set into a train set and a test set, the values of AUROC for the popularity-based algorithm are almost identical for both the cases of $x = 0.2$ and $x = 0.8$, what additionally confirms the reliability of the AUROC measure.

5 Conclusions

The PRI method generalizes RRI by enabling to scale the extent to which each context vector training step is performed as we have shown, such a scaling may be effectively driven by an information-theoretic measure representing the 'informational richness' of a set of context vectors. On the other hand, another (graph-centric) criterion may be used to effectively optimize the number of training cycles. Although, in contrast to RI or RRI, the proposed algorithm assumes the use of orthogonal (i.e., high-dimensional) index vectors, Random Projection may be (optionally) applied to context vectors in cases when dimensionality reduction is necessary, e.g., for scalability issues [7].

The experiments presented in this paper show that in the case of recommending items from the long tail, the accuracy of PRI is significantly higher than the accuracy of the most widely referenced collaborative filtering methods based on SVD. Moreover, our observations indicate that - as long as the density of a train set allows for reasonably accurate collaborative filtering - the more heavily-tailed a dataset is, the more visible the quality advantage of PRI (over the methods proposed by other authors) is.

Based on our novel method of modeling a collaborative data set as a bipartite graph, we are able to demonstrate that the main reason for the advantage of reflective data processing over dimensionality reduction is its ability to explore multi-hop item-item and user-user correspondences.

The results show a dependence between the optimal number of PRI reflections and structural properties of the dataset. The maximum length of the shortest path between users and items roughly indicates the ability of PRI to outperform SVD-based recommendation methods. It may be observed that the highest AUROC-measured classification probability closely corresponds to the case in which the reflective data processing is continued until all indirect connections (seen as the shortest paths) between all pairs of nodes (of the bipartite graph representing the train set) are 'discovered' and any over-training of context vectors is avoided.

Acknowledgments. This work is partly supported by the Polish Ministry of Science and Higher Education under grant N N516 196737, and by Poznan University of Technology under grant DS-MK 45-095/11.

References

1. Adomavicius, G., Tuzhilin, A.: Toward the Next Generation of Recommender Systems: A Survey of the State-of-the-Art and Possible Extensions. IEEE Transactions on Knowledge and Data Engineering 17(6), 734–749 (2005)
2. Anderson, C.: The Long Tail: Why the Future of Business is Selling Less of More. Hyperion (2006)
3. Basu, C., Hirsh, H., Cohen, W.: Recommendation as Classification: Using Social and Content-Based Information in Recommendation. In: Fifteenth National Conference on Artificial Intelligence, pp. 714–720 (1998)
4. Brynjolfsson, E., Hu, Y.J., Simester, D.: Goodbye Pareto Principle, Hello Long Tail: The Effect of Search Costs on the Concentration of Product Sales. Management Science (2011)
5. Ciesielczyk, M., Szwabe, A.: RI-based Dimensionality Reduction for Recommender Systems. In: 3rd International Conference on Machine Learning and Computing. IEEE Press, Singapore (2011)
6. Ciesielczyk, M., Szwabe, A., Prus-Zajaczkowski, B.: Interactive Collaborative Filtering with RI-based Approximation of SVD. In: 3rd International Conference on Computational Intelligence and Industrial Application. IEEE Press (2010)
7. Cohen, T., Schaneveldt, R., Widdows, D.: Reflective Random Indexing and Indirect Inference: A Scalable Method for Discovery of Implicit Connections. Journal of Biomedical Informatics 43(2), 240–256 (2010)
8. Cremonesi, P., Koren, Y., Turrin, R.: Performance of Recommender Algorithms on Top-n Recommendation Tasks. In: Fourth ACM Conference on Recommender Systems (RecSys 2010), New York, NY, USA, pp. 39–46 (2010)
9. Desrosiers, C., Karypis, G.: A Comprehensive Survey of Neighborhood-based Recommendation Methods. In: Ricci, F., Rokach, L., Shapira, B., Kantor, P.B. (eds.) Recommender Systems Handbook, 1st edn., pp. 107–144 (2011)
10. Herlocker, J.L., Konstan, J.A., Terveen, L.G., Riedl, J.T.: Evaluating Collaborative Filtering Recommender Systems. ACM Trans. Information Systems 22(1), 5–53 (2004)
11. Park, Y.-J., Tuzhilin, A.: The Long Tail of Recommender Systems and How to Leverage It. In: 2008 ACM Conference on Recommender Systems (RecSys 2008), pp. 11–18 (2008)
12. Ricci, F., Rokach, L., Shapira, B., Kantor, P.B. (eds.): Recommender Systems Handbook, 1st edn. (2011)
13. Sarwar, B.M., Karypis, G., Konstan, J.A., Riedl, J.: Application of Dimensionality Reduction in Recommender System - A Case Study. In: ACM WebKDD 2000 Web Mining for E-Commerce Workshop, Boston, MA, USA (2000)
14. van Rijsbergen, C.J.: The Geometry of Information Retrieval. Cambridge University Press, New York (2004)

Author Name Disambiguation for Ranking and Clustering PubMed Data Using NetClus

Arvin Varadharajalu[1], Wei Liu[1,*], and Wilson Wong[2]

[1] School of Computer Science and Software Engineering
The University of Western Australia, Australia
[2] School of Computer Science and Information Technology,
RMIT University, Australia
wei@csse.uwa.edu.au

Abstract. The ranking and clustering of publication databases are often used to discover useful information about research areas. NetClus is an iterative algorithm for clustering heterogenous star-schema information network that incorporates the ranking information of individual data types. The algorithm has been evaluated using the DBLP database. In this paper, we apply NetClus on PubMed, a free database of articles on life sciences and biomedical topics to discover key aspects of cancer research. The absence of unique identifiers for authors in PubMed introduces additional challenges. To address this, we introduce an improved author disambiguation technique using affiliation string normalisation based on vector space model together with co-author networks. Our technique for disambiguating authors, which offers a higher accuracy than existing techniques, significantly improves NetClus clustering results.

Keywords: Author Disambiguation, Clustering, NetClus, Heterogeneous Information Network.

1 Introduction

Governments, businesses, pharmaceutical companies and individual researchers frequently search publication databases to find the leading experts and their research groups [3]. PubMed is a free database accessing primarily the MEDLINE database of references and abstracts on life sciences and biomedical topics. Clustering query results of such databases will uncover invaluable information that would otherwise not available. For example, publications about ongoing research on cancer can be retrieved, and cluster analysis can be applied to answer questions such as new research findings, causes of cancer, common types of cancer and their treatment, as well as who the leading researchers and institutions are. However, PubMed's data collection is driven by crawling. In other words, authors and institutions are not assigned with unique identifiers as in the case of curated databases such as DBLP. Moreover, the increase of biomedical research citations in PubMed in the recent years makes the task of manually converting the literature into structured data extremely difficult [3].

* Corresponding author.

D. Wang and M. Reynolds (Eds.): AI 2011, LNAI 7106, pp. 152–161, 2011.

Publication data is best represented as a heterogeneous information network, which is a data model representing relations among multiple types of objects. RankClus [5] and its improvement NetClus [6] are among the more popular algorithms that integrate ranking and clustering for bi-typed and multi-typed heterogenous information networks, respectively. Both algorithms recognise the important fact that the ranking and the clustering mutually enhance each other. Ranking objects without clustering will lead to incomplete results, e.g., ranking artificial intelligence and database conferences together may not make much sense. Similarly, clustering a vast number of data objects in one huge cluster without differentiation is not informative either. However, to apply NetClus on PubMed, the ambiguity in author and institution names has to be resolved.

In this paper, we propose a technique to disambiguate authors by normalising affiliation string using a vector space model and then combining with co-author networks to further improve the system performance. The problem of affiliation string normalisation, especially on the PubMed dataset, has not been adequately addressed. Author disambiguation relying only on the affiliation strings suffers from the problems of variation in affiliation information, non-standard representation of affiliations, multiple affiliations for the same author and so on. We apply NetClus on PubMed publication records after disambiguation using our technique to extract useful information to understand the research trend, the leading journals, and organisations in the field of cancer research. To achieve this, we (1) developed a software module to collect PubMed records; (2) designed and developed the technique for author disambiguation; (3) reviewed existing clustering algorithms to identify the most suitable for this task; and (4) adapted the NetClus algorithm to work with PubMed data.

The paper is organised as follows. Section 2 examines the field of author disambiguation to identify the special focus of our disambiguation techniques. Section 3 outlines our disambiguation technique. In section 4, our algorithm is evaluated and clustering results are presented. A comparison of NetClus clustering results with and without author disambiguation is also presented. The paper concludes in Section 5 with an outlook to future work.

2 Related Work

2.1 The Challenges of Author Name Disambiguation on PubMed

The author name *"Wei Zheng"* is very common in PubMed articles. More than 1,700 publications were retrieved when *"Zheng W"* [Author] was searched on 10th of May, 2011. According to the PubMed search interface, all these articles were published by a single author but in actual fact, all these articles are published by authors with same name. This ambiguity problem leads to poor clustering results and incorrect co-author networks. Unlike DBLP, there is no unique ID that can identify an author or an institution. More sophiscated interfaces for PubMed such as GoPubMed[1] is able to offer multi-faceted search

[1] `http://www.gopubmed.org/`

experience such retrieving the affiliations and basic statistics about an individual author. These interfaces, however, are far from accurate and reliable. Just like PubMed, GoPubMed suffers the same problem of author name ambiguity, which messes up all the otherwise useful statistics.

Affiliation is probably the simplest starting point for author disambiguation. However, no standard format is enforced when affiliations are attached to an article in PubMed. Therefore the huge variation in affiliation strings makes it difficult to classify authors. *"Baylor College of Medicine"*, in the example below, is taking on a variety of names. `s1` has author name in it. `s2` and `s3` have email address but `s3` is missing the department. `s4` contains three different departments.

```
s1: Texas Children's Cancer Center, Michael E DeBakey Department of
    Surgery, Baylor College of Medicine, Houston, Texas 77030, USA.
s2: Texas Children's Cancer Center/Baylor College of Medicine,
    Houston, Texas 77030, USA. tmhorton@txccc.org
s3: Baylor College of Medicine, Houston, Texas 77030-2399, USA.
    dmetry@bcm.tmc.edu
s4: Texas Children's Cancer Center, Department of Pediatrics, Dan L.
    Duncan Cancer Center, Baylor College of Medicine, Houston, TX, USA.
```

To complicate matters, one author may have multiple affiliations at the same time, or at different times when they move between institutions, as illustrated by the author *"John M Maris"* below:

```
s1: Texas Tech University Health Sciences Center, Lubbock,
    Texas. min.kang@ttuhsc.edu.
s2: St. Jude Children's Research Hospital, Memphis, Tennessee.
```

2.2 Related Work on Disambiguation of PubMed Authors

Many techniques have been developed for disambiguating author names from PubMed articles. Torvik et al. [7], for instance, used a combination of features such as title words, journals, medical headings to calculate a probability to determine whether two articles have the same authors. This technique is based mainly on the assumption that authors publishing papers in similar research areas are more likely to be the same. However, the granularity and subjectivity of associating a publication with a research area may break this assumption.

Yu et al. [9] used the connections between organisations to articles and authors to disambiguate authors using affiliation strings. They disambiguate authors by extracting organisations and related entities from affiliation strings. There are, however, several problems with this technique. First, they assumed that an affiliation string follows this format: (address component, address component, country, email). This is never always true as shown in the example below:

```
s1: Integrated microRNA and mRNA expression profiling in a rat colon
    carcinogenesis model: Effect of a chemo-protective diet.[21406606]
    [Affiliation : Texas AM University]
s2: An algorithm to detect a center of pupil for extraction of point
    of gaze. [Dept. of Biomed. Eng., Inje Univ., Kimhae, South Korea.]
```

Second, dictionaries are required to differentiate geographical information from the names of organisations in the address component. Third, it requires that the organisation names contain only English words. The fact is, however, almost 10% of the affiliations contain words from the authors' native languages [3].

Jonnalagadda et al. [3] have taken these drawbacks into consideration and developed a technique called NEMO. They assumed that the information available in affiliation string is sufficient enough to disambiguate authors. In reality, many publications in PubMed use the affiliations of the first authors only. The affiliation strings therefore cannot always be used to disambiguate all authors.

3 A Multi-evidence Author Disambiguation System

There are two parts in our proposed technique as shown in Figure 1, namely, the extraction of different components from affiliation strings, and the 3-phase multi-evidence disambiguation process.

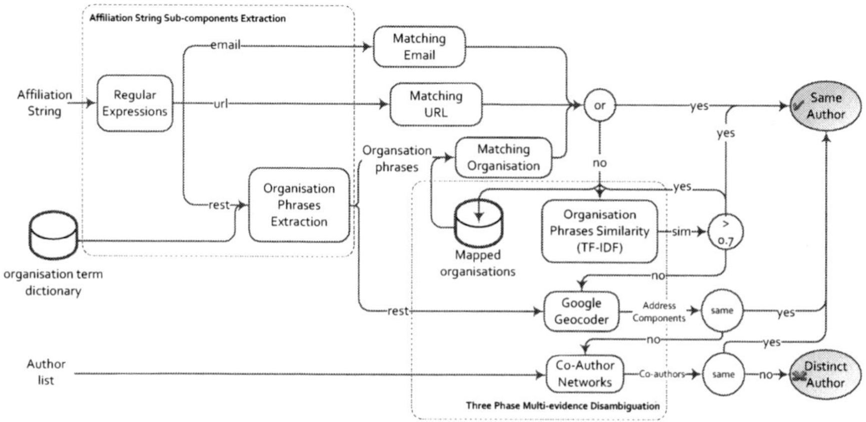

Fig. 1. Conceptual Overview of the Name Disambiguation System

In the first part, the affiliation strings of an ambiguous name (i.e. a name that can potentially refer to multiple individuals) are broken down into different components, namely, organisation names, organisation addresses, and email or homepage addresses, using a combination of regular expressions and other tools. The email address and URLs are first extracted using the following expressions:

```
/[\textbackslash.\_a-zA-Z0-9\-]+@[\textbackslash.\_a-z A-Z0-9-]+/i

((https?|ftp|gopher|telnet|file|notes|ms-help):((//)|(\textbackslash
\textbackslash\textbackslash\textbackslash))+[\textbackslash w
\textbackslash d:\#@\%/;\$()\~\_?\textbackslash+-=\textbackslash
\textbackslash\textbackslash.\&]*)
```

The remaining strings are then fed into a module which extracts the names of organisations using the following resources and tools:

1. Abbreviations and acronyms are disambiguated using the service provided by websites such as `Abbreviations.com`
2. Typographical mistakes are corrected using edit-distance [2] as in [8].
3. A table of organisation names as in [3].

The detected email addresses and URLs as well as organisations names are then removed from the affiliation strings. Finally, using Google's Geocoder[2], the address components, namely, countries, states, cities and street names, are extracted from the words that remain in the affiliation strings.

In the second part, the extracted components are used to determine if an author with multiple affiliation strings actually refer to one or multiple individuals. Initially, if the email addresses or URLs from two affiliation strings match, the two corresponding authors are considered to be the same. Otherwise, the technique proceeds further to the next two phases.

3.1 Disambiguation Using Organisation Names and Addresses

In this second phase, the organisation name components of the affiliation strings are compared using a vector space model [1]. Every term in the organisation name is represented as a dimension, and TF-IDF [4] is used to compute the weights of the individual terms. In our technique, the threshold value is set to 0.7. In other words, to be considered as identical organisations, the cosine similarity needs to be greater than 0.7. In the following example, the one author name has four different affiliation strings:

```
s1: Department of Molecular and Cellular Biology and Dan L. Duncan
    Cancer Center, Baylor College of Medicine
s2: Lester and Sue Smith Breast Center, Baylor College of Medicine
s3: Paris Breast Center, L'Institut Du Sein
s4: Clinical Research Division, Fred Hutchinson Cancer Research Center
```

All stop words are removed to improve the results. The term frequency, document frequency, inverse document frequency and term weights are then calculated. The word *"baylor"*, for instance, appears only once in the first organisation name. Therefore the term weight is assigned only to the first string. The word *"center"* appears in all the strings therefore its weight is reduced to zero using IDF. Similarly, the term weight for each and every term is calculated by multiplying the number of strings containing the term with the inverse document frequency. Next the dot products of all possible pairings of strings are calculated. If the cosine value is 1, then the strings within a pair are identical. In this example, the cosine values between $s1$, and $s2$, $s3$ and $s4$ are 0.6345, 0 and 0.1502, respectively. In other words, the second organisation name $s2$ is the most similar to $s1$. However, in this example, the cosine value is still less than 0.7. The chances

[2] `http://code.google.com/apis/maps/documentation/geocoding/`

of the authors of both strings being the same are very high. This problem is resolved by considering the address components in the affiliation strings. If three out of the four components (i.e. country, state, city, street name) match, the two organisation names and hence, the authors are considered to be the same. If this step fails, we proceed to the next step involving co-author networks.

3.2 Disambiguation Using Co-author Network

In this phase, the technique deals with a single author that has completely different email addresses, URLs, organisation names and organisation addresses. The author *"John M Maris"* mentioned at the end of Section 2.1 is an example. In cases like this, the technique will proceed to examining the co-author network to identify whether the multiple *"John M Maris"* are actually referring to the same or different individuals given these affiliation strings.

```
s1: Min H Kang, C Patrick Reynolds, Peter J Houghton, Denise Alexander,
    Christopher L Morton, EAnders Kolb, Richard Gorlick, Stephen T Keir,
    Hernan Carol, Richard Lock, John M Maris, Amy Wozniak, Malcolm A Smith
s2: Christopher L Morton, John M Maris, Stephen T Keir, Richard Gorlick,
    E Anders Kolb, Catherine A Billups, Jianrong Wu, Malcolm A Smith,
    Peter J Houghton
```

Using the co-author path from Microsoft Academic Search, the technique is able to determine that both *"John M Maris"* refer to the same author in the two publications above as *"John M Maris"*, *"Christopher L Morton"*, *"Malcom A Smith"* and *"Peter J Houghton"* have co-authored paper previously.

4 Evaluation of the Disambiguation Technique

For this evaluation, we queried PubMed using the word *"cancer"*. 12,707 results were retrieved between the year 2011 and 2010. From a total of 12,700 articles, our system has identified 80,042 paper-author combinations, and 71,810 unique authors. Table 1(a) lists the top 12 ambiguous names and the actual number of authors they represent. Out of the unique authors, 6,152 authors had more than two publications. Therefore, 65,658 authors have published only one paper in this dataset. From the 6,152 authors with multiple publications, 5,203 were identified using co-author network. The chance of these authors being different is therefore highly unlikely. Therefore, we only need to verify the correctness of the remaining 949 authors. These 949 authors' article titles and abstract were extracted. The entire code of this work can be downloaded from this link http://thesis.modusoperandi.com.au/dana.zip.

4.1 Accuracy of the Proposed Disambiguation Technique

It is labour-intensive and costly to obtain the ground truth of a dataset of such magnitude. For this experiment, we adopt the approach of finding similarities between the publications through the available abstracts. Along the line of Boyack et al. [1], we employ a combination of TF-IDF and PMRA (PubMed Related

Articles) to compute the similarity between PubMed abstracts of authors with the same names. All the stop words were firstly removed. If the similarity approaches 1, then these articles were considered to be from the same author.

The top author *"Kim Overvad"*'s publications can be taken as an example to illustrate how we evaluate the accuracy of the disambiguation algorithm. He has published 12 papers in the dataset we used. All the articles had a different affiliation but according to our algorithm the author is the same. To verify its correctness, the abstracts of all 12 articles demonstrating high level of similarity using TF-IDF and PMRA, confirming that the articles are from the same author.

Table 1(b) shows that the author disambiguation technique proposed in this paper demonstrates higher accuracy rate across the two similarity measures as ground truth, with the highest being 97.89% (929 out of 949)and the lowest being 94.9% (909 out of 949). Since PMRA has a higher coherence value [1], the string similarity measure of PMRA is considered as the accuracy of the system.

Table 1. Results and performance of our disambiguation system

Name	No.	Name	No.	Name	No.	Name	No.
Wei Wang	19	Ying Zhang	14	Li Wang	14	Wei Zhang	14
Ying Wang	13	Sang Lee	13	Ying Liu	13	Wei Li	13
Yan Li	13	Yan Wang	13	Wei Chen	12	Xiao Li	12

(a) Top 12 Ambiguious Names

Method	No.	Accuracy
PMRA	929	97.89%
TF-IDF	909	94.94%

(b) Accuracy

4.2 Evaluation of NetClus Results

In this evaluation, we look at the performance of NetClus on the dataset disambiguated using our technique. In Figure 2, the diamond shaped data points represent the objects in the *"Gastric Cancer"* cluster and the square shaped data points represent the objects in the *"Lung Cancer"* cluster. If the clusters are well separated, then the clustering results are considered as accurate. Partial dataset has been represented here in the scatter plots for improved readability.

In Figure 2(a), sixteen publications were represented in this scatter plot. The PMIDs related to lung cancer are placed on the extreme left of the graph and the PMIDs related to the gastric cancer are placed on the extreme right of the graph. Both the clusters are well separated indicating the clear demarcation of data objects based on the type of cancer. In Figure 2(b), the terms are well separated based on the type of cancer. All these terms are symptoms of cancer. Symptoms of lung cancer are placed on the top left part of the chart and the symptoms of gastric cancer are placed on the bottom right of the chart. Based on the component coefficients, the objects have been placed on the chart. For example, the term *"wheezing"* is a common lung cancer symptom, therefore it is placed in the lung cancer cluster while *"constipation"* is placed in the gastric cancer cluster. In Figure 2(c), treatment and sub-types of cancer are clustered. Lobectomy refers to the surgical excision of a lobe. This may refer to a lobe of the lung. Therefore, its position in the lung cancer cluster is correct.

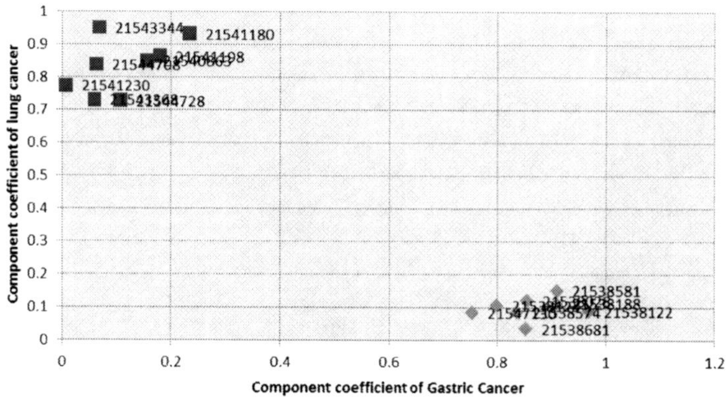

(a) PMID's scatter plot based on two component coefficients

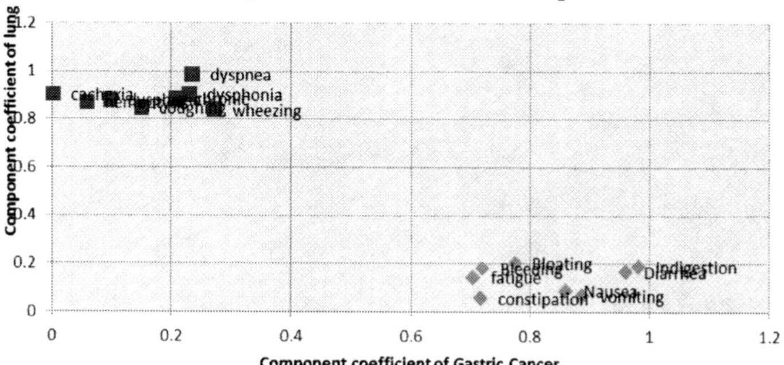

(b) Symptoms' scatter plot based on two component coefficients

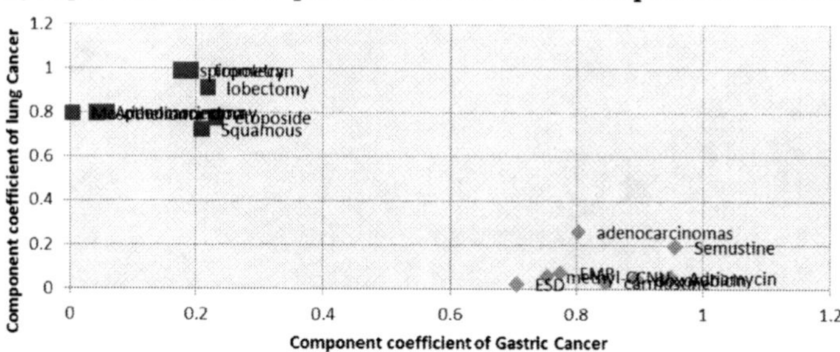

(c) Treatments' scatter plot based on two component coefficients

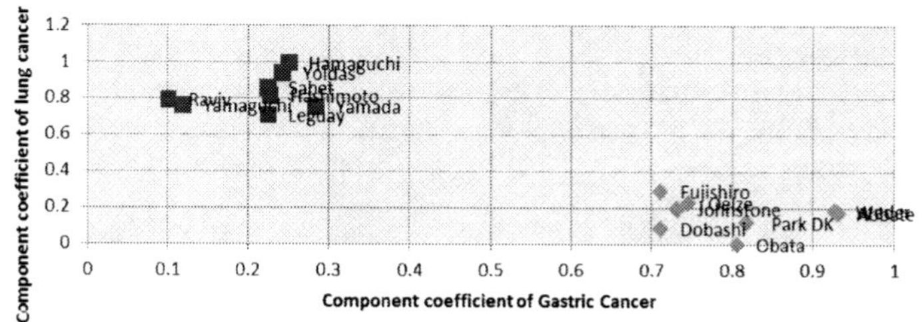

(d) authors' scatter plot based on two component coefficients

Fig. 2. Scatter Plot for articles, symptoms, treatments and authors

Similarly, since endoscopic submucosal dissection (ESD) is a mucosa resection, it is placed in gastric cancer cluster. Similarly, after author disambiguation, the authors were clustered based on the type of cancer they are researching on. The red square shaped objects represent the authors who have published more papers in lung cancer area, while the green diamond shaped objects for authors who have published more papers in gastric cancer area.

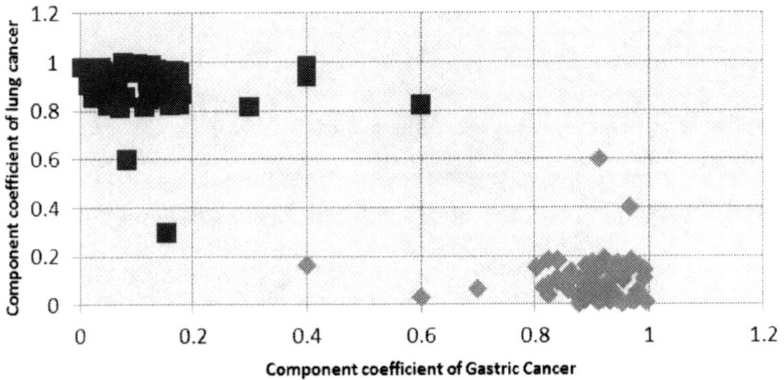

(a) Authors' scatter plot without disambiguation

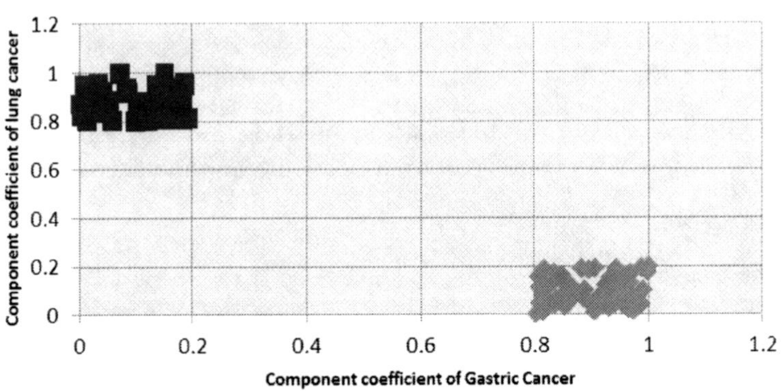

(b) Authors' scatter plot with disambiguation

Fig. 3. A comparison of clustering results with/without disambiguation

Figure 3(a) and (b) compares the clustering results of authors with and without disambiguation. In Figure 3(a), few authors are scattered throughout the chart. Because the author names were ambiguous, they are not properly placed in a cluster. In Figure 3(b), authors were disambiguated using our system. Since the ambiguity was resolved, the clusters are well separated.

5 Conclusion and Future Work

To cluster datasets represented as heterogeneous information networks such as PubMed, the ambiguity amongst the objects must first be resolved. The disambiguation technique proposed in this paper was used to mitigate the ambiguity of authors in the dataset. We have demonstrated that algorithms based on named entity recognition of affiliation string are not sufficient enough to disambiguate authors. In the proposed technique, a combination of vector space model based similarity measure, geographical information and co-author network was used to identify whether the authors with the same name are different or not.

The results of disambiguation were evaluated by comparing the PubMed abstracts of articles of disambiguated authors using text similarity algorithms such as PMRA and TF-IDF. The proposed technique showed a higher accuracy rate in both the methods. We also evaluated the NetClus algorithm with and without the disambiguation of authors. When the authors were not disambiguated, there were outliers and noise in the results. After disambiguation of authors, the clusters were clear and distinct. This disambiguation system and NetClus algorithm to identify interesting patterns that can be useful in bio-medical research. In future, we are planning to build a interface where users can use this multi-level normalisation system for author disambiguation of PubMed articles.

References

1. Boyack, K.W., Newman, D., Duhon, R.J., Klavans, R., Patek, M., Biberstine, J.R., Schijvenaars, B., Skupin, A., Ma, N., Brner, K.: Clustering more than two million biomedical publications: Comparing the accuracies of nine text-based similarity approaches. PLoS ONE 6(3), e18029 (2011)
2. Golic, J.D., Mihaljevic, M.J.: A generalized correlation attack on a class of stream ciphers based on the levenshtein distance. Journal of Cryptology 3, 201–212 (1991)
3. Jonnalagadda, S., Topham, P.: Nemo: Extraction and normalization of organization names from pubmed affiliation strings. J. Biomed. Discov. Collab. 5, 50–57 (2010)
4. Salton, G., Wong, A., Yang, C.S.: A vector space model for automatic indexing. Communications of ACM 18, 613–620 (1975)
5. Sun, Y., Han, J., Zhao, P., Yin, Z., Cheng, H., Wu, T.: RankClus: integrating clustering with ranking for heterogeneous information network analysis. In: EDBT 2009: Proceedings of the 12th International Conference on Extending Database Technology, pp. 565–576. ACM, New York (2009)
6. Sun, Y., Yu, Y., Han, J.: Ranking-based clustering of heterogeneous information networks with star network schema. In: Proceedings of the 15th ACM SIGKDD International Conference on Knowledge Discovery and Data Mining, KDD 2009, pp. 797–806. ACM, New York (2009)
7. Torvik, V.I., Weeber, M., Swanson, D.R., Smalheiser, N.R.: A probabilistic similarity metric for medline records: A model for author name disambiguation. Journal of the American Society for Information Science and Technology 56(2), 140–158 (2005)
8. Wong, W., Liu, W., Bennamoun, M.: Integrated scoring for spelling error correction, abbreviation expansion and case restoration in dirty text. In: Proceedings of the fifth Australasian Conference on Data Mining and Analystics - AusDM 2006, vol. 61, pp. 83–89. Australian Computer Society, Inc., Darlinghurst (2006)
9. Yu, W., Yesupriya, A., Wulf, A., Qu, J., Gwinn, M., Khoury, M.: An automatic method to generate domain-specific investigator networks using pubmed abstracts. BMC Medical Informatics and Decision Making 7(1), 17–26 (2007)

Self-Organizing Maps for Translating Health Care Knowledge: A Case Study in Diabetes Management

Kumari Wickramasinghe[1], Damminda Alahakoon[2], Peter Schattner[1], and Michael Georgeff[1]

[1] Department of General Practice, Faculty of Medicine, Nursing and Health Sciences
[2] Faculty of Information Technology
Monash University, Australia

Abstract. Chronic Disease Management (CDM) is an important area of health care where Health Knowledge Management can provide substantial benefits. A web-based chronic disease management service, called cdmNet, is accumulating detailed data on CDM as it is being rolled out across Australia. This paper presents the application of unsupervised neural networks to cdmNet data to: (1) identify interesting patterns in diabetes data; and (2) assist diabetes related policy-making at different levels. The work is distinct from existing research in: (1) the data; (2) the objectives; and (3) the techniques used. The data represents the diabetes population across the entire primary care sector. The objectives include diabetes related decision and policy making at different levels. The pattern recognition techniques combine a traditional approach to data mining,involving the Self-Organizing Map (SOM), with an extension to include the Growing Self-Organizing Map (GSOM).

1 Introduction

Health care Knowledge Management (HKM) [17] uses data analysis and mining techniques to provide high quality, well-informed and cost-effective patient-care decisions to health care stakeholders. Because of the increasing number of patients with chronic disease [8], such as diabetes, and the associated medical care costs [2], Chronic Disease Management (CDM) is emerging as one of the greatest challenges to health care systems worldwide. Applied to CDM, HKM has the potential to provide substantial benefits to stakeholders and patients.

The main components of CDM are described in the Chronic Care Model (CCM) developed by Wagner's group [19]. An Australian initiative, the chronic disease management network (cdmNet) [10] focuses on implementing the key components of the CCM. Since 2008, cdmNet has been gradually adopted by General Practitioners (GPs) to create and maintain care plans for patients with chronic disease. As cdmNet is beginning to be rolled out nationally, it is accumulating detailed data about CDM processes across the full breadth of primary care.

D. Wang and M. Reynolds (Eds.): AI 2011, LNAI 7106, pp. 162–171, 2011.

The tacit knowledge concealed within the cdmNet data has the potential to improve: (1) the understanding of chronic conditions; (2) development of treatments and guidelines for CDM; and (3) development of policies for managing and preventing chronic disease. A business intelligence module, called cdmNet-Business Intelligence (cdmNet-BI), is proposed to fulfil these objectives. cdmNet-BI uses business intelligence functionalities such as reporting, online analytical processing, analytics and data mining to systematically model, share and translate cdmNet data.

This paper describes the application of unsupervised neural networks, clustering in particular, within the cdmNet-BI module. Diabetes is chosen for the analysis as it: (1) is one of the most common chronic conditions; (2) is associated with a large number of complications (such as blindness, kidney failure and premature cardiovascular death); and (3) early detection and the proper management can make a significant difference to patient outcomes. Even though there is existing research on mining chronic disease related data [11,12,16], the uniqueness of our work is threefold: (1) the data; (2) the objectives; and (3) the combination of techniques used. Existing research is based on data collected from a specific health care organisation for a fixed time period. cdmNet data: (1) corresponds to national wide diabetes population; (2) covers the full breadth of the primary care environment; and (3) includes a comprehensive history over all time of key patient measurements such as body weight, blood pressure, and HbA1c (a measure used to diagnose diabetes). The objectives include: (1) identification of interesting patterns related to diabetes patients; and (2) assisting policy making and preventive care approaches to the management of diabetes at different levels. Two types of self-organizing neural networks: (1) Self-Organizing Map (SOM) [14]; and (2) Growing Self-Organizing Map (GSOM) [7] are used as the pattern recognition techniques. SOM is well-established and there are comprehensive tools available for the visualisation of the generated clusters. GSOM extends the capabilities of SOM through: (1) a flexible structure that self organises to accurately represent the structure of data; and (2) an inbuilt hierarchical clustering capability to start with a broader (generalised) view of data and drill down into specific details. Hierarchical clustering facilitates decision and policy making at different levels depending on the requirements. To our knowledge, hierarchical approaches have not been used to investigate and enhance decision-making in chronic disease.

The paper is organised as follows: Section 2 provides background on use of data mining techniques to diabetes data. Introductions to CDM, and the cdmNet system are carried out in Section 3. Section 4 presents the cdmNet-BI module. Data mining results are discussed in Section 5. Concluding remarks and future work are discussed in Section 6.

2 Background: Mining Diabetic Patient Data

A number of studies have used data mining techniques, particularly classification and clustering, to analyse diabetes data. A data mining tool [11] has been developed in Singapore for medical doctors, integrating classification with association rule mining (CBA) based on eight years of data. Huang and others [12] have applied feature selection via supervised model construction (FSSMC) [13] to determine individuals in the population with poor diabetes control. Patients' age, diagnosis duration, insulin treatment requirements, random blood glucose measurement and diet treatment were identified as the most influencing factors.

Use of Classification and Regression Trees on 10 predictors (age, sex, emergency department visits, office visits, comorbidity index, dyslipidemia, hypertension, cardiovascular disease, retinopathy, end-stage renal disease) has identified younger age as the determinant factor for bad glucose control [9]. The prediction model for diabetes constructed by Su and others [18] highlighted that the volume of trunk, left thigh circumference, right thigh circumference, waist circumference, volume of right leg, and age as being associated with the condition of diabetes. Porter and others [16] have applied clustering techniques to identify different diabetes sub groups. Their research findings are: (1) the best results can be obtained by clustering methods that use Euclidean distance as the cluster separation criteria; and (2) family history, body weight, and age are good indicators for identifying potential diabetic patients.

While the aforementioned existing analysis were based on patient data collected for a fixed time period, the data used in this paper is based on patients with diabetes across the whole primary care sector. Compared with Porters' finding above [16], both techniques are based on Euclidean distance. In addition, GSOM provides a hierarchical view of data, facilitating decision and policy making at different levels.

3 Application

3.1 Chronic Disease Management (CDM)

Chronic disease is the cause for 60% of all deaths worldwide and, by the year 2020, chronic disease will account for almost 75% all deaths [1]. Seven million Australians and 133 million Americans [8] have a chronic medical condition. In Australia, chronic disease accounts over 60% of health care costs ($60 billion per annum), and significantly impacts workforce productivity ($8 billion per annum) [2].

Diabetes is one of the most common chronic conditions worldwide. In Australia 700,000 people were diagnosed with diabetes in 2004-2005 [4] and the estimate for 2015 is 4.6 million, including both diagnosed and non-diagnosed patients [5]. In the united states, 7.8% of the population is diagnosed in 2007 [3]. Diabetes is associated with a large number of complications, such as blindness, kidney failure, leg amputations, and premature cardiovascular death [9].

Early detection and the proper management of diabetes can make a significant difference, for example, preventing up to 90% of blindness, and 50% of dialysis and amputations [9]. Therefore, much emphasis is on best practices, optimization of care and other management methods to improve outcomes. For example, in Australia, the government has introduced incentives to general practitioners (GPs) to undertake a more structured, planned and systematic approach to chronic disease management.

3.2 Chronic Disease Management Network (cdmNet)

The Chronic Care Model (CCM) developed by Wagner's group [19] emphasises collaboration among care providers and the patient in creating and maintaining a care plan for patients with chronic disease. cdmNet is focused on implementing the key processes of the CCM by creating best practice, personalised care plans, distributing care plans to the patients' care team and to the patients, continuously monitoring the care plan, facilitating collaborations, and supporting patient self-management.

cdmNet has been trialled in a number of regions in Victoria and Western Australia. Data corresponding to pre and post adoption of cdmNet indicate from 88% to 205% and from 80% to 201% increase in care plan generation and sharing respectively [10]. cdmNet is currently being implemented Australia wide.

4 Chronic Disease Management Network - Business Intelligence (cdmNet-BI) Module

The cdmNet-BI module aims to convert cdmNet transactional data to knowledge. It consists of three sub-modules: (1) pre-processing, (2) dashboard, and (3) data mining. The pre-processing sub-module converts transactional data to a format that can be used in the dashboard and the data mining sub-modules. The dashboard sub-module provides portals for the GPs presenting their use of cdmNet for care planning.

The data mining sub-module is the focus of this paper. It aims to identify interesting patterns in data and assist chronic care related policy making and preventive care approaches at different levels. Currently, the sub-module uses undirected knowledge discovery techniques, specifically clustering, as the data mining technique and applies them on diabetes data. The clustering techniques used, SOM and GSOM are briefly described in Sections 4.1 and 4.2.

4.1 The Self-Organizing Map (SOM)

The Self-Organizing Map (SOM) consists of a layer of input vectors and one dimensional (a row) or two dimensional (a lattice) output nodes [14]. All input and output nodes are connected through weight vectors. For a given input, certain regions of the array will fire strengthening the corresponding weights. If two input data items cause the same output neuron to fire, the similarity measure indicates that these two input data items belong to the same cluster. The concept of a neighbourhood is used in updating weights. That is, when updating the weights, the weights of the neurons in the neighbourhood of the winning neuron are also updated (strengthened). Once the clusters are formed, SOM provides a two-dimensional visualisation of the resulting clusters. Viscovery SOMine [6], which is a widely used commercial tool for the SOM algorithm, is used as one of the cluster visualisation tools within the data mining sub-module.

4.2 The Growing Self-Organizing Map (GSOM)

The Growing Self-Organizing Map (GSOM) [7] is a self-generating unsupervised neural network algorithm, which is an extended version of the SOM. It has a minimum number of starting nodes (usually four) and generates new nodes only when it is required using a heuristic to identify such a need, that is, GSOM generates new nodes as the data is input to the network. New nodes are created if and only if the nodes already present in the network are insufficient to represent the input data set.

The control of the spread of the map is achieved using a concept called Spread Factor (SF). A lower SF results in a smaller size map with the most significant clusters. A larger map with a higher number of nodes can be obtained with a higher SF. The larger map contains finer clusters or sub clusters. That is, SF provides a hierarchical clustering capabilities to the GSOM by linking GSOMs generated using high to low SFs. The data mining sub-module utilises different values of SF to obtain cluster hierarchies to visualise data at different levels.

5 Patterns in Diabetes Management

This section describes the use of clustering techniques to: (1) identify diabetes related interesting patterns; and (2) assist diabetes related policy making and preventive care approaches at different levels. The analysis is divided into two parts as:

(1) analysis of features common to any individual; and
(2) analysis of common features with diabetes specific medical features.

As diabetes is a condition that exists within the general population in a country, the former analysis helps to understand any common features that may contribute towards diabetes. The latter analysis aims at identifying any correlations among common features and diabetes specific features. We considered the following common and medical features:

1. common features: (a) demographics (gender, age, marital status); and (b) lifestyle (drinking status, smoking status).
2. medical features: (a) metabolic measure (HbA1c); and (b) clinical measure (BMI).

HbA1c or haemoglobin A1C is a metabolic measure used to diagnose diabetes. A normal non-diabetic has a HbA1c of 3.5-5.5%. Achieving HbA1c $< 7\%$ is considered as a diabetes management goal. Body Mass Index (BMI) indicates whether patients have a healthy weight to their height and is observed regularly for diabetes patients. Achieving BMI < 25 is considered as a diabetes management goal.

5.1 Analysis of Features Common to Any Individual

The GSOM tool with varying spread factor, SF, was used to obtain the hierarchies in data. The GSOM clusters corresponding to $SF = 0.1$ and $SF = 0.4$ are shown in Figures 1 and 2. Clusters in Figure 1 provide a generalised view of data. Figure 2 represents specific clusters by hierarchically expanding the clusters formed in Figure 1.

To illustrate the usefulness of hierarchical clustering in decision making, we use two clusters from Figure 1 (denoted by A1 and A2) and their corresponding clusters from Figure 2 (denoted by X1, X12, X2, and X14). The prominent features and the percentage of patients in each cluster are as follows. Cluster A1 consists of ex-smoking males indicating that within the diabetes population there is a specific group of patients who are males and ex-smokers. Cluster X1 represents divorced men, age between 60 and 70 whose drinking status is not known. Cluster X12 represents married men age between 70 and 80 who are drinkers. Clusters X1 and X12 indicate within the ex-smoking male group (Cluster A1), there are two sub groups which have different marital status, drinking status and they belong to different age groups. In terms of policy making, stakeholders can determine either general policies based on the most prominent feature values, ex-smoking and males; or specific policies based on marital status, drinking statuses and age. X12 consists of 85% of the patients from A1 leaving only a 15% to X1. Therefore, a general policy or campaign focused on ex-smoking males may be more cost effective than developing two specific policies, one for ex-smoking married males within the age of 70 and 80 who consume alcohol, and another policy for ex-smoking divorced males within the age of 60 and 70. On the other hand, X14 consists of 65% of the patients from A2 leaving 35% to X2. For this, a specific policy based on gender, marital status, drinking status and age may be more cost effective compared to a generalised solution.

5.2 Analysis of Common Features with Diabetes Specific Medical Features

Common features and diabetes specific medical features were clustered individually (Method 1) and as a combination (Method 2) as described below.
Method 1: A three step analysis:

Step 1: Cluster common features;
Step 2: Cluster medical features; and
Step 3: Identify correlations among Steps 1 and 2 clusters.

When $SF = 0.1$, Step 1 resulted in 11 clusters (say B0-B10) and Step 2 resulted in 4 clusters (say L0-L3). L0-L3 indicate different HbA1c and BMI combinations as:

Cluster L0: HbA1c = target, BMI = obese;
Cluster L1: HbA1c = very high, BMI = overweight;
Cluster L2: HbA1c = very high, BMI = obese; and
Cluster L3: HbA1c = target, BMI = overweight.

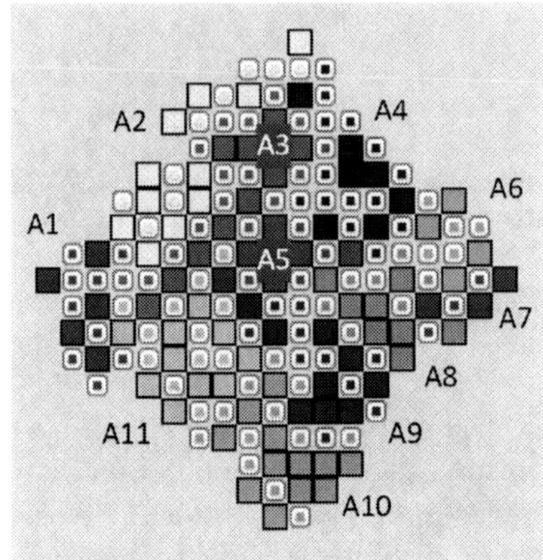

Fig. 1. The GSOM map corresponding to $SF = 0.1$

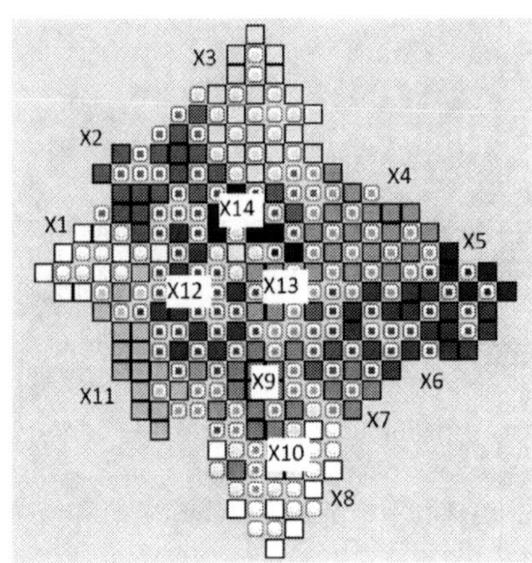

Fig. 2. The GSOM map corresponding to $SF = 0.4$

The patient identifier based cross analysis is carried out in Step 3 to determine the cluster a given patient is belonging to in Steps 1 and 2. Following correlations among features were identified:

(a) smoking and HbA1c; and
(b) alcohol consumption and BMI.

Smoking and HbA1c: To highlight this correlation, Step 3 output corresponding to smokers and non-smokers is analysed in Table 1. A row in Table 1 represents the percentage of patients, from a common feature cluster, that map to each medical feature cluster. As highlighted it indicates:

(a) Smokers (C1 and C2) have very high HbA1c (L1 and L2); and
(b) Non-smokers (C3 to C8) have target HbA1c (L0 and L3).

Alcohol consumption and BMI: To highlight this correlation, Step 3 output that corresponds to drinkers and non-drinkers is analysed in Table 2. A row in Table 2 represents the percentage of patients, from a common feature cluster, that map to each medical feature cluster. The bold faced text in the table indicate drinkers (C1, C4, C8 and C9) map to obese (L0).

Method 2: Common and medical features were clustered together. The resulting clusters indicate:

(1) Non-smoking and non-drinking lead to better controlled diabetes as a distinct cluster is formed with "non-smoking" , "non-drinking" patients whose HbA1c is maintained within the target;

(2) An unfavourable marital status can lead to poor controlled diabetes as two distinct clusters are formed, one with "divorced" patients with "too high" HbA1c and the other with "separated" patients with "too high" HbA1c;

(3) Alcohol consumption and "divorced" marital status can be the determinants of obesity: Clusters with patients having either "separated" or "unknown" marital status whose drinking status is unknown had BMI "obese" category; and

Table 1. Effect of smoking on HbA1c

Common feature clusters	Demographics and lifestyle values	L0(%)	L1(%)	L2(%)	L3(%)
C1	female, divorced, drinker, **smoker**, age between 50 and 60	47.6	0	**47.6**	4.8
C2	male and female, marital status known and defacto, drinking status unknown, **smoker**, age between 50 and 60	24.1	**41.4**	3.4	31
C3	male and female, widowed, non-drinker, **no-smoker**, age between 70 and 80	22.7	2.7	1.3	**73.3**
C4	female, widowed, drinker, **non-smoker**, age between 70 and 80	**48.4**	3.2	3.2	45.2
C5	female, marital status known and married, drinking status unknown, **non-smoker**, age between 60 and 70	**35.5**	23.7	21	19.7
C6	female, widowed, drinking status unknown, **non-smoker**, age between 70 and 80	**57.1**	28.6	0	14.3
C7	male, marital status unknown and married, drinking status unknown, **non-smoker**, age between 60 and 70	**38.5**	30.8	7.7	23.1
C8	male, divorced, drinker, **non-smoker**, age between 60 and 70	**55.6**	24.1	1.9	18.5

(4) The majority of diabetes patients (70.1%) belong to HbA1c "high" and BMI "obese" category.

5.3 Patterns Recognised from Diabetes Data: Outcomes

In relation to diabetes care policy development, the hierarchical clusters provided a basis to determine: (1) the approach (that is, general or specific policies); and (2) the features (parameters) of a policy. In addition following correlations were identified:

(1) Smoking increases HbA1c;
(2) Alcohol consumption increases BMI;
(3) Unfavourable marital statuses can lead to poor controlled diabetes;
(4) Alcohol consumption and "divorced" marital status can be the determinants of obesity; and
(5) Diabetes management is poor among married males, around 60 years of age who are heavy drinkers and ex-smokers.

There are existing studies that were carried out to determine the effect of smoking on HbA1c [15]. Item 1 above reinforces the high correlation between smoking and poor control in diabetes. One of the expected outcomes of unsupervised learning approach is to identify hypothesis for further testing. Items 2-5 can be considered as hypothesis to be tested.

Table 2. Effect of alcohol consumption on BMI

Common feature clusters	Demographics and lifestyle values	L0(%)	L1(%)	L2(%)	L3(%)
C1	female, divorced, **drinker**, smoker, age between 50 and 60	**47.6**	0	47.6	4.8
C4	female, widowed, **drinker**, non-smoker, age between 70 and 80	**48.4**	3.2	3.2	45.2
C8	male, married and marital status unknown, **drinker**, ex-smoker, age between 70 and 80	**60**	7.7	3.1	29.2
C9	male, divorced, **drinker**, non-smoker, age between 60 and 70	**55.6**	24.1	1.9	18.5
C3	male and female, widowed, **non-drinker**, no-smoker, age between 70 and 80	22.7	2.7	1.3	**73.3**

6 Conclusions and Future Work

cdmNet is developed based on a well-established organisational framework for CDM. Therefore, cdmNet data provides an evidence base for informed health policy for patients with chronic disease. This paper investigated the application of self-organizing maps for translating cdmNet diabetes patients' data into knowledge. The paper is distinct from existing research in: (1) data; (2) objective; and (3) approach. Data correspond to the current diabetes population, up to date and comprehensive. Objectives include identification of interesting patterns and decision and policy making at different levels. The clustering techniques, SOM and GSOM provide techniques to achieve the objectives by providing the advantage of a matured technique and representing data at different abstract levels.

The findings described in the paper: (1) provided insights into the identification of general and specific features in developing diabetes patient care policies; (2) established hypothesis for further testing; and (3) reinforce existing research findings. Our future work aims to enhance the data mining sub-module to utilise all data collected from all the functionalities available in cdmNet. As the effectiveness of existing treatments and guidelines for CDM is concealed within this data, the findings will indicate the effective treatments and guidelines and provide a basis for the development of CDM guidelines.

Acknowledgements. This work is supported by funding from the Australian Government under the Digital Regions Initiative and by the Victorian Government under the Victorian Science Agenda program. We are also grateful to Mrs Upuli Gunasinghe of CCSL, Monash University for the GSOM cluster visualisation tool and Dr Kay Jones and Dr Akuh Adaji of the Department of General Practice, Monash University for their advice.

References

1. WHO Report on Chronic Disease (2005), http://www.slideshare.net/Hfoodsupplements/who-report-on-chronic-disease (accessed: May 12, 2007)
2. National chronic disease strategy (2006), http://www.health.gov.au/internet/main/publishing.nsf/content/7E7E9140A3D3A3BCCA257140007AB32B/File/strata13.pdf (accessed: March 04, 2010)
3. Diabetes statistics (2007), http://www.diabetes.org/diabetes-basics/diabetes-statistics/ (accessed: November 12, 2010)
4. Diabetes australian facts (2008), http://www.aihw.gov.au/publication-detail/?id=6442468075 (accessed: January 10, 2010)
5. Quick facts (2010), http://www.australiandiabetescouncil.com/About-Diabetes/Quick-facts.aspx (accessed: October 12, 2010)
6. Viscovery (2010), http://www.viscovery.net/publications/books
7. Alahakoon, D., Halgamuge, S., Srinivasan, B.: Dynamic self-organizing maps with controlled growth for knowledge discovery. IEEE Transactions on Neural Networks 11(3), 601–614 (2002)
8. Anderson, G., Wilson, K.: Chronic disease in california: Facts and figures (2006), http://www.chcf.org/~/media/MEDIA%20LIBRARY%20Files/PDF/C/PDF%20ChronicDiseaseFactsFigures06.pdf
9. Breault, J., Goodall, C., Fos, P.: Data mining a diabetic data warehouse. Artificial Intelligence in Medicine 26(1-2), 37–54 (2002)
10. Georgeff, M.: Cdm-net:a broadband health network for transforming chronic disease management. Final report, Precedence Health Care (2010)
11. Hsu, W., Lee, M., Liu, B., Ling, T.: Exploration mining in diabetic patients databases: findings and conclusions. In: Proceedings of the Sixth ACM SIGKDD Conf. on KD and DM, p. 436. ACM (2000)
12. Huang, Y., McCullagh, P., Black, N., Harper, R.: Feature selection and classification model construction on type 2 diabetic patients' data. Artificial Intelligence in Medicine 41(3), 251–262 (2007)
13. Huang, Y., McCullagh, P., Black, N.: Feature selection via supervised model construction. In: Fourth IEEE International Conference on Data Mining, ICDM 2004, pp. 411–414. IEEE (2005)
14. Kohonen, T.: Self-organized formation of topologically correct feature maps. Biological Cybernetics 43(1), 59–69 (1982)
15. Nilsson, P., Gudbjornsdottir, S., Eliasson, B., Cederholm, J., et al.: Smoking is associated with increased HbA1c values and microalbuminuria in patients with diabetes–data from the National Diabetes Register in Sweden. Diabetes & Metabolism 30(3), 261–268 (2004)
16. Porter, T., Green, B.: Identifying Diabetic Patients: A Data Mining Approach. In: Proceedings of the 15th Americas Conference on Information Systems (AMCIS), San Francisco, California (2009)
17. Sibte, S., Abidi, R.: Healthcare Knowledge Management: The Art of the Possible. In: Riaño, D. (ed.) K4CARE 2007. LNCS (LNAI), vol. 4924, pp. 1–20. Springer, Heidelberg (2008)
18. Su, C., Yang, C., Hsu, K., Chiu, W.: Data mining for the diagnosis of type II diabetes from three-dimensional body surface anthropometrical scanning data*. Computers & Mathematics with Applications 51(6-7), 1075–1092 (2006)
19. Wagner, E.H., Austin, B.T., Davis, C., Hindmarsh, M., Schaefer, J., Bonomi, A.: Improving chronic illness care: translating evidence into action. Health Affairs 20(6), 64 (2001)

Distance-Based Feature Selection on Classification of Uncertain Objects

Lei Xu and Edward Hung

Department of Computing, The Hong Kong Polytechnic University
{cslxu,csehung}@comp.polyu.edu.hk

Abstract. We study the problem of classification on uncertain objects whose locations are uncertain and described by probability density functions (pdf). We propose a novel supervised UK-means algorithm for classifying uncertain objects to overcome the computation bottleneck of existing algorithms. Additionally, we consider to select features that can capture the relevant properties of uncertain data. We experimentally demonstrate that our proposed approaches are more efficient than existing algorithms and can attain comparatively accurate results on non-overlapping data sets.

Keywords: classification, feature selection, uncertain data.

1 Introduction

Classification is a classical problem in machine learning and data mining [4]. Numerous classification algorithms have been proposed and used in real applications, such as multiple instance learning [19], ensemble learning [18], transfer learning [7], etc. However, few algorithms can handle uncertain information. Our task is to construct a model that is able to predict the label of uncertain object correctly.

UK-means is proposed for clustering uncertain objects. However, the classification task is predicting the labels of unlabeled objects based on given labeled objects. In general, the difference between classification and clustering is that classification obtains initial K class representatives from labeled training objects. In this paper, we focus on non-overlapping data sets. Thus, assume $C(o_i) = c_j$ represents that o_i is assigned to class c_j, the goal of classification is to find the K cluster representatives such that the objective function $\sum_{i=1}^{n} EED(o_i, p_{C(o_i)}) = \sum_{i=1}^{n}(\int f_i(x) ED(x, p_{C(o_i)})dx)$ is minimized where ED is the Euclidean distance function, n is the number of testing objects, and $p_{C(o_i)}$ is the location of cluster representative of class c_j. Additionally, we select features out of the feature set to capture the relevant properties of uncertain data. Following previous work in [15,17], we propose *Averaging* Approach and *Distribution-based* Approach to extend supervised UK-means algorithm to handle uncertain objects. Our contributions of this work include 1) we build a classifier based on UK-means to handle uncertain data, 2) we experimentally show that the supervised UK-means algorithm can classify uncertain objects more efficiently

D. Wang and M. Reynolds (Eds.): AI 2011, LNAI 7106, pp. 172–181, 2011.

than existing algorithms, and 3) considering the relevant properties of uncertain data, we combine feature selection with supervised UK-means to attain a higher accuracy compared with other approaches.

In the rest of this paper, we first give some related work briefly in Section 2. Section 3 introduces UK-means and formally describes the problem definition. Two approaches (*Averaging* and *Distribution-based*) are described in details in Section 4. Section 5 shows experimental evaluation on the performance of the algorithms. Finally we conclude our work in Section 6.

2 Related Work

In this paper, we focus on the value uncertainty. A large number of classification algorithms have been proposed in the literature, such as multiple instance learning [12], support vector machine [10]. However, few researches focus on the problem of uncertain data classification. In [6], support vector machine is used to classify uncertain data. In the method, the uncertain object is assumed as a simple bounded geometric model. Support vector machine creates margins by using uncertain objects which lie on the boundary. In [16,17], an uncertain object is associated with a probability density function (pdf) and a finite region. The decision tree classifier is extended to handle uncertain data by using averaging or distribution-based approach. Similarly, Naive Bayes is extended to classify uncertain data in [15]. uRule is proposed in [14] to classify uncertain information. The key idea in uRule is that the algorithm computes which proportion of the instances is covered by a rule based on the uncertain attribute interval and probabilistic function. Though some algorithms have been extended to classify uncertain information, the problem of building classifiers on uncertain data is still a challenge. The algorithms take quite a long time to process uncertain data because of intensive computation bottleneck.

There are also researches on clustering on uncertain objects. UK-means is a generalization of the traditional K-means algorithm to handle uncertain objects whose locations are represented by pdfs. Another related area of research is fuzzy clustering. In fuzzy clustering, a cluster is represented by a fuzzy subset of objects. Each object has a degree of belongingness for each cluster. The fuzzy C-means algorithm is one of the most widely used fuzzy clustering methods [5].

3 Problem Definition

3.1 UK-Means

Assume there are n uncertain objects, and each object o_i is associated with a probability density function (pdf), $f_i(x)$, which is the probability density of object o_i at the possible location x. The goal of clustering is to group n these objects into K clusters so that the sum of *expected Euclidean distances* (EED) [8] between the uncertain objects and their cluster centers is minimized. Thus, suppose $C(o_i) = c_j$ represents that object o_i is assigned to cluster c_j, and $p_{C(o_i)}$

is the cluster's representative point, we want to find the K cluster representatives such that the objective function (Equation (1)) is minimized where ED is the distance function based on a metric d (i.e. Euclidean distance).

$$\sum_{i=1}^{n} EED(o_i, p_{C(o_i)}) = \sum_{i=1}^{n} (\int f_i(x) ED(x, p_{C(o_i)}) dx). \quad (1)$$

Expected Distance Calculation. As Figure 1 shows, the uncertain domain is

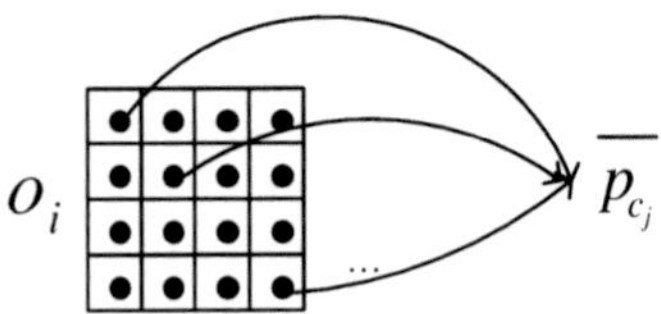

Fig. 1. Expected distance calculation from o_i to p_{c_j} in [13,11,8]

divided into a number of grid cells. Each grid cell represents a possible location of the uncertain object o_i. The expected Euclidean distance (EED) from object o_i (represented by a pdf f_i) to the cluster representative p_{c_j} is the weighted average of the distances between the samples in o_i and $\overline{p_{c_j}}$ (the mean vector of p_{c_j}), i.e. $EED(o_i, p_{c_j}) = \sum_{t=1}^{T} F_i(s_{i,t}) ED(s_{i,t}, \overline{p_{c_j}})$, where T is the number of samples in o_i, $s_{i,t}$ is the location (vector) of the t-th sample of o_i, p_{c_j} is the location (vector) of the cluster representative of cluster c_j, $F_i(s_{i,t}) = \int_{x \in cell_t} f_i(x)dx$ (F_i is a discrete probability distribution function over T grid cells, $cell_t$ is the grid cell that sample $s_{i,t}$ represents, x is the possible location of sample $s_{i,t}$ in $cell_t$), and the metric *ED* is Euclidean distance used in [13,11,8].

In traditional UK-means, n uncertain objects is assigned into K clusters by calculating expected distance between objects and cluster representatives and assign object o_i to the nearest cluster c_k with minimum expected distance between o_i and cluster representative. In UK-means, the mean vector $\overline{o_i}$ (expected value) of o_i is the weighted mean of all T samples (or possible locations) and is calculated as $\overline{o_i} = \sum_{t=1}^{T} s_{i,t} \times F_i(s_{i,t})$. The mean vector $\overline{p_{c_j}}$ of cluster representative p_{c_j} is obtained by $\overline{p_{c_j}} = \frac{1}{|c_j|} \sum_{i=1}^{|c_j|} \overline{o_i}$, where $|c_j|$ is the number of objects assigned to cluster c_j.

3.2 Supervised UK-Means

In supervised model, there are a set of N training objects $o_1, o_2, ..., o_N$, and m numerical (real-valued) feature attributes $A_1, ..., A_m$. The domain of attribute $A_u (1 \leq u \leq m)$ is $dom(A_u)$. Each o_i is associated with a probability density function (pdf $f_i(x)$) and a class label c_j ($c_j \in L$, where L is the set of all class labels), where x is a possible location of o_i, and $UD(o_i)$ is uncertain domain of o_i. Each tuple x is associated with a feature vector $x = (\tilde{x_1}, \tilde{x_2}, ..., \tilde{x_m})$, where

Algorithm 1. Supervised UK-means Algorithm

for i=0; $i < N$; i++ **do**
  compute $\overline{o_i}$ of training object by $\overline{o_i} = \sum_{t=1}^{T} s_{i,t} \times F_i(s_{i,t})$;
end for
for j=0; $j < K$; j++ **do**
  calculate all class representatives' mean vectors $\overline{p_{c_j}}$ by $\overline{p_{c_j}} = \frac{1}{|c_j|} \sum_{i=1}^{|c_j|} \overline{o_i}$;
end for
for i=0; $i < n$; i++ **do**
  compute $\overline{o_i}$ of testing object by $\overline{o_i} = \sum_{t=1}^{T} s_{i,t} \times F_i(s_{i,t})$;
end for
repeat
  for i=0; $i < n$; i++ **do**
    for j=0; $j < K$; j++ **do**
      compute expected distance by $EED(o_i, p_{c_j}) = \int_{UD(o_i)} f_i(x) ED(o_i, p_{c_j})\, dx$;
    end for
    assign object o_i to the nearest class c_k;
  end for
  update all cluster representatives by $\overline{p_{c_j}} = \frac{1}{|c_j|} \sum_{i=1}^{|c_j|} \overline{o_i}$;
until all cluster representatives converge

$\tilde{x_u} \in dom(A_u)(1 \leq u \leq m)$. The goal of supervised UK-means is to find K class representatives which can predict a testing object o_{test} to class c_k with the minimum expected Euclidean distance. In supervised UK-means, the initial class representative is obtained from the mean vectors of training objects associated with class labels. Then, we predict the labels of testing objects with the minimum expected Euclidean distance between objects and class centers. To obtain more accurate classification results, we repeat the testing process and update the class representatives until the algorithm converges. Algorithm 1 shows the generalized supervised UK-means, where N is the number of training objects, n is the number of testing objects, and K is the number of class labels.

Weighted Expected Euclidean Distance. ($||.||_w$) is calculated instead of expected Euclidean distance. The weighted distance between $s_{i,t}$ and p_{c_j} is calculated as $||s_{i,t} - p_{c_j}||_w = \sqrt{\sum_{u=1}^{m} w_u^2 \times (s_{i,t,u} - \overline{p_{c_{j,u}}})^2}$, where $s_{i,t,u}$ is the u-th dimension of t-th sample of object o_i, and w_u is the weight factor w on the u-th dimension. Furthermore, $||.||_w$ is calculated as $||o_i - p_{c_j}||_w = \sum_{t=1}^{T} F_i(s_{i,t}) ||s_{i,t} - p_{c_j}||_w$, which is the weighted average of weighted Euclidean distance between sample $s_{i,t}$ of object o_i and class center p_{c_j}.

4 Algorithms

In [15,17], *Averaging* Approach and *Distribution-based* Approach are proposed to modify decision tree algorithm to handle uncertain data. Similarly, we also use these approaches to handle uncertain objects in supervised UK-means. In addition, we select features to capture relevant properties of uncertain data.

4.1 Averaging Approach

A straightforward method to deal with uncertain object is to replace each pdf with its expected value [15,17]. Then the object is converted into exact value object, which reduces the problem back to that for certain data. Originally, object o_i is represented by T grid cells with pdf as Figure 1. In Averaging approach, the expected distance between o_i and class representative p_{c_j} is the exact distance between $\overline{o_i}$ and $\overline{p_{c_j}}$. The calculations of $\overline{o_i}$ and $\overline{p_{c_j}}$ have been illustrated in traditional UK-means. In Averaging approach, line 13 of Algorithm 1 is changed to $EED(o_i, p_{c_j}) = ED(\overline{o_i}, \overline{p_{c_j}})$ (where ED is Euclidean Distance).

4.2 Distribution-Based Approach

The difference between Averaging approach and Distribution-based approach is the calculation of expected distance between o_i and p_{c_j}. In this algorithm, line 13 of Algorithm 1 is $EED(o_i, p_{c_j}) = \sum_{t=1}^{T} F_i(s_{i,t}) ED(s_{i,t}, \overline{p_{c_j}})$, which is the same as that in UK-means. Algorithm 1 includes two parts: training process and testing process. In training process, the algorithm calculates the mean vectors of objects and the mean vectors of class representatives. The time complexity of training process is $O(NT)$,where N is the number of training objects, and T is the number of samples per object. In testing process, testing objects are predicted by calculating minimum expected Euclidean distance. The time complexity of Distribution-based approach in testing process is $O(nTT_1K)$, where n is the number of testing objects, T_1 is the number of iterations, and K is the number of class labels. If $N \geq nT_1K$, the computational complexity of Distribution-based approach is $O(NT)$. Otherwise, the computational complexity is $O(nTT_1K)$. In Averaging approach, the time complexity is $O(NT)$, because NT is usually larger than nT_1K. The time of training process of Averaging approach is the same as that of Distribution-based approach, but the time of testing process of Distribution-based approach is T times slower than that of Averaging approach.

Feature Selection. In previous work, all features are considered to be equally important. To build a classifier with high accuracy, it is necessary to select features from feature set to capture the relevant properties of uncertain data. [9] selects features which maximize the margins between objects from different classes.

Distance Based Evaluation Function. Existing algorithms on feature selection focus on exact value data, and we can see Averaging approach converting objects into deterministic point objects, so the existing algorithms can be readily used. Here we just extend the feature selection on Distribution-based approach. First we formulate the distance function of the selected set of features.

Definition 1. *Let Q be a set of uncertain objects and $o_q \in Q$. Let w be a weight vector over the feature set A, then the distance function of o_q is*

$$\theta_{o_q}^{w(A)} = \Big(\sum_{j=1, j \neq C(o_q)}^{K} ||o_q - p_{c_j}||_w \Big) - ||o_q - p_{C(o_q)}||_w. \tag{2}$$

where $||.||_w$ is weighted distance and has been described in details in Section 3.2.

Algorithm 2. Distance Based Feature Selection Algorithm (DBFS)

```
initialize weight vector w = (1,1,...,1);
pick randomly Q ⊆ N when N is training set;
for q = 1...|Q| do
    pick an object o_q from Q;
    for u = 0; u < m; u++ do
        ∇_u = 1/2 Σ_{o_q∈Q}((Σ_{j=1,j≠C((o_q))}^K ||o_q − p_{c_j}||^2 / ||o_q − p_{c_j}||_w) − ||o_q − p_{C(o_q)}||^2 / ||o_q − p_{C(o_q)}||_w) w_u;
        w_u = w_u + ∇_u;
    end for
end for
w = w^2 / w_max^2;
```

Definition 1 defines $w(A)$ to indicate the weight values on the feature set A. $\theta_{o_q}^{w(A)}$ can also be written as $\theta_{o_q}^{w}$. To make $\theta_{o_q}^{\lambda w} = |\lambda|\theta_{o_q}^{w}$ for any scalar λ, w is normalized in the way that max $w_u^2 = 1 (1 \leq u \leq m)$ (where w_u is the u-th value of $w(A)$, m is the number of attributes) to guarantee that $||.||_w^2 \leq ||.||^2$, where $||.||$ is Euclidean distance when $w = (1, ..., 1)$.

Definition 2. *Given a training set N $(Q \subseteq N)$ and a weight vector w, the distance-based evaluation function is* $e(w) = \sum_{o_i \in N} \theta_{o_i}^{w}$.

Definition 2 gives the evaluation function of feature selection. In the function, we aim to make all the objects in the training set N nearest to the class that they are labeled and farthest to other classes. The task of feature selection is to find $w(A)$ that can maximize the evaluation function $e(w)$.

Distance Based Feature Selection Algorithm (DBFS). DBFS tries to find the feature weight $w(A)$ to maximize the distance evaluation function $e(w)$. We use gradient ascent to maximize $e(w)$, since the evaluation function $e(w)$ can be seen smooth almost everywhere (there is $e(w) = e(w + \Delta)(if \Delta \to 0)$, but we will not demonstrate it because of space limitation) [9]. The gradient of $e(w)$ is shown as follows (Formula (3)) when it is evaluated on a sample o_q:

$$\bigtriangledown e(w)_u = \frac{\partial e(w)}{\partial w_u} = \sum_{o_q \in Q} \frac{\partial \theta_{o_q}^{w}}{\partial w_u} = \frac{1}{2} \sum_{o_q \in Q} ((\sum_{j=1, j \neq C(o_q)}^{K} \frac{||o_q - p_{c_j}||^2}{||o_q - p_{c_j}||_w}) - \frac{||o_q - p_{C(o_q)}||^2}{||o_q - p_{C(o_q)}||_w}) w_u. \tag{3}$$

In DBFS (algorithm 2), we use gradient over $e(w)$ to obtain $\bigtriangledown e(w)_u$ (we write as $\bigtriangledown_u$ for simplicity). We use a subset Q randomly picked from training set N to evaluate $e(w)$. In each iteration we use one object to calculate one term of the vector $\bigtriangledown$ and add it to the weight vector w. We have illustrated that the evaluation of $\bigtriangledown$ is invariant (i.e. $\bigtriangledown_u = \bigtriangledown e(\lambda w_u) \forall \lambda \geq 0$, see Proof 1). Therefore, since w_u increases by adding $\bigtriangledown_u$ during each iteration, the relative effect of the term $\bigtriangledown_u$ decreases (divided by increasing w) and the algorithm typically converges. The computational complexity of DBFS is $O(m|Q|TK)$ where m is the number of features, $|Q|$ is the number of iterations (usually 20 epochs), T is the number of samples of object o_q and K is the number of class labels. In Algorithm 1, we use DBFS in the training process to evaluate the weight values on feature set A. Then, testing objects are predicted by weighted expected Euclidean distance. Thus, Distribution-based approach (DBA) which uses DBFS to select features is weighted Distribution-based approach (Weighted DBA).

Proof 1. In Formula (4), because $||.||_{\lambda w} = \lambda||.||_w$ which has been described in Section 3.2, Formula (4) is equal to Formula (3).

$$\begin{aligned}\bigtriangledown(\lambda w_u) &= \frac{\partial e(\lambda w)}{\partial w_u} = \sum_{o_q \in Q} \frac{\partial \theta_{o_q}^{\lambda w}}{\partial w_u} \\ &= \frac{1}{2}\sum_{o_q \in Q}\Big(\Big(\sum_{j=1, j\neq C(o_q)}^{K} \frac{||o_q - p_{c_j}||^2}{||o_q - p_{c_j}||_{\lambda w}}\Big) - \frac{||o_q - p_{C(o_q)}||^2}{||o_q - p_{C(o_q)}||_{\lambda w}}\Big)\lambda w_u .\end{aligned} \tag{4}$$

5 Experimental Results

All codes were written in Java and were run on a Windows machine with an Intel 2.66GHz Pentium(R) Dual-Core processor and 4GB of main memory.

5.1 Data Sets

We run experiments on 3 UCI [3] non-overlapping data sets to study the performance of our algorithms and compare with the work in [17]. To compare with the work in [17], we did not show results on more other data sets. The parameters of the chosen data sets used for the experiments are summarized in Table 1. The attributes of all the 3 data sets are numerical obtained from measurements. Classifiers are built on the numerical attributes and their "class label" attributes. The 3 data sets contain "point values" without uncertainty. Thus, we generate the uncertain information following [17]. The point-value data become uncertain when we apply appropriate error models for them. For each object o_i on the u-th dimension (i.e. the attribute A_u), the point value $v_{i,u}$ reported in a data set is used as the mean of a pdf $f_{i,u}$, defined over an interval $[a_{i,u}, b_{i,u}]$. The range of values for A_u (over the whole data set) is noted and the width of $[a_{i,u}, b_{i,u}]$ is set to $un \times |A_u|$, where $|A_u|$ denotes the width of the range for A_u and un is a parameter to control the uncertainty of data set. We use two methods to generate pdf $f_{i,u}$. One is uniform distribution, which implies the pdf to be $f_{i,u} = (b_{i,u} - a_{i,u})^{-1}$. The other is Gaussian distribution, which the standard deviation is set to be $\frac{1}{4} \times (b_{i,u} - a_{i,u})$ (the same as that in [17]). In the above two cases, we use T samples to generate pdf over the interval. The point value is transformed into uncertain samples on Gaussian or uniform distribution by using the controlled parameter un and T samples. To compare with the work in [17], T is set to be 100 and un is from 1% to 20%.

Table 1. Selected Data Sets from the UCI Machine Learning Repository

Data Set	Training Tuples	No. of Attributes	No. of Classes	Test Tuples
Iris	150	4	3	10-fold
Ionosphere	351	32	2	10-fold
BreastCancer	569	30	2	10-fold

5.2 Performance Evaluation

Execution Time. Table 2 shows the execution time of Averaging (AVG) approach, Distribution-based approach (DBA), weighted DBA, and uncertain decision tree (UDT-ES) in [17] on 3 data sets. AVG can be seen as supervised UK-means for point value data while DBA and weighted DBA are supervised UK-means to handle uncertain objects. In Table 2, weighted DBA is at least 10 times faster than UDT-ES. DBA is at least 30 times faster than UDT-ES while AVG is 400 times faster than UDT-ES. DBA and weighted DBA are slower than AVG because of the expected distance calculation between uncertain objects and class centers. Weighted DBA has to calculate the weight values over feature set sample by sample which is a bit time consuming. However, the time used to evaluate weights is much shorter than that used in information gain in UDT-ES. The distribution of samples (uniform or Gaussian distribution) does not affect the execution time of DBA and weighted DBA. Thus, here we just presents the execution time of DBA and weighted DBA on Gaussian distribution. Figure 2(a) and (b) show the effects of increasing T samples on DBA and weighted DBA, respectively. Both our algorithms are 10 times faster than UDT-ES at least, and the time of our algorithms does not increase as fast as that of UDT-ES [17].

Table 2. Execution Time (Milliseconds)

Data Set	AVG	DBA	Weighted DBA	UDT-ES
Iris	10.8	42.5	76.7	5444.4
Ionosphere	18.8	261.1	881.2	9500
BreastCancer	28	370.3	912.4	11944.4

Accuracy. We use 10-fold cross validation on the three non-overlapping data sets to measure the accuracy. Figure 2(c) shows the accuracy with changing uncertainty un under different values of weighted DBA. To compare the effect of un, we put the accuracy of AVG which is exact point value algorithm at $un = 0$. In AVG, the point value on each dimension of an instance is the original data from the data sets. The accuracy is improved if uncertainty is taken into account. Table 3 shows the accuracy of our algorithms compared with UDT-ES in [17]. Table 3 chooses the accuracy of DBA (Gaussian pdf and uniform pdf) and weighted DBA (Gaussian pdf) from Figure 2(c) by the best results on un. From Table 3, we can see that DBA improves the accuracy from 0.3% to 2% while

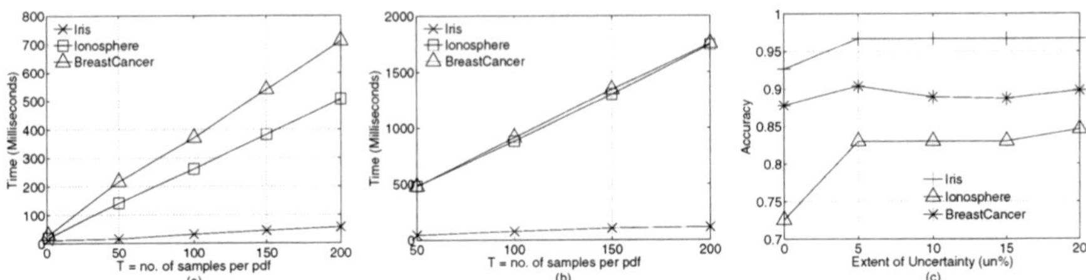

Fig. 2. (a) Effects of increasing T samples on Distribution Based Approach (b) Effects of increasing T samples on Weighted Distribution Based Approach (c) Weighted DBA Accuracy with controlled parameter un (Gaussian pdf)

Table 3. Accuracy

Data Set	AVG	DBA (Gaussian pdf)	DBA (Uniform pdf)	Weighted DBA	UDT-ES
Iris	92.67%	94%	94.67%	96.67%	96.13%
Ionosphere	72.6%	72.9%	72.86%	84.6%	91.69%
BreastCancer	87.86%	89.3%	88.04%	90.4%	95.93%

weighted DBA improves the accuracy from 2.5% to 12% compared with AVG. Our algorithms can attain comparatively accuracy compared with UDT-ES [17] by saving nearly 90%-95% time on non-overlapping data sets. We just illustrate the weights of field on Iris because of space limitation. The third feature is more important than other features on Iris with the weight value being 1.0 while other feature weight values are nearly 0. The dimensionality can be reduced since the weight values of some fields are nearly **zero**. Above all, weighted DBA is more accurate than DBA and AVG.

6 Conclusion

In this paper, we study the problem of classification on uncertain objects whose locations are presented by probability density functions (pdf). We have shown that supervised UK-means which overcomes the computation bottleneck is more efficient than existing algorithms with comparatively accurate classifying results by feature selection on non-overlapping data sets. Moreover, the accuracy is improved if uncertainty is taken into account.

Acknowledgement. The work described in this paper was partially supported by grants from the Research Grants Council of the Hong Kong Special Administrative Region, China (PolyU 5182/08E, PolyU 5191/09E).

References

1. Advances in Neural Information Processing Systems 17, Neural Information Processing Systems, NIPS 2004, Vancouver, British Columbia, Canada, December 13-18 (2004)
2. Proceedings of the 25th International Conference on Data Engineering, ICDE 2009, March 29-April 2, Shanghai, China. IEEE (2009)
3. Newman, D.J., Asuncion, A.: UCI machine learning repository (2007)
4. Agrawal, R., Imielinski, T., Swami, A.N.: Database mining: A performance perspective. IEEE Trans. Knowl. Data Eng. 5(6), 914–925 (1993)
5. Bezdek, J.C.: Pattern Recognition with Fuzzy Objective Function Algorithms. Kluwer Academic Publishers, Norwell (1981)
6. Bi, J.B., Zhang, T.: Support vector classification with input data uncertainty. In: NIPS [1]
7. Cao, B., Pan, S.J., Zhang, Y., Yeung, D.-Y., Yang, Q.: Adaptive transfer learning. In: Fox, M., Poole, D. (eds.) AAAI. AAAI Press (2010)
8. Chau, M., Cheng, R., Kao, B., Ng, J.: Uncertain Data Mining: An Example in Clustering Location Data. In: Ng, W.-K., Kitsuregawa, M., Li, J., Chang, K. (eds.) PAKDD 2006. LNCS (LNAI), vol. 3918, pp. 199–204. Springer, Heidelberg (2006)
9. Bachrach, R.G., Navot, A., Tishby, N.: Margin based feature selection - theory and algorithms. In: Brodley, C.E. (ed.) ICML. ACM International Conference Proceeding Series, vol. 69. ACM (2004)
10. Graf, H.P., Cosatto, E., Bottou, L., Durdanovic, I., Vapnik, V.: Parallel support vector machines: The cascade svm. In: NIPS [1]
11. Kao, B., Lee, S.D., Cheung, D.W., Ho, W.-S., Chan, K.F.: Clustering uncertain data using voronoi diagrams. In: ICDM, pp. 333–342. IEEE Computer Society (2008)
12. Maron, O., Pérez, T.L.: A framework for multiple-instance learning. In: Jordan, M.I., Kearns, M.J., Solla, S.A. (eds.) NIPS. The MIT Press (1997)
13. Ngai, W.K., Kao, B., Chui, C.K., Cheng, R., Chau, M., Yip, K.Y.: Efficient clustering of uncertain data. In: ICDM, pp. 436–445. IEEE Computer Society (2006)
14. Qin, B., Xia, Y.N., Prabhakar, S., Tu, Y.-C.: A rule-based classification algorithm for uncertain data. In: ICDE [2], pp. 1633–1640
15. Ren, J.T., Lee, S.D., Chen, X.L., Kao, B., Cheng, R., Cheung, D.W.-L.: Naive bayes classification of uncertain data. In: Wang, W., Kargupta, H., Ranka, S., Yu, P.S., Wu, X. (eds.) ICDM, pp. 944–949. IEEE Computer Society (2009)
16. Tsang, S., Kao, B., Yip, K.Y., Ho, W.-S., Lee, S.D.: Decision trees for uncertain data. In: ICDE [2], pp. 441–444
17. Tsang, S., Kao, B., Yip, K.Y., Ho, W.-S., Lee, S.D.: Decision trees for uncertain data. IEEE Trans. Knowl. Data Eng. 23(1), 64–78 (2011)
18. Zhang, D.Q., Chen, S.C., Zhou, Z.-H., Yang, Q.: Constraint projections for ensemble learning. In: Fox, D., Gomes, C.P. (eds.) AAAI, pp. 758–763. AAAI Press (2008)
19. Zhou, Z.-H., Sun, Y.-Y., Li, Y.-F.: Multi-instance learning by treating instances as non-i.i.d. samples. In: Danyluk, A.P., Bottou, L., Littman, M.L. (eds.) ICML. ACM International Conference Proceeding Series, vol. 382, page 157. ACM (2009)

Closure Spaces of Isotone Galois Connections and Their Morphisms

Radim Belohlavek and Jan Konecny

Palacky University, Olomouc, Czech Republic
radim.belohlavek@acm.org, jan.konecny@upol.cz

Abstract. We present results on closure spaces induced by isotone fuzzy Galois connections. Such spaces play a fundamental role in the analysis of relational data such as formal concept analysis or relational factor analysis. We provide a characterization of such closure spaces and study their morphisms. The results contribute to foundations of a matrix calculus over relational data.

1 Introduction

Closure structures are among the fundamental mathematical structures that naturally appear in many areas of pure and applied mathematics. In particular, closure structures are the fundamental structures behind formal concept analysis and other data analysis methods that are based on attribute sharing (rather than attribute distance). The results in this paper are motivated by the recent results on decompositions of matrices over residuated lattices and factor analysis of relational data described by such matrices, see e.g. [3–5]. These results reveal a fundamental role of closure and interior structures for the decompositions and motivate us to further investigate the calculus of matrices over residuated lattices. Such matrices include Boolean matrices as a particular case but have much richer structure. An important concept, studied in this paper, is that of a closure space of isotone and antitone Galois connections induced by such matrices. Such spaces are in fact the spaces of optimal factors for matrix decompositions [3, 4]. In the setting of Boolean matrices, there exists a natural bijective mapping between the spaces of isotone and antitone Galois connections. Moreover, these spaces exhaust all closure spaces. This is no longer true in the setting of matrices over residuated lattices. While it is known from the previous results that the closure spaces of antitone fuzzy Galois connections exhaust all fuzzy closure spaces, we show in this paper that the closure spaces of isotone fuzzy Galois connections are particular fuzzy closure spaces. We provide a characterization of such spaces. Moreover, we study morphisms of such spaces and show a correspondence between such morphisms and matrices (matrices induce morphisms and vice versa). The results contribute to the foundations of analysis of qualitative data, namely to the development of a matrix calculus for such data.

2 Preliminaries: Matrices, Decompositions, Concept Lattices

Matrices. We deal with matrices whose degrees are elements of residuated lattices. Note that instead of matrices, we could consider fuzzy relations (with degrees in complete residuated lattices) between possibly infinite sets. The results would then be more

D. Wang and M. Reynolds (Eds.): AI 2011, LNAI 7106, pp. 182–191, 2011.

general (matrices correspond to relations between finite sets). Recall that a (complete) residuated lattice [1, 10, 16] is a structure $\mathbf{L} = \langle L, \wedge, \vee, \otimes, \rightarrow, 0, 1\rangle$ such that

(i) $\langle L, \wedge, \vee, 0, 1\rangle$ is a (complete) lattice, i.e. a partially ordered set in which arbitrary infima and suprema exist (the lattice order is denoted by $\leq$; 0 and 1 denote the least and greatest element, respectively);
(ii) $\langle L, \otimes, 1\rangle$ is a commutative monoid, i.e. $\otimes$ is a binary operation which is commutative, associative, and $a \otimes 1 = a$ for each $a \in L$;
(iii) $\otimes$ and $\rightarrow$ satisfy adjointness, i.e. $a \otimes b \leq c$ iff $a \leq b \rightarrow c$.

Throughout the paper, $\mathbf{L}$ denotes an arbitrary (complete) residuated lattice. Common examples of complete residuated lattices include those defined on the real unit interval, i.e. $L = [0, 1]$, or on a finite chain in a unit interval, e.g. $L = \{0, \frac{1}{n}, \ldots, \frac{n-1}{n}, 1\}$. For instance, for $L = [0, 1]$, we can use any left-continuous t-norm for $\otimes$, such as minimum, product, or Łukasewicz, and the corresponding residuum $\rightarrow$ [1, 10, 16]. Residuated lattices are commonly used in fuzzy logic [1, 9, 10]. Elements $a \in L$ are called grades (degrees of truth). Operations $\otimes$ (multiplication) and $\rightarrow$ (residuum) play the role of a (truth function of) conjunction and implication, respectively.

We deal with compositions $I = A * B$ which involve an $n \times m$ matrix I, an $n \times k$ matrix A, and a $k \times m$ matrix B. We assume that $I_{ij}, A_{il}, B_{lj} \in L$. That is, all the matrix entries are elements of a given residuated lattice $\mathbf{L}$. Therefore, examples of matrices I which are subject to the decomposition are

$$\begin{pmatrix} 1.0 & 1.0 & 0.0 & 0.0 & 0.6 & 0.4 \\ 1.0 & 0.9 & 0.0 & 0.0 & 1.0 & 0.8 \\ 1.0 & 1.0 & 0.0 & 1.0 & 0.0 & 0.0 \\ 1.0 & 0.5 & 0.0 & 0.7 & 1.0 & 0.4 \end{pmatrix} \quad \text{or} \quad \begin{pmatrix} 0 & 0 & 1 & 1 & 1 \\ 0 & 0 & 1 & 1 & 0 \\ 0 & 0 & 0 & 0 & 1 \\ 0 & 1 & 1 & 1 & 0 \end{pmatrix}.$$

The second matrix makes it apparent that binary matrices are a particular case for $L = \{0, 1\}$.

For convenience and since there is no danger of misunderstanding, we take the advantage of identifying $n \times m$ matrices over residuated lattices (the set of all such matrices is denoted by $L^{n\times m}$) with binary fuzzy relations between X and Y (the set of all such relations is denoted by $L^{X\times Y}$). Also, we identify vectors with n components over residuated lattices (the set of all such vectors is denoted by L^n) with fuzzy sets in X (the set of all such fuzzy sets is denoted by L^X). As usual, we identify vectors with n components with $1 \times n$ matrices.

Composition Operators. We use three matrix composition operators, $\circ$, $\triangleleft$, and $\triangleright$. In the decompositions $I = A * B$, I_{ij} is interpreted as the degree to which the object i has the attribute j; A_{il} as the degree to which the factor l applies to the object i; B_{lj} as the degree to which the attribute j is a manifestation (one of possibly several manifestations) of the factor l. The composition operators are defined by

$$(A \circ B)_{ij} = \textstyle\bigvee_{l=1}^{k} A_{il} \otimes B_{lj}, \tag{1}$$

$$(A \triangleleft B)_{ij} = \textstyle\bigwedge_{l=1}^{k} A_{il} \rightarrow B_{lj}, \tag{2}$$

$$(A \triangleright B)_{ij} = \textstyle\bigwedge_{l=1}^{k} B_{lj} \rightarrow A_{il}. \tag{3}$$

Note that these operators were extensively studied by Bandler and Kohout, see e.g. [12] to which we refer for an overview of knowledge processing applications. The operators have natural verbal descriptions. For instance, $(A \triangleleft B)_{ij}$ is the truth degree of "for every factor l, if l applies to object i then attribute j is a manifestation of l". One may easily see that $\triangleright$ can be defined in terms of $\triangleleft$ and vice versa. Note also that for $L = \{0, 1\}$, $A \circ B$ coincides with the well-known Boolean product of matrices [11].

Concept Lattices Associated to I. For a positive integer n, we denote

$$\hat{n} = \{1, \ldots, n\}.$$

In addition, we put

$$X = \{1, \ldots, n\}, \quad Y = \{1, \ldots, m\}.$$

Recall that L^U denotes the set of all L-sets in U, i.e. all mappings from U to L. Consider the following pairs of operators between L^X and L^Y induced by matrix $I \in L^{n \times m}$:

$$C^{\uparrow}(j) = \bigwedge_{i=1}^{n}(C(i) \to I_{ij}), \quad D^{\downarrow}(i) = \bigwedge_{j=1}^{m}(D(j) \to I_{ij}), \tag{4}$$

$$C^{\cap}(j) = \bigvee_{i=1}^{n}(C(i) \otimes I_{ij}), \quad D^{\cup}(i) = \bigwedge_{j=1}^{m}(I_{ij} \to D(j)), \tag{5}$$

$$C^{\wedge}(j) = \bigwedge_{i=1}^{n}(I_{ij} \to C(i)), \quad D^{\vee}(i) = \bigvee_{j=1}^{m}(D(j) \otimes I_{ij}), \tag{6}$$

for $C \in L^X$, $D \in L^Y$, $j \in \{1, \ldots, m\}$, and $i \in \{1, \ldots, n\}$. Furthermore, denote the corresponding sets of fixpoints by $\mathcal{B}(X^{\uparrow}, Y^{\downarrow}, I)$, $\mathcal{B}(X^{\cap}, Y^{\cup}, I)$, and $\mathcal{B}(X^{\wedge}, Y^{\vee}, I)$, i.e.

$$\begin{aligned}
\mathcal{B}(X^{\uparrow}, Y^{\downarrow}, I) &= \{\langle C, D\rangle \mid C^{\uparrow} = D,\, D^{\downarrow} = C\},\\
\mathcal{B}(X^{\cap}, Y^{\cup}, I) &= \{\langle C, D\rangle \mid C^{\cap} = D,\, D^{\cup} = C\},\\
\mathcal{B}(X^{\wedge}, Y^{\vee}, I) &= \{\langle C, D\rangle \mid C^{\wedge} = D,\, D^{\vee} = C\}.
\end{aligned}$$

The sets of fixpoints are complete lattices, called concept lattices associated to I, and their elements are called formal concepts. These structures are the fundamental structures of formal concept analysis [6]. For a formal concept $\langle C, D\rangle$, C and D are called the extent and the intent and they represent the collection of objects and attributes to which the formal concept applies. The sets of all extents and intents of the respective concept lattices are denoted by $\mathrm{Ext}(X^{\uparrow}, Y^{\downarrow}, I)$, $\mathrm{Int}(X^{\uparrow}, Y^{\downarrow}, I)$, $\mathrm{Ext}(X^{\cap}, Y^{\cup}, I)$, $\mathrm{Int}(X^{\cap}, Y^{\cup}, I)$, $\mathrm{Ext}(X^{\wedge}, Y^{\vee}, I)$, and $\mathrm{Int}(X^{\wedge}, Y^{\vee}, I)$. It may be shown that

$$\begin{aligned}
\mathrm{Ext}(X^{\uparrow}, Y^{\downarrow}, I) &= \{C \in L^X \mid C = C^{\uparrow\downarrow}\},\\
\mathrm{Int}(X^{\uparrow}, Y^{\downarrow}, I) &= \{D \in L^Y \mid D = D^{\downarrow\uparrow}\},
\end{aligned}$$

and the same for the other cases.

The above-defined operators and their sets of fixpoints have extensively been studied, see e.g. [2, 7, 14]. Clearly, $\langle C, D\rangle \in \mathcal{B}(X^{\cap}, Y^{\cup}, I)$ iff $\langle D, C\rangle \in \mathcal{B}(Y^{\wedge}, X^{\vee}, I^{\mathrm{T}})$, where I^{T} denotes the transpose of I; so one could consider only one pair, $\langle^{\cap}, ^{\cup}\rangle$ or $\langle^{\wedge}, ^{\vee}\rangle$, and obtain the properties of the other pair by a simple translation. Note that if $L = \{0,1\}$, $\mathcal{B}(X^{\uparrow}, Y^{\downarrow}, I)$ coincides with the ordinary concept lattice of the formal context consisting of X, Y, and the binary relation (represented by) I; and that $\mathcal{B}(X^{\uparrow}, Y^{\downarrow}, I)$ is isomorphic to $\mathcal{B}(X^{\cap}, Y^{\cup}, \overline{I})$ with $\langle A, B\rangle \mapsto \langle A, \overline{B}\rangle$ being an isomorphism ($\overline{U}$ denotes the complement of U). Therefore, as is well known, for $L = \{0,1\}$, each of the three operators is definable by any of the remaining two. The mutual definability fails for general L because it is based on the law of double negation which does not hold for general residuated lattices. A simple framework that enables us to consider all the three operators as particular types of a more general operator is provided in [4], cf. also [7] for another possibility. For simplicity, we do not work with the general approach and use the three operators because they are well known.

3 Closure Spaces Induced by $\langle {}^{\wedge}, {}^{\vee}\rangle$

The following results are well known [1]. $\langle {}^{\uparrow}, {}^{\downarrow}\rangle$ forms an (antitone) **L**-Galois connection [1], ${}^{\uparrow\downarrow}$ and ${}^{\downarrow\uparrow}$ are **L**-closure operators in X and Y, and $\mathrm{Ext}(X^{\uparrow}, Y^{\downarrow}, I)$ and $\mathrm{Int}(X^{\uparrow}, Y^{\downarrow}, I)$ are **L**-closure systems in X and Y, respectively. Moreover, any **L**-closure system in X is in the form of $\mathrm{Ext}(X^{\uparrow}, Y^{\downarrow}, I)$ (same for Y).

Recall $V \subseteq L^U$ is called an **L**-*closure system* (in the context of fuzzy sets; or *c-subspace*, in the context of matrices) if

- V is closed under *left $\rightarrow$-multiplications*, i.e. $a \rightarrow C \in V$ for each $a \in L$ and $C \in V$ (here, $a \rightarrow C$ is defined by $(a \rightarrow C)(i) = a \rightarrow C(i)$ for $i = 1, \ldots, n$);
- V is closed under $\bigwedge$-*intersections*, i.e. for $C_j \in V$ ($j \in J$) we have $\bigwedge_{j\in J} C_j \in V$ (here, $\bigwedge_{j\in J} C_j$ is defined by $(\bigwedge_{j\in J} C_j)(i) = \bigwedge_{j\in J} C_j(i)$).

For $\langle {}^{\wedge}, {}^{\vee}\rangle$, it is known that $\langle {}^{\wedge}, {}^{\vee}\rangle$ forms an isotone **L**-Galois connection [7], ${}^{\wedge\vee}$ and ${}^{\vee\wedge}$ are **L**-interior and **L**-closure operators in X and Y, and $\mathrm{Ext}(X^{\wedge}, Y^{\vee}, I)$ and $\mathrm{Int}(X^{\wedge}, Y^{\vee}, I)$ are **L**-interior and **L**-closure systems in X and Y, respectively. The situation might seem completely dual to that of $\langle {}^{\uparrow}, {}^{\downarrow}\rangle$ (which is the case when $L = \{0,1\}$, see above). However, as the next example shows, it is not. Namely, there exist **L**-closure systems that are not of the form $\mathrm{Int}(X^{\wedge}, Y^{\vee}, I)$.

Example 1. Let **L** be the standard Gödel algebra, $U = \{u\}$, $\mathcal{S} = \{\{{}^{0.5}/u\}, \{{}^{1}/u\}\}$. Therefore, $L = [0,1]$ and $a \rightarrow b = 1$ if $a \leq b$ and $a \rightarrow b = b$ of $a > b$. Clearly, $\mathcal{S}$ is closed under intersections and $\rightarrow$-shifts, hence it is an **L**-closure system. However, $\mathcal{S}$ is not of the form $\mathcal{S} = \mathrm{Int}(X^{\wedge}, Y^{\vee}, I)$. (This claim is justified at the end of this section.)

Therefore, **L**-closure systems that are of the form $\mathrm{Int}(X^{\wedge}, Y^{\vee}, I)$ are just particular **L**-closure systmes. Below, we provide their characterization. For a system $\mathcal{S} \subseteq L^U$, put

$$
\begin{aligned}
[\mathcal{S}]_{\wedge} &= \{\textstyle\bigwedge \mathcal{T} \mid \mathcal{T} \subseteq \mathcal{S}\},\\
[\mathcal{S}]_{\rightarrow} &= \{a \rightarrow A \mid a \in L,\ A \in \mathcal{S}\},\\
[\mathcal{S}]^{\rightarrow} &= \{A \rightarrow a \mid a \in L,\ A \in \mathcal{S}\}.
\end{aligned}
$$

Note that $A \rightarrow a$ is defined by $(A \rightarrow a)(u) = A(u) \rightarrow a$ and call $A \rightarrow a$ the *right $\rightarrow$-multiple* of A by a. Therefore, $[\mathcal{S}]_{\wedge}$ is the system of all intersections of fuzzy sets from

$\mathcal{S}$, $[\mathcal{S}]_{\rightarrow}$ is the system of all left $\rightarrow$-multiplications of fuzzy sets from $\mathcal{S}$, and $[\mathcal{S}]^{\rightarrow}$ is the system of all right $\rightarrow$-multiplications of fuzzy sets from $\mathcal{S}$. It is known that for any $\mathcal{S} \subseteq L^U$, $[[\mathcal{S}]_{\rightarrow}]_{\bigwedge}$ is the least, w.r.t. inclusion, **L**-closure system containing $\mathcal{S}$. $[[\mathcal{S}]_{\rightarrow}]_{\bigwedge}$ is called the **L**-closure system generated by $\mathcal{S}$, or the *c-span* of $\mathcal{S}$.

Note that in fuzzy logic, $b \rightarrow 0$ is called the negation of the truth degree b. Correspondingly, the fuzzy set $A \rightarrow 0$ is called the complement of A. Clearly, in the above terms, $A \rightarrow 0$ is the right multiple of A by 0. From this point of view, the right multiples $A \rightarrow a$ generalize the concept of a complement of a fuzzy set. $A \rightarrow a$ could naturally be called the *a-complement* of A.

In the classical case ($L = \{0, 1\}$), every A is a complement of some B; namely, of $B = A \rightarrow 0$. This is no longer true for the general setting of residuated lattices (not even for $a = 0$). We only have:

Lemma 1. *A is an a-complement of some fuzzy set if and only if* $A = (A \rightarrow a) \rightarrow a$.

Proof. Easy, follows from $((b \rightarrow a) \rightarrow a) \rightarrow a = b \rightarrow a$. □

This lemma is, in a sense, the key observation in characterizing the **L**-closure systems $\mathrm{Int}(X^{\wedge}, Y^{\vee}, I)$. We are going to show that $\mathrm{Int}(X^{\wedge}, Y^{\vee}, I)$ are just the **L**-closure systems that are generated by a-complements of some collection $\mathcal{T}$ of fuzzy sets. Such systems are conveniently characterized by the following theorem.

Theorem 1. *For any* $\mathcal{T} \subseteq L^U$, $[[\mathcal{T}]^{\rightarrow}]_{\bigwedge}$ *is an* **L***-closure system. It is the least, w.r.t. inclusion,* **L***-closure system containing all a-complements (i.e., right* $\rightarrow$*-multiplications) of fuzzy sets from* $\mathcal{T}$.

Proof. Sketch: Clearly, $[[\mathcal{T}]^{\rightarrow}]_{\bigwedge}$ contains all a-complements of fuzzy sets from $\mathcal{T}$. Essential to the proof is to check that $[[\mathcal{T}]^{\rightarrow}]_{\bigwedge}$ is closed under left $\rightarrow$-multiplications (this follows from $a \rightarrow (b \rightarrow c) = b \rightarrow (a \rightarrow c)$). The rest is by standard arguments. □

The following theorem provides our characterization.

Theorem 2. *For any* $\mathcal{S} \subseteq L^U$, $\mathcal{S} = \mathrm{Int}(X^{\wedge}, Y^{\vee}, I)$ *for some I if and only if* $\mathcal{S} = [[\mathcal{T}]^{\rightarrow}]_{\bigwedge}$ *for some* $\mathcal{T} \subseteq L^U$*, i.e.* $\mathcal{S}$ *is an* **L***-closure system generated by a system of all a-complements of fuzzy sets from* $\mathcal{T}$.

Proof. Sketch: "$\Rightarrow$" is done by checking the conditions and using standard properties of residuated lattice.

"$\Leftarrow$": Let $X = \mathcal{T}$, $Y = U$, $I(A, u) = A(u)$ for $A \in \mathcal{S}$, $u \in U$. One can show that $\mathcal{S} = \mathrm{Int}(X^{\wedge}, Y^{\vee}, I)$. □

Definition 1. *We call the systems* $\mathcal{S}$ *satisfying the condition of Theorem 2* c-closure spaces *("c" for "complement").*

Example 1 (continued). Suppose, by contradiction, that $\mathcal{S} = \mathrm{Int}(X^{\wedge}, Y^{\vee}, I)$. Then $U = X$ and by Theorem 2, $\mathcal{S}$ is a system generated by a system of all a-complements of fuzzy sets from some $\mathcal{T}$. According to Theorem 1, $[[\mathcal{T}]^{\rightarrow}]_{\bigwedge} = \{\{^{0.5}/u\}, \{^{1}/u\}\}$. Then, $\{^{0.5}/u\}$ needs to be an intersection of other fuzzy sets from $[\mathcal{T}]^{\rightarrow}$ or $\{^{0.5}/u\} \in [\mathcal{T}]^{\rightarrow}$. Clearly, $\{^{0.5}/u\} \in [\mathcal{T}]^{\rightarrow}$ must be the case. Therefore, $\{^{0.5}/u\} = \{^{a}/u\} \rightarrow b$ for some b. Clearly, $a > b = 0.5$ must be the case. But then, we also have $\{^{a}/u\} \rightarrow 0.4 = \{^{0.4}/u\} \in [\mathcal{T}]^{\rightarrow}$, a contradiction to $[\mathcal{T}]^{\rightarrow} \subseteq [[\mathcal{T}]^{\rightarrow}]_{\bigwedge} = \{\{^{0.5}/u\}, \{^{1}/u\}\}$.

4 Morphisms of c-Closure Spaces

In this section we define morphisms of c-closure spaces, i.e. the particular **L**-closure spaces characterized in Section 3, and show that they are induced by matrices over residuated lattices via the $\triangleright$-product.

Definition 2. *A mapping $h : V \to W$ from a c-closure space $V \subseteq L^p$ into a c-closure space $W \subseteq L^q$ is called a* complement-preserving c-morphism *if*

- *h is an c-morphism, i.e. $h(a \to C) = a \to h(C)$ and $h(\bigwedge_{k\in K} C_k) = \bigwedge_{k\in K} h(C_k)$ for any $a \in L, C, C_k \in L^p$;*
- *if C is an a-complement then $h(C)$ is an a-complement.*

A complement-preserving c-morphism $h : V \to W$ from a c-subspace $V \subseteq L^p$ into a c-subspace $W \subseteq L^q$ is called an extendable *if there is a complement-preserving c-morphism $h' : L^p \to L^q$ such that $h'(C) = h(C)$ for each $C \in V$.*

A complement-preserving c-morphism h is called a complement-preserving c-isomorphism *if h is bijective and both h and h^{-1} are extendable complement-preserving c-morphisms.*

In what follows we assume only extendable complement-preserving c-morphisms.

First, every matrix induces a morphism:

Lemma 2. *For every matrix $A \in L^{p\times q}$, the mapping $h_A : L^p \to L^q$ defined by*

$$h_A(C) = C \triangleright A \quad (= C^{\wedge_A})$$

is a complement-preserving c-morphism.

Proof. Sketch: Being a c-morphism follows easily from the properties of residuated lattices. Let $C = D \to a$, then

$$\begin{aligned}[(D \to a) \triangleright A](j) &= \bigwedge_i A_{ij} \to (D(i) \to a) = \bigwedge_i ((A_{ij} \otimes D(i)) \to a) = \\ &= \bigvee_i (A_{ij} \otimes D(i)) \to a = (D \circ A)(j) \to a\end{aligned}$$

Whence, if C is an a-complement then $C \triangleright A$ is a-complement. □

Second, every morphism is induced by some matrix.

Lemma 3. *If $h : V \to L^q$ is a complement-preserving c-morphism of a c-closure space V, then there exists a matrix $A_h \in L^{p\times q}$ such that $h(C) = C \triangleright A$ for every $C \in V$.*

Proof. Let $A \in L^{p\times q}$ be defined by

$$A_{ij} = \textstyle\bigwedge_{C\in V}((h(C))(j) \to C(i)).$$

That is, $A_{i\_} = \bigwedge_{C\in V}(h(C) \to C(i))$, i.e. the row $A_{i\_}$ contains a vector of degrees that can be interpreted as the intersection of images of those vectors C from V for which the corresponding fuzzy set contains i (in Boolean case: for which the ith component is 1).

We now check $h(C) = C \triangleright A$ for every $C \in L^p$. First,

$$(C \triangleright A)(j) = \bigwedge_{i=1}^{p} [A_{ij} \to C(i)] =$$
$$= \bigwedge_{i=1}^{p} [(\bigwedge_{C' \in V} (h(C'))(j) \to C'(i))) \to C(i)] \geq (h(C))(j).$$

We omit the second part $((C \triangleright A)(j) \leq (h(C))(j))$, which is technically more involved, due limited space. □

As a corollary, we get the following characterization of morphisms:

Theorem 3. *$h : V \to L^q$ is a complement-preserving c-morphism of a c-closure space V if and only if there exists a matrix $A_h \in L^{p\times q}$ such that $h(C) = C \triangleright A$ for every $C \in L^p$.*

Proof. Directly from Lemma 2 and Lemma 3. □

5 Isomorphic c-Closure Spaces

The aim of this section is to provide a criterion of isomorphism of c-closure spaces.

Lemma 4. *Let $I, J \in L^{p\times q}$. We have $B^{\wedge_I} = B^{\wedge_J}$ for each $B \in L^p$ iff $I = J$.*

Proof. "⇒": Suppose $B^{\wedge_I} = B^{\wedge_J}$. Assume $I_{ij} \neq J_{ij}$ for some i, j. Without loss of generality, we may assume $I_{ij} \nleq J_{ij}$. Let

$$B(l) = \begin{cases} J_{ij} & \text{if } i = l, \\ 1 & \text{otherwise}. \end{cases}$$

Then $B^{\wedge_I}(j) = \bigwedge_l I_{lj} \to B(l) = I_{ij} \to J_{ij} \neq 1$, and $B^{\wedge_J}(j) = \bigwedge_l J_{lj} \to B(l) = J_{ij} \to J_{ij} = 1$, which is a contradiction.

"⇐": Obvious. □

We need to recall the following notions. $V \subseteq L^n$ is called an *i-subspace* if

- V is closed under $\otimes$-multiplication, i.e. for every $a \in L$ and $C \in V$, $a \otimes C \in V$ (here, $a \otimes C$ is defined by $(a \otimes C)(i) = a \otimes C(i)$ for $i = 1, \ldots, n$); and
- V is closed under $\bigvee$-union, i.e. for $C_j \in V$ $(j \in J)$ we have $\bigvee_{j\in J} C_j \in V$ (here, $\bigvee_{j\in J} C_j$ is defined by $(\bigvee_{j\in J} C_j)(i) = \bigvee_{j\in J} C_j(i)$).

A mapping $h : V \to W$ from an i-subspace $V \subseteq L^p$ into an i-subspace $W \subseteq L^q$ is called an *i-morphism* if it is a $\otimes$- and $\bigvee$-morphism, that is, $h(a \otimes C) = a \otimes h(C)$ and $h(\bigvee_{k\in K} C_k) = \bigvee_{k\in K} h(C_k)$ for any $a \in L$, $C, C_k \in L^p$. An i-morphism $V \to W$ is called

- an *extendable i-morphism* if h can be extended to an i-morphism of $L^p \to L^q$.
- an *i-isomorphism* if h is bijective and both h and h^{-1} are extendable i-morphisms.

Theorem 4. *Let $I \in L^{n\times m}$ and $J \in L^{p\times r}$ be matrices. Then there exist a complement-preserving c-isomorphism*

$$h : \mathrm{Int}(\hat{n}^\wedge, \hat{m}^\vee, I) \to \mathrm{Int}(\hat{p}^\wedge, \hat{r}^\vee, J)$$

if and only if there exists a matrix $K \in L^{p\times m}$ such that $\mathrm{Int}(\hat{n}^\wedge, \hat{m}^\vee, I) = \mathrm{Int}(\hat{p}^\wedge, \hat{m}^\vee, K)$ *and* $\mathrm{Ext}(\hat{p}^\wedge, \hat{r}^\vee, J) = \mathrm{Ext}(\hat{p}^\wedge, \hat{m}^\vee, K)$.

Proof. "⇒": Let $h : \mathrm{Int}(\hat{p}^\wedge, \hat{r}^\vee, J) \to \mathrm{Int}(\hat{n}^\wedge, \hat{m}^\vee, I)$ be a complement-preserving c-isomorphism. According to Lemma 3, there exist matrices $X \in L^{r\times m}$ and $Y \in L^{m\times r}$ such that

$$h(C) = C \triangleright X \text{ and } h^{-1}(D) = D \triangleright Y$$

for every $C \in \mathrm{Int}(\hat{p}^\wedge, \hat{r}^\vee, J)$ and $D \in \mathrm{Int}(\hat{n}^\wedge, \hat{m}^\vee, I)$.

Thus we have $C = h(h^{-1}(C)) = (C \triangleright X) \triangleright Y$. Now, since $(C \triangleright X) \triangleright Y = C \triangleright (X \circ Y)$ and since $B^{\wedge_J} \in \mathrm{Int}(\hat{p}^\wedge, \hat{r}^\vee, J)$ for every $B \in L^{\hat{p}}$,

$$B^{\wedge_J} = B^{\wedge_J \wedge_{(X\circ Y)}} = B^{\wedge_{J\circ X\circ Y}}$$

for every $B \in L^{\hat{p}}$. $J = J \circ X \circ Y$ now follows from Lemma 4. From that we have $\mathrm{Ext}(\hat{p}^\wedge, \hat{r}^\vee, J) = \mathrm{Ext}(\hat{p}^\wedge, \hat{m}^\vee, K)$. Furthermore, if $D \in \mathrm{Int}(\hat{n}^\wedge, \hat{m}^\vee, I)$, then $D \triangleright Y = h^D \in \mathrm{Int}(\hat{n}^\wedge, \hat{m}^\vee, I)$. Since $D = (D \triangleright Y) \triangleright X$, we get $D = (C \triangleright J) \triangleright X = C \triangleright (J \circ X)$ showing $D \in \mathrm{Int}(\hat{p}^\wedge, \hat{r}^\vee, J \circ X)$. We established $\mathrm{Int}(\hat{n}^\wedge, \hat{m}^\vee, I) \subseteq \mathrm{Int}(\hat{p}^\wedge, \hat{r}^\vee, J \circ X)$. If $D \in \mathrm{Int}(\hat{p}^\wedge, \hat{r}^\vee, J \circ X)$ then $D = C \triangleright (J \circ X) = (C \triangleright J) \triangleright X$ for some $C \in L^p$. Since $C \triangleright J \in \mathrm{Int}(\hat{p}^\wedge, \hat{r}^\vee, J)$, we get

$$D = (C \triangleright J) \triangleright X = h(C \circ J) \in \mathrm{Int}(\hat{p}^\wedge, \hat{r}^\vee, I),$$

proving $\mathrm{Int}(\hat{p}^\wedge, \hat{r}^\vee, J \circ X) \subseteq \mathrm{Int}(\hat{n}^\wedge, \hat{m}^\vee, I)$. Summing up, we proved $\mathrm{Int}(\hat{p}^\wedge, \hat{r}^\vee, J \circ X) = \mathrm{Int}(\hat{n}^\wedge, \hat{m}^\vee, I)$. Now, $J \circ X$ yields the required matrix K.

"⇐": Since $\mathrm{Ext}(\hat{p}^\wedge, \hat{r}^\vee, J) = \mathrm{Ext}(\hat{p}^\wedge, \hat{m}^\vee, K)$, there exists a matrix $S \in L^{m\times r}$ for which $K \circ S = J$ and a matrix $T \in L^{m\times r}$ for which $J \circ T = K$, respectively. Consider now mappings $f : \mathrm{Int}(\hat{p}^\wedge, \hat{m}^\vee, K) \to \mathrm{Int}(\hat{p}^\wedge, \hat{r}^\vee, J)$ and $g : \mathrm{Int}(\hat{p}^\wedge, \hat{r}^\vee, J) \to \mathrm{Int}(\hat{p}^\wedge, \hat{m}^\vee, K)$ defined for $D \in \mathrm{Int}(\hat{p}^\wedge, \hat{m}^\vee, K)$ and $F \in \mathrm{Int}(\hat{p}^\wedge, \hat{r}^\vee, J)$ by

$$f(D) = D \triangleright S \qquad \text{and} \qquad g(F) = F \triangleright T.$$

Notice that every $D \in \mathrm{Int}(\hat{p}^\wedge, \hat{m}^\vee, K)$ is in the form $D = C \triangleright K$ for some $C \in L^p$ and that every $F \in \mathrm{Int}(\hat{p}^\wedge, \hat{r}^\vee, J)$ is in the form $F = E \triangleright J$ for some $E \in L^p$. The mappings f and g are defined correctly. Indeed,

$$f(D) = D \triangleright S = (C \triangleright K) \triangleright S = C \triangleright (K \circ S) = C \triangleright J$$

for some C, and because $C \circ J \in \mathrm{Int}(\hat{p}^\wedge, \hat{r}^\vee, J)$, we have $f(D) \in \mathrm{Int}(\hat{p}^\wedge, \hat{r}^\vee, J)$. In a similar way one obtains $g(F) \in \mathrm{Int}(\hat{p}^\wedge, \hat{m}^\vee, K)$.

Next, observe that for D, which is the form $D = C \triangleright K$ for some C,

$$g(f(D)) = ((C \triangleright K) \triangleright S) \circ T = (C \triangleright (K \circ S)) \triangleright T = \\ = (C \triangleright J) \triangleright T = C \triangleright (J \circ T) = C \triangleright K = D$$

and, similarly, $f(g(F)) = F$, proving that f and g are mutually inverse bijections. Finally, due to Lemma 2, f (and g) is a complement-preserving c-morphism. This shows that $\mathrm{Int}(\hat{p}^\wedge, \hat{m}^\vee, K) \cong \mathrm{Int}(\hat{p}^\wedge, \hat{r}^\vee, J)$, and hence $\mathrm{Int}(\hat{n}^\wedge, \hat{m}^\vee, I) \cong \mathrm{Int}(\hat{p}^\wedge, \hat{r}^\vee, J)$. □

Note that switching h for its inverse h^{-1} in Theorem 4 brings a matrix $K' \in L^{p\times m}$ such that $\mathrm{Ext}(\hat{n}^\wedge, \hat{m}^\vee, I) = \mathrm{Ext}(\hat{p}^\wedge, \hat{m}^\vee, K')$ and $\mathrm{Int}(\hat{p}^\wedge, \hat{r}^\vee, J) = \mathrm{Int}(\hat{p}^\wedge, \hat{m}^\vee, K')$. The matrix K and K' does not need to be equal. As an counterexample consider L being a chain $0 < a < b < 1$ with $\otimes$ defined as follows

$$x \otimes y = \begin{cases} x \wedge y & \text{if } x = 1 \text{ or } y = 1, \\ 0 & \text{otherwise,} \end{cases}$$

for each $x, y \in L$. One can easily see that $x \otimes \bigvee_j y_j = \bigvee_j (x \otimes y_j)$ and thus an adjoint operation $\rightarrow$ exists such that $\langle L, \wedge, \vee, \otimes, \rightarrow, 0, 1\rangle$ is a complete residuated lattice. Namely, $\rightarrow$ is given as follows:

$$x \rightarrow y = \begin{cases} 1 & \text{if } x \leq y, \\ y & \text{if } x = 1, \\ b & \text{otherwise,} \end{cases}$$

for each $x, y \in L$. Now, matrices $I = (a), J = (b) \in L^{1\times 1}$ have the same set of intents, namely $\{[b], [1]\}$. It is easy to check, that the identity on L^1 is complement-preserving c-isomorphism. We get that $K = I$ and $K' = J$; on the other hand, there is no such matrix which could stand for both K and K'. This is contrary to analogous theorem for extendable i-morphisms.

The following theorem shows that i-isomorphism between extents of two concept lattices defines concept-preserving c-isomorphism between intents of the concept lattices.

Theorem 5. *If* $h_{\mathrm{Ext}} : \mathrm{Ext}(X_1^\wedge, Y_1^\vee, I_1) \rightarrow \mathrm{Ext}(X_2^\wedge, Y_2^\vee, I_2)$ *is i-isomorphism then corresponding mapping* $h_{\mathrm{Int}} : \mathrm{Int}(X_1^\wedge, Y_1^\vee, I_1) \rightarrow \mathrm{Int}(X_2^\wedge, Y_2^\vee, I_2)$ *is complement-preserving c-isomorphism.*

We omit the proof of Theorem 5 because of lack of space.

An analogy of Theorem 5 which would read that complement-preserving c-isomorphism between intents defines an i-isomorphisms between extents does not hold. The example following Theorem 4 can be used as the counterexample.

6 Conclusions

We investigated the closure spaces induced by isotone Galois connections, i.e. mappings induced by a matrix describing a graded relationship between objects and attributes. Such mappings naturally appear in analysis of relational data. We showed that unlike the bivalent case, these spaces are just particular closure spaces, we called c-closure spaces. We provided a characterization of such closure spaces: they are exactly the closure spaces generated by a-complements of fuzzy sets. Furthermore, we defined the notion of a morphism between such closure spaces and showed that these morphisms are just the mappings generated by matrices over residuated lattices by triangular product projections. In addition, we provided a criterion of isomorphism of two c-closure spaces in terms of row and column spaces of matrices over residuated lattices. The results show that behind the methods of relational data analysis, there is a reasonable calculus of matrices over residuated lattices. The role of this calculus is analogous to the role of ordinary matrix calculus for the analysis of real-valued data using the methods based on linear algebra.

Acknowledgment. Supported by Grant No. 202/10/0262 of the Czech Science Foundation.

References

1. Belohlavek, R.: Fuzzy Relational Systems: Foundations and Principles. Kluwer, Academic/Plenum Publishers, New York (2002)
2. Belohlavek, R.: Concept lattices and order in fuzzy logic. Annals of Pure and Applied Logic 128(1-3), 277–298 (2004)
3. Belohlavek, R.: Optimal triangular decompositions of matrices with entries from residuated lattices. Int. J. Approximate Reasoning 50(8), 1250–1258 (2009)
4. Belohlavek, R.: Optimal decompositions of matrices with entries from residuated lattices. J. Logic and Computation (2011), doi:10.1093/logcom/exr023
5. Belohlavek, R., Vychodil, V.: Factor Analysis of Incidence Data via Novel Decomposition of Matrices. In: Ferré, S., Rudolph, S. (eds.) ICFCA 2009. LNCS, vol. 5548, pp. 83–97. Springer, Heidelberg (2009)
6. Ganter, B., Wille, R.: Formal Concept Analysis. Mathematical Foundations. Springer, Berlin (1999)
7. Georgescu, G., Popescu, A.: Non-dual fuzzy connections. Archive for Mathematical Logic 43, 1009–1039 (2004)
8. Green, J.A.: On the structure of semigroups. Annals of Mathematics 54(1), 163–172 (1951)
9. Gottwald, S.: A Treatise on Many-Valued Logics. Research Studies Press, Baldock (2001)
10. Hájek, P.: Metamathematics of Fuzzy Logic. Kluwer, Dordrecht (1998)
11. Kim, K.H.: Boolean Matrix Theory and Applications. M. Dekker (1982)
12. Kohout, L.J., Bandler, W.: Relational-product architectures for information processing. Information Sciences 37(1-3), 25–37 (1985)
13. Markowsky, G.: The factorization and representation of lattices. Transactions of the AMS 203, 185–200 (1975)
14. Pollandt, S.: Fuzzy Begriffe. Springer, Heidelberg (1997)
15. Popescu, A.: A general approach to fuzzy concepts. Mathematical Logic Quarterly 50(3), 1–17 (2004)
16. Ward, M., Dilworth, R.P.: Residuated lattices. Trans. Amer. Math. Soc. 45, 335–354 (1939)

Ensemble Learning and Pruning in Multi-Objective Genetic Programming for Classification with Unbalanced Data

Urvesh Bhowan, Mark Johnston, and Mengjie Zhang

Evolutionary Computation Research Group,
Victoria University of Wellington, New Zealand

Abstract. Machine learning algorithms can suffer a performance bias when data sets are unbalanced. This paper develops a multi-objective genetic programming approach to evolving accurate and diverse ensembles of non-dominated solutions where members vote on class membership. We explore why the ensembles can also be vulnerable to the learning bias using a range of unbalanced data sets. Based on the notion that smaller ensembles can be better than larger ensembles, we develop a new evolutionary-based pruning method to find groups of highly-cooperative individuals that can improve accuracy on the important minority class.

1 Introduction

Classification with unbalanced data is an important problem in machine learning (ML) [1][2][3][4]. Data sets are unbalanced when the learning examples from one class are *rare* (the minority class), while the larger class makes up the rest (the majority class). Genetic Programming (GP) is an evolutionary ML technique based on the principles of Darwinian evolution and natural selection [5], which has been successful in building reliable and accurate classifiers to solve a range of classification problems [3][6][7]. However, GP, like other ML techniques, can evolve "biased" classifiers when data is unbalanced, i.e., classifiers with strong majority class accuracy but poor minority class accuracy. As the minority class usually represents the main class in many real-world problems, building classifiers with good accuracy on both classes is an important area of research [1][2][4][6].

The learning bias can occur because typical training criteria can be influenced by the larger majority class [1]. Addressing this issue either involves sampling the data set to artificially re-balance the class distributions during the learning process [2][4], or adapting the training criteria for class-specific cost adjustment, e.g., using a weighted average of the minority and majority class accuracies in the cost function [6]. This paper focuses on cost adjustment techniques within the learning algorithm. However, as the minority and majority class accuracies are usually in conflict, selecting suitable costs for the two classes *a priori* can be problem-specific and require a lengthy trial and error process. Evolutionary multi-objective optimisation (EMO) is a useful alternative where a Pareto frontier of the best trade-off solutions can be found in a single optimisation run [8][9].

D. Wang and M. Reynolds (Eds.): AI 2011, LNAI 7106, pp. 192–202, 2011.

Another advantage to EMO is that the combined knowledge of evolved solutions along the Pareto frontier can then be utilised cooperatively in an *ensemble* of classifiers to further improve generalisation ability [9][10]. An ensemble can be more accurate than any of its individual members if the members are accurate and diverse, i.e., make different errors in different inputs [10].

This paper develops a multi-objective GP (MOGP) approach using both the accuracy and diversity of solutions along the two classes as the learning objectives. This MOGP approach uses Pairwise Failure Crediting (PFC) [10] for diversity to negatively correlate the predictions of the frontier solutions. Our first research goal evaluates the effectiveness of the ensemble when the full Pareto-front of evolved classifiers works together to predict unseen instances for five real-world (binary) class imbalance tasks. We show that the learned ensembles can be vulnerable to the learning bias due to the influence of biased Pareto-front classifiers. To address this, our second research objective develops a new ensemble-pruning method using a second evolutionary search to find small subsets of highly-cooperative individuals. This approach is shown to improve ensemble performances on the important minority class. We also compare our MOGP results to another popular ensemble learning approach, namely, Naive Bayes with bagging and balanced bootstrap sampling.

2 Related Work: Ensemble Learning for Class Imbalance

Ensemble learning for class imbalance is typically used in conjunction with sampling to either create balanced bootstrap samples in bagging approaches [2][3] or re-balance the training data in EMO-based approaches using diversity measures in fitness [4]. However, sampling can incur a computational overhead, particularly in large data sets with high levels of imbalance, and some sampling techniques (such as under-samping) can potentially exclude useful training examples from the learning process. We use the multi-objective component for cost adjustment and the original unbalanced training data "as is" during learning.

Recent EMO-based approaches to evolving ensembles use Negative Correlation Learning (NCL) for ensemble diversity [4][7]. In [4], NCL is only applied to minority class instances (majority class instances are ignored) to evolve diverse neural network ensembles; while in [7], NCL serves as the secondary fitness measure in an MOGP approach where the Pareto front is determined using only the accuracy of the GP classifiers. This paper is different as a population-based diversity measure is used in fitness (PFC) which allows for equal selection preference between accurate and diverse solutions, potentially creating better diversity in the population. The PFC measure is also applied to both the minority and majority class separately where each contributes equally in fitness to ensure the ensemble members are equally diverse on both classes. Although we use PFC for ensemble diversity, sampling (such as over or under-sampling) may also be incorporated into the learning approach for ensemble diversity (such as [3]).

In [2] and [11], different ensemble pruning techniques are explored. In [2], a genetic algorithm evolves a set of weights to specify the contribution of individuals in the ensemble, using a separate validation set to learn these weights (in addition to the training set to generate the ensemble). In [11], an expectation propagation algorithm using Bayesian inference models is used to concurrently learn the optimal set of weights while also training the ensemble members. Although both of these are effective, a limitation of weight-based ensemble pruning is that suitable weights must be configured for *all* ensemble members. In contrast, this paper develops a GP-based pruning method that quickly explores different combinations of small subsets of individual only, using the original training set.

3 Multi-Objective GP (MOGP) for Evolving Ensembles

This paper develops a multi-objective GP approach to simultaneously evolving a *Pareto frontier* of GP classifiers along the objectives (minority and majority class accuracy) in a single optimisation run. An advantage of evolving a front of the best trade-off solutions is that the combined knowledge of these classifiers on the objectives can then be shared and used co-operatively in an *ensemble*. In an ensemble of classifiers, each member votes on the class label to assign to a given data instance, where the class label with the most votes determines the final ensemble prediction. Ensembles can have good generalisation ability and perform *better* than all of its individual members provided that the individuals are both accurate and diverse, i.e., generate different errors on different inputs [9][10]. However, if the individual members are not sufficiently accurate and diverse then the ensemble risks misclassifying all the same inputs together. For this reason, an explicit diversity measure in fitness is used to improve diversity between solutions so that if one individual generates an error for a given input, the other members do not also make the same error.

3.1 GP Framework for Classification

A tree-based structure is used to represent the genetic program solutions [5]. We use feature terminals (example features) and constant terminals (randomly generated floating point numbers), and a function set comprising of the four standard arithmetic operators, $+, -, \%$, and $\times$, and the conditional operator `if`. The $+,-$ and $\times$ operators have their usual meanings (addition, subtraction and multiplication) while % is *protected* division (usual division except that a divide by zero returns zero). The conditional `if` function takes three arguments and returns either the second argument if the first is negative, or the third argument otherwise. Each GP solution represents a mathematical expression that outputs a (floating-point) number for a given input (data example to be classified). This number is mapped to the class labels using zero as the threshold, i.e., *minority* class if the classifier output is zero or positive, or *majority* class otherwise.

3.2 MOGP Fitness

The objective performances of an evolved solution reflects both the accuracy and diversity of the solution on each of the two classes, minority and majority class. This is expressed by Eq. (1) for solution S_i on class c, where $Err_{c,i}$ is the total number of incorrect predictions in class c (by solution i) and N_c is the number of training examples in class c. An incorrect prediction occurs when the predicted and actual class labels differ for a given input. Weighting coefficient W specifies the trade-off between accuracy and diversity, where $W = 0.5$ is used to treat these two measures as equally important in the evolution. This gives equal selection preference to accurate and diverse solutions. The diversity estimate for solution i is represented by $PFC_{c,i}$ for all examples in class c, calculated using Pairwise Failure Crediting (PFC) [10]. PFC represents a penalty function in fitness to reduce the overlap of common errors between solutions in the population. In the PFC measure, T is population size and the indicator function $I(\cdot)$ returns 1 if the class labels returned from two solutions, i and j, are the same for the given input instance p, or 0 otherwise. In Eq. (1), both the objective performance S_i and PFC measure return values between 0 and 1; for PFC, the higher the value the better the diversity, and likewise for S_i where higher objective performances imply better accuracy and diversity.

$$\begin{aligned} (S_i)_c &= W\left(\tfrac{1-Err_{c,i}}{N_c}\right) + (1-W)PFC_{c,i} \\ PFC_{c,i} &= \tfrac{1}{T-1}\sum_{j=1,j\neq i}^{T} \tfrac{\sum_{p=1}^{N_c} I(i,j,p)}{Err_{c,i}+Err_{c,j}} \end{aligned} \quad (1)$$

Ranking the Objectives. Pareto dominance in fitness *ranks* the solutions in the population according to objective performances. This ranking is important as it affects the way selection is performed if the different objectives are to be treated separately in the evolution. Pareto dominance between two solutions, expressed by Eq. (2), states that a solution will *dominate* another solution if it is at least as good as the other solution on all the objectives and better on at least one. Solutions are *non-dominated* if they are not dominated by any other solution in the population.

$$S_i \succ S_j \longleftrightarrow \forall c[(S_i)_c \geq (S_j)_c] \wedge \exists k[(S_i)_k > (S_j)_k] \quad (2)$$

Our MOGP approach uses the popular and effective Pareto dominance-based EMO algorithm SPEA2 [8]. This algorithm is shown to evolve an accurate fronts of classifiers along the minority and majority class trade-off surface in these class imbalance tasks [7]. In SPEA2, each solution in the population is first assigned a *strength* value D based on the number of other solutions it dominates in the population. The final SPEA2 fitness value, Eq. (3) for solution S_i, is the sum of the strength values of all S_i's dominators, i.e., all other solutions in the population that dominate S_i. The lower the fitness value returned by Eq. 3, the better the solution on the objectives where non-dominated solutions in the population have the best fitness value of 0 (these solutions have no dominators).

$$\begin{aligned} fitness(S_i) &= \textstyle\sum_{j \in Pop, S_i \succ S_j} D(S_j) \\ D(S_i) &= |\{j | j \in Pop \wedge S_i \succ S_j\}| \end{aligned} \quad (3)$$

Table 1. Unbalanced classification tasks used in the experiments

Name	Classes (Minority/Majority)	Examples Total	Examples Minority	Imb. Ratio	Features No.	Features Type
Ion	Good/bad (ionosphere radar signal)	351	126 (35.8%)	1:3	34	Real
Spt	Abnormal/normal (tomography scan)	267	55 (20.6%)	1:4	22	Binary
Yst$_1$	*mit*/non-target (protein sequence)	1482	244 (16.5%)	1:6	8	Real
Yst$_2$	*me3*/non-target (protein sequence)	1482	163 (10.9%)	1:9	8	Real
Bal	Balanced/unbalanced (balance scale)	625	49 (7.8%)	1:12	4	Integer

3.3 MOGP Search

In SPEA2, the parent and offspring populations are merged together at every generation [8]. This combined parent-child population is sorted by fitness values where the fittest individuals are copied into a new population, called the *archive population.* The archive serves as the parent population in the next generation, and preserves elitism in the population over generations. The offspring population at every generation is generated using the traditional crossover and mutation genetic operators using binary tournament selection. At the end of the evolutionary cycle, the set of non-dominated solutions in the population represents the evolved Pareto-approximated front of classifiers. A *majority vote* of the class labels returned from the evolved set of non-dominated solutions (for a given input instance) determines the final ensemble output.

4 MOGP Ensemble Performance

In this section we outline the evolutionary parameters and unbalanced data sets used in the experiments, and evaluate the MOGP ensemble performances.

4.1 Evolutionary Parameters and Unbalanced Data Sets

The population size was 500, crossover and mutation rates were 60% and 40%, respectively, and the maximum program depth was 8 to restrict very large programs in the population. The evolution ran for 50 generations. Five benchmark binary classification problems taken from the *UCI Repository of Machine Learning Databases* [12], summarised in Table 1, are used in the experiments. These reflect classification tasks with varying levels of complexity and class imbalance. Half of the examples in each class are randomly chosen for the *training* and the *test* sets, to ensure that both sets preserve the same class imbalance ratio. While it is possible that the class distributions in the training set and test set are different, we only consider tasks with similar distributions in both sets.

4.2 MOGP Ensemble Results

Table 2 shows the average minority and majority class accuracies (± standard deviation) of the evolved ensembles, and the average ensembles sizes, on the *test*

Table 2. Average MOGP ensemble performances and sizes over 50 runs, and Naive Bayes (NB) ensemble using bagging (with balanced bootstrap sampling)

	MOGP Full-Front Ensemble			MOGP Pruned Ensemble			NB (Bagging)		
	Size	Minority	Majority	Size	Minority	Majority	Size	Minority	Majority
Ion	28.1	84.9 ± 5.1	92.4 ± 6.4	22.3	81.7 ± 5.8	95.8 ± 3.8	25	88.9	62.5
Spt	27.3	44.6 ± 5.4	90.8 ± 2.3	12.1	62.1 ± 8.0	80.5 ± 4.8	25	70.4	77.4
Yst_1	39.7	64.6 ± 4.8	82.5 ± 4.3	16.5	71.0 ± 4.4	75.5 ± 5.4	25	73.8	78.7
Yst_2	27.9	81.2 ± 4.9	95.5 ± 1.5	20.6	89.2 ± 3.2	92.3 ± 1.8	25	87.7	92.6
Bal	20.8	51.7 ± 18.2	95.4 ± 3.5	10.1	83.6 ± 9.4	79.5 ± 10.3	25	29.2	50.7

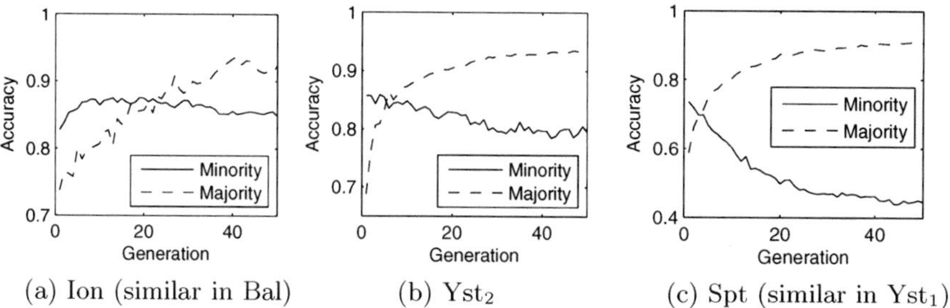

(a) Ion (similar in Bal) (b) Yst_2 (c) Spt (similar in Yst_1)

Fig. 1. MOGP ensemble performances using full Pareto front over generations

sets over 50 runs. Also included are the ensemble results for (a single run of) Naive Bayes (NB) using bagging with 25 balanced bootstrap samples [13]. The "full-front" ensemble results use *all* evolved non-dominated solutions (from a MOGP run) in the voting process; these results show that majority class accuracy is always higher than minority class accuracy in all tasks. The corresponding minority class accuracies are still reasonably good in some tasks (Ion and Yst_2), while in the others (Spt and Bal) these are poor. This shows that the evolved fronts can contain more solutions biased toward the majority class than the opposite case (solutions with good minority accuracy or middle-region solutions), as these solutions influence the ensemble vote in most tasks.

Analysis of the ensemble performances during the evolution reveals that this may be due to genetic drift in the population, toward non-dominated solutions biased toward the majority class objective. As the evolution advances over generations, *more* solutions with strong majority class accuracies achieve non-dominated status than solutions with good minority accuracies or middle-region solutions. This effect can be seen in Figure 1 to varying degrees. These figures show the average minority and majority class performance of the ensemble for 50 generations (over 50 runs on the test sets for three tasks). Figure 1 clearly shows that *more* solutions with stronger majority class accuracy (than solutions with stronger minority accuracy) are included in the ensemble over generations, as the ensemble accuracy simply reflects which class receives the most votes from the different members. In the remaining tasks (omitted for space constraints), Bal shows similar behaviour to Ion (Figure 1.a) and Yst_1 to Spt (Figure 1.c).

Pruning the Ensemble. To address the biased ensemble behaviour, a simple accuracy-based selection strategy is used to prune the MOGP ensembles to reduce the influence of biased non-dominated solutions on the ensemble vote. This strategy only selects non-dominated solutions with at least 50% accuracy on both objectives for the ensembles. The pruned ensemble performances and sizes, reported in Table 2, show that more balanced class performances are achieved, with noticeably better minority class accuracies, compared to the full-front results in all tasks (except Ion). The trade-off in majority class accuracy is relatively small in some tasks (Yst_1 and Yst_2) compared to others (Spt and Bal). These results show that the full ensembles are vulnerable to the learning bias in the unbalanced tasks, while the pruned ensembles can be better for the important minority class. The pruned MOGP ensembles also compare well to NB with bagging, outperforming NB in the two most unbalanced tasks (Yst_2 and Bal).

5 Ensemble Pruning

As the pruned ensembles show more balanced class performances and better minority class accuracies, in this section we investigate the effects of further pruning to create smaller ensembles. We develop two pruning methods to investigate whether smaller ensembles can be better for the unbalanced tasks. Although these pruning methods are developed in the context of the MOGP approach, they are not restricted to MOGP ensembles and can also be used in conjunction with any underlying ensemble learning algorithm.

5.1 Fitness-Based Pruning

In this pruning method, the non-dominated solutions are sorted according to their raw fitness values on the *training objectives* (from Eq. 1) and only the *best* (fittest) N are selected for the ensemble. Configuring N controls the pruned ensemble's size. As there are two objectives (minority and majority class accuracy), the average of these objective values is used as the final fitness value, to include only highly accurate and diverse solutions in the ensemble.

5.2 GP for Evolving Composite Voting Trees

As the fitness-based pruning strategy uses a linear ordering of the fittest N solutions for ensemble selection, this method does not guarantee that the overlap of common errors between the fittest N solutions are minimal with respect to *each other* only. A more robust ensemble-selection method explores different *combinations* of solution-subsets which are highly-diverse with respect to each other only. Let $X = \{p_1, p_2, ..., p_m\}$ be a set of m non-dominated individuals. The function $div(Y)$ calculates the diversity (i.e. overlap of common errors) between individuals in subset $Y \subseteq X$. In order to find the solution-subset with the best diversity we must compare the $div(Y)$ values for all possible subsets of

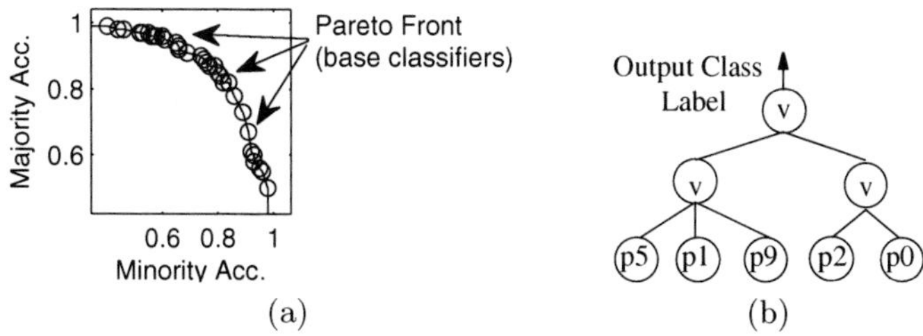

(a) (b)

Fig. 2. (a) Evolved MOGP Pareto front (small circles are classifiers) and (b) evolved CVT solution (depth 3) which uses a subset of MOGP classifiers

X, i.e., $\{p_1, p_2\}$, $\{p_1, p_3\}$,$\{p_1, p_2, p_3\}$ etc. Exploring all possible combinations of subsets of X is a computationally expensive and time-consuming combinatorial problem, particularly for large ensembles and data sets, as each $div(Y)$ estimate uses at least one pass through the full training set. To address this, we develop a GP-based search to efficiently explore this space of possible combinations, to quickly find diverse subsets of non-dominated solutions for the ensemble which are maximally diverse with respect to each other. The GP approach for ensemble pruning takes as input the evolved set of non-dominated classifiers returned from the MOGP search (called the base classifiers) as shown in Figure 2(a), and evolves a *composite voting tree* (CVT) representing a small subset of base classifiers that are highly-diverse and accurate when combined together in the ensemble voting process, as shown in Figure 2(b).

Representation. Tree-based GP is used to represent a CVT solution as shown in Figure 2(b). Each terminal node P_n represents a link to the nth base classifier in the input set (non-dominated MOGP classifiers), similar to feature terminals in MOGP. The root node of a CVT solution outputs a class label, determined by a majority vote of the *predictions* of each base classifier (terminal node) in the tree; this is the only component in a CVT tree which computes a value. Recall that the prediction of a MOGP base classifier will be *minority class* if the base classifier's output is non-negative or *majority class* otherwise. The new function v serves no purpose other than to join terminal nodes to the root node or other v nodes, where v can take any number of arguments between 1 and 3; this allows different CVT solutions to contain varying numbers of base classifiers.

Fitness Function. The output of a CVT solution (when evaluated on a given input instance) corresponds to the pruned ensemble output (class label) whose members are represented in the CVT solution. The fitness function calculates the average classification accuracy of the minority and majority class when each CVT solution is evaluated on the training set, aimed to evolve CVTs with good classification accuracy on both classes.

Evolutionary Search. The search process is akin to canonical (single-objective) GP where the fittest CVT solution in the population is returned from the evolution. Crossover, mutation and elitism rates are 60%, 35% and 5%, respectively, and the tournament selection size is 7. The evolution is limited to 50 generations

Table 3. Average MOGP ensemble performances and sizes for two pruning methods (∘ symbol shows the *dominating* pruning method for similar-sized ensembles)

	Fitness-based Pruning			CVT-based Pruning			
	Size	Minority	Majority	Size	Minority	Majority	
Ion	3	70.5 ± 8.4	91.3 ± 8.9	3.0	77.7 ± 6.1	91.9 ± 4.6	∘
	9	75.7 ± 7.2	94.8 ± 4.4	8.9	80.4 ± 5.5	94.3 ± 4.6	∘
Spt	3	68.1 ± 8.6	70.8 ± 5.4	3.0	72.7 ± 9.6	64.7 ± 10.6	
	9	64.2 ± 7.9	77.3 ± 4.7	7.5	58.4 ± 7.8	82.1 ± 4.1	
Yst_1	3	79.8 ± 15.2	53.8 ± 21.1	3.0	79.1 ± 15.4	59.2 ± 16.2	∘
	9	76.3 ± 13.7	61.4 ± 17.0	9.0	77.8 ± 6.1	66.7 ± 6.4	∘
Yst_2	3	95.3 ± 3.9	74.9 ± 6.5	3.0	95.3 ± 2.2	83.2 ± 4.4	∘
	9	93.0 ± 2.5	81.3 ± 4.6	9.0	93.5 ± 3.0	86.3 ± 3.9	∘
Bal	3	78.3 ± 14.4	76.6 ± 13.1	3.0	84.5 ± 10.3	76.3 ± 13.4	∘
	9	76.0 ± 14.4	81.6 ± 11.4	7.9	78.8 ± 10.6	85.5 ± 7.9	∘

unless a CVT solution with 100% accuracy on both classes on the training set is evolved, at which point the evolution is stopped. A population size of 1000 is used. To focus the evolution toward discovery of small but highly-effective CVT solutions, two maximum tree depths are compared, 2 and 3, to restrict the number of base classifiers in each solution. When tree depth is limited to 2, an evolved CVT solution represents a pruned ensemble of at most 3 members; similarly, a tree depth of 3 represents a pruned ensemble of at most 9 members.

5.3 Performance of Ensembles Using Puning Methods

For a fair comparison between these two ensemble-selection methods, we compare ensemble performances when a *similar* number of base classifiers is returned by the different selection methods, i.e., pruned ensembles limited to (at most) 3 and 9 members (odd-numbered ensemble sizes are preferred as *no draws* can occur in the voting process). This allows for a comparison of which selection method finds *more effective* (more accurate) subsets of base classifiers in the pruned ensemble, as well as an investigation of ensemble behaviour when fewer base classifiers are used in the ensemble voting (compared to the initial ensemble results from Table 2). To generate pruned ensembles limited to (at most) 3 and 9 members, the CVT-based pruning method uses a maximum CVT tree-depth of 2 and 3, respectively. Table 3 reports the performances and sizes of the pruned ensembles using the two pruning methods, i.e., fitness-based and CVT-based pruning, on the *test* sets over 50 independent runs; these correspond to the initial 50 MOGP experiments to generate the full ensembles (from Table 2).

Table 3 shows that the CVT-pruned ensembles outperform (i.e. dominate) the fitness-pruned ensembles for both the smallest (at most 3 members) and intermediate-sized (at most 9 members) ensembles in nearly all tasks (except Spt). This suggests that the *quality* of the base classifiers found using the CVT method is better than the fitness-based selection method, as these base classifiers improve the predictive ability of the pruned ensembles due to better cooperation between individuals. The evolutionary search to discover good CVTs is reasonably fast, taking between 0.2 and 5 seconds on the tasks (2–3% of the training time to evolve a MOGP front).

Comparing the pruned ensembles in Table 3 to the ensemble results from Table 2 shows that in nearly all tasks (except Ion for the MOGP ensemble and Yst_2 for NB), the smaller the ensemble, the *better* the minority class accuracies but the poorer the majority class accuracies. The smallest MOGP ensemble is dominated by the larger ensembles in only one task, Ion, also the least unbalanced task. This suggests that in these unbalanced tasks, the pruned ensembles are better than the larger ensembles but only for the minority class. The poorer majority class accuracies may be due to over-fitting from the secondary training phase. However, further investigation is required for future work.

6 Conclusions

The main goal of this paper develops a MOGP approach to classification with unbalanced data to evolve an accurate and diverse ensemble of non-dominated solutions along the minority and majority class trade-off frontier. We also compare ensemble behaviour using the full non-dominated set of solutions to smaller pruned ensembles, and develop a new pruning method to find small subsets of highly-cooperative individuals. Our goals were achieved by examining the classification performance of the full and pruned MOGP-evolved ensembles on five unbalanced (binary) tasks.

We show that the full MOGP ensembles is vulnerable to the learning bias due to the influence of more Pareto front solutions with stronger majority class accuracies (than solutions with good minority class accuracies). As the ensembles sizes are reduced, the pruned MOGP ensembles show better accuracies on the important minority class but not the majority class in the unbalanced tasks. The new GP-based ensemble pruning method finds highly-cooperative individuals for the pruned MOGP ensembles, as these have better accuracy on both classes compared to a fitness-based selection method for pruning on these tasks.

For future work we will investigate these methods on more unbalanced data sets and compare our results to canonical (single-objective) GP with different fitness functions for classification with unbalanced data. We also will investigate how the two new pruning techniques treat diversity in the pruned ensembles.

References

1. Weiss, G.M., Provost, F.: Learning when training data are costly: The effect of class distribution on tree induction. Journal of Artificial Intelligence Research 19, 315–354 (2003)
2. Chawla, N.V., Sylvester, J.: Exploiting Diversity in Ensembles: Improving the Performance on Unbalanced Datasets. In: Haindl, M., Kittler, J., Roli, F. (eds.) MCS 2007. LNCS, vol. 4472, pp. 397–406. Springer, Heidelberg (2007)

3. McIntyre, A., Heywood, M.: Multi-objective competitive coevolution for efficient GP classifier problem decomposition. In: IEEE International Conference on Systems, Man and Cybernetics, pp. 1930–1937 (2007)
4. Wang, S., Tang, K., Yao, X.: Diversity exploration and negative correlation learning on imbalanced data sets. In: International Joint Conference on Neural Networks, pp. 3259–3266 (2009)
5. Koza, J.R.: Genetic Programming: On the Programming of Computers by Means of Natural Selection. MIT Press (1992)
6. Holmes, J.H.: Differential negative reinforcement improves classifier system learning rate in two-class problems with unequal base rates. In: Koza, J.R., Banzhaf, W., Chellapilla, K., et al. (eds.) Genetic Programming 1998: Proceedings of the Third Annual Conference, pp. 635–644 (1998)
7. Bhowan, U., Zhang, M., Johnston, M.: Evolving ensembles in multi-objective genetic programming for classification with unbalanced data. In: Proceedings of 2011 Genetic and Evolutionary Computation Conference, pp. 1331–1339. ACM (2011)
8. Zitzler, E., Laumanns, M., Thiele, L.: Spea2: Improving the strength pareto evolutionary algorithm for multiobjective optimization. Technical report (2001), TIK-Report 103, Department of Electrical Engineering, Swiss Federal Institute of Technology
9. Jin, Y., Sendhoff, B.: Pareto-based multiobjective machine learning: An overview and case studies. IEEE Transactions on Systems, Man, and Cybernetics, Part C: Applications and Reviews 38, 397–415 (2008)
10. Chandra, A., Yao, X.: Ensemble learning using multi-objective evolutionary algorithms. Journal of Mathematical Modelling and Algorithms 5, 417–445 (2006)
11. Chen, H., Tino, P., Yao, X.: Predictive ensemble pruning by expectation propagation. IEEE Transactions on Knowledge and Data Engineering 21, 999–1013 (2009)
12. Asuncion, A., Newman, D.: UCI Machine Learning Repository, University of California, Irvine, School of Information and Computer Sciences (2007), `http://www.ics.uci.edu/~mlearn/MLRepository.html`
13. Hall, M., Frank, E., Holmes, G., Pfahringer, B., Reutemann, P., Witten, I.H.: The WEKA data mining software: An update. SIGKDD Explorations 11 (1) (2009)

Compiling Bayesian Networks for Parameter Learning Based on Shared BDDs

Masakazu Ishihata[1], Taisuke Sato[1], and Shin-ichi Minato[2]

[1] Tokyo Institute of Technology, Tokyo152-8552, Japan
{ishihata,sato}@mi.cs.titech.ac.jp
[2] Hokkaido University, Sapporo 060-0814, Japan
minato@ist.hokudai.ac.jp

Abstract. Compiling Bayesian networks (BNs) is one of the most effective ways to exact inference because a logical approach enables the exploitation of local structures in BNs (i.e., determinism and context-specific independence). In this paper, a new parameter learning method based on compiling BNs is proposed. Firstly, a target BN with multiple evidence sets are compiled into a single *shared binary decision diagram* (SBDD) which shares common sub-graphs in multiple BDDs. Secondly, all conditional expectations which are required for executing the EM algorithm are simultaneously computed on the SBDD while their common local probabilities and expectations are shared. Due to these two types of sharing, the computation efficiency of the proposed method is higher than that of an EM algorithm which naively uses an existing BN compiler for exact inference.

1 Introduction

Bayesian networks (BNs) are directed acyclic graphs which represent a joint distribution over random variables and have established and used for representing uncertain knowledge across a number of fields. Recently, compiling BNs has been attracting much attention as one of the most effective approaches to exact inference. BNs can be characterized by multi-linear functions (MLFs) [5] and logic-based BN compilation approaches can factorize them exploiting their local structures [3] (i.e., determinism and context-specific independence [1]). Also, compiling BNs and exploiting their local structures are effective for parameter learning [3,2]. For example, the EM algorithm [6] which is a popular parameter learning algorithm from incomplete data requires the computation of a large number of conditional probabilities conditioned on *evidence sets* (observations). Compiling a target BN into a structure which is convenient for exact inference speeds up the computation. Exploiting local structures reduces the number of parameters that have to be learned. However, existing BN compilation approaches to inference still have redundancy from the viewpoint of parameter learning because they compute multiple conditional probabilities separately despite the fact that there are common local probabilities. In addition, they do not consider multiple evidence sets when they compile a target BN even if those are given beforehand in learning unlike in inference. As one way of utilizing the information of given evidence sets, an approach which compiles a target BN with a *single* evidence set is proposed [2]. However, since

D. Wang and M. Reynolds (Eds.): AI 2011, LNAI 7106, pp. 203–212, 2011.

it is not designed for dealing simultaneously with *multiple* evidence sets, it is unable to share local probabilities even if they have common ones.

To fully utilize evidence sets given as observations, we propose a new parameter learning method based on BN compilation. The proposed method compiles a target BN with given evidence sets into a single *shared binary decision diagram* (SBDD) [8] and then executes the EM algorithm in a manner of dynamic programming on the SBDD. The advantages of our method are twofold. Firstly, it compresses multiple evidence sets into an SBDD while their common sub-graphs are shared, i.e., it never builds up the same sub-graph twice. Secondly, it simultaneously computes all *conditional expectations* (which are required for executing the EM algorithm) on the SBDD while their common local probabilities and expectations are shared, i.e., it never computes the same quantities over and over again. These advantages become especially apparent when the EM algorithm itself is repeated multiple times with different initial parameters, because the SBDD is constructed only once in the beginning of the whole learning process, i.e., it is never reconstructed even if parameters are changed.

In this paper, we compared our proposed method with an EM algorithm which naively uses the state-of-the-art BN compiler ACE [3]. We applied them to two types of randomly generated BNs, one without local structures and the other one with those. The result shows that the computational speed of our method is superior to the naive method in both types of BNs. It supports our claim that sharing common sub-graph and local probabilities is an effective approach to the parameter learning problem for BNs.

The rest of the paper is organized as follows: In Section 2, we briefly review BNs and formalize the parameter learning problem for them. Section 3 describes a new parameter learning method based on BN compilation. Experimental results are presented in Section 4. Finally, we describe conclusion and related work in Section 5.

2 Preliminary

In this section, we briefly review Bayesian networks (BNs) and formalize the parameter learning problem for them.

2.1 Bayesian Networks

A Bayesian network (BN) defines a joint distribution over a set of random variables $\boldsymbol{X} \equiv \{X_1, \ldots, X_N\}$ where X_i has a discrete finite domain. A BN consists of a *directed acyclic graph* (DAG) G representing conditional independence and a set of *conditional probability distributions* (CPDs). Each node in G is labeled by a random variable X_i and has a CPD $P(X_i \mid \Pi_i)$ where Π_i is its parent variables. Let x_i and π_i be a value of X_i and a value vector of Π_i, respectively. Then, a joint probability $P(x_1, \ldots, x_N)$ can be computed as a product of $P(x_i \mid \pi_i)$ $(1 \leq i \leq N)$ as $P(x_1, \ldots, x_N) \equiv \prod_{i=1}^{N} P(x_i \mid \pi_i)$, where $P(x_1, \ldots, x_N)$ and $P(x_i \mid \pi_i)$ are shorthands for $P(X_1 = x_1, \ldots, X_N = x_N)$ and $P(X_i = x_i \mid \Pi_i = \pi_i)$, respectively.

For example, Figure 1 shows a BN representing a joint distribution over $\boldsymbol{X} \equiv \{X_1, X_2, X_3\}$ where $X_i \in \boldsymbol{X}$ is a binary random variable. In the example, CPDs are expressed as *conditional probability tables* (CPTs) whose elements are called *parameters*.

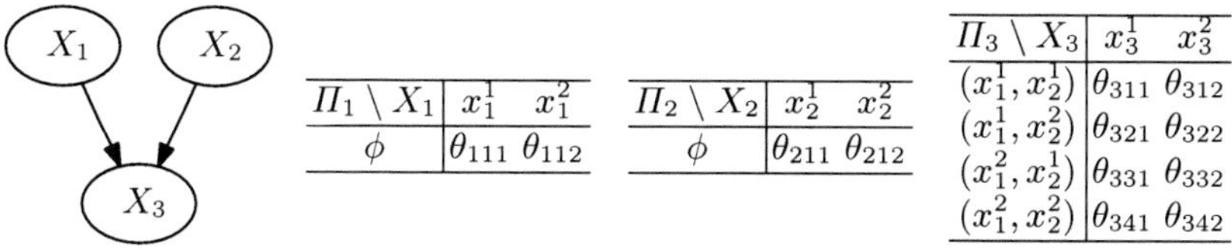

$\Pi_1 \setminus X_1$	x_1^1	x_1^2
ϕ	θ_{111}	θ_{112}

$\Pi_2 \setminus X_2$	x_2^1	x_2^2
ϕ	θ_{211}	θ_{212}

$\Pi_3 \setminus X_3$	x_3^1	x_3^2
(x_1^1, x_2^1)	θ_{311}	θ_{312}
(x_1^1, x_2^2)	θ_{321}	θ_{322}
(x_1^2, x_2^1)	θ_{331}	θ_{332}
(x_1^2, x_2^2)	θ_{341}	θ_{342}

Fig. 1. A BN for X_1, X_2 and X_3

A parameter θ_{ijk} corresponds to a conditional probability $P\left(x_i^k \mid \pi_i^j\right)$, where x_i^k is the k-th value of X_i and π_i^j is the j-th value vector of Π_i. An *evidence set* is a (partial) assignment of $\boldsymbol{X}$ and its probability is computed from parameters $\boldsymbol{\theta} \equiv \{\theta_{ijk}\}_{ijk}$. For example, the probability of an evidence set $e_1 \equiv (x_1^1, x_2^1, x_3^1)$ parameterized by $\boldsymbol{\theta}$ is computed as follows:

$$
\begin{aligned}
P(e_1; \boldsymbol{\theta}) &= P(x_1^1, x_2^1, x_3^1; \boldsymbol{\theta}) \\
&= P(x_1^1 \mid \phi)\, P(x_2^1 \mid \phi)\, P(x_3^1 \mid x_1^1, x_2^1) \\
&= \theta_{111}\theta_{211}\theta_{311}.
\end{aligned}
$$

According to [5], a BN can be characterized by a multi-linear function (MLF) which consists of two types of variables, parameters $\boldsymbol{\theta}$ and *indicators* $\boldsymbol{\lambda} \equiv \{\lambda_{ik}\}_{ik}$, where λ_{ik} represents that X_i takes its k-th value x_i^k. For example, the MLF for the BN in Figure 1 is defined as follows:

$$
\begin{aligned}
MLF \equiv\ & \lambda_{11}\lambda_{21}\lambda_{31}\theta_{111}\theta_{211}\theta_{311} + \lambda_{11}\lambda_{21}\lambda_{32}\theta_{111}\theta_{211}\theta_{312} \\
&+ \lambda_{11}\lambda_{22}\lambda_{31}\theta_{111}\theta_{212}\theta_{321} + \lambda_{11}\lambda_{22}\lambda_{32}\theta_{111}\theta_{212}\theta_{322} \\
&+ \lambda_{12}\lambda_{21}\lambda_{31}\theta_{111}\theta_{211}\theta_{331} + \lambda_{12}\lambda_{21}\lambda_{32}\theta_{111}\theta_{211}\theta_{332} \\
&+ \lambda_{12}\lambda_{22}\lambda_{31}\theta_{111}\theta_{212}\theta_{341} + \lambda_{12}\lambda_{22}\lambda_{32}\theta_{111}\theta_{212}\theta_{342}.
\end{aligned}
$$

Using the above MLF, the probability of an evidence set e is computed by setting indicators which contradict e to 0 and the others to 1. For example, the probability of e_1 is computed by setting $\lambda_{11}, \lambda_{21}$ and λ_{31} to 1 and the others to 0.

Our task in this paper is to learn parameters $\boldsymbol{\theta}$ of a target BN from evidence sets (observations) when its DAG structure G is given.

2.2 Parameter Learning Problem for BNs

We here formally define the parameter learning problem for BNs. We assume that a DAG structure G of a target BN and a date set $\boldsymbol{E}$ are given. $\boldsymbol{E}$ is a vector $\{e_t\}_{t=1}^{T}$ where e_t is the t-th evidence set observed from the target BN independently. The problem is to find a *maximum likelihood estimate* $\boldsymbol{\theta}^* \equiv \mathrm{argmax}_{\boldsymbol{\theta}}\, \mathcal{L}(\boldsymbol{\theta}|\boldsymbol{E})$, where $\mathcal{L}(\boldsymbol{\theta}|\boldsymbol{E})$ is the log likelihood defined as $\mathcal{L}(\boldsymbol{\theta}|\boldsymbol{E}) \equiv \log P(\boldsymbol{E}; \boldsymbol{\theta}) = \sum_{t=1}^{T} \log P(e_t; \boldsymbol{\theta})$. If each evidence set in $\boldsymbol{E}$ is a complete instantiation of $\boldsymbol{X}$, $\boldsymbol{E}$ is called *complete data*, otherwise called *incomplete data*. In the case of complete data, $\boldsymbol{\theta}^*$ is readily computed as follows:

$$
\boldsymbol{\theta}^* \equiv \{\theta_{ijk}^*\}_{ijk}, \qquad\qquad \theta_{ijk}^* \equiv N_{ijk} / \textstyle\sum_{k'} N_{ijk'},
$$

where N_{ijk} is the number of evidence sets in $\boldsymbol{E}$ which satisfy a proposition $A_{ijk} \equiv$ "$X_i = x_i^k, \Pi_i = \pi_i^j$".

On the other hand, it is not so easy to find $\boldsymbol{\theta}^*$ from incomplete data because the exact N_{ijk} is not computable. Instead of N_{ijk}, the EM algorithm [6], which is a way to find $\boldsymbol{\theta}^*$ from incomplete data, uses the expectation of N_{ijk}. It first randomly initializes parameters $\boldsymbol{\theta}^{(0)}$. Then, it updates parameters by the following E- and M-step.

E-step: Computes *conditional expectations* $E_{ijk}^{(m)}$ for all θ_{ijk} defined as:

$$E_{ijk}^{(m)} \equiv \mathrm{E}[N_{ijk}]_{P(\boldsymbol{X}|\boldsymbol{E};\boldsymbol{\theta}^{(m)})} = \sum_{e_t \in \boldsymbol{E}} P\left(x_i^k, \pi_i^j \mid e_t; \boldsymbol{\theta}^{(m)}\right). \tag{1}$$

M-step: Updates $\boldsymbol{\theta}^{(m)}$ to $\boldsymbol{\theta}^{(m+1)}$ by

$$\theta_{ijk}^{(m+1)} \equiv E_{ijk}^{(m)} / \textstyle\sum_{k'} E_{ijk'}^{(m)}. \tag{2}$$

The EM algorithm iterates the above two steps until parameters converge and then outputs converged $\boldsymbol{\theta}^{(m)}$ as an estimate of $\boldsymbol{\theta}^*$. It is ensured that the parameters estimated by the EM algorithm give a local maximum. To find the global maximum, the EM algorithm itself is usually repeated several times changing the initial parameters $\boldsymbol{\theta}^{(0)}$. According to the definition of the E-step, as many as $|\boldsymbol{\theta}| \times |\boldsymbol{E}|$ conditional probabilities would be computed in every E-step. As a matter of course, these probabilities are computable by existing BN compilation approaches to exact inference, e.g., ACE [3]. However, computing them separately incurs redundancy because they have common local probabilities. For example, $P(x_i^k, \pi_i^j | e_t; \boldsymbol{\theta}^{(m)})$ and $P(x_{i'}^{k'}, \pi_{i'}^{j'} | e_t; \boldsymbol{\theta}^{(m)})$ have those as they are conditioned by the same evidence set e_t. Also, $P(x_i^k, \pi_i^j | e_t; \boldsymbol{\theta}^{(m)})$ and $P(x_i^k, \pi_i^j | e_{t'}; \boldsymbol{\theta}^{(m)})$ have those since they correspond to the same probabilistic event "$X_i = x_i, \Pi_i^j = \pi_i^j$". To avoid such computational redundancy, we propose a new parameter learning method based on BN compilation. The method directly computes conditional expectations $E_{ijk}^{(m)}$ for all parameters θ_{ijk} while their common local probabilities and expectations are shared. Specially, two types of sharing are possible. One is over evidence sets $\boldsymbol{E}$ and the other is over parameters $\boldsymbol{\theta}$. The detail of our method will be described in the following section.

3 Proposed Method

In this section, we propose an efficient EM algorithm for BNs based on BN compilation. Firstly, it encodes a data set $\boldsymbol{E}$ into boolean formulas and compiles them into a single *shared binary decision diagram* (SBDD). Then, it learns parameters $\boldsymbol{\theta}$ in a manner of dynamic programming on the SBDD. The following is the detail of each step.

3.1 Encoding and Compiling

An evidence set $e_t \in \boldsymbol{E}$ observed from a target BN can be described as a boolean formula in propositional variables $X_{ik} \equiv$ "$X_i = x_i^k$". Let F_t, a boolean formula corresponding to

e_t, be defined as $F_t \equiv \bigwedge_{x_i^k \in e_t} X_{ik}$, where X_{ik} is also described as a boolean formula in $X_{ijk} \equiv$ "$X_i = x_i^k \mid \Pi_i = \pi_i^j$" recursively as follows:

$$X_{ik} \equiv \bigvee_{\pi_i^j} \left(X_{ijk} \wedge \bigwedge_{x_{i'}^{k'} \in \pi_i^j} X_{i'k'} \right).$$

For example, let's assume that an evidence set $e_1 \equiv (x_1^1, x_2^1, x_3^1)$ is observed from the BN in Figure 1. Then, its boolean formula F_1 is defined as $F_1 \equiv X_{11} \wedge X_{21} \wedge X_{31}$ where X_{11}, X_{21} and X_{31} are given as follows:

$$\begin{aligned} X_{11} &= X_{111}, \quad X_{21} = X_{211}, \\ X_{31} &= (X_{311} \wedge X_{11} \wedge X_{21}) \vee (X_{321} \wedge X_{11} \wedge X_{22}) \\ &\quad \vee (X_{331} \wedge X_{12} \wedge X_{21}) \vee (X_{341} \wedge X_{12} \wedge X_{22}). \end{aligned}$$

By substituting the above formulas into the formula of F_1, we finally obtain $F_1 = X_{111} \wedge X_{211} \wedge X_{311}$. We consider a propositional variable X_{ijk} as a random boolean variable which takes *true* with probability θ_{ijk} (i.e., $P(X_{ijk}; \boldsymbol{\theta}) \equiv \theta_{ijk}$). Here, we assume that X_{ijk} and $X_{i'j'k'}$ $(i \neq i')$ are independent, and also that X_{ijk} and $X_{ijk'}$ $(k \neq k')$ are exclusive. Due to these assumptions, the probability of F_1 is computed as follows:

$$\begin{aligned} P(F_1; \boldsymbol{\theta}) &\equiv P(X_{111} \wedge X_{211} \wedge X_{311}; \boldsymbol{\theta}) \\ &= P(X_{111}; \boldsymbol{\theta})\, P(X_{211}; \boldsymbol{\theta})\, P(X_{311}; \boldsymbol{\theta}) \\ &= \theta_{111}\theta_{211}\theta_{311} \\ &= P(e_1; \boldsymbol{\theta}). \end{aligned}$$

As the above example shows, the probability of an evidence set e_t can be computed as the one of F_t. Thus, a joint distribution of a target BN can be expressed as a distribution over random boolean variables $\{X_{ijk}\}_{ijk}$ which has the same parameters $\boldsymbol{\theta}$. The distribution is called a *logic-based probabilistic model* whose parameters $\boldsymbol{\theta}$ can be learned by the *BO-EM algorithm* [7] explained in 3.2.

There are two advantages of using the above encoding for parameter learning. The first one is that logical expressions can exploit local structures, i.e., determinism and context-specific independence (CSI), as shown in [3]. These local structures are expressed in our encoding as follows:

Determinism: If $\theta_{ijk} = 1$ (0), then set X_{ijk} to *true* (*false*).

CSI: If $\theta_{ijk} = \theta_{ij'k}$, then use X_{ijk} instead of $X_{ij'k}$.

For example, let's assume we know in advance that $\theta_{341} = 0, \theta_{342} = 1$ (determinism) and $\theta_{31k} = \theta_{32k}$ (CSI) in the CPT for X_3 in Figure 1. Then, the boolean formula of X_{31} is revised as $X_{31} = (X_{311} \wedge X_{11}) \vee (X_{331} \wedge X_{12} \wedge X_{21})$. This example shows that our encoding exploits local structures and consequently the length of formulas and the number of parameters are reduced.

The second advantage is that we can introduce state-of-the-art compression techniques for boolean formulas. In this paper, we introduce a *shared binary decision diagram* (SBDD) [8] to compress boolean formulas for $e_t \in \boldsymbol{E}$. An SBDD is a directed acyclic graph which compactly represents multiple boolean functions. For example, Figure 2 shows an SBDD representing boolean formulas for the following five evidence sets with the same local structures as mentioned above.

$$e_1 \equiv (x_1^1, x_2^1, x_3^1)\,, \quad e_2 \equiv (x_3^1)\,, \quad e_3 \equiv (x_3^2)\,, \quad e_4 \equiv (x_1^1, x_3^1)\,, \quad e_5 \equiv (x_2^1, x_3^1)\,.$$

An SBDD consists of two types of nodes, *variable nodes* and *terminal nodes*. A variable node is labeled by a boolean variable X_{ijk} and has exactly two outgoing edges, *1-edges* (solid edges) and *0-edges* (dashed edges). The 1-edge (resp. 0-edge) of X_{ijk} indicates that X_{ijk} takes *true* (resp. *false*). A terminal node is labeled by 1 or 0 and has no outgoing edge. An SBDD representing T boolean functions $\{F_t\}_{t=1}^T$ has T *input edges* labeled by a boolean function F_t. A path from an input edge F_t to the 1-terminal (resp. 0-terminal) indicates a (partial) instantiation of $\{X_{ijk}\}_{ijk}$ which makes F_t *true* (resp. *false*). For example, the SBDD shown in Figure 2 has a path from F_1 to the 1-terminal ($F_1 \to X_{111} \to X_{211} \to X_{311} \to 1$), and the path means that F_1 takes *true* if X_{111}, X_{211} and X_{311} take *true*. SBDDs sometimes have special edges called *negative edges* (dotted edges) corresponding to *negation*. For example, the SBDD has a path from F_3 to the 1-terminal ($F_3 \cdots\!\rightarrow X_{111} \dashrightarrow X_{211} \to X_{311} \to 1$) which includes a negative edge. The path indicates that F_3 takes *false* even if it points the 1-terminal. Due to negative edges, evidence sets which complement each other (e.g., F_2 and F_3) are completely shared in SBDDs.

As Figure 2 shows, multiple evidence sets $F_1, \ldots, F_5$ are compiled into an SBDD while their common sub-graphs are shared. Using an SBDD reduces not only memory but also time for parameter learning because our proposed method computes conditional expectations in a manner of dynamic programming on the SBDD. Consequently, sharing common sub-graphs in the SBDD implies sharing common local probabilities and expectations in the E-step.

Various encoding methods have been proposed for factorizing MLFs into *arithmetic circuits* (ACs) [4], and they employ two types of boolean variables, indicators $\boldsymbol{\lambda} \equiv \{\lambda_{ik}\}_{ik}$ and parameters $\boldsymbol{\theta} \equiv \{\theta_{ijk}\}_{ijk}$. On the other hand, our encoding method employs two types of random boolean variables, $\{X_{ik}\}_{ik}$ and $\{X_{ijk}\}_{ijk}$, where X_{ik} corresponds to an indicator λ_{ik} and also X_{ijk} corresponds to a parameter θ_{ijk}. However, the SBDD shown in Figure 2 consists only of $\{X_{ijk}\}_{ijk}$ (i.e., parameters). This is because, in parameter learning setting, observations are given before the SBDD is constructed. That is, the SBDD has already taken in the information of observations. Consequently, once the SBDD is constructed, is is never revised through the whole learning process. This is the key difference between our encoding and existing ones.

3.2 Learning

After encoding a data set $\boldsymbol{E}$ in boolean formulas and compiling them into an SBDD, we learn its parameters $\boldsymbol{\theta}$ by the *BO-EM algorithm* [7], which is an EM algorithm for *logic-based probabilistic models* (LBPMs). A LBPM is defined as a joint distribution over a set of random boolean variables. A probabilistic event on a LBPM is described as a boolean formula in them. As shown in 3.1, observations from a target BN can be described as boolean formulas in $\{X_{ijk}\}_{ijk}$, where X_{ijk} is a random boolean variable

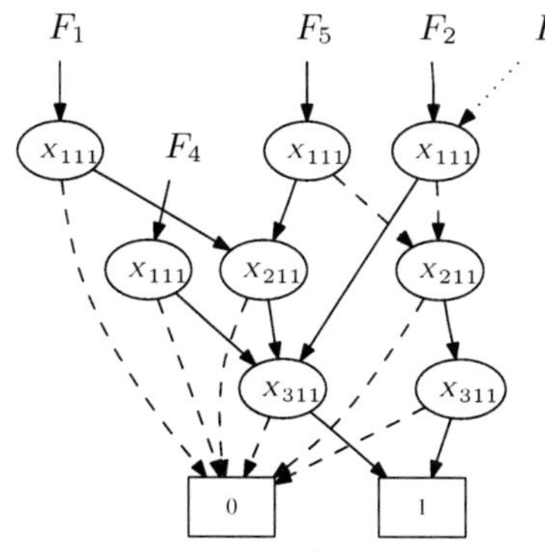

0. Compile $\boldsymbol{E}$ into an SBDD $\Delta_{\boldsymbol{E}}$.
1. Randomly initialize $\theta_{ijk}^{(0)}$ for each i, j, k.
2. Compute *backward probabilities* $B_{nx}^{(m)}$ for each node $n \in \Delta_{\boldsymbol{E}}$ from bottom-up ($x \in \{0, 1\}$).
3. Compute *forward expectations* $F_{nx}^{(m)}$ for each node $n \in \Delta_{\boldsymbol{E}}$ from top-down ($x \in \{0, 1\}$).
4. Compute *conditional expectations* $E_{ijk}^{(m)}$ for each parameter θ_{ijk} using $B_{nx}^{(m)}$ and $F_{nx}^{(m)}$.
5. Update $\theta_{ijk}^{(m)}$ to $\theta_{ijk}^{(m+1)}$ by Eq. (2).
6. Repeat 2.-5. until parameters converge.

Fig. 2. An SBDD for $F_1, \ldots, F_5$ and the overview of the BO-EM algorithm

with a probability θ_{ijk}. Consequently, the parameter learning problem for BNs can be formalized as the one for LBPMs. As a result, we can use the BO-EM algorithm to learn parameters of BNs. The brief overview of the BO-EM algorithm is in Fig. 2. After initializing parameters, the BO-EM algorithm computes *backward probabilities* and *forward expectations*, which are similar to backward and forward probabilities of the *Baum-Welch algorithm* (i.e., the EM algorithm specialized for *hidden Markov models* (HMMs)). Then, the BO-EM computes conditional expectations $E_{ijk}^{(m)}$ in linear time in the SBDD size. In particular, its time and space complexity for HMMs are the same as those of the Baum-Welch algorithm [7]. Consequently, our proposed method is a kind of a generalization of the Baum-Welch algorithm to BNs in general. The important point here is that our method exploits implicit common local patterns in the calculation of conditional expectations and shares them through an SBDD even if a target BN has no explicitly repeated patterns. In general, the EM algorithm itself is repeated several times with different initial parameters $\boldsymbol{\theta}^{(0)}$ to avoid local maximums. However, in our method, the SBDD is constructed only once in the beginning of the whole learning process. Therefore, the benefit of using our method gets bigger as the EM algorithm is repeated over and over again.

4 Experiments

In this section, we experimentally compare our proposed method with an *naive* EM algorithm for BNs, which computes conditional expectations $E_{ijk}^{(m)}$ as the sum of conditional probabilities $P(x_i^k, \pi_i^j \mid e_t; \boldsymbol{\theta}^{(m)})$. For computing these conditional probabilities, we use ACE[1] which is the state-of-the-art BN compiler package. ACE compiles a BN into an AC and computes probabilities of queries in linear time in the AC size. The key difference between ACE and our method is that ACE computes one probability at a time by tracing the AC whereas our method simultaneously computes all conditional expectations $E_{ijk}^{(m)}$ by tracing an SBDD representing evidence sets. For instance, suppose

[1] Available for download at http://reasoning.cs.ucla.edu/ace.

there is a BN which has 50 parameters and our task is to learn them from 100 evidence sets by the EM algorithm. The EM algorithm requires the computation of as many as 50 conditional expectations in every E-step. In the task, the naive method would compile a target BN with each evidence set. As a result, 100 ACs are constructed and 50 conditional probabilities are separately computed on each AC. Eventually, 5,000 conditional probabilities are independently computed and summed up to get conditional expectations in every E-step. The reason why the naive method compiles a BN with an evidence set is that exploiting evidence sets in ACs accelerates inference time [2]. Contrastingly, our method compiles 100 evidence sets into a single SBDD and directly computes 50 conditional expectations on the SBDD while common sub-graphs and local probabilities (or expectations) are shared.

For evaluating our method, we applied the naive method and our method to the parameter learning problem of randomly generated BNs, and compared their size (edge count), compile time and learning time. Here, the learning time is the time to execute one E-step (i.e., the time to compute all conditional expectations). Firstly, we randomly generated 100 BNs for binary random variables $\{X_n\}_{n=1}^{N}$ with $N \in \{10, 20, 30, 40, 50\}$. Then, we learned their parameters from a date set $\boldsymbol{E}$ with $|\boldsymbol{E}| \in \{10, 50, 100\}$. In this experiment, we generated two types of BNs, one without local structures and the other one with those. The results of applying the methods to the two types BNs are shown in Table 1 and Table 2. Each table shows averages of the edge count, the compile time and the learning time of the naive method and our method.

The experiments serve to demonstrate three points. The first point is that our method makes a significant improvement in learning time in all cases. This improvement is especially significant when the E-step is repeated numerous times. For example, suppose we iterate the E-step and the M-step 100 times in an EM execution and repeat it 100 times changing initial parameters in the bottom case in Table 1. Eventually, the E-step is repeated 10,000 times in total. This setting seems natural considering that the estimated parameters by the EM algorithm give a local maximum. To execute 10,000 E-steps, the naive method would take more than a whole day whereas our method takes only about 40 minutes. The experiments also show that the compile time is improved too. However, in the parameter learning setting, it matters less because the compilation is done only once at the beginning of learning and it takes much less time than the whole learning process.

The second point regards the effect of sharing evidence sets. In the result of the naive method, all of the edge count, the compile time and the learning time increase almost ten times when the number of evidence sets increases ten times. However, those of our method increase less than ten times. The result supports our claim that sharing sub-graphs between evidence sets improves on the parameter learning for BNs. The improvement is particularly remarkable on BNs with less parameters because SBDDs become less complex with less parameters and common sub-graphs often appear in such simple structures.

The final point shown in the results concerns the effect of local structures in learning. Local structures assumed in the experiments are the following context-specific independence (CSI) for each X_i:

$$X_i \perp\!\!\!\perp X_j \mid X_k = x_k^1 \qquad (\forall X_j, X_k \in \Pi_n, j > k).$$

Generally, a CPT size for X_i is $2^{|\Pi_i|+1}$, however, the above CSI decreases it to $2(|\Pi_n|+1)$, and that is surely confirmed in Table 2. As a result, the learning time is improved in the result of the both method. Especially in the case of $N=50$, improvement of our method is significant. This is because, as described in the second point, the effect of sharing evidence becomes prominent with less parameters. Consequently, our method deeply exploits local structures and the effect is much more significant than that of BN compilation approaches for exact inference.

Table 1. Comparison random BNs without local structures

Target BN property				Edge count		Compile Time [s]		learning Time [s]	
Nodes	Edges	Parameters	Evidence sets	ACs	SBDD	ACs	SBDD	ACE	Proposed
10	12.5	68.0	10	1.11e+03	4.59e+02	3.85e+00	4.61e−03	3.69e−01	9.69e−05
			50	5.50e+03	1.82e+03	1.92e+01	6.16e−03	1.85e+00	3.09e−04
			100	1.10e+04	3.34e+03	3.85e+01	8.35e−03	3.71e+00	5.43e−04
20	27.8	167.2	10	2.59e+03	3.33e+03	4.04e+00	1.48e−02	4.37e−01	4.89e−04
			50	1.33e+04	1.42e+04	2.02e+01	2.70e−02	2.21e+00	1.72e−03
			100	2.66e+04	2.79e+04	4.05e+01	4.29e−02	4.42e+00	3.03e−03
30	42.2	265.3	10	4.32e+03	1.72e+04	4.32e+00	9.22e−02	6.58e−01	1.92e−03
			50	2.16e+04	7.33e+04	2.18e+01	2.12e−01	3.29e+00	7.82e−03
			100	4.30e+04	1.43e+05	4.36e+01	3.53e−01	6.59e+00	1.53e−02
40	57.5	378.3	10	6.21e+03	8.76e+04	4.45e+00	8.38e−01	7.84e−01	9.45e−03
			50	3.09e+04	3.74e+05	2.22e+01	1.67e+00	3.93e+00	4.03e−02
			100	6.18e+04	7.10e+05	4.46e+01	2.52e+00	7.88e+00	7.65e−02
50	71.9	488.7	10	7.90e+03	2.24e+05	4.81e+00	1.40e+00	9.28e−01	2.44e−02
			50	4.00e+04	1.02e+06	2.37e+01	4.28e+00	4.69e+00	1.09e−01
			100	8.06e+04	2.04e+06	4.75e+01	7.84e+00	9.39e+00	2.25e−01

Table 2. Comparison for random BNs with local structures

Target BN property				Edge count		Compile Time [s]		learning Time [s]	
Nodes	Edges	Parameters	Evidence sets	ACs	SBDD	ACs	SBDD	ACE	Proposed
10	12.6	19.4	10	4.22e+02	6.24e+01	3.86e+00	3.12e−03	2.77e−01	3.48e−05
			50	2.12e+03	1.99e+02	1.94e+01	3.80e−03	1.39e+00	6.18e−05
			100	4.27e+03	3.10e+02	3.87e+01	5.00e−03	2.78e+00	8.73e−05
20	27.2	40.1	10	8.93e+02	1.85e+02	4.03e+00	4.03e−03	3.26e−01	5.15e−05
			50	4.45e+03	7.63e+02	2.02e+01	5.66e−03	1.63e+00	1.42e−04
			100	8.89e+03	1.41e+03	4.03e+01	7.94e−03	3.25e+00	2.49e−04
30	42.8	63.2	10	1.37e+03	3.60e+02	4.19e+00	5.16e−03	4.68e−01	7.44e−05
			50	6.87e+03	1.59e+03	2.10e+01	7.88e−03	2.38e+00	2.59e−04
			100	1.37e+04	3.09e+03	4.19e+01	1.18e−02	4.75e+00	4.65e−04
40	57.6	84.2	10	1.85e+03	5.74e+02	4.36e+00	6.20e−03	5.52e−01	1.05e−04
			50	9.17e+03	2.63e+03	2.21e+01	1.07e−02	2.73e+00	4.01e−04
			100	1.84e+04	5.14e+03	4.42e+01	1.66e−02	5.44e+00	7.19e−04
50	72.8	103.6	10	2.32e+03	9.74e+02	4.67e+00	7.87e−03	6.32e−01	1.59e−04
			50	1.17e+04	4.44e+03	2.35e+01	1.46e−02	3.17e+00	6.31e−04
			100	2.31e+04	8.72e+03	4.65e+01	2.34e−02	6.23e+00	1.08e−03

5 Conclusion and Related Work

In this paper, we considered the parameter learning problem for BNs and pointed out that the EM algorithm based on existing BN compilation approaches for inference has redundancy. Then, we proposed a new parameter learning method based on BN compilation. Our method is a kind of a propositionalized generalization of the Baum-Welch algorithm to any BNs, i.e., it detects implicit common local patterns in the calculation of conditional expectations and shares them through an SBDD even if a target BN has no explicitly repeated patterns. Finally, we experimentally showed that it makes a significant improvement in parameter learning for randomly generated BNs.

An approach to exploit a given evidence set in exact inference has been proposed by [2]. The key difference between their method and our method is that the former compiles a target BN with a single evidence set into an AC, whereas the latter does with multiple evidence sets into an SBDD. It follows that our method enables to share common sub-graphs and probabilities (or expectations) between evidence sets, and also that it is effective for the parameter learning problem because a massive number of observations are usually given in the problem.

The idea of sharing sub-graphs in compilation approaches has been proposed by [9]. Their method employes the same encoding as [5] and factorizes an MLF of a target BN using a *shared ZDD* (SZDD) which is similar to SBDDs. The point is that the method firstly compiles partial MLFs, which correspond to nodes in the BN, into an SZDD. Then, it constructs an ZDD representing the whole MLF depending on given query. Consequently, it is unable to share common local probabilities between evidence sets.

References

1. Boutilier, C., Friedman, N., Goldszmidt, M., Koller, D.: Context-Specific Independence in Bayesian Networks. In: Proc. of UAI 1996 (1996)
2. Chavira, M., Allen, D., Darwiche, A.: Exploiting Evidence in Probabilistic Inference. In: Proc. of UAI 2005 (2005)
3. Chavira, M., Darwiche, A.: Compiling Bayesian Networks with Local Structure. In: Proc. of IJCAI 2005 (2005)
4. Chavira, M., Darwiche, A.: On probabilistic inference by weighted model counting. Artificial Intelligence 172, 772–799 (2008)
5. Darwiche, A.: A Differential Approach to Inference in Bayesian Networks. Journal of the ACM 50(3), 123–132 (2003)
6. Dempster, A., Laird, N., Rubin, D.: Maximum likelihood from incomplete data via the EM algorithm. Journal of the Royal Statistical Society, Series B 39, 1–38 (1977)
7. Ishihata, M., Kameya, Y., Sato, T., Minato, S.: An EM algorithm on BDDs with order encoding for logic-based probabilistic models. In: Proc. of ACML 2010 (2010)
8. Minato, S., Ishiura, N., Yajima, S.: Shared Binary Decision Diagram with Attributed Edges for Efficient Boolean Function Manipulation. In: Proc. of DAC 1990 (1990)
9. Minato, S., Satoh, K., Sato, T.: Compiling Bayesian Networks by Symbolic Probability Calculation Based on Zero-suppressed BDDs. In: Proc. of IJCAI 2007 (2007)

An Empirical Study of Bagging Predictors for Imbalanced Data with Different Levels of Class Distribution

Guohua Liang, Xingquan Zhu, and Chengqi Zhang

The Centre for Quantum Computation & Intelligent Systems,
FEIT, University of Technology, Sydney NSW 2007 Australia
{gliang,xqzhu,chengqi}@it.uts.edu.au

Abstract. Research into learning from imbalanced data has increasingly captured the attention of both academia and industry, especially when the class distribution is highly skewed. This paper compares the Area Under the Receiver Operating Characteristic Curve (*AUC*) performance of bagging in the context of learning from different imbalanced levels of class distribution. Despite the popularity of bagging in many real-world applications, some questions have not been clearly answered in the existing research, e.g., which bagging predictors may achieve the best performance for applications, and whether bagging is superior to single learners when the levels of class distribution change. We perform a comprehensive evaluation of the *AUC* performance of bagging predictors with 12 base learners at different imbalanced levels of class distribution by using a sampling technique on 14 imbalanced data-sets. Our experimental results indicate that Decision Table (DTable) and RepTree are the learning algorithms with the best bagging *AUC* performance. Most *AUC* performances of bagging predictors are statistically superior to single learners, except for Support Vector Machines (SVM) and Decision Stump (DStump).

Keywords: imbalanced class distribution, AUC performance, bagging.

1 Introduction

Research into learning from imbalanced data has increasingly captured the attention of both academia and industry, especially when the applications involve a highly skewed class distribution: for instance, in fraud detection [1], text classification [2] and medical diagnostics [3-5]. In these extremely imbalanced situations, the minority class is more important than the majority class; however, most traditional learning algorithms generate high accuracy on the majority class and perform poorly on the minority class [2, 6-8]; the evaluation metrics, such as the overall accuracy or error rate, are ineffective for evaluating the performance of classifiers in extremely imbalanced data [8-12]. We have therefore adopted *AUC* of the Receiver Operating Characteristic (ROC) curve as an evaluation metric to investigate the effect of different imbalanced levels of class distribution on bagging performance over multiple imbalanced data-sets. This investigation utilizes a sampling technique to alter

D. Wang and M. Reynolds (Eds.): AI 2011, LNAI 7106, pp. 213–222, 2011.

the class distribution at different imbalanced levels. The statistical analyses [13] performed instills confidence in the validity of the conclusions of this research.

Bagging (bootstrap aggregating), one of the most popular and effective ensemble learning methods, was introduced to reduce the variance of an unstable base learner by Breiman in 1996 [14]. Many theories have been proposed regarding the effectiveness of bagging for classifications based on bias and variance decomposition [15]. Theoretical investigations into why bagging works have been described by previous researchers [14, 16, 17]. Existing studies demonstrate the effectiveness of the bagging predictor; however, a comprehensive study of the bagging predictors' AUC performance with respect to different imbalanced levels of class distribution has not been undertaken. Bagging has been widely applied in many real world applications, but some practical questions have not been clearly answered; e.g., which bagging predictors are the best learning algorithms on their average performance, when the imbalanced levels of class distribution change, and in such situations, whether bagging is superior to single learners. Answering these questions presents the following research challenges: (1) how to evaluate bagging performance at different imbalanced levels of class distribution, and (2) how to conduct a fair and rigorous study to evaluate multiple algorithms over multiple data-sets.

Our main contribution is twofold: we (1) conduct statistical comparisons to investigate when bagging significantly improves the predictive performance of the single learner with respect to different levels of class distribution, and (2) provide ranks of the AUC performance of all bagging predictors. As a result, our research provides a full comparison of the AUC performance of bagging predictors with underlying 12 base learners and with different levels of class distribution. The experimental results provide a useful guide for data mining practitioners to choose proper learners when using bagging predictors for imbalanced applications.

The paper is organized as follows. Section 2 presents the detail of the designed framework. Section 3 presents evaluation metrics and Section 4 provides the experimental setting. Section 5 presents the experimental results analysis to compare the performance of bagging with each of the single learners and to rank all bagging predictors with respect to different imbalanced levels of class distribution. Section 6 concludes the paper.

2 Designed Framework

The designed framework is presented in Fig. 1, and the evaluation is divided into tasks as follows: (1) utilize the sampling technique to alter each original imbalanced data-set into nine new data-sets with different imbalanced levels of class distribution; (2) perform 10-trial 10-fold cross-validation evaluation on the altered *nine* data-sets to obtain nine pair of (*FPR, TPR*) for each learner to form a ROC; (3) compare bagging with each of the single learners: the comparisons of the *AUC* performances are used to determine whether bagging is superior to each of the single learners at different imbalanced levels of class distribution over multiple imbalanced data-sets; and (4) use both average ranks of *AUC* performance and average *AUC* performance of bagging predictors to compare all the bagging predictors to determine which predictors have the best performance.

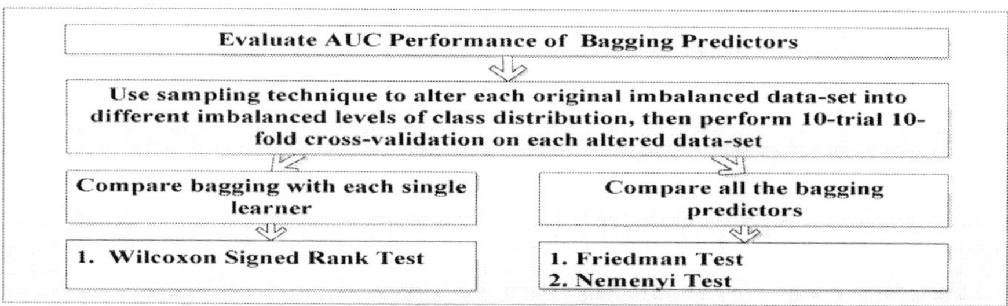

Fig. 1. Designed Framework

3 Evaluation Metrics

Mis-classification error rate is a popular choice for evaluating the performance of a classifier; however, it might not be a good metric for measuring the predictive performance of imbalanced data-sets. As in real world applications, the minority class is more important than the majority class; normally a high prediction accuracy is required in a minority class and therefore, a simple estimated error rate has limitations in evaluating the performance of a classifier on a minority class [18].

Table 1 presents the confusion matrix for a binary classification problem. Table 2 presents the formulas of True Positive Rate (*TPR*) and False Positive Rate (*FPR*).

The ROC is a well known performance metric for evaluating and comparing algorithms. We utilize the calculated *AUC* of ROC curves as an evaluation metric to compare bagging and single learners over multiple imbalanced data-sets.

A ROC graph [12] is a two-dimensional plot where the x-axis denotes the *FPR* of a classifier and the y-axis denotes the *TPR* of a classifier. In the ROC plot, the upper left point (0,1) is the most desired point, known as "ROC Heaven", presenting 100% true positive and zero false positives, while the point (1,0) is the least desired point, called "ROC Hell".

AUC [19, 20] is not biased against the minority class and it has an important statistical property [21], so it is commonly used as an evaluation criterion to assess the average performance of classifiers on data with imbalanced class distribution [18, 20-22].

Table 1. Confusion matrix for a binary classification problem

	Predicted Positives	Predicted Negatives
Positive Instances (*P*)	True Positive (*TP*)	False Negatives (*FN*)
Negative Instances (*N*)	False Positive (*FP*)	True Negatives (*TN*)

Table 2. True Positive Rate and False Positive Rate

$TPR = \frac{TP}{TP+FN}$	$FPR = \frac{FP}{FP+TN}$

4 Experimental Setting

We implement bagging prediction model in Java, and use WEKA implementations of the 12 algorithms with default parameter settings in this empirical study [23]. We investigate the *AUC* performance of bagging predictors with respect to *nine* difference imbalanced levels of class distribution on 14 imbalanced binary-class data-sets which are collected from the UCI Machine Learning Repository [24].

Firstly, we utilize the sampling technique to alter each original imbalanced data-set into *nine* new data-sets with *nine* different imbalanced levels of class distribution, i.e., we regard the original minority class (sample size *P*) as the positive class:

$$P = 10\%\ M_1 = 20\%\ M_2 = \ldots = 90\%\ M_9$$

Secondly, we randomly select samples from the majority class without a replacement as a negative class (sample sizes as $90\%\ M_1$, $80\%\ M_2 \ldots 10\%\ M_9$)

Thirdly, the *nine* new data-sets, D_1, $D_2 \ldots D_9$ *(sample size* M_1, $M_2 \ldots M_9$, *respectively)*:

- $M_1 = 10\%$ *positive class* + 90% *negative class*
- $M_2 = 20\%$ *positive class* + 80% *negative class*
- ...
- $M_9 = 90\%$ *positive class* + 10% *negative class.*

Finally, we perform a 10-trial 10-fold cross-validation evaluation on the altered *nine* data-sets for each learning algorithm to obtain *nine* pairs of (*FPR, TPR*) for the single learner and *nine* pairs of (*FPR, TPR*) for bagging to form ROC curves.

Overall, for each learning algorithm on each original data-set, we build 18 models to form two ROC curves, one for the single learner and one for the bagging predictor. We investigate the *AUC* performance of bagging predictors with 12 algorithms at *nine* levels of sample distributions on 14 data-sets. As a result, we have built 3024 models in total to evaluate the *AUC* performance of bagging predictors.

In order to reduce uncertainty and obtain reliable experimental results, all the evaluations are assessed under the same test conditions by using the same randomly selected bootstrap samples (with replacements) in each fold of the 10-trial 10-fold cross-validation on each data-set.

4.1 Selection of Base Learners

The 12 most common learning algorithms have been selected for this study from the Weka implementation. They are as follows: (1) C4.5 decision tree learner (*J48*) proposed by Quinlan, which is based on gain ratio to select the splitting attribute; (2) Naïve Bayes (*NB*) learner based on Bayes theorem, a simple, yet effective learner for large data-sets; (3) Support Vector Machines (SVM), a complex model for the classification, which uses mapping to transform the original training data into a higher dimension and the decision boundary is determined by finding the optimal separating hyper-planes, and SMO is selected from the Weka implementation for this study; (4) K-Nearest Neighbors (*KNN*), an IBK lazy learner in the Weka implementation is used for this study with the default setting; (5) Multi-layer Perceptron (MLP), a neural network learner; (6) PART, (7) Decision Table (DTable), and (8) OneR are rule learners; (9) Decision Stump (DStump), (10) Random Tree (RTree), (11) REPTree, and (12) Naïve-Bayes-Trees (NBTree) are tree family learners.

4.2 Data-Sets

A summary of the characteristics of the 14 imbalanced data-sets is displayed in Table 3. The data-sets were employed using different criteria, such as the number of instances from 57 up to 3772, the number of attributes from 7 up to 61, and the frequency of each class from almost balanced to extremely imbalanced.

Table 3. Imbalanced Data-Sets

Data-sets		Information Data		Class Data	
Index	Name	Attribute	Instances	Frequency	Classes
1	breastc	10	286	201, 85	2
2	bupa	7	345	145, 200	2
3	crx	16	690	307,383	2
4	Crx-g	21	1000	700,300	2
5	diabetes	9	768	500, 268	2
6	ionosphere	35	351	126,225	2
7	Kr-vs-kp	37	3196	1669,1527	2
8	laour	17	57	20,37	2
9	staloghear	14	270	120, 150	2
10	sick	30	3772	3541, 231	2
11	sonar	61	208	97,111	2
12	Tic-tac-toe	10	958	626,332	2
13	wbreastc	10	699	458, 241	2
14	WDBC	31	569	212,357	2

5 Experimental Results Analysis

The experimental results analysis includes the following three sub-sections: A. statistical test, B. compare bagging with each of single learners, and C. compare bagging predictors with one other.

5.1 Statistical Test

In order to conduct a rigorous and fair analysis, non-parametric tests were performed for the statistical comparison of learners: the Wilcoxon signed-rank test for comparison of two learners, and the Friedman test with the corresponding post-hoc Nemenyi test for comparison of multiple learners [13].

The Wilcoxon Signed-Rank Test. This is a non-parametric statistical hypothesis test which is considered to be an alternative to the paired t-test. The main difference from a t-test is that this test does not require assumptions about the populations of a normal distribution. This test is the most accurate non-parametric test for paired data to determine whether there is a difference between paired samples. The Wilcoxon signed-rank test is considered to be safe from a statistical point of view and is more powerful than the t-test when test conditions cannot meet the assumption requirements of a parametric test [13]. We therefore performed this test to determine whether there really is an improvement of performance between the two learners, bagging and single learner.

Friedman Test and Post-hoc Nemenyi Test. Both tests are non-parametric for comparing multiple algorithms over multiple datasets. Firstly, all the algorithms are ranked on each data-set, giving the best performing algorithm the rank of 1, the second best rank 2, and so on. If there are ties, average values are assigned.

Secondly, the average rank of the algorithm, $R_j = \frac{1}{N}\sum_i r_i^j$ is obtained, where r_i^j is the rank of the *j*-th of *d* algorithms on the *i*-th of *N* data-sets.

Finally, the Friedman test compares the average ranks of algorithms and checks whether there is a significant difference between the mean ranks. The Friedman statistic is calculated as:

$$\chi_F^2 = \frac{12N}{d(d+1)}\left[\sum_j R_j^2 - \frac{d(d+1)^2}{4}\right]$$

where *N* is the number of data-sets, *d* is the number of compared algorithms, and R_j is the average rank of algorithms. This statistic is χ_F^2 distributed with *k-1* degrees of freedom.

The Null Hypothesis of this test states that the performances of all algorithms are equivalent. If the Null Hypothesis is rejected, it does not determine which particular algorithms differ from one another. A post-hoc Nemenyi test is needed for additional exploration of the differences between mean ranks to provide specific information on which mean ranks are significantly different to each other to identify them. The critical difference is calculated as:

$$CD = q_\alpha\sqrt{\frac{d(d+1)}{6N}}$$

The critical values q_α are based on the Studentized range statistic divided by $\sqrt{2}$ [13]. If the mean ranks are different by at least the critical difference, the performance of two learners is significantly different.

5.2 Comparison between Bagging and Single Learners

The Wilcoxon Signed Rank Test is used to determine whether there really is an improvement of *AUC* performance between two learners, i.e., bagging SVM and single learner SVM. The Null Hypothesis shows that the median of differences between Bagging and each single learner equals 0.

Rule: Reject the Null Hypothesis if the p-value Test Statistic W is less than $\alpha = .05$ at the 95% confidence level of significance.

Table 4 presents the summarized results of the Wilcoxon signed-rank test for the difference *AUC* performance between bagging and single learners. If a calculated p-value is greater than α value, 0.05, then the P-values are highlighted and we accept the Null Hypothesis, for example SVM, and DStump. For all other cases, we reject the Null Hypothesis. In addition, we observe that the majority of bagging predictors' *AUC* values are larger than single learners' *AUC* values. The experimental results therefore demonstrate that the average performances of bagging are better than most single learners, except for SVM and DStump learners.

Table 4. The comparison of the *AUC* performance of bagging and single learners. The significance level is .05.

Wilcoxon Signed Rank Test to compare the AUC performance of Bagging and Single Learners						
Learners	J48	RepTree	RandTree	NB	SVM	DStump
p-values	.004	.035	.004	.001	**.096**	**.074**
Learners	OneR	DTable	PART	KNN	NBTree	MLP
p-values	.001	.001	.004	.008	.026	.004

5.3 Comparison between Bagging Predictors

The Friedman test is used for comparison of multiple learners over multiple data-sets. we first of all rank the bagging predictors on each data-set according to their *AUC* metric from 1 to 12, respectively, e.g., the best performance of the bagging predictor with the largest value of the *AUC* is signed as ranking 1, with the second largest value is signed as ranking 2, and so on; if there are ties, the averaged value of their ranking orders is signed as their ranking. We then perform the Friedman test to compare the *AUC* ranking of 12 bagging predictors and to obtain mean ranks in Table 5. Table 5 reports the average *AUC* ranks of the bagging predictors in the third and the last rows.

As the Null Hypothesis is rejected, the Friedman test indicates there is at least a difference between the mean ranks of bagging predictors. Therefore, the corresponding post-hoc Nemenyi test for additional exploration of the differences between mean ranks provides specific information on which mean ranks are significantly different from one another.

Table 5. Mean rank of Friedman Test for *AUC* performance of Bagging Predictors

Mean Rank of Friedman Test for AUC Performance of Bagging Predictors						
Predictors	DTable	RepTree	OneR	RandTree	J48	PART
Mean Ranks	3.00	4.14	4.71	5.36	5.43	6.36
Predictors	NBTree	DStump	SVM	KNN	MLP	NB
MeanRanks	6.50	7.07	7.64	9.00	9.21	9.57

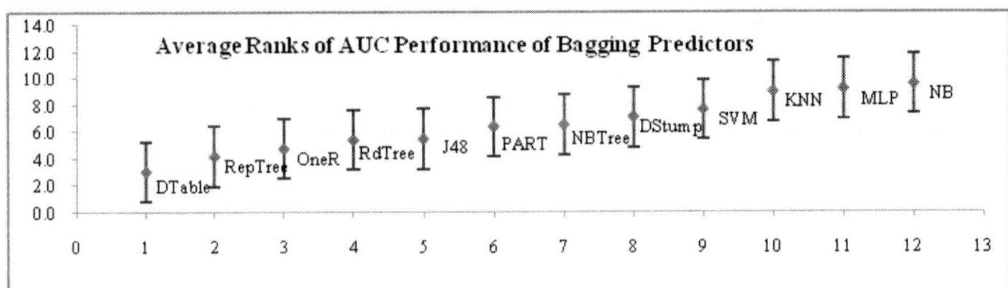

Fig. 2. Average ranks of *AUC* performance for 12 bagging predictors, where the *x-axis* denotes the ranking order of the bagging predictors, while the *y-axis* denotes the average rank of the bagging predictors' *AUC* performance. The error bars present the "critical difference" of the Nemenyi test.

Fig. 2 presents an empirical comparison of the *AUC* performances of 12 bagging predictors, where the *x-axis* denotes the ranking order of the bagging predictors, and the *y-axis* denotes the average rank of the bagging predictors' *AUC* values. The error bars indicate the "critical difference" of the Nemenyi test. The *AUC* performance of two bagging predictors is significantly different if the corresponding error bars do not overlap. Overall, the group of Bagging DTable and RepTree has the best average ranks of *AUC* performance, while the group of bagging NB, MLP and KNN has the worst average ranks of *AUC* performance. In addition, there are statistically significant differences between the two groups' average ranks of *AUC* performance. The authors [25] introduced a new method using two dimensional robustness and stability decomposition to categorize base learners into different categories. According to their results, DTable and RepTree are categorized as unstable base learners, while NB, MLP and KNN are categorized as stable learners. We therefore demonstrate that the unstable base learners, DTable and RepTree contribute to the best bagging predictors' *AUC* performance, while the stable base learners, NB, MLP and KNN lead to the worst bagging predictors' *AUC* performance when the imbalanced levels of class distribution changed at *nine* different levels on each data-set, over all 14 data-sets.

Table 6. Average *AUC* performance of Bagging Predictors on 14 imbalanced data-sets

Average *AUC* Performance of Bagging Predictors						
Predictors	DTable	RepTree	OneR	RandTree	J48	PART
Mean	.668	.598	.585	.561	.552	.530
Variance	.048	.097	.037	.065	.082	.071
Predictors	NBTree	DStump	SVM	KNN	MLP	NB
Mean	.519	.511	.510	.453	.435	.420
Variance	.064	.057	.084	.042	.073	.055

Table 6 presents the average *AUC* performance of bagging predictors on 14 data-sets. Mean (at the third and fifth rows) indicates the average value of the *AUC* performance of bagging predictors on 14 imbalanced data-sets in Fig. 3. Variance (at the fourth and last rows) indicates the corresponding value of error bars in Fig. 3.

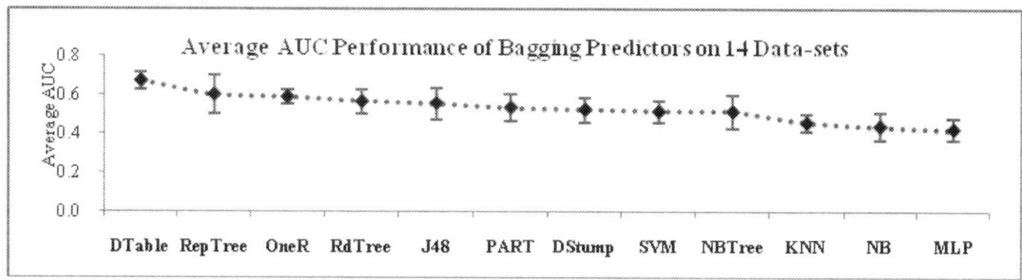

Fig. 3. The average *AUC* performance of bagging predictors over 14 data-sets, where the *x-axis* indicates the name of the bagging predictors, where the *y-axis* indicates the average value of AUC and the *error bar* indicates the variance value

Fig. 3 presents the summary of the observed average *AUC* performance of bagging predictors over 14 imbalanced data-sets in table 6. In this plot, the vertical axis indicates the average value of the *AUC* performance of bagging predictors, while the horizontal axis indicates the sorted average *AUC* performance of bagging predictors in descending order over the total benchmark of imbalanced data-sets, while the error bar indicates the variance of the observed average *AUC* performance. We note that Fig. 2 and Fig. 3 present a similar ranking order of bagging predictors, with the exception of NBTree.

6 Conclusion

We empirically investigated the *AUC* performances of 12 bagging predictors with respect to different levels of class distribution on 14 imbalanced data-sets. The under-sampling technique was utilized to alter the class distribution at different imbalanced levels. This research provided a full comparison of the *AUC* performances of bagging predictors with underlying 12 base learners at different levels of class distribution. The experimental results indicated that the *AUC* performances of bagging are statistically superior to single learners, except for SVM and DStump. Moreover, comparing the *AUC* performances of bagging, the group of the unstable learners, DeciTable and RepTree are the learning algorithms with the best bagging average performances, while the group of the stable learners, NB, MLP and KNN lead to the worst bagging predictors. In addition, there are significant differences between the two groups.

Acknowledgement. This research is supported by Australian Research Council Discovery Project (DP1093762).

References

1. Phua, C., Alahakoon, D., Lee, V.: Minority report in fraud detection: classification of skewed data. ACM SIGKDD Explorations Newsletter 6, 50–59 (2004)
2. Chawla, N.V., Bowyer, K.W., Hall, L.O., Kegelmeyer, W.P.: SMOTE: synthetic minority over-sampling technique. Journal of Artificial Intelligence Research 16, 321–357 (2002)

3. Mena, L., Gonzalez, J.: Machine learning for imbalanced datasets: application in medical diagnostic. In: Proceedings of the 19th International FLAIRS Conference (2006)
4. Rao, R.B., Krishnan, S., Niculescu, R.S.: Data mining for improved cardiac care. ACM SIGKDD Explorations Newsletter 8, 3–10 (2006)
5. Mazurowski, M.A., Habas, P.A., Zurada, J.M., Lo, J.Y., Baker, J.A., Tourassi, G.D.: Training neural network classifiers for medical decision making: The effects of imbalanced datasets on classification performance. Neural Networks 21, 427–436 (2008)
6. Koknar-Tezel, S., Latecki, L.J.: Improving SVM Classification on Imbalanced Data Sets in Distance Spaces. In: Proceedings of ICDM 2009, pp. 259–267 (2009)
7. Su, C.T., Hsiao, Y.H.: An evaluation of the robustness of MTS for imbalanced data. IEEE Transactions on Knowledge and Data Engineering, 1321–1332 (2007)
8. Maloof, M.: Learning when data sets are imbalanced and when costs are unequal and unknown. In: ICML 2003 Workshop on Learning from Imbalanced Data Sets II, Washington, DC (2003)
9. Weiss, G.M.: Mining with rarity: a unifying framework. ACM SIGKDD Explorations Newsletter 6, 7–19 (2004)
10. Chawla, N.V., Lazarevic, A., Hall, L.O., Bowyer, K.W.: SMOTEBoost: Improving Prediction of the Minority Class in Boosting. In: Lavrač, N., Gamberger, D., Todorovski, L., Blockeel, H. (eds.) PKDD 2003. LNCS (LNAI), vol. 2838, pp. 107–119. Springer, Heidelberg (2003)
11. Sun, Y., Kamel, M., Wong, A., Wang, Y.: Cost-sensitive boosting for classification of imbalanced data. Pattern Recognition 40, 3358–3378 (2007)
12. Zeng-Chang, Q.: ROC analysis for predictions made by probabilistic classifiers. In: Proceedings of ICMLC 2005, pp. 3119–3124 (2005)
13. Demšar, J.: Statistical comparisons of classifiers over multiple data sets. The Journal of Machine Learning Research 7, 1–30 (2006)
14. Breiman, L.: Bagging predictors. Machine Learning 24, 123–140 (1996)
15. Opitz, D., Maclin, R.: Popular ensemble methods: an empirical study. Journal of Artificial Intelligence Research 11, 169–198 (1999)
16. Büchlmann, P., Yu, B.: Analyzing bagging. Annals of Statistics 30, 927–961 (2002)
17. Buja, A., Stuetzle, W.: Observations on bagging. Statistica Sinica 16, 323 (2006)
18. Fawcett, T.: An introduction to ROC analysis. Pattern Recognition Letters 27, 861–874 (2006)
19. Bradley, A.P.: The use of the area under the ROC curve in the evaluation of machine learning algorithms. Pattern Recognition 30, 1145–1159 (1997)
20. Kotsiantis, S., Kanellopoulos, D., Pintelas, P.: Handling imbalanced datasets: a review. GESTS International Transactions on Computer Science and Engineering 30, 25–36 (2006)
21. Fawcett, T.: ROC graphs: Notes and practical considerations for researchers. Machine Learning 31, 1–38 (2004)
22. He, H., Garcia, A.E.: Learning from Imbalanced Data. IEEE Transactions on Knowledge and Data Engineering 21, 1263–1284 (2009)
23. Witten, I.H., Frank, E.: Data mining: practical machine learning tools and techniques. Morgan Kaufmann, San Francisco (2005)
24. Merz, C., Murphy, P.: UCI Repository of Machine Learning Databases (2006)
25. Liang, G., Zhu, X., Zhang, C.: An Empirical Study of Bagging Predictors for Different Learning Algorithms. In: Twenty-Fifth AAAI Conference on Artificial Intelligence, AAAI 2011. AAAI Press, San Francisco (2011)

A Simple Bayesian Algorithm for Feature Ranking in High Dimensional Regression Problems

Enes Makalic and Daniel F. Schmidt

Centre for MEGA Epidemiology, The University of Melbourne,
Carlton, VIC 3053, Australia
{emakalic,dschmidt}@unimelb.edu.au

Abstract. Variable selection or feature ranking is a problem of fundamental importance in modern scientific research where data sets comprising hundreds of thousands of potential predictor features and only a few hundred samples are not uncommon. This paper introduces a novel Bayesian algorithm for feature ranking (BFR) which does not require any user specified parameters. The BFR algorithm is very general and can be applied to both parametric regression and classification problems. An empirical comparison of BFR against random forests and marginal covariate screening demonstrates promising performance in both real and artificial experiments.

1 Introduction

Variable selection or feature ranking is a problem of fundamental importance in modern scientific research where data sets comprising hundreds of thousands of potential features and only a few hundred samples are not uncommon. In this setting, popular methods for importance ranking of features include the non-negative garotte [1], the least angle shrinkage and selection operator (LASSO) [2] and variants [3–5] as well as algorithms based on independence screening [6, 7]. The availability of computationally efficient learning algorithms for LASSO-type methods [8, 9] has made this approach particularly common in the literature. In addition, the LASSO and its variants fit all the covariates simultaneously, taking into account the correlation between the covariates, in contrast to marginal methods that examine each covariate in isolation.

An important issue with the application of LASSO-type methods for variable selection is how to specify the regularization or shrinkage parameter which determines the actual ranking of variables [10]. This is a highly challenging problem where a model selection method such as cross validation (CV) can lead to inconsistent results [11]. The problem may be circumvented by framing the LASSO in a Bayesian setting [12, 13] where the regularization parameter is automatically determined by posterior sampling. However, Bayesian LASSO-type algorithms cannot *fully* exclude any particular variable and thus do not provide an automatic importance ranking for the candidate features.

D. Wang and M. Reynolds (Eds.): AI 2011, LNAI 7106, pp. 223–230, 2011.

This paper presents a novel Bayesian algorithm, henceforth referred to as BFR, for variable selection in any parametric model where samples from the posterior distribution of the parameters are available. The new algorithm (see Section 2) computes an importance ranking of all observed features as well as credible intervals for these feature rankings. The credible intervals can then be used to remove features from further analysis that contribute little to explaining the data. The algorithm is very general, requires no user specified parameters and is applicable to both parametric regression and classification problems. The BFR algorithm is compared against random forests [14] and independence screening by generalized correlation [7] in Section 3. Empirical tests using artificially generated data as well as real data demonstrate excellent performance of the BFR algorithm.

2 Bayesian Feature Ranking (BFR) Algorithm

Given a data set comprising n samples

$$\mathcal{D} = \{(\mathbf{x}_1, y_1), (\mathbf{x}_2, y_2), \ldots, (\mathbf{x}_n, y_n)\}, \tag{1}$$

where $\mathbf{x}_i \in \mathbb{R}^p$ and $y_i \in \mathbb{R}$, the task is to select which of the p covariates, if any, are relevant to explaining the target $\mathbf{y} = (y_1, \ldots, y_n)'$. The target variable is assumed to be either real (regression task) or m-ary (classification task, $m \geq 2$), and to belong to the generalized linear family of statistical models with coefficients $\boldsymbol{\theta} \in \Theta$. Arrange the covariates into an $(n \times p)$ matrix $\mathbf{X} = (\mathbf{x}_1', \ldots, \mathbf{x}_n')'$. Without any loss of generality, we assume that the covariates are standardised to have zero mean and unit length, that is,

$$\sum_{i=1}^{n} X_{ij} = 0, \quad \sum_{i=1}^{n} X_{ij}^2 = 1. \tag{2}$$

Furthermore, we assume that there exists $B > 0$ samples $\{\boldsymbol{\theta}_1, \boldsymbol{\theta}_2, \ldots, \boldsymbol{\theta}_B\}$ from the posterior distribution $p(\boldsymbol{\theta}|\mathbf{X}, \mathbf{y})$ of the coefficients given the data.

The BFR algorithm proceeds by ranking the p covariates based on the absolute magnitude of the parameters in each posterior sample. That is, given a posterior sample $\boldsymbol{\theta}_i$, the parameters are ranked in descending order of $|\theta_{ij}|$ for $j = 1, 2, \ldots, p$. This process requires that the covariates are standardised as in (2) so that the absolute magnitude of some parameter θ_{ij} is an indication of the amount of variance explained by the corresponding column of the design matrix $(X_{kj}, k = 1, \ldots, n)$. The motivation for this comes from the fact that in a linear regression model, the amount of variance explained by covariate j with associated parameter θ_j is

$$\theta_j^2 \left(\sum_{k=1}^{n} X_{kj}^2 \right).$$

Due to the fact that we have standardised the covariates to have unit length, the amount of variance explained reduces simply to θ_j^2. This implies that ranking covariates in decreasing order of absolute magnitude of their associated coefficients is equivalent to ranking them in descending order of variance explained. The ranking process is repeated in turn for each of the posterior samples, resulting in B possible rankings of the p covariates. The final ranking of the covariates is determined from the complete set of rankings based on the empirical 75th percentile of each of the B possible rankings. Furthermore, the set of rankings can also be used to compute Bayesian credible intervals for the inclusion of each covariate. The BFR procedure is formally described in Algorithm 1.

Algorithm 1. BFR algorithm for feature ranking

Input: standardised feature matrix $\mathbf{X} \in \mathbb{R}^{n \times p}$, standardised target vector $\mathbf{y} \in \mathbb{R}^n$
Output: feature ranking $\tilde{\mathbf{r}} = (\tilde{r}_1, \tilde{r}_2, \ldots, \tilde{r}_p)$ $(1 \leq \mathbf{r}_i \leq p)$, credible intervals

1: Obtain B samples $\{\boldsymbol{\theta}_1, \boldsymbol{\theta}_2, \ldots, \boldsymbol{\theta}_B\}$ from the posterior distribution, $\boldsymbol{\theta}|\mathbf{X}, \mathbf{y}$

$$\boldsymbol{\theta}_i \sim \boldsymbol{\theta}|\mathbf{X}, \mathbf{y} \tag{3}$$

2: b $\leftarrow \lfloor B/10 \rfloor$ {number of burnin samples}
3: t $\leftarrow$ 5 {tempering step}
4: Initialise ranking matrix $\mathbf{R} = (\mathbf{r}_1, \mathbf{r}_2, \ldots, \mathbf{r}_p) = \mathbf{0}_{p \times B}$, $\mathbf{r}_i \in \mathbb{R}^p$
5: **for** $i = b$ to B step t **do**
6: Sort $\boldsymbol{\theta}_i = (\theta_{i1}, \theta_{i2}, \ldots, \theta_{ip})'$ by absolute magnitude $|\boldsymbol{\theta}_{ij}|$ in descending order
  1. Denote the sorted parameter vector

$$\boldsymbol{\theta}_i^* = (\theta_{i1}^*, \theta_{i2}^*, \ldots, \theta_{ip}^*)'$$

7: Compute ranking $\mathbf{r}_i = (r_{i1}, r_{i2}, \ldots, r_{ip})'$ from $\boldsymbol{\theta}_i^*$ for all p features
  1. The rank of feature j is r_{ij}, where $1 \leq r_{ij} \leq p$
  2. Absolute value of $|\theta_{ij}^*|$ determines rank of feature j
  3. If $r_{ij'} = 1$ then $|\theta_{ij'}^*| \geq |\theta_{ij}^*|$, $\forall j \neq j'$

8: **end for**
9: For each feature, compute the 25th and 75th rank percentiles using rank matrix $\mathbf{R}$
10: Compute the final feature ranking, $\tilde{\mathbf{r}}$, using $\mathbf{R}$
  1. Sort the 75th percentiles for each feature in ascending order
  2. Final rank of feature j is $\tilde{r}_j$, where $1 \leq \tilde{r}_j \leq p$
  3. If $\tilde{r}_{j'} = 1$ then 75th percentile for feature j is smaller than all $j \neq j'$

11: Compute 95% CI for features from 2.5th and 97.5th percentiles of $\mathbf{R}$

The algorithm begins by choosing which of the posterior samples will be used for ranking of the p covariates. The first ten percent of the initial posterior samples are discarded as *burnin* and then every $t = 5$th sample is used for the ranking method (Lines 1–3). The covariates are ranked based on the absolute magnitude of the parameters for each accepted sample, resulting in a set of possible rankings for the p covariates (Lines 5–8). The final ranking of the covariates, based on the 75-th percentile, and the 95% credible interval are computed in Line 10 and Line 11, respectively. The algorithm does not specify the type of sampler that should be used to generate the B posterior samples since, in theory, any reasonable sampling approach should result in a sensible covariate ranking procedure. This is furthed discussed in Section 3.

3 Discussion and Results

The BFR algorithm is now empirically compared against two popular feature selection methods: (i) random forests (RF) with default parameters [14], and (ii) independence screening by generalized correlation (HM) [7]. As this is a preliminary investigation of BFR, the empirical comparison will concentrate on the problem of covariate selection in the linear regression model. A Bayesian ridge regression sampler was chosen for the implementation of the BFR algorithm as it is: (i) similar to the commonly used method of least squares, and (ii) applicable when the number of covariates is greater than the sample size. The hierarchy depicting the Bayesian ridge regression [13] is

$$\begin{aligned}
\mathbf{y}|\mathbf{X},\boldsymbol{\beta},\sigma^2 &\sim \mathrm{N}_n(\mathbf{X}\boldsymbol{\beta},\sigma^2\mathbf{I}_n),\\
\boldsymbol{\beta} &\sim \mathrm{N}_p(\mathbf{0}_p,\sigma^2/\lambda^2\mathbf{I}_p),\\
\sigma^2 &\sim \sigma^{-2}d\sigma^2,\\
\lambda &\sim \mathrm{Gamma}(1,0.01),
\end{aligned}$$

where $\boldsymbol{\beta} \in \Theta \subset \mathbb{R}^p$ are the regression parameters, σ^2 is a normally distributed noise variable and λ is the ridge regularization parameter.

The three ranking methods will be compared on simulated data, where the true covariate set is known in advance, as well as two real data sets. The complete simulation code was written on the MATLAB numerical computing platform and is available for download from `www.emakalic.org/blog`.

3.1 Simulated Data

The BFR, RF and HM ranking methods are now compared on three linear regression functions borrowed from the simulation setup in [6]. For each of the three functions, we generated 100 data sets, with each data set comprising $n = 50$ samples and $p = 100$ covariates. All the generated data sets were standardised such that each covariate had a mean of zero and unit length. Noise was added to the target variables such that the signal-to-noise ratio (SNR) was in the set $\{1, 8\}$. The functions used for testing are detailed below.

(**Function** I). The generating regression coefficients were

$$\boldsymbol{\beta}^* = (1.24, -1.34, -1.35, -1.80, -1.58, -1.60, \mathbf{0}'_{p-6})',$$

where $\mathbf{0}_k$ is a k-dimensional zero vector. All predictors $\mathbf{x}_i$ ($i = 1, 2, \ldots, p$) were generated from the standard Gaussian distribution, $\mathbf{x}_i \sim \mathrm{N}_n(0, 1)$.

(**Function** II). The generating regression coefficients were

$$\boldsymbol{\beta}^* = (4, 4, 4, -6\sqrt{2}, \mathbf{0}'_{p-4})'.$$

The predictors were marginally distributed as per a standard Gaussian distribution, $\mathbf{x}_i \sim \mathrm{N}_n(0, 1)$; the correlation between predictors was $\mathrm{corr}(X_i, X_4) = 1/\sqrt{2}$ for all $i \neq 4$; $\mathrm{corr}(X_i, X_j) = 1/2$ if i and j were distinct elements in $\{1, 2, \ldots, p\}\backslash\{4\}$.

(**Function** III). The generating regression coefficients were

$$\boldsymbol{\beta}^* = (4, 4, 4, -6\sqrt{2}, 4/3, \mathbf{0}'_{p-5})'.$$

The predictors were marginally distributed as per a standard Gaussian distribution, $\mathbf{x}_i \sim \mathrm{N}_n(0, 1)$; the correlation between predictors was $\mathrm{corr}(X_i, X_5) = 0$ for all $i \neq 5$, $\mathrm{corr}(X_i, X_4) = 1/\sqrt{2}$ for all $i \notin \{4, 5\}$ and $\mathrm{corr}(X_i, X_4) = 1/2$ if both i and j were distinct elements in $\{1, 2, \ldots, p\}\backslash\{4, 5\}$.

Function I consists of independently generated covariates, while functions II and III contain varying levels of correlation. Feature selection is therefore expected to be somewhat more difficult for Functions II and III in contrast to function I. The ranking methods were compared on the `TopX` metric: the rank below which all the true features are included. For example, for Function I, a `TopX` of 15 indicates that the true six features are included among the first 15 selected covariates; the minimum possible `TopX` values for the three examples are six, four and five respectively. Box-and-whisker plots of the `TopX` metric for each method on the three test functions are depicted in Figure 1.

For all the tests functions, the BFR algorithm exhibited the smallest value of the median `TopX` metric of all the ranking methods considered. This was especially evident when the signal-to-noise ratio was larger, indicating that BFR is able to adapt well to varying levels of noise. Unsurprisingly, all three ranking methods performed better on function I, especially when SNR=8, in contrast to functions II and III. The HM algorithm performed better than random forests and slightly worse than the BFR method on all three test functions considered. As the amount of noise was decreased (SNR $\gg$ 8), the performance of the three methods became indistinguishable.

3.2 Real Data

The performance of the three methods was also examined on two real data sets: (i) the diabetes data set ($n = 442, p = 10$) downloaded from Trevor Hastie's homepage and analysed in [8], and (ii) the communities and crime data set ($n = 319, p = 123$) obtained from the UCI machine learning repository. Each data set was standardised similarly to the simulation data in Section 3.1. As the second data set contained a number of missing attributes, rows where one or more variables had missing entries were removed before analysis. The HM, RF and BH ranking of the $p = 10$ features for the diabetes data set is shown in Table 1.

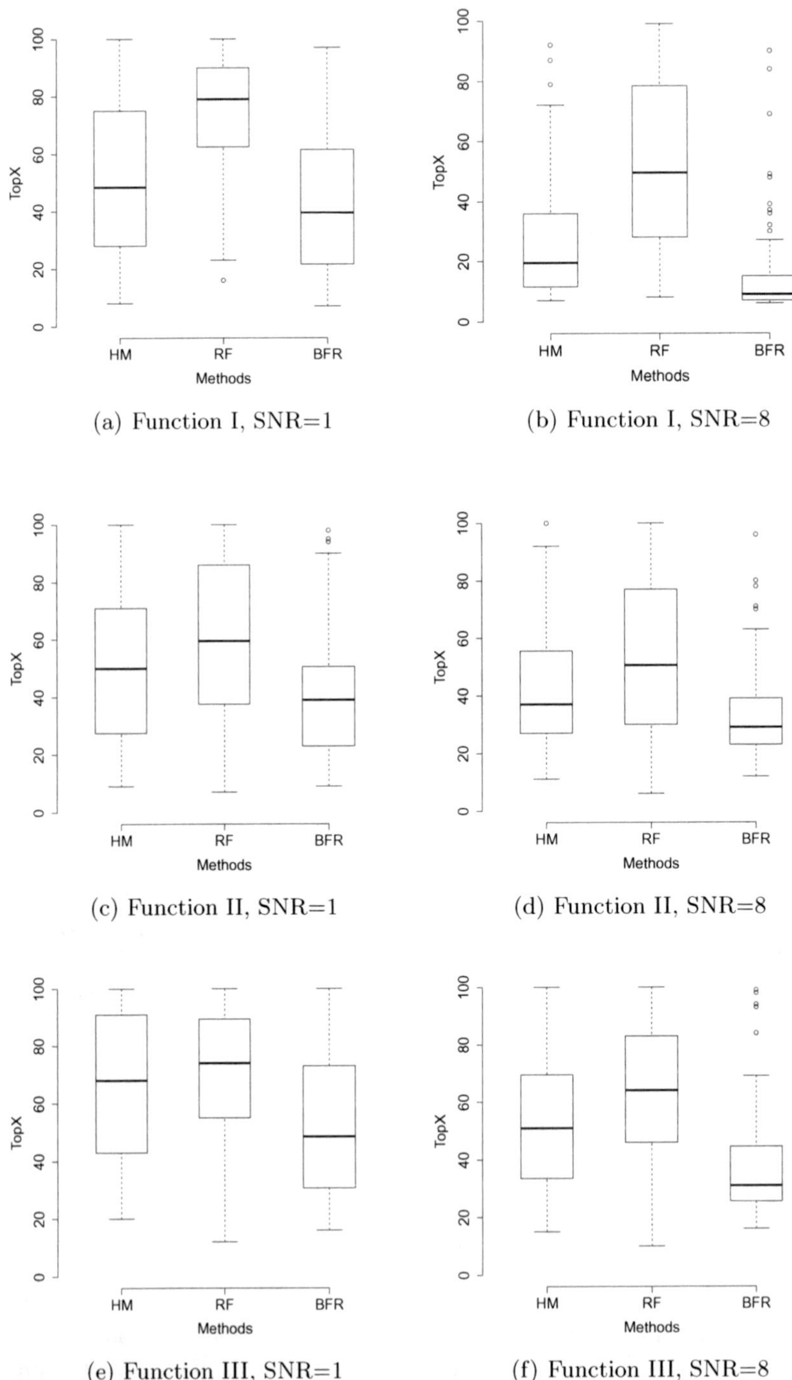

(a) Function I, SNR=1

(b) Function I, SNR=8

(c) Function II, SNR=1

(d) Function II, SNR=8

(e) Function III, SNR=1

(f) Function III, SNR=8

Fig. 1. Comparison of feature ranking methods on three test functions using the `TopX` metric

Table 1. HM, RF and BFR ranking of the ten features in the diabetes data set

Method	Feature rank									
	age	sex	bmi	map	tc	ldl	hdl	tch	ltg	glu
HM	3	9	4	7	8	10	5	1	6	2
RF	9	3	4	8	7	10	5	2	6	1
BFR	3	9	2	4	7	8	5	6	10	1

The top seven covariates selected by HM and RF were identical though with a slightly different ordering. All three ranking methods selected the `bmi` and `ltg` variables as the two most important features in terms of explanatory power. The BFR ranking is mostly similar to both HM and RF with one significant exception; BFR ranked the `sex` covariate much higher than the other ranking algorithms. Similarly, the BFR procedure ranked `glu` much lower in contrast to both HM and RF.

The performance of HM, RF and BFR was also examined on the communities and crime data set. Here, a five-fold cross validation procedure was used to estimate the generalisation error for each of the three methods over 100 test iterations. The mean squared prediction error for the BFR algorithm is shown in Figure 2. We notice that the generalisation error decreases sharply as the first few features are added to the model. The generalisation error begins to increase after approximately 30 features are included and smoothly rises until all $p = 123$

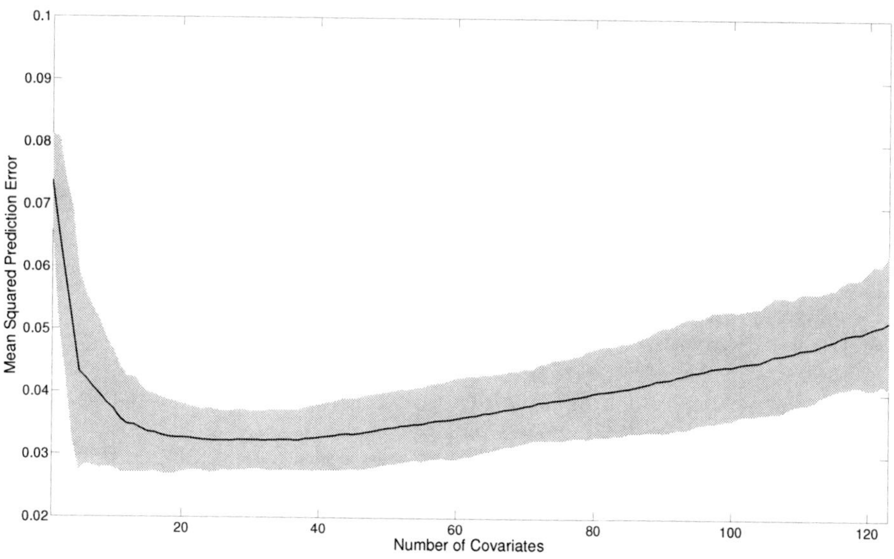

Fig. 2. Bayesian feature ranking for the communities and crime data set; standard errors represented by the shaded area

features are in the model. Importantly, the generalisation error does not drop after the first 30 features were included which indicates that BFR has included all the important features in the first 30 covariates. For this data set, both HM and RF algorithms were virtually indistinguishable from BFR and hence omitted from the plot for reasons of clarity.

4 Conclusion

This paper has presented a new Bayesian algorithm for feature ranking based on sampling from the posterior distribution of the parameters given the data. The new algorithm was applied to the linear regression model using both simulated and real data sets. BFR resulted in reasonable feature ranking in all empirical simulations, often outperforming random forests and feature ranking by generalised correlation. The excellent performance of BFR suggests that the idea is worthy of further exploration. Future work includes empirical examination of the sensitivity of BFR to the choice of Bayesian hierarchy, as well as application of BFR to feature ranking in classification problems.

References

1. Breiman, L.: Better subset regression using the nonnegative garrote. Technometrics 37, 373–384 (1995)
2. Tibshirani, R.: Regression shrinkage and selection via the lasso. Journal of the Royal Statistical Society (Series B) 58(1), 267–288 (1996)
3. Zou, H., Hastie, T.: Regularization and variable selection via the elastic net. Journal of the Royal Statistical Society (Series B) 67(2), 301–320 (2005)
4. Zou, H.: The adaptive lasso and its oracle properties. Journal of the American Statistical Association 101(476), 1418–1429 (2006)
5. James, G.M., Radchenko, P.: A generalized Dantzig selector with shrinkage tuning. Biometrika 96(2), 323–337 (2009)
6. Fan, J., Samworth, R., Wu, Y.: Ultrahigh dimensional feature selection: Beyond the linear model. Journal of Machine Learning Research 10, 2013–2038 (2009)
7. Hall, P., Miller, H.: Using generalized correlation to effect variable selection in very high dimensional problems. Journal of Computational and Graphical Statistics 18(3), 533–550 (2009)
8. Efron, B., Hastie, T., Johnstone, I., Tibshirani, R.: Least angle regression. The Annals of Statistics 32(2), 407–451 (2004)
9. Friedman, J., Hastie, T., Höfling, H., Tibshirani, R.: Pathwise coordinate optimization. The Annals of Applied Statistics 1(2), 302–332 (2007)
10. Zou, H., Hastie, T., Tibshirani, R.: On the "degrees of freedom" of the lasso. The Annals of Statistics 35(5), 2173–2192 (2007)
11. Leng, C., Lin, Y., Wahba, G.: A note on the lasso and related procedures in model selection. Statistica Sinica 16(4), 1273–1284 (2006)
12. Park, T., Casella, G.: The Bayesian lasso. Journal of the American Statistical Association 103(482), 681–686 (2008)
13. Kyung, M., Gill, J., Ghosh, M., Casella, G.: Penalized regression, standard errors, and Bayesian lassos. Bayesian Analysis 5(2), 369–412 (2010)
14. Breiman, L.: Random forests. Machine Learning 45(1), 5–32 (2001)

Semi-random Model Tree Ensembles: An Effective and Scalable Regression Method

Bernhard Pfahringer

University of Waikato, New Zealand
bernhard@cs.waikato.ac.nz
http://www.cs.waikato.ac.nz/~bernhard

Abstract. We present and investigate ensembles of semi-random model trees as a novel regression method. Such ensembles combine the scalability of tree-based methods with predictive performance rivalling the state of the art in numeric prediction. An empirical investigation shows that Semi-Random Model Trees produce predictive performance which is competitive with state-of-the-art methods like Gaussian Processes Regression or Additive Groves of Regression Trees. The training and optimization of Random Model Trees scales better than Gaussian Processes Regression to larger datasets, and enjoys a constant advantage over Additive Groves of the order of one to two orders of magnitude.

Keywords: regression, ensembles, supervised learning, randomization.

1 Introduction

Simple linear regression can work very well for arbitrary regression problems, especially when sample sizes are small and when good attributes are provided. Its performance usually breaks down when the relationship between the input and output contains significant non-linearity, and also in linear cases when there is strong collinearity present between pairs of inputs. Non-linear regression algorithms try to overcome these issues of simple linear regression. Still, on small data samples the main effect to be predicted is usually well modeled by a single global linear regression model, even if that effect or relationship in itself is not completely linear. Only when more data becomes available can non-linearity be extracted in a reliable and robust way. Samples sizes of 500 or even several thousand samples might be necessary. Non-linear methods include neural networks, support vector regression, gaussian process regression (GP) [9], and Additive Groves (AG) [11], among others. In this paper we introduce a new tree-based algorithm called Random Model Trees (RMT). We will compare RMTs with linear regression, GPs and AGs. We do not include results for additive regression over regression stumps (often also called "boosted stumps"), for space reasons, as they were not competitive. We also do not include support vector regression or neural networks in this comparison, as we have found GPs using Radial Basis function kernels to perform as well, and their parameter optimization is simpler involving only two tuning parameters. We also only include AGs and no other

D. Wang and M. Reynolds (Eds.): AI 2011, LNAI 7106, pp. 231–240, 2011.

tree-based methods like e.g. boosting, Random Forests [3], or Random Decision Trees [4], as AGs have been shown to perform at least as good as these tree-based alternatives. We show that RMTs are competitive in terms of predictive performance to both GPs and AGs, but that they can be an order of magnitude faster than AGs. Furthermore, being tree-based, RMTs scale with $O(NlogN)$, where N is the number of training examples. Therefore they can be applied to much larger problems than GPs.

In the following we will describe the new algorithm in Section 2, in Section 3 we will discuss the algorithms used for comparison, and will focus on parameter optimization. Parameter optimization is important, as the optimal values vary strongly depending on the specific datasets, and therefore no good default values exist. Section 4 will present results for more than 20 datasets. The final section summarizes and present directions for future research.

2 Random Model Trees

Random Model Trees are essentially the combination of two existing algorithms in Machine Learning: single model trees [8] are combined with Random Forest [3] ideas. Model trees are decision trees where every single leaf holds a linear model which is optimised for the local subspace described by this leaf. This works well in practise, as piece-wise linear regression can approximate arbitrary functions as long as the single pieces are small enough. For differentiable functions piece-wise linear regression can also be viewed as a crude one-step Taylor series expansion of such a function. Decision trees split the data into a number of small axis-parallel hypercubes, each of which will have its own local linear model. Issues with learning model trees include high training times searching for optimal splits and optimizing local linear models, potentially strong discontinuities in prediction at the borders between hypercubes, and erroneously overshooting extrapolation in sparse areas inside the hypercubes. Smoothing [14] inside a single tree and bagging of multiple trees [2] are standard ways to address some of these shortcomings.

Trees are also unstable, meaning that small changes in the training data can lead to the construction of trees that differ greatly in structure. While this may be problematic for a single tree, it is possible to take advantage of this effect in an ensemble. Random Forests [3] have shown to improve the performance of single decision trees considerably: tree diversity is generated by two ways of randomization. First the training data is sampled with replacement for each single tree like in Bagging. Secondly, when growing a tree, instead of always computing the best possible split for each node only a random subset of all attributes is considered at every node, and the best split for that subset is computed. Such trees have been used both for classification and for regression, but in the regression setting so far only trees with constant leaf prediction were used, i.e. regression trees were generated, but not model trees. Random model tree ensembles (RMT) for the first time combine model trees and random forests.

The success and efficiency of Random Model Trees critically depends on some specific engineering features. Determining the best split point for an attribute is

```
buildEnsemble(data, numTrees, k)
for i = 1 to numTrees
    do randomly split data into two:
        train + validate
        buildTree(train, validate, k)

buildTree(train, validate, k)
min ← minTargetValue(train)
max ← maxTargetValue(train)
localSSE ← linReg(train, validate)
▷
if |train| > 10 & |validate| > 10
    do split ← randomSplit(train, k)
        ▷
        smT ← smaller(train, split)
        smV ← smaller(validate, split)
        smaller ← buildTree(smT, smV, k)
        ▷
        laT ← larger(train, split)
        laV ← larger(validate, split)
        larger ← buildTree(laT, laV, k)
        ▷
        subSSE ← sse(smaller, larger, validate)
        ▷
        if localSSE < subSSE
            do smaller ← null
                larger ← null
          else
                localModel ← null

linReg(train, validate)
for ridge in 10^-8, 10^-4, 0.01, 0.1, 1, 10
    do model_r ← ridgeRegress(train, ridge)
        sse_r ← sse(model_r, validate)
if bestModel == model_10
    do build models for ridge = 100, 1000, ...
        and so on while improving
localModel ← bestModel
return minimum-sse-on-validation-data

randomSplit(train, k)
for i = 1 to k
    do splitAttr ← random_choice(allAttrs)
        stump ← stump(approx_median(splitAttr))
        compute sse(stump, train)
return minimum-sse stump
```

Fig. 1. Random model tree ensemble generator, defaults are $numTrees = 50$ and $k = 0.5 * data.numAttributes$

expensive: the data must be sorted according to this attribute, and then a linear scan can determine the best split for minimizing the weighted sum squared error (or a similar numeric loss function). Furthermore best splits are usually not balanced thus leading to potentially very skewed trees, i.e. trees where leaves can have vastly different numbers of examples. This in turn causes issues for the local model generation: to prevent against overfitting some form of regularization is needed. We use ridge regression [6], which like all such regularization methods depends on a user parameter, in this case the ridge value. The problem with large differences in leafsizes is that such regularization parameters strongly depend on the number of training examples. Thus no single good value exists that would work well for such skewed trees. Therefore they would need separate independent optimization at every single leaf, which is expensive.

Random model trees use an alternative approach: trees are approximately balanced by only splitting on the (estimated) median of some attribute. An approximate procedure for median computation was recently described in [12]. This procedure performs only two linear scans over the data to approximate the median. Random model trees employ this procedure for split selection and thus induce reasonably balanced trees, which guarantees that the tree generation complexity remains $O(NlogN)$ instead of approaching $O(N^2)$ for extremely skewed trees. Furthermore the approximate median computation itself is extremely fast, as it does not need to sort, and it also does not have to move around data in memory like QuickSelect or similar linear methods.

Regularization is tackled in a different way: for every tree generation iteration the full data is split randomly into two halves: a train set and a validation set. Similar to reduced error pruning [7] the validation set is used to prune back the final model tree. The validation set is also used to determine a good ridge value for the ridge regression procedure when computing local linear models.

Additionally, to prevent against extreme cases of extrapolation, each leaf (or hypercube) records the local minimum and maximum value for the target. Predictions from the local model are then compared to these thresholds and capped, if necessary. This simple procedure has proven very effective, as single extreme values can have a large influence on measures like root mean squared error, even after averaging multiple predictions from an ensemble of model trees.

Finally, as the trees are semi-random and therefore definitely not optimal in isolation, averaging an appropriate number of such trees is essential for good predictive performance. At least 30 trees should always be computed, and computing more (and sometimes a lot more) trees does further improve performance. Of course, due to the random nature of the process, adding more trees to an ensemble will never significantly degrade performance, but as for most ensemble methods any improvements diminish eventually.

The full ensemble generation procedure is depicted in Figure 1. As noted there, the procedure has only two user-settable parameters: the number of trees to generate, and the number of attributes to consider at each split decision. The default values of $numTrees = 50$ and $k = 0.5 * data.numAttributes$ seem to form a good compromise between speed and accuracy, and have been used in all experiments reported here. Regarding possible values of k we have noted that the extreme value of $k = 1$ usually generates more skewed trees, which take longer to build and also usually perform slightly worse. Values around the default k usually generate a good set of diverse and well-performing trees. This might be surprising, as in classification Random Forests usually use smaller numbers. As the random model trees do not search for the best possible splitting threshold, but instead use an approximate median split to ensure reasonably balanced trees, they can afford to test a larger number of attributes, without sacrificing diversity nor exhibiting overfitting behaviour.

3 Experiments

Random model trees are compared to linear regression, gaussian process regression, and additive groves. This comparison comprises datasets provided by Luis Torgo [13] and the UCI repository. Only datasets with at least 950 examples are included.

The different algorithms deal differently with either categorical or missing values, therefore all data was preprocessed by replacing categorical values with multiple binary indicator attributes; missing values were imputed using the respective attribute's mean value. This preprocessing should ensure that different algorithm-internal procedures do not impact the comparison.

All of the five algorithms have one or more tuning parameters which need careful optimization for good performance. Additive groves need an explicit validation set for tuning parameters, and the other three algorithms can be paired with some optimization procedure based on a validation set. Therefore the experiments reported here are 10 runs of three-fold cross-validation, where half of each training set was used as the "build" set and the other half was used as the "validation" set for internal parameter optimization. The results reported in the next section are averages of accuracies and total runtimes including time needed for internal parameter optimization. We view this as a fair procedure that reflects well the process that is necessary when deploying algorithms in real-world applications where tuning usually is essential for success. More details are listed below for every algorithm.

3.1 Linear Regression

We use the Weka [5] implementation of Linear (Ridge) Regression, perform no internal attribute selection, but vary the ridge parameter from 10^{-8} to 10^{10} in exponential steps of 10, selecting the value with the lowest squared error sum (SSE) on the validation set. Note that the complexity of linear regression is $O(N * K^2)$, provided $K < N$, where N is the number of examples, and K the number of attributes.

3.2 Gaussian Process Regression

For Gaussian Process Regression we have replaced the generic Weka implementation by a specialized version which firstly hard-codes a Radial Basis function kernel, and secondly uses a conjugate gradient descent solver [10]. These changes result in substantial speedups, usually between one and two orders of magnitude when compared to the standard Weka version. The implementation comprises two tuning parameters: the bandwidth of the kernel, and again a ridge value for the regression. Optimization employs a hill-climbing procedure over a grid of possible pairs of values: starting from some initial point all neighboring grid points are explored until a plateau is reached in terms of SSE on the validation set. The factor defining the grid for both the bandwidth and the ridge parameter is 2.0, i.e. doubles for going up, and halves for going down are used. Note that the complexity of the conjugate gradient descent solver is only $O(K * N^2)$, i.e. quadratic in the number of examples and linear in the number of attributes, as it is limited to at most 100 iterations. Usually 100 iterations are enough for full convergence, or at least for getting very close to full convergence. The real problem that GP faces is its memory consumption: as the kernel matrix needs to be precomputed, it takes $O(N^2)$ memory. Thus GPs become infeasible for the largest dataset employed, and also need on the order of about 28 gigabytes of memory for the second largest dataset.

3.3 Additive Groves

We use the C++ implementation as supplied by the authors. That implementation supplies both a "fast" and a "slow" training mode, as well as a Python script for iterative training improvements. As the results in the next section show, this implementation is substantially slower than the other three Weka-based algorithms. Therefore only the "fast" training mode was used. Better prediction would be achieved with more training, but is not feasible here. Even "fast" mode could not finish on the largest dataset. [11] do not discuss the theoretical computational complexity of additive groves, but one would expect $O(K * N * logN)$ behavior from a tree-based algorithm. The results further down seem to confirm this hypothesis, but also show a high constant factor when comparing to the new algorithm, random model trees.

3.4 Random Model Trees

A Weka-based Java implementation is used here. The number of randomly chosen attributes to evaluate is set to 50%. Ensemble size is fixed to 50 trees for parameter optimization search. The two parameters that would need optimizing in a standard model tree are tree-depth and the ridge value for the local linear regression models in the leaves of the tree. Due to the splitting of the full data into a train and a validation subset for every single tree these parameters are optimized in a dynamic fashion described above in Section 2. These optimizations happen internally and fully automatic without needing any user specified parameter values.

The complexity of building random model trees is $O(N * logN + N * K^2)$. The first term accounts for tree construction, which is independent of the number of attributes, as only a constant number of attributes will be considered for each split. The second term accounts for training of the local linear models and indicates a potential weakness of random model trees: large number of attributes can slow down training considerably. In the results reported below this was not an issue as all data had less than 100 attributes. This important issue will also be discussed further in the last section under future directions for research.

4 Results

In this section we compare all four algorithms both with respect to predictive performance as well as to efficiency, measured as build time. Results for the largest dataset are incomplete, showcasing the high resource needs and lack of scalability of some of the algorithms included in this comparison.

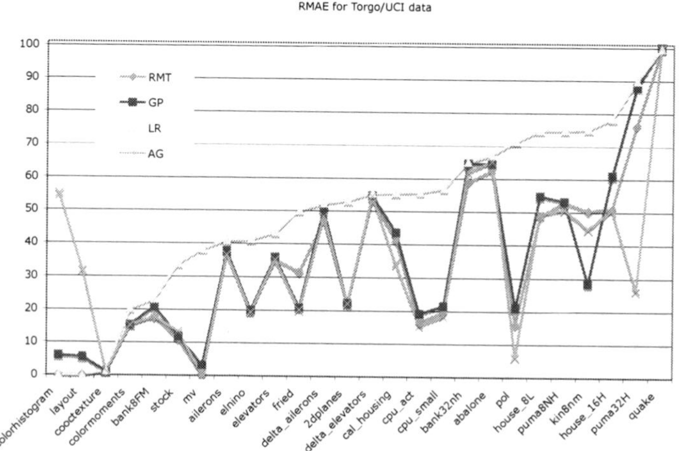

Fig. 2. RMAE for Torgo/UCI datasets, sorted by the linear regression result

4.1 Relative Mean Absolute Error

Relative mean absolute error (RMAE) is used here as a measure of accuracy, as it is a meaningful way of comparing across different datasets. In addition, a value of 100% indicates performance equal to simply always predicting the global mean, whereas a value of 0% indicates perfect prediction.

Figure 2 shows RMAE for all four algorithms over the Torgo/UCI datasets. The datasets have been sorted by the RMAE value of linear regression from low to high to facilitate comparison. All three non-linear algorithms usually improve over linear regression or are at least as good. This is absolutely true for Gaussian Processes regression, and also for Random Model Trees, which generally seem to perform very similar to GPs.

Additive Groves present a more varied picture, including both occasional large improvements over all other methods, e.g. on the "puma32H" data, as well as

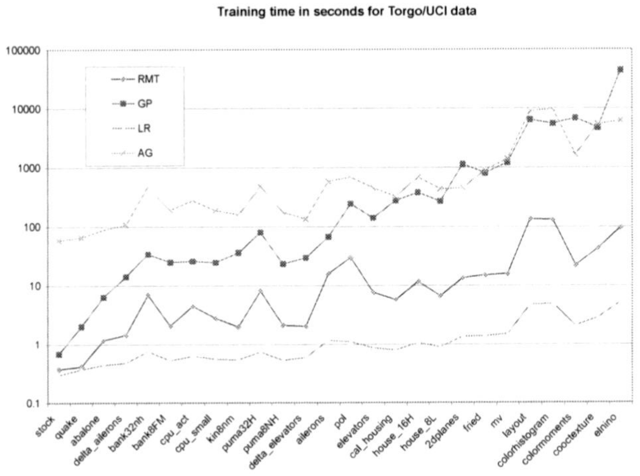

Fig. 3. Training time in seconds for Torgo/UCI datasets, sorted by the number of instances in each dataset; note the use of a logarithmic y-scale

occasional catastrophic failure, namely on the "colorhistogram", and the "layout" data. We suppose that using more complete search (using the "slow" setting or the iterative method) for AG on these latter datasets might improve performance to the level of GP or RMT.

The one dataset missing from Figure 2 is the largest set "census". Both GPs and AGs cannot process this data within reasonable amounts of resources. Table 1 displays the results on this dataset for linear regression, and for random model trees. RMT provides some improvement over linear regression, and does so within about 5.5 hours. Regarding AG, given a partial training run it is estimated that AG in "fast mode" would need about 25 days to complete. Note however, that AG also supports some parallelism, which was not used here. Similarly, search and ensemble construction of random model trees could be parallelized as well, so parallelism is really an orthogonal issue, which can safely be ignored in this comparison. GPs, or at least our implementation, cannot be employed for this dataset, as it would need about five terabytes of main memory to store the kernel matrix for the 800000 examples training set.

Figure 3 plots logarithmic build time in seconds for all algorithms over the Torgo/UCI datasets being sorted by number of instances, as this is the main complexity factor for most of the algorithms. Linear regression is the fastest for two reasons: all these datasets have 67 or fewer attributes, so the dominating factor is still the number of instances, and its influence is only linear; furthermore optimization concerns only one tuning parameter, looking at only 19 different values. Both tree-based methods RMT and AG show the expected $O(NlogN)$ behavior, but RMTs are consistently one to two order of magnitude faster. The variations visible are explainable by two factors: different numbers of attributes, but more importantly different number of steps in the hill-climbing

Table 1. Partial results for the UCI Census dataset, 2458285 examples in total, therefore about 800000 in the training fold

Method	RMAE	Time (secs)
LR	15.96	1205
RMT	9.78	19811
GP	?	? (would need 5 Tb RAM)
AG	?	? (estimated 2000000)

search, which at times terminates very quickly, and at other times explores a lot more parameter combinations. GPs start very fast for the smallest datasets, but their quadratic complexity is very apparent for larger datasets. For the "elnino" dataset a GP is already four orders of magnitude slower than linear regression.

5 Conclusions

We have introduced a new general regression method that combines model trees with random forests and some engineering detail in a novel way. A comparison to linear regression and to two other state-of-the-art regression algorithms over a substantial set of datasets of a wide range of properties has shown that the new algorithm can be competitive to the state of the art regarding predictive performance, but that it is considerably more efficient on datasets with relatively few attributes, and that it can scale reasonably to datasets of hundreds of thousands of examples. Still, when utmost predictive performance is needed in an application, an ensemble of well-tuned GPs, AGs, and RMTs would be the method of choice, provided enough computing resources are available.

There are a number of promising directions for future research. The most important one for random model trees is the issue of its complexity in the number of attributes. Either some form of local feature space reduction at each leaf in isolation, or some more global form of feature space reduction either per single tree or for the full ensemble can be explored. Local feature space reduction will have to be very careful with regard to runtime, but also potential loss of information. Another interesting direction will be investigating the possibility of a hybrid of the random model trees and additive groves ideas. And last but not least investigating efficient gaussian process regression for large datasets is a very challenging endeavour. Sparsification and gradient descent methods are potential candidates, but in regression settings they seem to trade off too much of the GP's predictive power for speed and memory savings. Again maybe a hybrid between Gaussian process regression and random model trees might provide a viable alternative.

References

1. Asuncion, A., Newman, D.J.: UCI Machine Learning Repository. University of California, School of Information and Computer Science, Irvine, CA (2007)
2. Breiman, L.: Bagging predictors. Machine Learning 24(2), 123–140 (1996)
3. Breiman, L.: Random Forests. Machine Learning 45(1), 5–32 (2001)
4. Fan, W., McClosky, J., Yu, P.S.: A General Framework for Accuracy and Fast Regression by Data Summarization in Random Decision Trees. In: KDD 2006, Philadelphia (August 2006)
5. Hall, M., Frank, E., Holmes, G., Pfahringer, B., Reutemann, P., Witten, I.H.: The WEKA Data Mining Software: An Update. SIGKDD Explorations 11(1) (2009)
6. Hoerl, A.E., Kennard, R.W.: Ridge regression: Biased estimation for nonorthogonal problems. Technometrics 12(3), 55–67 (1970)
7. Quinlan, J.R.: Simplifying decision trees. International Journal of Man-Machine Studies 27, 221–234 (1987)
8. Quinlan, J.R.: Learning with continuous classes. In: Proceedings Australian Joint Conference on Artificial Intelligence, pp. 343–348 (1992)
9. Rasmussen, C.E., Williams, C.K.I.: Gaussian Processes for Machine Learning. MIT Press (2006)
10. Saad, Y.: Iterative Methods for Sparse Linear Systems. Society for Industrial and Applied Mathematics (2003)
11. Sorokina, D., Caruana, R., Riedewald, M.: Additive Groves of Regression Trees. In: Kok, J.N., Koronacki, J., Lopez de Mantaras, R., Matwin, S., Mladenič, D., Skowron, A. (eds.) ECML 2007. LNCS (LNAI), vol. 4701, pp. 323–334. Springer, Heidelberg (2007)
12. Tibshirani, R.J.: Fast Computation of the Median by Successive Binning (2008) (unpublished manuscript), `http://stat.stanford.edu/~ryantibs/median/`
13. Torgo, L.: Regression data sets, `http://www.liaad.up.pt/~ltorgo/Regression/DataSets.html`
14. Wang, Y., Witten, I.H.: Induction of Model Trees for Predicting Continuous Classes. In: van Someren, M., Widmer, G. (eds.) ECML 1997. LNCS, vol. 1224, pp. 128–137. Springer, Heidelberg (1997)
15. Wold, H.: Soft Modeling by Latent Variables; the Nonlinear Iterative Partial Least Squares Approach. In: Gani, J. (ed.) Perspectives in Probability and Statistics, Papers in Honour of M.S. Bartlett, pp. 520–540. Academic Press (1975)

Supervised Subspace Learning with Multi-class Lagrangian SVM on the Grassmann Manifold

Duc-Son Pham and Svetha Venkatesh

Institute for Multi-sensor Processing and Content Analysis,
Curtin University, Perth, Western Australia

Abstract. Learning robust subspaces to maximize class discrimination is challenging, and most current works consider a weak connection between dimensionality reduction and classifier design. We propose an alternate framework wherein these two steps are combined in a joint formulation to exploit the direct connection between dimensionality reduction and classification. Specifically, we learn an optimal subspace on the Grassmann manifold jointly minimizing the classification error of an SVM classifier. We minimize the regularized empirical risk over both the hypothesis space of functions that underlies this new generalized multi-class Lagrangian SVM and the Grassmann manifold such that a linear projection is to be found. We propose an iterative algorithm to meet the dual goal of optimizing both the classifier and projection. Extensive numerical studies on challenging datasets show robust performance of the proposed scheme over other alternatives in contexts wherein limited training data is used, verifying the advantage of the joint formulation.

1 Introduction

Linear subspace learning has been an active area of research, especially in face recognition. Even though the dimensionality of the input data is high, previous works show that useful information for discriminating between classes can be found in a much lower dimensional space. Learning a suitable projection to this space not only removes noise to enhance classification performance but also helps uncover semantic meaning of discriminative information.

Supervised subspace learning algorithms vary widely in terms of the link between dimensionality reduction and classification. Most of the supervised subspace learning algorithms only consider weak connection between this two stages of the problem because the dimensionality reduction is not explicitly embedded in the classification formulation. Many supervised subspace learning algorithms implicitly assume a nearest neighbor classification and hence dimensionality reduction often relies on the Fisher-like criterion. For example, Fisherfaces (LDA) [1,5] maximizes the discrimination ratio of the between-class and within-class variances in the reduced subspace. Locality preserving projection (LPP) [8,4] on the other hand is formulated to preserve the local structure of the face. These can be generally unified under a graph embedding framework with different choices of graph configurations [16]. The Fisher criterion can also be replaced with other

D. Wang and M. Reynolds (Eds.): AI 2011, LNAI 7106, pp. 241–250, 2011.

statistical criterion, such as effective information [11]. However, the subsequent classification still uses nearest neighbor methods and this makes the connection weak. Some recent works, such as kernel dimensionality reduction (KDR) [6] and MLASSO [14], make the connection slightly more explicitly but the formulation is still in the regression framework and a direct classification error is still not used in dimensionality reduction. The exception is MLSVM [9] that considers a direct embedding of dimensionality reduction in a well-known learning framework of support vector machines (SVMs). The advantage and drawback of MLSVM is that it uses a simple linear multi-class SVM for classification, from which the manifold learning step is easily carried out using results from linear algebra. However, it also restricts the classification to a small hypothesis space of learning functions, leading to unsatisfactory performance and high computational cost.

To make the direct connection between dimensionality reduction and classification, we propose a new algorithm that learns an optimal subspace on the Grassmann manifold specifically for the SVM, embedding the classification error explicitly in the formulation. Our joint formulations differs from MLSVM [9] in two key aspects; first, we remove the restriction to linear SVMs and consider more general nonlinear SVMs with radial basis function (RBF) and polynomial kernels. Second, we use the Lagrangian SVM formulation [12] and deliver computational advantage. This is done through explicit derivation of the dual solution, removing the need to solve quadratic programming problems as customary in other SVM formulations. Like MLSVM [9], we consider the *all-versus-all* setting and derive the generalization of the basic binary Lagrangian SVM formulation.

The paper is organized as follows. In Section II, we explain the motivation for the proposed method, derive the new multi-class Lagrangian SVM in the *all-versus-all* setting, and show how the subspace can be jointly learned as an optimization on the Grassmann manifold in this new SVM formulation. Section III contains experimental results on challenging datasets, demonstrating the proposed method achieved lower error rates than recent subspace learning algorithms. Finally, Section IV concludes.

2 Proposed Method

To fix the notation, denote the original input data as $\mathbf{z} \in \mathbb{R}^N$. We seek a projection to a lower K-dimensional space $\mathbf{x} = \mathbf{P}^T\mathbf{z}$ via a linear projection matrix[1] $\mathbf{P} \in \mathbb{R}^{N\times K}$ which satisfies $\mathbf{P}^T\mathbf{P} = \mathbf{I}_K$. Hereinafter, we assume that K is already specified and we select K equal to the number of classes minus 1 for comparison with many subspace learning algorithms such as LDA. We propose to learn a functional relationship f between the input data $\mathbf{x} \in \mathcal{X}$ and the real-valued output data $y \in \mathcal{Y}$.

Denote as $R_{\mathrm{emp}}[f]$ the functional form of the empirical risk, then a standard formulation of the statistical learning theory can be equivalently expressed as

[1] $\mathbf{P}$ is actually an orthonormal transformation matrix and not a projection matrix in a strict linear algebra sense. However, we use the term *projection* for $\mathbf{P}$ in this work.

$\hat{f} = \arg\min_{f \in \mathcal{F}} R_{\text{emp}}[f], \quad C(f) \leq t$, where $C(f)$ represents the complexity of the learning function f.

In practice, a Lagrangian approach is often used to bring it to a more convenient form

$$\hat{g} = \hat{f} \circ \hat{\mathbf{h}} = \arg\min_{f,\mathbf{h}} \; R_{\text{emp}}[f, \mathbf{h}] + \lambda R_{\text{reg}}[f, \mathbf{h}], \tag{1}$$

where R_{reg} is the functional form for the regularization. In the SVM setting, it controls the (inverse) margin in the reproducing kernel Hilbert space (RKHS). Similar to many SVM settings, we restrict our attention to the RKHS for f. As $\mathbf{h}(\mathbf{z}) = \mathbf{P}^T\mathbf{z}$ and because of the restriction of orthonormal transformations $\mathbf{P}^T\mathbf{P} = \mathbf{I}_K$, finding $\mathbf{h}$ in its domain is equivalent to finding $\mathbf{P}$. Hereinafter, we shall refer to $\mathbf{P}$ rather than $\mathbf{h}$. To solve (1) we propose the following

The main algorithm
Step 1: Fix $\mathbf{P}$, solve the SVM problem
$f = \arg\min_{f \in \mathcal{F}} R_{\text{emp}}[f, \mathbf{P}] + \lambda R_{\text{reg}}[f, \mathbf{P}]$
Step 2: Fix f, solve
$\mathbf{P} = \arg\min_{\mathbf{P}} R_{\text{emp}}[f, \mathbf{P}] + \lambda R_{\text{reg}}[f, \mathbf{P}]$
Step 3: Check for convergence, else go back to Step 1.

The above procedure generates a sequence $\{f, \mathbf{P}\}$ which converges to a local minimum. At each step, a better solution for the projection and classifier is found. Note that this does not necessarily imply both the empirical risk and the regularized risk are reduced. However, their compromised linear combination via the regularization parameter λ is decreasing implying better generalization performance. We shall assume that λ is specified via standard cross-validation.

2.1 Multi-class Lagrangian SVM

We proposed to extend the basic Lagrangian SVM [12] to a multivariate version using the framework proposed in [10]. Denote the multivariate label for the jth class as a vector $\mathbf{y}_j \in \mathbb{R}^k$ of $-1/(k-1)$ except the jth position where it is 1. Assume that mis-classification is treated equally between other classes, the weighting for the the ith training sample which belongs to class j is $\mathbf{l}_i = \mathbf{l}(\mathbf{y}_i) \in \mathbb{R}^k$, which is a vector of 1's except the ith position where it is 0. Accordingly, the classification function is multivariate $\mathbf{f}(\mathbf{x}) = \left[f_1(\mathbf{x}) \; f_2(\mathbf{x}) \ldots f_\kappa(\mathbf{x})\right]^T \in \prod_{j=1}^k (\{1\} + \mathcal{H}_K)$ where $\mathcal{H}_{K_j} = \mathcal{H}_K, \;\; \forall j$. This implies the RKHS of each $f_j(\mathbf{x})$ are the same, i.e. they share the same kernel. The extra $\{1\}$ to the standard RKHS is due to the offset constant in the classification function. To avoid ambiguity, a sum-to-zero (simplex) constraint is needed $\sum_{j=1}^k f_j(\mathbf{x}) = 0$. Given these constraints, the generalized representer theorem in [10] states that the solution of standard SVM formulation admits the following representation

$$f_j(\mathbf{x}) = b_j + \sum_{i=1}^{n} c_{ji}\kappa(\mathbf{x}_i, \mathbf{x}), \tag{2}$$

where $\kappa(\bullet, \bullet)$ is the kernel associated with the RKHS $\mathcal{H}_K$.

The primal and dual problems. Extending the Lagrangian SVM to the multivariate case, we propose the following formulation

$$\arg \min_{f_j \in \mathcal{H}_K + \{1\}} \left\{ \frac{1}{2n} R_{\text{emp}}(\mathbf{f}) + \frac{1}{2}\lambda \sum_{j=1}^{k} \|f_j\|^2_{\mathcal{H}_K + \{1\}} \right\}, \tag{3}$$

where $R_{\text{emp}}(\mathbf{f}) = \sum_{i=1}^{n} (\mathbf{f}(\mathbf{x}_i) - \mathbf{y}_i)^T \mathsf{diag}(\mathbf{l}_i)(\mathbf{f}(\mathbf{x}_i) - \mathbf{y}_i)$ and n is the number of training samples. To simplify the notation, we swap the index of the sum of the first term from over training examples (i) to over classes (j). Denote the slack variable as $\boldsymbol{\xi}_j$ for $[f_j(\mathbf{x}_1), \ldots, f_j(\mathbf{x}_n)]^T - [y_{1j}, \ldots, y_{nj}]^T$. Let $\mathbf{L} = [\mathbf{l}_1, \ldots, \mathbf{l}_n]$ and denote $\mathbf{D}_j$ the diagonal matrix from the jth row of $\mathbf{L}$. Let $\mathbf{K} = [n \times n]$ be the kernel matrix and assume that $\mathbf{K}$ is positive definite $\mathbf{K} \succ \mathbf{0}$. Denote as $\mathbf{1}$ a vector of all 1's. Let $\mathbf{C} = [\mathbf{c}_1, \ldots \mathbf{c}_k]$ and denote $\mathbf{c}_j$ the jth column of $\mathbf{C}$. Let $\mathbf{Y} = [\mathbf{y}_1, \ldots, \mathbf{y}_n]$ and denote $\mathbf{v}_j^T$ the jth row of $\mathbf{Y}$. The *primal problem* is

$$\arg \min_{\boldsymbol{\xi}_j, \mathbf{c}_j, b_j} \quad \frac{1}{2} \sum_{j=1}^{k} \left\{ \boldsymbol{\xi}_j^T \mathbf{D}_j \boldsymbol{\xi}_j + n\lambda \left(\mathbf{c}_j^T \mathbf{K} \mathbf{c}_j + b_j^2 \right) \right\} \tag{4}$$

$$\text{s.t.} \sum_{j=1}^{k} \mathbf{1} b_j + \mathbf{K}\mathbf{c}_j = \mathbf{0}, \quad \mathbf{1} b_j + \mathbf{K}\mathbf{c}_j - \mathbf{v}_j \preccurlyeq \boldsymbol{\xi}_j \; j = 1, \ldots, k$$

Before solving this, we have an important remark that the constraints $\boldsymbol{\xi}_j \succcurlyeq \mathbf{0}$, $j = 1, \ldots, k$ which typically appear in standard SVM formulations, are not required due to the new formulation.

We introduce Lagrangian multipliers $\boldsymbol{\alpha}_j \succcurlyeq \mathbf{0}, \boldsymbol{\alpha}_j \in \mathbb{R}^n$, $j = 1, \ldots, k$ for the inequality constraints and $\boldsymbol{\delta} \in \mathbb{R}^n$ for the sum-to-zero constraint.

The *dual problem* can be written as

$$\arg \max_{\boldsymbol{\alpha}_j, \boldsymbol{\delta}} L_P + \sum_{j=1}^{k} \boldsymbol{\alpha}_j^T (\mathbf{1} b_j + \mathbf{K}\mathbf{c}_j - \mathbf{v}_j - \boldsymbol{\xi}_j) + \boldsymbol{\delta}^T \left(\sum_{j=1}^{k} \mathbf{1} b_j + \mathbf{K}\mathbf{c}_j \right),$$

$$\text{s.t.} \begin{cases} \frac{\partial L_D}{\partial \boldsymbol{\xi}_j} = \mathbf{D}_j \boldsymbol{\xi}_j - \boldsymbol{\alpha}_j = \mathbf{0}, & \frac{\partial L_D}{\partial \mathbf{c}_j} = n\lambda \mathbf{K}\mathbf{c}_j + \mathbf{K}(\boldsymbol{\alpha}_j + \boldsymbol{\delta}) = \mathbf{0}, \\ \frac{\partial L_D}{\partial b_j} = n\lambda b_j + \mathbf{1}^T(\boldsymbol{\alpha}_j + \boldsymbol{\delta}) = \mathbf{0}, \; \boldsymbol{\alpha}_j \succcurlyeq \mathbf{0} \end{cases} \tag{5}$$

for $j = 1, \ldots, k$, where L_P is the primal objective function in (4). This yields

$$\mathbf{c}_j = -(1/n\lambda)(\boldsymbol{\alpha}_j + \boldsymbol{\delta}), \quad b_j = -(1/n\lambda)\mathbf{1}^T(\boldsymbol{\alpha}_j + \boldsymbol{\delta}). \tag{6}$$

As $\mathbf{D}_j$ is a diagonal matrix of 0's and 1's hereinafter we assume that $\mathbf{D}_j$ is re-arranged so that the non-zero diagonal elements a collected to the top left block of $\mathbf{D}_j$ to simplify the notation. Likewise, we also assume that the entries of $\boldsymbol{\xi}_j$ and $\boldsymbol{\alpha}_j$ are also re-arranged to match with this rearrangement of $\mathbf{D}$ in the sense that they are consistent with the constraints

$$\mathbf{D}_j \boldsymbol{\xi}_j - \boldsymbol{\alpha}_j = \mathbf{0}, \quad j = 1, \ldots, k,$$

or equivalently in the new rearrangement

$$\begin{bmatrix} \mathbf{I}_{n_j} & \mathbf{0} \\ \mathbf{0} & \mathbf{0} \end{bmatrix} \begin{bmatrix} \tilde{\boldsymbol{\xi}}_j \\ \bar{\boldsymbol{\xi}}_j \end{bmatrix} = \begin{bmatrix} \tilde{\boldsymbol{\alpha}}_j \\ \bar{\boldsymbol{\alpha}}_j \end{bmatrix} \tag{7}$$

It follows that $\tilde{\boldsymbol{\xi}}_j = \tilde{\boldsymbol{\alpha}}_j$ and $\bar{\boldsymbol{\alpha}}_j = \mathbf{0}$ whilst $\bar{\boldsymbol{\xi}}_j$ can be arbitrary (because the weighting coefficients are zero so that whatever value it won't change the cost function). The dual problem can now be written more explicitly in terms of the dual variables

$$\arg\min_{\tilde{\boldsymbol{\alpha}}_j \succcurlyeq \mathbf{0}, \boldsymbol{\delta}} \left\{ \sum_{j=1}^{k} \frac{1}{2} \tilde{\boldsymbol{\alpha}}_j^T \tilde{\boldsymbol{\alpha}}_j + \mathbf{v}_j^T \boldsymbol{\alpha}_j + \frac{1}{2n\lambda} (\boldsymbol{\alpha}_j + \boldsymbol{\delta})^T \left(\mathbf{K} + \mathbf{1}\mathbf{1}^T\right) (\boldsymbol{\alpha}_j + \boldsymbol{\delta}) \right\} \tag{8}$$

with $\boldsymbol{\alpha}_j = [\tilde{\boldsymbol{\alpha}}_j^T \quad \mathbf{0}^T]^T$ as previously discussed. Let $\mathbf{H} = (1/n\lambda)(\mathbf{K} + \mathbf{1}\mathbf{1}^T)$, $\boldsymbol{\alpha} = [\boldsymbol{\alpha}_1^T, \ldots, \boldsymbol{\alpha}_k^T]^T$, $\mathbf{v} = [\mathbf{v}_1^T, \ldots, \mathbf{v}_k^T]^T$, then the *dual problem* is

$$\arg\min_{\tilde{\boldsymbol{\alpha}}_j \succcurlyeq \mathbf{0}, \boldsymbol{\delta}} \left\{ \frac{1}{2} \boldsymbol{\alpha}^T \mathbf{Q}_{11} \boldsymbol{\alpha} + \frac{k}{2} \boldsymbol{\delta}^T \mathbf{H} \boldsymbol{\delta} + \boldsymbol{\alpha}^T \mathbf{Q}_{12} \boldsymbol{\delta} + \mathbf{v}^T \boldsymbol{\alpha} \right\} \tag{9}$$

where $\mathbf{Q}_{11}$ is a block diagonal matrix of $\mathbf{D}_j + \mathbf{H}$ and $\mathbf{Q}_{12} = \mathbf{H} \otimes \mathbf{1}_k$ where $\otimes$ denotes the Kronecker product.

The dual solution. Consider the dual problem (9). If we fix $\boldsymbol{\alpha}$ then

$$\boldsymbol{\delta} = -(1/k)\mathbf{H}^{-1}\mathbf{Q}_{12}^T \boldsymbol{\alpha} = -(1/k) \sum_{j=1}^{k} \boldsymbol{\alpha}_j \tag{10}$$

where we have made use of the fact that there is no constraint on $\boldsymbol{\delta}$ and $\mathbf{H}$ is symmetric. On the other hand, if we fix $\boldsymbol{\delta}$ and solve for $\boldsymbol{\alpha}$, (9) becomes

$$\arg\min_{\tilde{\boldsymbol{\alpha}} \succcurlyeq \mathbf{0}} \left\{ (1/2) \boldsymbol{\alpha}^T \mathbf{Q}_{11} \boldsymbol{\alpha} + \boldsymbol{\alpha}^T \left(\mathbf{Q}_{12} \boldsymbol{\delta} + \mathbf{v}\right) \right\}. \tag{11}$$

In what follows, we denote as $\tilde{\boldsymbol{\beta}} = \tilde{\mathbf{v}} + \tilde{\mathbf{Q}}_{12}\boldsymbol{\delta}$ where $\tilde{\mathbf{Q}}_{12}$ and $\tilde{\mathbf{v}}$ are the reduced version of $\mathbf{Q}_{12}$ and $\mathbf{v}$ respectively that match with $\tilde{\boldsymbol{\alpha}}_j$ as indicated in (7). To solve (11), we rely on the following two results (Lemma 1 can be proved similarly using the approach in [12]).

Lemma 1. *For a strictly convex optimization problem,*

$$\arg\min_{\tilde{\boldsymbol{\alpha}} \succcurlyeq \mathbf{0}} \quad (1/2) \tilde{\boldsymbol{\alpha}}^T \tilde{\mathbf{Q}} \tilde{\boldsymbol{\alpha}} + \tilde{\boldsymbol{\beta}}^T \tilde{\boldsymbol{\alpha}}, \tag{12}$$

the solution satisfies $\tilde{\boldsymbol{\alpha}}^T(\tilde{\mathbf{Q}}\tilde{\boldsymbol{\alpha}} + \tilde{\boldsymbol{\beta}}) = 0$.

Lemma 2 (*Orthogonality and Clipping[12]*).

$$\{\mathbf{a}, \mathbf{b} \succcurlyeq \mathbf{0} \text{ and } \mathbf{a}^T\mathbf{b} = 0\} \Leftrightarrow \{\mathbf{a} = (\mathbf{a} - \gamma\mathbf{b})_+ \quad \forall \gamma > 0\}$$

where $(\bullet)_+$ *denotes the nonnegative operator.*

Now we discuss the solution for (11) using the KKT conditions[2, p.243]. From Lemmas 1 and 2, and f the fact that $\tilde{\mathbf{Q}}_{11}$ is block diagonal, (11) can be decomposed into k independent problems

$$\arg\min_{\tilde{\boldsymbol{\alpha}}_j \succcurlyeq \mathbf{0}_{n_j}} \left\{ \frac{1}{2} \tilde{\boldsymbol{\alpha}}_j^T (\mathbf{I}_{n_j} + \check{\mathbf{H}}_j) \tilde{\boldsymbol{\alpha}}_j + \tilde{\boldsymbol{\alpha}}_j^T (\tilde{\mathbf{H}}_j \boldsymbol{\delta} + \tilde{\mathbf{v}}_j) \right\}. \tag{13}$$

So actually the update for block $j = 1, \ldots, k$, becomes

$$\tilde{\boldsymbol{\alpha}}_j^{(i+1)} = \tilde{\mathbf{Q}}_j^{-1} \left[\left((\tilde{\mathbf{Q}}_j \tilde{\boldsymbol{\alpha}}_j^{(i)} + \tilde{\boldsymbol{\beta}}) - \rho_j \tilde{\boldsymbol{\alpha}}_j^{(i)} \right)_+ - \tilde{\boldsymbol{\beta}}_j \right], \tag{14}$$

where $\tilde{\mathbf{Q}}_j = \mathbf{I}_{n_j} + \check{\mathbf{H}}_j$ and $\tilde{\boldsymbol{\beta}}_j = \tilde{\mathbf{v}}_j + \tilde{\mathbf{H}}_j \boldsymbol{\delta}$ ($\tilde{\mathbf{H}}_j$ is the row-reduced version of $\mathbf{H}$ which matches with $\tilde{\boldsymbol{\alpha}}_j$. Similarly, $\check{\mathbf{H}}_j$ is both column-and-row-reduced). The convergence of this update rule is governed by the following result.

Theorem 1. *Provided that $0 < \rho_j \leq 2$ the update procedure for $\tilde{\boldsymbol{\alpha}}_j^{(i)}$ converges to the true global solution $\tilde{\boldsymbol{\alpha}}_j^*$ as*

$$\|\tilde{\mathbf{Q}}_j \tilde{\boldsymbol{\alpha}}_j^{(i+1)} - \tilde{\mathbf{Q}}_j \tilde{\boldsymbol{\alpha}}_j^*\|_2 \leq \|\mathbf{I}_{n_j} - \rho_j \tilde{\mathbf{Q}}_j^{-1}\|_2 \|\tilde{\mathbf{Q}}_j \tilde{\boldsymbol{\alpha}}_j^{(i)} - \tilde{\mathbf{Q}}_j \tilde{\boldsymbol{\alpha}}_j^*\|_2.$$

In summary, to solve for the dual variables of the optimization problem (9), one needs to iterate between $\boldsymbol{\delta}$ and $\boldsymbol{\alpha}$ using (10) and (14) until the dual objective function of (9) is converged (note: the step (14) to find $\boldsymbol{\alpha}_j$ is also iterative by itself). When the *dual* variables are determined, the *primal* variables can be easily deduced from (6). These primal variables constitute the SVM function.

2.2 Learning the Projection

We consider two popular kernels used in the SVM literature: the radial basis function (RBF) kernel: $K_{im}(\mathbf{P}) = \exp(-\mu\|\mathbf{P}^T\mathbf{z}_m - \mathbf{P}^T\mathbf{z}_i\|_2^2)$, with the smoothing parameter μ, and the polynomial kernel with degree d: $K_{im}(\mathbf{P}) = (1 + \mathbf{z}_m^T \mathbf{P}\mathbf{P}^T \mathbf{z}_i)^d$. As $\mathbf{c}_j$ and b_j are fixed, the objective function in terms of $\mathbf{P}$ is

$$g(\mathbf{P}) = R_{\text{emp}}(\mathbf{f}, \mathbf{P}) + \lambda \sum_{j=1}^{k} \mathbf{c}_j^T \mathbf{K}(\mathbf{P}) \mathbf{c}_j, \tag{15}$$

where we have written f as a function of $\mathbf{x}$ which implicitly depends on $\mathbf{P}$ as well. As $\mathbf{K}(\mathbf{P})$ is *invariant* to a right orthogonal transformation of $\mathbf{P}$, i.e. $\mathbf{K}(\mathbf{P}) = \mathbf{K}(\mathbf{P}\mathbf{R}), \mathbf{R}^T\mathbf{R} = \mathbf{I}$, the minimization of $g(\mathbf{P})$ is thus performed over the Grassmann manifold $\mathcal{G}_{N,K}$, which is the set of all k-dimensional subspaces:

$$\hat{\mathbf{P}} = \arg\min_{\mathbf{P} \in \mathcal{G}_{N,K}} g(\mathbf{P}). \tag{16}$$

To solve this optimization problem on the Grassmann manifold, we propose to use the manifold optimization framework in [13] for simpler computations.

The crucial step of this framework requires the gradient of the objective function $\mathbf{G} = \nabla_{\mathbf{P}} g(\mathbf{P})$ to be evaluated (see [13] for detail). Here, we explicitly derive the gradients for both cases. Denote the matrix function as $\mathbf{F}(\mathbf{P}^T\mathbf{Z}) = \left[\mathbf{f}(\mathbf{P}^T\mathbf{z}_1), \ldots, \mathbf{f}(\mathbf{P}^T\mathbf{z}_n)\right]$ and the residual matrix as $\boldsymbol{\Omega}(\mathbf{P}) = \mathbf{F}(\mathbf{P}^T\mathbf{Z}) - \mathbf{Y} = \mathbf{C}^T\mathbf{K}(\mathbf{P}) + \mathbf{b}\mathbf{1}^T - \mathbf{Y}$. Then, it can be easily seen that

$$g(\mathbf{P}) = (1/n)\|\boldsymbol{\Omega}(\mathbf{P}) \odot \mathbf{L}\|_F^2 + \lambda \mathsf{tr}\left[\mathbf{C}^T\mathbf{K}\mathbf{C}\right] \tag{17}$$

where $\odot$ denotes the element-wise product of two matrices.

Then we obtain the following results for simplified gradients on the Grassmann manifold:

Lemma 3. *For RBF kernel* $\nabla_{\mathbf{P}} g(\mathbf{P}) = -2\mu\mathbf{Z}\left(\mathsf{diag}\left[\bar{\mathbf{A}}\mathbf{1}\right] - \bar{\mathbf{A}}\right)\mathbf{Z}^T\mathbf{P}$ *where* $\mathbf{A} = (2/n)\mathbf{C}\boldsymbol{\Psi} \odot \mathbf{K}$, *and for polynomial kernel:* $\nabla_{\mathbf{P}} g(\mathbf{P}) = d\mathbf{Z}\left(\bar{\mathbf{A}}\right)\mathbf{Z}^T\mathbf{P}\mathbf{Z}^T$ *where* $\mathbf{A} = (2/n)\mathbf{C}\boldsymbol{\Psi} \odot \mathbf{K}^{\frac{d-1}{d}}$, *and* $\bar{\mathbf{A}} = \mathbf{A} + \mathbf{A}^T$.

In the above expressions, $\mathbf{K}$ is a matrix function of $\mathbf{P}$, and the notation $\mathbf{K}^{\frac{d-1}{d}}$ means each entry of $\tilde{\mathbf{K}}$ is raised with the power $\frac{d-1}{d}$ (element-wise).

Table 1. Error rates on the YaleB dataset

Train	2	3	4	5	6
PCA	0.84±0.02	0.80±0.01	0.77±0.01	0.74±0.01	0.72±0.01
R-LDA	0.60±0.03	0.48±0.02	0.40±0.02	0.35±0.02	0.30±0.02
R-LPP	0.63±0.02	0.56±0.02	0.51±0.02	0.46±0.02	0.42±0.02
OLPP	0.57±0.03	0.45±0.03	0.39±0.03	0.33±0.03	0.30±0.03
EI	0.56±0.01	0.48±0.03	0.38±0.02	0.34±0.02	0.31±0.02
KDR	0.57±0.02	0.47±0.02	0.36±0.03	0.34±0.01	0.30±0.01
MLSVM	0.79±0.02	0.71±0.02	0.64±0.01	0.54±0.02	0.51±0.04
SVM	0.53±0.03	0.44±0.02	0.40±0.02	0.36±0.02	0.33±0.02
GLSVM	0.53±0.03	0.40±0.02	0.33±0.02	0.28±0.02	0.24±0.02

3 Experiments

First, we illustrate the numerical properties of the proposed method, denoted as GLSVM, via a toy example with $k = 4$ classes, each with 4 samples whose (x, y) coordinates are distributed symmetrically over four quadrants (whilst their z coordinates are sampled randomly from the normal distribution). Here, we aim at learning a projection from 3D to 2D. We set $\mu = 0.1$ and $\lambda = 0.25$ for this experiment. Fig. 1 shows a typical reduction of the objective function versus the iteration number and the decision boundary for the 4 classes as projected onto the (x, y) plane (the actual subspace learned deviates from the (x, y) plane slightly due to noise). In this toy example, we found that the main algorithm converged in few iterations and that the decision boundary projected on the (x, y) plane is as expected. Next, we compare GLSVM with other

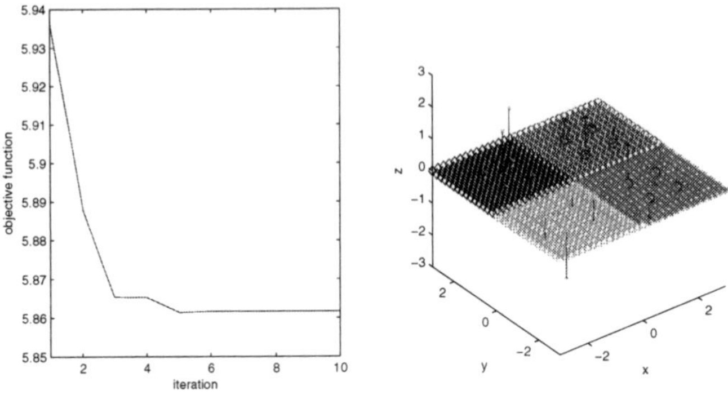

Fig. 1. A toy example (best seen in color)

recent subspace learning methods, including the regularized linear discriminant analysis (R-LDA), regularized locality preserving projection (R-LPP) [5], orthogonal locality preserving projection (OLPP) [4], effective information (EI) [11], kernel dimensionality reduction (KDR) [6], and the MLSVM of [9]. To demonstrate the advantage of the joint formulation, we also compare with the approach where dimensionality reduction and SVM (with RBF kernel) are learned separately (denoted as SVM).

Table 2. Error rates on the PIE dataset

Train	2	3	4	5	6
PCA	0.86±0.01	0.83±0.01	0.80±0.01	0.77±0.01	0.74±0.01
R-LDA	0.60±0.02	0.48±0.02	0.40±0.02	0.34±0.01	0.29±0.01
R-LPP	0.71±0.02	0.66±0.02	0.62±0.02	0.58±0.02	0.54±0.02
OLPP	0.66±0.03	0.56±0.02	0.49±0.02	0.45±0.02	0.42±0.02
EI	0.61±0.02	0.49±0.02	0.43±0.02	0.40±0.01	0.35±0.03
KDR	0.60±0.01	0.48±0.03	0.44±0.01	0.38±0.02	0.36±0.01
MLSVM	0.85±0.01	0.80±0.01	0.77±0.01	0.74±0.01	0.71±0.01
SVM	0.53±0.03	0.44±0.02	0.40±0.02	0.36±0.02	0.33±0.02
GLSVM	0.53±0.02	0.40±0.02	0.34±0.02	0.30±0.01	0.28±0.01

The datasets used in this experiment are well-known datasets. The original PIE database from Carnegie Mellon University [15] consists of 68 individuals with 41,368 images. This experiment uses a near-frontal subset of the PIE database obtained from [3]. In this subset, there are approximately 170 images per individual over about five different poses. We also select the YaleB face database [7] with 38 individuals, each having a total of 64 near-frontal images [3]. They are known to be difficult datasets in face recognition. With these two datasets, we generate 20 random splits. In each split, the images are randomly selected from each class for training, and the rest is used for testing. Then, we report the best average and standard deviation of the measured error rate.

In all cases, the pre-processing step involves cropping and resizing the faces to 32×32 gray-scale images, then centralizing about the mean, and finally normalizing each vector to unit norm. As we are only interested in small training sizes, we select 2-6 images for training. The parameters for R-LDA and LPP are the suggested values from the authors (in particular the choice of weighting matrix and the regularization $\alpha = 0.1$). To make it comparable to R-LDA, we set the projection onto a subspace with the dimension being the number of classes minus one. The reported PCA method is also based on the assumption of the same dimension. For all Grassmann based methods (EI, KDR, MLSVM, GLSVM), we use the LDA solution as an initialized dimensionality reduction. For methods with regularization and hyperparameters, it is possible to select the parameters via cross-validation for example. Here, we report the best performance as the number of training samples is small. For EI and KDR, we need to use subsampling (for EI maximum 100 terms for within and between distributions, for KDR: up to 5 neighbors) to evaluate the gradient as suggested by the authors because of prohibitive complexity with exact formulation.

Tables 1 and 2 show the average classification error rates on the YaleB and PIE datasets respectively. As can be shown, the proposed method outperforms other compared methods overall. When there are more training samples the joint formulation (GLSVM) does in fact improve over the disjoint alternative (SVM). Among other compared methods, R-LDA becomes more comparable when there are more training samples. We also found that the EI and KDR formulation perform similarly as their objective function does not change much with the LDA initialization. Additionally, the performance of MLSVM is not satisfactory, especially with PIE dataset, despite the fact that it was initialized with a reasonable LDA solution. It suggests that linear classifier is perhaps not suitable for this problem as the nonlinear classifier (SVM) performs much better.

As with other Grassmann manifold learning methods, e.g. EI and KDR, the manifold learning step takes the most computation. Due to explicit derivation of the gradient, we found that the manifold learning step of our method is still faster than that of EI and KDR for the aforementioned subsampling setting. The proposed method is suitable for the case where the number of training samples is small. In many practical situations, this assumption may be satisfied.

4 Conclusion

A new framework for robust learning of a discriminative projection has been proposed. The essence of the new approach is a joint formulation of the projection in a multi-class Lagrangian SVM problem. Intuitively, the algorithm selects the projection on the Grassmann manifold that yields the largest margin sum (defined in the corresponding augmented RKHS). By doing so, we have directly linked the two stages, i.e. dimensionality reduction and classification, which have been weakly made in most previous works on subspace learning. Thus, the meaning of "discriminative" for the projection is closely attached with the chosen classifier. Empirical results on challenging face datasets demonstrate the advantage of the new formulation over other alternatives in terms of reducing the classification errors when the number of training samples is small. Future work should address the current limitations of the approach in terms of computational complexity and the possibility of a semi-supervised setting where the projection is further constrained to yield a more robust performance.

References

1. Belhumeur, P., Hespanha, J., Kriegman, D.: Eigenfaces vs. Fisherfaces: Recognition using class specific linear projection. IEEE Transactions on Pattern Analysis and Machine Intelligence 19(7), 711–720 (1997)
2. Boyd, S., Vandenberghe, L.: Convex Optimization. Cambridge University Press (2004)
3. Cai, D.: Codes and datasets for subspace learning, `http://www.cs.uiuc.edu/homes/dengcai2/Data/data.html` (as retrieved on October 2007)
4. Cai, D., He, X., Han, J., Zhang, H.J.: Orthogonal Laplacianfaces for Face Recognition. IEEE Transactions on Image Processing 5(11), 3608–3614 (2006)
5. Cai, D., He, X., Hu, Y., Han, J., Huang, T.: Learning a spatially smooth subspace for face recognition. In: Proceedings CVPR (2007)
6. Fukumizu, K., Bach, F.R., Jordan, M.I.: Kernel dimension reduction in regression. Annals of Statistics 37(4), 1871–1905 (2009)
7. Georghiades, A., Belhumeur, P., Kriegman, D.: From few to many: Illumination cone models for face recognition under variablelighting and pose. IEEE Transactions on Pattern Analysis and Machine Intelligence 23(6), 643–660 (2001)
8. He, X., Niyogi, P.: Locality preserving projection. In: Proceedings NIPS (2003)
9. Ji, S., Ye, J.: Linear dimensionality reduction for multi-label classification. In: Proc. IJCAI (2009)
10. Lee, Y., Lin, Y., Wahba, G.: Multicategory support vector machines. Technical report, Department of Statistics, University of Wisconsin-Madison (2001)
11. Lin, D., Yan, S., Tang, X.: Pursuing informative projection on Grassman manifold. In: Proceedings CVPR (2006)
12. Mangasarian, O., Musicant, D.: Lagrangian support vector machines. Journal of Machine Learning Research 1, 161–177 (2001)
13. Manton, J.H.: Optimization algorithms exploiting unitary constraints. IEEE Transactions on Signal Processing 50(3), 4311–4322 (2002)
14. Pham, D.S., Venkatesh, S.: Robust learning of discriminative projection for multicategory classification on the stiefel manifold. In: Proceedings of the IEEE International Conference on Computer Vision and Pattern Recognition (CVPR 2008), Anchorage, Alaska, June 24-26 (2008)
15. Sim, T., Baker, S., Bsat, M.: The CMU pose, illumination, and expression (PIE) databse. IEEE Transactions on Pattern Analysis and Machine Intelligence 25(12), 1615–1618 (2003)
16. Yan, S., Xu, D., Zhang, B., Zhang, H.-J., Yang, Q., Lin, S.: Graph embedding and extensions: a general framework for dimensionality reduction. IEEE Transactions on Pattern Analysis and Machine Intelligence 29(1), 40–51 (2007)

Bagging Ensemble Selection

Quan Sun and Bernhard Pfahringer

Department of Computer Science,
The University of Waikato,
Hamilton, New Zealand
{qs12,bernhard}@cs.waikato.ac.nz

Abstract. Ensemble selection has recently appeared as a popular ensemble learning method, not only because its implementation is fairly straightforward, but also due to its excellent predictive performance on practical problems. The method has been highlighted in winning solutions of many data mining competitions, such as the Netflix competition, the KDD Cup 2009 and 2010, the UCSD FICO contest 2010, and a number of data mining competitions on the Kaggle platform. In this paper we present a novel variant: bagging ensemble selection. Three variations of the proposed algorithm are compared to the original ensemble selection algorithm and other ensemble algorithms. Experiments with ten real world problems from diverse domains demonstrate the benefit of the bagging ensemble selection algorithm.

1 Introduction

The problem of constructing an ensemble of classifiers from a library of base classifiers has always been of interest to the data mining community. Usually, compared with individual classifiers, ensemble methods are more accurate and stable. We here reproduce the mathematical expression used in [8] to illustrate the idea of ensemble learning: let x be an instance and $m_i, i = 1...k$, a set of base classifiers that output probability distributions $m_i(x, c_j)$ for each class label $c_j, j = 1...n$. The output of the final classifier ensemble $y(x)$ for instance x can be expressed as:

$$y(x) = \arg\max_{c_j} \sum_{i=0}^{k} w_i m_i(x, c_j), \qquad (1)$$

where w_i is the weight of base classifier m_i. In this particular form, ensemble learning strategies can be seen as methods for calculating optimal weights for each base classifier in terms of a classification goal. Since the mid-90's, many ensemble methods have been proposed. For a more detailed review of recent developments please refer to [2,9].

Before introducing the new methods, we briefly review bagging (bootstrap aggregating) [3] and the ensemble selection algorithm proposed in [5]. Bagging is based on the instability of base classifiers, which can be exploited to improve the predictive performance of such unstable base classifiers. The basic idea is

D. Wang and M. Reynolds (Eds.): AI 2011, LNAI 7106, pp. 251–260, 2011.

that, given a training set T of size n and a classifier A, bagging generates m new training sets with replacement, T_i, each of size $n' \leq n$. Then, bagging applies A to each T_i to build m models. The final output of bagging is based on simple voting [2].

Ensemble selection is a method for constructing ensembles from a library of base classifiers [5]. Firstly, base models are built using many different machine learning algorithms. Then a construction strategy such as forward stepwise selection, guided by some scoring function, extracts a well performing subset of all models. The simple forward model selection based procedure proposed in [5] works as follows: (1) start with an empty ensemble; (2) add to the ensemble the model in the library that maximizes the ensemble's performance to the error metric on a hillclimb set; (3) repeat Step 2 until all models have been examined; (4) return that subset of models that yields maximum performance on the hillclimb set. One advantage of ensemble selection is that it can be optimised for many common performance metrics or a combination of metrics. For variants of the ensemble selection algorithm, the reader is referred to [4,5]. In the next section, we will describe the proposed bagging ensemble selection algorithms and explain the motivation of combining bagging and ensemble selection.

2 Bagging Ensemble Selection

Based on the data sets and comparison results from [5], the simple forward model selection based ensemble selection algorithm is superior to many other well-known ensemble learning algorithms, such as stacking with linear regression at the meta-level, bagging decision trees, and boosting decision stumps. However, sometimes ensemble selection overfits the hillclimbing set, reducing the performance of the final ensemble. Figure 1(a) shows the hillclimb and test set learning curves of running ensemble selection on a data set. The red curve is the hillclimb set performance and the blue curve is the test set performance. It demonstrates that as the number of models in the model library increases, the performance (in terms of AUC) of ensemble selection on the hillclimb set gradually increases. However, the corresponding performance on the test set does not

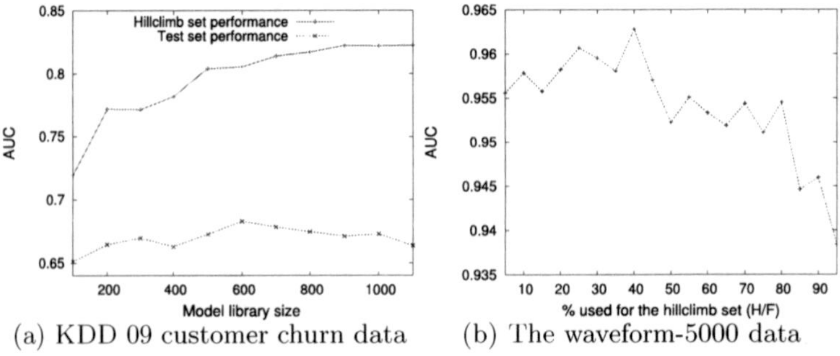

(a) KDD 09 customer churn data (b) The waveform-5000 data

Fig. 1. Ensemble selection hillclimb and test set learning curves

always increase; it may reach a peak (local or global) and then gradually decline. Also, as indicated in [5], for certain data sets, the root-mean-squared-error metric sometimes can decline very quickly. To overcome this problem, the authors of [5] proposed three additions to the simple forward selection procedure to reduce the chance of hillclimb set overfitting. The proposed additions are: (1) selection with replacement, where each individual classifier can be selected multiple times, which means some classifiers get larger weights than others; (2) sorted ensemble initialization, where instead of starting with an empty ensemble, models in the library are sorted by their performance, and the best N models are put into the initial ensemble; (3) "bagged" ensemble selection, where K groups (bags) of models are randomly selected from the model library, and ensemble selection is done inside each bag; the final ensemble is the union of the subsets selected for each of the bags. All three procedures also introduce additional parameters to the simple ensemble selection algorithm.

Furthermore, there is one more issue: how much data should be used for the hillclimb set? Figure 1(b) shows a typical test set learning curve for running ensemble selection with hillclimb sets of varying sizes. Assume the training set is F, and the hillclimb set H is a subset of F. Here, the x-axis shows the ratio H/F and indicates the percentage of F that is used for the hillclimb set. Based on the learning curve, we can see that the performance of ensemble selection is not stable, and is related to how much data is used for H. In the figure, there is a performance peak at $x = 40\%$, but performance starts to drop from $x = 50\%$. Different data sets may have different optimal ratios, which usually can be found only by using cross-validation. Therefore, this parameter indirectly increases the complexity of ensemble selection. Based on these observations, we propose a new ensemble learning algorithm called bagging ensemble selection: if we view the simple forward ensemble selection algorithm as an unstable base classifier, then we can apply the bagging idea to construct an ensemble of simple ensemble selection classifiers, which should be more robust than an individual ensemble selection classifier. In addition, the respective out-of-bag samples can be used as the hillclimb set. Specifically we will use the following three variations of bagging ensemble selection.

The **BaggingES-Simple** algorithm is the straightforward application of bagging to ensemble selection, with ensemble selection being the base classifier inside bagging. In this algorithm, the amount of data used for the hillclimb set is still a user-specified parameter (with a default of 30%). Each bootstrap sample is split into a train and a hillclimbing set according to this parameter.

The **BaggingES-OOB** algorithm uses the full bootstrap sample for model generation, and the respective out-of-bag instances as the hillclimb set for selection. The bootstrap sample is expected to contain about $1 - 1/e \approx 63.2\%$ of the unique examples of the training set [1,3]. Therefore the hillclimb set (out-of-bag sample) is expected to have about $1/e \approx 36.8\%$ unique examples of the training set for each bagging iteration. An advantage of BaggingES-OOB is that the user does not need to choose the size of the hillclimb set. Figure 2 shows the pseudocode for training the BaggingES-OOB ensemble.

Inputs:
Training set S; Ensemble Selection classifier E; Integer T (number of bootstrap samples)

Basic procedure:
```
for i = 1 to T {
    S_b = bootstrap sample from S (i.i.d. sample with replacement)
    S_oob = out of bag sample
    train base classifiers (can be a diverse model library) in E on S_b
    E_i = do ensemble selection based on base classifiers' performance on S_oob
}
```

Fig. 2. Pseudocode of the BaggingES-OOB algorithm

The **BaggingES-OOB-EX** algorithm is an extreme case of BaggingES-OOB, where in each bagging iteration only the single best classifier (in terms of performance on the hillclimb set) is selected. Therefore, if the number of bagging iterations is set M, then the final ensemble size will be exactly M as well.

3 Experimental Results

We experiment with ten classification problems. All of them are real world data sets which can be downloaded from the UCI repository [6], the UCSD FICO data mining contest website[1] and the KDD Cup 2009 website[2]. These data sets were selected because they are large enough, and they come from very different research and industrial areas. Table 1 shows the basic properties of these data sets. To make experiments possible for large model libraries, selecting from thousands of base classifiers, all five multiclass data sets were converted to binary problems by keeping only the two largest classes each. After this conversion to binary problems, for data sets that are larger than 10,000 instances, a subset of 10,000 instances is randomly selected for our experiments. Table 1 (in the rightmost column) shows the basic properties of the final data sets.

Ensemble selection is not restricted by the type of base classifiers used. Theoretically, any classifier can be used as a base classifier for ensemble selection. In this paper, the WEKA [7] implementation of the random tree classifier is used as the base classifier for all experiments. There are two reasons for focussing solely on random trees as base classifiers. The first one is simplicity: just by varying a single parameter, the random seed, we can obtain a large and relatively diverse model library. The second one is fair comparsion: most other ensemble methods are limited to uniform base classifiers. To speed up our experiments, parameter K of the random tree, the number of random attributes, is always set to 5, and the minimum number of instances at each leaf node is set to 50. In [5], the

[1] The University of California, San Diego and FICO 2010 data mining contest, http://mil.ucsd.edu/

[2] The KDD Cup 2009, http://www.kddcup-orange.com/

Table 1. Data sets: basic characteristics

Data set with release year	#Insts	Atts:Classes	Class distribution (#Insts)
Adult 96	48,842	14:2	23% vs 77% (10,000)
Chess 94	28,056	6:18	48% vs 52% (8,747)
Connect-4 95	67,557	42:3	26% vs 74% (10,000)
Covtype 98	581,012	54:7	43% vs 57% (10,000)
KDD09 Customer Churn 09	50,000	190:2	8% vs 92% (10,000)
Localization Person Activity 10	164,860	8:11	37% vs 63% (10,000)
MAGIC Gamma Telescope 07	19,020	11:2	35% vs 65% (10,000)
MiniBooNE Particle 10	130,065	50:2	28% vs 72% (10,000)
Poker Hand 07	1,025,010	11:10	45% vs 55% (10,000)
UCSD FICO Contest 10	130,475	334:2	9% vs 91% (10,000)
Original data sets			Final binary data sets

authors have shown that ensemble selection can be optimised to many common evaluation metrics. Bagging ensemble selection inherits this very useful feature; the goal metric is therefore a user-specified parameter. In this paper, the AUC (area under the ROC curve) metric is used for all experiments.

The following sections present two sets of results. One shows the results from comparing the three bagging ensemble selection algorithms to the simple forward ensemble selection algorithm (ES) and the ES++ algorithm, which is the improved version of ES with the three additions, as described in the introduction. This is followed by an analysis of the final ensemble sizes for these algorithms. The other set of results shows a comparison between bagging ensemble selection and other ensemble learning algorithms.

3.1 Comparison of Bagging Ensemble Selection Algorithms to the Forward Ensemble Selection Algorithms

In this experiment the following setup is used: the number of bags (bagging iterations) for BaggingES-Simple, BaggingES-OOB and BaggingES-OOB-EX is set to 50. For each data set, we run 10 experiments per algorithm, increasing the size of the model library per bag by 10 for each successive experiment: from 10 to 20, then to 30 and so on until 100 for the tenth experiment. For example, when the size of the model library is 100, then, in total, 5,000 base classifiers (random trees) are trained. Accordingly, we run 10 experiments on each data set for the ES algorithm and the ES++ algorithm (hillclimb ratio is set to 30% for both ES and ES++) that we want to compare. The size of the model library increases by 500 in each successive experiment, from a base 500 to 1,000, then 1,500 until it reaches 5,000 in the tenth experiment, which means all five algorithms in the comparison use the same number of base classifiers in each individual experiment. Also, for the ES++ algorithm, the number of subgroups is set to 50.

Figure 4 shows the test set learning curves of the ES algorithm, the ES++ algorithm, and the three bagging ensemble selection algorithms based on 500 individual experiments (5 algorithms, 10 data sets, 10 different model library sizes per data set). For each experiment, the algorithms are trained on 66% of the data set and evaluated on the other 34%. We repeated each experiment five times and the mean values were used for generating the figures and comparison. Based on Figure 4, we can see that ES and ES++ outperform bagging ensemble selection when the size of the model library is greater than 1,000 on the Adult-96 data set. For all other nine data sets, bagging ensemble selection, particularly BaggingES-OOB (blue curves) and BaggingES-OOB-EX (green curves), clearly outperform the ES algorithm and the ES++ algorithm. For data sets Chess-94, KDD-09 and Localization-10, BaggingES-OOB and BaggingES-OOB-EX gave similar performance.

An interesting pattern is that, for data sets Connect-4-95, Magic-07 and UCSD-10, the test performance of BaggingES-OOB-EX declines as the size of the model library increases. This is probably due to the fact that model diversity is more important for these data sets than for others. Thus, as the model library gets larger and larger, the best base classifier of each of the 50 bags of BaggingES-OOB-EX might become more similar to each other, thus losing model diversity.

For 6 out of 10 model library sizes, the BaggingES-Simple algorithm outperforms all other algorithms on the UCSD-10 data set. The ES++ algorithm outperforms other algorithms on the UCSD-10 data set when model library sizes are 500 and 5,000, but had a relatively poor performance when model library size is 1,000. Again, we can see that, for Covtype-98, KDD-09, MiniBooNe-10 and UCSD-10, the learning curves of the ES algorithm are not very stable. Figure 3 (left panel) shows the histogram presentation of the performance in terms of the number of wins for each algorithm over the ten data sets. We can see that BaggingES-OOB and BaggingES-OOB-EX are the top two winners.

Next, we look at the final ensemble sizes of ES, ES++, BaggingES-OOB, BaggingES-OOB-EX and BaggingES-Simple. Figure 5 shows the relationship

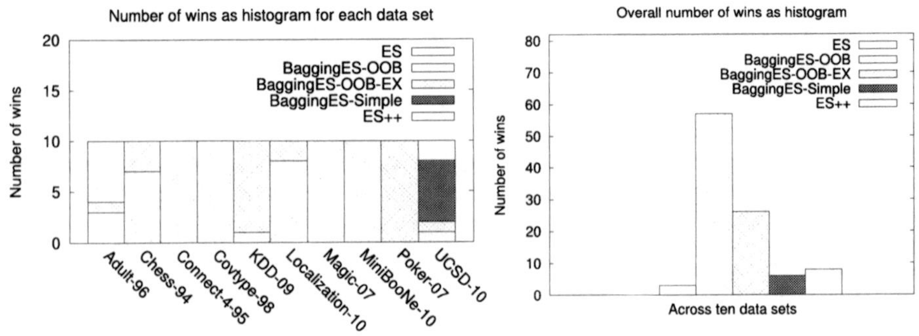

Fig. 3. Histogram presentation for counting number of wins for each algorithm

between model library size and the final ensemble size for these algorithms on the ten data sets. Please note that the final ensemble size of BaggingES-OOB-EX is always 50 because the number of bagging iterations is set to 50. Except for the BaggingES-OOB-EX algorithm, we can see that the final ensemble size of the other four ensemble algorithms increases linearly or sublinearly as the size of the model library increases (note that the y-axis is logarithmic). The final ensemble size of BaggingES-OOB, ES, and ES++ grows relatively faster than BaggingES-Simple's ensemble size. One possible reason is that in Bagging-OOB-Simple, the size of the build set (training set excluding the hillclimb set) is relatively small compared to BaggingES-OOB. Theoretically, for BaggingES-OOB, the hillclimb set (out-of-bag sample) has 36.8% unique instances of the training set, and the training set has 63.2% unique instances; however, BaggingES-Simple uses the bootstrap sample for both training and hillclimbing. For this experiment, the hillclimb ratio for BaggingES-Simple is set to 30%, thus its hillclimb set has fewer unique instances than BaggingES-OOB's hillclimb set. Therefore adding more base classifiers to BaggingES-Simple's model library may not necessarily improve the hillclimb performance since the hillclimb set might be too simple and the local hillclimb performance maximum could be achieved quickly.

Another interesting pattern is that ES has a much smaller ensemble size than BaggingES-OOB and BaggingES-Simple have. This could be because the local performance maximum of ES on the hillclimb set can be achieved more quickly compared to bagging ensemble selection. Again, adding more base classifiers to ES's model library may not necessarily improve the hillclimb performance.

Based on those observations, it seems that one reason for the good performance of BaggingES-OOB is that it usually has a larger final ensemble compared to all other algorithms. However, this does not imply that a larger final ensemble always yields better predictive performance. Refer to the learning curves in Figure 4, for data sets Chess-94, KDD-09 and Poker-07: BaggingES-OOB-EX's performance is competitive with BaggingES-OOB even though its final ensemble size is only 50. Therefore, whenever final ensemble size is crucial, for example, when an application requires fast real-time prediction, then the BaggingES-OOB-EX algorithm should be considered.

To sum up, we conclude that the advantage of the BaggingES-OOB algorithm and the BaggingES-OOB-EX algorithm over ES/ES++ is that their ensembles are evaluated on diverse hillclimb sets generated by the bagging procedure, and therefore are more robust and stable.

3.2 Comparison of Bagging Ensemble Selection Algorithms to Other Ensemble Learning Algorithms

In this experiment, we compare BaggingES-OOB (the most successful variant of the bagging ensemble selection based algorithms) to other popular ensemble learning methods. The following algorithms (WEKA [7] implementations)

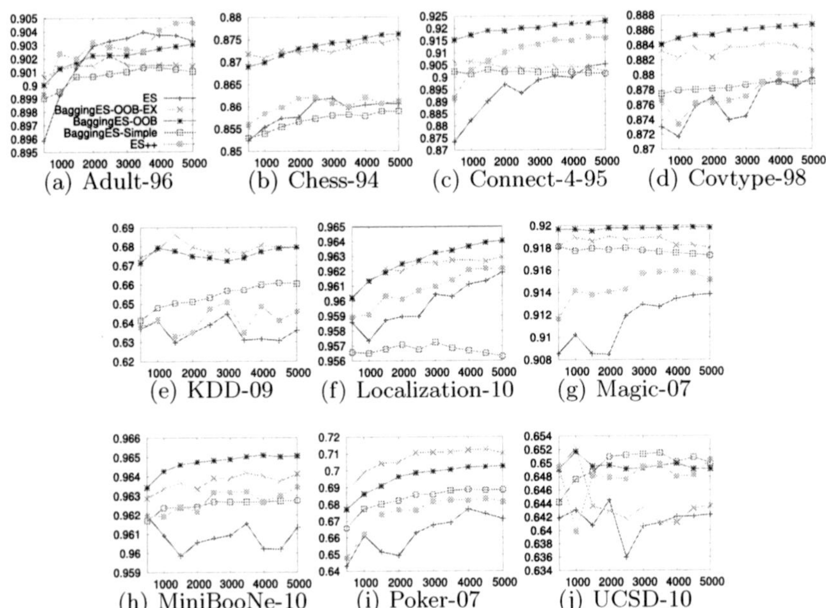

Fig. 4. Learning curves of ES, ES++ and the three bagging ensemble selection algorithms. X-axis is the model library size; y-axis is the AUC performance.

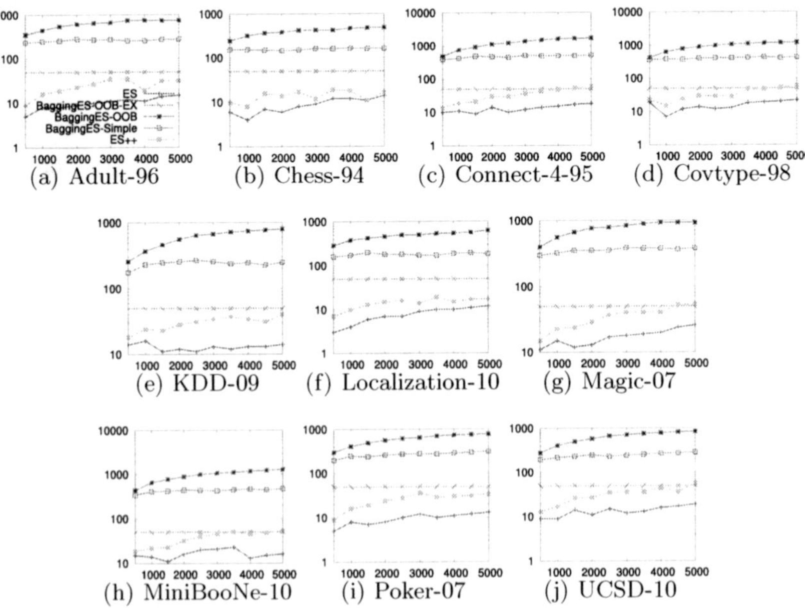

Fig. 5. Final ensemble sizes of ES, ES++ and the three bagging ES based algorithms. X-axis is the model library size; y-axis is the final ensemble size in logarithmic scale.

Table 2. Mean and standard deviation of the AUC performance of BaggingES-OOB and five other popular ensemble learning methods

Data set	BES-OOB	Voting	Stacking	AdaBst.M1	RandomFrst	ES++
Adult-96	0.905±0.001	0.902±0.002*	0.892±0.004*	0.783±0.008*	0.902±0.002*	0.906±0.002
Chess-94	0.875±0.004	0.859±0.003*	0.841±0.011*	0.971±0.002∘	0.862±0.004*	0.866±0.003*
Connt-4-95	0.918±0.006	0.911±0.006*	0.897±0.007*	0.905±0.005*	0.912±0.006*	0.916±0.005
Covtype-98	0.884±0.002	0.882±0.002*	0.875±0.004*	0.878±0.003*	0.882±0.002*	0.881±0.001*
KDD-09	0.678±0.029	0.678±0.027	0.656±0.031*	0.580±0.011*	0.675±0.029	0.669±0.029
Localiz-10	0.966±0.002	0.957±0.002*	0.940±0.006*	0.938±0.004*	0.960±0.002*	0.963±0.003*
Magic-07	0.920±0.004	0.916±0.004*	0.910±0.004*	0.868±0.005*	0.919±0.004*	0.913±0.002*
MiniB-10	0.964±0.002	0.963±0.002*	0.959±0.002*	0.928±0.006*	0.963±0.002*	0.963±0.001*
Poker-07	0.697±0.018	0.660±0.022*	0.620±0.041*	0.740±0.007∘	0.674±0.018*	0.671±0.020*
UCSD-10	0.649±0.011	0.648±0.008	0.612±0.016*	0.632±0.010*	0.646±0.008	0.646±0.007*
(win/tie/loss)		(0/2/8)	(0/0/10)	(2/0/8)	(0/2/8)	(0/3/7)

"*" BaggingES-OOB is significantly better, "∘" BaggingES-OOB is significantly worse, level of significance 0.05

are evaluated: Voting with probability averaging, stacking with linear regression at the meta-level (Stacking), AdaBoostM1, and RandomForest. ES++ is also included for comparison. All ensemble algorithms use the random tree as the base classifier. The total number of base classifiers allowed to be trained for each ensemble algorithm is equal. For bagging ensemble selection the number of bags is set to 50, and the number of base classifiers of individual ensemble selection in each bag is set to 100; thus in total 5,000 base classifiers (random trees) are trained. For other ensemble algorithms, the number of base classifiers is set to 5,000. The training complexity of random tree is $O(nlogn)$, where n is the size of the training set. In this experiment, all ensemble algorithms train on the same number of random trees, therefore the training costs for the model library of each ensemble algorithm in this comparison are roughly the same.

Table 2 shows the performance of each algorithm on the ten data sets. Standard deviations and significant test results were calculated from five independent runs of 66% (training) versus 34% (testing) split validation. The results for which a significant difference with BaggingES-OOB was found, are marked with a "*" or "∘" next to them. An asterisk "*" next to a result indicates that BaggingES-OOB was significantly better than the respective method (column) for the respective data set (row). A circle "∘" next to a result indicates that BaggingES-OOB was significantly worse than the respective method. We can see that AdaBoost.M1 significantly outperforms BaggingES-OOB on the Chess-94 and the Poker-07 data sets. On the other eight data sets, BaggingES-OOB is competitive (7 ties) to or superior (41 significant wins) to all other ensemble algorithms.

4 Conclusions

Ensemble selection is a popular ensemble learning method. Over the past several years, ensemble selection has been empirically examined and has proven to be a very effective and accurate ensemble learning strategy. One disadvantage of ensemble selection is that it is unstable and sometimes overfits the hillclimb set. In this paper, to further improve ensemble selection we proposed using the bagging strategy, which utilises the unstable property, to reduce the variance of a single ensemble selection. Our experiments on ten real world problems show that the bagging ensemble selection, especially BaggingES-OOB, which uses the out-of-bag sample as the hillclimb set, yields a robust and more accurate classifier ensemble than the original ensemble selection.

When the underlying problem requires fast prediction, we suggest using BaggingES-OOB-EX instead, because the user can control the size of the final ensemble. In terms of predictive performance, bagging ensemble selection is also competitive (in many cases, superior) to other state-of-art ensemble learning algorithms, such as voting, random forest, stacking and boosting. Again, bagging ensemble selection is not restricted by the type of base classifiers.

We experimented with only one type of base classifier in this paper, but to get the best out of the algorithm, we suggest using a more diverse model library. The bagging ensemble selection idea can be easily generalised to regression problems, since bagging is applicable to both classification and regression. In future research, we will compare bagging ensemble selection to other ensemble methods for regression problems. The success of the proposed methods on the diverse data sets selected for the study strongly suggests the applicability of the bagging ensemble selection algorithm to a wide range of problems.

References

1. Bauer, E., Kohavi, R.: An empirical comparison of voting classification algorithms: bagging, boosting, and variants. Machine Learning 1(38) (1998)
2. Brazdil, P., Giraud-Carrier, C., Soares, C., Vilalta, R.: Metalearning: Application to Data Mining. Springer, Heidelberg (2009)
3. Breiman, L.: Bagging predictors. Machine Learning 24(2), 123–140 (1996)
4. Caruana, R., Munson, A., Niculescu-Mizil, A.: Getting the most out of ensemble selection. In: Proceedings of the Sixth International Conference on Data Mining, ICDM 2006 (2006)
5. Caruana, R., Niculescu-Mizil, A., Crew, G., Ksikes, A.: Ensemble selection from libraries of models. In: Proceedings of the Twenty-First International Conference on Machine Learning, ICML 2004 (2004)
6. Frank, A., Asuncion, A.: UCI machine learning repository (2010), `http://archive.ics.uci.edu/ml`
7. Hall, M., Frank, E., Holmes, G., Pfahringer, B., Reutemann, P., Witten, I.: The weka data mining software: An update. SIGKDD Explorations 11(1) (2009)
8. Partalas, I., Tsoumakas, G., Vlahavas, I.: An ensemble uncertainty aware measure for direct hill climbing ensemble pruning. Machine Learning 81(3) (2010)
9. Rokach, L.: Ensemble-based classifiers. Artificial Intelligence Review 33, 1–39 (2010)

Augmented Spatial Pooling

John Thornton, Andrew Srbic, Linda Main, and Mahsa Chitsaz

Institute for Integrated and Intelligent Systems, Griffith University, QLD, Australia
j.thornton@griffith.edu.au,
{andrew.srbic,linda.main,mahsa.chitsaz}@griffithuni.edu.au

Abstract. It is a widely held view in contemporary computational neuroscience that the brain responds to sensory input by producing sparse distributed representations. In this paper we investigate a brain-inspired spatial pooling algorithm that produces such sparse distributed representations by modelling the formation of proximal dendrites associated with neocortical minicolumns. In this approach, distributed representations are formed out of a competitive process of inter-column inhibition and subsequent learning. Specifically, we evaluate the performance of a recently proposed binary spatial pooling algorithm on a well-known benchmark of greyscale natural images. Our main contribution is to augment the algorithm to handle greyscale images, and to produce better quality encodings of binary images. We also show that the augmented algorithm produces superior population and lifetime kurtosis measures in comparison to a number of other well-known coding schemes.

1 Introduction

Advances in computational neuroscience over the last twenty years have produced increasingly realistic and viable models of the functioning of the mammalian neocortex. These advances provide a compelling evidence-based picture of the kinds of physical processes and structures that underpin natural intelligence – a picture that suggests various computational realisations. In the current paper, we investigate a computational model of the neocortex proposed by Jeff Hawkins, known as *hierarchical temporal memory* (HTM) [3]. Hawkins first published his ideas in 2004, but only recently developed a practical computational description of its low-level functioning [8]. Our task is to evaluate the spatial pooling component of this algorithm in terms of its ability to robustly and efficiently encode Willmore and Tolhurst's well-known benchmark of greyscale natural scene images [11].

Research suggests that the neocortex uses a *sparse coding* strategy to represent information within a hierarchal structure of layers [9]. A sparse code is one where a relatively small proportion of code elements are active at any one time. If we take the cortical column to be the basic unit of neocortical activation [7], this implies that only a small proportion of columns connected to a given input will be active when the input is present. In addition, for a code to be representationally useful, differing inputs must activate different subsets of columns. This can be

D. Wang and M. Reynolds (Eds.): AI 2011, LNAI 7106, pp. 261–270, 2011.

achieved by maximising the *statistical independence* of the generated codes [4]. The advantage of sparse coding over local coding is that sparse codes have greater representational capacity while still being able to encode simultaneous inputs without interference.

The HTM model extends existing work by explicitly handling temporal sequences of input within a hierarchical Bayesian framework [2]. This is achieved by localised collections of cortical minicolumns learning to predict sequences of feed-forward input arriving either from sensory receptors or from other regions of the neocortex, and having the entire hierarchy learn and exchange inferences about temporal sequences (i.e. events) rather than spatial patterns. It is this temporal *predictive* function of groups of minicolumns that sets the HTM model apart from other hierarchical Bayesian approaches (e.g. [1,6]).

However, it is only recently that the computational details of the HTM model have explicitly incorporated a sparse coding strategy into the *spatial pooler* component of the architecture [8]. To date there has been no published evidence evaluating the performance of the new spatial pooler, or of the new cortical column architecture. To address this, we implemented the HTM spatial pooler and evaluated it on a set of static image benchmarks. In the remainder of the paper we provide a description of this implementation and explain the principles upon which it works. We then introduce a number of modifications that were necessary to make the pooler operate efficiently on our benchmark problems and present an empirical study comparing the HTM spatial pooler with our modified pooler and with the various techniques presented in Willmore and Tolhurst's paper on characterising the sparseness of neural codes [11]. As part of this empirical study we investigate a range of measures to capture the important dimensions of the spatial pooler's behaviour.

2 Spatial Pooling

Basic Principles: The latest HTM architecture [8] introduces a more sophisticated and biologically plausible neural model than is typically employed in artificial neural network research. This model is structured as a hierarchy of regions, where each region consists of a set of columns and each column consists of a set of neurons and their associated dendrites and synapses. According to HTM theory, these neurons control which columns in a region are currently active, and which are currently *predicting* they will be active. The first function is determined by a procedure known as *spatial pooling* and the second by a procedure known as *temporal pooling.*

The basic task of the spatial pooler is to form a sparse distributed representation of the input. This is required by the temporal pooler in order to learn and predict the sequential order of particular input streams. However, to be biologically plausible as well as practically useful, the spatial pooler must also be able to *efficiently* form a relatively *stable* representation of a *continuous* stream of input. These requirements rule out existing solutions, such as independent components analysis [4], as these lack the flexibility and efficiency to adjust to online data streams.

As the internal structure of an HTM column and its associated neurons is only relevant to the implementation of temporal pooling, we shall not discuss these details further. To understand spatial pooling, we need only consider a column as a unified entity with an associated set of proximal dendrites that synapse directly with the input (see [8]). These synapses are *not* associated with weights that multiplicatively determine the strength of the signal. Instead, each dendrite is associated with a *potential* synapse and each synapse is associated with a *permanence value*. If the permanence value of a synapse passes a certain threshold then the synapse is connected and the dendrite will transmit the input to which it is connected, otherwise the synapse remains potential and inactive.

To justify the use of potential synapses, Hawkins argues that the traditional artificial neural network approach of learning by adjusting the strength or weight of individual synapses is not biologically realistic. He acknowledges that synapses have differing strengths, but argues that the synaptic release of neurotransmitters is too stochastic to explain the fine distinctions that are made between differing inputs. Instead, he points to recent research that shows how synapses can rapidly form and un-form [10] and argues that this provides a better mechanism for synaptic learning.

A second important aspect of the operation of the spatial pooler is the use of inhibition between columns to produce sparse distributed representations. It is this feature that produces the *self-organising* capacity of the system to adjust itself to the structure of the input data. As with Kohonen's self-organising maps [5], the spatial pooler performs learning on the basis of how well the synapses from a particular column match (or overlap) the input to which the synapses are connected. However, instead of altering the relative weights of the synapses of neighbouring columns, a strongly activated column will compete with and *inhibit* its less active neighbours [10]. At the end of this process, only the potential synapses belonging to the winning columns that best represent the current input will be able to learn. Here learning entails increasing the permanence values of potential synapses that are connected to active input and decreasing the permanence values of those connected to inactive input. This implements the forming and un-forming of synaptic connections discussed above.

Binary Spatial Pooling: The basic functioning of spatial pooling is described in [8]. Here we only provide a brief outline of the algorithm and explain those areas which deviate from or extend the original proposal. Firstly, each HTM column has a set of potential synapses that are randomly connected with probability $P(connect)$ to each input coordinate. Every potential synapse s is then initialised with a randomly generated permanence value $perm(s)$ bounded within a small range of a threshold $permThreshold$, such that the probability of connection varies inversely and linearly with the distance of the column from the input coordinate. Potential synapses with a permanence value greater than $permThreshold$ are now defined as *connected*.

Algorithm 1. performLearning($columns$)

for each potential synapse s in each active column c **do**
  if s has active input **then** $perm(s) = min(perm(s) + pInc, 1)$
  else $perm(s) = max(perm(s) - pDec, 0)$
end for
for each c in $columns$ **do**
  if $activity(c) < minActivity(c)$ **then** $boost(c) = boost(c) + bInc$
  else $boost(c) = 1$
  if $overlapSum(c) < minActivity(c)$ **then**
    for each potential synapse s in c **do** $perm(s) = min(perm(s) \times pMult, 1)$
  end if
end for

The system then calculates the activity level of each column's response to each *image*. For binary images, this column activity (or overlap) is a simple count of the number of connected synapses that are receiving active input. However, each column's overlap ($overlap(c)$) must also exceed a $minOverlap$ threshold, in which case $overlap(c)$ is multiplicatively boosted by a factor $boost(c)$ (determined by the learning procedure), otherwise it is set to zero. A column c then becomes active if it is one of the n most active columns within the $meanInhibitionArea$ of c, where $meanInhibitionArea$ is the mean size of the receptive fields of all columns and n is set by the parameter $desiredActivity$.

Algorithm 1 ($performLearning$) implements the basic learning strategy. Firstly, the permanence values of all synapses belonging to active columns are adjusted by incrementing those connected to active input and decrementing those connected to inactive input. Then two strategies are used to increase the activity of insufficiently active columns. This first involves counting how often a column c has been active over the last i iterations ($activity(c)$) and how often it has exceeded the $minOverlap$ threshold ($overlapSum(c)$). These values are compared with $minActivity(c)$, calculated by $maxActivity \times minActivityThreshold$, where $maxActivity$ is the maximum activity of any column falling within the $meanInhibitionArea$ of c and $minActivityThreshold$ is a user defined parameter. If column c's $activity(c)$ falls below $minActivity(c)$ then $boost(c)$ is incremented by $bInc$ and if $overlapSum(c)$ falls below $minActivity(c)$ then the permanence values of all c's potential synapses are increased by a factor of $pMult$. The first strategy ensures all columns maintain a minimum level of activity and the second ensures they maintain a minimum level of synapse connectivity.

Finally, the system converges when no changes have been made to the permanence value of any synapse since the last iteration through the entire set of images. The end result is a sparse, distributed encoding of each image presented to the pooler, comprising of the set of columns that are active when an image is present. The sparse distributed nature of the encoding is produced by the self-organising interaction of inhibition, which focuses activity on a small subset of columns (sparsifying), and learning, which ensures all columns become at least minimally active (distributing).

3 Modifications

Handling Greyscale Images: The HTM specifications only handle binary input. Hence we term the original algorithm *binary spatial pooling* (BSP). Our first extension was to redefine the notion of overlap so that synapse inputs can take on integer values. To achieve this, a column's overlap becomes the sum of the integer input values at each connected synapse rather than a simple count of active bits. The main alteration occurs in the updating of the permanence values of potential synapses of active columns (compare lines 1–4 of Algorithm 1 with lines 1–9 of Algorithm 2). Previously a permanence value was incremented whenever a potential synapse is associated with an active input. Now we redefine the notion of an active input to be an input that is greater than the mean activation level of the current image. In addition, $minOverlap$ is adjusted by being multiplied by the mean value of all non-zero pixels in the current image. This preserves the original value of $minOverlap$ for binary images (as the mean value of non-zero binary pixels is one) while ensuring that (on average) for each column at least $minOverlap$ synapses are connected to active (above mean) non-binary inputs.

Accelerating Convergence: The convergence behaviour of the pooler can be accelerated by switching off the basic learning function in lines 1–4 of Algorithm 1 at the point where the two boosting strategies become inactive. This is achieved by keeping count of the number of columns that are either boosted or have their potential synapses incremented during a single iteration through the entire set of images (or over a sufficiently long period of time). If this count is zero then all columns will have attained a sufficient level of activation over the entire data set and there is no further need to adjust the synapse connections. The advantage of this approach is that the pooler can have its main learning function suspended and yet still remain responsive to new input, i.e. if new input cannot be represented by the existing pattern of synapses, some form of boosting will occur and learning will be resumed. The system is still not considered to have finally converged until a complete iteration through all images has occurred such that the permanence values of all synapses remain the same at the end of the iteration as they were at the start.

Augmented Spatial Pooling: The main contribution of the paper, aside from evaluating the current HTM spatial pooler, is the development of a more robust learning strategy. This strategy was suggested by observing that the existing boosting strategy often fails to sufficiently alter the pattern of connected synapses: although boosting succeeds in elevating an inactive column into activity, because the boost value is then immediately reset to one, the column does not remain active long enough for any of its currently inactive synapses to become connected. If no new connections are made in the first iteration of activity, the column can immediately fall into inactivity and again have to wait for its boost value to increment to a point where it becomes active. If a large number of columns are in this position, then an escalating boosting competition

Algorithm 2. performAugmentedLearning(*image*, *columns*)

```
for each potential synapse s in each active column c do
  if input(s) > meanInput(image) and perm(s) >= connectThreshold then
    perm(s) = min(perm(s) + pInc, 1)
  else if input(s) > meanInput(image) and perm(s) < connectThreshold then
    perm(s) = min(perm(s) + pInc, connectThreshold − pInc)
  else
    perm(s) = max(perm(s) − pDec, 0)
  end if
end for
for each c in columns do
  if activity(c) < minActivity(c) and (boost(c) = boost(c) + bInc) > bMax then
    boost(c) = 1
    for each disconnected synapse s in c in ascending distance order from c do
      if perm(s) > maxPerm then maxPerm = perm(s) and maxS = s
    end for
    perm(maxS) = connectThreshold + pInc
  end if
end for
```

can occur where, although each column is slowly gaining new connections, so are its competitors, meaning none remain active long enough to form stable representations. The end result is that the pooler can fail to converge, especially on complex (high entropy) images.

Algorithm 2 details the augmented learning procedure. Here the updating of active column synapses is altered so that disconnected synapses only have their permanence values incremented to a point just below *connectThreshold* (lines 4–5). Now, the only place where synapses can become connected is in the boosting procedure (line 16). As before, if a column's activity is below the $minActivity(c)$ threshold its boost value is increased (line 11). However, if $boost(c)$ exceeds a $bMax$ threshold then the closest synapse to c ($maxS$) is selected from the set of *disconnected* synapses with the greatest permanence value ($maxPerm$) and this synapse has its permanence value set so that it is connected. In this way the connection of synapses is controlled entirely within the boosting procedure and the earlier ineffective escalating boosting behaviour is remedied.

4 Experimental Evaluation and Discussion

In order to evaluate the HTM spatial pooler we used the 64 greyscale images from Willmore and Tolhurst's influential study on measures of sparsity [11], and similarly generated ten sets of $10,000$ 16×16 image patches selected randomly from the 64 full images. In the original paper, sparseness was characterised according to two statistics: population kurtosis and lifetime kurtosis. Population sparseness is defined as the average kurtosis of the distribution of the activities of the complete set of N columns for each *image* in the set of input images. The population kurtosis of a single image i is given by:

$$populationKurtosis_i = \left\{ \frac{1}{N} \sum_{c=1}^{N} \left[\frac{a_c - \bar{a}}{\sigma_a} \right]^4 \right\} - 3 \quad (1)$$

where $a_1 \ldots a_N$ are the post-inhibition overlap activities of columns $1 \ldots N$ for image i, and $\bar{a}$ and σ_a are the mean and standard deviation of these activities. To allow for comparison between methods, and again following [11], we standardised the activities to have a mean of zero and a standard deviation of one, giving an averaged population kurtosis of $\frac{1}{M} \sum_{i=1}^{M} populationKurtosis_i$ for an entire set of M images. This statistic measures the infrequency or *sparseness* of collective column activity in response to individual images but fails to measure how the responses are *distributed* between columns, i.e. how infrequently individual columns become active. To capture this second dimension, we need the average *lifetime* kurtosis of the columns, which measures the averaged kurtosis of each column's responses to an entire set of input images. Lifetime kurtosis is defined in the same way as population kurtosis except that $a_1 \ldots a_N$ are now the activities of column c in response to the entire set of N images and the average kurtosis is taken over $i = 1 \ldots M$ columns.

Comparisons with Greyscale Images: Given these measures we can now compare the sparseness and distribution of the spatial pooler representations with the results reported in [11] (see Table 1). Here, following [11], we generated a spatial pooler column for each image pixel (256 in all), and then set the pooler parameters as follows: $P(connect) = 0.15$, $connectThreshold = 0.2$, $pInc = pDec = 0.02$, $bInc = 0.005$, $bMax = 4$, $minActivityThrehold = 0.01$, $desiredActivity = 0.05 \times meanInhibitionArea$, and decay $d = 100$. In addition, $minOverlap$ was dynamically set to be the product of the mean pixel intensity of the current image and the mean number of connected synapses for an individual column. These values proved fairly robust for the augmented spatial pooler (ASP) and were subsequently used as ASP defaults.

In contrast, despite extensive parameter tuning, the original binary spatial pooler (BSP) was unable to converge to a stable representation on any of the $10 \times 10{,}000$ greyscale image sets (after allowing 500 cycles through each image set). This reflects that BSP was not developed to process greyscale images. If we binarise the input by setting each pixel with an intensity greater than the mean intensity for a given image to one and all others to zero, then BSP can successfully converge. However, as the Willmore study was concerned with greyscale coding schemes, we cannot fairly compare BSP with the other coding schemes, and so we only report statistics for ASP in Table 1.

Overall, the results show that ASP significantly outperforms all the coding schemes considered in [11], having a lifetime kurtosis on the raw images 4.6 times greater than the best alternative (ICA) and 3.9 times greater than Gabor on the whitened images. The population kurtosis improvements were less pronounced on the raw images (but still 1.5 times greater than PCA) but even more pronounced on the whitened images (8.31 times greater than Gabor).

Table 1. Comparison of augmented spatial pooling (ASP) averaged lifetime and population kurtosis measures with results published in [11] where Gabor = Gabor filters, ICA = independent components filters, O&F = Olshausen-Field bases, PCA = principal components filters, Sinu = sinusoids, Walsh = Walsh functions, Gaus = Gaussian filters, Pixel = single pixel, Raw = unprocessed images, and White = whitened images.

Kurtosis Measure	Image Process	Coding Scheme								
		ASP	Gabor	ICA	O&F	PCA	Sinu	Walsh	Gaus	Pixel
Lifetime	Raw	87.10	18.50	18.74		8.24	10.33	10.69	7.37	6.76
	White	71.23	18.47		17.21	8.13	10.05	10.91	8.93	11.13
Population	Raw	50.54	21.66	6.42		32.64	27.12	27.75	0.21	1.66
	White	44.64	5.37		2.17	3.07	4.62	4.01	0.52	2.68

Table 2. Comparison of augmented spatial pooling (ASP) with binary spatial pooling (BSP) on the complete set of binary scaled natural images taken from [11]

	Algorithm		
	ASP	BSP	BSP
minOverlap setting	auto	4.00	3.00
Converge Time (secs)	8.48	27.88	2.73
Converge Cycles	15.70	56.40	4.70
% Duplicates	10.96	14.94	8.19
% Zero Length	0.00	9.22	4.40
Lifetime Kurtosis	61.94	47.35	26.98
Population Kurtosis	36.95	27.98	20.37
Mean Code Length	15.59	19.02	22.64

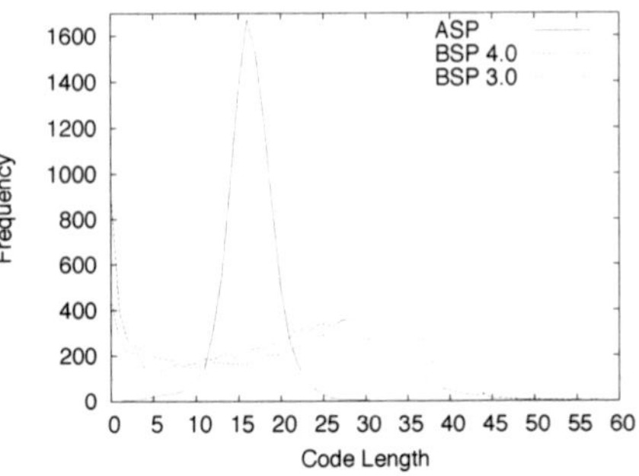

ASP was also able to *efficiently* converge on stable representations, requiring, on average, 13.62 cycles through each of the ten raw data sets (where each cycle processes all 10,000 images in a set) and 10.67 cycles through the whitened data. This took an average 8.01 seconds per convergence on the raw data and 6.98 seconds on the whitened data (all ASP and BSP experiments were run on an Apple MacBook Pro 2.93 GHz Intel Core 2 Duo processor with 4 GB of 1067 MHz DDR3 RAM and running Mac OS X version 10.6.7).

Comparisons with Binary Images: As the original spatial pooling algorithm (BSP) was unable to converge on the greyscale images, we ran a separate experiment using the same set of natural image patches but after performing a binary conversion. BSP still found these binary images challenging in comparison to simpler binary encodings and was unable to converge using ASP's default parameter settings. We found that BSP will only converge on these images if the effect of decrementing the permanence value of a synapse is much stronger than the effect of an increment, making it easier for a synapse to become disconnected than for it to become connected. ASP achieves a similar effect by not incrementing a permanence value past *connectThreshold* unless the associated column is sufficiently inactive (see Algorithm 2). To similarly influence BSP we set $pInc$ (= 0.0005) to be five times weaker than $pDec$ (= 0.0025). In addition, to enable BSP to *reliably* converge within 500 cycles we reduced $P(connect)$ to 0.1 and limited $minOverlap$ to range between 3.0 to 4.0.

Table 2 compares BSP at these adjusted settings with ASP using the standard defaults. The results first show the significant effect of altering $minOverlap$ on the convergence behaviour of BSP, i.e. a reduction of from 4.0 to 3.0 causes a tenfold speedup in convergence, making BSP 3.0 the fastest of the three algorithms. However, this superior convergence is bought at the cost of longer codes, as shown by the mean code length and the distribution of code lengths in the graph. Here a code is the set of columns C_i that become active when an image patch i is present, and a distribution of code lengths is the set of code lengths $|C_i|$ for each image patch $i = 1 \ldots M$. All else being equal, shorter codes are preferred over longer codes because they are more efficient. On this measure, and on the measures of lifetime and population kurtosis, ASP is clearly better than either BSP 3.0 or 4.0. This means ASP reliably produces shorter codes that involve fewer columns and that are more evenly distributed across all columns (as shown by the sharp peak for ASP on the graph in Table 2).

However, an additional dimension is the degree to which a code can distinguish between different inputs. Again, all else being equal, a code that produces finer distinctions is to be preferred. To measure this, we looked at the proportion of image patches that were encoded using common sets of columns (% duplicates and % zero length in Table 2). We used two measures because the duplicate percentage cannot represent the difference between an encoding that captures 1000 images using one set of columns and one that captures 1000 images using 500 sets of columns, where each column set encodes a pair images. In practice, the majority of duplicates only involved column sets encoding image pairs, except for zero length encodings. Such encodings occur when an image fails to make any column active, i.e. the image is ignored or remains unencoded. Clearly, duplicates involving a high proportion of zero length codes (BSP 3.0 and 4.0) make poorer distinctions than encodings where all duplicates are made up of column sets encoding pairs of images (ASP). We can therefore conclude that ASP produces better encodings, both in terms of efficiency, and in terms of making finer distinctions. The price is that ASP converges more slowly than BSP 3.00. However, if speed of convergence is an issue, the ASP parameter defaults can be altered to produce results equivalent to BSP 3.0, whereas we could find no BSP settings that could improve upon the BSP 4.0 encodings.

5 Conclusions

Firstly, we have shown that augmented spatial pooling significantly outperforms the coding schemes presented in Willmore and Tolhurst's original study, both in terms of population and lifetime kurtosis. Secondly, we can conclude that augmented spatial pooling is better than binary spatial pooling for encoding the natural images in the Willmore and Tolhurst data set. The results hold most strongly for the greyscale encodings of the images, where BSP is unable to converge on any of the data sets. It also holds on the binary encodings, where ASP produces better quality sparse representations, both in terms of efficiency (code lengths) and discrimination (duplicates and zero length codes). More generally, we conjecture that the reason BSP performs poorly on natural images is because it forms synapses too easily. This behaviour comes out in relation to natural images because such images have relatively poorly defined structure (i.e. they have high entropy), meaning synapses will tend to form uniformly across the entire image. ASP controls this behaviour by more tightly constraining the situations where new synapses will form.

In future work, we intend to compare ASP and BSP on a wider range of images to confirm our conjecture concerning the complexity of the encodings. We also intend to investigate greyscale spatial pooling using two forms of synapse, one responsive to darker shades and the other responsive to light.

Acknowledgments. We thank David Tolhurst and Ben Willmore for supplying the images used in their original paper.

References

1. Chikkerur, S., Serre, T., Tan, C., Poggio, T.: What and where: A Bayesian inference theory of attention. Vision Research 50(22), 2233–2247 (2010)
2. George, D., Hawkins, J.: A hierarchical Bayesian model of invariant pattern recognition in the visual cortex. In: Proceedings of the International Joint Conference on Neural Networks (IJCNN 2005), pp. 1812–1817 (2005)
3. Hawkins, J., Blakeslee, S.: On intelligence. Henry Holt, New York (2004)
4. Hyvärinen, A., Karhunen, J., Oja, E.: Independent Components Analysis. John Wiley and Sons, Inc., New York (2001)
5. Kohonen, T.: Self-organization and associative memory. Springer, Berlin (1989)
6. Lee, T.S., Mumford, D.: Hierarchical Bayesian inference in visual cortex. Journal of the Optical Society of America A 20(7), 1434–1448 (2003)
7. Mountcastle, V.B.: Introduction to the special issue on computation in cortical columns. Cerebral Cortex 13(1), 2–4 (2003)
8. Numenta Inc.: Hierarchical temporal memory including HTM cortical learning algorithms. Tech. rep., Numenta, Inc, Palto Alto (2010), `http://www.numenta.com/htm-overview/education/HTM_CorticalLearningAlgorithms.pdf`
9. Olshausen, B.A.: Sparse codes and spikes. In: Rao, R.P.N., Olshausen, B.A., Lewicki, M.S. (eds.) Probabilistic Models of the Brain: Perception and Neural Function, pp. 257–272. MIT Press, Cambridge (2002)
10. Stuart, G., Spruston, N., Häusser, M.: Dendrites. OUP, New York (2008)
11. Willmore, B., Tolhurst, D.J.: Characterizing the sparseness of neural codes. Network: Computational Neural Systems 12, 255–270 (2001)

An Analysis of Sub-swarms in Multi-swarm Systems

Antonio Bolufé Röhler[1] and Stephen Chen[2]

[1] University of Havana, Havana, Cuba
bolufe@matcom.uh.cu
[2] School of Information Technology, York University 4700 Keele Street, Toronto, Ontario M3J 1P3
sychen@yorku.ca

Abstract. Particle swarm optimization cannot guarantee convergence to the global optimum on multi-modal functions, so multiple swarms can be useful. One means to coordinate these swarms is to use a separate search mechanism to identify different regions of the solution space for each swarm to explore. The expectation is that these independent sub-swarms can each perform an effective search around the region where it is initialized. This regional focus means that sub-swarms will have different goals and features when compared to standard (single) swarms. A comprehensive study of these differences leads to a new set of general guidelines for the configuration of sub-swarms in multi-swarm systems.

Keywords: Particle swarm optimization, exploration-exploitation, multi-swarm system, multi-modal search spaces.

1 Introduction

Particle Swarm Optimization (PSO) is an effective search technique for optimization problems in continuous domains [10]. Inspired by the principles that influence the flocking of birds and the schooling of fish, the main idea is the combination of personal experience from the individual and social experience from the group. In PSO, this experience is represented as an attraction to the best position found by a given individual (particle) and to the best one found by a set of individuals. Together with a particle's momentum, these attraction forces define the movement of each particle in the swarm.

The movement of particles in a swarm is naturally convergent. The convergence rate can be slower or faster depending on the communication topology (e.g. ring or star)[1], but eventually the swarm will focus its search efforts around the best-found solution(s). The convergent nature of this search process is not ideally suited to multi-modal search spaces where it is important to achieve an effective balance between exploration and exploitation. Compared to modifications which seek to improve the balance between exploration and exploitation in PSO (e.g. [4][7]), an alternative approach is to separate these processes into distinct phases which focus primarily on either exploration or exploitation. Multi-swarms systems (e.g. [2][9]) use multiple sub-swarms to search in different regions of the

D. Wang and M. Reynolds (Eds.): AI 2011, LNAI 7106, pp. 271–280, 2011.

solution space, and this two-phase organization supports an exploration around a diverse number of "best positions".

A considerable amount of research (e.g. [5]) has been dedicated to study the optimal way to configure particle swarms, and a large amount of this research is summarized in the definition for standard PSO [1]. However, it is not expected that standard (single) swarms and smaller sub-swarms will have the same optimal configuration. For example, sub-swarms that use fewer iterations/function evaluations (FEs) can require more constriction to ensure convergence [3].

In this paper we investigate how different features of PSO – the number of particles, constriction factor, initial velocities, initial positions, and function evaluations – influence the behaviour of sub-swarms in comparison with standard swarms. As a reference point, all experiments in this paper use a novel method based on Estimation Distribution Algorithms (EDA) for selecting the initial positions of the sub-swarms. This initialization simulates the exploratory phase of a hypothetical multi-swarm system. A study of sub-swarms is then carried out with the purpose of providing general considerations and good design features for multi-swarm systems.

This analysis of sub-swarm behaviour begins in Section 2 with some background on different examples of multi-swarm systems. A brief description of the benchmark functions and the experimental design is given in Section 3. In Section 4, the influence of initial velocities on standard swarms in comparison to multi-swarm systems is analyzed. Sections 5 and 6 present some considerations about the constriction factor and the population size, respectively. In Section 7, results are combined into an overall recommendation for sub-swarm parameters. The discussion in Section 8 puts previous results in context, and a brief summary is presented in Section 9.

2 Multi-swarm Systems

The design of multi-swarm systems divides the processes of exploration and exploitation into two distinct phases. Each individual swarm focuses on exploitation in a specific region, and a separate mechanism which chooses these regions focuses on exploration. This exploratory mechanism can be considered as the essential part that differentiates one multi-swarm system from another.

For example, Waves of Swarm Particles (WoSP) [9] bases its diversification mechanism on the "collision" of particles. When particles get too close, a repulsive force expels the particles into new waves/sub-swarms, and this avoids a complete convergence. A key feature of the new sub-swarms is that their initial positions are not randomly selected as in normal swarms. Instead, they maintain some information from the previous trajectories of the particles. A similar relationship exists with initial velocities. In WoSP, the initial search direction after the ejection is based on the previous velocity of the particle.

The significance of initial positions and velocities is much clearer in locust swarms [2]. This multi-swarm system bases its diversification mechanism on a "devour and move on" strategy. Once a sub-swarm has devoured a region (intensive search) the swarm is ready to move on to another promising region. The initial positions of the new sub-swarm are selected using a scouting process around the best position found by the previous sub-swarm. The initial velocities are directed away from this previous optimum to further push the subsequent sub-swarms away from previously devoured parts of the search space.

Although the design of multi-swarm systems tends to focus on the selection of initial positions and initial velocities for the particles of a new sub-swarm, additional design considerations are also required. For example, a multi-swarm system that uses the same overall number of FEs as a standard swarm will require each sub-swarm to use a highly reduced number of iterations/function evaluations. With fewer iterations per particle, and considering each sub-swarm as an exploitation mechanism, it may be necessary to increase the convergence rate by decreasing the constriction factor. To increase the iterations for each particle (given a fixed number of total function evaluations), the swarm size can be reduced. These previously under-studied aspects of sub-swarm design are the focus of this paper.

3 Experimental Design

The experiments presented in this paper have been performed using set 3 (uni-modal functions with high conditioning – functions 10-14) and set 4 (multi-modal functions with adequate global structure – functions 15-19) of the Black-Box Optimization Benchmarking (BBOB) functions [8]. To provide some consistency with other results (e.g. [3][4]) five trials were run on the first five instances of each benchmark function for a total of 25 trials per function. This previous work also used a fixed number of function evaluations based on the dimensions D (i.e. $FEs = 5000 * D$), and they focused on a problem size of $D = 20$ dimensions.

The following experiments require a set of initial positions which are of high quality, but that are not completely converged. These initial positions can then be used to simulate the result of the exploratory phase in a multi-swarm system. Estimation Distribution Algorithms (EDAs) [11] are a promising candidate for selecting the initial positions. Among EDAs, the UMDA algorithm was chosen because it is a simple and methodical way to explore a search space.

The exploratory phase in a multi-swarm system doesn't use 100% of the available function evaluations. For each sub-swarm, a small number of FEs are used for exploration (to find initial positions), and the sub-swarm then tries to find the best possible solution from there (e.g. a nearby local optimum). The following analysis of sub-swarm behaviour uses initial positions selected by an UMDA algorithm after 20,000 FEs (UMDA 20).

The benchmark PSO for the current experiments is a constricted, ring topology version (i.e. standard PSO [1]) developed from the source code published in El-Abd and Kamel [6]. The published implementation uses a swarm size of $p = 40$ particles and a constriction factor $\chi = 0.792$. Together with random initial velocities and UMDA 20 initial positions (i.e. 20,000 FEs for a standard implementation as proposed in [11] with a population of 1,000 individuals and a selection coefficient of 0.2), these will be the default parameters for the experiments in this paper.

4 Effects of Initial Velocities

The effectiveness of a selection method for initial velocities based on Differential Evolution (DE) [12] has been reported by Chen and Montgomery in [3]. In Differential Evolution, a new solution is created by applying a difference vector to a base solution. This update equation (1) uses three unique solutions x_1, x_2 and x_3 drawn from the population, and a scaling factor F.

$$x = x_1 + F(x_2 - x_3) \tag{1}$$

To determine if this technique (DE-velocities) can improve any PSO algorithm regardless of the number of function evaluations, the following experiment compares the performance of swarms with random initial velocities and swarms with initial velocities selected with difference vectors (using F = 1.0). The initial positions are the p best UMDA 20 solutions, and the first results are for a high number of FEs (100,000, 80,000, 60,000, and 40,000). All other "sub-swarm" parameters are from the benchmark [10] (e.g. $\chi = 0.792$ and $p = 40$).

The relative improvement (%-diff) achieved by DE-velocities versus random ones is reported in Table 1. The last row shows the mean improvement over all 10 functions for a given amount of FEs. In these results, the DE-velocities do not show any meaningful improvement compared to the random velocities. A possible explanation is that over the course of a long run, the swarm will conduct a thorough exploration of the search space regardless of the initial velocities.

Table 1. Comparison of DE Velocities vs Random Velocities

fn	100,000 FEs	80,000 FEs	60,000 FEs	40,000 FEs
10	2.1%	−7.8%	5.6%	−7.9%
11	−1.3%	−0.2%	−9.7%	−1.4%
12	0.4%	−6.6%	−5.3%	2.3%
13	−3.0%	−3.1%	2.9%	3.8%
14	1.3%	−1.8%	5.4%	17.6%
15	19.5%	−5.0%	−11.1%	6.0%
16	0.6%	−1.4%	−4.6%	9.1%
17	2.7%	9.0%	15.7%	8.3%
18	3.8%	−13.0%	−4.9%	−0.4%
19	3.2%	4.8%	−3.1%	0.8%
mean	2.9%	−2.5%	−0.9%	3.8%

In multi-swarm systems, the number of FEs is drastically reduced for each sub-swarm. To determine how this decrement influences the effects of initial velocities, the previous experiment is repeated with much fewer function evaluations. In the results shown in Table 2, each column corresponds to swarms with

10,000, 5,000, 2,000, and 1,000 FEs, i.e. 10, 5, 2, and 1 percent respectively of the total FEs of a standard swarm.

The relative improvement obtained by DE-velocities versus random velocities shows that the selection of the initial velocities can be beneficial for swarms which must converge in a small number of function evaluations. The total improvement presented in the last row demonstrates a clear trend – the importance of selected initial velocities increases as the number of FEs used by the sub-swarms decreases. These results replicate the benefit shown in [3] and demonstrate that initial velocities are an important design consideration in sub-swarms that does not exist for standard swarms.

Table 2. Comparison of DE Velocities vs Random Velocities

fn	10,000 FEs	5,000 FEs	2,000 FEs	1,000 FEs
10	7.6%	9.0%	32.2%	54.7%
11	−6.5%	−1.2%	9.9%	−7.1%
12	63.6%	63.8%	92.6%	99.2%
13	29.3%	28.9%	48.9%	79.7%
14	19.0%	35.0%	67.6%	86.8%
15	1.0%	5.5%	16.7%	18.7%
16	4.7%	7.4%	−7.3%	−2.9%
17	7.2%	15.4%	42.7%	69.9%
18	3.2%	4.3%	34.5%	55.1%
19	−0.9%	4.0%	3.1%	9.0%
mean	12.8%	17.2%	34.1%	46.3%

5 Effects of Smaller Constriction Factors

With fewer iterations per particle, it may be beneficial to increase the convergence rate of the sub-swarms (i.e. decrease the constriction factor χ). In standard PSO [1] the velocities of each particle are updated by

$$v_d = \chi(v_d + c_1\epsilon_1(pbest_d - x_d) + c_2\epsilon_2(gbest_d - x_d)) \tag{2}$$

In (2), v is the particle's velocity, x is the position of the particle, and d is a given dimension. The variables ϵ_1 and ϵ_2 are random values, which together with the weights c_1 and c_2 determine the contribution of attractions to the personal and global bests $pbest_d$ and $gbest_d$, respectively. The constriction factor is represented by χ, the specific value used for the constriction factor in [6] is $\chi = 0.792$. By changing the value of this parameter, it is possible to modify the particle's momentum, and therefore to either promote a more exploratory or a more exploitative behaviour.

The following experiments examine the effect of reducing the constriction factor on the performance of sub-swarms. The reported results (see Figure 1) are the relative improvement (%-diff) achieved with a reduced constriction factor versus the value used in the benchmark PSO (i.e. $\chi = 0.792$). The constriction χ was decreased by multiplying it by an additional reduction factor α ($\alpha \leq 1$). The initial positions were selected using UMDA 20, and all other sub-swarm parameters are from the benchmark (e.g. random initial velocities and a swarm size of $p = 40$).

Figure 1 shows the relationship between the improvement in performance and the reduction of the constriction factor for sub-swarms with different amounts of FEs (10,000, 5,000, 2,000 and 1,000). For example, the largest improvement of 67.4% is achieved for sub-swarms with 1,000 FEs by multiplying the original constriction factor with $\alpha = 0.55$ (i.e. $\chi = 0.401$) – see the highlighted tick-mark. The %-diff values are averages for all of the benchmark functions in BBOB sets 3 and 4 (e.g. the mean value in Tables 1 and 2).

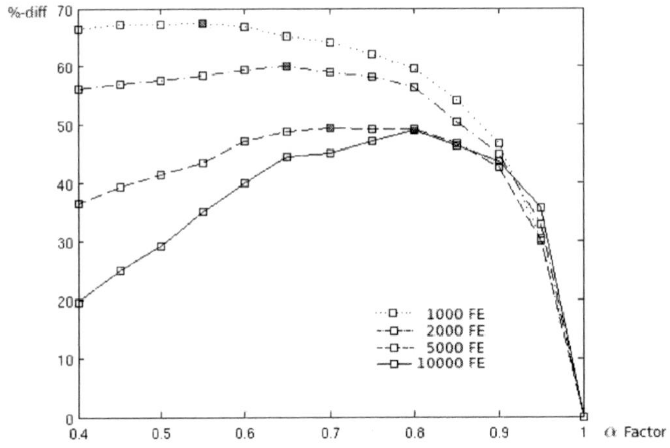

Fig. 1. Relationship between constriction factor and sub-swarm improvement

By analyzing Figure 1, two observations can be made. First, as the number of FEs decreases the (relative) improvement achieved by reducing the constriction factor increases. Second, the best value for the constriction factor gets smaller as the amount of function evaluations is reduced. A smaller constriction factor decreases the particles' momentum which gives the swarm a less exploratory (and more exploitative) behaviour. More exploitation allows sub-swarms to converge, but constriction values that are too small can cause premature convergence which can again decrease the performance of the sub-swarm.

6 Effects of Swarm Size

With a fixed quantity of function evaluations, the number of iterations can be altered by adjusting the swarm size: $FEs = popsize * iterations$. If the swarm size is decreased then it is possible to execute more iterations. On the other hand, a small population may affect the ability of the swarm to explore different regions of the solution space. Thus, the number of individuals causes a trade-off between exploration and exploitation.

In the recent definition for standard PSO [1], Bratton and Kennedy analyzed the influence that the number of particles can have on PSO performance. They report that "no swarm size between 20 – 100 particles produced results that were clearly superior or inferior to any other value for a majority of the tested problems". The best value for a swarm size may depend on problem-specific features like the number of dimensions, constraints, and other characteristics of the objective function. To provide some consistency with other results (e.g. [6]), a swarm size of $p = 40$ particles has been used in the previous experiments.

The purpose of the following set of experiments is to observe the effects of population size in sub-swarms which use a highly limited number of FEs. The reported results (see Figure 2) show the relative improvement (mean %-diff over the 10 functions in BBOB sets 3 and 4) achieved with smaller swarm sizes versus a swarm with $p = 40$ particles. The results in Figure 2 correspond to sub-swarms with different amounts of FEs (10,000, 5,000, 2,000 and 1,000). The initial positions were selected using UMDA 20, and all other sub-swarm parameters are from the benchmark (e.g. random initial velocities and a constriction factor of $\chi = 0.792$).

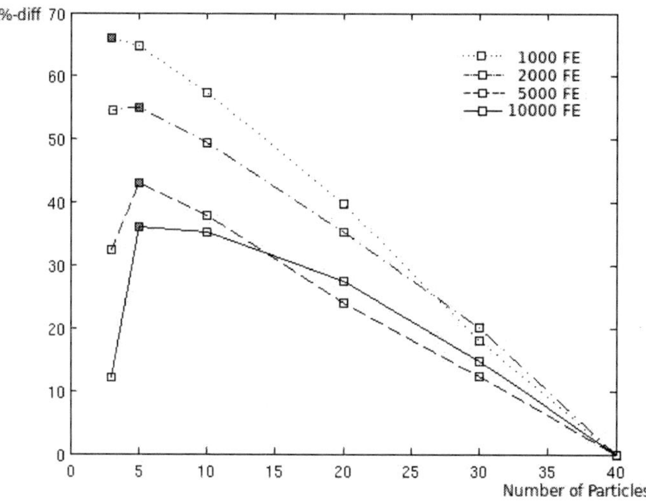

Fig. 2. Relation between number of particles and sub-swarm improvement

In Figure 2, a considerable improvement in performance can be observed when the swarm size is decreased to values far below those suggested by Bratton and Kennedy for a standard swarm [1]. With fewer function evaluations, sub-swarms benefit from smaller populations which allow more iterations and a subsequent increase in their ability to adapt to the function's landscape.

7 Recommended Parameters

So far, the different parameters have been analysed separately with the aim of better understanding the effects that each of them has on sub-swarm behaviour. In this section, the best found combination of parameters for sub-swarms is reported. The focus is on swarm parameters given a set of initial positions (i.e. UMDA 20). Similar to the experiments in Sections 4–6, all of the swarms start with the same initial positions.

The selection of initial velocities is a binary decision, whether to use random or non-random velocities, and the results reported in Section 4 support the use of (non-random) DE-based initial velocities. The selection of the two other parameters, i.e. constriction factor and sub-swarm size, depends on different characteristics of the sub-swarm. Extensive tests (partially shown in Sections 4–6) have led to suggested values of $\alpha = 0.8$ ($\chi = 0.634$) and a swarm size of $p = 15$ particles. In Table 3, the total improvement achieved with these parameters is presented. The results represent the relative improvement (%-diff) achieved with the recommended parameters for sub-swarms versus standard parameters (i.e. random initial velocities, a constriction factor of $\chi = 0.792$, and $p = 40$ particles).

Table 3. Improvement of well parametrized sub-swarms vs. standard parameters

fn	10,000 FEs	5,000 FEs	2,000 FEs	1,000 FEs
10	62.4%	61.3%	66.4%	78.0%
11	23.8%	22.7%	99.9%	99.9%
12	100%	100%	100%	100%
13	88.9%	92.3%	100%	100%
14	92.4%	94.2%	100%	100%
15	36.8%	26.8%	99.9%	100%
16	−0.68%	−1.82%	100%	100%
17	55.9%	67.6%	100%	100%
18	47.8%	50.6%	100%	100%
19	0.26%	25.8%	100%	100%
mean	50.7%	53.9%	96.6%	97.8%

The first two columns in Table 4 show the difference between the initial positions (UMDA 20) and the final results that can be achieved by UMDA in 100,000 FEs (i.e. UMDA 100). This difference represents an initial target for the performance of a multi-swarm system. Starting from the UMDA 20 positions, a sub-swarm using the recommended parameters (i.e. column 1 from Table 3) can achieve results comparable with those from UMDA 100. The UMDA 20 + PSO 10 system uses a total of 30,000 FEs, and it is already more effective than UMDA on 6 of the 10 functions. On the remaining 4 functions, a large amount of the gap between UMDA 20 and UMDA 100 has been covered. Future work will attempt to use the remaining 70,000 FEs to build a multi-swarm system that is more effective than either UMDA or PSO alone.

Table 4. UMDA vs. PSO Sub-swarm

fn	UMDA 100	UMDA 20	UMDA 20+PSO 10
10	$1.68e+04$	$4.24e+04$	$\mathbf{8.81e+03}$
11	$7.80e+01$	$9.03e+01$	$\mathbf{5.86e+01}$
12	$7.47e-01$	$9.61e+03$	$\mathbf{5.70e-01}$
13	$6.44e+00$	$4.84e+01$	$\mathbf{2.19e+00}$
14	$2.63e-03$	$9.77e-02$	$\mathbf{1.39e-03}$
15	$3.06e+00$	$1.03e+02$	$5.39e+01$
16	$1.42e+01$	$2.02e+01$	$\mathbf{1.41e+01}$
17	$2.38e-03$	$3.24e-01$	$7.65e-02$
18	$1.53e-01$	$2.33e+00$	$8.08e-01$
19	$2.84e+00$	$4.26e+00$	$3.56e+00$

8 Discussion

Multi-swarm systems do not base their search process on standard swarms, but on sub-swarms which have a more regional search focus. Two main issues differentiate sub-swarms from standard (single) swarms: the considerable difference in FEs and their non-random initial positions (previously selected by a separate search mechanism that guides the multi-swarm system). Both conditions are reflected in the tests performed in this paper – the later is recreated through the initialization of sub-swarms at the best solutions provided by a relatively short UMDA search.

These differences between standard swarms and sub-swarms imply that different design decisions and parameter values are necessary in multi-swarm systems. In particular, the use of non-random initial velocities leads to large improvements in sub-swarm performance, but they provide no benefits in standard swarms. The optimal swarm size also changes in sub-swarms with the reported results showing that sub-swarms benefit from smaller populations. When the overall number of function evaluations is greatly reduced, fewer particles lead to more iterations, and this allows the sub-swarm to better adapt to the function's landscape.

Sub-swarms usually start in good positions of a specific sub-region, so there is less need to boost exploration as in standard swarms. Subsequently, sub-swarms also benefit from a reduced constriction factor that promotes a more exploitative behaviour. However, it should be noted that the recommended constriction factor has a direct relation with the initial magnitude of the particle velocities, and thus the method used to select the initial velocities [3].

9 Summary

Standard PSO recommends a set of parameters and design decisions such as random initial velocities, a constriction factor of $\chi = 0.729$, and swarms with 20 – 100 particles. These values lead to optimal performance in standard (single) swarms. Sub-swarms have been shown to perform better with features like (non-random) DE-based initial velocities, a constriction factor of $\chi = 0.634$, and $p = 15$ particles. Future work will use these new design recommendations in the development of a multi-swarm system that uses UMDA during the exploratory phase.

References

1. Bratton, D., Kennedy, J.: Defining a Standard for Particle Swarm Optimization. In: Proceedings of the 2007 IEEE Swarm Intelligence Symposium, pp. 120–127 (2007)
2. Chen, S.: Locust Swarms – A New Multi–Optima Search Technique. In: Proceedings of the 2009 IEEE Congress on Evolutionary Computation, pp. 1942–1948 (2009)
3. Chen., S., Montgomery, J.: Selection Strategies for Initial Positions and Initial Velocities in Multi–optima Particle Swarms. In: Proceedings of the 2011 Genetic and Evolutionary Computation Conference, pp. 53–60 (2011)
4. Chen., S., Noa Vargas, Y.: Improving the Performance of Particle Swarms through Dimension Reductions – A Case Study with Locust Swarms. In: Proceedings of the 2010 IEEE Congress on Evolutionary Computation, pp. 211–220 (2010)
5. Clerc, M., Kennedy, J.: The particle swarm – explosion, stability, and convergence in a multidimensional complex space. IEEE Transactions on Evolutionary Computation 6(1), 58–73 (2002)
6. El–Abd, M., Kamel, M.S.: Black–Box Optimization Benchmarking for Noiseless Function Testbed using Particle Swarm Optimization. In: Proceedings of the 2009 Genetic and Evolutionary Computation Conference, pp. 2269–2273 (2009)
7. Engelbrecht, A.P.: Heterogeneous particle swarm optimization. In: Proceedings of 7th International Conference on Swarm Intelligence, pp. 191–202 (2010)
8. Hansen, N., Fick, S., Ros, R., Auger, A.: Real–Parameter Black–Box Optimization Benchmarking 2009: Noiseless Functions Definitions. INRIA Technical Report, RR–6829 (2009)
9. Hendtlass, T.: WoSP: A Multi–Optima Particle Swarm Algorithm. In: Proceedings of the 2005 IEEE Congress on Evolutionary Computation, pp. 727–734 (2005)
10. Kennedy, J., Eberhart, R.: Particle Swarm Optimization. In: Proceedings of the 1995 IEEE International Conference on Neural Networks, pp. 1942–1948 (1995)
11. Larrañaga, P., Lozano, J.A.: Estimation of Distribution Algorithms, A New Tool for Evolutionary Computation. Kluwer Academic Publishers (2011)
12. Storn, R., Price, K.: Differential Evolution – A Simple and Efficient Heuristic for Global Optimization over Continuous Spaces. Journal of Global Optimization 11, 341–359 (1995)

A Simple Strategy to Maintain Diversity and Reduce Crowding in Particle Swarm Optimization

Stephen Chen[1] and James Montgomery[2]

[1] School of Information Technology, York University
4700 Keele Street, Toronto, Ontario M3J 1P3
sychen@yorku.ca

[2] College of Engineering and Computer Science, Australian National University
Canberra, ACT 0200, Australia
james.montgomery@anu.edu.au

Abstract. Each particle in a swarm maintains its current position and its personal best position. It is useful to think of these personal best positions as a population of attractors – updates to current positions are based on attractions to these personal best positions. If the population of attractors has high diversity, it will encourage a broad exploration of the search space with particles being drawn in many different directions. However, the population of attractors can converge quickly – attractors can draw other particles towards them, and these particles can update their own personal bests to be near the first attractor. This convergence of attractors can be reduced by having a particle update the attractor it has approached rather than its own attractor/personal best. This simple change to the update procedure in particle swarm optimization incurs minimal computational cost, and it can lead to large performance improvements in multi-modal search spaces.

Keywords: Particle swarm optimization, crowding, niching, population diversity, multi-modal search spaces.

1 Introduction

The development of particle swarm optimization (PSO) includes inspirations from "bird flocking, fish schooling, and swarming theory in particular" [11]. Each particle (e.g. a simulated bird) is attracted to its personal best position and the best position of a neighbouring member of the swarm. In original PSO [11], the neighbourhood for all particles is the entire swarm (i.e. a star topology) – the global best position attracts all of the other particles towards it. This concentration of search around a single attractor can work well in unimodal search spaces, but this level of convergence can also lead to poor performance in multi-modal search spaces.

To improve the balance between exploration and exploitation, standard PSO [1] recommends a ring topology – each particle communicates with only two neighbours. With this reduced communication, a single good position will not immediately attract all of the other particles in the swarm. Specifically, several different positions can each act as the attractor for a small subset of particles, and the overall swarm can

D. Wang and M. Reynolds (Eds.): AI 2011, LNAI 7106, pp. 281–290, 2011.

subsequently explore many regions of the search space. This increased exploration generally improves PSO performance in multi-modal search spaces [1].

The use of a ring topology can lead to local behaviours that are similar to sub-swarms. For example, if particle 2 is the attractor for particles 1 and 3, and particle 5 is the attractor for particles 4 and 6, then this six particle swarm could temporarily behave like two independent swarms of three particles each. Many multi-swarm techniques exist which use sub-swarms (in sequence or in parallel) to explore multiple local optima (e.g. [3][10][12]). Compared to standard PSO, these multi-swarm techniques tend to have their most consistent performance improvements in multi-modal search spaces (e.g. as shown for locust swarms in [4]).

In population search techniques, another way to explore multiple local optima is niching (e.g. [2]). The effect of niching is to cause the overall population to divide into several sub-populations that each explores the area around a distinct local optimum. The intention of niching can be to simultaneously explore multiple local optima with the goal of finding many or all of the local optima in a search space.

A related idea that ultimately allows the population to converge is to reduce crowding (e.g. [6][13]). Crowding occurs when two or more population members are too close to each other. As crowds gather, population diversity is reduced and the explorative capacity of the search technique is similarly reduced. To prevent crowds, a new candidate solution should replace a similar solution in the population. This replacement strategy ensures that these two solutions will not be able to form a crowd.

If the personal best positions are viewed as a population of attractors, it can be seen that the basic operation of PSO promotes crowding. An attractor draws another particle towards it with the explicit purpose of having that other particle search in the nearby area. If the attracted particle subsequently finds a new personal best position near this local best attractor, it will update its own personal best attractor to be near the first attractor – these two attractors have now formed a crowd.

To reduce crowding, the standard procedure is to compare the new candidate solution with several existing members of the population. Among these solutions, the minimal loss of diversity occurs if the most similar solution is replaced [6]. Transferring this idea to PSO, a new personal best position should be compared with the nearest/most similar member in the population of personal best attractors. Subsequently, a new update strategy is proposed which allows particles to update the personal bests of other particles.

A modified PSO that implements the above strategy to reduce crowding has been developed. Starting with standard PSO [1] (and its ring topology), the procedure to update personal best (*pbest*) positions is changed to first check if the new position is close to its local best (*lbest*) attractor. If the new position is within a threshold distance to its *lbest* attractor, it is compared with and potentially updates this attractor. Outside of the threshold distance, normal PSO comparisons and updates occur. The effectiveness of this strategy is tested across a broad range of benchmark functions.

The proposed new strategy to maintain diversity in particle swarm optimization draws inspiration from crowding techniques which are reviewed in Section 2. A brief introduction to PSO is given in Section 3 before the details of the new update strategy are presented in Section 4. Experiments on a broad range of standard benchmark problems are performed in Sections 5 and 6. The results of these experiments are discussed in Section 7 before the paper is summarized in Section 8.

2 Background

The balance between exploration and exploitation is a recurring theme in many heuristic search techniques. For example, selection pressure in genetic algorithms will increase the proportion of "fit" schemata in a population [8]. However, the unselected schemata can be eliminated from the population which will lead to decreased diversity. In general, maintaining diversity will reduce the rate of convergence which will reduce the likelihood of stagnation in a poor local optimum, and this ability to continue progress can ultimately lead to the discovery of a better final solution. However, slower convergence also tends to increase the time required by the search process to produce a final result, so the balance between exploration and exploitation is important for both the efficient and effective performance of many heuristic search techniques.

One method to maintain diversity in a population is to reduce crowding. The basic technique is to compare each new candidate solution with its most similar individual in a subset of the overall population. The fitter of these two solutions survives as a member of the population. The size of the subset to find a neighbour for comparison can be small [6], which can cause "replacement errors", or it can be large, which can cause significant increases to the required computational effort [13].

The basic crowding technique always replaces the most similar individual in the examined subset, but there is no guarantee (especially at early stages of the search process) that it would not be beneficial to have both of these solutions survive. This effect is related to a replacement error – one effect of a replacement error is that an unexamined solution is more similar and that its survival allows a crowd to form, and another effect of a replacement error is that a relatively diverse and potentially useful solution is unnecessarily removed from the population. This second effect can occur even if the crowd size is the entire population.

These two effects highlight the key objectives of crowd control: maintain a diverse set of promising solutions and reduce (premature) convergence. Similar goals are useful for the population of personal best attractors in particle swarm optimization. Specifically, a particle with crowded personal best and local best attractors will be drawn/constrained to this small region of the search space. Since this particle is not immediately affected by the position of other attractors, they can be (temporarily) ignored. Thus, there are only two attractors of concern, and the new strategy becomes similar to crowding with a subset of size two: the personal best and the local best for each particle.

3 Particle Swarm Optimization

The benchmark and baseline PSO for the current experiments is a constricted LBest version (i.e. standard PSO [1]) developed from the published source code for the constricted GBest version by El-Abd and Kamel [7]. In a constricted PSO, each dimension d of a particle's velocity v is updated for the next iteration i+1 by

$$v_{i+1,d} = \chi\left(v_{i,d} + c_1\varepsilon_1\left(pbest_{i,d} - x_{i,d}\right) + c_2\varepsilon_2\left(lbest_{i,d} - x_{i,d}\right)\right) \tag{1}$$

where χ is the constriction factor, c_1 and c_2 are weights which vary the contributions of personal best and local best attractors, ε_1 and ε_2 are independent uniform random numbers in the range of [0,1], x is the position of the particle, *pbest* is the best position found by the current particle, and *lbest* is the best position found by any particle communicating with the current particle (e.g. all particles in the GBest star topology and only two neighbours in an LBest ring topology). The key parameters used in [7] are $\chi = 0.792, \chi * c_1 = \chi * c_2 = 1.4944$, i.e. $c_1 = c_2 \approx 1.887$, and $p = 40$ particles.

The following experiments use a fixed number of function evaluations (*FE*) based on the number of dimensions *D*. The chosen limit of *FE* = 5000 * *D* promotes consistency with previous results. In particular, results for the original GBest version of this benchmark PSO are reported in [5][7], and results for the constricted LBest version are available in [4].

4 A New Update Strategy for PSO

In PSO, the update of a particle's velocity shown in (1) is based on three distinct components: a momentum term (m), an attraction to *pbest* (f_p), and an attraction to *lbest* (f_l). An example of how these three component vectors might combine to create the new velocity v_i at iteration i is shown in Fig. 1. Applying this new velocity to the previous position leads to a new position x_i.

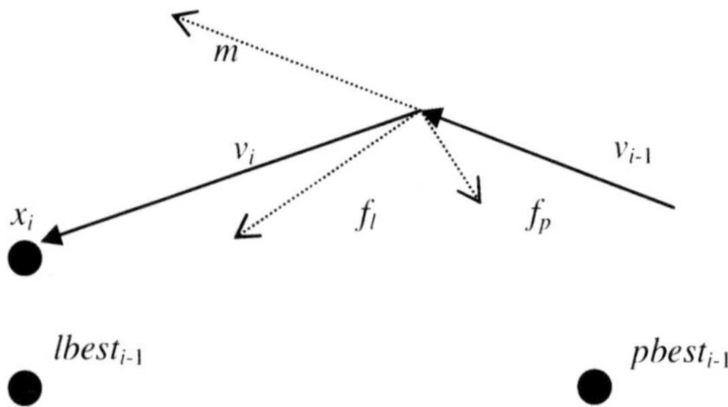

Fig. 1. A particle's path is influenced by attractions to *pbest* and *lbest* positions. In this example, the new particle position has been drawn close to its *lbest* attractor.

After determining the new position, the fitness is calculated and the personal best position is updated if necessary.

Pseudo code for the standard update procedure used in PSO

```
if f(x_i) < f(pbest_i-1) then
  pbest_i = x_i
```

Starting from the example in Fig. 1, assume that $f(x_i) < f(pbest_{i-1})$. The standard update procedure will then make x_i the new position for $pbest_i$, and this will cause the two attractors to become very close (i.e. form a crowd). During the next iteration after the standard update shown in Fig. 2, the closeness of the attractors $pbest_i$ and $lbest_i$ will help to constrain the future search path of this particle to a small area of the search space around these two points. Low diversity in the population of attractors leads to reduced explorative behaviour in the flight paths of a swarm's particles.

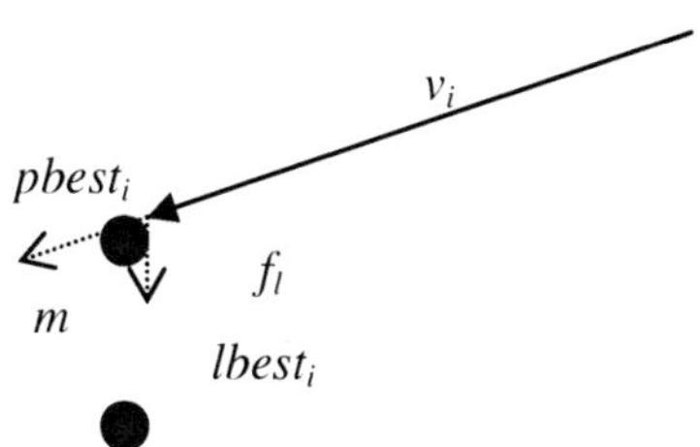

Fig. 2. If $f(x_i) < f(pbest_{i-1})$ in the example from Fig. 1, then the standard update procedure will update *pbest* to be next to *lbest*

A small number of converged particles might not be too damaging, but the convergence of attractors can have a cascading effect (even with a ring topology). For example, assume that the *pbest* for particle 1 is the *lbest* attractor for particle 2. After an update like the one shown in Fig. 2, it is possible that the new *pbest* for particle 2 can become the *lbest* attractor for particle 3. This third particle will now be drawn towards this area with a high concentration of *pbest* attractors. If it also finds a new *pbest* in this area, then this cascade of convergence in the population of *pbest* attractors can continue until all particles have been drawn into this area.

Focusing on this population of *pbest* attractors, the key concept from crowding is that a new solution should replace the most similar member in the existing population. Therefore, instead of replacing *pbest* in Fig. 2, the new position x should replace *lbest*.

Pseudo code for the new update strategy

```
if ||xi - lbesti-1|| < threshold then
   if f(xi) < f(lbesti-1)
      lbesti = xi
else if f(xi) < f(pbesti-1) then
   pbesti = xi
```

In crowding [6], a "crowding factor" specifies the size of the (randomly selected) subset from the overall population which can undergo replacement. The new solution is compared to the members of this subset, and it replaces the most similar solution (if the new solution is fitter). The new update strategy is similar to crowding with a crowding factor or two. However, these two solutions are not selected randomly – they are the *pbest* and the *lbest* for the current particle. Further, the closer of these two points is not automatically replaced. The new update strategy also uses a threshold function to control the minimum required diversity. As discussed in Section 2, the most similar solution in a population can still represent a useful area for further exploration.

If the new update strategy is applied to the example in Fig. 1, it will prevent the creation of a crowd between x_i (which becomes $pbest_i$ in Fig. 2) and $lbest_i$. Instead x_i will replace $lbest_{i-1}$ (see Fig. 3). With this update of *lbest*, the new strategy separates the two roles of *pbest*: store the best known position and act as an attractor in the search space. The swarm as a whole still remembers the best known position (which is stored in *lbest*), but greater diversity is maintained in the population of *pbest* attractors. The effect of reduced crowding is to maintain diversity in the attractors and subsequently to encourage a greater exploration of the overall search space.[1]

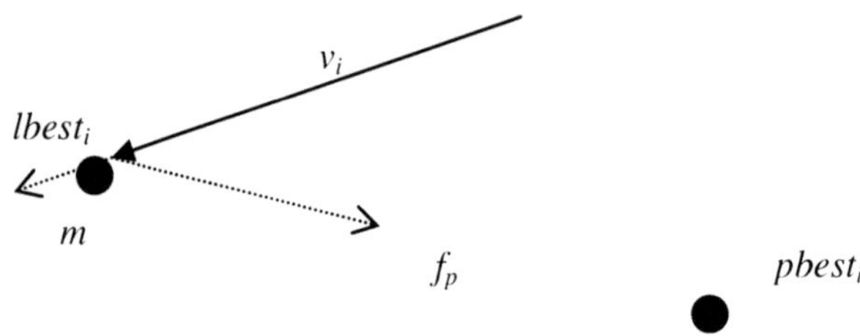

Fig. 3. Compared to the standard update procedure, the new update strategy will update *lbest* instead. This will help maintain diversity in the *pbest* and *lbest* attractors.

5 Results on Multi-modal Functions

The value of increased diversity is to lessen the risk of premature convergence to a poor local optimum. Since local optima do not exist on unimodal functions, the new strategy is not expected to provide benefits on these functions. The following experiments compare the performance of standard PSO [1] based on the benchmark implementation of [7] with a modified version which replaces the "standard update procedure" with the "new update strategy".

The functions (with their ranges) for the following experiments are Fletcher-Powell $[-\pi,\pi]$, Langerman (with m = 7) $[0,10]$, Rastrigin $[-5.12,5.12]$, Schwefel $[-500,500]$, and Shubert $[-10,10]$, and all functions are in D = 20 dimensions. The details for the benchmark PSO are available in the published source of [7], and the key features and parameters are repeated in Section 3. Preliminary experiments with this modified PSO determined that a "threshold" parameter was required to properly calibrate the new balance between exploration and exploitation.

The parameter tuning experiments revealed that the *threshold* should decay over time (to allow the swarm to converge), and that the threshold should only be applied

[1] If the distances between the new position x and the previous *pbest* and *lbest* positions are both less than the threshold, these two distances should both be measured to ensure that x replaces its nearest attractor. Without the extra distance calculation, approximately 1% of the updates under the new strategy can replace a more distant *lbest* attractor. However, since *pbest* and *lbest* must already be quite close for this event to occur, it is not expected to have a large effect on the overall performance.

to *lbest*. If the new update strategy is applied to all attractors/*pbests* in the population, the computational effort is much larger and the performance is much worse – a result presumably caused by a complete lack of convergence. The *threshold* used in the following experiments starts with an initial value of 10% of the search space diagonal (i.e. $\alpha = 0.10$), and it decays with a cubic function (i.e. $\gamma = 3$) – in (2), n is the total number of iterations and i is the current iteration.

$$threshold = (\alpha * diagonal) * ([n-i]/n)^{\gamma} \tag{2}$$

The experiments involve 50 independent runs started with different random seeds. The final solution from each technique is collected after 100,000 function evaluations (i.e. 5,000 * D). For these 50 runs, the minimum (min), mean, maximum (max), and standard deviation (std dev) are presented in Table 1. Except for the maximum and the standard deviation on Rastrigin, PSO with the modified update strategy (Mod) leads to better results (or same for minimum on Langerman) when compared to standard PSO (Std). The p-value for a one-tailed *t*-tests show that the differences in performance have some variability – since the p-values are not all much less than 5%, these results represent more of a promising trend than a strongly significant result.

Table 1. Results for the new update strategy on several benchmark multi-modal functions

Function	PSO	min	mean	max	std dev	t-test
Fletcher-	Std	1,245	10,460	34,181	8,173	1.8%
Powell	Mod	997	7,258	18,828	4,387	
Langerman	Std	–0.513	–0.399	–0.100	0.118	5.2%
m = 7	Mod	–0.513	–0.440	–0.272	0.086	
Rastrigin	Std	7.96	28.77	46.76	8.01	3.0%
	Mod	7.39	23.75	47.96	8.23	
Schwefel	Std	890	1,605	2,360	347	0.0%
	Mod	594	1,139	1,780	272	
Shubert	Std	–3.14e+22	–3.77e+21	–6.34e+19	5.16e+21	5.0%
	Mod	–5.75e+22	–7.34e+21	–2.50e+20	1.15e+21	

6 Results on Other Functions

The modified update strategy is designed explicitly for multi-modal search spaces, but it is still useful to observe its effects across a board range of search spaces. The following experiments use the Black-Box Optimization Benchmarking (BBOB) functions [9]. The BBOB problems are broken into five sets – (1) separable functions, (2) functions with low or moderate conditioning, (3) unimodal functions with high conditioning, (4) multi-modal functions with adequate global structure, and (5) multi-modal functions with weak global structure.

The results for standard PSO on the BBOB functions with dimension $D = 20$ are taken from previous work by the authors [4]. These results (means and standard deviations) are for 25 independent trials of 100,000 function evaluations each (i.e. 5 trials on each of the first 5 instances of each BBOB function). On these functions, standard PSO is able to get within 1e–8 of the optimal solution on every trial of BBOB fn 1, 2, and 5. Errors of this size are considered negligible on the BBOB, so these functions are considered as fully solved. The following experiments only consider the remaining 21 BBOB functions which cannot be solved by standard PSO.

Several sets of parameters for the modified PSO were tried. Preliminary experiments determined that values of 1 and 4 for γ never led to the best-overall results. Thus, the results in Table 2 represent the best performance by the modified PSO across a total of eight parameter pairs – 0.01, 0.04, 0.10, or 0.33 for α and 2 or 3 for γ. For the best set of parameters as shown, the mean errors from optimum (mean), standard deviations (std dev), percent improvement in the mean for the results of modified PSO compared to the results of standard PSO (%-diff), and the p-value for a one-tailed *t*-test are reported.

Table 2. Results for the new update strategy on Black-Box Optimization Benchmarking functions

Set	fn	Standard PSO		Modified PSO		Parameters		%-diff	*t*-test
		mean	std dev	mean	std dev	α	γ		
1	1	0.00e+0	0.00e+0						
	2	0.00e+0	0.00e+0						
	3	2.56e+1	4.99e+0	2.18e+1	6.17e+0	0.10	3	**14.9%**	**1.0%**
	4	3.23e+1	8.55e+0	2.80e+1	6.492+0	0.04	3	**13.4%**	**2.5%**
	5	0.00e+0	0.00e+0						
2	6	8.53e–1	8.89e–1	7.77e–1	5.20e–1	0.10	3	8.9%	35.8%
	7	7.04e+0	2.68e+0	5.40e+0	2.27e+0	0.04	2	23.4%	1.2%
	8	1.22e+1	3.67e+0	1.07e+1	5.00e+0	0.01	3	11.8%	12.6%
	9	1.55e+1	2.24e+0	1.51e+1	2.95e+0	0.01	3	3.0%	26.7%
3	10	6.85e+3	3.39e+3	8.54e+3	3.36e+3	0.01	3	–24.6%	4.2%
	11	6.54e+1	1.71e+1	5.72e+1	1.50e+1	0.01	3	12.6%	3.8%
	12	1.53e+0	4.23e+0	7.38e–1	9.04e–1	0.01	3	51.7%	18.5%
	13	1.50e+0	1.99e+0	1.09e+0	6.16e–1	0.04	3	27.4%	16.6%
	14	1.34e–3	2.66e–4	2.28e–3	4.84e–4	0.01	3	–70.7%	0.0%
4	15	6.05e+1	1.46e+1	4.89e+1	1.37e+1	0.01	3	**19.2%**	**0.3%**
	16	5.37e+0	1.53e+0	4.42e+0	1.21e+0	0.01	3	**17.6%**	**1.0%**
	17	6.61e–1	2.64e–1	4.30e–1	1.49e–1	0.04	2	**34.9%**	**0.0%**
	18	2.87e+0	1.28e+0	2.33e+0	8.00e–1	0.10	3	**18.9%**	**4.0%**
	19	3.61e+0	4.32e–1	3.50e+0	5.11e–1	0.01	2	3.1%	20.7%
5	20	1.14e+0	1.38e–1	9.07e–1	1.46e–1	0.01	2	**20.1%**	**0.0%**
	21	1.41e+0	1.21e+0	5.68e–1	7.70e–1	0.33	2	**59.8%**	**0.3%**
	22	1.69e+0	1.51e+0	1.05e+0	6.44e–1	0.33	3	**38.1%**	**2.9%**
	23	1.33e+0	2.49e–1	1.26e+0	3.02e–1	0.01	2	5.2%	19.3%
	24	1.13e+2	1.12e+1	1.10e+2	1.54e+1	0.01	2	2.6%	22.5%

Although only the best results are shown, it is worth mentioning that the unreported results for non-multi-modal functions are highly inconsistent. On many of the functions, the modified PSO was able to produce an improvement for only the reported parameter set, and none of the eight parameter sets led to an improvement on BBOB fn 10 and 14. Conversely, the modified PSO performed much more consistently on the multi-modal functions for which it was designed. On these functions, the bold values represent statistically significant improvements of more than 10%. Further, there is some robustness to these results as each of these functions had at least one additional parameter set that also led to an improvement of more than 10%. From these observations, it is hypothesized that matching the α parameter to the spacing of the local optima in the search space will lead to the best performance for the proposed strategy, and that the best value for the γ parameter may depend on the contour of the fitness landscape around each local optimum.

7 Discussion

Standard particle swarm optimization shows broad improvements over original PSO across a diverse range of problems [1], but it is still only a starting point for the design of a practical application. In accordance with "no free lunch" [14], there is no single set of parameters that can be expected to lead to the best possible performance of a technique on multiple problems. Therefore, parameter tuning and other modifications are a necessary part of achieving the best possible results for any specific application of a heuristic search technique. Given the large performance improvements that can be achieved with the new strategy, the addition of a new threshold function should not be unduly cumbersome.

The proposed modification to the update strategy in PSO is generally ineffective outside of the targeted multi-modal functions. This is not a major concern since multi-modal functions are the primary application for heuristic search techniques like PSO – gradient descent methods tend to be much more effective than heuristic search techniques on unimodal functions (e.g. BBOB set 3). The underlying mechanisms of the new update strategy attempt to maintain diversity by reducing crowding, and the value of this increased diversity is primarily realized in multi-modal search spaces where it can help prevent premature convergence to a poor local optimum.

The new modification is also simple and computationally efficient. To change from the "standard update procedure" to the "new update strategy", only a distance calculation between two specific points is required – the position of a particle and the position of its *lbest* attractor. In comparison, other diversification strategies based on niching and crowding are either computationally expensive (as distances between a new solution and all existing population members must be calculated) or prone to "replacement errors" (if only a subset of the population is compared against) [13]. In PSO, it is possible to identify the most likely population member that a new candidate solution might form a crowd with – its *lbest* attractor. This insight allows the proposed modification to achieve many of the benefits of niching and crowding at a fraction of the computational cost.

8 Summary

Particle swarm optimization must find the proper balance between exploration and exploitation to maximize its performance. The proposed modification to improve exploration by maintaining diversity is simple and computationally efficient. The reduction in crowding achieved by the new update strategy leads to significant performance improvements in the targeted multi-modal search spaces. The key insight in the current research is the ability to identify with which existing population member a new solution might form a crowd. Future work will attempt to apply this insight to other population search techniques.

References

1. Bratton, D., Kennedy, J.: Defining a Standard for Particle Swarm Optimization. In: Proceedings of the 2007 IEEE Swarm Intelligence Symposium, pp. 120–127. IEEE Press, Los Alamitos (2007)
2. Brits, R., Engelbrecht, A.P., Van den Bergh, F.: A Niching Particle Swarm Optimizer. In: Proceedings of the 4th Asia-Pacific Conference on Simulated Evolution and Learning, pp. 692–696. IEEE Press, Los Alamitos (2002)
3. Chen, S.: Locust Swarms – A New Multi-Optima Search Technique. In: Proceedings of the 2009 IEEE Congress on Evolutionary Computation, pp. 1745–1752. IEEE Press, Los Alamitos (2009)
4. Chen, S., Montgomery, J.: Selection Strategies for Initial Positions and Initial Velocities in Multi-optima Particle Swarms. In: Proceedings of the 2011 Genetic and Evolutionary Computation Conference, pp. 53–60. ACM, New York (2011)
5. Chen, S., Noa Vargas, Y.: Improving the Performance of Particle Swarms through Dimension Reductions – A Case Study with Locust Swarms. In: Proceedings of the 2010 IEEE Congress on Evolutionary Computation, pp. 2950–2957. IEEE Press, Los Alamitos (2010)
6. De Jong, K.A.: An Analysis of the Behavior of a Class of Genetic Adaptive Systems, PhD thesis. Dept. of Computer and Communication Sciences, University of Michigan (1975)
7. El-Abd, M., Kamel, M.S.: Black-Box Optimization Benchmarking for Noiseless Function Testbed using Particle Swarm Optimization. In: Proceedings of the 2009 Genetic and Evolutionary Computation Conference, pp. 2269–2273. ACM, New York (2009)
8. Goldberg, D.E.: Genetic Algorithms in Search, Optimization and Machine Learning. Addison Wesley, Reading (1989)
9. Hansen, N., Finck, S., Ros, R., Auger, A.: Real-Parameter Black-Box Optimization Benchmarking 2009: Noiseless Functions Definitions. INRIA Technical Report RR-6829 (2009)
10. Hendtlass, T.: WoSP: A Multi-Optima Particle Swarm Algorithm. In: Proceedings of the 2005 IEEE Congress on Evolutionary Computation, pp. 727–734. IEEE Press, Los Alamitos (2005)
11. Kennedy, J., Eberhart, R.C.: Particle Swarm Optimization. In: Proceedings of the IEEE International Conference on Neural Networks, vol. 4, pp. 1942–1948. IEEE Press, Los Alamitos (1995)
12. Liang, J.J., Suganthan, P.N.: Dynamic Multi-Swarm Particle Swarm Optimizer. In: Proceedings of the 2005 IEEE Swarm Intelligence Symposium, pp. 124–129. IEEE Press, Los Alamitos (2005)
13. Thomsen, R.: Multimodal Optimization using Crowding-based Differential Evolution. In: Proceedings of the 2004 IEEE Congress on Evolutionary Computation, pp. 1382–1389. IEEE Press, Los Alamitos (2004)
14. Wolpert, D.H., Macready, W.G.: No free lunch theorems for optimization. IEEE Transactions on Evolutionary Computation 1, 67–82 (1997)

Weighted Preferences in Evolutionary Multi-objective Optimization

Tobias Friedrich[1], Trent Kroeger[2], and Frank Neumann[2]

[1] Max-Planck-Institut für Informatik, Saarbrücken, Germany
[2] School of Computer Science, University of Adelaide, Adelaide, Australia

Abstract. Evolutionary algorithms have been widely used to tackle multi-objective optimization problems. Incorporating preference information into the search of evolutionary algorithms for multi-objective optimization is of great importance as it allows one to focus on interesting regions in the objective space. Zitzler et al. have shown how to use a weight distribution function on the objective space to incorporate preference information into hypervolume-based algorithms. We show that this weighted information can easily be used in other popular EMO algorithms as well. Our results for NSGA-II and SPEA2 show that this yields similar results to the hypervolume approach and requires less computational effort.

1 Introduction

Evolutionary algorithms are very powerful problem solvers especially when dealing with multi-objective optimization problems [5, 6]. Many successful methods have been developed in evolutionary multi-objective optimization (EMO) during recent years. Popular evolutionary algorithms for multi-objective optimization are (among many others) NSGA-II [8] and SPEA2 [17], as well as hypervolume-based approaches such as SMS-EMOA [3] and MO-CMA-ES [11, 12].

Recently, there has been significant interest in strategies for introducing user preferences into EMO methods. The goal is to give specific regions of the objective space a higher priority. Consequently, more solutions should be computed and maintained in the population for highly preferred regions of the objective space. For NSGA-II a reference point based approach has been proposed by [7]. They give the example that in the problem of maximizing throughput and minimizing latency, a decision maker may have a clue that throughput should be about 99.9%. Several authors [1, 9, 14] also used reference points to guide multi-objective particle swarm algorithms. Wickramasinghe and Li [15] use preferred areas. Auger et al. [2] present an approach of sampling the weighted hypervolume to incorporate user-defined preferences into the search for problems with many objectives. Thiele et al. [13] extended this such that at each iteration, a decision maker is asked to give preference information in terms of his reference point. Furthermore, Hu et al. [10] changed the crowding distance assignment in NSGA-II in order to achieve a non even spread of the points along the Pareto front.

D. Wang and M. Reynolds (Eds.): AI 2011, LNAI 7106, pp. 291–300, 2011.

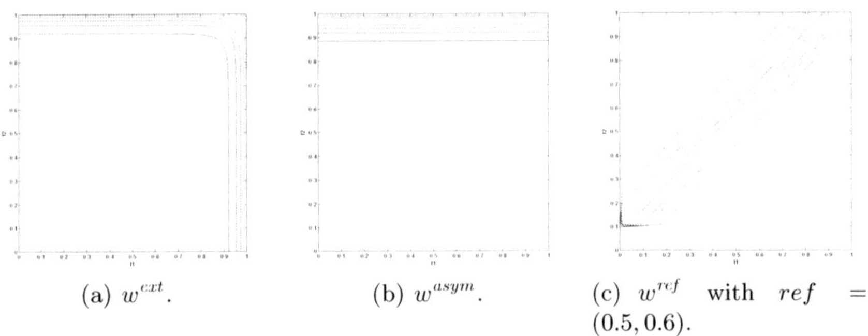

(a) w^{ext}.

(b) w^{asym}.

(c) w^{ref} with $ref = (0.5, 0.6)$.

Fig. 1. Contour plots of the weight distribution functions

We consider the case where there is a weight function on the objective space. This preference information can be used by the weighted hypervolume indicator instead of the standard hypervolume indicator. Zitzler et al. [18] show that this weight integration is very well suited for incorporating user preferences. Furthermore, they compare their results to the ones obtained by NSGA-II and SPEA2 and show that these two algorithms do not perform well as they do not take into account the preference information.

In this paper, we present a neat and simple approach to use the preference information given by weightings on the objective space in classical algorithms such as NSGA-II and SPEA2. Up to now, this weight information has only been used by hypervolume-based approaches that have the drawback of needing a runtime exponential in the number of dimensions [4]. We present very simple approaches to incorporate the weight information on the objective space into a wide range of EMO algorithms. We exemplify this by using NSGA-II and SPEA2 and show that this leads to results similar to the ones of the weighted hypervolume indicator presented in [18]. Furthermore, our algorithms are as efficient as the original implementation of NSGA-II and SPEA2 and do not have to deal with the expensive computations that hypervolume-based algorithms have to face [4].

The outline is as follows. In Section 2, we introduce some basic concepts of multi-objective optimization and the weight functions used. Section 3 shows how to incorporate weight information into NSGA-II and SPEA2. We report our experimental results in Section 4 and finish with some concluding remarks.

2 Preliminaries

A multi-objective optimization problem is given by a vector-valued objective function

$$f = (f_1, \ldots, f_d) \colon S \to \mathbb{R}^d$$

on a search space S. W.l.o.g. we assume that each function f_i, $1 \leqslant i \leqslant d$, should be minimized. We first define a partial order on the objective space. An objective vector $x = (x_1, \ldots, x_d) \in \mathbb{R}^d$ weakly dominates an objective vector $y = (y_1, \ldots, y_d) \in \mathbb{R}^d$ ($x \preceq y$) if it is not worse in any objective, i. e., $x \preceq y :\Leftrightarrow x_i \leqslant y_i$ for $1 \leqslant i \leqslant d$.

Let $f(A)$ be the set of objective vectors of the search points in A, i. e., $f(A) = \{f(a) \colon a \in A\}$. Then, we denote by $\mathrm{Min}(f(A), \preceq)$ the set of minimal objective vectors in $f(A)$ with respect to the partial order $\preceq$ on $f(A)$. The goal in multi-objective optimization is to compute a set X^* with $f(X^*) = \mathrm{Min}(f(S), \preceq)$, where S is the considered search space. $f(X^*)$ is called the Pareto front of the given problem.

Often the size of the Pareto front is large, i. e., exponential with respect to the given input or even infinite in the case of continuous functions. In this case, it is not possible to compute the whole set of minimal elements of $f(S)$ efficiently and $f(X^*)$ should be a smaller subset of them.

So far, there has not been any preference between incomparable solutions. Having to cope with a large Pareto front, we have to decide between incomparable solutions. Basically, all successful evolutionary algorithms have certain diversity mechanisms to deal with this issue. Our goal is to investigate user preference in evolutionary multi-objective optimization. These user preferences give additional information for the search process and distinguish between sets of incomparable solutions.

We assume that we have access to a weight function $w \colon \mathbb{R}^d \mapsto \mathbb{R}$ which describes the preferences of the decision maker. In principle, w can be an arbitrary function that gives preferences to certain regions of the objective space. We will use the following weight distribution functions on the objective space which have been introduced and investigated by [18] in the context of hypervolume-based algorithms:

- Uniform weight: $w^{uni}(x) = 1$
- Sum of two exponential functions in the direction of the axes:

$$w^{ext}(x) = (e^{20 \cdot x_1} + e^{20 \cdot x_2})/(2 \cdot e^{20})$$

- Exponential function in the f_2-direction:

$$w^{asym}(x) = e^{20 \cdot x_2}/e^{20}$$

- Weighted depending on a reference point $ref = (a, b)$:

$$w^{ref}(x) = \begin{cases} c + \frac{(2-((2(x_1-a))^2+(2(x_2-b))^2))}{(0.001+(2(x_1-a)-2(x_2-b))^2)} \\ \quad \text{if } |x_1 - a| < 0.5 \wedge |x_2 - b| < 0.5 \\ c \quad \text{otherwise} \end{cases}$$

Note, that the uniform weight does not imply any preferences on the objective space. For NSGA-II and SPEA2 this will imply that we are just running the original versions of these algorithms. The weight distribution for the other three functions are illustrated in Figure 1 for the objective space $[0, 1]^2$ and we will show in the following sections how to generalize NSGA-II and SPEA2 such that they can make use of this preference information.

Algorithm 1. Crowding assignment for weighted NSGA-II

$\ell \leftarrow |X|$;
foreach $x^j \in X$ **do**
    $d(x^j) \leftarrow 0$;
for $i \leftarrow 1$ **to** d **do**
    Sort the solutions in P such that $f_i(x^1) \leqslant f_i(x^2) \leqslant \ldots \leqslant f_i(x^\ell)$ holds;
    $d(x^1) \leftarrow d(x^\ell) \leftarrow \infty$;
    for $j \leftarrow 2$ **to** $\ell - 1$ **do**
        $d(x^j) \leftarrow d(x^j) + \frac{f_i(x^{j+1}) - f_i(x^{j-1})}{f_j^{\max} - f_j^{\min}}$;
foreach $x^j \in P$ **do**
    $d(x^j) \leftarrow w(x^j) \cdot d(x^j)$;

3 Algorithms

In this section, we show how to integrate the user preferences given by weightings on the objective space into NSGA-II [8] and SPEA2 [17]. Both algorithms are based on the Pareto dominance relation and use diversity mechanisms to decide between incomparable solutions. We transfer the diversity mechanisms for incomparable solutions to the weighted case and adjust them such that they can make use of the weight information provided on the objective space.

3.1 Weighted NSGA-II

NSGA-II is a very popular evolutionary multi-objective algorithm. It is based on dominance ranking which ensures that non-dominated solutions are preferred over dominated ones. Furthermore, the algorithm has a diversity mechanism which distinguishes between incomparable solutions by a crowding distance measure. This measure prefers solutions of less crowded regions in the objective space.

We will keep the dominance ranking as in the original algorithm. To incorporate the user preferences, we will change the crowding distance assignment according to the weight information on the objective space. Let X be a set of incomparable solutions then the assignment of a crowding distance, taking into account the weighting on the objective space, is given in Algorithm 1.

As the original NSGA-II, it iterates over all objectives. For each objective f_i, the solutions are sorted in increasing order and the distance of a solution is changed according to its neighboring points for that objective. Solutions that are maximal (or minimal) with respect to one objective obtain an infinite distance which gives strong preference to the extreme points of the Pareto front. Our weighted crowding distance assignment differs from the original crowding distance assignment by the last for-loop. In this loop the crowding distance of each solution is multiplied by the weight that its objective vector has according to the weight distribution on the objective space. Note, that this does not change the crowding distance assignment of the extreme points as they have already obtained an infinite distance. However, it changes the assignment and the preferences for the other points and gives preferences based on the weighting of the objective space.

3.2 Weighted SPEA2

SPEA2 is another very popular approach. It sorts individuals of a population based on a fitness assignment strategy that incorporates both a coarse-grain evaluation of Pareto dominance that results in an integer raw fitness value and a fine-grained evaluation of density that allows the algorithm to distinguish between solutions with the same raw fitness. The sum of the density and raw fitness values yields the overall fitness of a solution, which is used within both the mating selection and environmental selection functions of the algorithm. The density $D(i)$ of a solution i is calculated as follows:

$$D(i) = \frac{1}{\sigma_i^k + 2}$$

Where σ_i^k is the distance within the objective space from the solution i to its k-th nearest neighbour in the population. Solutions in less crowded regions of the objective space will be assigned lower density values and will be preferred when compared to solutions in more crowded regions of the same Pareto front. In this way, SPEA2 maximises the diversity of solutions within the population. Note that the density function is constructed such that $D(i) \leqslant 0.5$ and as such, the density cannot affect Pareto dominance relationships between solutions, which have a raw fitness value of integer type.

Our weighted density measure incorporates information from a weight distribution function as follows:

$$D(i) = \frac{1}{(w_i \cdot \sigma_i^k) + 2}$$

Where w_i is the value of the weight distribution function at the point in the objective space corresponding to solution i. Solutions corresponding to higher values of w_i will have a lower density $D(i)$ and will be favoured by the SPEA2 selection functions. Conversely, solutions corresponding to lower values of w_i will be less favoured by the SPEA2 selection functions. It is important to note that for any weight distribution function $w \colon \mathbb{R}^d \mapsto \mathbb{R}_+$, it still holds that $D(i) \leqslant 0.5$ and so, Pareto compliance is maintained.

For each iteration of the SPEA2 main loop, the environmental selection function involves copying non-dominated individuals from the archive and population at the previous iteration into a new archive. If the nondominated front fits exactly into the archive then the environmental selection step is completed. If there are not enough non-dominated individuals to fill the archive, the remaining places in the archive are filled with dominated individuals according to fitness.

If there are too many non-dominated solutions to fit into the archive, then a truncation procedure is invoked to iteratively remove individuals until the non-dominated solutions fit within the archive. At each iteration of the truncation procedure, a solution i is chosen for removal that has the minimum distance to another individual. If there are several individuals with minimum distance then the second and, if necessary subsequent, smallest distances are considered to break the tie. Our weighted version of SPEA2 incorporates user preferences within the truncation procedure by multiplying the calculated distance from a solution i to its k-th nearest neighbor by w_i to yield a weighted distance. The result of this modification is that solutions that are situated in highly-weighted regions of the objective space will have a relatively high weighted distance and so will be less likely to be removed by the truncation procedure.

4 Experimental Results

In this section, we report on our experimental results for the weight integration into NSGA-II and SPEA2. We use the same setting as [18] for the weighted hypervolume indicator and examine the classical benchmark functions ZDT1, ZDT3, and ZDT6 [16] and the weight distribution functions defined in Section 2.

We now examine how the three weight distribution functions defined above influence the search process of NSGA-II and SPEA2 for the three test problems. The functions ZDT1, ZDT3, and ZDT6, are optimized by NSGA-II runs with population size 100 for 25000 generations and SPEA2 runs with population and archive sizes of 100 for 25000 generations. For both algorithms, a crossover probability setting of 0.90 is used and mutation probability is set to 0.03. Each of the test problems was constructed with 30 decision variables.

Figures 2 and 3 show the computed Pareto front approximations for NSGA-II and SPEA2 after 25000 generations for the three ZDT functions and the three weight distribution functions w^{ext}, w^{asym} and w^{ref}. The reference point for w^{ref} is chosen as $ref = (0.5, 0.6)$ for ZDT1 and ZDT6 and as $ref = (0.5, 1.2)$ for ZDT3 which is the same as in [18]. The difference in reference point position is due to the fact that the ZDT3 function has a larger range of values in the f_2-direction. The computed Pareto front approximation for a uniform weighting scheme w^{uni} is also shown to allow comparisons to the results of the original NSGA-II and SPEA2 variants.

Charts of the experiments show that for test functions ZDT1 and ZDT6, the weighted versions of NSGA-II and SPEA2 were highly successful in directing solutions towards regions of the objective space in accordance with all three of the weight distribution functions trialled. For the weighting scheme w^{ext}, both algorithms yielded a set of solutions that were concentrated near the boundary regions of the Pareto front. Similarly, the use of the weighting scheme w^{asym} resulted in a set of solutions concentrated near the boundary of the Pareto front in the f_2-direction. When the w^{ref} weight distribution function was used, both algorithms yielded results that were concentrated in a the region of the Pareto front that was closest to the specified reference point.

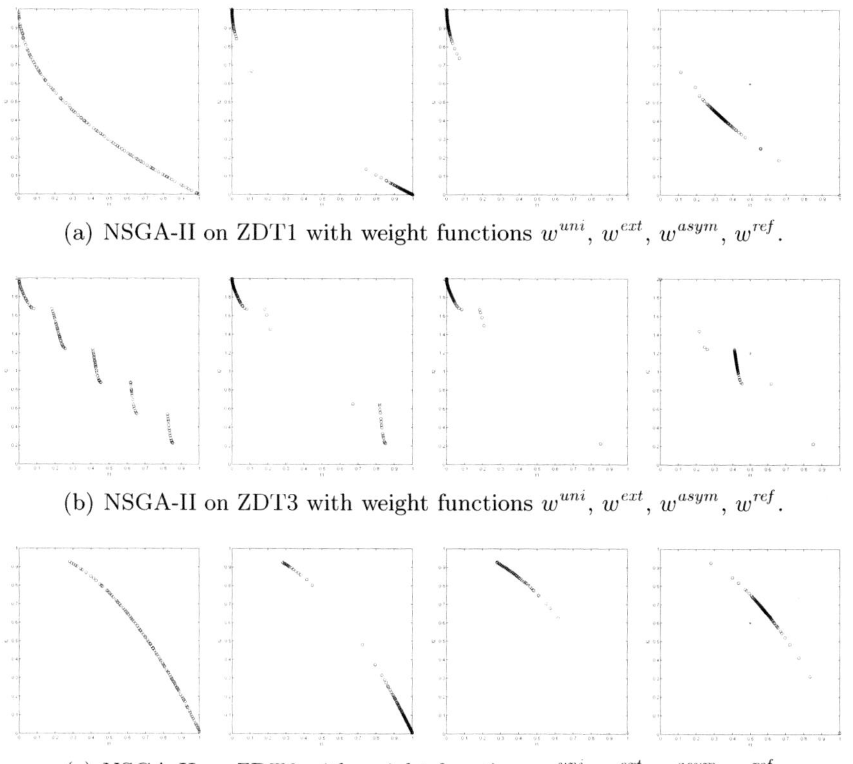

(a) NSGA-II on ZDT1 with weight functions w^{uni}, w^{ext}, w^{asym}, w^{ref}.

(b) NSGA-II on ZDT3 with weight functions w^{uni}, w^{ext}, w^{asym}, w^{ref}.

(c) NSGA-II on ZDT6 with weight functions w^{uni}, w^{ext}, w^{asym}, w^{ref}.

Fig. 2. Experimental results for NSGA-II

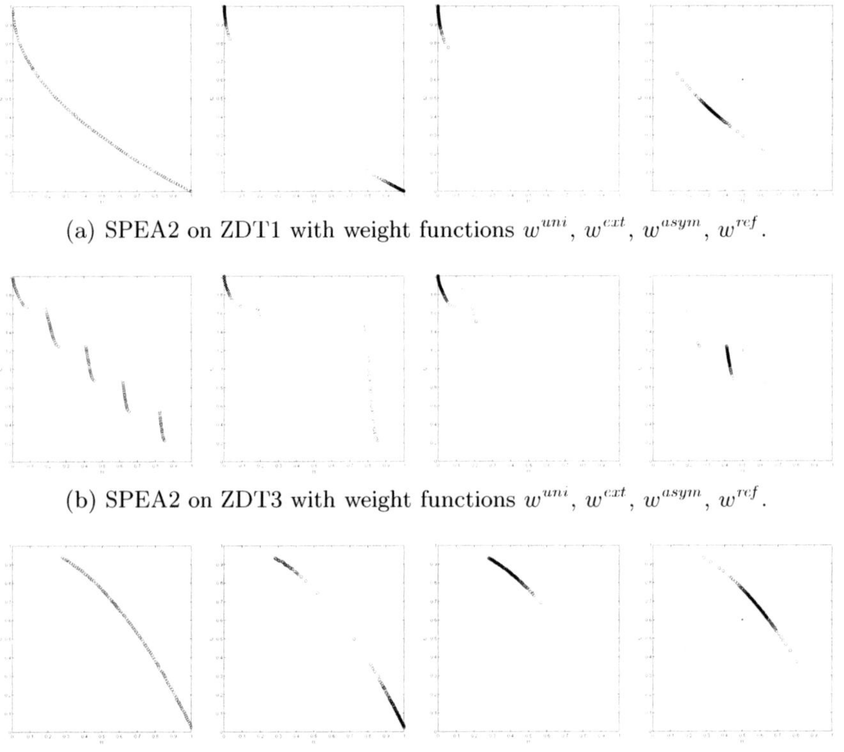

(a) SPEA2 on ZDT1 with weight functions w^{uni}, w^{ext}, w^{asym}, w^{ref}.

(b) SPEA2 on ZDT3 with weight functions w^{uni}, w^{ext}, w^{asym}, w^{ref}.

(c) SPEA2 on ZDT6 with weight functions w^{uni}, w^{ext}, w^{asym}, w^{ref}.

Fig. 3. Experimental results for SPEA2

Similar behaviour was observed for the ZDT3 test function when the weighted NSGA-II and SPEA2 algorithms were executed using weighting schemes w^{asym} and w^{ref}. However, both algorithms produced less definitive results when using the ZDT3 test function and the w^{ext} weighting scheme. In particular, it appears that for the weighted SPEA2 algorithm, application of the w^{ext} weight distribution function led to a poor approximation of the true Pareto front in some regions.

Despite this, the experimental results clearly demonstrate that the proposed weighted versions of NSGA-II and SPEA2 can be used successfully to guide solutions towards areas of the objective space according to an arbitrary weight distribution function. Importantly, these approaches are implemented in such a way that they can be used to incorporate user preferences without compromising Pareto compliance of the algorithms. Furthermore, weights are introduced to the diversity measures of each algorithm in such a way as to modify, but not completely destroy its diversity characteristics.

5 Conclusions

The integration of user preference into EMO methods is an important research topic as it allows the user of an EMO algorithm to focus on interesting regions of the objective space. Different models for the integration of user preference have been proposed. Incorporating user preferences by weight information on the objective space has been shown to work very well for hypervolume algorithms [18]. We have presented a simple and very effective alternative way to use these user preferences in other state-of-the-art approaches such as NSGA-II and SPEA2. Our experimental results show that the weight integration into these algorithms performs very well, produces similar results as the weighted hypervolume indicator, and requires less computational effort than the hypervolume approach.

References

[1] Allmendinger, R., Li, X., Branke, J.: Reference Point-Based Particle Swarm Optimization Using a Steady-State Approach. In: Li, X., Kirley, M., Zhang, M., Green, D., Ciesielski, V., Abbass, H.A., Michalewicz, Z., Hendtlass, T., Deb, K., Tan, K.C., Branke, J., Shi, Y. (eds.) SEAL 2008. LNCS, vol. 5361, pp. 200–209. Springer, Heidelberg (2008)

[2] Auger, A., Bader, J., Brockhoff, D., Zitzler, E.: Articulating user preferences in many-objective problems by sampling the weighted hypervolume. In: Proc. 11th Annual Conference on Genetic and Evolutionary Computation, pp. 555–562 (2009)

[3] Beume, N., Naujoks, B., Emmerich, M.T.M.: SMS-EMOA: Multiobjective selection based on dominated hypervolume. European Journal of Operational Research 181, 1653–1669 (2007)

[4] Bringmann, K., Friedrich, T.: Approximating the volume of unions and intersections of high-dimensional geometric objects. Computational Geometry: Theory and Applications 43, 601–610 (2010)

[5] Coello Coello, C.A., Van Veldhuizen, D.A., Lamont, G.B.: Evolutionary Algorithms for Solving Multi-Objective Problems. Kluwer Academic Publishers, New York (2002)
[6] Deb, K.: Multi-objective optimization using evolutionary algorithms. Wiley, Chichester (2001)
[7] Deb, K., Sundar, J.: Reference point based multi-objective optimization using evolutionary algorithms. In: Proc. 8th Annual Conference on Genetic and Evolutionary Computation Conference (GECCO 2006), pp. 635–642 (2006)
[8] Deb, K., Agrawal, S., Pratap, A., Meyarivan, T.: A fast and elitist multiobjective genetic algorithm: NSGA-II. IEEE Trans. Evolutionary Computation 6(2), 182–197 (2002)
[9] Ho, S.-l., Yang, S., Ni, G.: Incorporating a priori preferences in a vector pso algorithm to find arbitrary fractions of the pareto front of multiobjective design problems. IEEE Trans. Magnetics 44, 1038–1041 (2008)
[10] Hu, Q., Xu, L., Goodman, E.D.: Non-even spread nsga-ii and its application to conflicting multi-objective compatible control. In: Proc. Genetic and Evolutionary Computation Conference Summit (GEC 2009), pp. 223–230 (2009)
[11] Igel, C., Hansen, N., Roth, S.: Covariance matrix adaptation for multi-objective optimization. Evolutionary Computation 15, 1–28 (2007)
[12] Suttorp, T., Hansen, N., Igel, C.: Efficient covariance matrix update for variable metric evolution strategies. Machine Learning 75, 167–197 (2009)
[13] Thiele, L., Miettinen, K., Korhonen, P.J., Luque, J.M.: A preference-based evolutionary algorithm for multi-objective optimization. Evolutionary Computation 17(3), 411–436 (2009)
[14] Wickramasinghe, U.K., Li, X.: Integrating user preferences with particle swarms for multi-objective optimization. In: Proc. 10th Annual Conference on Genetic and Evolutionary Computation, pp. 745–752 (2008)
[15] Wickramasinghe, U.K., Li, X.: Using a distance metric to guide pso algorithms for many-objective optimization. In: Proc. 11th Annual Conference on Genetic and Evolutionary Computation, pp. 667–674 (2009)
[16] Zitzler, E., Deb, K., Thiele, L.: Comparison of Multiobjective Evolutionary Algorithms: Empirical Results. Evolutionary Computation 8(2), 173–195 (2000)
[17] Zitzler, E., Laumanns, M., Thiele, L.: SPEA2: Improving the strength Pareto evolutionary algorithm for multiobjective optimization. In: Proc. Evolutionary Methods for Design, Optimisation and Control with Application to Industrial Problems (EUROGEN 2001), pp. 95–100 (2002)
[18] Zitzler, E., Brockhoff, D., Thiele, L.: The Hypervolume Indicator Revisited: On the Design of Pareto-compliant Indicators via Weighted Integration. In: Obayashi, S., Deb, K., Poloni, C., Hiroyasu, T., Murata, T. (eds.) EMO 2007. LNCS, vol. 4403, pp. 862–876. Springer, Heidelberg (2007)

Genetic Programming for Edge Detection Based on Accuracy of Each Training Image

Wenlong Fu[1], Mark Johnston[1], and Mengjie Zhang[2]

[1] School of Mathematics, Statistics and Operations Research
[2] School of Engineering and Computer Science
Victoria University of Wellington, PO Box 600, Wellington, New Zealand

Abstract. This paper investigates fitness functions based on the detecting accuracy of each training image. In general, machine learning algorithms for edge detection only focus on the accuracy based on all training pixels treated equally, but the accuracy based on every training image is not investigated. We employ genetic programming to evolve detectors with fitness functions based on the accuracy of every training image. Here, average (arithmetic mean) and geometric mean are used as fitness functions for normal natural images. The experimental results show fitness functions based on the accuracy of each training image obtain better performance, compared with the Sobel detector, and there is no obvious difference between the fitness functions with average and geometric mean.

Keywords: Genetic Programming, Edge Detection, Image Analysis.

1 Introduction

Edge detection is a well developed area of image analysis. Many different techniques for edge detection have been developed based on window filters [1]. Based on some existing features extracted from a fixed window, such as gradient, different methods are used to detect edges [2]. Generally, the accuracy of a method is over all training examples, not taking into account the accuracy of every training image. It is possible that the good accuracy over all pixels has good detection in many images but poor detection for a few images. The accuracy based on every training image is worth investigating. It has not so far been addressed what performance methods based on the accuracy of each training image can obtain and what difference exists in different measure methods based on the accuracy of each training image.

Genetic programming (GP) has been employed for object detection and image analysis since the 1990s [3], but there are only a few reports for edge detection. Almost all existing methods require prior or domain specific knowledge. Similarly to other machine learning approaches, GP usually evolves image classifiers or detectors using a reasonable number of actual images as the training set. The evaluation only takes into account the detecting results for all sampling pixels. However, GP can create detectors without using a window, extract edge

D. Wang and M. Reynolds (Eds.): AI 2011, LNAI 7106, pp. 301–310, 2011.

information automatically, and directly output the whole detecting result of one image using a whole image as input [4]. The output based on one whole image is well-suited for evaluating a detector based on the accuracy of each image, but what difference measures based on the accuracy of every training image will make needs to be investigated.

Goals. The overall goal of this paper is to investigate fitness functions based on the accuracy of each training image. Average (arithmetic mean) and geometric mean are used as fitness functions. The purpose of using the geometric mean is to penalise the worse detection from the training images. Based on a normal natural image experiment, we will analyse the influence from evaluation methods with average and the geometric mean. Specifically, we would like to investigate the following research objectives.

- Whether the fitness function with average of the accuracy of every training image can achieve reasonably good performance.
- Whether the fitness function with the geometric mean of the accuracy of every training image also can achieve reasonably good performance, and whether it can compete with average.
- Whether a combination of the geometric mean and average can further improve performance.

The remainder of this paper is organised as follows. Section 2 briefly describes the background. Section 3 develops fitness functions based on the accuracy of each training image for edge detection using GP. After presenting the experimental design in Section 4, Section 5 describes the results with discussions. Section 6 gives conclusions and future work directions.

2 Background

2.1 Edge Detection

Edge detection is one of the most essential tasks in image processing and computer vision. For example, it is very useful for feature detection and extraction. The purpose of edge detection is to identify points in an image at which the pixel intensities change sharply or irregularly. Edge detection in untextured images aims at finding these interesting points based on local discontinuities. Edge detection in textured images is more complex, where these edge points mark the boundary of regular changes or irregular changes.

Almost all methods use features extracted based on discontinuities, such as gradients. Detectors in the traditional methods based on derivatives are, e.g., Sobel detectors based on first derivatives, Laplacian detectors based on the zero crossings in the second derivative and other detectors based on differentiation [5]. From approximation to the shape of spatial receptive fields, Gaussian filters along with the Laplacian detectors are very similar to the difference of Gaussians (DOG) [1]. Canny detectors [6] are derived from the optimal filter leading to a Finite Impulse Response filter which turns out to be well approximated by the derivative of a Gaussian function [1]. A popular strategy is to design a local window to filter non-edge points with a threshold; the window size is usually 3×3 for the sake of speed.

From the traditional way to extract edge information, machine learning methods are employed to search for special detectors for one image dataset. Edge information extracted based on a fixed window are used as input variables. However, features extracted based on a window are sensitive to the size of the window, e.g., a small size may be affected by noise and a larger size may lead to wrong edge localisation. Our previous work [4] employed GP as a global method to evolve edge detectors without using a window, and showed good performance for finding good detectors. However, differences between fitness functions based on the accuracy of each training image were not addressed.

2.2 Related Work to Genetic Programming for Edge Detection

There is little existing work on GP for edge detection. Harris and Buxton [7] designed approximate response detectors in one-dimensional signals by GP, but it is based on the theoretical analysis of the ideal edge detector and the corresponding properties. Poli [9] suggested to use four macros for searching a pixel's neighbours using GP. Ebner [8] used four shift functions and other functions to approximate the Canny detector. The Sobel detector is approximated by hardware design [10] with the relationship between a pixel and its neighbourhood as terminals. Bolis et al [11] simulate an artificial ant to search edges in image. Zhang and Rockett [12] evolved a 13×13 window filter for comparison with the Canny edge detector. Wang and Tan [13] used linear GP to find binary image edges, inspired by morphological operators, erosion and dilation, as terminals [14] for binary images. A 4×4 window is employed to evolve digital transfer functions (combination of bit operators or gates) for edge detection by GP [15]. Our previous work [4] used GP for edge detection based on ground truth and without using windows.

3 The New Approach

3.1 Main Idea

To measure the performance of an edge detector over a number of training examples, the traditional approach is to adopt the classification-based measure of the number of pixels correctly classified as a proportion of all actual edge pixels. However, edge detection is different from traditional classification problems; the test performance should be based on each image, not just the set of all pixels from all testing images. In general, machine learning algorithms do not contain any indication for the detecting result of each training image. Therefore it is worth investigating what result different fitness function based on the accuracy of every training image will make.

3.2 Sets of Terminals and Primitive Functions

We use the same terminal set and function set as in our previous work [4]. The terminal set contains one input image x and random constants rnd in the range of $[-10, 10]$. The function set is $\{+, -, *, \div, shift_{n,m}, abs, sqrt, square\}$. Here, function "$shift_{n,m}$" is a main operator, meaning that the image will shift n columns and m rows. If n is negative, it shifts to left, otherwise shifts to right. If m is negative, it means shifting up, otherwise down. For example, "$shift_{1,0}$" and "$shift_{-1,0}$" mean that the input image will be shifted right and left by 1 column, respectively. If a shift direction is out of the range, the nearest row (column) will be used as the result for the row (column) shifting. All these functions are able to operate on the input image matrix. Here n and m are randomly generated from 0 to 2, which means, for an input matrix, each element has equal probability to choose a neighbour in a 5×5 window. For the four direction shifting functions in [8], neighbours of each element have different probability, e.g., the probability for the top left neighbour for each element is lower than the left neighbour because this neighbour needs the left shifting function and the top shifting function. However, neighbours in a small window should have the same importance for determining the pixel as an edge point or not. The "$\{+, -, *, abs, square\}$" have their usual meanings. The square root function "*sqrt*" is protected, which produces a result of 0 for negative inputs. Division "$\div$" is also protected, producing a result of 1 for a 0 divisor.

A classical filter can be expressed as an individual program based on this function set. For example, the 2×2 window Robert filter [5] can be shown as

$$(sqrt(+(square(-x\,(shift_{1,1}\,x)))(square(-(shift_{1,0}\,x)(shift_{0,1}\,x))))).$$

3.3 Fitness Function

We treat the edge detection task as a binary classification task (with the edge pixels as the main class) in the evolutionary training process. For the output of a program, we do not use the threshold ratio for marking edge points; rather we simply use zero as the threshold and all images use the pixel intensities. For the negative values, we always classify them as non-edge points. The output is directly evaluated without post-processing, following the suggestion from [16].

Assume we have recall r_i (the number of pixels on the edges correctly detected as a proportion of the total number of pixels on the edges in image i), and precision p_i (the number of pixels on the edges correctly detected as a proportion of the total number of pixels detected as edges in images i). The *F-measure* [2] is employed to evaluate the accuracy of a detector for one image. We use f_i (see (1) below) to indicate the fitness for image i. Higher f_i indicates the better performance for detecting image i. Given the training data contains N images, the normal way is to use the (arithmetic) average of all images f_i, $i = 1, 2, \ldots, N$, and the fitness function is shown in (2). Here we minimize $1 - f_i$ rather than maximize f_i. F_{avg} focuses on the overall result for every training image. For penalising bad performance from some training images, the geometric mean is employed to evaluate the performance of all images and the fitness function F_{gm} is shown in (3). Taking into account average and the geometric mean, we combine F_{avg} and F_{gm} together and design a new fitness function F_{com} (see (4)) with a weight $0 \leq \alpha \leq 1$.

$$f_i = \frac{2r_ip_i}{r_i + p_i} \tag{1}$$

$$F_{avg} = 1 - \frac{1}{N}\sum_{i=1}^{N}(f_i) \tag{2}$$

$$F_{gm} = 1 - (\prod_{i=1}^{N} f_i)^{\frac{1}{N}} \tag{3}$$

$$F_{com} = (1-\alpha)F_{avg} + \alpha F_{gm} \tag{4}$$

4 Experimental Design

The Berkeley Segmentation Dataset (BSD) [2] comes from natural images (of size 481×321 pixels) with ground truth provided. This image dataset is not polluted by additional noise. For simplicity, we select some images from the BSD training dataset as our training dataset, and some images from the BSD test dataset as our test dataset. For the fairness of the judgment of edges, the ground truth are combined from five to seven persons as grey-level images. Fig. 1 shows the training images and the ground truth. In order to decrease the computation time, we sample five different sub-images of size 41×41 from each image as training sub-images, and use one whole sub-image as an input. The size 41×41 is enough for covering the edge features, such as texture gradients [2]. Ten additional images (Fig. 2) are used as the test dataset.

The parameter values for GP are: population size 500; maximum generations 200; maximum depth (of a program) 10; and probabilities for mutation 0.15, crossover 0.80 and elitism 0.05. These values are chosen based on common settings and initial experiments. For analysing the weight α, a set of $\{0.1, 0.2, 0.35, 0.5, 0.65, 0.8, 0.9\}$ is used and a variant α_t based on (5):

$$\alpha_{t+1} = 0.01 + 0.99\alpha_t \tag{5}$$

where t is the generation and $\alpha_0 = 0$. In the initial stages of evolution, all individuals may have poor detection for some training images so that the geometric mean does not work well. Therefore the weight α focuses on average at the beginning and then increases the weight to the geometric mean. The GP experiment is repeated for 30 independent runs with different random seeds. We use 30 fixed and different random seeds so that we have same initial population when different fitness functions are used. For checking the performance of the fitness functions, the related results will be compared with the 3×3 Sobel detector. We use 11 threshold levels to find the minimum value $1 - f_i$ for the Sobel detector and these threshold values for the 11 levels are $\frac{i}{11} * maxO$, where $i = 0, 1, 2, \ldots, 10$.

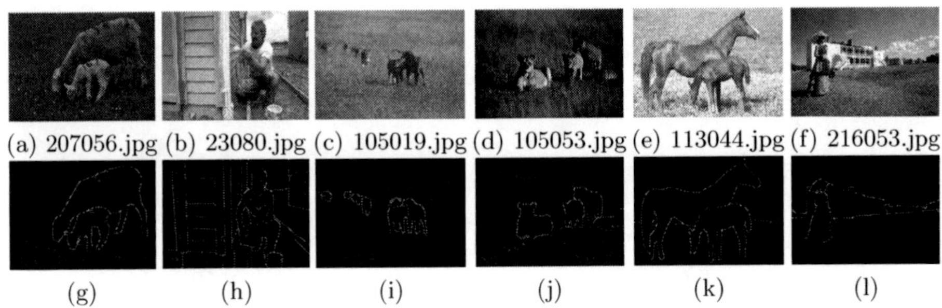

(a) 207056.jpg (b) 23080.jpg (c) 105019.jpg (d) 105053.jpg (e) 113044.jpg (f) 216053.jpg

(g) (h) (i) (j) (k) (l)

Fig. 1. Training images from BSD dataset and the ground truth

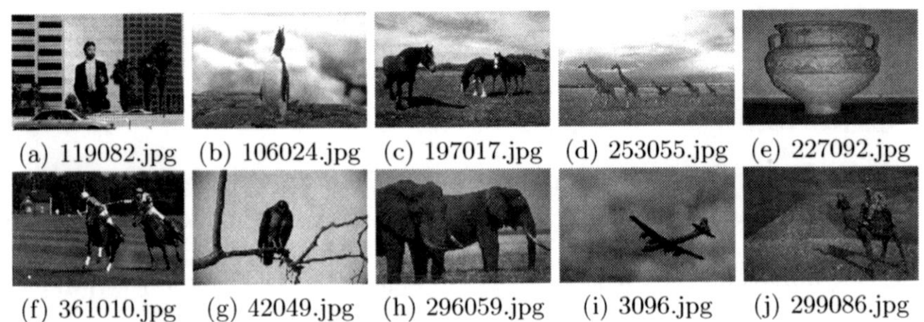

(a) 119082.jpg (b) 106024.jpg (c) 197017.jpg (d) 253055.jpg (e) 227092.jpg

(f) 361010.jpg (g) 42049.jpg (h) 296059.jpg (i) 3096.jpg (j) 299086.jpg

Fig. 2. Test images from BSD dataset

5 Experimental Results and Discussion

Training Performance. Table 1 shows the training results: mean and standard deviation over 30 replications of the *mean*, *minimum* or *maximum* (respectively) over all training sub-images i of $1 - f_i$. The columns feature the best detector evolved using each of F_{avg}, F_{gm} and $F_{0.5}$ (this is F_{com} with $\alpha = 0.5$), respectively. The lower the value, the better the performance. Here "*" indicates that there is a statistically significant difference between the mean of *mean*, *min* or *max* using fitness function F_{gm} and using fitness function $F_{0.5}$ with significance level 0.05 (in a t-test). Comparing F_{avg} with F_{gm} and $F_{0.5}$, there is no significant difference for the means. From Table 1, we can see that *mean* and *min* for $F_{0.5}$ are significantly lower than the values in F_{gm}. For F_{avg} and F_{gm}, there is no significant difference, even though *mean* and *min* in F_{avg} are smaller than the values in $F_{0.5}$; this is because the standard deviations in F_{avg} are relatively high compared to the standard deviations of $F_{0.5}$. Comparing standard deviations, the values in F_{avg} are also higher than F_{gm}, which indicates that detectors evolved with F_{gm} are more stable than those evolved with F_{avg} on the training images. For the combination of the geometric mean and average, $F_{0.5}$ is more stable than F_{avg} and has better F-measure than F_{gm} for the training images.

Test Performance. Table 2 shows the test results (mean ± standard deviation) with F_{avg}, F_{gm}, $F_{0.5}$ and the Sobel edge detector. Here "*" has the same meaning

Table 1. Comparison of $1 - f_i$ among detectors evolved with fitness functions F_{avg}, F_{gm} and $F_{0.5}$ on training sub-images

Fitness	F_{avg}	F_{gm}	$F_{0.5}$
mean	0.5581 ± 0.0382	0.5713 ± 0.0214	$0.5591^* \pm 0.0225$
min	0.2707 ± 0.0445	0.2897 ± 0.0329	$0.2726^* \pm 0.0246$
max	0.8020 ± 0.0498	0.7892 ± 0.0342	0.7832 ± 0.0226

Table 2. Results $1 - f_i$ for F_{avg}, F_{gm} and F_{com} on ten test images

Image i	F_{avg}	F_{gm}	$F_{0.5}$	Sobel
(a) 119082	0.6808 ± 0.0219	$0.6868^* \pm 0.0201$	0.6777 ± 0.0153	$\mathbf{0.7608}^+$
(b) 106024	0.7149 ± 0.0390	$0.7137^\diamond \pm 0.0230$	0.7068 ± 0.0264	$\mathbf{0.6875}^-$
(c) 197017	0.7045 ± 0.0377	$0.7027^\diamond \pm 0.0215$	0.6966 ± 0.0261	0.6930
(d) 253055	0.6033 ± 0.0458	0.6007 ± 0.0274	0.5951 ± 0.0224	$\mathbf{0.7359}^+$
(e) 361010	0.6482 ± 0.0359	0.6499 ± 0.0199	0.6409 ± 0.0217	$\mathbf{0.7598}^+$
(f) 227092	0.6978 ± 0.0299	0.6946 ± 0.0288	0.6933 ± 0.0172	$\mathbf{0.7130}^+$
(g) 42049	0.4609 ± 0.0339	0.4711 ± 0.0245	0.4596 ± 0.0275	$\mathbf{0.4964}^+$
(h) 296059	0.6107 ± 0.0590	$0.6114^\diamond \pm 0.0324$	0.6009 ± 0.0351	0.5930
(i) 3096	0.4851 ± 0.0548	$0.4900^\diamond \pm 0.0336$	0.4782 ± 0.0417	0.4766
(j) 299086	0.6962 ± 0.0301	0.6875 ± 0.0131	0.6857 ± 0.0146	$\mathbf{0.8533}^+$

as in Table 1; "$\diamond$" indicates that the performance of F_{gm} is significantly higher (worse) than the performance of the Sobel detector; "+" means the performance for the Sobel detector is significantly worse than all other fitness functions; and "−" means the performance for the Sobel detector is significantly better than them. Overall, the performances on the ten test images show no significant differences among F_{avg}, F_{gm} and $F_{0.5}$, but they are generally significantly better

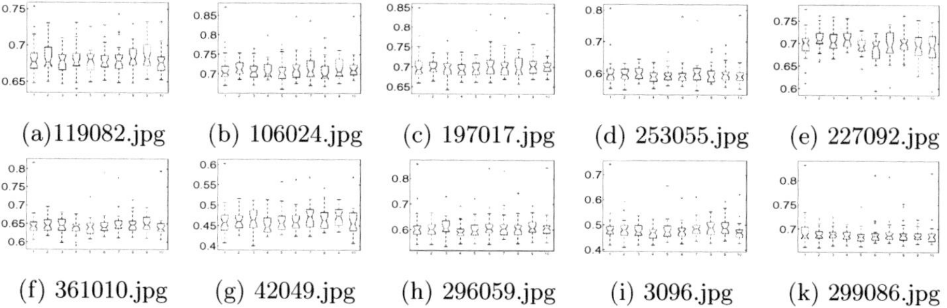

(a)119082.jpg (b) 106024.jpg (c) 197017.jpg (d) 253055.jpg (e) 227092.jpg

(f) 361010.jpg (g) 42049.jpg (h) 296059.jpg (i) 3096.jpg (k) 299086.jpg

Fig. 3. Comparison among F_{avg}, F_{gm} and F_{com} for the ten test images. The test results $(1 - f_i)$ for each subfigure, from left to right, are: F_{avg}, F_{com} with $\alpha \in \{0.1, 0.2, 0.35, 0.5, 0.65, 0.8, 0.9\}$, F_{gm} and the variant α_t.

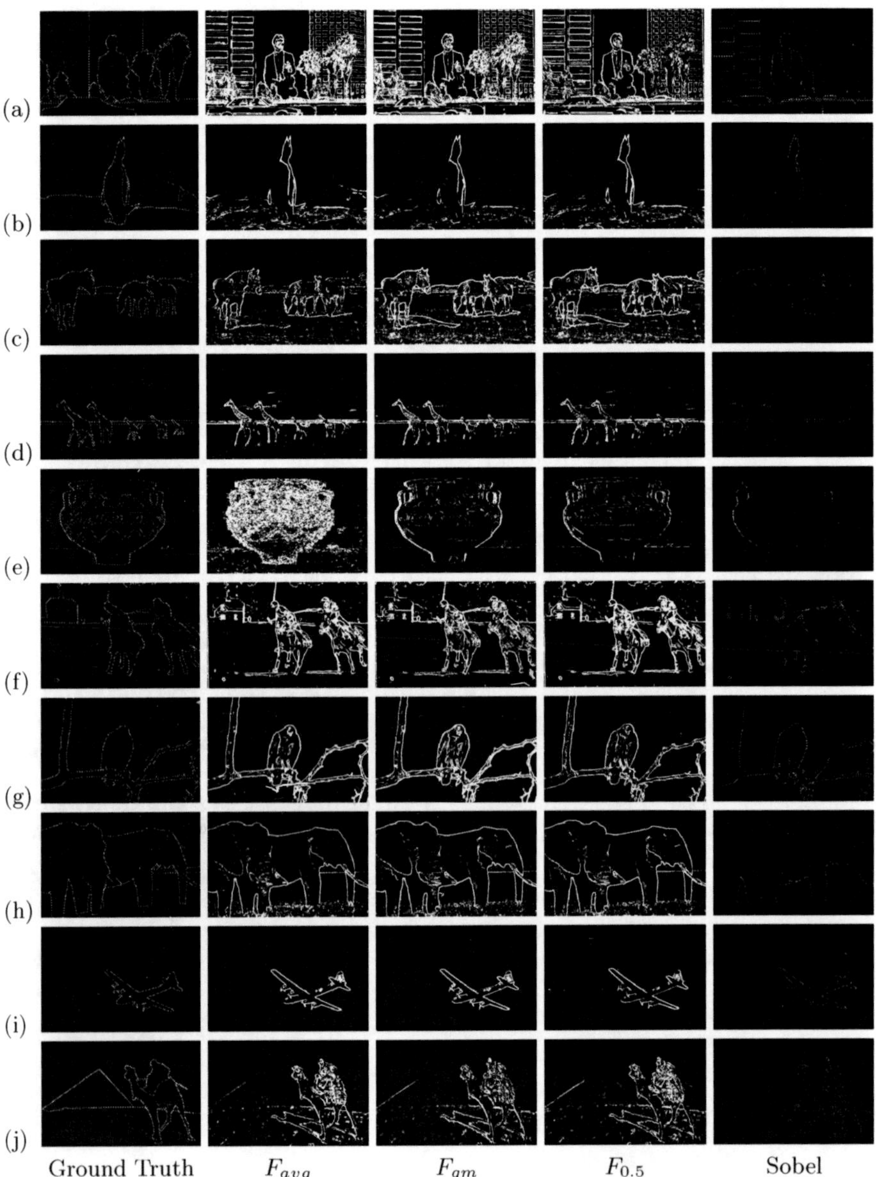

Fig. 4. Best results for the ten test images for each method

than the Sobel detector. The only significant difference between F_{gm} and $F_{0.5}$ is that $F_{0.5}$ is better than F_{gm} for image (a). For F_{avg} and $F_{0.5}$, only test image (b) has significantly worse performance than the Sobel detector, but six of the ten test images have better performances for all fitness functions compared to Sobel. For F_{gm} compared to Sobel, interestingly when it is not significantly better it is always significantly worse. The standard deviations of the ten test images in

F_{gm} and $F_{0.5}$ are lower than the values in F_{avg} and in seven of the ten images F_{gm} has slightly lower standard deviation than $F_{0.5}$. Therefore, F_{gm} and $F_{0.5}$ appear to be more stable than F_{avg}.

Comparison of α Values. For the further analysis, Fig. 3 shows the test results using boxplots. Comparing medians, lower and upper quartiles, there are no distinct differences among these results with different fitness functions. But the details for outliers are different between F_{avg} and F_{gm}; F_{gm} only contains them in four test images and they are closer to the largest observation than F_{avg} (the minimum distance to the largest observation or the smallest observation). All ten test images in F_{avg} contain outliers, which means that at least one of the 30 detecting results for each test image in F_{avg} is unusual. Referring to the standard deviations of the ten test images in Table 2, it is a possible reason that F_{avg} has larger deviations in the ten test images. From the comparison among different F_{com}, α does not affect the detecting result of each test image.

Examples of Detected Edges and Further Analysis. The best detecting results from F_{avg}, F_{gm} and $F_{0.5}$ are shown in Fig. 4. These results from GP are binary boundaries, and the detecting results from the Sobel detector are soft boundaries. These results from GP are hard to distinguish, except image (e). For image (e), F_{avg} has the best detecting result, but high recall with low precision. For image (g), F_{gm} has a higher false alarm, compared with F_{avg}. Compared with the results from the Sobel detector, the detectors evolved by GP appear to find more details.

Summary. In summary, F_{avg}, F_{gm} and F_{com} have no obvious differences for the normal natural images based on mean and they outperform the Sobel detector based on the ten test images. However, F_{gm} and F_{com} are more stable than F_{avg}; it seems that the geometric mean can slightly restrict to choose the results of the training images during the training progress so that the stability for test images are better than average. A possible reason is that average only takes into account the overall result of every training image, and does not know the differentiation from training images, but geometric mean can indicate the bad detection result from images (too low f_i in one training image will make the product of all f_i very low).

6 Conclusions

The goal of this paper was to investigate fitness functions based on the accuracy of each training image. The geometric mean and (arithmetic) average were employed as fitness functions based on the accuracy of each training image. The fitness functions based on average and geometric mean can be used to evolve detectors outperforming the Sobel detector. The combination of the geometric mean and average as fitness functions, improves the accuracy, compared with the fitness function with the geometric mean, and is more stable than the fitness function with average.

For future work, we will test this technique on further natural images, especially image datasets containing a few special images and compare results from the fitness function based on overall pixels. We will also compare our algorithm with traditional window-based methods.

References

1. Basu, M.: Gaussian-based edge-detection methods: a survey. IEEE Transactions on Systems, Man, and Cybernetics—Part C: Applications and Reviews 32(3), 252–260 (2002)
2. Martin, D.R., Fowlkes, C.C., Malik, J.: Learning to detect natural image boundaries using local brightness, color, and texture cues. IEEE Transactions on Pattern Analysis and Machine Intelligence 26(5), 530–549 (2004)
3. Krawiec, K., Howard, D., Zhang, M.: Overview of object detection and image analysis by means of genetic programming techniques. In: Frontiers in the Convergence of Bioscience and Information Technologies, pp. 779–784 (2007)
4. Fu, W., Johnston, M., Zhang, M.: Genetic programming for edge detection: a global approach. In: Proceedings of the IEEE Congress on Evolutionary Computation, pp. 254–261 (2011)
5. Ganesan, L., Bhattacharyya, P.: Edge detection in untextured and textured images: a common computational framework. IEEE Transactions on Systems, Man, and Cybernetics, Part B: Cybernetics 27(5), 823–834 (1997)
6. Canny, J.: A computational approach to edge detection. IEEE Transactions on Pattern Analysis and Machine Intelligence 8(6), 679–698 (1986)
7. Harris, C., Buxton, B.: Evolving edge detection with genetic programming. In: GECCO 1996, pp. 309–314 (1996)
8. Ebner, M.: On the edge detectors for robot vision using genetic programming. In: Horst-Michael Groβ, Workshop SOAVE 97- Selbstorganisation von Adaptivem Verhalten, pp. 127–134 (1997)
9. Poli, R.: Genetic programming for image analysis. In: Proceedings of the First Annual Conference on Genetic Programming, pp. 363–368 (1996)
10. Hollingworth, G.S., Smith, S.L., Tyrrell, A.M.: Design of highly parallel edge detection nodes using evolutionary techniques. In: Proceedings of the Seventh Euromicro Workshop on Parallel and Distributed Processing, pp. 35–42 (1999)
11. Bolis, E., Zerbi, C., Collet, P., Louchet, J., Lutton, E.: A GP Artificial Ant for Image Processing: Preliminary Experiments with EASEA. In: Miller, J., Tomassini, M., Lanzi, P.L., Ryan, C., Tetamanzi, A.G.B., Langdon, W.B. (eds.) EuroGP 2001. LNCS, vol. 2038, pp. 246–255. Springer, Heidelberg (2001)
12. Zhang, Y., Rockett, P.: Evolving optimal feature extraction using multi-objective genetic programming: a methodology and preliminary study on edge detection. In: GECCO 2005, pp. 795–802 (2005)
13. Wang, J., Tan, Y.: A novel genetic programming based morphological image analysis algorithm. In: GECCO 2010, pp. 979–980 (2010)
14. Quintata, M., Poli, R., Claridge, E.: Morphological algorithm design for binary images using genetic programming. Genetic Programming and Evolvable Machines 7(1), 81–102 (2006)
15. Golonek, T., Grzechca, D., Rutkowski, J.: Application of genetic programming to edge detector design. In: International Symposium on Circuits and Systems, pp. 4683–4686 (2006)
16. Moreno, R., Puig, D., Julia, C., Garcia, M.A.: A new methodology for evaluation of edge detectors. In: 16th IEEE International Conference on Image Processing (ICIP), pp. 2157–2160 (2009)

Improving Robustness of Multiple-Objective Genetic Programming for Object Detection

Rachel Hunt[1], Mark Johnston[1], and Mengjie Zhang[2]

[1] School of Mathematics, Statistics and Operations Research
[2] School of Engineering and Computer Science
Victoria University of Wellington, PO Box 600, Wellington, New Zealand
huntrach1@myvuw.ac.nz,
{mark.johnston,mengjie.zhang}@vuw.ac.nz

Abstract. Object detection in images is inherently imbalanced and prone to overfitting on the training set. This work investigates the use of a validation set and sampling methods in Multi-Objective Genetic Programming (MOGP) to improve the effectiveness and robustness of object detection in images. Results show that sampling methods decrease runtimes substantially and increase robustness of detectors at higher detection rates, and that a combination of validation together with sampling improves upon a validation-only approach in effectiveness and efficiency.

1 Introduction

Object detection is the task of correctly locating and classifying objects of interest inside a larger image [9]. Object classification requires differentiating between different kinds of objects and object localisation consists of identifying the positions of all objects of interest in a large image. A detector's ability to detect objects in an image is primarily measured in terms of *detection rate*, DR, the number of correctly located objects as a proportion of the total number of objects in the image, and the *false alarm rate*, FAR, the number of falsely reported objects as a proportion of the total number of objects in the image. Here $0 \leq DR \leq 1$ but $0 \leq FAR$ as it is possible for the detector to report more objects than there are in the image. The aim is to maximise DR and minimise FAR. There is certainly a trade-off between these two objectives. A classifier could have 0% FAR by not reporting any objects, hence DR is also 0%. Similarly a classifier could correctly detect every object by reporting an object at each location in the image, in which case FAR would be very large.

Genetic Programming (GP) is a form of evolutionary computation based on the principles of biological evolution [12]. In tree-based GP [12], the genetic programs (solutions) are represented as trees, and then evolved using Darwinian evolutionary principles. Natural selection, recombination and mutation evolve the population towards a solution for the given problem; this is survival of the fittest. GP has been used widely for a variety of tasks, including classification [2,4,6], regression and object detection [10,12,13].

D. Wang and M. Reynolds (Eds.): AI 2011, LNAI 7106, pp. 311–320, 2011.

Object detection is inherently imbalanced, and prone to overfitting on the training set. Our previous work [6] has shown that more accurate classifiers under imbalanced datasets can be built by dynamic sampling of the training set during training. Also, validation is a common way for guarding against overfitting. Hence the goal of this paper is to determine whether sampling and validation can improve the robustness of object detectors and compare the effectiveness and efficiency of both.

The remainder of this paper is organised as follows. Section 2 describes the background work relating to this paper. Section 3 describes the new validation approach and sampling methods used for object detection tasks. Section 4 describes the experimental design and Section 5 discusses the results. Section 6 concludes the paper.

2 Background

Class Imbalance and Sampling. In object detection there are a large number of non-object instances and only a few object pixels in each image, an example of class imbalance in which there are a large number of instances of one class and only a small number of instances in the other [2]. Re-sampling can be used to artificially balance the data set, often through under and over sampling. Under-sampling uses fewer than the total number of majority class instances and over-sampling replicates minority class examples. Gathercole and Ross [5] compare three methods of subset selection, Dynamic Subset Selection (DSS), Historical Subset Selection (HSS) and Random Subset Selection (RSS), on a large unbalanced dataset. DSS randomly selects a sample from the training set of size N biased towards those instances which are often misclassified or have not been included in the subset for several generations. HSS uses standard GP runs on the training dataset to determine the 'difficulty' of each instance, by counting the number of times it is mis-classified by the best population member in each run. The cases with greater 'difficulty' are then used as the sample subset for HSS runs. In RSS each training instance has equal probability of being selected in the subset, and a new subset is taken at each generation.

Hunt et al. [6] establish that the variation in training performance introduced by sampling examples from the training set is no worse than the variation between GP runs already accepted. On binary classification tasks, results show that the use of sampling methods during the training process improves minority class classification accuracy and the robustness of classifiers evolved.

SOGP and MOGP for Object Detection. Zhang [12] propose a two-phase training method and the use of False Alarm Area (FAA) in Single Objective GP (SOGP). The first phase uses object cutouts from the full training set using a fitness function which maximises classification accuracy; this is purely object classification. In the second phase, initialised with the population from the first phase, programs are now trained on the full training images, using a moving window approach. The minimisation fitness function of this stage is

$fitness_1 = K_1 \cdot (1 - DR) + K_2 \cdot FAR + K_3 \cdot FAA + K_4 \cdot size$, where K_i are constants, FAA is the number of positive classifications less the number of objects in the image, and *size* is the size of the program. A clustering approach is used so false alarms are not reported if the false alarm is within *tolerance* of a previously detected object or false alarm. Results showed the two-phase approach outperformed single-phase GP, with a greater detection rate (higher number of identified objects) and shorter training time on a coins dataset (described in Section 4).

Pixel statistics are a domain independent approach to image analysis. Pixel values are a measure of overall brightness/intensity and contrast, pixel statistics frequently used are the mean and variance of regions of images. Zhang et al. [13] investigated the use of three different terminal sets: rectilinear features; circular features; and pixels. These terminal sets were used with SOGP with the fitness function $fitness_2 = K_1 \cdot (1 - DR) + K_2 \cdot FAR$. Three image sets were used: shapes, coins and retina images. Results suggest that rectilinear features are more effective for these problems than circular features.

Liddle et al. [10] extend the two-phase training method of [12] into Multiple-Objective GP (MOGP) based on NSGA-II [3]. The first phase, classification, used true positive rate and true negative rate as two objectives. The second phase used DR and FAR as two objectives with FAA used as a tie-breaking measure. Summary attainment surfaces [8] were used to visualise the MOGP results. Results on the shapes and coins datasets show that MOGP has significant promise in evolving a larger set of more diverse classifiers in the same or less CPU time as SOGP. However there was a high FAR, with the best classifier having 5000% FAR and 90% DR which is not desirable. There was also a large presence of classifiers with worst-best performance, i.e., those than have the worst performance on the first objective, DR, and attain the best performance on the second objective, FAR. These classifiers are essentially useless, as they classify all pixels as non-object.

The outcome of any one run of SOGP is a single best genetic program for object detection. The outcome of a single run of MOGP is a Pareto-front of non-dominated genetic program solutions. A genetic program is *non-dominated* if it weakly dominates all other genetic programs in the current population. A genetic program, x, *weakly dominates* another genetic program, y, if for all i, $x_i \leq y_i$ and there exists a j such that $x_j < y_j$ for i, j in objectives (all objectives to be minimised) [10]. Bhowan et al. [2] use MOGP for classification of unbalanced data. They use P-dominance, where a program x weakly dominates another program y, if x weakly dominates y, as defined above, and x achieves P, a minimum performance level, on each objective. P-dominance was used to limit the number of one sided classifiers in the population. The use of P-dominance was shown to improve front diversity, and reduce the number of one-sided programs in all of the datasets used.

Validation. The dataset is split into training, validation and testing sets, which are distinct. The main idea of the validation set is to determine when to terminate training, by considering when independent performance on the validation set begins to get worse [7]. The training set is used to drive the evolutionary process, the performance of each program is recorded as observed over this set. The validation set is used to measure the generalization ability of the current population of classifiers. When the classification performance decreases, it is a possible indicator that the classifiers are overfitting to the training data, so evolution is stopped.

3 Validation and Sampling Methods

Following Zhang [12] and Liddle et al. [10] we use the same two phase approach. Phase one still uses true positive and true negative rates as the two objectives, both to be maximised, and phase two uses DR and FAR as the two objectives to be maximised and minimised respectively.

Benchmark. The benchmark for each fitness scheme evolves the population for 20 generations in the first phase (classification) and 40 generations in the second phase (detection). The training set for the first phase is the 480 cutouts and in phase two the full training set of size 128800 is used to evaluate the current population of genetic programs at each generation.

Sampling Methods. We leave the first phase of the evolution process as is, the genetic programs being trained on the same 480 image cutouts for the full 20 generations. In the second phase of evolution we sampled with uniform probability from the training set, of size 128800, until we had a sample of the desired size. We used two different sample sizes: 500 and 5000. Each object centre is included in the sample, and the sample is taken without replacement. This is a form of under-sampling, as not all of the non-object instances are included in the sample. A new sample is taken at each generation of evolution and used in training to evaluate the entire current population of genetic programs at that generation. Phase one is effectively a static sample from the training set which is taken before training begins, and phase two is dynamic sampling during training. *Sampling 500*, like the benchmark, evolves the population for 20 generations in the first phase and 40 generations in the second phase. In the second phase a sample of size 500 of training instances is taken from the training set at each generation. The current population is evaluated on this subset of training instances. *Sampling 5000* uses the same methods as *Sampling 500* but with a sample size of 5000.

Validation Methods. Validation is performed every two generations. The population is first evaluated on the validation set, and the best front found. Generalization ability is monitored by the use of two measures: hyperarea and distance. Hyperarea is the area under the best Pareto-front (as a staircase as in Figure 1):

$$hyperarea = \sum_{i=2}^{n-1} (S_{i,0} - S_{i-1,0}) \times S_{i,1}$$

where n is the number of classifiers in the Pareto-front and $S_{i,j}$ is the performance of the ith classifier on the jth objective. Distance is a measure of how much each classifier's performance differs on the validation set in comparison to the training set, i.e., how much further each classifier's performance, $(1 - DR, \text{FAR})$, is from the best possible performance, 100% DR and 0% FAR, on the validation set, in comparison to performance on the current training sample.

$$distance = \sum_{i=1}^{N} \left(\sqrt{(S_{i,0}^{val})^2 + (S_{i,1}^{val})^2} - \sqrt{(S_{i,0}^{tr})^2 + (S_{i,1}^{tr})^2} \right)$$

where N is the size of the population, S^{tr} is the performance on the training set and S^{val} is the performance on the validation set. A moving average of three validations is used, and if the moving average of both hyperarea and distance increase between validations, evolution is terminated, otherwise evolution runs for the set number of generations. *Validation* evolves the population for 20 generations in the first phase and for up to 40 generations in the second phase. In phase two the full training set is used for evaluating the population. In phase two the validation set is used to monitor the generalisability of the population of genetic programs to determine if overfitting is occurring. If validation does not indicate that overfitting is occurring, then evolution terminates after generation 40, and then the best front is evaluated on the test set. *Validation-Sampling* combines aspects of *Sampling 5000* and *Validation*. The population evolves for 20 generations in the first phase, with no sampling or validation. The second phase uses both sampling and validation. A sample of size 5000 is taken at every generation from the full training set and this is used to evaluate the performance of each program in the population. Validation again occurs every two generations, and evolution is terminated under the same conditions as *Validation*.

4 Experimental Design

Dataset. The *coins* dataset consists of seven photographs of New Zealand 5 cent coins, either heads up or tails up, against a noisy background. Each of the photographs in this dataset contains 16 coins with diameter approximately 63 pixels, and size approximately 500 × 500 pixels. The dataset was split so five photographs were used for training, one for validation, and one for testing. For the first phase (classification) 480 cutouts were selected from the training images: 80 objects; 50 partial; and 350 randomly selected. In the second training phase (detection) a sweeping window moved by three pixels each time. In both validation and testing, the sweeping window moved by one pixel each time.

Functions and Terminals. The terminal set represents the input into the GP system: features extracted from the images and random real number nodes in the range $[-10, 10]$. For each window cutout of the image, the window cutout is divided into five regions equal in size. These regions are the four quarters, and the square of the same size as these quarters centred in the middle of the window cutout. For each of these five regions, the mean and variance of the pixel values were calculated, giving 10 features per window cutout. The use of pixel statistics, which are domain independent, rather than individual pixel values increases the generalisability of our classifiers. The function set is $\{+, -, \times, \%,$ `if`$\}$. The arithmetic operators take two arguments. The first three arithmetic operators, $+$, $-$, $\times$, have their usual meanings. The % is as usual division except when dividing by zero where the value returned is zero. The `if` function takes three arguments, if the first argument is positive then it returns the second, else the third is returned.

Evolutionary Parameters. A population of 500 programs was used; the initial population was generated using the ramped half-and-half method. For benchmark MOGP the population was evolved for 20 generations in the first phase, and 40 generations in the second phase. For our modified MOGP the population was evolved for a maximum of 20 generations in the first phase, and a maximum of 40 generations in the second phase. Minimum tree depth 2, maximum tree depth 6, crossover rate 70%, mutation rate 30%, and tournament selection with tournament size of 2. We used the same 40 randomly generated seeds for both MOGP and our modified versions of MOGP. A *tolerance* value of 3 pixels was used, meaning that detected object centres reported within 3 pixels of an already reported object centre are not considered a new object centre.

5 Experimental Results

Summary attainment surfaces [8] are used to present the experimental results. The attainment value of a particular solution is the probability that the MOGP system will evolve a solution which is better than the current solution, i.e., weakly dominates the solution [8]. The number of attainment surfaces is equal to the number of MOGP runs, for this project this means there are 40 attainment surfaces. The best attainment surface, therefore, represents the solutions which have attainment value of zero. These are the solutions that are non-dominated in comparison to the solutions from all 40 Pareto-fronts. Similarly the median attainment surface is made up of solutions which have attainment value of 50% in comparison to the solutions from the 40 Pareto-fronts.

These attainment surfaces give a way of comparing performance and distribution of results over two objectives similar to the way a boxplot represents the distribution over one objective. In analysing the comparative performance of the seven methods, five attainment surfaces will be used to summarise the results: the best, upper quartile, median, lower quartile and worst summary attainment surfaces, representing attainment of 0%, 25%, 50%, 75% and 100% respectively. DR is to be maximised and FAR is to be minimised, as ideally a solution will correctly report all object centres without reporting any non-object centres. The attainment surfaces which are lower and more to the right are better. Therefore the region below and to the right of the best summary attainment surface represents the area which has not been attained in any of the runs, and the region above and to the left of the worst summary attainment surface represents the area which has been attained by all of the runs. Further the area below and to the right of, for example, the median summary attainment surface, represents the area which 50% of GP runs attain. Hence when comparing two median summary attainments surfaces, the better one, is the one which is lower and more to the right.

Results for Benchmark MOGP. The benchmark test results are shown in Figure 1(a). The average runtime of the benchmark method over 40 replications was 26420 seconds (see Table 1). The results show that over 50%, but less than 75%, of GP runs have a Pareto-front which has the full range of detection rates. All GP runs attain a detection rate of at least 0.9375 on the test set.

Results for Sampling. The test results of *Sampling 500* and *Sampling 5000* are shown in Figures 1(b) and (c) respectively. *Sampling 5000* outperforms *Sampling 500.* The median attainment surface of *Sampling 5000* is better than upper quartile attainment surface of *Sampling 500* for DR $\in (0, 0.8375)$. *Sampling 5000* attainment surfaces are all lower through lower detection rates (0 to 0.40), particularly when $0.2 \leq DR \leq 0.4$ where all summary attainment surfaces of *Sampling 5000* are below the median attainment surface of *Sampling 500. Sampling 5000* attainment surfaces have lower FAR than *Benchmark* for the highest DR for the best summary attainment surface. For both *Sampling 500* and *Sampling 5000* at least 75% of GP runs have a Pareto-front which attains the full range of detection rates, this is better than the benchmark. Like the benchmark test results, all GP runs have a Pareto-front which contains a detection rate of at least 0.9375.

Results for Validation. The test results of *Validation* and *Validation-Sampling* are shown in Figures 1(e) and (d) respectively. *Validation-Sampling* uses a sample size of 5000. For both *Validation* and *Validation-Sampling* less than 25% of Pareto fronts evolved in the 40 runs have the full DR range of 0 to 1. *Validation-Sampling* outperforms *Validation* for the high detection rates. Table 1 shows that the runtime in seconds of *Sampling 5000* is 20 times faster than that of the benchmark, and *Sampling 500* is 60 times faster than the benchmark. *Validation-Sampling* has almost ten times longer runtimes than just Sampling. The worst summary attainment surface for the four new methods are all considerably worse than the benchmark. *Validation* and *Validation-Sampling* reach 100% detection rate on less than 25% of evolutionary runs. Best and upper quartile summary attainment surfaces are better in MOGP with *Validation-Sampling* than MOGP with *Validation*, however the median, lower-quartile and worst fronts are better in MOGP with only validation. Best and upper quartile summary attainment surfaces of MOGP with *Validation-Sampling* is better than *Validation* for $DR < 0.9375$, after which *Validation* is better. The median summary attainment surface of *Validation-Sampling* is better than *Validation.* Comparing the test results of *Validation* and *Validation-Sampling* to the test results of the benchmark, there is not much improvement offered by the use of the validation set. The worse results than the benchmark can be attributed partially to the earlier termination of some program's evolution. Although the aim in this project is to terminate

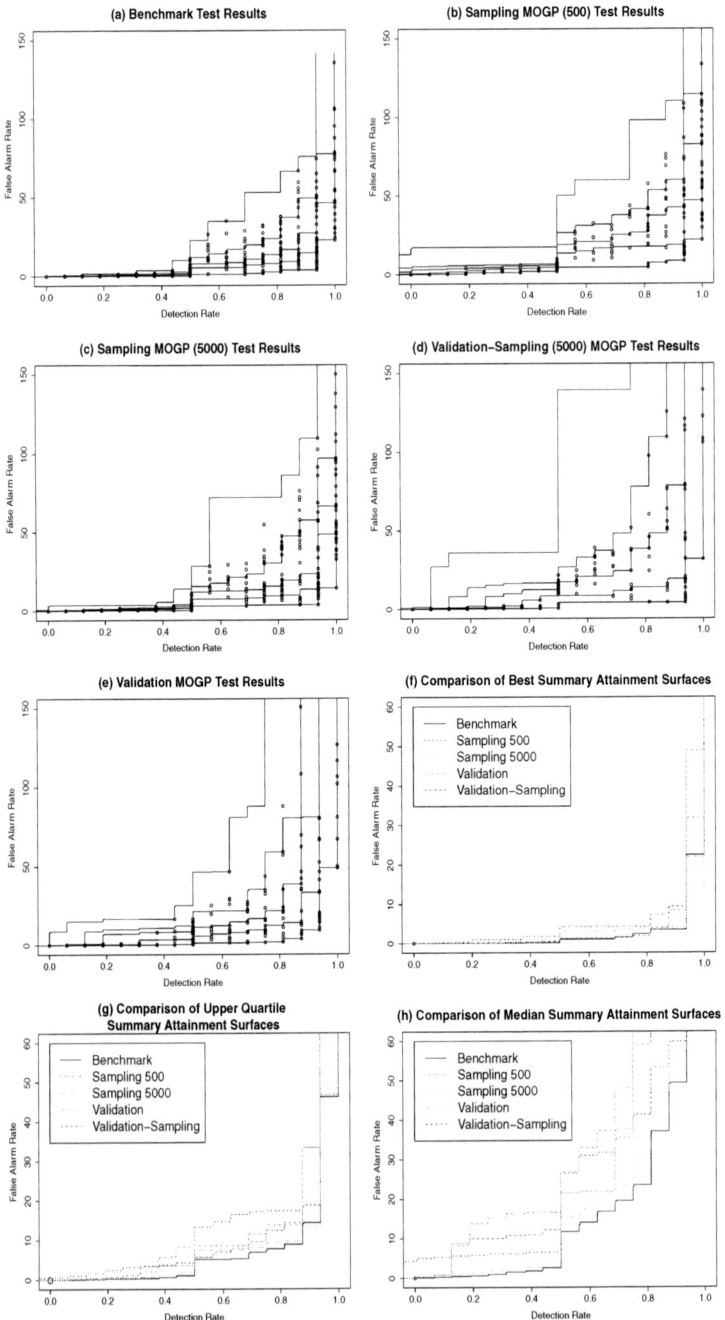

Fig. 1. Graphs showing the summary attainment surfaces for the benchmark and each of the four methods, Figures (a) to (e), and comparisons of best, upper quartile and median summary attainment surfaces between all methods, Figures (f) to (h), on MOGP (coins)

Table 1. Average runtimes in seconds, with standard deviations, of 40 runs of each MOGP method

MOGP	Runtime (s)
Benchmark	26420±4480
Sampling (500)	411 ± 220
Sampling (5000)	1203± 316
Validation	30025±9734
Sampling Validation (5000)	12232±24780

evolution before overfitting occurs, it is difficult to ascertain whether the decrease in validation set performance detected is due to having passed a global minimum (in which case evolution should be stopped) or a local minimum (in which case evolution should continue). Many programs terminated evolution before reaching even 15 second phase generations. The high number of programs that do not terminate before the end of the 40 second phase generations suggests the number of generations for evolution could be increased, as overfitting does not appear to be occurring in the majority of the population.

Comparison. The best summary attainment surfaces of each method are shown in Figure 1(f). The best summary attainment surfaces of all methods except *Sampling 500* are very similar for $DR < 0.5$. *Sampling 5000* is equal to or better than the benchmark for $DR > 0.8$. *Validation-Sampling* is better than *Validation* except for at $DR = 1.0$. The upper quartile summary attainment surfaces of each method are shown in Figure 1(g). *Sampling 5000* is very similar, but not quite as good as the benchmark and is better for $DR = 0.9375$. The median summary attainment surfaces of each method are shown in Figure 1(h). The benchmark has the best median summary attainment surface for $DR < 0.9375$ (which is the highest DR reached), however *Sampling 5000* has the lower FAR at $DR = 0.9375$, and is very similar to the benchmark front through $0 \leq DR \leq 0.5$. *Validation* has a worse median summary attainment surface than *Validation-Sampling.*

6 Conclusions

The goal of this paper was to increase the robustness of classifiers evolved in MOGP for object detection in images. Sampling methods decrease the run time of the evolutionary runs by a very substantial amount, and increase robustness at higher detection rates. Further, with the use of sampling the proportion of GP runs which have a Pareto-front spanning the full range of detection rates increases. This shows that the use of sampling does improve the robustness of Pareto-fronts evolved in MOGP. The comparison of the test results of *Validation* and *Validation-Sampling* show that *Validation-Sampling* has a considerably better median than *Validation.* This shows that in this situation the use of sampling in combination with validation also improves the robustness of the Pareto-fronts evolved. However in comparing the test results of *Validation* and *Validation-*

Sampling to those of *Sampling 500* and *Sampling 5000* it is clear that the use of the validation set does not provide an increase in robustness.

In future work, NSGAII will be compared with SPEA2 [11], and we will use more difficult datasets. More work is needed on MOGP with validation and sampling, e.g., changing the frequency of validation so it is only every 10 generations, and introducing a minimum number of generations that evolution must go through in the second phase before evolution can terminate. It would then be interesting to increase the upper limit on number of generations evolved to be much higher, say 200, and see how results compare. If evolution is allowed to take up to 200 generations then the time taken will be very long, this is when the sampling methods will be very useful, as they decrease the time taken hugely, making validation and longer evolutionary periods feasible.

References

1. Asuncion, A., Newman, D.J.: UCI Machine Learning Repository (2007), `http://www.ics.uci.edu/~mlearn/MLRepository.html`
2. Bhowan, U., Zhang, M., Johnston, M.: Multi-objective genetic programming for classification with unbalanced data. In: Proceedings of the 22nd Australasian Joint Conference on Artificial Intelligence, pp. 370–380 (2009)
3. Deb, K., Pratap, A., Agarwal, S., Meyarivan, T.: A fast and elitist muliobjective genetic algorithm: NSGA-II. IEEE Transactions on Evolutionary Computation, 182–197 (2002)
4. Doucette, J., Heywood, M.I.: GP Classification under Imbalanced Data Sets: Active Sub-sampling and AUC Approximation. In: O'Neill, M., Vanneschi, L., Gustafson, S., Esparcia Alcázar, A.I., De Falco, I., Della Cioppa, A., Tarantino, E. (eds.) EuroGP 2008. LNCS, vol. 4971, pp. 266–277. Springer, Heidelberg (2008)
5. Gathercole, C., Ross, P.: Dynamic Training Subset Selection for Supervised Learning in Genetic Programming. In: Davidor, Y., Männer, R., Schwefel, H.-P. (eds.) PPSN 1994. LNCS, vol. 866, pp. 312–321. Springer, Heidelberg (1994)
6. Hunt, R., Johnston, M., Browne, W., Zhang, M.: Sampling methods in genetic programming for classification with unbalanced data. In: Proceedings of the 23rd Australasian Joint Conference on Artificial Intelligence, pp. 273–282 (2010)
7. Iba, H.: Bagging, boosting, and bloating in genetic programming. In: GECCO, vol. 2, pp. 1053–1060 (1999)
8. Knowles, J.: A summary-attainment-surface plotting method for visualizing the performance of stochastic multiobjective optimizers. In: Proceedings of the 5th International Conference on Intelligent Systems Design and Applications, pp. 552–557 (2005)
9. Krawiec, K., Howard, D., Zhang, M.: Overview of object detection and image analysis by means of genetic programming techniques. In: Proceedings of the International Conference Frontiers in the Convergence of Bioscience and Information Technologies, pp. 779–784 (2007)
10. Liddle, T., Johnston, M., Zhang, M.: Multi-objective genetic programming for object detection. In: CEC, pp. 3345–3352 (2010)
11. Sitzler, E., Laumanns, M., Thiele, L., Talbi, E.: SPEA2: Improving the strength pareto evolutionary algorithm. In: USENIX Technical Conference (2001)
12. Zhang, M.: Improving object detection performance with genetic programming. International Journal on Artificial Intelligence Tools 16(5), 849–873 (2003)
13. Zhang, M., Ciesielski, V., Andreae, P.: A domain-independent window approach to multiclass object detection using genetic programming. EURASIP Journal on Applied Signal Processing 2003(8), 841–859 (2003)

DE/isolated/1: A New Mutation Operator for Multimodal Optimization with Differential Evolution

Takahiro Otani*, Reiji Suzuki, and Takaya Arita

Graduate School of Information Science, Nagoya University
Furo-cho, Chikusa-ku, Nagoya 464-8601, Japan
t-otani@alife.cs.is.nagoya-u.ac.jp,
{reiji,arita}@nagoya-u.jp

Abstract. This paper proposes a new variant of differential evolution for multimodal optimization termed DE/isolated/1. It generates new individuals close to an isolated individual in a current population as a niching scheme. This mechanism will evenly allocate search resources for each optimum. The proposed method was evaluated along with the existing methods through computational experiments using eight two-dimensional multimodal functions as benchmarks. Experimental results show that the proposed method shows better performance for several functions which are not effectively solved by existing algorithms.

Keywords: differential evolution, multimodal optimization, niching.

1 Introduction

Differential evolution (DE) [1] is a very powerful population-based algorithm for function optimization. Despite its very good performance on various problems, it is very easy to implement since it consists of only four simple operations, initialization, mutation, crossover and selection. This paper deals with extending DE in order to solve multimodal optimization problems. The goal of a multimodal optimization process is to find all or most of the multiple optima in a solution space. Real-world problems, such as classification problems in machine learning [2] and the inversion of teleseismic waves [3], are considered to be highly multimodal problems. They are likely to have several global and/or local optima, and in many cases it is desirable to accurately find as many as possible [4].

Some DE-variants for solving multimodal optimization problems have been proposed so far. Thomsen proposed two variants, CrowdingDE and SharingDE [5]. He showed the CrowdingDE outperformed the SharingDE on fourteen commonly used benchmark problems. DE with local selection [6] generates new individuals close to a current vector in a population. The local selection mechanism partitions the population into some niches which evolve in isolation. Epitropakis

* Research Fellow of the Japan Society for the Promotion of Science.

D. Wang and M. Reynolds (Eds.): AI 2011, LNAI 7106, pp. 321–330, 2011.

et al. proposed DE/nrand/1 and DE/nrand/2 in which a new vector is generated close to the nearest neighbor of a current vector [4]. Furthermore, DE with restricted tournament selection [7] and species-based DE [8] have been proposed.

This paper proposes a new DE-variant termed DE/isolated/1. This algorithm selects an isolated vector in a current population as a base of newly generated vectors. In this setting, a new candidate vector is generated close to the isolated vector in the current population. Then the current vector will be replaced by the generated vector if the generated vector has a better fitness value. Thus we could expect each candidate solution actively migrates to the isolated area in each generation, and will evenly allocate search resources for each optimum in the search space.

The rest of the paper is arranged as follows. Section 2 briefly explains the conventional differential evolution algorithm. Section 3 shows the proposed method and explains design concepts. Section 4 shows performance evaluation experiments. The performance is compared with existing DE-variants. Finally, Section 5 concludes the paper.

2 Differential Evolution

This section briefly explains the conventional differential evolution algorithm according to [9]. A pseudocode of the algorithm is described in Fig. 1. The algorithm sequentially chooses a vector $x_{i,c}$ in a population X_c. The chosen vector is called *target* vector, and a new vector is generated according to this by mutation and crossover operator. Then the target vector will be replaced by the newly generated one if it has a better fitness value than the target.

We adopt conventional terminology for some vectors used in DE as follows:

- Target vector $\boldsymbol{x}_{i,c}$: A vector chosen from the current population X_c.
- Donor vector $\boldsymbol{v}_i$: A mutant vector obtained through the differential mutation operation.
- Base vector: A vector to be perturbed by the differential vector through the mutation operation.
- Trial vector $\boldsymbol{u}_i$: An offspring formed by recombining the donor with the target vector.

2.1 Initialization

First, the algorithm generates an initial population of N-dimensional vectors by uniformly randomizing individuals within the search space constrained by the specified minimum and maximum bounds. The j-th component of a vector $\boldsymbol{x}_{i,c}$ in X_c is initialized as

$$x_{i,j,c} = x_{j,\min} + rand_{i,j}\left[0,1\right] \cdot \left(x_{j,\max} - x_{j,\min}\right), \tag{1}$$

where $x_{j,\min}$ is the minimum bound of the j-th element, and $x_{j,\max}$ is the maximum bound. $rand_{i,j}[0,1]$ is a uniformly distributed random number lying between 0 and 1, and is instantiated independently for each component of the i-th vector.

```
input:
  f(x) - The function to be minimized
  F - Scaling factor
  Cr - Crossover rate
  NP - Population size
procedure:
  X_c ← Randomly initialized vectors [x_{1,c}, x_{2,c}, ··· , x_{NP,c}]
  while (the stopping condition is not satisfied) {
    X_n ← X_c
    for (i ← 1 to NP) {
      v_i ← Mutation(i, X_c, F)
      u_i ← Crossover(x_{i,c}, v_{i,c}, Cr)
      Selection(i, X_c, u_i, X_n, f)
    }
    X_c ← X_n
  }
  Output the best vector in X_c
```

Fig. 1. The pseudocode of DE algorithm

2.2 Mutation

The mutation operator generates a donor vector $\boldsymbol{v}_i$ using vectors randomly selected from the current population. In the conventional DE, the following operator is used:

$$\boldsymbol{v}_i = \boldsymbol{x}_{r_1^i,c} + F \cdot \left(\boldsymbol{x}_{r_2^i,c} - \boldsymbol{x}_{r_3^i,c} \right) . \tag{2}$$

The indices r_1^i, r_2^i and r_3^i are mutually exclusive integers randomly chosen from $\{1, 2, \ldots, NP\} \setminus \{i\}$. F is a scaling factor which has to be determined by the user.

2.3 Crossover

The crossover operator generates a trial vector $\boldsymbol{u}_i$ by recombining the target vector $\boldsymbol{x}_{i,c}$ and the donor vector $\boldsymbol{v}_i$. In the conventional DE, two types of operators, *exponential* and *binomial*, are used. We use the *binomial* operator in this study. In the *binomial* operator, each element in the trial vector inherits the donor vector's value with probability Cr which is specified by the user, and inherits the target vector's value with probability $1 - Cr$. The trial vector $\boldsymbol{u}_i$ is obtained as

$$u_{j,i} = \begin{cases} v_{j,i} & rand_{i,j}[0,1] \leq Cr \text{ or } j = j_{rand} \\ x_{j,i,c} & \text{otherwise,} \end{cases} \tag{3}$$

where $j_{rand} \in [1, 2, \ldots, D]$ is a randomly chosen index which ensures that $\boldsymbol{u}_i$ inherits at least one component from the trial vector $\boldsymbol{v}_i$.

2.4 Selection

The selection operator determines whether the target or the trial vector survives to the next generation. If the new trial vector yields an equal or lower value of the objective function, it replaces the corresponding target vector. The target vector in the next generation $\boldsymbol{x}_{i,n}$ is obtained as

$$\boldsymbol{x}_{i,n} = \begin{cases} \boldsymbol{u}_i & f(\boldsymbol{u}_i) \leq f(\boldsymbol{x}_{i,c}) \\ \boldsymbol{x}_{i,c} & f(\boldsymbol{u}_i) > f(\boldsymbol{x}_{i,c}). \end{cases} \tag{4}$$

3 Proposed Method: DE/isolated/1

In this section, we propose a new DE-variant for multimodal optimization termed DE/isolated/1. The main feature of the variant is that it selects an isolated vector in the current population as a base vector of the mutation operator. This generates a trial vector close to the isolated vector. Thus, the target vector may migrate to the isolated area by the selection operator.

Fig. 2 describes behavioral feature of typical niching methods and the proposed method in a one-dimensional problem. In typical niching methods, individuals in a same valley tend to gather together by descending the valley. In the proposed method, individuals actively migrate to other valleys.

Fig. 3 describes a pseudocode of the proposed method. In the next subsections, we explain design concepts of the algorithm.

3.1 Isolated Vector-Based Mutation

The mutation operator used in DE/isolated/1 selects an isolated vector in the current population as a base vector. The operator is described as follows:

$$\boldsymbol{v}_i = \boldsymbol{x}_{isolated} + F \cdot \left(\boldsymbol{x}_{r_1^i} - \boldsymbol{x}'_{r_1^i} \right), \tag{5}$$

where *isolated* is an index of an isolated vector obtained as follows:

$$isolated = \arg\max_{1 \leq j \leq NP} \left\{ \min_{1 \leq k \leq NP, j \neq k} ||\boldsymbol{x}_j - \boldsymbol{x}_k|| \right\}. \tag{6}$$

This means the isolated vector has the maximum distance from its nearest neighbor vector.

The second term of the differential vector $\boldsymbol{x}'_{r_1^i}$ is chosen randomly from N_d nearest neighbors of $\boldsymbol{x}_{r_1^i}$. $N_d \in [1, 2, \ldots, NP-1]$ is a control parameter specified by the user. The case with $N_d = NP - 1$ corresponds to that with the original differential vector in the conventional DE except that the target vector will be

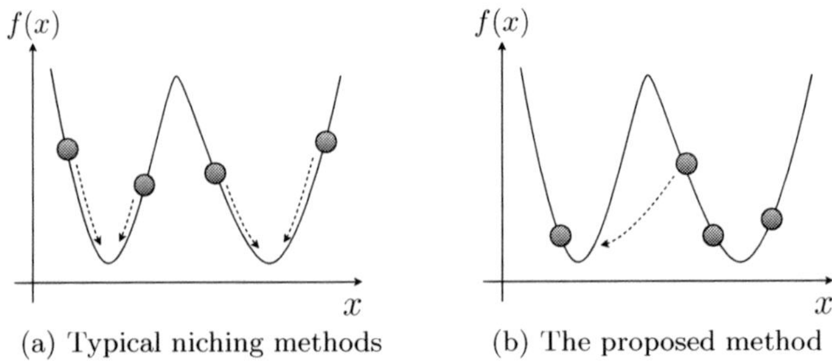

Fig. 2. Behavioral feature of typical niching methods and the proposed method

```
input:
  f(x) - The function to be minimized
  F - Scaling factor
  Cr - Crossover rate
  NP - Population size
  N_d - The number of candidate differential vectors
  N_w - Threshold of rejected trials
procedure:
  X ← Randomly initialized vectors [x_1, x_2, ··· , x_NP]
  while (the stopping condition is not satisfied) {
    for (i ← 1 to NP) {
```

$$v_i \leftarrow \begin{cases} x_{r_1^i} + F \cdot (x_{r_2^i} - x_{r_3^i}) & isolated = i \wedge n_w \geq N_w \\ x_{isolated} + F \cdot (x_{r_1^i} - x'_{r_1^i}) & \text{otherwise} \end{cases}$$

$$u_i \leftarrow \texttt{Crossover}(x_i,\ v_i,\ Cr)$$

$$x_i \leftarrow \begin{cases} u_i & f(u_i) \leq f(x_i) \\ x_i & f(u_i) > f(x_i). \end{cases}$$

```
    }
  }
  Output X
```

Fig. 3. The pseudocode of DE/isolated/1

chosen. The design intention of the second term is to generate trial vectors close to the isolated vector absolutely. In conventional DE, differential vectors are chosen randomly from the current population. In this setting differential vectors which step over mountains will be generated if two chosen vectors are not in a same valley. Therefore, a trial vector will not be close to the isolated vector if this setting is used in the proposed method.

3.2 Immediate Update of the Population

The conventional DE updates the current population after generating new vectors for each target vector in the population, i.e. newly generated vectors are stored in another memory X_n during executing the three steps, then these are copied to the current population memory X_c (or the reference is changed). If DE/isolated/1 is used under the update mechanism, the next population tends to converge to the isolated vector. This causes premature convergence.

To avoid this situation, we adopt another update mechanism, *immediate update*. In this mechanism, if the selection operator determines to replace the target vector by the newly generated one, the target in the current population is replaced immediately. i.e. we use only one memory X for storing the population, and all operations manipulate vectors in the memory. Therefore, the information of the isolated vector will be changed immediately after the update.

3.3 Escape from Local Optima

If we use the isolated vector-based mutation, a bad situation could happen. When an isolated vector is in a local optimal area and others are in a better area, all trial vectors will be rejected by the selection operator since all such vectors are close to the isolated vector and show bad quality than target vectors.

To overcome this situation, the isolated vector needs to escape the local optima and move to another area. However, the isolated vector cannot escape since all trial vectors will be generated close to the isolated vector even if the isolated vector is chosen as a target vector.

As an escape mechanism, conventional mutation operator is used if the trials with an isolated vector are rejected for a specified times. The improved mutation operator is described as follows:

$$\boldsymbol{v}_i = \begin{cases} \boldsymbol{x}_{r_1^i} + F \cdot \left(\boldsymbol{x}_{r_2^i} - \boldsymbol{x}_{r_3^i}\right) & isolated = i \wedge n_w \geq N_w \\ \boldsymbol{x}_{isolated} + F \cdot \left(\boldsymbol{x}_{r_1^i} - \boldsymbol{x}'_{r_1^i}\right) & \text{otherwise}, \end{cases} \tag{7}$$

where $n_w (\geq 0)$ is the number of rejected trials during using the current isolated vector, and is reinitialized to 0 when a current vector is replaced by a newly generated one. N_w is a threshold specified by the user. To set $N_w = \infty$ corresponds to using the original mutation operator described in the equation (5).

4 Experiments

To evaluate the performance of the proposed algorithm, we conducted computational experiments using two-dimensional multimodal functions as benchmarks. The performance was compared with existing DE-variants, CrowdingDE [5], DE with local selection (DELS) [6] and DE/nrand/{1,2} [4]. We used the literature [4] as a reference to design the experiments.

4.1 Benchmark Functions

We used eight benchmark multimodal functions to be minimized, that are the same ones adopted in the literature [4]. Table 1 shows mathematical formulas and domains of these functions.

The functions F_1 and F_2 have a low number of irregularly spaced global optima and no local optima. The number of global optima are 3 and 4 respectively. The function F_3 has 18 global optima and a high number (742) of local optima. The function F_5 has 6^2 global optima without local optima and is partially irregular, with the differences between optima to increase along the value of x. The function F_6 contains 5^2 evenly spaced global optima and does not have any local optima. Similarly, the function F_7 has the same number of optima, but the distances between each global optimum decrease towards the origin. Finally, the function F_8 is a modified version of the well-known Rastrigin function, having 4 evenly spaced global optima and 96 local optima.

4.2 Performance Measure

The following two measures, *peak ratio (PR)* and *success ratio (SR)*, were used to evaluate the performance of each algorithm. The peak ratio is defined as follows:

$$Peak\ ratio = \frac{\text{the number of global optima found}}{\text{the number of total global optima}}. \tag{8}$$

Table 1. Benchmark functions

Function	Mathematical formula	Domain
Branin	$F_1(\boldsymbol{x}) = \left(x_2 - \frac{5.1}{4\pi^2}x_1^2 + \frac{5}{\pi}x_1 - 6\right)^2$ $+10\left(1 - \frac{1}{8\pi}\right)\cos(x_1) + 10$	$x_1 \in [-5, 10]$ $x_2 \in [0, 15]$
Himmelblau	$F_2(\boldsymbol{x}) = \left(x_1^2 + x_2 - 11\right)^2 + \left(x_1 + x_2^2 - 7\right)^2$	$\boldsymbol{x} \in [-6, 6]^2$
Shubert	$F_3(\boldsymbol{x}) = \sum_{i=1}^{5} i\cos((i+1)x_1 + i)$ $\cdot\sum_{i=1}^{5} i\cos((i+1)x_2 + i)$	$\boldsymbol{x} \in [-10, 10]^2$
Six-hump camel back	$F_4(\boldsymbol{x}) = \left(4 - 2.1x_1^2 + \frac{x_1^4}{3}\right)x_1^2$ $+x_1x_2 + \left(-4 - 4x_2^2\right)x_2^2$	$x_1 \in [-1.9, 1.9]$ $x_2 \in [-1.1, 1.1]$
Vincent	$F_5(\boldsymbol{x}) = -\frac{1}{2}\sum_{i=1}^{2}\sin(10\log(x_i))$	$\boldsymbol{x} \in [0.25, 10]^2$
Deb 1	$F_6(\boldsymbol{x}) = -\frac{1}{2}\sum_{i=1}^{2}\sin^6(5\pi x_i)$	$\boldsymbol{x} \in [0, 1]^2$
Deb 3	$F_7(\boldsymbol{x}) = -\frac{1}{2}\sum_{i=1}^{2}\sin^6\left(5\pi\left(y_i^{\frac{3}{4}} - 0.05\right)\right)$	$\boldsymbol{x} \in [0, 1]^2$
Modified Rastrigin	$F_8(\boldsymbol{x}) = 20 + \sum_{i=1}^{2}\left(x_i^2 + 10\cos(2\pi x_i)\right)$	$\boldsymbol{x} \in [-5.12, 5.12]^2$

An optimum is considered to be found when a vector in the population is within a specified Euclidean distance ε(*accuracy level*) from the optimum.

The success ratio is defined as follows:

$$Success\ ratio = \frac{\text{the number of trials all global optima was found}}{\text{the number of trials}}. \tag{9}$$

Finding all global optima in a trial corresponds to that the peak ratio at the last generation of the trial equals to 1.

According to these definitions, we checked out the average peak ratio at the last generation and the success ratio for all algorithms.

4.3 Parameter Setup

The experiment was conducted 100 times for each function and for each accuracy level. Accuracy levels were $\varepsilon = \{10^{-3}, 10^{-4}, \ldots, 10^{-8}\}$. Control parameters of DE were set as follows. The population size NP and the number of generations were set to 100 and 1000 respectively. For CrowdingDE, DELS, DE/nrand/1 and DE/nrand/2, the common setting of $F = 0.5$ and $Cr = 0.9$ were used, which are the same as the literature [4]. The binomial crossover operator was used in all algorithms.

For the proposed method DE/isolated/1, $F = 0.9$ and $Cr = 0.9$ were used. The reason of setting comparatively a high value of F is that the neighbors-based choice of differential vectors described in section 3.1 tends to generate short differential vectors. The number of candidate differential vectors N_d is set to 2. The threshold of rejected trials $N_w = NP - 1$ was used. This means the escaping mechanism is allowed if the current population is not updated during one generation. We determined these parameters according to the results of preliminary experiments. These showed good performance on the experiments, but may not be an optimal one.

Table 2. Peak ratio and success ratio measures for the multimodal function F_1 – F_4

(a) Branin (F_1)

ε	DE/isolated/1 PR	DE/isolated/1 SR	CrowdingDE PR	CrowdingDE SR	DELS PR	DELS SR	DE/nrand/1 PR	DE/nrand/1 SR	DE/nrand/2 PR	DE/nrand/2 SR
10^{-3}	0.870	0.62	0.993	0.98	**1.000**	**1.00**	**1.000**	**1.00**	**1.000**	**1.00**
10^{-4}	0.893	0.69	0.157	0.01	**1.000**	**1.00**	**1.000**	**1.00**	**1.000**	**1.00**
10^{-5}	0.900	0.70	0.003	0.00	**1.000**	**1.00**	**1.000**	**1.00**	**1.000**	**1.00**
10^{-6}	0.903	0.71	0.000	0.00	**1.000**	**1.00**	**1.000**	**1.00**	**1.000**	**1.00**
10^{-7}	0.880	0.64	0.000	0.00	**1.000**	**1.00**	**1.000**	**1.00**	**1.000**	**1.00**
10^{-8}	0.837	0.53	0.000	0.00	0.820	0.55	**1.000**	**1.00**	**1.000**	**1.00**

(b) Himmelblau (F_2)

ε	DE/isolated/1 PR	DE/isolated/1 SR	CrowdingDE PR	CrowdingDE SR	DELS PR	DELS SR	DE/nrand/1 PR	DE/nrand/1 SR	DE/nrand/2 PR	DE/nrand/2 SR
10^{-3}	**1.000**	**1.00**	**1.000**	**1.00**	**1.000**	**1.00**	**1.000**	**1.00**	**1.000**	**1.00**
10^{-4}	**1.000**	**1.00**	**1.000**	**1.00**	**1.000**	**1.00**	**1.000**	**1.00**	**1.000**	**1.00**
10^{-5}	**1.000**	**1.00**	0.395	0.07	**1.000**	**1.00**	**1.000**	**1.00**	**1.000**	**1.00**
10^{-6}	**1.000**	**1.00**	0.003	0.00	**1.000**	**1.00**	**1.000**	**1.00**	**1.000**	**1.00**
10^{-7}	**1.000**	**1.00**	0.000	0.00	0.990	0.96	**1.000**	**1.00**	0.995	0.98
10^{-8}	**1.000**	**1.00**	0.000	0.00	0.173	0.02	**1.000**	**1.00**	0.883	0.76

(c) Shubert (F_3)

ε	DE/isolated/1 PR	DE/isolated/1 SR	CrowdingDE PR	CrowdingDE SR	DELS PR	DELS SR	DE/nrand/1 PR	DE/nrand/1 SR	DE/nrand/2 PR	DE/nrand/2 SR
10^{-3}	0.988	0.86	0.164	0.00	0.953	0.41	0.718	0.01	**0.998**	**0.97**
10^{-4}	0.977	0.76	0.004	0.00	0.489	0.00	0.727	0.00	**0.999**	**0.99**
10^{-5}	0.982	0.80	0.000	0.00	0.014	0.00	0.733	0.01	**0.997**	**0.96**
10^{-6}	**0.978**	**0.81**	0.000	0.00	0.000	0.00	0.714	0.01	0.921	0.54
10^{-7}	**0.981**	**0.76**	0.000	0.00	0.000	0.00	0.712	0.01	0.166	0.00
10^{-8}	**0.980**	**0.80**	0.000	0.00	0.000	0.00	0.708	0.00	0.002	0.00

(d) Six-hump came back (F_4)

ε	DE/isolated/1 PR	DE/isolated/1 SR	CrowdingDE PR	CrowdingDE SR	DELS PR	DELS SR	DE/nrand/1 PR	DE/nrand/1 SR	DE/nrand/2 PR	DE/nrand/2 SR
10^{-3}	**1.000**	**1.00**	**1.000**	**1.00**	**1.000**	**1.00**	**1.000**	**1.00**	**1.000**	**1.00**
10^{-4}	**1.000**	**1.00**	**1.000**	**1.00**	**1.000**	**1.00**	**1.000**	**1.00**	**1.000**	**1.00**
10^{-5}	**1.000**	**1.00**	**1.000**	**1.00**	**1.000**	**1.00**	**1.000**	**1.00**	**1.000**	**1.00**
10^{-6}	**1.000**	**1.00**	**1.000**	**1.00**	**1.000**	**1.00**	**1.000**	**1.00**	**1.000**	**1.00**
10^{-7}	**1.000**	**1.00**	0.215	0.07	**1.000**	**1.00**	**1.000**	**1.00**	**1.000**	**1.00**
10^{-8}	**1.000**	**1.00**	0.000	0.00	**1.000**	**1.00**	**1.000**	**1.00**	**1.000**	**1.00**

4.4 Results

Table 2 and Table 3 show experimental results of all algorithms over all benchmark functions. The proposed method showed good performance when the accuracy level ε is set to tight settings. When the level was set to $\varepsilon \leq 10^{-6}$, the proposed method showed the best performance on 6 functions (F_2 – F_7). These results suggest that the proposed method has better convergence speed than others.

The proposed method showed higher performance than others especially on Vincent function and Deb 3 function. These functions are partially irregular; some optima's basins are wide-spreading over the search space and others are very narrow. Existing methods can not easily find the latter optima since almost all search resources tend to be allocated to the former.

On Branin function, the performance of the proposed method was worse than some existing methods since necessary vectors are moved by the conventional mutation operator. We examined by further experiments that it was solved by

Table 3. Peak ratio and success ratio measures for the multimodal function F_5 – F_8

(a) F_5: Vincent (F_5)

	DE/isolated/1		CrowdingDE		DELS		DE/nrand/1		DE/nrand/2	
ε	PR	SR	PR	SR	PR	SR	PR	SR	PR	SR
10^{-3}	**0.945**	**0.55**	0.587	0.00	0.508	0.00	0.359	0.00	0.339	0.00
10^{-4}	**0.943**	**0.44**	0.065	0.00	0.431	0.00	0.331	0.00	0.240	0.00
10^{-5}	**0.940**	**0.49**	0.001	0.00	0.239	0.00	0.304	0.00	0.062	0.00
10^{-6}	**0.930**	**0.40**	0.000	0.00	0.029	0.00	0.268	0.00	0.001	0.00
10^{-7}	**0.941**	**0.46**	0.000	0.00	0.001	0.00	0.211	0.00	0.000	0.00
10^{-8}	**0.940**	**0.35**	0.000	0.00	0.000	0.00	0.084	0.00	0.000	0.00

(b) F_6: Deb 1 (F_6)

	DE/isolated/1		CrowdingDE		DELS		DE/nrand/1		DE/nrand/2	
ε	PR	SR	PR	SR	PR	SR	PR	SR	PR	SR
10^{-3}	**1.000**	0.99	**1.000**	**1.00**	0.980	0.58	0.984	0.63	0.988	0.73
10^{-4}	**1.000**	0.99	**1.000**	**1.00**	0.975	0.52	0.986	0.69	0.981	0.61
10^{-5}	0.998	0.96	**1.000**	**1.00**	0.965	0.36	0.984	0.66	0.985	0.69
10^{-6}	**1.000**	**0.99**	0.551	0.02	0.949	0.18	0.983	0.64	0.984	0.65
10^{-7}	**1.000**	**1.00**	0.008	0.00	0.893	0.06	0.979	0.61	0.984	0.64
10^{-8}	**1.000**	**1.00**	0.000	0.00	0.435	0.00	0.972	0.44	0.983	0.67

(c) F_7: Deb 3 (F_7)

	DE/isolated/1		CrowdingDE		DELS		DE/nrand/1		DE/nrand/2	
ε	PR	SR	PR	SR	PR	SR	PR	SR	PR	SR
10^{-3}	0.998	0.96	**1.000**	**1.00**	0.957	0.29	0.804	0.00	0.840	0.01
10^{-4}	**0.999**	**0.98**	0.638	0.00	0.898	0.03	0.749	0.00	0.286	0.00
10^{-5}	**0.999**	**0.97**	0.018	0.00	0.384	0.00	0.654	0.00	0.004	0.00
10^{-6}	**0.999**	**0.98**	0.000	0.00	0.008	0.00	0.478	0.00	0.000	0.00
10^{-7}	**1.000**	**1.00**	0.000	0.00	0.000	0.00	0.110	0.00	0.000	0.00
10^{-8}	**1.000**	**0.99**	0.000	0.00	0.000	0.00	0.004	0.00	0.000	0.00

(d) F_8: Modified Rastrigin (F_8)

	DE/isolated/1		CrowdingDE		DELS		DE/nrand/1		DE/nrand/2	
ε	PR	SR	PR	SR	PR	SR	PR	SR	PR	SR
10^{-3}	0.988	0.98	**1.000**	**1.00**	**1.000**	**1.00**	**1.000**	**1.00**	**1.000**	**1.00**
10^{-4}	0.978	0.96	**1.000**	**1.00**	**1.000**	**1.00**	**1.000**	**1.00**	**1.000**	**1.00**
10^{-5}	0.963	0.93	**1.000**	**1.00**	**1.000**	**1.00**	**1.000**	**1.00**	**1.000**	**1.00**
10^{-6}	0.980	0.96	0.963	0.85	**1.000**	**1.00**	**1.000**	**1.00**	**1.000**	**1.00**
10^{-7}	0.975	0.95	0.053	0.00	**1.000**	**1.00**	**1.000**	**1.00**	**1.000**	**1.00**
10^{-8}	0.975	0.95	0.000	0.00	0.940	0.82	**1.000**	**1.00**	**1.000**	**1.00**

setting the threshold N_w to a high value, but then the method showed poor performance on Shubert function (which has many local optima). Therefore, introduction of an adaptive threshold value N_w would further improve the performance of the proposed algorithm.

5 Conclusion

In this paper, we proposed a new DE-variant, DE/isolated/1, for solving multimodal optimization problems, and evaluated the performance of the algorithm along with the existing methods through computational experiments using eight multimodal benchmark functions. The experimental results showed that the proposed method has better performance than the existing algorithms on the whole.

Future work includes to conduct further experiments using high-dimensional problems and practical problems, and performance comparison with other existing DE-variants, DE with restricted tournament selection [7] and species-based

DE [8]. Through these experiments, we would understand the properties of the introduced mechanisms and identify the strengths and weaknesses of the proposed algorithm corresponding to the problems to be solved.

Another direction would be to improve the proposed algorithm. A promising candidate is to develop an adaptation mechanism of the N_w parameter in order to solve a broad range of problems efficiently. In addition, it would be important to reduce the amount of calculation. Determination of an isolated vector described in equation (6) is equivalent to do nearest neighbor search, and has high run-time complexity. For example, the run-time of the proposed algorithm will be improved by adopting an efficient search algorithm like kd-tree [10]. Further speed-up will also be achieved by using an approximate algorithm like locality sensitive hashing [11] although we should mind the influences of approximation error.

References

1. Storn, R., Price, K.: Differential Evolution – A Simple and Efficient Heuristic for Global Optimization over Continuous Spaces. Journal of Global Optimization 11(4), 341–359 (1997)
2. Mahfoud, S.W.: Niching Methods for Genetic Algorithms. Ph.D. dissertation, Urbana, IL, USA (1995)
3. Koper, K., Wysession, M., Wiens, D.: Multimodal Function Optimization with a Niching Genetic Algorithm: A Seismological Example. Bulletin of the Seismological Society of America 89(4), 978–988 (1999)
4. Epitropakis, M., Plagianakos, V., Vrahatis, M.: Finding Multiple Global Optima Exploiting Differential Evolution's Niching Capability. In: 2011 IEEE Symposium on Differential Evolution, pp. 80–87 (2011)
5. Thomsen, R.: Multimodal Optimization using Crowding-based Differential Evolution. In: Congress on Evolutionary Computation 2004, pp. 1382–1389 (2004)
6. Lampinen, J.: An Extended Mutation Concept for the Local Selection Based Differential Evolution Algorithm. In: The 2007 Conference on Genetic and Evolutionary Computation, pp. 689–696 (2007)
7. Qu, B., Suganthan, P.N.: Novel Multimodal Problems and Differential Evolution with Ensemble of Restricted Tournament Selection. In: Congress on Evolutionary Computation 2010, pp. 1–7 (2010)
8. Li, X.: Efficient Differential Evolution using Speciation for Multimodal Function Optimization. In: The 2005 Conference on Genetic and Evolutionary Computation, pp. 873–880 (2005)
9. Das, S., Suganthan, P.N.: Differential Evolution: A Survey of the State-of-the-Art. IEEE Transactions on Evolutionary Computation 15(1), 4–31 (2011)
10. Bentley, J.L.: Multidimensional Binary Search Trees used for Associative Searching. Communications of the ACM 18(9), 517–590 (1975)
11. Indyk, P., Motwani, R., Raghavan, P., Vempala, S.: Localit-Preserving Hashing in Multidimensional Spaces. In: 29th ACM Symposium on Theory of Computing, pp. 618–625 (1997)

The Effects of Diversity Maintenance on Coevolution for an Intransitive Numbers Problem

Tirtha R. Ranjeet, Martin Masek, Philip Hingston, and Chiou-Peng Lam

School of Computer and Security Science,
Edith Cowan University
{t.ranjeet,m.masek,p.hingston,p.lam}@ecu.edu.au

Abstract. In this paper, we investigate the effectiveness of several techniques commonly recommended for overcoming convergence problems with coevolutionary algorithms. In particular, we investigate effects of the Hall of Fame, and of several diversity maintenance methods, on a problem designed to test the ability of coevolutionary algorithms to deal with an intransitive superiority relation between solutions. We measure and analyse the effects of these methods on population diversity and on solution quality.

Keywords: coevolution, diversity maintenance, HOF, fitness sharing.

1 Introduction

Evolutionary algorithms are population-based, stochastic search algorithms modelled on evolutionary processes in nature. Potential solutions to a problem are assigned a *fitness* that reflects how well they solve the problem, and these values guide the search. In a coevolutionary algorithm (CEA), this fitness value depends on interactions with other potential solutions. CEAs offer advantages over ordinary evolutionary algorithms in certain situations: when there is no objective function to measure fitness of a solution; in a large search space when there are two or more interacting subspaces and in certain complex problem domains [1-7]. However, CEAs can also suffer from pathologies which interfere with convergence. Many techniques have been proposed to address these pathologies. One approach is to use an *archive* of high quality solutions - the Hall of Fame is a well-known of example [8]. Another idea is to use a *diversity maintenance* mechanism, such as fitness sharing [9-13].

This work is an empirical study, using a recent method for estimating solution set quality [9,14], to investigate how diversity maintenance techniques can improve the effectiveness of CEAs, both with and without the additional use of an archive. More specifically, we empirically test variants of a standard CEA with different mutation rates, with and without competitive fitness sharing, and with and without a Hall of Fame, on a test problem designed to challenge CEAs. We examine how solution set diversity and quality is affected in the variants.

The aim of fitness sharing and HOF is to improve the quality of solutions found by the CEA, yet for many problems, there is no predefined quality metric– rather quality

D. Wang and M. Reynolds (Eds.): AI 2011, LNAI 7106, pp. 331–340, 2011.

can only be judged based on how evolved solutions interact with other solutions. In [9,14], Chong et al. proposed that the appropriate quality measure for CEAs is *generalization performance*, and introduced a set of methods for estimating it. They explored the relationship between diversity and quality, using various implicit and explicit diversity methods, and concluded that appropriate diversity improves quality.

In our paper, we have adapted the methods of Chong et al. to a different kind of problem. They used a problem with a single population of interacting agents, Iterated Prisoner's Dilemma (IPD), whereas our intransitive number test problem uses two competing populations, as is suitable when evolving competing sets of solutions in an asymmetric domain. As well as diversity and quality, we also investigate the effect of HOF, and its interaction with diversity maintenance.

The remainder of this paper is structured as follows. In Section 2, we review the basics of CEAs and the Hall of Fame and some common diversity maintenance methods, as well as describing methods for measuring diversity and quality in CEAs. In Section 3, we give a description of the design of our experiments. In the final two sections, we describe our results and conclude.

2 Coevolutionary Algorithms

Evolutionary algorithms (EAs) are stochastic search methods inspired by biological evolution. EAs work with populations of solutions (individuals). Each individual's *fitness* depends on its performance against a criterion. Individuals with high fitness are selected preferentially to produce "offspring" individuals for the next generation. Two selected parents produce several offspring by exchanging genes (crossover). Then, each offspring alters its gene structure with some probability (mutation) and becomes a new individual in the next generation. This process of variation and selection is repeated until some stopping condition is met.

A coevolutionary algorithm (CEA) is an evolutionary algorithm in which the fitness of each individual depends on interactions between it and other individuals [1]. In CEAs, individuals are organised into sub-populations which coevolve [2,3,15,16]. The fitness calculation in CEAs is *subjective*: each individual interacts with individuals from another population. Unlike *objective* fitness, subjective fitness is dependent on the composition of the populations. A typical subjective fitness calculates the average score of an individual in interactions with opposing individuals in the current populations.

2.1 Hall of Fame

The Hall of Fame (HOF) is a technique that allows the population to interact with a set of the best individuals from previous generations of the opponent population. The best individuals from both populations in every generation are collected and stored in an archive, which interacts with the populations during the fitness evaluation. The functionality of the HOF is to preserve some old individuals to avoid the cycling and forgetting pathologies. When the HOF is used, subjective fitness is modified to be the average score of an individual in interactions with opposing individuals in the current populations and also in the Hall of Fame [8].

2.2 Diversity

Too much selective pressure and/or not enough exploration in an evolutionary process can cause premature convergence [9]. Maintaining diversity in the population has proved to avoid premature convergence [12] in many instances. Chong, et al. [9,14] categorize diversity maintenance methods into two types, implicit and explicit:

Implicit diversity maintenance methods use the selection process. A typical implicit method is competitive fitness sharing (FS), where diversity is maintained in the population by discouraging individuals with similar characteristics. Fitness values are reduced for individuals with common gene structures. The shared fitness of an individual f'_i is calculated by dividing simple fitness by the *niche count*:

$$f_i' = \frac{f_i}{c_i} \tag{1}$$

The symbol c_i is a niche count, which is calculated on the basis of the individual's gene structure variation (d_j) in the population. The following formulas are used to calculate gene variation and niche count respectively.

$$d_j = \sqrt{\sum_{m=1}^{u} (x_m - y_{j,m})^2} \qquad c_i = \sum_{j=1}^{n} \begin{cases} 1 - \left(\frac{d_j}{n_r}\right)^{\tau}, & if\ d_j \le n_r \\ 0, Otherwise \end{cases}$$

The symbol u is the genome length, x is an individual and y_j is an individual from the same population, and x_m and $y_{j,m}$ are their m^{th} gene values. The symbol τ is a constant. The symbol n_r is a constant niche radius and n is a population size.

Explicit diversity maintenance methods achieve diversity through variation. A simple method is to increase the mutation rate.

Two types of diversity are *genotypic* and *phenotypic* diversity. Genotypic diversity in a population is a measure of the gene structure variation, calculated as the average gene variation over the population. Phenotypic diversity is calculated based on the entropy [11,12] of the distribution of fitness values. The fitness values present in the population are divided among N equal sized buckets, and then equation (2) is applied.

$$E(P) = -\sum_{k=1}^{N} p_k . log p_k \tag{2}$$

2.3 Quality

We adopt the approach of Chong et al. to measure quality, i.e. we use a statistical estimate of the generalization performance of a solution, but we modify it slightly to account for the fact that we are using two populations. Chong et al. begin by defining generalization performance as the mean score of a solution in all possible test cases. This intuitively appealing idea is usually impractical to calculate. Therefore, they propose a statistical approximation approach, in which a mean score is computed for a suitable sample of test cases. In many cases, scores against "high quality" test cases might be considered more important. They therefore propose two different methods for sampling the space of test cases: unbiased sampling (purely random) and biased sampling (favours higher quality). In the present study, due to space limitations, we report only on results using biased sampling. To obtain a biased test set, we follow the procedure in Chong et al., using a sample size of 200. Once we have generated test sets, we can use them to estimate the quality of each solution as its mean score against the test set solutions, and we can combine these in various ways to obtain an overall quality measure for an evolved population of solutions.

Estimated Average Quality In an evolutionary algorithm, we are usually most interested in the top few evolved solutions. Thus, we first sort the population according to internal fitness, and then consider only the top few. Average quality is then estimated as

$$E_i = \frac{1}{nTest}\sum_{j=1}^{nTest} score_{i,j} \qquad Est.average\ quality = \frac{1}{nBest}\sum_{i=1}^{nBest} E_i \tag{3}$$

where E_i is the estimated quality of solution *i*, *nTest* is the size of the test set, and *nBest* is the number used in the estimate (i.e. we use only the best *nBest*).

Estimated Best Quality This is the quality of the best solution amongst the top *nBest* solutions in the population, when they are sorted on internal fitness:

$$Estimated\ best\ quality = \ max_{i<nBest}E_i \tag{4}$$

3 Experiments

In this section, we describe our experimental design. We describe the test problem we have chosen to study, the algorithm variants that we test, and the measurements that we gather during the testing.

As our test problem, we chose an intransitive number problem which was introduced by Watson and Pollack [17]. It has advantages over the test problem used by Chong et al, the IPD. IPD is an important problem and widely studied. It is an extremely difficult problem for a CEA, with complex evolutionary dynamics, an enormous search space (in fact researchers always restrict their search to solutions that can be represented using some restricted representation). The intransitive number problem has one specific feature that makes it difficult (intransitive superiority) and a simple representation, as well as a known objective quality criterion, making it very suitable for testing.

Watson and Pollack [17] introduced intransitive number test problems to test the functionality of CEAs. We pose a version with two populations. Individual solutions in both populations consist of pairs of real numbers in (0, 100), which we call x and y. The score when solution a from one population meets solution b from the other population is given in Equation (5):

$$score\left((a_x, a_y), (b_x, b_y)\right) = \begin{cases} score(a_x, b_x), if |a_x - b_x| < |a_y - b_y| \\ score(a_y, b_y), if |a_x - b_x| > |a_y - b_y| \\ random\ choice\ if |a_x - b_x| = |a_y - b_y| \end{cases}$$

$$where, score(a, b) = \begin{cases} 1, if\ a > b \\ 0, if\ a < b \\ random\ choice\ if\ a = b \end{cases} \quad (5)$$

Consider three solutions: A =<10;90>, B =<11;88> and C =<8;89>. Now score (A, B) is 0 (B beats A), because 10 and 11 are closer than 90 and 88, so the score is determined by which solution has the larger x value. Similarly, C beats B (based on a larger y), and yet A beats C. Thus the superiority relation between solutions is intransitive. Although this is problematic, generally speaking, the closer the solution is to <100;100>, i.e. the larger both x and y values are, the higher quality the solutions is. We define the *actual quality* of solution i as $A_i = (x+y)/2$, the average of the solutions x and y values. We can then define measures for the actual quality of a population, in a similar way as for estimated quality.

3.1 Algorithms Tested

For this experiment, four algorithms, naïve CEA, CEA with fitness sharing (CEAFS), CEA with HOF (CEAHOF) and combination of FS and HOF (CEAFH) were considered. For each, the mutation rate was varied from 5% to 25% with 5 intervals.

In all algorithms tested, single point crossover [20] and polynomial mutation [18] were used for the reproduction process. Parents were selected using a stochastic universal sampling method [19] and an elite individual is copied to the next generation. Initial gene values were randomly generated between 0 and 100. Population size (25) and crossover rate (60%) are as recommended by Watson and Pollack, and we chose 300 generations based on initial testing that showed algorithms has stabilised well before this. Each run of an algorithm was repeated 60 times to account for variation.

4 Results and Analysis

In this section, we review the results of our experiments by examining quality and diversity in the evolved populations produced using each algorithm. First we examine the quality. In Fig. 1, a convergence plot for the CEA naïve algorithm is shown. Each data point is an average across 60 runs of the algorithm for a specific generation. The y-axis is the estimated best quality. By about 100 generations, the algorithm has converged, except in the case of 5% mutation, which needs around 200 generations. The best mutation rate in terms of estimated quality appears 25%. The actual best quality plot is similar except that the mutation rate has little effect.

In order to quantify this visual impression, we compute average figures over the last 60 generations (as the algorithms appear to have converged by then) and all 60 runs (i.e. an average of 3600 data values) for each mutation rate. These averages are presented in Table 1 (along with diversity data).From the table we can see that, in the case of the naïve algorithm CEA, higher mutation rates tend to give higher best quality (both estimated and actual), and that there is little effect on average quality. Convergence plots for average quality are qualitatively similar to those for best quality, and are omitted.

Looking at CEAFS, we see that best quality is not sensitive to mutation rate, and that estimated best quality is high when compared with CEA, while actual quality is improved compared with CEA. Thus, fitness sharing is effective in increasing the performance of the algorithm (higher best quality). Average quality is reduced when compared with CEA, and decreases with higher mutation rates. The reduction in average quality is due at least in part to the increased diversity of the population, as expected. Convergence plots are quite similar to those for CEA, apart from the final quality levels being different.

CEAHOF has improved quality compared to CEA, with estimated best quality very similar to CEAFS, and the actual best quality also similar, but more sensitive to mutation rate. In fact the best performance over all the algorithms on this measure was CEAHOF with 25% mutation. However average quality levels are actually higher than for CEA, suggesting that the improved performance is not due to an increase in diversity.

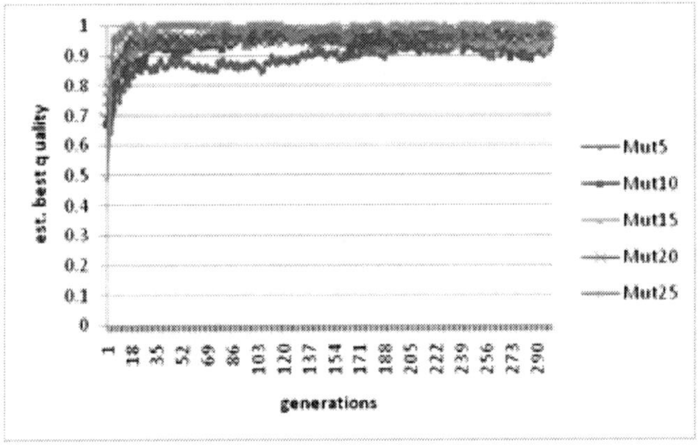

Fig. 1. Convergence plot for CEANaive with different mutation rates, showing average estimated best quality over 60 runs

Finally, the performance of CEAFH is rather erratic, with best quality levels similar to the naïve algorithm, along with a lower average quality. We conjecture that this is because the mechanism of HOF and diversity maintenance methods interfere and conflict with each other, rendering both ineffective.

As well as solution quality, we also focus on the role of diversity. Following Chong et al., we measured both genotypic and phenotypic diversity. Fig. 2 is a generational plot showing the progress of genotypic diversity for CEA - diversity drops swiftly, with a slight recover in phenotypic diversity, before levelling out. Phenotypic diversity is similar. This low diversity might be expected to cause problems such as premature convergence. Higher mutation rates reduce the loss of diversity.

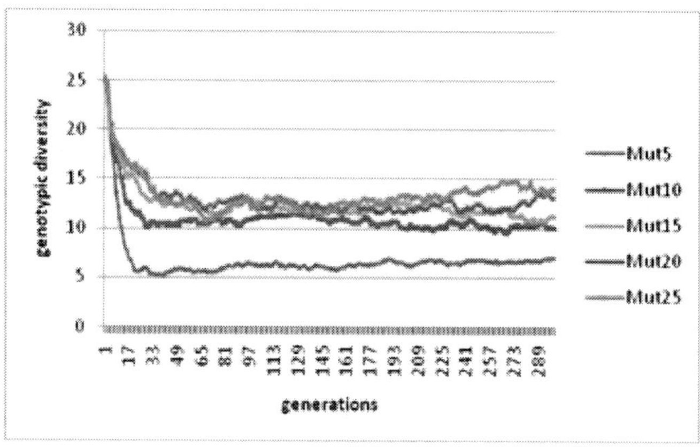

Fig. 2. Generational plot of genotypic diversity with CEANaive. Data values are averaged over 60 runs.

Table 1. Population quality and diversity figures for all algorithm variants. Each column shows the mean for the last 60 generations, over 60 runs of the algorithm.

Algorithm	Est.Average	Est.Best	Act.Average	Act.Best	Geno	Pheno
CEANaive05	0.85	0.93	75.45	75.36	6.72	0.98
CEANaive10	0.82	0.92	74.95	83.47	10.25	1.12
CEANaive15	0.84	0.95	74.50	84.16	11.42	1.17
CEANaive20	0.82	0.96	72.93	83.79	12.50	1.32
CEANaive25	0.84	0.98	74.67	84.74	14.18	1.25
CEAFS05	0.69	0.95	71.11	91.74	25.99	1.51
CEAFS10	0.68	0.96	70.04	91.39	26.11	1.55
CEAFS15	0.66	0.96	70.00	91.12	26.08	1.64
CEAFS20	0.63	0.95	68.89	91.61	27.17	1.70
CEAFS25	0.63	0.96	69.03	91.41	27.19	1.71
CEAHOF05	0.88	0.95	82.15	88.58	6.32	0.88
CEAHOF10	0.88	0.95	84.76	90.95	8.58	0.95
CEAHOF15	0.86	0.95	83.94	91.81	9.62	0.98
CEAHOF20	0.84	0.96	83.31	91.14	10.79	1.09
CEAHOF25	0.85	0.96	88.09	95.51	11.20	0.93
CEAFH05	0.49	0.95	61.98	82.78	23.18	1.16
CEAFH10	0.50	0.90	65.06	85.95	23.41	1.62
CEAFH15	0.48	0.88	63.95	83.97	23.72	1.67
CEAFH20	0.46	0.87	63.82	83.60	23.56	1.71
CEAFH25	0.45	0.87	65.20	85.77	24.12	1.71

The last two columns of Table 1 summarise diversity values for variants of each algorithm. It can be seen that higher mutation rates increase diversity, as expected, and that this effect is much smaller when fitness sharing is used, as diversity is already effectively maintained. Also, the level of diversity is much higher in every case when fitness sharing is used than in any case where fitness sharing is not used. The effect of HOF is to reduce diversity, again emphasising that the improvement in quality when HOF is used is due to a different mechanism.

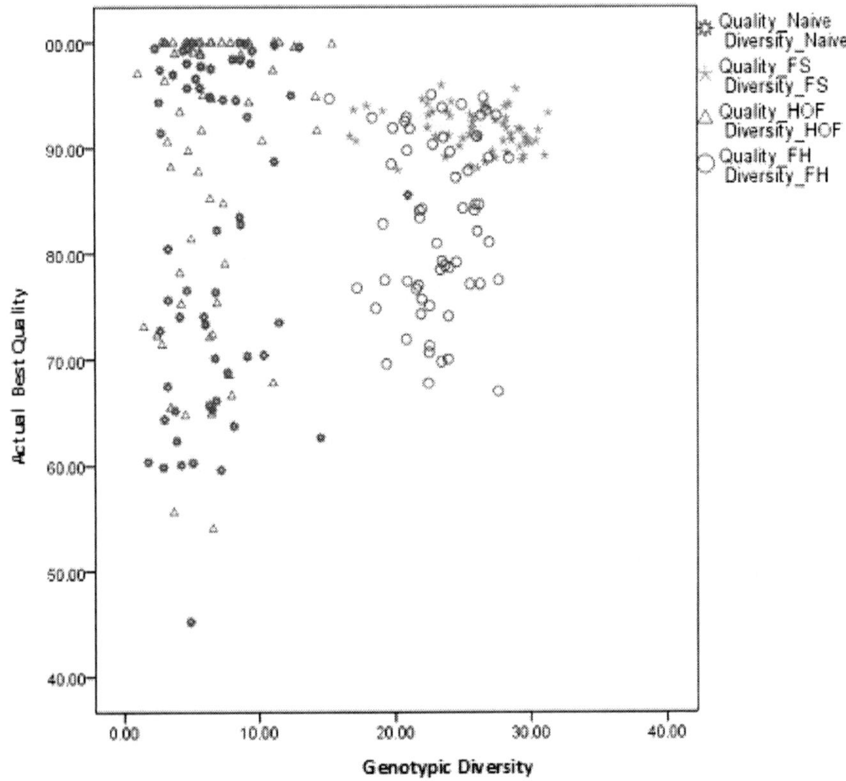

Fig. 3. Scatter plot of diversity versus quality for each of the four algorithms, with a mutation rate of 5%. For each data point, the x value is the mean value of genotypic diversity over the last 60 generations in one run of the particular algorithm, while the y value is the corresponding mean of the actual best quality measure.

Due to space restrictions, we have omitted generational diversity plots for CEAFS, CEAHOF and CEAFH, but we can provide a qualitative description of them as follows: For CEAFS, the plots show a small but rapid rise in genotypic diversity, after which the level remains steady. There is an initial small increase in phenotypic diversity then a quick drop and a leveling out at about the initial diversity level.

The overall shape of the plots for CEAHOF is similar to those for CEA, except that the final diversity levels are a little lower. CEAFH is similar to CEAFS, with genotypic diversity levels slightly lower. The fact that the performance of CEAFH is so poor, even though diversity is only slightly reduced, again suggests that HOF and diversity maintenance are interfering with each other.

To further scrutinize the relationship between diversity and quality, we present Fig. 3, a scatter plot of genotypic diversity versus actual best quality, for all algorithms, with a mutation rate of 5%. It is clear that the naïve algorithm and CEAHOF provide all the points on the left of the plot, i.e. those with lower diversity, and that their quality values are widely spread, i.e. the algorithm is unreliable (though it sometimes converges on very high quality). In contrast, the two algorithms with fitness sharing contribute all the higher diversity points, and reliable quality, with CEAFS being more consistent than CEAFH.

5 Conclusion

In this paper, we have described our experiments with different variations on a naïve CEA, introducing combinations of fitness sharing, Hall of Fame, and a range of mutation rates. We have tested these variations on a test problem designed to be difficult for CEAs due to an intransitive superiority relationship between solutions. We have measured the effects of these variations on the performance of the algorithm in terms of population diversity and solution quality. With regards to diversity, our results are in broad agreement with those found by Chong et al. on a different problem: Iterated Prisoner's Dilemma: fitness sharing is an effective way to maintain population diversity in a CEA, and a moderate amount of diversity helps to ensure that high quality solutions are reliably found. In addition, we found that the Hall of Fame method can also improve quality, but not as reliably as fitness sharing, and that the diversity maintenance methods that we tested do not combine well with Hall of Fame.

In future, we intend to carry out similar tests on further test problems having different characteristics, such as multi-modal problems, to try to improve understanding of which methods are most effective for which kinds of problems. We would also like to investigate whether there are ways to combine diversity maintenance with HOF effectively.

References

1. Axelrod, R.: The evolution of strategies in the iterated Prisoner's Dilemma. Genetic Algorithms and Simulated Annealing, 32–41 (1987)
2. de Jong, E., Stanley, K., Wiegand, P.: Introductory tutorial on coevolution. In: Proceedings of the 2007 Genetic and Evolutionary Computation Conference (GECCO 2007), pp. 3133–3157. ACM, New York (2007)
3. Ficici, S.G.: Solution concepts in coevolutionary algorithms. Ph.D. Dissertation. Brandeis University (2004)
4. Hillis, W.D.: Coevolving parasites improve simulated evolution as an optimization procedure. Physica D: Nonlinear Phenomena 42, 228–234 (1990)

5. Porter, M.A., de Jong, K.A.: A Cooperative Coevolutionary Approach to Function Optimization. In: Davidor, Y., Männer, R., Schwefel, H.-P. (eds.) PPSN 1994. LNCS, vol. 866, pp. 249–257. Springer, Heidelberg (1994)
6. Rosin, C.D.: Coevolutionary search among adversaries. Ph.D. Dissertation. University of California, San Diego (1997)
7. Wiegand, R.P.: An analysis of cooperative coevolutionary algorithms. George Mason University, Virginia (2003)
8. Rosin, C.D., Belew, R.K.: New methods for competitive coevolution. Evolutionary Computation 5, 1–29 (1997)
9. Chong, S.Y., Tino, P., Yao, X.: Relationship between generalization and diversity in coevolutionary learning. IEEE Transactions on Computational Intelligence and AI in Games 1, 214–232 (2009)
10. Mckay, R.I.: Fitness sharing in genetic programming. In: Proceedings of the Proceedings of the Genetic and Evolutionary Computation Conference, Las Vegas (2000)
11. Ray, T.S.: Evolution, complexity, entropy and artificial reality. Physica D: Nonlinear Phenomena, 239–263 (1993)
12. Rosca, J.P.: Entropy-driven adaptive representation. In: Proceedings of the Workshop on Genetic Programming: From Theory to Real-World Applications, pp. 23–32 (1995)
13. Yao, X., Liu, Y.: How to Make Best Use of Evolutionary Learning. Complex Systems - From Local Interactions to Global Phenomena, 229–242 (1996)
14. Chong, S.Y., Tino, P., Yao, X.: Measuring Generalization Performance in Coevolutionary Learning. IEEE Transactions on Evolutionary Computation 12, 479–505 (2008)
15. Casillas, J., Cordon, O., Herrera, F., Merelo, J.J.: A cooperative coevolutionary algorithm for jointly learning fuzzy rule bases and membership functions. Artificial Evolution, 1075–1105 (2002)
16. Ficici, S.G., Pollack, J.B.: Pareto Optimality in Coevolutionary Learning. In: Kelemen, J., Sosík, P. (eds.) ECAL 2001. LNCS (LNAI), vol. 2159, pp. 316–325. Springer, Heidelberg (2001)
17. Watson, R.A., Pollack, J.B.: Coevolutionary dynamics in a minimal substrate. In: Proceedings of the Proceedings of the Genetic and Evolutionary Computation Conference, GECCO 2001. Morgan Kaufmann, San Francisco (2001)
18. Deb, K., Goyal, M.: A combined genetic adaptive search (gene AS) for Engineering Design. Computer Science and Informatics 26, 30–45 (1996)
19. Barker, J.E.: Adaptive Selection Methods for Genetic Algorithms. In: Proceedings of the 1st International Conference on Genetic Algorithms, Hillsdale, NJ, pp. 101–111 (1985)
20. Poli, R., Langdon, W.B.: A new schema theorem for genetic programming with one-point crossover and point mutation. Evolutionary Computation 6, 231–252 (1998)

Eliminating Useless Object Detectors Evolved in Multiple-Objective Genetic Programming

Aaron Scoble[1], Mark Johnston[1], and Mengjie Zhang[2]

[1] School of Mathematics, Statistics and Operations Research
[2] School of Engineering and Computer Science
Victoria University of Wellington, PO Box 600, Wellington, New Zealand

Abstract. Object detection is the task of correctly identifying and locating objects of interest within a larger image. An ideal object detector would maximise the number of correctly located objects and minimise the number of false-alarms. Previous work, following the traditional multiple-objective paradigm of finding Pareto-optimal tradeoffs between these objectives, suffers from an abundance of useless detectors that either detect nothing (but with no false-alarms) or mark every pixel as an object (perfect detection performance with but a very large number of false-alarms); these are very often Pareto-optimal and hence inadvertently rewarded. We propose and compare a number of improvements to eliminate useless detectors during evolution. The most successful improvements are generally more inefficient than the benchmark MOGP approach due to the often vast numbers of additional crossover and mutation operations required, but as a result the archive populations generally include a much higher number of Pareto-fronts.

1 Introduction

Object detection is the task of correctly identifying and locating objects of interest within a larger image [4]. A sweeping window slightly larger than the target objects passes over the full image, classifying each pixel as object or background, and object centres are reported. There is a need to detect patterns quickly (often in real time, hence with constraints on program complexity) and reliably (implying tight constraints on tolerable error rates). Further desirable, but conflicting, objectives include: detecting all objects of interest, minimizing the number of false-alarms, and ensuring proposed object positions are close to the centres of the true target objects.

This paper concentrates on improving Genetic Programming (GP) approaches to object detection. Zhang [4] proposed a two-phase single-objective GP approach. In the first phase, object cutouts from the full set of training images are used to train object classifiers based only on maximising classification accuracy. In the second phase, object detectors are trained on the full training images using a linear combination of detection rate, false alarm rate, false alarm area and size of GP tree as the fitness function. Liddle et al [3] extend the approach of Zhang [4] to a multiple-objective context using Multiple-Objective Genetic Programming

D. Wang and M. Reynolds (Eds.): AI 2011, LNAI 7106, pp. 341–350, 2011.

(MOGP) based on NSGAII [1]. These objectives are evolved into a set of best-performing detectors along the objective trade-off surface (Pareto front), which allows the decision maker to select the most suitable detector after the process rather than weighting the objectives, or otherwise making a priori assumptions. Although this MOGP approach has shown considerable promise, comparing very favourably in terms of execution time and program size, it has several major limitations: (1) using Pareto-based fitness, as in NSGAII [1], MOGP suffers from an abundance of useless detectors that either detect nothing (with no false-alarms) or mark every pixel as an object (with perfect detection performance but a very large number of false-alarms) and these useless detectors are very often Pareto-optimal and hence inadvertently propagated to the next generation of evolution; (2) many different detectors report the same (good) performance on the objectives; (3) there is a danger that over-trained detectors perform poorly on the test set; and (4) some systematic bias in the prediction of object centres is observed in practice.

The goal of this paper is to compare a number of improvements to MOGP in order to eliminate useless detectors during evolution. We wish to gain some understanding of the cost of these improvements in terms of: (1) efficiency; (2) effectiveness; (3) control over the balance between crossover rate and mutation rate parameters; and (4) the inclusion of detectors well below the Pareto-front.

The rest of this paper is divided up as follows. Section 2 provides some further background on single-objective and MOGP approaches to object detection. Section 3 details our response to some of the problems that have been encountered. Section 4 outlines the experimental design and analyses the results. Section 5 concludes the paper.

2 Background

Objectives for Object Detectors. In the two-phase training approach proposed by Zhang [4] and adopted by Liddle et al [3], the objectives in the first training phase (object classification only for the first 20 generations) are to maximize the two objectives of *true positive rate* and *true negative rate*. In the second phase (object detection for the next 40 generations), the detectors are refined on the complete images, using *detection rate* (DR) and *false alarm rate* (FAR), to be maximised and minimised, respectively:

$$DR = \frac{\text{number of correctly located objects}}{\text{total number of objects in the image}}$$

$$FAR = \frac{\text{number of falsely reported objects}}{\text{total number of objects in the image}}$$

Secondary objectives (tie-breakers) used were Crowding Distance (CD) and False Alarm Area (FAA). Crowding distance is the Manhattan distance between any two programs in objective space; the idea is to preserve the spread of programs by rejecting those that are crowded together. False alarm area is the count of all pixels that are reported as positive classifications before clustering, except for those that are reported as the correctly detected centre of an object.

NSGA-II Algorithm. The Nondominated Sorting Genetic Algorithm II (NSGA-II) [1] is a popular multiple-objective evolutionary algorithm. A program is said to (weakly) *dominate* another if it is at least as good as the other program on all objectives, and better on at least one. For each solution two entities are calculated, domination count n_p (the number of solutions that dominate the solution p), and S_p, the set of solutions that p dominates. All solutions in the first nondominated front (rank 0) have $n_p = 0$. At each iteration of the process, the nondominated front is stored, and for each member p with $n_p = 0$, we visit each member $q \in S_p$ and reduce its domination count by one. If for any member q the domination count becomes zero, we add it to the list Q and the next nondominated front (rank i) is established.

The main evolutionary process begins by initially creating a random parent population P_0 of size N which is sorted according to nondomination. Binary tournament selection, crossover, and mutation operators are applied to create an offspring population Q_0 of size N, and $R_0 = P_0 \cup Q_0$ of size $2N$ is sorted by nondomination. As the programs in the first nondominated front are the best in R_t, they will be selected for the new parent population before any others. The algorithm then works through the nondominated fronts, until it reaches one which cannot fit fully into the population of size N. It is at this point where the secondary objectives are used to select the best solutions in the front to fill the population P_{t+1} to size N.

Worst-Best Detectors. An under-reporting detector does not detect all the objects in the image; an extreme case is a detector that reports no objects has $DR = 0$ and $FAR = 0$, and hence is very likely to be Pareto-optimal. An extreme over-reporting detector with $DR = 1$ and very large FAR is also likely to be Pareto-optimal. We call these effectively useless detectors “worst-best” detectors. Liddle et al [3] showed that approximately 94% of the evolved classifiers were worst-best classifiers, implying that the MOGP process is grossly inefficient.

Preliminary investigation discovered that within the first 10 generations, the evolutionary process generally tended to settle to a point where each parent population was selected from the Pareto-optimal front only. This would, in turn, create an offspring population from a process that has selected from what is effectively an unfit set of parents. Once this pattern had been established, it was uncommon for the algorithm to search deeper into the Pareto ranks again, until the beginning of the second phase (due to the shift of objectives). After an initial bump, the process tended to settle on individuals from the Pareto-optimal front again, to the exclusion of all others. We have measured this effect by introducing a metric we have called *reach* (see Section 4). The effect of crowding distance is that the worst-best programs will be selected only after all other programs in that front.

The large number of worst-best classifiers produced at the end of the evolutionary process, indicated that the NSGA-II algorithm was retaining the worst-best classifiers, and leaving little (if any) room in the parent population for

Table 1. Summary of Improvement Methods

Method	(0)	(1)	(2)	(3)	(4)	(5)	(6)	(7)	(8)
Initial population P_0		■	■	■	■				
Single node		■	■	■	■				
Zero objective						■	■	■	■
Merged population R_t			■			■			■
Child population Q_t				■	■		■	■	
Population size N	500	500	500	500	100	500	500	100	100
Legend colour	black	red	green	blue	cyan	magenta	orange	grey	yellow

other programs. The effect, therefore, is to divert computing resources away from other, more suitable programs by preventing them from being selected.

3 Improvements to the MOGP Approach

We observed the effect of worst-best classifiers on the evolutionary process, and introduced several improvements to better observe and isolate the processes that were allowing and propagating the useless detectors, and then to eliminate them. The eight methods proposed are summarised in Table 1, in addition to the original MOGP which we call method (0).

Targeting single node programs. A GP program with only one node cannot perform any calculation and any fitness can only be coincidental. As well as being unable to perform the task, and therefore unsuitable for selection, the lack of branches in their structure also makes them unsuitable for the evolutionary processes. This nature also makes them ideal candidates for worst-best programs. Analysis of tree size [3] suggested that most programs of tree size 0–10 were worst-best performers. Three approaches to eliminating single node programs are proposed:

(1) Modify the creation of the initial population P_0 by regenerating any single node programs in-place as they were created, and not affecting the creation of such programs during the evolutionary process.
(2) Regenerate single node programs in both P_0 and the combined population R_t, which would allow the creation of single node programs but then substitute them out of the combined population during the next nondominated sorting step.
(3) Regenerate single node programs in P_0 and intercept the generation of the child population Q_t, regenerating any single node child that was created in-place. Where a crossover produces an unfit child, both parents are randomly

selected again and process is repeated until both resulting children are fit. This is repeated for mutation, although for a single child.

Although the latter two approaches are similar in that they simulated an infinite population from which the first N fit children were selected, the difference between them is whether the parent selection process is repeated for an unfit child. In all cases, the method of regenerating a program was to replace it with an entirely new randomly created GP program that consisted of more than one node.

Targeting worst objective fitness. As it may be more effective to encourage exploration around the middle of the Pareto front rather than the edges, we have explored an approach in which we simply discard any individual that obtains a result of zero on either objective. In the first phase, both TPR and TNR were evaluated. A non-zero objective value was required in both objectives to consider the program fit. In the second phase there is no fixed worst fitness for FAR, so it was not assessed, i.e., any program with a non-zero detection rate was considered fit, regardless of the FAR. Simply *removing* individuals that had a worst objective fitness could not be implemented due to the implications on the population, i.e., removing the vast number of worst-best programs quickly reduced the population to a fraction of its required size N. The two approaches we consider are to target either the combined population R_t, or the child population Q_t, both designed to maintain correct population size.

(5) Addressing R_t, we randomly regenerate the target programs in-place when they are discovered by the NS process, effectively removing useless programs from the parent population P_{t+1}. Unfit children are allowed to be produced in Q_t, but are substituted at the NS phase. A side effect of this approach is to tip the balance between mutation and crossover in favour of mutation, removing the control of this balance from the setup parameters. Although not ideal in this respect, we were able to observe the effect on the evolutionary process in a somewhat less constrained state.

(6) When Q_t is addressed, we intercept any unfit children during the creation of a new child population. Upon creation of an unfit child, the selection process for parent(s) is repeated, as well as the crossover or mutation operation. This simulates the creation of a population of infinite size, where unfit programs are discarded and the parametrised balance between mutation and crossover is maintained.

Small populations. Once we had found effective approaches, we reduced the size of the population under the hypothesis that the more effective process would be capable of producing comparable results using a smaller set of programs, whilst also offsetting the added overhead of the modifications. Methods (4), (7) and (8) are identical to methods (3), (6) and (5) except that the population size is reduced from $N = 500$ to $N = 100$.

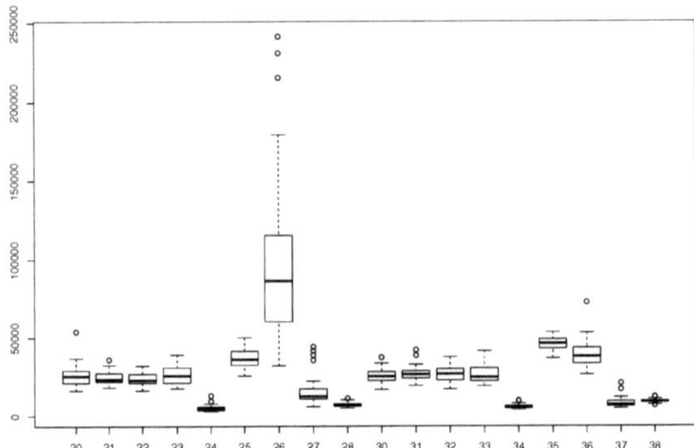

Fig. 1. Boxplots of CPU time (seconds) for each method. Here 20–28 are methods (0)–(8) applied to the shapes dataset and 30–38 are methods (0)–(8) applied to the coins dataset.

4 Experimental Design and Results

Datasets. We have used the same datasets as Liddle et al [3]. The *shapes* dataset consists of equally sized circles and squares against a uniform background in 10 images of size 200×200 pixels containing approximately 15 objects of size 25×25 pixels. The *coins* dataset consists of photographs of New Zealand 5 cent coins [5] either heads or tails up against a noisy background in 10 images of size approximately 500×500 pixels containing 16 objects of size approximately 63×63 pixels. The dataset was split half-and-half for training and testing, and the window cutout size (in phase one) is the same as the object size.

Function and Terminal Sets. The functions $\{\times, \%, +, -, \texttt{if}\}$ are used, where % stands for protected division and the `if` operator takes three arguments and returns the second argument if the first argument is positive, otherwise it returns the third argument. The terminals are features extracted from the images (mean and standard deviation of five equally-sized square regions, consisting of the four quarters and a middle "quarter") and random constants in $[-10, 10]$.

GP System Parameters. We have used the RMIT-GP package with the following parameters: tree depth minimum 2 and maximum 6; crossover 70%; mutation 30%; tournament size 2; 20 first phase generations and 40 second phase generations; and population size 500 or 100. We ran all the experiments on the same 39 randomly generated seeds, so all replications with the same N share the same initial population of trees, thereby reducing bias in the results.

Efficiency Results. Figure 1 shows boxplots of CPU time (seconds) for each method across 39 replications. Method (6) is very inefficient on the shapes dataset and (4), (7) and (8) (all with $N = 100$) are reliably fast on both datasets. Figure 2 shows counts of *additional* crossover and mutation operations for the

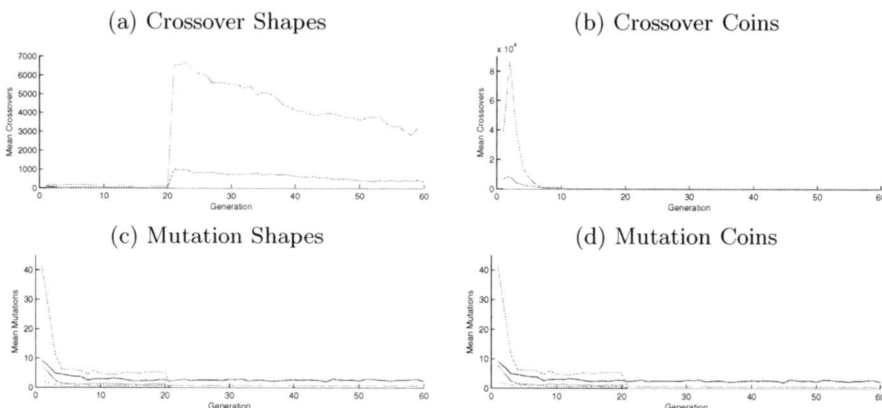

Fig. 2. Frequency of additional crossover and mutation operations for methods that operate on the child population Q_t, i.e., methods (3),(4),(6),(7)

methods that operate on the child population Q_t, i.e., $\{(3),(4),(6),(7)\}$ only. On the shapes dataset, (6, orange) has a high number of mutations in phase one and a high number of crossovers in phase two, which explains the very high CPU time. On the coins dataset, (6) and (7, grey) have high number of crossovers and mutations in phase one. Method (7) with $N = 100$ has proportionately similar additional operations as (6) but its CPU time is much more stable. Note that (6) has $N = 500$ and (7) has $N = 100$ so these figures are excessively high, i.e., alot of additional crossover or mutation operations are required to produce just one non-useless GP tree.

Effectiveness Results. A point on the pth *summary attainment surface* (SAS) [2] is weakly dominated by some solution in each of a proportion p of all 39 replications. For example, the median ($p = 0.5$) SAS is a curve consisting of all points which are weakly dominated in half of the replications. A SAS estimates the probability of an additional replication obtaining better results. Figure 3 shows the $p \in \{0.25, 0.5, 0.75\}$ SAS for both datasets, with legend colours from Table 1; the best SAS are very similar for all methods. On the shapes dataset, for the remaining SAS, method (5, magenta) is clearly the most effective, followed by (6, orange) and (8, yellow), and these are the only ones distinguishably better than the benchmark (0, black). Of these, (5) has a very high reach (see Figure 4) but reasonable CPU time; (8) is the $N = 100$ sized population version of (5); (6) is very inefficient, with very high CPU time due to extremely large number of additional crossovers and mutations; and (7, grey), the $N = 100$ sized population version of (6), performs about the same as the benchmark but is much quicker. On the coins dataset, the SAS are much more tightly packed. Method (6) stands out as clearly the best; its CPU time (see Figure 1) is much better behaved than for the shapes dataset and notice that it again has a high number of additional crossovers and mutations. Methods (5) and (1, red) are the next best methods

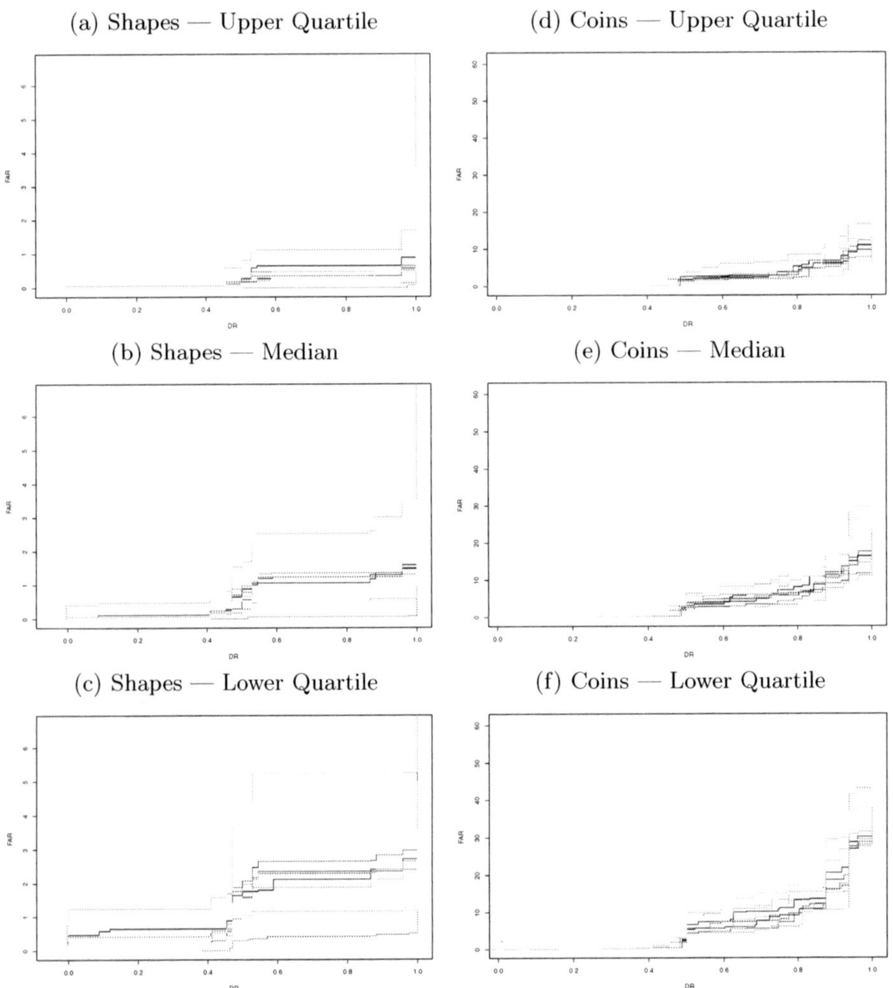

Fig. 3. Summary attainment surfaces for shapes and coins datasets

and the only other methods that are better than (0) are (2, green) and (3, blue). In particular, all the $N = 100$ methods perform worse than (0) on the coins dataset; this indicates that a small population is not sufficient for this more difficult object detection task.

Further Analysis. At each generation, *reach* measures the number of Pareto fronts that are represented in the archive population P_t, e.g., reach is two when the archive consists only of solutions from the best and second-best Pareto fronts in the combined parent and child populations. Figure 4 shows the reach at each generation averaged over the 39 replications. There is a clear jump in reach for all methods at generation 21 at which phase two begins. The peak for shapes is

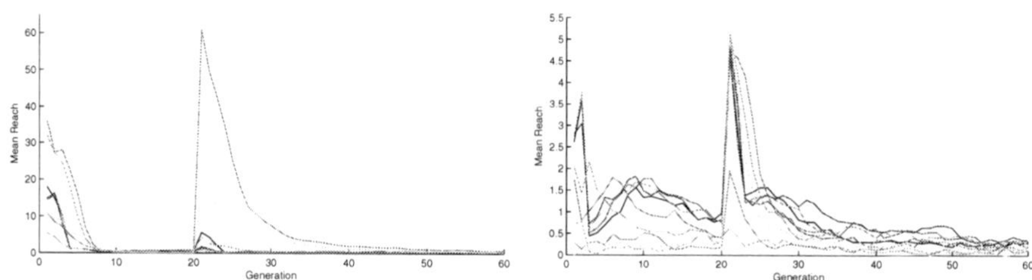

Fig. 4. Reach for shapes dataset (left) and coins dataset (right) averaged across replications for 60 generations

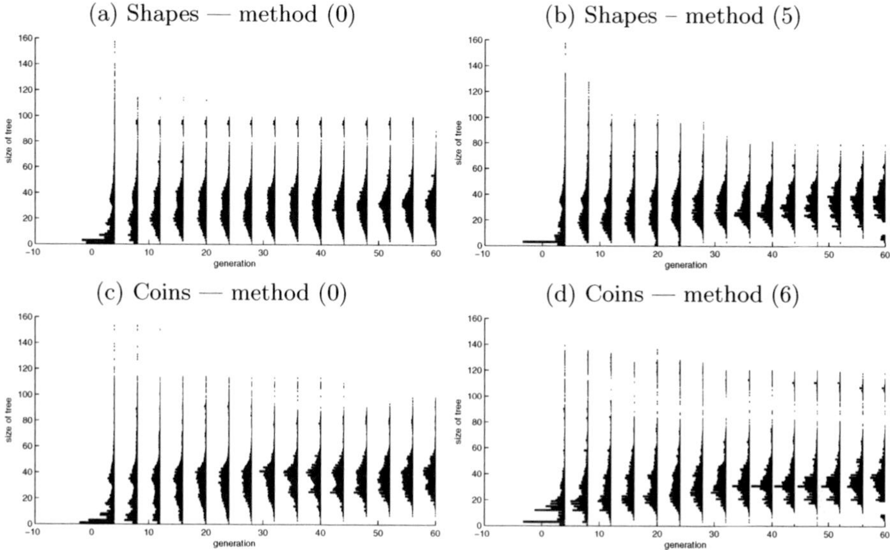

Fig. 5. Frequency of sizes of GP programs at each generation, averaged over 39 replications and averaged over groups of 4 generations

61 whereas the peak for coins is only 5.15. Method (5, magenta) is extreme, but methods (1)–(4) are very well behaved and have lower reach than method (0) after the jump at generation 21. For the coins dataset, the methods are more similar.

Figure 5 estimates the distribution of *sizes* of GP programs (number of nodes) across the generations of evolution for the most effective methods in each dataset to compare against the benchmark. The initial populations in generation 1 are identical. Each distribution plotted is a sum over 39 replications and over 4 generations, hence 60 generations is presented by 15 distributions with $500 \times 4 \times 39 = 78000$ trees in each distribution. Initially there are a large number of small trees. On the shapes dataset, we see that method (5) slowly increases the median size of programs and concentrates the distribution into a narrower range than the benchmark. On the coins dataset, method (6) appears to produce a very similar distribution of sizes to the benchmark at the end of evolution but the intermediate steps move more quickly away from small program sizes.

5 Conclusions

Zitzler, Deb and Thiele [6] suggest three measurable goals of multiple-objective evolutionary algorithms (MOEA): (1) distance of resulting solutions to the Pareto front should be minimised; (2) extent (spread) of solutions should be maximised; and (3) a uniform distribution of solutions found is desirable. We conclude that these goals are not the most effective goals to pursue when evolving object detectors using MOEA. Proximity to the Pareto-optimal front is perhaps less desirable if the solution does not achieve its real objective (detecting objects), and a classifier that is ranked on the second or higher front may be a better solution than a nondominated, under-reporting classifier. Maximising the spread of solutions, while preserving diversity, may also encourage results that are outside of a desirable area, i.e., worst-best classifiers. Hence, perhaps, classical Pareto-based fitness is not a good approach to evolving object detectors in particular.

Improving on the benchmark MOGP results is possible in terms of effectiveness, largely due to establishing deeper reach (as in method (5) for the shapes dataset), but often comes at the expense of vast numbers of additional crossover and mutation operators (as in method (6) for the coins dataset) which eats up CPU time. Smaller population sizes than $N = 500$ may prove effective in combating increase in CPU time but investigating this thoroughly remains future work.

References

1. Deb, K., Pratap, A., Agarwal, S.: A fast and elitist multiobjective genetic algorithm: NSGA-II. IEEE Transactions on Evolutionary Computation 6(2), 182–197 (2002)
2. Knowles, J.: A summary-attainment-surface plotting method for visualizing the performance of stochastic multiobjective optimizers. In: Proceedings of the 5th International Conference on Intelligent Systems Design and Applications, pp. 552–557 (2005)
3. Liddle, T., Johnston, M., Zhang, M.: Multi-objective genetic programming for object detection. In: Proceedings of 2010 IEEE Congress on Evolutionary Computation, pp. 3345–3352 (2010)
4. Zhang, M.: Improving object detection performance with genetic programming. International Journal on Artificial Intelligence Tools 16(5), 849 (2007)
5. Zhang, M., Andreae, P., Pritchard, M.: Pixel Statistics and False Alarm Area in Genetic Programming for Object Detection. In: Raidl, G.R., Cagnoni, S., Cardalda, J.J.R., Corne, D.W., Gottlieb, J., Guillot, A., Hart, E., Johnson, C.G., Marchiori, E., Meyer, J.-A., Middendorf, M. (eds.) EvoIASP 2003, EvoWorkshops 2003, EvoSTIM 2003, EvoROB/EvoRobot 2003, EvoCOP 2003, EvoBIO 2003, and EvoMUSART 2003. LNCS, vol. 2611, pp. 455–466. Springer, Heidelberg (2003)
6. Zitzler, E., Deb, K., Thiele, L.: Comparison of multiobjective evolutionary algorithms: Empirical results. Evolutionary Computation 8(2), 173–195 (2000)

Asymmetric Pareto-adaptive Scheme for Multiobjective Optimization

Siwei Jiang, Jie Zhang, and Yew Soon Ong

School of Computer Engineering, Nanyang Technology University, Singapore
{sjiang1,zhangj,asysong}@ntu.edu.sg

Abstract. A core challenge of Multiobjective Evolutionary Algorithms (MOEAs) is to attain evenly distributed Pareto optimal solutions along the Pareto front. In this paper, we propose a novel asymmetric Pareto-adaptive (apa) scheme for the identification of well distributed Pareto optimal solutions based on the geometrical characteristics of the Pareto front. The apa scheme applies to problem with symmetric and asymmetric Pareto fronts. Evaluation on multiobjective problems with Pareto fronts of different forms confirms that apa improves both convergence and diversity of the classical decomposition-based (MOEA/D) and Pareto dominance-based MOEAs ($pa\epsilon$-MyDE).

Keywords: Multiobjective Optimization, Hypervolume, Pareto-adaptive.

1 Introduction

Multiobjective optimization problems (MOPs) involve several conflicting objectives to be optimized simultaneously. For Pareto optimal solutions, improvement on one objective leads to the decrement of at least one other objective. Multiobjective Evolutionary Algorithms (MOEAs) have been well established as efficient approaches to deal with various MOPs [1].

MOEAs can be generally categorized into two major classes, namely decomposition-based (MOEA/D) and Pareto dominance-based MOEAs [2]. MOEA/D decomposes MOPs into a number of scalar subproblems and optimizes them simultaneously. The assigned weight vectors of classical MOEA/D, however, may not always suit different *Pareto front* (PF). Pareto dominance-based MOEAs use the Pareto dominance definition with the crowding distance or neighbor density estimator to evaluate individuals. However, both of them are less effective to deal with MOPs with asymmetric PFs.

In this paper, we propose a novel asymmetric Pareto-adaptive (apa) scheme. Driven by the hypervolume [3–5] , apa is designed to evenly distribute Pareto optimal solutions along both asymmetric and symmetric PFs. Experimental results on different shapes of 2-dimensional MOPs showed that MOEA/D and $pa\epsilon$-MyDE (one from each category of MOEAs) using apa, labeled here as $apa\lambda$-MOEA/D and $apa\epsilon$-MyDE respectively, lead to higher hypervolume, better convergence and more evenly distributed solutions.

D. Wang and M. Reynolds (Eds.): AI 2011, LNAI 7106, pp. 351–360, 2011.

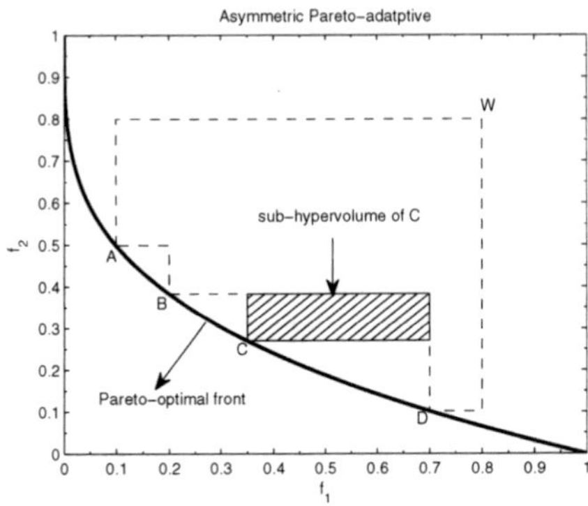

Fig. 1. Asymmetric Pareto-adaptive Scheme

2 *apa*: Asymmetric Pareto-adaptive

2.1 *apa* for 2-Dimensional Pareto Front

The new asymmetric Pareto-adaptive (apa) scheme is driven by the hypervolume. When minimizing bi-objectives for instance, hypervolume is the area enclosed within the discontinuous dash line *WABCDW* (Figure 1). $X = \{A, B, C, D\}$ denotes the set of non-dominated solutions and W is the reference point constructed by the worst objective function values.

Assume points $\{A, B, D\}$ are fixed and point C moves along curve BD. The hypervolume of X is then decided by the sub-hypervolume of point C, which is indicated by the shaded rectangle. In general, we define the normalized asymmetric Pareto optimal front as $f_1^{p_1} + f_2^{p_2} = 1$, where $p_1 \neq p_2$. The points on the curve are $B(x_1, y_1), C(x, y)$, $D(x_2, y_2)$, and the sub-hypervolume of point C is calculated as:

$$\vartheta(x) = (x_2 - x)(y_1 - y) = xy - y_1 x - x_2 y + x_2 y_1 \tag{1}$$

To maximize sub-hypervolume, the optimal position of point C is $(\hat{x}, \hat{y} = (1 - \hat{x}^{p_1})^{\frac{1}{p_2}})$, which can be calculated by the Newton Iterative method defined as:

$$x_{k+1} = x_k - \vartheta'(x_k)/\vartheta''(x_k) \tag{2}$$

where $\vartheta'(x)$ and $\vartheta''(x)$ are the first and second order of $\vartheta(x)$, respectively. The initial value $x_0 = (x_1 + x_2)/2$. When stopping criteria $x_{k+1} - x_k < \xi$ is satisfied, $\hat{x} = x_{k+1}$. The maximum sub-hypervolume is calculated as $\vartheta(\hat{x})$.

Algorithm 1 presents the details of the apa scheme. The N initial points $X = \{(x_1, y_1), \cdots, (x_N, y_N)\}$ along PF are constructed by equally dividing the f_1 axis (Line 1). In Line 9, the point (x_{i_m}, y_{i_m}) is replaced by $(\hat{x}_{i_m}, \hat{y}_{i_m})$, which makes the maximum increment to the hypervolume. In Lines 10-14, the movement of point i_m only impact the neighborhood points. We update the sub-hypervolume and the maximum sub-hypervolume of the neighborhood points $(i_m - 1, i_m, i_m + 1)$. When the hypervolume increment is less than ξ, the algorithm terminates and outputs X.

Algorithm 1. Asymmetric Pareto-adaptive scheme

Initialize N points X along Pareto optimal front
Sort the N points ascending by the first objective
for $i = 2, \cdots, N-1$ **do**
  Calculate sub-hypervolume $\vartheta(x_i)$
  Calculate maximum sub-hypervolume $\vartheta(\hat{x}_i)$
  $\triangle\vartheta(x_i) = \vartheta(x_i) - \vartheta(\hat{x}_i)$
Find $i_m = \max\{\triangle\vartheta(x_i) : i \in 2, \cdots, N-1\}$
while $\triangle\vartheta(x_{i_m}) > \xi$ **do**
  Move the point (x_{i_m}, y_{i_m}) to $(\hat{x}_{i_m}, \hat{y}_{i_m})$
  Set $\vartheta(x_{i_m}) = \vartheta(\hat{x}_{i_m})$, $\triangle\vartheta(x_{i_m}) = 0$
  Update $\vartheta(x_{i_m-1})$ and $\vartheta(x_{i_m+1})$
  Update $\vartheta(\hat{x}_{i_m-1})$ and $\vartheta(\hat{x}_{i_m+1})$
  $\triangle\vartheta(x_{i_m-1}) = \vartheta(x_{i_m-1}) - \vartheta(\hat{x}_{i_m-1})$
  $\triangle\vartheta(x_{i_m+1}) = \vartheta(x_{i_m+1}) - \vartheta(\hat{x}_{i_m-1})$
  Find $i_m = \max\{\triangle\vartheta(x_i) : i \in 2, \cdots, N-1\}$
Output X

2.2 Curve Function for Asymmetric Pareto Front

Upon generating a set of Pareto optimal solutions (points), we estimate a curve function to represent the PF based on the available points. Define 2D Pareto optimal solutions as $F = \{(x_i, y_i) : i = 1, \cdots, |F|\}$ ($|F|$ is the number of points, normalized into $[0, 1]$). To estimate the asymmetric PF $f_1^{p_1} + f_2^{p_2} = 1$, we define the Sum of Square Error as:

$$SSE(F) = \sum\nolimits_i^{|F|} (x_i^{p_1} + y_i^{p_2} - 1 + \theta)^2 \qquad (3)$$

where θ denotes the relaxing parameter. A small $SSE(F)$ implies that the curve function approaches the Pareto optimal solutions better. In reality, it is not easy for all solutions to fall exactly on the true PF, thus, it is natural to set the relaxing parameter with a small value to estimate the curve function.

3 *apa* for Decomposition and ϵ-Dominance

The well established two major categories of MOEAs includes the decomposition-based and Pareto dominance-based MOEAs. In this section, we describe how the *apa* scheme enhances the performances of MOEAs.

3.1 $apa\lambda$: *apa* for Decomposition

MOEA based on decomposition (MOEA/D) transforms the PF into a number of scalar optimization subproblems and optimizes them simultaneously. Three major approaches of the MOEA/D are weighted sum, Tchebycheff and Boundary intersection (BI) [2].

Define $\lambda = (\lambda^1, \cdots, \lambda^m)^T$ as a weight vector for m objectives, and $\sum_i^m \lambda^i = 1$. Classical MOEA/D produces N weight vectors in 2-dimensional objective spaces as: $(\frac{0}{H}, \frac{H}{H}), (\frac{1}{H}, \frac{H-1}{H}), \cdots, (\frac{H}{H}, \frac{0}{H})$, where $H = N - 1$. The gradients of weight vectors are represented by λ lines (see Figures 2, 3). These λ lines produce N intersection points along PF. Such weight vectors are perfectly distributed only when PF is $f_1 + f_2 = 1$ (i.e. a linear line), but not suitable for $f_1^{p_1} + f_2^{p_2} = 1, p_1, p_2 \neq 1$ (i.e. non-linear PF).

Asymmetric Pareto-adaptive weight vectors ($apa\lambda$) is formed by applying the apa scheme to MOEA/D. Since the apa scheme (Algorithm 1) can obtain evenly distributed intersection points $\{(x_i, y_i), i = 1, \cdots, N\}$ along different shapes of Pareto optimal front, the weight vectors along asymmetric Pareto optimal front can be adjusted as:

$$apa\lambda_i = (\frac{x_i}{x_i + y_i}, \frac{y_i}{x_i + y_i}) \tag{4}$$

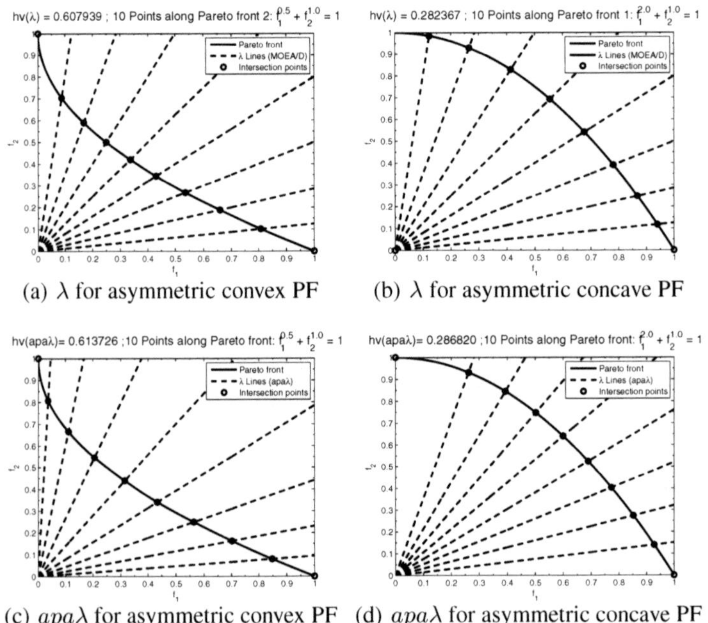

(a) λ for asymmetric convex PF (b) λ for asymmetric concave PF

(c) $apa\lambda$ for asymmetric convex PF (d) $apa\lambda$ for asymmetric concave PF

Fig. 2. 10 Points along 2-dimensional *asymmetric* Pareto Fronts by λ and $apa\lambda$

Figure 2 shows an example of 10 intersection points along 2-dimensional asymmetric PFs generated by λ (MOEA/D) and the $apa\lambda$ scheme, respectively. When p_1=0.5, p_2=1.0, the hypervolume of the intersection points is $hv(\lambda) = 0.607939$ in MOEA/D (Figure 2(a)). From Figure 2(c), $apa\lambda$ obtains a larger $hv(apa\lambda) = 0.613726$, and the λ lines are scattered to the two endpoints of PF and distributed more evenly. When $p_1 = 2.0$ and $p_2 = 1.0$, the hypervolume of intersection points is $hv(\lambda) = 0.282367$ in MOEA/D (Figure 2(b)). $apa\lambda$ obtains a larger $hv(apa\lambda) = 0.286820$ (Figure 2(d)), and the λ lines are well assembled and divide the objective space more uniformly.

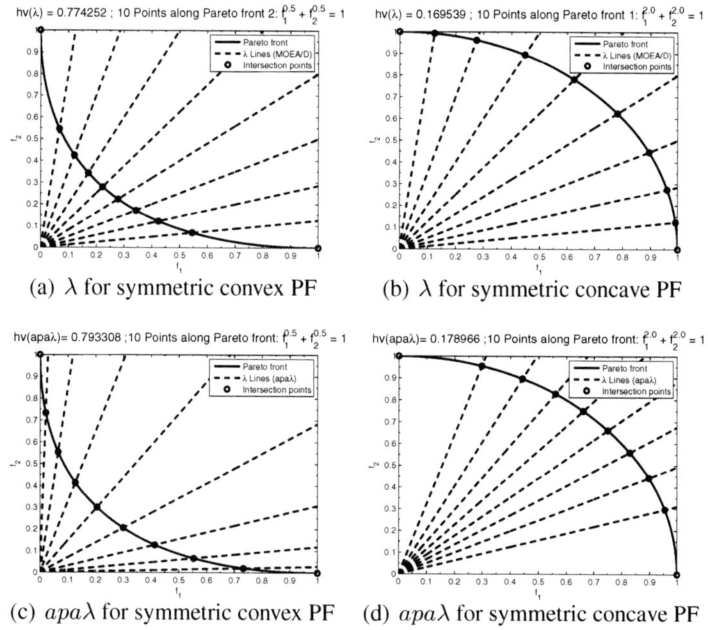

(a) λ for symmetric convex PF (b) λ for symmetric concave PF

(c) $apa\lambda$ for symmetric convex PF (d) $apa\lambda$ for symmetric concave PF

Fig. 3. 10 Points along 2-dimensional *symmetric* Pareto Fronts by λ and $apa\lambda$

In addition, further studies on MOEA/D also assert that $apa\lambda$ obtained higher hypervolume on both symmetric convex and symmetric concave PFs (Figure 3). The apa scheme is shown to significantly improve the performance of classical MOEA/D.

3.2 $apa\epsilon$-Dominance: apa for ϵ-Dominance

The ϵ-dominance is an advanced dominance concept that includes additive and multiplicative schemes [6]. It divides the m objective spaces into equal-sized hyper-boxes and only one solution can survive in a hyper-box. When two solutions exist in the same hyper-box and non-dominates each other, ϵ-dominance remains the one which is nearer to the corner of hyper-box. When minimizing MOPs, the additive scheme f is said to ϵ-dominate g, if $\forall i \in \{1, \cdots, m\}, f_i - \epsilon \leq g_i$.

In ϵ-dominance, the parameter ϵ is user-specific. Pareto-adaptive ϵ-dominance ($pa\epsilon$-dominance) is a new ϵ-dominance, which calculates $\epsilon^j = (\epsilon^1, \epsilon^2, \cdots, \epsilon^N)$ (N is population size) depending on the geometric characteristics of PFs [7]. When minimizing MOPs, f is said to $pa\epsilon$-dominate g in j-th hyper-box, if $\forall i \in \{1, \cdots, m\}, f_i - \epsilon^j \leq g_i$.

$pa\epsilon$-dominance handles asymmetric PFs by approximating $f_1^{p1} + f_2^{p2} = 1, p1 \neq p2$ as $f_1^p + f_2^p = 1$. Asymmetric Pareto-adaptive ϵ-dominance concept ($apa\epsilon$-dominance), on the other hand, applies the proposed apa scheme to ϵ-dominance. $\epsilon_i^j = (\epsilon_i^1, \epsilon_i^2, \cdots, \epsilon_i^N)$ in i-th objective is calculated for different shapes of PFs. When minimizing MOPs, f is said to $apa\epsilon$-dominate g in the j-th hyper-box, if $\forall i \in \{1, \cdots, m\}, f_i - \epsilon_i^j \leq g_i$.

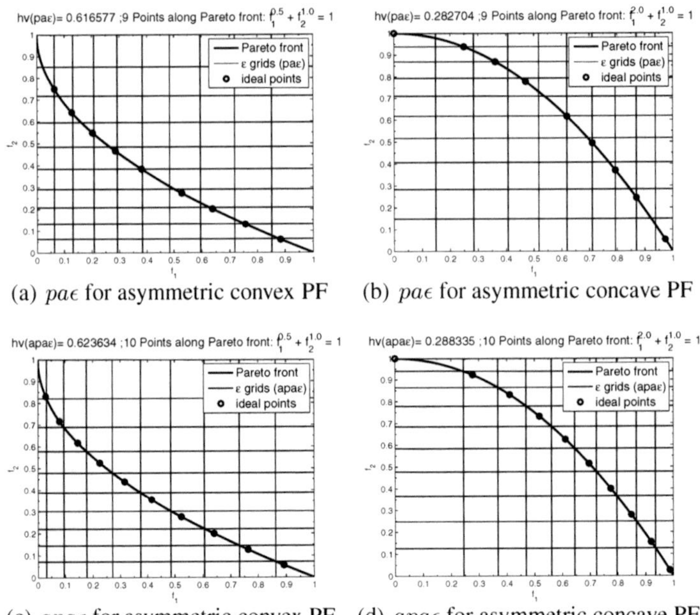

(a) $pa\epsilon$ for asymmetric convex PF (b) $pa\epsilon$ for asymmetric concave PF

(c) $apa\epsilon$ for asymmetric convex PF (d) $apa\epsilon$ for asymmetric concave PF

Fig. 4. Ideal Points along 2-dimensional *asymmetric* Pareto Fronts by $pa\epsilon$ and $apa\epsilon$

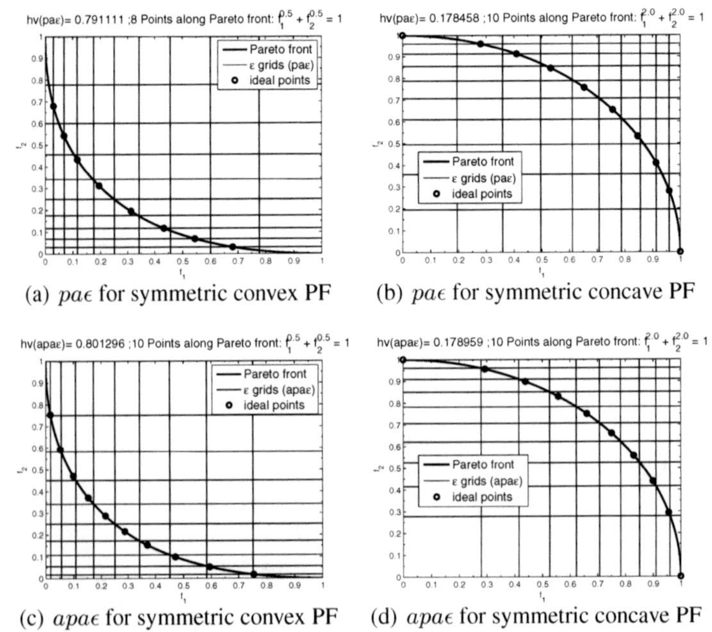

(a) $pa\epsilon$ for symmetric convex PF (b) $pa\epsilon$ for symmetric concave PF

(c) $apa\epsilon$ for symmetric convex PF (d) $apa\epsilon$ for symmetric concave PF

Fig. 5. Ideal Points along 2-dimensional *symmetric* Pareto Fronts by $pa\epsilon$ and $apa\epsilon$

The $apa\epsilon$-dominance divides the i-th objective space into N non-equal segments (They can be equal only if $f_1 + f_2 = 1$). To begin, the apa scheme (Algorithm 1) distributes $N + 1$ solutions $\{(x_i, y_i), i = 1, \cdots, N + 1\}$ along the PF. Then ϵ_i^j ($i = 1, 2; j = 1, \cdots, N$) can be calculated as follows:

$$\begin{cases} \epsilon_1^j = x_{j+1} \\ \epsilon_2^j = y_{j+1} \end{cases} \tag{5}$$

Minimizing bi-objective problems, Figures 4- 5 show $pa\epsilon$ and $apa\epsilon$-dominance distribute 10 points along PFs. In some cases, $pa\epsilon$-dominance cannot get 10 points. Each ideal point is carefully drawn under ϵ-dominance concept. Only one point can survive in a hyper-box, and the point has the minimum distance to the left bottom corner of the hyper-box. For asymmetric PF, $pa\epsilon$-dominance approximates it as an symmetric PF, which has the same hypervolume as the original asymmetric PF[1]. Figures 4(a-b) show that the region near the origin $(0, 0)$ is a *square*, which is obviously unsuitable for asymmetric PFs. In contrast, $apa\epsilon$-dominance arrive at *rectangle* shape (Figures 4(c-d)).

Focusing on the number of points for different PFs, $pa\epsilon$-dominance produces 9 points on the two asymmetric PFs (Figure 4(a-b)), 8 points on $f_1^{0.5} + f_2^{0.5} = 1$ and 10 points on $f_1^2 + f_2^2 = 1$ (Figure 5(a-b)), while $ap\epsilon$-dominance produces 10 points on all PFs. Comparing the hypervolume on asymmetric PFs, $pa\epsilon$-dominance obtains $hv(pa\epsilon) = 0.616577$ for $f_1^{0.5} + f_2 = 1$ and $hv(pa\epsilon) = 0.282704$ for $f_1^2 + f_2 = 1$ (Figure 4(a-b)). However, $apa\epsilon$-dominance is able to obtain higher values of $hv(apa\epsilon) = 0.623634$ and $hv(apa\epsilon) = 0.288335$, respectively (Figure 4(c-d)). In addition, for symmetric PFs, Figure 5 shows that $apa\epsilon$-dominance also obtains the higher hypervolume than $pa\epsilon$-dominance. The apa scheme is thus shown to successfully enhance the performance of Pareto dominance-based MOEAs.

4 Experimental Results and Discussion

4.1 Benchmark Problems and Experimental Setting

The experiments are performed on jMetal 3.0 [4], which is a Java-based framework that is aimed at facilitating the development of metaheuristics for solving MOPs[2]. Testing MOPs include 4 with asymmetric PFs: $f_1^{0.5} + f_2 = 1$ (ZDT1, ZDT4) and $f_1^2 + f_2 = 1$ (ZDT2, ZDT6), 4 with symmetric PFs: $f_1^{0.5} + f_2^{0.5} = 1$ (ZDT1.1, ZDT4.1[3]) and $f_1^2 + f_2^2 = 1$ (WFG4, DTLZ2.2D), and 2 discrete PFs as ZDT3 and Kursawe [4].

The classical MOEA/D - MOEA/D with the Tchebycheff approach [2] and $pa\epsilon$-MyDE [7] are included for comparison. Two instances of the apa scheme are considered here: MOEA/D with asymmetric Pareto-adaptive weight vectors ($apa\lambda$-MOEA/D) and differential evolution with asymmetric Pareto-adaptive ϵ-dominance ($apa\epsilon$-MyDE).

---

[1] For $f_1^{0.5} + f_2 = 1$, $pa\epsilon$ approximates it as $f_1^{0.723} + f_2^{0.723} = 1$; for $f_1^{2.0} + f_2 = 1$, $pa\epsilon$ approximates it as $f_1^{1.445} + f_2^{1.445} = 1$.

[2] `http://jmetal.sourceforge.net`

[3] ZDT1.1 and ZDT4.1 are symmetric PF by modifying ZDT1 and ZDT4 respectively. The true PS is formed by equally dividing circle into 200 sections in term of angle.

The experimental settings are outlined as follows. The population size is 25 and the maximum number of fitness function evaluations is $25,000$. Every algorithm runs 100 times independently for each test problem, to obtain statistically significant results. In MOEA/D, the number of neighborhoods is $T = 20$. For DE (Differential Evolution) operator, $CR = 0.25$ and $F = 0.5$. For polynomial mutation, $\eta = 20$ and $p_m = 1/n$ (n is the number of decisional variables). For apa scheme, termination condition is $\xi = 1e - 10$ and relaxing parameter to estimate PF is $\theta = 0.01$.

Five performance metrics are reported: Hypervolume (HV), Inverted Generational Distance (IGD), Generational Distance (GD), Unary Additive Epsilon Indicator ($I^1_{\epsilon+}$) and Spread. The higher Hypervolume and lower IGD, GD, $I^1_{\epsilon+}$ and Spread, the better is the algorithm's performance. The obtained results are compared using median values and the superior results of test problems are highlighted by grey background.

Table 1. Median of Hypervolume (**HV**)

	MOEA/D	$apa\lambda$-MOEA/D	$pa\epsilon$-MyDE	$apa\epsilon$-MyDE
ZDT1	$6.4496e-01$	$6.4721e-01$	$6.4576e-01$	$6.4752e-01$
ZDT4	$6.4460e-01$	$6.4690e-01$	$6.4678e-01$	$6.4828e-01$
ZDT2	$3.1346e-01$	$3.1505e-01$	$3.1417e-01$	$3.1589e-01$
ZDT6	$3.8570e-01$	$3.8558e-01$	$3.8667e-01$	$3.8713e-01$
ZDT1.1	$8.1150e-01$	$8.1937e-01$	$8.1781e-01$	$8.1880e-01$
ZDT4.1	$8.1043e-01$	$8.1879e-01$	$8.1836e-01$	$8.1896e-01$
WFG4	$2.0373e-01$	$2.0760e-01$	$2.0694e-01$	$2.0782e-01$
DTLZ2.2D	$1.9816e-01$	$2.0190e-01$	$2.0008e-01$	$2.0138e-01$
ZDT3	$5.0189e-01$	$5.0555e-01$	$4.9971e-01$	$4.9945e-01$
Kursawe	$3.8594e-01$	$3.8607e-01$	$3.8479e-01$	$3.8484e-01$

4.2 Discussions on Statistical Results

Table 1 shows the performance of the MOEAs on hypervolume. For decomposition-based algorithms, $apa\lambda$-MOEA/D reported superior HV values on 9 problems. MOEA/D fares better only on ZDT6. Both ZDT2 and ZDT6 share the same true PF of $f_1^2 + f_2 = 1$. The PF of ZDT2 exists in $f_1, f_2 \in [0.0, 1.0]$, but that of ZDT6 is in $f_1 \in [0.2809, 1.0]$ and $f_2 \in [0.0, 0.9211]$. The results indicate that Tchebycheff approach is unsuitable for dealing with disparately scaled objectives problems. To solve such problems, Zhang suggest to use the *Objective Normalization* [2].

On Pareto dominance-based algorithms, $apa\epsilon$-MyDE reported superior HV values on 9 problems, while $pa\epsilon$-MyDE fares better only on ZDT3, which has a discrete PF. The lower HV of $apa\epsilon$-MyDE on ZDT3 indicates that it is less suitable for discrete PF.

Table 2. Median of Inverted Genetic Distance (**IGD**)

	MOEA/D	$apa\lambda$-MOEA/D	$pa\epsilon$-MyDE	$apa\epsilon$-MyDE
ZDT1	$6.3658e-04$	$5.5097e-04$	$6.7638e-04$	$5.6731e-04$
ZDT4	$6.3606e-04$	$5.5161e-04$	$6.1060e-04$	$5.3733e-04$
ZDT2	$5.7958e-04$	$6.2226e-04$	$7.6548e-04$	$5.8589e-04$
ZDT6	$3.6948e-04$	$4.0058e-04$	$9.7866e-04$	$3.9867e-04$
ZDT1.1	$4.7732e-03$	$3.0518e-03$	$4.5052e-03$	$3.2133e-03$
ZDT4.1	$4.7694e-03$	$3.0493e-03$	$4.5169e-03$	$3.4396e-03$
WFG4	$4.3374e-04$	$3.8385e-04$	$4.0915e-04$	$3.9105e-04$
DTLZ2.2D	$1.4219e-03$	$1.7491e-03$	$1.7322e-03$	$1.7056e-03$
ZDT3	$2.0143e-03$	$1.5427e-03$	$2.8749e-03$	$2.7623e-03$
Kursawe	$6.5540e-04$	$6.8078e-04$	$9.7884e-04$	$8.7020e-04$

Table 3. Median of Genetic Distance (**GD**)

	MOEA/D	$apa\lambda$-MOEA/D	$pa\epsilon$-MyDE	$apa\epsilon$-MyDE
ZDT1	$1.0730e-04$	$1.3747e-04$	$9.5069e-05$	$1.0304e-04$
ZDT4	$1.8162e-04$	$1.7120e-04$	$1.0070e-04$	$9.8641e-05$
ZDT2	$9.3999e-05$	$1.0670e-04$	$9.5799e-05$	$9.2231e-05$
ZDT6	$9.6881e-04$	$8.9086e-04$	$1.0054e-03$	$1.1717e-03$
ZDT1.1	$9.5795e-03$	$8.2646e-03$	$9.3162e-03$	$8.5521e-03$
ZDT4.1	$9.4369e-03$	$8.2199e-03$	$9.1691e-03$	$8.6565e-03$
WFG4	$1.5613e-03$	$1.5306e-03$	$8.2842e-04$	$7.0187e-04$
DTLZ2.2D	$5.1184e-04$	$5.3937e-04$	$5.6100e-04$	$5.4583e-04$
ZDT3	$3.9503e-04$	$3.5437e-04$	$3.1126e-04$	$2.7977e-04$
Kursawe	$1.9408e-04$	$2.3288e-04$	$3.1429e-04$	$3.3285e-04$

Table 2 shows the performance on Inverted Genetic Distance metric. On decomposition based algorithms, $apa\lambda$-MOEA/D reported superior results on 6 problems, while MOEA/D fares better on ZDT2, ZDT6, DTLZ2.2D and Kursawe, respectively. Except the discrete PF (kursawe), these problems have concave true PFs as $f_1^2 + f_2 = 1$ (ZDT2, ZDT6) and $f_1^2 + f_2^2 = 1$ (DTLZ2.2D). The results indicate that the apa scheme can lead to reduce IGD performance on concave PFs.

On Pareto dominance-based MOEAs, $apa\epsilon$-MyDE got better IGD on all problems.

Tables 3 and 4 tabulated the performance of the Genetic Distance and epsilon metrics, respectively. Among the 10 test problems, $apa\lambda$-MOEA/D and $apa\epsilon$-MyDE reported superior results for these two metrics on most of the problems.

Table 4. Median of **epsilon** ($I^1_{\epsilon+}$) Metric

	MOEA/D	$apa\lambda$-MOEA/D	$pa\epsilon$-MyDE	$apa\epsilon$-MyDE
ZDT1	$3.3184e-02$	$2.0204e-02$	$3.2003e-02$	$2.6870e-02$
ZDT4	$3.3184e-02$	$2.0865e-02$	$3.1665e-02$	$2.3536e-02$
ZDT2	$2.5697e-02$	$1.9412e-02$	$3.0058e-02$	$2.0158e-02$
ZDT6	$1.9300e-02$	$2.5540e-02$	$4.3798e-02$	$1.5601e-02$
ZDT1.1	$2.9057e-02$	$5.7351e-03$	$1.2688e-02$	$7.8903e-03$
ZDT4.1	$2.9275e-02$	$6.0491e-03$	$1.1796e-02$	$1.3107e-02$
WFG4	$7.8637e-02$	$5.0718e-02$	$7.7030e-02$	$7.2197e-02$
DTLZ2.2D	$2.8321e-02$	$1.6760e-02$	$2.8834e-02$	$1.7828e-02$
ZDT3	$5.8978e-02$	$3.6942e-02$	$4.0652e-02$	$4.5659e-02$
Kursawe	$2.8456e-01$	$3.4308e-01$	$2.4865e-01$	$2.4459e-01$

Table 5 presents the performance on Spread metric, which evaluates the distribution of non-dominated solutions. On decomposition-based algorithms, $apa\lambda$-MOEA/D reported better Spread values on 6 problems, while MOEA/D on 4 namely ZDT6, WFG4, DTLZ2.2D and Kursawe. Similar to the IGD metric, the results indicate that the apa scheme may not favor Spread metric on concave PFs.

Among Pareto dominance-based algorithms, $apa\epsilon$-MyDE reported smaller Spread values on all problems. For the asymmetric problems including $f_1^{0.5} + f_2 = 1$ (ZDT1, ZDT4) and $f_1^2 + f_2 = 1$ (ZDT2, ZDT6), it is worth mentioning that $apa\epsilon$-MyDE arrives at significantly better spread value than $pa\epsilon$-MyDE.

To summarize, experimental results highlight that $apa\lambda$-MOEA/D and $apa\epsilon$-MyDE are able to obtain consistently higher hypervolume, better convergence and more evenly distributed solutions than classical MOEA/D and $pa\epsilon$-MyDE on majority of the benchmark problems. The apa scheme is validated by demonstrating empirically performance improvement of both decomposition-based and Pareto dominance-based MOEAs.

Table 5. Median of **Spread** Metric

	MOEA/D	$apa\,\lambda$-MOEA/D	$pa\epsilon$-MyDE	$apa\epsilon$-MyDE
ZDT1	$2.8126e-01$	$8.9546e-02$	$2.4730e-01$	$1.5013e-01$
ZDT4	$2.8085e-01$	$9.0885e-02$	$2.4298e-01$	$1.3152e-01$
ZDT2	$1.3611e-01$	$1.0650e-01$	$1.9993e-01$	$9.4430e-02$
ZDT6	$1.4975e-01$	$1.6385e-01$	$1.0006e-01$	$3.8206e-02$
ZDT1.1	$6.8130e-01$	$2.9226e-01$	$4.2275e-01$	$3.3306e-01$
ZDT4.1	$6.8115e-01$	$2.9534e-01$	$4.2975e-01$	$3.6701e-01$
WFG4	$2.0255e-01$	$3.0316e-01$	$2.8014e-01$	$2.7552e-01$
DTLZ2.2D	$1.8513e-01$	$2.8223e-01$	$3.1004e-01$	$2.6857e-01$
ZDT3	$7.6543e-01$	$7.3368e-01$	$5.8611e-01$	$5.4899e-01$
Kursawe	$6.0004e-01$	$6.0182e-01$	$4.1732e-01$	$3.9657e-01$

5 Conclusion and Future Research

In this paper, we have proposed a novel Asymmetric Pareto-adaptive (apa) scheme, which automatically adjusts the position of Pareto optimal solutions according to the geometric characteristics of 2-dimensional Pareto optimal front. The new scheme is shown to work well on both symmetric and asymmetric PFs and improves the performance of general MOEAs such as decomposition-based and Pareto dominance-based MOEAs. Experimental results further confirm that the apa scheme led to significant improvements on the performance of MOEA/D and $pa\epsilon$-MyDE.

The apa scheme is showed to be efficient and for dealing with 2-dimensional MOPs. Future research is to extend the scheme to higher dimensional MOPs. Another potential further research would be to improve the MOP search based on the paradigm of Memetic Computation [8–11].

References

1. Coello, C., et al.: Evolutionary multi-objective optimization: a historical view of the field. IEEE Comput. Intell. Mag. 1(1), 28–36 (2006)
2. Li, H., Zhang, Q.: Multiobjective optimization problems with complicated Pareto sets, MOEA/D and NSGA-II. IEEE Trans. Evol. Comput. 13(2), 284–302 (2009)
3. Zitzler, E., Thiele, L.: Multiobjective evolutionary algorithms: A comparative case study and the strength pareto approach. IEEE Trans. Evol. Comput. 3(4), 257–271 (1999)
4. Durillo, J.J., Nebro, A.J., Alba, E.: The jmetal framework for multi-objective optimization: Design and architecture. IEEE Congr. Evol. Comput., 1–8 (2010)
5. Bader, J., Zitzler, E.: HypE: An algorithm for fast hypervolume-based many-objective optimization. Evol. Comput. 19(1), 45–76 (2011)
6. Laumanns, M., Thiele, L., Deb, K., Zitzler, E.: Combining convergence and diversity in evolutionary multiobjective optimization. Evol. Comput. 10(3), 263–282 (2002)
7. Hernlandez-Dlaz, A.G., Santana-Quintero, L.V., Coello Coello, C.A., Molina, J.: Paretoadaptive ϵ-dominance. Evol. Comput. 15(4), 493–517 (2007)
8. Lim, D., Jin, Y., Ong, Y.S., Sendhoff, B.: Generalizing surrogate-assisted evolutionary computation. IEEE Trans. Evol. Comput. 14(3), 329–355 (2010)
9. Ong, Y.S., Lim, M.H., Zhu, N., Wong, K.W.: Classification of adaptive memetic algorithms: A comparative study. IEEE Trans. Syst., Man, Cybern., Part B: Cybernetics 36(1), 141–152 (2006)
10. Ong, Y.S., Lim, M.H., Chen, X.S.: Research frontier: memetic computation-past, present & future. IEEE Comput. Intell. Mag. 5(2), 24–31 (2010)
11. Acampora, G., Loia, V., Gaeta, M.: Exploring e-learning knowledge through ontological memetic agents. IEEE Comput. Intell. Mag. 5(2), 66–77 (2010)

Convergence of a Recombination-Based Elitist Evolutionary Algorithm on the Royal Roads Test Function

Aram Ter-Sarkisov and Stephen Marsland

Department of Computer Science
Massey University, New Zealand
{a.ter-sarkisov,s.r.marsland}@massey.ac.nz
http://www.massey.ac.nz/seat

Abstract. We present analysis of performance of an elitist Evolutionary algorithm using a recombination operator 1-Bit-Swap on the Royal Roads test function. We derive complete, approximate and asymptotic convergence rates. Both complete and approximate models show the benefit of the size of the population and recombination pool when they are small and leveling out of this effect when limit conditions are applied. Numerical results confirm our findings.

Keywords: Evolutionary algorithms, computational complexity, probabilistic models.

1 Introduction

Evolutionary Algorithms (EA) are a set of heuristic optimization tools, that are well-suited to solve problems with poorly-understood landscape (black-box optimization). Despite rich history in application, theoretical analysis has been lagging behind. Although in the past few years a large amount of research in this area has evolved, it is mostly restricted to single-parent algorithms.

We analyze an elitist (hence +) $(\mu{+}\lambda)$ algorithm that operates on a population of solutions size μ and recombination pool size λ using a genetic operator called 1-Bit-Swap (1BS) and tournament selection function. Each generation a subset of best species α is saved (hence elitism).

We find three expressions for the expected runtime of $(\mu{+}\lambda)\text{EA}_{1BS}$: one exact, one approximate and the third one asymptotic. Asymptotic expression does not contain the variables for the size of the population and recombination pool due to cancellation, which is the leveling out effect: as the size of the population grows large enough, its effect relaxes. Approximate and asymptotic expressions are necessary, since the complete one doesn't seem to exist in the closed form.

An important idea, on which the derivations in this article are based, is the distribution of elite species in the population, α, which is assumed Uniform. It does not seem to be the case, that this approach was used in EA literature before.

D. Wang and M. Reynolds (Eds.): AI 2011, LNAI 7106, pp. 361–371, 2011.

1.1 Royal Roads Function

Royal Roads (RR) is a test function introduced in [1] and analyzed in [2], where a population-based Evolutionary Algorithm (EA) was found to have underperformed a simpler heuristic Randomized Local Search (RLS), which contradicted the theoretical findings in the same article. An interesting feature of RR is a plateau of fitness (i.e. same fitness for a large subset of genotype).

The notation we use in this article is different from the one in [1,2]: a string length n is broken down into K consecutive bins numbered $\kappa_1, \kappa_2, \ldots, \kappa_K$ and the size/length of each bin is M (therefore $n = KM$). Fitness of each bin is equal to 0 if at least 1 bit is 0 and M if all bits are equal to 1.

Therefore, fitness of the string/parent is $0, M, 2M ... n$. Additional notation is presented in Section 3.

1.2 Past Work

Unlike OneMax (Counting Ones), RR has seen less attention in EA literature, though in [3,4] a variant of RR was analyzed and the upper bound of $O(n^6)$ was found for a version of RR in [4]. In [2] the bounds on convergence for RR were found to be $O(2^K \log N)$, N being the length of the string, K the length of schemata (length of the bin), up to a linear term tighter than for RLS ($O(2^K N \log N)$), although numerically RLS outperformed EA. Besides, this result is somewhat vague, since it doesn't involve population or recombination pool size in any way.

Most research in EA literature is focused on mutation-based single-species algorithms solving pseudo-boolean type functions, that includes OneMax and RR with some very sharp bounds derived for OneMax problem (e.g. $0.982 n \log n$ in [5]). Recently a number of recombination-based algorithms were analyzed in [6,7] for some cases where crossover can be provably effective.

Research on population-based algorithms (including EA) is more numerous, including that on OneMax test function, although $(1+1)$, $(\mu+1)$ and $(1+\lambda)$ setups are still more widespread. In [8] it was proven that the effect of population is problem-specific, i.e., increase in population size may not improve performance at all. Very recently, in [9], it was shown that population size $O(\log n)$ boosts performance and size $\Omega(\frac{n}{\log n})$ impairs the progress of the algorithm (on TrapZeros multimodal function) and reduces the probability of global convergence.

2 Analyzed Algorithm: $(\mu + \lambda)\mathrm{EA}_{1BS}$

k-Bit-Swap genetic operator (KBS) was introduced in [10]. It contains some features of both mutation and uniform crossover and recombines information between two parents in a random manner. In this article we use 1-Bit-Swap (1BS), which picks exactly 1 bit from each parent uniformly at random.

Pseudocode for the analyzed algorithm is presented in Table 1 and is very simple both in outline and implementation.

Table 1. $(\mu + \lambda)\mathrm{EA}_{1BS}$

1	Initialize population size μ
	repeat for t generations:
2	select $\frac{\lambda}{2}$ pairs of parents from the population using Tournament selection
	repeat $\frac{\lambda}{2}$ times:
3a	select a bit at random in Parent 1
3b	select a bit at random in Parent 2
3c	swap values in the selected bits
4	after the recombination, keep α best species in the population,
	replace the rest with the best species from the pool

We use Tournament selection detailed below because it is fairly straightforward both in implementation and analysis.

- Select two species x_i, x_j uniformly at random
- if $f(x_i) = f(x_j)$, either x_i or x_j enters the pool at random
- else the species with better fitness enters the pool.

3 Model Setup and Assumptions

The main quantity we analyze in this article is the first hitting time of the global solution of the test problem:

$$\tau_A^{RR} = \min\{t \geq 0 : f(\alpha) = n\}$$

where A is the set of all possible populations that include a global solution. We want to find $\mathbf{E}\tau^{RR}_{(\mu+\lambda)EA_{1BS}}$, the expectation of this time parameter.

3.1 Improvement Process

We assume that each bin κ starts with an equal number of 0's and 1's, which implies that the starting fitness of the population is 0. To measure the progress of the algorithm we introduce, in addition to the fitness function, the auxiliary function OneMax (or counting Ones, for further reference see e.g. [11]) that we denote V_κ. Due to the nature of 1BS bins evolve in a sequence, i.e. two different bins cannot evolve at the same generation, therefore κ can be viewed as the 'active bin'.

We also assume that the starting auxiliary function of each bin, $\min V_\kappa = \frac{M}{2}$.

The successful event G is defined as evolution of at least one more elite species in the population. To avoid confusion, the number of bits equal to 0 left to swap in a bin we use l (they are numbered 0 through $\frac{M}{2}-1$), and the bins left to fill we use κ numbered 0 through $K-1$. We restrict our attention to elite pairs in the recombination pool, i.e. pairs in which both parents are currently-elite species. Following the process described in greater details in Section 2, the probability to select such a pair into a recombination pool is

$$P_{\text{sel}}(\alpha) = \frac{\alpha^2(\alpha + 2(\mu - \alpha))^2}{\mu^4} = \frac{(\alpha(2\mu - \alpha))^2}{\mu^4}$$

Having selected the pair, the probability that as a results of swapping bits between them, a better species evolves is

$$P_{\text{swap}} = \frac{2(\frac{M}{2} - l)(\frac{n}{2} + \frac{\kappa M}{2} + l)}{n^2} = \frac{(M - 2l)(n + \kappa M + 2l)}{2n^2}$$

This probability comes from the fact that we want to select any 0 in bin κ in one of the parents and 1 anywhere in the other parent. Obviously as the number of 1's in both parents keeps growing, this probability grows too. In Section 4.1 we also extensively use the probability of failure:

$$P_{\text{F}} = 1 - P_{\text{swap}}$$

3.2 Population and Elitism Assumptions

This is a very important part of the paper. We assume that each generation currently-elite species in the population are distributed uniformly:

$$\alpha \sim Uniform\Big(\frac{1}{\mu}\Big)$$

This is a static model, i.e. this distribution does not change throughout the run of the algorithm. We also assume that the rate of elitism (number of species saved for the next generation) is *high enough*, that is, high enough to keep all elite species. We expect this result to yield a type of a lower bound, because this probability distribution assigns relatively high values to very high sizes of elite species. Say, in a real run the probability of have μ elite species in the population is much lower than $\frac{1}{\mu}$.

4 Derivation of the Expectation of Convergence Time

We present three main results: exact, approximate and asymptotic. The latter two are necessary, since, the complete one doe s not exist in he closed form.

4.1 Exact Expression

We start with introducing the probability of failure to improve V_κ (see also Appendix B):

$$P(G_0) = \sum_{j=0}^{\frac{\lambda}{2}} P(G_0|H_j) \sum_{\alpha=1}^{\mu} P(H_j|\alpha)P(\alpha) \tag{1}$$

where H_j is j'th elite pair in the recombination pool λ, α is the number of elite species in the population μ.

The probability to fail to improve a bit in a bin given l improvements so far is

$$
\begin{aligned}
P(G_{0l}) &= \frac{1}{\mu}\sum_{j=0}^{\frac{\lambda}{2}}\Big(\frac{2n^2-(M-2l)(n+\kappa M+2l)}{2n^2}\Big)^j\binom{\frac{\lambda}{2}}{j} \\
&\cdot\sum_{\alpha=1}^{\mu}\Big(\frac{(\alpha(\alpha+2\mu(\mu-\alpha)))^2}{\mu^4}\Big)^j\Big(1-\frac{(\alpha(\alpha+2\mu(\mu-\alpha)))^2}{\mu^4}\Big)^{\frac{\lambda}{2}-j} \\
&= \frac{1}{\mu}\sum_{j=0}^{\frac{\lambda}{2}}P_{\mathrm{F}}^{j}\binom{\frac{\lambda}{2}}{j}\sum_{\alpha=1}^{\mu}(P_{\mathrm{sel}}(\alpha))^j(1-P_{\mathrm{sel}}(\alpha))^{\frac{\lambda}{2}-j} \\
&= \frac{1}{\mu}\sum_{\alpha=1}^{\mu}\sum_{j=0}^{\frac{\lambda}{2}}\binom{\frac{\lambda}{2}}{j}(P_{\mathrm{F}}P_{\mathrm{sel}}(\alpha))^j(1-P_{\mathrm{sel}}(\alpha))^{\frac{\lambda}{2}-j} \\
&= \frac{1}{\mu}\sum_{\alpha=1}^{\mu}(1-P_{\mathrm{sel}}(\alpha)P_{\mathrm{swap}})^{\frac{\lambda}{2}} \qquad (2)
\end{aligned}
$$

The last step is due to the Binomial expansion: $(a+b)^n = \sum_{k=0}^{n}\binom{n}{k}a^k b^{n-k}$. Therefore,

$$
P(G_l) = 1 - P(G_{0l}) = 1 - \frac{1}{\mu}\sum_{\alpha=1}^{\mu}(1-P_{\mathrm{sel}}(\alpha)P_{\mathrm{swap}})^{\frac{\lambda}{2}}
$$

Expected time until improving the fitness of a bin κ is the sum of improvements over all values of the auxiliary function:

$$
\mathbf{E}T_\kappa = \sum_{l=0}^{\frac{M}{2}-1}\frac{1}{P(G_l)} \qquad (3)
$$

and, finally, summing over all κ from 0 to $K-1$ we obtain (since G depends on both l and κ)

$$
\begin{aligned}
\mathbf{E}\tau_{(\mu+\lambda)EA_{1BS}} &= \sum_{\kappa=0}^{K-1}\sum_{l=0}^{\frac{M}{2}-1}\frac{1}{P(G_{l,\kappa})} = \sum_{\kappa=0}^{K-1}\sum_{l=0}^{\frac{M}{2}-1}\frac{1}{1-\frac{1}{\mu}\sum_{\alpha=1}^{\mu}(1-(\alpha(2\mu-\alpha))^2P_{swap})^{\frac{\lambda}{2}}} \\
&= \sum_{\kappa=0}^{K-1}\sum_{l=0}^{\frac{M}{2}-1}\frac{1}{1-\frac{1}{\mu}\sum_{\alpha=1}^{\mu}(1-\frac{(\alpha((2\mu-\alpha)))^2(M-2l)(n+\kappa M+2l)}{2\mu^4n^2})^{\frac{\lambda}{2}}} \qquad (4)
\end{aligned}
$$

This derivation quite clearly shows the benefit of the population size due to $\frac{1}{\mu}$ term in front of the sum over α and μ^4 in the denominator of this sum. Also, increase in the size of λ leads to reduction the probability of failure.

We test this expression numerically for different values of n, μ, λ (see Appendix C). Unfortunately, this expression does not seem to exist in closed form, so we instead go ahead with finding an approximation to it in the next subsection.

4.2 Approximate and Asymptotic Expressions

$$P(G_{0l}) = \frac{1}{\mu}\sum_{\alpha=1}^{\mu}\Big(1-\frac{(\alpha(\alpha+2(\mu-\alpha)))^2(M-2l)(n+\kappa M+2l)}{2\mu^4 n^2}\Big)^{\frac{\lambda}{2}}$$
$$= \frac{1}{\mu}\sum_{\alpha=1}^{\mu}\Big(1-\frac{(\alpha(2\mu-\alpha))^2(M-2l)(n+\kappa M+2l)}{2\mu^4 n^2}\Big)^{\frac{\lambda}{2}}$$
$$\leq \frac{1}{\mu}\sum_{\alpha=1}^{\mu} e^{-\frac{\lambda(\alpha(2\mu-\alpha))^2(M-2l)(n+\kappa M+2l)}{4\mu^4 n^2}} \approx \frac{1}{\mu}\int_1^{\mu} e^{-\left(\frac{\alpha(2\mu-\alpha)}{\sqrt{\gamma}}\right)^2} d\alpha \quad (5)$$

The last two steps in the summand were due to $\lim_{n\to\infty}(1-\frac{k}{n}) = e^{-k}$ and the monotone nature of the summand. Note that $\gamma = \frac{4\mu^4 n^2}{\lambda(M-2l)(n+\kappa M+2l)}$, and, assuming that $\mu = \lambda$, the upper bound on γ is $\frac{4\mu^4 n^2}{\lambda(M-2l)(n+kM+2l)} < \frac{4\mu^4 n^2}{3\lambda n} = O(\mu^3 n)$. although for monotonically decreasing functions, like the one we have got, by the integral test the sum is larger than the corresponding integral, for $\mu << n$ the sum is closely approximated by the integral.

Denote $I_1 = \int_1^{\mu} f(\alpha)d\alpha = \int_1^{\mu} e^{-\left(\frac{\alpha(2\mu-\alpha)}{\sqrt{\gamma}}\right)^2} d\alpha$. Expanding the integrand in Taylor series around $\alpha_0 = 1$ up to the second term, we get (since $f'(\alpha_0) = -\frac{4(2\mu-1)(\mu-1)}{e^{(\frac{2\mu-1}{\sqrt{\gamma}})^2}\gamma}$)

$$f(\alpha) \approx e^{-(\frac{2\mu-1}{\sqrt{\gamma}})^2} - \frac{4(2\mu-1)(\mu-1)(\alpha-1)}{e^{(\frac{2\mu-1}{\sqrt{\gamma}})^2}\gamma} \quad (6)$$

Therefore the integral turns into

$$I_1 = \int_1^{\mu} f(\alpha)d\alpha \approx \int_1^{\mu}\left(e^{-(\frac{2\mu-1}{\sqrt{\gamma}})^2} - \frac{4(2\mu-1)(\mu-1)(\alpha-1)}{e^{(\frac{2\mu-1}{\sqrt{\gamma}})^2}\gamma}\right)d\alpha$$
$$= e^{-(\frac{2\mu-1}{\sqrt{\gamma}})^2}(\mu-1)\Big[1-\frac{2(2\mu-1)(\mu-1)^2}{\gamma}\Big] \quad (7)$$

The probability of failure is approximately (with the assumptions specified above)

$$P(G_{0l}) \approx \frac{e^{-(\frac{2\mu-1}{\sqrt{\gamma}})^2}(\mu-1)\Big[1-\frac{2(2\mu-1)(\mu-1)^2}{\gamma}\Big]}{\mu}$$

Accordingly, probability of a successful swap is

$$P(G_l) \approx 1 - \frac{e^{-(\frac{2\mu-1}{\sqrt{\gamma}})^2}(\mu-1)\Big[1-\frac{2(2\mu-1)(\mu-1)^2}{\gamma}\Big]}{\mu}$$

Using the sum of expectations of Geometric random variables with different parameters (Appendix B), the expected time until filling a bin, i.e. improvement of the fitness function, is (we keep the γ substitution to simplify the notation)

$$\mathbf{E}T_\kappa = \sum_{l=0}^{\frac{M}{2}-1} \frac{\gamma}{\gamma - \gamma e^{-(\frac{2\mu-1}{\sqrt{\gamma}})^2} + 2(2\mu-1)(\mu-1)^2 e^{-(\frac{2\mu-1}{\sqrt{\gamma}})^2}} \tag{8}$$

We do two approximations here, first the Riemannian sums approximation to obtain $[0,1]$ bounds on the integral, and then expand the integrand in Taylor series with 2 terms around midpoint to obtain a good approximation of the integral. The Riemannian sums approximation is defined by

$$\lim_{n\to\infty} \sum_{j=1}^{n} f(x_j) = n \int_0^1 f(nx)dx + o(n)$$

and γ is transformed accordingly:

$$\gamma = \frac{4\mu^4 n^2}{(M - 2(\frac{M}{2}-1)l)(n + \kappa M + (\frac{M}{2}-1)l)}$$

and

$$I_2 = \int_0^1 \frac{\gamma dl}{\gamma - \gamma e^{-(\frac{2\mu-1}{\sqrt{\gamma}})^2} + 2(2\mu-1)(\mu-1)^2 e^{-(\frac{2\mu-1}{\sqrt{\gamma}})^2}} \tag{9}$$

therefore, the expected first hitting time until the evolution of the bin κ is (Taylor series expansion of the integrand is in the Appendix A due to its length).

$$\begin{aligned}\mathbf{E}T_\kappa &\approx \left(\frac{M}{2}-1\right) \int_0^1 \frac{\gamma dl}{\gamma - \gamma e^{-(\frac{2\mu-1}{\sqrt{\gamma}})^2} + 2(2\mu-1)(\mu-1)^2 e^{-(\frac{2\mu-1}{\sqrt{\gamma}})^2}} \\ &\approx \frac{4\mu^4 n^2 (\frac{M}{2}-1)}{\lambda(\frac{M}{2}+1)(\frac{M}{2}+n+\kappa M-1)\left[\frac{2(2\mu-1)(\mu-1)^2}{\sigma_1} + \frac{4\mu^4 n^2}{\lambda(\frac{M}{2}+1)\sigma_2} - \frac{4\mu^4 n^2}{\lambda(\frac{M}{2}+1)\sigma_2\sigma_1}\right]}\end{aligned} \tag{10}$$

where

$$\sigma_1 = e^{\frac{\lambda(2\mu-1)^2(\frac{M}{2}+1)\sigma_2}{4\mu^4 n^2}}$$

$$\sigma_2 = \frac{M}{2} + n + \kappa M - 1$$

It's easy to notice that σ_1 has a very interesting property (given $\mu = \lambda$):

$$\lim_{n\to\infty} e^{\frac{\lambda(2\mu-1)^2(\frac{M}{2}+1)(\frac{M}{2}+n+\kappa M-1)}{4\mu^4 n^2}} = \lim_{n\to\infty} e^{\frac{M(M+n+KM)}{\mu n^2}} = \lim_{n\to\infty} e^{\frac{M}{\mu n}+O\left(\frac{M^2}{\mu n^2}\right)} = 1$$

which means, that for sufficiently large values of n and μ the second and the third terms in the square brackets cancel each other out, and the first term is just $2(2\mu-1)(\mu-1)^2$.

Finally, summing over all κ, the number of bins in the string, we get the approximation of the convergence time of the $(\mu+\lambda)$ algorithm on RR test function.

$$\begin{aligned}
\mathbf{E}\tau^{RR}_{(\mu+\lambda)EA_{1BS}} &\approx \frac{2\mu^4 n^2(M-2)}{\lambda(M+2)(2\mu-1)(\mu-1)^2}\sum_{\kappa=0}^{K-1}\frac{1}{\frac{M}{2}+n-1+\kappa M} \\
&= \frac{2\mu^4 n^2(M-2)}{\lambda M(M+2)(2\mu-1)(\mu-1)^2}\cdot \\
&\left[\psi_0\left(\frac{\frac{M}{2}+n-1+M+KM}{M}\right)-\psi_0\left(\frac{\frac{M}{2}+n-1}{M}\right)\right] \\
&\approx \frac{2\mu^4 n^2(M-2)}{\lambda M(M+2)(2\mu-1)(\mu-1)^2}\log\left(1+\frac{2KM}{M+2n}\right)
\end{aligned} \tag{11}$$

where ψ_0 is a Digamma function (see e.g. [12, 13]). In the derivation of the asymptotic expression for this bound, all population-related terms cancel out (since $\mu=\lambda$ and both numerator and denominator have the highest term μ^4), and the order of convergence is

$$\mathbf{E}\tau^{RR}_{(\mu+\lambda)EA_{1BS}} = O\left(\frac{n^2\log\left(1+\frac{KM}{M+n}\right)}{M}\right) \tag{12}$$

which seems to be a result comparable to those available in literature covering fitness functions with plateaus of fitness (e.g. [8, 11, 4]). Nevertheless for small μ we show in Appendix D that the effect of the population is beneficial, but converges to a constant as $\mu\to\infty$, thus the effect relaxes with the growth of the population size.

5 Conclusions and Future Work

We have derived three expressions for convergence of an elitist $(\mu+\lambda)\mathrm{EA}_{1BS}$ on Royal Roads test function: exact, approximate and asymptotic. Both the exact and approximate expressions for the expected convergence time clearly show the benefit of increase in the population when the population is relatively small, asymptotically population effect is $O(1)$, which means that as the size keeps growing its effect relaxes.

An important assumption for the approximation of $\mathbf{E}\tau^{RR}_{(\mu+\lambda)EA_{1BS}}$ was that $\mu << n$, but we never specified the relation, unlike in [9]. This is something to look at in the future. Since the effect of the population is known to be problem-specific, we will be able to get good insights into it for unimodal functions with plateaus, such as Royal Roads.

Numerical results are consistent with our findings, with the computational results lower-bounded by theoretical. Since lower ratios of $\frac{\mu}{n}$ give sharper bounds, this may shed more light on the optimal population size for problems with function plateaus.

We have performed our analysis assuming Uniform distribution of elite species in the population, something noone seems to have done in EA literature before. This is a static approach to convergence (i.e. the distribution assumption does not change throughout the run of the algorithm). We would like to look at the dynamics of the elite species and their effect on the probability of success, $P(G_l)$ and expected convergence time.

References

1. Mitchell, M., Forrest, S., Holland, J.H.: The Royal Road for Genetic Algorithms: Fitness Landscapes and GA Performance. In: European Conference on Artificial Life, vol. 1, pp. 245–254 (1992)
2. Mitchell, M.: Introduction to Genetic Algorithms. Kluwer Academic Publishers (1996)
3. Watson, R.A., Jansen, T.: A Building Block Royal Road Where Crossover is Provably Essential. In: Genetic and Evolutionary Computing Conference (GECCO), pp. 1452–1459 (2007)
4. Storch, T., Wegener, I.: Real Royal Road Function for Constant Population Size. In: Cantú-Paz, E., Foster, J.A., Deb, K., Davis, L., Roy, R., O'Reilly, U.-M., Beyer, H.-G., Kendall, G., Wilson, S.W., Harman, M., Wegener, J., Dasgupta, D., Potter, M.A., Schultz, A., Dowsland, K.A., Jonoska, N., Miller, J., Standish, R.K. (eds.) GECCO 2003. LNCS, vol. 2724, pp. 1406–1417. Springer, Heidelberg (2003)
5. Doerr, B., Fouz, M., Witt, C.: Sharp Bounds by Probability-Generating Functions and Variable Drift. In: Genetic and Evolutionary Computing Conference (GECCO), pp. 2083–2090 (2011)
6. Doerr, B., Happ, E., Klein, C.: Crossover Can Provably be Useful in Evolutionary Computation (in press, 2011)
7. Kótzing, T., Sudholt, D., Theile, M.: How Crossover Helps in Pseudo-Boolean Optimization. In: Genetic and Evolutionary Computing Conference (GECCO), pp. 705–712 (2011)
8. He, J., Yao, X.: From an Individual to a Population: An Analysis of the First Hitting Time of Population-Based Evolutionary Algorithms. IEEE Transactions on Evolutionary Computation 6(5), 495–511 (2002)
9. Chen, T., Tang, K., Chen, G., Yao, X.: A large population size can be unhelpful in evolutionary algorithms (in press, 2011)
10. Ter-Sarkisov, A., Marsland, S., Holland, B.: The k-Bit-Swap: A New Genetic Algorithm Operator. In: Genetic and Evolutionary Computing Conference (GECCO), pp. 815–816 (2010)
11. Chen, T., He, J., Sun, G., Chen, G., Yao, X.: A New Approach for Analyzing Average Time Complexity of Population-Based Evolutionary Algorithms on Unimodal Problems. IEEE Transactions on Systems, Man, and Cybernetics B 39(5), 1092–1106 (2009)
12. Abramowitz, M., Stegun, I.: Handbook of Mathematical Functions. United States Government Printing Office (1965)
13. Graham, R.L., Knuth, D.E., Patashnik, O.: Concrete Mathematics: A Foundation for Computer Science. Addison-Wesley Publishing Company (1995)

A Taylor Series Approximation of the Integrand

We give the expression for the Equation 10 here due to its length. It's Taylor series expansion of the integrand around midpoint of the interval (0.5)

$$\phi(l) \approx \frac{4\mu^2 n^2}{\lambda(\frac{M}{2}+1)\sigma_1\sigma_3} + s_3\left(\frac{s_3\left\{\frac{\frac{s_1}{\sigma_2\sigma_4} - \frac{s_2}{\mu^4 n^2 \sigma_2(\frac{M}{2}+1)}}{\sigma_3} - \frac{16(M-2)}{\sigma_2\sigma_3^2(\frac{M}{2}+1)}\right\} - s_3\frac{(16M-32)\sigma_5}{\sigma_4\sigma_3^2} + \varphi_1}{(\frac{M}{2}+1)\sigma_1^2\sigma_3} \cdot \right.$$
$$\left. + \frac{4(M-2)\sigma_5}{\sigma_1\sigma_3^2\sigma_4}\right)(l - \frac{1}{2})$$

where

$$\sigma_1 = \frac{2(\mu-1)(\mu-1)^2}{\sigma_2} + \frac{4\mu^2 n^2}{\lambda(\frac{M}{2}+1)\sigma_3} - \frac{4\mu^4 n^2}{\lambda\sigma_2\sigma_3(\frac{M}{2}+1)}$$

$$\sigma_2 = e^{\frac{\lambda(2\mu-1)(\frac{M}{2}+1)\sigma_3}{4\mu^4 n^2}},\ \sigma_3 = \frac{M}{2} + n + kM - 1,\ \sigma_4 = \left(\frac{M}{2}+1\right)^2,\ \sigma_5 = n + kM - 2$$

$$s_1 = 16(M-2),\ s_2 = 4\lambda(2\mu-1)^2((M-2)(\frac{M}{2}+1) - \sigma_3(M-2)),\ s_3 = \frac{\mu^4 n^2}{\lambda}$$

$$\varphi_1 = \frac{2(2\mu-1)^3(\mu-1)^2(2M + 2n + 2kM - nM - kM^2 - 4)}{s_3\sigma_2}$$

B Probability Theory

To derive expressions in Section 4.1, we extensively used properties of independent Geometric RVs that are not identically distributed, which is also known as Coupon collector's problem (see e.g. [13]): if $X_i \sim Geom(p_i)$ it expectation is $\mathbf{E}[X_i] = \frac{1}{p_i}$. Therefore, if $Y = \sum_{i=1}^n X_i$, $\mathbf{E}Y = \sum_{i=1}^n \mathbf{E}[X_i] = \frac{1}{p_1} + \frac{1}{p_2} + \ldots + \frac{1}{p_n}$

For Equation 1 we use the Law of total probability twice: first, conditioning on H_j, then on α:

$$P(A) = \sum_{i=1}^m P(A|B_i)P(B_i) = \sum_{i=1}^m P(A|B_i)\sum_{j=1}^n P(B_i|C_j)P(C_j)$$

C Numerical Results to Verify Equation 4

Column $\tilde{\tau}_{(\mu+\lambda)EA_{1BS}}$ was obtained by running the algorithm with different parameters 20 times, each run was 2000 generation each. The earliest achievement of the global minimum for each run was saved and then averaged over.

The results are very consistent in terms of exposing the effect of the population growth and are sharper for smaller ratios of $\frac{\mu}{n}$. Like we expected in the Assumptions section, theoretical bounds obtained are optimistic due to higher probabilities of observing high numbers of elite species.

Table 2. Theoretical and computational bounds for $(\mu+\lambda)EA^{RR}_{1BS}$

n	K	M	μ	λ	$\mathbf{E}\tau^{RR}_{(\mu+\lambda)EA_{1BS}}$	$\tilde{\tau}_{(\mu+\lambda)EA_{1BS}}$
32	4	8	4	4	145	315.3077
			10	10	72.4	268.2195
			20	20	44.2	192.2917
			30	30	34.5	173.5625
64	8	8	4	4	570.625	612.46
			10	10	279.88	497.93
			20	20	153.46	454.4681
			30	30	112.297	372.04
128	16	8	4	4	2264.36	1365
			10	10	1048	1239
			20	20	570.44	1091.5
			30	30	401.99	949.4

D Effect of the Population

We rewrite Equation 11 in order to factor out terms involving μ, taking for simplicity $\mu=\lambda$.

$$\mathbf{E}\tau_{(\mu+\lambda)EA_{1BS}} = \varphi(n,M,K)\frac{\mu^4}{\lambda\mu(2\mu-1)(\mu-1)^2} = \varphi(n,M,K)\frac{\mu^3}{2\mu^3-5\mu^2+4\mu-1}$$

For small values of μ this expression lies between 0.5 and 1 and quickly converges to 0.5, so $\mathbf{E}\tau_{(\mu+\lambda)EA_{1BS}} = O(\varphi(n,M,K))$ because asymptotically

$$\frac{\mu^3}{2\mu^3-5\mu^2+4\mu-1} = O(1)$$

This explains the benefit of the growth of the population when it is small and its leveling out for larger values

Simplex Model Based Evolutionary Algorithm for Dynamic Multi-Objective Optimization

Jingxuan Wei[1,2] and Mengjie Zhang[2]

[1] School of Computer Science and Technology, Xidian University, Xi' an, China
[2] School of Engineering and Computer Science, Victoria University of Wellington, New Zealand
{Jingxuan.Wei,Mengjie.Zhang}@ecs.vuw.ac.nz

Abstract. Most real-world problems involve objectives, constraints and parameters which constantly change with time. Treating such problems as static problems requires knowledge of the prior time but the computational cost is still high. In this paper, a simplex model based evolutionary algorithm is proposed for dynamic multi-objective optimization, which uses a modified simplex model to predict the optimal solutions (in variable space) of the next time step. Thereafter, a modified evolutionary algorithm which borrows ideas from particle swarm optimization is applied to solve multi-objective problems when the time step is fixed. This method is tested and compared on a set of benchmarks. The results show that the method can effectively track varying Pareto fronts over time.

Keywords: Simplex Model, PSO, Multi-objective Optimization.

1 Introduction

Dynamic multi-objective optimization problems (DMOPs) usually involve objective functions, constraints and parameters which change with time. These problems often arise from real-world problems solving, particularly in optimal control problems or problems requiring on-line optimization. The nature of a time-changing based dynamic optimization problem is concerned with tracking the moving optima. Two basic concepts of this kind of problem are the frequency and severity of a change [2]. If the severity of a change is large, the problem will become less relevant to the next time step and it requires a re-start of the algorithm. Frequency is a constant throughout a fixed time step, but it varies with different problems.

The performance of EAs in such dynamic optimization problems has been studied mostly in the past decade, and many enhancement techniques have been proposed. A large part of the existing algorithms can be classified into two categories. The first category handles two basic functions of an algorithm: diversity for exploring the search space for the locations of the next optimal solutions and convergence to the current global optima. These two functions usually compete with each other, and a balance is needed between them. The multi-population scheme [8] is an example for the diversity control. Other approaches, such as

D. Wang and M. Reynolds (Eds.): AI 2011, LNAI 7106, pp. 372–381, 2011.

elitism based immigrants method [9] are for the convergence. The Second category of approaches is concerned with exploiting the past information which might be useful for the problem solving. A typical example is the memory-based approach [5], which employs an extra memory to store useful information to guide further search.

Recently, Deb et al. proposed two versions DNSGAII-A and DNSGAII-B [2] to deal with DMOPs. In the former, some randomly generated individuals are inserted into the population of the next time step when a change happens, while in the latter version, a small portion of population is replaced with mutated solutions. The main limitations of the two versions are: how close or how far of the next optimal solutions to the current optimal solutions is not considered, and the two algorithms become less effective if some complex constraints are added to the dynamic problems.

To address the above issues, the goal of the paper is to develop a simplex model based evolutionary algorithm for dynamic multi-objective problems (DMOPs). We expect the proposed algorithm to track varying Pareto fronts effectively and the optimal solutions obtained at each time step to be widely distributed along the true Pareto front.

2 Background

2.1 Problem Definition

Without loss of generality, a dynamic constrained multi-objective optimization problem can be described as Eq.1:

$$\begin{cases} \min & f(x,t) = \{f_1(x,t), f_2(x,t), ..., f_m(x,t)\} \\ \text{s. t.} & g_j(x,t) \leq 0, j = 1, \cdots, p \\ & h_j(x,t) = 0, j = 1, ..., s \\ & x \in [L, U] \end{cases} \tag{1}$$

where $t \in [a,b]$ is time variable, $x = (x_1, x_2, ...x_n)$ is the decision vector, $g_j(x,t)(j = 1,2,...,p)$ are inequality constraints. All equality $h(x,t)$ can be converted into inequality constraints. All of these constraints depend on time variable t. $[L,U] = \{x = (x_1, x_2, ...x_n) | l_i \leq x_i \leq u_i, i = 1, 2, ..., n\}$ is the search space. Constraint violation at time t is defined as:

$\Phi(x,t) = \sum_{j=1}^{p} \max(0, g_j(x,t))$. A vector $\mu = (\mu_1, \mu_2, ...\mu_n)$ is said to dominate a vector $\nu = (\nu_1, \nu_2, ...\nu_n)$ (denoted as $\mu \prec \nu$) if: $\forall i \in \{1, 2, ...m\}, f_i(\mu, t) \leq f_i(\nu, t) \bigwedge \exists j \in \{1, 2, ...m\}, f_j(\mu, t) < f_j(\nu, t)$. A solution x is called a Pareto optimal solution for problem (1) at a fixed time t if $\Phi(x,t) = 0$ and $\sim \exists \widetilde{x} \in [L,U]$ such that $\Phi(\widetilde{x}, t) = 0$ and $\widetilde{x} \prec x$.

2.2 Performance Criteria

To compare different approaches in a dynamic environment, it is not sufficient to only compare the best solutions found so far, because the optimal solutions change with time. A reasonable alternative is to use a set of offline performance criteria, which average over the best solutions' performance at each time step. We use modified C-metric, FS-metric, and S-metric to evaluate the performance of different approaches in convergence, diversity and distribution. All these criteria are well-known and designed by Zitzler and Thiele [3]. If two algorithms A and B are executed, two series of Pareto optimal solution sets are acquired, $\{A^t\}_{t=1}^T$ and $\{B^t\}_{t=1}^T$, where T is the total numer of the time steps.

C-metric: $C(A,B) = \frac{1}{T}\sum_{t=1}^{T} \frac{|\{y \in B^t : \exists x \in A^t, s.t. x \prec y\}|}{|B^t|}$, which is used to measure the average convergence rate. The equation shows that the quality of the Pareto optimal solution sets by algorithm B has better convergence than those by A if $C(A,B) < C(B,A)$.

FS-metric: $FS = \frac{1}{T}\sum_{t=1}^{T} \sqrt{\sum_{i=1}^{m} \max_{(x_0,\, x_1) \in A^t \times A^t} \{(f_i(x_0) - f_i(x_1))^2\}}$, which is used to measure the average coverage range of the Pareto fronts.The larger the value of FS, the better diversity of solutions on the Pareto front will be.

S-metric: $S = \frac{1}{T}\sum_{t=1}^{T}[\frac{1}{n_{PF}}\sum_{i=1}^{n_{PF}}(d_i^{'} - \overline{d^{'}})^2]^{1/2}$, $\overline{d^{'}} = \frac{1}{n_{PF}}\sum_{i=1}^{n_{PF}} d_i^{'}$, which is used to measure the average distribution of the Pareto optimal solutions. n_{PF} is the number of the solutions in the found Pareto front at time step t, $d_i^{'}$ is the distance (in the objective space) between the member y_i and its nearest member in the found Pareto front at a fixed time step t . The smaller the value of S-metric, the more uniformity the Pareto front will be.

2.3 Previous Work

During the last decade, a number of different EAs were suggested to solve dynamic optimization problems. This section reviews the approaches for DMOPs.

The challenge for DMOPs is to track a set of Pareto optimal solutions rather than one solution only, and we expect the solutions to achieve a better performance in convergence, diversity and distribution. In [4], an artificial immune system is proposed to solve DMOPs. In [7], a direction based algorithm is used to solve DMOPs, which is a direct extension of static MOPs. In this algorithm, each objective function is minimized using a hybrid evolutionary-deterministic strategy, a change detecting strategy is added to the direction based algorithm, and the Pareto optimal solutions from previous time are directly used as the initial population of the next time. Although these algorithms can solve the DMOPs successfully to some extent, there still are some limitations: how far the current optimal solutions to the next optimal solutions is not considered, and how to handle constraints in DMOPs is not considered.

3 A New Simplex Model Based Evolutionary Algorithm for DMOPs

The new strategies proposed in this work consist of using a modified simplex model to predict the locations of the next optimal solutions and using a PSO-based algorithm to find the exact optimal solutions of the next time step. Before we describe the two new strategies, we first outline the overall algorithm.

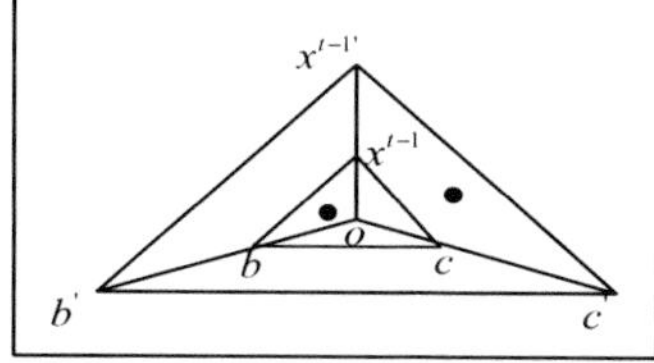

● optimal solution of the next time step

x^{t-1} : one optimal solution at t-1

b, c : are two solutions generated from x^{t-1} :

$x^{t-1'}, b', c'$: are solutions generated from x^{t-1}, b, c

o : is the center of the both triangles

Fig. 1. Relationship between current optimal solution x^{t-1} and the optimal solution of the next time, 2D space

At the end of time step $t-1$, two optimal solutions x_1^{t-1}, x_2^{t-1} which are located in the boundary of the Pareto front are used as inputs.

Step1. Use a modified simplex model to generate solutions around the input optimal solutions. All of these solutions constitute the prediction solution sets, (section 3.1).
Step2. Insert the prediction solution sets to the population of the next time step and run the PSO-based evolutionary algorithm for a fixed function evaluations and get the new optimal solutions for the next time step (section 3.2).
Step3. Update the information of the current optima, then go to step1.

Note that each input solution is used to construct a simplex model and get a prediction solution set for the next time step. Thus, two prediction solution sets are obtained. The total population at the beginning of time step t is composed of two parts: prediction solution sets for placing a team of solutions in the neighborhood of the next optimal solutions to achieve a fast convergence, and the randomly generated solutions to aid the discovery of the optimal solutions in case the prediction is unsuccessful.

3.1 A Modified Simplex Model for Prediction

The use of the modified simplex model is to make a prediction for the optimal solutions of the next time step. The main differences between the modified model and the existing one are: (1) instead of using a simplex model as a crossover operator [11], we use it to approximate the locations of the next optimal solutions; (2) in [11], some random solutions (or solutions chosen according to a probability) are used to form a simplex model, while we use the gradient information of objective functions to generate new points.

In a general case, there are two relationships between the optimal solution at $t-1$ and the optimal solution of the next time step, as shown in Fig. 1: (1) the optimal solution of the next time step is within the inner triangle $x^{t-1}bc$, which is created by the optimal solution x^{t-1}; (2)the optimal solution of the next time step is within the inner triangle $x^{t-1'}b'c'$. The modified simplex model is described in the following algorithm, where one of the optimal solutions, denoted as x^{t-1} from the previous time step, is used as the input.

Step1. Generate b and c from the optimal solution x^{t-1} according to the following rules: for each objective function $f_i(x, t-1)$, $i = 1, 2, ..., m$, we calculate the approximate gradient value of $f_i(x, t-1)$ at x^{t-1} and denote it as:

$$\nabla f_i(x^{t-1}, t-1) = (\nabla f_{i1}(x^{t-1}, t-1), ..., \nabla f_{in}(x^{t-1}, t-1)),$$

where $\nabla f_{ij} = \frac{f_i(x^{t-1}+\delta e_j, t-1) - f_i(x^{t-1}, t-1)}{\delta}$, and e_j is unit vector that the jth variable equals one and the other variables equal zero, $j = 1, 2, ..., n$. n is the number of dimensions of vector x. δ is a small positive number.
Step2. Let $d_i = -\nabla f_i(x^{t-1}, t-1)$, $i = 1, 2, ..., m$. We search along d_i from solution x^{t-1} with a small step ξ, then two solutions b, c are generated according to the following equations: $b = x^{t-1} + \xi d_1$ and $c = x^{t-1} + \xi d_2$.
Step3. Calculate the center of the three solutions, $o = \frac{x^{t-1}+b+c}{3}$.
Step4. Generate the points $x^{t-1'}, b', c'$. For example, $b' = (b-o) \times (1+\varepsilon)$, where $\varepsilon > 0$ is the expanding rate. $x^{t-1'}$ and c' are created in the same way.
Step5. Output the prediction solution set, which is composed of $x^{t-1'}, b', c'$ and x^{t-1}, b, c.

It is worth noting that totally $2 \times (m+1)$ solutions will be created by the above strategies, namely the size of each prediction set is $2 \times (m+1)$. We let two parameters $\xi = \varepsilon = 0.5$.

3.2 PSO Based Evolutionary Algorithm for DMOPs

The PSO based evolutionary algorithm is aiming at making a balance between diversity and convergence and we expect it to aid the simplex model to find the optimal solutions of the next time step. The main difference between our algorithm and the commonly used EAs is that we use a PSO based crossover operator to generate offsprings, aiming at handling constraints appeared in DMOPs.

PSO Based Crossover Operator. Crossover operator is often used to generate the offsprings that inherit the best information from parents. The arithmetic crossover operator, uniform crossover operator and simplex crossover operator [11] are often used due to their effectiveness in keeping population's diversity. However, most crossover operators can only deal with the aspect of how to keep diversity or how to make the algorithm converge faster, but rarely consider both aspects. Since developed by Kennedy and Eberhart in 1995 [6], PSO has attracted a high level of interest due to the easy implementation. In essence, the trajectory of each particle is updated according to its own best position *pbest* and the global best position *gbest*. If we use a good principle to choose *pbest* and *gbest*, a balance between diversity and convergence can be maintained. The reason is that *gbest* can be seen as a factor related to convergence and *pbest* can be seen as a factor related to diversity. Based on the idea, we modify the PSO velocity updating equation and use it as a crossover operator. The difference between the standard PSO velocity equation and the proposed crossover operator is that we add one item to the standard velocity equation, which is used to handle constraints. Suppose x_c is the solution that is chosen to undergo crossover, and x_c' is the offspring, then:

$$x_c' = \omega x_c + c_1 rand_1(pbest_c(k) - x_c(k)) + c_2 rand_2(gbest(k) - x_c(k)) + (1 - \frac{k}{k_{max}})(pbest_f - x_c(k)) \quad (2)$$

where, ω, c_1, c_2 are the parameters that are the same as the standard PSO, k is the generation number for a fixed time step, $gbest(k)$ is a solution randomly selected from the current Pareto optimal solution set at generation k for a fixed time step, and $pbest_c(k)$ is the best solution of x_c found so far (up to generation k at a fixed time step). For comparing two solutions, Pareto constraint dominance principle [1] is directly used to choose *pbest*. The last part is the most important, which makes the algorithm search a larger region so that the diversity is kept. $pbest_f$ is chosen from a predefined infeasible solution set [10], which preserves the solutions that are infeasible but have better rank values than the current optimal solutions (see section 2.1 for the concept of infeasible solution and rank value). k_{max} is the maximum generation for a fixed time step. The weight parameter for the last item $(1 - \frac{k}{k_{max}})$ is a function of current generation number k. At the beginning of evolution, the weight value is large so that a large region is explored, while at the late stage of evolution, the weight value is close to zero, then more offsprings are located in a small region so that convergence is guaranteed.

PSO Based Evolutionary Algorithm for DMOPs. At time step t, we use a PSO based EA to evolve the Pareto front, which is outlined as follows.

Step1. Use prediction solution sets from section 3.1 and some randomly generated solutions to constitute the initial population of time step t. The Pareto optimal solution set at generation k for this time step is denoted as PF_k. Let $k = 1$.
Step2. Select individuals for crossover according to crossover probability p_c and use the PSO based crossover operator to generate offsprings.
Step3. Select individuals from the crossover offsprings for mutation according to mutation probability p_m. For each selected individual, say $X = (x_1, x_2, ..., x_n)$ randomly change it to another individual $X = X + \Delta X$, where each component of ΔX is generated by a random number generator using Gaussian distribution.
Step4. Update the Pareto optimal solution set PF_k by crossover and mutation.
Step5. Select the next generation population by using the constraint Pareto dominance principle [1].
Step6. If the maximum generation number k_{max} is reached, then stop and output PF_k as the optimal solutions at time step t; otherwise, go to step2.

4 Experiment Design

To examine the performance of the proposed algorithm, we compare it with two well known algorithms DNSGAII-A and DNSGAII-B on three benchmarks DCTP1, DCTP2 and DCTP3 [4]. The parameter settings refer to [4].

4.1 Test Functions

$$DCTP1: \begin{cases} \min\ f_1(x,t) = x_1 \\ \min\ f_2(x,t) = c(x,t)exp(-f_1(x,t)/c(x,t)) \\ s.t.\ g_1(x,t) = f_2(x,t) - 0.858\exp(-0.541 f_1(x,t)) \geq 0 \\ s.t.\ g_2(x,t) = f_2(x,t) - 0.728\exp(-0.295 f_1(x,t)) \geq 0 \end{cases}$$

$x = (x_1, x_2, ..., x_5)$, $0 \le x_1 \le 1$, $-5 \le x_i \le 5$, $i = 2, 3, 4, 5$.

$$DCTP2 \sim DCTP3: \begin{cases} \min\ f_1(x,t) = x_1 \\ \min\ f_2(x,t) = c(x,t)(1 - \frac{f_1(x,t)}{c(x,t)}) \\ s.\,t.\ g_1(x,t) = \cos(\theta)(f_2(x,t) - e) - \sin(\theta) f_1(x,t) \ge \\ a|\sin(b\pi(\sin(\theta)(f_2(x,t) - e) + \cos(\theta) f_1(x,t))^c)|^d \end{cases}$$

$x = (x_1, x_2, ..., x_5)$, $0 \le x_1 \le 1$, $-5 \le x_i \le 5$, $i = 2, 3, 4, 5$. For $DCTP1 \sim DCTP3$, $c(x,t) = 1 + \sum_{i=1}^{5}(x_i - sin(0.05\pi t))^2$. We set $t \in [1, 4]$ and divide it into four equal time steps, namely, $t_1 = 1, t_2 = 2, t_3 = 3$ and $t_4 = 4$.

For DCTP1, since the Pareto front at each time step is a part of the constraint boundary, it becomes more difficult to find a considerable number of optimal solutions. DCTP2 is an extremely complex problem, as in each time step, the Pareto front consists of a number of disconnected optimal fronts. DCTP3 is even more difficult because each of the disconnected feasible regions only contains one Pareto optimal solution for each time step.

4.2 Parameter Settings

For the proposed algorithm, we record the results of the four time steps for each function. Each algorithm is run for a maximal 200 generations at each time step. All of these algorithms are required to execute 30 independent runs for each test function. For a single run, each algorithm performs once at each of the four time steps. For the proposed algorithm, the crossover probability is 0.9, and the mutation probability is 0.2 (same as DNSGAII); inertia weight $\omega = 0.4$, c_1 and c_2 are two random numbers in the range $[0, 1]$. These parameter values are the same as the standard MOPSO [6].

5 Simulation Results

5.1 Overall Results

Table 1 summarizes the statistic results of the Pareto optimal solutions obtained by the three algorithms. For brevity, our algorithm is denoted as PEA, DNSB and DNSA denote the algorithms DNSGAII-B and DNSGAII-A respectively and NA denotes the value is unavailable. AC(A,B) is the mean of the 30 coverage rates C(A,B). FS denotes the average coverage range of the Pareto fronts obtained by an algorithm at different time steps. S is used to measure the average distribution of the Pareto optimal solutions gained at different time steps (see section 2.2).

For DCTP1, all of those three algorithms almost have the same convergence performance as the AC values have a little distinction among them. For example, AC(DNSGAII-B,PEA)=27 while AC(PEA,DNSGAII-B)=25, which means that 27 Pareto optimal solutions in PEA are dominated by DNSGAII-B while 25 optimal solutions in DNSGAII-B are dominated by PEA. By comparing FS-metric, we can find that the FS-metric value found by our algorithm is bigger than those of DNSGAII-B and DNSGAII-A. For example, the mean value of FS for the proposed algorithm is 1.1766, while that corresponding value is 1.1333 for

Table 1. Comparisons of the statistic results of the Pareto optimal solution sets

Prob.	AC(A,B) %			FS				S			
	PEA	DNSB	DNSA	Best	Worst	Mean	Std	Best	Worst	Mean	Std
DCTP1											
PEA	0	25	21	1.2645	1.0804	1.1766	0.0658	0.0053	0.0240	0.0097	0.0080
DNSB	27	0	NA	1.1681	1.0892	1.1333	0.0283	0.00534	0.0084	0.0064	0.0012
DNSA	29	NA	0	1.1962	1.0745	1.1344	0.0625	0.0054	0.0068	0.0062	0.0006
DCTP2											
PEA	0	31	29	1.4653	1.3428	1.4073	0.0483	0.0079	0.0464	0.0168	0.0166
DNSB	20	0	NA	1.4164	1.2372	1.3287	0.0682	0.0061	0.1534	0.0358	0.0657
DNSA	19	NA	0	1.3722	1.3451	1.3610	0.0114	0.0051	0.0170	0.0091	0.0056
DCTP3											
PEA	0	52	72	1.3318	1.0591	1.2236	0.1011	0.0727	0.1288	0.0868	0.0240
DNSB	39	0	NA	1.3119	0.8261	1.0034	0.1967	0.0607	0.0963	0.0806	0.0140
DNSA	17	NA	0	1.0150	0.9123	0.9788	0.0534	0.0423	0.0830	0.0571	0.0183

DNSGAII-B and 1.1344 for DNSGAII-A. This suggests that the new algorithm can find good solutions that are more widely spread along the Pareto front than the other two methods. Regarding S-metric, the mean value of the proposed algorithm is bigger than the corresponding value for DNSGAII-B and DNSGAII-A. The results suggest that the uniformity of the proposed algorithm is not as good as those by DNSGAII-B and DNSGAII-A.

DCTP2 is a more difficult example than DCTP1, as the Pareto fronts are disconnected at different time steps. The proposed PEA has a better convergence to the true Pareto front compared with DNSGAII-B and DNSGAII-A as only a small number of optimal solutions obtained by PEA are dominated by DNSGAII-B and DNSGAII-A, respectively. In addition, PEA has a better performance than DNSGAII-B and DNSGAII-A in terms of the widespread of the Pareto optimal solutions due to the big mean value of FS-metric. Furthermore, the optimal solutions obtained by PEA are more uniformly distributed than those by DNSGAII-B due to the small mean S-metric value.

DCTP3 is the most difficult example because each of its disconnected Pareto front only contains one solution. In this case, our algorithm has the best performance in the aspect of convergence, diversity and distribution. For example, the proposed algorithm has the biggest FS-metric value, suggesting that the optimal solutions are widespread along the Pareto front. Overall, the results suggest that our algorithm can find a widespread Pareto front regardless of the shape of the Pareto fronts and the obtained Pareto front has a better convergence to the true Pareto front, particularly for difficult cases. In addition, we find that the proposed algorithm is significantly better than the compared ones in terms of FS-metric based on a standard T-test.

5.2 Evolved Pareto Fronts

Fig.2 shows some example Pareto fronts evolved by the different algorithms for the three test functions at a fixed time step. It can be seen that the proposed PEA can almost converge to the true Pareto front and evolve a diverse Pareto front at the fixed time step for all test functions.

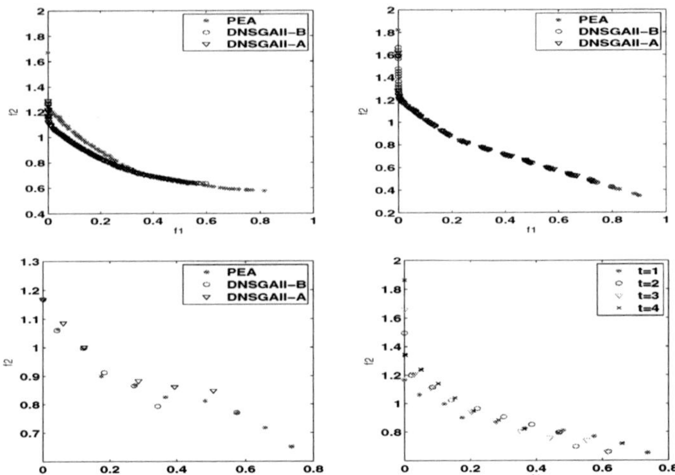

Fig. 2. Evolved Pareto fronts for (1)DCTP1: at $t = 2$ (first row left), (2)DCTP2: at $t = 3$ (first row right), (3)DCTP3: at $t = 1$ (second row left), (4)DCTP3: Pareto fronts at all time steps obtained by PEA (second row right). [in colour, see PDF]

For DCTP1, the proposed PEA can find solutions distributed in the region $f_1 \in [0, 0.8]$ and $f_2 \in [0.60, 1.6]$, while both DNSGAII-B and DNSGAII-A can only find a small region $f_1 \in [0, 0.6]$ and $f_2 \in [0.65, 1.2]$. It is worth noting that the convergence of DNSGAII-B and DNSGAII-A is better than PEA at $t = 2$, however the average convergence rate AC of PEA is almost the same as the compared ones (see Table.1).

For DCTP2, PEA can find the optimal solutions in the larger range $f_1 \in [0, 1]$ and $f_2 \in [0.4, 1.8]$. However, DNSGAII-A and DNSGAII-B can not find the solutions in $f_1 \in [0.8, 1]$. In addition, from Table 1, we can see that the convergence of PEA is much better than DNSGAII-A and DNSGAII-B. Hence, the optimal solutions found by PEA are highly competitive in terms of convergence and diversity at $t = 3$.

For DCTP3 (the most difficult example, only a few solutions can be found), the proposed PEA can find 10 optimal solutions, while DNSGAII-B can only find 7 solutions and the solutions found by DNSGAII-A are worse than those of PEA in the aspect of convergence (second row of Fig.2, left). We can also see that for DCTP3, the proposed PEA can track the varying Pareto fronts and the optimal solutions are widely spread along the Pareto front at each time step (second row of Fig.2, right).

Overall, Fig.2 shows that the proposed PEA can evolve a diverse optimal solution set at the fixed time steps and those solutions have better convergence than DNSGAII-A and DNSGAII-B.

6 Conclusions and Future Work

The goal of this paper was to investigate a simplex model based evolutionary algorithm for DMOPs. The goal was successfully achieved by developing a modified simplex model to approximate the optimal solutions of the next time step and a PSO based crossover operator to handle constraints in a dynamic environment. The results show that the proposed PEA has better performance than DNSGAII-A and DNSGAII-B in terms of the average coverage rate, average coverage range and the average distribution.

In the future work, we will prove the convergence of the proposed algorithm and perform more experiments to analyze the two new developments.

Acknowledgments. The work was supported in part by the Fundamental Research Funds for the Central Universities (No.72005502), China and by the URF (101154/2880) at Victoria University of Wellington.

References

1. Deb, K., Pratap, A., Agarwal, S., Meyarivan, T.: A fast and elitist multi-objective genetic algorithm: NSGAII. IEEE Transactions on Evolutionary Computation 6, 182–197 (2003)
2. Deb, K., Rao, U.B.N., Karthik, S.: Dynamic Multi-objective Optimization and Decision-Making Using Modified NSGA-II: A Case Study on Hydro-thermal Power Scheduling. In: Obayashi, S., Deb, K., Poloni, C., Hiroyasu, T., Murata, T. (eds.) EMO 2007. LNCS, vol. 4403, pp. 803–817. Springer, Heidelberg (2007)
3. Zitzler, E., Thiele, L.: Multiobjective evolutionary algorithms: a comparative case study and the strength Pareto approach. Evolutionary Computation 3, 257–271 (1999)
4. Zhang, Z.H., Qian, S.Q.: Artificial immune system in dynamic environments solving time-varying non-linear constrained multi-objective problems. Soft Computing 7, 1333–1349 (2011)
5. Hatzakis, I., Wallace, D.: Dynamic multi-objective optimization with evolutionary algorithms: a forward-looking approach. In: Proceedings of the Genetic and Evolutioanry Computation Conference, GECCO, pp. 1201–1208. ACM Press (2006)
6. Kennedy, J., Eberhart, R.: Particle swarm optimization. In: Proceedings of the IEEE International Conference on Neural Networks, pp. 1941–1948. IEEE Press (1995)
7. Farina, M., Deb, K., Amato, P.: Dynamic mutiobjective optimization problems: test cases, approximations, and applications. IEEE Transactions on Evolutionary Computation 8, 425–442 (2004)
8. Bui, L.T., Abbass, H.A., Brabke, J.: Multiobjective optimization for dynamic environments. In: Proceedings of the IEEE Congress on Evolutionary Computation, CEC, pp. 2349–2356. IEEE Press (2005)
9. Yang, S.: Genetic algorithms with memory and elitism-based immigrants in dynamic environments. Evolutionary Computation 16, 385–416 (2008)
10. Wei, J.X., Zhang, M.J.: A memetic particle swarm optimization for costrained multi-objective optimization problems. In: Proceedings of the IEEE Congress on Evolutionary Computation, CEC, pp. 45–53. IEEE Press (2011)
11. Tsutsui, S., Yamamura, M., Higuchi, T.: Multi-parent recombination with simplex crossover in real coded genetic algorithm. In: Proceedings of the Genetic and Evolutioanry Computation Conference, GECCO, pp. 657–664. ACM Press (1999)

Color Quantization Using Modified Artificial Fish Swarm Algorithm

Danial Yazdani[1], Hadi Nabizadeh[1],
Elyas Mohamadzadeh Kosari[2], and Adel Nadjaran Toosi[3]

[1] Department of Electronic, Computer and Information Technology,
Azad University of Qazvin, Iran
{d_yazdani,h_nabizadeh}@qiau.ac.ir
[2] Department of Computer Engineering, Faculty of Engineering,
Ferdowsi University of Mashhad, Mashhad, Iran
elyas.kosari@stu-mail.um.ac.ir
[3] Department of Computer Science and Software Engineering,
University of Melbourne, Melbourne, Australia
adeln@csse.unimelb.edu.au

Abstract. Color quantization (CQ) is one of the most important techniques in image compression and processing. Most of quantization methods are based on clustering algorithms. Data clustering is an unsupervised classification technique and belongs to NP-hard problems. One of the methods for solving NP-hard problems is applying swarm intelligence algorithms. Artificial fish swarm algorithm (AFSA) fits in the swarm intelligence algorithms. In this paper, a modified AFSA is proposed for performing CQ. In the proposed algorithm, to improve the AFSA's efficiency and remove its weaknesses, some modifications are done on behaviors, parameters and the algorithm procedure. The proposed algorithm along with other multiple known algorithms has been used on four well known images for doing CQ. Experimental results comparison shows that the proposed algorithm has acceptable efficiency.

Keywords: Color quantization, compression, artificial fish swarm algorithm, data clustering.

1 Introduction

One of the available challenges in image processing is high color variety in pixels. Therefore, usually a decreasing technique of color variety is used as a preprocessing for different works in graphic and image processing applications. By decreasing the number of colors, it can decrease image file size to conserve storage space, reduce time for transmission, and reduce computation. Color Quantization (CQ) is one of the most famous techniques of decreasing the numbers of colors which is applied in image compression [1], graphic [2] and image processing [3,4].

CQ process is done in two steps [5]. In the first step, a codebook is constructed. In this step, it has to be determined how many colors have to be decreased at first. In

D. Wang and M. Reynolds (Eds.): AI 2011, LNAI 7106, pp. 382–391, 2011.

fact, the number of considered colors is the number of codewords in the codebook. Each codeword represents a color and its index in the codebook that each of these codewords is representative of multiple colors on the original image. After constructing the codebook by means of a CQ method, the image is encoded by that and every pixel just would possess index number of its representative color in the codebook. On the second step, each image is decoded by its corresponding codebook. From the view of compression, CQ is taken into account as a lossy compression method in which some information are lost. Indeed, after performing CQ on an image, most of pixels cannot have their primary colors anymore [5].

Natural images often have many colors and determining a proper codebook is a challenging problem for decreasing the colors. Generally, CQ techniques could be categorized into two classes: first category consists of those methods independent from image, which specify a comprehensive codebook regardless of any specific image. In these methods, first, a fixed codebook is produced by using a training set, and then all images are encoded and decoded by means of this codebook. Second category contains dependent techniques to image, in which for every image a codebook is built. Usually, in these methods, every codebook is built based on color distribution in a specific image. Thereafter, to use it for decoding the image, the codebook with encoded image is transferred. Dependent image techniques are slower than independent image techniques, but obtained results from the former have higher quality than the latter [6].

In CQ, the main goal is to obtain an appropriate codebook. If the codebook is not proper, the resulted image from CQ has much disharmony with the original image and distortion increases between the original image and decoded one. One of the applied approaches for producing codebook based on the color distribution is using clustering algorithms like k-means [6] and FCM [7]. Clustering is an unsupervised classification technique in which datasets that are usually data vectors in multi dimensional space. Data vectors are divided into some clusters based on a similarity criterion. After performing a clustering algorithm, each of data vectors of dataset is assigned to one of clusters. Clustering process is done with respect to a specific similarity criterion such that assigned data to a cluster are more similar than other data in other clusters. The way of using clustering algorithms in CQ is such that, first, colors histogram is produced for the original image and after that, clustering according to the color distribution among pixels is done. The number of cluster centers in clustering algorithms is determined equal to the number of decreased colors in the codebook. In clustering process, cluster centers contain smaller set of colors. Other colors with respect to difference between their color numbers and the numbers of cluster center colors become a member of one of the clusters. That is, each of colors becomes a member of a cluster that its center color is more similar than other cluster centers. One of the applied methods for clustering is use of the swarm intelligence algorithms such as particle swarm optimization [8], and artificial fish swarm algorithm [9].

Artificial fish swarm algorithm (AFSA) was presented by Li Xiao Lei in 2002 [10]. This algorithm is a technique based on swarm behaviors that was inspired from social behaviors of fish swarm in nature. AFSA works based on population, random search, and behaviorism. This algorithm has been used in optimization applications, such as clustering [11, 12], machine learning [13, 14], PID control [15], data mining [16], image segmentation [17].

In this paper, a modified AFSA is proposed as a color quantizer. In the modified AFSA, we try to remove weaknesses of standard AFSA to increase the algorithm efficiency. Then, this algorithm is configured to perform CQ. In order to comparison, the proposed algorithm along with three other clustering algorithms is used for performing CQ on 4 well-known images. Comparing qualitative efficiency of the algorithms confirms the competence of the proposed algorithm. The remainder of the paper is organized as follows: in section 2, standard AFSA will be described and in section 3, the proposed algorithm will be presented. Section 4 studies the experiments and analyzes the results. Final section concludes the paper and outlines possibilities of the future works.

2 Artificial Fish Swarm Algorithm

In the underwater world, fish can find areas that have more foods rather than their current area, which is done with individual or swarm search by fishes. According to this characteristic, artificial fish (AF) model is represented by prey, free_move, swarm and follow behaviors. AFs search the problem space by those behaviors. The environment, which AF lives in, substantially is solution space and other AF's domain. Food consistence degree in water area is AFSA objective function. Finally, AFs reach to a point, which its food consistence degree is maxima (global optimum).

In AFSA, AF perceives external concepts with sense of sight. Current position of AF is shown by vector $X=(x_1, x_2, \ldots, x_n)$. The *visual* is equal to length of sight field of AF in each dimension and X_v is a position in *visual* where the AF wants to go. Then if X_v has better food consistence than current position of AF, it goes one step toward X_v which causes change in AF position from X to X_{next}, but if the current position of AF is better than X_v, it continues searching in its *visual* area. *Food consistence* in position X is fitness value of this position and is shown with $f(X)$. The step is equal to maximum length of the movement. The distance between two AFs which are in X_i and X_j positions is shown by $Dis_{ij}=\| X_i-X_j\|$ (Euclidean distance). AF model consists of two parts of variables and functions. Variables include X (current AF position), *step* (maximum length step), *visual* (length of sight field), *try-number* (the maximum test interactions and tries) and crowd factor δ ($0<\delta<1$). Also functions consist of prey behavior, free move behavior, swarm behavior and follow behavior. In each step of optimization process, AF looks for locations with better fitness values in problem search space by performing these four behaviors based on algorithm procedure [10, 14, 15].

3 Proposed Algorithm

In this section, a modified artificial fish swarm algorithm called MAFSA is presented. Then, MAFSA is configured as a color quantizer. Generally, modifications which are imposed on standard AFSA structure include: removing two parameters step and crowd_factor, adding contraction coefficient parameter to the algorithm, removing blackboard, changing visual parameter value during algorithm execution, changing in follow and prey behaviors, removing swarm behavior and changing in the procedure of algorithm execution. In the following, modified AFSA algorithm is described. First, modified AFSA behaviors are explained:

3.1 Prey Behavior

This behavior is an individual behavior that each AF performs independently and performs a local search around itself. Every AF by performing this behavior attempts try-number times to move to a new position with better fitness. Here, it is supposed that AF *i* is in position $\vec{X}_i$ and wants to perform prey behavior. In prey behavior, following steps are done:

1) AF *i* considers a goal position $\vec{X}_T$ in its *visual* by means of Eq. (1), then evaluates its fitness. *d* shows dimension number and Rand generates a random number by uniform distribution in [-1, 1].

$$X_{T,d} = X_{i,d} + Visual \times Rand_d(-1,1) \tag{1}$$

2) If fitness value of position $\vec{X}_T$ is better than fitness value of the current position of AF *i*, position of AF *i* is updated by Eq. (2).

$$\vec{X}_i(t+1) = \vec{X}_i(t) + \left(\vec{X}_T - \vec{X}_i(t)\right) \times Rand(0,1) \tag{2}$$

Steps 1 and 2 are repeated *try-number* times. By executing above steps, an AF can update its position at most try-number times.

AF moves as a random percentage of the distance between its current position and goal position at each movement. Also, it is possible that none of its attempts for finding better positions is efficacious. If AF *i* couldn't move toward better positions by performing two mentioned steps (*try_number* times), it moves with a random step in its *visual* by means of Eq. (3):

$$X_{i,d}(t+1) = X_{i,d}(t) + Visual \times Rand_d(-1,1) \tag{3}$$

In MAFSA, by executing Eq. (3) on AF, it is attempted to preserve swarm diversity, but it wouldn't be used for the best AF of swarm because this behavior may result in worse position for an AF. Therefore, the best AF of swarm wouldn't lose its position even when it doesn't find better position in its neighborhood and just displaces when it could find better position in its visual. Thereafter, in this condition, the best found position during previous iterations by swarm is the best AF's position since at each iteration, the best AF of swarm changes its position when it moves toward a better position. Consequently, MAFSA doesn't require to blackboard anymore.

3.2 Follow Behavior

As it was mentioned in subsection 3-1, the best AF of swarm locates in the best found position so far by swarm. In follow behavior, each of AF moves one step toward the best AF of swarm by Eq. (4):

$$\vec{X}_i(t+1) = \vec{X}_i(t) + \frac{\vec{X}_{Best} - \vec{X}_i(t)}{Dis_{i,Best}} \times [Visual \times Rand(0,1)] \tag{4}$$

Where $\vec{X}_i$ is position vector of AF *i* which performs follow behavior and $\vec{X}_{Best}$ is the position vector of the best AF of swarm. Hence, AF *i* moves as a random percentage of *visual* in each dimension toward the best AF of swarm. Indeed, after that an AF finds more food, other members follow it to reach more food, too. Performing follow behavior of the best AF of swarm causes increase in convergence rate of swarm and helps to keep integrity of AF of a swarm. This behavior is a group behavior and interactions between members of swarm are done globally among them. Thus, this behavior can also perform the duty of swarm move (keeping swarm integrity) in standard AFSA since it can keep AF in a swarm and prevent from swarm splitting in problem space. As a result, swarm behavior is eliminated in MAFSA. To execute follow behavior in standard AFSA, at each iteration of algorithm execution, it has to calculate Euclidean distance between all AF with each other (for detecting neighbors of each AF), which is of heavy computational load. But follow behavior in MAFSA has lower computational load while it is very effective in increasing the convergence rate of the algorithm.

3.3 Modified AFSA Procedure

In MAFSA, each of prey and follow behaviors are done for each of AF at every iteration. In MAFSA, first all AF perform prey behavior and their position are updated based on prey behavior execution procedure. Then, follow behavior is performed and all members except the best AF of swarm move to a new position in direction of moving toward the best found position by swarm. At the end of each of iteration of MAFSA algorithm, *visual* value is updated for AF to make a balance between global search and local search abilities [13, 14]. To reach this goal, *visual* has to be large at first, such that AF converge to their goals fast and perform global search well. Simultaneously with swarm convergence toward goal, *visual* decreases gradually until AF with small *visual* could get better results by doing an acceptable local search around goal.

For this purpose, *visual* is multiplied by a positive number less than one at each of iterations, which this number can be determined with different approaches [14]. In this paper, to decrease *visual*, a random number generator with uniform distribution is applied in the considered interval that is given in Eq. (5):

$$Visual(t+1) = Visual(t) \times \left(L_{Low} + \left(Rand \times \left(L_{High} - L_{Low}\right)\right)\right) \tag{5}$$

In Eq. (5), at each of iterations, *visual* is obtained randomly with respect to this parameter in previous iteration. L_{low} and L_{high} are lower bound and upper bound of change percentage of *visual* to previous iteration respectively and Rand is random number generator with uniform distribution in interval [0, 1]. Therefore, *visual* is a random percentage of its value in previous iteration between L_{low} and L_{high}. For this reason, L_{high} should be considered a number less than one. Pseudo code of MAFSA is represented in figure 1.

```
MAFSA:
for each Artificial Fish i • [1 .. N]
    initialize x_i
endfor
repeat:
    for each Artificial Fish i • [1 .. N]
        flag[i]=0;
        for counter=1 to try_number
            Obtain X_T with Eq. 1 and Calculate f(X_T)
            if f(X_T) • f(X_i) then
                apply Eq. 2
                flag[i]=1;
            endif
        endfor
        if flag[i]==0 then
            apply Eq. 3
        endif
    endfor
   for each Artificial Fish i • [1 .. N]
        apply Eq. 4
   endfor
   Update Visual according Eq. 5
until stopping criterion is met
```

Fig. 1. MAFSA pseudo code

3.4 MAFSA Configuration for CQ

In this section, MAFSA configuration is discussed as a color quantizer. Application of images is usually for observing by an individual. The eye is very good at interpolation, that is, the eye can tolerate some distortion. The eye has more acuity for luminance (gray scale) than chrominance (color). This is why we will concentrate on compressing gray scale (8 bits per pixel) image.

As mentioned before, the goal of this paper is to solve CQ problem as a clustering problem. First, a dataset has to be determined that clustering has to be done on it. In this problem, dataset consists of all pixels' values. In gray scale images, every pixel has an 8 bit color characteristic, so data are one dimensional. Then, it has to be specified a fitness function for clustering which MAFSA should optimize. In this paper, to find optimal values of cluster centers which their number has been predetermined, one of the most known clustering criteria called sum of intra cluster distances is used [18]. Eq. (6) is a function which calculates sum of intra cluster distances that according to it, the best clustering is the one when this function's value is minimum.

$$J(C_1,C_2,\ldots,C_K)=\sum_{i=1}^{K}\left(\sum_{X_j\in C_i}\left\| Z_i - X_j \right\|\right) \tag{6}$$

This function is used as a fitness function for MAFSA algorithm and is considered as a minimizing problem. In Eq. (6), the Euclidean distance between each data vector in a cluster and the centroid of that cluster is calculated and summed up. Here, we have K clusters C_i ($1 \le i \le K$) that each of N data vectors X_j ($1 \le j \le N$) are clustered on the basis of distance from each of these cluster centers Z_i ($1 \le i \le K$). Data vectors belong to a cluster that their Euclidean distance from its cluster center is less than their Euclidean distance from other cluster centers. Therefore, MAFSA goal is to determine cluster centers which minimize Eq. (6), and consequently optimal cluster centers are determined. In fact, according to clustering conditions of CQ, Eq. (6) shows the sum of differences between original image's pixel color numbers and decoded image's. Codebook is determined by using final result of clustering. Indeed, the codebook contains cluster centers and their indices. Each cluster center is a one dimensional vector that consists of gray scale color values. Hence, after completing the codebook, every pixel of the original image is transformed into cluster center index in the codebook that it belongs to (encoding). To represent the image again, each pixel takes its corresponding cluster center color values with respect to its encoded value (decoding).

Since data and cluster centers are one dimensional and there are K clusters (the number of codewords in the codebook), so every AF has to represent K cluster centers. As a result, each AF is K dimensional or has K components in its vector. As a matter of fact, each of components includes one of colors which are supposed to be considered as the replacement of some more similar colors to it.

In MAFSA, first, AFs are initialized randomly in the problem space. Therefore, every AF consists of K initial random cluster centers which displace these cluster centers in the problem space by means of MAFSA behaviors and their goal is to determine cluster centers in a way that Eq. (6) to be minimizes as a fitness function. At last, the codebook would be the same as obtained cluster centers from MAFSA.

4 Experiments

Experiments are done on 4 well-known images which are mostly used for measuring the efficiency of CQ algorithms. These images are *Barbara*, *Boat*, *Lenna* and *Pepper* that their size is 512*512 pixels. Figure 2 shows applied images in this paper.

Fig. 2. Applied images in this paper

The most important measurement criteria for CQ algorithms efficiency include mean squared error (*MSE*) and peak signal to noise ratio (*PSNR*) [5,6]. *MSE* is usually used for assessing distortion between the original image and resulted image from CQ. Let the original image x have n pixels. *MSE* is computed by Eq. (7):

$$MSE = \frac{1}{n}\sum_{i=1}^{n}\left(x_i - \hat{x}_i\right)^2 \tag{7}$$

Where, $\hat{x}$ is the obtained image after performing CQ. *MSE* represents the average distortion and lesser value of it shows better efficiency of CQ algorithm. *PSNR* is a standard way for evaluating fidelity between the original image and the obtained image from CQ. *PSNR* is calculated by Eq. (8):

$$PSNR = 10\log_{10}\left(\frac{m^2}{MSE}\right) \tag{8}$$

Where, m is the largest amount which a pixel can take that is 255 in gray scale images. *PSNR* is measured in decibels (dB) and the larger value of it shows better efficiency of CQ method. The proposed algorithm along with standard AFSA, PSO and k-means is used for performing CQ on 4 mentioned images. PSO parameters are adjusted with respect to [18] and Forgy initializing method is used for k-means [6]. Population size is considered 5 times the number of problem space dimensions for standard AFSA and MAFSA [18, 9]. That is, population size is 5 times the number of codebooks' colors. Based on multiple experiments which have been done, visual, try-number, L_{high} and L_{low} are 10, 10, 1 and 0.95, respectively. Standard AFSA's parameters are adjusted according to [9]. Experiments are repeated 50 times and average of obtained *PSNR* and *MSE* from 4 algorithms on 4 images are represented in table 1. In this table, each image has been compressed with rates 8:3, 8:4 and 8:5 which their colors have been decreased to 8, 16 and 32, respectively. The best result is shown by bold face for each case. As it is observed, MAFSA has achieved better results in all cases. MAFSA has achieved better results than standard AFSA because of not having the weaknesses of standard AFSA specially imbalance between global search and local search [14]. In fact, AF perform global and local search well in MAFSA and generate a codebook by decreasing the sum of intra cluster distances which decreases distortion in decoded image. Therefore, obtained images from the proposed algorithm would have more fidelity with the original image.

According to results of table 1, generally, standard AFSA has less efficiency than PSO. But MAFSA has achieved better efficiency than PSO by improving different parts of standard AFSA. Figure 3 shows Lenna and Peppers images whose colors have been decreased to 8 colors and have been compacted by rate 8:3.

On the whole, experimental results show that compressed images by means of generated codebook by the proposed algorithm are of higher quality than other algorithms.

Fig. 3. Two decoded images with 8 colors

Table 1. MSE and PSNR comparison of the quantization methods

Image	Compression Ratio	Criteria	Std-AFSA	K-means	PSO	MAFSA
Lenna	8:3	MSE	29.81	28.16	26.04	**22.48**
		PSNR	33.38	33.69	33.97	**34.61**
	8:4	MSE	8.93	8.29	8.69	**6.27**
		PSNR	38.62	39.03	38.76	**40.16**
	8:5	MSE	3.71	2.78	3.34	**1.71**
		PSNR	42.43	43.70	42.93	**45.80**
Barbara	8:3	MSE	25.48	24.94	25.81	**22.78**
		PSNR	34.07	34.19	34.02	**34.55**
	8:4	MSE	9.99	8.37	9.87	**6.48**
		PSNR	38.14	38.95	38.20	**40.01**
	8:5	MSE	3.83	3.50	3.01	**1.61**
		PSNR	42.29	42.97	43.39	**46.05**
Boat	8:3	MSE	26.79	23.99	24.91	**23.39**
		PSNR	33.85	34.32	34.22	**34.44**
	8:4	MSE	10.14	9.11	10.61	**6.78**
		PSNR	38.15	38.64	37.93	**39.82**
	8:5	MSE	4.18	2.62	3.90	**2.06**
		PSNR	42.17	44.05	42.28	**44.97**
Pepper	8:3	MSE	29.58	30.04	32.55	**28.90**
		PSNR	33.42	33.38	33.01	**33.52**
	8:4	MSE	10.43	9.84	10.93	**7.10**
		PSNR	37.95	38.22	37.77	**39.61**
	8:5	MSE	4.56	2.83	3.11	**1.86**
		PSNR	41.57	43.66	43.21	**45.43**

5 Conclusion

In this paper, a modified artificial fish swarm algorithm was proposed. In the proposed algorithm, it has been attempted to remove standard AFSA's weaknesses and algorithm to be able to reach acceptable and good results. The proposed algorithm is utilized in CQ application and its efficiency is compared qualitatively with efficiency of standard AFSA, PSO and k-means. In this study, images are compressed only with respect to the number of their colors. Experimental results show that obtained images from the proposed algorithm are of higher quality than obtained results from other tested algorithms. However, the proposed algorithm has more complexity than other tested algorithms. Reducing complexity is issue that merits further research.

References

1. Yang, C.K., Tsai, W.H.: Color Image Compression Using Quantization, thresholding, and Edge Detection Techniques all Based on the Moment-Preserving Principle. Pattern Recognition Letters 19, 205–215 (1998)
2. Wang, S., Cai, K., Lu, J., Liu, X., Wu, E.: Real-time coherent stylization for augmented reality. The Visual Computer 26, 445–455 (2010)
3. Deng, Y., Manjunath, B.: Unsupervised Segmentation of Color–Texture Regions in Images and Video. IEEE Transactions on Pattern Analysis and Machine Intelligence 23, 800–810 (2001)
4. Sertel, O., Kong, J., Catalyurek, U.V., Lozanski, G., Saltz, J.H., Gurcan, M.N.: Histopathological Image Analysis Using Model-Based Intermediate Representations and Color Texture: Follicular Lymphoma Grading. Journal of Signal Processing Systems 55, 169–183 (2009)
5. Sayood, K.: Introduction to Data Compression, 3rd edn. Morgan Kaufmann (2006)
6. Celebi, M.E.: Improving the Performance of K-means for Color Quantization. Journal of Image and Vision Computing 29, 26–271 (2011)
7. Schaefer, G., Zhou, H.: Fuzzy Clustering for Color Reduction in Images. Telecommunication Systems 40, 17–25 (2009)
8. Tsai, C.Y., Kao, I.W.: Particle Swarm Optimization with Selective Particle Regeneration for Data Clustering. Journal of Expert Systems with Applications 38, 6565–6576 (2011)
9. Yazdani, D., Golyari, S., Meybodi, M.R.: A New Hybrid Approach for Data Clustering. In: 5th International Symposium on Telecommunication (IST), Tehran, pp. 932–937 (2010)
10. Li, L.X., Shao, Z.J., Qian, J.X.: An Optimizing Method Based on Autonomous Animate: Fish Swarm Algorithm. In: Proceeding of System Engineering Theory and Practice, pp. 32–38 (2002)
11. Hi, S., Belacel, N., Hamam, H., Bouslimani, Y.: Fuzzy Clustering with Improved Artificial Fish Swarm Algorithm. In: International Joint Conference on Computational Sciences and Optimization 2009, Hainan, pp. 317–321 (2009)
12. Xiao, L.: A Clustering Algorithm Based on Artificial Fish school. In: 2nd International Conference on Computer Engineering and Technology, Chengdu, pp. 766–769 (2010)
13. Yazdani, D., Golyari, S., Meybodi, M.R.: A New Hybrid Algorithm for Optimization Based on Artificial Fish Swarm Algorithm and Cellular Learning Automata. In: 5th International Symposium on Telecommunication (IST), Tehran, pp. 932–937 (2010)
14. Yazdani, D., Nadjaran Toosi, A., Meybodi, M.R.: Fuzzy Adaptive Artificial Fish Swarm Algorithm. In: 23th Australian Conference on Artificial Intelligent, Adelaide (2010)
15. Luo, Y., Zhang, J., Li, X.: The Optimization of PID Controller Parameters Based on Artificial Fish Swarm Algorithm. In: IEEE International Conference on Automation and Logistics, Jinan, pp. 1058–1062 (2007)
16. Zhang, M., Shao, C., Li, M., Sun, J.: Mining Classification Rule with Artificial Fish Swarm. In: 6th World Congress on Intelligent Control and Automation, Dalian, pp. 5877–5881 (2006)
17. Li, C.X., Ying, Z., JunTao, S., Qing, S.J.: Method of Image Segmentation Based on Fuzzy C-means Clustering Algorithm and Artificial Fish Swarm Algorithm. In: International Conference on Intelligent Computing and Integrated Systems (ICISS), Guilin (2010)
18. Kao, Y.T., Zahara, E., Kao, I.W.: A Hybridized Approach to Data Clustering. Journal on Expert System with Applications 34, 1754–1762 (2008)

Coordinated Learning for Loosely Coupled Agents with Sparse Interactions

Chao Yu, Minjie Zhang, and Fenghui Ren

School of Computer Science and Software Engineering,
University of Wollongong, Wollongong, 2500, NSW, Australia
cy496@uowmail.edu.au, {minjie,fren}@uow.edu.au

Abstract. Multiagent learning is a challenging problem in the area of multiagent systems because of the non-stationary environment caused by the interdependencies between agents. Learning for coordination becomes more difficult when agents do not know the structure of the environment and have only local observability. In this paper, an approach is proposed to enable autonomous agents to learn where and how to coordinate their behaviours in an environment where the interactions between agents are sparse. Our approach firstly adopts a statistical method to detect those states where coordination is most necessary. A Q-learning based coordination mechanism is then applied to coordinate agents' behaviours based on their local observability of the environment. We test our approach in grid world domains to show its good performance.

Keywords: Multiagent Reinforcement Learning, Coordination.

1 Introduction

Multiagent learning is one of the most important issues in the research area of multiagent systems (MASs), finding increasing applications in a variety of domains such as robotics, distributed control, resource management and economics, etc.. A significant part of the research on multiagent learning focuses on reinforcement learning techniques. In multiagent reinforcement learning, agents can carry out the learning processes concurrently or simultaneously such that the reward each agent received can be impacted by other agents' actions. These interdependencies between agents make the learning environment non-stationary. Agents need to communicate and/or coordinate with each other in this non-stationary learning environment to improve their performance.

In many MASs, the interactions between agents are sparse, which means the agents are loosely coupled and do not need to coordinate with each other frequently. For example, two robots are navigating in a building. Most of the time they can move around independently according to their own decisions. Only when both robots come around the doorway should they coordinate their behaviours in case of colliding with each other. In such type of MASs, coordination is required among agents only when it is necessary. Agents should learn from experiences to determine in which situations coordination is most beneficial and

D. Wang and M. Reynolds (Eds.): AI 2011, LNAI 7106, pp. 392–401, 2011.

how to coordinate their behaviours after these situations are determined. However, due to agents' local observability, learning where and how to coordinate agents' behaviours in loosely coupled MASs is a very challenging problem.

Some approaches have been developed to deal with the coordinated learning problem with sparse interactions in recent years, using techniques such as coordination graphs [4][6], statistical learning [1][4] and learning automata [2], etc. However, these approaches are based on some assumptions, which are to (1) predefine the specific states where coordination is necessary [6], (2) require learners to have prior knowledge about their optimal policies [1], and (3) assume agents to have full observability of joint-states and/or joint-actions [4][6]. These assumptions heavily limit these approaches in real world applications.

In this paper, a new approach is proposed to enable agents to learn where and how to coordinate their behaviours by using local information during sparse interactions. Our approach starts with a statistical learning process to detect the possible states where coordination is required. After that, a Q-learning based reinforcement learning approach is used to coordinate the agents' behaviours based on their local observability of the environment. Our approach does not require the agents to have any prior knowledge or full observability of the environment, thus solving more realistic problems than most of the state-of-the-art works do.

The problem description and definitions, as well as the proposed approach are illustrated in Section 2. Section 3 presents the experimental results and analysis to show the performance of our approach. Section 4 compares our work with some related works. Finally, Section 5 concludes this paper and lays out some directions for future research.

2 Learning to Coordinate

2.1 Problem Description and Definitions

Consider two very simple problems in which two robots are navigating in an environment, each trying to reach its own goal as illustrated in Fig. 1, where R_1, R_2 represent two robots and G_1 and G_2 are their goals (In Fig. 1 (a), G_1 and G_2 are in the same grid denoted by G). In these multiagent domains, each robot can be modeled as an independent learner and the optimal policy of each robot can be learnt by using a single-agent learning approach. However if both robots choose its own optimal policy in order to achieve their individual

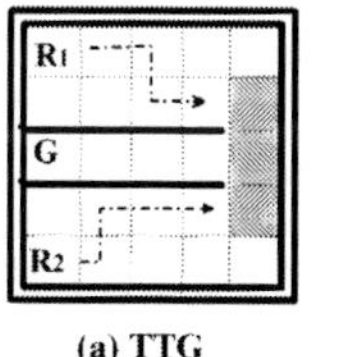

(a) TTG

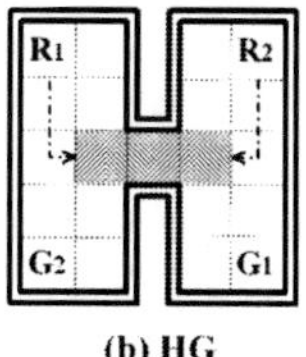

(b) HG

Fig. 1. Two domains where agents need to coordinate in the shadowy states [3]

goals and do not coordinate with each other when they come to the conflicting states (shadowy areas in Fig. 1(a) and Fig. 1 (b)), they may bounce into each other and get stuck there. In many applications, this kind of conflicts may affect agents (robots) to achieve their goals or even are not allowed to happen in some domains such as agent-based disaster management, emergency rescue systems. How to coordinate agent's behaviours to decrease the probability of conflicts during learning in this kind of loosely coupled MASs is a challenging issue.

As can be seen from Fig. 1, there are two types of states in this kind of loosely-coupled MASs, which are *coordinated states* and *uncoordinated states*. Let S be the state space of the domain. The coordinated states and uncoordinated states are formally defined by the following two definitions.

Definition 1. *Coordinated States* are the states where agents need to coordinate with each other, which are defined as a set $S^c = \{s_i^c | s_i^c \in S(1 \leq i \leq m)\}$.

Definition 2. *Uncoordinated States* are the states where agents can act independently, which are defined as a set $S^{\bar{c}} = \{s_j^{\bar{c}} | s_j^{\bar{c}} \in S(1 \leq j \leq n)\}$.

An approach is proposed in this paper to solve the coordinated learning problem represented by the domains in Fig. 1. The main idea of our approach is to (1) dynamically identify the coordinated states during agent learning; and (2) develop a coordinated learning approach to adapt agent's behaviours after the coordinated states are determined. Each part of this approach is introduced in detail by Subsection 2.2 and Subsection 2.3, respectively.

2.2 Learning the Coordinated States

In a reinforcement learning setting, the only feedback from the environment is the reward. When agents have received severely penalized reward in a state, they are notified by the environment that coordination should be considered in this conflicting state. However, at the beginning of learning, agents are exploring the environment, which to some extent makes the learning a stochastic process. As such, a conflicting state is not sufficient to reveal the true structure of the environment. However, from a statistical point of view, more frequent conflicts in a state indicate that this state is more likely to be one of the coordinated states. Further more, if agents conflict in a certain state, it means that agents are also likely to conflict in the neighboring states.

Based on the considerations stated ahead, we choose *kernel density estimation* (KDE) approach to detect the coordinated states. The basis of KDE can be represented by the *kernel* function F satisfying $\int F(x)dx = 1$. A simple illustration of KDE is given in Fig. 2, where x-axis stands for the one-dimensional variable space, y-axis stands for the density, the dashed lines represent the kernels and the solid line is the overall estimation. The overall estimation gives a belief of how the corresponding observation (denoted as a cross in Fig. 2) is likely to be the real point (the dot in Fig. 2) of the estimated variable.

In our problem, however, the variable space is two-dimensional because the state space is a plane. The observation in this variable space means that agents

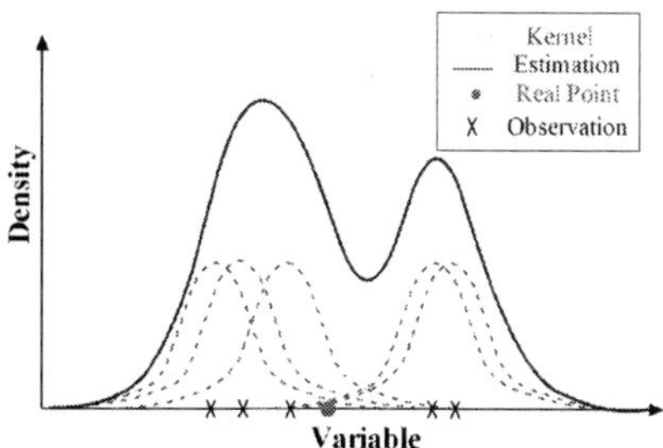

Fig. 2. An illustration of KDE

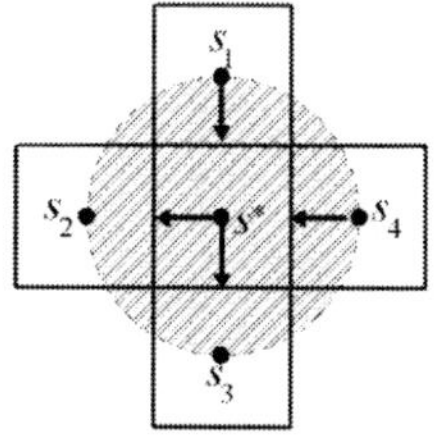

Fig. 3. An example of determining the coordinated states

conflict with each other in a state. An observation with the highest density signifies that the corresponding state is a location where coordination is most required. Let s represent a state with the central point of $P_s(x_s, y_s)$ and let $F_P(x, y)$ be the kernel function centralized at point P. The overall estimation for state s is calculated by summing up all the overlapping kernels given by $\sum_P F_P(x_s, y_s)$. After the statistics collecting period, agents can determine the coordinated states S^c according to Algorithm 1.

Algorithm 1. Determining S^c

Input: P_{s^*}, R;
Output: S^c;
1 **for** *each agent* k **do**
2 $\quad \widetilde{S^c_k} \leftarrow \varnothing$;
3 $\quad$ **for** *each state* $s \in S$ **do**
4 $\quad\quad$ **if** $|P_s - P_{s^*}| \leq R/2$ **then**
5 $\quad\quad\quad \widetilde{S^c_k} \leftarrow \widetilde{S^c_k} \cup \{s\}$;
6 $\quad\quad$ **end**
7 $\quad$ **end**
8 $\quad \widetilde{S^c_k} \leftarrow eliminate(\widetilde{S^c_k})$;
9 **end**
10 $S^c \leftarrow \bigcup_{k=1}^{N} \widetilde{S^c_k}$;

Algorithm 2. Elimination Mechanism

$sort(\widetilde{S^c_k})$ according to the density;
$S^{c\prime}_k \leftarrow \varnothing$;
for *each state* $s^c_k \in \widetilde{S^c_k}$ **do**
$\quad$ **for** *each neighboring state* $\widehat{s^c_k} \in \widetilde{S^c_k}$ *of* s^c_k *and* $\widehat{s^c_k} \notin S^{c\prime}_k$ **do**
$\quad\quad$ **if** $n_{\widehat{s^c_k} \rightarrow s^c_k} < n_{s^c_k \rightarrow \widehat{s^c_k}}$ **then**
$\quad\quad\quad \widetilde{S^c_k} \leftarrow \widetilde{S^c_k} \setminus \{\widehat{s^c_k}\}$;
$\quad\quad$ **end**
$\quad$ **end**
$\quad S^{c\prime}_k \leftarrow S^{c\prime}_k \cup \{s^c_k\}$;
end
return $\widetilde{S^c_k}$;

In Algorithm 1, P_{s^*} is the central point with the highest density and R is the scanning distance of the agent. $\widetilde{S^c_k}$ represents the coordinated states of agent k and can be computed by involving the states that are located in the scanning distance of the agent (Lines 3-7). Fig. 3 gives an illustration of determining the coordinated states with a scanning distance twice longer than the side length of each grid. However, not all the states in $\widetilde{S^c_k}$ are causes of the conflict in central state s^*. As shown in Fig. 3, an agent transits from s_1, s_4 to s^*, causing the conflict in s^*, and transits from s^* to s_2, s_3. It is obvious that s_2 and s_3 are not the causes of the conflict in s^* such that they should be eliminated from $\widetilde{S^c_k}$.

An elimination mechanism (given in detail by Algorithm 2) is applied to eliminate this kind of states (Line 8). Finally, the overall coordinated states S^c are the union of the coordinated states $\widetilde{S_k^c}$ of all the agents (Line 10).

Algorithm 2 illustrates the process of the elimination mechanism. The states in $\widetilde{S_k^c}$ are sorted in a descending order according to the density derived from KDE process(Line 1). For each sorted state s_k^c, the agent determines whether its neighboring state (e.g. $\widehat{s_k^c}$) is the cause for the conflict in state s_k^c. This can be done by collecting statistics of transitions between states s_k^c and $\widehat{s_k^c}$ (Lines 3-10). If the agent transits from central state s_k^c to state $\widehat{s_k^c}$ more often than the reserve (Line 5, where $n_{s \to s'}$ represents the times of transitions from s to s'), state $\widehat{s_k^c}$ is not the cause of conflict in s_k^c and should be eliminated from $\widetilde{S_k^c}$ (Line 6).

2.3 Learning for Coordination

After determining coordinated states S^c and uncoordinated states $S^{\bar{c}}$, agents should learn how to coordinate their behaviours by taking other agents into account. At the beginning of learning, each agent maintains a single-state-action Q-value table denoted by $Q_k(s_k, a_k)$ for all states, where s_k is the state of agent k and a_k is its action. After the KDE detects the coordinated states, a joint-state-action Q-value table for the coordinated states is created by combining all the state-action information from the single learning process. Suppose there are total N agents in the environment. Let S_k^c and A_k be the coordinated state space and the action space of agent k. The joint-state space of all agents in coordinated states can be given by $JS = \times_{k=1}^{N} S_k^c$, and the joint-action space of all agents is $JA = \times_{k=1}^{N} A_k$. This joint Q-value $Q_c(js, ja)$ can be initialized by adding the single Q-values $Q_k(s_k, a_k)$ of each agent, which can be given by Equation 1.

$$Q_c(js, ja) = \sum_{k=1}^{N} Q_k(s_k, a_k) \quad s_k \in S^c, js \in JS, ja \in JA \tag{1}$$

After adding the joint Q-value $Q_c(js, ja)$, agents can coordinate their behaviours according to this Q-value when in the coordinated states. The basic idea of our coordinated learning approach is to let agents act optimistically when facing uncertainties caused by their local observability. In more detail, when there are agents out of the coordinated states, the agents that are in the coordinated states cannot receive the joint-state-action information of all the agents to determine their joint actions from the joint Q-value $Q_c(js, ja)$. In this case, agents can act optimistically by giving a best estimation of those agents that cannot be observed, which means it will act according to the highest Q-value only based on the available state-action information.

Let s_k be the current state of the agent at step t and s_k' be the state in the next step $t+1$. JA_m and JS_m denotes the joint action space and joint state space of m agents in the coordinated states. $ja_m \in JA_m$ and $js_m \in JS_m$ are their joint action and joint state, respectively. There are mainly two scenarios according to the current state and the transition situations of agent k.

(1) $s_k \in S^{\bar{c}}$. In this scenario, agent k is in an uncoordinated state. It looks up its own single Q-value table $Q_k(s_k, a_k)$ and takes an action a_k that has the highest Q-value to jump into a new state. If the new state s'_k is still in the uncoordinated states, a normal single Q-learning can be applied to update the Q-value. This updating is given by Equation 2, where $R_k(s_k, a_k)$ is the immediate reward, $\alpha \in (0, 1]$ is the learning rate and $\gamma \in [0, 1)$ is the discount factor.

$$Q_k(s_k, a_k) \leftarrow Q_k(s_k, a_k) + \alpha[R_k(s_k, a_k) + \gamma \max_{a'_k} Q_k(s'_k, a'_k) - Q_k(s_k, a_k)] \quad (2)$$

However, if the new state s'_k is in the coordinated states, the agent needs to back up its Q-value by adding the expected reward from the coordinated state s'_k. Note that in the coordinated states, agents only maintain a joint Q-value table $Q_c(js, ja)$ which represents the overall expected reward when all the agents are in the coordinates states with a joint state js and joint action ja. However agent k only has a local observability of the coordinated states and cannot observe the agents that are out of the coordinated states, thus the joint state js cannot be determined to choose a joint action ja that maximizes the Q-value. Suppose there are m agents in the coordinated states and n agents in the uncoordinated states at step $t + 1$. Agent k observes the joint state js_m of the m agents and chooses the highest $Q_c(js, ja)$ based on this information by giving an optimistic estimation of the unobserved n agents. The value of $Q_c(js, ja)$ represents the overall expected reward and can be averaged by the total number of all agents N. The Q-value updating rule is formally given by Equation 3.

$$Q_k(s_k, a_k) \leftarrow Q_k(s_k, a_k) + \alpha[R_k(s_k, a_k) + \gamma \frac{1}{N} \max_{ja'} Q_c(js', ja') - Q_k(s_k, a_k)] \quad (3)$$

where ja' is selected as follows,

$$\forall js_n \in JS_n, \forall ja_n \in JA_n, \exists js' \in JS, \exists ja' \in JA \Rightarrow \max_{ja'} Q_c(js', ja'). \quad (4)$$

(2) $s_k \in S^c$. In this scenario, agent k is in the coordinated state at step t. It observes the whole coordinated states to gain the state-action information of other agents that are in the coordinated states at current time. Assume there are now $m (m \leq N)$ agents existing in the coordinated states with the joint state $js_m \in JS_m$ and other n agents in the uncoordinated states. The m agents will look up the joint Q-value table Q_c and choose ja_m according to Equation 5.

$$\forall js_n \in JS_n, \forall ja_n \in JA_n, \exists js \in JS, \exists ja_m \in JA_m \Rightarrow \max_{ja_m} Q_c(js, ja). \quad (5)$$

After taking the joint action ja_m, each agent jumps to a new state. Suppose among the m agents, there are p $(p \leq m)$ agents still in the coordinated states and other $q = (m - p)$ agents moving out to uncoordinated states. The m agents should back up the future rewards from Q_c according to the joint-state of the p agents and from Q_k according to the state of each agent that jumps out of the coordinated states. The joint Q-value can be updated by Equation 6.

$$Q_c(js, ja) \leftarrow Q_c(js, ja) + \alpha[\frac{N}{m}(R_m + \gamma(\frac{p}{N}\max_{ja'} Q_c(js', ja')$$
$$+ \sum_{k=1}^{q} \max_{a'_k} Q_k(s'_k, a'_k))) - Q_c(js, ja)], \forall js_n \in JS_n, \forall ja_n \in JA_n. \quad (6)$$

where $R_m = \sum_{k=1}^{m} R_k(s_k, a_k)$ is the sum of the reward of the m agents. In Equation 6, (1) $\max_{ja'} Q_c(js', ja')$ is the expected reward of all the N agents based on the information of the p agents. This value multiplied by $\frac{p}{N}$ represents the expected reward of the p agents; (2) $\max_{a'_k} Q_k(s'_k, a'_k)$ is the expected reward of each agent that moves out of the coordinated states. Summing up these value represents all the expected reward of q agents; (3)$R_m + \gamma(\frac{p}{N}\max_{ja'} Q_c(js', ja') + \sum_{k=1}^{q} \max_{a'_k} Q_k(s'_k, a'_k))$ is the expected reward of the m agents. This value multiplied by $\frac{N}{m}$ represents the expected reward of all the N agents; (4) $\forall js_n \in JS_n, \forall ja_n \in JA_n$ means that this Q-value updating is applied for all the possible joint-state-action of the n agents. In this way, the joint Q_c value can be updated by using the available information among the m agents and giving an optimistic estimation of the unobserved n agents.

3 Experiment

In this section, experiments are carried out to demonstrate the performance of our approach, denoted as *CL* (*Coordinated Learning*). To give a benchmark, we compare our approach to other two approaches. The first one is to let each agent learn its policy independently, which is denoted as *IL* (*Independent Learning*). The second is called *JL* (*Joint Learning*), which is a centralized learning approach that agents have a full observability of the environment and receive the joint-state-action information of all the agents to control the learning process.

3.1 Experimental Setting

We test our approach in the domains given in Fig. 1, where each robot has 4 actions, “Move East”, “Move South”, “Move West” and “Move North”. Each action moves the robot to the corresponding direction deterministically. When robots collide into the wall, they will rebound back. If they collide into each other, both are transferred back to the original states. The exploration policy is the fixed $\varepsilon - greedy$ policy with $\varepsilon = 0.1$. The learning rate $\alpha = 0.05$, discount factor $\gamma = 0.95$ and rewards are given as follows: +20 for reaching the goal state, -1 for colliding into the wall, -10 for colliding into the other robot. To use our approach, we choose a two-dimensional normal distribution function $N(\mu_1, \mu_2, \sigma_1^2, \sigma_2^2, \rho)$ as the kernel function just because of the simplicity of implementation. We set the side length of each grid be 1, the kernel function be $N(0, 0, 1, 1, 0)$ and $R = 2$. We run the robots for $10,000$ episodes with the first 1000 episodes to determine the coordinated states and the last 2000 episodes averaged to compute the overall performance. All results are averaged over 10 runs.

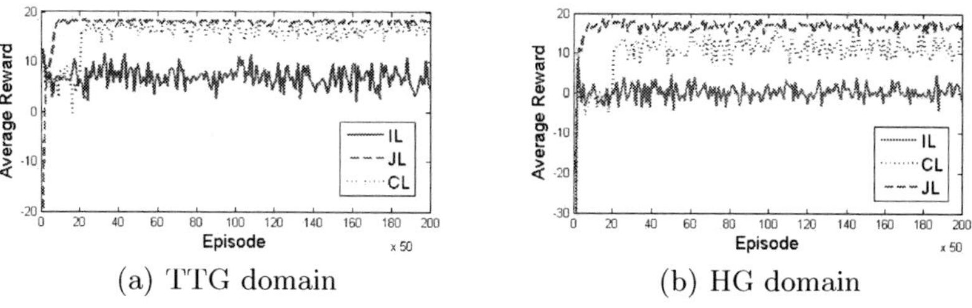

(a) TTG domain (b) HG domain

Fig. 4. Average reward of the three approaches

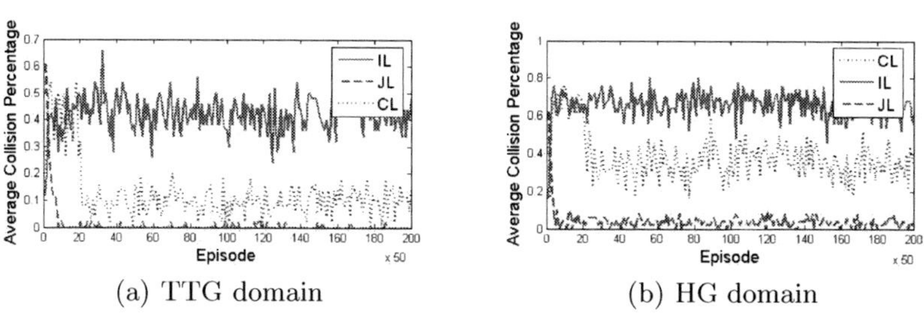

(a) TTG domain (b) HG domain

Fig. 5. Collision percentage of the three approaches

3.2 Results and Analysis

The learning processes in terms of the average rewards gained by both robots are given in Fig. 4. As can be seen in the figure, the *JL* approach converges to the optimal value because it can receive joint state-action information about both robots. The performance of *CL* is almost the same with *IL* during the first 1000 episodes but quickly outpaces *IL* after the coordination mechanism is added. As for the results of collision percentage given in Fig. 5, *JL* can learn a collision-free path in both domains while *IL* has a high probability of collision because both robots are learning independently and do not take the other robot into account. Although our approach cannot acquire a totally collision-free path due to local observability of the robots, it decreases the probability of collision dramatically compared with the uncoordinated *IL*.

Table. 1 gives the overall performance of these three approaches. To give a comparison, the state and action space are also laid out to show the computational complexity. In TTG domain, there are 3 coordinated states among the whole 25 states. The state space each agent keeps in *CL* thus can be calculated as $22 + 3^2/2 = 26.5$ and the action space is $4 \times 22/25 + 4^2 \times 3/25 = 5.44$. As can be seen in the results, *CL* reduces the computational complexity a lot compared with *JL*. This reduction is more desirable in larger scale domains where the computational complexity is too high to be implemented. Another important aspect showing the performance of these three approaches is the step number for both robots to reach their own goals. The results show that robots in *IL* always find the shortest paths to their goals. This in turn causes the high probability

Table 1. Performance of difference learning approaches in the grid world domains

Domain	Approach	State	Action	Reward	Collision(%)	Step
TTG	*IL*	25	4	6.925	0.420	12.544
	CL	26.5	5.44	16.582	0.104	16.794
	JL	25^2	4^2	18.210	0.002	22.653
HG	*IL*	21	4	0.2176	0.649	12.446
	CL	22.5	5.71	11.193	0.338	17.793
	JL	21^2	4^2	17.246	0.041	23.384

of collision because they do not coordinate with each other when both come to the coordinated states. In *JL*, a central controller receives the joint-state-action information of both robots. As a result, a safe detour strategy will be adopted by the robots to reduce the probability of collision, which accordingly increases the steps to the goals. However, our *CL* approach combines the merits of both *IL* and *JL*, allowing robots to find the shortest path to the goals while only making a small detour around the coordinated states. This is why the step number to goals in *CL* is higher than that in *IL* but much lower than that in *JL*.

In conclusion, the experimental results show that our approach *CL* outperforms the uncoordinated approach *IL* by considering coordinations when necessary. On the other hand, by removing the assumption of centralized controller, *CL* reduces the state-action space considerably and enables robots to learn a shorter path to the goal than centralized approach *JL*.

4 Related Work

Much attention has been paid to the problems of learning from sparse interactions for coordination in recent multiagent research. In [5], Kok and Vlassis proposed an approach called sparse tabular Q-learning to learn joint action values on those states where coordination is beneficial. However, these coordination states are specified beforehand and assumed to be prior knowledge to the agents. Their approach was extended to enable the agents to coordinate their actions when there exist more complicated dependencies between agents [6]. In later work [4], Kok et al. used statistical information about the obtained rewards to learn these dependencies. All these approaches are based on the agents' full observability of the joint-state space and confined to fully cooperative MASs.

In [3], Spann and Melo introduced a model for solving the learning problem in loosely coupled MASs called interaction-driven Markov Games (IDMG). They specified in advance the states where agents should coordinate with each other. Then a fully cooperative Markov Game is defined in these coordinated states such that agents can compute the game structure and Nash equilibria to choose their actions correspondingly. Our work differs from [3] in that our approach uses statistics to learn the coordinated states other than predefines them.

In [1], an algorithm called CQ-learning was proposed to enable agents to adapt the state representation in order to coordinate with other agents. This approach however depends on the assumption that agents have already learnt optimal single policies such that every agent has a model of its expected rewards. In our work, the coordinated states are detected by collecting statistics while agents are learning. This merit renders our approach more feasible in applications where agents have no prior knowledge about the structure of the environment.

5 Conclusion and Future Work

In this paper, we proposed a coordinated learning approach that enables agents to learn where and how to coordinate their behaviours with sparse interactions. Our approach does not require agents to have full observability of the whole environment, thus modeling more realistic problems than the centralized approach. The experimental results show that our approach improves the performance considerably than the uncoordinated learning approach. For future work, it is possible to improve our learning process to be totally dynamic and online. It is also necessary to test our approach in larger scale grid world domains and extend the approach to continuous environments.

References

1. De Hauwere, Y.M., Vrancx, P., Nowé, A.: Learning multi-agent state space representations. In: Proceedings of the 9th International Conference on Autonomous Agents and Multiagent Systems, pp. 715–722. IFAAMAS, Richland (2010)
2. De Hauwere, Y.M., Vrancx, P., Nowé, A.: Learning what to observe in multi-agent systems. In: Proceedings of the 20th Belgian-Netherlands Conference on Artificial Intelligence, pp. 83–90. University of Twente Press, Enschede (2009)
3. Spaan, M., Melo, F.S.: Interaction-driven Markov games for decentralized multiagent planning under uncertainty. In: Proceedings of the 7th International Conference on Autonomous Agents and Multiagent Systems, pp. 525–532. IFAAMAS, Richland (2008)
4. Kok, J.R., Hoen, P., Bakker, B., Vlassis, N.: Utile coordination: Learning interdependencies among cooperative agents. In: Proceedings of the Symposium on Computational Intelligence and Games, pp. 29–36. IEEE Press, New York (2005)
5. Kok, J.R., Vlassis, N.: Sparse tabular multiagent Q-learning. In: Annual Machine Learning Conference of Belgium and the Netherlands, pp. 65–71. Universiteit Twente Press, Enschede (2004)
6. Kok, J.R., Vlassis, N.: Sparse cooperative Q-learning. In: Proceedings of the 21st International Conference on Machine Learning, pp. 61–68. ACM Press, New York (2004)

TATM: A Trust Mechanism for Social Traders in Double Auctions

Jacob Dumesny[1], Tim Miller[1,*], Michael Kirley[1], and Liz Sonenberg[2]

[1] Dept. of Computer Science & Software Engineering, University of Melbourne
[2] Dept. of Information Systems, University of Melbourne
j.dumesny@ugrad.unimelb.edu.au,
{tmiller,mkirley,l.sonenberg}@unimelb.edu.au

Abstract. Traders that operate in markets with multiple competing marketplaces can use learning to choose in which marketplace they will trade, and how much they will shout in that marketplace. If traders are able to share information with each other about their shout price and market choice over a social network, they can trend towards the market equilibrium more quickly, leading to higher profits for individual traders, and a more efficient market overall. However, if some traders share false information, profit and market efficiency can suffer as a result of traders acting on incorrect information. We present the *Trading Agent Trust Model* (TATM) that individual traders employ to detect deceptive traders and mitigate their influence on the individual's actions. Using the JCAT double-auction simulator, we assess TATM by performing an experimental evaluation of traders sharing information about their actions over a social network in the presence of deceptive traders. Results indicate that TATM is effective at mitigating traders sharing false information, and can increase the profit of TATM traders relative to non-TATM traders.

1 Introduction

Niu et al. [8] demonstrate that competition between marketplaces is reflected directly by the migration of traders between those marketplaces. Traders migrate based on estimates of expected profits, derived from the trader's past experience with that specialist.

Individual traders can improve their strategies based on shared information. Intra-marginal traders — those sellers (buyers) whose shout price is below (above) the market clearing price, and are therefore successfully matched – could communicate to fellow intra-marginal traders about marketplaces that are highly efficient, which would lead to an increase in the number of intra-marginal traders in that marketplace, thus increasing profits for both trader and marketplace. Furthermore, intra-marginal traders can communicate to extra-marginal traders — those sellers (buyers) whose shout price is above (below) the marking clearing price, and are therefore not matched — which provides the extra-marginal traders with some bounds on the market clearing price in a given marketplace.

* Corresponding author.

D. Wang and M. Reynolds (Eds.): AI 2011, LNAI 7106, pp. 402–411, 2011.

However, traders can share false information. For example, deceitful sellers can communicate that a successfully matched trade had a higher price than is true, encouraging extra-marginal buyers and sellers to increase their price to obtain a match. In addition, deceitful sellers can falsely claim a high price was obtained in another marketplace, thus encouraging other intra-marginal sellers in its marketplace to migrate away, leaving less competition. Such false information has the potential to disrupt a market by increasing the profit of deceitful traders at the expense of other traders and market specialists themselves.

In this paper, we present the *Trading Agent Trust Model* (TATM) that individual traders employ to detect deceptive traders and mitigate their influence on the individual's actions. TATM is a simple trust mechanism based in part on the FIRE model [4]. Traders employing TATM receive information from their neighbours on a social network that outlines their shout information from the previous trading day, such as price and marketplace choice. TATM traders will mimic their neighbours on some trading days, and use their own success to judge whether their neighbour is truthful or deceitful.

Using the JCAT double-auction simulator [8], experimental evaluate TATM traders sharing information about their actions over a social network in the presence of deceptive traders. Results indicate that TATM is effective at mitigating traders sharing false information. The profit of deceitful traders is reduced in the presence of TATM traders, but in most cases, still remains higher than truthful traders, and the profit of TATM traders is increased compared to naïve traders that employ no trust model.

2 Related Work: Trust and Reputation Mechanisms

Enhancing decision making in trading markets by mimicing successful peers has the potential to improve both individual agent performance and market efficiency. However, notions of *trust* and *reputation* must be considered if reliable estimates of peer ability are to be constructed. In this section, we review key trust and reputation models from the multi-agent systems literature.

McKnight and Chervany [6] identify four primary categories of trust: *competence*, *integrity*, *benevolence*, and *predictability*, which can be used to facilitate effective interactions and cooperation between agents. Typically, trust models consider a variety of information sources that are combined to determine a measure of trust according to the specific preferences of the agent. Closely related to trust is an agent's reputation, which can be thought of as an assessment based on the history of interactions with, or observations of, other agents [4,11].

Recently, Castelfranchi and Falcone [1] have provided an elaborate analysis of the role of trust in agent-based systems, with a focus on autonomous cognitive agents, but including cultural, institutional, technical, and normative dimensions. This endeavour to provide the foundations of a general theory of trust has attracted some critique, for example [2].

In an earlier trust model, Marsh [5] considered a local trust dimension derived directly from agent interactions. The trust value was a probability value based on independent criteria with additional ad-hoc factors associated with *risk* and *importance*. Mui et al. [7] proposed a model that extended this idea by incorporating a multi-part reputation metric derived from components embedded in a social network. In this model, individual reputation could be derived from direct observation of other agents, or from inferences based on information gathered from a social network. In subsequent work, Smith and Desjardins [9] incorporated aspects of *competence* and *integrity* into a formal framework for decision making based on trust and reputation. Two key phases were used in their framework: (i) assessing the capability of an agent to fulfil its stated commitments; and (ii) applying this knowledge to make effective decisions when interacting with other agents.

The FIRE model [4] integrates four different information sources to produce a comprehensive assessment of an agent's likely performance. FIRE uses a single composite trust-reputation value derived from: *interaction trust, role-based trust, witness reputation* and *certified reputation.* Both direct interactions and social network interactions (witness and certified) are used in trust calculations.

A notable trust model from the recommender systems domain is proposed by Walter et al. [11]. In their model, agents use their social network to gather information and use trust relationships to filter the collected information. Recommendations from neighbours may be received directly or indirectly via the larger pool of connected agents in the network. The trust values provide a ranking of the recommendations received. A probabilistic selection mechanism is then used for decision making.

3 The Trading Agent Trust Model

TATM employs a trust and reputation mechanism that aims to detect deceptive traders on a social network used for sharing trade information. Fundamentally, TATM is a reinforcement learning-based model, consisting of three components:

1. a return updating policy for estimating the trustworthiness of its neighbours, based on the interactions it has had with these neighbours;
2. an action choosing policy for deciding which neighbour on the social network is to be imitated in the next round; and
3. a decision-making strategy for mimicking the marketplace selection and last shout placed by a neighbour.

3.1 Returning Updating Policy

The return updating policy employed in TATM is based on a component of the FIRE model [4], presented in Section 2. Interaction trust is built from the direct experience of an agent. Specifically, each agent rates its partner's performance after every transaction and stores its ratings. When an agent requires the trust value of another agent, it calculates this based on the past ratings using a rate weighting function that favours more recent interactions. The rating recency function is given by the formula:

$$w(r_i) = e^{\frac{-\Delta t(r_i)}{\lambda}} \tag{1}$$

in which r_i the rating of a particular interaction, $\Delta t(r_i)$ is the time that has elapsed since that interaction, and λ is a parameter used to modify the decay of a rating (a lower value of λ means that older ratings are weighted lower).

The return updating policy must be customised for a particular domain. In JCAT, traders are given an upper bound on the number of commodities they can trade each day. On a given day, each trader attempts to trade as many of these as possible at its shout price in a specified marketplace. The shout price, marketplace, and number of trades made is shared on the social network after each trading day, as well how many successful trades were made.

In TATM, a trader mimics a neighbour's shout (see Section 3.2 for a discussion of neighbour selection) by using the same shout price and marketplace (all traders have the same maximum number of trades). After playing this strategy, the trader updates its feedback for that neighbour using a parameter ϵ:

1. $r_i = -\epsilon$ if the trade resulted in a *smaller* numbers of trades than specified by the neighbour. We refer to this as a deceptive case.
2. $r_i = \epsilon$ if the trade resulted in *more than or equal to* the number of trades specific by the neighbour. We refer to this as a non-deceptive case.

However, in a double auction, traders can take advantage of their private information to *preemptively* determine deceit. For example, if a seller receives information that a neighbour had sold its allocation at price q in market m, and the seller itself placed shouts at price p where $p < q$, also in market m, that were not matched, then it is highly likely that the neighbour is attempting to deceive; otherwise it is likely that the seller would also have received successful matches. As a result, TATM uses the following preemptive rules:

1. $r_i = -2\epsilon$ if a seller (buyer) indicates a successful shout $\langle q, m\rangle$, and the trader's own shout $\langle p, m\rangle$, where $p < q$ ($p > q$), was not matched.
2. $r_i = -\epsilon$ if a seller (buyer) and the trader's own shout $\langle p, n\rangle$, where $p < q$ ($p > q$), was not successfully matched. The traders are in different markets (m and n), so we are less sure that the neighbour is deceptive.

In each of these cases, the neighbours trade is not mimicked, and the trader reverts to its underlying strategy. This feedback policy is simple and does not detect the *degree* to which a neighbouring trader is deceptive by, for example, measuring the difference in the number of trades.

3.2 Action-Choosing and Decision-Making Policies

To choose which neighbour to mimic, a trader calculates a score for each neighbour as a function of its trust value for that neighbour and the neighbour's claim of its performance on the previous trading day. We consider only the previous day for simplicity. The trust of each neighbour, a, which we will call the neighbour's *Q-value*, is given by the average of all interactions (Equation 1):

$$Q(a) = \frac{\sum_{i=1}^{n} w_a(r_i)}{n} \tag{2}$$

in which n is the number of interactions between the trader and its neighbour.

Each trader should explore its neighbours to obtain recent feedback, but should also exploit the knowledge it has built up over time. This *exploit-vs-explore* dilemma is addressed using a softmax action selection method that uses a Boltzmann distribution [10]. Using this strategy, an agent explores its environment first and then gradually moves its stance towards exploitation when it learns more about the environment. Thus, the trading agent chooses an action a with the probability of:

$$P(a) = \frac{e^{Q(a)/\tau}}{\sum_{b=1}^{n} e^{Q(b)/\tau}}, \tag{3}$$

in which $Q(a)$ is the value of an action based on feedback from previous applications of that action (in our case, interactions with a neighbour), and τ is a positive number that dictates how much of an influence the past data has on the decision. A high τ value specifies a low influence, while a low value causes them to be close to their $Q(a)$ values. A parameter, $\alpha \in (0..1]$, specifies a rate of decay such that after each action, the value of τ becomes $\tau_0 \cdot \alpha$, in which τ_0 is the value of τ in the previous round.

The selection of a neighbour is governed by the following formula:

$$A = \begin{cases} \max\{a \mid profit(a) \times P(a)\} & \text{(where the trader is a seller)} \\ \min\{a \mid profit(a) \times P(a)\} & \text{(where the trader is a buyer)} \end{cases} \tag{4}$$

in which $profit(a)$ is trader a's profit from the previous day (see Equation 7).

Therefore, the score for each neighbour is the multiple of its shout price from the previous day and its trust value relative to other neighbours. A seller (buyer) chooses the neighbour with the highest (lowest) score. If and only the neighbour's shout price is greater than the traders, the trader will mimic the neighbour.

4 Experimental Setup

We use JCAT 0.17[1] to run CAT simulations to examine the effectiveness of the TATM model. We measure that average daily profit for each type of trader, as well as the global allocative efficiency, which is a measure of social welfare.

4.1 Traders

We implemented four different types of trading agents: 1) a naïve truthful trader (no trust model); 2) a naïve deceptive trader; 3) a TATM truthful trader; and 4) a TATM deceptive trader.

Deceptive traders. Deceptive traders deceive their neighbours by modifying their shout information before sharing it with their neighbours. There are two pieces of information that are modified: shout price, and the number of matches achieved. A parameter, $\delta > 0$, specifies the amount by which this information is modified. Given a shout of p in which the trader received n matches, a deceptive trader will share the false information:

[1] `http://jcat.sourceforge.net/`.

$$\begin{array}{ll} \langle p \times (1+\delta), n \times (1+\delta)\rangle & \text{(where the trader is a seller)} \\ \langle p \times (1-\delta), n \times (1+\delta)\rangle & \text{(where the trader is a buyer)} \end{array} \quad (5)$$

That is, a seller will attempt to raise the general price level of the market, while a buyer will attempt to lower it, thus pushing intra-marginal traders of the same type to be extra-marginal, and inducing traders of the opposite type into the intra-marginal range.

Naïve traders. Naïve traders employ a system in which they simply choose the neighbour with the best offering. That is, they have no learning mechanism to determine the trustworthiness of neighbours, and they simply choose a neighbour to mimic using the following:

$$A = \begin{cases} \max\{a \mid profit(a)\} & \text{(where the trader is a seller)} \\ \min\{a \mid profit(a)\} & \text{(where the trader is a buyer)} \end{cases} \quad (6)$$

Underlying strategies. In our experiments, when a trader chooses not to mimic a neighbour, it employs its own *underlying* strategies, which are the *zero-intelligence constrained* strategy for shout prices, which chooses a random value between the minimum and maximum range, provided that this does not result in a loss for the trader, and random market selection.

4.2 Experiment Variables and Parameters

The independent variables of the experiment are the type of trader. We run two sets of experiments: one in which all traders are naïve (the non-TATM markets), and one in which all traders employ TATM (the TATM markets). To help generalise the results, we vary other parameters in the experiment. For both sets of experiments, we modify the following two parameters:

1. **Number of deceptive traders** (ξ) — We vary the ratio of deceptive traders to non-deceptive in the market from 0.1–0.9, in intervals of 0.1.
2. **Deceit level of deceptive traders** (δ) – We vary the degree to which deceptive traders exaggerate their success from 0.1–1.0, in intervals of 0.1.

We run all pairwise combinations of these parameters, resulting in 90 different configurations in each experiment. Each configuration is run 30 times and each game lasts 400 days. The results to be presented in the next section are averaged over the total 12,000 days. Traders are each allowed to trade three units of goods each day and their private values are drawn from the uniform distribution between 50 and 100. Other parameters are held constant. Each marketplace runs a *continuous double auction* [3]. We run five marketplaces in each experimental run, in which each marketplace charges at a different level on the profit of traders: 0%, 20%, 40%, 60%, and 80% respectively. Traders operate on a 14 × 14 toroidal grid social network, evenly divided between sellers and buyers, and with neighbours randomly assigned; that is, on aggregate, sellers are connected to an even number of buyers and sellers.

Measures. In these experiments, we record two measures. First, we measure the *mean trader type profit*, which is the mean daily profit over all simulations of each type of trader (random, TATM, and deceptive). The daily profit for a trader i is:

$$profit(a) = \begin{cases} (n \times |v_a - p_a|) - f_a & \text{(where } p_a > 0) \\ -f_a & \text{(where } p_a = 0) \end{cases} \tag{7}$$

in which v_i is the private valuation of trader i, p_i is the price of the trade made by trader i, n is the number of successful trades, and f_i are the fees paid by trader i. In the case that a trader does not make a successful trade that day, they lose the fees charged by the marketplace.

The mean daily profit of a trader type on a single day is:

$$P = \frac{\sum_{i=m}^{n} pr_i}{N} \tag{8}$$

in which traders $m..n$ are the traders of a particular type.

Second, we measure the *global allocative efficiency*, which measures how close the entire market is to trading at the equilibrium price, where the *equilibrium price* is defined as the price at which demand equals supply when all traders offer to buy or sell at their private value, assuming that all traders in the market can trade with each other. The global allocative efficiency is calculated using:

$$E = \frac{\sum_j \sum_i |v_i^j - p_i^j|}{\sum_j \sum_i |v_i^j - p_0|} \tag{9}$$

in which p_0 is the equilibrium price of the market, v_i^j is the private value of trader i in marketplace j, and p_i^j is the price paid by trader i in marketplace j.

5 Results

Figure 1 plots the mean of daily trader profit (Equation 8) for all 90 configurations of the experiment over the 30 iterations for the non-TATM market. These plots are included to illustrate the effect of the experiment parameters. The plots for global efficiency and for the TATM-market look similar: a clear downward trend as the deceit level increases, so these plots are omitted for brevity.

From these figures, we can see a clear downward trend in profit as the deceit level of the deceptive traders increases. Surprisingly, the number of deceptive traders has little impact on either trader profit or efficiency. This minimal impact can be explained by the fact that the number of deceptive sellers is in balance with the number of deceptive buyers, and on aggregate, each trader is connected to an equal number of buyers and sellers. As a result, when a seller (buyer) deceives another trader by increasing (decreasing) their previous shout by the specified deceit level, the receiving traders' new shout is likely to be matched by a trader of the opposite type. The trend downwards as deceit level increases is expected. The probability of getting a match reduces, first, as the range of shouts starts to increase, and second, as traders' shout values move around the range instead of moving towards the market equilibrium.

A more important result is the effect of the trust model. Figure 2 shows the difference in mean profit (expressed as a percentage) between deceptive and truthful in the non-TATM market (Figure 2a) and in the TATM market (Figure 2b). The horizontal plane shows 0%, making it easier to see the distinction between a negative and positive change.

From Figure 2a, we can see that deceptive traders perform better than naïve truthful traders, except in the cases in which the deceit level is 0.9. This sharp spike is likely due to the fact that profit obtained by the deceptive traders themselves becomes so poor that they will mimic naïve truthful traders. Figure 2b indicates that the impact of deceit can be mitigated using TATM. Truthful TATM traders outperform their deceptive counterparts for deceit levels $0.7-0.9$, and the difference between the two for other parameters is significantly lower, bottoming at just above 2% compared with almost 6% for the naïve traders.

Figure 3 shows the inter-market comparison of deceptive traders and truthful traders. It is important to note the different ranges on the Z axes between Figures 3a and 3b. Figure 3a shows the percentage change in mean trader profit for truthful agents between the non-TATM markets and the TATM markets respectively. This figure demonstrates that employing the TATM model results in a higher trader profit for all of the parameters, and that the higher the level of deceit, the larger the change. This upward trend is because as deceptive agents increase their deceit level, deceit becomes easier to identify.

Figure 3b plots the same data for the deceptive traders, showing some interesting results. First, even in the presence of the TATM model, deceit can be beneficial. However, this only holds if there are few other deceptive traders in the market. We attribute this increase in profit to the fact that the deceptive traders themselves are employing the TATM model, so are less likely to mimic other deceptive traders. As the level of deceptive traders increases, being deceptive becomes less profitable.

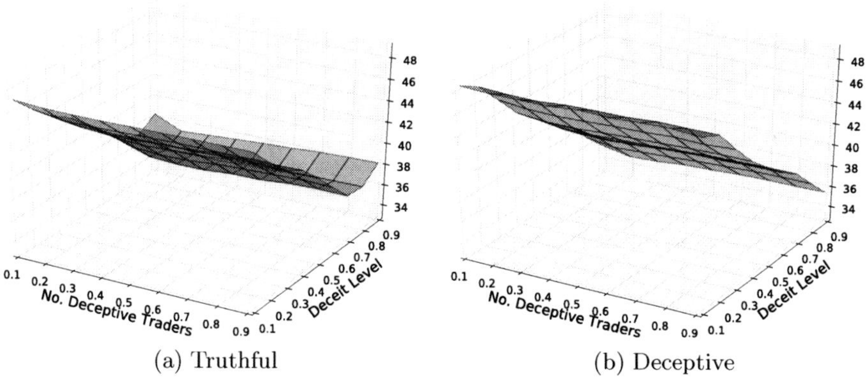

(a) Truthful (b) Deceptive

Fig. 1. Mean trader profit per type (truthful or deceptive) for the non-TATM market

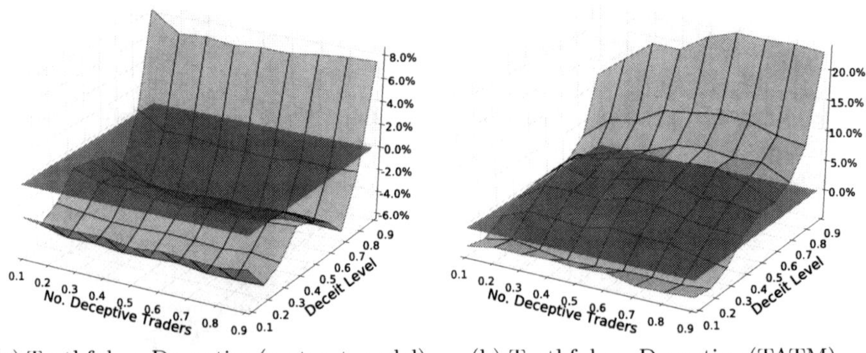

(a) Truthful vs. Deceptive (no trust model). (b) Truthful vs. Deceptive (TATM).

Fig. 2. Plots of the difference in mean profit between deceptive and truthful agents in the two experiments respectively (intra-market comparison), expressed as a percentage. This is calculated as $(P_B - P_A)/P_A$, where P_A and P_B are mean trader profit Equation 8), for A vs. B. Note the different limits on the Z axis. The gray plane is 0%.

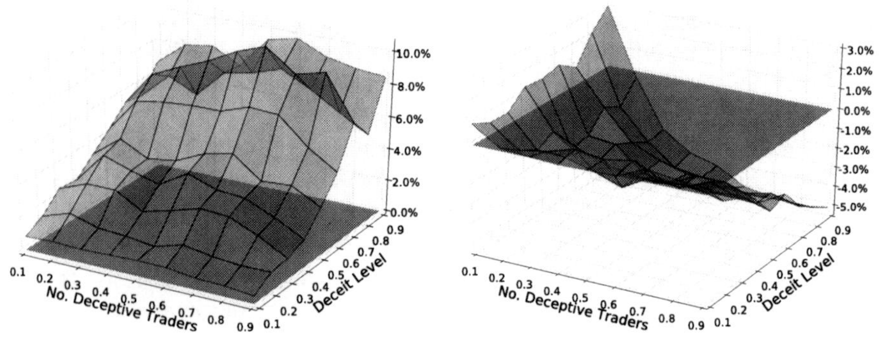

(a) Truthful(no trust model) vs. Truthful (TATM). (b) Deceptive(no trust model) vs. Deceptive(TATM).

Fig. 3. Plots of the percentage change of mean profit between deceptive traders in each experiment, and truthful traders in each experiment (inter-market comparison)

6 Discussion and Conclusions

Our results demonstrate that employing TATM is always preferably to a base-line "no trust" model, as the mean daily profit achieved by traders is higher than their naïve counterparts for all experiment configurations.

The TATM model reduces the effects of deceptive traders, but these effects cannot be completely eliminated. The TATM model also helps to mitigate the differences between truthful and deceptive traders. While deceptive traders increased their profit in some experimental runs of the TATM market, this is attributed to themselves employing the TATM model. However, the difference between the truthful and deceptive traders is smaller in the TATM markets. Market efficiency also improves in the TATM model, except when there are a high number of deceptive traders with a high deceit level. In these particular cases, the deceptive traders perform worse themselves.

Overall, the conclusions support our hypothesis that a simple trust model such as TATM can mitigate the problems of deception in markets.

In future work, we plan to investigate indirect information sharing within a social network and extending the TATM model to handle this. We also plan to investigate how TATM can be improved to further mitigate the effects of deceptive traders.

Acknowledgements. The authors thank Peter McBurney of King's College London for his insight into this work, and the University of Melbourne Visiting Scholar's Scheme for funding Peter's visit to Melbourne.

References

1. Castelfranchi, C., Falcone, R.: Trust Theory: A Socio-Cognitive and Computational Model. John Wiley & Sons, Ltd. (2010)
2. Elsenbroich, C.: Review of Trust Theory: A socio-cognitive and computational model: Castelfranchi, Cristiano and Falcone, Rino. Journal of Artificial Societies and Social Simulation 14(2) (2011)
3. Friedman, D.: The double auction institution: A survey. In: Friedman, D., Rust, J. (eds.) The Double Auction Market: Institutions, Theories and Evidence, ch. 1, pp. 3–25 (1993)
4. Huynh, T.D., Jennings, N.R., Shadbolt, N.R.: An integrated trust and reputation model for open multi-agent systems. Autonomous Agents and Multi-Agent Systems 13(2), 119–154 (2006)
5. Marsh, S.: Formalising Trust as a Computational Concept. PhD thesis, University of Stirling (1994)
6. McKnight, D.H., Chervany, N.L.: Trust and Distrust Definitions: One Bite at a Time. In: Falcone, R., Singh, M., Tan, Y.-H. (eds.) AA-WS 2000. LNCS (LNAI), vol. 2246, pp. 27–54. Springer, Heidelberg (2001)
7. Mui, L., Mohtashemi, M., Halberstadt, A.: A computational model of trust and reputation. In: Proceedings of the 35th Annual Hawaii International Conference on System Sciences, pp. 2431–2439. IEEE (2002)
8. Niu, J., Cai, K., Gerding, E., McBurney, P., Parsons, S.: JCAT: A platform for the TAC market design competition. In: Proc. of 7th Int. Conf. on Autonomous Agents and Multiagent Systems, pp. 1649–1650. IFAAMAS (2008)
9. Smith, M.J., Desjardins, M.: Learning to trust in the competence and commitment of agents. Autonomous Agents and Multi-Agent Systems 18(1), 36–82 (2009)
10. Sutton, R.S., Barto, A.G.: Reinforcement Learning: An Introduction (1998)
11. Walter, F.E., Battiston, S., Schweitzer, F.: A model of a trust-based recommendation system on a social network. Autonomous Agents and Multi-Agent Systems 16(1), 57–74 (2008)

Sequential Single-Cluster Auctions for Robot Task Allocation

Bradford Heap and Maurice Pagnucco

ARC Centre of Excellence in Autonomous Systems,
School of Computer Science and Engineering, UNSW,
Sydney, NSW, 2052, Australia
{bradfordh,morri}@cse.unsw.edu.au

Abstract. Multi-robot task allocation research has focused on *sequential single-item auctions* and various extensions as quick methods for allocating tasks to robots with small overall team costs. In this paper we outline the benefits of grouping tasks with positive synergies together and auctioning clusters of tasks rather than individual tasks. We show that with task-clustering the winner determination costs remain the same as sequential single-item auctions and that auctioning task-clusters can result in overall smaller team costs.

1 Introduction

Consider a team of autonomous mobile robots operating in an office-like environment. These robots may be required to deliver documents between departments, clean up spillages, or act as tour guides to visitors. In many situations there will be a set of tasks to be completed and we wish for the robots to distribute these tasks amongst themselves in a manner that satisfies a global team objective. Recently, multi-robot cooperative auctions have become a popular approach for solving task-allocation problems [3].

We can achieve an optimal allocation of a set of tasks to robots using a *single-round combinatorial auction*. However, in most situations where there are many tasks, combinatorial auctions fail to perform efficiently due to high communication and winner determination costs [1]. As an alternative, much of the research focus has been on the use of *sequential single-item auctions* (SSI auctions) for task allocation over multi-round auctions [6]. Although SSI auctions produce a team cost that is at least as large as combinatorial auctions, they have much lower communication and winner determination costs which results in a much quicker allocation of tasks. To lower the team cost in SSI auctions researchers have looked at improvements and extensions to the bidding phases of SSI auctions through the use of techniques like *rollouts*, *regret clearing* and *bundle-bids* (the interested reader is referred to [5,7]).

SSI auctions with bundles are an interesting hybrid of standard SSI auctions and combinatorial auctions in which each robot can bid on dynamic combinations of up to k tasks and, during the winner determination phase, a robot can

D. Wang and M. Reynolds (Eds.): AI 2011, LNAI 7106, pp. 412–421, 2011.

be allocated between $0 - k$ tasks. In general, this approach results in lower team costs as each bundle bid takes into account more synergies between tasks, however, because of the additional calculations involved in the bidding and winner determination phases it performs a lot slower than standard SSI auctions.

In this paper we extend the idea of bidding on a collection of tasks and allow robots to bid on fixed clusters of tasks where a robot will either win all items in the cluster or none. We show empirically that this method results in lower team costs than standard SSI auctions and performs much faster than SSI auctions with bundle bids. More specifically, we demonstrate that SSC auctions result in lower MiniMax distances than SSI auctions when the number of robots is greater than 2. Moreover, for the MiniSum team objective, SSC auctions perform well when the capacity constraint is small.

2 Multi-robot Task-Allocation

We formalise the definition of the task-allocation problem in the same manner as Koenig *et al.* [7]. Given a set of robots $R = \{r_1, \ldots, r_m\}$ and a set of tasks $T = \{t_1, \ldots, t_n\}$, any tuple $\langle T_{r_1}, \ldots, T_{r_m} \rangle$ of pairwise disjoint bundles $T_{r_i} \subseteq T$ and $T_{r_i} \neq T_{r_j}$ for $i \neq j$, for all $i = 1, \ldots, m$, is a partial solution of the task-allocation problem. This means that robot r_i performs the tasks T_{r_i}, and no task is assigned to more than one robot. To determine a complete solution to the task-allocation problem we need to find a partial solution $\langle T_{r_1} \ldots T_{r_m} \rangle$ with $\cup_{r_i \in R} T_{r_i} = T$, that is, where every task is assigned to exactly one robot.

The standard testbed of the task-allocation problem is multi-robot routing. The tasks represent locations to visit. Robots know their locations and can calculate the costs between locations. We assume costs are symmetric, $\lambda(i,j) = \lambda(j,i)$ and are the same for all robots. The robot cost $\lambda_{r_i}(T_{r_i})$ is the minimum cost for an individual robot r_i to visit all locations T_{r_i} assigned to it. There can be synergies between tasks, such that, $\lambda_{r_i}(T_{r'}) + \lambda_{r_i}(T_{r''})$ may not equal $\lambda_{r_i}(T_{r'} \cup T_{r''})$. A positive synergy is when $\lambda_{r_i}(T_{r'} \cup T_{r''}) < \lambda_{r_i}(T_{r'}) + \lambda_{r_i}(T_{r''})$. Robots can also have capacity constraints where they can have at most a fixed number of tasks. We wish to find a solution to the task-allocation problem that achieves a team objective. In this paper we study two common team objectives:

MiniMax. $\max_{r_i \in R} \lambda_{r_i}(T_{r_i})$, that is to minimise the maximum distance each individual robot travels.

MiniSum. $\sum_{r_i \in R} \lambda_{r_i}(T_{r_i})$, that is to minimise the sum of the paths of all robots in visiting all their assigned locations.

These two team objective result in different allocations of tasks due to how each robot calculates their bids incorporating synergies between tasks. Lagoudakis *et al.* [8] explores these differences in more detail.

3 Sequential Auctions with Clusters

Auction-based methods for task allocation have become increasingly popular in the recent literature. An auction is composed of three separate phases: the *initial phase* in which an auctioneer sends a request to all robots indicating the tasks up for auction; a *bidding phase* in which each robot evaluates the tasks up for auction and responds with a bid for those in which it is interested; and, a *winner determination* phase in which the auctioneer determines the winner for each task. Common auction types include *combinatorial auctions*, *parallel auctions* and *sequential auctions*. In combinatorial auctions each robot bids on all subsets of the tasks on offer. This yields optimal results but the computation tends to be intractable and is certainly not feasible for any but the smallest scenarios. In parallel auctions the robots develop a bid for each task and the auctioneer then allocates the tasks all at once. The computational complexity is minimal but solutions are likely to be sub-optimal. Sequential auctions represent a compromise between these two extremes. They progress over several rounds in which a subset of tasks is auctioned in each round. In the case of SSI auctions, one item (i.e., task) is auctioned in each round.

We now develop an extension to SSI auctions in which individual tasks are organised into clusters taking into account positive synergies between tasks. Robots bid on these clusters to solve the task-allocation problem. We call this *sequential single-cluster auctions* (SSC auctions). An SSC auction consists of three phases: clustering phase, bidding phase, and winner determination phase. Initially, all tasks are unassigned. Before the auction, a clustering algorithm is used to allocate all individual tasks into a cluster with the goal of maximising the positive synergy between tasks in each cluster (clustering phase). Each task can be assigned to one, and only one cluster. Clusters can be of varying sizes. During each round, all robots bid on all unassigned task clusters (bidding phase), the auctioneer then determines the winner and assigns the winning cluster to the winning robot (winner determination phase). The winning robot must then complete all tasks in that cluster.

Clustering Phase: Expanding upon our definition of the task-allocation problem given in Section 2 we introduce the set of clusters $C = \{c_1, \ldots, c_o\}$. We now need to allocate all tasks to one and only one cluster. This is achieved by taking any tuple $\langle T_{c_1}, \ldots, T_{c_o} \rangle$ of pairwise disjoint bundles $T_{c_j} \subseteq T$ for all $j = 1, \ldots, o$ that satisfies $\cup_{c_j \in C} T_{c_j} = T$. For multi-robot routing the synergy between tasks is represented by the distance between them. Tasks with a large distance separating them have a low synergy, whereas, tasks with a small distance have a high positive synergy. In this paper, we use the standard k-means algorithm [4] for clustering tasks during the empirical experimentation. However, our proposal does not depend upon k-means and other clustering methods that satisfy these properties may produce better results.

Once we have organised all tasks into clusters we must ensure that all clusters are allocated to one and only one robot. We do this by taking any tuple $\langle C_{r_1}, \ldots, C_{r_m} \rangle$ of pairwise disjoint bundles $C_{r_i} \subseteq C$ for all $i = 1, \ldots, m$ that satisfy $\cup_{r_i \in R} C_{r_i} = C$. As a result of this we have now allocated all tasks into clusters, and assigned all clusters to robots and therefore it holds that we still have a valid solution to the task-allocation problem of all tasks being allocated such that each task is allocated to one and only one robot.

Now we consider a single round of a SSC auction. We assume that robot $r_i \in R$ has already been assigned the set of task clusters $C_{r_i} \subseteq C$ in previous rounds for all $r_i \in R$. Therefore $U = C \setminus \cup_{r_i \in R} C_{r_i}$ is the set of unassigned task clusters. Let a bid b be a triple of a robot b_r, a task cluster b_c and a bid cost b_λ, such that, $b = \langle b_r, b_c, b_\lambda \rangle$.

Bidding Phase: The set of submitted bids $B = \{b_1, \ldots, b_m\}$ satisfies: 1) for all $b \in B$, it holds that $b_r \in R$ and $b_c \in U$; and 2) for all $r_i \in R$ and $c' \in U$ there exists exactly one bid $b \in B$ with $b_r = r_i$ and $b_c = c'$. That is each robot submits one bid on each task cluster. For the MiniMax team objective, $b_\lambda = \lambda_{b_r}(C_{b_r} \cup \{b_c\})$. That is the robot bids the costs to do all tasks assigned to it plus the tasks in the cluster it is bidding on. For the MiniSum team objective, $b_\lambda = \lambda_{b_r}(C_{b_r} \cup \{b_c\}) - \lambda_{b_r}(C_{b_r})$. That is the robot bids the increase in its costs for doing all of its currently allocated tasks plus the tasks in the cluster it is bidding on.

Winner Determination Phase: Once all bids have been received, the auctioneer evaluates a potentially winning bid $b' \in B$ according to the value b'_λ. The winning bid for both the MiniMax and MiniSum team objective is the bid b' with the smallest b'_λ. The auctioneer then assigns all tasks in the cluster b'_c to the robot b'_r.

4 Properties

We now describe the unique behavioural properties of SSC auctions. These properties allow SSC auctions to operate in an efficent manner and generally result in a small team cost.

1. The number of rounds in a SSC auction is no more than the number of rounds in a SSI auction.

   Proof: We define an SSI auction as the tuple $A_{ssi} = \langle R, T \rangle$ where R represents the set of available robots and T the set of tasks. The number of rounds in A_{ssi} is equal to the number of tasks, $N_{ssi} = |T|$, as only one task is allocated per round. We define an SSC auction is the tuple $A_{ssc} = \langle R, T, C \rangle$. The number of rounds in A_{ssc} is equal to the number of clusters, $N_{ssc} = |C|$, as one cluster is allocated per round. Each cluster can have one or more tasks, therefore, $|C| \leq |T|$, and as a result of this $N_{ssc} \leq N_{ssi}$.
2. Winner determination time in a SSC auction is equal to winner determination time in a SSI auction.

   Proof: In an SSI auction each bid b_s consists of a robot b_r, a task b_t, and a cost b_λ. In an SSC auction the structure of a bid remains the same, with the exception that b_t is replaced by b_c (as defined in Section 3). For winner determination, we have a set of bids B and the value of each b_λ is compared in the same manner in both auction frameworks and $|B|$ does not change. Therefore the winner determination time does not change.

   N.B. SSC winner determination time is much faster than SSI with bundles. This is because in SSI with bundles each bid must include b_λ for each combination of the k tasks that is being bid on. To determine the winner in SSI with bundles each b_λ for each combination needs to be compared to all other bids and combinations to determine the winner.

3. When clusters employ positive synergies between tasks the resultant team cost in a SSC auction is less than in a SSI auction.

   Take for example, the same task-allocation problem as *Exploration Task 4* in Koenig *et al.* [6] (Figure 1). In this problem an SSI auction fails to consider enough synergies between tasks and results in a less than optimal solution. For the MiniSum team objective the overall distance sum is 20 and the resultant paths for each robot to traverse are $r_1 \rightarrow t_2 \rightarrow t_1$ and $r_2 \rightarrow t_4 \rightarrow t_3$. For an SSC auction we define our clusters $c_1 = \{t_1, t_3\}$ and $c_2 = \{t_2, t_4\}$. Auctioning with the MiniSum team objective results in an allocation of c_1 to r_2 and c_2 to r_1 with the resultant paths $r_1 \rightarrow t_2 \rightarrow t_4$ and $r_2 \rightarrow t_3 \rightarrow t_1$. The overall distance sum is 15. However, it should be noted that if a cluster fails to employ synergies correctly SSC auctions may result in team costs that are worse than SSI auctions.

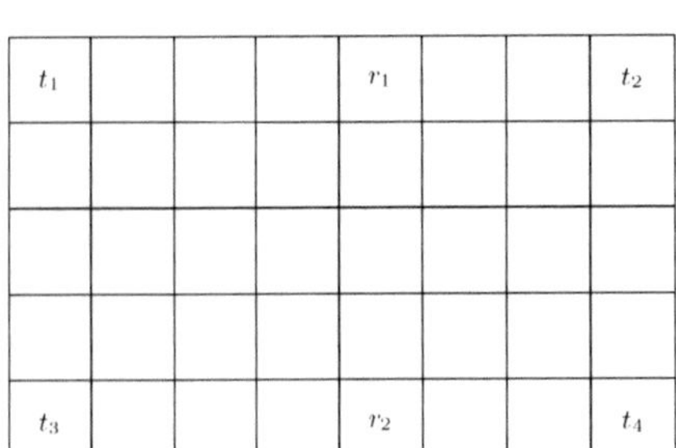

Fig. 1. Exploration Task 4 (Koenig *et al.* [6])

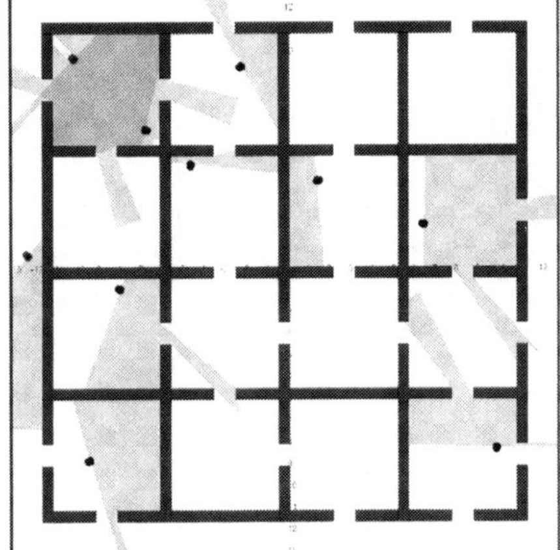

Fig. 2. Simulation of an office-like environment (cf. Koenig *et al.* [6])

5 Experiment Setup

To test SSC auctions we simulate an office-like environment (Figure 2) as in Koenig *et al.* [6]. For each experiment, doors between different rooms and the hallway are either opened or closed. We tested on 25 different randomly generated configurations of opened and closed doors with each robot in each configuration starting in a different random location which is standard in the literature and therefore provides a common setting for comparison. Robots can only travel between rooms through open doors and cannot open or close doors. In each experiment robots are set a fixed task-capacity constraint of the ratio of the number of tasks to the number of robots. Robots stop being allocated additional tasks once these capacities are met. For each configuration we test with $|R| \in \{2, 4, 6, 8, 10\}$ and $|T| \in \{6, 7, \ldots, 60\}$.

We use standard k-means clustering to quickly create clusters of geographically close tasks to be auctioned. It is important to note that k-means clustering does not take into account walls and closed doors. This means that it is possible for tasks to be clustered together that may have a large navigational distance between them (low synergy). However, this approach best represents a real world situation where it would be extremely complex to always create an optimal grouping of tasks. For our experiments we test two different total numbers of task-clusters. Our first experiment uses a cluster count of half the number of tasks, and the second uses a cluster count of two-thirds the number of tasks.

For each auction round robots bid on the cluster that will result in the lowest increase to the team objective. To determine their bid cost each robot needs to solve a version of the travelling salesperson problem (TSP) where it needs to travel to all tasks allocated to it but does not return to its initial location. Solving the TSP is an NP-Hard problem so we need to approximate the true cost. We do this by using the cheapest-insertion heuristic to add new tasks into our path and then use the two-opt heuristic [2] to improve our solution.

To compare the effectiveness of SSC auctions we also run parallel, SSI, and SSI with bundles auctions on the same 25 configurations. For SSI with bundles we test $k = 2$ and $k = 3$ with a *non-cautious auctioneer*, that is, all k tasks are allocated in each round. Furthermore, we test *hard* and *soft* capacity constraints for SSI with bundles. Hard capacity constraints ensure that all robots are allocated exactly their capacity of tasks. Soft capacity constraints allow robots to go slightly over their capacity, provided they are under their capacity before the round winner determination and allocation. This comparison of capacity constraints is necessary because SSC auctions may result in allocations where robots are slightly over their capacities because of the requirement that all tasks in a cluster are allocated to the same robot.

6 Results

We begin our analysis with the MiniMax Team Objective with the mean experimental results shown in Table 1. We observe that in all Robot/Task combinations tested that SSC auctions result in a lower mean MiniMax result than SSI auctions. Overall there is an average MiniMax distance reduction of 20% where the number of clusters $|C| = \frac{1}{2}|T|$ and a reduction of 25% where the number of clusters $|C| = \frac{2}{3}|T|$. However, $|C| = \frac{2}{3}|T|$ does not result in lower mean MiniMax distances than $|C| = \frac{1}{2}|T|$ in all Robot/Task combinations. We also note that SSI auctions with bundles also result in lower mean MiniMax distances than both standard SSI auctions and SSC auctions. Interestingly SSI auctions with bundles where $k = 3$ do not always result in lower results than SSI auctions with bundles where $k = 2$ for all Robot/Task combinations. This result, however,

Table 1. Mean MiniMax Experimental Results

			Standard		SSC		SSI bundles $k = 2$		SSI bundles $k = 3$	
Capacity	Robots	Tasks	Parallel	SSI	$\lvert C\rvert = \frac{1}{2}\lvert T\rvert$	$\lvert C\rvert = \frac{2}{3}\lvert T\rvert$	Hard-Cap	Soft-Cap	Hard-Cap	Soft-Cap
3	2	6	1039	1130	1085	944	823	811	607	613
3	4	12	1094	1138	880	946	828	808	762	755
3	6	18	1060	1156	899	833	743	704	730	675
3	8	24	1199	1112	853	760	668	680	763	706
3	10	30	1092	1159	802	733	656	651	670	636
4	2	8	1318	1284	1242	1108	965	950	1060	1194
4	4	16	1430	1239	1034	1042	880	851	1038	969
4	6	24	1301	1352	1030	868	779	781	762	679
4	8	32	1310	1299	857	856	747	767	789	999
4	10	40	1438	1249	889	758	704	687	821	855
5	2	10	1464	1364	1260	1257	1132	1101	1326	1248
5	4	20	1545	1297	1138	1142	928	905	1001	1119
5	6	30	1485	1289	1087	1003	850	835	915	853
5	8	40	1506	1341	989	952	819	797	974	891
5	10	50	1574	1347	933	872	773	732	850	1051
6	2	12	1699	1690	1421	1459	1231	1197	1092	1117
6	4	24	1711	1457	1274	1142	1039	1010	972	923
6	6	36	1782	1409	1129	1051	840	884	1076	1061
6	8	48	1713	1463	1132	1012	907	812	894	964
6	10	60	1736	1492	957	909	836	813	928	856
Overall Mean:			1425	1313	1045	983	857	839	901	908

is consistent with Koenig's prior results for SSI auctions with bundles where a non-cautious auctioneer has been used [7]. Despite, SSI with bundles producing lower results than SSC the computational overhead is significantly higher and the consequences of this are discussed further below.

To confirm the validity of our results we perform *two-sample independent one-tailed t tests* comparing the SSC auction results to the SSI auction results for each Robot/Task combination. We define our null hypothesis as $H_0 : \mu A_{ssc} \geq \mu A_{ssi}$ and our alternative hypothesis as $H_a : \mu A_{ssc} < \mu A_{ssi}$, that is, we wish to prove that the mean result for SSC auctions are lower than SSI auctions. We declare any result a significant difference if the result of the t test P is less than 0.05, that is, the probability of the decrease between the mean results of SSC auctions compared to SSI auctions being a result of random variation is less than 5%.

The significance tests show that in all but three Robot/Task combinations we have a statistically significant reduction in the MiniMax distance, that is, we accept the alternative hypothesis. The non-significant results occur, in both $|C|$ sizes, when there are only 2 robots with total tasks $\{6, 8, 10\}$. However, in these scenarios we can expect that clustering will not perform well due to the low numbers of robots and tasks.

Finally we perform *two-sample independent two-tailed t tests* for the difference between the cluster sizes for all Robots/Tasks combinations ($H_0 : \mu A_{|C|=\frac{1}{2}|T|} = \mu A_{|C|=\frac{2}{3}|T|}$, $H_a : \mu A_{|C|=\frac{1}{2}|T|} \neq \mu A_{|C|=\frac{2}{3}|T|}$). Only two combinations, $\langle |R| = 6, |T| = 24\rangle$ and $\langle |R| = 10, |T| = 40\rangle$, result in a significant difference between the two cluster sizes, in which $|C| = \frac{2}{3}|T|$ produces the smallest distances. Overall we can conclude that SSC auctions result in lower MiniMax distances than SSI auctions when the number of robots is greater than 2.

Table 2. Mean MiniSum Experimental Results

			Standard		SSC		SSI bundles $k=2$		SSI bundles $k=3$	
Capacity	Robots	Tasks	Parallel	SSI	$\|C\|=\frac{1}{2}\|T\|$	$\|C\|=\frac{2}{3}\|T\|$	Hard-Cap	Soft-Cap	Hard-Cap	Soft-Cap
3	2	6	1653	1819	1589	1615	1661	1617	1398	1398
3	4	12	2757	2867	2331	2411	2243	2378	1997	1984
3	6	18	3580	3542	2864	2982	2643	2628	2284	2285
3	8	24	4723	4395	3366	3596	3191	3281	2489	2462
3	10	30	5057	4928	3764	3869	3394	3408	2751	2663
4	2	8	2085	1941	1889	1971	1857	1796	1850	1844
4	4	16	3564	3180	2783	2892	2641	2637	2514	2497
4	6	24	4417	4033	3448	3718	3268	3301	2612	2652
4	8	32	5428	4780	3749	4144	3607	3703	3125	3391
4	10	40	6370	5391	4371	4605	3998	4181	3183	3442
5	2	10	2378	2149	2202	2198	2154	2127	2444	2145
5	4	20	4026	3029	3170	3372	2981	3038	3019	2933
5	6	30	5129	4086	4044	4078	3677	3825	3025	2842
5	8	40	6334	4741	4549	4637	4221	4320	3529	3403
5	10	50	7087	5353	4745	5078	4526	4848	3850	3947
6	2	12	2834	2628	2397	2417	2482	2464	2402	2478
6	4	24	4435	3537	3512	3498	3360	3407	3014	2859
6	6	36	5941	4475	4302	4139	4207	4081	3377	3593
6	8	48	7234	5268	5028	5022	4753	4961	3791	3824
6	10	60	8059	5805	5523	5731	5289	5093	4007	4170
Overall Mean:			4654	3897	3481	3599	3308	3355	2833	2841

The mean results of the MiniSum Team Objective is shown in Table 2. Overall there is a mean MiniSum distance reduction of 12% where the number of clusters $|C| = \frac{1}{2}|T|$ and a reduction of 8% where the number of clusters $|C| = \frac{2}{3}|T|$ when compared to SSI auctions. However, in contrast to the MiniMax results, there is not a mean distance reduction in every Robot/Task combination. In particular, the combination $\langle |R| = 4, |T| = 20\rangle$ shows a substantial increase in the MiniSum distances in both cluster sizes. The results for our experiments using SSI with bundles show that they result in lower distances than SSI and SSC auctions. We observe that in experiments with bundle size $k = 3$ the mean distance is consistently lower than experiments with bundle size $k = 2$. This is in line with, and validates, Koenig's previous work on SSI with bundles.

We perform *two-sample independent one-tailed t tests* comparing the SSC auction results to the SSI auction results for those Robot/Task combinations where the SSC result is less than the SSI result ($H_0 : \mu A_{ssc} \geq \mu A_{ssi}$, $H_a : \mu A_{ssc} < \mu A_{ssi}$), that is, we test for a statistically significant decrease in the mean results of SSC compared to SSI. When the SSC result is greater than the SSI result we perform *two-tailed t tests* for a difference between the two samples ($H_0 : \mu A_{ssc} \neq \mu A_{ssi}$, $H_a : \mu A_{ssc} = \mu A_{ssi}$), that is, we test for no statistically significant difference between the mean results.

The results of these tests give an interesting partition of the data. In experiments where the robot capacity is 3 or 4 we confirm a significant result in the reduction of the mean MiniSum distances for all combinations except those where $|R| = 2$. However, in all cases where the robot capacity is 5 or 6 we get no significant difference between the SSI and SSC auctions, except, in the previously mentioned combination $\langle |R| = 4, |T| = 20\rangle$ with $|C| = \frac{2}{3}|T|$ which, in the two-tailed t tests, confirmed a significant increase in distance.

The MiniSum results are not in line with our predictions. The raw data appears to show SSC auctions mostly performing better than SSI auctions. However, our statistical testing does not confirm this. We can conclude that when the capacity constraint is small SSC auctions perform well. However, more experiments are needed to examine situations where robots are allocated many tasks. For instance using a different clustering algorithm, such as potential fields or graph partitioning, may produce a significant reduction in the distance.

Table 3. Mean Total Task-Allocation Determination Time *(seconds)*

			Standard		SSC		SSI bundles $k = 2$		SSI bundles $k = 3$	
Capacity	Robots	Tasks	Parallel	SSI	$\lvert C\rvert = \frac{1}{2}\lvert T\rvert$	$\lvert C\rvert = \frac{2}{3}\lvert T\rvert$	Hard-Cap	Soft-Cap	Hard-Cap	Soft-Cap
3	2	6	1.9	2.4	2.6	2.5	3.3	2.4	2.9	2.9
3	4	12	4.0	7.8	8.2	8.1	14.4	14.9	14.9	15.2
3	6	18	7.3	16.3	16.4	16.1	46.9	47.2	47.4	47.5
3	8	24	12.7	29.6	32.1	29.5	117.8	119.0	121.3	120.8
3	10	30	20.0	47.2	49.3	46.4	245.3	244.8	244.9	242.3
4	2	8	2.3	3.4	3.5	3.4	4.1	4.0	4.4	4.4
4	4	16	5.1	11.7	11.7	11.5	23.4	23.9	24.9	24.9
4	6	24	9.4	25.4	26.2	24.8	80.2	80.3	82.9	81.0
4	8	32	16.1	45.2	46.8	45.8	201.0	200.4	207.5	207.7
4	10	40	29.8	72.9	73.4	72.5	415.0	422.8	419.9	431.4
5	2	10	2.7	4.4	4.6	4.6	5.5	5.5	5.5	5.4
5	4	20	6.1	16.1	16.2	16.3	34.2	34.2	35.3	35.6
5	6	30	11.4	35.9	36.5	35.6	120.8	123.1	124.7	125.4
5	8	40	19.9	65.3	66.1	65.5	312.5	326.3	320.5	315.2
5	10	50	31.4	104.3	105.0	102.4	649.5	649.4	659.9	661.9
6	2	12	3.2	5.7	6.9	5.7	7.1	7.1	7.2	8.1
6	4	24	7.2	21.4	22.1	25.5	48.3	48.4	50.5	50.6
6	6	36	14.1	47.1	50.5	49.0	171.7	170.5	180.4	178.7
6	8	48	24.3	87.1	87.8	87.0	445.7	432.6	445.4	458.4
6	10	60	37.9	138.4	146.4	140.0	908.4	890.0	929.5	1082.3
Overall Mean:			8.9	26.3	27.1	26.4	128.5	128.2	131.0	136.7

Table 3 shows the mean time to run auctions and allocate all tasks for each Robot/Task combination. For all auctions except SSC we begin timing when the robots are informed of the tasks to bid on and stop timing when all tasks have been allocated. For the SSC auctions we begin timing when the clustering algorithm begins and stop when all tasks have been allocated.

Parallel auctions are always the quickest auction to finish, however, they produce the most sub-optimal distance results. Standard SSI are on average around three times slower than Parallel auctions. SSC auctions run in a comparable time to SSI auctions. This is an important point because SSC auctions need to generate the task clusters before auctions can begin which can take considerable time. However, once the auctioning phases begin they are quicker than SSI auctions because they have fewer auction rounds. This result validates our properties from Section 4 and analysing both the mean distance results and the timing results empirically demonstrates that SSC auctions can result in a lower team objective distance in a similar time to SSI auctions.

Finally, SSI auctions with bundles perform around five times slower than SSI and SSC auctions and 13 times slower than Parallel auctions. Although SSI auctions with bundles produce the lowest team objective distances the performance trade-off cost is very high.

7 Conclusions and Further Work

In this paper we have shown the benefits of SSC auctions as an alternative to SSI auctions for the allocation of tasks to robots. We developed the theoretical foundations of SSC auctions and outlined their unique behavioural properties. Using the standard multi-robot routing test-bed we demonstrated empirically that SSC auctions can produce smaller team objective results than SSI auctions. We also compared these results to another extension of SSI auctions which involves grouping tasks, *SSI auctions with bundles*, and showed that SSC auctions perform much quicker.

This paper provides scope for further investigation of SSC auctions. For instance, a comparison of the effectiveness of different clustering algorithms could provide an insight into the trade-off between run-time speed and the optimality of the final allocation. Applying SSC auctions to dynamic task allocation and reallocation in a manner similar to [9] can also be considered. Finally, clustering non-homogeneous tasks could be advantageous in the quick allocation of complex task sets.

References

1. Berhault, M., Huang, H., Keskinocak, P., Koenig, S., Elmaghraby, W., Griffin, P., Kleywegt, A.: Robot exploration with combinatorial auctions. In: Proceedings of the 2003 IEEE/RSJ International Conference on Intelligent Robots and Systems (IROS 2003), vol. 2, pp. 1957–1962. IEEE (2003)
2. Croes, G.: A method for solving traveling-salesman problems. Operations Research 6(6), 791–812 (1958)
3. Dias, M.B., Zlot, R., Kalra, N., Stentz, A.: Market-based multirobot coordination: A survey and analysis. Proceedings of the IEEE 94(7), 1257–1270 (2006)
4. Hartigan, J., Wong, M.: A k-means clustering algorithm. Journal of the Royal Statistical Society C 28(1), 100–108 (1979)
5. Koenig, S., Keskinocak, P., Tovey, C.: Progress on agent coordination with cooperative auctions. In: Proceedings of the AAAI Conference on Artificial Intelligence (2010)
6. Koenig, S., Tovey, C., Lagoudakis, M., Markakis, V., Kempe, D., Keskinocak, P., Kleywegt, A., Meyerson, A., Jain, S.: The power of sequential single-item auctions for agent coordination. In: Proc. AAAI 2006 (2006)
7. Koenig, S., Tovey, C., Zheng, X., Sungur, I.: Sequential bundle-bid single-sale auction algorithms for decentralized control. In: Proc. IJCAI 2007, pp. 1359–1365 (2007)
8. Lagoudakis, M., Markakis, E., Kempe, D., Keskinocak, P., Kleywegt, A., Koenig, S., Tovey, C., Meyerson, A., Jain, S.: Auction-based multi-robot routing. In: Proc. Int. Conf. on Robotics: Science and Systems, pp. 343–350 (2005)
9. Schoenig, A., Pagnucco, M.: Evaluating Sequential Single-Item Auctions for Dynamic Task Allocation. In: Li, J. (ed.) AI 2010. LNCS, vol. 6464, pp. 506–515. Springer, Heidelberg (2010)

Correlation between Genetic Diversity and Fitness in a Predator-Prey Ecosystem Simulation

Marwa Khater, Elham Salehi, and Robin Gras

School of Computer Science, University of Windsor, ON, Canada
{khater,salehie,rgras}@uwindsor.ca

Abstract. Biologists are interested in studying the relation between the genetic diversity of a population and its fitness. We adopt the notion of entropy as a measure of genetic diversity and correlate it with fitness of an evolutionary ecosystem simulation. EcoSim is a predator-prey individual based simulation which models co-evolving sexual individuals evolving in a dynamic environment. The correlation values between entropy and fitness of all the species that ever existed during the whole simulation are presented. We show how entropy strongly correlates with fitness and investigate the factors behind this result using machine learning techniques. We build a classifier based on different species' features and successfully predict the resulting correlation value between entropy and fitness. The best features affecting the quality of classification are also being investigated.

Keywords: artificial life modeling, individual-based modeling, genetic diversity, entropy, fitness.

1 Introduction

Genetic diversity serves as a way for populations to adapt to changing environments. With more variation, it is more likely that some individuals in a population will possess variations of alleles that are suited for the environment. Those individuals are more likely to survive to produce offspring bearing that allele. The population will continue for more generations because of the success of these individuals. In summary, genetic diversity strengthens a population by increasing the likelihood that at least some of the individuals will be able to survive major disturbances, and by making the group less susceptible to inherited disorders. Many biological studies showed that decreased population genetic diversity can be associated with declines in population fitness [10] [7] [16]. However, populations also learn from their environment by selecting the individuals with highest fitness. This driving force is opposite to the previous one and leads to unstable equilibrium value for genetic diversity. Because overall population diversity affects both short-term individual fitness and long-term population adaptive capacity, there is a need to develop an empirical quantitative understanding of the relationship between population genetic diversity and population viability.

D. Wang and M. Reynolds (Eds.): AI 2011, LNAI 7106, pp. 422–431, 2011.

Like in many disciplines, simulation modeling played a great role in studying evolutionary processes. In this paper we investigate the relation between species fitness and species genetic diversity using EcoSim; an Individual based predator-prey ecosystem simulation. We use the Shannon entropy, as a measure of genetic diversity and study its correlation with species fitness. We present the different correlation values obtained between entropy and fitness and investigate the factors behind these values. The rest of the paper is organized as follows: A brief description of our model used is presented in Section 2. Section 3 depicts the details of the entropy as a genetic diversity measure. The correlation results between entropy and fitness are presented in Section 4. Furthermore, building a classifier for inference, and feature selection is illustrated in Section 5, followed by a summed up conclusion in Section6.

2 The Model

In order to investigate several open theoretic ecological questions we have designed the individual-based evolving predator-prey ecosystem simulation platform EcoSim introduced by Gras et al. [4] [5] [3]. Our objective is to study how individuals and local events can affect high level mechanisms such as community formation, speciation or evolution. In this paper, we have used EcoSim, to compute and study the relation between genetic diversity and fitness. EcoSim uses Fuzzy Cognitive Map as a behavior model [6] which allows a combination of compactness with a very low computational requirement while having the capacity to represent complex high level notions. The complex adaptive agents (or individuals) of this simulation are either prey or predators which act in a dynamic environment of 1000 x 1000 cells. Each cell may contain several individuals and some amount of food from which individuals gain energy. Preys consume grass which is dynamically distributed, whereas predators predate on prey individuals. An individual consumes some energy each time it performs an action such as evasion, search for food, eating and breeding. Each individual performs one action during a time step based on its perception of the environment.

Fuzzy Cognitive Map (FCM) [6] is used to model the individual's behavior and to compute the next action to be performed. The FCM is coded in the individual's genome through which evolution acts. Each agent possesses its unique proper FCM, and the system can still manage several hundreds of thousands of such agents simultaneously into the world with reasonable computational requirements. A typical run lasts several tens of thousands of time steps, during which, several hundreds of millions of agents will be born and several thousands of species [1] will be generated, allowing evolutionary process to take place and new behaviors to emerge to react to a constantly changing environment. A FCM is a graph which contains a set of nodes, each node being a concept, and a set of edges, each edge representing the influence of one concept on another. In each FCM, three kinds of concepts are defined: sensitive (such as distance to foe or food, amount of energy, etc), internal (fear, hunger, curiosity, satisfaction, etc) and motor (evasion, socialization, exploration, breeding, etc.). The FCM serves as a genome for each individual. The genome length is fixed to 390 sites, where each site corresponds to an edge between two concepts of the FCM. In a breeding event, the FCM of the two parents is combined and is transmitted to their unique offspring after the possible addition of some mutations.

3 Entropy as a Measure of Genetic Diversity

Depending on the specific problem or representation being used, ranging from biological domain to genetic programming, numerous diversity measures and methods exist. The use of information theoretic measures such as Shannon entropy [13] or mutual information was controversial in many of the areas of biology that aim to understand how organisms have evolved to deal with information, including behavioral biology, evolutionary ecology and genetics. Sherwin [14] showed that Shannon entropy proves its ability in measuring diversity in ecological community and genetics. He also highlighted the advantages of using entropy based genetic diversity measures, along with surveying these diversity measures. A close relationship between; biological concepts of Darwinian fitness and information-theoretic measures such as Shannon entropy or mutual information, was found. Furthermore, it was shown that in evolving biological systems, the fitness value of information is bounded above by the Shannon entropy [2]. Shannon Information theory defines uncertainty (entropy) as the number of bits needed to fully specify a situation, given a set of probabilities. These probabilities can be estimated by simply counting the abundance of each genotype (site) in the population. Therefore, these probabilities are only meaningful when calculated with respect to population of individuals. The entropy content of the whole sequence (genome) is approximated by summing the per-site entropy and then summing over all sites in the sequence. This is only an approximation because it ignores interactions between sites (epistasis).

The lower the entropy, the less diverse are the genomes of a population and vice versa. There is a limit in the desired values of entropy. When it approaches its maximum, it indicates a completely non-uniform population close to randomness. On the other hand very low entropy (close to 0) means too much similarity between individual genomes which need to diverge more in order to learn and survive in their dynamic environment. When the entropy values are within an intermediate range, it could be considered as a desirable diversity indication. So a good balance between learning from the environment (low genetic diversity) and increasing the diversity (high genetic diversity) should be met in order to ensure the well being of species. Initially all prey and predator individuals are given the same value for their genome respectively. Step after step as more individuals are created, changes in their genomes occur. In each time step we compute the entropy for all existing species. We also calculate the fitness for every species as the average fitness of its individuals. We define fitness of an individual as the age of death of the individual plus the age of death of its entire offspring population. Accordingly, the fitness value mirrors the individual's capability to survive longer and produce high number of strong adaptive offspring.

4 Measuring Correlation between Entropy and Fitness

EcoSim gives us the chance to study the relation between species genetic diversity and species fitness, not only in certain environmental conditions and at specific time like done in biological studies [7] [16] [8], but also through evolution. In EcoSim the environment changes from one place to another and from a time step to another. Individuals that evolve in different parts of the world have different information stored in their genome about the environment they evolve in. Furthermore, as we model a predator-prey system, we have co-evolution. This means that the strategies (behaviors) of each kind are continuously changing trying to adapt to the other kind. Thus there are many factors affecting the genetic diversity and fitness and controlling values of correlation between them. At every time step we calculate entropy and fitness for all existing species. In order to investigate their possible correlations, we first begin by calculating the Spearman's cross correlation [15], for all prey species, between their genetic diversity and their fitness. The Spearman measure ranks two sets of variables and tests for a linear relationship between the variables' ranks. A perfect Spearman correlation of +1 or -1 occurs when each of the variables is a perfect monotone function of the other.

In our evolutionary ecosystem the effect of the diversity measure on fitness is not immediate. There must be a time shift between the variation in genetic diversity and its effect on fitness. Also because we did not determine which attribute is the cause of the other we calculate the correlation in both shift directions. We compute the Spearman correlation coefficient, between these two time series for every possible shift between -s and +s time steps. Basically we correlate the entropy at time t with fitness at time t + s where s ranges from -s to +s.

Although there are many factors that might affect fitness beside genetic diversity, we managed to find strong correlation between entropy and fitness for all prey species. We present the cross-correlation charts for some prey species in Fig.1. The x-axis in these charts represents the different shifts for the time series. The y-axis represents the cross-correlation value at the corresponding shift. From the figure we see that not only different species have different cross-correlation values, but also the same species correlate differently based on the time shift. Note that the dynamic environment, co-evolution and changing parameters with time, all affect species behavior. Thus, correlation values for the same species might vary with time and through the course of evolution, a fact that is feasible to study in our model but not in biological experiments. This fact encouraged us to add a time frame to the two series and measure correlation within the specific time frame. Consequently, we split these time series into sliding windows of 200 time steps centered at every time step. Within each window we calculate all possible correlations with different shifts +-s. Then we choose the highest correlation value (whether positive or negative) and assign it to the species at that time step.

We present the results of 5 different runs of the simulation each one containing 16,000 time steps and generating around 110,000 instances in average.

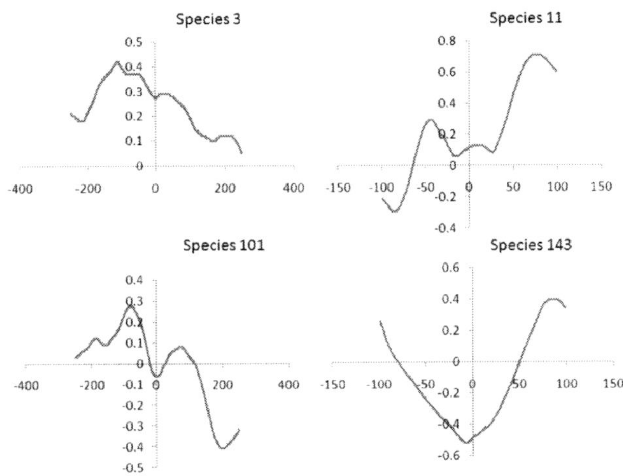

Fig. 1. Different prey species correlation values between entropy and fitness. x-axis represents the different time shifts. Y-axis represents the correlation values.

We assign three different classes to the correlation values. Correlation with values between -0.5 and 0.5 are class WEEK CORR which shows the situation where there is either no or weak correlation. Correlation values above 0.5 are high positive (HIGHP) and correlation values below -0.5 are high negative (HIGHN) respectively. We calculate these correlation classes for all instances (which are each species at every time step) in every run and present the percentage of each class with a window of 200 and maximum shift of 25 in both directions. In average of 5 runs there is 26.8%, 38.4%, 34.6% for classes HIGHP, HIGHN and WEEK CORR respectively.

We investigate variations in window and shift values to better tune our model. Having a window of 200 and a maximum shift of 20 in both directions gave in average of 5 runs 17%, 29.6% and 53.4% for HIGHP, HIGHN and WEEK CORR correlation classes respectively. Increasing the window and maximum shift to 400 and 50 was also tested. The average percentages were 23.7%, 27.5% and 48.8% for HIGHP, HIGHN and WEEK CORR classes respectively. Increasing the shift values increase the percentage of high correlation instances, as more time is needed to detect an increase in fitness after an increase in genetic diversity. Also note that increasing the window does not necessary increase the high correlation values as some fluctuations in the entropy or fitness time series could exist. The values of shift that leads to the highest correlation values were also examined. We found that 37.7% of instances in 5 runs obtained highest correlations from a positive shift between 10 and 25. In addition, 38.7% of instances in average of 5 runs found highest correlation in negative shift between -10 and -25. It shows that for more than 76% of the cases it need between 10 to 25 time steps to see the effect of genetic diversity on the fitness or vice-versa. These values correspond roughly to 1 to 3 'biological generations' which seems a reasonable time to observe the effect of genetic variations in a population.

From the above discussion we observe high values for both negative and positive correlations. These results support the claim of the great influence the genetic diversity has on the well being of species. High positive correlation values mean that an increase in the genetic diversity, results in an increase in species fitness. There are many ways to interpret these results. A newly forming species with a small population would gradually tend to increase its genetic diversity and subsequently positively correlates with the fitness. It is worth mentioning that individuals in EcoSim adapts to their constantly changing environment. This adaptation could be mirrored in the increase of similarity of the species FCM (and thus a decrease in entropy), as new interesting behavior for the current environment has been discovered and then diffuse in the population. Negative correlations imply the fact that a species decreases diversity in order to reach stability by learning from its environment. Our motivation to validate these results and further investigate the reason behind these correlation values encouraged us to build a classifier. The interest of building this classifier is first to see if some specific species properties can predict the current evolutionary behavior of a species (that is if it is learning from the environment or increasing its diversity to be able to react to a future change in the environment). It can also help to understand what are the factors and conditions that affect the evolutionary behavior. Therefore, we try to infer the correlation value knowing some features about the species. If we are able to correctly classify unknown instances based on a trained classifier, it would validate our correlation results.

5 Building Classifier for Inference

In order to validate our high correlation values found between entropy and fitness we make use of machine learning classifiers. We built a classifier to infer the class correlation using decision trees. We use the C4.5 algorithm [9] with pruning, implemented in the WEKA [17]. The C4.5 is a powerful tool which also provides decision rules that can help in the interpretation of the classifier. In order to build a classifier we had to choose the features that would best describe the species and has direct effect on the species fitness. We choose features from both internal and physical concepts. These features are: the number of individuals in species, the average age of individuals in species, the average speed of the individuals and their average energy level. The average number of reproduction events, average number of reproduction failing events, average activation level of reproduction and the spatial dispersal are also included. In addition we also include the average activation level of fear, hunger, satisfaction, nuisance, curiosity (which encourage individuals to move). Finally, we include the entropy and fitness for each species. In total we have 16 features including the class variable which is the correlation with values HIGHN, HIGHP and WEEK CORR. Our next step was to try to select the best features from these 16 features in order to both simplify the model and discover the most important features.

5.1 Feature Selection

To increase the quality of the classifier we use feature selection in order to extract the most important features from the above list. This step will provide more semantics about which features most influence the value of correlation. We use a wrapper feature selection method [18] [11] based on an estimation of distribution algorithm (EDA) called CMSS-EDA [12]. This feature selection method is particularly efficient for problems with high level of interdependency between features. We search for the subset of variables which maximizes AUC (Area Under ROC Curve) obtained by a Bayesian network classifier.

The best chosen features are population size, entropy, fitness, spatial dispersal, age and reproduction fail. We ran the feature selection algorithm on all the 5 runs and they all found the same best features. This fact shows the stability of the simulation which is important to be able to discover meaningful generic rules. Clearly, entropy and fitness are chosen among the best features as they are the two features being correlated and subsequently have a direct effect on the correlation class variable. But also fitness and entropy values determines the sign of correlation being either positive or negative. But this is not a bias in our analysis as what is measured here is how a specific value of either entropy or fitness, at a given time step, affects the future (or is affected by the past) correlation between fitness and entropy. Studying the effect the population size, which was among the selected features, has on fitness is a major study in biology. Some studies showed that population size and genetic variation are strongly positively correlated with fitness [8]. Also, loss in fitness and genetic diversity was accompanied by a drop in population size in [10]. Furthermore, positive correlation between genetic diversity fitness, and population size was shown in [16]. Another selected feature was spatial dispersal. It was also discovered that spatial dispersal is a very important factor maintaining genetic diversity and subsequently fitness [16]. The last two selected features are the average age and the average reproduction fail. From the fitness definition we used,clearly these two features have a direct effect on the fitness value as the higher the average age of species population the higher its fitness. Also, the decrease in the reproduction failure is accompanied by the increase in the fitness. The similarity between the best features discovered by our system and the most significant biological features affecting the genetic diversity and fitness is noticeable. Furthermore, the significance of the best features chosen highlights the validity of our calculations and the founding of the strong correlation between genetic diversity and fitness in our system.

5.2 Classification Results

We build a classifier using the C4.5 algorithm implemented in WEKA environment. We use the window of 400 and fix the shift to 25 time steps for calculating the correlations. The reason behind that is to have all instances on the same scale and thus comparable. Also, increasing the window for more than 400, subject the fitness and entropy series to fluctuations. Furthermore, decreasing the window

Table 1. Percentage of high positive, high negative, week correlation and high correlation prey instances for five different runs

Run	Percentage HIGHP	Percentage HIGHN	Percentage WEEK CORR	Percentage HIGH CORR
Run 1	13%	15.5%	71.5%	28.5%
Run 2	13.3%	17.6%	69.1%	30.9%
Run 3	11.3%	15.9%	72.8%	27.2%
Run 4	9.8%	11.4%	78.8%	21.2%
Run 5	11.8%	14.9%	73.3%	26.7%
Average	11.8%	14.9%	73.3%	26.7%

Table 2. Accuracy percentages for training and testing with the C4.5 classifier for 5 runs of the simulation

Run	Train Accuracy	Test Accuracy on same run	Average Test Accuracy on other 4 runs	STD test accuracy on other runs	Number of rules
Run 1	79.3%	80.3%	60.1%	4.9	294
Run 2	74.7%	75.3%	66.8%	0.8	307
Run 3	77.2%	78.1%	63.2%	3.2	280
Run 4	80.2%	80.2%	69.1%	2.6	181
Run 5	78%	78%	66.9%	3.8	263
Average	77.9%	78.4%	65.2%	3.1	265

tend to influence the correlation results to higher correlations. We choose 25 as a shift value based on the analysis of which shift leads to the highest correlations. Table 1 presents the percentages for HIGHP, HIGHN, WEEK CORR and the sum of HIGHP and HIGHN called HIGH CORR, for the 5 runs. The 6 features used for the model are the ones selected from the feature selection process. We split the instances for each 5 runs into 80% for training the classifier using 10-fold cross validation and 20% for testing with C4.5 pruning model. Table 2 presents training and testing accuracy on data set from the same run. We also tested training the classifier on data set from one run and testing on another data set from the other runs to infer generality of the model. The confusion matrix showed high true positive results for training and testing on the same run. The results from testing on another run showed only reasonably high true positive values when accuracies is above 65%. This is due to the fact that each run has variations in terms of attributes values and ranges and also to possible overfitting. However, the model was able to discover some rules that can make good prediction on unclassified instances. The good classification accuracy on the test set of the same run shows the validity of our calculations of entropy as genetic diversity and its high correlation with fitness. It also shows that there exist specific conditions of the species that lead to a positive or negative correlation between fitness and genetic diversity.

6 Conclusion

In this paper we introduced the use of Shannon entropy as a measure of genetic diversity of an individual based evolutionary ecosystem simulation. We found very high correlation both negative and positive between entropy and fitness. In order to validate our correlation results and further understand the reasons behind these results we built a classifier to predict the correlation class variable based on training and testing sets. We found high accuracy for classification which proves the interest of our genetic diversity measure and its correlation with fitness. In addition, we used feature selection to find the best features affecting the correlation values. We showed how these extracted features are similar to the factors affecting genetic diversity and fitness in community ecology. The similarity between results of five different runs of the simulation proves the stability of the simulation and the generality of our findings. This study allows us to show that the relation between genetic diversity and fitness changes based on time and other features such as reproduction rate, population size and spatial dispersal. In the future we will work on our classification problem to try to reduce the overfitting effect. We will also investigate more about the values of the features and which values lead to negative or positive correlation which would have a great impact on community ecology domain.

Acknowledgments. This work is supported by the NSERC grant ORGPIN 341854, the CRC grant 950-2-3617 and the CFI grant 203617 and is made possible by the facilities of the Shared Hierarchical Academic Research Computing Network (SHARCNET:www.sharcnet.ca).

References

1. Aspinal, A., Gras, R.: K-Means Clustering as a Speciation Mechanism within an Individual-Based Evolving Predator-Prey Ecosystem Simulation. In: An, A., Lingras, P., Petty, S., Huang, R. (eds.) AMT 2010. LNCS, vol. 6335, pp. 318–329. Springer, Heidelberg (2010)
2. Bergstrom, C., Lachmann, M.: Shannon information and biological fitness. In: IEEE Information Theory Workshop, pp. 50–54 (2004)
3. EcoSim: An Ecosystem Simulation, `http://sites.google.com/site/ecosimgroup/research/ecosystem-simulation`
4. Gras, R., Devaurs, D., Wozniak, A., Aspinall, A.: An individual-based evolving predator-prey ecosystem simulation using fuzzy cognitive map as behavior model. Artificial Life 15(4), 423–463 (2009)
5. Gras, R., Golestani, A., Hosseini, M., Khater, M., Farahani, Y.M., Mashayekhi, M., Ibne, S.M., Sajadi, A., Salehi, E., Scott, R.: Ecosim: an individual-based platform for studying evolution. In: European Conference on Artificial Life (in press, 2011)
6. Kosko, B.: Fuzzy cognitive maps. Int. Journal of Man-Machine Studies, 65–75 (1986)
7. Markert, J., Champlin, D., Gutjahr-Gobell, R., Grear, J., Kuhn, A., McGreevy, T., Roth, A., Bagley, M., Nacci, D.: Population genetic diversity and fitness in multiple environments. BMC Evolutionary Biology 10, 1471–2148–10–205 (2010)

8. Oostermeijer, J., van Eijck, M., den Nijs, J.: Offspring fitness in relation to population size and genetic variation in the rare perennial plant species gentiana pneumonanthe (gentianceae). Oecologia 97, 289–296 (1994)
9. Quinlan, R.: C4.5: Programs for Machine Learning. Morgan Kaufmann Publishers (1993)
10. Reed, D., Frankham, R.: Correlation between fitness and genetic diversity. Conserv. Biology 17, 230–237 (2003)
11. Saeys, Y., Inza, I., Larraaga, P.: A review of feature selection techniques in bioinformatics. Bioinformatics 23(19), 2507–2517 (2007)
12. Salehi, E., Gras, R.: Efficient eda for large optimization problem via constraining the search space of models. In: Genetic and Evolutionary Computation Conference, Dublin, Ireland (in press, 2011)
13. Shannon, C.: A mathematical theory of communication. Bell Systems Technical Journal, 379–423 (1948)
14. Sherwin, W.B.: Entropy and information approaches to genetic diversity and its expression: Genomic geography. Entropy 12, 1765–1798 (2010)
15. Siegel, S.: Nonparametric Statistics for the Behavioral Sciences. McGraw-Hill, New York (1956)
16. Vandewoestijne, S., Schtickzelle, N., Baguette, M.: Positive correlation between genetic diversity and fitness in a large, well-connected metapopulation. BMC Biology 6, 1741–7007–6–46 (2008)
17. Witten, I., Frank, E.: Data Mining- Practical Machine Learning Tools and Techniques with Java Implementations. Morgan Kaufmann, USA (2000)
18. Yang, Q., Salehi, E., Gras, R.: Using Feature Selection Approaches to Find the Dependent Features. In: Rutkowski, L., Scherer, R., Tadeusiewicz, R., Zadeh, L.A., Zurada, J.M. (eds.) ICAISC 2010. LNCS (LNAI), vol. 6113, pp. 487–494. Springer, Heidelberg (2010)

Is Revision a Special Kind of Update?

Abhaya C. Nayak*

Department of Computing
Macquarie University
Sydney, Australia 2109

Abstract. It is widely acknowledged that belief revision and belief update are two very different types of processes – one is appropriate to model belief change in a static environment, the other in a dynamic environment. Technically speaking, the former is constructed with the aid of a global preference ordering over possible worlds as the selection mechanism, and the latter with the aid of a family of such local orderings. It has been argued that update can be defined via revision. In this paper I argue that indeed revision can be defined via update in a restricted sense if a distance function is used as the selection mechanism.

1 Introduction

How a rational agent ought to change its beliefs has been a focal point for AI researchers for the better part of last two and half decades. The story goes that artificial agents such as robots must be provided with the ability to update their beliefs as they receive and process new information, perceptual or otherwise. There are two major threads of research that have been carried out under this umbrella concept of belief change. One is called *belief revision*, the central operation in the AGM framework named after its founding fathers [1]. The other is called *belief update* that was put forward by Katsuno and Mendelzon as an alternative to belief revision [7]. The received wisdom has it that belief revision is appropriate for processing new information in a static world, where the new information either supplements the current information, or rectifies existing error in the current knowledge. On the other hand, belief update is appropriate for processing new information in a dynamic world, where the new information indicates that the world has changed and the current knowledge is dated. The purpose of this paper is to explore if this nice storyline is actually as good as it appears to be. I will show that if we use a distance measure as the underlying mechanism for driving both revision as update, as has been advocated by Lehmann, Magidor and Schlechta [8], then revision by some information of one belief body reduces to update by the same information of a different body of beliefs. It is, as it were, one person's revision is another person's update.

In what follows, I will try to tell the story of this somewhat interesting result in a rather informal manner. As usual, I will assume a propositional language

* The author was a visitor at the School of CSE, University of New South Wales, Australia during the period this paper was written.

D. Wang and M. Reynolds (Eds.): AI 2011, LNAI 7106, pp. 432–441, 2011.

generated from a finite set of atomic sentences. Given the finitary nature of the language, the body of beliefs of an agent can be represented as a single sentence; I typically denote it by k, and variously call it a belief base, belief set or a body of beliefs ignoring the standard distinction drawn between these names. I denote beliefs by lower case Roman letters, and sets of beliefs by upper case Roman letters. By a world, interpretation (or even model) I mean a function that tells for each sentence whether it is true or false. Worlds are denoted by lower case Greek letters, sets of worlds by upper case Greek letters, and the set of all worlds by Ω. Given a set of sentences X, by $[X]$ I denote the set of all models of X; and I drop curly brackets if X is a singleton (thus $[\{x\}]$ is simplified to $[x]$). As standard, I use $*$ and $\diamond$ for revision and update operations, thus $k * x$ and $k \diamond x$ are sentences, respectively representing the result of revising and updating k by x. I will use $\sqsubseteq$ to denote a preference relation (plausibility preorder) over worlds; its strict part is denoted by $\sqsubset$, and the equivalence part by $\approx$.

2 Need for Selection Mechanisms

It is not always obvious how a piece of new information can be accommodated into one's body of beliefs, or a piece of old information can be removed from it, if the process in question is to be deemed rational. Consider for instance a situation where an agent believes:

1. a : *the moon is made of cheese,*
2. b : *cheese is green*

and everything else that logically follows from them, such as,

3. $a \wedge b$: *the moon is made of green cheese*[1].

Now, suppose the agent in question reliably learned that

4. $a \rightarrow \neg b$: *the moon is made of cheese only if it (cheese) is not green.*

Arguably a rational agent must maintain consistency among its beliefs, and hence cannot accommodate this new piece of knowledge while believing both a and b and their logical consequences. Accordingly it must purge some offending beliefs before accommodating $a \rightarrow \neg b$. Theoretically, there are different choices available to the agent. For instance, it might want to discard all its current beliefs, to the extent it is possible,[2] before accommodating the new piece of information. In this example, then, the agent would only be believing $a \rightarrow \neg b$ (and whatever else follows from it). However, this is hardly a rational course of action, considering that it entails massive loss of information. How the agent should proceed in this case would depend on one's favoured conception of rationality.

[1] The *green* in the proverb, *the moon is made of green cheese*, indeed denotes the quality of being young or fresh, not the colour green.

[2] The agent cannot discard beliefs that are willy nilly true, namely logical truths such as $a \rightarrow a$.

A very simple approach towards getting a handle on rationality in this context is to assume that the agent must *minimally* contract its body of beliefs such that the result is consistent with $a \rightarrow \neg b$ (that is, $a \wedge b$ is no longer a consequence of its beliefs), and then accommodate the new piece of information $a \rightarrow \neg b$. It is well known in the literature that this proposal, dubbed *maxichoice* revision (respectively contraction) in general is not acceptable [1,4]. In particular, there are too many ways of carrying out such minimal change to one's body of knowledge,[3] and the information removed from one's body of knowledge is so little that after incorporating the new piece of information, the agent would behave like an omniscient agent, with an opinion on the truth of every proposition expressible in its object language.

One way out of this choice problem – of choosing exactly one from the numerous maxichoice options available – and consequently avoid its undesirable consequences, is to consider all the available options as equally good, and believe only what is commonly supported by all of them. This is called the *full meet* revision (respectively contraction) [1,4]. It turns out that this leads to massive loss of information in that after a nontrivial revision by some sentence x, the agent would lose all its old knowledge and end up believing only x and whatever else logically follows from it.

Clearly then the rational option is to choose some middle path between these to extremes. The agent needs access to some mechanism which would guide it to choose the best among all the maxichoice options available, and the agent can then believe in only those propositions commonly supported by them. Hence what is required is an appropriate selection mechanism. There are different such selection mechanisms suggested and studied in the literature, starting from a selection function over remainder sets [1], through a preference ordering over interpretations (modulo the object language) [6], to *epistemic entrenchment*, a relational measure over sentences indicating which belief is how hard to discard [5]. In the next two sections, I briefly describe two different types of selection mechanism, both semantic in character. The first type, called the *global selection mechanism*, is used to deal with belief change in static environment, and as the name suggests, a single such mechanism is adequate for this purpose. The second type, called the *local selection mechanism*, is appropriate to model belief change in a dynamic environment – and typically a family of such mechanisms would be required, one for each possible state of the environment in which change is being effected.

3 Global Selection Mechanisms and Belief Revision

Let us briefly look at a very popular selection mechanism designed to handle belief change in static environments. It was proposed by Adam Grove, and best known as a *system of spheres* [6]. The idea is, we envisage the interpretations (or

[3] In the nontrivial case, when a sentence x is being removed from one's body of knowledge, there are as many ways of carrying out such minimal mutilation as there are models of the sentence $\neg x$.

worlds) forming a number of spheres, the bigger ones enclosing the smaller ones. The innermost sphere in this conception, the very core of this system of spheres, consists of the set of worlds that the agent considers most plausible, namely exactly the set of worlds that model its current knowledge. The system of sphere in question is said to be *centred* on this core set of worlds. The next sphere consists of this core, together with the set of next most plausible worlds, and so on.... The largest sphere consists of all the possible worlds. Hence the system of spheres is really a visualisation of the plausibility ordering in question. This ordering helps in tie breaking among the worlds, and indirectly select among competing maxichoice options. Let us assume that the current knowledge of the agent is represented by a sentence k. Then, according to this conception, the agent's *belief state* is represented by a single selection mechanism, a global system of sphere, as follows:

- The global system of spheres, a total pre-order $\sqsubseteq_{[k]}$, also called a plausibility ranking, over Ω is centred on the set of worlds $[k]$.[4]
- The result of revising k by a piece of new information represented by a sentence x is determined by the set of most plausible worlds where x holds:

$$[k * x] = \{\omega \in [x] \mid \omega \sqsubseteq_{[k]} \omega' \text{ for all } \omega' \models x\}$$

Let us consider an example originally due to Katsuno and Mendelzon [7], now considered a classic.

Example 1. All that our agent believes is that exactly one of the two items, a book and a magazine, is on the desk, the other being on the floor. In other words, the agent's knowledge k can be represented by the sentence $b \leftrightarrow \neg m$ where b (respectively m) denotes the book (magazine) is on the desk. A satellite image just retrieved by the agent shows that the magazine is on the desk (that is, m). What should the agent do?

Assuming that b and m are the only two atoms in the language, let us denote the four worlds (with obvious understanding) as bm, $b\overline{m}$, $\overline{b}m$ and $\overline{bm}$. Let the plausibility preorder $\sqsubseteq_{[k]}$ be as follows:

$$\sqsubseteq_{[k]}:\ b\overline{m} \approx \overline{b}m \sqsubset \overline{bm} \approx bm$$

There is only one $\sqsubseteq_{[k]}$-minimal world among the two models of the new information m, namely $\overline{b}m$. Hence $[k * m] = \{\overline{b}m\}$ whereby $k * m$ can be represented by the sentence $\neg b \wedge m$. In other words, the agent should believe that the book is on the floor and the magazine on the desk.

This example accords well with our intuition. Note, however, that this example is a rather trivial case of revision, since the new evidence m is consistent with the current knowledge $b \leftrightarrow \neg m$, and the outcome in this case can be directly computed without recourse to a selection mechanism, as $(b \leftrightarrow \neg m) \wedge m$. But this simple example contrasts well with the corresponding situation in a dynamic environment, as discussed in the next section.

[4] A total preorder is a binary relation that is reflective, transitive and total. On occasion we will drop the subscript $[k]$ when no confusion is imminent.

4 Local Selection Mechanisms and Belief Update

The belief revision process discussed in the last section is not always appropriate. Arguably, it works only when the new piece of information received informs the agent only of what she did not previously know, that is, it adds to the agent's body of knowledge, or alternatively corrects some erroneous beliefs of the agent. In Example 1 that we discussed in the last section, the agent had no particular opinion as to whether the magazine was on the desk or on the floor; it was a piece of information that supplemented the agent's (rather incomplete) knowledge. The presumption here is that the domain in question is not undergoing change with respect to the new piece of information. As noted by Gärdenfors, it is akin to Bayesian conditionalisation [4]. What is undergoing change here is the knowledge of the agent, not the world that the knowledge is about. Hence a global selection mechanism was adequate for the purpose at hand.

However, on occasion, the new information received may inform the agent of change that the world has undergone. In such a case, a global selection mechanism is inadequate – a mechanism needs to be attached with each world in order that the agent can calculate the state this world would be in after undergoing the change in question. All such resultant worlds will jointly determine the updated knowledge of the agent. This view was propounded by Katsuno and Mendelzon [7], and is known to be the non-probabilistic counterpart of the account of *imaging* propounded by David Lewis in order to develop a theory of conditionals [9]. The basic idea is as follows:

- A number of local systems of spheres over Ω, one centred on each world $\omega \in \Omega$ is assumed. A system of spheres centred on a world ω is represented by a total pre-order $\sqsubseteq_\omega$. One may take $\omega' \sqsubset_\omega \omega''$ to mean that from the vantage point of ω, the world ω' is considered to be more plausible than ω''.
- The result of the update of k by x is determined by a big union, namely, that of the sets of most plausible (from the perspective of each world $\omega \in [k]$) worlds in which x is true:

$$[k \diamond x] = \bigcup_{\omega \in [k]} \{\omega' \in [x] \mid \omega' \sqsubseteq_\omega \omega'' \text{ for all } \omega'' \models x\}$$

.

Let us continue with the example by Katsuno and Mendelzon [7] that we started in the previous section:

Example 2. Our agent believes that exactly one of the two items, a book and a magazine, is on the desk, the other being on the floor. It commands its housekeeping robot to ensure that the magazine is on the desk (m), and soon afterwards the robot intimates its owner that the mission has been successful. What is the agent's new knowledge base?

As before, let us assume that b and m are the only two atoms in the language, and the corresponding four worlds are bm, $b\overline{m}$, $\overline{b}m$ and $\overline{bm}$. We will also refer to these worlds by the corresponding truth vectors,

11, 10, 01 and 00. Accordingly, there are four local plausibility preorders, respectively denoted: $\sqsubseteq_{11}$, $\sqsubseteq_{10}$, $\sqsubseteq_{01}$ and $\sqsubseteq_{00}$. Let them be:

1. $\sqsubseteq_{11}$: $bm \sqsubset b\overline{m} \approx \overline{b}m \sqsubset \overline{bm}$
2. $\sqsubseteq_{10}$: $b\overline{m} \sqsubset \overline{bm} \approx bm \sqsubset \overline{b}m$
3. $\sqsubseteq_{01}$: $\overline{b}m \sqsubset bm \approx \overline{bm} \sqsubset b\overline{m}$
4. $\sqsubseteq_{00}$: $\overline{bm} \sqsubset \overline{b}m \approx b\overline{m} \sqsubset bm$

Now, there are only two models of the agent's initial knowledge, namely $b\overline{m}$ and $\overline{b}m$. Accordingly we accumulate the $\sqsubseteq_{10}$-minimal and $\sqsubseteq_{01}$-minimal worlds of $[m]$ in order to obtain $[k \diamond m]$. Thus we get

$$[k \diamond m] \;=\; \{bm\} \cup \{\overline{b}m\} \;= \{bm,\; \overline{b}m\}$$

whereby all the agent knows after the update, denoted $k \diamond m$, can be represented by m. The agent now knows that the magazine is on the desk, but has no opinion as to the location of the book.

Again, this accords well with our intuition. The robot would ensure that the magazine is on the desk. But it will not interfere with the location of the book. Since the agent allowed both the possibilities of the book being on the floor as well as on the desk, it would not know the location of the book post event; as far as the agent is concerned, it could be at either of the locations.

5 Distance Function as a Selection Mechanism

In the last two sections we have observed that belief change in static domain can be modelled via a global selection mechanism, and that in a dynamic domain can be modelled via a family of local selection mechanisms. It is worth noting that the second approach – via a family of selection mechanisms – is much richer than the first. This is easily seen in the context of iterated belief change. In case of belief revision, after one's knowledge (set or base) changes from k to $k * x$ with the help of the preorder $\sqsubseteq_{[k]}$, assuming that k and $k * x$ are different, the old mechanism is no longer useful, and must be replaced by a new mechanism $\sqsubseteq_{[k*x]}$ centred on the set of worlds $[k * x]$. In case of update, however, no such modification in the selection mechanism(s) is necessary since we still have a selection mechanism centred on every world $\omega \in [k * x]$. A natural question is whether a mechanism can be devised that can help with belief change both in that static and dynamic domains, and would be rich enough so that it would not need modification after each case of belief change. Schlechta, Magidor and Lehmann [8] have observed that a distance function will precisely fit the bill.[5]

Definition 1. *A pseudodistance function $d : \Omega \times \Omega \;\rightarrow\; Z$ satisfies the following four conditions: for all worlds $\omega, \omega', \omega'' \in \Omega$,*

[5] Schlechta, et. al. define a pseudo-distance measure via a total preorder. We will slightly deviate from their account. For further works on distance functions, see [2,3].

1. $d(\omega, \omega') \geq 0$	(non-negativity)
2. $d(\omega, \omega) = 0$	(Identity)
3. $d(\omega, \omega') = d(\omega', \omega)$	(Symmetry)
4. $d(\omega, \omega') + d(\omega', \omega'') \geq d(\omega, \omega'')$	(Triangular Inequality)

One may wonder where the pseudo distance between different models is obtained from. The pseudo distance may be taken to be analogous to the notion of revealed preference in social choice theory – we assume it to be a theoretical construct, revealed by the belief-change behaviour of an agent.

For operational purpose, we can assume the distance function to be anything that satisfies the above constraints. For instance, we can take the distance between any two models to be their Hamming distance, that is the number of atoms that are satisfied by one but not the other. Continuing our earlier example:

Example 3. We have two atoms, namely b and m in our language. The corresponding four worlds are $11 : bm$, $10 : b\overline{m}$, $01 : \bar{b}m$ and $00 : \overline{bm}$. The Hamming distance d^H between different pairs of these worlds is as follows:[6]

1. $d(11, 11) = d(01, 01) = d(10, 10) = d(00, 00) = 0$
2. $d(11, 10) = d(10, 11) = d(11, 01) = d(01, 11) = 1$
3. $d(10, 00) = d(00, 10) = d(01, 00) = d(00, 01) = 1$
4. $d(11, 00) = d(00, 11) = d(10, 01) = d(01, 10) = 2$

Now that we have the distance between every pair of worlds, how would it be useful in belief revision or update? Belief update is performed relatively easily. Given a distance function d and a world ω, we define a total preorder $\sqsubseteq^d_\omega$ centered on ω in the obvious manner:

Definition 2. *$\omega' \sqsubseteq^d_\omega \omega''$ if and only if $d(\omega, \omega') \leq d(\omega, \omega'')$, for any two worlds ω' and ω'' in Ω.*

Thus the distance function easily lends to define the family of local selection mechanisms required for belief update. To continue our running example,

Example 4. We know that d is symmetric. Now,

1. Since $d(11, 10) = d(11, 01) = 1$, and $d(11, 00) = 2$, we get $\sqsubseteq_{11}$: $bm \sqsubset b\overline{m} \approx \bar{b}m \sqsubset \overline{bm}$.
2. Since $d(10, 00) = d(10, 11) = 1$ and $d(10, 01) = 2$ we get $\sqsubseteq_{10}$: $b\overline{m} \sqsubset \overline{bm} \approx bm \sqsubset \bar{b}m$.
3. Since $d(01, 11) = d(01, 00) = 1$ and $d(01, 10) = 2$ we get $\sqsubseteq_{01}$: $\bar{b}m \sqsubset bm \approx \overline{bm} \sqsubset b\overline{m}$.
4. Since $d(00, 01) = d(00, 10) = 1$ and $d(00, 11) = 2$ we get $\sqsubseteq_{00}$: $\overline{bm} \sqsubset \bar{b}m \approx b\overline{m} \sqsubset bm$.

Thus we get exactly the family of total preorders as in Example 2, and the belief update can be carried our exactly as in that example.

---

[6] We drop the superscript H when it is clear from the context that we are talking of Hamming distance, not any arbitrary distance.

Now let us see how a distance function can be effectively used to carry out belief revision. For this purpose, we generalise the distance function d in order to obtain a distance, say d' between any set of worlds Δ and any world ω. It is standard to define it as the minimal distance between any world in Δ and ω.

Definition 3. *For every world ω and set of worlds Δ:*

1. $d'(\Delta, \omega) = min\{d(\omega', \omega) \mid \omega' \in \Delta\}$
2. $d'(\omega, \Delta) = min\{d(\omega, \omega') \mid \omega' \in \Delta\}$

It is easily observed that the symmetry of the distance between worlds is transferred to the symmetry of the distance between a world and a set of worlds:

Observation 1. $d'(\Delta, \omega) = d'(\omega, \Delta)$ *for any world ω and any set of worlds Δ.*

Furthermore, the following result shows that the function d' can be effectively used to do belief update as well:

Observation 2. *Given a world ω and a set of worlds Δ, the set of worlds in Δ that are minimally away from ω, namely, $\{\omega' \in \Delta \mid d(\omega, \omega') \leq d(\omega, \omega'')$ for all $\omega'' \in \Delta\}$ is exactly the set of worlds $\{\omega' \in \Delta \mid d(\omega, \omega') = d'(\omega, \Delta)\}$.*

We are now in a position to carry out belief revision.

Example 5. We recall that the total knowledge k of the agent is represented by the sentence $b \leftrightarrow \neg m$. We need to construct a global system of sphere centred on the set of worlds $[k] = \{10, 01\}$. Now,

1. $d'(\{10, 01\},\ 10) = d'(\{10, 01\},\ 01) = 0$, and
2. $d'(\{10, 01\},\ 00) = d'(\{10, 01\},\ 11) = 1$

whereby $\sqsubseteq_{\{10,01\}}$: $b\overline{m} \approx \bar{b}m \sqsubset \overline{bm} \approx bm$, reducing it to Example 1.

Since d' gives the distance between a world and a set of worlds, we can define a distance function d'' that returns the distance between two sets of worlds:

Definition 4. *For any two sets of worlds Δ and Δ':*

$$d''(\Delta, \Delta') = min\{d'(\omega, \Delta') \mid \omega \in \Delta\}$$

Again, the symmetry of the distance functions d and d' is carried over to the symmetry of the distance between two set of worlds:

Observation 3. $d''(\Delta, \Delta') = d''(\Delta', \Delta)$ *for any sets of worlds Δ and Δ'.*

Indeed, the function d'' can be effectively used to do belief revision:

Observation 4. *Given two sets of worlds Δ and Δ', the set of worlds in Δ' that are minimally away from Δ, namely, $\{\omega' \in \Delta' \mid d'(\Delta, \omega') \leq d'(\Delta, \omega'')$ for all $\omega'' \in \Delta'\}$ is exactly the set of worlds $\{\omega' \in \Delta' \mid d'(\Delta, \omega') = d''(\Delta, \Delta')\}$.*

6 Belief Revision as a Special Type of Belief Update

We have so far seen how a given distance function can be used to do both belief revision and belief update. Now, a world can be considered as a complete theory (a maximally consistent set of sentences). Hence, *each* local system of sphere (centred on a world) can be viewed as a mechanism to *revise* complete theories. In this sense, belief update can be defined via a family of belief revisions. Now we consider the converse issue.

Suppose we want to revise a body of beliefs k by some received information x. We need to identify the set $[k * x]$ of worlds in $[x]$ which are at minimal distance from $[k]$. In other words, we are interested in identifying exactly those worlds in $[x]$ which are precisely $d''([k],[x])$ distance away from $[k]$.

Let us assume that $\omega' \in [x]$ is such that $d'([k],\omega') = d''([k],[x])$. It follows then that there is at least one world $\omega \in [k]$ such that $d(\omega,\omega') = d''(([k],[x])$. In fact, it can be shown that all the worlds in $[x]$ that are closest to this world ω are exactly $d''(([k],[x])$ distance away from it, and will be members of the set $[k * x]$ if the revision operation $*$ is based on the distance function d. More formally:

Lemma 1. *Consider a belief base k and new information x. Let d be a distance function between the worlds, and d' and d'' be defined from d as appropriate. Let the revision operation $*$ be based on these distance functions.*

1. *If $\omega \in [k]$ and $\omega' \in [x]$ are such that $d(\omega,\omega') = d''([k],[x])$, then $\omega' \in [k * x]$.*
2. *Let ω' be a world in $[k * x]$. Then there exists a world $\omega \in [k]$ such that $d(\omega,\omega') = d''([k],[x])$.*

What Lemma 1 says is very simple: the k-worlds that are minimally away from $[x]$ are closest to those x-worlds that are minimally away from $[k]$. Let us denote by k' a sentence such that $[k']$ exactly consists of the worlds in $[k]$ that are minimally away from $[x]$, that is: $[k'] = \{\omega \in [k] \mid d'(\omega,[x]) = d''([k],[x])\}$. Then, with the help of Lemma 1 it can be easily established that $k' \diamond x = k * x$.

Theorem 1. *Let sentences k, x and k' be such that $[k']$ consists of exactly those members of $[k]$ that are at minimal distance from $[x]$. Then $k' \diamond x = k * x$.*

Thus there is a restricted sense in which revision is reducible to belief update – given a finitary language and a distance function, we can construct a sentence k' such that update of k' by x is exactly the revision of k by x. It is interesting to observe that since $[k']$ consists of those members of $[k]$ that are minimally away from $[x]$, we can replace k' by $x * k$. Hence our final result:

Theorem 2. *Given two sentences k and x, and the relevant revision and update functions determined by an appropriate distance function, $(x * k) \diamond x = k * x$.*

I conclude this section with the last scene of our running example:

Example 6. We recall that the current knowledge k is $b \leftrightarrow \neg m$, and x, the new information received, is m.

1. First we compute $[m * (b \leftrightarrow \neg m)]$. Note that $[m] = \{11, 01\}$ and $[b \leftrightarrow \neg m] = \{10, 01\}$. Noting that $d''(\{11, 01\}, \{10, 01\}) = 0$, we get: $[m * (b \leftrightarrow \neg m)] = \{01\} = \{\bar{b}m\}$.
2. Now we compute $[(m * (b \leftrightarrow \neg m)) \diamond m]$. As we saw in Example 4, the local selection mechanism centred on the world $\{\bar{b}m\}$ is given by: $\sqsubseteq_{01}$: $\bar{b}m \sqsubset bm \approx \overline{bm} \sqsubset b\overline{m}$. Accordingly the set of $\sqsubseteq_{01}$-minimal worlds in m are exactly $\{\bar{b}m\}$.

Thus we see that $(x * k) \diamond x = k * x \equiv (\neg b \wedge m)$. The book is on the floor, and the magazine is on the desk, as desired.

7 Concluding Remarks

We started with the intention of exploring if revision can be defined via update. We have seen that if both revision and update are driven by the same distance function, then revision can be seen as update in a rather restricted sense: revision of x by y can be seen as the update by y of the result of revising y by x.

Received wisdom suggests that revision is appropriate for belief change in static domains, and update in dynamic domains. Now that revision of x by y is seen to be equivalent to the update of some other sentence x' (namely $y * x$) by the same received information y, that nice storyline that cleanly demarcates revision from update appears not to be such a good story after all. It will be interesting to see how best we can iron out this unforeseen wrinkle.

Since revision can be viewed as a form of update, it has further possible ramifications. Preliminary investigation shows that it can help develop a new theory of probabilistic belief revision/contraction (as opposed to update/erasure) based on imaging. This is a topic beyond the scope of the current paper.

References

1. Alchourrón, C.E., Gärdenfors, P., Makinson, D.: On the logic of theory change: Partial meet contraction and revision functions. J. Symb. Log. 50, 510–530 (1985)
2. Audibert, L., Lhoussaine, C., Schlechta, K.: Distance based revision of preferential logics. Logic Journal of the IGPL 7(4), 429–446 (1999)
3. Ben-Naim, J.: Lack of finite characterizations for the distance-based revision. In: Doherty, P., Mylopoulos, J., Welty, C.A. (eds.) KR, pp. 239–248. AAAI Press (2006)
4. Gärdenfors, P.: Knowledge in Flux: Modeling the Dynamics of Epistemic States. Bradford Books, MIT Press, Cambridge (1988)
5. Gärdenfors, P., Makinson, D.: Revisions of knowledge systems using epistemic entrenchment. In: Procs. of the 2nd TARK, pp. 83–96 (1988)
6. Grove, A.: Two modellings for theory change. Journal of Philosophical Logic 17, 157–170 (1988)
7. Katsuno, H., Mendelzon, A.: On the difference between updating a knowledge base and revising it. In: Gärdenfors, P. (ed.) Belief Revision, pp. 183–203. CUP (1992)
8. Lehmann, D.J., Magidor, M., Schlechta, K.: Distance semantics for belief revision. J. Symb. Log. 66(1), 295–317 (2001)
9. Lewis, D.: Probabilities of conditionals and conditional probabilities. The Philosophical Review 85(3), 297–315 (1976)

A Parallel, Multi-issue Negotiation Model in Dynamic E-Markets

Fenghui Ren[1], Minjie Zhang[1], Xudong Luo[3], and Danny Soetanto[2]

[1] School of Computer Science and Software Engineering
[2] School of Electrical, Computer and Telecom Engineering
University of Wollongong, Australia
[3] School of Computer Engineering
Nanyang Technological University, Singapore

Abstract. Negotiating agents play a key role in e-markets and become more popular. However, in much existing work, the e-markets are assumed to be closed and static, which is unrealistic. To address the issue, this paper developed negotiating agents that can adapt their negotiation strategies, outcome expectations, offer evaluations, and counter-offers generations in dynamic, open e-markets. Also, the proposed agents can generate multiple counter-offers according to different preferences so as to further improve their negotiation outcomes. Finally, the experimental results show the improvements on agents' profits by employing our negotiation model.

1 Introduction

In recent years, electronic marketplace (e-market) has changed the traditional ways of doing business and intelligent agents make the business processes in e-market more efficient. In an e-market, people can easily publish information, retrieve items of interest, and negotiate with opponents concurrently. In such a frequently changing environment, agents' expectations on negotiation outcomes may not be achieved successfully without considering the impacts from changes of e-markets. For example, when a market changes from a buyer's market to a seller's market, if buyers fail to be aware of such a change and insist on their original expectations, the negotiation could fail due to buyers' original expectations being hard to be satisfied in a seller's market. To the contrary, if sellers fail to be aware of such a change and insist on their original expectations, the sellers may loss the chance to maximize their benefits. Therefore, in order to be successful in such highly dynamic e-market, negotiation agents should adapt their negotiation strategies accordingly.

Many multi-issue negotiation models have been proposed. For example, the model in [5] can achieve optimal negotiation outcomes, but it works only in the environment with fixed number of agents. The model in [8] is also an one for multi-attribute negotiations between two agents. However, impacts on agents' strategies from outside options are still not taken into account. In model of [7], a multilateral multi-issue negotiation protocol is proposed, in a cooperative scenario, by employing a mediator agent. However, when the number of negotiation participators fluctuates, the mediator can hardly make an unbiased and accurate response to all agents. The work in [4] studied multi-issue

D. Wang and M. Reynolds (Eds.): AI 2011, LNAI 7106, pp. 442–451, 2011.

negotiation models in incomplete information settings and illustrated equilibrium solutions for different negotiation agendas and procedures. However, their model is only applicable in static negotiation environments. To remove the limitation, this paper will propose a parallel, multi-issue negotiation model for open, dynamic e-markets.

The rest of this paper is organized as follows. Section 2 introduces a new model to represent the e-markets and the negotiation issues. Section 3 proposes an offer evaluation approach. Section 4 discusses our counter-offer generation approach. Section 5 carries out experimental analysis of our model in different e-markets. Section 6 compares our work with related work. Finally, Section 7 concludes this paper.

2 E-Market and Negotiation Issue

2.1 E-Market Change and Agent's Response

Circumstance plays a crucial role in negotiations [11]. Especially in a dynamic e-market, the market change will impact agents' behaviours during negotiations.

Let α ($\alpha = 1$ for seller, and $\alpha = -1$ for buyer) denote an agent's role, s denote the number of sellers in an e-market, and b denote the number of buyers. Then the market's situation by considering the relationship between supply and demand can be defined as:

$$\Phi(s, b, \alpha) = \frac{b - s}{b + s} \times \alpha \tag{1}$$

Clearly, the value of Φ is in $(-1, 1)$. Intuitively, if $0 < \Phi < 1$, the e-market is a beneficial market (i.e., agents in role α have advantages in such a market); if $-1 < \Phi < 0$, the e-market is an inferior market (i.e., agents in role α have disadvantages); if $\Phi = 0$, the e-market is an equitable market (i.e., $b = s$ and agents play fairly).

However, Formula (1) only reflects the objective status of an e-market, but does not take agents' subjectiveness into account. In real world markets, different people behave differently in the same market situation. Therefore, we need a mapping from objective market situations to subjective responses of agents. Let β denote an agent's attitude the market changes. In general, it can be seen that the agent has three typical attitudes when the e-market's situation changes, i.e., calm ($\beta > 1$), excited ($1 > \beta > 0$), and normal ($\beta = 1$). Formally, by considering attitudes, agents' responses to the e-market changes are defined as:

$$\Psi(s, b, \alpha, \beta) = \begin{cases} (\Phi(s, b, \alpha))^{\beta}, & \text{if } \Phi(s, b, \alpha) \geq 0 \\ -(-\Phi(s, b, \alpha))^{\beta}, & \text{otherwise} \end{cases} \tag{2}$$

In fact, the formula reflects well the agent's subjective responses to e-market changes: (1) $1 > \Psi > 0$ indicates a positive response; (2) $\Psi = 0$ indicates a normal response; and (3) $-1 < \Psi < 0$ indicates a negative response.

2.2 Issue's Significance and Relationship

In multi-issue negotiation, the significance of each issue and the relationship between issues play important roles for the offer evaluation, the counter-offer generation, and the negotiation outcome [6]. Usually, the significance of issues are represented by weights, and there are no logical relationships between issues [4]. To put into consideration the logic relationships among the issues, here we propose alternative approach.

Definition 1. An agent's significance on a negotiation issue m is represented by tag $\kappa_m \in \{N, L, I, V, E\}$, which means not-important (N), little-important (L), important (I), very-important (V), and extremely-important (E), respectively. $\Gamma : \{N, L, I, V, E\} \rightarrow \{1, 2, 3, 4, 5\}$ maps a significance tag into its corresponding order. Atomic Preference (AP) is a collection of all issues' significance tags i.e., $AP = (\kappa_1, \ldots, \kappa_M)$, and Complete Preferences (CP) contains all APs the agent has, i.e., $CP = (AP_1, \ldots, AP_J)$.

Definition 2. The relationship between two issues or two APs is represented by a unique relationship tag $\xi \in \{\cap, \cup\}$. Tag $\cap$ indicates a union relationship of two connected parts (i.e., an agent's expectation on two parts connected by $\cap$ must be satisfied). Tag $\cup$ indicates an alternative relationship of two connected parts (i.e., an agent's expectation on only one part connected by $\cup$ must be satisfied).

3 Offer Evaluation

In this paper, we employ the package deal negotiation procedure [3], but do not use multi-attribute theory [1] to evaluate offer package because (1) quantitative representation of issues' significance does not always accord with human's ways of thinking [9]; and (2) The multi-attribute utility approach cannot model the logical relationship between negotiation issues. However, a human often assigns different logic relationships between their concerned issues. In this section, we propose a novel approach to perform the evaluation process in multi-issue negotiation by using significance tags and relationship tags. Let $\boldsymbol{O}_{i,t} = (o_{i,t,1}, \ldots, o_{i,t,M})$ denote an offer package from opponent i at round t, where $o_{i,t,m}$ denotes opponent i's proposal on issue m. Let $\boldsymbol{O}_{ini} = (o_{ini,1}, \ldots, o_{ini,M})$ denote an agent's initial offer package, and $\gamma = (\gamma_1, \ldots, \gamma_M)$ ($\gamma_m \in \{-1, 1\}$) indicate the agent's preference on each negotiating issue. That is, if the agent prefers a higher value than $o_{ini,m}$, $\gamma_m = 1$, otherwise $\gamma_m = -1$. Firstly, when the agent receives opponent i's proposal $o_{i,t,m}$, the proposal is evaluated as follows:

$$\Lambda(o_{i,t,m}, o_{ini,m}, \gamma_m) = \mathbf{th}\left(\frac{o_{i,t,m} - o_{ini,m}}{o_{ini,m}} \times \gamma_m\right) + 1 \tag{3}$$

where $\mathbf{th}(x) = \frac{e^x - e^{-x}}{e^x + e^{-x}}$ is the Hyperbolic Tangent function.[1] Formula (3) ($\Lambda \in (0, 2)$) can model well how an agent's initial offer is satisfied by an opponent's proposal. For example, suppose an agent prefers a lower value than the initial value, then γ_m equals -1. When $o_{i,t,m} = o_{ini,m}$, $\Lambda = 1$ and the buyer's expectation is fully satisfied. When $o_{i,t,m} > o_{ini,m}$, $0 \leq \Lambda < 1$ and the buyer's expectation is partially achieved. And when $o_{i,t,m} < o_{ini,m}$, $1 < \Lambda < 2$ and the buyer's expectation is overachieved. However, formula (3) only evaluates the proposal $o_{i,t,m}$ based on an agent's initial offer without considering the market situations. To remove the limitation, we update the formula as follows:

$$\Theta(o_{i,t,m}, o_{ini,m}, \gamma_m, s, b, \alpha, \beta) = \frac{\Lambda(o_{i,t,m}, o_{ini,m}, \gamma_m)}{\Psi(s, b, \alpha, \beta) + 1} \tag{4}$$

[1] That is, $\mathbf{th}(0) = 0$, $\lim_{x \to \infty} \mathbf{th}(x) = 1$, and $\lim_{x \to -\infty} \mathbf{th}(x) = -1$.

Θ ($\in [0,1]$) can well evaluates the opponent's proposal $o_{i,t,m}$ according to both the agent's and the market's situations. In an equitable market, the proposal is evaluated unbiasedly (i.e., when $\Psi = 0$, $\Theta = \Lambda$). In a beneficial market, the proposal is undervalued (i.e., when $0 < \Psi < 1$, $\Theta < \Lambda$). In an inferior market, the proposal is overvalued (i.e., when $-1 < \Psi < 0$, $\Theta > \Lambda$).

Let $\Theta(\boldsymbol{O}_{i,t}) = (\Theta(o_{i,t,1}), \ldots, \Theta(o_{i,t,M}))$ indicate the evaluation results on opponent i's all proposals. Then, these results need to be combined in order to get an overall evaluation result on opponent i's proposals. The traditional combination approach is the weighted sum approach. However, we propose a non-linear combination approach by using the significance tags, and agents' multiple preferences are also considered. We firstly calculate the relative importance between issues according to their significance orders as below:

$$d_m = \Gamma(\kappa_m) / \sum_{m=1}^{M} \Gamma(\kappa_m) \tag{5}$$

Then the combined evaluation result by considering all negotiation issues based on the agent preference AP is:

$$T^{AP}_{prod}(\Theta(\boldsymbol{O}_{i,t})) = \prod_{m=1}^{M} \Theta(o_{i,t,m})^{d_m} \tag{6}$$

If an agent has more than one preferences, since the agent might have different significance orders on issues in different preferences, when it evaluates a proposal package, it should consider all preferences, and select one as the final result. Different selection criterion can be employed by agents to indicate their attitudes, i.e., a pessimistic agent will select the minimal one; a optimistic agent will select the maximal one; while other agents will select between these two extreme values. In this paper, we choose the maximal one as the final result. That is because a satisfaction on any one of the agent's preferences can lead to an agreement, and thus select the maximal one accelerate the process to the agreement. The selection result in this paper is:

$$T^{CP}_{prod}(\Theta(\boldsymbol{O}_{i,t})) = \max_{AP_j \in CP} \{T^{AP_j}_{prod}(\Theta(\boldsymbol{O}_{i,t}))\} \tag{7}$$

4 Counter-Offer Generation

In this section, we present a novel counter-offer generation approach by considering issues' significance and e-market situations. Let $\boldsymbol{O}_{m,t} = (o_{1,t,m}, \ldots, o_{I,t,m})$ denote all offers from all available opponents on issue m at round t. In Figure 1, the x-axis indicates opponents' proposals on issue m, and the y-axis represents the occurrence density of these proposals. The solid curve indicates the distribution of $\boldsymbol{O}_{m,t}$, and the dotted line is the estimated distribution of $\boldsymbol{O}_{m,t+1}$ in the next round. The distribution of $\boldsymbol{O}_{m,t}$ may be different from case to case. However, without losing any generality, we make the assumption that the shape of the distribution curve of set $\boldsymbol{O}_{m,t+1}$ is similar to that of $\boldsymbol{O}_{m,t}$, but the range of span is changed. Suppose a buyer agent negotiates a car's price, then the market displayed in Figure 1 is a beneficial market for the buyer agent. In a beneficial market, opponents' proposals are estimated to become smaller in average

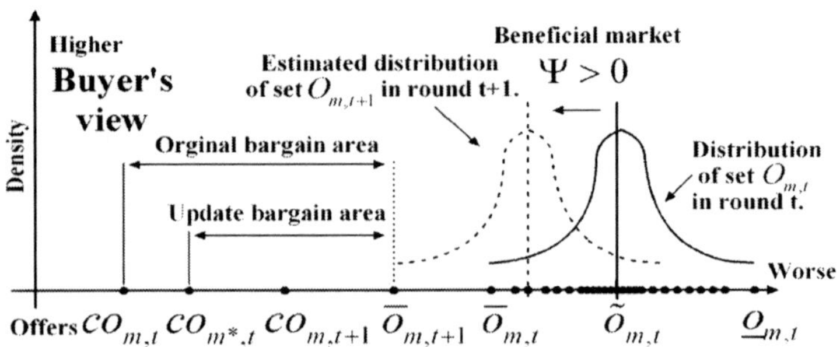

Fig. 1. Counter-offer generation in a beneficial market for multiple issues

in the next round (i.e., $\tilde{O}_{m,t+1} < \tilde{O}_{m,t}$). The distance between the buyer's current offer $co_{m,t}$ and the expected best offer $\bar{o}_{m,t+1}$ in the next round is the bargain area. Then the buyer's new counter-offer $co_{m,t+1}$ is generated within this area according to the buyer's negotiation strategies, the remaining rounds, the distribution of $\boldsymbol{O}_{m,t}$, and the issue's significance.

Firstly, we estimate the expected best offer $\bar{o}_{m,t+1}$ in the next round. Such an estimation is based on an assumption that all agents are self-interested and rational and thus they are trying to balance their profits and negotiation success. In a beneficial market, they will look for more profits; while in an inferior market, they will sacrifice profits in order to increase the negotiation success. Formally, $\bar{o}_{m,t+1}$ can be calculated as:

$$\bar{o}_{m,t+1} = \bar{o}_{m,t} + \gamma_m \cdot \Psi(s, c, \alpha, \beta) \cdot 3\sqrt{D(\boldsymbol{O}_t)} \tag{8}$$

$$D(\boldsymbol{O}_{m,t}) = \sum_{i=1}^{I} (o_{i,t,m} - E(\boldsymbol{O}_{m,t}))^2 \cdot p_{i,m} \tag{9}$$

where $D(\boldsymbol{O}_{m,t})$ indicates the variance of $\boldsymbol{O}_{m,t}$, $E(\boldsymbol{O}_{m,t})$ indicates the mathematical expectation of $\boldsymbol{O}_{m,t}$, and $p_{i,m}$ indicates the distribution probability of $o_{i,t,m}$. We set the maximal possible change of the expected best offer to $3\sqrt{D(\boldsymbol{O}_t)}$ because 99% of observed value locates in interval $[-3\sqrt{D(\boldsymbol{O}_t)}, 3\sqrt{D(\boldsymbol{O}_t)}]$ in mathematics. Usually, when the distribution of $\boldsymbol{O}_{m,t}$ is a normal distribution, $E(\boldsymbol{O}_{m,t}) = \frac{1}{I}\sum_{i=1}^{I} o_{i,t,m}$ and $p_{i,m} = 1/I$.

Secondly, we modify the bargain area. As shown in Figure 1, the bargain area is originally between $co_{m,t}$ and $\bar{o}_{m,t+1}$. However, when agents assign different significance tags on issues, their expectations on negotiation outcomes are different. Intuitively, an agent should concede less at more significant issues, but more at less significant issues. In order to represent such a reality, the starting point of the bargain area needs to be updated as follows:

$$co_{m*,t} = co_{m,t} + (\bar{o}_{m,t+1} - co_{m,t})(\Gamma(E) - \Gamma(\kappa_m))/(\Gamma(E) - \Gamma(N)) \tag{10}$$

For example, if issue m's significance tag is E, then the updated stating point equals to the original starting point (i.e., $co_{m*,t} = co_{m,t}$), and the agent does not shrink its bargain area. If issue m's significance tag is N, the updated stating point equals to the expected best offer (i.e., $co_{m*,t} = \bar{o}_{m,t+1}$), and the agent minimizes its bargain area. For issues with other significance tags, the changes of the bargain area are between these two extreme cases. Then the counter-offer $co_{m,t+1}$ is generated as follows:

$$co_{m,t+1} = \left\{ \begin{array}{ll} o_{ini,m}, & \text{if } t = 0 \\ co_{m*,t} + (\bar{o}_{m,t+1} - co_{m,t})(\frac{t}{\tau})^{\lambda}, & \text{if } t \leq \tau \end{array} \right\} \tag{11}$$

Where τ is the agent's deadline and λ is the negotiation strategy [6]. Finally, by applying formula (11) on all issues, a counter-offer package is generated as follows.

$$\Upsilon(AP, t) = (co_{1,t+1}, \ldots, co_{M,t+1}) \tag{12}$$

According to Formula (10), the starting point of bargain area for issues with different significance tag will be updated differently. If an agent has several preferences and assigns different significance on an issue in different preferences, different counter-offers should be generated. In our proposed counter-offer generation approach, the number of counter-offers generated by the agent in each negotiation round equals the number of the agent's preferences. An agent generates parallel counter-offers by considering all preferences as follows:

$$\Upsilon(CP, t) = (\Upsilon(AP_1, t), \ldots, \Upsilon(AP_J, t))^T \tag{13}$$

By comparison with the traditional single counter-offer approach, the proposed approach will increase the negotiation efficiency and overall profit. Because parallel counter-offers will result an agreement quicker and decrease the lose of overall profit by considering time constraints.

5 Experiment

5.1 Experimental Settings

We set the total of agents to 100 (50 buyers and 50 sellers) and two negotiating issues (the price and warranty of a car). For the buyer agents, their initial prices, reservation prices, initial warranty, and reservation warranty are randomly selected between \$1500 and \$4500, \$5000 and \$15000, 5 years and 10 years, and 1.5 years and 4.5 years, respectively. For the seller agents, their initial prices, reservation prices, initial warranty, and reservation warranty are randomly selected between \$5000 and \$15000, \$1500 and \$4500, 1.5 years and 4.5 years, and 5 years and 10 years. For all negotiating agents, the parameters of their negotiation strategy are randomly selected in interval $[0, 2]$, their negotiation deadlines are randomly selected in interval $[15, 25]$. For the classic negotiating agents [3], their weighting on the two issues are randomly selected in interval $[0, 1]$; and for our negotiation agents, their significance tags on the issues are randomly selected in $\{N, L, I, V, E\}$.

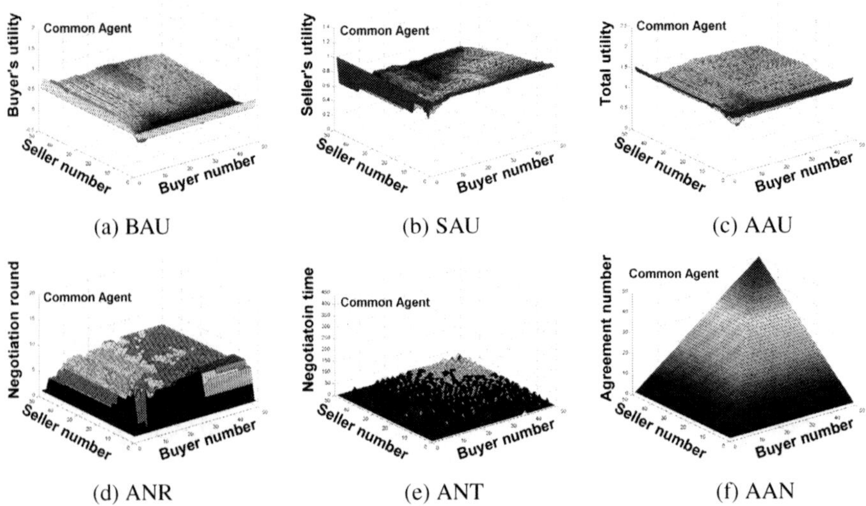

(a) BAU (b) SAU (c) AAU

(d) ANR (e) ANT (f) AAN

Fig. 2. Negotiation results with the classic model

5.2 Experimental Results

To show the performance of our negotiation model, two experiments are carried out. In the first experiment, both buyer and seller agents employ the classic negotiation model [3] (that cannot handle the dynamics of e-markets). Both the buyer and the seller agents' number are started from 1, and gradually increased to 50. We analyze the experimental results, in terms of the buyers' average utility (*BAU*), the sellers' average utility (*SAU*), all negotiating agents' average utility (*AAU*), average negotiation rounds (*ANR*), average negotiation time (*ANT*, in millisecond), and average agreement number (*AAN*). In Figure 2, the experimental results by using the classic negotiation model are displayed, we can see that except for *AAN*, the other five results almost has no changes when the number of negotiating agents changes. Both *BAUs* and *SAUs* are around 0.5 for different market situations, and *AAUs* are round 1.0. All negotiations are finished in-between 5 and 8 rounds, and spend around $10ms$. That is because the classic negotiation model does not consider the impacts from the e-market changes, and thus the agents cannot adapt their behaviours when the market changes.

In the second experiment, all buyer agents employ our negotiation model, and all seller agents employ the classic negotiation model. To simplify the experiment, we set the buyer agents' attitudes on market changes to normal (i.e., $\beta = 1$). The experimental results are shown in Figure 3. It can be seen that *BAUs* (see Figure 3(a)) show different values in different market situations. When it is an equitable market, *BAUs* are similar as the values gained in the first experiment. However, when the market becomes more beneficial to the buyers, *BAUs* increase gradually. The maximum *BAU* is around 1.5, and appears when there are 50 sellers but only 1 buyer in the market. On the other hand, when the market becomes inferior to the buyers, *BAUs* decreases gradually. In the extreme case, when the market contains only 1 seller but 50 buyers, *BAU* is minimized and almost equals to 0. The reason for such differences is because the buyer agents adapt their negotiation behaviours when the market situation changes, and try to enlarge profits in beneficial markets, and to guarantee success in inferior markets.

Even though the seller agents cannot adapt their negotiation behaviours initiatively, *SAUs* (see Figure 3(b)) are also varied in different market situations with the changes of the buyer agents' behaviours. In general, *SAU* is increased from 0 to 1 when the market changes from an extreme buyer's market to an extreme seller's market. In a buyer's market, since the buyer agents know their advantages, they would not make big concessions, and so the seller agents have to. By contrast, in order to beat other competitors, the buyer agents have to make great concessions in a seller's market, so the seller agents' profits are increased. However, since the seller agents employ the classic negotiation model and cannot be aware of their advantages in a seller's market, they cannot further enlarge their profits subjectively as the buyer agents do in a buyer's market, but immediately accept offers when their initial expectations are satisfied. That is why the maximum *BAU* is 1.5, but the maximum *SAU* is only 1.

AAUs (see Figure 3(c)) are increased around 0.2 on average by comparison with the outcomes of classic negotiation model (see Figure 2(c)). Such an increment implies that our negotiation model can improve the outcome of the whole market in different market situations. In a buyer's market, the buyer agents would like to spend more time on bargaining to maximize their profits, and so *ANRs* (see Figure 3(d)) are increased in the second experiment. However, when the buyer agents try to prevent profits loss and to guarantee successes in a seller's market, they would like to reach agreements as quick as possible, and so *ANRs* are decreased. Nevertheless, no matter in a buyer's market or in a seller's market, because the buyer agents need extra time to analyze the market situation and accordingly select their following actions in each negotiation round, our

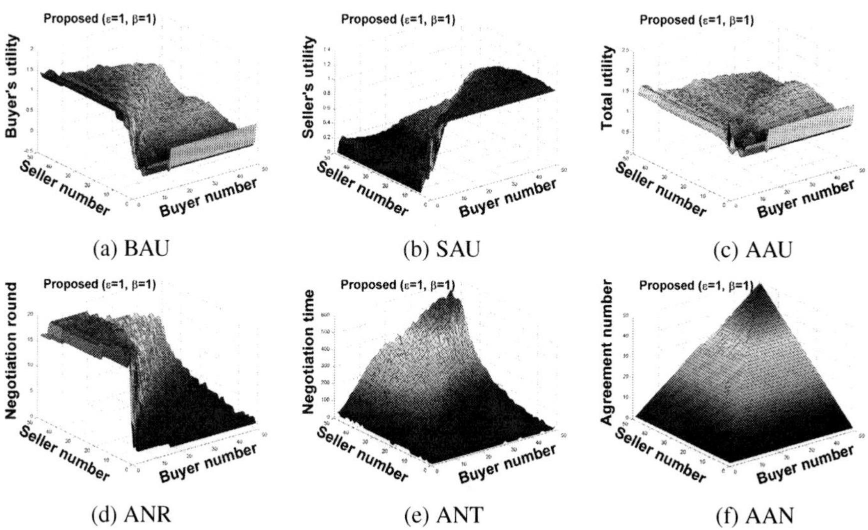

(a) BAU (b) SAU (c) AAU

(d) ANR (e) ANT (f) AAN

Fig. 3. Negotiation results with our model

negotiation model could spend more time than the classic one. It can be seen that in the most complex market (50 buyer agents and 50 seller agents), our model spends only 0.5 second more than the classic model in finishing all negotiations. However, compared with the benefits bringed to the whole market, such a little delay is acceptable.

According to the experimental settings, since the bargain areas exist between all buyer and seller agents, the classic negotiation model can reach agreements between all buyer and seller agents. However, by employing our negotiation model, the buyer agents will adapt their offer evaluation results in different market situations. Especially, in a buyer's market, the seller agents' offers are usually under-valued by the buyer agents, so the buyer agents' requirements are not easy to be satisfied. That is the reason for the slightly decrement on *AAN* in a buyer's market (see Figure 3(e)).

6 Related Work

Dasgupta and Hashimoto proposed an approach to address the problem of dynamic pricing in a competitive online economy where a product is differentiated by buyers and sellers on multi-issue [2]. Agents may have incomplete knowledge of the negotiation parameters. A seller employs a collaborative filtering algorithm to determine a temporary consumer's purchase preferences and a dynamic pricing algorithm to determine a competitive price for the product. However, their approach pays attention only to sellers, and our negotiation model considers both negotiation sides.

Nguyen and Jennings proposed a concurrent bilateral negotiation model to handle multi-lateral negotiations [10]. When an agent negotiates with more than one opponents, this model treats the negotiation between the agent and each opponent as an 1-to-1 negotiating thread, and a coordinator is employed to control all negotiating threads. The coordinator will select a suitable negotiation strategy from predefined strategies for each thread. However, their model did not consider the impacts from negotiation environment changes (i.e., the change of sellers and buyers). In contrast, our negotiation model captures the dynamic changes of a negotiation environment.

Ren proposed a Market-Driven Agents (MDAs) model to model relationship between agents' negotiation strategies and the negotiation environment [11]. In the MDAs model, agents are guided by four concession factors, and these factors determine how much concession agents can give during the negotiation based on the environment. However, the MDAs model does not take into account the situation when the negotiation environment becomes open and dynamic, and our model address well how a dynamic changing e-market be handled in negotiation.

By comparison with the above related work, our negotiation model has features. It models negotiations in e-markets by considering (i) both e-markets' and agents' situations, (ii) both current and future possible situations of e-markets, and (iii) multi-attribute and multi-preference. However, we also recognize that the performance of our model still can be improved in aspects of time cost and negotiation success rate.

7 Conclusion and Future Work

This paper proposed a parallel, multi-issue negotiation model for dynamic e-markets. This model describes an e-market as beneficial, inferior or normal according to the supply and demand. Agents evaluate offers or generate counter-offers based on the e-market situation and themselves attitudes. The experimental results showed clearly that our negotiation model capture the e-market changes well and adapt agents' negotiation behaviours dynamically and accordingly. Our future work will improve the e-market model by considering the number of goods in supply and demand relationship and opponent agents' reputations in proposal evaluations.

Acknowledgement. The authors would like to acknowledge the financial support from the Australian Research Council (ARC) Linkage Scheme LP0991428 and Transgrid Australia for this project.

References

1. Barbuceanu, M., Lo, W.: Multi-attribute Utility Theoretic Negotiation for Electronic Commerce. Agent-Mediated Electronic Commerce III, 15–30 (2001)
2. Dasgupta, P., Hashimoto, Y.: Multi-Attribute Dynamic Pricing for Online Markets Using Intelligent Agents. In: Proc. of 3rd Int. Conf. on Autonomous Agents and Multiagent Systems, pp. 277–284 (2004)
3. Fatima, S., Wooldridge, M., Jennings, N.: Multi-Issue Negotiation Under Time Constraints. In: Proc. of 1st Int. Conf. on Autonomous Agents and Multi-Agent Systems, pp. 143–150 (2002)
4. Fatima, S., Wooldridge, M., Jennings, N.: An Agenda-Based Framework for Multi-Issue Negotiation. Artificial Intelligence 152(1), 1–45 (2004)
5. Fatima, S., Wooldridge, M., Jennings, N.: Approximate and Online Multi-Issue Negotiation. In: Proc. of 6th Int. Conf. on Autonomous Agents and Multi-Agent Systems, pp. 947–954 (2007)
6. Fatima, S., Wooldridge, M., Jennings, N.: An Analysis of Feasible Solutions for Multi-Issue Negotiation Involving Nonlinear Utility Functions. In: Proc. of 8th Int. Conf. on Autonomous Agents and Multiagent Systems, pp. 1041–1048 (2009)
7. Hemaissia, M., Seghrouchni, A., Labreuche, C., Mattioli, J.: A Multilateral Multi-Issue Negotiation Protocol. In: Proc. of 6th Int. Conf. on Autonomous Agents and Multiagent Systems, pp. 939–946 (2007)
8. Lai, G., Sycara, K., Li, C.: A Pareto Optimal Model for Automated Multi-attribute Negotiations. In: Proc. of 6th Int. Conf. on Autonomous Agents and Multi-Agent Systems, pp. 1040–1042 (2007)
9. Luo, X., Zhang, C., Jennings, N.: A Hybrid Model for Sharing Information between Fuzzy, Uncertain and Default Reasoning Models in Multi-Agent Systems. Int.l J. of Uncertainty, Fuzziness and Knowledge-Based Systems 10(4), 421–450 (2007)
10. Nguyen, T., Jennings, N.: Coordinating Multiple Concurrent Negotiations. In: 3rd Int. Joint Conf. on Autonomous Agents and Multiagent Systems, pp. 1064–1071 (2004)
11. Ren, F., Zhang, M., Sim, K.: Adaptive Conceding Strategies for Automated Trading Agents in Dynamic, Open Markets. Decision Support Systems 46(3), 704–716 (2009)

Parallel Monte Carlo Tree Search Scalability Discussion

Kamil Rocki and Reiji Suda

The University of Tokyo,
Department of Computer Science,
7-3-1, Hongo, Bunkyo-ku, Tokyo, Japan
{kamil.rocki,reiji}@is.s.u-tokyo.ac.jp

Abstract. In this paper we are discussing which factors affect the scalability of the parallel Monte Carlo Tree Search algorithm. We have run the algorithm on CPUs and GPUs in Reversi game and SameGame puzzle on the TSUBAME supercomputer. We are showing that the most likely cause of the scaling bottleneck is the problem size. Therefore we are showing that the MCTS is a weak-scaling algorithm. We are not focusing on the relative scaling when compared to a single-threaded MCTS, but rather on the absolute scaling of the parallel MCTS algorithm.

Keywords: Monte Carlo, Scalability, Reversi, SameGame.

1 Introduction

Monte Carlo Tree Search (MCTS)[1][2][3] is a method for making optimal decisions in artificial intelligence (AI) problems, typically move planning in combinatorial games. It combines the generality of random simulation with the precision of tree search. In this paper we are focusing on the parallel MCTS usage and its scaling limitations. First we will very briefly explain how the MCTS algorithm works.

1.1 MCTS Algorithm Overview

A simulation is defined as a series of random moves which are performed until the end of a game is reached (until neither of the players can move). The result of this simulation can be successful, when there was a win in the end or unsuccessful otherwise. So, let every node i in the tree store the number of simulations t_i (visits) and the number of successful simulations S_i. First the algorithm starts only with the root node. The general MCTS algorithm comprises 4 steps (Figure 1) which are repeated until a particular condition is met (i.e. no possible move or time limit is reached).

1.2 MCTS Iteration Steps

Selection - a node from the game tree is chosen based on the specified criteria. The value of each node is calculated and the best one is selected. In this paper,

D. Wang and M. Reynolds (Eds.): AI 2011, LNAI 7106, pp. 452–461, 2011.

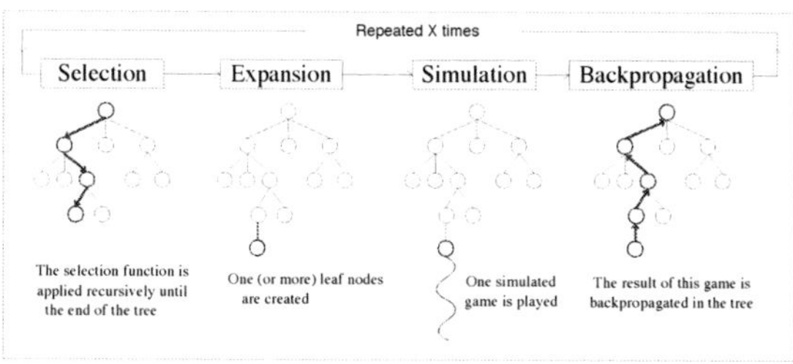

Fig. 1. A single MCTS algorithm iteration's steps (from [1])

the formula used to calculate the node value is the Upper Confidence bound applied to Trees (UCT)[2].

$$UCB_i = \frac{S_i}{t_i} + C * \sqrt{\frac{logT_i}{t_i}}$$

Where:
T_i - total number of simulations for the parent of node i
C - a parameter to be adjusted (low - exploitation, high - exploration).

Supposed that some simulations have been performed for a node, first the average node value is taken and then the second term which includes the total number of simulations for that node and its parent. The first one provides the best possible node in the analyzed tree (exploitation), while the second one is responsible for the tree exploration. That means that a node which has been rarely visited is more likely to be chosen, because the value of the second terms is greater. The C parameter adjusts the exploitation/exploration ratio.

Expansion - one or more successors of the selected node are added to the tree depending on the strategy.

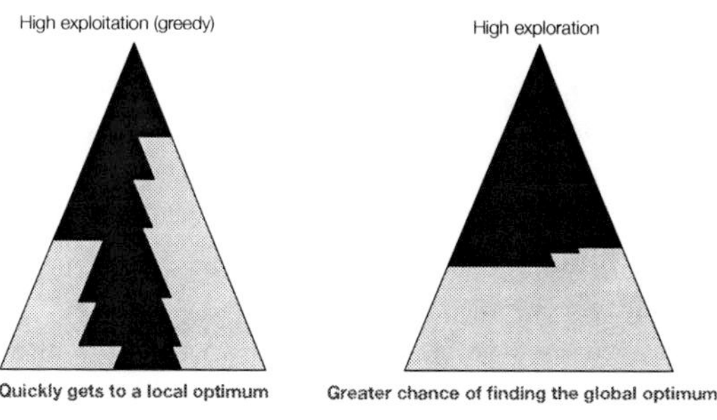

Fig. 2. Explotation and Exploration illustrated)

Simulation - for the added node(s) perform simulation(s) and update the node(s) values (successes, total).

Backpropagation - update the parents' values up to the root nodes.

2 Parallelization of Monte-Carlo Tree Search

2.1 Some of the Existing Parallel Approaches

In 2007 Cazenave and Jouandeau[4] propose 2 methods of parallelization of MCTS and later in 2008 Chaslot et al.[3] propose another one an analyze 3 approaches(Figure 3):

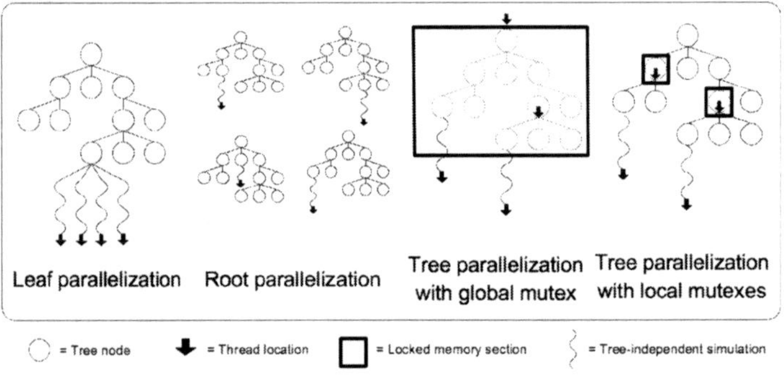

Fig. 3. Parallel MCTS as in [3])

1. **Leaf Parallelization.** It is one of the easiest ways to parallelize MCTS. Only one thread traverses the tree and adds one of more nodes to the tree when a leaf node is reached (*Selection* and *Expansion* phases). Next, starting from the leaf node, independent simulated games are played for each available thread (*Simulation* phase). When all games are finished, the result of all these simulated games is propagated backwards through the tree by one single thread (*Backpropagation* phase).
2. **Root Parallelization.** Cazenave[4] proposed a second parallelization called *single-run* parallelization. It is also referred to as *root parallelization*[3]. It consists of building multiple MCTS trees in parallel, with one thread per tree. The threads do not share information with each other. When the available time is spent, all the root children of the separate MCTS trees are merged with their corresponding clones. For each group of clones, the scores of all games played are added. The best move is selected based on this grand total. This parallelization method only requires a minimal amount of communication between the threads.It is more efficient than simple leaf parallelization[3], because building more trees diminishes the effect of being stuck in a local extremum and increases the chances of finding the true global maximum.

The goal of this paper is to determine which parameters affect the scalability of the algorithm when running on multiple CPUs/GPUs using root parallelization.

3 Methodology

3.1 MPI Usage

In order to run the simulations on more machines the application has been modified in the way that communication through MPI is possible. This allows to take advantage of systems such as the TSUBAME supercomputer or smaller heterogenous clusters. The implemented scheme (Figures 4) defines one master process (with id 0) which controls the game and I/O operations, whereas other processes are active only during the MCTS phases. The master process broadcast the input data (current state/node) to the other processes. Then each process performs an independent Monte Carlo search and stores the result. After this phase the master process collects the data (through the *reduce* MPI operation which sums the results.

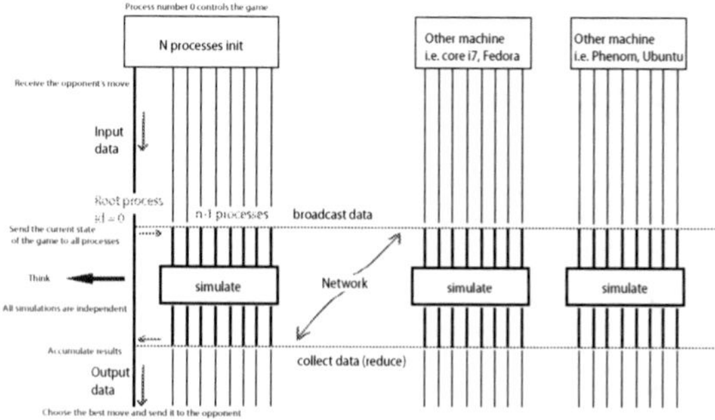

Fig. 4. MPI Processing scheme

3.2 Reversi

In general there are some basic features of graphs presenting the results which need to be explained, it is important to understand how to read them as they may seem to be complicated.

1. **Number of simulation per second** in regard to the number of CPU threads, GPU threads or other factors - (i.e. Figure 5). By this I mean the average number of random playouts performed (MCTS algorithm - step 3) during the game.
2. **Score** (or average score) in regard to the number of CPU threads or GPU threads. Score means the point difference between player A and player B who play against each other. The higher score, the better. If score is greater than 0, it means a winning situation, 0 means a draw, otherwise a loss.

3. **A game step** is a particular game stage starting from the initial Reversi position until the moment when no move is possible. There can be up to 60 game steps in Reversi. As the game progresses, the average complexity of the search tree changes, the way the algorithm behaves changes as well. Instead of showing the score or the speed of the algorithm in regard to the number of cores, I also show the performance considering the game stage.
4. **Win ratio** - Another type of measuring the strength of an algorithm. It means the proportion of the games won to the total number of games played. The higher, the better.

3.3 Samegame

A modified version of the MCTS algorithm is used[7] to be able to solve SameGame puzzle in order to get as many points as possible. The first reason to do this was to test if MCTS algorithm can be easily applied to other domains. And the second one to check the scalability limitations, since it is easier to adjust fewer parameters in SameGame rather than having a 2-player game and to see if there are any similarities in 2 problems. Before applying MCTS to SameGame it was not sure if the problem itself may cause limitations in scalability.

4 Results and Analysis

4.1 CPU MPI Implementation

Figure 5 shows the average number of simulations per second in total depending on the number of cpu cores. The very little overhead is observed during the MPI communication. The number of simulations per second increases almost linearly and for 1024 threads the speedup is around 1020-fold (around 8 million simulations/sec).

Figure 4.1B shows the average score and win ratio of MCTS parallel algorithm playing against sequential MCTS agent depending on the number of cpu cores. From this graph it can be seen that obviously when 1 root-parallel MCTS thread plays against the sequential MCTS, they are equal (the winning percentage of around 48% for each of them - the missing 4% are the draws). In this graph I present the absolute score (not the score difference between players, so it is not so tightly associated with the winning ratio, but still the more, the stronger the algorithm). When the number of cores doubles, the parallel algorithm wins in more than 60% of the cases, and when the number of threads equals 16, it reaches 90%, to get to the level of around 98% for 64 threads. Here, a high increase in the algorithm's strength is observed when the thread number is increased up to 16, later the strength increase is not as significant, which can lead to a conclusion that the root parallel MCTS algorithm can have some limitations regarding the strength scalability given the constant increase in the speed shown in Figure 5A. Actually this has been studied ([4][3][6]) and some conclusions have been formed that such a limit exists and that the root-parallel algorithm performs well only up to several cores.

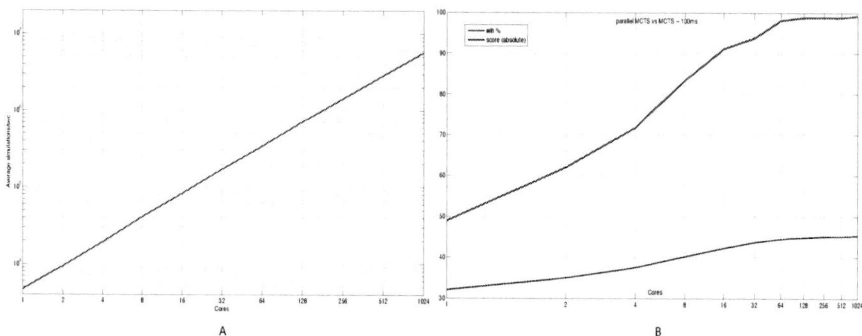

Fig. 5. (A) Average number of simulations per second in total depending on the number of cpu cores (multi-node), (B) - Average score and win ratio of MCTS parallel algorithm playing against sequential MCTS agent depending on the number of cpu cores

The first and main conclusion of the results obtained is that the root parallelization method is very scalable in terms of multi-CPU communication and number of simulations performed in given time increases significantly. Another one is that there is a point when raising the number of trees in the root parallel MCTS does not give a significant strength improvement.

4.2 Multi GPU Implementation

Multi-GPU implementation follows the same pattern as the multi-CPU scheme. Each TSUBAME node has 3 GPUs and using more than 3 GPUs requires MPI utilization. In Figure 6 it can be seen that just like with the multiple CPUs, there is no inter-node communication bottleneck and the raw simulation speed

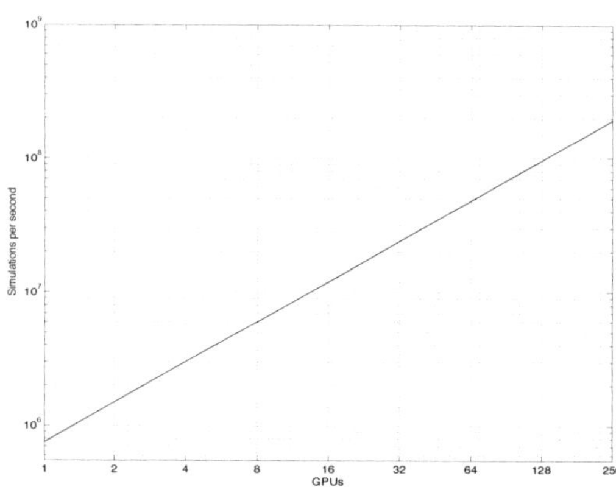

Fig. 6. Number of simulations per second depending on the GPU number, 56x112 threads per GPU

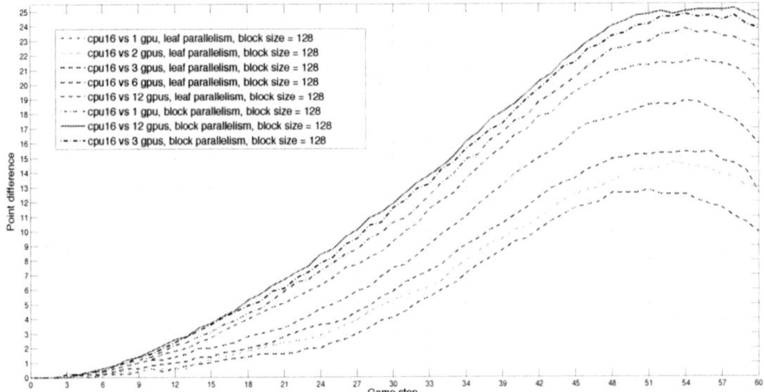

Fig. 7. 16 CPU cores vs different GPU configurations at particular game stages

increases linearly reaching approximately $2 * 10^8$ simulations/second when 256 GPUs (128 nodes, 2 GPUs per node) are used. Another graph (Figure 7 shows that just like in the case of multiple CPUs, even when the simulation speed increases linearly for multiple devices, the strength of the algorithm does not.

4.3 Scalability Analysis

The next illustrations (Figure 9) shows results of changing the MCTS environment to analyze the scalability affecting parameters. First in figures 9ABC we see how changing the sigma(C - exploitation/exploration ratio) affects it and Figure 9 presents how changing the problem size impacts the strength increase of the parallel approach. There are 2 things to be observed which are important. First, the sigma parameter affects the scalability in the way, that when the exploitation of a single tree is promoted, the overall strength of the algorithm improves better with the thread increase. In my opinion, this is due to the deeper tree search for each of the separate trees. When the exploration ratio is higher, then

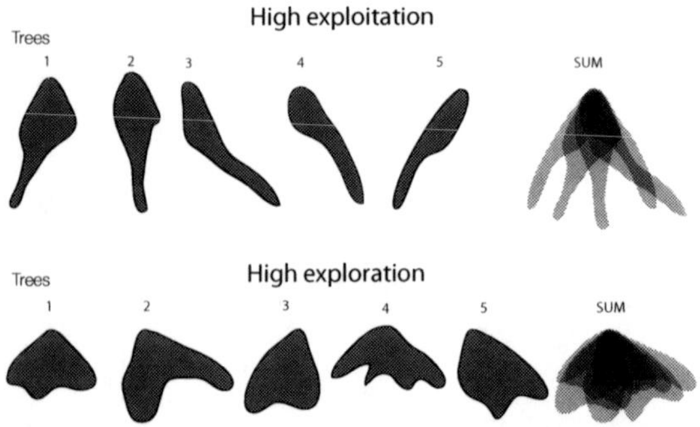

Fig. 8. Explanation of performance change caused by sigma constant adjustment

the trees basically form similar trees and reducing it propagates more diversified tree structures among the threads. Then in the next figure we see that as the problem size decreases, the improvement while parallelizing the algorithm also diminishes. It means that parallel MCTS algorithm presents *weak parallelism*, so it would also mean as long as we increase the problem size, the number of threads can grow and the results improve.

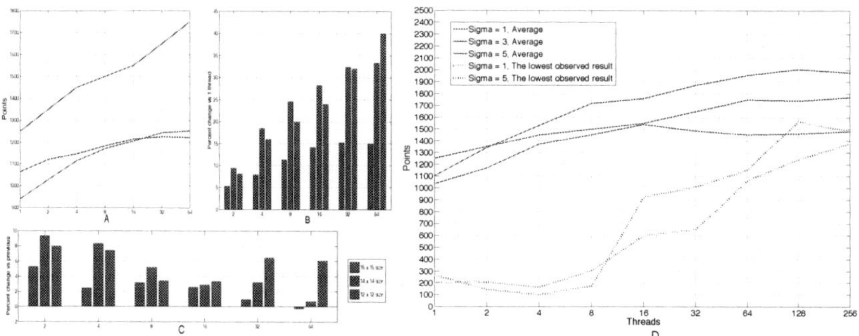

Fig. 9. Scalability of the algorithm when the problem size changes, (A) - absolute score, (B) - relative change in regard to 1 thread score, (C) - relative change when number of threads is doubled, sigma = 3, Root parallelism, (D) - Differences in scores during increasing number of threads and changing sigma constant, Root parallelismMax nodes = 10k

4.4 Random Sampling with Replacement

What is interesting in this case, when large number of threads is simulating at random is to obtain the probability of having exactly x distinct simulations after n simulations. Then, assuming that m is the total number of possible combinations (without order), $D(m, n)$ can be calculated, which is the expected number of distinct samples.

$$P(m, n, x) = \frac{\binom{m}{n}\binom{n-1}{n-x}}{\binom{m+n-1}{n}}$$

$P(m, n, x)$ is the probability of having exactly x distinct simulations after n simulations, where m is the total number of possible combinations (according the theorems of combinatorics, sampling with replacement). Then:

$$D(m, n) = \sum_{x=1}^{n} x * P(m, n, x)$$

It is hard to calculate D(m,n) for big numbers because of the factorials, but according to [5] an approximation can be used. Let $r = \frac{n}{m}$. Then:

$$D(m,n) \sim m(1-e^{-r}) + \frac{r}{2}e^{-r} - \frac{O(r(1+r)e^{-r}}{n}$$

The first term is the most important. Having this I was able to analyze the impact of having large number of samples in regard to the state space size and check how many of those samples would repeat (theoretically). For an instance if:

$m = 10^8$, $n_1 = 10^4$, $n_2 = 10^7$

In the first case (n_1): $\frac{D(m,n_1)}{n_1} \sim 99.5\%$ - almost no repeating samples.

Then if I consider (n_2): $\frac{D(m,n_2)}{n_2} \sim 95.1\%$ - around 4.9% samples are repeated.

This means that the scalability clearly depends on the space state size/number of samples relation. As the number of samples approaches the number of possible paths in a tree, the algorithm will lose its parallel properties and even finding the exact solution is not guaranteed, since we would have to consider infinite number of samples. It can be concluded that as the tree gets smaller (the solution is closer) the number of repeated samples increases (the higher the line the more repeated samples there are). When the tree is shallow enough (depth is lower than 10) it is very significant. If the state-space is small, the impact of the parallelism will be diminished (Figure 10). Low problem complexity may be caused by the problem is simple itself.

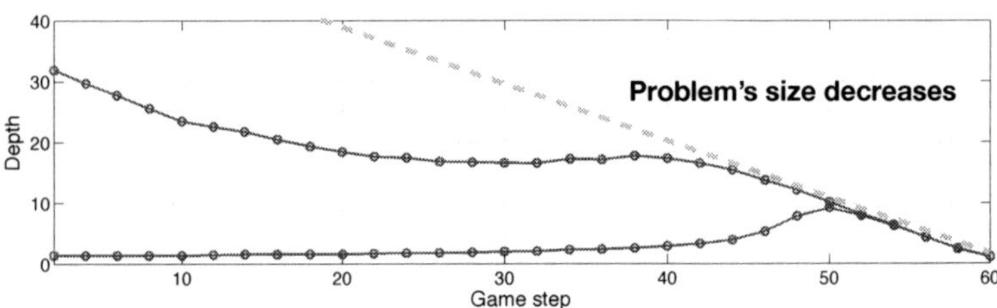

Fig. 10. Explanation of performance change caused by the problem size change

5 Conclusions

Results from Reversi and SameGame show the same problem. Both in case of CPU and GPU usage an improvement limit exists. The communication is not the problem as the simulation speed increases linearly. The most likely causes of this issue is the problem size and therefore repeating samples. We were able to gain performance improvement in both problems by increasing the thread number. Therefore at a certain point a scaling problem arose. We showed that the scaling is affected by:

1. The problem complexity - MCTS shows weak scaling
2. Random number generation, bounds for the number of unique random sequences, repeating samples
3. Exploitation/exploration ratio - the higher exploitation, the better scaling, more unique trees in root parallelism
4. The implementation itself - i.e. leaf parallelism/root parallelism.

References

1. Coulom, R.: Efficient Selectivity and Backup Operators in Monte-Carlo Tree Search. In: van den Herik, H.J., Ciancarini, P., Donkers, H.H.L.M(J.) (eds.) CG 2006. LNCS, vol. 4630, pp. 72–83. Springer, Heidelberg (2007)
2. Kocsis, L., Szepesvari, C.: Bandit based Monte-Carlo Planning. In: Proceedings of the EMCL 2006, pp. 282–293 (2006)
3. Chaslot, G.M.J.-B., Winands, M.H.M., van den Herik, H.J.: Parallel Monte-Carlo Tree Search. In: van den Herik, H.J., Xu, X., Ma, Z., Winands, M.H.M. (eds.) CG 2008. LNCS, vol. 5131, pp. 60–71. Springer, Heidelberg (2008)
4. Cazenave, T., Jouandeau, N.: On the parallelization of UCT. In: Proceedings of the Computer Games Workshop, pp. 93–101 (2007)
5. Kolchin, V.F., Sevastyanov, B.A., Chistyakov, V.P.: Random Allocations (1976)
6. Segal, R.B.: On the Scalability of Parallel UCT. In: van den Herik, H.J., Iida, H., Plaat, A. (eds.) CG 2010. LNCS, vol. 6515, pp. 36–47. Springer, Heidelberg (2011)
7. Schadd, M.P.D., Winands, M.H.M., van den Herik, H.J., Chaslot, G.M.J.-B., Uiterwijk, J.W.H.M.: Single-Player Monte-Carlo Tree Search. In: van den Herik, H.J., Xu, X., Ma, Z., Winands, M.H.M. (eds.) CG 2008. LNCS, vol. 5131, pp. 1–12. Springer, Heidelberg (2008)

How Questions Guide Choices: A Preliminary Logical Investigation

Zuojun Xiong[1,⋆] and Jeremy Seligman[2]

[1] Institute of Logic and Intelligence
Southwest University, Chongqing, China
xiongzuojun@gmail.com
[2] Department of Philosophy
University of Auckland, New Zealand
j.m.seligman@gmail.com

1 How Questions Guide Choices

We will be concerned with decisions that involve some sort of action. You have an apple and can exchange it for an orange. If you prefer the orange then you exchange; if not, you don't. The concept of preference here is a practical one, which presupposes that it is within your capability to move from one thing to another only if it is at least as good. Yet it is also important to be aware of what your capabilities are, and this is where questions are useful.

> Alice is considering moving house. She is unhappy with the fact that her house is far from the bus stop. She searches the listings for a house that is better located and sees several that she likes better. She goes to visit the one of them with Betty, her good friend. When Betty sees the house, she says 'what about a garden?' This is not a question that Alice had considered before. Her own house doesn't have one, but she is influenced by Betty to go back to the listings and check out houses with gardens. Eventually, she finds a house and moves. It has a nice big garden. But a few months later, she visits Chandra, a friend of Betty's who lived in a concrete house. Alice finds it quite charming. Her new house is timber-framed, like her old house and every house she has ever lives in. That night, she goes back to the listings...

The story illustrates how the process of practical decision making is guided by the questions one asks. Alice may well have asked very direct questions, such as 'does it have a roof?' and 'can I afford it?' but often it is the more open ended questions such as 'what about a garden?' that helped her to enlarge the options available to her, and it is these questions that we will focus on.

Here is a simple model of Alice's situation, with the full facts revealed. There is a twist that we will come to shortly.

⋆ Supported by the National Social Science Foundation of China (09CZX033), the Foundation for Humanities and Social Sciences by the Ministry of Education of China (08JC72040002) and the Fundamental Research Funds of Southwest University (SWU0909512).

D. Wang and M. Reynolds (Eds.): AI 2011, LNAI 7106, pp. 462–471, 2011.

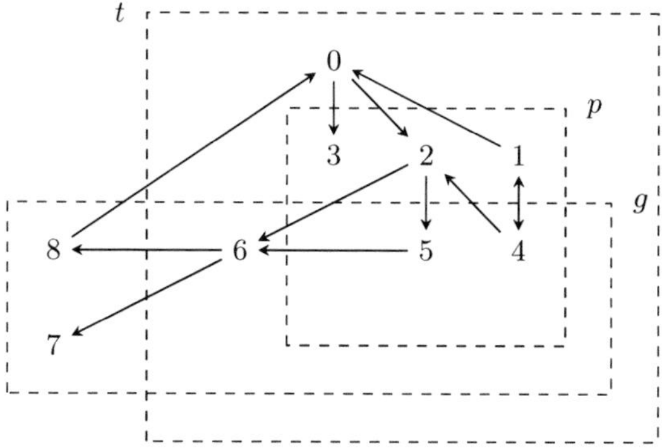

Fig. 1. Alice's Story

where the numbers represent the various houses, with 0 being Alice's original house and an arrow from n to m indicated that Alice regards m as at least as good as n. The rectangles enclose the houses that are p close to public transport, g have gardens and c are built of concrete. Initially, before Alice considers moving, she doesn't even consider other houses. When she asks about public transport, the rectangle marked p fades away and she has a range of options: 0, 1, 2 and 3. Of these, she rejects 1, which is worse than her own house. Houses 2 and 3 are both better but incomparable. But when Betty asks about gardens, the g rectangle also fades and she has another three houses to choose from: 4, 5 and 6. Of these, 5 and 6 are both better than any other house she has looked at. Eventually, she opts for 6. Later, at Chandra's house, she considers houses that are not built from timber and 7 and 8 become available.

We will develop a language and logic for reasoning in this way in the following sections. For now, consider a further development in the story:

> Alice is so impressed with the houses she looks at. Why had she never thought of houses being made out of anything but wood? Next month, she has moved into her new plaster house, which looks very modern and stylish. But after a while she happens to walk past her old house - the first one. Taken by it's quaint charm and worn woodwork, she realises that she prefers it to her new house.

Something has clearly gone wrong with Alice's decision making but it is an all-too-familiar situation. In the excitement of the search, something was missing. An easy answer is that Alice's preferences are clearly not transitive. She is therefore irrational, in some sense. But the apparent plausibility of each of her decisions suggests there is more to be said. We will return to this example in Section 4 below.

Our second example comes from a the Warring States period in ancient China, which we have altered a little to suit our purposes.

> Racing horses was the most popular entertainment in the ancient state of Qi. Despite having a fine stable, the minister Tian Ji's horses habitually

> lost to the King of Qi, who had the fastest horses in the state. The famous military strategist Sun Bin offered his help. Each day there are three races: the blue race, the yellow race and the red race for increasingly fast horses. The horses wear ribbons of the colour of the race they will be in, so that the handlers can bring them to the starting position. Sun Bin observed that Tian Ji's horses only lost by a small amount, and quickly devised a strategy to make Tian Ji win. The next day, Tian Ji went to the races as usual, but with one difference. His slowest horse wore a red ribbon, his fastest horse wore the yellow ribbon and the middle horse wore a blue ribbon. And that day he beat the King in two out of the three.

Here the actions involved are races between two horses, which is only a decision in the metaphorical sense, but as practical decisions go, it is among the more dramatic. Sun Bin achieves his subterfuge by swapping the horses ribbons without the other side knowing. The race order is therefore established with the question that the handlers have to ask at the beginning of each race, namely: 'what colour?'

The model looks like this (with the first day shown on the left and the second on the right):

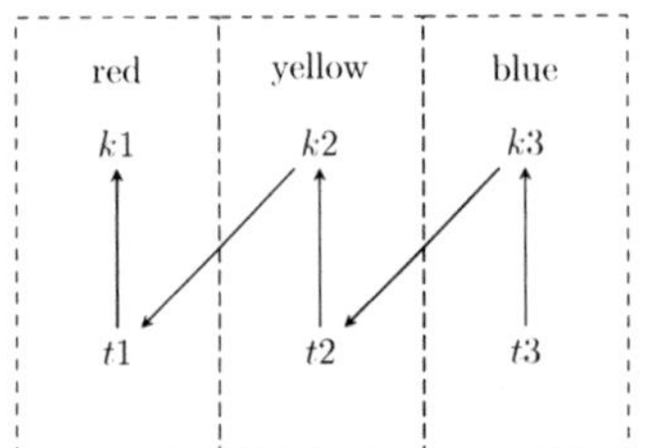

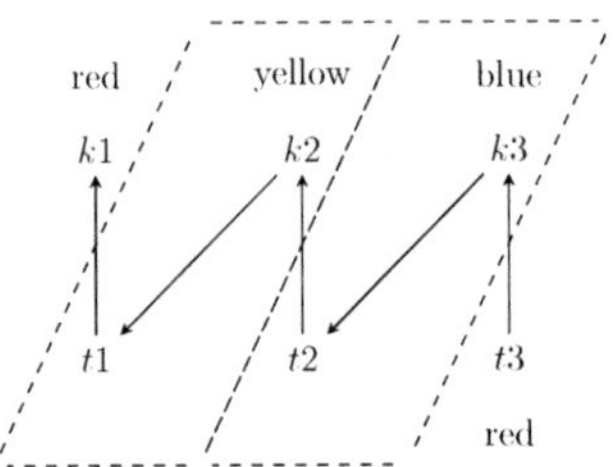

In this case, the effect of the question 'what colour?' is to look for an answer, and so impose the dashed boundaries on the space of actions. When these are as shown on the left (represented the first day), each race is decided in favour of the King. After the switching of ribbons, on the second day, the boundaries given by the same question are as shown on the right, and two out of the three races are decided in favour of Tian Ji.

2 A Simple Logic of Rational Choice

We model an agent's preferences by a binary relation $\leq$ on which we initial impose no constraints. $u \leq v$ means that the agent regards v as at least as good as u and is capable of deciding to move to v when in state u, other things being equal. As discussed in the introduction, this is not a theoretical preference but a practical one. If $u \leq v$ and $v \not\leq u$ we write $u < v$ and say that the agent *prefers* v to u. This latter relation is clearly asymmetric (and so irreflexive) but beyond that there are no constraints. A *choice frame*, then, is a structure $\langle W, \leq \rangle$.

We reason about choice frames using the basic hybrid modal language with a universal modal operator U

$$i \mid p \mid \neg \mid \wedge \mid \Box \mid U$$

where $i \in \text{NOM}$, a set of *nominals*, $p \in \mathsf{Prop}$, a set of *propositional variables* and $\Box$ is the unary modal operators for the $\leq$ relation. The nominals are included in the language so that we can reason about particular agent's preferences concerning particular states/entities. We abbreviate $\Diamond = \neg\Box\neg$and $E = \neg U \neg$ to get the corresponding existential modalities. As always, we can express the fact that φ holds at state i by defining $@_i\varphi \quad = \quad U(i \to \varphi)$.

A *choice model* $M = \langle F, V\rangle$ consists of a choice frame F together with a valuation function $V\colon \mathsf{Prop} \cup \text{NOM} \to \mathcal{P}(W)$, satisfying the usual restriction that $V(i) = \{\underline{i}\}$ for each $i \in \text{NOM}$. The semantic conditions of all the operators are perfectly standard, so we will not reproduce them here. Moreover, the set of valid formulas has a well-known axiomatisation $\mathcal{H}$ together with a single axiom for the universal modality and the reflexivity of $\leq$:[1]

$$\textsf{Universal } Ei \qquad\qquad \textsf{Reflexivity } i \to \Diamond i$$

How, then, do we reason about *rational* choice on the basis of the available choices given by a choice frame? Let us introduce a new operator D with the following interpretation:

$D\varphi$ holds if after any rational choice that the agent can make, φ holds.

To analyse this in our model, we have to say, for each state u whether or not changing to state v is not merely a possible choice but a rational one. As we are not making the assumption that $\leq$ is transitive, standard accounts of rational decision making cannot be directly applied.[2]

Instead, we say that a choice is rational if it is at least as good as the the present state and is stable, in the sense that any move away can be followed by a move back. So we define:

$$u \text{ is } \textit{stable} \text{ iff } v \leq u \text{ for every } v \text{ such that } u \leq v$$

Reformulating this slightly, note that u is stable just in case there is no $v > u$. Any such v would permit an irreversible move away from u. When Alice is considering moving house, it is sensible for her to make a stable choice, one that she will not abandon, at least not without changing her preferences. Stability can now be used to give the semantic condition for rational choice:

---

1 For details, see [7], p.87.

2 Intransitive preferences have been considered in the foundations of rational choice theory, most notable by Anand in [1], who argues for an alternative to the standard account in which intransitive preferences are permitted. His view remains controversial as a normative account but there is little doubt that transitivity of preference is simply false as a descriptive account of preference. Our present concern is to focus on the role of questions in guiding choices. As Alice's story indicates, and we will later explain, this role is illuminated more clearly when we do not make the idealizing assumption that the agent has transitive preferences.

$$M, u \models D\varphi \quad \text{iff} \quad M, v \models \varphi \text{ for all stable } v \geq u.$$

The operator D cannot be defined in the basic language but it can be characterized on frames with a pure formula:

$$\textsf{Rational Choice} \quad \langle D \rangle i \quad \leftrightarrow \quad \Diamond(i \wedge \Box\Diamond i)$$

where $\langle D \rangle = \neg D \neg$, the dual of D. Thus, adding this as an axiom to $\mathcal{H}$ and Universal gives us a complete axiomatisation.

In the considerable literature on preference logic, starting in philosophy with [9] and in AI with [5], preferences are assumed to be transitive, making any direct comparison with the current approach a little difficult.

Firstly, a common approach to say that φ is preferred if it is true is all the 'best' states: those that are maximal with respect to the preference order $\leq$. A state u is maximal if there is no state v that is strictly better than it; i.e., no v such that $u \leq v$ and not $v \leq u$. This of course, is exactly our definition of 'stable', which we claim is the natural generalisation of the concept of maximality to the intransitive setting.

Secondly, a rather odd consequence of our definition of rational choice is that an agent may strictly prefer v to u and not be (rationally) able to choose v because there is a w that is strictly better than v, making v unstable. Yet because of a failure of transitivity, the agent may also be unable to choose w over u. This kind of quandry is exactly what wish to analyse. But so far we lack the resources to do so. A crucial element is missing: context change.

3 *Ceteris Paribus* Preference Logic: The Atomic Case

So far we have assumed that the space W of alternatives from which the agent must choose is fixed. If she is currently in state u, all of the other states are available as possible choices. But we are interested in the possibility that asking questions can open up different possibilities. Suppose, for example, that p is true in state u. If the agent presupposes (consciously or not) that p is a fact of life that cannot be altered, then she will not consider those states in which p is false. For example, suppose you are thinking of changing your job as a waitress. You look through the situations vacant ads in the local newspaper, perhaps even search online for openings in the hospitality industry, but you may fail to consider the possibility of training as a high school teacher. When someone asks you about this, or you ask yourself, the space of options expands.

This phenomenon is closely related to the common observation that preference judgements are *ceteris paribus*, 'other things being equal'. From Von Wright's pioneering work [9], logicians have attempted to incorporate the *ceteris paribus* assumption into logical systems. Recent work ([8] and [6]) has provided a general framework for logics of *ceteris paribus* reasoning, in which the modal operator are restricted by an equivalence relation $\approx$ which holds between states that are equivalent 'other things being equal'. The focus of that work is to include a description of the conditions under which 'things are equal' in the language itself, as a parameter of the modal operators. Here we take a slightly different approach.[3] We will suppose that by default two states are equivalent *ceteris paribus* if and only if the same 'facts' hold in the two states. By a *fact* we mean a non-modal proposition. A necessary and sufficient condition for this to be the case is that the two states satisfy the same propositional variables. The effect asking the question 'what about p?' is to allow more states to be equivalent, specifically, after asking the question, two states are equivalent iff they satisfy the same propositional variables *other than* p.

Considering atomic questions will give us enough to be getting on with. So let us define an *atomic ceteris paribus choice model* $M = \langle F, V, P \rangle$ to consist of a choice model $\langle F, V \rangle$ together with a set P of propositional variables, which we interpret as the atomic facts that are held constant. That is, we define for $u, v \in W$

$$u \approx v \quad \text{iff} \quad \text{for each } p \in P,\ u \in V(p) \text{ iff } v \in V(p)$$

We require P to be cofinite, so that only a finite number of propositional variables are allowed to vary. Then the semantic conditions for our language must be restricted to observe the dominion of $\approx$:[4]

$$\begin{array}{lll} M, w \models \Box\varphi & \text{iff} & M, v \models \varphi \text{ for each } v \approx w \text{ such that } w \leq v \\ M, w \models U\varphi & \text{iff} & M, v \models \varphi \text{ for each } v \approx w \\ M, w \models D\varphi & \text{iff} & M, v \models \varphi \text{ for each } v \approx w \text{ such that } v \text{ is also } \leq \cap \approx\text{-stable.} \end{array}$$

Recall the example of Tian Ji's horse racing success and the diagram presented earlier on p.464. Call the frame depicted here, F. The pictures represent two models $M_1 = \langle F, V_1, P \rangle$ and $M_2 = \langle F, V_2, P \rangle$, where the valuations for the propositional variables 'red', 'yellow' and 'blue' are given by V_1 and V_2. Another two variables 'king' and 'tianji' have values $\{k1, k2, k3\}$ (the King's horses) and $\{t1, t2, t3\}$ (Tian Ji's horses).

The formula that expresses that the King always wins is $D(\text{king})$. If this is evaluated with the semantics of a choice model, it is valid in both, because the King's horse $k1$ is the fastest in the land. But rules of the race state that only horses of the same colour can race. We represent this by taking $P = \{\text{red, yellow, blue}\}$. The resulting equivalence classes for $\approx$ are shown with dashed boxes. They are different in the two models, representing the different ways of matching horses in the race. The conventions of the race ensure that only horses of the same colour are compared. Then we can see that $M_1 \models D(\text{king})$ but $M_2 \not\models D(\text{king})$. In fact, we can see the complete race record in M_2:

$$\begin{array}{ll} M_2 \models (\text{red} \rightarrow D(\text{king})) & \text{the King wins} \\ M_2 \models (\text{yellow} \rightarrow D(\text{tianji})) & \text{Tian wins} \\ M_2 \models (\text{blue} \rightarrow D(\text{tianji})) & \text{Tian wins} \end{array}$$

---

[3] A detailed comparison is given in [10].

[4] Note that we also liberalise D from the $\leq$ relation, relying instead on the $\approx$ relation to provide the range of choices from which the agent must make a selection. This removes the odd consequences of intransitive preferences that we mentioned a the end of Section 2, at least in the local context. The relation $\approx$ *is* transitive and so all stable states within the current context are reachable.

And so, finally, Tian Ji wins! Races are not literally decisions, of course, although the vocabulary of decision making is often applied metaphorically. Likewise, this ancient tale provides a vivid metaphor for how the outcome of our decisions can be crucially effected by those contextual factors that we take to be constant and those that we permit to vary. This is the role that questioning plays in the rational process, we claim, and it is to this that we turn in the next section. But first, we have the following quick result about the logic of atomic *ceteris paribus* choice:

Theorem 1. *The set of formulas valid over the class of atomic* ceteris paribus *preference models is axiomatized by basic hybrid logic, $\mathcal{H}$, together with*

Universal Equivalence	$i \rightarrow Ei$	Reflexivity	$i \rightarrow \Diamond i$
	$i \rightarrow UEi$	Inclusion	$\Diamond i \rightarrow Ei$
	$EEi \rightarrow Ei$	Preferential Choice	$\langle D\rangle i \leftrightarrow E(i \wedge \Box\Diamond i)$

Call this the logic $\mathcal{PC}$ of Preferential Choice.

Proof. Firstly, each of these axioms is valid in any atomic *ceteris paribus* model, so the axiomatisation is sound. Secondly, for completeness, say that $F = \langle W, \leq, R_U, C\rangle$ is a *quasi-preference frame* iff

1. R_U is an equivalence relation,
2. $\leq$ is reflexive and $\leq \subseteq R_U$, and
3. Cuv iff $R_U uv$ and v is $\leq$-maximal.

For a *quasi-preference model* $M = \langle F, \leq, R_U, C, V\rangle$, define

$$M, w \models U\varphi \quad \text{iff} \quad M, v \models \varphi \text{ for all } v \in W \text{ such that } R_U wv$$

Then the set of formulas valid in all quasi-preference models is axiomatised as shown above. This is because (1) is characterised by Universal Equivalence, (2) is characterised by Inclusion and Reflexivity, and (3) by Preferential Choice. These are all pure formulas, so completeness is automatic (by [7] again) . So suppose that φ is a consistent formula and let P be the set of propositional variables that do not occur in φ, noting that this set is cofinite. Then the set $\{\varphi\} \cup \{Up \mid p \in P\}$ is also consistent and so is satisfied at a state w of a quasi-preference model $M = \langle W, \leq, R_U, C, V\rangle$. Now let

$$W' = \bigcup\{[\underline{i}] \mid i \text{ occurs in } \varphi\}$$

where $[\underline{i}]$ is the R_U-equivalence class of $\underline{i}$. For each nominal i occurring in φ, assign a distinct propositional variable $p_i \in P$. Pick a state $w_0 \in W$ and let V' be the valuation defined as follows for $x \in \mathsf{Prop} \cup \textsc{Nom}$,

$$V'(x) = \begin{cases} \{w_0\} & \text{if } x \in \textsc{Nom} \text{ but } x \text{ does not occur in } \varphi \\ [\underline{i}] & \text{if } x = p_i \\ V(x) \cap W' & \text{otherwise} \end{cases}$$

and let $M' = \langle W', \leq, R_U, C, V' \rangle$. Since V' agrees with V on all variables in φ, this formulas is also satisfied at w in M'. Now define an atomic *ceteris paribus* preference model $M'' = \langle W, \leq, V', P \rangle$. Then the following are equivalent:

$R_U uv$	
$v \in V'(p_i)$	for some nominal i such that $R_U u\underline{i}$
$u \in V'(p_i)$ iff $v \in V'(p_i)$	for all i occurring in φ
$u \in V'(p)$ iff $v \in V'(p)$	for all $p \in P$
$u \approx v$	in M''

This is the crucial step in an inductive proof that for all formulas ψ and all $u \in W$, $M', u \models \psi$ iff $M'', u \models \psi$. The rest of the argument is routine. Hence ψ, in particular, is satisfied at w in the atomic *ceteris paribus* model M'', and so the axiomatisation is complete. ♣

Although this logic uses the semantics of *ceteris paribus* preferences, it differs from recent approaches, such as [8], [6] and [4], in not including a description of the context in the language. Instead, we will focus on the dynamics of *changing* the context using questions.

4 Adding Questions as Dynamic Operators

With the semantics of atomic *ceteris paribus* preference models in place, we can start to model the effect of asking questions on the *ceteris paribus* relation.[5] The idea is very simple. When we ask the question 'what about p?' we move from the model $M = \langle F, V, P \rangle$ to the model $[?p]M = \langle F, V, P \setminus \{p\} \rangle$ in which p has been subtracted from the set of propositional variables that must be held constant when we search for alternatives. More generally, for a finite set Q of propositional variables, we define

$$[?Q]M = \langle F, V, P \setminus Q \rangle$$

This operation on models will be expressed by a new operator $[?Q]$ in our language, with the obvious semantic condition

$$M, w \models [?Q]\varphi \quad \text{iff} \quad [?Q]M, w \models \varphi$$

With reference to Figure 1, we model M Alice's decisions about moving from house 0 using the frame show F with nominals for each of the houses. t, g, p are propositional variables. Initially, before Alice considers moving, all of these are fixed, so we take $P = \mathsf{Prop}$. When Alice is unhappy about living far from the bus stop, she searches for houses and considers $1, 2$, and 3. These options are available because she has asked the question $[?p]$. She rejects 1 and we have

[5] Although there have been various proposals for developing a dynamic logic of questions, especially Johan van Benthem and Ştefan Minică's [2], our focus on decision making and the effect of questioning on *ceteris paribus* conditions, is a different project. In an earlier paper, [3], one of us developed an account of what we now call 'closed' questions using a similar framework.

$M, 0 \models [?p]D(2 \vee 3)$ She decides to move to house 2, but her friend Betty asks 'what about a garden?' $[?g]$. Then the listings for houses is enlarged and since $M, 2 \models [?p][?g]D6$ she moves to house 6. We'll return to the example when we can say more about what happens next.

We have interpreted the question 'what about p?' as an *open* question, inviting not an answer but a further investigation of situations in which p may or may not be the case. But one can also interpret the question in a *closed* way, as a request for an answer. Closure is achieved by finding out whether p is the case and then sticking to the answer. We can easily add an operation $[!p]$ (more generally $[!Q]$) to our language that has this effect. Let L^c be the extension of L to include closed question operators. Given a model $M = \langle F, V, P \rangle$, let

$$[!Q]M = \langle F, V, P \cup Q \rangle$$

and define $M, u \models [!Q]\varphi$ iff $[!Q]M, u \models \varphi$.[6]

Closed questions allow us to give a model of decisions that are actually taken. If we ask about Q and make a decisive choice, the options opened by $[?Q]$ are typically closed. We represent this by

$$[?Q]D[!Q]$$

The closing of the question marks the end of the decision-making process. This is no clearer than with examples in which the decision involves a clear action, such as moving house. Alice deciding to move to a house with a garden and then actually moving is represented as

$$[?g]D[!g]$$

After this point, she no longer considers houses without a garden. If Alice kept the garden question open, she would not have moved the second time (move to the house 8), which would have not been a stable option. That is

$$M, 0 \models [?p]\langle D \rangle [?g]\langle D \rangle [!g][?t]\langle D \rangle 8 \text{ but not } M, 0 \models [?p]\langle D \rangle [?g]\langle D \rangle [?t]\langle D \rangle 8$$

Finally, we can model the last part of Alice's story, when she discovers that she prefers her old house 0 to the new one 8. That is

$$M, 0 \models [?p]\langle D \rangle [?g]\langle D \rangle [!g][?t]\langle D \rangle [!p][?g]\langle D \rangle 0$$

5 Closing Remarks

Inspired by a simple example of decision making and how it is guided by asking questions, we introduced a rational choice operator D to model the act of making a decision which applies even to agents with intransitive preferences. This gave us a logic of rational choice, formulated in a basic hybrid modal language using the axiom of Rational Choice. We then transformed this into a *ceteris paribus*

[6] Note that if P is cofinite and Q is finite then $P \cup Q$ is also cofinite.

logic in which choices are allowed to vary over a range of contextually supplied alternatives. This was axiomatised as the logic $\mathcal{PC}$ of Preferential Choice, and illustrated by giving an analysis of the strategic wisdom of Sun Bin.

The logic $\mathcal{PC}$ provided the background for a novel approach of modelling *ceteris paribus* context change using dynamic operators. In particular, we suggested that the role of questions in decision making is to alter those factors that are held 'equal' when evaluating *ceteris paribus* preferences. This allowed us to give an account of our main example, in which an agent's intransitive preferences allow her to make a series of locally rational choices, which eventually lead her around in a circle. A full logical investigation of these operators is beyond the scope of this short paper, and we refer the reader to the more technical treatment in [10], which also includes application to Condorcet's Voting Paradox and the extension to compound questions.

References

1. Anand, P.: Foundations of Rational Choice Under Risk. Oxford University Press (1993)
2. van Benthem, J., Minică, Ş.: Toward a dynamic logic of questions. In: Pacuit, E., He, X., Horty, J. (eds.) Logic Rationality and Interaction, Chongqing, China, pp. 27–41. Springer, Heidelberg (2009)
3. Guo, M., Xiong, Z.: A dynamic preference logic with issue-management. Studies in Logic 1 (Spring 2011)
4. Bienvenu, M., Lang, J., Wilson, N.: From preference logics to preference languages, and back. In: Proceedings of KR 2010 (2010)
5. Boutilier, C.: Toward a logic for qualitative decision theory. In: Proceedings of the KR 1994, pp. 75–86. Morgan Kaufmann (1992)
6. Seligman, J., Girard, P.: Being flexible about ceteris paribus. Australasian Journal of Logic (2011)
7. ten Cate, B.: Model Theory for Extended Modal languages. PhD thesis, Institute for logic, Language and Computation, Universiteit van Amsterdam, Amsterdam, The Netherlands, ILLC Dissertation series DS-2005-01 (January 2005)
8. van Benthem, J., Girard, P., Roy, O.: Everything else being equal: a modal logic for ceteris paribus preferences. Journal of Philosophical Logic (August 2008), `http://www.springerlink.com/content/p756008882505667/`
9. von Wright, G.H.: The Logic of Preference. Edinburgh (1963)
10. Xiong, Z., Seligman, J.: Open and closed questions in decision-making. In: van Ditmarsch, H. (ed.) Proceedings of M4M 7: 7th International Workshop on Methods for Modalities. ENTCS, Elsevier Science Inc. (2011)

A Defeasible Logic for Clauses

David Billington

School of Information and Communication Technology, Nathan Campus,
Griffith University, Brisbane, Queensland 4111, Australia
d.billington@griffith.edu.au

Abstract. A new non-monotonic logic called clausal defeasible logic (CDL) is defined and explained. CDL is the latest in the family of defeasible logics, which, it is argued, is important for knowledge representation and reasoning. CDL increases the expressive power of defeasible logic by allowing clauses where previous defeasible logics only allowed literals. This greater expressiveness allows the representation of the Lottery Paradox, for example. CDL is well-defined, consistent, and has other desirable properties.

Keywords: Defeasible logic, Non-monotonic reasoning, Knowledge representation and reasoning, Artificial intelligence.

1 Introduction

Non-monotonic reasoning systems represent and reason with incomplete information where the degree of incompleteness is not quantified. A very simple and natural way to represent such incomplete information is with a defeasible rule of the form "antecedent $\Rightarrow$ consequent"; with the meaning that provided there is no evidence against the consequent, the antecedent is sufficient evidence for concluding the consequent. Creating such rules is made easier for the knowledge engineer as each rule need only be considered in isolation. The interaction between the rules is the concern of the logic designer.

Reiter's normal defaults [23] have this form, with the meaning that if the antecedent is accepted and the consequent is consistent with our knowledge so far then accept the consequent. Of course the consequent could be consistent with current knowledge and yet there be evidence against the consequent. This results in multiple extensions. However multiple extensions are avoided by interpreting a defeasible rule as "if the antecedent is accepted and all the evidence against the consequent has been nullified then accept the consequent". This interpretation forms the foundation of a family of non-monotonic logics all based on Nute's original defeasible logic [21]. (Different formal definitions of "accepted", "evidence against", and "nullified" have been used by different defeasible logics.)

Unlike other non-monotonic reasoning systems, these defeasible logics use Nute's very simple and natural "defeasible arrow" to represent incomplete information. This simplicity and naturalness is important when explaining an implementation to a client. All defeasible logics have a priority relation on rules and use classical negation rather than negation-as-failure.

D. Wang and M. Reynolds (Eds.): AI 2011, LNAI 7106, pp. 472–480, 2011.

A key feature of these defeasible logics is that they all have efficient easily implementable deduction algorithms [10,20,22]. Indeed defeasible logics have been used in an expert system, for learning and planning [22], in a robotic dog which plays soccer [4,5], in a robotic poker player [6], to improve the accuracy of radio frequency identification [11], to model the behaviour of autonomous robots [9], and to facilitate the encoding of software requirements [8] so they can be automatically translated into a programming language [7]. Defeasible logics have been advocated for various applications including modelling regulations and business rules [1], agent negotiations [13], the semantic web [2,3,25], modelling agents [16], modelling intentions [15], modelling dynamic resource allocation [17], modelling contracts [12], legal reasoning [18], modelling deadlines [14], and modelling dialogue games [24]. Moreover, defeasible theories, describing policies of business activities, can be mined efficiently from appropriate datasets [19].

The unique features and diverse range of practical applications show that defeasible logics are useful and their language is important for knowledge representation and reasoning. Using defeasible logic as the inference engine in an expert system is obvious. But it is less obvious to use defeasible logic to deal with the error-prone output of sensors (possibly in a robot), because this can be done using classical logic. The advantages of using defeasible logic are that the system can be developed incrementally, there are fewer rules, and the rules are simpler [4,5,11].

In this article we shall define a new defeasible logic, called clausal defeasible logic (CDL) and explain how it works. The main purpose of CDL is to increase the expressive power of defeasible logic by allowing clauses where previous defeasible logics only allowed literals.

The rest of the paper has the following organisation. Section 2 gives an overview of CDL. The formal definitions of CDL are in Section 3. An example is considered in Section 4. The results in Section 5 show that CDL is well-defined, consistent, and has other desirable properties. A summary forms Section 6.

2 Overview of Clausal Defeasible Logic (CDL)

CDL reasons with both factual and plausible information. The factual information is represented by strict rules of the form $A \rightarrow c$ where A is a finite set of literals and c is a clause. If all the literals in A are proved then c can be deduced, no matter what the evidence against c is.

The plausible information is represented by defeasible rules, warning rules, and a priority relation, $>$, on these rules.

Defeasible rules have the form $A \Rightarrow c$ where A is a finite set of clauses and c is a clause. If all the clauses in A are proved and all the evidence against c has been nullified then c can be deduced. For example, "Birds usually fly." can be represented by $\{b\} \Rightarrow f$.

Warning rules, for example $A \leadsto \neg l$, warn against concluding usually l, but do not support usually $\neg l$. For example, "Sick birds might not fly." can be represented by $\{s, b\} \leadsto \neg f$. The idea is that a bird being sick is not sufficient evidence to conclude that it usually does not fly; it is only evidence against the conclusion that it usually flies.

There is an acyclic priority relation between non-strict rules; $r_2 > r_1$ means that r_2 is preferred over r_1.

CDL has four major proof algorithms, μ, ρ, π, and β, which cater for different intuitions about what should follow from a reasoning situation. The μ algorithm is monotonic and uses only strict rules. CDL restricted to μ is essentially classical propositional logic. The ambiguity blocking algorithm is denoted by β. There are two ambiguity propagating algorithms denoted by ρ and π, ρ being more reliable than π.

The task of proving a formula is done by a recursive proof function P. The input to P is the proof algorithm to be used, the formula to be proved, and the background. The background is an initially empty storage bin into which is put all the clauses that are currently being proved as P recursively calls itself. The purpose of this background is to detect loops. The output of P is one of the following proof-values $+1$, 0, or -1. The $+1$ means that the formula is proved in a finite number of steps, 0 means that the proof got into a loop which was detected in a finite number of steps, and -1 means that in a finite number of steps it has been demonstrated that there is no proof of the formula and that this demonstration does not get into a loop.

3 Clausal Defeasible Logic (CDL)

Atm is a non-empty countable set of (propositional) atoms, and $Lit = Atm \cup \{\neg a : a \in Atm\}$ is the set of all literals. If C is a clause or a set of clauses then $Lit(C)$ denotes the set of literals in C. The set of all tautologies is denoted by $Taut$. The set of all clauses which are resolution-derivable from the set C of clauses is denoted by $Res(C)$.

The **complement**, $\sim l$, of a literal l is defined as follows. If a is an atom then $\sim a$ is $\neg a$, and $\sim\neg a$ is a. If L is a set of literals then $\sim L = \{\sim l : l \in L\}$.

Definition 1. A **rule**, r, is any triple $(A(r), arrow(r), c(r))$, such that $A(r)$ is a finite set of non-empty clauses called the set of **antecedents** of r, $arrow(r) \in \{\rightarrow, \Rightarrow, \leadsto\}$, $c(r)$ is a non-empty clause called the **consequent** of r, and if $arrow(r)$ is $\rightarrow$ then $A(r)$ is a finite set of literals.

Strict rules use the **strict arrow**, $\rightarrow$, and are written $A(r) \rightarrow c(r)$. **Defeasible rules** use the **defeasible arrow**, $\Rightarrow$, and are written $A(r) \Rightarrow c(r)$. **Warning rules** use the **warning arrow**, $\leadsto$, and are written $A(r) \leadsto c(r)$.

Definition 2. Let R be any set of rules, R_s be any set of strict rules, C be any set of clauses, L be any set of literals, c be any clause, and l be any literal.

$R_s = \{r \in R : r \text{ is a strict rule}\}$. $R[\vee L] = \{r \in R : c(r) = \vee L\}$.

$R_d = \{r \in R : r \text{ is a defeasible rule}\}$. $R[c; 1] = \{r \in R[c] : |A(r)| \leq 1\}$.

$R_w = \{r \in R : r \text{ is a warning rule}\}$. $R[C] = \{r \in R : c(r) \in C\}$.

$R_{dw} = R_d \cup R_w$. $Cl(R_s) = \{\vee(L \cup \sim A) : A \rightarrow \vee L \in R_s\}$.

$Ru(C) = \{\sim(L - K) \rightarrow \vee K : \vee L \in C \text{ and } \{\} \subset K \subseteq L\}$.

Cl converts a strict rule to a clause. Conversely, *Ru* converts a clause with n literals to $2^n - 1$ strict rules.

Definition 3. If R is a set of rules then $>$ is a **priority relation** on R iff $>$ is an acyclic binary relation on R_{dw}. $R[l; s] = \{t \in R[l] : t > s\}$.

Let C be a set of clauses. We want to remove from C all the clauses which are empty, or tautologies, or are strict superclauses of other clauses. The result is called the reduct of C, $Red(C)$. A set of literals such that each clause in C contributes exactly one literal to the set is called a transversal of C.

To help prove clauses we "move literals from the antecedent to the consequent". For instance the rule $\{a, b\} \Rightarrow c$ generates the rules $\{a\} \Rightarrow \vee\{\sim b, c\}$ and $\{\} \Rightarrow \vee\{\sim a, \sim b, c\}$.

Definition 4. Let R be any set of rules. The set, $Gen(R)$, of **rules generated from** R is defined by $Gen(R) = \{(A(r) - C)\ arrow(r) \vee(L \cup \sim T) : r \in R,\ C \subseteq A(r),$ $c(r) = \vee L$, and T is a transversal of $C\}$.

Definition 5. Let R be a finite set of rules. The ordered pair $(R, >)$ is called a **clausal defeasible theory (cdt)** iff DT1, DT2, and DT3 all hold.
DT1) $R_s = Ru(Red(Res(Cl(R_s))))$.
DT2) $R_{dw} = Gen(R_{dw})$.
DT3) $>$ is a priority relation on R_{dw}.

Definition 6. Let $\Theta = (R, >)$ be a cdt. The set $Ax(\Theta)$ of **axioms** of Θ is defined by $Ax(\Theta) = Cl(R_s)$. Define $A^*(R_s[l; 1]) = \{A(r) : r \in R_s[l; 1]\} \cup \{\{l\}\}$.

To cater for various intuitions we will introduce the following proof algorithms: μ, ρ, π, β, π', and ρ', which are explained after their formal definition.

Definition 7. Suppose $\lambda \in \{\rho, \pi, \beta\}$. Then λ' is the **co-algorithm** of λ, where $\beta' = \beta$. Moreover we define $(\lambda')' = \lambda$.

A **clausal defeasible logic** consists of a clausal defeasible theory Θ and its **proof function**, P. To define P we shall define some auxiliary functions and the proof algorithms $\mu, \rho, \pi, \beta, \pi', \rho'$. For non-empty sets max and min have their usual meaning. But we also define $\max\{\} = -1$, and $\min\{\} = +1$. We now define P, its auxiliary functions, and the proof algorithms.

Definition 8. Suppose $\lambda \in \{\mu, \rho, \pi, \beta, \pi', \rho'\}$, C is a finite set of clauses, and L is a finite set of literals.

$P\lambda$set) $P(\lambda, C, B) = \min\{P(\lambda, c, B) : c \in C\}$.
$P\lambda$taut) If $\vee L$ is a tautology then $P(\lambda, \vee L, B) = +1$.
$P\lambda$mt) $P(\lambda, \vee\{\}, B) = -1$.

Definition 9. Suppose $\lambda \in \{\rho, \pi, \beta, \pi', \rho'\}$, L is a non-empty finite set of literals, $\vee L$ is not a tautology, and l is a literal.

$P\mu$cl) If $\{\} \subset K \subseteq L$ and $\vee K \in Cl(R_s)$ then $P(\mu, \vee L, B) = +1$;
else $P(\mu, \vee L, B) = -1$.

$P\lambda$0) If $\vee L \in B$ then $P(\lambda, \vee L, B) = 0$.

$P\lambda$pc) If $\vee L \notin B$ and $|L| \geq 2$ then $P(\lambda, \vee L, B) = \max(\{P(\lambda, l, B) : l \in L\} \cup \{P(\lambda, A(r), \{\vee K\} \cup B) : K \subseteq L,\ |K| \geq 2, \text{ and } r \in R_s[\vee K; 1] \cup R_d[\vee K]\})$.

$P\lambda$lit) If $l \notin B$ then $P(\lambda, l, B) = \max(\{P(\lambda, A(r), \{l\} \cup B) : r \in R_s[l; 1]\} \cup \{Plaus(\lambda, l, B)\})$.

Definition 10. Suppose $\lambda \in \{\rho, \pi, \beta\}$, l is a literal, B is a finite set of clauses, and $l \notin B$.

Plausρ') $Plaus(\rho', l, B) = \max\{P(\rho', A(r), \{l\} \cup B) : r \in R_d[l]\}$.

Plausλ) $Plaus(\lambda, l, B) = \min(\{For(\lambda, l, B)\} \cup \{Nulld(\lambda, l, B, I) : I \in A^*(R_s[\sim l; 1])\})$.

Plausπ') $Plaus(\pi', l, B) = \max\{Evid(\pi', l, B, r) : r \in R_d[l]\}$.

Forλ) $For(\lambda, l, B) = \max\{P(\lambda, A(r), \{l\} \cup B) : r \in R_d[l]\}$.

Evidπ') $Evid(\pi', l, B, r) = \min(\{P(\pi', A(r), \{l\} \cup B)\} \cup \{Nulldr(\pi', l, B, r, I) : I \in A^*(R_s[\sim l; 1])\})$.

Nulldλ) $Nulld(\lambda, l, B, I) = \max\{Discred(\lambda, l, B, q) : q \in I\}$.

Nulldrπ') $Nulldr(\pi', l, B, r, I) = \max\{Discredr(\pi', l, B, r, q) : q \in I\}$.

Discredλ) $Discred(\lambda, l, B, q) = \min\{Dftd(\lambda, l, B, s) : s \in R_s[q; 1] \cup R_{dw}[q]\}$.

Discredrπ') $Discredr(\pi', l, B, r, q) = \min\{Dftd(\pi', l, B, s) : s \in R[q; r]\}$.

Dftdλ) Now suppose $\lambda \in \{\rho, \pi, \beta, \pi'\}$. Then $Dftd(\lambda, l, B, s) = \max(\{P(\lambda, A(t), \{l\} \cup B) : t \in R_d[l; s]\} \cup \{-P(\lambda', A(s), \{l\} \cup B)\})$.

Definition 11. Suppose Θ is a cdt, P is the proof function of Θ, c is a clause, and $\lambda \in \{\mu, \rho, \pi, \beta, \pi', \rho'\}$. We define $\Theta(\lambda+) = \{c : P(\lambda, c, \{\}) = +1\}$, and $\Theta(\lambda-) = \{c : P(\lambda, c, \{\}) = -1\}$.

We shall now give some insight into the above proof algorithms. Note that min and max behave like quantifiers. Suppose $\lambda \in \{\mu, \rho, \pi, \beta, \pi', \rho'\}$.

A set of clauses, C, is proved by proving each clause in C. So for $P(\lambda, C, B)$ to be $+1$ each $P(\lambda, c, B)$ must be $+1$, where $c \in C$. Hence $P\lambda$set.

Now consider proving a clause. If a clause, $\vee L$, is a tautology then we declare it proved, whether or not it is in the background. Hence $P\lambda$taut. The empty clause is disproved, whether or not it is in the background. Hence $P\lambda$mt. So suppose $\vee L$ is not a tautology and $L \neq \{\}$.

Apart from the cases considered above, a clause $\vee L$ is proved by proving any non-empty subclause of $\vee L$.

The μ algorithm declares $\vee L$ to be proved if a non-empty subclause of $\vee L$ is an axiom; otherwise it is disproved. Hence $P\mu$cl. So suppose $\lambda \in \{\rho, \pi, \beta, \pi', \rho'\}$.

If $\vee L$ is in the background B then we are already in the process of trying to prove $\vee L$. So we are now in a loop. Hence $P(\lambda, \vee L, B) = 0$ and so $P\lambda$0. So suppose $\vee L$ is not in B.

To prove a proper clause, $\vee L$, we must either prove a literal which occurs in L, (hence $\{P(\lambda, l, B) : l \in L\}$), or prove a subclause $\vee K$ of $\vee L$ which is proper.

To prove $\bigvee K$ we need to prove every clause in the set of antecedents of a strict or defeasible rule whose consequent is $\bigvee K$, provided that the set of antecedents of the strict rule contains at most 1 literal. The need for this restriction is shown by Example 1. We have the strict rule r_7: $\{\neg s_1, \neg s_2\} \rightarrow s_3$, and also by Evaluations $E1$ and $E2$ we can prove both $\neg s_1$ and $\neg s_2$. But the fact that $\neg s_1$ and $\neg s_2$ can be independently proved is not evidence for s_3. Hence $P\lambda$pc.

To prove a literal l we must prove every clause in the set of antecedents of a strict or defeasible rule whose consequent is l. If a strict rule is used then evidence against l need not be considered, and its set of antecedents must contain at most 1 literal, $\max\{P(\lambda, A(r), \{l\}\cup B) : r \in R_s[l;1]\}$. If a defeasible rule is used then evidence against l must be considered, $\mathit{Plaus}(\lambda, l, B)$. Hence $P\lambda$lit.

The ρ' algorithm ignores all evidence against l. Hence $\mathit{Plaus}\rho'$.

Now consider the ρ, π, and β algorithms. Suppose $\lambda \in \{\rho, \pi, \beta\}$. These algorithms must consider evidence for l, $\mathit{For}(\lambda, l, B)$ and $\mathit{For}\lambda$, and nullify the evidence against l, $\{\mathit{Nulld}(\lambda, l, B, I) : I \in A^*(R_s[\sim l;1])\}$. Hence $\mathit{Plaus}\lambda$.

The evidence against l consists of any set of literals which is inconsistent with l. The set of all such sets is $A^*(R_s[\sim l;1])$. The evidence against l is nullified by, for each I in $A^*(R_s[\sim l;1])$, finding a literal, q, in I such that every rule, s, whose consequent is q, is defeated. The only restriction is that if s is a strict rule then its set of antecedents must contain at most 1 literal. Hence $\mathit{Nulld}\lambda$ and $\mathit{Discred}\lambda$. A rule, s, is defeated by either using the λ' algorithm to disprove a clause in its set of antecedents, $-P(\lambda', A(s), \{l\}\cup B)$, or by using team defeat. That is by finding a defeasible rule t whose consequent is l (a member of the team for l) and which is superior to s, and then proving every clause in the set of antecedents of t, $\{P(\lambda, A(t), \{l\}\cup B) : t \in R_d[l;s]\}$. Hence $\mathit{Dftd}\lambda$.

Finally the π' proof algorithm is similar to the π algorithm, except that we choose one defeasible rule r supporting l and only consider rules s which are superior to r as evidence against l, and hence need defeating.

4 Example

The following example shows how the lottery example can be represented.

Example 1. (The 3-Lottery example)
Consider a fair 3-sided die. Let s_i denote side i. Then for each i in $\{1,2,3\}$ it is plausible that the outcome of a roll of this die is not s_i. So for each i in $\{1,2,3\}$ we have $\{\} \Rightarrow \neg s_i$. Moreover for each i in $\{1,2,3\}$ it is plausible that the outcome of a roll of this die is in $\{s_1, s_2, s_3\} - \{s_i\}$. So for each i in $\{1,2,3\}$ we have $\{\} \Rightarrow \bigvee(\{s_1, s_2, s_3\} - \{s_i\})$. Furthermore the outcome of a roll of this die is exactly one of s_1, s_2, or s_3. So we have $\bigvee\{s_1, s_2, s_3\}$, $\neg\bigwedge\{s_1, s_2\}$, $\neg\bigwedge\{s_1, s_3\}$, $\neg\bigwedge\{s_2, s_3\}$. Converting these facts to clauses gives: $\bigvee\{s_1, s_2, s_3\}$, $\bigvee\{\neg s_1, \neg s_2\}$, $\bigvee\{\neg s_1, \neg s_3\}$, $\bigvee\{\neg s_2, \neg s_3\}$.

The cdt $\Theta = (R, >)$ which captures this situation is defined as follows.

The strict rules are: r_1: $\{\} \rightarrow \bigvee\{s_1, s_2, s_3\}$, r_2: $\{\neg s_1\} \rightarrow \bigvee\{s_2, s_3\}$, r_3: $\{\neg s_2\} \rightarrow \bigvee\{s_1, s_3\}$, r_4: $\{\neg s_3\} \rightarrow \bigvee\{s_1, s_2\}$, r_5: $\{\neg s_2, \neg s_3\} \rightarrow s_1$, r_6: $\{\neg s_1, \neg s_3\} \rightarrow s_2$, r_7: $\{\neg s_1, \neg s_2\} \rightarrow s_3$, r_8: $\{\} \rightarrow \bigvee\{\neg s_1, \neg s_2\}$,

r_9: $\{s_1\} \rightarrow \neg s_2$, r_{10}: $\{s_2\} \rightarrow \neg s_1$, r_{11}: $\{\} \rightarrow \vee\{\neg s_1, \neg s_3\}$, r_{12}: $\{s_1\} \rightarrow \neg s_3$, r_{13}: $\{s_3\} \rightarrow \neg s_1$, r_{14}: $\{\} \rightarrow \vee\{\neg s_2, \neg s_3\}$, r_{15}: $\{s_2\} \rightarrow \neg s_3$, r_{16}: $\{s_3\} \rightarrow \neg s_2$.

The defeasible rules are: r_{17}: $\{\} \Rightarrow \neg s_1$, r_{18}: $\{\} \Rightarrow \neg s_2$, r_{19}: $\{\} \Rightarrow \neg s_3$, r_{20}: $\{\} \Rightarrow \vee\{s_1, s_2\}$, r_{21}: $\{\} \Rightarrow \vee\{s_1, s_3\}$, r_{22}: $\{\} \Rightarrow \vee\{s_2, s_3\}$.

There are no warning rules and the priority relation, $>$, is empty.

We show that if λ is in $\{\rho, \pi, \beta\}$ then λ is able to prove every clause in $U(3) = \{\neg s_1, \neg s_2, \vee\{s_1, s_2\}\}$. We note $A^*(R_s[s_1; 1]) = \{\{s_1\}\}$.

Evaluation $E1$

1) $P(\lambda, \neg s_1, \{\}) = \max\{P(\lambda, \{s_2\}, \{\neg s_1\}), P(\lambda, \{s_3\}, \{\neg s_1\}), Plaus(\lambda, \neg s_1, \{\})\}$, by $P\lambda$lit

2) $Plaus(\lambda, \neg s_1, \{\}) = \min\{For(\lambda, \neg s_1, \{\}), Nulld(\lambda, \neg s_1, \{\}, \{s_1\})\}$, by $Plaus\lambda$

3) $For(\lambda, \neg s_1, \{\}) = P(\lambda, \{\}, \{\neg s_1\})$, by $For\lambda$

4) $= \min\{\} = +1$, by $P\lambda$set.

5) $\therefore Plaus(\lambda, \neg s_1, \{\}) = Nulld(\lambda, \neg s_1, \{\}, \{s_1\})\}$, by lines 4, 3, and 2 of $E1$.

6) $= Discred(\lambda, \neg s_1, \{\}, s_1)$, by $Nulld\lambda$

7) $= \min\{\} = +1$, by $Discred\lambda$

8) $\therefore P(\lambda, \neg s_1, \{\}) = +1$, by lines 7, 6, 5, and 1 of $E1$.

A similar evaluation, say $E2$, shows that $P(\lambda, \neg s_2, \{\}) = +1$. Finally we show that $\vee\{s_1, s_2\}$ is provable by any λ in $\{\rho, \pi, \beta\}$.

Evaluation $E3$

1) $P(\lambda, \vee\{s_1, s_2\}, \{\}) = \max\{P(\lambda, \{s_1\}, \{\}), P(\lambda, \{s_2\}, \{\}),$
$P(\lambda, \{\neg s_3\}, \{\vee\{s_1, s_2\}\}), P(\lambda, \{\}, \{\vee\{s_1, s_2\}\})\}$, by $P\lambda$pc

2) $P(\lambda, \{\}, \{\vee\{s_1, s_2\}\}) = \min\{\} = +1$, by $P\lambda$set

3) $\therefore P(\lambda, \vee\{s_1, s_2\}, \{\}) = +1$, by lines 2 and 1 of $E3$.

EndExample1

5 Results

Our first result shows that the proof function P really is a function. (The proofs of these and other results are in the full article available from the author.)

Theorem 1. Let $\Theta = (R, >)$ be a clausal defeasible theory. Then P is a function with co-domain $\{+1, 0, -1\}$.

The next theorem says that CDL has a linear proof hierarchy, and also the reverse for disproof. This shows that proofs with different levels of confidence can be achieved without using numbers.

Theorem 2. Suppose Θ is a clausal defeasible theory.
(1) $\Theta(\mu+) \subseteq \Theta(\rho+) \subseteq \Theta(\pi+) \subseteq \Theta(\beta+) \subseteq \Theta(\pi'+) \subseteq \Theta(\rho'+)$.
(2) $\Theta(\rho'-) \subseteq \Theta(\pi'-) \subseteq \Theta(\beta-) \subseteq \Theta(\pi-) \subseteq \Theta(\rho-) \subseteq \Theta(\mu-)$.

Our final result is that CDL is consistent; that is, each pair of provable clauses is satisfiable, which as the lottery example shows is the desired result.

Theorem 3. Suppose $\Theta = (R, >)$ is a clausal defeasible theory, $Ax(\Theta)$ is consistent, and $\lambda \in \{\mu, \rho, \pi, \beta\}$. If $\{c_1, c_2\} \subseteq \Theta(\lambda+)$ then $\{c_1, c_2\}$ is satisfiable.

6 Summary

In Section 1 we argued that, among non-monotonic logics, the family of defeasible logics is important for knowledge representation and reasoning. Defeasible logics are powerful enough for a diverse range of practical applications, and yet their language has a unique combination of expressiveness, simplicity, and naturalness.

The rules of previous defeasible logics were constructed from literals, but the non-strict rules of CDL are constructed from clauses. This greater expressiveness allows the representation of the Lottery Paradox. The results listed in Section 5 are some of the more important results required to show that CDL is well-defined and consistent.

Acknowledgement. The author wishes to thank Andrew Rock, René Hexel, and Vladimir Estivill-Castro for their assistance and advice; and Michael Maher and Kewen Wang for their comments on a draft of an earlier version of this article.

References

1. Antoniou, G., Billington, D., Maher, M.J.: On the analysis of regulations using defeasible rules. In: Proceedings of the 32nd Hawaii International Conference on Systems Science. IEEE Press (January 1999)
2. Antoniou, G.: Nonmonotonic Rule Systems on Top of Ontology Layers. In: Horrocks, I., Hendler, J. (eds.) ISWC 2002. LNCS, vol. 2342, pp. 394–398. Springer, Heidelberg (2002)
3. Bassiliades, N., Antoniou, G., Vlahavas, I.P.: DR-DEVICE: A Defeasible Logic System for the Semantic Web. In: Ohlbach, H.J., Schaffert, S. (eds.) PPSWR 2004. LNCS, vol. 3208, pp. 134–148. Springer, Heidelberg (2004)
4. Billington, D., Estivill-Castro, V., Hexel, R., Rock, A.: Non-monotonic reasoning for localisation in robocup. In: Proceedings of the 2005 Australasian Conference on Robotics and Automation (2005)
5. Billington, D., Estivill-Castro, V., Hexel, R., Rock, A.: Using Temporal Consistency to Improve Robot Localisation. In: Lakemeyer, G., Sklar, E., Sorrenti, D.G., Takahashi, T. (eds.) RoboCup 2006: Robot Soccer World Cup X. LNCS (LNAI), vol. 4434, pp. 232–244. Springer, Heidelberg (2007)
6. Billington, D., Estivill-Castro, V., Hexel, R., Rock, A.: Architecture for Hybrid Robotic Behavior. In: Corchado, E., Wu, X., Oja, E., Herrero, Á., Baruque, B. (eds.) HAIS 2009. LNCS, vol. 5572, pp. 145–156. Springer, Heidelberg (2009)
7. Billington, D., Estivill-Castro, V., Hexel, R., Rock, A.: Modelling Behaviour Requirements for Automatic Interpretation, Simulation and Deployment. In: Ando, N., Balakirsky, S., Hemker, T., Reggiani, M., von Stryk, O. (eds.) SIMPAR 2010. LNCS, vol. 6472, pp. 204–216. Springer, Heidelberg (2010)
8. Billington, D., Estivill-Castro, V., Hexel, R., Rock, A.: Requirements Engineering via Non-monotonic Logics and State Diagrams. In: Maciaszek, L.A., Loucopoulos, P. (eds.) ENASE 2010. CCIS, vol. 230, pp. 121–135. Springer, Heidelberg (2011)

9. Billington, D., Estivill-Castro, V., Hexel, R., Rock, A.: Plausible logic facilitates engineering the behaviour of autonomous robots. In: Fox, R., Golubski, W. (eds.) The IASTED International Conference on Software Engineering 2010, pp. 41–48. ACTA Press (February 2010)
10. Billington, D., Rock, A.: Propositional plausible logic: Introduction and implementation. Studia Logica 67(2), 243–269 (2001)
11. Darcy, P., Stantic, B., Derakhshan, R.: Correcting stored rfid data with non-monotonic reasoning. International Journal of Principles and Applications of Information Science and Technology 1(1), 65–77 (2007)
12. Governatori, G.: Representing business contracts in ruleml. International Journal of Cooperative Information Systems 14(2-3), 181–216 (2005)
13. Governatori, G., Dumas, M., ter Hofstede, A.H., Oaks, P.: A formal approach to protocols and strategies for (legal) negotiation. In: Proceedings of the 8th International Conference on Artificial Intelligence and Law, pp. 168–177. ACM Press (2001)
14. Governatori, G., Hulstijn, J., Riveret, R., Rotolo, A.: Characterising Deadlines in Temporal Modal Defeasible Logic. In: Orgun, M.A., Thornton, J. (eds.) AI 2007. LNCS (LNAI), vol. 4830, pp. 486–496. Springer, Heidelberg (2007)
15. Governatori, G., Padmanabhan, V.: A Defeasible Logic of Policy-Based Intention. In: Gedeon, T.D., Fung, L.C.C. (eds.) AI 2003. LNCS (LNAI), vol. 2903, pp. 414–426. Springer, Heidelberg (2003)
16. Governatori, G., Rotolo, A.: Defeasible Logic: Agency, Intention and Obligation. In: Lomuscio, A., Nute, D. (eds.) DEON 2004. LNCS (LNAI), vol. 3065, pp. 114–128. Springer, Heidelberg (2004)
17. Governatori, G., Rotolo, A., Sadiq, S.: A model of dynamic resource allocation in workflow systems. In: Schewe, K.D., Williams, H.E. (eds.) Conference Research and Practice of Information Technology Database Technology 2004, vol. 27, pp. 197–206. Australian Computer Science Association, ACS (2004)
18. Governatori, G., Rotolo, A., Sartor, G.: Temporalised normative positions in defeasible logic. In: Proceedings of the 10th International Conference on Artificial Intelligence and Law, pp. 25–34. ACM Press (2005)
19. Johnston, B., Governatori, G.: An algorithm for the induction of defeasible logic theories from databases. In: Schewe, K.D., Zhou, X. (eds.) Conference Research and Practice of Information Technology Database Technology, vol. 17, pp. 75–83. Australian Computer Science Association (2003)
20. Maher, M.J., Rock, A., Antoniou, G., Billington, D., Miller, T.: Efficient defeasible reasoning systems. International Journal of Artificial Intelligence Tools 10(4), 483–501 (2001)
21. Nute, D.: Defeasible reasoning. In: Proceedings of the 20th Hawaii International Conference on System Science, pp. 470–477. University of Hawaii (1987)
22. Nute, D.: Defeasible Logic. In: Bartenstein, O., Geske, U., Hannebauer, M., Yoshie, O. (eds.) INAP 2001. LNCS (LNAI), vol. 2543, pp. 151–169. Springer, Heidelberg (2003)
23. Reiter, R.: A logic for default reasoning. Artificial Intelligence 13, 81–132 (1980)
24. Thakur, S.S., Governatori, G., Padmanabhan, V., Eriksson Lundström, J.: Dialogue Games in Defeasible Logic. In: Orgun, M.A., Thornton, J. (eds.) AI 2007. LNCS (LNAI), vol. 4830, pp. 497–506. Springer, Heidelberg (2007)
25. Wang, K., Billington, D., Blee, J., Antoniou, G.: Combining Description Logic and Defeasible Logic for the Semantic Web. In: Antoniou, G., Boley, H. (eds.) RuleML 2004. LNCS, vol. 3323, pp. 170–181. Springer, Heidelberg (2004)

Applying Multiple Classification Ripple Round Rules to a Complex Configuration Task

Ivan Bindoff[1] and Byeong Ho Kang[2]

[1] University of Tasmania, School of Computing and Information Systems & School of Pharmacy,
[2] University of Tasmania, School of Computing and Information Systems
{Ivan.Bindoff,bhkang}@utas.edu.au

Abstract. A new expert systems methodology was developed, building on existing work on the Ripple Down Rules (RDR) method. RDR methods offer a solution to the maintenance problem which has otherwise plagued traditional rule-based expert systems. However, they are, in their classic form, unable to support rules which use existing classifications in their rule conditions. The new method outlined in this paper is suited to multiple classification tasks, and maintains all the significant advantages of previous RDR offerings, while also allowing the creation of rules which use classifications in their conditions. It improves on previous offerings in this field by having fewer restrictions regarding where and how these rules may be used. The method has undergone initial testing on a complex configuration task, which would be practically unsolvable with traditional multiple classification RDR methods, and has performed well, reaching an accuracy in the 90th percentile after being trained with 1073 rules over the course of classifying 1000 cases, taking ~12 expert hours.

Keywords: ripple, down, rules, multiple, classification, round, configuration, knowledge acquisition.

1 Introduction and Previous Work

Expert systems development and research underwent a surge of popularity during the 1970s and 80s, as they were one of the first examples of a successful and practical artificial intelligence method [1]. However, the approach later fell out of favor, in a large part due to maintenance issues. Knowledge-engineers working on traditional rule-based systems found that, although initially these systems allowed easy rule creation, as the complexity grew it became more and more difficult to define new rules that did not interfere with the existing rules adversely [2, 3].

These maintenance concerns lead to the development of tools, and validation methods which were intended to help the knowledge engineer validate and verify their knowledge base after the addition of new rules. However, these methods did not offer complete proof against errors being introduced into the knowledge base, and the maintenance concerns remained, albeit reduced in severity [3, 4].

After experiencing these issues through work with the GARVAN-ES1 thyroid expert system [3], Compton & Jansen made some important observations about the

D. Wang and M. Reynolds (Eds.): AI 2011, LNAI 7106, pp. 481–490, 2011.

nature of the knowledge which was provided by experts. They identified that experts tend to provide knowledge which justifies their classification in the current context, rather than expressing the complete knowledge path which leads to their classification. That is, they express *why* their classification is correct, rather than *how* [5]. This observation offered an alternative viewpoint to the conventional wisdom of the day, which was that experts intrinsically found it difficult to communicate their knowledge [6]. Furthering these observations, Compton & Jansen also became aware that experts were very good at describing the differences between similar cases, and precisely why the classifications should vary between them [5].

In light of these discoveries, they proposed RDR as an alternative approach to traditional rule-based expert systems.

1.1 Ripple Down Rules

The RDR methodology makes use of a true-false binary tree structure in order to ensure that rules are always added in context [5, 7].

With the RDR approach the problem of maintainability was relatively simple to solve. When creating a new rule, the system would store the current case against that new rule as a "cornerstone", which stored the entire context that the rule was created under. When creating a rule the system could detect if this new rule caused a conflict with the past cornerstone. If so, the expert was required to select a relevant difference between the current and cornerstone cases. When performing an inference with this approach, the knowledge tree would be traversed until a leaf node was reached, and then the last known true node would fire [8, 9].

The RDR method described above is unsuitable for multiple classification tasks, since it would require the use of either multiple knowledge bases or compound classifications, which can cause an undesirable explosion in the amount of knowledge required [10]. As such, to extend this method to multiple classification tasks, Kang altered the underlying knowledge representation structure to that of an n-tree, altered the cornerstone case approach such that multiple cornerstone cases might apply to a single case, and modified the inference strategy accordingly. Contrasting it with RDR, multiple classification RDR (MCRDR) explores every node that is attached to the root node, and adds only the deepest satisfied node of any branch to the result set [11]. An example of a simple MCRDR knowledge base is shown in Figure 1.

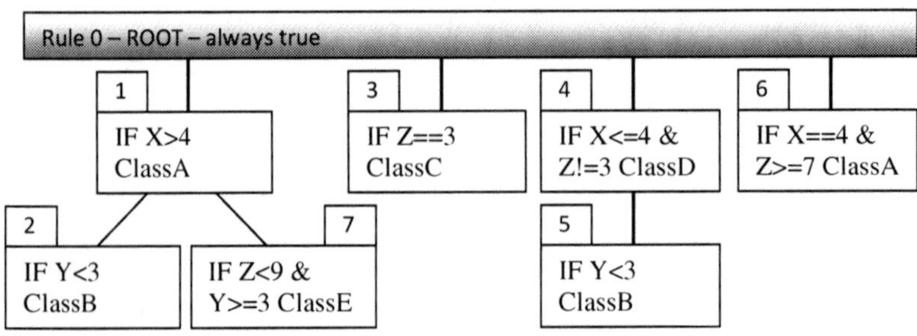

Fig. 1. A simple MCRDR knowledge base

To facilitate understanding of the MCRDR inference approach, consider the knowledge base in Figure 1 being applied to a case [X=4, Y=2, Z=7]. Rule 1 will not fire, so that branch will be disregarded. Rule 3 will be similarly skipped. Rule 4 will fire, but then so will the exception rule 5, which will supersede it, adding ClassB to the result list. Rule 6 will also fire, adding ClassA to the result list.

RDR methods have proven to be unusually valuable in real world tasks, with one very successful commercial system for pathology interpretation [12], a business rules system for Tesco [13], and several commercial applications under development, in areas including high volume call centre management [14] and medication review [15].

1.2 Using Classifications as Conditions

One of the compromises that were made when developing the RDR methodology was that the ability to create rules which used classifications as conditions was lost. This ability had previously been integral to success in some domains, and is of particular value in complex configuration tasks where the positioning of each module may influence the positioning of the other modules.

During early RDR research this shortcoming was recognized. Mulholland developed an RDR based system to solve Ion Chromatography configuration tasks, although this solution was highly specialized and somewhat unreliable, since it did not cope well with the introduction of cyclic knowledge [16]. After this, Beydoun & Hoffmann developed Nested RDR, which was a single classification RDR approach which allowed the creation of "intermediate classifications", as stepping stones towards the end classification [17]. This method was targeted at single classification problem domains, although it may be extended to multiple classification problems. Later, a proposition was made for a more generalized version of RDR, which included provision for a Repeat Inference MCRDR (RIMCRDR) approach whereby the existing MCRDR method was augmented with the ability to use classifications as conditions [18, 19]. However, this approach included restrictions as to when and how these types of rules could be used, and about how the knowledge base must be inferred and interpreted. This was done as an attempt to eliminate the potential for cyclic rules – rules which depend on the existence of a classification, yet upon firing, either directly or indirectly cause that classification to be retracted. Particularly, RIMCRDR asserts that rules must be evaluated in strict chronological order, and that no retractions are allowed. By not allowing retractions you are removing the expert's ability to create exception rules which use classifications as conditions [19]. In making these restrictions, it is felt that the RIMCRDR method will alienate some experts, who may be frustrated at being denied the ability to define rules which make perfect sense, just on the chance that it may result in a cyclical rule definition at some point.

In light of these concerns, it was attempted to develop a new MCRDR based method which would preserve all the essential benefits and strategies of the RDR method, while augmenting it with the ability to create rules using classifications as conditions. It should do this while offering minimal restrictions as to when and where the expert may define these rules.

2 Method

The approach that was decided upon, in order to achieve the goals outlined above, involved changes to the knowledge representation approach as well as the inference strategy. These changes in turn necessitated some revisions to the knowledge acquisition process, particularly provisions for the detection of cyclic rule definitions and an update to the cornerstone case mechanism.

2.1 Knowledge Representation

It was considered desirable to maintain all the essential benefits of the traditional MCRDR knowledge structure, but necessary to extend it, so as to facilitate the detection of potential cycles. As such, the n-tree structure was altered to become a type of directed graph. The underlying n-tree structure was still present, with each rule being a node with one or more conditions and a classification; however, nodes were given the added ability to store one or more classifications which must be satisfied in order for the rule to fire. These classifications could each be defined as classifications which must be present, or must not be present in order for the rule to fire, and were termed switches. Each switch represented a classification, and maintained a counter. Whenever a particular classification was added to the result set during the inference strategy, every switch concerning that classification was incremented by one. If a particular classification was removed from the result set, the corresponding counters were decremented by one. The counter behavior is necessary, since the same classification can be reached through potentially multiple paths.

To facilitate the inference strategy, a store of dependencies which indicated which rule nodes are dependent on which classifications, was also necessary. The list of dependencies for each classification ensures that it was always known which nodes must be revisited and re-inferred whenever a classification was added or removed from the result set, and to enable the switch counters to be adjusted efficiently.

2.2 Inference

The typical MCRDR inference strategy is quite simple, being similar to that of a depth first search where the deepest satisfied node is added to the result set, but where the search does not stop until every node at the first level has been traversed.

The new inference strategy must be substantially more complex since there is now the possibility of nodes being revisited and of results being removed. The new inference algorithm is shown in a simplified pseudo-code form here.

```
infer(Node, Case) {
      clearResult(Node, Case)
      If (Node's rule is satisfied
      AND all of its children's rules aren't)
             If (All Node's parent rules are satisfied)
                    Add Node to result list
                    Mark Node as having fired
                    Activate all dependents of Node's class
                           For each node that changed state
```

```
                            infer(ActivatedNode, Case)

        If (Node has a non-root parent rule)
                clearResult(Node's parent)
        For each Child that isn't marked as avoid
                Clear avoid markers
                infer(Child, Case)
}
clearResult(Node, Case) {
        If (Node is marked as having fired)
                Remove it from the result set
                Clear its fired flag
                Deactivate all dependents of Nodes class
                For each node that changed state
                        infer(DeactivatedNode, Case)
                If (Node has a non-root parent rule
                AND all Node's parent rules are satisfied
                AND no siblings are satisfied)
                        Mark Node to avoid
                        infer(Node's parent, Case)
}
```

2.3 Knowledge Acquisition

From the expert's perspective as the user of the system, the knowledge acquisition process remains largely unchanged from MCRDR. They are presented with a case, and the system's current belief as to which classifications apply to that case. If the expert believes a classification is missing, or that a classification is incorrectly provided by the system, they indicate as such and enter the rule creation process. During this process the expert may select what the classification should be and which valid conditions of the case are relevant to this classification, as normal. The only difference here is that the expert is also able to select any of the classifications the system is currently aware of as conditions of the case. If the classification is currently present on the case, then it can be added to reflect that the classification must be present in order for this rule to fire. If it is not currently present, the condition will be added to reflect that the classification must *not* be present in order for this rule to fire.

Cycles. Where the knowledge acquisition does change is largely behind the scenes. Whenever the expert creates a rule, the system must check that their new rule does not have the potential to cause a cycle in the knowledge base. An example of a cycle can be seen in Figure 2, where each node is represented by 3 boxes, the top left being the switches which must be on for the rule to fire, the bottom left being the conditions which must be satisfied, and the bottom right being the classification that the rule will add to the result set if it fires. It can be seen that in a case where X, Y and Z are all satisfied, the inference algorithm would fire on R1 but fail on the exception R2, thus adding ClassA. It would then consider the next rule, R3, adding ClassB. As a consequence of adding ClassB R2 would be re-evaluated and this time fire, which would add ClassC to the result list. However, R2 is an exception to R1, and thus supersedes R1, necessitating the removal of ClassA from the result set. Because ClassA was removed, R3 which depends on ClassA would have to be re-evaluated, which would in turn remove ClassB and so forth, with no termination possible.

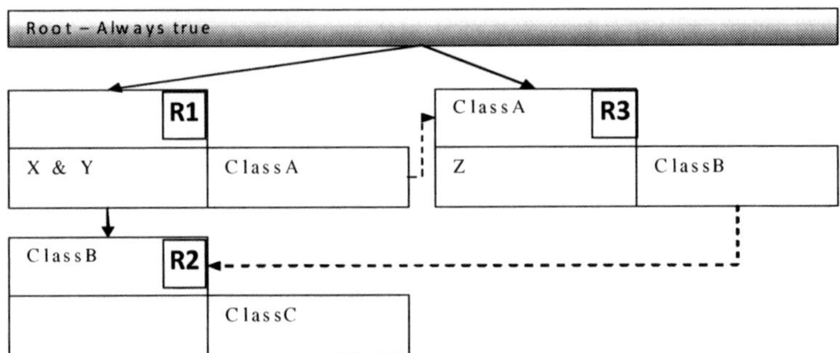

Fig. 2. A simple example of a cyclic rule set. Solid arrows represent exception links, while dotted arrows represent dependency links.

Perhaps the most common method of detecting cycles in a directed graph structure is to perform a topological sort on the graph, as described by Kahn [20]. If a sorted topology cannot be found then there is a cycle. This method is efficient when considering whole graphs, however it was noted that in the context of this method it is only necessary to check the dependencies that are relevant to the new rule. As such, a method was used where the dependencies of all the classifications present in the new rule were examined in turn to see if they lead to a condition where the classification was no longer valid. That is, is there a dependency chain such that the classification might ultimately be dependent on itself being false. If a potential cycle was identified, the expert was informed that their new rule could cause conflicts, and was asked to revise the rule until no further conflicts were found. Considering again the example in Figure 2 we can see that due to R3 ClassB depends on ClassA, due to R2 ClassC depends on ClassB, and due to the R1/R2 exception ClassA depends on ClassB *not* being present (!ClassB). The chain in this example is ClassB depends on ClassA which depends on !ClassB. This paradoxical dependency chain would have been detected when the expert attempted to make the last of these three rules, and disallowed, thus ensuring the cycle was never created.

Cornerstone cases. It was also necessary to adjust the cornerstone case strategy. In MCRDR a cornerstone case can most robustly be defined as being any case which has been previously approved by the expert, but which would be altered by the addition of the new rule that the expert is trying to create [11]. From the user's perspective, the cornerstone case process remains essentially unchanged. When they define a rule and attempt to add it the system will prompt them with a list of any past cases which would be altered by this new rule, and they must either select differences which will eliminate the cornerstone cases, or they must accept that the new rule should in fact alter the past case. However, behind the scenes, there is a loss of efficiency. MCRDR was able to interrogate cornerstone cases very efficiently, as a reference to each cornerstone case could be stored against the nodes they relate to. With the new method this approach becomes challenging, since it is now necessary to store each classification of each case in addition to the attributes. These may change many times for each case, thus requiring many different cornerstone cases to be created as their contexts shift. When the complexities of attempting to maintain this library of cornerstone cases were fully considered, it was deemed simpler to just re-infer all past cases with a temporary version of the knowledge base that included the new rule, and flag each one which had the new rule in its result set after inference as being a cornerstone case. This will cause efficiency concerns as the number of past cases and the size of the knowledge base grows, but many domains are sufficiently small (less than thousands of rules/cases) to be managed with this approach. However, since performing this experiment, a more computationally efficient approach to finding cornerstone cases has been proposed [21].

2.4 Task

The method was initially tested on a small scale pizza preferences domain, a task at which MCRDR might otherwise have been used [22]. However, it was felt that a configuration task would be appropriate for a more robust test, since this type of task would make more use of the added features of the method. Further to this, it was desirable to test the method on a domain which would stress the areas of the method which were identified as potential weaknesses, particularly it was sought to stress the cycle detection and management processes, by designing a domain in which cycles were very likely to occur. To this end, a configuration task was designed which would require a large number of rules which used classifications as conditions, and would have a fundamentally limited number of total classifications.

A blocks placement task was devised, where each case was represented with a 5x5 grid which had between 0 and 4 cells randomly unavailable for block placement (25 boolean attributes per case, each corresponding to a cell in the grid). For a case to be correctly solved, the block modules shown in Figure 3 must be placed into appropriate places in the grid, with no overlaps. Of the blocks in the grayed area, *only* 3 of the 4 were allowed to be placed in the grid. The modules were designed such that modules are likely to be placed relative to other modules (i.e. in interlocking formations), or unavailable cells, as this would tend to be more efficient. One thousand cases were generated randomly and the expert was tasked with training the system such that these 1000 cases were correctly classified. Each placement of a block was represented by a classification, such as “Block A at location X,Y”. As such, each case required exactly 7 unique classifications to be correctly classified.

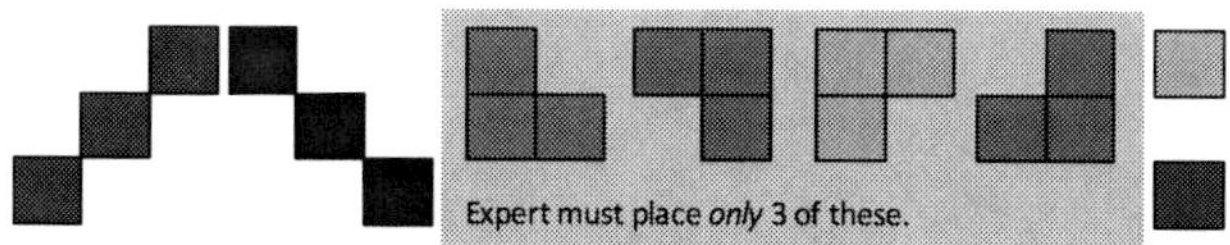

Fig. 3. The block modules which must be placed in the grid

3 Results and Discussion

As the system was trained by the expert, all important actions were recorded. Through analysis of these logs, and of the knowledge base, evaluations could be performed.

3.1 Nature of the Knowledge Base

Perhaps the most indicative figure in determining if the system is working as intended is the growth of the knowledge base. The number of rules will typically grow initially fast, but gradually plateau as the system reaches higher levels of accuracy and thus requires less refinement. This result was seen in this system, with a gradual decline in the growth rate of rules, before the experiment was ended with 1073 rules input.

Of particular interest in this evaluation is the number of classifications used as conditions, as this figure gives some indication of how often these types of rules were deemed necessary by the expert. The substantial majority of rules in this system used at least one classification as a condition (72%), while there were a maximum of 4 classifications used as conditions in a single rule. On average there were 1.20 classifications used as conditions per rule, and 25% of all conditions used were classifications. There were 298 rules which had no classifications as conditions, 339 rules which had one, 364 rules which had two, 65 which had three, and 8 with four.

Perhaps the most telling result is the accuracy of the system. This can be calculated quite effectively in MCRDR systems by determining the ratio of how many *correct* classifications were provided when the expert first loads a case vs. the number of classifications on the case when the expert has finished assessing it [15, 22, 23]. This result is seen in Figure 4 where it can be seen that although the accuracy of any particular case was quite unpredictable, the overall trend when averaged was increasing into the 90th percentile towards the end of the training period.

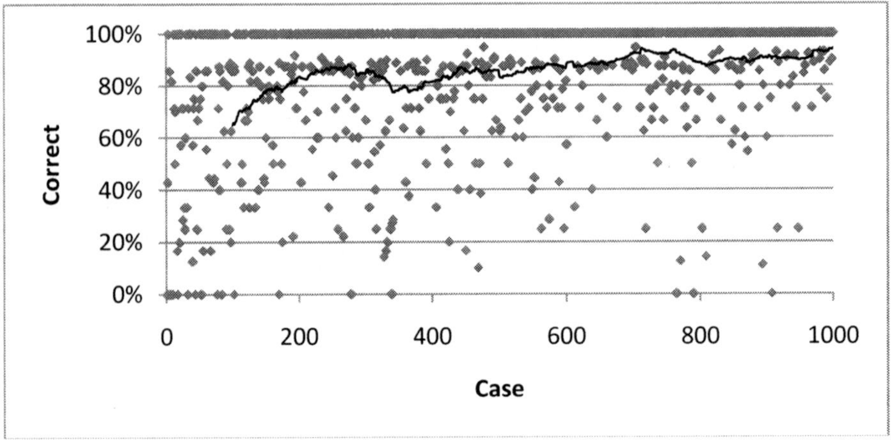

Fig. 4. The accuracy of the system as each case was seen. Shown with a moving average with a period of 100.

3.2 Time Taken

It is an important claim of RDR systems that the average time taken to define a rule is very low, usually in the order of only a few minutes, and that this time does not significantly increase as the knowledge base is populated with more rules. This claim is upheld with this system with an average time taken per rule of 35.9 seconds, and this average was maintained with high consistency throughout training. This indicates that the expert did not find it challenging to create rules, although this figure may be somewhat lower than usual since the expert in this instance was the author of the method, and may have had a better understanding than a typical expert.

3.3 Cycles

It was considered important that the expert was not restricted from creating the rule they wanted to too often. As such, the number of times the expert was told their rule was invalid due to a cycle was measured.

Despite having a very high level of rules which used classifications as conditions, and having a fundamentally limited number of possible classifications, the expert was refused a rule due to a potential cycle only 76 times in total. Of these instances only 11 rules were rejected on the second attempt (after the expert was prompted to refine it), and only one rule on the third. This suggests that it was quite easy for the expert to adjust their rule such that the cycle would be avoided. Furthermore the frequency of these events did not appear to significantly increase as the experiment progressed, even though one might expect it to as the level of inter-dependency increased. This is likely to be due to the fact that the expert was tending not to define cyclic rules, since they do not make sense by definition. However, there may again be a certain level of bias in this result, as the expert in this experiment was the author of the method.

4 Conclusions and Further Work

An initial test of the new method, Multiple Classification Ripple Round Rules (MCRRR) has been undertaken on a complex configuration task. Advantages over previous attempts at offering a similar functionality have been discussed and the work presented shows promising results that indicate the method is suitable for configuration style tasks.

However, the experiment is unfortunately biased, partly because a custom designed problem was used – although a genuine attempt was made to design the task such that it would challenge the method, rather than suit it – and partly because the author of the method also acted as the expert for the purposes of the experiment. This means he would be likely to have insights into the nature of the method that another expert using such a system might not. To overcome these concerns further experimentation is needed with a range of problem domains, a range of unbiased experts, and a range of potential methodological approaches. However, similar limitations are common within research in this field, as expert time is both valuable and scarce, so it is felt that this work is still a significant research contribution.

References

1. Buchanan, B.: Expert systems: working systems and the research literature. Expert Systems 3, 32–50 (1986)
2. Bachant, J., McDermott, J.: R1 Revisited: four years in the trenches. American Association for Artificial Intelligence Readings from the AI Magazine, 177–188 (1988)

3. Compton, P., Horn, K., Quinlan, R., Lazarus, L., Ho, K.: Maintaining an expert system. Applications of Expert Systems 2, 366–384 (1989)
4. Suwa, M., Scott, A., Shortliffe, E.: An approach to verifying completeness and consistency in a rule-based expert system. AI Magazine 3, 16–21 (1982)
5. Compton, P., Jansen, R.: A philosophical basis for knowledge acquisition. In: European Knowledge Acquisition for Knowledge-Based Systems, pp. 75–89 (1989)
6. Boose, J.H., Bradshaw, J.M.: Expertise transfer and complex problems: using AQUINAS as a knowledge-acquisition workbench for knowledge-based systems. International Journal of Man-Machine Studies 26, 3–28 (1987)
7. Compton, P., Kang, B.H., Preston, P., Mulholland, M.: Knowledge Acquisition without Analysis. In: Knowledge Acquisition for Knowledge-Based Systems, pp. 278–299 (1993)
8. Kang, B., Compton, P.: A Maintenance Approach to Case Based Reasoning. In: Haton, J.-P., Keane, M., Manago, M. (eds.) European Workshop on Advances in Case-Based Reasoning, vol. 984, pp. 226–239. Springer, Chantilly (1994)
9. Preston, P., Edwards, G., Compton, P.: A 2000 Rule Expert System Without a Knowledge Engineer. In: AIII-Sponsored Banff Knowledge Acquisition for Knowledge-Based Systems (1994)
10. Kang, B., Compton, P., Preston, P.: Multiple Classification Ripple Down Rules: Evaluation and Possibilities. In: AIII-Sponsored Banff Knowledge Acquisition for Knowledge-Based Systems (1995)
11. Kang, B.H.: Validating knowledge acquisition: multiple classification ripple-down rules. University of New South Wales, Sydney (1995)
12. Compton, P., Peters, L., Edwards, G., Lavers, T.G.: Experience with ripple-down rules. Knowledge Based Systems 19, 356–362 (2006)
13. Sarraf, Q., Ellis, G.: Business Rules in Retail: The Tesco.com Story. 2010 (2007)
14. Vazey, M., Richards, D.: Troubleshooting at the Call Centre: A Knowledge-based Approach. In: Hamza, M.H. (ed.) International Conference on Artificial Intelligence and Applications, vol. 23, pp. 721–726. IASTED/ACTA Press, Innsbruck (2005)
15. Bindoff, I., Tenni, P., Peterson, G., Kang, B.H., Jackson, S.: Development of an intelligent decision support system for medication review. Journal of Clinical Pharmacy and Therapeutics 32, 81–88 (2007)
16. Mulholland, M.: The Evaluation of the Applicability of Artificial Intelligence Software to Solving Problems in Ion Chromatography. University of New South Wales (1995)
17. Beydoun, G., Hoffmann, A.: NRDR for the Acquisition of Search Knowledge. In: Sattar, A. (ed.) Canadian AI 1997. LNCS, vol. 1342, pp. 177–186. Springer, Heidelberg (1997)
18. Compton, P., Richards, D.: Extending ripple down rules. In: International Conference on Knowledge Engineering and Knowledge Management (1999)
19. Compton, P., Richards, D.: Generalising ripple-down rules. In: Knowledge Engineering and Knowledge Management: Methods, Models, Tools, pp. 2–6 (2000)
20. Kahn, A.B.: Topological sorting of large networks. Communications of the ACM 5, 558–562 (1962)
21. Bindoff, I.K.: Multiple Classification Ripple Round Rules: Classifications as Conditions. University of Tasmania, Hobart (2010)
22. Bindoff, I., Ling, T., Kang, B.H.: Multiple Classification Ripple Round Rules: A Preliminary Study. In: Richards, D., Kang, B.-H. (eds.) PKAW 2008. LNCS, vol. 5465, pp. 76–90. Springer, Heidelberg (2009)
23. Bindoff, I., Kang, B., Ling, T., Tenni, P., Peterson, G.: Applying MCRDR to a Multidisciplinary Domain. In: Orgun, M.A., Thornton, J. (eds.) AI 2007. LNCS (LNAI), vol. 4830, pp. 519–528. Springer, Heidelberg (2007)

Semantic Foundation for Preferential Description Logics

Katarina Britz, Thomas Meyer, and Ivan Varzinczak

Centre for Artificial Intelligence Research
CSIR Meraka Institute and University of KwaZulu-Natal, South Africa
{arina.britz,tommie.meyer,ivan.varzinczak}@meraka.org.za

Abstract. Description logics are a well-established family of knowledge representation formalisms in Artificial Intelligence. Enriching description logics with non-monotonic reasoning capabilities, especially preferential reasoning as developed by Lehmann and colleagues in the 90's, would therefore constitute a natural extension of such KR formalisms. Nevertheless, there is at present no generally accepted semantics, with corresponding syntactic characterization, for preferential consequence in description logics. In this paper we fill this gap by providing a natural and intuitive semantics for defeasible subsumption in the description logic $\mathcal{ALC}$. Our semantics replaces the propositional valuations used in the models of Lehmann et al. with structures we refer to as *concept models*. We present representation results for the description logic $\mathcal{ALC}$ for both preferential and rational consequence relations. We argue that our semantics paves the way for extending preferential and rational consequence, and therefore also rational closure, to a whole class of logics that have a semantics defined in terms of first-order relational structures.

1 Introduction

The preferential and rational consequence relations first studied by Lehmann and colleagues [8,10] play a central role in non-monotonic reasoning, not least because they provide the foundation for the determination of the important notion of rational closure. Although they can be applied directly to a large variety of knowledge representation languages, these constructions suffer from the limitation that they are largely propositional in nature, whereas many logics of interest for Artificial Intelligence have more structure.

One of the main obstacles in moving beyond the propositional setting has been the lack of a formal semantics which appropriately generalizes the preferential and ranked models of Lehmann et al. The first tentative exploration of preferential predicate logics by Lehmann et al. didn't fly primarily because propositional logic was sufficiently expressive for the non-monotonic reasoning community at the time, and first-order logic introduced too much complexity [9]. But this changed with the surge of interest in description logics as knowledge representation formalism and their many applications in AI.

D. Wang and M. Reynolds (Eds.): AI 2011, LNAI 7106, pp. 491–500, 2011.

Description logics (DLs) [1] are decidable fragments of first-order logic, and are ideal candidates for the kind of extension to preferential reasoning we have in mind: the notion of subsumption present in all DLs is a natural candidate for defeasibility, while at the same time, the restricted expressivity of DLs ensures that attempts to introduce preferential reasoning are not hampered by the complexity of full first-order logic. The aim of this paper is therefore to extend the work of Lehmann et al. [8,10] beyond propositional logic without moving to full first-order logic. We restrict our attention to the description logic $\mathcal{ALC}$ here, but the results are broadly applicable to other DLs, as well as other similarly structured logics such as logics of action and logics of knowledge and belief [3].

The rest of the paper is structured as follows. After some DL preliminaries (Section 2), we give a brief account of preferential and rational consequence in the propositional case (Section 3). In Section 4, which is the heart of the paper, we define the semantics for both preferential and rational subsumption for $\mathcal{ALC}$ and present representation results for both. Importantly, these are with respect to the corresponding *propositional* properties. From this we conclude that our semantics forms the foundation of a semantics for preferential and rational consequence for a whole class of DLs and related logics. In Section 5 we show that the notions of propositional preferential entailment and rational closure can be 'lifted' to the case for DLs, specifically $\mathcal{ALC}$. In Section 6 we discuss related results. We conclude with Section 7 in which we also discuss future work.

2 Description Logics

The language of $\mathcal{ALC}$ is built upon a finite set of atomic *concept names* $\mathsf{N}_{\mathscr{C}}$ (together with the distinguished concept $\top$), and a finite set of *role names* $\mathsf{N}_{\mathscr{R}}$, using the constructors $\sqcap$ (concept conjunction), $\neg$ (complement), and $\exists$ (existential restriction). An atomic concept is denoted by A, possibly with subscripts, and a role name by r, possibly with subscripts. Complex concepts are denoted by $C, D, \ldots$ and are constructed according to the rule

$$C ::= A \mid \top \mid C \sqcap C \mid \neg C \mid \exists r.C$$

Concepts built with the constructors $\sqcup$ and $\forall$, and the special concept $\bot$ are defined in terms of the others in the usual way. We let $\mathcal{L}$ denote the set of all $\mathcal{ALC}$ concepts.

The semantics of $\mathcal{ALC}$ is the standard set theoretic Tarskian semantics. An *interpretation* is a structure $\mathcal{I} = \langle \Delta^{\mathcal{I}}, \cdot^{\mathcal{I}} \rangle$, where $\Delta^{\mathcal{I}}$ is a non-empty set called the *domain*, and $\cdot^{\mathcal{I}}$ is an *interpretation function* mapping concept names A to subsets $A^{\mathcal{I}}$ of $\Delta^{\mathcal{I}}$, and mapping role names r to binary relations $r^{\mathcal{I}}$ over $\Delta^{\mathcal{I}} \times \Delta^{\mathcal{I}}$:

$$A^{\mathcal{I}} \subseteq \Delta^{\mathcal{I}},\ r^{\mathcal{I}} \subseteq \Delta^{\mathcal{I}} \times \Delta^{\mathcal{I}}, \top^{\mathcal{I}} = \Delta^{\mathcal{I}},\ \bot^{\mathcal{I}} = \emptyset$$

Given an interpretation $\mathcal{I} = \langle \Delta^{\mathcal{I}}, \cdot^{\mathcal{I}} \rangle$, $\cdot^{\mathcal{I}}$ is extended to interpret complex concepts in the following way:

$$(\neg C)^{\mathcal{I}} = \Delta^{\mathcal{I}} \setminus C^{\mathcal{I}},\ \ (C \sqcap D)^{\mathcal{I}} = C^{\mathcal{I}} \cap D^{\mathcal{I}},$$

$$(\exists r.C)^{\mathcal{I}} = \{a \in \Delta^{\mathcal{I}} \mid \text{for some } b, (a, b) \in r^{\mathcal{I}} \text{ and } b \in C^{\mathcal{I}}\}$$

Given $C, D \in \mathcal{L}$, $C \sqsubseteq D$ is a *subsumption statement*, and it is read as "C is subsumed by D". $C \equiv D$ is an abbreviation for both $C \sqsubseteq D$ and $D \sqsubseteq C$. An ($\mathcal{ALC}$) *TBox* $\mathcal{T}$ is a finite set of subsumption statements.

An interpretation $\mathcal{I}$ *satisfies* $C \sqsubseteq D$ (denoted $\mathcal{I} \Vdash C \sqsubseteq D$) if and only if $C^{\mathcal{I}} \subseteq D^{\mathcal{I}}$. $\mathcal{I} \Vdash C \equiv D$ if and only if $C^{\mathcal{I}} = D^{\mathcal{I}}$. $C \sqsubseteq D$ is (classically) *entailed* by a TBox $\mathcal{T}$, denoted $\mathcal{T} \models C \sqsubseteq D$, if and only if every interpretation $\mathcal{I}$ which satisfies all elements of $\mathcal{T}$, also satisfies $C \sqsubseteq D$.

For more details on description logics in general, and the description logic $\mathcal{ALC}$ in particular, the reader is referred to the DL handbook [1].

3 Propositional Preferential and Rational Consequence

In this section we give a brief introduction to propositional preferential and rational consequence, as initially defined by Kraus et al. [8]. A propositional defeasible consequence relation $\mid\sim$ is defined as a binary relation on formulas $\alpha, \beta, \gamma, \ldots$ of an underlying (possibly infinitely generated) propositional logic equipped with a standard propositional entailment relation $\models$. $\mid\sim$ is said to be *preferential* if it satisfies the following set of properties:

$$\text{(Ref)}\ \alpha \mid\sim \alpha \qquad \text{(LLE)}\ \frac{\alpha \equiv \beta,\ \ \alpha \mid\sim \gamma}{\beta \mid\sim \gamma} \qquad \text{(And)}\ \frac{\alpha \mid\sim \beta,\ \ \alpha \mid\sim \gamma}{\alpha \mid\sim \beta \wedge \gamma}$$

$$\text{(RW)}\ \frac{\alpha \mid\sim \beta,\ \ \beta \models \gamma}{\alpha \mid\sim \gamma} \qquad \text{(Or)}\ \frac{\alpha \mid\sim \gamma,\ \ \beta \mid\sim \gamma}{\alpha \vee \beta \mid\sim \gamma} \qquad \text{(CM)}\ \frac{\alpha \mid\sim \beta,\ \ \alpha \mid\sim \gamma}{\alpha \wedge \beta \mid\sim \gamma}$$

The semantics of (propositional) preferential consequence relations is in terms of *preferential models*; these are partially ordered structures with states labeled by propositional valuations. We shall make this terminology more precise in Section 4, but it essentially allows for a partial order on states, with states lower down in the order being more preferred than those higher up. Given a preferential model $\mathscr{P}$, a pair $\alpha \mid\sim \beta$ is in the consequence relation defined by $\mathscr{P}$ if and only if the minimal states (according to the partial order) of all those states labeled by valuations that are propositional models of α, are also labeled by propositional models of β. The representation theorem for preferential consequence relations then states [8]:

Theorem 1 (Kraus et al.). *A defeasible consequence relation is a preferential consequence relation if and only if it is defined by some preferential model.*

If, in addition to the preferential properties, $\mid\sim$ also satisfies the following Rational Monotony property, it is said to be a *rational* consequence relation:

$$\text{(RM)}\ \frac{\alpha \mid\sim \beta,\ \ \alpha \not\mid\sim \neg\gamma}{\alpha \wedge \gamma \mid\sim \beta}$$

The semantics of rational consequence relations is in terms of *ranked* preferential models, i.e., preferential models in which the preference order is *modular*:

Definition 1. *Given a set* S, $\prec \subseteq S \times S$ *is modular if and only if* $\prec$ *is a partial order on* S, *and there is a ranking function* $rk : S \mapsto \mathbb{N}$ *such that for every* $s, s' \in S$, $s \prec s'$ *if and only if* $rk(s) < rk(s')$.

The representation theorem for rational consequence relations then states [10]:

Theorem 2 (Lehmann and Magidor). *A defeasible consequence relation is a rational consequence relation if and only if it is defined by some ranked model.*

4 Semantics for DL Preferential Subsumption

Description logics are ideal candidates for extending propositional preferential consequence since the notion of subsumption in DLs lends itself naturally to defeasibility [2,7,4]. The basic idea is to reinterpret defeasible consequence of the form $\alpha \mid\!\sim \beta$ as *defeasible subsumption* of the form $C \precsim D$, and classical entailment $\models$ as DL subsumption $\sqsubseteq$. For example, if $M \sqsubseteq \neg F$ is read as "meningitis is not fatal", then $M \precsim \neg F$ can be read as "meningitis is usually not fatal". The above properties of preferential consequence are then immediately applicable.

Definition 2. *A subsumption relation* $\precsim \subseteq \mathcal{L} \times \mathcal{L}$ *is a preferential subsumption relation if and only if it satisfies the properties (Ref), (LLE), (And), (RW), (Or), and (CM), with propositional entailment replaced by classical DL subsumption.* $\precsim$ *is a rational subsumption relation if and only if in addition to being a preferential subsumption relation, it also satisfies the property (RM).*

Since DLs have a standard first-order semantics, the obvious generalization from a technical perspective is to replace the propositional valuations in preferential models with first-order interpretations. Intuitively, this also turns out to be a natural generalization of the propositional setting, with the notion of normal first-order interpretation characterizing a given concept replacing the propositional notion of normal worlds satisfying a given proposition. Formally, our semantics is based on the notion of a *concept model*, which is analogous to that of a Kripke model in modal logic [5]:

Definition 3 (Concept Model). *A concept model is a tuple* $\mathscr{M} = \langle W, R, V \rangle$ *where* W *is a set of possible worlds,* $R = \langle R_1, \ldots, R_n \rangle$, *where each* $R_i \subseteq W \times W$, $1 \leq i \leq |\mathsf{N}_{\mathscr{R}}|$, *and* $V \colon W \mapsto 2^{\mathsf{N}_{\mathscr{C}}}$ *is a valuation function.*

Observe that the valuation function V can be viewed as a propositional valuation with propositional atoms replaced by concept names. From the definition of satisfaction in a concept model below it is then clear that, within the context of a concept model, a world occurring in that concept model is a proper generalization of a propositional valuation.

Definition 4 (Satisfaction). *Given* $\mathscr{M} = \langle W, R, V \rangle$ *and* $w \in W$:

- $\mathscr{M}, w \Vdash \top$;
- $\mathscr{M}, w \Vdash A$ *iff* $A \in V(w)$;

- *$\mathscr{M}, w \Vdash C \sqcap D$ iff $\mathscr{M}, w \Vdash C$ and $\mathscr{M}, w \Vdash D$;*
- *$\mathscr{M}, w \Vdash \neg C$ iff $\mathscr{M}, w \not\Vdash C$;*
- *$\mathscr{M}, w \Vdash \exists r_i.C$ iff there is $w' \in W$ s.t. $(w, w') \in R_i$ and $\mathscr{M}, w' \Vdash C$.*

Let $\mathscr{U}$ denote the set of all pairs $(\mathscr{M}, w)$ where $\mathscr{M} = \langle W, R, V\rangle$ is a concept model and $w \in W$. Worlds are, loosely speaking, interpreted DL objects. And while this correspondence holds technically (from the correspondence between $\mathcal{ALC}$ and multimodal logic K [14]), a possible worlds reading of the meaning of a concept is also more intuitive in the current context, since this leads to a preference order on rich first-order structures, rather than on interpreted objects. This is made precise below.

Let S be a set, the elements of which are called *states*. Let $\ell : S \mapsto \mathscr{U}$ be a *labeling function* mapping every state to a pair $(\mathscr{M}, w)$ where $\mathscr{M} = \langle W, R, V\rangle$ is a concept model such that $w \in W$. Let $\prec$ be a binary relation on S. Given $C \in \mathcal{L}$, we say that $s \in S$ *satisfies* C (written $s \models C$) if and only if $\ell(s) \Vdash C$, i.e., $\mathscr{M}, w \Vdash C$. We define $\widehat{C} = \{s \in S \mid s \models C\}$. $\widehat{C}$ is *smooth* if and only if each $s \in \widehat{C}$ is either $\prec$-minimal in $\widehat{C}$, or there is $s' \in \widehat{C}$ such that $s' \prec s$ and s' is $\prec$-minimal in $\widehat{C}$. We say that S satisfies the smoothness condition if and only if for every $C \in \mathcal{L}$, $\widehat{C}$ is smooth.

We are now ready for our definition of preferential model.

Definition 5 (Preferential Model). *A preferential model is a triple $\mathscr{P} = \langle S, \ell, \prec\rangle$ where S is a set of states satisfying the smoothness condition, ℓ is a labeling function mapping states to elements of $\mathscr{U}$, and $\prec$ is a strict partial order on S, i.e., $\prec$ is irreflexive and transitive.*

These formal constructions closely resemble those of Kraus et al. [8] and of Lehmann and Magidor [10], the difference being that propositional valuations are replaced with elements of the set $\mathscr{U}$.

Definition 6 (Preferential Subsumption). *Let $C, D \in \mathcal{L}$ and $\mathscr{P} = \langle S, \ell, \prec\rangle$ be a preferential model. C is preferentially subsumed by D in $\mathscr{P}$ (noted $C \sqsubseteq\!\!\!\sim_{\mathscr{P}} D$) if and only if every $\prec$-minimal state $s \in \widehat{C}$ is such that $s \in \widehat{D}$.*

We are now in a position to prove one of the central results of this paper.

Theorem 3. *A defeasible subsumption relation is a preferential subsumption relation if and only if it is defined by some preferential model.*

The significance of this is that the representation result is proved with respect to the same set of properties used to characterize propositional preferential consequence. We therefore argue that preferential models, as we have defined them, provide the foundation for a semantics for preferential (and rational) subsumption for a whole class of DLs and related logics. We do not claim that this is *the* appropriate notion of preferential subsumption for $\mathcal{ALC}$, but rather that it describes the basic framework within which to investigate such a notion. In order to obtain a similar result for rational subsumption, we restrict ourselves to those preferential models in which $\prec$ is a modular order on states (cf. Definition 1):

Definition 7 (Ranked Model). *A ranked model* $\mathscr{P}_r$ *is a preferential model* $\langle S, \ell, \prec \rangle$ *in which* $\prec$ *is modular.*

Since ranked models are preferential models, the notion of rational subsumption is as in Definition 6. We can then state the following result:

Theorem 4. *A defeasible subsumption relation is a rational subsumption relation if and only if it is defined by some ranked model.*

5 Rational Closure

One of the primary reasons for defining non-monotonic consequence relations of the kind we have presented above is to get at a notion of *defeasible entailment*: Given a set of subsumption statements of the form $C \precsim D$ or $C \sqsubseteq D$, which other subsumption statements, defeasible and classical, should one be able to derive from this? It can be shown that hard subsumption statements $C \sqsubseteq D$ can be encoded as defeasible subsumptions of the form $C \sqcap \neg D \precsim \bot$ [10, Section 2]. For the remainder of this paper we shall therefore concern ourselves only with *finite* sets of defeasible subsumption statements, and refer to these as *defeasible TBoxes*, denoted $\mathcal{T}$. We permit ourselves the freedom to include classical subsumption statements of the form $C \sqsubseteq D$ in a defeasible TBox, with the understanding that it is an encoding of the defeasible subsumption statement $C \sqcap \neg D \precsim \bot$.

Our aim in this section is to show that the results for the propositional case [10] with respect to the question above can be 'lifted' to $\mathcal{ALC}$. We provide here appropriate notions of preferential entailment and rational closure. It must be emphasized that the results obtained in this section rely heavily on similar results obtained by Lehmann and Magidor [10] for the propositional case, and the semantics for preferential and rational subsumption presented in Section 4. Similar to the results of that section, our claim is not that the versions of preferential and rational closure here are *the* appropriate ones for $\mathcal{ALC}$. In fact, our conjecture is that they are *not*, due to their propositional nature. However, we claim that they provide the appropriate springboard from which to investigate more appropriate versions, for $\mathcal{ALC}$, as well as for other DLs and related logics.

The version of rational closure defined here provides us with a strict generalization of classical entailment for $\mathcal{ALC}$ TBoxes in which the expressivity of $\mathcal{ALC}$ is enriched with the ability to make defeasible subsumption statements. For example, consider the defeasible $\mathcal{ALC}$ TBox:

$$\mathcal{T} = \{BM \sqsubseteq M, VM \sqsubseteq M, M \precsim \neg F, BM \precsim F\}, \tag{1}$$

where BM abbreviates the concept *BacterialMeningitis*, M stands for *Meningitis*, VM for *viralMeningitis*, and F abbreviates *FatalDisease*. One should be able to conclude that viral meningitis is usually non-fatal ($VM \precsim \neg F$). On the other hand, we should not conclude that fatal versions of meningitis are usually bacterial ($F \sqcap M \precsim BM$), nor, for that matter, that fatal versions of meningitis are usually *not* bacterial ones ($F \sqcap M \precsim \neg BM$).

Armed with the notion of a preferential model (cf. Section 4) we define preferential entailment for $\mathcal{ALC}$ as follows.

Definition 8. *$C \precsim D$ is preferentially entailed by a defeasible TBox $\mathcal{T}$ if and only if for every preferential model $\mathscr{P}$ in which $E \precsim_{\mathscr{P}} F$ for every $E \precsim F \in \mathcal{T}$, it is also the case that $C \precsim_{\mathscr{P}} D$.*

Firstly, we can show that preferential entailment is well-behaved and coincides with *preferential closure* under the properties of preferential subsumption (i.e., the intersection of all preferential subsumption relations containing a defeasible TBox). More precisely, if $\mathcal{T}$ is a defeasible TBox, the set of defeasible subsumption statements preferentially entailed by $\mathcal{T}$, viewed as a binary relation on $\mathcal{L}$, is a preferential subsumption relation. Furthermore, a defeasible subsumption statement is preferentially entailed by $\mathcal{T}$ if and only if it is in the preferential closure of $\mathcal{T}$.

From this it follows that if we use preferential entailment, the meningitis example can be formalized by letting $\mathcal{T}$ be as in Equation 1. However, $VM \precsim \neg F$ is *not* preferentially entailed by $\mathcal{T}$ above (we cannot conclude that viral meningitis is usually not fatal) and preferential entailment is thus too weak. Hence we move to rational subsumption relations.

The first attempt to do so is to use a definition similar to that employed for preferential entailment: $C \precsim D$ is rationally entailed by a defeasible TBox $\mathcal{T}$ if and only if for every ranked model $\mathscr{P}_r$ in which $E \precsim_{\mathscr{P}_r} F$ for every $E \precsim F \in \mathcal{T}$, it is also the case that $C \precsim_{\mathscr{P}_r} D$. However, this turns out to be *exactly* equivalent to preferential entailment [10, Section 4.2]. Therefore, if the set of defeasible subsumption statements obtained as such is viewed as a binary relation on concepts, the result is a preferential subsumption relation and is not, in general, a rational consequence relation.

The above attempt to define rational entailment is thus not acceptable, as shown by Lehmann and Magidor. Instead, in order to arrive at an appropriate notion of (rational) entailment we first define a preference ordering on rational subsumption relations, with relations further down in the ordering interpreted as more preferred.

Definition 9. *Let $\precsim_0$ and $\precsim_1$ be rational subsumption relations. $\precsim_0$ is preferable to $\precsim_1$ (written $\precsim_0 \ll \precsim_1$) if and only if*

- *there is $C \precsim D \in \precsim_1 \setminus \precsim_0$ s.t. for all E s.t. $E \sqcup C \precsim_0 \neg C$ and for all F s.t. $E \precsim_0 F$, we also have $E \precsim_1 F$; and*
- *for every $E, F \in \mathcal{L}$, if $E \precsim F$ is in $\precsim_0 \setminus \precsim_1$, then there is an assertion $G \precsim H$ in $\precsim_1 \setminus \precsim_0$ s.t. $G \sqcup E \precsim_1 \neg E$.*

Space considerations prevent us from giving a detailed motivation for $\ll$ here, but it is essentially the motivation for the same ordering for the propositional case provided by Lehmann and Magidor [10]. Given a defeasible TBox $\mathcal{T}$, the idea is now to define rational entailment as the most preferred (with respect to $\ll$) of all those rational subsumption relations which include $\mathcal{T}$.

Lemma 1. *Let $\mathcal{T}$ be a finite defeasible TBox and let $\mathcal{R}$ be the class of all rational subsumption relations which include $\mathcal{T}$. There is a unique rational subsumption relation in $\mathcal{R}$ which is preferable to all other elements of $\mathcal{R}$ with respect to $\ll$.*

This puts us in a position to define an appropriate form of (rational) entailment for defeasible TBoxes:

Definition 10. *Let $\mathcal{T}$ be a defeasible TBox. The rational closure of $\mathcal{T}$ is the (unique) rational subsumption relation which includes $\mathcal{T}$ and is preferable (with respect to $\ll$) to all other rational subsumption relations including $\mathcal{T}$.*

It can be shown that $VM \precsim \neg F$ is in the rational closure of $\mathcal{T}$ (we can conclude viral meningitis is usually not fatal), but that neither $F \sqcap M \precsim BM$ nor $F \sqcap M \precsim \neg BM$ is.

We conclude this section with a result which can be used to define an algorithm for computing the rational closure of a defeasible TBox $\mathcal{T}$. For this we first need to define a ranking of concepts with respect to $\mathcal{T}$ which, in turn, is based on a notion of exceptionality. A concept C is said to be *exceptional* for a defeasible TBox $\mathcal{T}$ if and only if $\mathcal{T}$ *preferentially* entails $\top \precsim \neg C$. A defeasible subsumption statement $C \precsim D$ is exceptional for $\mathcal{T}$ if and only if its antecedent C is exceptional for $\mathcal{T}$.

It turns out that checking for exceptionality can be reduced to classical subsumption checking.

Lemma 2. *Given a defeasible TBox $\mathcal{T}$, let $\mathcal{T}^{\sqsubseteq}$ be its classical counterpart in which every defeasible subsumption of the form $D \precsim E$ in $\mathcal{T}$ is replaced by $D \sqsubseteq E$. C is exceptional for $\mathcal{T}$ if and only if $\top \sqsubseteq \neg C$ is classically entailed by $\mathcal{T}^{\sqsubseteq}$.*

Let $E(\mathcal{T})$ denote the subset of $\mathcal{T}$ containing statements that are exceptional for $\mathcal{T}$. We define a non-increasing sequence of subsets of $\mathcal{T}$ as follows: $\mathcal{E}_0 = \mathcal{T}$, and for $i > 0$, $\mathcal{E}_i = E(\mathcal{E}_{i-1})$. Clearly there is a smallest integer k such that for all $j \geq k$, $\mathcal{E}_j = \mathcal{E}_{j+1}$. From this we define the *rank* of a concept with respect to $\mathcal{T}$: $r_{\mathcal{T}}(C) = k - i$, where i is the smallest integer such that C is not exceptional for $\mathcal{E}_i$. If C is exceptional for $\mathcal{E}_k$ (and therefore exceptional for all $\mathcal{E}$s), then $r_{\mathcal{T}}(C) = 0$. Intuitively, the lower the rank of a concept, the more exceptional it is with respect to the TBox $\mathcal{T}$.

Theorem 5. *Let $\mathcal{T}$ be a defeasible TBox. The* rational closure *of $\mathcal{T}$ is the set of defeasible subsumption statements $C \precsim D$ such that either $r_{\mathcal{T}}(C) > r_{\mathcal{T}}(C \sqcap \neg D)$, or $r_{\mathcal{T}}(C) = 0$ (in which case $r_{\mathcal{T}}(C \sqcap \neg D) = 0$ as well).*

From this result one can construct a (naïve) decidable algorithm to check whether a given defeasible subsumption statement is in the rational closure of a defeasible TBox $\mathcal{T}$. Also, if checking for exceptionality is assumed to take constant time, the algorithm is quadratic in the size of $\mathcal{T}$. Given that exceptionality reduces to subsumption checking in $\mathcal{ALC}$ which is ExpTime-complete, it immediately follows that checking whether a given defeasible subsumption is in the rational closure of $\mathcal{T}$ is an ExpTime-complete problem. This result relates to a result by Casini and Straccia [4] which we refer to again in the next section.

6 Related Work

Quantz and Ryan [12,13] were probably the first to consider the lifting of non-monotonic reasoning formalisms to a DL setting. They propose a general framework for Preferential Default Description Logics (PDDL) based on an $\mathcal{ALC}$-like language by introducing a version of default subsumption and proposing a semantics for it. Their semantics is based on a simplified version of standard DL interpretations in which all domains are assumed to be finite and the unique name assumption holds for object names. In that sense, their framework is much more restrictive than ours, as we do not make these assumptions here. They focus on a version of entailment which they refer to as preferential entailment, but which is to be distinguished from the version of preferential entailment that we have presented in this paper. In what follows, we shall refer to their version as *Q-preferential entailment.*

Q-preferential entailment is concerned with what ought to follow from a set of classical DL statements, together with a set of default subsumption statements, and is parameterised by a fixed partial order on (simplified) DL interpretations. They prove that any Q-preferential entailment satisfies the properties of a preferential consequence relation and, with some restrictions on the partial order, satisfies Rational Monotony as well. Q-preferential entailment can therefore be viewed as something in between the notions of preferential consequence and preferential entailment we have defined for DLs. It is also worth noting that although the Q-preferential entailments satisfy the properties of a preferential consequence relation, Quantz and Ryan do not prove that Q-preferential entailment provides a *characterisation* of preferential consequence.

Britz et al. [2] and Giordano et al. [7] use typicality orderings on *objects* in first-order domains to define versions of defeasible subsumption for $\mathcal{ALC}$. Both approaches propose specific non-monotonic consequence relations, and hence their semantic constructions are special cases of the more general framework we have provided here. In contrast, we provide a general semantic framework which is relevant to all logics with a possible worlds semantics. This is because our preference semantics is not defined in terms of orders on interpreted DL objects relative to given concepts, but rather in terms of a single order on relational structures. Our semantics for defeasible subsumption yields a single order at the meta level, rather than ad hoc relativized orders at the object level.

Casini and Straccia [4] recently proposed a syntactic operational characterization of rational closure in the context of description logics, based on classical entailment tests only, and thus amenable to implementation. Their work is based on that of Lehmann and Magidor [10], Freund [6] and Poole [11], and represents an important building block in the extension of preferential consequence to description logics. However, this work lacks a semantics, and we can only at present conjecture that the rational closure produced by their algorithm coincides with the notion of the rational closure of a defeasible TBox presented in this paper.

Finally, Britz et al. [3] present the modal counterpart of our notions of preferential reasoning and rational closure, illustrated by examples from epistemic reasoning and reasoning about actions.

7 Conclusion and Future Work

The main contribution of this paper is the provision of a natural and intuitive formal semantics for preferential and rational subsumption for the description logic $\mathcal{ALC}$. We claim that our semantics provides the foundation for extending preferential reasoning in at least three ways. Firstly, as we have seen in Section 5, it allows for the 'lifting' of preferential entailment and rational closure from the propositional case to the case for $\mathcal{ALC}$. Without the semantics such a lifting may be possible in principle, but will be very hard to prove formally. Secondly, it paves the way for defining similar results for other DLs, as well as other similarly structured logics, such as logics of action and belief [3]. And thirdly, it provides the tools to tighten up the versions of preferential and rational subsumption for $\mathcal{ALC}$ presented in this paper in order to truly move beyond the propositional. The latter point is the obvious one to pursue first when it comes to future work.

Acknowledgments. This work was partially funded by Project number 247601, Net2: Network for Enabling Networked Knowledge, from the FP7-PEOPLE-2009-IRSES call.

References

1. Baader, F., Calvanese, D., McGuinness, D., Nardi, D., Patel-Schneider, P.: The Description Logic Handbook, 2nd edn. Cambridge (2007)
2. Britz, K., Heidema, J., Meyer, T.: Semantic preferential subsumption. In: Proc. of KR, pp. 476–484. AAAI Press/MIT Press (2008)
3. Britz, K., Meyer, T., Varzinczak, I.: Preferential reasoning for modal logic. In: Proc. of Methods for Modalities (2011)
4. Casini, G., Straccia, U.: Rational Closure for Defeasible Description Logics. In: Janhunen, T., Niemelä, I. (eds.) JELIA 2010. LNCS, vol. 6341, pp. 77–90. Springer, Heidelberg (2010)
5. Chellas, B.: Modal logic: An introduction. Cambridge University Press (1980)
6. Freund, M.: Preferential reasoning in the perspective of Poole default logic. Artificial Intelligence 98, 209–235 (1998)
7. Giordano, L., Olivetti, N., Gliozzi, V., Pozzato, G.L.: $\mathcal{ALC} + T$: a preferential extension of description logics. Fundamenta Informaticae 96(3), 341–372 (2009)
8. Kraus, S., Lehmann, D., Magidor, M.: Nonmonotonic reasoning, preferential models and cumulative logics. Artificial Intelligence 44, 167–207 (1990)
9. Lehmann, D., Magidor, M.: Preferential logics: the predicate calculus case. In: Proc. of TARK, pp. 57–72 (1990)
10. Lehmann, D., Magidor, M.: What does a conditional knowledge base entail? Artificial Intelligence 55, 1–60 (1992)
11. Poole, D.: A logical framework for default reasoning. Artificial Intelligence 36, 27–47 (1988)
12. Quantz, J.: A Preference Semantics for Defaults in Terminological Logics. In: Proc. of KR, pp. 294–305 (1992)
13. Quantz, J., Ryan, M.: Preferential Default Description Logics. Tech. rep., TU Berlin (1993), `http://www.tu-berlin.de/fileadmin/fg53/KIT-Reports/r110.pdf`
14. Schild, K.: A correspondence theory for terminological logics: Preliminary report. In: Proc. of IJCAI, pp. 466–471 (1991)

From Approximate Clausal Reasoning to Problem Hardness

David Rajaratnam and Maurice Pagnucco

ARC Centre of Excellence for Autonomous Systems, School of Computer Science and Engineering, The University of New South Wales, Sydney, Australia
{daver,morri}@cse.unsw.edu.au

Abstract. Approximate propositional logics provide a response to the intractability of classical inference for the modelling and construction of resource-bounded agents. They allow the degree of logical soundness (or completeness) to be balanced against the agent's resource limitations.

We develop a logical semantics, based on a restriction to Finger's *logics of limited bivalence* [5], and establish the adequacy of a clausal tableau based proof theory with respect to this semantics. This system is shown to characterise DPLL with restricted branching, providing a clear path for the adaptation of DPLL-based satisfiability solvers to approximate reasoning. Furthermore it provides insights into the traditional notion of problem hardness, as we show that the parameter set of these logics correspond to the *strong backdoor* for an unsatisfiable problem.

1 Introduction

Research into propositional satisfiability (SAT) and approximate propositional logics represent two distinct but related areas of research. SAT is generally focused on the development of efficient reasoning techniques and understanding the characteristics of computationally hard problems. In contrast, accepting that some problems are inherently intractable (unless $\mathbf{P} = \mathbf{NP}$), approximate logics provide techniques for modelling and constructing resource-bounded agents.

This difference is highlighted in how the two approaches deal with a logical problem that exceeds the available computational resources of a reasoner. Because a SAT problem can only be (classically) satisfiable or unsatisfiable, when a traditional SAT reasoner fails to yield a definite result, for example due to exceeeding some predetermined time constraint, there is little semantic information that can be gleaned from this failure. In contrast, approximate logics are parameterised such that the failure to determine a classical result can still yield useful semantic information. For example, it can provide local knowledge-base consistency when modelling the beliefs of resource-bounded agents [2].

A detailed motivation for approximate logics is beyond the scope of this paper, and the interested reader is referred to [1] for a more comprehensive discussion. Rather the intention here is to show that the two research areas should be viewed as complementary and that there exist close theoretical linkages that have remained largely unexplored.

D. Wang and M. Reynolds (Eds.): AI 2011, LNAI 7106, pp. 501–510, 2011.

To develop these linkages we consider two families of sound approximate logics: Cadoli and Schaerf's S-3 logics [1], and Finger's *logics of limited bivalence* [5]. As SAT research is largely focused on clausal reasoning, the developments of this paper are similarly focused towards providing a detailed examination of approximate reasoning in the clausal case.

The main contribution of this paper is to show that a sub-class of the logics of limited bivalence provides a logical characterisation of the well-known DPLL algorithm [3], when branching is restricted to a subset of the propositional variables (RDPLL). Furthermore, we show that the parameter set of a sound approximate logic corresponds exactly to the notion of a *strong backdoor* for an unsatisfiable problem.

These results are of dual significance. Firstly, they provide a path for translating SAT research to the approximate reasoning domain. For example, allowing for the adaptation of SAT solvers as approximate reasoners. Secondly, it provides insight into the notion of problem hardness. *Strong backdoors* have previously been shown to be strongly correlated to problem hardness. Consequently, their correspondence to approximate logic parameter sets provides for a generalisation of the notion of problem hardness established in terms of logical semantics and not just an algorithmic property dependent on particular proof methods.

2 Background

We consider an underlying propositional language $\mathcal{L}$ over a set of propositional variables (atoms) $\mathcal{P}$. We consider the full propositional language, as well as negation normal form (NNF) and conjunctive normal form (CNF) restrictions. The positive and negative literals for a variable are referred to as *conjugates*, with $\bar{l}$ denoting the conjugate of l. This notion is extended to arbitrary formulae in the obvious manner. Furthermore, we use $atoms(\alpha)$ to denote the set of propositional variables occurring in the formula α.

2.1 S-3 Logics

We consider the S-3 logics [1] primarily as a point of reference. Consequently, we provide only a brief overview of their construction. An S-3 logic is defined in terms of a *parameter set* S of atoms, and mappings from literals to truth values *true* and *false*. The most salient aspect of the semantics is that an atom and its conjugate are forced to map to opposite truth values only for atoms in the parameter set, while literals for atoms outside the parameter set can both map to *true* at the same time. This is the basis of an *S-3-interpretation*, which is then extended to arbitrary NNF formulae, and the concepts of *S-3-satisfiable* and *S-3-entailment* (written $\models^3_S$) are defined in the expected manner.

Example 1. Consider a set of formulae $\Gamma = \{\neg c \vee g, \neg g \vee m, \neg m \vee v\}$ and $S = \{g\}$. Now $\Gamma \models^3_S \neg c \vee m$, as $g \in S$ and hence g and $\neg g$ must map to opposite truth values. However $\Gamma \not\models^3_S \neg c \vee v$ (although $\Gamma \models \neg c \vee v$), as m and $\neg m$ can be simultaneously *true* under an S-3-interpretation where $m \notin S$.

A significant property of these logics is the monotonicity of S-3-entailment with respect to increasing parameter sets. In particular, when the parameter set includes all atoms the resulting S-3 logic becomes classical.

Theorem 1 ([1]). *For CNF formula Γ and clause γ, and sets of atoms S and S' s.t. $S \subseteq S' \subseteq \mathcal{P}$, if $\Gamma \models_S^3 \gamma$ then $\Gamma \models_{S'}^3 \gamma$ (hence $\Gamma \models \gamma$) .*

2.2 Logics of Limited Bivalence

The intuition for the logics of limited bivalence ($LB(\Sigma)$) comes from restricting the branching rule (i.e., the *principle of bivalence* (PB)) of a KE-tableau [11]. Historically, the development of these logics has proceeded in two parts. An initial presentation provided only a semantics [5], while a later version provided a (different) semantics and proof theory [4]. We only consider the earlier semantics, as the latter does not provide the necessary clausal correspondence.

The parameter set for the $LB(\Sigma)$ logics consists of a set Σ of propositional formulae. This contrasts with the use of propositional variables for the S-3 logics. Furthermore, Σ must be *closed under formula formation*; meaning that if $\alpha \in \Sigma$ then $\neg\alpha \in \Sigma$, and if $\alpha, \beta \in \Sigma$ then $\alpha \vee \beta, \alpha \wedge \beta, \alpha \rightarrow \beta \in \Sigma$.

The $LB(\Sigma)$ logics are based on a three-level lattice $(L, \sqcap, \sqcup, 0, 1)$, such that L is the countable set of elements $\{0, 1, \epsilon_0, \epsilon_1, \epsilon_2, \ldots\}$, $\sqcup$ is the *least upper bound*, and $\sqcap$ is the *greatest lower bound* (Figure 1 (a)). The partial ordering relation $\sqsubseteq$ is defined as follows: $a \sqsubseteq b$ iff $a \sqcup b = b$ iff $a \sqcap b = a$. $\sqsubseteq$ is then a partial ordering over elements in L such that $0 \sqsubseteq \epsilon_i \sqsubseteq 1$ for every $i < \omega$ and $\epsilon_i \not\sqsubseteq \epsilon_j$ for $i \neq j$. The values 1 and 0 can be referred to as *true* and *false* respectively, while the ϵ_i values are called *neutral*. Furthermore, a *converse operator* $\sim$ is defined over the lattice such that $\sim 0 = 1$, $\sim 1 = 0$, and $\sim \epsilon_i = \epsilon_i$ for all $i < \omega$ (Figure 1 (b)).

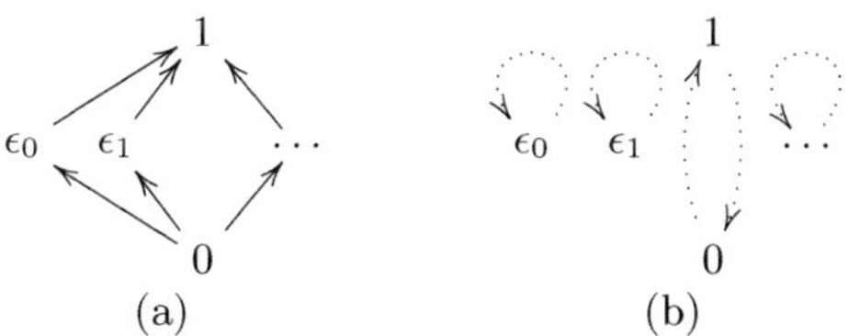

Fig. 1. Lattice semantics (a) and converse operation (b)

An unlimited valuation is a function $v_\Sigma : \mathcal{P} \rightarrow L$ that maps atoms to elements of the lattice. This is extended to arbitrary formulae, $v_\Sigma : \mathcal{L} \rightarrow L$ as follows:

$$\begin{aligned} v_\Sigma(\neg\alpha) &= \sim v_\Sigma(\alpha) \\ v_\Sigma(\alpha \wedge \beta) &= v_\Sigma(\alpha) \sqcap v_\Sigma(\beta) \\ v_\Sigma(\alpha \vee \beta) &= v_\Sigma(\alpha) \sqcup v_\Sigma(\beta) \\ v_\Sigma(\alpha \rightarrow \beta) &= (\sim v_\Sigma(\alpha)) \sqcup v_\Sigma(\beta) \end{aligned}$$

Using the notion of an unlimited valuation, a *limited valuation* is defined by adding the restriction that for every formula α:

$$v_\Sigma(\alpha) = 0 \text{ or } v_\Sigma(\alpha) = 1, \text{ if } \alpha \in \Sigma$$

Example 2. Consider a paramater set Σ such that $\{a, \neg a \vee b\} \in \Sigma$ but $b \notin \Sigma$. We can construct a limited valuation v_Σ by setting $v_\Sigma(a) = 0$ and $v_\Sigma(b) = \epsilon_0$. Applying the rules of an unlimited valuation $v_\Sigma(\neg a) = 1$ and hence $v_\Sigma(\neg a \vee b) = 1$. Note that as b is not in the parameter set it can be assigned a neutral value, but only so long as $\neg a \vee b$ results in a classical value (0 or 1).

Notions of satisfiability and entailment are defined in terms of limited valuations. A valuation v_Σ *satisfies* α if and only if $v_\Sigma(\alpha) = 1$; α is said to be *satisfiable.* A set of formulae Γ is satisfied by v_Σ if and only if all its formulae are satisfied by v_Σ. A valuation v_Σ *contradicts* α if and only if $v_\Sigma(\alpha) = 0$. If a formula is neither satisfied nor contradicted by a valuation then it is *neutral* with respect to that valuation. A set of formulae Γ is said to $LB(\Sigma)$-*entail* a formula α (written $\Gamma \models^{LB}_{\Sigma} \alpha$) if and only if for all limited valuations v_Σ that satisfy Γ, v_Σ does not contradict α (i.e., if $v_\Sigma(\Gamma) = 1$ then $v_\Sigma(\alpha) \neq 0$). Note: v_Σ may be neutral with respect to α.

Theorem 2 ([5]). *Given a knowledge base Γ, formula α, and parameter sets of formulae Σ and Σ' s.t. $\Sigma \subseteq \Sigma' \subseteq \mathcal{L}$ and both Σ and Σ' are closed under formula formation, if $\Gamma \models^{LB}_{\Sigma} \alpha$ then $\Gamma \models^{LB}_{\Sigma'} \alpha$ (hence $\Gamma \models \alpha$).*

Theorem 2 establishes the classical soundness of $LB(\Sigma)$ logics, and shows that the approximation to classical logic becomes more accurate as the parameter set approaches the full language. This mirrors the case for the S-3 logics.

2.3 DPLL and Restricted Branching

The Davis-Putnam-Logemann-Loveland (DPLL) algorithm [3] forms the basis of most modern complete SAT solvers. It is commonly formalised with a branching rule (*analytic cut*), and a deterministic rule for *unit propagation*:

$$\frac{\begin{array}{l} A \cup \{l\} \\ \bar{l} \end{array}}{A} \text{(Unit Propagation)} \qquad \frac{}{l \mid \bar{l}} \text{(Analytic Cut)}$$

In DPLL branching can occur on any variable. In contrast, DPLL with *restricted branching* (RDPLL) allows branching only on a restricted set of variables [7,10]. This can be defined by replacing the analytic cut rule with a *restricted cut* rule:

$$\frac{p \in S}{p \mid \neg p} \text{(Restricted Cut)}$$

Importantly, logical completeness is guaranteed when the truth values of the non-branching variables are uniquely determined by the branching variables.

While RDPLL can provide performance improvements through a reduction in the size of the problem search space, theoretical and practical results have shown that the proof system for RDPLL cannot polynomially simulate DPLL [7]. In short, DPLL with restricted branching can in some cases result in an exponentially larger proof than would be needed for unrestricted branching.

2.4 Problem Hardness and Strong Backdoors

Problem hardness is concerned with identifying the characteristics of hard and easy SAT instances. A *strong backdoor* is a set of variables such that under any truth value assignment the resulting formula can be solved in polynomial time. Empirical studies have shown that the size of the minimal strong backdoor is strongly correlated to problem hardness [13,8].

We adopt the formalisation of a strong backdoor from [15]. For a set of clauses Γ, let $\Gamma[v/x]$ denote the set obtained from Γ by setting the value of variable v to x. Let $a_S : S \subseteq \mathcal{P} \to \{1, 0\}$ be a partial assignment, and let $\Gamma[a_S]$ denote the set of clauses obtained by setting the variables defined in a_S for the set Γ. A *sub-solver* is used to solve tractable sub-cases, subject to the conditions:

(i) it either rejects the input or determines Γ correctly.
(ii) it runs in polynomial time.
(iii) it determines if Γ is trivially *true* or trivially *false*.
(iv) if it determines Γ then it determines $\Gamma[v/x]$ for any variable v and Boolean value x.

Definition 1 (strong backdoor). *Let A be a sub-solver and Γ be a set of clauses. Then, a set $S \subseteq \mathcal{P}$ is a* strong backdoor *in Γ for A if for all $a_S : S \to \{1, 0\}$, A returns a satisfying assignment or concludes unsatisfiability of $\Gamma[a_S]$.*

We highlight two particular sub-solvers. The *trivial sub-solver* performs no problem simplification; if Γ contains an empty clause then Γ is unsatisfiable, if Γ is empty then Γ is satisfiable, otherwise the problem is rejected. In contrast, the *unit propagation sub-solver* either determines the satisfiability or unsatisfiability of Γ through unit propagation, or rejects Γ. This latter sub-solver is the most widely used for studying backdoors [8,9].

3 Clausal $LB(\Sigma)$ Reasoning

In this section we consider the $LB(\Sigma)$ logics in the context of clausal reasoning. We define a logical sub-class based on an added restriction to the parameter set. A clausal tableau based proof theory is then defined and shown to be sound and complete for clausal reasoning with respect to the restricted semantics.

3.1 $LB(\Sigma^*)$ Semantics

We consider the case where the parameter set Σ is not only closed under formula formation, but is also *downwardly saturated* under formula formation. That is, for every formula in the parameter set, every subformulae of that formula is also in the parameter set (i.e., if $\alpha \in \Sigma$ then for every subformula α' of α, $\alpha' \in \Sigma$). Hence, every atom of every formula in the parameter set will itself be a member of the parameter set. This sub-class is referred to as $LB(\Sigma^*)$ semantics.

This restricted semantics allows for a reduction theorem that shows that $LB(\Sigma^*)$ entailment for an arbitrary parameter set is equivalent to $LB(\Sigma^*)$ entailment with an empty parameter set provided that a number of tautological clauses of the form $p \vee \neg p$ are added to the knowledge base.

Theorem 3 (reduction). *Given a parameter set Σ^*, closed under formula formation and downwardly saturated, a knowledge base Γ, and a query α, let Σ^T denote the set of disjunctive formulae constructed from the conjugate pairs of atoms in Σ that are also atoms in either Γ or α; that is $\Sigma^T = \{p \vee \neg p \mid p \in \Sigma^*$ and either $p \in atoms(\Gamma)$ or $p \in atoms(\alpha)\}$. For the empty parameter set $\emptyset$ then $\Gamma \models^{LB}_{\Sigma^*} \alpha$ iff $\Sigma^T \cup \Gamma \models^{LB}_{\emptyset} \alpha$.*

Proof. Follows from observing that every $LB(\Sigma^*)$-valuation that maps Γ to 1 is also an $LB(\emptyset)$-valuation that maps $\Sigma^T \cup \Gamma$ to 1 and vice-versa.

Theorem 3 is particularly relevant to clausal reasoning. Given a clausal knowledge base, a query, and an arbitrary Σ^* parameter set, the reduction theorem establishes that we only need to consider the propositional atoms in Σ^* in order to perform reasoning. And since only clauses are being added to the knowledge base (i.e., $p \vee \neg p$ for each $p \in \Sigma^*$) therefore pure clausal reasoning is maintained.

3.2 Clausal Tableau

As a convention, we use the uppercase Roman characters, A and B, with or without subscripts, to represent clauses, and lower case letters k and l to represent literals. As is common practice, we extend the notion of a *conjugate* to *signed* formulae to refer to the opposite sign for some formula (i.e., $\mathbf{T}\alpha$ is the conjugate of $\mathbf{F}\alpha$ and vice-versa).

$$\frac{\mathbf{T}\ A \cup \{l\} \quad \mathbf{F}\ l}{\mathbf{T}\ A}(\mathrm{T}\vee) \qquad \frac{\mathbf{F}\ A}{\mathbf{F}\ l,\ \forall l \in A}(\mathrm{F}\vee)$$

$$\frac{\mathbf{T}\ l}{\mathbf{F}\ \bar{l}}(\mathrm{T}^-) \qquad \frac{\mathbf{F}\ l}{\mathbf{T}\ \bar{l}}(\mathrm{F}^-)$$

$$\frac{\mathbf{T}\ \{l, \bar{l}\}}{\mathbf{T}\ l \mid \mathbf{F}\ l}(\mathrm{T}_{l\vee\bar{l}}) \qquad \frac{p \in S}{\mathbf{T}\ p \mid \mathbf{F}\ p}(\mathrm{LPB(S)})$$

Fig. 2. $CKELB(S)$ clausal tableau rules

Figure 2 provides the signed clausal tableau rules for the system $CKELB(S)$ (the name representing that it provides Clausal KE-tableau based rules for the Limited Bivalence semantics with a parameter set S). The system consists of four linear expansion rules and two branching rules. (LPB(S)) and $(\mathrm{T}_{l\vee\bar{l}})$ are the only rules that allow the splitting of a branch. For the (LPB(S)) rule to be applied for some atom p two conditions must be met: p must be a member of S, and the subformula property must be satisfied (i.e., p must be a subformula of a formula further up the branch of the tableau).

Tableau properties are defined in the usual manner; a branch is *closed* if it contains a pair of conjugate literals, and *open* otherwise. It is *saturated* if no rule can be applied to produce a new formula in the branch. A tableau is closed if all its branches are closed and saturated if all its branches are saturated.

Definition 2. *Let S be a set of atoms, and $A_1, \ldots, A_n$ and B be clauses. A proof of consequence $A_1, \ldots, A_n \vdash_S^{CKELB} B$ is a closed $CKELB(S)$ tableau with the items $TA_1, \ldots, TA_n, FB$ at the head of the tableau.*

Soundness and completeness of the $CKELB(S)$-tableau with respect to clausal $LB(\Sigma^*)$ semantics is established in a similar manner to [4]. This is encapsulated in the following adequacy theorem.

Theorem 4 (adequacy). *Let Σ^* be a parameter set that is closed under formula formation and downwardly saturated, and S be a set of atoms s.t. $p \in S$ iff $p \in \Sigma^*$. For any set of clauses Γ and clause γ then $\Gamma \vdash_S^{CKELB} \gamma$ iff $\Gamma \models_{\Sigma^*}^{LB} \gamma$.*

Proof. The notion of a valuation is extended to signed formulae, from which soundness of each tableau rule can be observed. Completeness is proven by considering the contrapositive. Namely, that if there is an open saturated branch of the tableau, then the set of clauses must be satisfiable.

Complexity of Reasoning. Importantly, when tautological clauses are allowed entailment checking for $LB(\Sigma)$ is no better than the classical case. We can see this by first establishing the following classical equivalence theorem.

Theorem 5 (classical equivalence). *Given a knowledge base Γ, let Γ^T denote the set of conjugate paired tautological clauses of all atoms of formulae in Γ; that is $\Gamma^T = \{p \vee \neg p \mid p \in atoms(\beta) \text{ where } \beta \in \Gamma\}$. For any parameter set Σ, and any formula α then $\Gamma \models \alpha$ iff $\Gamma^T \cup \Gamma \models_{\Sigma}^{LB} \alpha$.*

Proof. Follows from observing that every classical valuation that maps Γ to 1 is a $LB(\Sigma)$-valuation that maps $\Gamma^T \cup \Gamma$ to 1 and vice-versa.

From Theorem 5 classical entailment can be transformed to determining $LB(\Sigma)$ entailment through the addition of a set of tautological clauses. Furthermore, constructing this set can be trivially undertaken in linear time in the size of the knowledge base. This establishes the **coNP**-hardness of $LB(\Sigma)$ entailment, showing that it is theoretically no easier than classical entailment.

In the context of clausal $LB(\Sigma^*)$ reasoning this result can be seen to follow from the unrestricted branching of the ($T_{l \vee \bar{l}}$) rule. Potentially, the tableau may have to split on every atom, resulting in an exponential number of branches.

However, where a knowledge base is restricted to non-tautological clauses then it is simple to observe that the ($T_{l \vee \bar{l}}$) rule can never be triggered. Consequently, the only possibility for branching of the tableau is when the (LPB(S)) rule is applied. In this case, the potential number of branches is exponentially bounded by the size of the parameter set. This offers the possibility of balancing the complexity of reasoning against the completeness of the deductive process.

4 Characterising RDPLL

As shown earlier, RDPLL tableau consists of two rules: unit propagation, and restricted cut. Both have a clear correspondence to the $CKELB(S)$-tableau (Figure 2). The (T$\vee$) rule is simply a signed version of unit propagation, while (LPB(S)) is restricted cut. Similarly, the (T$^-$) and (F$^-$) rules ensure the existence of the signed positive and negative versions of a literal and therefore fit the RDPLL framework. Finally, when used for testing entailment, the classical relationship between satisfiability and entailment is employed to convert the negation of the query into a conjunction of negated literals. Consequently, an unsigned version of (F$\vee$) is implicitly assumed in the practical use of RDPLL.

The only clear difference between the RDPLL and $CKELB(S)$ is the ($\mathrm{T}_{l \vee \bar{l}}$) rule. However, this difference is of little practical importance, as most SAT solvers parse the input to remove tautological clauses. Consequently, for such cases $LB(\Sigma^*)$ semantics does indeed provide a logical characterisation of the soundness of the RDPLL algorithm.

Proof Limitations. RDPLL has primarily been considered, and abandoned, as an optimisation strategy for SAT solvers. Nevertheless, the results from this research are still highly relevant. For example, proof complexity results showing that RDPLL cannot polynomially simulate DPLL [7,10] apply equally to approximate reasoning. In practice this means that there may be cases where an approximate reasoner provides a result more slowly than a classical reasoner.

Despite this observation, the goal of approximate reasoning is not to find a shortest proof but is instead to provide computationally bounded, well-defined logical behaviour. Determining the degree of any practical loss in performance from the optimal would be a consideration for future research.

5 Approximate Logics to Problem Hardness

In this section we examine the relationship between clausal approximate logics and problem hardness. We show that the notion of a strong backdoor of an unsatisfiable problem corresponds precisely to the parameter set for the $LB(\Sigma^*)$ and S-3 families of approximate logics. The relationship between parameter sets and strong backdoors is established separately for each logical family.

Theorem 6. *Given a classically unsatisfiable set of non-tautological clauses Γ and set $S \subseteq \mathcal{P}$, then Γ is S-3-unsatisfiable iff S is a strong backdoor for Γ using the trivial sub-solver.*

Proof. The proof follows from two observations. Firstly, that Γ is 3-unsatisfiable (i.e., the S-3-logic where $S = \emptyset$) iff the trivial sub-solver determines that Γ is unsatisfiable. Secondly, for any assignment a_S of variables in S then the S-3-unsatisfiability of $\Gamma[a_S]$ reduces to determining the 3-unsatisfiability of $\Gamma[a_S]$.

Theorem 7. *Given a parameter set Σ^* that is closed under formula formation and downwardly saturated, let $S \subseteq \mathcal{P}$ contain all the atoms in Σ^*. A classically* *unsatisfiable set of non-tautological clauses Γ is unsatisfiable under $LB(\Sigma^*)$ semantics iff S is a strong backdoor for Γ using the unit propagation sub-solver.*

Proof. Both proof directions follow from constructing a $CKELB(S)$ tableau where the (LPB(S)) rule is applied before any other. There will be exactly $2^{|S|}$ branches, each corresponding to an assignment a_S over all elements of S. The tableau will be closed iff every $\Gamma[a_S]$ is determined to be unsatisfiable by the unit propagation sub-solver.

Theorems 6 and 7 establish an equivalence between the parameter set of an approximate logic and the strong backdoors for the corresponding sub-solvers. This is significant for a number of reasons. Firstly, it is particularly important to the study of approximate logics as it allows existing research into backdoors and problem hardness to be applied directly to approximate reasoning. For example, existing results establishing the intractability of finding minimal strong backdoors [8,14] means that finding minimal parameter sets is also intractable. Furthermore, heuristics for finding backdoors can be translated directly into the approximate logic domain as heuristics for finding parameter sets. As a consequence, these heuristics can be contrasted against existing parameter set construction strategies, such as the syntactic relevance, graph, and incremental tableau based methods proposed in [2], [12], and [6] respectively.

Finally, these results are also of interest for providing a logical characterisation of the notion of a strong backdoor for unsatisfiable problems. Such a characterisation provides for a greater degree of generalisation of the concept, allowing equivalence to be established between backdoors defined in terms of different proof methods, provided the proof methods share a common semantics.

6 Summary and Future Research

In this paper we examined the properties of sound approximate reasoning. The $LB(\Sigma^*)$ logics were considered for clausal reasoning and a clausal tableau based proof system was established. Importantly, $LB(\Sigma^*)$ was shown to logically characterise the behaviour of the well-known DPLL algorithm when restrictions are placed on the allowable branching variables (RDPLL).

In characterising sound approximate clausal reasoning a number of important linkages were made to existing SAT research. Firstly, the resource-bounded nature of RDPLL was established, with a worst-case complexity exponentially bounded by the size of the parameter set. Secondly, the set of atoms in the parameter sets of $LB(\Sigma^*)$ and S-3 were shown to logically characterise the notion of a strong backdoor of an unsatisfiable problem. This result allows existing SAT research to be translated to the approximate logic domain. Furthermore, empirical results establishing the strong correlation between minimal strong backdoors and problem hardness can now be understood in terms of a strong correlation between the size of the minimal parameter set and problem hardness.

This research provides a number of avenues for future work. Complete approximate logics should be considered. As the parameter set of a sound approximate logic corresponds to a strong backdoor of an unsatisfiable problem, it would be expected that the parameter set of a complete approximate logic would correspond to a strong backdoor of a satisfiable problem. Together, this would capture the entire notion of a backdoor within an approximate logic setting. Finally, this research opens up practical possibilities for the implementation of resource-bounded agents based on SAT technologies.

Acknowledgements. This material is based on research sponsored by the Air Force Research Laboratory, under agreement number FA2386-10-1-4122. The U.S. Government is authorized to reproduce and distribute reprints for Governmental purposes notwithstanding any copyright notation thereon.[1]

References

1. Cadoli, M.: Tractable Reasoning in Aritificial Intelligence. LNCS (LNAI), vol. 941. Springer, Heidelberg (1995)
2. Chopra, S., Parikh, R., Wassermann, R.: Approximate belief revision. In: Proc. of the Workshop on Language, Logic and Information, WoLLIC (2000)
3. Davis, M., Logemann, G., Loveland, D.: A machine program for theorem-proving. CACM 5(7), 394–397 (1962)
4. Finger, M.: Polynomial approximations of full propositional logic via limited bivalence. In: Proc. 9th European Conf. on Logics in AI, pp. 526–538 (2004)
5. Finger, M.: Towards polynomial approximations of full propositional logic. In: 17th Brazilian Symposium on AI, pp. 11–20 (2004)
6. Finger, M., Wassermann, R.: Tableaux for approximate reasoning. In: Proceedings of the IJCAI 2001 Workshop on Inconsistency in Data and Knowledge, Seattle, WA, USA, August 6-10, pp. 71–79 (2001)
7. Järvisalo, M., Junttila, T.A., Niemelä, I.: Unrestricted vs restricted cut in a tableau method for Boolean circuits. Annals of Math. and AI 44(4), 373–399 (2005)
8. Kilby, P., Slaney, J.K., Thiébaux, S., Walsh, T.: Backbones and backdoors in satisfiability. In: Proc. 20th Nat. Conf. on AI, pp. 1368–1373 (2005)
9. Li, C.M.: A constraint-based approach to narrow search trees for satisfiability. Information Processing Letters 71(2), 75–80 (1999)
10. Liberatore, P.: Complexity results on DPLL and resolution. ACM Trans. on Computational Logic 7(1), 84–107 (2006)
11. Mondadori, M.: Classical analytical deduction. Tech. rep., Annali dell' Università di Ferrara, Nuova Serie, sezione III, Filosofia, Paper, n. 1 (1988)
12. Riani, J., Wassermann, R.: Using relevance to speed up inference. some empirical results. In: 17th Brazilian Symposium on AI, pp. 21–30 (2004)
13. Ruan, Y., Kautz, H.A., Horvitz, E.: The backdoor key: A path to understanding problem hardness. In: Proc. 19th Nat. Conf. on AI, pp. 124–130 (2004)
14. Szeider, S.: Backdoor sets for DLL subsolvers. J. of Automated Reasoning 35(1-3), 73–88 (2005)
15. Williams, R., Gomes, C.P., Selman, B.: Backdoors to typical case complexity. In: Proc. 18th Int. Joint Conf. on AI, pp. 1173–1178 (2003)

[1] The views and conclusions contained herein are those of the authors and should not be interpreted as necessarily representing the official policies or endorsements, either expressed or implied, of the Air Force Research Laboratory or the U.S. Government.

A Logic for Knowledge Flow in Social Networks

Ji Ruan and Michael Thielscher

School of Computer Science & Engineering, The University of New South Wales
{jiruan,mit}@cse.unsw.edu.au

Abstract. In this paper, we develop a formal framework for analysing the flow of information and knowledge through social networks. Specifically, we propose a multi-agent epistemic logic in which we can represent and reason about communicative actions based on social networks and the resulting knowledge and ignorance of agents. We apply this logic to formally analyse the "Revolt or Stay-at-home" problem known from the literature, where social networks play an important role in agents' knowledge acquisition and decision-making. We evaluate our work by proving some mathematical properties of our new logic, including the fact that it generalises the existing Logic of Public Announcement.

1 Introduction

The emergence of online social networks such as Facebook and Twitter has enabled richer and easier interactions among people globally. Research on social networks has a long history and is a very interdisciplinary area with important links to sociology, economics, epidemiology, computer science, and mathematics (see [8,9,1]). It deals with topics such as exchange of information, spread of diseases, trade of goods and services and diffusion of patterns of social behaviours.

This paper focuses on the modelling of knowledge and ignorance within social networks, and its crucial role in decision-making. As a motivating example, consider the two alternative social networks depicted in Fig. 1, involving four agents, Alice, Bob, Cath and Dave. Suppose they are unhappy about their dictatorial government and consider a revolt. Suppose further that each of them thinks that "I will revolt on condition that I know for sure that at least two others in my social network will also revolt; otherwise I will stay at home." Next, everyone posts their thought to their social network (assuming no government agent is watching and the network structure is common knowledge). Under these assumptions, will these four people actually revolt in these two different social networks? In this scenario, which is an adapted version of an example introduced in [5], the social network structures play a key role in spreading people's intentions and their knowledge about other people's knowledge of their intentions. These intentions can be seen as social interaction protocols, which are specifications for carrying out tasks with specific social goals, such as fair division of desirable goods, rational decision making in groups, voting, and so on.

In this paper, we develop a formal framework for analysing the flow of information and knowledge through social networks. More specifically, we propose a

D. Wang and M. Reynolds (Eds.): AI 2011, LNAI 7106, pp. 511–520, 2011.

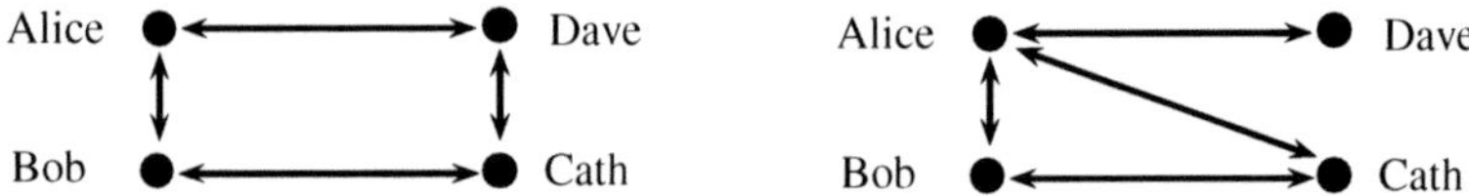

Fig. 1. Two networks structures for the "Revolt or Stay-at-home" scenario

multi-agent epistemic logic in which we can represent and reason about communicative actions based on social networks and the resulting knowledge and ignorance of agents.

The remainder of the paper is organised as follows. Section 2 recalls an existing basic version of Dynamic Epistemic Logic: Public Announcement Logic. Section 3 extends this logic with a social network component. The resulting Social Network Logic is then applied to formally analyse the "Revolt or Stay-at-home" problem. In Section 4 we evaluate our new logic by proving some theoretical results. We conclude in Section 5 with a discussion of related work.

2 Preliminaries

Dynamic Epistemic Logic [3,4,6] studies how actions affect knowledge in a multi-agent setting. Public Announcement Logic (PAL) is an example of such a logic and is an extension of standard multi-agent epistemic logic. We give a concise overview of this logic; intuitive explanations of the epistemic part of the semantics can be found in [7].

Definition 1 (The language of Public Announcement Logic). *Given are a set of agents Ag and a set of atoms At. The* language of Public Announcement Logic $\mathcal{L}_{\mathsf{PAL}}$ *is defined as follows:*

$$\phi \equiv p \mid \neg\phi \mid \phi \wedge \psi \mid K_i\phi \mid C_G\phi \mid \langle\phi\rangle\psi$$

where $p \in \mathsf{At}, i \in Ag, G \subseteq Ag$. For $K_i\phi$, read 'agent i knows ϕ.' For $C_G\phi$, read 'ϕ is common knowledge for the group of agents G.' For $\langle\phi\rangle\psi$, read 'after truthful public announcement of ϕ, formula ψ holds'.

Definition 2 (Epistemic Model). *An* epistemic model *M is a structure $\langle W, \{\sim_i : i \in Ag\}, V\rangle$, where Ag is a set of* agents, *W is a set of* possible worlds, *each $\sim_i \subseteq W \times W$ is an equivalence relation (the* accessibility relation*) for each agent $i \in Ag$, and $V : \mathsf{At} \mapsto 2^W$ is a* valuation function *that assigns each atomic proposition a set of worlds (said to be* true *in those worlds). For model M and world $w \in W$, entailment is defined as follows:*

$M, w \models p$ *iff* $w \in V(p)$;
$M, w \models \neg\phi$ *iff* $M, w \not\models \phi$;
$M, w \models \phi \wedge \psi$ *iff* $M, w \models \phi$ *and* $M, w \models \psi$;
$M, w \models K_i\phi$ *iff for all* w', *if* $w \sim_i w'$ *then* $M, w' \models \phi$;
$M, w \models C_G\phi$ *iff for all* w', *if* $w \sim_G w'$ *then* $M, w' \models \phi$;
$M, w \models \langle\phi\rangle\psi$ *iff* $M, w \models \phi$ *and* $M|\phi, w \models \psi$.

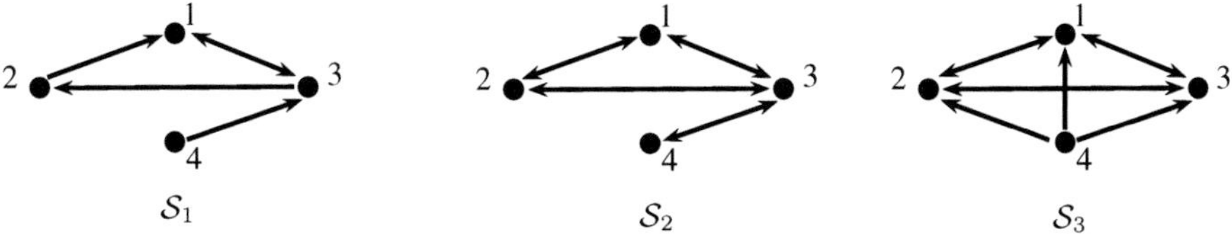

Fig. 2. Example Social Networks

where the group accessibility relation $\sim_G$ *is the transitive and reflexive closure of the union of all accessibility relations for the individuals in* G*:* $\sim_G \equiv (\bigcup_{i \in G} \sim_i)^*$*; and the* restricted *model* $M|\phi = \langle W', \{\sim'_i : i \in Ag\}, V'\rangle$ *is given by*

$$W' = \{w' \in W \mid M, w' \models \phi\};$$
$$\sim'_i = \sim_i \cap (W' \times W');$$
$$V(p') = V(p) \cap W'$$

The modal operator $\langle\phi\rangle$ ('after publicly announcing ϕ') is interpreted as an epistemic state transformer: the model $M|\phi$ is the model M restricted so as to only contain worlds in which ϕ is true. Validity and logical consequence are defined in the standard way. For a proof system, see [6].

3 Social Network Logic

In Public Announcement Logic, all agents have the same source of information, making the logic suitable for modelling epistemic problems such as the famous Muddy Children example [7]. However, the language lacks the possibility to explicitly indicate who made the announcement, or to model the case of informing a subgroup of agents with prior dependencies. We address these shortcomings by extending epistemic models with social networks, and the language with more subtle communication actions.

Definition 3 (Social Network). *A social network* $\mathcal{S}$ *is a tuple* $\langle Ag, F\rangle$*, where* Ag *is a set of agents and* $F \subseteq Ag \times Ag \setminus \{(i,i) \mid i \in Ag\}$ *a binary relation on* Ag *(indicating a specific social relation) among agents.*

Of all the various types of social relations such as friendship, kinship, common interest, or dislike, we are interested in modelling the relations that influence agents' knowledge acquisition. In particular, a social relation F defines information flow among agents as follows: iFj (or $(i,j) \in F$) means that agent i gets information from j. Fig. 2 shows three social networks. The second one is symmetric and the third one is transitive. In the following, we will sometimes use the notation $F_{\mathcal{S}}$ to indicate the social relation belonging to social network $\mathcal{S}$.

We extend an epistemic model with a social network component as follows.

Definition 4 (Social Epistemic Model). *A* social epistemic model E *is a structure* $\langle Ag, \mathcal{S}, M\rangle$*, where* Ag *is a set of* agents*,* $\mathcal{S}$ *is a social network with* Ag*,* M *is a multi-agent epistemic model with agents* Ag*.*

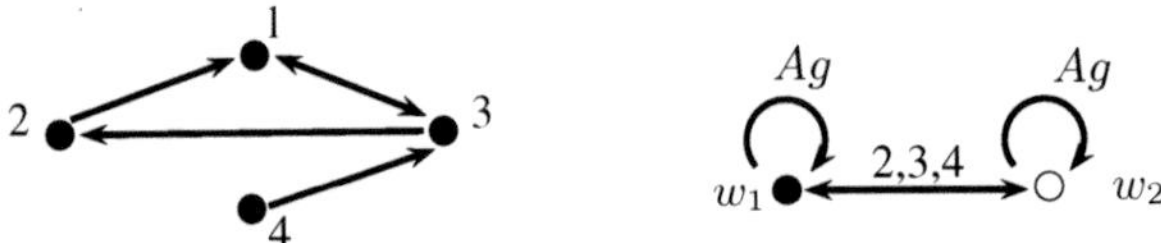

Fig. 3. Example social epistemic mode. Property p is true in w_1 and false in w_2.

The common part shared by $\mathcal{S}$ and M is Ag. In this paper, we assume that $\mathcal{S}$ is common knowledge among Ag (see Section 5 for a discussion). Fig. 3 shows a social epistemic model where $Ag = \{1, 2, 3, 4\}$, the social network is $\mathcal{S}_1$ from Fig. 2, atomic proposition p is true in world w_1 and false in world w_2, and only agent 1 can distinguish w_1 from w_2 (as indicated by the link between w_1, w_2 being labelled '2,3,4', that is, these three agents cannot distinguish w_1 from w_2.)

The social network and knowledge of agents can be changed by their actions. We want to capture two kinds of actions:

1. **Network Actions:** *Follow* an agent, *Unfollow* an agent.
   Such actions change the social network structure. Suppose $(i, j) \notin F$. If agent i acts to follow agent j, then the result will be $(i, j) \in F$. Take, say, $\mathcal{S}_1$ from Fig. 2, then executing the three actions '1: Follow 2', '2: Follow 3' and '3: Follow 4' results in $\mathcal{S}_2$ in Fig. 2. Action 'Unfollow' has the reverse effect.
2. **Message Actions:** *Post* a message ϕ.
   Such actions change agents' knowledge. The effect of agent i posting a message ϕ is that all the agents that follow i will know ϕ. We assume that agents post only messages known to them, that is, both ϕ and $K_i\phi$ are true.

Definition 5 (Social Epistemic Language). *The language of Social Network Logic $\mathcal{L}_{\mathsf{SNL}}$ is defined as follows:*

$$\phi \equiv p \mid f_{(i,j)} \mid \neg\phi \mid \phi \wedge \psi \mid K_i\phi \mid C_G\phi \mid \langle\pi\rangle\phi$$
$$\pi \equiv i : \mathsf{Fo}(j) \mid i : \mathsf{uFo}(j) \mid i : \phi$$

where $p \in \mathsf{At}$; $f_{(i,j)} \in \mathsf{SAt}$; $i, j \in Ag$; and $G \subseteq Ag$.

Unlike the language for Public Announcement Logic, our language has a special kind of atomic propositions $f_{(i,j)}$, indicating that agent i follows j, and dynamic modalities π for social network changing and message posting. Action '$i : \mathsf{Fo}(j)$' means that agent i acts to follow agent j; action '$i : \mathsf{uFo}(j)$' means that agent i acts to unfollow j; and action '$i : \phi$' means that agent i posts message ϕ. In terms of the epistemic model, the effect of the last action is to limit all i's followers' access to worlds in which ϕ is true. To define the precise meaning of $i : \phi$, we adapt the concept of action models from Dynamic Epistemic Logic [6] as follows.

Definition 6 (Action Model). *Given a social network $\mathcal{S}$, an action model $A^{\mathcal{S}}_{(k,\phi)}$ for agent k posting message ϕ in $\mathcal{S}$ is a structure*

$$\langle \{a_1, a_2\}, \{\sim_i : i \in Ag\}, pre\rangle$$

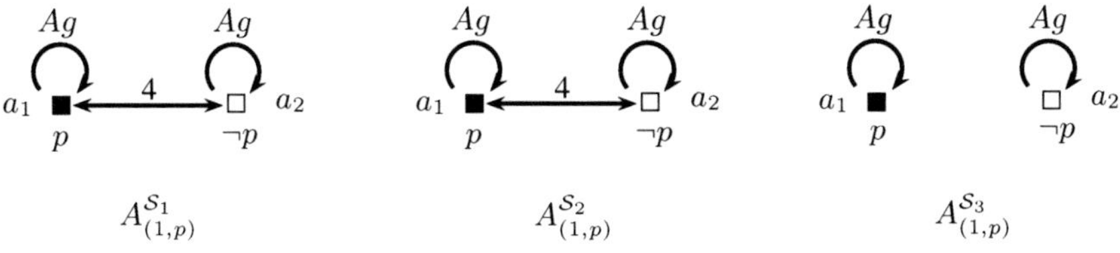

Fig. 4. Example Action Models. $\mathcal{S}_1, \mathcal{S}_2$ and $\mathcal{S}_3$ correspond to those in Fig. 2.

where a_1, a_2 are two atomic actions; $pre(a_1) = \phi$ and $pre(a_2) = \neg\phi$; and $\sim_i \subseteq \{a_1, a_2\}^2$, for which we distinguish two cases: if $i = k$ or $iF_{\mathcal{S}}k$, then $\sim_i = \{(a_1, a_1), (a_2, a_2)\}$; otherwise, $\sim_i = \{(a_1, a_1), (a_1, a_2), (a_2, a_1), (a_2, a_2)\}$.

An action model is somewhat similar to an epistemic model: an atomic action represents a possible action that can be executed by agents, and the accessibility relation $\sim_i$ expresses the uncertainty of agent i about which action has been executed. But instead of having a valuation function, an action model features a function *pre* that assigns a precondition to each atomic action. In order to be executable in an epistemic state w, an action's precondition must be satisfied in that state, which in our case means that if an agent wants to send message ϕ, then the agent must know ϕ. Fig. 4 shows three examples of action models derived from agent 1 announcing p in $\mathcal{S}_1$, $\mathcal{S}_2$, and $\mathcal{S}_3$, respectively, of Fig. 2. It is interesting to note that $A^{\mathcal{S}_1}_{(1,p)}$ and $A^{\mathcal{S}_2}_{(1,p)}$ are identical, while $A^{\mathcal{S}_3}_{(1,p)}$ is equivalent to a public announcement of p, since all other agents follow agent 1.

We can now formally define entailment for $\mathcal{L}_{\mathsf{SNL}}$ wrt. social epistemic models.

Definition 7 (Semantics). *Given a social epistemic model $E = \langle Ag, \mathcal{S}, M\rangle$ and a social epistemic formula $\phi \in \mathcal{L}_{\mathsf{SNL}}$, the entailment relation $\models$ is defined by*

$E, w \models p$ *iff* $w \in V(p)$;
$E, w \models f_{(i,j)}$ *iff* $iF_{\mathcal{S}}j$;
$E, w \models \neg\phi$ *iff* $E, w \not\models \phi$;
$E, w \models \phi \wedge \psi$ *iff* $E, w \models \phi$ *and* $E, w \models \psi$;
$E, w \models K_i\phi$ *iff for all* w', *if* $w \sim_i w'$ *then* $E, w' \models \phi$;
$E, w \models C_G\phi$ *iff for all* w', *if* $w \sim_G w'$ *then* $E, w' \models \phi$;
$E, w \models \langle i : \mathsf{Fo}(j)\rangle\phi$ *iff* $E', w \models \phi$ *where* $E' = \langle Ag, \mathcal{S}', M\rangle$ *and* $F'_{\mathcal{S}'} = F_{\mathcal{S}} \cup \{(i, j)\}$;
$E, w \models \langle i : \mathsf{uFo}(j)\rangle\phi$ *iff* $E', w \models \phi$ *where* $E' = \langle Ag, \mathcal{S}', M\rangle$ *and* $F'_{\mathcal{S}'} = F_{\mathcal{S}} \setminus \{(i, j)\}$;
$E, w \models \langle i : \phi\rangle\psi$ *iff* $E, w \models K_i\phi$ *and* $E', (w, a_1) \models \psi$ *where* $E' = \langle Ag, \mathcal{S}, M\rangle \otimes A^{\mathcal{S}}_{(i,\phi)}$.

The update operation $\otimes$ is given as $\langle Ag, \mathcal{S}, M\rangle \otimes A^{\mathcal{S}}_{(i,\phi)} \equiv \langle Ag, \mathcal{S}, M'\rangle$, where $W_{M'} = \{(w, a) \mid M, w \models pre(a), a \in \{a_1, a_2\}\}$; $(w, a) \sim_i (w', a')$ iff $w \sim_i w'$ and $a \sim_i a'$; and $(w, a) \in V_{M'}(p)$ iff $w \in V_M(p)$.

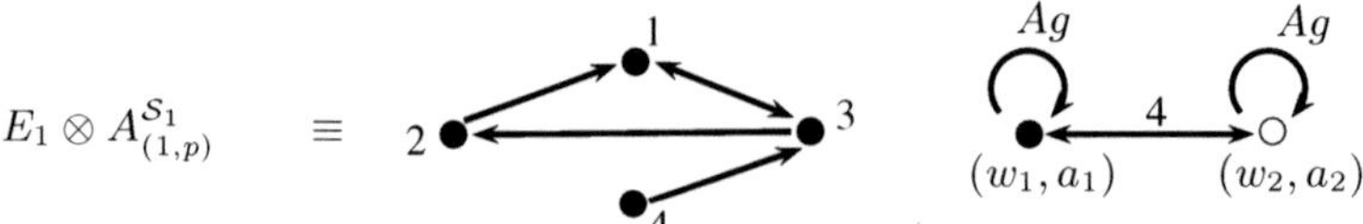

Fig. 5. Example Update Result

The definition of the update operation $\otimes$ essentially follows from [3]. The main difference is that the action models in our approach are constructed based on social networks. Intuitively $\otimes$ takes a social epistemic model and an action mode to produce a new social epistemic model. Agent i cannot distinguish two new worlds (w, a) and (w', a') if i cannot distinguish the world w from w' nor the action a from a'.

The other logical connectives $\vee, \rightarrow$ can be defined as usual. If $E, w \models \phi$ for all E, w, then ϕ is valid, written as $\models \phi$. To illustrate the update operation, Fig. 5 gives the result of updating E_1 from Fig. 3 with $A^{S_1}_{(1,p)}$ from Fig. 4. In the resulting model, the social network remains the same and p becomes common knowledge for agents 1,2,3, but agent 4 is still ignorant about p. Formally,

$$E_1, w_1 \models \langle 1 : p \rangle (C_{\{1,2,3\}} p \wedge \neg K_4 p).$$

Example: Revolt or Stay-at-home. Our new Social Network Logic provides a framework for formally analysing the motivating example from the introduction. Let $p_{(A,x)}$ represent the proposition "Agent $A(lice)$ will revolt if at least x of her friends intend to revolt." It suffices to consider $x \in \{0, 1, 2, 3, 4\}$, where $x = 0$ means A will definitely revolt and $x = 4$ means A will definitely stay at home, given that there are only three other agents in this example. Each possible collection of the thresholds for every agent constitutes a possible world. We describe a world by tuples like 0123, which means that agent A's threshold is 0, B's is 1, C's is 2, and D's is 3. There are $5^4 = 625$ different possible worlds in total. Initially, each agent only knows about his or her own threshold. The accessibility relation for agent A is defined as $(ijkl \sim_A i'j'k'l')$ iff $i = i'$, for agent B as $(ijkl \sim_B i'j'k'l')$ iff $j = j'$, and similar for agents C and D. Valuation V is defined naturally as $V(p_{(A,x)}) \equiv \{ijkl \mid i = x\}$ and similar for the other atomic propositions. Let E^L_{init} denote the initial social epistemic model with the network on the left-hand side of Fig. 1; E^R_{init} the initial social epistemic model with the network on the right-hand side. We use E_{init} to refer to either of them. Assume that our agents all have threshold 2 (i.e., world 2222) and only know their own threshold initially. The following (about everyone's knowledge of agent A's threshold) can be formally concluded for both social network structures:

$$E_{init}, 2222 \models K_A p_{(A,2)} \wedge \neg K_B p_{(A,2)} \wedge \neg K_C p_{(A,2)} \wedge \neg K_D p_{(A,2)}.$$

After A posts her threshold, what agents know about $p_{(A,2)}$ starts to diverge:

$$E^L_{init}, 2222 \models \langle A : p_{(A,2)} \rangle (K_A p_{(A,2)} \wedge K_B p_{(A,2)} \wedge \neg K_C p_{(A,2)} \wedge K_D p_{(A,2)}),$$

$$E^R_{init}, 2222 \models \langle A : p_{(A,2)}\rangle(K_A p_{(A,2)} \wedge K_B p_{(A,2)} \wedge K_C p_{(A,2)} \wedge K_D p_{(A,2)}).$$

After all agents have posted their thresholds, in the epistemic model resulting from $(E^L_{init}, 2222)$, agent A knows that C considers it possible that $p_{(A,4)}$ is true:

$$E^L_{init}, 2222 \models \langle A : p_{(A,2)}\rangle\langle B : p_{(B,2)}\rangle\langle C : p_{(C,2)}\rangle\langle D : p_{(D,2)}\rangle(K_A \neg K_C \neg p_{(A,4)}).$$

Hence, agent C considers it possible that A will stay at home. A similar analysis applies to B's ignorance about D's threshold. So from C's perspective, agent B may think it is possible that both A and D will stay at home. Therefore, A will also choose to stay at home. The key condition for them to revolt, namely, that they should have common knowledge about their thresholds among a group of at least three, is not given. Formally, $(E^L_{init}, 2222)$ does *not* entail

$$\langle A : p_{(A,2)}\rangle\langle B : p_{(B,2)}\rangle\langle C : p_{(C,2)}\rangle\langle D : p_{(D,2)}\rangle(C_{\{A,B,C\}}(p_{(A,2)} \wedge p_{(B,2)} \wedge p_{(C,2)})).$$

The situation is different for $(E^R_{init}, 2222)$, where agents A, B, C manage to achieve common knowledge about their thresholds. Formally, $(E^R_{init}, 2222)$ entails

$$\langle A : p_{(A,2)}\rangle\langle B : p_{(B,2)}\rangle\langle C : p_{(C,2)}\rangle\langle D : p_{(D,2)}\rangle(C_{\{A,B,C\}}(p_{(A,2)} \wedge p_{(B,2)} \wedge p_{(C,2)})).$$

Here, agents A, B, C will all revolt while agent D chooses to stay at home.

4 Theoretical Results

Having illustrated how our new logic can be used to formally analyse the flow of knowledge in a social network, we now present some general results. First, we will formally prove that the existing PAL can be obtained as a special case of our SNL by introducing a special agent who knows everything, and a social network in which every agent follows this special agent.

Proposition 1. *Given an epistemic model $M = \langle W, \{\sim_i : i \in Ag\}, V\rangle$, there is a corresponding social epistemic model E s.t. for all public announcements of ϕ,*

$$M, w \models \langle \phi \rangle \psi \quad \textit{iff} \quad E, w \models \langle Announcer : \phi \rangle \psi$$

where Announcer is a special agent in E and ϕ, ψ are formulas in $\mathcal{L}_{\mathsf{PAL}}$.

Proof. We construct $E = \langle Ag', \mathcal{S}, M'\rangle$ from M as follows:

- $Ag' = Ag \cup \{Announcer\}$; special atoms $f_{(i,j)}$ are introduced for $i, j \in Ag'$;
- $\mathcal{S} = \langle Ag', F\rangle$ with $(i, Announcer) \in F$ for all $i \in Ag$;
- M' is exactly the same as M except that $\sim_{Announcer} = \{(w, w) \mid w \in W\}$.

From left to right: Assume $M, w \models \langle \phi \rangle \psi$, then $M, w \models \phi$ and $M|\phi, w \models \psi$. Since $\sim_{Announcer}$ is the identity relation, we have $E, w \models K_{Announcer}\phi$. Action model $A^{\mathcal{S}}_{(Announcer,\phi)}$ has only identity pairs since all agents in Ag follow $Announcer$ (the model is similar to $A^{\mathcal{S}_3}_{(1,p)}$ in Fig. 4). It is easy to show that $M|\phi$ is isomorphic to the epistemic part of $E \otimes A^{\mathcal{S}}_{(Announcer,\phi)}$. By structural induction on ψ, we have $E \otimes A^{\mathcal{S}}_{(Announcer,\phi)}, (w, a_1) \models \psi$. Therefore $E, w \models \langle Announcer : \phi \rangle \psi$ holds as desired. Since this line of argument is reversible, we immediately have the other direction. □

The above result concerns model equivalence, but we can also show equivalence of formula validity.

Proposition 2. *Let a syntactic translation* trs *from* $\mathcal{L}_{\mathsf{PAL}}$ *to* $\mathcal{L}_{\mathsf{SNL}}$ *be given as:*

$$\begin{array}{llll} \mathsf{trs}(p) & \equiv p & \mathsf{trs}(K_i\phi) & \equiv K_i\mathsf{trs}(\phi) \\ \mathsf{trs}(\phi \wedge \psi) & \equiv \mathsf{trs}(\phi) \wedge \mathsf{trs}(\psi) & \mathsf{trs}(C_G\phi) & \equiv C_G\mathsf{trs}(\phi) \\ \mathsf{trs}(\neg\phi) & \equiv \neg\mathsf{trs}(\phi) & \mathsf{trs}(\langle\phi\rangle\psi) & \equiv \langle Announcer : \mathsf{trs}(\phi)\rangle\mathsf{trs}(\psi) \end{array}$$

then $\models_{\mathsf{PAL}} \phi$ *iff* $\models_{\mathsf{SNL}} \bigwedge_{i\in Ag} f_{(i,Announcer)} \rightarrow \mathsf{trs}(\phi)$.

Proof. Assume $\models_{\mathsf{PAL}} \phi$. Take an arbitrary social epistemic model $E = \langle Ag \cup \{Announcer\}, \mathcal{S}, M\rangle$ and a world w. Assume $E, w \models_{\mathsf{SNL}} \bigwedge_{i\in Ag} f_{(i,Announcer)}$. $\mathcal{S}$ is a social network where every agent follows the $Announcer$, and $M, w \models_{\mathsf{PAL}} \phi$ follows from the assumption. By induction on ϕ and using similar reasoning as in Proposition 1, $E, w \models_{\mathsf{SNL}} \mathsf{trs}(\phi)$. Similar for the other direction. □

Our next results are about whether two actions can be executed in a different order and still result in epistemic situations that satisfy the same formulas, that is, $\models \langle\pi_1\rangle\langle\pi_2\rangle\phi \leftrightarrow \langle\pi_2\rangle\langle\pi_1\rangle\phi$. Clearly, this will depend on the type of actions. For the network actions alone, if π_1, π_2 are both of the type of $i : \mathsf{Fo}(j)$ or $i : \mathsf{uFo}(j)$, then the above principle indeed holds: It is easy to verify, for instance, that

$$\models \langle i : \mathsf{Fo}(j)\rangle\langle k : \mathsf{Fo}(l)\rangle\phi \leftrightarrow \langle k : \mathsf{Fo}(l)\rangle\langle i : \mathsf{Fo}(j)\rangle\phi.$$

But clearly we cannot in general mix different structural actions, that is,

$$\not\models \langle i : \mathsf{Fo}(j)\rangle\langle i : \mathsf{uFo}(j)\rangle\phi \leftrightarrow \langle i : \mathsf{uFo}(j)\rangle\langle i : \mathsf{Fo}(j)\rangle\phi.$$

More interesting are cases which also involve message actions:

$$\not\models \langle j : \phi\rangle\langle i : \mathsf{uFo}(j)\rangle\psi \leftrightarrow \langle i : \mathsf{uFo}(j)\rangle\langle j : \phi\rangle\psi.$$

However, for a special class of models where i does not follow j, we do have

$$\models \neg f_{(i,j)} \rightarrow (\langle j : \phi\rangle\langle i : \mathsf{uFo}(j)\rangle\psi \leftrightarrow \langle i : \mathsf{uFo}(j)\rangle\langle j : \phi\rangle\psi).$$

The next result says that if two agents know some propositional facts initially, then no matter in what order they post these facts, the resulting models will satisfy the same formulas (i.e., no formulas can distinguish them).

Proposition 3. *Given propositional formulas* ϕ_1, ϕ_2,

$$\models K_i\phi_1 \wedge K_j\phi_2 \rightarrow (\langle i : \phi_1\rangle\langle j : \phi_2\rangle\psi \leftrightarrow \langle j : \phi_2\rangle\langle i : \phi_1\rangle\psi).$$

Proof. Given an arbitrary epistemic model E and a world w, assume $E, w \models K_i\phi_1 \wedge K_j\phi_2$. From left to right: Assume $E, w \models \langle i : \phi_1\rangle\langle j : \phi_2\rangle\psi$. It follows that $E, w \models K_i\phi_1$ and $E \otimes A_{(i,\phi_1)}, (w, a_1) \models \langle j : \phi_2\rangle\psi$; hence, $E \otimes A_{(i,\phi_1)}, (w, a_1) \models K_j\phi_2$ and $E \otimes A_{(i,\phi_1)} \otimes A_{(j,\phi_2)}, ((w, a_1), a_1) \models \psi$. There is an isomorphism between $E \otimes A_{(i,\phi_1)} \otimes A_{(j,\phi_2)}$ and $E \otimes A_{(j,\phi_2)} \otimes A_{(i,\phi_1)}$ by mapping every $((w, a), b)$ to $((w, b), a)$. It is easy to show that $E \otimes A_{(j,\phi_2)} \otimes A_{(i,\phi_1)}, ((w, a_1), a_1) \models \psi$, and $E \otimes A_{(j,\phi_2)}, (w, a_1) \models K_i\phi_1$ (as $E, w \models K_i\phi_1$ and ϕ_1, ϕ_2 are propositional). We then have $E, w \models \langle j : \phi_2\rangle\langle i : \phi_1\rangle\psi$. The other direction is similar. □

It is worth noting that this may not hold if we lift the restriction on ϕ_1, ϕ_2 to be propositional (i.e., without modal operator). Suppose, for example, a situation E, w where jFi and hFi along with $K_i p$ and $K_j \neg K_h p$ hold. We then have $E, w \models \langle j : \neg K_h p\rangle\langle i : p\rangle\top$, but not $E, w \models \langle i : p\rangle\langle j : \neg K_h p\rangle\top$.

Finally, we consider the question whether two message actions have the same effect as one. Again this depends on the social network. While it is not true in general it does hold under certain conditions. Let us consider one result for transitive social networks, i.e., where $\bigwedge_{i,j,h\in Ag}(f_{(i,j)} \wedge f_{(j,h)} \to f_{(i,h)})$.

Proposition 4. *Given propositional formula ϕ and two agents $m, n \in Ag$:*

$$\models f_{(m,n)} \wedge \bigwedge_{i,j,h\in Ag} (f_{(i,j)} \wedge f_{(j,h)} \to f_{(i,h)}) \to (\langle n : \phi\rangle\psi \leftrightarrow \langle n : \phi\rangle\langle m : \phi\rangle\psi).$$

Proof. Given a social epistemic model $E = \langle Ag, \mathcal{S}, M\rangle$, assume $E, w \models f_{(m,n)} \wedge \bigwedge_{i,j,h\in Ag}(f_{(i,j)} \wedge f_{(j,h)} \to f_{(i,h)})$. This ensures that $\mathcal{S}$ is transitive and mFn. From left to right: Suppose $E, w \models \langle n : \phi\rangle\psi$. So it is the case that $E, w \models K_n\phi$ and $E \otimes A_{(n,\phi)}, (w, a_1) \models \psi$. From transitivity and mFn, we know that $\{h \mid hFm\} \subseteq \{h \mid hFn\}$, that is, all followers of m are also followers of n. Since ϕ is propositional, we can show that $E \otimes A_{(n,\phi)}$ and $E \otimes A_{(n,\phi)} \otimes A_{(m,\phi)}$ are isomorphic. It follows that $E \otimes A_{(n,\phi)} \otimes A_{(m,\phi)}, ((w, a_1), a_1) \models \psi$, and hence $E, w \models \langle n : \phi\rangle\langle m : \phi\rangle\psi$. The other direction follows by reverse reasoning. □

5 Conclusion

We have introduced a logic for reasoning about knowledge and change in social networks that generalises Public Announcement Logic by an information flow network. For further research, we intend to find an axiomatic system to characterise communications in social networks, and to study the computational complexity of verifying formulas for given social epistemic models.

We conclude with a short discussion of related work. Ruan [12] gives a logic of private message passing in which a message expression $CC_G\phi$ denotes a private message ϕ being sent to group G (similar to email Carbon Copying). Roelofsen [11] proposes a more general logic by introducing the notion of communication channels among groups G_1 and G_2, and message actions by which group G_1 sends a message ϕ to group G_2. Seligman et al. [13] propose a Facebook Logic that has an explicit social network as part of a possible world and where the social relations are assumed to be symmetric. All of these frameworks use a dynamic epistemic semantics based on [3] or [4].

Pacuit and Parikh [10] give a logic of communication graphs by introducing a commonly known, static, directed graph, which explicitly represents the communication links between individual agents, and a temporal expression $\Diamond\phi$, which represents that "after some communications, ϕ becomes true". They use a history-based semantics rather than dynamic epistemic semantics of [3,4]. Apt et al. [2] and Wang et al. [15] also use a history-based semantics, but the communication structure (named hypergraph) is formed on groups: A message from an agent is received by all members of the same group. Both [10] and [2] limit the message contents to atomic propositions, and message actions are specified only in the history model, while [15] has a richer language to represent both message contents and actions. Sietsma and van Eijck in their recent paper [14] propose a framework for message passing that combines the dynamic epistemic semantics and history-based approaches.

Using a dynamic epistemic semantics, our work is in line with [12,11,13]. Different from [12,11], our communication channels are explicitly represented in social networks, and we link individual agents rather than groups. Different from [13], our social network does not need to be symmetric and is not part of a possible world; in addition, we have actions that changes network structures. Our work shares with [10] the assumption that the social network relations are commonly known by the agents. This assumption might be too strong for real-life social networks in which the social relationships between agents are quite complex and highly context-dependent. However, we can generalise our approach in a similar style as [13] by making a social network as part of a possible world.

References

1. Abraham, A., Hassanien, A.E., Snasel, V. (eds.): Computational Social Network Analysis. Springer, London (2010)
2. Apt, K.R., Witzel, A., Zvesper, J.A.: Common knowledge in interaction structures. In: Heifetz, A. (ed.) Proc. of TARK, pp. 4–13 (2009)
3. Baltag, A., Moss, L.: Logics for epistemic programs. Synthese 139, 165–224 (2004), Knowledge, Rationality & Action, 1–60
4. van Benthem, J., van Eijck, J., Kooi, B.P.: Logics of communication and change. Inf. Comput. 204(11), 1620–1662 (2006)
5. Chwe, M.S.Y.: Structure and strategy in collective action. American Journal of Sociology 105(1), 128–156 (1999)
6. van Ditmarsch, H., van der Hoek, W., Kooi, B.: Dynamic Epistemic Logic, Synthese Library, vol. 337. Springer, Heidelberg (2007)
7. Fagin, R., Halpern, J.Y., Moses, Y., Vardi, M.Y.: Reasoning About Knowledge. MIT Press (1995)
8. Freeman, L.: The Development of Social Network Analysis: A Study in the Sociology of Science. Empirical Press, Vancouver (2004)
9. Jackson, M.: Social and Economic Networks. Princeton University Press (2008)
10. Pacuit, E., Parikh, R.: Reasoning about communication graphs. In: van Benthem, J., Loewe, B., Gabbay, D. (eds.) Interactive Logic: Games and Social Software (2007)
11. Roelofsen, F.: Exploring Logical Perspectives on Distributed Information and its Dynamics. Master's thesis, ILLC, University of Amsterdam (2005)
12. Ruan, J.: Exploring the Update Universe. Master's thesis, ILLC Publications, Master of Logic Thesis Series, University of Amsterdam (2004)
13. Seligman, J., Liu, F., Girard, P.: Logic in the Community. In: Banerjee, M., Seth, A. (eds.) Logic and Its Applications. LNCS, vol. 6521, pp. 178–188. Springer, Heidelberg (2011)
14. Sietsma, F., van Eijck, J.: Message passing in a dynamic epistemic logic setting. In: Apt, K.R. (ed.) Proc. of TARK, pp. 212–220. ACM (2011)
15. Wang, Y., Sietsma, F., van Eijck, J.: Logic of Information Flow on Communication Channels. In: Omicini, A., Sardina, S., Vasconcelos, W. (eds.) DALT 2010. LNCS, vol. 6619, pp. 130–147. Springer, Heidelberg (2011)

Object Detection by Admissible Region Search

Xiaoming Chen[1], Senjian An[1], Wanquan Liu[1], and Wanqing Li[2]

[1] Department of Computing, Curtin University, Perth, Australia
[2] Information and Communication Technology (ICT) Research Institute
University of Wollongong, NSW 2522, Australia

Abstract. Efficient Subwindow Search(ESS) is an effective method for object detection and localization, which adopts a scheme of branch-and-bound to find the global optimum of a quality function from all the possible subimages. Since the number of possible subimage is $O(n^4)$ for an images with $n \times n$ resolution, the time complexity of ESS ranges from $O(n^2)$ to $O(n^4)$. In other words, ESS is equivalent to the exhaustive search in the worst case. In this paper, we propose a new method named Adimissible Region Search(ARS) for detecting and localizing the object with arbitrary shape in an image. Compared with the sliding window methods using ESS, ARS has two advantages: firstly, the time complexity is quadratic and stable so that it is more suitable to process large resolution images; secondly, the admissible region is adaptable to match the real shape of the target object and thus more suitable to represent the object. The experimental results on *PASCAL VOC 2006* demonstrate that the proposed method is much faster than the ESS method on average.

1 Introduction

Object detection is a highly active research topic in computer vision society due to an increasing number of practical applications. The task is to find objects in an image automatically and separate them from the background [6][7][9][10][11]. The *Sliding window* approaches [12][13][14]domonstrate the outstanding effectiveness of finding objects from an image. In these approaches, we first train a classifier as a quality function based on the extracted features of training images such as SIFT[1], SURF[5] and so on, next we apply the quality function to all possible subimages within an image and find out the one with maximum value. However, the number of possible subimages rises quadratically with the number of image pixels, so the time cost of the exhaustive search is $O(n^4)$ for the images with $n \times n$ resolution. Recently, a iterative brand and bound method for finding the optimal subwindow is developed, called *Efficient Subwindow Search*(*ESS*)[3]. If the object's quality value is much higher than those of other subimages, ESS can find the globally optimal bounding box quickly; otherwise, the iteration number can be large and ESS can be very slow. In [17], two faster subwindow search methods called *I-ESS* and *A-ESS* were proposed and the upper bound of time complexity was reduced to $O(n^3)$. A-ESS is the faster one but it can not guarantee that the global optimal bounding box can be found.

D. Wang and M. Reynolds (Eds.): AI 2011, LNAI 7106, pp. 521–530, 2011.

In this paper, we propose to detect object by searching for the optimal admissible region, which maximizes the quality value among all possible admissible regions instead of the bounding box in the image. An admissible region is a connected *x-monotone region*, which requires that the whole area is connected and the columns are vertically connected but not require the rows are connected. Therefore, the admissible region is a natural description of the shape of objects whose parts are vertically connected but may not be horizontally connected. For example, in the image of a cow standing on the ground, the legs are vertically connected but not horizontally connected. Comparing with the bounding box which are both vertically and horizontally connected, the admissible region is more natural and less restrictive for description of objects in images. In most images, the object is usually standing on the ground and the parts of the object in each column are usually connected. However, the parts of object may not be connected in each row. Since the bounding box is a special case of admissible region, we have more possible admissible regions in the image and one may worry that the searching is too computationally expensive. Fortunately, in the case of linear quality function, the object detection problem by admissible region search turns into how to find an admissible region in a matrix with the largest sum of entries and this problem can be solved in $O(n^2)$ time for $n \times n$ matrices [16].

The rest of the paper is organized as follows. In section 2, we overview the method for finding the optimal admissible region in a matrix and analyze its time complexity. Section 3 addresses the Admissible Region Search (ARS) method for object detection and localization. The experimental results are shown in Section 4 in comparison with the ESS method and Section 5 concludes the paper with some discussions.

2 Overview of Admissible Region Search

First of all, we introduce the definition of the admissible region briefly. A two-dimensional region is called an *x-monotone region*, if its intersection with any vertical line is connected. And a region is called an *admissible region* if it is a connected *x-monotone region*. We illustrate the difference between the *admissible region* and the *x-monotone region* in Fig.1. Since each column of the shade areas in both Fig.1(a) and Fig.1(b), is an interval and thus connected, both the two shade areas are *x-monotone regions*. In Fig.1(a), for any two points in the shade, we can find a connective path to link them, so the shade area in Fig.1(a) is an admissible region. In Fig.1(b), the shade area is not connected and thus not an admissible region.

Next, we will overview the admissible search problem on matrices and the quadratic time algorithm to solve this problem.

2.1 Admissible Region Search on Matrices

Given a matrix, the admissible Region Search problem is to find an admissible region in the matrix so that the sum of entries are maximized among all possible

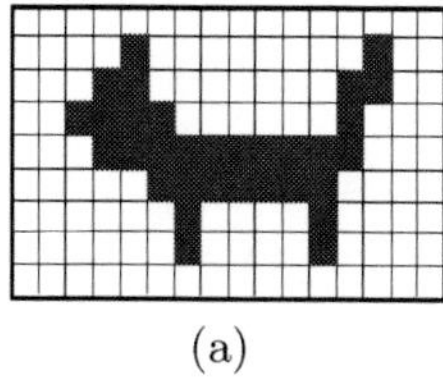

(a)

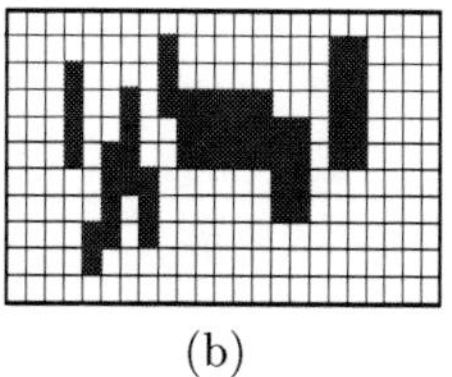

(b)

Fig. 1. Demonstration of admissible region and x-monotone region: (a) an admissible region (b) an x-monotone region but not an admissible region

admissible regions in the matrix. In [16], an quadratic time algorithm was proposed to solve the admissible region search problem . Now we review the main ideas of this algorithm.

Let $A = \{a[i,j]\}$ be an $N \times M$ matrix with real values. $cover_m(i,i')$ is defined as the largest sum of the entries of the mth column $a[\cdot, m]$ amongst all its subarrays that contains the ith and i'^{th} entries, that is

$$cover_m(i,i') \triangleq \max\left\{\sum_{k=i_1}^{i_2} a[k,m] : 1 \leq i_1 \leq i_2 \leq N, i \in [i_1,i_2], i' \in [i_1,i_2]\right\}. \tag{1}$$

$F(i, \leq m)$ is defined as the largest sum of entries of A amongst all admissible regions which includes the position $[i,m]$ and is within the first m columns. More precisely,

$$F(i,\leq m) \triangleq \max\left\{\sum_{[i,j]\in R} a[i,j] : R \text{ is admissible, } R \sqsubset [1:N, 1:m], [i,m] \in R\right\}. \tag{2}$$

The optimal admissible region must be one of the solutions of the above maximization problem for some i and m. Next, We will show that all these maximization problems can be solved in $O(NM)$ time.

It is shown in [16] and is easy to check that

$$F(i,1) = \max_{1\leq j\leq N}\{cover_1(i,j)\} \tag{3}$$

and for $m \geq 1$,

$$F(i,m+1) = \max_{1\leq j\leq N}\{\max(F(j,m),0) + cover_{m+1}(i,j)\}. \tag{4}$$

Define matrix B_1 as

$$B_1[i,j] \triangleq \max_{1\leq j\leq N}\{cover_1(i,j)\}, \tag{5}$$

and define B_m, for $m > 1$, as

$$B_m[i,j] \triangleq \max_{1\leq j\leq N}\{\max(F(j,m),0) + cover_{m+1}(i,j)\}. \tag{6}$$

Then $F(i, m+1)$ is the maximum of B_m's i^{th} row. Fortunately, B_m has the following nice property and its row maxima can be computed in $O(N)$ time.

Lemma 1. *[16] Let B_m be defined as in (5) for $m = 1$ and defined as in (6) for $m > 1$. Then both the upper-triangular and lower-triangular parts of B are totally monotone matrices and all its row maxima can be computed in $O(N+M)$ time by the well-known SMAWK algorithm [2], provided that each entry of B_m can be computed in constant time.*

Next, we show that how to compute $cover_m(i, j)$ in constant time. By introducing the following two indices,

$$bottom_m(i) \triangleq \arg\max_j \{\sum_{k=i}^{j} a[k, m], j \in [i, N]\} \tag{7}$$

$$top_m(i) \triangleq \arg\max_j \{\sum_{k=j}^{i} a[k, m], j \in [1, i]\} \tag{8}$$

$cover_m(i, i')$ can be computed as below [16]

$$\begin{aligned} cover_m(i, i') &= \begin{cases} \sum_{k=top_m(i)}^{bottom_m(i')} a[k, m] & if \quad i \leq i' \\ \sum_{k=top_m(i')}^{buttom_m(i)} a[k, m] & if \quad i' < i \end{cases} \\ &= \begin{cases} \hat{a}[bottom_m(i'), m] - \hat{a}[top_m(i), m] & if \quad i \leq i' \\ \hat{a}[bottom_m(i), m] - \hat{a}[top_m(i'), m] & if \quad i' < i \end{cases} \end{aligned} \tag{9}$$

where $\hat{a}[\cdot, m]$ is the integral of $a[\cdot, m]$, i.e.,

$$\hat{a}[i, m] \triangleq \sum_{k=1}^{i} a[k, m]. \tag{10}$$

Hence $cover_m(i, i')$ can be computed in constant time if $\hat{a}[k, m]$, $bottom_m(i)$ and $top_m(i)$ are obtained in advance for all $i = 1, 2, \cdots, N$. From (10), we have $\hat{a}[1, m] = a[1, m]$ and $\hat{a}[k+1, m] = \hat{a}[k, m] + a[k+1, m]$ for $k > 1$, and therefore $\hat{a}[k, m]$ can be computed in $O(N)$ time for all $k = 1, 2, \cdots, N$. Furthermore, the indices $bottom_m(i)$ and $top_m(i)$, for all $i = 1, 2, \cdots, N$, can also be computed in $O(N)$ time as shown in [16]. In summary, we have.

Lemma 2. *The indices $top_m(i)$, $bottom_m(i)$ and $\hat{a}[i, m]$ for all $i = 1, 2, \ldots, N, m = 1, 2, \cdots, M$ can be computed in $O(NM)$ time.*

So the algorithm proposed in [16] first compute the indices $top_m(i)$, $bottom_m(i)$ and $\hat{a}[i, m]$ for all $i = 1, 2, \ldots, N, m = 1, 2, \cdots, M$. Then apply SMAWK algorithm [2] to compute $F(i, \leq 1)$ for all $i = 1, 2, \cdots, N$. And then, for any $m > 1$, apply SMAWK algorithm to compute recursively $F(i, \leq m+1)$ for all $i = 1, 2, \cdots, N$ based on $F(i, \leq m)$ for all $i = 1, 2, \cdots, N$. All these steps can be done in $O(NM)$ time.

Fig. 2. Object Localization by Admissible Region Search

3 Object Detection by Admissible Region Search

In this paper, we restrict the detected area in which the object is included to be an admissible regions. One of the motivations of using admissible region is that it is less restrictive and more natural for representation of objects in images comparing with rectangle regions and circles. For example, in a picture of a cow standing on the grass as shown as in Fig.2, the cow's front legs and back legs are vertically connected but not horizontally connected, and therefore admissible regions are more suitable to cover the cow than rectangle regions. Another motivation of using admissible region for object detection is that an quadratic time complexity algorithm exists to find the maximum sum of entries of a matrix amongst all possible admissible regions in the matrix as we discussed in Section 2. Later, we will show that when the quality function is linear on the image histograms, the object detection problem can be transformed into an admissible region search problem on matrices and therefore can be solved by the efficient algorithm introduced in Section 2.

Consider the quality function:$f: \mathcal{X} \times \mathcal{Y} \rightarrow I\!R$, where $\mathcal{X}$ is the set of all images and $\mathcal{Y}$ is the set of admissible regions. $f(x,y)$ denotes the quality value of the prediction that an object of the searching class is at the position y in image x. Under the condition of fixed image x, we can simplify $f(x, y)$ by $f(y)$. Then the task of predicting the best location of the object is transformed into an optimization problem as below:

$$y_{opt} = \arg\max_{y \in \mathcal{Y}} f(y) \tag{11}$$

In computer vision community, one usually use the histograms of the extracted features to represent images. First, one needs to extract sufficient large number of features from the training images and cluster them into K bins to make a dictionary. Then, for any image, we count the number of its features in each bin and form a vector which is called a histogram. In object detection by admissible region search, we treat each admissible region of the image as a subimage and this subimage is also represented by its feature histogram. More precisely, for a given admissible region y in an image x, we count the number, denoted by

$h_y(k)$, of the features that is extracted in the area y and belongs to k^{th} cluster bin. The histogram $h_y = (h_y(1), h_y(2), \ldots, h_y(K))^T$ is then used to represent the subimage located at y and the optimization problem (12) for object detection. In this case,

$$y_{opt} = \arg\max_{y \in \mathcal{Y}} f(h_y) \tag{12}$$

Let $H[i,j,k]$ denote the number of k-th bin features extracted at pixel(i,j) $1 \leq i \leq N, 1 \leq j \leq M$. Then $h_y(k) = \sum_{(i,j)\in y} H[i,j,k]$.

Note that $h_y = \sum_{k=1}^{K} e_k h_y(k)$ where e_k is a K-dimensional vector whose k^{th} entry is 1 and other entries are zeros. If $f(h_y)$ is a linear function of h_y, then

$$f(h_y) = f\left(\sum_{k=1}^{K} e_k h_y(k)\right) = \sum_{(i,j)\in y} A[i,j] \tag{13}$$

where matrix A is defined as $A[i,j] = \sum_{k=1}^{K} f(e_k H(i,j,k))$. Hence $f(h_y)$ is the sum of the entries of A within the admissible region y and therefore the optimization problem (13) is transformed to be an admissible search problem on matrices, i.e.,

$$y_{opt} = \arg\max_{y \in \mathcal{Y}} \sum_{(i,j)\in y} A[i,j]. \tag{14}$$

The efficient algorithm for admissible the search problem on matrices can be applied for object detection if the quality function is linear on the image histograms. In a typical support vector machine with linear kernel $f(h_y) = b + \sum_i \alpha_i \langle h_y, h^i \rangle$, where $\langle \cdot, \cdot \rangle$ denotes the scalar product and h^i are histograms of training images. Since removing the constant b does not affect the optimal solution, one can remove b and f is then a linear function of h_y and the proposed algorithm can be applied.

The proposed method is of $O(n^2)$ time complexity for $n \times n$ images. Table 1 shows the time and memory complexity of the proposed method with comparisons to ESS and I-ESS methods.

Table 1. Time Complexity and Memory Requirements for $n \times n$ Images

	Time Complexity		Memory Requirement	
Method	Best Case	Worst Case	Best Case	Worst Case
ARS	$O(n^2)$	$O(n^2)$	$O(n^2)$	$O(n^2)$
ESS	$O(n^2)$	$O(n^4)$	$O(n^2)$	$O(n^4)$
I-ESS	$O(n^2)$	$O(n^3)$	$O(n^2)$	$O(n^2)$

4 Experimental Results

In this section, we report the performance of the proposed method on two data sets with comparison to ESS and I-ESS methods. The first data set is a toy data set that are generated randomly, and the second one is **PASCAL VOC 2006**[1] which contains 5304 nature images from 10 categories in different resolutions. For ESS and I-ESS methods, we used the codes downloaded from the authors' websites. The experiments are conducted on a standard desktop personal computer with Windows XP operating system. The CPU is Intel Xeon E5345 2.33GHZ and the compiler is Visual Studio.NET 2008.

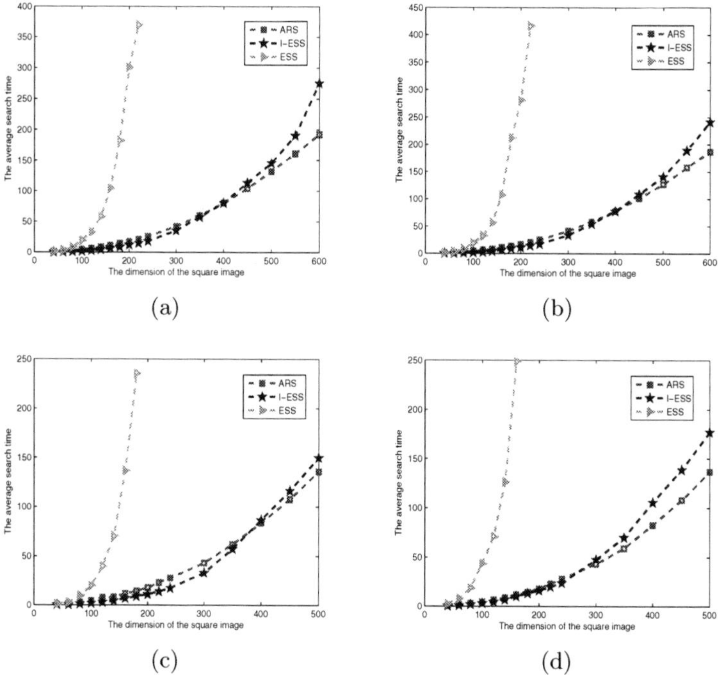

Fig. 3. The average searching time on different image size $n \times n$ and different value of parameter α: (a) $\alpha = 0.001$ (b) $\alpha = 0.01$ (c) $\alpha = 0.1$ (d) $\alpha = 1$. n varies from 40 to 600.

4.1 A Toy Data

The experiments in this section aims to test the scale of the time complexity with varying matrix sizes. We generate 540 random matrices with 18 resolutions of $n \times n$ where n varys from 40 to 600. In each resolution we created 30 matrices each matrix C being the sum of two matrices A and B. A represents the object area of the image. The nonzero area in A represents the position of the object and

[1] `http://pascallin.ecs.soton.ac.uk/challenges/VOC/voc2006/`

the entry in this area is randomly selected in [0, 1]. On the other hand, matrix B represents the background noise and its entries are selected randomly with zero mean and unit variance under Gaussian distribution. We use a weighting coefficient α to control the noise level, so the matrix $C = A + \alpha B$. We selected four values, 0.001, 0,01, 0.1 and 1, for α. At each value, we test ARS method, ESS method and I-ESS method and calculate the average search time on 30 matrices for each resolution. The results are shown in Fig.3. The ARS method scales much better with matrix sizes n than ESS and I-ESS methods.

4.2 PASCAL VOC 2006

In this section, we test the proposed method on *PASCAL VOC 2006* standard data set[2] The data set contains 5304 images and in some images there are more than one object, so there are 9507 objects from 10 categories in all. In the step of feature extraction, we inherit the representation method of images in [17] and use the same feature data and weights for each class of objects.

We compare the performance of the ARS method with methods ESS and I-ESS. There are more than 100 different resolutions in all the 5304 images but only 2 contains more than a thousand of images. Since the variation of performances of ESS and I-ESS are quite large on different images, in order to make the search time distribution reliable, we choose the images of these two resolutions, 489×363 and 626×468, to report the search time distribution and the scales with matrix dimensions.

The average, minimal and maximal search time of the three methods are reported in Table 2. The scales of the average and maximal search time with image dimension are reported in Fig. 4-5. Table 2 shows that ARS is faster than ESS on average but slower than I-ESS. However, from Fig. 4-5, the scale of ARS with image dimension is much better than I-ESS and ESS. For higher resolution images, ARS can be better than I-ESS on average.

In Figures 4-5, we choose the search time for low resolution 489×363 and its dimension, approximated as the square root of 489×363 since it is not square, as references and compute the ratio between the approximate dimensions of the two resolutions and the ratio between their search times. In order to check whether search time scales quadratic or cubic, we also draw the quadratic and cubic curves. For example, if the time complexity of the method is quadratic, the point will be close to the quadratic line. From Fig. 4(a), one can see that the growing rate of ESS's average search time is higher than cubic line while those of ARS and I-ESS are close to quadratic line. From Fig. 4(b), one can see that the growing rate of I-ESS is nearly cubic and that of ESS is more expensive than cubic while that of ARS is still close to the quadratic line. ARS scales the best and is most stable for different images with same sizes.

[2] http://pascallin.ecs.soton.ac.uk/challenges/VOC/voc2006/..

Table 2. Search Time Comparison (L represent 489 × 363 resolution, H represent 626 × 468 resolution)

	CPU Time Cost (millionseconds)					
	Average		Minimum		Maximum	
Method	L	H	L	H	L	H
ARS	98.27	164.78	91.40	154.73	142.40	217.37
ESS	368.19	867.08	10.17	13.62	8691.10	25440
I-ESS	14.90	25.39	4.03	6.29	62.87	128.75

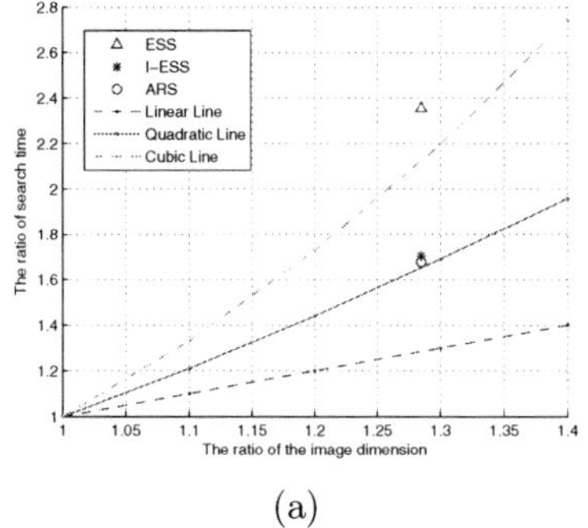

(a)

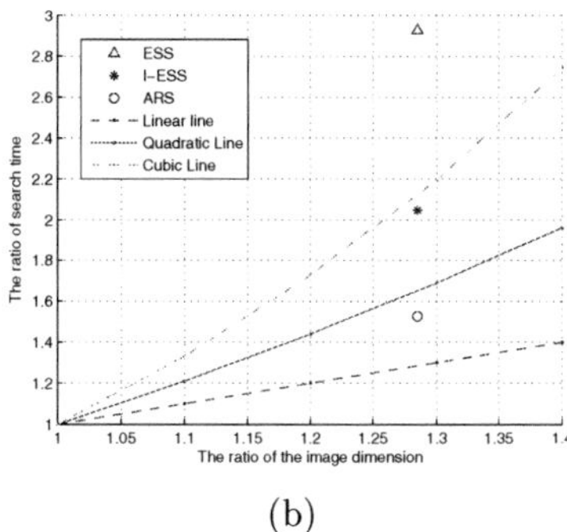

(b)

Fig. 4. (a) is the Average Time Cost Comparison on Images of Two Resolutions 489 × 363 and 626 × 468. (b)is the Maximum Time Cost Comparison on Images of Two Resolutions 489 × 363 and 626 × 468.

5 Conclusion and Discussion

In this paper, we have proposed a new object detection method by using admissible region search. Comparing with the well-known sliding window methods using ESS or I-ESS, the proposed method's running time scales better with matrix sizes and thus faster for high resolution images. Also, admissible region is closer to the shape of real objects in images and has potential to find the shape of the object if the extracted features and quality function are accurate enough.

References

1. Lowe, D.G.: Distinctive image features from scale-invariant keypoints. International Journal of Computer Vision 60(2), 91–110 (2004)
2. Aggarwal, A., Klawe, M.M., Moran, S., Shor, P., Wilber, R.: Geometric applications of a matrix-searching algorithm. Algorithmica 2(1), 195–208 (2007)

3. Lampert, C.H., Blaschko, M.B., Hofmann, T.: Beyond sliding windows: Object localization by efficient subwindow search. In: IEEE Conference on Computer Vision and Pattern Recognition, pp. 1–8 (2008)
4. Lampert, C.H., Blaschko, M.B., Hofmann, T.: Efficient subwindow search: A branch and bound framework for object localization. IEEE Transactions on Pattern Analysis and Machine Intelligence 31(12), 2129–2142 (2009)
5. Bay, H., Tuytelaars, T., Van Gool, L.: SURF: Speeded Up Robust Features. In: Leonardis, A., Bischof, H., Pinz, A. (eds.) ECCV 2006. LNCS, vol. 3951, pp. 404–417. Springer, Heidelberg (2006)
6. Bosch, A., Zisserman, A., Munoz, X.: Representing shape with a spatial pyramid kernel. In: 6th ACM International Conference on Image and Video Retrieval, pp. 401–408 (2007)
7. Ferrari, V., Fevrier, L., Jurie, F., Schmid, C.: Groups of adjacent contour segments for object detection. IEEE Transactions on Pattern Analysis and Machine Intelligence 30(1), 36–51 (2008)
8. Fulkerson, B., Vedaldi, A., Soatto, S.: Localizing Objects with Smart Dictionaries. In: Forsyth, D., Torr, P., Zisserman, A. (eds.) ECCV 2008, Part I. LNCS, vol. 5302, pp. 179–192. Springer, Heidelberg (2008)
9. Rowley, H.A., Baluja, S., Kanade, T.: Human face detection in visual scenes. In: Advances in Neural Information Processing Systems, pp. 875–881 (1996)
10. Fritz, M., Schiele, B.: Decomposition, discovery and detection of visual categories using topic models. In: IEEE Conference on Computer Vision and Pattern Recognition (2008)
11. Opelt, A., Pinz, A., Zisserman, A.: A Boundary-Fragment-Model for Object Detection. In: Leonardis, A., Bischof, H., Pinz, A. (eds.) ECCV 2006. LNCS, vol. 3952, pp. 575–588. Springer, Heidelberg (2006)
12. Chum, O., Zisserman, A.: An exemplar model for learning object classes. In: IEEE Conference on Computer Vision and Pattern Recognition, pp. 1–8 (2007)
13. Dalal, N., Triggs, B.: Histograms of oriented gradients for human detection. In: IEEE Conference on Computer Vision and Pattern Recognition (2005)
14. Viola, P., Jones, M.: Rapid object detection using a boosted cascade of simple features. In: European Conference on Computer Vision (2001)
15. Asano, T., Chen, D.Z., Katoh, N., Tokuyama, T.: Polynomial-time solutions to image segmentation. In: 7th ACM-SIAM Symposium on Discrete Algorithms, pp. 104–113 (1996)
16. Fukuda, T., Morimoto, Y., Morishita, S., Tokuyama, T.: Data mining with optimized two-dimensional association rules. ACM Transactions on Database Systems 26(2), 179–213 (2001)
17. An, S., Peursum, P., Liu, W., Venkatesh, S.: Efficient algorithm for subwindow search in object detection and localization. In: IEEE Conference on Computer Vision and Pattern Recognition (2009)

Structured Light-Based Shape Measurement System of Human Body

Jun Cheng[1,2], Shiguang Zheng[1,2], and Xinyu Wu[1,2]

[1] Shenzhen Institutes of Advanced Technology, Chinese Academy of Sciences, China
{sg.zheng,jun.cheng,xy.wu}@siat.ac.cn
[2] The Chinese University of Hong Kong, Hong Kong, China

Abstract. Recent progress in stereo matching algorithms performance especially belief propagation algorithm encourages us greatly. However, it is difficult to develop a stable, real time and low cost shape measurement system. In this paper, we propose a novel laser-scanning based 3D measurement system to obtain depth information in real time. Efficient Belief Propagation algorithm is adopted to match the correspondent points of laser stripe between calibration image and measurement image. When laser beam is projected onto scene, occlusion problem will occur and influence 3D reconstruction accuracy. Outliers are detected by the ratio of the second highest peak over the highest peak. Experimental results demonstrate the feasibility of proposed approach.

Keywords: Efficient belief propagation, outlier, 3D reconstruction.

1 Introduction

Laser triangulation scanners usually project a stripe (or a single spot) onto the scene, and then use a sensor usually a CCD camera to view the scene. There are many kinds of active triangulation technologies just differ in forming the laser beam [2].Occlusion is one of the biggest challenges in stereo vision. A number of approaches have been developed to deal with it. Egnal and Wildes [3] gave empirical comparisons of five Binocular Half-Occlusion detecting approaches, BMD (Bimodality), MGJ (Match Goodness Jumps), LRC (Left-Right Checking), ORD (Ordering constraint), and OCC (Occlusion constraint). Brown and Burschka [4] found the above distinction fails for many applications when determining whether occlusion is caused by error or other factors (e.g. image noise, lack texture). They classified algorithms for handling occlusion into three categories: methods that detect occlusion [5], methods that reduce sensitivity to occlusion [6] and methods that model the occlusion geometry [7]. Another method of detecting occlusion regions [8] was proposed to exploit multiple cameras. As more images are added, the amount of occlusion will increase since each pixel of reference image will be invisible in more than one supporting camera. In laser triangulation, it is impossible for the stripe to be imaged because of occlusion. Some researchers [9] used another sensor to obtain a second view or project another stripe [10] to reduce the influence of occlusion. A narrower triangulation angel was used to make the system more robust to occlusion [11].

D. Wang and M. Reynolds (Eds.): AI 2011, LNAI 7106, pp. 531–539, 2011.

In order to develop a low cost system that can measure depth information of human body in indoor environments in real time and stably, we project laser strip onto the scene vertically and sweep human body from left to right in temporal succession for a dense reconstruction. We adopt the efficient belief propagation algorithm [1] to extract center of laser lines in the laser image. The spread of intensity values across the stripe conforms to a Gaussian distribution. Because of ambient noise and sensor noise (electrical noise, quantization noise), the camera images a laser stripe with a certain amount of unexpected lighting peak superimposed to it. When the surface exhibits a low reflection index (e.g. hairs of human), occlusion problem occur or the distance between the laser range system and the scene exceeds the maximum range of the system, the noises are significant and laser stripe discontinue. We assume that there are no highlights or specularities in the scene. The peak ratio of intensity value is used to determine whether a point is an outlier or not. The peak ratio is the ratio of the second highest peak over the highest peak. If laser stripe discontinue at one point, the peak ratio of intensity value in row will be very close to one. Otherwise, it will be less than a threshold. Experiments show the effective of proposed method.

We observe that the number of points whose belief vector exhibits unimodal profile starts to fluctuate up and down with increase of iterations. Meanwhile, the depth map is over-smoothed and foreground-fattening effect occurs. It is necessary and reasonable to stop iteration adaptively by counting the number of points whose belief vector conforms to unimodal distribution.

The remainder of this paper is organized as follows. In section 2, we give a general description of our system. Section 3 details the approaches we proposed including stripe detection, outlier detection and our adaptively iteration of belief propagation. Experiment results are given in section 4 and we summarize the results finally.

2 System Overview

The system adopts active triangulation technology to measure depth information of the human body. It consists of a positioning device controlling the movement of laser strip, a CCD camera, a laser line generator and a computer. A single laser stripe is projected onto the scene one time, and then reflected and imaged by a CCD camera. To acquire a dense 3D map of the scene, a positioning device is used to control the horizontal shift of the laser line. The laser sheet sweep the scene from left to right and the next stripe generally appears on the right of the previous one. If it shifts 64 times, a full 3D image of the scene is formed. The picture captured by the CCD camera is 40*384 pixels in one scanning time. After 64 times shift, we acquire 64 frames and synthesis them into one laser image with size of 2560*384. The coordinates of points in each frame can be represented as $I(x, y, t)$. x, y are the horizontal coordinate and vertical coordinate respectively and t is the frame index. All of this above is finished by a pre-processing procedure, shown in Fig. 1.

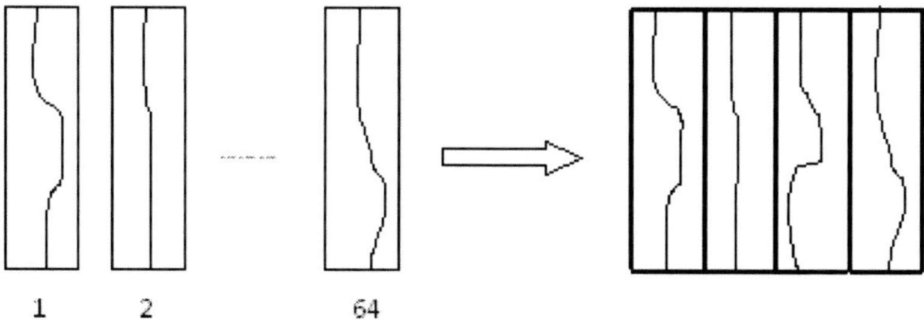

Fig. 1. The pre-processing procedure

3 Proposed Approach

3.1 Calibration

In our system, the focal lens of camera and the distance between the laser and the image sensor are 550 pixels and 70 pixels respectively. These parameters are calculated using approach proposed by Tsai [12] at two meters away from camera. We let a laser line sweep a smooth wall at two meters place from left to right without other object in front of it to extract center of background laser line. Then, we adopt efficient belief propagation algorithm to detect laser stripes and save offset of laser line correspond to the first column in each frame as $bk(y,t)$ for subsequent process where y represents the row and t is the laser line index. The algorithm that we adopt to calculate offsets of background laser lines just differs in cost function compared to subsequent stripe detection. Here, the cost function is defined as follows:

$$C(p, f_p) = 255 - I_P(x), \tag{1}$$

and

$$f_p = x. \tag{2}$$

Where, x is the horizontal coordinate of each pixel $I(x, y, t)$ in the same row with laser point p. $I_p(x)$ is the intensity value of pixel $I(x, y, t)$. $C(p, f_p)$ is the cost of assigning f_p to horizontal position of laser point p in time-succession frames. The efficient belief propagation algorithm will be described later.

3.2 Stripe Detection

Most of the researches adopt peak detection to locate the position of laser stripe. This method fails when signal to noise is low. To obtain accuracy 3D reconstruction result, we use efficient belief propagation algorithm to detect stripes in raw laser image. Global optimization algorithms such as graph cuts and belief propagation perform far superior than local methods, but they are too slow for practical use. Fortunately, Pedro and Daniel [1] presented three technologies to substantially reduce the running time of loopy belief propagation approach. First, they assume each message update is similar to min convolution, and can be computed in linear time. Second, they spit the nodes of grid graph into two sets and the messages of two sets are updated in turn. Finally, they present a multi-grid technology for performing BP in a coarse-to-fine manner. In this paper, we benefit from the first two technologies. The stereo match is formulated as an energy minimization problem. The energy function includes a data term and a smooth term. The data term is the cost of assigning a label to the node and the smooth term measures the discontinuity penalty.

In our system, the frames in Fig. 1 are 40 pixels width and 384 pixels height. For each laser point, $C(p, f_p)$ is the cost of assigning f_p to its horizontal offset corresponding to calibration image and it is defined as follows:

$$C(p,f_p)=\begin{cases}255-I_p(x) & if\ x \geq bk(y,t)\ and\ x<40 \\ 255 & if\ x<bk(y,t)\ and\ x\geq 0\end{cases} \quad (3)$$

and

$$f_p=(x-bk(y,t)+40) \bmod 40 . \quad (4)$$

x is the horizontal coordinate of each pixel $I(x,y,t)$ in time-succession frames. $bk(y,t)$ is the horizontal position of laser point corresponding to p in calibration image. $I_p(x)$ is the intensity value of each pixel $I(x,y,t)$.

We assume that laser strip distorts continuously and a discontinuity penalty function is defined linearly as

$$W(f_p,f_q)=k^* | f_p-f_q |, k=5 . \quad (5)$$

The factor k is determined empirically. Belief propagation is an iterative inference algorithm that propagates messages in parallel. Before message passing, each message $m^0_{p->q}$ is initialized to zero and we update it in the following way,

$$m^t_{p->q}=\min_{f_p}(k^* | f_p-f_q |+h(f_p)) \quad (6)$$

and

$$h(f_p)=C(p,f_p)+\sum_{s\in N(p)\backslash p} m^{t-1}_{s->p}(f_p) . \quad (7)$$

$N(p)\setminus p$ denotes the neighbors of p other than p. The form of Equation (6) is commonly referred to as a min convolution and can be computed in $o(n)$ time using the simple two-pass algorithm detailed described in [1].

We propagate the message in checkerboard pattern rather the standard way. Checkerboard is a board on which the squares are of alternating dark and light color. Thus, you can think the nodes of grid graph are painted with black or white. When iteration number t is odd, we update the messages sent from nodes colored with black and keep the old value for the messages sent from nodes colored with white. When iteration number t is even, we update the message sent from nodes colored with white and keep the old value for the messages sent from nodes colored with black. Thus, we reduce the memory to half as the normal belief propagation and obtain nearly the same performance. After t iterations, a belief vector is computed for each laser point in such a way,

$$b_q(f_q) = C(q, f_q) + \sum_{p \in N(q)} m^t_{p->q}(f_q), \tag{8}$$

and

$$f_q^* = \min_{f_q}(b_q(f_q)). \tag{9}$$

Finally, the label f_q^* that minimizes $b_q(f_q)$ individually at each node is selected and the locations of laser stripe centers are determined.

3.3 Outlier Detection

When we project a plane of light into the scene, the intersection of this plane with the visible surfaces in the scene forms an illuminated light stripe. In theory, the energy pattern of a stripe corresponds to a Gaussian profile. However, ambient noise and sensor noise (electrical noise, quantization noise) are inevitable in practical use so that the camera images a laser stripe with a certain amount of undesired lighting peak superimposed to it. When the object surface exhibits a low reflection index (e.g. hairs of human), occlusion occur or the distance between laser range system and the scene exceeds the maximum range of the system, the reviewing system cannot observe the projected laser light, with no depth data gathered in these regions, these points can be represented as outliers. Meanwhile, the noises are very significant and laser stripe disconnect. Large areas of outliers within depth image will have adverse affects on subsequent processing such as clustering or recognition.

We assume that there are no highlights or specularities in the scene. The peak ratio of intensity value is chose to determine whether a laser point is an outlier or not. The peak ratio is the ratio of the second highest peak over the highest peak. In each frame, if laser stripe discontinue, the peak ratio of intensity value in row will be very close to one. Otherwise, it will be less than a fixed threshold T. This algorithm is expressed as

$$outlier = \begin{cases} true & \dfrac{I_2}{I_1} > T \\ false & \dfrac{I_2}{I_1} < T \end{cases} \tag{10}$$

where I_2 and I_1 are the second highest peak and highest peak of intensity value in row. In our system, the threshold T is determined as 1/2 empirically.

3.4 Adaptively Iteration of Belief Propagation

We observed that the number of points whose belief vector conformed to unimodal distribution started to fluctuate up and down with increase of the iterations. Meanwhile, the depth map is over-smoothed and fatting effect occurs. The iterations algorithm used in different environment is different. A fixed iteration number is not suitable, so it is necessary and reasonable to stop iteration adaptively by counting the numbers of points whose belief vector exhibits unimodal distribution. In our adaptively stop iteration algorithm, outliers are not considered and one more iteration is done at the most with a small number of points are over-smoothed compared to optimal iterations. We set a fixed value 10 as the number max iterations for the system. If the iteration number exceeds this max value, the algorithm will exit loop. We count the number of points whose belief vector exhibits single peak distribution for each iteration. If it meets Equation (11), we stop iteration

$$PeakNum(t) - PeakNum(t-1) < PeakNum(t) * 0.1 \tag{11}$$

Where $PeakNum(t)$ represents the number of points whose belief vector conforms to single peak distribution for current iteration and $PeakNum(t-1)$ is the number of points whose belief vector exhibits unimodal distribution in last iteration. In addition, t is the iteration index.

4 Experiment Results

We arranged BMD (Bimodality) [3] and disparity crosscheck experiments to evaluate our outlier detection algorithm. The test laser raw images are shown in Fig. 2. First, the search window size of BMD is 3*3, which produces a little effect to the result and the peak similarity threshold in this paper is 0.8. Second, you can think crosscheck as a variant of LRC (Left-Right Checking). Here, we simply describe this algorithm. Because the same laser raw image is processed by stripe peak detection algorithm and efficient belief propagation algorithm, the disparity value may differ only in sign except at the points that not be illuminated by laser stripe. If the difference of disparity value at corresponding points exceeds one pixel, this point is an outlier.

Table.1 shows the performance summary of outlier detection result. Of the detected outliers, 81.6% were indeed outliers and 43.7% of the true outliers were found by our proposed approach. Moreover, computation speed of our approach is much faster than that of others. Although the percentage of true outliers found by crosscheck is a little higher than our algorithm, the ratio of correct outliers to detected outliers in our approach is apparently higher than crosscheck. The BMD algorithm performs very well at occlusion border but fails when occlusion region is of primary interests. The BMD and crosscheck algorithm are all processed after obtaining the disparities by efficient belief propagation algorithm. Thus, a large amount of time is spent in computing the disparities.

Fig. 3 shows our depth computation result for the test images shown in Fig.2. The first row is the depth computing result of peak detection. The second row represents the result of our adaptively efficient belief propagation. We give the classic belief propagation result in the last row with three iterations. The text in the image indicates the depth of the central point in the image with mm level precision and the text below the image is the iteration numbers. And the farthest point is colored with blue, following with green, yellow, red and the nearest point is painted with black. It can be observed that our approach can obtain nearly the same performance compared to classic belief propagation when a small number of points are over-smoothed and perform far superior than peak detection especially in noisy area.

Fig. 2. Four laser raw images PC obtained from pre-processor

Table 1. The average outlier detection result

Approach	%Outlier Found	%Outlier Correct	Time (ms)
Our Approach	43.7	81.6	31
BMD	4.8	61.2	94
Crosscheck	44.6	50.8	188

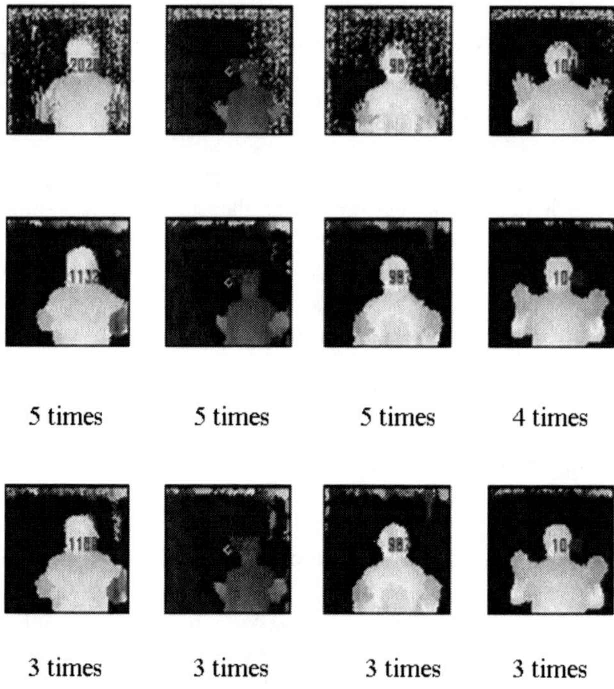

Fig. 3. The result of depth computation

5 Conclusions

A real time laser-scanning based shape measurement system is developed. We adopt efficient belief propagation algorithm to calculate the offsets of laser line between calibration image and measurement image. Experiments show that it performs superior compared to traditional peak detection algorithm. In addition, the outlier detection algorithm proposed in this paper is more effective than BMD and disparity crosscheck when there are no highlights or specularities in the scene. We make Efficient Belief Propagation algorithm stop iteration adaptively by counting the number of points whose belief vector conforms to single peak distribution.

Although this work corroborates that such systems are feasible, there are also some limitations in our system. First, too few point cloud data is obtained in our system. Second, a single stripe is projected one time; it severely limits the data acquisition speed. Third, if there are some specular surfaces or translucid surfaces in the scene, our outlier detection algorithm fails. In future, we will adopt structure light coding technology to speed up the data acquisition speed and project more laser lines to obtain more point cloud for higher resolution. Our outlier detection algorithm directly used the raw laser image obtained from the pre-processor; we can make a post-processing of scanned 3D surface data for stable outlier detection.

Acknowledgment. Special thanks to Key Laboratory of Robotics and Intelligent System of Guangdong Province (2009A060800016), the National Natural Science Foundation of China(60806050), Shenzhen Key Laboratory of Precision Engineering (CXB201005250018A), CAS and Locality Cooperation Projects (ZNGZ-2011-012).

References

1. Felzenszwalb, P.F., Huttenlocher, D.P.: Efficient belief propagation for early vision. In: Proceedings of the IEEE International Conference on Computer Vision and Pattern Recognition, vol. 1, pp. 261–268 (2004)
2. Besl, P.J.: Active, optical range imaging sensors. Machine Vision and Applications 1(2), 127–152 (1988)
3. Egnal, G., Wildes, R.: Detecting binocular half-occlusions: empirical comparisons of five approaches. IEEE Transactions on Pattern Analysis and Machine Intelligence 24(8), 1127–1133 (2002)
4. Brown, M.Z., Burschka, D., Hager, G.D.: Advances in computational stereo. IEEE Transactions on Pattern Analysis and Machine Intelligence 25(8), 993–1008 (2003)
5. Sun, J., Li, Y., Kang, S.B., Shum, H.-Y.: Symmetric stereo matching for occlusion handling. In: Proceedings of the IEEE International Conference on Computer Vision and Pattern Recognition, vol. 2, pp. 399–406 (2005)
6. Sorgi, L., Neri, A.: Bidirectional dynamic programming for stereo matching. In: Proceedings of the IEEE International Conference on Image Processing, pp. 1013–1016 (2006)
7. Min, D.B., Sohn, K.: Stereo matching with asymmetric occlusion handling in weighted least square framework. In: Proceedings of the IEEE International Conference on Acoustics, Speech and Signal Processing, pp. 1061–1064 (2008)
8. Lin, H.Y., Subbarao, M.: Multiple base-angle rotational stereo for accurate 3D model reconstruction. In: Proceedings of the 3rd International Symposium on Image and Signal Processing and Analysis, vol. 2, pp. 931–935 (2003)
9. Wang, L., Bo, M., Gao, J., Ou, C.S.: A novel double triangulation 3D camera design. In: Proceedings of the IEEE International Conference on Information Acquisition, pp. 877–882 (2006)
10. Mavrinac, A., Chen, X., Denzinger, P., Sirizzotti, M.: Calibration of dual laser-based range cameras for reduced occlusion in 3D imaging. In: Proceedings of the IEEE/ASME International Conference on Advanced Intelligent Mechatronics, pp. 79–83 (2010)
11. Fu, W.Y., Qiao, A.K., Fu, P.B.: Boundary identification and triangulation of STL model. In: Proceedings of the International Conference on BioMedical Engineering and Informatics, pp. 500–504 (2008)
12. Tsai, R.Y.: A versatile camera calibration technique for high-accuracy 3D machine vision metrology using off-the-shelf TV cameras and lenses. IEEE Journal of Robotics and Automation 3(4), 323–344 (1987)

Fast Sub-window Search with Square Shape

Antoni Liang, Senjian An, and Wanquan Liu

Department of Computing
Curtin University, WA 6102
Australia

Abstract. Research in this paper is focused to make a change on variety of Efficient Sub-window Search algorithms. A restriction is applied on the sub-window shape from rectangle into square in order to reduce the number of possible sub-windows with an expectation to improve the computation speed. However, this may come with a consequence of accuracy loss. The experiment results on the proposed algorithms were analysed and compared with the performance of the original algorithms to determine whether the speed improvement is significantly large to make the accuracy loss acceptable. It was found that some new algorithms show a good speed improvement while maintaining small accuracy loss. Furthermore, there is an algorithm designed from a combination of a new algorithm and an original algorithm which gains the benefit from both algorithms and produces the best performance among all new algorithms.

Keywords: Sub-window, Object detection, Branch and Bound Search.

1 Introduction

There are many applications which involve sub-window search in order to track the locations of particular objects in an image. For instance, images from surveillance camera can be used to find face position prior to face recognition process. It is expected that the object in the image can be tracked in acceptable time and accuracy. Because of this reason, many investigations have been conducted in order to find efficient approaches to do sub-window search. The idea of sub-window search is to find the location of target object (e.g human) in an image by drawing a bounding box (sub-window) around it (Fig. 1). Sub-window search is really useful in a large number of images because it is done automatically by computers, so it will save time and effort compared to doing it manually by human.

One simple approach to do object detection is the sliding window approach (exhaustive search). The idea is to explore all possibilities of sub-windows in an image and choose the one with the highest score. However, the complexity will be $O(n^4)$ for an image with size n x n in terms of the number of sub-windows. This is not applicable since the computation speed is too slow for a large number of images. There are other approaches which are more efficient to find the optimal sub-window. One of the well-known approaches is called Efficient Sub-window Search (ESS) [4], [5] which uses the branch and bound search to explore only sub-windows with high

D. Wang and M. Reynolds (Eds.): AI 2011, LNAI 7106, pp. 540–549, 2011.

scores. Furthermore, there are some extensions of ESS such as χ^2-ESS (Chi-square ESS) and A-ESS (Alternating ESS) [1-3]. All of these approaches are using rectangle sub-window to show the location of a target object in an image.

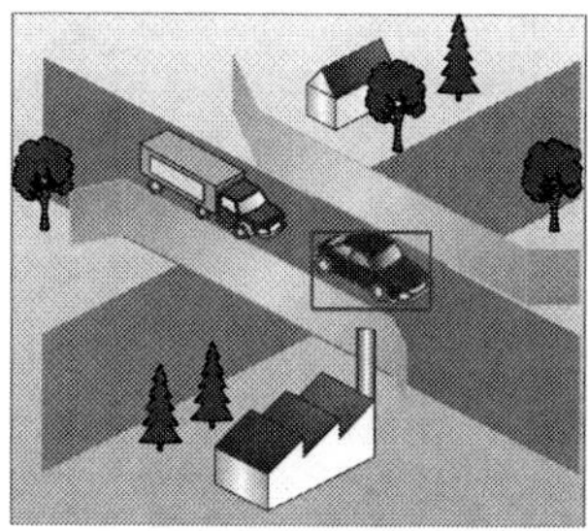

Fig. 1. Bounding box sample on red car in image

A rectangle sub-window can be represented with 4 variables: L, R, T, and B which refer to left, right, top, and bottom coordinates respectively. As a consequence, the complexity of computing the number of possible sub-windows is about $O(n^4)$ for an image with size n x n. However, if the shape of a sub-window has a restriction (e.g square shape), the number of possible sub-windows can be reduced, hence may improve the computation speed. For example, square sub-window can be represented with only 3 variables: X, Y, and S. X and Y are the top left coordinate of the sub-window. S is side length of the sub-window. This paper will discuss the potential of square-shaped sub-window used in sub-window search.

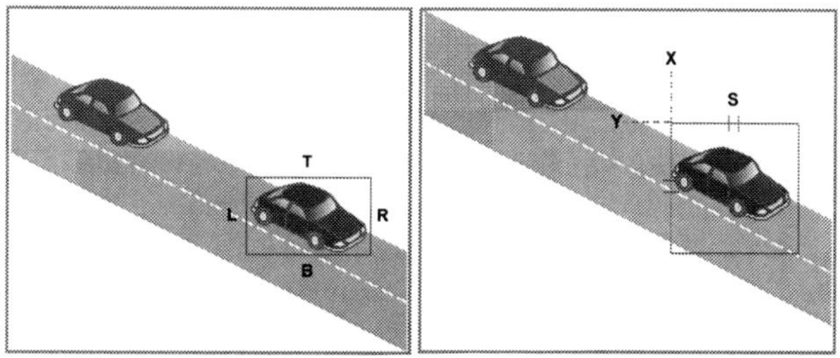

Fig. 2. (a) Rectangle Sub-window (b) Square Sub-window

There are three proposed algorithms discussed in this paper: square ESS, square χ^2-ESS, and combination of square ESS and A-ESS. All these algorithms are using the square sub-window in order to improve the computation speed. However, the challenge is to maintain high accuracy of sub-window search. The contributions of presented in this paper are as following.

1) Square sub-window reduces the number of possible sub-windows in images, hence might reduce the time needed to find the best sub-window. **2)** Square sub-window can simplify the model of sub-window, so only 3 variables are needed to represent it. **3)** The worst complexity of sub-window search can be reduced into $O(n^3)$

instead of O(n^4). The organization of this paper is as follows. Section 2 contains explanations on the existing sub-window search algorithms such as ESS, χ^2-ESS, and A-ESS in more detail. These algorithms are the foundation of the proposed algorithms. Section 3 explains the detail of proposed algorithms based on square-shaped sub-window. The proposed algorithms are square ESS, square χ^2-ESS and combination of square ESS and A-ESS. Section 4 provides brief analysis on results of the experiments on the proposed algorithms. The measurement will be based on the time improvement and accuracy loss. Lastly, section 5 contains the summary of the major contributions and conclusions on this research.

2 Efficient Sub-window Search (ESS) Algorithms

Prior to sub-window search, some pre-processing need to be done. SIFT (Scale Invariant Feature Transform) features [6] are extracted and used in the simulation presented in this paper. In order to reduce the number of classes of features, these features need to be clustered. Each feature has a cluster ID. Based on the cluster ID in a sub-window, a histogram is created. Histogram contains the amount of each cluster. The histograms are classified by using SVM (Support Vector Machine). The idea of classification is to create a function *f(x)* to determine which histogram belongs to target object [7]. The function *f(x)* will have the output of positive score if the input is the histogram that belongs to the target object. On the other hand, the function *f(x)* will have an output of negative score if the input is the histogram which belongs to the non-target object or background. The purpose of sub-window search approach such as ESS is to find the sub-window with the highest score in an image.

2.1 ESS

ESS is designed to find the global optimal sub-window based on a rectangle sub-window [4], [5]. This algorithm uses a branch and bound approach in order to find the globally maximum sub-window. The sub-window range is defined with four parameters: L, R, T, and B to represent left, right, top and bottom positions respectively. The full set of sub-windows is defined as following set.

$$\boldsymbol{W}_{\text{rectangle}} \triangleq \{[\boldsymbol{L},\boldsymbol{R},\boldsymbol{T},\boldsymbol{B}] \mid \boldsymbol{L} \in [\mathbf{1},\boldsymbol{column}], \boldsymbol{R} \in [\mathbf{1},\boldsymbol{column}], \\ \boldsymbol{T} \in [\mathbf{1},\boldsymbol{row}], \boldsymbol{B} \in [\mathbf{1},\boldsymbol{row}], \boldsymbol{L} \leq \boldsymbol{R}, \boldsymbol{T} \leq \boldsymbol{B}\}$$

The idea is to split the sets (L, R, T, B) that has the highest range into half (Fig. 3). As a result, there will be two window sets for each split range (the first half and second half). Each set will have the upper bound computed and put into a priority queue. The set with the highest bound will be put in the head of the priority queue, so it will be split further in the next iteration. This process is repeated until there is a set which only contains one sub-window, i.e., the highest bound (branch and bound search). The complexity of ESS is O(n^4), however it is still fast on average since it only explores window sets with higher score.

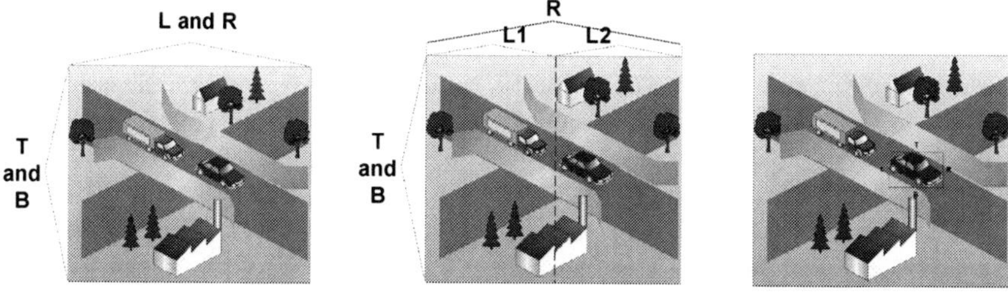

Fig. 3. ESS simulation

2.2 χ^2-ESS (Chi-Square ESS)

This is a non-linear ESS algorithm [2], [4], [5]. Only some special non-linear functions can be used with ESS. In this study, we use the χ^2 distance, between the histogram from sub-window and the histogram of the target object, as the score function. The χ^2 distance is defined as below:

$$\sum_{i=1}^{n} \frac{(H_{target}(i) - H_{window}(i))^2}{H_{target}(i) + H_{window}(i)}$$

n = amount of different cluster indexes

χ^2 distance is used to measure the difference between the histogram from the sub-window and the target object. The difference is scaled down by the cluster size of both sub-window and target object in order to remove the effect of larger histograms. The procedure is the same as ESS where it uses the branch and bound search to find the optimal sub-window. As a result, the complexity of this algorithm is also $O(n^4)$.

2.3 A-ESS (Alternating ESS)

A-ESS (Alternating ESS) is an approximation algorithm to find a *locally* optimal sub-window [1]. The result might not be globally optimal. However, the computation speed is very fast.

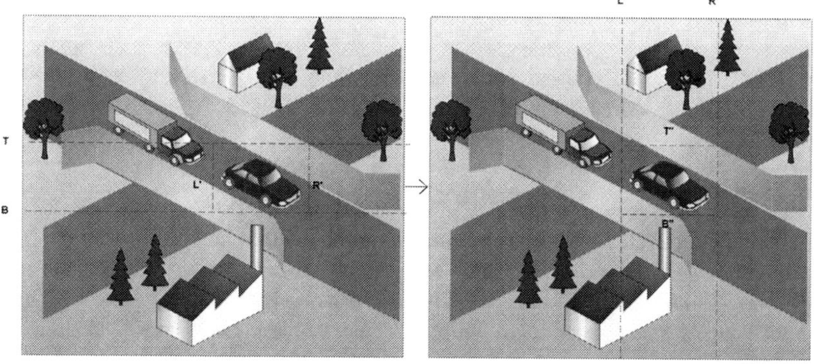

Fig. 4. A-ESS simulation

A-ESS can start from two possibilities: row interval or column interval. The idea of A-ESS is to keep improving the score starting from the initial row interval and then finding the best column interval (L' and R') given that row interval. The next step is to find the best row interval (T'' and B'') given column interval L' and R'. If the score improvement is still significantly large after getting row interval T'' and B'' then the process is repeated (start from T'' and B'') until there is no score improvement (or very small improvement). The complexity of this algorithm is $O(n^3)$.

3 Proposed Algorithms

3.1 Square ESS

This new algorithm has the same basic logic of using a branch and bound method as the original rectangle-based Efficient Sub-window Search (ESS). The change is very explicit with clear intuition. The change is on the set variables from **L, R, T, B** (left, right, top, bottom) to **X, Y, and S** (coordinate (X,Y) and side length) that represents all possible sub-windows.

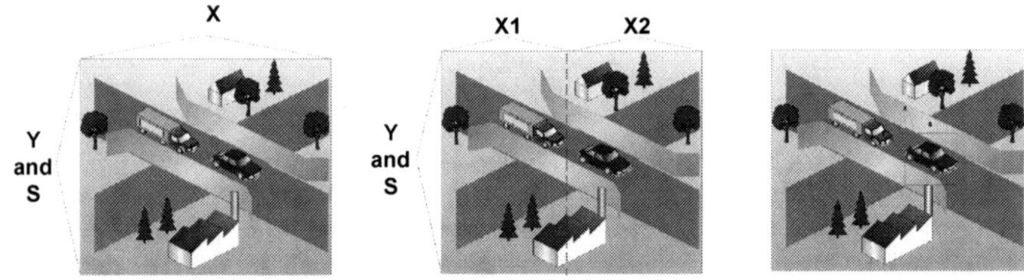

Fig. 5. Square ESS simulation

The idea is to create a *window set* that contains the range of coordinate X, Y and side length S. By having only three variables, there are only three ranges to be split, hence reducing the number of iterations in the loop. Because of the square-shape restriction, a window set that cannot fit any square in its range after splitting will not be put in to the priority queue. Otherwise, calculate the score of the sub-window on the *window set* and put it into the priority queue. This procedure is repeated for the *window set* that has the highest score until only one sub-window fits in the window set range. So, square ESS is guaranteed to find the optimal sub-window for the square-shaped sub-window. (But perhaps little bit less accurate than sub-window from ESS)

3.2 Square χ^2-ESS

The change proposed for this algorithm is similar to the change for square ESS. It does the same procedure with branch and bound search strategy to find the optimal sub-window. The computation of the distance is done by calculating the histogram of the sub-window and compares it with the target histogram using χ^2 distance. This algorithm tries to find the optimal sub-window with the shortest distance with target histogram.

3.3 Combination Square ESS and A-ESS

This is the algorithm to combine the advantages from both high computation speed (expected) from square ESS and accuracy from original A-ESS (and it is also fast) in order to achieve best performance. At the end of square ESS algorithm, the position of the sub-window (left, right, top, and bottom) is passed into the A-ESS algorithm. When A-ESS starts from row interval first, it will use the T (top) and B (bottom) position for the row interval initialisation. It is similar when A-ESS starts from column interval first. The only difference is it uses L (left) and R (right) parameters as the column interval initialisation.

The main idea of this approach is to apply square ESS algorithm on an image to find the approximate location of a target object (because the object might not be covered tight enough) with high computation speed. Square sub-window will be created as a result. The next step is to improve the accuracy of the square sub-window by covering the object tighter by applying A-ESS algorithm on it and produce the final sub-window (Fig. 6).

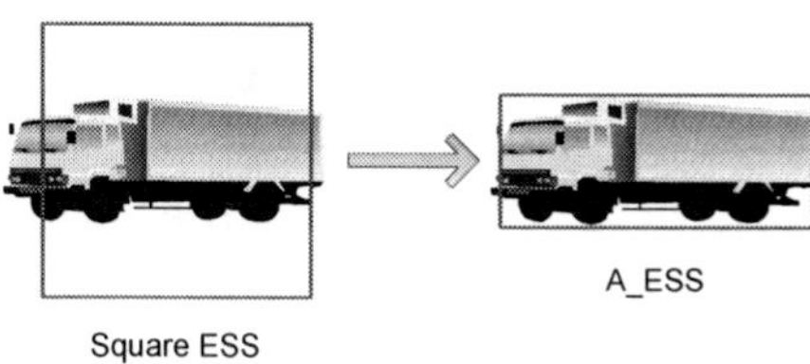

Fig. 6. Square ESS + A-ESS simulation

4 Experiments

All the image sets used in this experiment are from PASCAL Visual Object Classes Challenges 2006 (VOC) [8]. In total, there are 5,304 different images containing 9,507 objects with 10 different class objects. The class objects include bicycle, bus, car, cat, cow, dog, horse, motorbike, person, and sheep. These images will be tested on each original and proposed algorithm to measure the computation speed and score on the best sub-window.

4.1 Square ESS

Table 1. Statistic of ESS and Square ESS

Time statistic			
	Average	**Min**	**Max**
ESS	108.1865 ms	1.5700 ms	4489.0 ms
Square ESS	10.4813 ms	0.6600 ms	102.9560 ms
Relative Error = 14%			

From the information in Table 1, it can be seen that computation speed of square ESS is about 10 times faster than original ESS (on average) regardless of the existence of the target object. However, the relative error is about 14% worse, on average. This is as expected since there are some objects that cannot be covered tightly with a square (for instance, long and narrow object). Fig. 7 shows some image samples to see the result from both original ESS and square ESS. The red box represents the ground truth. Yellow box is the optimal sub-window from the square ESS. Blue box is the optimal sub-window from the original ESS. The first one is an image of a cat which is reasonably large and has the shape of rectangle with ratio between width and height around 4:1. The second image contains a motorbike of large size and has the shape near to square (ratio about 3:2).

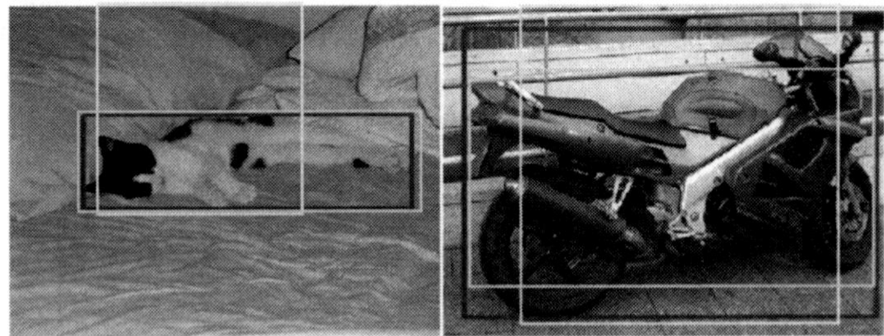

Fig. 7. Square ESS sample for (a) cat image (b) motorbike image

4.2 Square χ^2-ESS

Table 2. Statistic of χ2-ESS and Square χ2-ESS

Time statistic			
	Average	**Min**	**Max**
χ^2-ESS	7775.1 ms	261.9 ms	107412 ms
Square χ^2-ESS	2231.8 ms	251.325 ms	7992.6 ms
Relative error = 1.12%			

From the information on Table 2, square χ^2-ESS has a relative error 1.12% in average with the distance from original χ^2-ESS which is a good achievement since it does not lose too much performance. The computation speed on square χ^2-ESS is about 3.5 times faster in average. The speed improvement is not as high compared to the square ESS. This is as expected because both algorithms have to do a data pre-processing (integral image) which takes quite long time in order to make it easier to calculate Chi-Square distance which is not affected by sub-window shape. In addition, calculating Chi-Square distance also takes longer time than calculating sum of scores in linear cases.

Fig. 8 shows some image samples to see the result from both original χ^2-ESS and square χ^2-ESS. The red box represents the ground truth. Yellow box is the optimal sub-window from the square χ^2-ESS. Blue box is the optimal sub-window from the original χ^2-ESS.

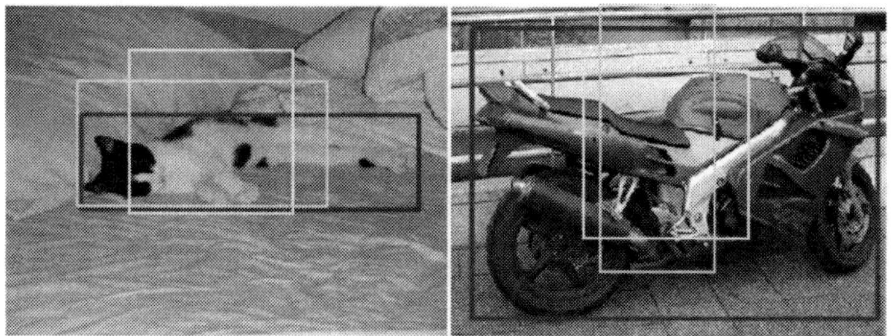

Fig. 8. Square χ^2-ESS sample for (a) cat image (b) motorbike image

4.3 Combination Square ESS and A-ESS

Table 3. Statistic of ESS, Square ESS, and Square ESS + A-ESS

	Average Time	**Relative Error (with ESS)**
ESS	108.1865 ms	-
Square ESS	10.4813 ms	14%
Square ESS + A-ESS	11.1691 ms	1%

After applying this approach, the score relative error is reduced to 1 % compared to score from the original ESS. This is a good achievement since the accuracy is improved significantly (better than 14% relative error with only square ESS). However, the computation speed is only 9.5 times faster (square ESS alone can perform about 10 times faster than the original ESS) since there is a bit overhead while applying A-ESS. Overall, this approach is worth to use since the performance increases a lot just by sacrificing a bit of the computation speed.

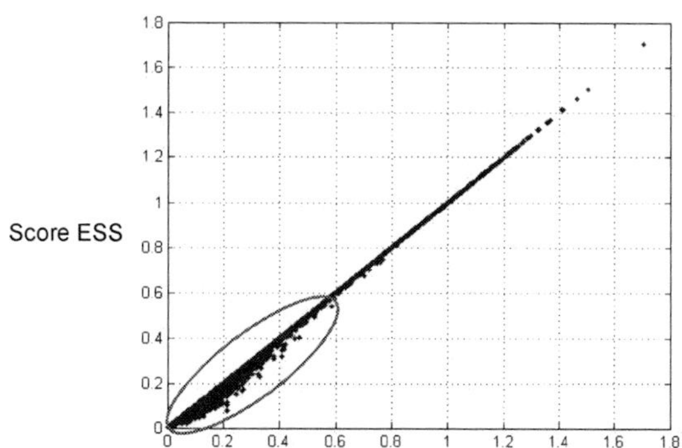

Fig. 9. Score comparison between square ESS + original A-ESS with original ESS

Fig. 9 shows that when the score is low (less than 0.6), there are many cases where the score of square ESS combined with A-ESS is lower than the original ESS. Most likely, the objects are too small so the coverage is not tight enough with approximation algorithm. On the other hand, when the score is large, the score between both approaches are quite similar or perfectly identical. This performance is much better than applying sub-window search only with the square ESS.

Fig. 10 shows some image samples to see the result from both original ESS and combination of square ESS and A-ESS. The red box represents the ground truth. Yellow box is the optimal sub-window from the combination of square ESS and A-ESS. Blue box is the optimal sub-window from the original ESS.

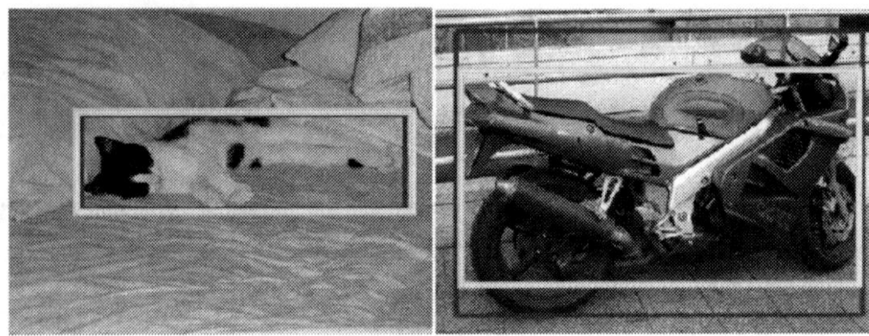

Fig. 10. Square ESS + A-ESS sample for (a) cat image (b) motorbike image

Both sub-windows are perfectly identical for these two image samples. Obviously, it will not happen for all images. These two samples show the great improvement done by combining square ESS and original A-ESS. The good accuracy still can be obtained with higher computation speed.

5 Conclusions

The main purpose of this research is to improve the sub-window search processing speed of the well-known approach called Efficient Sub-window Search (ESS). This is achieved by modifying the shape of the sub-window to be a square. By having square shape, only 3 variables (X coordinate, Y coordinate, and side length S) are needed to represent a single sub-window. Therefore, the iteration will be reduced significantly to achieve the optimal sub-window. The worst case complexity can be reduced from $O(n^4)$ into $O(n^3)$. This experiment is used to introduce the potential of fixed-shape (square-shaped in this research) sub-window in order to improve the computation speed while maintaining good accuracy.

Some proposed algorithms in this research show a good improvement in terms of computation speed. The first one is square ESS which improves the speed up to 10 times faster. However, the accuracy loss is quite large. The second one is square χ^2-ESS which also has a good accuracy compared to original χ^2-ESS, but the speed improvement is not as high as square ESS (only 3.5 times faster). Finally, the last approach is the combination of square ESS and original A-ESS. The result of this experiment shows that the accuracy of the sub-window is really high (very close to original ESS) and the computation speed is faster than original ESS. This approach is the best algorithm among all the proposed algorithms.

References

1. An, S., Peursum, P., Liu, W., Venkatesh, S.: Efficient Algorithms for Subwindow Search in Object Detection and Localization. In: IEEE Conferences on Computer Vision and Pattern Recognition, Miami, pp. 264–271 (2009)
2. An, S., Peursum, P., Liu, W., Venkatesh, S., Chen, X.: Exploiting Monge Structures in Optimum Subwindow Search. In: IEEE Conferences on Computer Vision and Pattern Recognition, San Fransisco, pp. 926–933 (2010)
3. Bentley, J.: Programming Pearls: Perspective in Performance. Commun. ACM 27, 1087–1092 (1984)
4. Lampert, C.H., Blaschko, M.B., Hofmann, T.: Beyond Sliding Windows: Object Localization by Efficient Subwindow Search. In: IEEE Conferences on Computer Vision and Pattern Recognition, Anchorage, pp. 1–8 (2008)
5. Lampert, C.H., Blaschko, M.B., Hofmann, T.: Efficient Subwindow Search: A Branch and Bound Framework for Object Localization. IEEE Transactions on Pattern Analysis and Machine Intelligence 31, 2129–2142 (2009)
6. Lowe, D.G.: Distinctive Image Features from Scale-Invariant Keypoints. International Journal of Computer Vision 60, 91–110 (2004)
7. Scholkopf, B., Smola, A.J.: Learning with Kernels: Support Vector Machines, Regularization, Optimization, and Beyond. MIT Press, Cambridge (2001)
8. The PASCAL Visual Object Classes Challenge 2006 (VOC 2006) Results, `http://www.pascal-network.org`

Binocular Structured Light Stereo Matching Approach for Dense Facial Disparity Map

Shiwei Ma, Yujie Shen, Junfen Qian, Hui Chen, Zhonghua Hao, and Lei Yang

School of Mechatronic Engineering & Automation, Shanghai Key Laboratory of Power Station Automation Technology, Shanghai University, NO.149, Yanchang Rd. 200072 Shanghai, China
syjman@gmail.com

Abstract. Binocular stereo vision technology shows a particular interesting for face recognition, in which the accurate stereo matching is the key issue for obtaining dense disparity map used for exploiting 3D shape information of object. This paper proposed a binocular structured light stereo matching approach to deal with the challenge of stereo matching to objects having large disparity and low texture, such as facial image. By introducing global system to coordinate the binocular camera and projector, a projector cast structured light pattern which added texture to the face scene. Binocular epipolar constraint and semi-global stereo matching algorithm were applied. The experiments showed that the accuracy had improved compared to that of purely binocular vision for getting dense facial disparity map.

Keywords: facial disparity map, binocular, structured light, stereo matching.

1 Introduction

Exploiting 3D shape information for face recognition is attracting more attention in recent years. However, the face stereo pair has a larger disparity and lowly textured skin regions, so it is still not an easy task to obtain facial nuances or oddity in a 3D virtual implementation. Though some model-based methods can be used to overcome these obstacles [1], it is hard to achieve accurate reconstruction of the true effect automatically due to these approaches require fully cooperation between operator and identifier. Hence, many research works focus on model-free approaches and vision based methods for 3D facial reconstruction due to their general low-cost usage of off the shelf hardware.

In this filed, binocular stereo vision technology shows a particular interesting in recent years in exploiting 3D shape information since it is adaptive and flexible. In binocular stereo vision technology, the disparity map is commonly used for obtaining 3D shape information of object, and stereo matching is the basis to compute dense disparity map. However, since the inherent uncertainty of stereo matching, it has no uniform solution up to now. In our study, we combined the structured light and binocular stereo vision to overcome the challenge in stereo matching for obtaining

D. Wang and M. Reynolds (Eds.): AI 2011, LNAI 7106, pp. 550–559, 2011.

dense facial disparity map. In this paper, the key problem and previous work are investigated firstly. Then, the proposed binocular structured light approach and two typical real-time matching algorithms are introduced in detail. Finally, performance of the method is validated by simulation experiment and concluding remarks are presented.

2 Problem Formulation and Previous Work

The main process of making disparity map by binocular stereo vision is shown as Figure 1. The process of rectification removes lens distortions and turns the camera stereo pair in standard form by using the results obtained from calibration. The disparity maps can be obtained after the stereo matching was done.

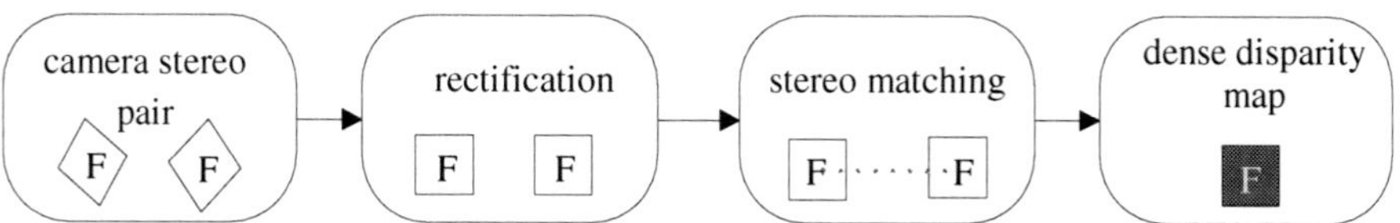

Fig. 1. Main process of binocular stereo vision

In the whole process of binocular stereo vision technology, stereo matching is the key step and attracted lots of research attentions. Various techniques have been proposed for estimating depth or disparity from image pairs, in which existing stereo matching methods can be roughly divided into two categories [2]. The first solves the correspondence problem by using different local matching methods, while the others minimize energy functions in a global sense, they are global matching methods.

In order to properly deal with ambiguity in stereo matching, the local (window-based) methods generally use statistical relationship between color or intensity patterns in the local support window. In this category, there are several stereo matching algorithms [3-5], such as adaptive window matching, variable window methods, multiple-window methods, multi-view stereo non-linear diffusion and mutual information based matching.

The global matching algorithm used global optimization to estimate disparity. By establishing and minimizing the global energy function, it obtains the best visual difference. In the established energy function, there are smooth items except for data items. Data items measure the similarity between pixels, while smoothing items ensure the smoothness of disparity between the adjacent pixels. Global optimization algorithms such as Belief Propagation algorithm [6, 7] and Graph Cut [8] algorithm have attracted much attention. Dynamic programming-based methods [9] are based on assumptions of "uniqueness constraint" and "ordering constraint". Generally, global matching algorithm can achieve a better matching result than local method, but its time complexity is also obvious. For example, top-performing Graph Cut based global algorithm has notably higher time complexity.

In fact, stereo matching has been one of the core challenges in computer vision for decades. One of main problems in stereo matching is the tradeoff between the quality and the time to compute depth map. Many stereo methods focus on high quality results instead of fast computation times. These high quality methods need at least several seconds to compute a single depth map from one stereo pair images [10]. However, for most of the real applications, real-time computation is essential.

Considering that, with active structured light, the quality of the disparity map will improve significantly while the computation time unchanged essentially. Therefore, in our study, we combined the structured light and binocular stereo to overcome the challenge in stereo matching for obtaining dense facial disparity map. However, we are not the first one who combines binocular and active lighting. In Woodward's review literature [11], three kinds of 3D facial reconstruction methods were compared, including binocular stereo, structured lighting and photometric stereo. And comparative experiments show that the combination of structured lighting with symmetric dynamic programming based binocular stereo vision has good prospects due to its reasonable processing time and sufficient accuracy. And Yang's work [12] improved the accuracy of data acquisition using data redundancy in binocular structured light system, which does not require that the optical axes of two camera lenses be coplanar and the structured light sheets be perpendicular to that plane.

3 Binocular Structured Light Approach for Facial Disparity Map

The feature areas of human face such as eyes, mouth, nose and cheek carry most of the visual information expressed by humans, they are especially important for 3D face reconstruction. In order to obtain facial details, face region should be close enough to binocular camera, which will result in two harmful factors to the stereo matching. The first one is large disparity, and the second is low texture area. For example, comparing the two images in Figure 2, we can see that the face model shows the property of low texture (the color is simplex) and large disparity. Obviously, low texture can cause serious ambiguity which will increase the difficulty for matching, or even cause match disaster. These two factors are not isolated. When the face is enlarged, the difficulty for matching in low texture regions, such as cheeks, will seriously affects the quality of disparity map generated. Therefore, purely binocular approach cannot match facial stereo pair well. And in dark scenes, the purely binocular vision will lose its role, while active illumination will perform well.

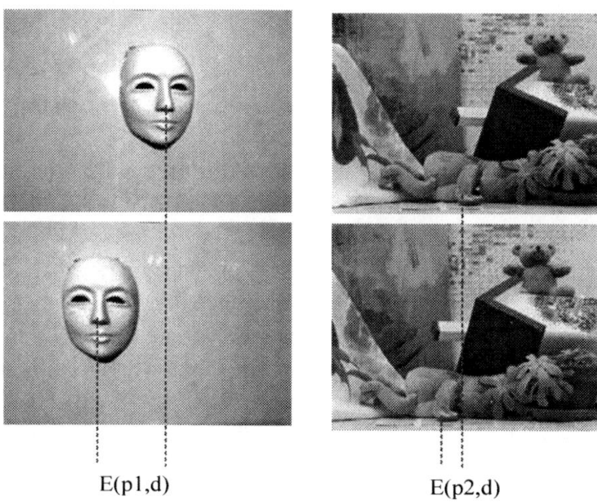

Fig. 2. Face model shows low texture and large disparity

Figure 3 shows the experiment setup of our binocular structured light stereo vision system and the schematic diagram of our proposed approach. In this setup, binocular structured light stereo system is specially designed for making accurate dense disparity map. The structured light from the LCD projector are illustrated as following. Let $I(x; y)$ be an image of a scene without active illumination used (only with ambient illumination), and $I_c(x; y)$ be an image of same scene with active illumination used (plus the same ambient illumination). Then, the image $I_a(x; y) = I_c(x; y) - I(x; y)$ will be the color pattern in visible light spectrum.

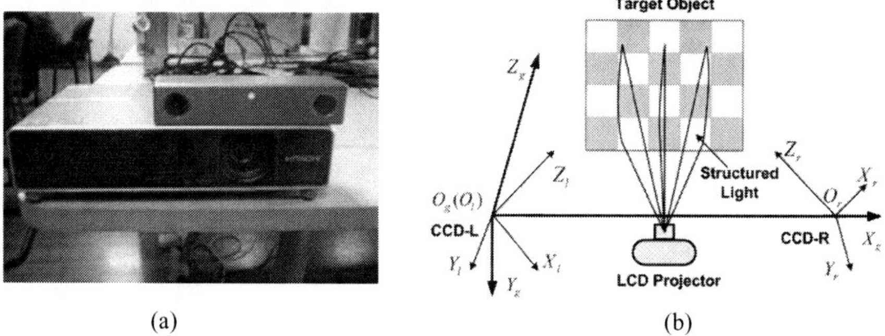

Fig. 3. (a) Setup of binocular cameras and projector; (b) Schematic diagram of binocular structured light stereo vision system

In the implementation of the binocular structured light system, a global coordinate system is established by using binocular cameras and projector. The projector casts structured light pattern onto object, which makes the large smooth area in face be full of texture. Binocular epipolar constraint is used to achieve exact match result. Hence, large disparity and low-texture regions is acceptable, and the accuracy may be improved compared to purely binocular vision. This stereo setup with active illumination can benefit from the improved local scene textures. Hence, better correspondences and more accurate construction of dense disparity maps will be allowed.

In the application of this method, one should employ the best suited projection pattern for specific object. Figure 4 gives three kind of commonly used projection patterns. A number of researchers have worked on optimizing the employed patters that are mostly based on vertical stripes [13, 14]. The main idea of these works is to adapt the pattern in such a way, that interference by the scene is minimized. In traditional structured light approach, a combination of geometric coding, color coding and tracking over time is used. However, the binocular structured light approach does not need to identify the pattern to reconstruct the disparity map. That gives following advantages [15]. 1) One does not need to know the exact position of projector. Therefore, the complex calibrating to structured light can be avoided. 2) One does not need prior knowledge about the structure of the projected light. These two advantages make binocular structured light approach much easier than the traditional structured light approach. Therefore, most of the dense structure light patterns are suitable for the binocular structured light approach and one don't need to pay much attention to encode or decode the light pattern.

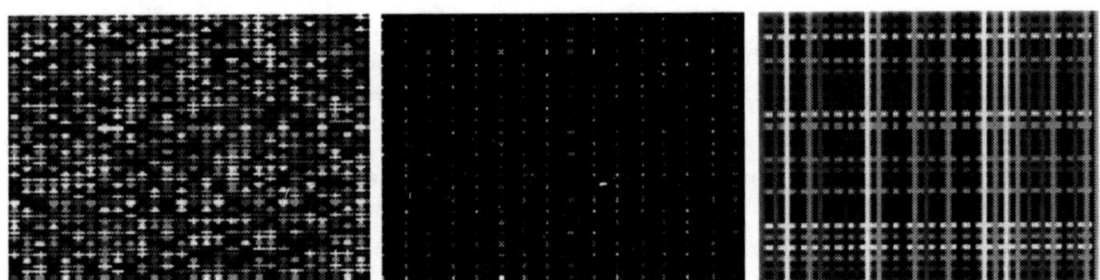

Fig. 4. Instances of common used structured light pattern

4 Selection of Stereo Matching Algorithm

Since structured light significantly reduces the ambiguity of the low texture regions, it can improve the accuracy of the result of stereo matching. In this case, considering the time cost, the complex global minimization method isn't suitable [7-9]. In rencent year, two stereo matching algorithm are proposed. One is local algorithm [16], which is based on correlation and can be implemented very efficiently. The other is semi-global stereo matching (SGM) algorithm [17], which is based on the idea of pixelwise matching and 2D smoothness constraint by combining some 1D constraint, is suitable for increasing the accuracy of stereo matching.

In the SGM algorithm, the pixelwise cost and the smoothness constraints are expressed by defining an energy function $E(D)$ that depends on the disparity image D, i.e.:

$$E(D)=\sum_{P}C(p,D_P)+\sum_{q\in N_P}P_1T\left[\left|D_p-D_q\right|=1\right]+\sum_{q\in N_P}P_2T\left[\left|D_p-D_q\right|>1\right] \quad (1)$$

Where, in the right side, the first term is the sum of all pixel matching costs for the disparities of D. The second term adds a constant penalty P_1 for all pixels q in the neighborhood N_p of p, for which the disparity changes a little bit. The third term adds a larger constant penalty P_2, for all larger disparity changes. The function T [] equal to one only when its argument is true, otherwise it is zero.

For 2D images, looking for global minimum of function $E(D)$ has been shown to be NP-complete problem, and energy minimization in one-dimensional path can be implemented with DP (dynamic programming) method. This leads to an idea that combining the results of multiple 1D path to approximate the 2D situation. Therefore, the disparity map can be calculated by a DP method along the direction of 8 or 16 one-dimensional paths, as following:

$$\begin{aligned}L_r(p,d)=C(p,d)+\min(L_r(p-r,d),L_r(p-r,d-1)+P_1,\\ L_r(p-r,d+1)+P_1,\min_i L_r(p-r,i)+P_2)-\min_k L_r(p-r,k)\end{aligned} \quad (2)$$

In the right side of Equation (2), the first item defines the matching cost of pixel p in depth d. The second item minimize matching cost of point p_r in the current path r containing the penalty coefficient. The third item does not change the actual path through disparity space. Since the subtracted value is constant for all disparities of a pixel p, the optimal path of the minimum does not change. However, the upper limit can now be given as $L\leq C_{max}+P_2$. The cost L_r is summed over paths in all direction r. For example, when 16 paths is accepted, the upper limit of of all pixel matching costs $S(p,d)$, which is defined by

$$S(p,d)=\sum_{r}L_r(p,d) \quad (3)$$

can be easily determined as $S\leq 16\ (C_{max}+P_2)$. The disparity image D_b that corresponds to the base image I_b can be obtained similar as that in the local algorithm by selecting each pixel p the disparity d that corresponds to the minimum cost, i.e. $\min_d S(p, d)$.

It had been manifested that the SGM algorithm can perform better matching effects than that of the local methods and is almost as accurate as global methods, however, the time cost of SGM is less than global methods [17]. Therefore, in the implementation of our approach, we employ the SGM algorithm to acquire a accurate disparity map.

5 Simulated Experimental Results

In order to validate the binocular structured light stereo matching approach for dense facial disparity map, both proposed approach and purely binocular approach were implemented in our experiment to compare their matching results. The experimental devices consist of a binocular camera (Bumblebee@2, 7.4um square pixels, 12cm baseline), a projector (Hitachi-HCP-600X, 1024*768 solution), and a personal computer (1.6GHz CPU and 2GB memory). In the experiment, designed structured light color pattern as shown in the left picture of Figure 4 was projected onto experimental scenes in visible light spectrum. To manifest different matching results, experiments on simulated low-texture scene, large disparity scenes and facial model scene are conducted respectively. The stereo image pairs, left image and right image, were captured by using binocular vision system. The semi-global stereo matching algorithm SGM were used to calculate disparity map.

First, a simple low-texture scene, a flat white plate placed in front of a white background board, was arranged in room environment. From the obtained results shown in Figure 5 in which only the left image are presented in the left column, the performances of disparity map on low-texture scene can be observed and compared intuitively in the right column. For purely binocular approach, since lack of texture on the blackboard, it is difficult to match the pixels accurately regardless of what kind of real-time stereo matching algorithm used. On the contrary, since it used active structured light illumination to make the blackboard full of color texture, the proposed binocular structured light stereo matching approach shows a good matching result.

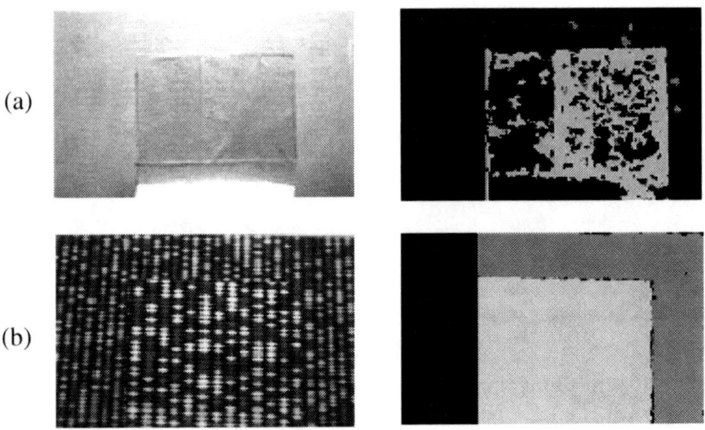

Fig. 5. The results of purely binocular approach (a) and the proposed approach (b) on a low texture scene

Second, a large disparity scene, including chairs, tea carton, barrels, a model facial mask, was arranged in room environment as show in Figure 6. Both near field and far field stereo image pairs were captured to compare performances of proposed approach and purely binocular approach on different disparity scenes. From the obtained results in Figure 6, in which only the left image are presented in the left column, the performances of disparity map on large disparity scene can be observed and compared intuitively in the right column. For purely binocular approach, since the disparity will get larger as the binocular cameras getting closer to the scene, it can only match lesser pixels and result in a significant decline in the matching quality, which lead discontinuous region in Fig.6 (a). What's more, the level of disparity map in Fig.6 (a) is blurred, and the nuances between range disparities cannot be reflected. On the contrary, the results of proposed binocular structured light approach showed a robust performance in Fig.6 (a). Even at near field, as projection light becomes clear, a good matching result is obtained.

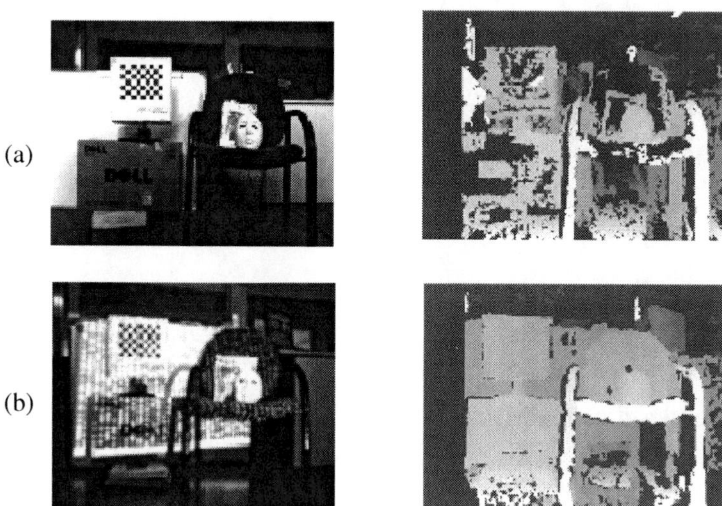

Fig. 6. The results of purely binocular approach (a) and the proposed approach (b) on a large disparity scene

Finally, a model facial mask was selected to conduct experiment on facial scene in room environment as show in Figure 7. First collected face stereo image pairs for both approaches, and then apply GSM algorithm to obtain the disparity map. From the obtained results, the performances can be observed and compared intuitively. Compared to purely binocular approach, the proposed binocular structured light stereo matching approach can get improved quantity of matching pixels for stereo facial scene which has larger disparity and lower texture.

The performances on running time was also emphasized in the proposed method.For an 800 * 700 pixels stereo pairs conducted on a PC with 1.6GHz CPU and 2GB memory, average time cost of Semi-Global Matching algorithms is 1.7s. And with smaller pixels, i.e. 320*240 pixels, they will approach real-time performance.

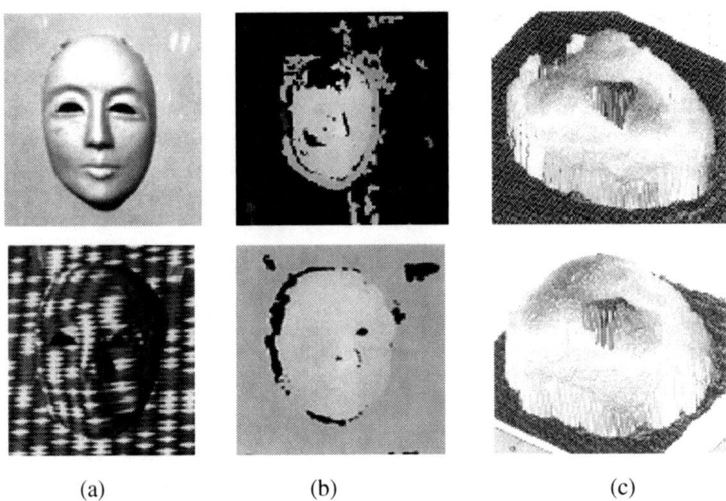

(a) (b) (c)

Fig. 7. (a) The left view of the object scene; (b) The result of obtained disparity map; (c) Three-dimensional view of disparity map

6 Conclusion and Future Work

The proposed method combined active structured light illumination with passive binocular stereo vision to deal with the challenge of stereo matching to objects having large disparity and low texture, such as facial image. With the help of epipolar constraint, this approach don't need to know the exact position of the projector and prior knowledge about the projected patterns in obtaining disparity map. By employing existing semi-global algorithm to optimize the time cost and accuracy for stereo matching, the approach is easily implemented. Simulated experimental results show that the approach performs better than purely binocular vision method in the aspects of large disparity, low-texture, and is suitable for getting dense facial disparity map.

However, the cognitive behavior and natural scenes is the complexity for current stereo matching approach including proposed approach. Other problems in stereo matching such as foreshortening, perspective distortions, occlusions and discontinuities, transparent objects, which will affect the results seriously in some situations, still exist in proposed approach. There is no perfect algorithm conquers these flaws, therefore it is a feasible way to overcome these problems by combining the advantages of various approach. And limited to a fixed binocular system, we didn't use the vary baseline,which will help to reconstruct the near field scene. Since illumination in active approach was under control, which can make the object scene match the Lambertian surface model. The Lambertian surface model can recover three-dimensional shapes of the object scene by SFS (shpae from shading) method. If SFS method is blend into binocular structured light approach, the data will be redundancy and a more accurate stereo matching is possible. This will be left for our future investigation.

Acknoledgements. This work was supported by Shanghai University Graduate Students Innovation Foundation No.SHUCX112172, Shanghai Science and Technology Research Foundation No.09dz2273400 and 08dz2272400.

References

1. Zhang, Y., Pralmeh, E.C., Sung, E.: Hierarchical modeling of a personalized face for realistic expression animation. In: Proceedings of the IEEE International Conference on Multimedia and Expo., pp. 457–460. IEEE, Lausanne (2002)
2. Scharstein, D., Szeliski, R.: A taxonomy and evaluation of dense two-frame stereo correspondence algorithms. IJCV 47(1), 7–42 (2002)
3. Veksler, O.: Fast variable window for stereo correspondence using integral images. In: CVPR 2003, pp. I: 556–561 (2003)
4. Fusiello, A., Roberto, V., Trucco, E.: Efficeint stereo with multiple windowing. In: CVPR 1997, pp. 858–863 (1997)
5. Kim, J., Kolmogorov, V., Zabih, R.: Visual correspondence using energy minimization and mutual information. In: CVPR 2003, pp. 1033–1040 (2003)
6. Li, G., Zucker, S.W.: Surface geometric constraints for stereo in belief propagation. In: IEEE Computer Society Conference, vol. 2, pp. 2355–2362 (2006)
7. Felzenszwalb, P.F., Huttenlocher, D.P.: Efficient belief propagation for early vision. Int. J. Comput. Vis. 70(1) (2006)
8. Kolmogorov, V., Zabih, R.: Computing Visual Correspondence with Occlusions using Graph Cuts. In: Proc. Int. Conf. Computer Vision (2001)
9. Cox, I.J., Hingorani, S.L., Rao, S.B., Maggs, B.M.: A maximum-likelihood stereo algorithm. CVIU 63(3), 542–567 (1996)
10. Denker, K., Umlauf, G.: Accurate Real-Time Multi-Camera Stereo-Matching on the GPU for 3D Reconstruction. Journal of WSCG 19(1), 9–16 (2011)
11. Alexander, W., Da, A., Georgy, G., Patrice, D.: A Comparison of Three 3-D Facial Reconstruction Approaches. In: IEEE International Conference on Multimedia and Expo, pp. 2057–2060 (2006)
12. Rongqian, Y., Sheng, C., Yazhu, C.: Flexible and accurate implementation of a binocular structured light system. Optics and Lasers in Engineering 46(5), 373–379 (2008)
13. Blake, A., McCowen, D., Lo, H.R., Lindsey, P.J.: Trinocular active range-sensing. IEEE Trans. Pattern Anal. Mach. Intell. 15, 477–483 (1993)
14. Koninckx, T., Gool, L.V.: Real-time range acquisition by adaptive structured light. IEEE Transac. on Pattern Analysis & Machine Intelligence 28, 432–445 (2006)
15. Kosov, S., Thormählen, T., Seidel, H.-P.: Using Active Illumination for Accurate Variational Space-Time Stereo. In: Heyden, A., Kahl, F. (eds.) SCIA 2011. LNCS, vol. 6688, pp. 752–763. Springer, Heidelberg (2011)
16. Veksler, O.: Fast variable window for stereo correspondence using integral images. In: CVPR, vol. I, pp. 556–561 (2003)
17. Hirschmuller, H.: Accurate and efficient stereo processing by semi-global matching and mutual information. In: IEEE Computer Society Conference on Computer Vision and Pattern Recognition, CVPR 2005, vol. (2), pp. 807–814 (2005)

Fast Object Detection with Foveated Imaging and Virtual Saccades on Resource Limited Robots

Adrian Ratter, David Claridge,
Jayen Ashar, and Bernhard Hengst

School of Computer Science and Engineering,
University of New South Wales, UNSW Sydney 2052 Australia

Abstract. This paper describes the use of foveated imaging and virtual saccades to identify visual objects using both colour and edge features. Vision processing is a resource hungry operation at the best of times. When the demands require robust, real time performance with a limited embedded processor, the challenge is significant. Our domain of application is the RoboCup Standard Platform League soccer competition using the Aldebaran Nao robot. We describe algorithms that use a combination of down-sampled colour images and high resolution edge detection to identify objects in varying lighting conditions. Optimised to run in real time on autonomous robots, these techniques can potentially be applied in other resource limited domains.

1 Introduction

Real time identification of objects in a video feed is a significant research area in robotics, and forms the major component of many perception systems. For the rich environments we encounter in everyday life this is still an open research problem. RoboCup Soccer[1] is an international research and education initiative that constrains the environment to a soccer field with a limited number of objects, namely a ball, field, goal posts, and other robots. Vision algorithms are able to exploit these constraints, but face significant challenges.

Small, mobile robots are limited in their processing power. Vision needs to share this limited resource with other functions such as world modelling and behaviour generation. Success in soccer also depends on the speed at which robots can react. A major challenge is for the vision system to deliver real time object recognition at the full frame rate and still leave resources for the other functions.

Colour cameras provide a high native pixel resolution in a three dimensional colour space. It is taxing on resources to process the image in its full resolution. When objects are relatively far away, and appear small in the visual field, we would like to take advantage of the higher resolution.

The human eye has a region with maximum acuity in the centre of the macula known as the fovea. Motivated by this physiology the above dilemma can be

D. Wang and M. Reynolds (Eds.): AI 2011, LNAI 7106, pp. 560–569, 2011.

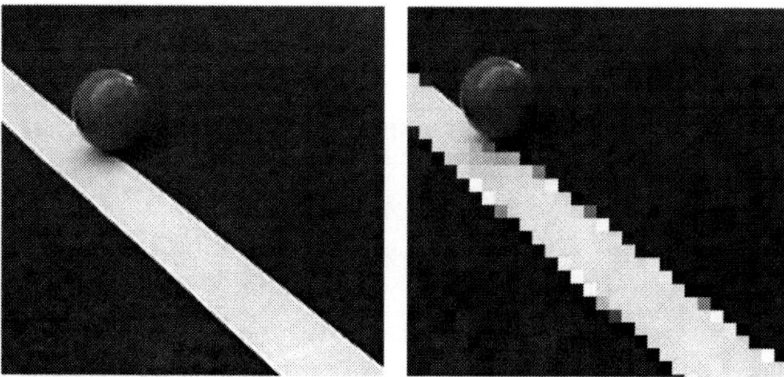

Fig. 1. Foveated imaging. Original Image (left). Virtual fovea with a coarse resolution of the field around the higher resolution fovea area centred on the ball (right).

addressed by varying the resolution and processing across the image according to one or more points of fixation. This technique is called *foveated imaging*. The fovea provides a high resolution image, but a very narrow field of view. *Peripheral vision* is provided by the image outside the foveal regions at lower resolution. These ideas have been used in computer vision inspiring both software and hardware solutions [3]. Figure 1 shows an image at full camera resolution on the left and a foveated image on the right, with the fovea saccaded and fixed on the ball. The ball has a smooth round edge, in contrast to the jagged edges of the field line in the peripheral parts of the image.

The RoboCup soccer environments are characterised by objects with distinct colours. It is not surprising that algorithms to date have largely used colour to identify objects. Organisers have gradually increased the vision challenge by progressively removing crutches such as walls, beacons and coloured goal posts. In particular, the practice of providing special high luminescent and uniform lighting has been discontinued and robots need to cope with whatever lighting is provided by the venue. Lighting often changes during games as audience numbers fluctuate creating varying overshadowing conditions during the game. One solution is for vision to rely less on colour and more on shape cues.

The contribution of this paper is a vision system that addresses the above needs with the following characterstics:

1. A peripheral vision system to locate salient features. A novelty is the detection of field edges for localisation using a subsampled and colour classified image alone.
2. Employing foveated imaging techniques to limit resource usage.
3. Relying more on edges and reducing the dependence on colour.
4. Meeting real time requirements running close to the full frame rate.

The application of these methods have broad applicability. We describe them in the context of the Standard Platform League that uses the small humanoid Nao robot from Aldebaran Robotics. The rules of the league disallow external processing or any modification to the machine. The robots' embedded computer is limited to an AMD Geode LX 800 processor running at a modest 500MHz.

The playing area of the soccer field is currently 4 by 6 meters with colour coded open goals. A team size of three robots was used in 2010 and this will increase to four in 2011. The ball is a standard orange coloured street hockey ball. Each robot has two CCD 640 × 480 pixel cameras in its head (although only one can be used at a time).

In the rest of this paper we will describe the down-sampled *saliency* frames that are used to identify possible locations of various objects on the field. The saliency image is used to find field edges to aid in the localisation of the robot. We next show how the saliency image leads to virtual saccades to multiple points of fixation representing interest regions corresponding to the ball and goals. Multimodal colour and edge data at high resolution is used at these foveal points in the image, achieving both high accuracy and high efficiency.

This approach was implemented by the UNSW team *rUNSWift* for the Standard Platform League in RoboCup World Competition in Singapore in 2010, for both the technical challenges and the soccer tournament. The University of New South Wales (UNSW) placed first in the technical challenges and second in the soccer competition against 23 other international teams.

2 Saliency Scan

In order to achieve our goal of identifying areas of interest in the image as fast as possible, the first step of the vision pipeline is to subsample the image in a regular grid pattern. We reduce the image size by a factor of n for each of the two image dimensions, reducing the number of pixels to be processed by a factor of n^2. We have chosen $n = 4$ for the 2010 competition to provide a substantial reduction in image size, but this could be reduced further to $n = 8$ if required, at the potential cost of increasing the number of false negatives of small, far away objects. By mapping every 4th pixel in the raw 640 × 480 image we derive a 160 × 120 pixel resolution image giving a 16 fold reduction in image size. Figure 1 (right) shows a down-sampled part-image of the green field, field line, and ball for $n = 8$. The immediate area around the ball shows a virtual fovea region at the original raw resolution, constrasting with the low resolution in the rest of the image.

Each pixel in the subsampled image is colour classified before it is stored in the *saliency image*. Figure 2 (left) shows an example saliency image. This colour coding is performed offline using a weighted kernel classification algorithm developed for previous RoboCup competitions [6], where each training sample increases a weighting for a particular YUV value toward the classified colour. The neighbouring colours in YUV space to the sample also have their weights increased, but at a normally decreasing amount the further they are from the sample. The kernel file is used to generate a constant time lookup table on the robot at runtime. The colours classified are orange (the ball), green (the field), white (the field lines and parts of the robots), yellow (the yellow goals), red (the pink band worn by robots on the red team), blue (the blue goals and the blue band worn by robots on the blue team) and background (to remove background objects with a similar colour to the ones used on a RoboCup field).

Fig. 2. Colour classified saliency image at 160×120 resolution ie. $n = 4$ (left). Goal and robot regions identified during the region detection process (right).

As the saliency image is generated for every frame at 30 fps, any further optimisation is desirable. We analysed the compiler generated assembly code to find optimisation opportunities. The main optimisations are as follows:

- Reducing the amount of memory used
- Reducing the number of memory accesses
- Reducing the number of local variables used (thereby reducing the number of variables in memory)
- Reducing the number of calculations required for image access by using pointer arithmetic instead of array indexing.

In the following sections we will describe how the colour classified saliency scan can be used to rapidly identify objects of interest in the image.

3 Field Edge Detection Using the Saliency Scan

To further reduce the amount of the image that has to be processed for object identification, and to assist with localisation, the edges of the green field are detected using the saliency scan image. In 2009 B-Human used a convex hull algorithm to exclude areas above the field edge[9], which achieves the first goal of reducing the area of the image to be processed. In 2010 rUNSWift used a similar method of vertical scanning to detect points on the edge of the field, but rather than find an arbitrary convex hull, multiple iterations of the RANSAC algorithm[4] were used to find straight lines. When two field edge lines are detected, the possible positions of the robot are reduced to four hypotheses; one for each corner of the field.

Initially, the first green pixel in each column of the saliency scan is recorded, by scanning vertically from the horizon down (the horizon is found by using kinematics; by using the robot dimensions and joint angles to calculate what pixels in the image correspond to the horizon [7]) - Figure 3 (top-left). Secondly, the pixels are fit to a line of the form $t_1x + t_2y + t_3 = 0$ with the RANSAC algorithm, to maximise the number of points that fit a line - Figure 3 (top-right).

Fig. 3. Candidate points for a field edge line (top-left). Line found by performing RANSAC on candidate points (top-right). Lines found by performing RANSAC twice on the candidate points (bottom-left). False positive field edge (bottom-right).

Finally, the consensus set of the first line is removed from the candidate points, and RANSAC is repeated, possibly finding a second line - Figure 3 (bottom-left). Figure 3 (bottom-right) shows a false positive for one of the field edges caused by the triangular goal post support. Its effect is rapidly filtered out with goal post localisation information.

In addition to reducing the amount of the image to be scanned for objects to the parts of the image below the field edge, these field edge observations were able to be used to provide useful updates to the robot's estimated position on the field.[2]

4 Multi-modal Object Analysis in Foveated Regions

We scan the colour classified pixels underneath the field edge to identify potential areas, or regions, that could represent important features, such as the ball, other robots, or field lines. The contents of each of these regions are analysed to determine what objects they may represent. By only examining small areas of interest at the full resolution, this method of virtual saccades enabled us to greatly increase the run time speed of the vision processing system. An example of this region detection is shown in Figure 2 (right).

An alternative to the use of colour is to use edges to find the outline of objects. Unfortunately, common edge detection methods to identify all the edges in the image, such as Canny, are computationally too expensive to run in real time on the Nao, before considering the additional challenge of complex shape identification. A foveated image hybrid solution of these two methods was used to combine the accuracy and robustness of edge detection with the computational speed of colour coding to identify both balls and goal posts. The hybrid solution involved firstly using colours in the lower resolution peripheral vision to quickly identify salient locations, and then edge detection to perform accurate and reliable identification at the higher resolution foveated points. While other robots may have faster processors than available on the Nao, the increasing reolution of video cameras and the increasing computational demands in other areas of robotics extend the potential application of these techniques outside the RoboCup competition.

4.1 Ball Detection

For ball detection, edge detection is used in the full resolution image only around the region that has been identified as a probable ball. The objective is to find a set of points on the edge of the ball. A circle is then fitted to these points to allow the location of the ball to be accurately determined. Rows and columns of the full resolution image are scanned outwards from the region until the V channel of adjacent pixels differs by more than a certain threshold. The best value for this threshold changes depending on the lighting conditions and camera settings, so it is set experimentally according to the current conditions. Only the V channel was used in the ball edge detection as this chromatic dimension of the ball tends to change quite markedly near the edge of the ball. Edges are often detected inside the ball when a combination of the Y, U and V channels are used.

In order to further increase the efficiency of this method, the space between rows and columns scanned for edges was adjusted according to the size of the region to ensure that balls close to the robot did not take too long to process, but balls far away from the robot could still be properly identified.

Once points around the edge of a ball have been identified, a circle can be quickly fitted to these points by randomly selecting three edge points, and finding the intersection of the perpendicular bisectors of the lines joining the three points. The intersection gives the centre of the ball, and the distance between the intersection and any of the three points gives the radius of the ball. When this process is repeated several times and the median of the centre and radius measurements is taken, any small errors in the edge detection are greatly reduced.

Figure 4 shows an example of the edge detection being used to accurately identify a ball. The image on the left shows the colour classified image. It can be seen that a substantial part of the ball is unclassified (note that unclassified colours appear as light blue in the screenshot). The image on the right shows that the edge detection has enabled the edge of the ball to be precisely located. This is particuarly important for ball detection as kicks need to be lined up very precisely for them to work well.

Fig. 4. A screenshot of the ball detection. The left image shows the colour classified image, while the right shows the edge points identified and the circle fitted to the edge points.

4.2 Goal Post Detection

As only the very bottom part of each goal post appears below the field edge, goals are not identified during region building. Instead, histograms are generated while the saliency scan is being built, and are used to identify the likely approximate positions of the goals. Edge detection is then used to find the exact position of the goal posts, or to remove false positives from the histogram stage.

This is achieved by firstly finding the maximum value in the y-axis histogram for one of the goal colours - the row that contains the most number of goal coloured pixels. Only one y coordinate is used because if there are two goal posts in the image, they will occupy approximately the same y coordinate range, and the maximum in the histogram will most likely occur at a y coordinate occupied by both posts. The x-axis histogram is then scanned to find local maximums above a certain threshold for the goal colour. To avoid several local maximums being detected in the same goal post, the histogram value of the goal post colour has to decrease to be at least three times less than the maximum value before another local maximum can be recorded. The same procedure is used for both goal colours.

Several horizontal and vertical scan lines are used around each pair of x and y coordinates identified using the histograms. Each scan starts around the pair of x and y coordinates, and continues outwards until an edge is detected. For goal detection, an edge is found when the two pixels differ in the sum of the differences in the Y, U and V values by more than a certain threshold. All channels are used as the colour of the background around the goal posts cannot be controlled, so any significant change in any channel needs to be registered as an edge. These scan lines result in a rectangle representing the goal post, which can then be used by the localisation algorithms.

Figure 5 (left) shows a very deteriorated colour classified image of the blue goal. Despite the poor quality, the foveated higher resolution edge detection approach is able to clearly identify both goal posts, as shown in Figure 5 (right).

Fig. 5. Poor colour classified image of goal posts (left). Correctly identified goal post using edge information (right).

5 Performance in RoboCup

The set of algorithms presented in this paper form the cornerstone of UNSW's visual object identification for the 2010 RoboCup competition. In particular, the foveated vision algorithm was able to successfully handle the difficult conditions of a final game without noticeable degradation in performance where people crowded around the field creating significant challenges for vision by affecting the lighting. In testing before the competition, we found that vision was able to run at approximately 30 frames per second during game conditions.

As the region builder uses the field edge detection to only scan the image below the field edge, and field edges are used for localisation, field edge detection is a vital part of our vision system. We found that when the field edge(s) could be seen clearly, or with a few small obstructions, the field edge detection worked consistently and accurately. However, when there was a lot of obstruction, such as several robots, or a referee, the field lines were often misplaced. At times this caused a noticeable deterioration in the localisation while lining up to make a kick for goals.

Using first the foveated image and virtual saccade approach and then accurate edge detection proved to be very beneficial to the performance of both the goal detection and the ball detection. In following this method, only a very small number of pixels in the saliency scan needed to be the correct colour for the edge detection to give an accurate match. This allowed us to consistently and accurately detect the balls and goals, even from the opposite side of the field, despite the large amount of colour variation due to the curved surfaces of the goals and the ball, and various shadows on the goals.

6 Related Work

A number of alternate methods have been devised to solve the complex task of object identification in the resource limited environment of RoboCup.

In order to limit the amount of interference from the background, it is often a useful first step to identify the edge of the field in the image. Any item above this edge can therefore be eliminated. The method used by B-Human [9] to find the edge of the field is to scan down each column in the image to find a green segment of a minimum length, and fit a convex hull to the start of the green segments. As RANSAC's performance deteriorates significantly with noise around the field edge, this method could likely make the position of the field edge more accurate when there are a lot of non green objects around the edge. However, fitting lines to the field edge has the advantage of being easily used to localise the robot on the field.

Due to the limited processing power available on the Nao, it is not possible to scan every pixel in the image fast enough to run in real time. An interesting approach is taken by North[5], where the density of horizontal and vertical scan lines is changed depending on how close the lines are to the horizon. This uses the theory that objects close to the camera will be large enough to be seen using extremely low resolution scan lines, but objects further away, near the horizon, will appear much smaller, and therefore need a much higher density of scan lines in order to be detected. The drawback to this approach is that shape identification and repeated accesses are harder and slower. This approach is also inappropriate for robots with a higher camera, such as our 58cm Nao. In an alternate approach by Von Hundelshausen and Rojas[10], regions are grown from the green field, with the white field lines, robots and balls separating the green regions. The authors propose that, as the robot moves, the regions can be incrementally grown and shrunk, resulting in far fewer pixels needing to be processed and updated each frame. This idea of using previous frames to help lower the computation time of the current frame, while not explored in our 2010 vision system, is a worthwhile avenue for future research, and in line with the concept of foveated imaging.

One of the most difficult parts of the object identification for RoboCup is the distinction between field lines and robots, as many parts of the robots are white or close to white. This means that some kind of processing, other than colour, has to be used to separate field lines and robots. The method used by B-Human[9] to achieve this is to first create a series of small white coloured regions that could represent either parts of a line or parts of a robot. These regions are then analysed in terms of their shape, and ones that more likely represent robots are marked. Finally, areas of the images where there is a cluster of these marked regions are considered to most likely contain robots, and every region in this area is thus removed.

Röfer and Jüngel[8] propose a different approach of edges and colour to achieve fast object recognition. In this method, a grid of horizontal and vertical scan lines is used to search for pixels where there is a significant drop in the y channel compared to the previous pixels searched. As the field is generally darker than the field lines and the robots, this can indicate an edge between an object and the field.

7 Conclusion

A vision processing system must be highly efficient, robust, and accurate to enable it to perform reliably in the dynamic world of a soccer game. We have presented a foveated imaging approach using colour CCD cameras that can perform the vision task in real time. We have also presented several processor optimisations to help improve code for low powered embedded systems. By utilising the hybrid methodologies of colour classification and edge detection, we are able to reliably identify robots, goals, field lines and balls in the RoboCup environment. Our approach of using virtual saccades to points of fixation of high resolution foveal areas in the image allowed us to reduce the processing of redundant data, and achieve processing speeds of approximately 30 frames per second in changing lighting conditions.

References

1. Board of trustees: Robocup, http://www.robocup.org/
2. Ashar, J., Claridge, D., Hall, B., Hengst, B., Nguyen, H., Pagnucco, M., Ratter, A., Robinson, S., Sammut, C., Vance, B., White, B., Zhu, Y.: RoboCup Standard Platform League - rUNSWift 2010. In: Australasian Conference on Robotics and Automation (2010)
3. Camacho, P., Arrebola, F., Sandoval, F.: Multiresolution Sensors With Adaptive Structure. In: Proceedings of the 24th Annual Conference of the IEEE Industrial Electronics Society, IECON 1998 (1998)
4. Fischler, M.A., Bolles, R.C.: Random Sample Consensus: A Paradigm For Model Fitting With Applications to Image Analysis and Automated Cartography. Commun. ACM 24, 381–395 (1981), http://doi.acm.org/10.1145/358669.358692
5. North, A.: Object Recognition From Sub-Sampled Image Processing. Honours thesis, The University of New South Wales (2005)
6. Pham, K.C.: Incremental Learning of Vision Recognition Using Ripple Down Rules. Honours thesis, The University of New South Wales (2005)
7. Ratter, A., Hengst, B., Hall, B., White, B., Vance, B., Claridge, D., Nguyen, H., Ashar, J., Robinson, S., Zhu, Y.: rUNSWift Team Report 2010 Robocup Standard Platform League. Tech. rep., School of Computer Science and Engineering, University of New South Wales (2010)
8. Röfer, T., Jüngel, M.: Fast and Robust Edge-Based Localization in the Sony Four-Legged Robot League. In: Polani, D., Browning, B., Bonarini, A., Yoshida, K. (eds.) RoboCup 2003. LNCS (LNAI), vol. 3020, pp. 262–273. Springer, Heidelberg (2004)
9. Röfer, T., Laue, T., Müller, J., Bösche, O., Burchardt, A., Damrose, E., Gillmann, K., Graf, C., ry de Haas, T.J., Härtl, A., Rieskamp, A., Schreck, A., Sieverdingbeck, I., Worch, J.H.: B-Human Team Report and Code Release (2009), http://www.b-human.de/index.php?s=publications
10. von Hundelshausen, F., Rojas, R.: Tracking Regions. In: Polani, D., Browning, B., Bonarini, A., Yoshida, K. (eds.) RoboCup 2003. LNCS (LNAI), vol. 3020, pp. 250–261. Springer, Heidelberg (2004)

Fast Multi-view Graph Kernels for Object Classification

Luming Zhang, Mingli Song, Jiajun Bu, and Chun Chen

College of Computer Science, Zhejiang University
{zglumg,brooksong,bjj,chenc}@cs.zju.edu.cn

Abstract. Object classification is an important problem in multimedia information retrieval. In order to better objects classification, we often employ a set of multi-view images to describe an object for classification. However, two issues remain unsolved: 1) exploiting the spatial relations of local features in the multi-view images for classification, and 2) accelerating the classification process. To solve them, Fast Multi-view Graph Kernel (FMGK), is proposed. Given a set of multi-view images for an object, we segment each view image into several regions. And inter- and intra- view linkage graphs are constructed to describe the spatial relations of the regions between and within each multi-view image respectively. Then, the inter- and intra- view graphs are integrated into a so-called multi-view region graph. Finally, the kernel between objects is computed by accumulating all matchings' of walk structures between corresponding multi-view region graphs. And a SVM [11] classifier is trained based on the computed kernels for object classification. The experimental results on different datasets validate the effectiveness of our FMGK.

1 Introduction

Object classification is an important issue for many multimedia applications, such as scene recognition and surveillance. To classify an object, two schemes can be adopted, we can either describe an object by either a single-view image and classify this image into an object category or, describe an object by a set of multi-view images and classify the set of images into an object category. Obviously, the second scheme is more robust because it contains richer information for an object. That is, in the first scheme, some discriminative information may be occluded, while in multi-view case, the occluded information is recovered. However, it is still a challenging task to deal with the multi-view image based object classification successfully due to two factors: on one hand, the components in the multi-view images and their spatial relations are complex and unstable, which makes it difficult to extract features discriminative enough for classification; on the other hand, the huge number of components and their bilateral relations bring challenges to computer to be processed efficiently. Therefore, more discriminative and efficient features are becoming more and more important for multi-view object classification.

In the evolution of image analysis, many features have been proposed and they can be categorized into two groups: global features and local features. Global feature, e.g., eigenspace [1], represent the entire image by singly vector and are hence tractable for conventional classifiers, e.g., Support Vector Machine(SVM) [11]. However, global

D. Wang and M. Reynolds (Eds.): AI 2011, LNAI 7106, pp. 570–579, 2011.

features are sensitive to occlusions and clutters, which results in poor classification accuracy. In contrast to global features, local features, e.g., Scale Invariant Feature Transformation (SIFT) [10], are extracted at interest points and are robust to image deformations. Different images may produce different number of local features. In order to be tractable for conventional classifier, these local features are often integrated into an orderless bag-of-features representation. Unfortunately, as a non-structural representation, the bag-of-features representation ignores the geometric property of images, i.e., the spatial of local features, which prevents it from being discriminative.

In order to enhance multi-view based object classification, several methods have been proposed. Lazebink et al. [12] developed the Spatial Pyramid Matching(SPM) by partitioning an image into increasingly fine grids and computing histograms of local features inside each grid cell. However, SPM requires nonlinear classifier, which is high time-consuming, to achieve good classification accuracy. In , Latent Dirichlet Allocation (LDA) [13] is used to model the geometric property of the scene images. Specifically, scene image is represented by a set of codewords, which are independently generated by corresponding latent topics. However, as empirically demonstrated in [14], LDA affects adversely on scene classification. In [2,3], each image is modeled as a tree and image matching is formulated into tree matching. Unfortunately, compared to general graphs, the capability of modeling regions' relations by trees is limited Felzenszwalb et al. [4] modeled the relation of different parts of an object as a spring; however, [4] relies heavily on optimal background subtraction. In [5], Hedau et al. defined a new measure of pairwise regions based on the overlaps between regions; but just region overlaps are too simple to capture the complicated spatial relations between regions. Keselaman et al. [6] defined a graph, called Least Common Abstraction(LCA), to represent the spatial relations of components of an object; however, LCA cannot be output to conventional classifier, e.g., SVM [11], directly. By exploring the complementary property of different types of features, Multi-view Spectral Embedding [18] and Multiview Stochastic Neighbor Embedding [19] obtains a physical meaningful embedding of the multimodal features. However, [18] and [19] fail to integrate the geometric information of image from each view.

To solve or at least reduce the aforementioned problems, an efficient kernel, called Fast Multi-view Graph Kernel (FMGK), which exploit the spatial relations of local features between and within multi-view images, is proposed. First of all, given a set of multi-view images for an object, we first segment each view image into a number of regions in terms of their color intensity distribution. And two types of graphs, the inter- and the intra- view linkage graph are constructed to model the set of multi-view images, i.e., the inter-view graph describes the spatial relations of regions from different view images while the intra-view graph describes the spatial relations of regions within each view image. Then, we integrate the two types of graphs into a so-called multi-view region graph. Finally, by constructing a product graph, the kernel between a pair of objects is computed efficiently by accumulating all matchings' of walk structures between the corresponding multi-view graphs. Based on the obtained kernel, a SVM [11] classifier is trained for object classification.

The contributions of this paper are as follows: 1). FMGK, a new method to build the representation of multi-view images, is presented for object classification; 2). inter- and intra- view linkage graphs to represent the spatial relations of regions between and within multi-view images; 3). product graph for efficient computation of the kernel between multi-view region graphs.

2 Inter- and Intra- view Linkage Graphs

2.1 Segmenting Each View Image

Image segmentation is essentially a clustering problem. Features are extracted from each pixel and pixels with similar features are clustered into a region. In our approach, normalized cut [15] is employed to segment each view image. Specifically, given M the number of segmented regions, normalized cut clusters pixels in each view image into a set of regions $\{r_1, r_2, \cdots, r_M\}$ with strongly intra- region connectivity and weak inter-regions connectivity.

In our approach, over-segment setting is applied by setting a large M (usually $M >$ 100). That is to say, components in an object may span more than one region, but very few regions span more than one components. This is because: 1). determining the number of segmented regions is heuristic-based, and 2). in contrast to deficient-segment setting, more regions are obtained in over-segment setting, so it is rarer for one region spans several components, fewer discriminative components are neglected. It is noticeable that, more regions implies higher time consumption in kernel computation, this again demonstrates the necessity of accelerating the kernel computation.

2.2 Constructing Inter- and Intra- view Linkage Graphs

By segmenting each view image into a set of regions$\{r_1, r_2, \cdots, r_M\}$, the intra-view linkage graph G^k ($1 \leq k \leq K$) is constructed to model the spatial relations of regions within the k-th intra-view image I^k, i.e.,

$$G^k = (V, E, H, h) \tag{1}$$

where $V = \{v_1, v_2, \cdots, v_M\}$ is a finite set of vertices, v_i represents region r_i; $h : V \rightarrow H$ is a function assigning a label to each $v \in V$, i.e., $h(v)$ is the RGB histogram of the region corresponding to v; $E = \{(v_i, v_j)|v_i, v_j \in V \wedge v_i \sim v_j\}$ is a set of edges, $v_i \sim v_j$ means two regions corresponding to v_i and v_j are spatial adjancent.

To model the spatial relations of regions between different intra-view images, the inter-view linkage graph $G^{'}$ is constructed, i.e.,

$$G^{'} = (V^{'}, E^{'}, H, h) \tag{2}$$

where $V^{'} = \{v_1, v_2, \cdots, v_{KM}\}$ is a finite set of vertices, v_{k*M+i} represents the i-th region from the k-th image; $E^{'} = \{(v_i, v_j)|v_i, v_j \in V^{'} \wedge ||h(v_i) - h(v_j|| < \delta\}$, where $|| \cdot ||$ is the Euclidean norm, δ is a small parameter making the regions corresponding to v_i and v_j have similar color intensity distribution.

3 Multi-view Region Graphs

As seen from (1) and (2), each object corresponds to $K + 1$ graphs, i.e., K intra-view linkage graphs and one inter-view linkage graph. Towards an efficient representation for an object, we combine all the $K + 1$ graphs into a multi-view region graph. Firstly, all regions from the first intra-view image are saved in a region list. Then the candidate regions of the next intra-view images are removed from the region list if the candidate regions are inter-view linked by the regions in the region list. Finally, inter-view linkage of the remaining regions are save and these regions are added to the region list. If no next intra-view graph is found, the region list becomes the multi-view region list. Details of large region graph construction are in Table 1:

Table 1. Multi-view region graph construction(Algorithm 1)

input: K intra-view graphs $\{G^1, G^2, \cdots, G^K\}$; one intra-view graph G';
output: A multi-view region graph G_{mv};

begin:
1.Put the K segmented region $\{r_1^1, r_1^2, \cdots, r_1^K\}$ from G^1 into region list $\mathcal{L}$ and retain the intra-region linkage E^1.
2. **for** $i = 2 : K$ **do**
  for $j = 1 : M$ **do**
    if region r_j^i are not linked by any $r \in \mathcal{L}$
      Save r_j^i into $\mathcal{L}$;
  end for;
  Retain the intra-region linkage E^k in $\mathcal{L}$;
end for;
3. Transfer $\mathcal{L}$ into G_{mv} ;
end

4 Fast Multi View Graph Kernel

Based on the multi-view region graphs obtained in Section 3, we present the efficient kernel computation of our FMGK.

4.1 Direct Walk Kernels

Given a multi-view region graph G_{mv}, a walk is a finite sequence of neighboring vertices which allows repetitive vertices. Let us denote $W_{G_{mv}}^{len}$ and $W_{G'_{mv}}^{len}$ the set of *len*-length walks in G_{mv} and G'_{mv} respectively. A *len*-length walk contains $len + 1$ vertices. The *len*- length walk kernel $k_w^{len}(G, G')$ sums matchings of all *len*-length walks between G_{mv} and G'_{mv}, i.e.,

$$k_w^{len}(G_{mv}, G'_{mv}) = \sum_{\substack{(r_1,\cdots,len)\in W_{G_{mv}}^{len} \\ (r_1,\cdots,len)\in W_{G'_{mv}}^{len}}} \prod_{i=1}^{len+1} k(h(r_i), h(s_i)) \tag{3}$$

where $k(\cdot, \cdot)$ is a pre-defined basis kernel between vertices in graph G_{mv} and graph G'_{mv}; $\prod_{i=1}^{len+1} k(h(r_i), h(s_i))$ denotes matching of a pair of len length walks between G_{mv} and G'_{mv}.

(3) cannot be computed directly as it is impossible to obtian $W^{len}_{G_{mv}}$ and $W^{len}_{G'_{mv}}$ explicitly. To compute $k^{len}_{w}(G_{mv}, G'_{mv})$, based on [7], a recursive scheme is applied, we call it direct walk kernel and the computation is as follows:

The *len*-length kernel with starting point of r in G_{mv} and s in G'_{mv} is defined from (4) to (7):

$$k^{len}_{w}(G_{mv}, G'_{mv}, r, s) = k(f_{G_{mv}}(r), f_{G'_{mv}}(s)) \cdot \sum_{\substack{r' \in N_{G_{mv}(r)} \\ s' \in N_{G_{mv}(s)}}} k^{len-1}_{w}(G_{mv}, G'_{mv}, r', s') \quad (4)$$

where $len \geq 1$; $N_{G_{mv}}(r)$ is the set of spatial adjacent regions to r in G_{mv}.

The recursion is initialized with

$$k^{0}_{w}(G_{mv}, G_m v', r, s) = k(f_{G_{mv}}(r), f_{G'_{mv}}(s)) \quad (5)$$

where $k(f_{G_{mv}}(r), f_{G'_{mv}}(s))$ is a base kernel defined as follows:

Given a pair of segmented regions r and s, the distance between region r and s is: $d(r, s) = ||h(r) - h(s)||$. Based on $d(r, s)$, the basis kernel defined as:

$$k(f_{G_{mv}}(r), f_{G'_{mv}}(s)) \propto \exp(-\lambda * d(r, s)) \quad (6)$$

where λ is a tuning parameter. It is noticeable that, (6) is positive semi-definite and thus can be used as basis kernel.

The final kernel is an accumulation of all kernel values with one start point in G_{mv} and the other start point in G'_{mv}:

$$k^{len}_{w}(G_{mv}, G'_{mv}) = \sum_{\substack{r \in V_{G_{mv}} \\ s \in V_{G'_{mv}}}} k^{len}_{w}(G_{mv}, G'_{mv}, r, s) \quad (7)$$

Denote average vertex of degree D, based on (4) and (5), the number of basis kernels need to compute is:

$$K^2 M^2 * (1 + D^2 + \cdots + D^{len}) = K^2 M^2 * \frac{D^{len+1} - 1}{D - 1} \quad (8)$$

As seen from (8), the computational complexity is exponential increasing with the length of walk, *len*, making it is computational intractable to obtain an expressive kernel. Therefore, we develop an scheme in Section 4.2 to accelerate the computation of (7).

4.2 Multi-view Region Product Graph

Given a pair of multi-view region graphs G_{mv} and G'_{mv}, the product graph G_p with respect to G_{mv} and G'_{mv} is defined as:

$$G_p = (V_p, E_p) \quad (9)$$

where $V_p = \{(v_i, v'_i) | v_i \in V \wedge v'_i \in V'\}$, $E_p = \{((v_i, v'_i), (v_j, v'_j)) | (v_i, v_j) \in E \wedge (v'_i, v'_j) \in E'\}$.

As shown in (9), G_p is a graph with KM vertex pairs, each representing a pair of vertices from G_{mv} and G'_{mv} respectively. An edge exists in G_p if an edge exists in both G_{mv} and G'_{mv}. Therefore, performing a simultaneously random walk on G_{mv} and G'_{mv} equals to performing a walk on G_p.

Based on (9), *len*-length walk kernel starts from (r, s) and end in (r', s') is defined as:

$$k_w^{len}(G_{mv}, G'_{mv}, r, s, r', s') = k(f_{G_{mv}}(r), f_{G'_{mv}}(s)) * \sum k_w^{len-1}(G_{mv}, G'_{mv}, r^{lin}, s^{lin}, r'^{lin}, s'^{lin}) * k(f_{G_{mv}}(r), f_{G'_{mv}}(s)) \tag{10}$$

where r^{lin} denote a spatial adjacent region of r . We sum all walks kernels from (r, s) to (r', s') of *len*-length. The final kernel value is obtained by summing all kernel values start and end in G_{mv},G'_{mv}, i.e.,

$$k_w^{len}(G_{mv}, G'_{mv}) = \sum_{\substack{r\in V, s\in V\\ r'\in V', s'\in V'}} w * k_w^{len}(G_{mv}, G'_{mv}, r, s, r', s') \tag{11}$$

where w is a weight on each $k_w^{len}(G_{mv}, G'_{mv}, r, s, r', s')$ as larger regions have more chance to become a start or end vertex. Eq. 11 can be rewrite as:

$$k_w^{len}(G_{mv}, G'_{mv}) = \sum_{\substack{r\in V, s\in V\\ r'\in V', s'\in V'}} Vk_{start}(ind(r) * (MK-1) + ind(s)) k_w^{len}(G, G', r, s, r', s') \; Vk_{end}(ind(r') * (MK-1) + ind(s')) \tag{12}$$

where $ind(r)$ denotes the index of region r, and $Vk_{start}$is a MK-length vector, each entity denotes the probability of choosing the start region pair. More weights will give regions with larger pixels, i.e.,

$$Vk_{start}(i(M-1)+j) = \frac{nPixel(r) * nPixel(s)}{nPixel(I) * nPixle(I')} \tag{13}$$

where $nPixel(\cdot)$ counts the number of pixels. In our approach, we set $Vk_{start} = Vk_{end} = V$ for simplicity.

We can rewritten (12) in a matrix form as:

$$k_w^{len}(G_{mv}, G'_{mv}) = (Vk_{start})' * \lambda^{len} W^{len} * Vk_{end} \tag{14}$$

where W is a matrix on $\mathbb{R}^{MK\times MK}$and each entity of W is defined as:

$$W_{in'+i', jn+j'} = k(l_{G_{mv}}(v_i), l_{G'_{mv}}(v'_i)) * k(l_(G_{mv})(v_j), l_(G'_{mv})(v'_j)) \tag{15}$$

where $W_{in'+i', jn+j'}$ denotes the kernel value between a 1-length walk (v_i, v_j) on G_{mv} and $(v_{i'}, v_{j'})$ on G'_{mv} and λ is a tuning parameter.

Based on the matrix W obtained in (15), the final kernel is obtained by summing all kernel value of length 1 to length infinity,i.e.,

$$\begin{aligned} k(G_{mv}, G'_{mv}) &= \sum_{p=1}^{\infty} (Vk_{start})' * \lambda^p W^p * Vk_{end} \\ &= (Vk_{start})' * lim(k \to \infty) \frac{(\lambda W)(I - (\lambda W)^{k-1})}{I - \lambda W} * Vk_{end} \end{aligned} \tag{16}$$

where I is an identity matrix; entity in W is between 0 and 1, $lim_{p\to\infty}(W)^{p-1} = 0$,thus (16) can be rewritten as:

$$k(G_{mv}, G_{mv}^{'}) = (Vk_{start})^{'} * \lambda W(I - \lambda W)^{-1} * Vk_{end} \quad (17)$$

In our approach, we develop an algorithm to speed up the calculation of $(I - \lambda W)^{-1} * Vk_{end}$ in (17). As shown in Table 2, we first solve equation:

$$(I - \lambda W)x = Vk_{end} \quad (18)$$

Let $x_1 = Vk_{end}$, denote x_t the value of x in the t-th iteration, we then need to compute $x_{t+1} = Vk_{end} + \lambda W x_t$ repeatedly until the stopping criterion is met. The stopping criterion can either based on the variation of x_t between two consecutive steps, or more simply, on a maximal number of iterations. In this paper, we use the first criterion. Iteration stopped when $||x_{t+1} - x_t|| < \delta$, δ a pre-defined threshold, in our approach, we set $\lambda < 1/\max(eig(W))$ to ensure converge, $\max(eig(W))$ is the maximal eigenvalue of W.

Table 2. Calculation of $(I - \lambda W)^{-1} * Vk_{end}$(Algorithm 2)

input: W and Vk_{end}; **output**: $(I - \lambda W)^{-1} * Vk_{end}$;
begin:
1. Set $x = Vk_{end} + \lambda W x$;
2. **do**
   $x^{'} = Vk_{end} + \lambda W x$; $x^{'} = x$;
   while $||x_{t+1} - x_t|| \geq \delta$
3. Return $(I - \lambda W)^{-1} * Vk_{end} = x$;
end

4.3 Analysis of Computational Complexity

In Algorithm 2, W and x are both $MK \times MK$ matrices. As shown in Table 2, in each iteration, calculation time complexity is $O(M^2K^2)$, therefore total time complexity $O(TM^2K^2)$ is needed to obtain the $MK \times MK$ matrix $(I - \lambda W)^{-1} * Vk_{end}$, where T is the number of iterations. Vk_{start} is a MK-length vector and time complexity of $Vk_{start} * x$ is $O(MK)$, thus time complexity to calculate our FMGK is $O(TM^2K^2)$, in our experiment, we found that number of iteration $T < 20$, therefore the computational complexity of FMGK is reduced to $O(M^2K^2)$ if the M and K is larger compared to T, which is common case as view images are over-segmented.

Compared with direct walk kernel whose time complexity $O(K^2M^2D^{len})$, which is exponential increasing with walk length,the time complexity of our FMGK is quadratically increasing with the region number M. Besides, all length walk kernel are calculated and summed, leading to a more expressive kernel.

5 Experimental Results and Analysis

The experiment is carried out on two datasets, i.e., ETH-80 and VOC2008. The experiment runs on a system equipped with Intel E8500 and 4GB RAM. And the algorithm of our FMGK is implemented on Matlab platform.

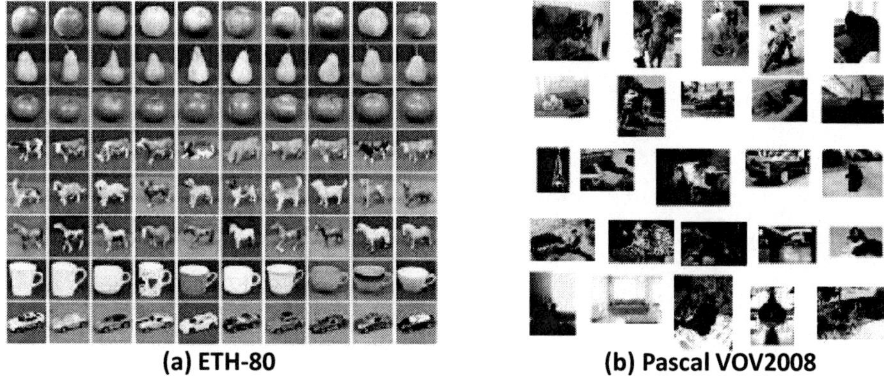

Fig. 1. Sample images from ETH-80(a) and Pascal VOC2008(b)

5.1 Classifying Multi View Images on ETH-80

In this experiment, we evaluate our FMGK on ETH-80. As shown in Fig. 1(a), ETH-80 image dataset [17] consists of color images of 80 objects from eight different categories: apples, tomatoes, pears, toy-cows, toy-horses, toy-dogs, toy-cars and cups. Each category contains 10 objects with 41views per object, spaced equally over the view hemisphere. The whole dataset contains 3280 128×128 images. A set of binary images, called segmentation mask, is provided for each color image. In each object category, we use 9 objects as training data and the rest 1 object as testing data. We evaluate objects describing with different number of view images, K. We set N as 41, 10, 4 and 1 respectively. The corresponding sample number is 80, 320, 800 and 3280. Notice that when $N = 10$ and 4, we use 40 view images per object, the last view image are abandoned. We use color histograms as features of each region.

Several other previous methods: i.e., global RGB histogram, global SIFT histogram, two rotation-variant descriptors, i.e., derivatives in x and y direction over three different scales (DxDy) and gradient magnitude and the Laplacian (Mag-Lap) over three different scales, PCA on raw segmentation masks(PCA mask), PCA on the segmented gray-value images(PCA gray), and a combination of the six above global features(Cmob6). We present the average classification over each category of ETH-80 in Table3.

5.2 Discussions the Time Consumption

To evaluate the time consumption of our FLMK on PASCAL VOC 2008 [16]. As shown in Fig. 1(b), PASCAL VOC 2008 consists of 10057 images from 20 categories, with 4340 images for training and 5717 image for testing. In this experiment, we fix the number of views, K, to 1. Then we implement the algorithm of FLWK [20], and present the comparison of time consumption with respect to the length of FLWK in Fig. 2. As expected, time consumption of FLWK increasing shapely with the value of len, which is consistent with the theoretical analysis in Section 4.1. Besides, the time consumption of our FMGK is much smaller than that of FLWK, which demonstrate the efficiency of our FMGK.

Table 3. The average recognition accuracy (%) on ETH-80 dataset

Cate.	RGB Hist.	SIFT Hist.	D_xD_y	Mag-Lap	PCA mask	PCA gray	Cmob6	FMGK
apple	57.56	77.34	85.37	80.24	78.78	88.29	97.10	98.6
pear	66.10	80.23	90.00	85.37	99.51	99.76	90.22	97.8
toma.	98.54	88.66	94.63	97.07	67.80	76.59	91.32	93.2
cow	86.59	70.32	82.68	94.39	75.12	62.44	70.24	77.4
dog	34.63	60.22	62.44	74.39	72.20	66.34	64.44	69.4
horse	32.68	71.02	58.78	70.98	77.80	77.32	63.26	74.3
cup	79.76	90.12	66.10	77.80	96.10	96.1	97.34	98.8
car	62.93	93.14	98.29	77.56	100.0	97.07	96.16	100.0

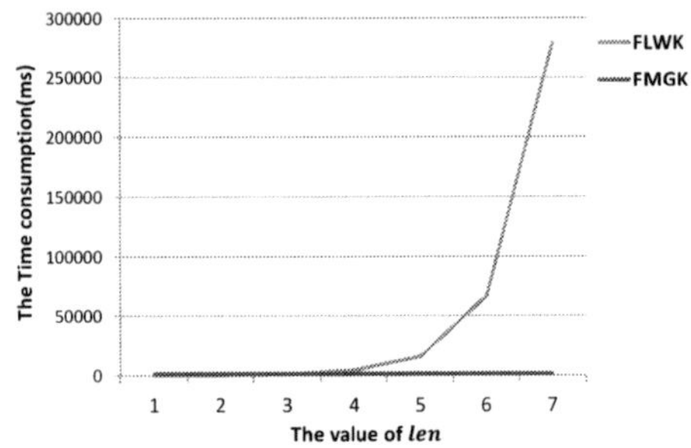

Fig. 2. Time consumption with respect to the length of FMGK

6 Conclusions

In this paper, we present FMGK,a new method to build the representation of multi-view images,for object classification. By constructing inter- and intra- view linkage graphs to represent the spatial relations of regions between and within each multi-view images,we integrate the two types of graphs into a multi-view region graph to describe an object. Towards an efficient kernel computation,we propose product graph between multi-view region graphs. Experimental results on two popular datasets validate the effectiveness of our FMGK.

References

1. Yuan, X.T., Zhu, H.W., Yang, S.T.: A Robust Framework For Eigenspace Image Reconstruction. In: IEEE Workshop on Motion and Video Computing, pp. 54–59 (2005)
2. Todorovic, S., Ahuja, N.: Region-based hierarchical image matching. In: IJCV (2007)
3. Fatih Demirci, M., Shokoufandeh, A., Keselman, Y., Bretzner, L., Dickinson, S.J.: Object recognition as many-to-many feature matching. In: IJCV (2006)
4. Felzenszwalb, P.F.: Pictorial structure for object recognition. In: IJCV (2005)
5. Hedau, V., Arora, H., Ahuja, N.: Matching images under unstable segmentations. In: Proc. of CVPR, pp. 1–8 (2008)
6. Keselman, Y., Dickinson, S.J.: Generic Model Abstraction from Examples. IEEE T-PAMI, 1141–1156 (2005)

7. Shervashidze, N., Vishwanathan, S.V.N., Petri, T., Mehlhorn, K., Borgwardt, K.M.: Efficient Graphlet Kernels for Large Graph Comparison. In: Proc. of ICAIS, pp. 488–495 (2009)
8. Lazebnik, S., Schmid, C., Ponce, J.: Beyond Bags of Features: Spatial Pyramid Matching for Recognizing Natural Scene Categories. In: Proc. of CVPR, pp. 2169–2178 (2006)
9. Cao, L., Fei-Fei, L.: Spatially Coherent Latent Topic Model for Concurrent Segmentation and Classification of Objects and Scenes. In: Proc. of ICCV, pp. 1–8 (2007)
10. Porway, J., Wang, K., Yao, B., Zhu, S.C.: Scale-invariant shape features for recognition of object categories. In: Proc. of ICCV, pp. 90–96 (2004)
11. Duda, R.O., Hart, P.E., Stork, D.G.: Pattern Classification. Wiley-Interscience (2000)
12. Lazebnik, S., Schmid, C., Ponce, J.: Beyond Bags of Features: Spatial Pyramid Matching for Recognizing Natural Scene Categories. In: Proc. of ICCV, pp. 2169–2178 (2006)
13. Cao, L., Fei-Fei, L.: Spatially Coherent Latent Topic Model for Concurrent Segmentation and Classification of Objects and Scenes. In: Proc. of ICCV, pp. 1–8 (2007)
14. Quelhas, P., Monay, F., Odobez, J.-M., Gatica-perez, D., Tuytelaars, T., Van Gool, L.: Modeling scenes with local descriptors and latent aspects. In: Proc. of ICCV, pp. 883–890 (2005)
15. Shi, J., Malik, J.: Normalized Cuts and Image Segmentation. IEEE T-PAMI (2000)
16. Everingham, M., Van Gool, L., Williams, C.K.I., Winn, J., Zisserman, A.: The PASCAL Visual Object Classes Challenge (VOC 2008) Results (2008), `http://www.pascal-network.org/challenges/VOC/voc2008/workshop/index.html`
17. Leibe, B., Schiele, B.: Analyzing Appearance and Contour Based Methods for Object Categorization. In: Proc. of CVPR, pp. 409–415 (2003)
18. Xia, T., Tao, D., Mei, T., Zhang, Y.: Multiview Spectral Embedding. IEEE T-SMCB, 1438–1446 (2010)
19. Xie, B., Mu, Y., Tao, D., Huang, K.: m-SNE: Multiview Stochastic Neighbor Embedding. IEEE T-SMCB, 338–346 (2010)
20. Harchaoui, Z., Bach, F.: Image Classification with Segmentation Graph Kernels. In: Proc. of CVPR (2007)

Image Feature Selection Based on Ant Colony Optimization

Ling Chen[1,2], Bolun Chen[1], and Yixin Chen[3]

[1] Department of Computer Science, Yangzhou University, Yangzhou, China
Chenbolun1986@gmail.com

[2] State Key Lab of Novel Software Tech, Nanjing University, Nanjing, China
lchen@yzu.edu.cn

[3] Department of Computer Science, Washington University in St Louis, USA
chen@cse.wustl.edu

Abstract. Image feature selection (FS) is an important task which can affect the performance of image classification and recognition. In this paper, we present a feature selection algorithm based on ant colony optimization (ACO). For n features, most ACO-based feature selection methods use a complete graph with $O(n^2)$ edges. However, the artificial ants in the proposed algorithm traverse on a directed graph with only $2n$ arcs. The algorithm adopts classifier performance and the number of the selected features as heuristic information, and selects the optimal feature subset in terms of feature set size and classification performance. Experimental results on various images show that our algorithm can obtain better classification accuracy with a smaller feature set comparing to other algorithms.

Keywords: ant colony optimization, dimensionality reduction, feature selection, image classification.

1 Introduction

Reduction of pattern dimensionality via feature extraction is one of the most important tasks for pattern recognition and classification. Feature selection has considerable importance in areas such as bioinformatics [1], signal processing [2], image processing [3], text categorization [4], data mining [5], pattern recognition [6], medical diagnosis [7], remote sensor image recognition [8]. The goal of feature selection is to choose a subset of available features by eliminating unnecessary features. To extract as much information as possible from a given image set while using the smallest number of features, we should eliminate the features with little or no predictive information, and ignore the redundant features that are strongly correlated. Quality of the results of feature selection can affect the performance of image classification and recognition [6].

Many feature selection algorithms involve heuristic or random search strategies in order to reduce the computing time. For a large number of features, heuristic search is often used to find the best subset of features. More recently, nature inspired algorithms are used for feature selection. Those population-based optimization algorithms for

D. Wang and M. Reynolds (Eds.): AI 2011, LNAI 7106, pp. 580–589, 2011.

feature selection such as genetic algorithm (GA) [3,9], ant colony optimization (ACO) [1,4], particle swarm optimization (PSO) [10] have been proposed. These methods are stochastic optimization techniques attempting to achieve better solutions by referencing the feedback and heuristic information.

Ant colony optimization (ACO) is an evolution simulation algorithm proposed by M. Dorigo et al. [11]. It has been successfully used for system fault detecting, job-shop scheduling, network load balancing, graph coloring, robotics and other combinational optimization problems.

In this paper, we present an ACO-based feature selection algorithm, ACOFS, to reduce the memory requirement and computation time. In this algorithm, the artificial ants traverse on a directed graph with only $2n$ arcs. The algorithm adopts classifier performance and the number of the selected features as heuristic information, and selects the optimal feature subset in terms of the feature set size and classifier performance. Experimental results on image data sets show that the proposed algorithm has superior performance. Comparing with other existing algorithms, our algorithm can obtain better classification accuracy with a smaller feature set from images.

2 The ACO Algorithm for Image Feature Selection

Given a feature set of size n, the feature selection problem is to find a minimal feature subset of size s ($s< n$) while maintaining a fairly high classification accuracy in representing the original features. Most of the ACO based algorithms for feature selecting use a complete graph, on which the ants try to construct a path with part of the nodes. In the ACO on such complete graph, ant on one node (feature) selects an edge connecting another node (feature) based on the pheromone and heuristic information assigned on this edge between the two nodes (features). Since the solution of feature selection is a subset of those selected features, there is no any ordering among the components of the solution. Therefore, it unnecessary to use a complete graph with $O(n^2)$ edges in the ACO algorithm.

To efficiently apply an ACO algorithm for feature selection, we must redefine the way that the representation graph is used. We proposed ant optimization algorithm on a discrete search space represented by a directed graph with only $O(n)$ arcs as shown in Figure 1, where the nodes represent features, and the arcs connecting two adjacent nodes indicating the choice of the next feature.

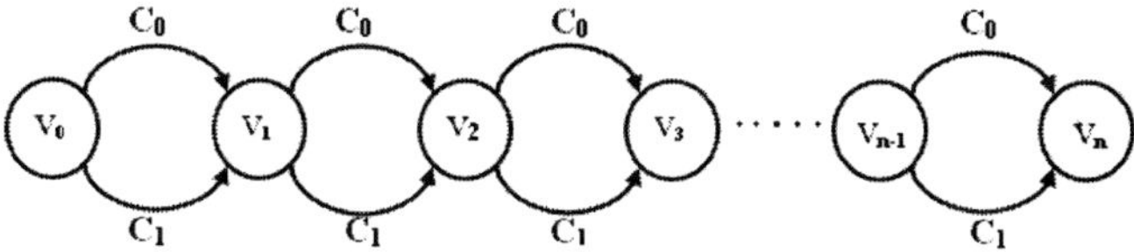

Fig. 1. The directed graph

Denote the n features as $f_1, f_2, \ldots, f_n$, the ith node v_i is used to represent feature f_i. An additional node v_0 is placed at the beginning of the graph where each ant starts its search. As shown in Figure 1, the ants travel on the directed graph from v_0 to v_1, and then to v_2, and so on. The ant terminates its tour and outputs this feature subset as it reaches the last node v_n. When an ant completes the search from v_0 to v_n, the arcs on its trace form a solution.

There are two arcs named C_j^0 and C_j^1 linking two adjacent nodes v_{j-1} and v_j. If an artificial ant at v_j selects arc C_j^0 (or C_j^1), the jth feature is selected (or not selected). On each arc C_i^j, virtual pheromone value τ_i^j is assigned as the feedback information to direct the ants' searching on the graph. We initialize pheromone matrix τ as $\tau_i^j=1$ for all i=1,2,…,n and j=0,1.

The search for the optimal feature subset is the procedure of the ants traversing through the graph. Suppose an ant is currently at node v_{i-1} and has to choose one path connecting v_i to pass through. A probabilistic function of transition, denoting the probability of an ant at node v_{i-1} to choose the path c_i^j to reach v_i is designed by combining the heuristic desirability and pheromone density of the arc. The probability of an ant at node v_{i-1} to choose the arc c_i^j at time t is:

$$p_i^j(t)=\frac{[\tau_i^j(t)]^\alpha(\eta_i^j)^\beta}{[\tau_i^0(t)]^\alpha(\eta_i^0)^\beta+[\tau_i^1(t)]^\alpha(\eta_i^1)^\beta} \quad (i=1,2,\ldots,n;\quad j=0,1) \tag{1}$$

Here, $\tau_i^j(t)$ is the pheromone on the arc c_i^j between nodes v_{i-1} and v_i at time t, which reflects the potential tend for ants to follow arc c_i^j (j=0,1). η_i^j is the heuristic information reflecting the desirability of choosing arc c_i^j. α and β are two parameters that determine the relative importance of the pheromone and the heuristic information.

From (1) we can see that the transition probability used by ACO depends on the pheromone intensity $\tau_i^j(t)$ and heuristic information η_i^j.To effectively balance the influences of positive feedback information from previous high-quality solutions and the desirability of the arc, we should chose proper values of the parameters α and β. When α= 0, no positive feedback information is used. Since the previous search experience is lost, the search degrades to a stochastic greedy search. When β=0, the potential benefit of arcs is neglected, and it becomes a entirely random search.

The heuristic information η_i^1 is the desirability of choosing the arc c_i^j between nodes v_{i-1} and v_i, which means the preference of ant to choose the feature f_i. There are many ways to define a suitable value of η_i^1. It could be any evaluation function on the discrimination ability of a feature f_i, such as rough set dependency measure, or entropy-based measure. We set the value of η_i^1 using F-score, which is a easy measurement to evaluate the discrimination ability of feature f_i, defined as follows:

$$\eta_i^1 = \frac{\sum_{k=1}^{m}\left(\overline{x_i}^k - \overline{x_i}\right)}{\sum_{k=1}^{m}\left(\frac{1}{N_i^k - 1}\sum_{j=1}^{N_i^k}\left(x_{ij}^k - \overline{x_i}^k\right)^2\right)} \quad (i = 1,...,n) \tag{2}$$

Here, m is the number of classes of the image set; n is the number of features; N_i^k is the number of samples of the feature f_i in class k, (k =1,2,…,m, i=1,2,…,n), x_{ij}^k is the jth training sample for the feature f_i of the images in class k, (j=1,2,…, N_i^k), $\overline{x_i}$ is the mean value of the feature f_i of all images , $\overline{x_i}^k$ is the mean of the feature f_i of the images in class k.

In (2), the numerator indicates the discrimination between the classes of the image set, and the denominator specifies the discrimination within each class. A larger η_i^1 value implies that the feature f_i has a greater discriminative ability.

For the value of η_i^0, we simply set $\eta_i^0 = \frac{\xi}{n}\sum_{i=1}^{n}\eta_i^1$, where $\xi\in(0,1)$ is a constant.

3 Implementation of the Algorithm

In an ACO based optimization method, the design of the pheromone update strategy, and the measurement of the quality of the solutions are critical.

3.1 Pheromone Updating

In each iteration, the algorithm ACOFS updates the pheromone value on each arc according to the pheromone and heuristic information on the arc.

Obviously, if an ant chooses the arc c_i^j, pheromone on this arc should be assigned more increment, and ants should select arc c_i^j with higher probability in the next iteration. This forms a positive feedback of the pheromone system. In each iteration, the pheromone on each arc is updated according to formulas (3),(4) and (5).

$$\tau_i^j(t+1) = \rho\cdot\tau_i^j(t) + \Delta\tau_i^j(t) + Q_i^j(t) \tag{3}$$

where

$$\Delta\tau_i^j(t) = \frac{1}{|S_i^j|}\sum_{s\in S_i^j} f(s) \tag{4}$$

and

$$Q_i^j(t)=\begin{cases} Q & c_i^j \in s_{best} \\ 0 & otherwise \end{cases} \tag{5}$$

In (4) , s_i^j is the set of solutions generated at the t-th iteration passing through c_i^j . In (5), S_{best} is the best solution found so far, and Q is a positive constant. To emphasize the influence of the best-so-far solution, we add an extra pheromone increment on the arcs included in S_{best} .

3.2 The Fitness Function

Based on the ant's solution, which is a selected feature subset, the solution quality in terms of classification accuracy is evaluated by classifying the training data sets using the selected features. The test accuracy measures the number of examples that are correctly classified. In addition, the number of features in the set is also considered in the quality function. The subset with less features could get higher quality function value. The quality function $f(s)$ of a solution s is defined as follows:

$$f(s)=\frac{N_{corr}}{1+\lambda N_{feat}} \tag{6}$$

where N_{corr} the number of examples that are correctly classified, N_{feat} is the number of features selected in s, λ is a constant to adjust the importance of the accuracy and the number of features selected. The scheme obtaining higher accuracy and with less features will get greater quality function value.

4 Experimental Results

To test the effectiveness and performance of our proposed feature selection algorithm ACOFS, we test it by a series of experiments. All experiments have been run on Pentium IV, Windows XP, P1.7G, using VC++ 6. 0, and the results are visualized on Matlab 6.0.

A set of images was tested to demonstrate the classificatory accuracy and determine whether the proposed algorithm can correctly select the relevant features. The data set contains 80 images in 4 classes. The data set has 19 features including first and second order origin moment, first and second order central moment, twist degree, peak values, entropy of the moments, and the statistical of the gray differential statistics, such as contrast, angle second-order moment (ASM), mean value, entropy etc.

On the image set, the ACFS algorithm is applied to select the relevant features and is compared to GA-based approach GAs [21] and the modified ACO algorithm for feature selection presented in [26] which is denoted as mACO.

For GA-based feature selector GAs, we set the length of chromosomes as the number of features. In a chromosome, each gene g_i corresponds to the ith feature.

If g_i=1, this means we select the ith feature. Otherwise, g_i=0 , which means the ith feature is ignored. By iterations of producing chromosomes for the new generation, crossover and mutation, the algorithm tries to find a chromosome with the smallest number of 1's and higher classifier accuracy. In order to select the individuals for the next generation, GA's roulette wheel selection method was used. We set the parameters of GAs as follows: probabilities of crossover and mutation are P_{croos}=0.9 and $P_{mutation}$ =0.25, the population size is m=50, and the maximum iterations k = 50.

For ACO-based algorithms ACOFS and mACO, we have applied them with the same population size as GA based algorithm GAs. Various parameters leading to a better convergence are tested and the best parameters that are obtained by simulations are as follows: α=1, β=0.5, evaporation rate ρ =0.951, the initial pheromone intensity of each arc is equal to 1, the number of ant in each iteration m = 50 and the maximum iterations k = 50. These values are chosen to justify the comparison with GAs. For each subset of the features obtained, its quality is measured by classifying the training image sets using SVM classifier. The number of the selected features and the quality of the classification results are considered for performance evaluation.

To evaluate the average classification accuracy of the selected feature subsets, 10-fold and 5-fold cross validation (CV) is used. For the three algorithms, the CV accuracy on the training and testing data of the best-so-far solution at each iteration are computed and recorded.

Table 1 shows the number of features selected in the best solution obtained by the three algorithms. From the table we can see that ACOFS selects the smallest number of features while maintain the high accuracy of classification.

Table 1. Number of features selected by the three algorithms

Algorithm	ACOFS	GAs	mACO
5-fold CV	7	8	10
10-fold CV	9	10	12

We measure the quality of the classification results in two criterions, namely, recall and precision of each class. The average recall and the precision of the classification of the ith class are defined as follows:

$$recall(i)=\frac{N_{TP}(i)}{N_c(i)} \qquad precision(i)=\frac{N_{TP}(i)}{N_{FP}(i)+N_{TP}(i)} \tag{7}$$

Here, $Nc(i)$ is the number of images in the ith class, $N_{TP}(i)$ is the number of images correctly classified into the ith class, $N_{FP}(i)$ is the number of images incorrectly classified into the ith class.

To obtain a more reliable result, 10 runs were conducted by 10-fold and 5-fold cross-validation with on each of the ten image sets. Tables 2 to 5 present the average recall on the first four training and testing image sets by 5-fold and 10-fols CV tests.

Table 2. Recall of the results in 5-fold CV on training sets

Class ID	ACOFS	GAs	mACO
1	100%	87.5%	93.75%
2	100%	100%	100%
3	100%	87.5%	81.25%
4	87.5%	81.25%	87.5%
Average	96.88%	89.06%	90.63%

Table 3. Recall of the results in 5-fold CV on testing sets

Class ID	ACOFS	GAs	mACO
1	75%	75%	75%
2	100%	100%	100%
3	100%	100%	100%
4	100%	75%	75%
Average	93.75%	87.5%s	87.5%

Table 4. Recall of the results in 10-fold CV on training sets

Class ID	ACOFS	GAs	mACO
1	100%	100%	94.44%
2	100%	100%	100%
3	94.44%	94.44%	100%
4	88.89%	88.89%	83.33%
Average	95.83%	95.83%	94.44%

Table 5. Recall of the results in 10-fold CV on testing sets

Class ID	ACOFS	GAs	mACO
1	100%	100%	94.44%
2	100%	100%	100%
3	100%	50%	100%
4	100%	100%	100%
Average	100%	87.5%	98.61%

Comparing the criterion of recall and the number of features selected, we can see from the tables that proposed ACOFS algorithm outperforms the mACO, GAs algorithms. The number of features selected by the ACOFS algorithm is 7 and 9 in the 5-fold and 10-fold CV test respectively, while 8 and 10 by the GAs, 10 and 12 by mACO. Furthermore, while using less features, the ACOFS algorithm gets higher recall than the mACO, GAs algorithms. For instance, in the 10-fold CV test on the testing data, the average recall of ACOFS is 100%, while that is 87.5% for GAs and 98.61% for mACO. This means the recall of the results by ACOFS are always better than those of mACO, GAs algorithms.

Tables 6 to 9 list the average precision of the training and testing image sets by 5-fold and 10-fols CV tests.

Table 6. Precision of the results in 5-fold CV on training sets

Class ID	ACOFS	GAs	mACO
1	88.89%	82.35%	78.95%
2	100%	94.12%	88.89%
3	100%	82.35%	100%
4	100%	100%	100%
Average	97.22%	89.71%	91.96%

Table 7. Precision of the results in 5-fold CV on testing sets

Class ID	ACOFS	GAs	mACO
1	100%	75%	75%
2	100%	100%	100%
3	80%	80%	100%
4	100%	100%	75%
Average	95%	88.75%	87.5%

Table 8. Precision of the results in 10-fold CV on training sets

Class ID	ACOFS	GAs	mACO
1	85.71%	85.71%	85%
2	100%	100%	100%
3	100%	100%	100%
4	100%	100%	93.75%
Average	96.43%	96.43%	94.69%

Table 9. Precision of the results in 10-fold CV on testing sets

Class ID	ACOFS	GAs	mACO
1	100%	66.67%	100%
2	100%	100%	100%
3	100%	100%	100%
4	100%	100%	94.74%
Average	100%	91.66%	98.69%

We can see from the tables that the proposed ACOFS algorithm has better precision than the mACO, GAs algorithms. Even using fewer features, the ACOFS algorithm still can obtain higher precision than the mACO, GAs algorithms. For instance, in the 5-fold CV test on the testing sets, the average precision of ACOFS is 95%, while that is 88.75% for GAs and 87.5% for mACO. This means the precision of the results by ACOFS are always better than those of algorithms mACO and GAs.

Comparison of the average recall and precision of the three algorithms on the ten data sets are shown in Figures 2 and 3 respectively. We can conclude from the figures and tables that the proposed ACOFS algorithm can successfully select subset of features which can obtain high classification accuracy. Compared with algorithms GAs and mACO in the tests using the same image set, ACOFS can obtain better classification accuracy but had a smaller feature set.

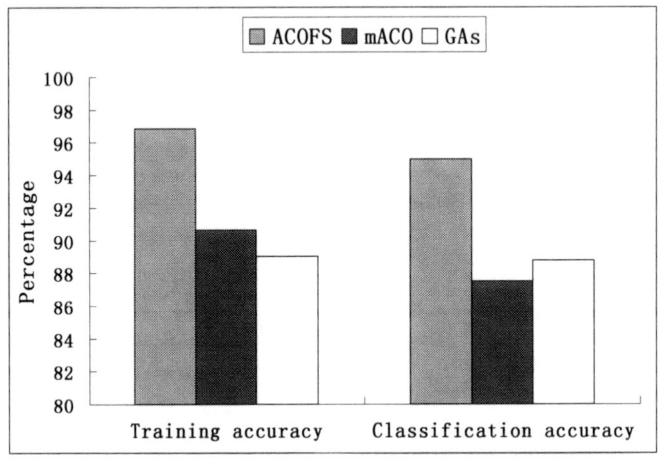

Fig. 2. Comparison of the average recall of the three algorithms

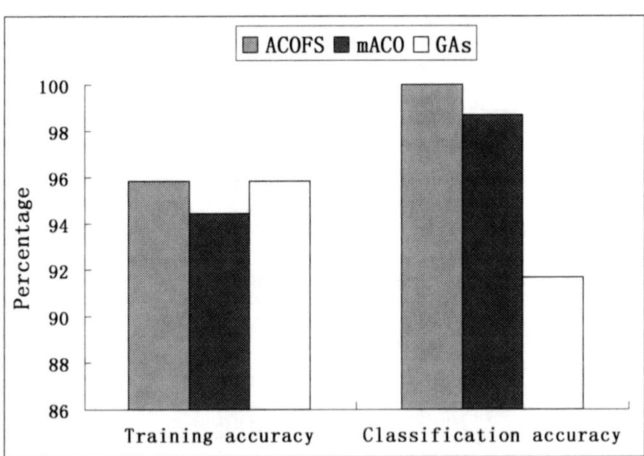

Fig. 3. Comparison of the average precision of three algorithms

5 Conclusions

We proposed an ACO-based feature selecting algorithm ACOFS. The algorithm adopts classifier performance and the number of the selected features as heuristic information, and selects the optimal feature subset in terms of smallest feature number and the best performance of classifier. The experimental results on image data sets show that the algorithm ACOFS can obtain better classification accuracy but had a smaller feature set than other similar methods.

Acknowledgments. This research was supported in part by the Chinese National Natural Science Foundation under grant No. 61070047,61070133 and 60773103, Natural Science Foundation of Jiangsu Province under contract BK21010134, and Natural Science Foundation of Education Department of Jiangsu Province under contract 08KJB520012 and 09KJB20013.

References

1. Basiri, M.E., Ghasem-Aghaee, N., Aghdam, M.H.: Using Ant Colony Optimization-Based Selected Features for Predicting Post-synaptic Activity in Proteins. In: Marchiori, E., Moore, J.H. (eds.) EvoBIO 2008. LNCS, vol. 4973, pp. 12–23. Springer, Heidelberg (2008)
2. Yeh, Y.-C., Wang, W.-J., Chiou, C.W.: Feature selection algorithm for ECG signals using Range-Overlaps Method. Expert Systems with Applications 37(4), 3499–3512 (2010)
3. Lu, J., Zhao, T., Zhang, Y.: Feature selection based-on genetic algorithm for image annotation. Knowledge-Based Systems 21(8), 887–891 (2008)
4. Aghdam, M.H., Ghasem-Aghaee, N., Basiri, M.E.: Application of ant colony optimization for feature selection in text categorization. In: Proc. CEC 2008, Proceeding of the Fifth IEEE Congress on Evolutionary Computation. IEEE Press, Hong Kong (2008)
5. Lutu, P.E.N., Engelbrecht, A.P.: A decision rule-based method for feature selection in predictive data mining. Expert Systems with Applications 37(1), 602–609 (2010)
6. Jensen, R.: Combining rough and fuzzy sets for feature selection. PhD thesis, University of Edinburgh (2005)
7. Polat, K., Güneş, S.: Medical decision support system based on artificial immune recognition immune system (AIRS), fuzzy weighted pre-processing and feature selection. Expert Systems with Applications 33(2), 484–490 (2007)
8. Gundimada, S., Asari, V.K., Gudur, N.: Face recognition in multi-sensor images based on a novel modular feature selection technique. Information Fusion 11(2), 124–132 (2010)
9. Oliveira, L.S., Sabourin, R., Bortolozzi, F., Suen, C.Y.: A methodology for feature selection using multi-objective genetic algorithms for hand written digit string recognition. International Journal of Pattern Recognition and Artificial Intelligence 17(6), 903–929 (2003)
10. Wang, X., Yang, J., Teng, X., Xia, W., Jensen, R.: Feature selection based on rough sets and particle swarm optimization. Pattern Recognition Letters 28, 459–471 (2007)
11. Dorigo, M., Birattari, M., Stützle, T.: Ant Colony Optimization:Artificial Ants as a Computational Intelligence Technique. IEEE Computational Intelligence Magazine (11), 28–29 (2006)

Realistic Smile Expression Recognition Using Biologically Inspired Features

Cong He, Huiyun Mao, and Lianwen Jin

School of Electronic and Information Engineering
South China University of Technology
386 Wushan Road, Guangzhou, 510641, China
{riverho1015,huiyun.mao01,Lianwen.jin}@gmail.com

Abstract. A robust smile recognition system could be widely used for many real-world applications. In this paper, we introduce biologically inspired model (BIM) into the building of realistic smile classification system for solving challenging realistic tasks. To improve the performance of BIM, we develop a modified BIM (MBIM), which utilizes a more efficient pooling operation and boosting feature selection. Experiments demonstrate the effectiveness of themodifications and adjustments of BIM. By testing on the challenging realistic database, GENKI, our method is proved to be superior to some other state-of-the-art smile classification algorithms.

Keywords: Facial expression classification, smile recognition, biologically inspired model.

1 Introduction

Automatic facial expression recognition have been studied world widely in the last two decades and has become a very active research area in computer vision and pattern recognition. Facial expression classification is a key component for human-computer interaction and related fields. Smile, an important part of facial expression, delivers a variety of emotional information, e.g., joy, happiness, attractiveness and kindness, so it plays an important role in human communication. Automatic smile classification technology has entered people's lives and become more and more popular with the widespread use of commercial digital products, such as digital cameras, digital videos and social robots.

The last few years have witnessed considerable progress in the field of automatic facial expression classification and a large number of classification algorithms have been proposed. Common approaches include appearance-based analysis[1,2], local feature based classifiers[3,4,5,6], bag-of-features[7], bag-of-keypoints[8] , Gabor feature [23], and so on. Littlewort et al. [23] proposed to select a subset of Gabor features using an AdaBoost method and thereafter train a support vector machine algorithm using the selected features. They reported very high accuracy in recognition of facial expressions. Bai et.al proposed a high-performance smile classification method using pyramid histogram of orientation gradients features and classifier with

D. Wang and M. Reynolds (Eds.): AI 2011, LNAI 7106, pp. 590–599, 2011.

Adaboost and SVM[24]. These algorithms perform well on some public standard databases, such as Cohn-Kanade AU-Coded Facial Expression Image (CKACFEID) Database [20], Japanese Female Facial Expression (JAFFE) Database [21], etc. However, the accuracies of most of these algorithms decline when they are used to more challenging and realistic problems of classifying spontaneous expressions with varieties of illumination, pose variations, geometric transformations, occlusion and clutter. One of the weakness is that they ignore high level information, such as relations of local orientations. Therefore, there is much room for developing more efficient algorithms addressing practical problems in real world.

As known to all, human and primate's eyes outperform the best machine vision systems by almost any measure. According to models of object classification in visual cortex, the brain uses a hierarchical approach in which simple, low-level features with high position and scale specificity are pooled and combined into more complex, higher-level features with greater location and scale invariance [9,15]. It has always been an attractive idea to build a system that emulates object classification in cortex. Strictly following the organization of the cortex, a biologically inspired model (BIM) for object classification was developed by Serre et al[10]. Experiments on different classification tasks such as object recognition [26], scene classification [27] has illustrated that BIM and its variants have achieved best state-of-the-art performance.

In this paper, we introduce BIM into smile recognition for addressing solving challenging realistic task. We develop a modified biologically inspired model (MBIM) that improves the performance of BIM. It improves the sensitivity and informativeness of the pooling operation model by applying a new pooling operation which averages the sum of the max response and its neighbors. Furthermore, a boosting feature selection is applied in order to retain effective biologically inspired features (BIF) and remove uninformative features. The final features, which are called modified biologically inspired features (MBIF), are sent to a SVM classifier for final classification. Experiments on a challenging database, GENKI [11], which contains pictures from thousands of different people taking in many different real-world imaging conditions, show that MBIF is more effective and efficient than BIF and some other state-of-the-art approaches. We also try some other modifications on BIM and obtain several benefits in reducing computational time and storage of features extraction.

The organization of this paper is as follows. In Section 2, a brief review of BIM is given. In Section 3, we present modified biologically inspired features (MBIF) and our smile classification system based on MBIF. Experimental results on GENKI dataset are given in section 4. Section 5 summarizes this paper.

2 Biologically Inspired Model

Biologically inspired model (BIM) was built by Serre et al. [10,18] basing on the "HMAX" model of Riesenhuber and Loggia [9]. It follows the standard model of object classification in primate cortex [9]. There are four layers of computational units: S1, C1, S2, and C2, where S and C are named by analogy with the V1 simple and complex cells discovered by Hubel and Wiesel [12,13]. The framework of BIM and MBIM is shown in figure 1.

S1 Units: The units in S1 layer correspond to simple cells in primate primary visual cortex, V1, i.e., the first visual cortical stage [12,13]. The S1 units are obtained by applying to the input image a group of Gabor filters, each of which can be described as the product of an elliptical Gaussian envelope and a complex plane wave [10,16,17,18],

$$G(x,y)=\exp(-\frac{X+\gamma^2Y^2}{2\sigma^2})\times\cos(\frac{2\pi}{\lambda}X) \tag{1}$$

where $X = x\cos\theta + y\sin\theta$, $Y = -x\sin\theta + y\cos\theta$.

The filter parameters, i.e. σ (effective width), θ (orientation), λ (wavelength), γ (aspect ratio), are adjusted so that the tuning profiles of S1 units match those of V1 simple cells. There are 16 filters at 4 orientations and different parameters [10].

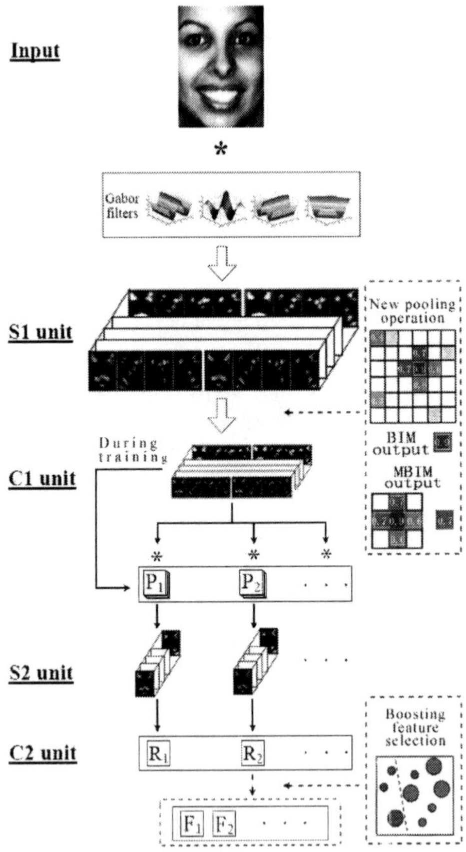

Fig. 1. Framework of BIM (do not include dashed parts) and MBIM

C1 Units: The C1 units correspond to complex cells which show some shift invariance and scale invariance [12,13]. C1 units pool over S1 units (of the same orientation) using the maximum operation of Riesenhuber and Poggio [9]. As a result, it can subsample S1 units to reduce the number of units and create position and scale tolerance (shown in Figure 2 [10,18]).

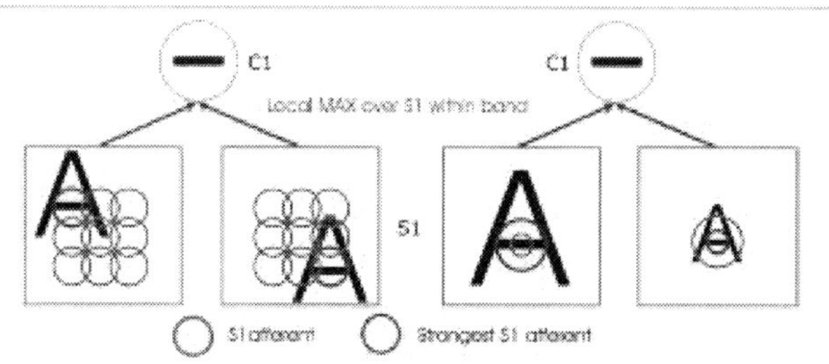

Fig. 2. Scale- and position-tolerance at C1 level

S2 Units: The S2 Units pool over C1 Units by performing a Gaussian-like template matching between afferent C1 units and each of the stored prototype patches [10,18],

$$Y = \exp(-\beta \|X - P_i\|^2) \tag{2}$$

where β defines the sharpness and P_i is one of the K prototype patches. The patches are randomly extracted from the positive training images at the C1 layers during the initial feature-learning state. This process is computed in all orientations, and it ends by setting each of the patches as prototypes of the S2 units which behave as radial basis function (RBF) units [19] during classification. Eq. (2) calculates the Euclidean distance between a new input patch and the stored prototype, and it reflects the similarity of them.

C2 Units: Finally, a global maximum over all scales and positions of each S2 units is taken to create a set of K shift- and scale-invariant C2 units. The size of the C2 feature vector depends only on the number (i.e., K) of stored prototypes but not on the size of the input image.

SVM Classifier: The C2 features (also known as BIF) are sent to a linear Support Vector Machine (SVM) for training and classification.

3 Modified BIM for Smile Classification

In this section we introduce a more effective polling operation and boosting feature selection for BIM. For challenging real-world smile classification tasks, we construct a classification system based on the proposed modified biologically inspired model (MBIM).

3.1 A More Effective Pooling Operation Method

As discussed in section 2, the original BIM simply uses the maximum model as the pooling model. The response of complex units is the maximum of the responses of all simple units over local area. There are two drawbacks in this maximum model: firstly, noises always appear in the high frequency band of images, especially in the images photographed in the real-world conditions, and simple maximum model may output the noises as the max responses of the simple units; secondly, the most strongly activated units strengthen responses of their neighbors [14], thus simple maximum model will lose some informative responses.

Therefore, in the MBIM,we introduce a more effective pooling operation between simple and complex cells. We first find the maximal response and its neighbors, then we remove all other weak responses, and finally, the average of all responses remained is computed. The pooling function (shown in Figure 3) in MBIM is given by

$$C = \frac{1}{N_I} \sum_{x_i, y_i \in I} S(x_i, y_i) \tag{3}$$

where $S(x_i, y_i)$ is the response of the ith simple unit, C is the response of complex unit, I is the neighborhood of the maximal response point in the local area, N is the number of responses in I.

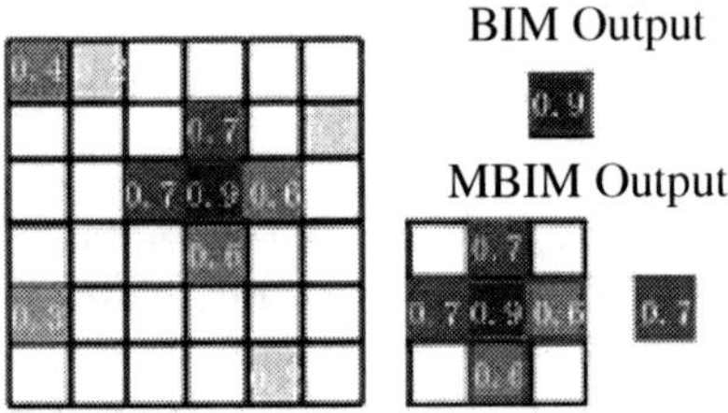

Fig. 3. The pooling operation in MBIM

It is expected that this new pooling function can not only smooth the noises, but also remain more information.

3.2 Feature Selection by AdaBoost

In the initial feature-learning stage of BIM, K prototype patches are randomly sampled from C1 layer of the training positive images. The total number of the final BIF is very large, this results in heavy computational cost and storage in classifier building stage, and part of the BIF is uninformative or even harmful to the accuracy of the classifier.

To solve these problems, we employ feature selection after C2 Units using the Adaboost algorithm[25]. In order to ensure fast and accurate classification, the feature selection process should exclude a large majority of BIF, and focus on a small set of critical features. In our AdaBoost procedure for BIM, the weak learner is restricted to only one single feature, so that each stage of the boosting process, which selects a new weak classifier, can be viewed as a feature selection process.

In MBIM, important and useful MBIFs are retained after the boosting feature selection and then they are sent to the SVM classifier for training and classification.

3.3 Other Strategies

In the empirical studies, we also try some other modifications on BIM to see what can improve the performance of BIM, through two strategies: (1) prototype patches are only extracted from the mouth or eyes area which is marked artificially; (2) prototype patches are clustered using K-means algorithm. The detailed performances of these strategies will be discussed in section 4.

3.4 Classification System Using Boosted MBIF

With the benefit of MBIF, we set up a system for smile classification, as shown in Figure 4. Firstly, we detect the face area in an image through the Haar-like detector of Viola and Jones [22]. Secondly we convert the face image to grayscale and normalize it to an image with size of 128 ×96 pixels. Then we extract the MBIF of the image and fed it to SVM classifier to produce final recognition result.

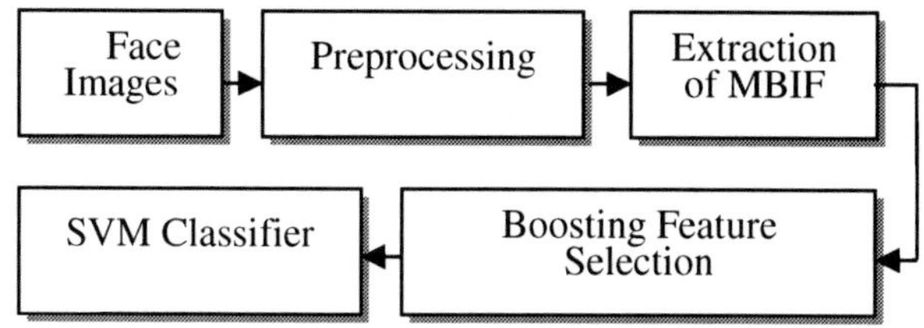

Fig. 4. Smile classification system using boosted MBIF

4 Experiments and Results

To evaluate the performances of the MBIF and our smile classification system, we make a series of empirical studies on smile classification on a challenging real-world database, GENKI [11].

4.1 GENKI Datasets

So far as we know, most current automatic facial expression classification studies have focused on databases that were collected under controlled conditions on a relatively small number of subjects. So, in order to validate the effectiveness of our system on real-world conditions, we choose GENKI [11] for our experiments. GENKI is collected from the Internet by the Machine Perception Laboratory of University of California. The GENKI dataset contains over 63,000 pictures, photographed by the subjects themselves, from tens of thousands of different people in many different real-world imaging conditions. It represents the diversity of illumination conditions, camera models, pose variations, personal difference and many other characters that are found in the real world. What we use here is GENKI-R2009a (some examples are shown in Figure 5), a subset of GENKI. We divide GENKI-R2009a into two parts: a training set which contains 2645 positive images (smile faces) and 2490 negative images (non-smile faces) for training, and a testing set of 597 positive images and 551 negative images.

4.2 Performance Comparison of MBIM against BIM

To test the effectiveness of each modification of MBIM, two set of experiments are designed: first, we justify the effectiveness of the new pooling operation by replacing the maximum operation with the new pooling operation in BIM, performance curves shown in Figure 6 demonstrate that the new pooling operation is valuable to improve accuracy and local optimal performance is obtained when K=1,000 (K is the number of features/patches); and then, we justify the effectiveness of the boosting feature selection by imposing it on BIM (using K=1,000), performance curves shown in Figure 7 demonstrate that the feature selection is useful to improve the classification accuracy and reduce the computational cost and storage, local optimal performance is obtained with 200 selected features, far less than original 1,000 features.

Fig. 5. Examples form GENKI-R2009a

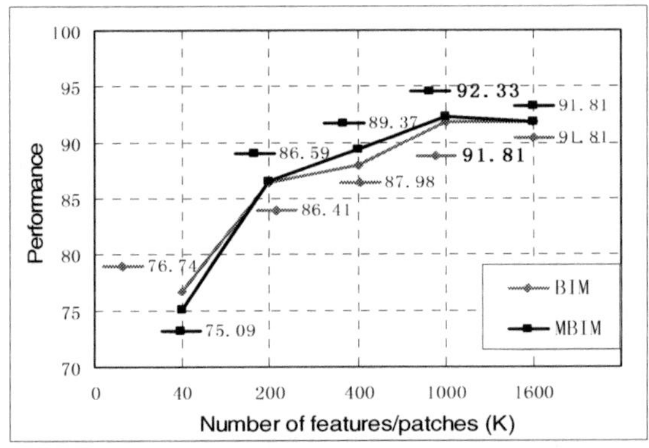

Fig. 6. Comparison between BIM and MBIM according to the new pooling operation

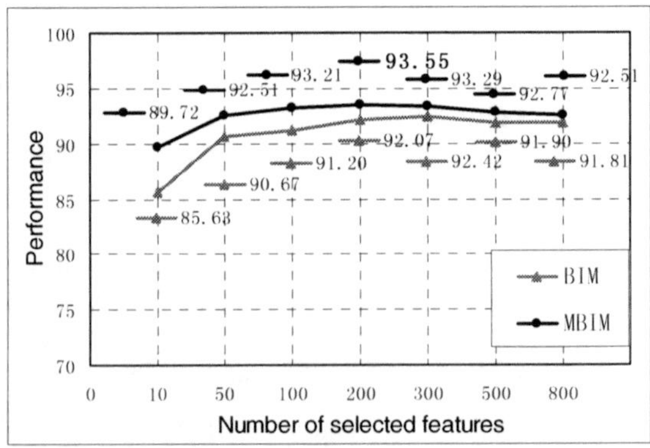

Fig. 7. Comparison between BIM and MBIM according to boosting feature selection

4.3 Performance of MBIF against BIF, Gabor Feature and PHOG Feature

We compare the proposed MBIF with the original BIF and two other state-of-the-art features for smile classification, namely the Gabor feature [23] and PHOG feature [24]. The recognition results are given in table 1. It can be seen that our method (MBIF) achieved the highest accuracy.

Table 1. Performance comparison among Gabor, PHOG, BIM and MBIM

Feature	GABOR	PHOG	BIF	MBIF
Recognition accuracy (%)	80.31	83.62	91.81	93.55

4.4 Performance When Extracting Patches from Specified Area

To compare the effectiveness of the patches extracted from different area, we artificially mark the mouth and eyes on the face (shown in Figure 8) and then extract K (=1,000) patches from each area respectively. The performance is shown in table 2. It can be seen that patches extracted from mouth area contain more information than those from eyes area, but both of them don't achieve high enough performance as patches extracted from all face area.

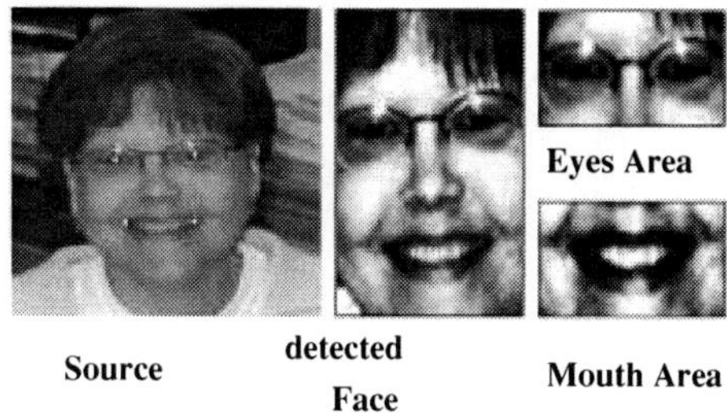

Fig. 8. Example of artificially marked mouth and eyes area

Table 2. Recognition accuracies (%) with patches extracted from face, eyes, mouth area

Area	BIM	MBIM
Whole Face	91.81	93.55
Eyes	85.71	86.06
Mouth	91.20	91.03

4.5 Performance of Clustering Patches

We extract K (=1,000) prototype patches, and then cluster them into k (k<<K) clusters though k-means algorithm. Then the k clusters are used as patches to compute BIF or MBIF. The performance is shown in Figure 9. It can be seen that the technique of clustering patches can reduce computational cost and feature storage and retain the representativeness of the prototype patches with little accuracy loss.

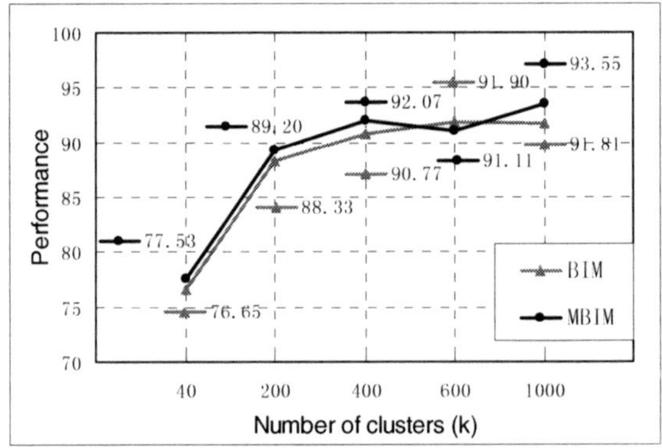

Fig. 9. Results of BIM and MBIM using clustered patche

5 Conclusion

In this paper, we have shown that a modified biologically inspired model (MBIM) is competitive with other state-of-the-art approaches for challenging real-world smile classification task. We have presented several modifications to improve the effectiveness of biologically inspired model (BIM). A biologically-motivated framework for robust real-world smile classification is introduced in the field of smile classification. Experimental results on GENKI, a challenging database with a variety of real-world smile faces show the effectiveness of the proposed approach. It has been shown from experiments that our MBIM is superior to BIM in terms of effectiveness with less number of features and computational cost.

Acknowledgments. This work is supported in part by NSFC (Grant no. U0735004 and 61075021), GDSTP(no.2008A050200004, 2010B09040039） and Fundamental Research Funds for the Central Universities, SCUT (No. 2009ZZ0014).

References

1. Niblack, W., Barber, R., Equitz, W., Fickner, M., Glasman, E., Petkovic, D., Yanker, P.: The QBIC Project: Querying Images By Content Using Color, Texture and Shape. In: Proc. Storage and Retrieval for Image and Video Databases, SPIE, vol. 1908, p. 173 (1993)
2. Schiele, B., Crowley, J.: Recognition Without Correspondence Using Multidimensional Receptive Field Histograms. International J. of Computer Vision 36(1), 31–35 (2000)
3. Lowe, D.G.: Distinctive Image Features from Dcale-Invariant Key-Points. International Journal of Computer Vision 60(2), 91–110 (2004)
4. Mikolajczyk, K., Schmid, C.: A Performance Evaluation of Local Descriptors. IEEE Trans. on PAMI 27(10), 1615–1630 (2005)
5. Lazebnik, S., Schmid, C., Ponce, J.: A Sparse Texture Representation Using Local Affine Regions. Technical Report, Beckman Institute, University of Illinois (2004)
6. Dalal, N., Triggs, B.: Histograms of Oriented Gradients for Human Detection. In: IEEE International Conference of CVPR, pp. 886–893. IEEE Press, California (2005)

7. Leung, T., Malik, J.: Representing and Recognizing the Visual Appearance of Materials Using Three-Dimensional Textons. International J. of Computer Vision 43(1), 29–44 (2001)
8. Csurka, G., Bray, C., Dance, C., Fan, L.: Visual Categorization with Bags of Keypoints. In: ECCV International Workshop on Statistical Learning in Computer Vision, Prague, pp. 1–22 (2004)
9. Riesenhuber, M., Poggio, T.: Hierarchical Models of Object Recognition in Cortex. Nature Neuroscience 2(11), 1019–1025 (1999)
10. Serre, T., Wolf, L., Poggio, T.: Object Recognition with Features Inspired By Visual Cortex. In: IEEE International Conference of Computer Vision and Pattern Recognition, pp. 994–1000. IEEE Press, California (2005)
11. The MPLab GENKI Dataset: GENKI-R2009 a Subset, `http://mplab.ucsd.edu`
12. Hubel, D., Wiesel, T.: Receptive Fields of Single Neurons in the Cat'S Striate Cortex. Journal of Physiology 148, 574–591 (1959)
13. Hubel, D., Wiesel, T.: Receptive Fields and Functional Architecture in Two Nonstriate Visual Areas (18 And 19) of the Cat. Neurophys. 28, 229–289 (1965)
14. Berkes, P., Wiskott, L.: Slow Feature Analysis Yields a Rich Repertoire of Complex Cell Properties. Journal of Vision 5, 579–602 (2005)
15. Afraz, D.A., Bondar, I.V., Giesem, A.: Norm-Based Face Encoding by Single Neurons in the Monkey Infer Temporal Cortex. Nature 442, 572–575 (2006)
16. Mutch, J., Lowe, D.G.: Multiclass Object Recognition with Sparse, Localized Features. In: IEEE International Conference on CVPR, pp. 11–18. IEEE Press, New York (2006)
17. Mutch, J., Lowe, D.G.: Object Class Recognition and Localization Using Sparse Features with Limited Receptive Fields. International Journal of Computer Vision 80(1), 45–57 (2008)
18. Serre, T., Wolf, L.: Robust Object Recognition with Cortex-Like Mechanisms. IEEE Trans. Pattern Analysis and Machine Intelligence 29(3), 411–426 (2007)
19. Poggio, T., Bizzi, E.: Generalization in Vision and Motor Control. Nature 431(7010), 768–774 (2004)
20. Kanade, T., Cohn, J., Tian, Y.: Comprehensive Database for Facial Expression Analysis. In: IEEE 4th International Conference on Face and Gesture Recognition, pp. 46–53. IEEE Press (2000)
21. Michael, J.L., Shigeru, A., Miyuki, K., Jiro, G.: Coding Facial Expressionswith Gabor Wavelets. In: 3rd IEEE International Conference on Automatic Face and Gesture Recognition, pp. 200–205. IEEE Press, Nara, Japan (1998)
22. Viola, P., Michael, J.J.: Robust Real-Time Face Detection. International Journal of Computer Vision 57(2), 137–154 (2004)
23. Littlewort, G., Bartlett, M.S., Fasel, I., Susskind, J., Movellan, J.: Dynamics of Facial Expression Extracted Automatically from Video. Image and Vision Computing 24(6), 615–625 (2006)
24. Bai, Y., Guo, L., et al.: A Novel Feature Extraction Method Using Pyramid Histogram of Orientation Gradients for Smile Recognition. In: Proceedings of IEEE ICIP. IEEE Press, Cairo (2009)
25. Yoav, F., Robert, E.S.: A Decision-Theoretic Generalization of On-Line Learning and an Application To Boosting. Journal of Computer and System Sciences 55, 119–139 (1997)
26. Song, D., Tao, D.C.: Biologically Inspired Feature Manifold for Scene Classification. IEEE Transactions on Image Processing 19(1), 174–184 (2010)
27. Huang, Y.Z., Huang, K.Q., Wang, L.G., Tao, D.C., Tan, T.N., Li, X.L.: Enhanced Biologically Inspired Model. In: IEEE International Conference on CVPR. IEEE Press, Alaska (2008)

Learning Colours from Textures by Sparse Manifold Embedding

Jun Li, Wei Bian, Dacheng Tao, and Chengqi Zhang

Center for Quantum Computation & Intelligent Systems,
University of Technology, Sydney

Abstract. The capability of inferring colours from the texture (grayscale contents) of an image is useful in many application areas, when the imaging device/environment is limited. Traditional colour assignment involves intensive human effort. Automatic methods have been proposed to establish relations between image textures and the corresponding colours. Existing research mainly focuses on linear relations.

In this paper, we employ sparse constraints in the model of texture-colour relationship. The technique is developed on a locally linear model, which assumes manifold assumption of the distribution of the image data. Given the texture of an image patch, learning the model transfers colours to the texture patch by combining known colours of similar texture patches. The sparse constraint checks the contributing factors in the model and helps improve the stability of the colour transfer. Experiments show that our method gives superior results to those of the previous work.

1 Introduction

Human vision percepts the world with colours. Colours do not only make images feel more vivid to us, they also contains important visual clues of the image. Although an inexpensive camera can now record colour images easily, there are many circumistances where colours needs to be inferred according to the texture of an image. For example, old monochrome photos may need to be colourised; pictures shot with severely wrong white balance settings can be rescued by keeping only the captured textures and transfer colours from another source. Colourisation is essential in areas of specialised imaging, where the sensors captures signals that are out of visible spectrum of light, eg X-ray, MRI, near infrared images. Pseudo colours for these images helps human experts interpret the information. But it is both important and difficult to assign colours consistently and efficiently.

A technique of transferring colours to images must be adaptive, because is no one-to-one map between the luminance and the chromatic (hue/saturation) value for an image pixel. Ie pixels of the same intensity may have different colours in an image and vice versa. Human can tell the colours given a monochrome picture, because human brain interprets the contents in an image and guesses the colours for the image components based on prior knowledge. Unfortunately,

D. Wang and M. Reynolds (Eds.): AI 2011, LNAI 7106, pp. 600–608, 2011.

to implement the guess-in-mind to values that represent colours involves arduous work and can be inconsistent. For a machine, the tedious work of assigning colours is a tractable. However, it can be difficult to recognise the context in the image and to retrieve background knowledge of the colours. To combine the advantage of both sides, a semi-automatic system [12] employs human to indicate essential clues for adding colours. However, this can become time-consuming and impractical when the size of the task grows large.

Recent techniques entail less human effort. Welsh et al [17] has proposed a scheme as follows. First, a user provides the system a colour image A of the similar contents to the image B needs colourisation. Second, the system learns the relations between the colours and the grayscale intensities of pixels in image A. Finally, grayscale pixels in image B are assigned colours according to the learned relations.

The scheme has been extended in [8], where they assume non-linear manifold distribution of the image data, and learn the texture-colour relations from a small number of relevant image patches. In the framework, after the reference image A has been selected, for a grayscale image B, they extract overlapping patches from A and B. The patches are assumed to have non-linear manifold distribution in their feature space ([1,6]). Then for each from image B, a neighbourhood in the patches from image A is retrieved. A linear system [10] is built from these patches neighbours to transfer the colours to the patch from B. However, the locally linear method in [8] may become instable when there are insufficient training patches; it can also mix colours from many contributing patches and cause artifical effects.

In this paper, we develop the semi-automatic scheme framework in [17] and [8] by adopting sparsity constraints in the colour prediction model. The processing steps of our system resembles that in [8], however, with the key difference in the core algorithm of predicting patch colours. We impose the linear model sparse constraint, which help check the model complexity in the case of insufficient training samples. The sparse model also limits the number of contributing components from the reference patches. This may prevent fusing colours from irrelevant sources and improve the results.

The rest of the paper is organised as follows. We will have a brief review of related techniques. We will refer the reader to more thorough treatment of the literature due to the space limit. The technical details of our system is introduced in Section 3. Section 4 demonstrates two examples of using our system to transfer colours. We conclude the paper in Section 5.

2 Related Works

Automatic systems for assigning colours to images has been developed in the cartoon industry since long ago. Early systems needs intensive human interaction [12]. Qu et al. proposed an method to alleviate human effort [9]; but the system needs an initial estimation of the colours for each components in the image, which may pose difficulties for users without corresponding expertise. Moreover, direct user interaction prevents batch processing: assigning colours to multiple grayscale images efficiently. Welsh et al. [17] have proposed an alternative scheme, which requires only a reference colour image. The algorithm computes colour for each pixel from the reference image as we have introduced above. Li and Hao [8] extends the algorithm in [17] by adopting patch-based computation and by accounting for the manifold distribution of the image data [1,11]. They use a linear prediction model [10].

Image manifold assumption and manifold learning techniques [7,13,18] have been exploited for several visual computation tasks Manifold learning techniques. Eg the patch-based model have been used for super-resolution [2,6].

Linear prediction models are popular data anlysis tools. Recent research of these models has shown desirable characteristics when sparse constraints are adopted [15]. The models tend to be more robust; and the results are more interpretable [19]. When constructing linear systems for the manifold neighbourhood, we impose sparsity constraints [5] to the model.

3 Colourising Images by Sparse Manifold Embedding

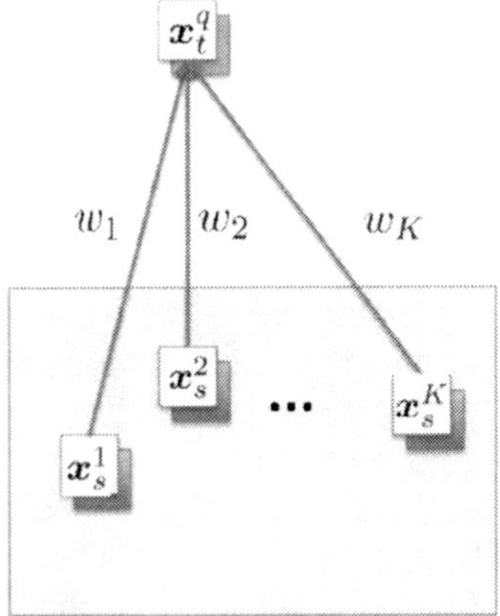

The white boxes indicates grayscale patches. The big enclosing gray box indicates the reference image. The blue boxes behind each grayscale patch represents the colour information of those patches. After the coefficients $w_1, \ldots, w_K$ have been computed, they are used to recover the colours of the query patch (shown in orange box).

Fig. 1. Flow Chart of Locally Linear Embedding with Sparsity Constraint

(a) input; (b) training; (c) – (g) features. (Courtesy of the authors of [8])

Fig. 2. Nearest neighbours

3.1 Problem Formulation

We firstly introduce frequently used symbols and formulate the problem of predicting colours from texture in mathematical form. We adopt the denotation convention in [8]. In the following, we will use the latin letter "X/x" for grayscale texture information and "Y/y" for the colour information; we will use subscript "t" (target) for the image where colour information needs to be predicted and subscript "s" (source) for the information from the reference image.

The task is to predict the colours for a monochrome image $\mathbf{X}_t$, given a reference image represented in $\mathbf{X}_s$ and $\mathbf{Y}_s$. $\mathbf{X}_s$ represents the grayscale intensities of the reference image; and $\mathbf{Y}_s$ represents the chromatic values. The model needs to learn how $\mathbf{X}_s$ relates to $\mathbf{Y}_s$ and utilise the knowledge to estimate the chromatic information of the input image, $\mathbf{Y}_t$, from its grayscale intensities $\mathbf{X}_t$.

An image is a big object consisting many components. Each pixel may only be related to some neighbouring pixels of the relevant component. For the purpose of transfering colours, a small part of the reference image $(\mathbf{X}_s, \mathbf{Y}_s)$ is sufficient to provide appropriate information for a counterpart in $\mathbf{X}_t$. Therefore the images are represented by overlapped patches as follows. $\mathbf{X}_t := \{\boldsymbol{x}_t^p\}_{p=1}^{N_t}$, $\mathbf{Y}_t := \{\boldsymbol{y}_t^p\}_{p=1}^{N_t}$, $\mathbf{X}_s := \{\boldsymbol{x}_t^q\}_{q=1}^{N_s}$ and $\mathbf{Y}_s := \{\boldsymbol{y}_s^q\}_{q=1}^{N_s}$. N_t and N_s are the number of patches in the input grayscale image and the reference image respectively.

3.2 Manifold of Patches

Rich research has pointed patches from an image lie in a manifold [4,16,14,2,6]. This is because image patches are not random. They consists of a small subspace in the feature space where a patch is represented; and the variation of the patches are controlled by a few meaningful factors. Some of the factors affects both the grayscale intensities (texture) and the colours in a patch in a similar manner, although the texture and the colours of the patches form difference manifolds embedded in distinct feature spaces.

The relations between the manifolds can be established by first locating related neighbourhoods on the manifolds and then building a linear system for the neighbourhoods. The underlying assumption is that the neighbourhoods have similar linear structures, thus the structure on one manifold can help prediction on the other manifold. We will introduce the linear system in details in the next subsection.

3.3 Estimating the Colours with Sparsity Constraints

For an input grayscale patch $\boldsymbol{x}_t^q \in \mathbf{X}_t$, the procedure of estimating its colours $\boldsymbol{y}_t^q$ are as follows.

1. Find the K nearest neighbours $\mathcal{N}^q$ of $\boldsymbol{x}_t^q$ in $\mathbf{X}_s$.
2. Compute K combination coefficients for each $\boldsymbol{x}_s^r$, $r \in \mathcal{N}_q$, such that the combination approximates $\boldsymbol{x}_t^q$.
3. Synthesize $\boldsymbol{y}_t^q$ by combining the corresponding neighbours $\boldsymbol{y}_s^r$, $r \in \mathcal{N}_q$ with the coefficients computed in Step 2.

In Step 1, the corresponding neighbourhoods related to the patch $\boldsymbol{x}_t^q$ of interest are found on the two manifolds (the manifold of grayscale patches and the one of colours). Step 2 represents the vital step to discover the linear structure of the corresponding neighbourhoods. Formally, the linear system is constructed as follows. The coefficients $\boldsymbol{w}^q$ is solved by minimising an objective function

$$\boldsymbol{w}^q = \arg\min_{\boldsymbol{w}} \|\boldsymbol{x}_t^q - \sum_{r=1}^{K} w_r \boldsymbol{x}_s^{\mathcal{N}_q(r)}\|_2^2 + \lambda\|\boldsymbol{w}\|_1, \qquad (1)$$

where w_r represents the r-th element of $\boldsymbol{w}$ and $\mathcal{N}_q(r)$ represnets the r-th neighbour in $\mathbf{X}_s$.

The regularisation term $\|\boldsymbol{w}\|_1$ encourages sparse solution of $\boldsymbol{w}$. The optimal $\boldsymbol{w}^q$ in (1) can be readily found by solving a linear system [10] using the LARS [5] algorithm. In Fig. 1, we draw a flowchart for our framework of colourising grayscale images.

4 Experiment

4.1 Representing Patches in Feature Vectors

The grayscale patches are represented by feature vectors, which is constructed as in [2,3]. The feature vector for a patch consists of three components: the average pixel intensity, the first and the second order intensity gradients at individual pixels.

The figure shows learning the relation between the colours and the texture of an image (a) and inferring the colours of a grayscale input image (b). In both images, there is a squirrel present against a background of natural scene. The methods of [17] and [8] generates the image in (c) and (d), respectively. Our sparse manifold embedding generates the image in (e). Image (f) is the ground truth colour image of (b). This figure is best viewed on screen with colours.

Fig. 3. Transfer colours to an image of a squirrel

Formally, the feature vector of an image patch is built as follows. An image can be considered as a function $\mathcal{I} : \mathbb{Z}^2 \to \mathbb{R}$. The horizontal and vertical differentiation operators are defined as

$$\begin{aligned}\nabla_x \mathcal{I}(x, y) &= \mathcal{I}(x + 1, y) - \mathcal{I}(x - 1, y) \\ \nabla_y \mathcal{I}(x, y) &= \mathcal{I}(x, y + 1) - \mathcal{I}(x, y - 1).\end{aligned} \tag{2}$$

The feature vector of a grayscale patch $\mathcal{P}$ is then defined as

$$\left[\lambda \overline{\mathcal{I}|_\mathcal{P}} \; \nabla_x \mathcal{I}|_\mathcal{P} \; \nabla_y \mathcal{I}|_\mathcal{P} \; \nabla_x^2 \mathcal{I}|_\mathcal{P} \; \nabla_y^2 \mathcal{I}|_\mathcal{P}\right]^T,$$

where the first element represent the average pixel intensity in that patch

$$\overline{\mathcal{I}|_\mathcal{P}} = \frac{\sum_{(x,y)\in\mathcal{P}} \mathcal{I}(x, y)}{|\mathcal{P}|},$$

and λ is the weight of the intensity. The weight should be chosen according to the patch size, which is to keep the balance between the influence of the average intensity (representing luminance) and the gradients (representing the texture details). The weight is necessary, because the entries of the gradients in the feature vector grows with respect to the size of the patch, but the average intensity is a scalar irrelevant to the size of a patch.

The colour patches are the hue/saturation values at individual pixels in the corresponding patches.

Fig. 2 demonstrate the construction of the feature vector for an input grayscale patch and that for its 3 nearest neighbours in the reference grayscale patches.

Fig. 2 (a) and (b) shows the input and the reference images, respectively. The blueish box in (a) indicates an input patch, whose colours are to be predicted. The first column in Fig. 2(c–g) represents the feature of the query patch: the intensity values, the horizontal graidents, the vertical gradients, the horizontal second order gradients and the vertical second order gradients, respectively.

Three patches with closest features to the query feature has been shown in Fig. 2(b) by greenish boxes. Their feature components are shown in the second, the third and the last column in Fig. 2(c–g).

4.2 Experiment Results

Figure 3 demonstrates the results of transfering colours to an image, where a subject of middle size (a squirrel) presents against a complex background. The background consists of both sharp components (the flowers and the woody stump) and out-of-focus backdrop. The result (c) shows that the non-adaptive

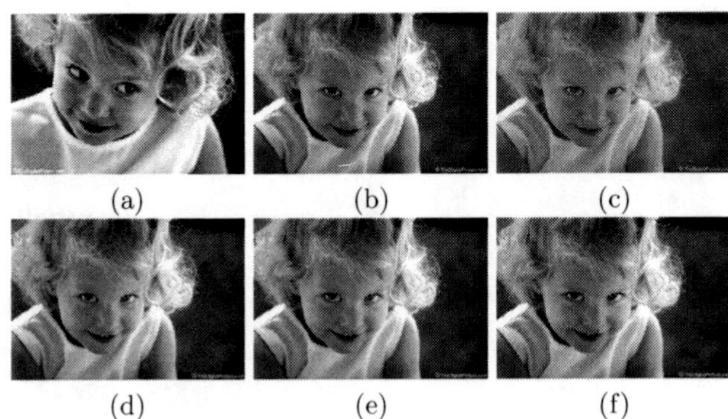

The figure shows learning the relation between the colours and the texture of an image (a) and inferring the colours of a grayscale input image (b). Both images consist of a portrait of a child against a texture-less background. The methods of [17] and [8] generates the image in (c) and (d), respectively. Our sparse manifold embedding generates the image in (e). Image (f) is the ground truth colour image of (b). This figure is best viewed on screen with colours.

Fig. 4. Transfer colours to an image of a portrait

method of [17] produces greenish colours for both the subjects and the background. Both results (d) (of the method of [8]) and (e) (ours) are satisfactory, as they apply correct colour tunes to the squirrel and the background respectively. However, result (e) demonstrates sharper contrast in colour tunes between the subject and the scenery. A possible reason lies in the spare method. For each patch in the input image, the sparse constraint limits the number of patches in the training image that can transfer their colours to the input patch. Therefore the constraint reduces the mix of colours and produce distinctive colours for different components in the image.

Figure 4 demonstrates the results of transfering colours to an image of a portrait of a child. As in the last experiment, images (c), (d) and (e) show the result of [17], [8] and our method, respectively. Image (f) represents the ground truth colour image. Compared to [8], our method is better in recovering colours in vital image components (see Figure 5).

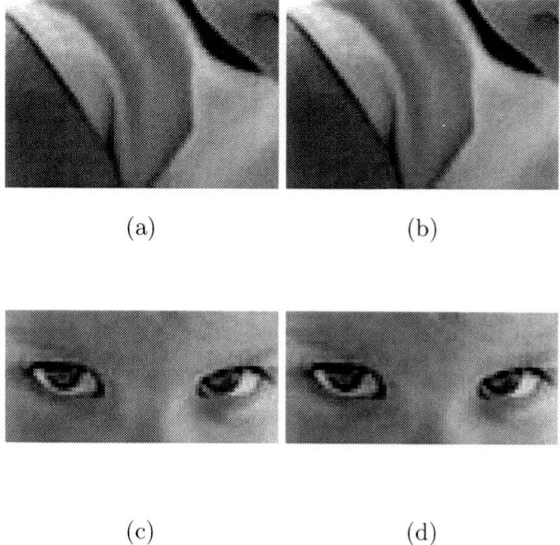

The figure compares parts of the image of interest in Figure 4. The images shown in (a) and (c) are produced by the method in [8] and that in (b) and (d) are resulted by our method. The sparse manifold embedding method correctly recovers the colours in the eyes and on the skin.

Fig. 5. Comparison of colours recovered by [8] and our method

5 Conclusion

In this paper, we have proposed an automatic method of learning the relations between colours and texture in an image, and transferring colours to a grayscale image by exploiting the learned relations. The method is based on the manifold assumption of the image data, and on the sparsity constraints on linear models.

Compared to existing research on the problem of colourisation of grayscale images, the proposed method has several advantages. It involves small amount of human effort. The model is robust when the reference information is insufficient. The resultant colours may also be more sharp than the previous locally linear model due to the sparsity constraint.

Future research may focus on the efficient construction of the manifold, as well as on more sophisticated techniques of treating the overlapping patches.

References

1. Beymer, D., Poggio, T.: Image representation for visual learning. Science (1996)
2. Chang, H., Yeung, D.Y., Xiong, Y.: Super-resolution through neighbor embedding. In: Proceedings of CVPR (2004)
3. Dalal, N., Triggs, B.: Histograms of oriented gradients for human detection. In: Proceedings of CVPR (2005)
4. Donoho, D.L., Grimes, C.: Image manifolds which are isometric to euclidean space. J. Math. Imaging Vis. 23(1), 5–24 (2005)
5. Efron, B., Hastie, T., Johnstone, I., Tibshirani, R.: Least angle regression. Annals of Statistics 32(2), 407–451 (2004)
6. Fan, W., Yeung, D.-Y.: Image hallucination using neighbor embedding over visual primitive manifolds. In: Proceedings of CVPR (2007)
7. Huo, X., Ni, X., Smith, A.K.: A survey of manifold-based learning methods. Technical report, Statistics Group, Georgia Institute of Technology (2006)
8. Li, J., Hao, P.: Transferring colours to grayscale images by locally linear embedding. In: BMVC (2008)
9. Qu, Y., Wong, T.-T., Heng, P.-A.: Manga colorization. In: Proceedings of Siggraph (2006)
10. Roweis, S.T., Saul, L.: Nonlinear dimensionality reduction by locally linear embedding. Science (2000)
11. Seung, H.S., Lee, D.D.: The manifold way of perception. Science, 2268–2269 (2000)
12. Silberg, J.: Cinesite press article (1998), `http://www.cinesite.com/core/press/articles/1998/10_00_98-team.html`
13. Song, M., Tao, D., Chen, C., Li, X., Chen, C.W.: Color to gray: Visual cue preservation. IEEE Trans. Pattern Anal. Mach. Intell. 21(9), 1537–1552 (2010)
14. Souvenir, R.: Manifold learning for natural image sets. Ph.D. thesis (2006)
15. Tibshirani, R.: Regression shrinkage and selection via the lasso. Journal of the Royal Statistical Society Series B (Methodological), 267–288 (1996)
16. Verbeek, J.: Learning non-linear image manifolds by combining local linear models. IEEE Transactions on Pattern Analysis & Machine Intelligence 28(8), 1236–1250 (2006)
17. Welsh, T., Ashikhmin, M., Mueller, K.: Transferring color to greyscale images. In: Proceedings of Siggraph (2002)
18. Zhang, T., Tao, D., Li, X., Yang, J.: Patch alignment for dimensionality reduction. IEEE Trans. Knowl. Data Eng. 21(9), 1299–1313 (2009)
19. Zou, H., Hastie, T., Tibshirani, R.: Sparse principal component analysis. Jcgs 15(2), 262–286 (2006)

Investigating Particle Swarm Optimisation Topologies for Edge Detection in Noisy Images

Mahdi Setayesh[1], Mengjie Zhang[1], and Mark Johnston[2]

[1] School of Engineering and Computer Science
{mahdi.setayesh,mengjie.zhang}@ecs.vuw.ac.nz
[2] School of Mathematics, Statistics and Operations Research
Victoria University of Wellington,
PO Box 600, Wellington, New Zealand
mark.johnston@msor.vuw.ac.nz

Abstract. This paper investigates the effects of applying different well-known static and dynamic neighbourhood topologies on the efficiency and effectiveness of a particle swarm optimisation-based edge detection algorithm. Our experiments show that the use of different topologies in a PSO-based edge detection algorithm does not have any significant effect on the accuracy of the algorithm for noisy images in most cases. That is in contrast to many reported results in the literature which claim that the selection of the neighbourhood topology affects the robustness of the algorithm to premature convergence and its accuracy. However, the fully connected topology in which all particles are connected to each other and exchange information performs more efficiently than other topologies in the PSO-based based edge detector.

Keywords: particle swarm optimisation, edge detection, noisy images, neighbourhood topology.

1 Introduction

Edges as low level features in an image contain important information that are utilised in image analysis and computer vision systems. Many algorithms have been proposed to detect edges for different applications using various different paradigms such as curve fitting [4], optimization of a criterion [3], statistical testing [9] and soft computing [2] to detect edges. The selection of an edge detection algorithm for a particular application depends on its performance in variant environmental conditions (such as illumination and noise) and the requirements of the system of interest (such as real time ability, continuity of edges, thinness of edges and scale insensitivity).

PSO as a meta-heuristic method has been used to successfully solve global optimisation problems and was introduced by Kennedy and Eberhart in 1995 [7]. The main general advantages of PSO in comparison with other heuristic methods such as genetic algorithms, are ease of its implementation, fewer

D. Wang and M. Reynolds (Eds.): AI 2011, LNAI 7106, pp. 609–618, 2011.

operators, a limited memory for each particle and high speed of convergence [1]. As PSO has a high capability to optimise noisy functions [12], it has been successfully applied to many problems in noisy environments, such as image segmentation and vision tracking [20].

We previously applied two PSO-based algorithms with different encoding schemes and fitness functions to noisy binary images containing simple shapes, such as rectangles, squares, circles, crosses and triangles [16]. Their performances were acceptable in the binary images but they were inefficient and did not operate well on non-binary images. We revised our PSO-based algorithm through developing a new encoding scheme and fitness function and examined it on real images corrupted by two different types of noise (Gaussian and impulsive) [19]. The main idea in this algorithm was to maximise **interset** distance between the average pixel intensities of two regions separated by a continuous edge represented by a particle and minimise **intraset** distances within both regions. Our experiments showed that this version could outperform the Canny algorithm as a Gaussian filter-based algorithm especially in the images with high levels of noise. However, it produced jagged edges and its overall performance was worse than robust-rank order (RRO) algorithm as a statistical-based edge detection algorithm and was slower than the Canny and RRO algorithms. We changed the fitness function of our PSO-based algorithm through considering a larger area around each single pixel on a continuous edge than the previous version of our algorithm and introducing a curvature cost of a continuous edge to reduce the effect of producing jagged edges [19]. The experiments showed that the new revised version could detect edges more accurate, more continuous, smoother than the older version introduced in [17], the Canny and RRO algorithms. But, the new algorithm was still slower than Canny and RRO. We introduced a discrete constrained PSO-based algorithm with two constraints and used a penalising method to handle these constraints to detect edges [18]. Our experiments showed that the new algorithm is faster than the algorithm presented in [19] and there was no significant difference between the localisation accuracy of the algorithms. In all experiments, we have utilised the fully connected graph as a neighbourhood structure in our PSO-based edge detection algorithm and never investigated the influence of the chosen topology on the performance of the algorithm.

In many cases, researchers use the same social topologies (fully connected and ring graph) in the PSO algorithm to solve an optimisation problem, but there is a strong relationship between the selection of the social topology and the robustness of the algorithm to premature convergence [5]. Therefore it is needed to investigate which topology is more efficient and more accurate for the PSO-based edge detection algorithm and how we can improve accuracy or speed of the algorithm. In this paper, the influence of the chosen topology on the accuracy and speed of the PSO-based edge detection algorithm will be investigated.

2 Background

2.1 Particle Swarm Optimisation

PSO as a branch of swarm intelligence was inspired by the social behavior of animals and simulated a simplified social model such as flocking of birds and schooling of fish.

In PSO, there is a population of m particles. The position of ith particle in an n-dimensional search space at time t is represented as the vector $\boldsymbol{X}_i(t) = (x_{i1}(t), x_{i2}(t), ..., x_{in}(t))$. The position of the particle is influenced by its own experience (particle and memory influence) and that of its neighbours (swarm influence). Each particle of the population has a velocity represented by $\boldsymbol{V}_i(t)$ that is used to update $\boldsymbol{X}_i(t)$ at each iteration of PSO as in equation (1).

$$\boldsymbol{X}_i(t+1) = \boldsymbol{X}_i(t) + \boldsymbol{V}_i(t+1). \tag{1}$$

The velocity is updated according to three components: current motion influence, particle memory influence, and swarm influence:

$$\boldsymbol{V}_i(t+1) = w\boldsymbol{V}_i(t) + C_1 r_1(\boldsymbol{X}_{pbest_i} - \boldsymbol{X}_i(t)) + C_2 r_2(\boldsymbol{X}_{leader} - \boldsymbol{X}_i(t)) \tag{2}$$

where r_1 and r_2 are uniform random variables between 0 and 1; w denotes the inertia weight that controls the impact of the previous velocity; C_1 and C_2 are the self and swarm confidence learning factors respectively; $\boldsymbol{X}_{pbest}$ represents the personal best position of each particle so far; and $\boldsymbol{X}_{leader}$ is the position of the leader which is the particle that is defined by a neighbourhood topology and guides other particles toward better regions of the search space.

2.2 Neighbourhood Topologies

An important feature of the PSO algorithm is the topology which defines how particles are connected to each other as an information sharing or exchanging mechanism [14]. A topology defines the social structure among a swarm's particles. The topology specifies the leader of each particle based on a typical neighbourhood graph. There are several typical neighbourhood topologies that have been proposed in the literature as follows:

- **Fully connected graph (FCG):** In this case, each particle is fully connected to the other particles (the opposite of the empty topology) [7]. In this topology, each particle is influenced by the best particle of the entire swarm (gbest), as well as its own past experience (pbest). In this case, the leader is global best particle ($leader = gbest$ in equation (2)). This topology is shown in Figure 1(a).
- **Local best graph (LBG):** There are k immediate neighbours for each particle in the graph [7]. It means each particle has a local best particle among k particles within its neighbourhood. In this topology, each particle is influenced by a leader in its local neighbourhood plus its own past experience (pbest). In this case, the leader is called the local best (lbest) particle. This topology is shown in Figure 1(b). This topology does not need to be symmetrical.

- **Ring topology (RT):** This topology is a special representation of the local best topology in which $k = 2$ [7]. It means, each particle has just two particles in its neighbourhood as shown in Figure 1(c).
- **Star graph (SG):** In this case, one particle is just connected to all other particles [14] as shown in Figure 1(d). It is called the focal particle. In this topology, particles are isolated from each other and they communicate through the focal particle. This topology is sometime called the wheel topology. In this topology, $leader = focal$ in equation (2).
- **Tree-based graph (TBG):** each particle corresponds to one node in a tree [14]. An example of this topology is shown in Figure 1(e). In this case, the leader of each particle is its parent in the tree. Whenever each child particle finds a solution better than the best particle found by its parent, the child and parent particle are exchanged. In this topology, $leader = pbest_{parent}$ in equation (2).
- **The von Neumann topology (VNT):** in this case, each particle has four neighbours within its neighbourhood and exchanges the information with them [14]. These particles are usually located in its four different directions. An example of this topology in 2-dimensional search space is shown in Figure 1(f).

In some papers (e.g., [10], [5]), it has been indicated that if the neighbourhood size of a particle increases, the performance of the swarm may deteriorate. On the other hand, if it decreases, the run time of the algorithm may be increased. In [11], it has been shown that there is a strong relation between the chosen topology for the PSO algorithm and its robustness to premature convergence to optimise some benchmark fitness functions. It has been pointed out that the main cause of premature convergence in the PSO algorithm is the kind of topology which is chosen for it [5]. In many applications of PSO algorithm, the fully connected or local best graph is mostly utilised.

2.3 PSO-Based Edge Detection Algorithm

The most important goals of the PSO-based edge detection algorithm is to detect continuous edges in noisy images. Therefore, to reduce broken edges, we proposed an encoding scheme for the particles where each particle represents the global structure of a continuous edge [19]. This edge partitions an area of an image into two regions, the light and dark regions as can be seen in Figure 2(b), such that it maximises **interset** distance between the average pixel intensities of two regions and minimises **intraset** distances within both regions.

A continuous edge is encoded into a particle as $\langle\langle o_1, o_2\rangle, \langle m_1, m_2, \ldots, m_{max/2}\rangle$ $, \langle m_{max/2+1}, \ldots, m_{max}\rangle\rangle$, where $max + 1$ is the number of pixels on the edge. The encoding scheme has three parts: the offsets of the closest edge to each pixel of the image ($\langle o_1, o_2\rangle$) and two sets of movement direction sequences from the pixel ($\langle m_1, m_2, \ldots, m_{max/2}\rangle$ and $\langle m_{max/2+1}, m_{max/2+2}, \ldots, m_{max}\rangle$). The values of two offsets (o_1 and o_2) are integers ranging from 0 to $SqrSize - 1$ and m_i ranging from 0 to 7. Here, m_i shows the movement direction from a pixel to one of

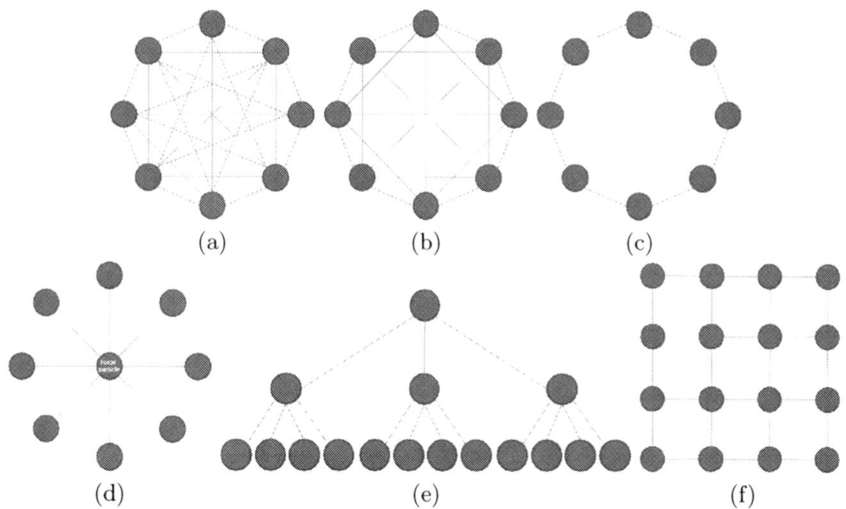

Fig. 1. Some well-known topologies used in PSO: (a) FCG (b) LBG, (c) RT, (d) ST, (e) TBG and (f) VNT

the eight possible adjacent pixels in its neighbourhood along the continuous edge as shown in Figure 2(a). For example, the particle encoding for the continuous edge in Figure 2(b) can be seen in Figure 2(c).

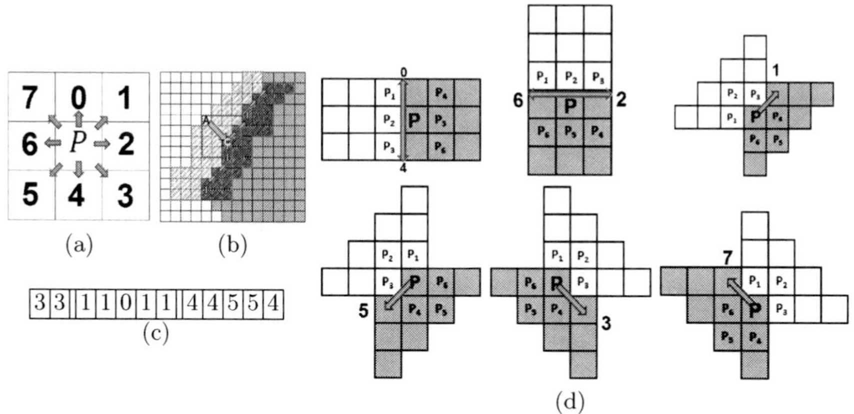

Fig. 2. The particle encoding scheme [18]: (a) eight movement directions from a pixel P; (b) an example of a curve with two regions; (c) the particle representing the curve with $max = 10$; (d) eight moving ways from pixel P to its neighbours

For evaluation of each particle at each generation of the PSO algorithm, at the first step, the intraset and interset distances of each single pixel on the continuous edge represented by the particle are calculated by the equations proposed in [18] according to eight moving ways from the pixel to its neighbours (see Figure 2(d)). Then the possibility score ($PScore$) and curvature cost factor ($CCost$) of the curve fitting on a continuous edge are computed as the proposed equations in [18]. The score in conjunction with the curvature cost factor as equation (3) is used for evaluation of each particle [18],

$$Fitness(C) = PScore(C) - CCost(C) \tag{3}$$

subject to two constraints:

$$Cross(C) = 0 \; and \; PScore(C) > HP$$

where C is the curve represented by a particle, $Cross(C)$ shows how many times the curve C crosses itself and HP is a threshold value which is defined by the user. The curves, represented by the particles, may sometimes intersect themselves, so we set a constraint $Cross(C) = 0$. On the other hand, $PScore(C)$ should be larger than HP to avoid false alarms. Therefore, $PScore(C) > HP$ as another constraint should be satisfied in the PSO algorithm. We proposed a non-stationary and multi-stage penalising method to handle these two constraints in [18]. In all experiments that we have arranged so far, the fully connected topology has been used in the PSO-based algorithm. Therefore, to evaluate the influence of using different topologies on the accuracy of the algorithm, we change the velocity equation in [18] as the equation (2) in order to specify the leader's position according to the chosen topology which defines the neighborhood structure of each particle in the PSO algorithm.

3 Experiment Design

We will compare the performance of PSO with six topologies for edge detection in noisy environment. We will describe the image set first and then the performance measure used in this paper.

To investigate the influence of chosen topology on the efficiency and effectiveness of the algorithms, we apply the algorithm on a set of benchmark images including four real images (Saturn, multi-cube, wall and road). The real images and their ground truth edge maps are available from [6]. The size of each image is 256×256 pixels and their resolution is 8 bits per pixel. These images are shown in Figure 3. All images are corrupted by two different types of noise. The noise probability for the impulsive noise ranges from 0.1 to 0.5 with a step size of 0.1. The peak signal-to-noise ratio (PSNR) value ranges from 0 to $22dB$ with a step size of $4dB$ for the Gaussian noise.

To compare the accuracy of the PSO-based edge detection algorithm with different described neighbourhood topologies, Pratt's Figure of Merit (PFOM)

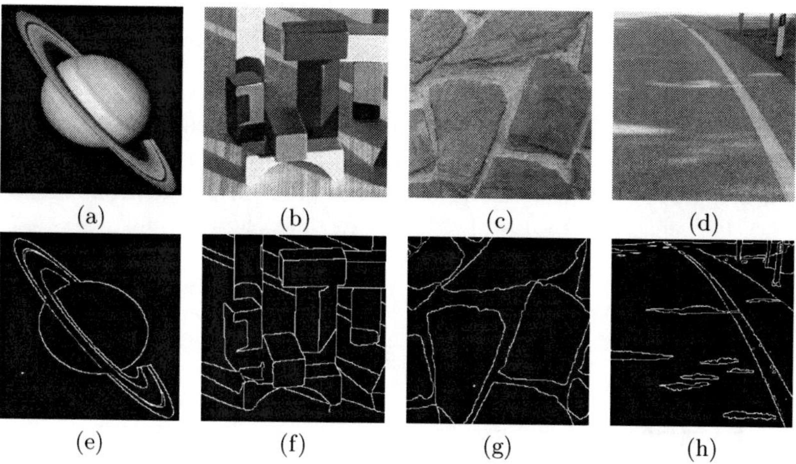

Fig. 3. (a)–(h) four real image from the UCO university and (e)–(h) their manual ground truth images [6]

is used as a quantitative measure. This measure is commonly utilised to compare the *localisation accuracy* of edge detection algorithms [13]. This measure is defined by equation (4),

$$R_{PFOM} = \frac{1}{max(I_I, I_A)} \sum_{i=1}^{I_A} \frac{1}{1 + \beta d(i)^2} \tag{4}$$

where I_I and I_A indicate the number of ideal and actual edge points in the ground truth and the generated edge map images, $d(i)$ is the distance between the pixel i in the generated edge map and the nearest ideal edge point in the ideal edge map, and β is a constant scale factor which is typically set to $\frac{1}{9}$. The ideal value of R_{PFOM} is 1.0 and the minimum could be very small. A larger value indicates stronger performance.

We use the values $w = 0.7298$, $C_1 = 1.4962$, $C_2 = 1.4962$ for PSO parameters in equation (2). The population size was set at 50 and the maximum number of iterations was 200. These values were chosen based on common settings [8]. In the PSO-based edge detection algorithm, the minimum length of a continuous edge, $max+1$ was set at 21, $SqrSize$ at 4, and HP at 0.5 [18]. For the tree-based topology, the branching factor was set at 3 [15].

4 Results

Table 1 shows PFOM estimated from the resulting images after applying the the PSO-based algorithm with different topologies. G6, G10, G14, G18 and G22 represent PSNR from 6dB to 22dB for Gaussian noise and N0.1, N0.2, N0.3, N0.4 and N0.5 represent noise probability from 0.1 to 0.5 for impulsive noise.

The columns FCG, LBG, RT, SG, TBG and VNT show the 95% confidence intervals for the localisation accuracy of the PSO-based algorithm with the fully connected, local best, ring, star, tree-based and von Neoman topologies after 30 runs for each image in each noise level. The Student two paired t-test was used to compare the pairwise accuracy means of the topologies. Alternative hypothesis was inequality of the means. The statistical analysis showed that the null hypothesis was accepted in most cases, i.e, there is often no significant difference between their means. This suggests that the topology does not have any influence on the accuracy of the algorithm.

Table 1. Comparison of accuracy of the PSO-based algorithm with different topologies

Image	Noise Level	95% Confidence Interval for Accuracy					
		FCG	LBG	RT	SG	TBG	VNT
Sat	G22	0.7728±0.0032	0.7688±0.0033	0.7684±0.0037	0.7649±0.0059	0.7568±0.0048	0.7533±0.0032
Sat	G18	0.8534±0.0026	0.8582±0.0022	0.8595±0.0018	0.8547±0.0022	0.8561±0.0040	0.8584±0.0021
Sat	G14	0.7846±0.0028	0.7867±0.0056	0.7924±0.0075	0.7953±0.0075	0.7918±0.0099	0.7976±0.0094
Sat	G10	0.8832±0.0032	0.8817±0.0023	0.8871±0.0073	0.8899±0.0050	0.8836±0.0071	0.8802±0.0068
Sat	G6	0.7674±0.0028	0.7668±0.0033	0.7630±0.0050	0.7660±0.0065	0.7707±0.0071	0.7670±0.0074
Cube	G22	0.6182±0.0032	0.6178±0.0020	0.6263±0.0008	0.6287±0.0013	0.6270±0.0011	0.6229±0.0011
Cube	G18	0.6466±0.0025	0.6399±0.0018	0.6359±0.0011	0.6384±0.0081	0.6414±0.0003	0.6440±0.0019
Cube	G14	0.5166±0.0030	0.5145±0.0018	0.5152±0.0024	0.5099±0.0086	0.5121±0.0042	0.5089±0.0062
Cube	G10	0.6333±0.0027	0.6316±0.0053	0.6346±0.0044	0.6344±0.0053	0.6368±0.0054	0.6301±0.0054
Cube	G6	0.5892±0.0027	0.5851±0.0036	0.5860±0.0015	0.5819±0.0040	0.5774±0.0016	0.5767±0.0014
Wall	G22	0.7466±0.0029	0.7585±0.0051	0.7602±0.0035	0.7453±0.0041	0.7659±0.0034	0.7649±0.0016
Wall	G18	0.7470±0.0030	0.7504±0.0033	0.7495±0.0008	0.7367±0.0008	0.7372±0.0094	0.7463±0.0034
Wall	G14	0.7913±0.0034	0.7933±0.0039	0.7977±0.0054	0.7922±0.0026	0.7939±0.0031	0.7939±0.0021
Wall	G10	0.8063±0.0030	0.8115±0.0000	0.8034±0.0003	0.7953±0.0002	0.7894±0.0006	0.7957±0.0004
Wall	G6	0.7805±0.0028	0.7820±0.0039	0.7756±0.0009	0.7796±0.0019	0.7772±0.0048	0.7789±0.0037
Street	G22	0.8091±0.0027	0.8086±0.0034	0.8046±0.0013	0.8101±0.0013	0.8208±0.0032	0.8158±0.0027
Street	G18	0.7440±0.0031	0.7510±0.0036	0.7515±0.0014	0.7442±0.0036	0.7607±0.0034	0.7514±0.0083
Street	G14	0.7468±0.0029	0.7480±0.0038	0.7562±0.0052	0.7524±0.0075	0.7532±0.0081	0.7537±0.0075
Street	G10	0.6412±0.0032	0.6388±0.0029	0.6318±0.0037	0.6337±0.0047	0.6352±0.0026	0.6347±0.0092
Street	G6	0.7502±0.0035	0.7539±0.0021	0.7504±0.0035	0.7448±0.0062	0.7353±0.0073	0.7314±0.0085
Sat	N0.1	0.4218±0.0027	0.4209±0.0063	0.4305±0.0045	0.4289±0.0073	0.4215±0.0065	0.4149±0.0080
Sat	N0.2	0.4701±0.0027	0.4693±0.0027	0.4677±0.0026	0.4656±0.0013	0.4712±0.0026	0.4746±0.0035
Sat	N0.3	0.4836±0.0029	0.4913±0.0028	0.4918±0.0030	0.4866±0.0038	0.4845±0.0021	0.4885±0.0055
Sat	N0.4	0.1912±0.0031	0.1886±0.0034	0.1873±0.0026	0.1934±0.0028	0.1932±0.0018	0.1933±0.0025
Sat	N0.5	0.1925±0.0027	0.1904±0.0013	0.1897±0.0005	0.1964±0.0026	0.1997±0.0012	0.2002±0.0018
Cube	N0.1	0.5698±0.0030	0.5707±0.0023	0.5706±0.0016	0.5699±0.0030	0.5700±0.0049	0.5685±0.0064
Cube	N0.2	0.5356±0.0029	0.5417±0.0022	0.5383±0.0023	0.5263±0.0094	0.5289±0.0002	0.5267±0.0011
Cube	N0.3	0.5344±0.0029	0.5463±0.0025	0.5515±0.0032	0.5457±0.0016	0.5369±0.0082	0.5427±0.0094
Cube	N0.4	0.4066±0.0025	0.4077±0.0010	0.4045±0.0022	0.4024±0.0045	0.3967±0.0024	0.3995±0.0027
Cube	N0.5	0.2914±0.0031	0.3035±0.0041	0.3028±0.0072	0.2909±0.0081	0.3061±0.0075	0.3012±0.0098
Wall	N0.1	0.4772±0.0025	0.4788±0.0040	0.4736±0.0075	0.4749±0.0046	0.4773±0.0009	0.4834±0.0014
Wall	N0.2	0.4887±0.0028	0.4922±0.0021	0.5005±0.0064	0.4881±0.0069	0.5002±0.0062	0.4964±0.0062
Wall	N0.3	0.5822±0.0030	0.5841±0.0027	0.5834±0.0037	0.5740±0.0040	0.5735±0.0095	0.5777±0.0065
Wall	N0.4	0.4400±0.0027	0.4389±0.0030	0.4322±0.0028	0.4374±0.0036	0.4431±0.0017	0.4433±0.0037
Wall	N0.5	0.2564±0.0030	0.2578±0.0007	0.2579±0.0033	0.2568±0.0022	0.2585±0.0006	0.2530±0.0017
Street	N0.1	0.5421±0.0030	0.5414±0.0019	0.5563±0.0020	0.5548±0.0003	0.5538±0.0005	0.5438±0.0010
Street	N0.2	0.3814±0.0030	0.3835±0.0020	0.3831±0.0026	0.3900±0.0028	0.3908±0.0085	0.3792±0.0039
Street	N0.3	0.4565±0.0028	0.4582±0.0039	0.4561±0.0023	0.4595±0.0010	0.4491±0.0001	0.4468±0.0046
Street	N0.4	0.4133±0.0022	0.4082±0.0017	0.4139±0.0027	0.4067±0.0035	0.3983±0.0002	0.3989±0.0001
Street	N0.5	0.2755±0.0033	0.2689±0.0046	0.2713±0.0046	0.2731±0.0053	0.2764±0.0037	0.2759±0.0035

Table 2 gives the 95% confidence intervals for the number of fitness function evaluations of the PSO-based algorithm with the described topologies after 30 runs for each image at each noise level. Statistic analysis showed that the number of fitness function evaluations of the algorithm with the fully connected graph is less than that of the PSO-based algorithm with the other topologies. This implies that the algorithm with the fully connected graph is faster than the algorithm with the other topologies. However, the accuracy of the algorithm with the fully connected graph topology does not have any significant difference with those of the algorithm with the other topologies. This suggests that the fully connected graph is the best topology for the PSO-based algorithm in terms of efficiency.

Table 2. Comparison of the number of the fitness function evaluations for PSO with different topologies

Image	Noise Level	Number of Fitness Function Evaluation					
		FCG	LBG	RT	SG	TBG	VNT
Sat	N0.1	354948±589	356011±571	358978±582	359835±585	358119±634	356935±654
Sat	N0.2	385045±1069	386010±1056	389105±1001	389909±1143	388017±969	387110±1055
Sat	G22	342871±414	343867±426	346829±476	347672±373	345831±412	344754±411
Sat	G14	367991±733	368995±746	372131±666	372950±651	371062±834	369955±736
Cube	N0.1	345031±361	346036±311	348974±323	350176±448	347902±393	347018±415
Cube	N0.2	374948±589	375972±582	379052±475	380050±701	377892±591	376899±639
Cube	G22	342654±297	343700±335	346492±353	347747±319	345544±245	344688±335
Cube	G14	362553±519	363527±497	366575±432	367584±522	365590±474	364609±508
Wall	N0.1	365010±1239	365996±1277	369095±1236	370103±1249	368229±1214	366920±1232
Wall	N0.2	394998±1472	395949±1451	399121±1473	400005±1493	397851±1515	396965±1476
Wall	G22	351359±918	352454±925	355419±817	356295±888	354417±994	353319±919
Wall	G14	378192±1319	379201±1338	382203±1383	383247±1327	381161±1372	380146±1317
Street	N0.1	294990±503	295954±477	299021±559	300025±541	297816±473	297039±500
Street	N0.2	324903±693	325845±680	328845±741	329780±801	327720±626	326989±753
Street	G22	271502±376	272557±382	275574±393	276416±359	274472±402	273500±375
Street	G14	310648±710	311589±662	314668±735	315693±728	313716±645	312486±709

5 Conclusions

For the PSO-based edge detection algorithm with two different constraints, it was demonstrated that the fully connected topology is the superior to the other described topologies in terms of efficiency. However the accuracy of the PSO-based edge detection algorithms was not influenced by the use of different topologies and there is no significant difference among their accuracies. These results are in contrast to the comments in the literature that the fully connected neighbourhood topology may converge to a local optima since all particles are connected together and they quickly communicate and share acquired information in the swarm. The results also showed that if the size of the particle neighbourhood is increased in the PSO-based edge detection algorithm, the algorithm speeds up meanwhile the accuracy of the algorithm is not significantly changed.

References

1. Rashidi, M.R.A., El-Hawary, M.E.: A survey of particle swarm optimization applications in electric power systems. Trans. Evol. Comp. 13(4), 913–918 (2009)

2. Baştürk, A., Günay, E.: Efficient edge detection in digital images using a cellular neural network optimized by differential evolution algorithm. Expert Syst. Appl. 36(2), 2645–2650 (2009)
3. Canny, J.: A computational approach to edge detection. IEEE Trans. Pattern Anal. Mach. Intell. 8(6), 679–698 (1986)
4. Chen, G., Hong Yang, Y.H.: Edge detection by regularized cubic b-spline fitting. IEEE Transactions on Systems, Man and Cybernetics 25(4), 636–643 (1995)
5. Czogalla, J., Fink, A.: Particle Swarm Topologies for Resource Constrained Project Scheduling. In: NICSO. SCI, vol. 236, pp. 61–73. Springer, Heidelberg (2009)
6. Fernández-García, N.L., Carmona-Poyato, A., Medina-Carnicer, R., Madrid-Cuevas, F.J.: Images from automatic generation of consensus ground truth for comparison of edge detection techniques, `http://www.uco.es/~ma1fegan/investigacion/imagenes/ground-truth.html`
7. Kennedy, F., Eberhart, R., Shi, Y.: Swarm Intelligence. Morgan Kaufmann, San Francisco (2001)
8. Laskari, E.C., Parsopoulos, K.E., Vrahatis, M.N.: Particle swarm optimization for integer programming. In: Proceedings of the 2002 Congress on Evolutionary Computation, pp. 1582–1587. IEEE Press (2002)
9. Lim, D.H.: Robust edge detection in noisy images. Comput. Stat. Data Anal. 50(3), 803–812 (2006)
10. Mendes, R., Kennedy, J., Neves, J.: The fully informed particle swarm: Simpler, maybe better. IEEE Transactions on Evolutionary Computation 8, 204–210 (2004)
11. Montes de Oca, M.A., Stützle, T.: Convergence behavior of the fully informed particle swarm optimization algorithm. In: GECCO, pp. 71–78. ACM Press, New York (2008)
12. Pan, H., Wang, L., Liu, B.: Particle swarm optimization for function optimization in noisy environment. Applied Math. and Compu. 181(2), 908–919 (2006)
13. Pratt, W.: Digital Image Processing. Wiley Interscience (2007)
14. Reyes-Sierra, M., Coello Coello, C.A.: Multi-objective particle swarm optimizers: A survey of the state-of-the-art. International Journal of Computational Intelligence Research 2(3), 287–308 (2006)
15. Schor, D., Kinsner, W., Anderson, J.: A study of optimal topologies in swarm intelligence. In: 23rd Canadian Conference on Electrical and Computer Engineering (CCECE), pp. 1–8 (2010)
16. Setayesh, M., Johnston, M., Zhang, M.: Edge and Corner Extraction Using Particle Swarm Optimisation. In: Li, J. (ed.) AI 2010. LNCS, vol. 6464, pp. 323–333. Springer, Heidelberg (2010)
17. Setayesh, M., Zhang, M., Johnston, M.: Improving edge detection using particle swarm optimisation. In: Proceedings of the 25th International Conference on Image and Vision Computing. IEEE Press, New Zealand (2010)
18. Setayesh, M., Zhang, M., Johnston, M.: Detection of continuous, smooth and thin edges in noisy images using constrained particle swarm optimisation. In: GECCO, pp. 45–52 (2011)
19. Setayesh, M., Zhang, M., Johnston, M.: Edge detection using constrained discrete particle swarm optimisation in noisy images. In: Proceedings of the 2011 IEEE Congress on Evolutionary Computation, pp. 246–253. IEEE Press (2011)
20. Zhao, J., Li, Z.: Particle filter based on particle swarm optimization resampling for vision tracking. Expert Systems with Applications 37(12), 8910–8914 (2010)

Adaptive Binarization Method for Enhancing Ancient Malay Manuscript Images

Sitti Rachmawati Yahya[1], Siti Norul Huda Sheikh Abdullah[1], Khairuddin Omar[1], and Choong-Yeun Liong[2]

[1] Center for Artificial Intelligence Technology, Faculty of Information Science and Technology, Universiti Kebangsaan Malaysia, 43600 UKM Bangi, Selangor, Malaysia
sitti.rachma@gmail.com, {mimi,ko}@ftsm.ukm.my

[2] School of Mathematical Sciences, Faculty of Science and Technology, Universiti Kebangsaan Malaysia, 43600 UKM Bangi, Selangor, Malaysia
lg@ukm.my

Abstract. In order to transform ancient Malay manuscript images to be cleaner and more readable, enhancement must be performed as the images have different qualities due to uneven background, ink bleed, or ink bleed and expansion of spots. The proposed method for image improvement in this experiment consists of several stages, which are Local Adaptive Equalization, Image Intensity Values, K-Means Clustering, Adaptive Thresholding, and Median Filtering. The proposed method produces an adaptive binarization image. We tested the proposed method on eleven ancient Malay manuscript images. The proposed method has the smallest average value of Relative Foreground Area Error compared to the other state of the art methods. At the same time, the proposed method have produced the better results and better readability compared to the other methods.

Keywords: Local Adaptive Equalization, Image Intensity Values, K-Means Clustering, Automatic Threshold, Median Filtering.

1 Introduction

Many researchers have successfully implemented image enhancement techniques for cleaning and separating background from the foreground in manuscripts or documents with a history of degraded or poor quality. A combination method is proposed to improve degraded images of the documents involving direct information of the detected edge images [1]. In line with that, a general threshold value can also separate the background and foreground on a shadow image [2, 3]. On the other hand, separation between foreground and background on carbon copied medical forms is done using the wave trajectory method [4]. Later, multiple threshold levels have been introduced to separate the text from the background by [5]. Besides that, an adaptive thresholding method based on adaptive window generation is used to separate textual content from the background in old Arabic documents [6]. This method begins with

D. Wang and M. Reynolds (Eds.): AI 2011, LNAI 7106, pp. 619–627, 2011.

text normalization and then separates the background using a 3×3 kernel which is used in the reading of the text block. They use edge direction matrixes and combination of projection profile to perform binarization of images.

Objective of this paper is to propose enhancement steps to overcome extreme major ink bleed surrounding textual manuscript images. This paper is divided into five sections. Section 1 introduces the background of the proposed method, including the researchers who had previously studied the expansion of the image, amendment of the threshold value, and adjustment in the binary image of the manuscript. Section 2 deals with the current methods which are the basis to this new proposed method. Section 3 explains the proposed method. Section 4 presents the results and discussion on the proposed research, while the last section presents the conclusions of this research.

2 State of the Art

Several methods proposed by previous researchers have been used as the basis of this experiment, which are:

- **Niblack's Method [7]:** The Niblack's Method is a simple and efficient method for adaptive thresholding. Niblack's Method can read the region of the image on a field that has less quality level. The local threshold used on the Niblack's method is set as follows:

$$t = \mu + w \tag{1}$$

where t is local threshold, μ and σ are a local mean and standard deviation which calculated over a local $(i \times j)$ window, w is the parameter to kernel window size.

– **Nick's Method [8]:** This method is proposed by Khurshid et al. [8]. The Nick's method was developed from the Niblack's method. It tried to solve low contrast problem by shifting down the thresholding value. The thresholding formula is as following:

$$T = m + k \times \sqrt{\frac{(\sum P_i^2 - m^2)}{NP}}, \tag{2}$$

where k is a control factor in the range of [–0.1, –0.2], P_i = the image pixel grey-scale value and NP = the total number of pixels in the image. The author suggested the $k = -0.1$ [7]. Kefali et al. [9] claimed that Nick's method gave the best performance compared to previous methods. However, problems of low contrast images still remained unsolved.

- **Bataineh's Method [6]:** This method suggests an adaptive threshold for low-contrast images and thin pen stroke problems. At first, the method only includes adaptive thresholding equation [10], then they extend the method [6] uses adaptive window generation and adaptive thresholding value towards repairing the image contrast based on global and local image information.

$$T_W = m_W - \frac{m_W^2 - \sigma_W}{(m_g + \sigma_W) \times (\sigma_{Adaptive} + \sigma_W)} \tag{3}$$

$$\sigma_{Adaptive} = \frac{\sigma_W - \sigma_{\min}}{\sigma_{max} - \sigma_{\min}} \tag{4}$$

where, T is the thresholding value, m_W is the mean value of the window's pixels, σ_W is the standard deviation of the window's pixels, m_g is the mean value of all pixels in the image and $\sigma_{Adaptive}$ is the adaptive standard deviation of the window.

Based on this T_W values, the binarization process is defined as follows:

$$I(x,y) = \begin{cases} black, & i(x,y) < T_W, \\ white, & Otherwise \end{cases} \tag{5}$$

where $I(x, y)$ is the binary image and $i(x, y)$ is the input pixel value of the image.

Alternatively, Bousellaa et al. [11] use iterative segmentation estimation approach to enhance Tunisian degraded manuscript images in Y channel. They perform iterative estimation by using Expectation Maximum (*EM*). They have also extended the *EM* by introducing maximum likelihood to approximate the probability that falls into either text or background classes.

Niblack's method produces images with characters of better shape but the thresholding value is not appropriate because the image is darker. The Nick's method almost produces good image, but the shape for images with ink-bleed expands-spots images is not so obvious and several results has lots of black regions. Bataineh's image is almost the same as the Nick's method, but Bataineh's method could further boost the image for more obvious characters shape for the low quality image. However, the results is also not clear for image that has damage around the characters.

All the methods above have been tested on different types of document image quality such multi-color image consisting different size fonts, spotted, low and very low-quality image, non-uniform illumination including thin pen stroke problems based on the DIBCO 2009 and 2011 benchmark image datasets. However, this dataset neglect document image that contains extremely major ink bleed around the textual information. This problem is found to be a major issue in preserving ancient Jawi-Malay handwritten manuscript in the Malaysian National Library. Figure 1 shows some examples of images of old Jawi-Malay Manuscript which have different levels of image quality.

In respond to the vital need from the Jawi manuscript reader community, we explore methods to overcome the above mentioned problems.

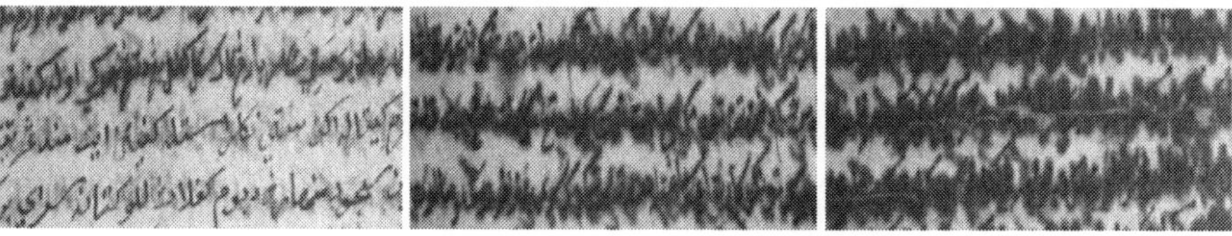

Fig. 1. Several examples of images of old Jawi-Malay manuscripts that have phases of different qualities that were used in this experiment. From the left: uneven background, ink bleed and ink bleed-expands spots images.

3 The Proposed Method

The Jawi-Malay manuscript images used in this research focus on the problem of ink bleed around the Jawi handwriting although there are some dirt or stain outside the Jawi handwriting. To solve these problems, we propose a method of binary adaptation to improve and enhance the quality of old Jawi Manuscript images. The method is illustrated in Figure 2 below:

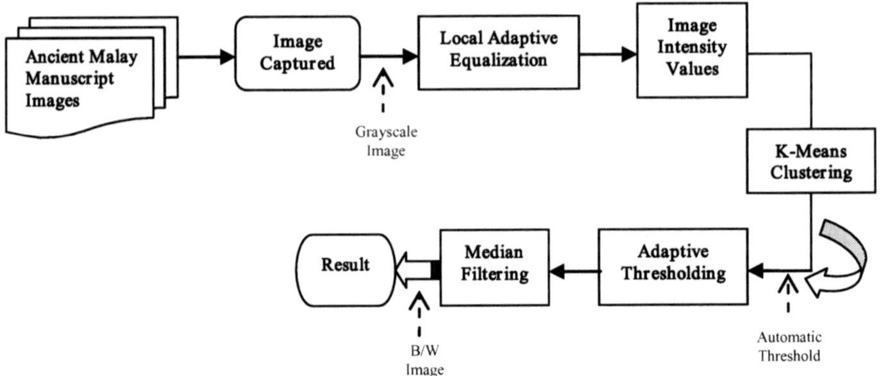

Fig. 2. The proposed flowcharts of adaptive binarization of ancient Malay manuscript images

Our proposed method are made up of the following steps: Local Adaptive Equalization *(LAE)* and image intensity values *(IIV)* process, K-Means Clustering to determine the automatic threshold, Adaptive Thresholding, and finally Median Filtering. Firstly, we improve and enhance the quality of the ancient images of Jawi-Malay Manuscripts using the *LAE* as follows:

$$\dot{g}_{(x,y)} = R \times \left[g_{(x,y)} - m\right] + m + M^2 \times m^2, \quad (6)$$

$$R = \left(k \frac{M}{\sigma + c}\right) \quad (7)$$

where $\dot{g}_{(x,y)}$ is the result of image transformation while $g_{(x,y)}$ is the input image, R is the coefficient as defined in Equation (7) with $k = 0.8$, m and σ are the mean and are the mean and standard deviation values of a fixed window subsequently, M is the average of the original image, and c is a constant. In this experiment, we apply 31×31 window size as proposed by Niblack [7]. Next, we perform Image Intensity Values (*IIV*) process as below:

$$\ddot{g}_{(x,y)} = \alpha \times (\dot{g}_{max} - \dot{g}_{min}) + \dot{g}_{min} + \dot{g}_{(x,y)} \quad (8)$$

where the value of $\alpha = 0.1$, $\dot{g}_{max}$ and $\dot{g}_{min}$ are the maximum and minimum values of pixels in $\dot{g}_{(x,y)}$, which are the resulting images of the *LAE* process. Consequently, the *IIV* process helps to reduce apparent background noise. However, this step is still insufficient for smaller noise or shadow noise.

In order to decide either a pixel belongs to the foreground or the background, we use K-Means Clustering technique as one of the steps. Then, we separate the shadow around the characters by proposing an adaptive threshold process to all clustering pixels. This proposed method, an extension to [12] and [13], searches for the adaptive threshold value based on a bi-level histogram. In [12], the threshold value searching is carried out by determining the balance point of the uncertain threshold value, then balancing it with weights closes to the uncertain threshold point until the actual threshold value is found.

We apply histogram graph by using Gaussian Windows to obtain smoother line graph before calculating the two highest peak values. We summarize the proposed automatic thresholding process as Algorithm 1 and the process is illustrated in Figure 3:

Algorithm 1. Proposed automatic thresholding process.

```
BEGIN
Let fi = Image histogram , P1 = First highest peak (background image), P2
= Second highest peak (foreground image), Ci = The cluster of i, Val_ti =
Minimum grey level of I, npi= Number of pixel of i, and tmp = The step
to next grey level;
  DIVIDE grey level to 5 clusters (Ci);{Each cluster = 50 pixels number}
  DETECT which cluster belongs to P2;
  IF P2 < C1max
     THEN r1 = C1max/2;
    IF P2 ≥ r1
     THEN r2 = (C2max + C1max)/2;
     W2 = [P2: r1];
     Val_t1 = minimal grey level value in W2;
       IF W3 = [1: r1];
        Val_t2 = minimum grey level value in W3;
       ELSE
        W4 = [C1max: r2];
        Val_t3 = minimum grey level value in W4;
       ENDIF
    ENDIF
  ENDIF
  IF Val_ti < 100
     THEN tmpi = tmp -1;
     Val_npi(tmpi) = npi(tmpi) - Val_ti;
  ELSE
     thresh = Val_ti;
  ENDIF
  IF Val_npi(tmpi) < 100
     THEN tmpi = tmp -1;
     Val_npi(tmpi) = Val_npi(tmpi) - Val_ti;
  ELSE
     thresh = Val_npi;
  ENDIF
```

where $r_1$dan r_2 are the range for the first and second cluster subsequently, C_1_max and C_2_max are the maximum gray level values of the first and second cluster in order, W_2, W_3, W_4 are the range limit set for r_1 and r_2, p_1 and p_2 are the first and second peak values in the histogram bi level correspondingly, and f_i is the histogram of the relevant image. Lastly, Val_t, Val_t$_l$, Val_t$_2$ are minimum limit values surrounding to a maximum value of the second peak.

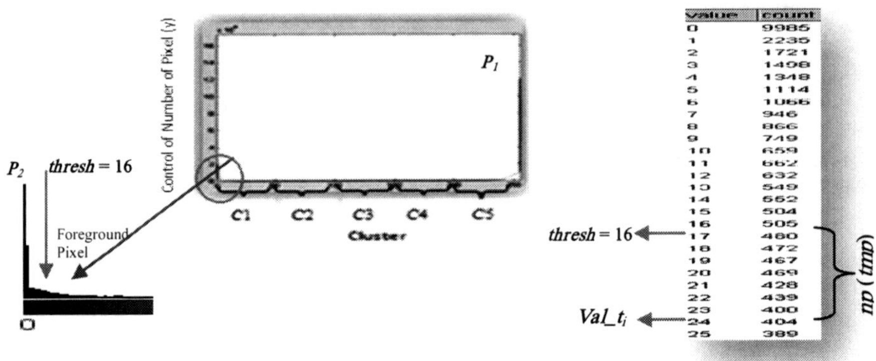

Fig. 3. The graph shows the distribution of the existing 5 groups (clusters) and the table on the right is to clarify the limits of pixel gray level to obtain the automatic threshold. Number 16 is an automatic threshold (thresh) value and it was taken from the total number of pixel-based gray level value of the second peak.

Next, we accomplish final step in adaptive threshold process as below:

$$\dddot{g}_{(x,y)} = \begin{cases} 1, & if\ \ddot{g}_{(x,y)} < thresh \\ 0, & otherwise \end{cases} \tag{9}$$

where $\dddot{g}_{(x,y)}$ is an output image after performing adaptive binarization, $\ddot{g}_{(x,y)}$ is an image result after applying *K-Means* clustering, and *thresh* is automatic threshold value based on bi-level histogram.

$$\breve{g}_{(x,y)} = \dddot{g}_{(x,y)} \times 2 - M, \tag{10}$$

where M is an average value of the variable $\ddot{g}$.

$$\hat{g}_{(x,y)} = \bar{g}_{(x,y)} - \breve{g}_{(x,y)} - C \tag{11}$$

where $\hat{g}_{(x,y)}$ is an output image of $\bar{g}_{(x,y)}$ after applying median filter with a 20×20 20×20 window size, and C is a constant with value of 0.03.

In order to remove unwanted noise, we reapply median filtering with a 3×3 kernel size onto $\hat{g}$ image.

$$\tilde{g}_{(x,y)} = \left[\bar{\bar{g}}_{(x,y)} \xleftarrow{median\ filter} \hat{g}_{(x,y)}\right]^2, \tag{12}$$

where $\tilde{g}_{(x,y)}$ is a sum product image of Median filtering process of image, $\hat{g}_{(x,y)}$.

4 The Experimental Results and Discussion

In this experiment, eleven ancient Hang Tuah Malay Manuscript images which were taken from the Malaysian National Library [14] have been used. The images were divided into three levels of quality of uneven background, ink bleed and ink bleed-expands spots images (the first row of Figure 4). We used 640 × 512 image size in grayscale format. Additionally, we compared our proposed method with other state of the art methods namely Niblack's [7], Nick's [8] and Bataineh's method [6]. The resulting images after applying the proposed and the other methods are given in Figure 4.

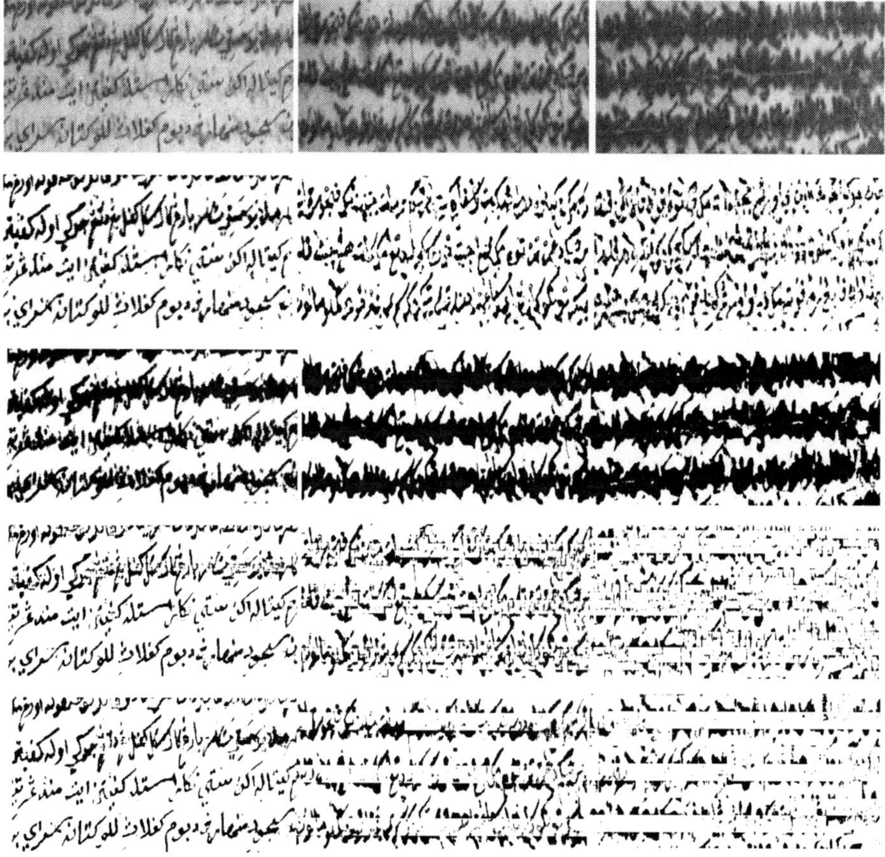

Fig. 4. Several images of the Hang Tuah Malay Manuscript which are divided into 3 levels of quality. From the left: uneven background, ink bleed, and ink bleed-expands spots images. The following rows are the resulting images after applying the proposed method, Niblack's Method [7], Nick's Method [8] and Bataineh's Method [6], respectively.

We measure performance of our proposed method based on the Relative Foreground Area Error (*RAE*) criterion proposed by Sezgin and Sankur [15]. This criterion calculates the expected values within [0, 1]. In all cases, the measure that is closer to zero corresponds to the best binarization result [10]. It can be expressed as below:

$$RAE = \begin{cases} \frac{A_o - A_k}{A_o}, & A_k < A_o \\ \frac{A_k - A_o}{A_k}, & A_k \geq A_o \end{cases} \tag{13}$$

where, A_o and A_k are the foreground areas in the reference image and the test data image. The analytical score values for each of the three types of degraded document images after binarization by the various methods are shown in Table 1.

Table 1. The *RAE* values and their averages for images of different quality levels taken from Hang Tuah Malay Manuscript after using the proposed, Niblack's [7], Bataineh's [6] and Nick's [8] methods

Quality Levels	Image	Proposed Method *RAE* Value	Niblack's Method [7] *RAE* Value	Bataineh's Method [6] *RAE* Value	Nick's Method [8] *RAE* Value
Uneven Background Images	Im63	0.0769	0.3258	0.1600	0.1311
	Im65	0.0787	0.3377	0.1748	0.1347
	Im67	0.0769	0.2965	0.1497	0.1360
	Im69	0.0782	0.1931	0.1570	0.1428
	Im77	0.0815	0.1356	0.1480	0.1331
Ink-Bleed Images	Im99	0.0764	0.3429	0.1855	0.1620
	Im101	0.0649	0.3905	0.1712	0.1766
	Im107	0.0678	0.3812	0.1553	0.1421
Ink-Bleed and Expansion Spot Images	Im61	0.0248	0.4025	0.1294	0.1177
	Im109	0.0467	0.4509	0.1432	0.1122
	Im111	0.0342	0.4117	0.1412	0.1224
AVERAGE		**0.0642**	**0.3335**	**0.1559**	**0.1373**

The smaller the value of *RAE* for an image, the better the quality of the image. Also, this means that the error of the pixels in the foreground area of the images is little. From Table 1, the proposed method achieved better *RAE* results of 0.0643 compared to Niblack's, Nick's and Bataineh's methods which achieved 0.3335, 0.1373, and 0.1559 respectively. Therefore the resulting images produced by our proposed method are better and more readable for all types of image studied, i.e. uneven background, ink bleed and ink bleed-expands spots images.

5 Conclusion

Most of the Hang Tuah Manuscript image datasets are suffering from extremely bad qualities that leads to inconvenience among readers. Therefore, the Pattern Recognition Research Group has been continuously put in effort to improve the existing image processing methods in order to dig up invaluable information from our local ancient manuscripts. In summary, the proposed method shows a smaller value of average *RAE* for all the three levels of image qualities in comparison to the other state of the art methods which are Niblack's [7], Nick's [8] and Bataineh's [6] methods.

Acknowledgments. Thanks to the National Library of Malaysia (PNM). This research project was funded by the research grants UKM-TT-03-FRGS0130 and UKM-TT-03-FRGS0129.

References

1. Gatos, B., Pratikakis, I., Perantonis, S.J.: Improved Document Image Binarization by Using a Combination of Multiple Binarization Techniques and Adapted Edge Information. In: 19th International Conference on Pattern Recognition (ICPR), Tampa, Florida, USA, pp. 1–4 (2008) ISBN: 978-1-4244-2175-6/08
2. Yosef, I.B., Beckman, I., Kedem, K., Dinstein, I.: Binarization, Character Extraction, and Writer Identification of Historical Hebrew Calligraphy Documents. IJDAR 9, 89–99 (2007)
3. Shafait, F., Keysers, D., Breuel, T.M.: Efficient Implementation of Local Adaptive Thresholding Techniques Using Integral Images. In: Proc. SPIE. Document Recognition and Retrieval XV (2008)
4. Milewski, R., Govindaraju, V.: Binarization and Cleanup of Handwritten Text from Carbon Copy Medical Form Images. Pattern Recognition 41, 1308–1315 (2008)
5. Arora, S., Acharya, J., Verma, A., Panigrahi, P.K.: Multilevel Thresholding for Image Segmentation through a Fast Statistical Recursive Algorithm. Pattern Recognition Letters 29, 119–125 (2008)
6. Bataineh, B., Abdullah, S.N.H.S., Omar, K.: An adaptive local binarization method for document images based on a novel thresholding method and dynamic windows. Journal of Pattern Recognition Letters 32, 1805–1813 (2011)
7. Niblack, W.: An Introduction to Digital Image Processing. Prentice Hall, Upper Saddle River (1985)
8. Khurshid, K., Siddiqi, I., Faure, C., Vincent, N.: Comparison of Niblack Inspired Binarization Methods for Ancient Documents. In: 16th International Conference on Document Recognition and Retrieval. SPIE, USA (2010)
9. Kefali, A., Sari, T., Sellami, M.: Evaluation of Several Binarization Techniques for Old Arabic Documents Images. In: The First International Symposium on Modeling and Implementing Complex Systems, MISC 2010, Constantine, Algeria, pp. 88–99 (2010)
10. Bataineh, B., Abdullah, S.N.H.S., Omar, K., Faidzul, M.: Adaptive Thresholding Methods for Documents Image Binarization. In: Martínez-Trinidad, J.F., Carrasco-Ochoa, J.A., Ben-Youssef Brants, C., Hancock, E.R. (eds.) MCPR 2011. LNCS, vol. 6718, pp. 230–239. Springer, Heidelberg (2011)
11. Boussellaa, W., Bougacha, A., Zahour, A., El Abed, H., Alimi, A.: Enhanced Text Extraction from Arabic Degraded Document Images using EM Algorithm. In: 10th International Conference on Document Analysis and Recognition, pp. 743–747 (2009)
12. António, A., Leite, R., Cancela, M.L., Shahbazkia, H.R.: MAQ – A Bioinformatics Tool for Automatic Macroarray Analysis. International Journal of Computer Applications 4, 51–58 (2010)
13. Atae-Allah, Z., Aroza, J.M.: A Filter to Remove Gaussian Noise by Clustering the Gray Scale. Journal of Mathematical Imaging and Vision 17(1), 15–25 (2002)
14. Manuscripts, National Library of Malaysia (Perpustakaan Negara Malaysia, PNM) (April 27, 2009), `http://www.pnm.gov.my/pnmv3/index.php?id=84`
15. Sezgin, M., Sankur, B.: Survey Over Image Thresholding Techniques and Quantitative Performance Evaluation. J. Electron Imaging 13(1), 146–165 (2004)

A New Approach to Use Concepts Definitions for Semantic Relatedness Measurement

Ehsan KhounSiavash and Kamran Zamanifar

Computer Engineering Department, Engineering Faculty, University of Isfahan, Isfahan, Iran
{Khounsiavash,Zamanifar}@eng.ui.ac.ir

Abstract. Semantic Relatedness Measurement (SRM) is one of the most important applications of reasoning by ontologies and different disciplines of AI, e.g. Information Retrieval, are firmly tied to it. The accuracy of SRM by lexical resources is largely determined by the quality of the knowledge modeling by the knowledge base. The limited types of relations modeled by ontologies have caused most of the SRM methods to be able to detect and measure only a few special types of semantic relationships that is very far from the concept of semantic relatedness in human brain. Concepts of lexical resources are usually accompanied with a plain text narratively defines the concept. The information included in the definition of concepts sound very promising for SRM. This paper intends to treat this information as formal relations to improve SRM by distance-base methods. In order to do so, concepts glosses are mined for the semantic relations that are not modeled by the ontology. Then, these relations are employed in combination with classic relations of the ontology for semantic relatedness measurement according to the shortest path between concepts. Our evaluation demonstrated qualitative and quantitative improvement in detection of previously unknown semantic relationships and also stronger correlation with human judgment in SRM.

Keywords: Semantic Relatedness Measurement, Ontology, Definition of Concepts, Gloss, and Path.

1 Introduction

Semantic Relatedness Measurement (SRM) is one of the most important and challenging applications of ontologies. It is widely used in Information Retrieval and Knowledge Extraction (e.g. semantic annotation, indexing, query expansion, word sense disambiguation, and named entity recognition), Machine Learning, Knowledge Management and Engineering, etc.

Human`s brain may recognize two things semantically related because of wide variety of semantic links between them. The most familiar types of semantic relationships consist of similarity, inclusion, interaction, bilaterality, or even opposition. For example, taxi and car are semantically related because of their common applications, car and wheel because wheel is part of car, car and fuel because car uses fuel, or even heat and cold because they are opposite. However, due

D. Wang and M. Reynolds (Eds.): AI 2011, LNAI 7106, pp. 628–637, 2011.

to limitations of ontologies to model a domain knowledge, existing ontology-based methods of SRM can not perceive and measure most of such semantic relations.

Research Challenge. Effective SRM needs a detailed model of the knowledge, whereas elaborating the model of domain knowledge is very expensive and complicates reasoning mechanisms [1]. Moreover, the knowledge of a domain, whatever restricted, is too complicated that a fully elaborated ontology that models all aspects of the domain knowledge is extremely elusive. Considering these trade-offs, ontology engineers usually prefer to connect the concepts by a small set of general relations and derive complicated relations from the general ones. Hence, most ontologies suffer from the lack of non-classical semantic relations, like those introduced by [2], because it is not possible to drive all of non-classical relations from the general ones.

SRM methods, like other applications of ontologies, are seriously affected by this limitation. This problem has been widely addressed in the past two decades by different approaches like using concepts glosses or employing external resources e.g. collection of documents. However, the lack of non-classical semantic relations introduces biases into any application that relies upon SRM.

Contribution. An ontology is an explicit, formal specification of a shared conceptualization of a domain of interest [3], where formal implies that the ontology should be machine-readable [4]. However, using natural language to express concepts definitions prevents the machine to use a great deal of information conveyed in this way.

In this paper, a new method of using the definition of concepts in SRM has been proposed. The method relies on the semantic relationships among the concepts cooperate to define a concept and the defined concept. This idea is employed in order to extract new semantic relations from concepts glosses and increase the connectivity of the semantic network. The richer connectivity in the knowledge base, the more accurate semantic relatedness measurement is possible.

Organization. In the next section, the literature review will be presented. In the third section, our solution to improve the accuracy of SRM methods, that is based on [1] contribution to WordNet will be proposed. The proposed method will be examined in the forth section. Section 5 concludes and draws some future work.

2 Related Work

The classic categories of semantic relatedness measurement methods using lexical resources consist of distance-based, information theoretic-based, and gloss-based approaches [5].

Distance-based methods, like [6, 7], rely on the notion that the shorter is the length of path between two concepts in a semantic network, the stronger the semantic relationship between them will be. This idea is seriously affected by common inconsistencies of semantic networks [8, 9] and different heuristics, like patterns for meaningful paths and weighting paths, have been proposed to alleviate the problem [10-13].

Information theoretic-based methods use information contents of concepts to measure their semantic relatedness. The information content of a concept is defined as *–log[p(c)]*, so that *p(c)* is the probability by which the concept c is used in a collection of documents [14]. In this approach, the semantic similarity is measured by the amount of shared information content by the two concepts. They generally use corpus statistics in order to calculate the *p(c)*.

The dependency of information theoretic-based methods motivated [15, 16] to calculate the information content of concepts without using external resources.

Gloss-based methods, the third category, intend to utilize the definition of concepts--the unused parts of the information provided by ontologies. Thus, these methods are able to reflect some kinds of semantic relationships that are not observable through classic relations. Reference [17] introduced this idea for Word Sense Disambiguation. It compared definitions of different meanings of a word to its neighbor words in a text in order to find the best meaning of an ambiguous word. This idea has been expanded by Pedersen *et al.*

Reference [18] used the comparison of concepts definitions in order to measure their semantic relatedness. Since definitions of concepts are generally short, they incorporated the definitions of directly connected concepts to the definition of the concept. Moreover, [18] valued longer phrases more than shorter ones when compared.

Pederson *et al.* used the definition of concepts in [19] too. In this method, the definitions of concepts are used as a corpus to constitute the co-occurrence matrix of the concepts. Using this matrix and the relations of the ontology, the second order co-occurrence matrix will be constituted that describes each concept as a vector in the semantic space of the ontology.

There are also hybrid methods that try to combine different approaches in order to overcome one`s disadvantages by the other one`s advantages [10, 14, 20].

In spite of different efforts and studies, comprehensive SRM that takes into consideration the most possible semantic relationships is still an open problem.

3 Proposed Method

The concepts of an ontology are usually accompanied with a narrative plain text which informally defines the concept and is called definition or gloss. The definition of each concept includes other concepts that implies important semantic relationships among the defined concept and the concepts used in the definition. Many times, these semantic relations are not formally modeled by the ontology because they do not conform to the classic and standard types of the semantic relations defined by the ontology. For instance, consider the way "gasoline" is shown in WordNet. (Fig. 1)

The gloss of "*gasoline*" in WordNet implies that there are strong semantic relationships among "*gasoline*" from one side and "*hydrocarbon*", "*hexane*", "*heptane*", "*octane*", "*petroleum*", "*fuel*", and "*internal-combustion engines*" from the other side. However, the non-classic relationship between "*gasoline*" and "*internal-combustion engine*" is not modeled because it does not conform to the standard semantic relations of WordNet like *is-a* and *is-part-of*.

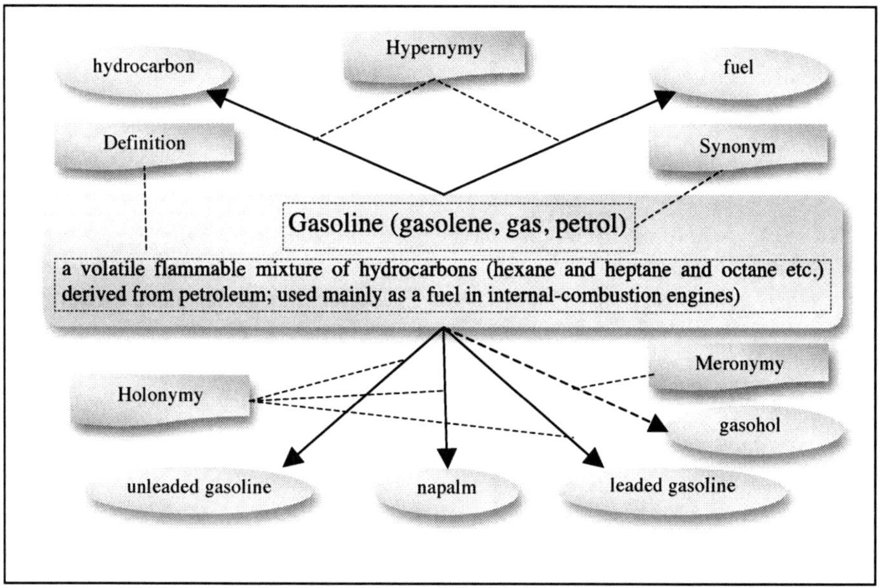

Fig. 1. The representation of WordNet for "Gasoline"

These types of semantic relations could play crucial roles in Information Retrieval and Natural Language Processing applications like Word Sense Disambiguation, text summarization, and spelling error correction [2, 21]. However, since they are not introduced as formal relations, they are not usually taken into account by the SRM methods that use the ontology. Therefore, in order to employ the information embedded in the definition of concepts, it should be explicitly extracted.

According to [1], Gloss and Attribute are the two types of direct semantic relations between noun concepts embedded in their definitions. Our study focuses on the semantic relations exist between noun concepts in order to avoid complicated reasoning rules. Since, for example, the degree of semantic relatedness between "automobile" from one side and "wheel", "fuel", "road", "SUV", or "cab" from another side is the same, although each pair is connected by different semantic relations. Thus, the semantic relations derived from definition of concepts could be generalized to definition-based relations.

3.1 Definition-Based Relations

Considering ontology O and the definition of each concept in O as a set of concepts of O that are cohered with stop words, the semantic relationships among the noun concepts used in a definition and the defined concept can be defined as the following:

Definition1. *Concept c_1 has Refer-To relationship with concept c_2, if c_2 is used in the definition of c_1. (1)*

$$ReferTo(c_1, c_2) \Leftrightarrow \forall c_1, c_2 \in O \wedge c_2 \in \{Definition\ of\ c_1\} \quad (1)$$

Definition2. *Concept c_1 has Referred-By relationship with concept c_2, if c_1 is used in the definition of c_2.(2)*

$$ReferredBy(c_1, c_2) \Leftrightarrow \forall c_1, c_2 \in O \wedge c_1 \in \{Definition\ of\ c_2\} \quad (2)$$

For example, according to WordNet (Fig. 1) *"gasoline"* has *Refer-To* relationship with *"internal combustion engine"*. Conversely, *"internal combustion engine"* has *Referred-By* relationship with *"gasoline"*. These relations could be very helpful in applications like Word Sense Disambiguation (WSD). For instance, the semantic relationship between *"car"(as "automobile")* and *"gas"*(as *"gasoline"*) can not be effectively measured through general relations of WordNet, but definition-based relations easily reveal their semantic relatedness by a 2-long-path (*"gasoline"*→ *"internal-combustion engine"*→ *"automobile"*).

Table 1 lists the concepts that are in definition-based relation with *"gasoline"* among which *"fuel"*, *"napalm"*, *"leaded gasoline"*, and *"unleaded gasoline"* are crossed because of the existing classic relationships defined by WordNet between *"gasoline"* and these concepts.

Table 1. Definition-based semantic relations for "Gasoline"

Relation Type	Related Concept		
Refer-To	1) Petroleum	2) Hexane	3) ~~Fuel~~
	4) Octane	5) Hydrocarbon	6) Heptane
	7) Internal-combustion engine		
Referred-By	1) mileage	10) put put	19) ethyl alcohol
	2) antiknock	11) gas pump	20) vapor lock
	3) miles per gallon	12) ~~napalm~~	21) gasoline gauge
	4) gasoline_engine	13) gas engine	22) power mower
	5) gasoline_station	14) isobutylene	23) tetraethyl lead
	6) gasohol	15) fuel line	24) carburetor
	7) gas tank	16) octane number	25) hydrocracking
	8) additive	17) ~~unleaded_gasoline~~	26) gasoline tax
	9) gas line	18) ~~leaded_gasoline~~	

3.2 Semantic Relatedness Measurement

Reference [22] is one of the earliest and most simple distance-based method of SRM. It defined the semantic distance of two concepts as the length of the shortest path between them. Using the same idea, equation (3) linearly distributes the semantic relatedness form *MaxVal* to *zero* according to the length of the shortest path between the two concepts.

$$SemRel(c_1, c_2) = MaxVal * \left(1 - \frac{len(c_1, c_2)}{MaxLen}\right), MaxLen = \max_{\forall c_1, c_2 \in O} len\,(c_1, c_2) \quad (3)$$

Where *len*(c_1, c_2) is the length of the shortest path between c_1 and c_2 and *MaxLen* is the maximum length of shortest path between two concepts in the semantic network. The shortest path is a mixture of all available relations in the ontology and there is no pattern to select or exclude relations. The farther are the two concepts, the less is their semantic similarity, and vice versa.

4 Evaluation

In this section, the extraction of definition-based relations from WordNet, as the most famous lexical ontology, will be described. Then, the effects of these relations on semantic relatedness measurement by equation (3) will be examined and discussed.

4.1 Extraction of the Relations

As it was stated in the previous section, other concepts than nouns are ignored in this study. In order to avoid the difficulties of automatic word sense disambiguation we used WordNet Gloss Corpus, which is a manually-tagged disambiguated gloss set of WordNet 3.0. Mining the glosses of noun concepts, 476026 links were found, among them there were two groups of redundant links.

The first group existed between two concepts that mutually include the other one in their glosses, i.e. when c_1 is used in the gloss of c_2 and c_2 is used in the gloss of c_1. There were13733 instances of such cases where Referred-By relations were omitted in favor of Refer-To relations.

The other redundant links are the relations previously modeled by classic relations of the ontology. These links (106248) were also ignored in favor of classic relations of the ontology.

After redundancy elimination 356047 links remained, that is more than 4 definition-based relationships for each noun synset in WordNet 3.0.

4.2 Evaluation Method and Results

Rubenstein and Goodenough (1965) asked 51 persons to judge about the semantic similarity of 65 pairs of words ranging from “highly synonymous” to “semantically unrelated”. Miller and Charles (1991) found similar results in a similar study using 30 pairs of Rubenstein and Goodenough`s and 31 subjects [21]. The results of their experiments made a baseline for the evaluation of semantic similarity methods. In the absence of appropriate dataset for the evaluation of semantic relatedness measurement methods, these experiments have been adopted as an independent preliminary method for the evaluation and comparison of SRM methods.

In order to evaluate the effects of definition-based relations, the semantic relatedness of Miller and Charles’ pairs were calculated by equation (3) with two different circumstances. At the first time, the shortest paths do not include definition-based relations. This run is called Run I and makes the baseline of the evaluation. At the Second time, called Run II, the shortest paths include classic relations of WordNet along with the definition-based relations.

The results of using equation (3) to calculate the semantic relatedness of Miller and Charles' pairs are depicted in Fig. 2. The series marked by ◆ indicates the human judgments, the series that marked by ■ indicates the results of the experiment using just the standard relations of WordNet, and the series marked by ▲ indicates the results of using equation (3) when the semantic network of WordNet is enriched by definition-based relations.

The results of Run I and Run II are compared with the human judgments obtained from Miller and Charles experiment. The Pearson correlation coefficient of the human judgment and the results of Run I and Run II are respectively 0.872 and 0.905.

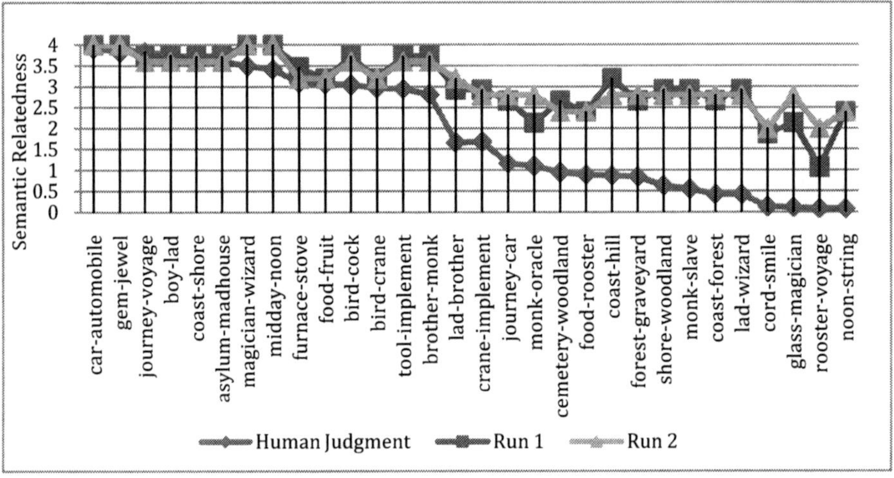

Fig. 2. The values of semantic relatedness for instances of Miller and Charles (1991), Run I, and Run II

4.3 Discussion

According to the statistics mentioned in section 4.1, about 25% of the links obtained through mining of concepts glosses have been formerly modeled by classic relations of WordNet. This fact supports the idea that there are strong semantic relationships between a concept and the other concepts which constitute its definition.

Furthermore, the more smooth descent of the curve of Run II and its higher correlation with human judgment, in comparison to Run I, confirm the positive contribution of definition-based relations to [22]'s distance-based SRM method.

The Pearson correlation coefficients of the two runs with the human judgment, along with the most prominent methods of semantic similarity/relatedness measurement have been listed in Table 2. The correlation of 0.905 for the Run II shows that using equation (3) and the shortest path consists of all various relations of the ontology perform as well as state-of-the-art distance-based SRM methods.

Table 2. The Pearson's correlation coefficients of some of semantic relatedness or similarity measurement methods [23]

Method	M&C	Path	IC	Gloss
Resnik	0.774	N	Y	N
Hirst and St–Onge	0.744	Y	N	N
Leacock and Chodorow	0.816	Y	N	N
Lin	0.829	N	Y	N
Jiang and Conrath	0.836	Y	Y	N
Li et al.	0.882	Y	Y	N
Patwardhan and Pedersen	0.910	N	N	Y
KhounSiavash and Barrani-Dastjerdi	0.910	N	N	Y
Yang and Powers	0.921	Y	N	N
Alvarez and Lim	0.913	Y	N	N
Pirró	0.912	N	Y	N
Shortest Path(Run I)	0.872	Y	N	N
Shortest Path(Run II)	0.905	Y	N	Y

However, comparing the 3 curves in Fig. 2 demonstrates that the values of semantic relatedness for the both runs decrease more slowly than the human judgments, so that the minimum SRM for Run I and Run II are, respectively, 1 and 2 that are more than our sense of semantic relatedness for the pairs at the button of the list. This inconsistency, which exists in both runs, probably originates from the two assumptions, namely linear behavior of equation (3) and using all possible combination of relations to find the shortest paths. In that, diversifying the types of relations and increasing the length of a path decrease the semantic information conveyed by the path.

Furthermore, the experiments of Rubenstein-Goodenoug and Miller-Charles suffer from two important flaws. First, they asked subjects to judge about semantic similarity of pairs and not semantic relatedness. It is noticeable by considering the scores of instances; for example, "*automobile-cushion*" in spite of having direct *is-part-of* relation is assigned a small score. Moreover, there are non-standard semantic relationships between concepts like those relate "*bird*" to "*woodland*" or "*car*" to "*journey*" which were not scored as much as similarity. Additionally, the limited number of instances does not contain a lot of important semantic relationships that should be considered when SRM methods are being evaluated.

Second, judging the intended meaning of a polysemous word is possible just by considering its application in the context. For this reason, some methods, like [18], in spite of having good results in application based evaluation, are not very successful in independent evaluation like Rubenstein-Goodenough or Miller- Charles.

Apart from that, we strongly believe that including the relations embedded in the definition of concepts expands the span of semantic relatedness measureable through the information provided by ontologies. The considerable improvement in the correlation of Run I by adding these relations confirms the value of information these relations provide for the simple distance-based method. Although, the dataset does not contain a lot of non-classical semantic relations which are easily detectable by definition-based relations.

5 Conclusion and Future Work

This study intended to improve semantic relatedness measurement by the information embedded in the definition of concepts. To this end, the narrative information of glosses was extracted as formal relations in order to enrich the connectivity of the semantic network. These relations enhance the poor ability of semantic network to reflect non-classical semantic relations and lead to more accurate semantic relatedness measurement.

In order to evaluate the effects of these relations, they were extracted from a manually sense tagged version of WordNet. Adding the definition-based relations to the semantic network of WordNet caused 3% improvement in SRM by a simple distance-based method in comparison to when it used just classic relations of WordNet.

As the future work, this study will be continued in two areas. The first issue to be solved is some small anomalies caused by adding definition-based relations. This can be done by pruning irrelevant relations, weighting relations, and changing the way of finding the shortest path between concepts. The second is to find a way to compare different SRM methods in measuring non-classical semantic relationships.

References

1. Harabagiu, S., Moldovan, D.I.: Knowledge Processing on an Extended WordNet. In: Fellbaum, C. (ed.) WordNet: An Electronic Lexical Database, vol. 305, pp. 381–405. MIT Press, Cambridge (1998)
2. Morris, J., Hirst, G.: Non-classical Lexical Semantic Relations. In: The HLT-NAACL Workshop on Computational Lexical Semantics, pp. 46–51. Association for Computational Linguistics, Stroudsburg (2004)
3. Gruber, T.R.: Toward Principles for the Design of Ontologies Used for Knowledge Sharing. International Journal of Human Computer Studies 43, 907–928 (1995)
4. Buitelaar, P., Cimiano, P., Magnini, B.: Ontology Learning from Text: An Overview. In: Buitelaar, P., Cimiano, P., Magnini, B. (eds.) Ontology Learning from Text: Methods, Evaluation and Applications, vol. 123, pp. 3–12. IOS Press (2005)
5. Varelas, G., Voutsakis, E., Raftopoulou, P., Petrakis, E.G.M., Milios, E.E.: Semantic Similarity Methods in WordNet and Their Application to Information Retrieval on the Web. In: 7th Annual ACM International Workshop on Web Information and Data Management, pp. 10–16. ACM Press, New York (2005)
6. Hirst, G., St-Onge, D.: Lexical Chains as Representations of Context for the Detection and Correction of Malapropisms. In: Fellbaum, C. (ed.) WordNet: An Electronic Lexical Database, vol. 305, pp. 305–332. The MIT Press, Cambridge (1998)
7. Wu, Z., Palmer, M.: Verbs Semantics and Lexical Selection. In: 32nd Annual Meeting on Association for Computational Linguistics, pp. 133–138. Association for Computational Linguistics, Stroudsburg (1994)

8. KhounSiavash, E., Baraani-Dastjerdi, A.: Using the Whole Structure of Ontology for Measuring Semantic Relatedness Measurement. In: 22th International Conference on Software Engineering and Knowledge Engineering, pp. 79–83. Knowledge Systems Institute Graduate School, IL, USA (2010)
9. KhounSiavash, E., Baraani-Dastjerdi, A.: Using the Density of Paths for Semantic Relatedness Measurement. In: 2010 Semantic Web and Web Services, pp. 18–24. CSREA Press, USA (2010)
10. Li, Y., Bandar, Z.A., McLean, D.: An Approach for Measuring Semantic Similarity between Words Using Multiple Information Sources. IEEE Transactions on Knowledge and Data Engineering 15, 871–882 (2003)
11. Mazuel, L., Sabouret, N.: Semantic Relatedness Measure Using Object Properties in an Ontology. In: Sheth, A.P., Staab, S., Dean, M., Paolucci, M., Maynard, D., Finin, T., Thirunarayan, K. (eds.) ISWC 2008. LNCS, vol. 5318, pp. 681–694. Springer, Heidelberg (2008)
12. Yang, D., Powers, D.M.W.: Measuring Semantic Similarity in the Taxonomy of WordNet. In: 28th Australasian Conference on Computer Science, vol. 38, pp. 315–322. Australian Computer Society, Inc., Newcastle (2005)
13. Rhee, S.K., Lee, J., Park, M.W.: Semantic Relevance Measure between Resources Based on a Graph Structure. In: International Multiconference on Computer Science and Information Technology, vol. 8, pp. 229–236. IEEE Press (2008)
14. Resnik, P.: Using Information Content to Evaluate Semantic Similarity in a Taxonomy. In: 14th International Joint Conference on Artificial Intelligence, Canada, Montreal, vol. 1, pp. 448–453 (1995)
15. Blanchard, E., Kuntz, P., Harzallah, M., Briand, H.: A Tree-based Similarity for Evaluating Concept Proximities in an Ontology. In: 10th Conference of the International Federation of Classification Societies, pp. 3–11. Springer, Heidelberg (2006)
16. Pirró, G.: A Semantic Similarity Metric Combining Features and Intrinsic Information Content. Data & Knowledge Engineering 68, 1289–1308 (2009)
17. Lesk, M.: Automatic Sense Disambiguation Using Machine Readable Dictionaries: How to Tell a Pine Cone from an Ice Cream Cone. In: 5th Annual International Conference on Systems Documentation (SIGDOC 1986), pp. 24–26. ACM Press, New York (1986)
18. Banerjee, S., Pedersen, T.: Extended Gloss Overlaps as a Measure of Semantic Relatedness. In: 18th International Joint Conference on Artificial Intelligence, vol. 18, pp. 805–810. Morgan Kaufmann Publishers Inc., San Francisco (2003)
19. Patwardhan, S., Pedersen, T.: Using WordNet-based Context Vectors to Estimate the Semantic Relatedness of Concepts. In: EACL 2006 Workshop Making Sense of Sense: Bringing Computational Linguistics and Psycholinguistics Together, Italy, pp. 1–8 (2006)
20. Jiang, J.J., Conrath, D.W.: Semantic Similarity Based on Corpus Statistics and Lexical Taxonomy. In: International Conference Research on Computational Linguistics (ROCLING X), Taiwan, pp. 19–33 (1997)
21. Budanitsky, A., Hirst, G.: Evaluating WordNet-based Measures of Lexical Semantic Relatedness. Computational Linguistics 32, 13–47 (2006)
22. Rada, M., Bicknell, E.: Ranking Documents with a Thesaurus. American Society for Information Science and Technology 40, 304–310 (1989)
23. Alvarez, M.A., Lim, S.J.: A Graph Modeling of Semantic Similarity between Words. In: International Conference on Semantic Computing (ICSC), pp. 355–362. IEEE Press, Washington, DC (2007)

Using a Lexical Dictionary and a Folksonomy to Automatically Construct Domain Ontologies

Daniel Macías-Galindo, Wilson Wong,
Lawrence Cavedon, and John Thangarajah

School of Computer Science and I.T., RMIT University, Melbourne, Australia
{daniel.macias,wilson.wong,lawrence.cavedon,john.thangarajah}@rmit.edu.au

Abstract. We present and evaluate *MKBUILD*, a tool for creating domain-specific ontologies. These ontologies, which we call Modular Knowledge Bases (MKBs), contain concepts and associations imported from existing large-scale knowledge resources, in particular WordNet and Wikipedia. The combination of WordNet's human-crafted taxonomy and Wikipedia's semantic associations between articles produces a highly connected resource. Our MKBs are used by a *conversational agent* operating in a small computational environment. We constructed several domains with our technique, and then conducted an evaluation by asking human subjects to rate the domain-relevance of the concepts included in each MKB on a 3-point scale. The proposed methodology achieved precision values between 71% and 88% and recall between 37% and 95% in the evaluation, depending on how the middle-score judgements are interpreted. The results are encouraging considering the cross-domain nature of the construction process and the difficulty of representing concepts as opposed to terms.

1 Introduction

Conventional approaches to building domain ontologies typically rely on collections of domain text (i.e., ontology learning from text) or expert-crafted structured knowledge resources (e.g., WordNet [6], Cyc [11]). Such centralised approaches require enormous effort from domain experts and knowledge engineers; hence, these resources are slow to keep up with new knowledge and have considerably smaller coverage. The realisation of these drawbacks has resulted in the rise of an ontology construction approach using collaboratively-maintained resources: e.g., Freebase [3], YAGO [20] and DBPedia [1]. Despite the advantages of collaboratively maintained resources, issues of trustworthiness and subjectiveness related to social tagging can translate to poorer quality categorisations. For this reason, a backbone provided by expert-crafted resources is still desirable.

In this paper we present a methodology for construcing modular knowledge bases (MKBs) using WordNet and Wikipedia. As both resources provide their own strengths and shortcomings, their amalgamation increases the coverage and reliability of the resulting knowledge bases [8]. These MKBs combine the strengths of both resources as follows. The developer of the MKB first defines a domain using a Wikipedia article. A set of relevant concepts are extracted based

D. Wang and M. Reynolds (Eds.): AI 2011, LNAI 7106, pp. 638–647, 2011.

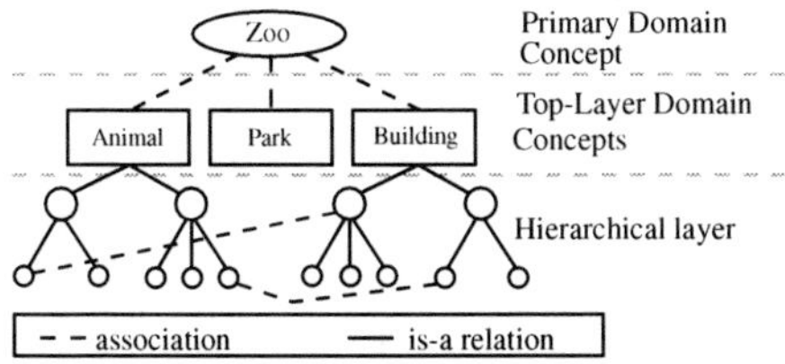

Fig. 1. Schema of an MKB

on being linked from the article. WordNet is then used to add parent and child concepts. Our methodology has been implemented as a tool called *MKBUILD* to construct MKBs for specific domains with minimal involvement from the developer. Our work on MKBs is motivated by the need to provide knowledge bases for a conversational agent designed to operate on a mobile platform with a small computational footprint. This agent is unable to accommodate large knowledge resources such as Cyc or DBPedia due to issues related to memory and storage size, and efficient access and processing. Our approach allows different MKBs to be loaded onto the platform as required depending on conversational flow.

The tasks of extracting domain-specific terms and of automatically constructing ontologies (typically using language processing techniques over Wikipedia or text corpora) have been widely studied: e.g., [12,15] for the former and [17,22] for the latter. To some degree, our approach combines these tasks. First, a set of "maximally general" concepts are extracted from Wikipedia and WordNet, which form the roots of the multiple sub-ontologies associated with a target domain. Second, the sub-ontologies rooted at each of these concepts are constructed, including *association* links between the concepts. These links form the basis of a generic *semantic relatedness* technique (not described here). The construction process of the MKBs is outlined in the following section, followed by the description of a user-based evaluation and a discussion.

2 Building Domain-Specific MKBs

In this section, we briefly describe the proposed methodology for building Modular Knowledge Bases[1]. An MKB is an ontology built around a main concept representative of a domain, and features a set of sub-taxonomies linked by the associations amongst its nodes (concepts). The target architecture of MKBs is shown in Figure 1. To build MKBs, we use two knowledge resources, namely, WordNet [6], a lexical dictionary that contains multiple word senses grouped by their meaning, and Wikipedia[2], an online encyclopaedia that operates like a collaborative wiki. WordNet features a taxonomy of concepts, but lacks relationships that are not lexical (e.g., ***Lion*** `lives in` ***Savannah***). On the other

[1] We omit some details, such as related work here for reasons of space: full details can be found in [13].

[2] `http://en.wikipedia.org/`

hand, Wikipedia does not have a taxonomical organisation; rather, we focus on the *wikilinks* featured in every article. A *wikilink* represents a concept that helps in the understanding of definitions[3]. Although *wikilinks* do not always describe a positive association, we are interested only in existence of such associations rather than their nature. The combination of Wikipedia *wikilinks* ("flexible" in the sense that humans themselves choose what to link in Wikipedia articles) and the WordNet hierarchy ("rigid" because property inheritance cannot be changed by humans) helps us to produce richer MKBs. *Wikilinks* have been previously analysed as a reliable set, though not absolute, of associations between articles [9,16]. For our approach, we use unidirectional *wikilinks* instead of mutual (from article a to b and vice versa) since we are interested in using such associations for conversational topic transitions. Thus, we are prepared to tolerate a more liberal notion of "relatedness".

Our process for constructing domain-specific MKBs consists of the following three stages. An overview of the process is shown in Figure 2:

1. **Define the domain**, i.e., select the *primary domain concept* by choosing a Wikipedia article that unambiguously reflects the main concept of the target domain;
2. **Build the top-layer** by extracting concepts to represent the most general and representative concepts associated with the *primary domain concept*;
3. **Extend the MKB** by adding sub-concepts to each top layer concept and analysing, for each concept's articles, the corresponding *wikilinks*.

The first stage of this process is performed manually, where the module designer chooses a Wikipedia entry that best matches the domain of the MKB. In this work, we refer to the selected entry's identifier (which may be qualified by a specific "sense" for ambiguous terms) as the *primary domain concept*.

The next two stages of the process are executed by the *MKBUILD* tool. *MKBUILD* performs all tasks necessary for those stages and produces an MKB *automatically*. *MKBUILD* has been developed in Java and uses the OWL-API Library[4] for handling the ontology. All these stages may be performed separately using *MKBUILD*, thus allowing intermediate manual modifications to the MKB in order to improve the coverage of the module. The rest of this section contains a brief description of the process. For full details, see [13].

Stage 2. Building the MKB Top Layer

The *primary domain concept* identified in the first stage is used as the input to *MKBUILD*, which performs Stages 2 and 3 automatically. In Stage 2, the concepts that form the top layer of the MKB are discovered. The tasks that comprise this stage are briefly described below.

2.1. Page link extraction. *MKBUILD* retrieves all terms that appear as *wikilinks* in the article referenced by the *primary domain concept*. This extraction

[3] See `http://en.wikipedia.org/wiki/Wikipedia:Manual_of_Style`

[4] `http://owlapi.sourceforge.net/`

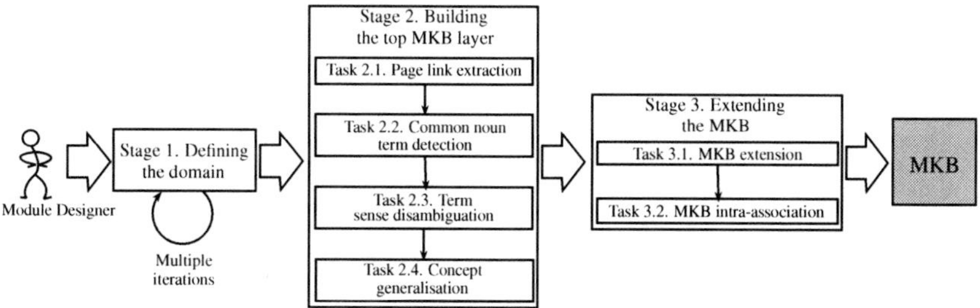

Fig. 2. An overview of the process to build MKBs

process is performed using DBPedia [1] (version 3.5.1), which contains Wikipedia links stored as triplets. MKB also extracts any *redirect links* that accompany each term, as these contain the original name of the Wikipedia article (i.e., *wikilinks* are proposed by authors; *redirect links* reconcile other concepts to point to the same article). In contrast to previous work that has considered the category structure provided by Wikipedia [7,10,18], we propose the use of *wikilinks* as the initial source of concepts directly related to a domain. We do, however, propose to leverage Wikipedia's category-based hierarchical *folksonomy* in future improvements, as discussed in the Evaluation section. The *wikilink* terms extracted are validated using a *named entity recognition (NER)* tool[5] and a "Wikipedia-to-WordNet" conversion table provided by DBPedia. At the end of this task, *MKBUILD* obtains a set of *preliminary concept terms.*

2.2. Common noun term detection. The preliminary terms may refer to either concepts or *instances* of concepts (e.g., specific people or places) as Wikipedia itself does not distinguish between the two [9]. This task performs a second detection and removal of terms that correspond to instances. These terms are detected using two tools: a Part-of-Speech (POS) tagger implemented in the Language Technology tool MorphAdorner[6], and (ii) WordNet word forms. Terms are retained for the next step as long as MorphAdorner determines that they contain at least one common noun and no proper nouns, proper adjectives nor non-English words. Additionally, WordNet helps with removing terms that start with a capital letter, as this has proven to be a sufficient heuristic to determine instances [14]. After this task is performed, a list of terms is obtained.

2.3. Term sense disambiguation. Terms retained in the above step may be ambiguous, in that they have multiple senses in WordNet. Consequently, to obtain concepts, a disambiguation process is required. This process finds the concepts that are related to the *primary domain concept* using semantic similarity measure of Lesk, adapted to WordNet glosses[7][2].

[5] The Stanford NER tool, that can be obtained from `http://nlp.stanford.edu/ner/`

[6] `http://morphadorner.northwestern.edu/`

[7] This value is obtained from the Java WordNet:Similarity Library, available in: `http://www.cogs.susx.ac.uk/users/drh21/`.

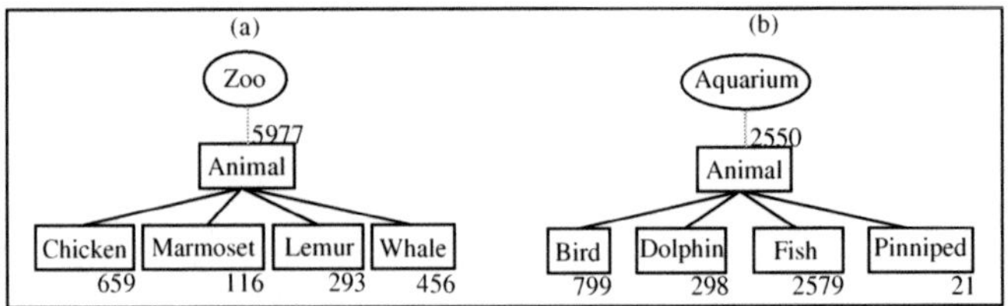

Fig. 3. Detection of more general classes via WordNet: (a) an accepted generalisation; and (b) a rejected generalisation. The top oval corresponds to the *primary domain concept*, and each number represents the co-occurrence between the *pdc* and a concept.

2.4. Concept generalisation. The concepts obtained in the previous task may not represent the level of generality required for the domain (i.e., the domain covers more general concepts than those identified). Concept generalisation requires extracting all WordNet *hypernyms* (super-classes) of the concepts obtained in step 2.3. This task is executed by the following two steps:
(i) *Generalisation using available concepts:* In this step, *MKBUILD* removes a concept if another concept in the list is its parent, as they will be later added as sub-concepts of the corresponding top-layer concept at a later stage.
(ii) *Generalisation using WordNet hierarchy: MKBUILD* detects if two or more concepts $sc_i, \ldots sc_j$ can be generalised using a common super-class h. If a super-class is detected, *MKBUILD* compares the co-occurrence of the *primary domain concept* (*pdc*) and h against the co-occurrence of *pdc* and each concept $sc_i, \ldots sc_j$ using Wikipedia articles as a corpus. If concept h is more commonly associated with the *pdc* than the sum of all *sc*, then the sub-concepts are replaced by h in the list of related concepts. An example of this is shown in Figure 3.

Stage 3. Building the Hierarchical Layer

With a top-layer of concepts obtained from Stage 2, two more tasks are performed before an initial version of the MKB is produced. In the first task, *MKBUILD* adds sub-classes from WordNet below each top-layer concept, which now become the root nodes of sub-ontologies. As in Stage 2, only WordNet senses that are common nouns are included. Finally, in the second task, *MKBUILD* adds association links between concepts that are not lexically based. These association links support a notion of *semantic relatedness* featuring more general links between concepts. These links are used by our conversational agent for concept-based topic transitions. *MKBUILD* inserts an association between two concepts if a *wikilink* between the articles corresponding to those concepts exists in DBPedia (as long as there is not already a lexical link from WordNet).

3 Evaluation

In this section, we describe an evaluation of the Stage 2 of *MKBUILD*, i.e., identifying the *top-layer domain concepts* of the sub-ontologies related to the specified domain.[8] We conducted a user study by asking subjects to judge whether the *top-layer domain concepts* extracted by *MKBUILD* were appropriate to the domain. We focus on evaluating *precision* and *recall* of the extraction process for *top-layer domain concepts* and not the hierarchy below, since concepts in the hierarchy below a *top-layer domain concept* are assumed to be related to it.

Setup. We used *MKBUILD* to construct MKBs for 14 domains, which are shown in the first column of Table 1. The total number of *top-layer domain concepts* (*tldc*) across all domains is 490 (set T). We extracted a subset of T, namely T', with the highest *idf* in each domain. T' was distributed across 6 different survey files. The breakdown of these concepts according to the different domains is summarised in columns 2 and 3 of Table 1. Each survey contains 3 domains, each domain comprises up to 10 concepts.

We asked 55 anonymous users to score how "related" each *tldc* is to a proposed domain D. Surveys were randomly assigned, following an even distribution across users. Users scored each domain-concept pair with an integer number of either 2, 1 or 0, where 2 indicates that the concept is highly related to D, 1 indicates it is related, and 0 for unrelated concepts. Users could also separately select *Unsure.* Users were also requested to add, for each domain, a set of up to five concepts that were not in the survey but what they considered to be highly related to the domain.

We obtained assessment scores from between 8 and 10 participants for each survey. We calculated the average Pearson correlation between subjects for each survey, obtaining values ranging from 0.28 (indicating medium low correlation) to 0.54 (strong correlation)[5]. Although these values indicate some agreement, these also show the difficulty of finding similarly scored participations.

Results. To determine users' agreement with the system for each *top-layer domain concept*, we calculated an aggregated value in three different ways, each representing a different assessment of relatedness. First, p_a (i.e., precision) was calculated by adding the number of participants scoring either 1 (i.e., "related") or 2 ("highly related") and subtracting the number of 0's ("unrelated") scored for each *tldc*. Second, the scores of 1's were changed to 0.5 to calculate p_b. Third, p_c took into consideration only the number of 0's and 2's, with the number of 1's used only to break any ties (i.e., the numbers of 0's and 2's were the same). The first criterion is standard according to the definition of our experiment, which is that both scores of 1 and 2 represent a certain degree of relatedness. The latter two criteria represent a less generous interpretation of the middle score (i.e., 1). These criteria bias against our system, hence we include them for comparison.

Using the total number of concepts together with the aggregated values obtained as per the three criteria, we calculated the *Precision* and *Recall* for all domains, as defined by [19]. We employ these measures as they reflect the coverage of the concepts with respect to the target domain. Our evaluation of the 36% of all the available 490 *top-layer domain concepts* resulted in the following

[8] Evaluating other stages would be effectively evaluating WordNet and DBpedia.

precision values, namely, $p_a = 0.88$, $p_b = 0.80$ and $p_c = 0.71$. These values reflect a high number of human participants agreeing with the *top-layer domain concepts* extracted by *MKBUILD*, particularly on the standard interpretation of the middle score.

Next, we estimated recall using the *top-layer domain concepts* deemed as related, plus the extra concepts provided by participants. Only 38 out of the 55 participants provided any extra concepts; a total of 366 extra concepts were provided, ranging from 12 to 46 per domain. Due to the lack of a gold standard, we artificially created one with these extra concepts and the scores obtained from the provided *top-layer domain concepts.* We analysed these extra concepts in two ways: first, assuming that it was due to a lack of coverage of WordNet or Wikipedia that such concepts were not added to the MKB (method d); and second, assuming that all the suggested extra concepts should be in the MKB (method e). These concepts are proposed as our *false negatives*, while the concepts with a positive score are the *true positives.* These criteria affected the results for recall, which are shown in Table 1 as $r_{m|i}$, where m is method d or e and i refers to the method for calculating precision, as described above.

Table 1. Sample domains with their evaluated precision and recall values

Domain(D)	T	T'	p_a	p_b	p_c	$r_{d\|a}$	$r_{d\|b}$	$r_{d\|c}$	$r_{e\|a}$	$r_{e\|b}$	$r_{e\|c}$
Amusement park	26	10	0.9	0.7	0.6	1	1	1	0.6	0.54	0.5
Association football	25	10	0.9	0.8	0.8	1	1	1	0.56	0.53	0.53
Automobile	41	20	0.85	0.4	0.4	0.89	0.8	0.8	0.35	0.2	0.2
Beach	28	10	1	1	1	1	1	1	0.38	0.38	0.38
Computer	73	20	0.95	0.9	0.8	0.86	0.85	0.84	0.39	0.38	0.35
Economy	56	20	0.85	0.85	0.75	1	1	1	0.47	0.47	0.44
Food	88	20	0.9	0.8	0.55	1	1	1	0.49	0.46	0.37
Museum	32	10	1	0.8	0.5	1	1	1	0.56	0.5	0.38
Music	37	10	0.9	0.9	0.9	0.82	0.82	0.82	0.45	0.45	0.45
Public aquarium	11	10	0.5	0.5	0.4	0.83	0.83	0.8	0.25	0.25	0.21
School	25	10	0.8	0.7	0.7	1	1	1	0.32	0.29	0.29
Sport	24	10	0.9	0.8	0.7	1	1	1	0.53	0.5	0.47
Theatre	18	10	0.9	0.9	0.9	1	1	1	0.36	0.36	0.36
Zoo	8	8	1	0.75	0.5	1	1	1	0.42	0.35	0.27
Total	490	178	0.88	0.8	0.71	0.95	0.95	0.94	0.42	0.4	0.37

We do not have a comparable task for direct comparison, but can compare to performance in domain term extraction; e.g., for this task, [12] reported values of precision and recall of 0.354 and 0.183 respectively. [15] obtained an F1 quality score of 0.25 in term extraction using the Web. On the other hand, our lowest F1 score reported is 0.486 for p_c and $r_{e|c}$. Some care has to be taken when interpreting these figures because there are clear differences between our approach and domain term extraction which makes them not comparable. First, we focus on extracting concepts, not just terms, so we have to resolve against concepts (which includes performing word sense disambiguation). Second, term extraction is commonly applied in closed environments using well-defined domain corpora, whereas we extract from a resource as broad as Wikipedia. Hence, in comparison to this (related) baseline task, we consider our results as encouraging.

Error Analysis. We can analyse the set of concepts suggested by participants to obtain insights into *MKBUILD*'s inability to extract certain concepts. The suggested concepts can be classified in four ways, namely, (A) they ambiguously refer to proper instead of common nouns (e.g., Shakespeare) or to other parts of speech besides noun (e.g., play), (B) they are in an MKB but were not shown to the user, and (C) they do not appear anywhere in the MKB. From the set of 366 suggested concepts, 8 concepts fall under category (A), 151 under (B), and 207 under (C).

From category (C), we can create three subgroups. Group (C1) contains those concepts that do not appear in WordNet. Analogously, group (C2) contains concepts that do not appear in the Wikipedia article of the domain as *wikilinks*. Group (C3) contains those suggested concepts appearing in both WordNet and Wikipedia which did not appear in the resulting MKBs. Concepts in group (C2) represent the largest limitation of our approach, showing that using only the *primary domain concept* article is not enough to find concepts associated with the domain. Earlier, we mentioned that *MKBUILD* does not currently use Wikipedia's *folksonomy*. Therefore, a broader, more systematic exploration of related articles, considering the article categorisation in Wikipedia, should be performed in future work.

Only 8 suggested concepts fall within group (C3). These suggested concepts missing from the MKBs are classified into four types. The first type of missing suggested concept, namely (C3-a) features those concepts with an ambiguous WordNet taxonomy. For example, the concept ***Dolphin*** has two different senses, where one corresponds to its meat, and the other defines a type of ***Mammal***. If two concepts have similar names and no other synonyms available, *MKBUILD* is unable to create a new concept, thus the concept referring to the second sense and its children concepts are not included. This issue can be resolved if by analysing the definitions of concepts according to WordNet. In cases where some definitions for different senses of a word are complimentary (e.g., Dolphin is an edible fish AND a mammal) we must merge both senses in our produced MKB.

The second type (C3-b) occurs with the NER tool (Step 2.1), which performs suboptimally due to the lack of context for terms. Therefore, some concepts that correspond to common nouns are treated as referring to instances, and are removed from the process. For example, the term ***Algorithm*** is recognised as an entity expressing a location.

Finally, the third type, (C3-c) occurs due to our heuristic to identify instances using WordNet. Our approach automatically eliminates a term if it contains a word form (a synonym) starting with a capitalised letter. This applies to concepts such as ***hydrogen***, which can be also represented with the letter "H".

Error type (C3-c) is the most frequent, occurring with four suggested concepts. Error (C3-b) was detected on three occasions and (C3-a) only once. This means that in order to improve entity recognition, we have to use longer texts rather than only terms. One possibility is to feed the NER tool with a sentence from the short abstract extracted from Wikipedia containing the analysed term.

4 Conclusions and Future Work

We have described a process for constructing domain-specific ontologies, called Modular Knowledge Bases, to be used by a conversational agent with a modular infrastructure. The process has been programmed as *MKBUILD*, a tool that allows the *automatic* extraction of concepts and relations specific to a given domain using large resources such as WordNet and Wikipedia/DBPedia. The ontology construction process we described saves developers a significant amount of effort in constructing an ontology specific to a conversational domain, while at the same time allowing the developers to easily intervene at any point in time to correct any egregious errors.

We have conducted an experiment involving human assessors to determine the precision and recall of a critical stage of the construction process, namely identifying the top-level concepts for the domain-specific ontologies. We obtained encouraging results considering the difficulty of cross-domain concept extraction. This experiment has also allowed us to determine that the exploration of only the Wikipedia article associated with the *primary domain concept* of the MKB is insufficient. Other related Wikipedia articles have to be considered in order to extract a broader range of domain-specific concepts. We also discussed limitations in the current extraction process and proposed solutions.

Our main application of the domain ontologies constructed using *MKBUILD* is to generate a *Topic Network* that can be used to link conversational fragments together into more coherent longer-running threads, using ontology-based semantic similarity measures. We are also conducting an evaluation to measure ontology-based semantic relatedness involving sets of relations that go beyond previously considered (e.g., [4,18]), and evaluate its efficacy in topic transitioning in conversational dialogue. Other future work includes extending the coverage of the concepts and relations in the MKBs through the use of other large knowledge bases constructed using information extraction techniques (e.g., [21]).

Acknowledgements. The first author acknowledges scholarship 201228 from the Consejo Nacional de Ciencia y Tecnologia (CONACYT), Mexico. This work was partially supported by Australian Research Council Linkage Projects LP0882013 and LP110100050. We acknowledge the support and collaboration of our partner, Realthing Entertainment Pty Ltd. Aidan Martin implemented the framework for data collection.

References

1. Auer, S., Bizer, C., Kobilarov, G., Lehmann, J., Cyganiak, R., Ives, Z.G.: DBpedia: A Nucleus for a Web of Open Data. In: Aberer, K., Choi, K.-S., Noy, N., Allemang, D., Lee, K.-I., Nixon, L.J.B., Golbeck, J., Mika, P., Maynard, D., Mizoguchi, R., Schreiber, G., Cudré-Mauroux, P. (eds.) ASWC 2007 and ISWC 2007. LNCS, vol. 4825, pp. 722–735. Springer, Heidelberg (2007)

2. Banerjee, S., Pedersen, T.: An Adapted Lesk Algorithm for Word Sense Disambiguation Using Wordnet. In: Gelbukh, A. (ed.) CICLing 2002. LNCS, vol. 2276, pp. 136–145. Springer, Heidelberg (2002)
3. Bollacker, K., Evans, C., Paritosh, P., Sturge, T., Taylor, J.: Freebase: a Collaboratively Created Graph Database for Structuring Human Knowledge. In: SIGMOD, pp. 1247–1250 (2008)
4. Budanitsky, A., Hirst, G.: Evaluating WordNet-based Measures of Lexical Semantic Relatedness. Computational Linguistics 32(1), 13–47 (2006)
5. Cohen, J.: Statistical power analysis for the behavioral sciences, 2nd rev. edn. Academic Press, London (1977)
6. Fellbaum, C.: WordNet: an electronic lexical database. The MIT Press (1998)
7. Grieser, K., Baldwin, T., Bohnert, F., Sonenberg, L.: Using Ontological and Document Similarity to Estimate Museum Exhibit Relatedness. J. Computing and Cultural Heritage 3(3), 10:1–10:20 (2011)
8. Gruber, T.: Ontology of Folksonomy: A Mash-up of Apples and Oranges. Semantic Web and Information Systems 3(2), 1–11 (2007)
9. Hepp, M., Siorpaes, K., Bachlechner, D.: Harvesting Wiki Consensus: Using Wikipedia Entries as Vocabulary for Knowledge Management. Internet Computing 11, 54–65 (2007)
10. Herbelot, A., Copestake, A.: Acquiring Ontological Relationships from Wikipedia Using RMRS. In: Web Content Mining with Human Language Technologies, pp. 1–10 (2006)
11. Lenat, D., Guha, R.V.: Building Large Knowledge-Based Systems; Representation and Inference in the Cyc Project. Addison-Wesley (1990)
12. Liu, Z., Huang, W., Zheng, Y., Sun, M.: Automatic Keyphrase Extraction via Topic Decomposition. In: EMNLP, pp. 366–376 (2010)
13. Macias-Galindo, D., Cavedon, L., Thangarajah, J.: Building Modular Knowledge Bases for Conversational Agents. In: Knowledge Representation and Reasoning for Practical Dialogue Systems, pp. 16–23 (2011)
14. Martin, P.: Correction and Extension of WordNet 1.7. In: Ganter, B., de Moor, A., Lex, W. (eds.) ICCS 2003. LNCS, vol. 2746, pp. 160–173. Springer, Heidelberg (2003)
15. Massey, L., Wong, W.: A Cognitive-Based Approach to Identify Topics in Text Using the Web as a Knowledge Source. In: Wong, W., Liu, W., Bennamoun, M. (eds.) Ontology Learning and Knowledge Discovery Using the Web. IGI Global (2011)
16. Milne, D., Medelyan, O., Witten, I.H.: Mining Domain-specific Thesauri from Wikipedia: A case study. In: Web Intelligence, pp. 442–448 (2006)
17. Ponzetto, S.P., Strube, M.: Deriving a Large Scale Taxonomy from Wikipedia. In: The National Conference on Artificial Intelligence, vol. 22, pp. 1440–1446 (2007)
18. Ponzetto, S.P., Strube, M.: Knowledge Derived From Wikipedia For Computing Semantic Relatedness. J. AI Research 30, 181–212 (2007)
19. Sabou, M., Wroe, C., Goble, C., Mishne, G.: Learning domain ontologies for web service descriptions: an experiment in bioinformatics. In: World Wide Web, pp. 190–198 (2005)
20. Suchanek, F.M., Kasneci, G., Weikum, G.: YAGO: A Large Ontology from Wikipedia and WordNet. Web Semantics 6(3), 203–217 (2008)
21. Yates, A., Cafarella, M., Banko, M., Etzioni, O., Broadhead, M., Soderland, S.: TextRunner: open information extraction on the web. In: NAACL-Demos, pp. 25–26 (2007)
22. Zirn, C., Nastase, V., Strube, M.: Distinguishing between Instances and Classes in the Wikipedia Taxonomy. In: Bechhofer, S., Hauswirth, M., Hoffmann, J., Koubarakis, M. (eds.) ESWC 2008. LNCS, vol. 5021, pp. 376–387. Springer, Heidelberg (2008)

Generality Evaluation of Automatically Generated Knowledge for the Japanese ConceptNet

Rafal Rzepka, Koichi Muramoto, and Kenji Araki

Graduate School of Information Science and Technology, Hokkaido University
Kita-ku, Kita 14, Nishi 8, Sapporo, Japan
{kabura,koin,araki}@media.eng.hokudai.ac.jp
http://arakilab.media.hokudai.ac.jp

Abstract. In this paper we introduce three methods for automatic generality evaluation of commonsense statements candidates generated for Open Mind Common Sense (OMCS), which is the basis of ConceptNet, a commonsense knowledge base. By using sister terms from Japanese WordNet, our system generates new statements which are automatically evaluated by using WWW co-occurrences and hit number retrieved by a Web search engine. These values are used in three generality judgment methods we propose. Evaluation experiments show that the best of them was "exact match ratio" which achieved accuracy of 62.6% when evaluating general sentences and "co-occurrences in snippets" method scored highest with 48.6% when judging unnatural phrases. Compared to the data without noise elimination, the "exact match ratio" achieved 38.2 points increase in accuracy.

Keywords: Common Sense Knowledge, Open Mind Common Sense, ConceptNet, WordNet, Automatic Generality Evaluation.

1 Introduction

To understand language, a machine needs knowledge that human beings gather from experience since the very beginning of their lives. This knowledge is obvious and general, and we call it common sense knowledge. Many AI researchers have tried and are still trying to collect it, usually input it by hand – by specialists (as in CyC[1]) or by amateur contributors (as in OMCS[2]). Also in Japan there are engineers using general knowledge in their research, however they limit their methods to, for instance, question answering, and they create their databases manually, making it much easier to use[4]. But such limitations of usage range of knowledge is contradictory to common sense which in our opinion has more universal and inter-conceptual usage. Our approach is directed toward as fully automatic as possible methods of acquiring wide range of various kinds of knowledge people usually share. As some entries for OMCS show, the volunteers entering commonsense descriptions of the world like to joke and "generality" of many entries is doubtful ([2] states that 15% of entries do not make sense). The same

D. Wang and M. Reynolds (Eds.): AI 2011, LNAI 7106, pp. 648–657, 2011.

tendencies are visible in the latest trend - common concepts acquisition through on-line games[1]. After a while, players get bored and start to be original rather than general. However, human contributors are an important part of systems such as ConceptNet[3] based on Open Mind Common Sense where "UsedFor" or "IsA" are examples of edges which denote relationship between concepts. Although ConceptNet has been used by different researchers for a decade since MIT Media Lab has developed it, most of the projects used the English language version (and lately Chinese), while other languages versions (as Japanese) produced much smaller scientific output. The reason is quite obvious since English OMCS has currently 1,035,681 registered statements expressing common sense knowledge and there are only 14,546 for Japanese. If we could increase the number of general sentences, the usability of this knowledge would also increase. For that reason we decided to tackle this problem. Our first idea was to use WordNet[5] and WWW search to harvest Japanese concepts to acquire new commonsense statements. The basic proposal of our ideas was introduced in [6], however erroneous statements generated from Internet search gave us low accuracy not allowing the system to be somehow useful. In this paper, we propose methods to improve our system by adding automatic generality evaluation of phrases retrieved from the Web.

2 Related Work

Trials on automatic retrieval, usually based on syntactic patterns, are not new [7][8][9]. Van Durme et. al have also tried to use the WordNet in their KNEXT[10] project. Hyponym-hypernym links between noun synsets were investigated to figure out how reliably hyponyms can be viewed as mutually exclusive. Their findings (summarized on the project site [2]) were that the hypernym links were only two-thirds correspondence to true subtypes, and that the hyponyms are about 70% truly exclusive. They studied many ways to improve the extraction process, but concluded that the causes were too diverse to enable large improvement by any automated means. Hanheide et al. prove usefulness of such data presenting a similar approach for combining OMCS statements with WWW search results to quantifying commonsense knowledge for intelligent robots[11].

3 Commonsense Knowledge Generation

3.1 Definition

We define Common Sense Knowledge as an experience-based general knowledge (e.g. "dogs walk") but also broaden it to more concrete information shared by users of a given language ("Todai-ji is a temple in Japan", "Madonna sings", "you can work at Sony". Such broader definition increases capability of non-task oriented dialog systems which we are also working on.

[1] http://nadia.jp

[2] http://www.cs.rochester.edu/ schubert/projects/world-knowledge-mining.html

3.2 System Overview

The idea is to use existing OMCS sentences and exchange nouns with sister terms from the WordNet dictionary to generate new similar statements and then use a Web search to determine how usual the generated knowledge is. Figure 1 provides an overview of our system. By "sister terms" we mean hyponyms under the same hypernyms. For example, "lions roar" can be transformed into "tigers roar". Then, to remove possible noise (untrue or unnatural statements) such a phrase becomes a query for search engine, which in this study is Yahoo! Japan[3]. Usualness (generality) calculation uses thresholds which will be described later in detail.

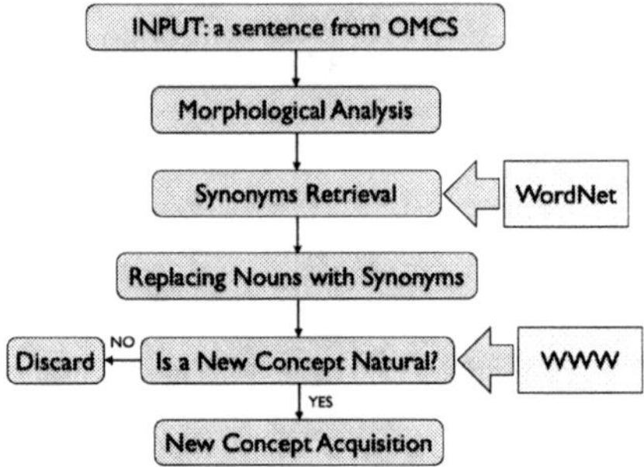

Fig. 1. Overview of our system for harvesting concepts by using WordNet sister terms and evaluating them by WWW search

3.3 Japanese WordNet

WordNet, developed at Princeton University, is a semantic lexicon consisting of concepts called synsets. Words that are similar are kept within the same synset. A synset is labeled as "number ID - part of speech" where, for instance, "n" means a "noun", and "v" indicates a verb. It is also connected to other synsets associated by relationships like hyponymy, hyperonymy or meronymy. In Japanese WordNet there are 57,238 synsets, 93,834 words and 158,058 pairs of synsets and words.

3.4 Retrieving Sister Terms and Generating Sentences

A sentence from OMCS set for Japanese language becomes an input to our system. Then its noun is replaced by a sister term from the WordNet, so, for example, "(one can) throw a ball" produces statements like "(one can) throw a fastball", "(one can) throw a Frisbee" or "(one can) throw a [playground] slide". As you can see from the last example, statements generated by nouns from a broad category as "toys" will not always produce a general, commonsense

[3] http://search.yahoo.co.jp

knowledge and such erroneous phrases cannot be added to the knowledge base. Therefore the noise elimination becomes crucial for newly generated data quality – in the next section we explain in details what methods we developed.

4 Noise Elimination Methods

To eliminate semantically erroneous generations we propose three shallow web-mining methods, which we named "co-occurrences in snippets", "exact match ratio" and "conjugated keywords hit ratio".

4.1 Co-occurrences in Snippets (a)

Verbs, nouns and (if they appear) adjectives are extracted from the input sentence, and the original particle[4] is used to form a search query "$NounParticle$" + "($Verb|Adjective$)". Web search using such a query outputs set of snippets (short summary passages output by a search engine) and the system counts how many times both queried phrases occurred. The condition is to be in the same sentence and in the same order and this type of results we call "co-occurrences in snippets". We define "sentence" here as a phrase between punctuation marks as dots, commas, exclamation marks, question marks, etc. We set a threshold for co-occurrences in snippets, and if their number falls below the threshold then queried sentence is determined as noise. Thresholds are explained later in the paper.

4.2 Exact Match Ratio (b)

Unlike the "co-occurrences in snippets" method, here noun, particle and verb (or adjective) create one exact match query (without OR operator): "$NounParticle$ ($Verb|Adjective$)". At the same time following additional queries are created: "$NounParticle$" + "$Verb|Adjective$", "$NounParticle$", and "$Verb|Adjective$". System uses search engine results for all these queries to calculate an "exact match ratio" with the Formula (1). Again thresholds are set to eliminate erroneous output.

$$Pp = \frac{N_p}{N_n + N_v - N_c} \tag{1}$$

- P_p: exact match ratio
- N_p number of hits for "$NounParticle(Verb|Adjective)$"
- N_n: number of hits for "$NounParticle$"
- N_v: number of hits for "$Verb|Adjective$"
- N_c: number of hits for "$NounParticle$" + "$Verb|Adjective$"

[4] Japanese particles are suffixes that immediately follow the modified noun, verb, adjective, or sentence. For example in *booru o nageru* (throw a ball, to throw a ball, throwing a ball, one throws a ball, etc.) *o* states that the noun it follows is a direct object of the action described by following verb.

4.3 Conjugated Keywords Hit Ratio

In this method we decided to add a natural language processing module for stemming as search engines ignore the fact that verbs conjugate. The main reason for adding this technique is to increase number of hits, which allows to get a better accuracy of the investigated data. Phrase “eat a cake” after stemming can find five or six other forms which may be Japanese equivalents of “eating a cake”,“ate a cake”, ’“will eat” or “will be eating”[5]. Calculations are similar (it is the sum of all stemmed keywords) to the previous method (see Formula (2)) and also here adequate thresholds are set.

$$Pc = \sum \frac{N_p}{N_n + N_v - N_c} \tag{2}$$

P_c: conjugated keywords hit ratio (see Formula 1 for the full description).

Table 1. Results of threshold setting experiment

(a) “co-occurrence in snippets”

Authors’ evaluation	Number of Sentences	Average Appearance in Snippets
0 points	545	7.3
1point	211	11.6
2points	244	16.4

(b) “exact match ratio”

Authors’ evaluation	Number of Sentences	Average Ratio of Exact Matching
0 points	723	0.00185
1point	56	0.00390
2points	221	0.00414

(c) “conjugated keywords hits ratio”

Authors’ evaluation	Number of Sentences	Average Ratio of Conjugated Keywords
0 points	709	0.000106
1point	60	0.000587
2points	231	0.00131

5 Preliminary Experiments for Setting Noise Elimination Thresholds

As mentioned in previous sections it was necessary to set thresholds to eliminate as much unnatural output as possible. Fifty sentences including nouns were

[5] They cover more than tenses but the examples show only this type for the sake of simplicity.

randomly selected from OMCS Japanese data and system used sister terms to harvest candidates. It produced 13,240 sentences and we randomly selected 1,000 of them, and then a manual evaluation was performed by a native speaker of Japanese. The following criteria were used in the evaluation: "unnatural knowledge = 0 points", "possible but not general knowledge = 1 point", "general knowledge = 2 points". Table 1 shows results for all three methods described in Section 4. In case of "co-occurrence in snippets" (a), more than half of the generated sentences appeared to be unnatural , while about 24% of the acquired phrases was evaluated as useful general knowledge. "Exact match ratio" and "conjugated keywords hit ratio" produced 22% and 23% common sense statements respectively. Accordingly to these results we have decided that in case of (a), threshold for unnatural sentences is less than 7 co-occurrences of queried phrases in snippets, for non-general is more than 7 and less than 11, and for general there must be more than 11. Scores for (b) were set to 0.00185, 0.00390, and 0.00414; while for (c) we set number of hits threshold: 775,849 as the unnaturalness borderline, 860,909 for "arguable zone" and 1,349,698 as a starting point for regarding outputs as natural.

Table 2. Automatic vs. manual evaluation ("co-occurrence in snippets")

System Evaluation Score	Evaluators' Score	Average Number of Sentences
2 points	2 points	27.5
	1 point	12.5
	0 points	10.0
	Ratio of Correct Answers	**55.0%**
1 point	2 points	21.0
	1 point	14.0
	0 points	15.0
	Ratio of Correct Answers	**28.0%**
0 points	2 points	14.7
	1 point	11.0
	0 points	24.3
	Ratio of Correct Answers	**48.6%**

6 Evaluation Experiment and Its Results

After setting thresholds described in the previous section, we have performed experiments in order to see how accurately our system eliminated noisy, non-general knowledge from harvested data and how confident it can be about correct output. The rating method was the same as in the preliminary experiment and 50 statements (after noise elimination) for each method were randomly chosen (150 sentences in total). The same sets were also evaluated (in the same 3 grade scale) by 6 subjects who were two male college students from the science department plus two male and two female students from the literature department.

Table 3. Automatic vs. manual evaluation ("ratio of exact matches")

System Evaluation Score	Evaluators' Score	Average Number of Sentences
2 points	2 points	31.3
	1 point	10.9
	0 points	7.8
	Ratio of Correct Answers	**62.6%**
0 points	2 points	25.5
	1 point	13.8
	0 points	10.7
	Ratio of Correct Answers	**21.4%**

Table 4. Automatic vs. manual evaluation ("conjugated keywords hit ratio")

System Evaluation Score	Evaluators' Score	Average Number of Sentences
2 points	2 points	31.0
	1 point	9.5
	0 points	9.5
	Ratio of Correct Answers	**62.0%**
0 points	2 points	28.3
	1 point	10.2
	0 points	11.5
	Ratio of Correct Answers	**23.0%**

Table 5. Evaluators agreement ("co-occurrence in snippets")

System Evaluation Score	Evaluators' Score	3 Evaluators	4 & More Evaluators
2 points	2 points	25	25
	1 point	5	5
	0 points	10	5
	Ratio of Correct Answers	**62.5%**	**71.4%**
1 point	2 points	22	20
	1 point	7	5
	0 points	15	13
	Ratio of Correct Answers	**15.9%**	**13.2%**
0 points	2 points	12	8
	1 point	8	2
	0 points	25	21
	Ratio of Correct Answers	**55.6%**	**67.7%**

The experimental results for method (a) are shown in Table 2. In 55.0% of the cases, system correctly estimated that knowledge is general, in 28.0% of the cases that it is non-general and in 48.6% that it was unnatural and should be discarded. As defining what is general and what is not is often difficult even for human evaluators, we also took into account the agreement between users.

Table 6. Evaluators agreement ("exact match ratio")

System Evaluation Score	Evaluators' Evaluation	3 Evaluators	4 & More Evaluators
2 points	2 points	36	31
	1 point	6	4
	0 points	7	6
	Ratio of Correct Answers	**73.5%**	**75.6%**
0 points	2 points	28	21
	1 point	8	6
	0 points	7	6
	Ratio of Correct Answers	**16.3%**	**18.2%**

Table 7. Evaluators agreement ("conjugated keywords hit ratio")

System Evaluation Score	Evaluators' Evaluation	3 Evaluators	4 & More Evaluators
2 points	2 points	35	33
	1 point	4	2
	0 points	7	7
	Ratio of Correct Answers	**76.1%**	**78.6%**
0 points	2 points	29	24
	1 point	8	7
	0 points	9	7
	Ratio of Correct Answers	**19.6%**	**22.6%**

Table 5 shows that in cases of less arguable knowledge (0 and 2 points, more than 4 evaluators agreed), the system's accuracy increases from 55.0% to 71.4% (general knowledge) and from 48.6% to 67.7% (unnatural knowledge). Because of this lack of agreement and the fact that system discovered too few[6] sentences that could be evaluated as not general, we decided to exclude it from the evaluation process. Tables 3 and 4 show experimental results for methods (b) and (c), Tables 6 and 7 indicate results where user agreement is considered. In case of "exact match ratio", a significant increase of accuracy (62.6% to 75.6%) can be observed for general knowledge but in discovering unnatural statements this method appeared worse (decreased from 21.4% to 18.2% when agreed by more than 4 evaluators). "Conjugated keywords hit ratio" method performed much better in case of common sense statements (62.0% to 78.6%) but again was slightly worse in discovering erroneous knowledge (23.0% to 22.6%). As shown in Table 1(a), without noise elimination, we could retrieve only 24.4% of usable general knowledge. Method (a) "Co-occurrence in snippets", after eliminating erroneous statements, allowed to correctly find 55.0% of such knowledge. In case of "exact match ratio" (b), 62.6% of the generations were correct and of "conjugated keywords hit ratio" (c), 62.0% were evaluated as proper automatic judgment. The highest accuracy was achieved by method (b) - compared to the results without noise removal there was 38.2 points improvement in accuracy.

[6] Too few to be statistically significant.

There were 7 sentences which were evaluated "0 points" by the system and "2 points" by more than 4 evaluators. Five of these statements were generated by the morphological analysis tool, which cuts off suffixes that are nouns but have different meaning when used separately. For example *-hen* is used as a "compilation suffix"; when added to novels or poems means "collection of novels" or "collection of poems", but by itself it sounds odd. As we decided to use one noun, not a noun phrase, this type of errors depending on third party tools was inevitable. Another problem was context dependency – one of the sentences that showed a significant difference in evaluation was "summer is cold". Depending on places and particular days, summers can be cold cold and such statements are not rare on the WWW.

7 Conclusions and Future Work

In this paper we introduced three methods for automatic generality evaluation of Japanese sentence candidates generated for Open Mind Common Sense (OMCS), which is the base for ConceptNet, a freely available commonsense knowledge base and NLP tool-kit developed by MIT. By using sister terms from Japanese WordNet, our system was able to generate new statements that possibly represent common sense knowledge, however only part of newly produced outputs are obviously general. Therefore we implemented a module using Yahoo! Japan search engine to retrieve co-occurrences and hit numbers, which became a base for three methods we proposed. Evaluation experiments showed that the best of them was "exact match ratio" method which achieved accuracy of 62.6% when evaluating general sentences. For judging unnatural (impossible) knowledge, "co-occurrences in snippets" method scored highest with 48.6%.

As we noticed that human contributors get bored soon after starting to type commonsense statements, we assume it would be much faster and efficient to let them choose if something indicates general knowledge or not. Using our methods would definitely decrease burden of the proper entry choice task by showing only statements which scored 2 points in the 0-1-2 scale of generality to an evaluator. However, to come closer to accuracy allowing fully automatic generation, there is still plenty of room for future work. During the development and experiments we noticed many tendencies that could allow improvements. The more examples are found, the wider coverage we could get. There is thus a need for extending queries, for example by alternating particles – Japanese topic indicating particle *wa* can be replaced with subject indicating particle *ga*. We will also add techniques for so called (in linguistics) "genericity" and use grammatical structures and words that often suggest generality of a sentence (e.g. adverbs like "usually"). We also noticed that context dependent errors can be reused with negations to find new knowledge and every arguable statement could be rewritten and processed again. Combinations of "usually", "not" and "but" could also bring interesting results, therefore we want to increase quality by widening the web-mining process by taking grammatical information and neighboring words (also noun phrases) into consideration. We are also planning to transfer proposed shallow methods to ConceptNet versions for other languages that suffer the same lack of OMCS sentences as Japanese.

References

1. Lenat, D., et al.: Common Sense Knowledge Database CYC (1995)
2. Singh, P., Lin, T., Mueller, E.T., Lim, G., Perkins, T., Zhu, W.: Open Mind Common Sense: Knowledge Acquisition from the General Public. In: Meersman, R., et al. (eds.) CoopIS 2002, DOA 2002, and ODBASE 2002. LNCS, vol. 2519, pp. 1223–1237. Springer, Heidelberg (2002)
3. Havasi, C., Speer, R., Alonso, J.: ConceptNet 3: a Flexible, Multilingual Semantic Network for Common Sense Knowledge. In: Proceedings of Recent Advances in Natural Languages Processing, pp. 277–293 (2007)
4. Oe, N., Watabe, H., Kawaoka, T.: The construction method of commonsense judgment system for understanding the conversation of geography (in Japanese). IPSJ SIG Notes. ICS (24), 163–168 (2005)
5. Fellbaum, C.: WordNet: An Electronic Lexical Database. MIT Press, Cambridge (1998)
6. Muramoto, K., Rzepka, R., Araki, K.: Generation method of sentences for common sense knowledge bases using WordNet and web search (in Japanese). Kotoba Kenkyuu-kai SIG Technical Report 35, 1–7 (2010)
7. Chklovski, T.: Learner: A system for acquiring commonsense knowledge by analogy. In: K-CAP 2003, pp. 4–12 (2003)
8. Clark, P., Harrison, P.: Large-scale extraction and use of knowledge from text. In: K-CAP 2009, pp. 153–160 (2009)
9. Yu, C., Chen, H.: Commonsense Knowledge Mining from the Web. In: Proceedings of the Twenty-Fourth AAAI Conference on Artificial Intelligence (AAAI 2010), pp. 1480–1485 (2010)
10. Van Durme, B., Michalak, P., Schubert, L.K.: Deriving generalized knowledge from corpora using WordNet abstraction. In: Proceedings of the 12th Conference of the European Chapter of the Association for Computational Linguistics, pp. 808–816 (2009)
11. Hanheide, M., Hawes, N., Gretton, C., Aydemir, A., Zender, H., Pronobis, A., Wyatt, J., Gobelbecker, M.: Exploiting probabilistic knowledge under uncertain sensing for efficient robot behaviour. In: Proc. of IJCAI 2011 (2011)

Processing Coordinated Structures in PENG Light

Rolf Schwitter

Centre for Language Technology,
Macquarie University,
Sydney NSW 2109, Australia
Rolf.Schwitter@mq.edu.au

Abstract. PENG Light is a controlled natural language designed to write unambiguous specifications that can be translated automatically via discourse representation structures into a formal target language. Instead of writing axioms in a formal language, an author writes a specification and the associated background axioms directly in controlled natural language. In this paper, we first review the controlled natural language PENG Light and show how a discourse representation structure is generated for sentences written in PENG Light. We then discuss two different solutions of how discourse representation structures can be implemented for coordinated structures. Finally, we show how an efficient implementation of coordinated structures combined with a suitable parsing strategy affects the parsing performance of the controlled natural language.

Keywords: controlled natural language, parsing, coordination, discourse representation structures.

1 Introduction

PENG Light is a controlled natural language designed for representing knowledge in an unambiguous way [10]. Specifications written in PENG Light can be translated into a formal target language for automated reasoning. The language of PENG Light covers a strict subset of standard English and is defined by a controlled grammar and a controlled lexicon [12]. The language processor of PENG Light uses a unification-based phrase structure grammar that is based on a definite-clause grammar (DCG) notation [8]. In order to avoid redundant analysis and for practical reasons that we will discuss later, the DCG notation is automatically transformed into a format that can be processed easier by a chart (= tabular) parser [1].

In general, the DCG notation can be used to write context-free as well as context-sensitive grammars and these grammars can be executed directly using Prolog's top-down, depth-first, backtrack search. If no left recursion is present in context-free grammars, then the worst-case parsing performance is exponential in the size of the input, and if tabulation is used, then the performance is – in theory – cubic for context-free grammars (but tabulation is expensive in Prolog).

D. Wang and M. Reynolds (Eds.): AI 2011, LNAI 7106, pp. 658–667, 2011.

The situation, however, is more complex in our case, since the parsing performance is less clear for context-sensitive grammars because of the wide range of expressiveness that is allowed in these formats. The grammar of PENG Light uses, for example, feature structures in the arguments of the grammar rules and interleaves syntactic, semantic, and pragmatic information. Furthermore, the anaphora resolution algorithm of PENG Light is directly embedded into the grammar and investigates the discourse representation structure during parsing to find possible antecedents. Another thing that adds to the complexity is that the chart parser collects look-ahead information during parsing to support the writing process of the user. In this paper, we show that the way the grammar of PENG Light is written is critical for the parsing performance. We demonstrate this by example of coordinated structures in PENG Light and investigate how discourse representation structures can be constructed efficiently for coordinated structures. The parsing performance is then evaluated in an empirical way for an entire specification text.

The rest of this paper is structured as follows: In the next section we review the controlled natural language PENG Light and provide an example specification that we then use for our performance analysis. In Section 3, we look at the language processor of PENG Light and explain how chart parsing works with a special focus on chart parsing in Prolog. In Section 4, we first show how simple and complex discourse representation structures are generated in PENG Light, and then discuss different ways how discourse representation structures for coordinated structures can be implemented. In Section 5, we use the example specification introduced in Section 2 and evaluate the performance of the chart parser taking various settings for the parser as well as the proposed solutions for coordinated structures into consideration.

2 PENG Light

PENG Light is a controlled natural languages that can be used as a high-level knowledge representation language [12]. By design, PENG Light eliminates both ambiguity and complexity of full natural language but adheres to the same rigorous principles as formal languages do. At first glance, a specification written in PENG Light **looks seemingly informal** and is therefore easy to read and understand by a human; nevertheless, the specification can be translated unambiguously via discourse representation structures into a first-order logic theory. Below is an example of a specification text written in PENG Light that describes knowledge in a dynamic domain. This specification can be written as a coherent piece of text, but we can distinguish four different forms of knowledge (A-D) in this dynamic domain:

A. Knowledge about Events and their Effects

1. If a person arrives at a location then the person will be at that location.
2. If a person is at a location and a vehicle is waiting at that location and the person gets on that vehicle then the person will be in that vehicle.

3. If a person is in a vehicle and the vehicle leaves a location then the person will no longer be at that location and the vehicle will no longer be waiting at that location.

B. Terminological Knowledge

4. Every airport is a location.
5. Everybody who is John is a person.
6. Every bus is a vehicle.
7. Every Burwood bus is a bus.

C. Initial Domain State

8. The Burwood bus is waiting at the airport.

D. Sequence of Domain Events

9. John arrives at 10:10 with Qantas Flight QF2 at the airport of Sydney.
10. John gets on the Burwood bus.
11. The bus leaves the airport at 11:00.

The language processor of PENG Light translates this specification incrementally into the input language of the Simplified Event Calculus [5,6,11], and the author can query the resulting theory in controlled natural language as the events (9-11) unfold.

As the specification text illustrates, the syntactic structures of PENG Light sentences can be simple or complex. Simple sentences such as (8-11) have the following functional structure: subject + verb + [complements] + {adjuncts}. Complex sentences (1-7) are built from simpler sentences with the help of coordination, subordination, quantification and negation. Questions are derived from declarative sentences via movement of constituents, usually referred to as filler-gap dependencies [8]. Only restricted forms of anaphoric references are allowed in PENG Light: anaphoric references can be establish via definite noun phrases, proper nouns, and variables. For example, sentence (8) introduces a new object (*Burwood bus*) into the discourse and the corresponding noun phrases in (10) and (11) refer to this object. It is important to note that the author of a specification text does not need to remember the restrictions of the controlled natural language since these restrictions are enforced by a predictive authoring tool [9]. This authoring tool guides the writing process and informs the author with the help of look-ahead information about the words and phrases that can follow the current input.

3 The Language Processor

The language processor of PENG Light uses a chart parser and a unification-based phrase structure grammar to process the input text incrementally. The chart parser resolves anaphoric references, builds a syntax tree, a discourse representation structure, a paraphrase, and extracts look-ahead information from the chart. The grammar itself is written in DCG notation and consists of about

200 grammar rules; however, we do not use the DCG directly. Instead we use term expansion [13], a source-to-source transformation technique, to transform the DCG notation into a format that is easier to process with the help of a chart parser. We do this, because we want to avoid redundant analysis and because we need a way to process the input incrementally on a word-by-word basis so that we can extract look-ahead information from the chart to guide the writing process of the author.

The chart parser is an agenda-based active chart parser [1,3,4] that stores active edges (hypotheses about constituents) and passive edges (complete constituents) in the knowledge base. That means the chart parser uses the Prolog knowledge base directly as agenda. Every active edge represents a hypothesis that needs to be further explored. The fundamental rule [1] of chart parsing combines active edges with passive edges to generate new edges. These new edges can be either active or passive. We use a top-down rule invocation strategy and regard the agenda as a queue; new edges are added to the queue (by asserting them to the end of the knowledge base); this implements a breadth-first search strategy (we will discuss the impact of the search strategy on the parsing performance in Section 5). The edges of the chart are stored in the following form in the knowledge base:

12. *edge(SNum, Mode, SVNum, EVNum, LHS, Found, ToFind)*

Here, the first argument `SNum` stands for the sentence number, the second argument `Mode` for the mode of the sentence, the third argument `SVNum` for the number of the start vertex, the fourth argument `EVNum` for the number of the end vertex, the fifth argument `LHS` for the category on the left-hand side of a grammar rule, the sixth argument `Found` for the categories that have been found so far, and finally the seventh argument `ToFind` for the categories that still have to be found in order to complete an edge. For efficiency reasons, we index the sentence number (`SNum`), the number of the start vertex (`SVNum`), the number of the end vertex (`EVNum`), and the category on the left-hand side of a grammar rule (`LHS`). In SWI Prolog [13] up to four arguments of a predicate can be indexed, compound terms such as the left-hand side category of a grammar rule are indexed on the combination of their name and arity. We will discuss the impact of argument indexing on the parsing performance in Section 5.

4 Discourse Representation Structures (DRSs)

The grammar of PENG Light uses feature structures to generate a discourse representation structure (DRS) during parsing. A DRS consists of a set of discourse referents `U` and a set of conditions `Con`. Conditions can be either basic or complex: basic conditions store information about discourse referents or relations between discourse referents, and complex conditions contain embedded DRSs. DRSs are defined recursively and their nesting predicts which discourse referents are accessible via anaphoric expressions and which ones are not accessible. In contrast to standard discourse representation theory [2], we use a neo-Davidsonian representation for events and thematic roles [7] to connect discourse referents that represent events with other discourse referents that are described in a sentence.

4.1 Building Basic and Complex DRSs

We represent a basic DRS in the grammar of PENG Light as a Prolog term of the form: `drs(U, Con)` where both `U` and `Con` are lists. This basic DRS is initially empty and processed with the help of a difference list that collects the discourse referents and relevant conditions during parsing. The following grammar rule `s0` takes the existing DRS `D1` as input and generates the DRS `D3` as output. The contribution of the verb phrase `v3` to the DRS `D3` is collected with the help of the difference list `D2-D3` and this information is combined with the contribution of the noun phrase `n3` (depending on the determiner of the noun phrase):

```
13. s0([ sem:[E], tree:[s0, T1, T2], drs:D1-D3, para:P1-P4,
         gap:G1-G3, snum:N ]) -->
      n3( [ syn:[agr:[Pers, Num, case:nom]], sem:[M], tree:T1,
            drs:D1-D3, sco:D2-D3, para:P1-P2, gap:G1-G2,
            snum:N ]),
      v3( [ crd:C, syn:[vform:fin, agr:[Pers, Num, case:nom]],
            sem:[E, M], tree:T2, drs:D2-D3, para:P3-P4,
            gap:G2-G3, snum:N ]).
```

Apart from the feature structures for the DRS (`drs, sco`), the grammar rule (13) contains feature structures for the processing of syntactic information (`crd, syn, tree, gap`), semantic information (`sem`), pragmatic information (`para`) and information for keeping track of the sentence number (`snum`). Let us illustrate how a simple DRS looks like. In PENG Light the processing of the sentence (14):

14. A bus leaves the airport at 11:00.

results in the following DRS:

```
15. drs([A, B, C, D],
        [object(A, bus), theta(B, agent, A),
         event(B, leaving),
         theta(B, theme, C), object(C, airport),
         theta(B, time, D), timex(D, '11:00')])
```

The single conditions that occur in this DRS are retrieved during parsing from the lexicon. For example, the lexical entry for the word form *leaves* contains among other information the three conditions: `theta(B, agent, A), event(B, leaving)`, and `theta(B, theme, C)`.

The contribution of the determiners to a DRS deserves closer investigation: determiners are the most important constituents to establish the DRS. Semantically, determiners have two arguments: a restrictor (`res`) and a scope (`sco`). The restrictor consists of the information in the noun phrase minus the determiner, and the scope is the information outside of the noun phrase. In the case of an indefinite noun phrase (*a bus*), the lexical entry (16) for the indefinite determiner (*a*) specifies – among other things – that the output variable of the restrictor (D2) is same as the input variable for the scope (D2):

16. `drs:D1-D3, res:D1-`**D2**`, sco:`**D2**`-D3.`

This has the consequence that the conditions derived from the verb phrase are simply added to the current DRS (D2) that contains the previous contextual information, inclusive the information derived from the noun phrase. The situation gets a bit more complex if other determiners such as *every* or *no* occur in a sentence. The determiner *every* triggers a complex DRS (17) where the DRS for the restrictor (`drs(U2, C2)`) is combined with the DRS for the scope (`drs(U3, C3)`) and the result is embedded into the conditions of the superordinate DRS (`drs(U1, C1)`) using the operator `=>` to denote material implication. The lexical entry of this determiner contains the following information:

```
17. drs:D1-[drs(U1,[drs(U2, C2) => drs(U3, C3)|C1])|T],
    res:[drs([],[])|D1]-D2,
    sco:[drs([],[])|D2]-[drs(U3, C3), drs(U2, C2), drs(U1, C1)|T].
```

The determiner *no* triggers a similar structure (18) like *every*. The only difference is that the DRS built up in the scope is negated `~drs(U3, C3)` and embedded into a DRS in the consequent of the implication:

```
18. drs:D1-[drs(U1,[drs(U2, C2) => drs([],[~drs(U3, C3)])|C1])|T],
    res:[drs([],[])|D1]-D2,
    sco:[drs([],[])|D2]-[drs(U3, C3), drs(U2, C2), drs(U1, C1)|T].
```

4.2 Building DRSs for Coordinated Structures

In PENG Light, we can coordinate sentences, relative clauses, verb phrases, adjective phrases, and adverbial phrases by means of *and* and *or*. Coordination is only possible between complete constituents of the same syntactic type and the standard binding order of logic applies (that means the coordinator *and* binds stronger than the coordinator *or*). Our example specification contains two complex sentences (2 and 3) where coordination occurs between simple sentences. Apart from these syndetic cases of coordination where a coordinator marks the coordinated constituents, sentence (9) shows an asyndetic case of coordination where the coordinator is absent. In sentence (9), a number of prepositional phrases occur in adjunct position of the sentence without an explicit coordinator between these constituents, but the implementation of asyndetic coordination requires a very similar grammar rule as for syndetic coordination. In the following, we present two solutions of how coordinated structures can be implemented in PENG Light and focus on the coordination of simple sentences, coordination for the other constituents follows similar rules.

First Solution. The first solution relies on two grammar rules: one for conjunction and one for disjunction. Recall that a DRS has the form `drs(U, Con)` and that this structure is updated during parsing. For example, we can implement sentence coordination using the following two grammar rules (19 and 20):

```
19. s1([ crd:yes, tree:[s1, T1, T2, T3], drs:D1-D3, para:P1-P4,
        snum:N ]) -->
      s1( [ crd:no, tree:T1, drs:D1-D2, para:P1-P2, snum:N ]),
      crd([ cat:conj, tree:T2, para:P2-P3 ]),
      s1( [ crd:C, tree:T3, drs:D2-D3, para:P3-P4, snum:N ]).

20. s1([ crd:yes, tree:[s1, T1, T2, T3],
        drs:D1-[drs(U1, [D3 v D4|Con1])|Top],
        para:P1-P4, snum:N ]) -->
      s1( [ crd:no, tree:T1,
            drs:[drs([], [])|D1]-[D3|D2],
            para:P1-P2, snum:N ]),
      crd([ cat:disj, tree:T2, para:P2-P3 ]),
      s1( [ crd:C, tree:T3,
            drs:[drs([], [])|D2]-[D4, drs(U1, Con1)|Top],
            para:P3-P4, snum:N ]).
```

In the case of a conjunction (19), the DRS is passed through the grammar rule with the help of a difference list, first from `D1` to `D2` and then from `D2` to `D3`. In the case of a disjunction (20), each disjunct uses its own DRS, in our case `D3` and `D4`, and these DRSs are finally embedded into the existing DRS `drs(U1, Con1)` resulting in: `drs(U1, [D3 v D4|Con1])`. It is important to note that both disjuncts start with an empty DRS `drs([], [])` followed by the superordinate DRS `D1` and `D2`, respectively. At first glance, this solution looks intuitive, but it has the disadvantage that it duplicates the grammar rules for coordination on each level and this increases the processing time dramatically, as we will see in Section 5.

Second Solution. The second solution uses only one grammar rule for conjunction and disjunction and delegates the work for coordination to the category `crd` for coordination. This makes the grammar less speculative, since we provide the relevant structure for the DRS only when the coordinator is processed:

```
21. s1([ crd:yes, tree:[s1, T1, T2, T3], drs:D1-D3, para:P1-P4,
        snum:N ]) -->
      s1( [ crd:no, tree:T1, drs:D1-D2, para:P1-P2, snum:N ]),
      crd([ cat:coord, tree:T2, drs:D2-D3, hld:D1, sco:D4,
            para:P2-P3 ]),
      s1( [ crd:C, tree:T3, drs:D4, para:P3-P4, snum:N ]).
```

Here, the category `crd` takes the DRS `D1` as well as the DRS `D2` as input. The former one contains the information before the first conjunct `s1` is processed, and the latter one contains the information after the first conjunct has been processed. The individual grammar rules for the category `crd` then take care of the further processing as we will see now.

In the case of a conjunction, the grammar rule (22) makes sure that the DRS in `D2` can flow from the first conjunct to the second one and returns the result in `D3`:

```
22. crd([ cat:coord, tree:[conj, [and]], drs:D2-D3, hld:_,
          sco:D2-D3, para:P1-[[and]|P1] ]) -->
        [and],
        { lexicon([cat:conj, wform:[and]]) }.
```

In the case of a disjunction, the grammar rule (23) makes sure that the DRSs for the disjuncts are properly embedded into the existing DRS. Recall that the DRS `D1` contains the information before processing the first disjunct `s1` and the DRS `D2` contains the information after processing this disjunct. The difference between these two DRSs is the DRS `drs(U3, Con3)` that contains the relevant information for the first disjunct. The DRS for the second disjunct is then built up in `D4`:

```
23. crd([ cat:coord, tree:[disj, [or]],
          drs:D2-[drs(U5, [drs(U3, Con3) v D4|Con5])|Top],
          hld:D1,
          sco:[drs([],[])|D3]-[D4, drs(U5, Con5)|Top]
          para: P1-[[or]|P1] ]) -->
        [or],
        { lexicon([ cat:disj, wform:[or],
                    drs:D2-[drs(U3, Con3)|D3], hld:D1 ]) } .
```

Note that the actual difference between the DRS `D1` and `D2` is calculated with the help of the lexical rule for the coordinator `or` whenever this coordinator is processed.

5 Evaluation

We evaluated the performance of the chart parser for the two presented solutions using the example specification introduced in Section 2. We processed this specification on a 2.4 GHz Intel Core 2 Duo Windows machine with 2 GB RAM running SWI-Prolog (32 bits, Version 5.11.23). The lexicon that we used for this evaluation consists of 2,326 entries and is of a realistic size for a controlled language application, most of these entries represent content words.

In Figure 1, we compare the processing times for each sentence of the example specification and take the two solutions for processing coordinated structures into consideration. We used the chart parser with a top-down, breadth-first parsing strategy and measured the processing times in milliseconds for the first solution with argument indexing (s1-a) and without argument indexing (s1-b), as well as for the second solution with argument indexing (s2-a) and without argument indexing (s2-b). Recall that the sentences are processed in context; that means the chart parser takes the DRS of the previous sentences into account while the current sentence is processed in order to resolve anaphoric references. The most efficient solution is the second solution with argument indexing (s2-a). This solution is on average 4.29 times faster than the first solution with argument indexing (s1-a). Argument indexing gives us a speed-up of about 15% for the second solution (s2-a) compared to the same solution without argument indexing (s2-b). In Figure 2, we compare the processing times for the best solution using a breadth-first parsing strategy (s2-a-bf) and a depth-first strategy (s2-a-df). The breadth-first strategy gives us an overall speed-up of about 12%, only sentence 4 was processed slightly faster using the depth-first strategy.

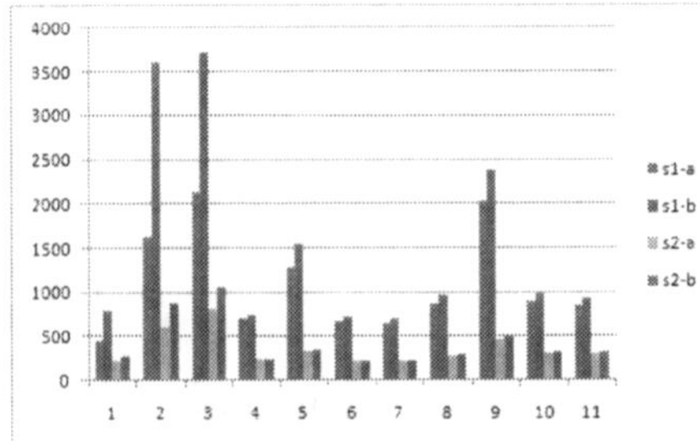

Fig. 1. Overall Comparison

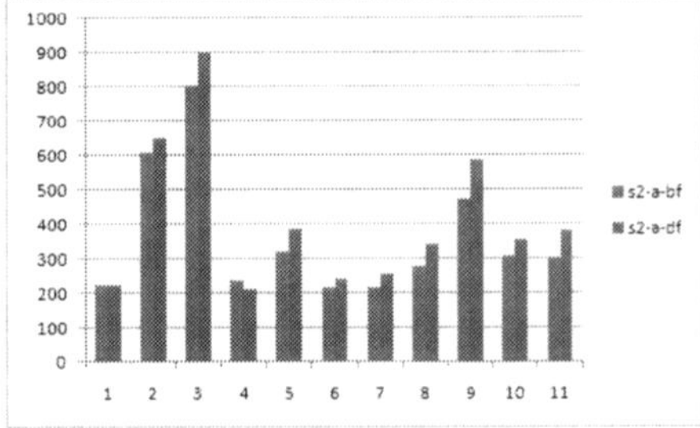

Fig. 2. Parsing Strategy

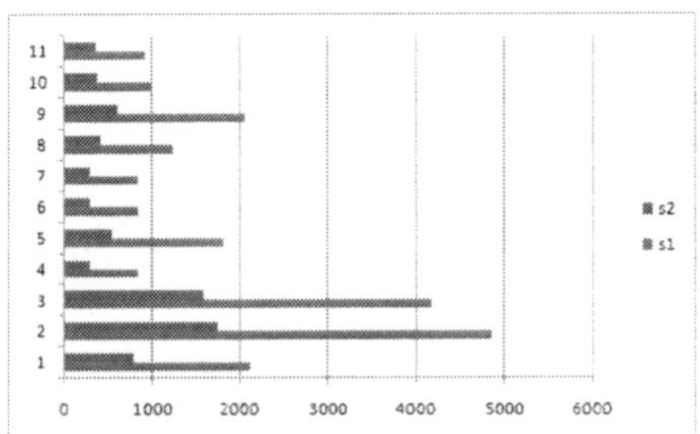

Fig. 3. Number of Edges

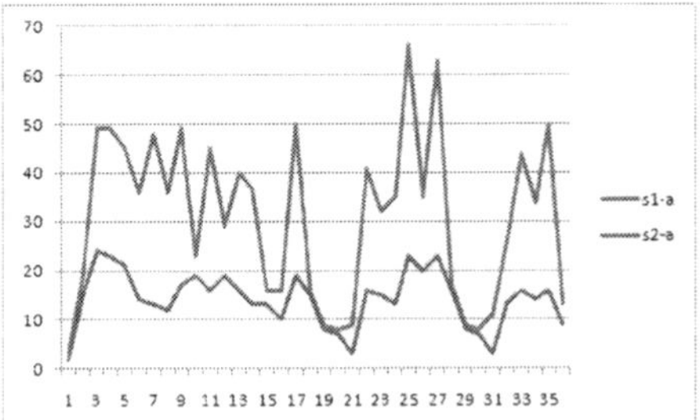

Fig. 4. Sentence 3

Figure 3 shows the number of edges that are generated for each sentence of the specification and compares the first solution for coordinated structures (s1) with the second solution (s2). The first solution creates 20.648 edges for the entire specification and the second solution 7.369 edges, that means the second solutions reduces the number of edges by a factor of 2.8. Since the specification text is usually written in an incremental fashion and look-ahead categories are generated for each word, it is interesting to compare the processing times for each word of a sentence. In Figure 4, we compare the processing times for each word of sentence 3 taking the first solution (s1-a) and the second solution (s2-a) into consideration, using argument indexing and the top-down, breadth-first parsing strategy. The interesting result here is that the processing times for both solutions are in the millisecond range, in the case of the second solution (s2-a), the processing times are well below 30 milliseconds, that means the author will not experience any delay while typing a text, since a delay of up to 100 milliseconds is generally not perceivable by a human.

6 Conclusions

In this paper, we showed how discourse representation structures are implemented for coordinated structures in PENG Light. The choice of the implementation for coordinated structures has a dramatic effect on the overall parsing performance. Argument indexing and a suitable parsing strategy (in our case a top-down, breadth first search strategy) can further speed up the processing of a specification. Furthermore, we showed that our approach is fast enough to support the incremental processing of a specification text, since there will be no perceivable delay when an author writes a text in PENG Light.

References

1. Gazdar, G., Mellish, C.: Natural Language Processing in PROLOG. Addison Wesley (1989)
2. Kamp, H., Reyle, U.: From Discourse to Logic. Kluwer, Dordrecht (1993)
3. Kay, M.: The MIND system. In: Rustin, R. (ed.) Natural Language Processing, pp. 155–188. Algorithmics Press, New York (1973)
4. Kay, M.: Algorithm schemata and data structures in syntactic processing. Technical Report CSL-80-12, Xerox PARC, Palo Alto, CA (October 1980)
5. Kowalski, R., Sergot, M.: Logic-Based Calculus of Events. New Generation Computing 4, 67–95 (1986)
6. Mueller, E.T.: Automating Commonsense Reasoning Using the Event Calculus. Communications of the ACM 52(1), 113–117 (2009)
7. Parsons, T.: Events in the Semantics of English: A Study in Subatomic Semantics. Current Studies in Linguistics. MIT Press (1994)
8. Pereira, F.C.N., Shieber, S.M.: Prolog and Natural-Language Analysis. CSLI Publications (1987)
9. Schwitter, R., Ljungberg, A., Hood, D.: ECOLE – A Look-ahead Editor for a Controlled Language. In: Proceedings of EAMT-CLAW 2003, Controlled Translation, pp. 141–150. Dublin City University, Ireland (2003)
10. Schwitter, R.: Controlled Natural Language for Knowledge Representation. In: Proceedings of COLING 2010, pp. 1113–1121 (2010)
11. Schwitter, R.: Specifying Events and their Effects in Controlled Natural Language. In: Proceedings of PACLING 2011 (forthcoming, 2011)
12. White, C., Schwitter, R.: An Update on PENG Light. In: Pizzato, L., Schwitter, R. (eds.) Proceedings of ALTA 2009, Sydney, Australia, pp. 80–88 (2009)
13. Wielemaker, J.: SWI-Prolog 5.10.4 Reference Manual. Department of Computer Science VU University Amsterdam (2010)

A Malay Stemmer for Jawi Characters

Suliana Sulaiman[1], Khairuddin Omar[2], Nazlia Omar[2], Mohd Zamri Murah[2], and Hamdan Abdul Rahman[2]

[1] Fakulti Seni, Komputeran dan Industri Kreatif, Universiti Pendidikan Sultan Idris, Tanjong Malim, Malaysia
suliana@fskik.upsi.edu.my

[2] Fakulti Teknologi dan Sains Maklumat, Universiti Kebangsaan Malaysia, Bangi, Malaysia
{ko,no,zamri,hamdan}@ftsm.ukm.my

Abstract. The Malay language may be written using either Roman or Jawi characters. Most Malay stemmers cover only Roman (*Rumi*) affixes. This paper proposes a stemmer for Jawi characters using two sets of rules in Jawi: one set of rules is used to stem various forms of derived words, and another set is used to replace the use of a dictionary by producing the root word for each derivative. This stemmer has been tested using 1185 derived words consisting of prefix, circumfix, suffix, and infix. The results show that 84.89% of Jawi root words have been successfully stemmed.

Keywords: Jawi, Stemmer, Malay Stemmer.

1 Introduction

A stemmer algorithm is used to find the morphological root word, known as the stem, of a word containing an affix [1]. A stemmer can be used not only for indexing and reducing vocabulary size but also to improve the performance of information retrieval [2]. Stemmer is very important in many languages and so does in Malay. The Malay language has two different types of scripts. A Jawi script resembles an Arabic character with the addition of six new characters that makes it suitable for use in the Malay language and a Roman script is distinguished as a Roman alphabet. The Malay language is spoken throughout many countries including Indonesia, Malaysia, Singapore and Brunei. The Jawi script is used in books, manuscripts, letters between kings and others [3].

Studies on Malay stemmers have been conducted by Asim [4], Fatimah [5], Idris [6], Sock [7] and Muhamad [2]. Most studies used morphological rules in their stemming algorithm, although Sock [7] used a combination method of an N-gram and stemming. Idris [6] identified the use of an additional dictionary, known as a local dictionary, which can reduce stemmer errors. Fatimah [5] and Muhamad [2] also used a root word dictionary to ensure that the root word was correct. However, these stemmers applied the root word dictionaries and are applicable only to Malay-derived words that are written in Roman characters. This paper proposes a stemmer for Jawi

D. Wang and M. Reynolds (Eds.): AI 2011, LNAI 7106, pp. 668–676, 2011.

characters to stem all possible Malay-derived words into their respective root words using two sets of rules without using a dictionary. This paper has been organized into six parts. Section 2 introduces an overview of related works. Section 3 reviews studies related to Malay affixes. Section 4 describes the Jawi stemmer. Section 5 discusses the experiment and the results. Finally, section 6 presents our conclusions.

2 Related Works

The first Malay stemmer was developed by Asim [4]. The aim of the stemmer is to stem words derived in the Roman script into their root words. Fatimah [5] proposed a stemming process called the 'Rule Application Order' and improved the rule previously used in Asim [4]. In the 'Rule Application Order', the common dictionary [5] was replaced with a root word dictionary. According to Idris [6], the use of a 'local dictionary' can improve the stemmer's result. The 'local dictionary' mainly holds a clear context of the root word and provides higher accuracy for the stemmer. Later, Muhamad [2] enhanced the technique used in Fatimah [5] and introduced the 'Rule Frequency Order' to stem Malay words in Roman script. The accuracy of Muhamad's [2] stemmer is higher as compared to Fatimah's [5].

Errors reported on Malay stemmers have been identified as understemming, overstemming, spelling exceptions and others [2]. Understemming is an error that occurs when the root word produced by the stemmer contains the root word along with other characters [8]. For example applying a stemmer to the derived word تمبهكن (*tmbhkn*) should produce the actual root word تمبه (*tmbh*); in an understemming error, the stemmer produces the stem تمبهک (*tmbhk*). Overstemming is said to occur when a root word is overstemmed such as the derived word تمبهكن (*tmbhkn*) is stemmed to مبه (*mbh*) [8]. In the Malay language, when the prefix is removed from the derived word, the character following the prefix sometimes needs to be replaced with another character. Failing to do so can lead to spelling exception errors, for example, when the word ڤڠلوار (pngluar) is stemmed to الوار (*aluar*) instead of كلوار (*kluar*) [2]. Other errors occur when the derived word does not overstem or understem, such as the word ڤربيذاٴن (*prbezaan*) stemmed to ڤربي (*prbe*) rather than بيذا (*beza*) [2].

The use of a Malay stemmer is not limited to information retrieval alone, but it can also be used in transliteration especially to transliterate Jawi words into Roman characters. In the 'Jawi-Malay Transliteration' [9], the stemming process is used to eliminate the affix from the root word. According to [9], the use of a root word dictionary is required to ensure that the root word is correct before the transliteration process can be done.

Most studies use morphological rules and dictionaries in their stemming algorithms. However, the use of a dictionary can lead to some issues such as the dictionary itself must be comprehensive to make sure that the derived word is correctly stemmed [10],[11]. Moreover, some maintenance needs to be performed to update newly-discovered words and the use of a large-sized dictionary will affect the storage space and processing time [11],[12].

3 Malay Affixes

A derived word can be described as a combination of a prefix, a circumfix, a suffix or an infix with a root word to form a new word [13]. Differences can be seen between affixes in a Roman script and affixes in Jawi regarding characters and spelling methods [14]. The differences are found in the effects of the circumfixes and suffixes. This occurs primarily because Jawi uses only three characters ا (*alif*), و (*wau*) and ي (*ya*) for six vowel sounds [15]. People may assume that two vowels share one common symbol, but that is not the case. Table 1 shows that each vowel may be spelled differently depending on its position in the word – initial, medial or final. Moreover, certain vowels cannot be spelled using any vowel letter.

Table 1. Different types of spelling using Jawi vowels, based on their positions

Vowel	Initial	Medial	Final
[ə]	< ا (*alif*) > : emak = امق (*emk*)	< ø >* : lemak = لمق (*lmk*)	< ى (*ya*) > : egoism = ايگوءيسمى (*egoism*)
[a]	< ا (*alif*) > : atas = اتس (*ats*)	< ا > : batas = باتس (*bts*)	< ا > : sila = سيلا (*sila*) ; < ø >* : tika = تيک (*tik*)
[i]	< ايـ (*i*) > : ikan = ايکن (*ikn*)	< ي (*ya*) > : lipas = ليڤس (*lps*)	< ي (*ya*) > : roti = روتي (*roti*)
[e]	< ايـ (*e*) > : elak = ايلق (*elk*)	< ي (*ya*) > : perak = ڤيرق (*perk*)	< ي (*ya*) > : sate = ساتي (*sate*)
[u]	< او (*u*) > : udang = اودڠ (*udng*)	< و (*u*) > : buta = بوتا (*buta*)	< و (*u*) > : biru = بيرو (*biru*)
[o]	< او (o) > : orang = اورڠ (*orng*)	< و (o) > : roda = رودا (*roda*)	< و (o) > : polo = ڤولو (*polo*)

< ø >* = Vowels [ə] in the medial position and [a] in the final position are not spelled with < ا (*alif*) > character.

To stem the words *binaan* and *bukaan* in the Roman script, we need to delete the suffix {+*an*} to form the root words *bina* and *buka*. In the same way, to stem the derived word بيناءن (*binaan*) in Jawi, the suffix ءن (*an*) need to be deleted to form the root word بينا (*bena*). However, in some situations, to stem a derived word such as بوکأن (*bukaan*), the suffix أن should be deleted to produce the root word بوک. (*buk*) Some words require ا (*alif*) at the end of the root word, but others do not. Table 2 shows some spelling examples in suffixes between the Roman and Jawi scripts.

Table 2. Spelling examples in suffixes between the Roman and Jawi scripts

Jawi	Roman Script
+أن	+an
+ن	+an
+ءن	+an
+ان	+an
+أي	+i
+ءي	+i
+ي	+i
+اي	+i

In Malay, when we want to remove a prefix from a derived word, we need to add another character to the root word after the prefix is removed. This is known as a spelling exception [2]. For example, after removing the prefix +ڤ (*pa*) from the word ڤمليه (*pmilih*), the character م (*mim*) should be replaced by character ڤ (*pa*) to form the word ڤيليه (*pilih*).

4 The Jawi Stemmer

The development of a deaffixation rule must consider vowel placement, spelling exceptions and the minimum length of the morphemes in a Jawi word to produce the best stemmed word. When using vowel placement, it is important to note that after removing the circumfix and suffix, ا (*alif*) should be used at the end of the word only if it represents the vowel [a] and not immediately preceded by the character ݢ (*ga*) or ك (*kaf*). We also need to consider the use of ا (*alif*) at the end of the word if the vowel in the preceding syllables is [a] and the second syllable begin with either د (*dal*), ل (*lam*), و (*wau*), ر (*ro*) or ڠ (*nga*). To avoid errors, these two rules should be applied every time a circumfix or a suffix is eliminated. In Jawi, spelling exceptions occur more often in a prefix and a circumfix. Example of spelling exceptions for the Jawi stemmer is shown in Table 3.

The minimum length of the Jawi stemmer can be defined in two parts. The first part is a root word with a minimum length of 2 and the second part is a root word with

Table 3. Example of spelling exceptions for prefixes and suffixes

Prefix	Character added at the beginning of the root word after removing the prefix	Circumfix	Character added at the beginning of the root word after removing the circumfix
+مم (*pm*) /+ ڤم (*pm*)	ڤ (*pa*)	كن (*kn*) +مم (*mm*)	ڤ (*pa*)
+من (*pn*) /+ڤن (*pn*)	ت (*ta*), ج (*jim*), چ (*cha*)	كن (*kn*) +من (*mn*)	ت (*ta*), ج (*jim*), چ (*cha*)
+مڠ (*mng*) /+ڤڠ (*png*)	ک (*kaf*) / ا (*alif*)	كن (*kn*) +مڠ (*mng*)	ک (*kaf*) / ا (*alif*)
+مڽ (*mny*) /+ڤڽ (*pny*)	س (*sin*)	كن (*kn*) +مڽ (*mny*)	س (*sin*)

a minimum length of 3. Prefixes, circumfixes and suffixes use a minimum length of 2 to prevent overstemming from occurring and infixes use a minimum length of 3. This is because, in Jawi, the use of two characters sometimes refers to a disyllable; for example, the affix براڤ (*brap*) will stem into the root word اڤ (*ap*).

4.1 The Algorithm for the Malay Stemmer for Jawi Characters

The deaffixation rules for the Malay Stemmer for Jawi Characters' algorithm have been developed based on the book '*Panduan Mengeja dan Menulis Jawi*' written by Hamdan Abdul Rahman [16]. These rules are being adapted with added capabilities appropriate for the Jawi stemmer. Instead of using the root word dictionary to check that the word is correctly stemmed, the proposed algorithm uses the Spelling Error Detector Rules (SEDR) [17] to ensure that the root word produced is spelled correctly. SEDR has been tested using 3018 Jawi words and produced 97.8% of accuracy as reported in [17]. The algorithm for the Malay Stemmer for Jawi Characters is described as follows:

```
BEGIN;

LOAD Deaffixation rule;

  IF Circumfix present in a word;

    THEN apply Circumfix rule;

    WRITE the result into R.Txt;

  END IF

  IF prefix present in a word;

    THEN apply prefix rule

    WRITE the result into R.Txt;              } Deaffixation
                                                Rule
  END IF

  IF suffix present in a word;

    THEN apply suffix rule;

    WRITE the result into R.Txt;

  END IF

  IF infix present in a word;

    THEN apply infix rule;
```

```
    WRITE the result into R.Txt;

    ELSE output the word as a stemmed word;

  END IF

OPEN R.txt;

    APPLY Spelling Error Detector Rule;        \
                                                |
      IF Spelling == Correct;                   |  Spelling Error
                                                 > Detector Rule
        OUTPUT the result as a root word;       |
                                                |
      END IF;                                   /

CLOSED R.txt;

END
```

For example, the word فرکولقن (*prgolkn*) contains the prefix +فر (*pr*), the circumfix ن (*an*) +فر (*pr*) and the suffix ن (*an*) +. After the deaffixation rule is applied, the result must be checked using the SEDR. This set of rules involves checking the vowel placement and spelling method for Jawi [16]. To avoid understemming, the SEDR [17] must detect the best root word for the Jawi disyllable and follows it with three or four syllables. This is because the amount of disyllabic words is larger compared to other words. If the rule detects more than one available results, then it will take the circumfix result, followed by the prefix, suffix and infix. After the word فرکولقن (*prgolkn*) has been stemmed by the deaffixation rule, it must be checked using the SEDR [17] for disyllables. In this case کولق (*golk*) will be given as the best root word of affix فرکولقن (*prgolkn*). This is because the result obtained after individually removing the prefix +فر and the suffix ن+ will give کولقن and فرکولق, which does not match the set of rule of SEDR [17] for disyllables.

5 Experimental and the Results

All of the data were taken from online newspaper websites [18],[19] and were transliterated using the Transliteration Engine for Rumi to Jawi [20]. An experiment was performed to determine which deaffixation rules are the best sequences to be used with the SEDR [17]. This experiment used 1185 unique derived words and was tested using six different tests, labeled as D1, D2, D3, D4, D5 and D6. In this experiment, the infix was always in the last sequence because the infix is a smaller part of a derived word. Table 4 shows the accuracy of each test.

Table 4. The accuracy of each test

	D1	D2	D3	D4	D5	D6
Accuracy	78.39%	73.67%	84.89%	80.42%	72.66%	73.76%

- D1: Prefix, circumfix, suffix and infix
- D2: Prefix, suffix, circumfix and infix
- D3: Circumfix, prefix, suffix and infix
- D4: Circumfix, suffix, prefix and infix
- D5: Suffix, prefix, circumfix and infix
- D6: Suffix, circumfix, prefix and infix

The results show that the highest accuracy for the Jawi stemmer was D3. Thus, to get the best results from the deaffixation rule, the sequence of stemming is to firstly stem the circumfix, followed by the prefix and the suffix, and finally the infix and then check the results using the SEDR [17].

Next, two more experiments were conducted. The first was a controlled experiment called TEST A. To perform this experiment, the work of Fatimah [5] and Muhamad [2] have been used to test the data in Jawi without using a root word dictionary and then compare the results with those from the Malay Stemmer for Jawi Characters. 1185 unique derived words and the same set of deaffixation rules for all sets have been used in TEST A. However, neither the root word dictionary nor the SEDR has been used here. The results were compared based on the accuracy.

For the second experiment, the SEDR [17] has been used to replace the use of a root word dictionary as were done in Fatimah [5] and Muhamad [2]. This experiment was called TEST B. Again, the same set of data and deaffixation rules have been used. The results were compared based on the accuracy of each set. Details of the errors are shown in Table 5.

Table 5. Types of errors in Test A and Test B

Types of Error	TEST A			TEST B		
	Fatimah	Muhamad	Jawi Stemmer	Fatimah	Muhamad	Jawi Stemmer
Understemming	537	495	30	147	138	34
Overstemming	5	3	70	30	29	61
Spelling Exception	20	16	37	21	31	40
Others	95	126	69	58	26	44

From Table 5 we see that in TEST A, Fatimah [5] shows the highest error, followed by Muhamad [2] and the Jawi stemmer if we use only deaffixation rules to stem the derived words. This error can be reduced as reported in TEST B if we use the SEDR to replace the use of a dictionary as in Fatimah [5] and Muhamad [2]. In TEST B, we see that Fatimah [5] and Muhamad [2] produced the most errors for understemming and the Jawi stemmer produced the most errors for overstemming. Figure 1 presents the accuracies for TEST A and TEST B.

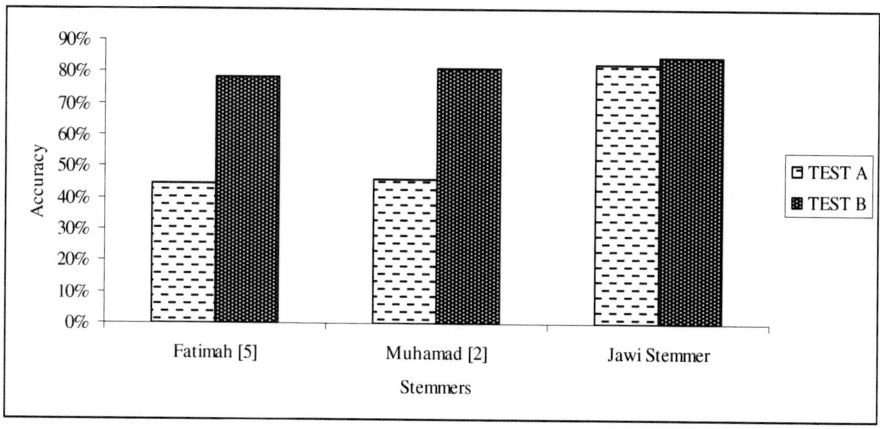

Fig. 1. Accuracy of the stemmers

Based on the graph in Figure 1, TEST A shows the control experiment for the Jawi stemmer and its accuracy compared to the studies of Fatimah [5] and Muhamad [2]. It is clear that the accuracy of the Jawi Stemmer is the highest. In TEST B, the use of a dictionary has been replaced by the SEDR [17] in the studies of Fatimah [5] and Muhamad [2]. The graph shows that the highest accuracy is 84.89% which is that of the Jawi stemmer. In conclusion, it is clear that the use of the SEDR [17] increased the accuracy of all stemmers and that stemmers achieve accuracies in the range of 78.39% to 84.89%.

6 Conclusion

This study shows that the use of both the deaffixation rules and the SEDR can improve the accuracy of the Jawi stemmer. These two sets of rules can replace the use of a root word dictionary in developing the Jawi stemmer. The use of these rules can also reduce the time required to update a new root word in the dictionary when a new stem is produced after the derived word has been stemmed. However, the limitation of the Jawi stemmer is that it does not apply to English or Arabic-derived words because the vowels used in English and Arabic-derived words are different compared to pure Malay words. Further work needs to be done to ensure that this stemmer can also stem these types of words correctly.

References

1. Melucci, M.: A Basis for Information Retrieval in Context. ACM Transaction on Information System (TOIS) 26, 14–41 (2008)
2. Muhamad, T., Fatimah, A., Ramlan, M., Tengku, M.T.S.: Rules Frequency Order Stemmer for Malay Language. International Journal of Computer Science and Network Security 9, 433–438 (2009)

3. Mohammad, F.N., Khairuddin, O., Mohamad, S., Liong, C.Y.: Handwritten Cursive Jawi Character Recognition: A Survey. In: 5th International Conference on Computer Graphics, Imaging and Visualisation, pp. 247–249. IEEE (2008)
4. Asim, O.: Pengakar Perkataan Melayu dan Sistem Capaian Dokumen. Universiti Kebangsaan Malaysia, Bangi (1993)
5. Fatimah, A.: A Malay Document Retrieval System: An Experimental Approach and Analysis. National University of Malaysia, Bangi (1995)
6. Idris, N., Syed Mustapha, S.M.F.D.: Stemming for Term Conflation in Malay Texts. In: International Conference of Artificial Intelligence, ICAI 2001, Las Vegas, pp. 1512–1517 (2001)
7. Sock, Y.T., Cheng, S.O., Abdullah, N.A.: On Designing an Automated Malaysia Stemmer for the Malay Language. In: Proceedings of the 5th International Workshop Information Retrieval with Asian Languages, pp. 201–208. ACM, New York (2000)
8. Vivian, M.O.: A Stemming Algorithm for the Portuguese. In: 8th International Symposiums on String Processing and Information Retrieval, pp. 186–193. IEEE (2001)
9. Roslan, A.G., Mohamad, S.Z., Khairuddin, O.: Jawi-Malay Transliteration. In: Proceedings of the International Conference on Electrical Engineering and Informatics, ICEEI, pp. 154–157. IEEE Selangor (2009)
10. Kazeem, T., Rania, E., Jeffrey, C.: Arabic Stemming Without a Root Dictionary. In: Proceedings of the International Conference on Information Technology: Coding and Computting, pp. 152–157. IEEE Computer Society, Washington (2005)
11. Mirna, A., Jelita, A., Bobby, N., Tahaghoghi, S.M.M., Hugh, E.W.: Stemming Indonesian: A Confix-Stripping Approach. ACM Transaction on Asian Language Information Processing (TALIP) 6, 22–33 (2007)
12. Pramod, P.S., Sanjay, K.D.: Advancement of Clinical Stemmer. Special Volume: Problems of Passing in India Languages, 45–50 (2011)
13. Nik, S., Farid, M.O., Hashim, H.M.: Tatabahasa Dewan Edisi Baharu. Dewan Bahasa dan Pustaka (1993)
14. Dewan, B.P.: Daftar Kata Bahasa Melayu Roman-Sebutan-Jawi. Dewan Bahasa dan Pustaka, Kuala Lumpur (2001)
15. Abdullah, H.: The Morphology of Malay. Dewan Bahasa dan Pustaka, Kuala Lumpur (1974)
16. Hamdan, A.R.: Panduan Menulis dan Mengeja Jawi. Dewan Bahasa dan Pustaka, Kuala Lumpur (1999)
17. Suliana, S., Khairuddin, O., Nazlia, O., Mohd, Z.M., Hamdan, A.R.: Spelling Error Detector for Jawi Stemmer. In: International Conference on Pattern Analysis and Intelligent Robotics ICPAIR, pp. 78–82. IEEE (2011)
18. Utusan Melayu Online, http://www.utusan.com.my
19. Berita Harian Online, http://www.bharian.com.my
20. Transliteration Engine for Roman to Jawi, http://www.jawi.ukm.my

Executability in the Situation Calculus

Timothy Cerexhe and Maurice Pagnucco

ARC Centre of Excellence in Autonomous Systems,
School of Computer Science and Engineering, UNSW,
Sydney, NSW, 2052, Australia
{timothyc,morri}@cse.unsw.edu.au

Abstract. This paper establishes a formal relationship between theories of computation and certain types of reasoning about action theories expressed in the situation calculus. In particular it establishes a formal correspondence between Deterministic Finite-State Automata (DFAs) and the 'literal-based' class of basic action theory, and identifies the special case of DFAs equivalent to 'context-free' action theories. These results formally describe the relative expressivity of different action theories. We intend to exploit these results to drive more efficient implementations for planning, legality checking, and modelling in the situation calculus.

Keywords: Reasoning about Action and Change, Situation Calculus, Theory of Computation.

1 Introduction

The *situation calculus* is a general-purpose dialect of first-order logic for reasoning about the effects of actions in dynamic domains. It allows us to model the state of a changing world, or determine which sequence of actions will achieve a goal. However not all sequences will be executable—the effects of one action may violate the preconditions of another. In this light, the set of executable sequences of actions for a given theory corresponds to the more general concept of a language whose strings are built from an alphabet of actions.

Executability in the situation calculus has been briefly defined and discussed by Reiter [5]. However there has been no attempt to describe the complexity of an action theory based on the language of sequences of actions it accepts. This is surprising since, compared with logic-based action calculi, automata theory provides a powerful and more thoroughly studied means of classifying the 'hardness' of different computational machinery—including those required for recognising or constructing action theories.

In this paper we identify several existing special cases of action theories in the situation calculus and prove their equivalence with classes of deterministic finite automata. This facilitates a greater understanding of the expressivity of situation calculus theories. We also expect it to lead to more efficient techniques for representing and manipulating action theories, such as in planning problems or in new interpreters for the cognitive robotics language Golog [1] whose semantics are based on the situation calculus. By the same token, our translations

D. Wang and M. Reynolds (Eds.): AI 2011, LNAI 7106, pp. 677–686, 2011.

allow AI behaviours encoded via finite state machines—such as XABSL [3] used for robotics—to be converted to the situation calculus where it may be formally reasoned about.

2 Background

The situation calculus is a sorted first-order logic for reasoning about actions in dynamic systems with several distinguished elements. *Situation* terms are histories (sequences of actions) composed of the binary $do(a, s)$ function which returns the situation that results from performing action a in situation s. Thus $do(a_n, \ldots, do(a_2, do(a_1, S_0)))$ represents the situation after performing the actions $a_1, a_2, \ldots, a_n$ in that order, starting from the distinguished *initial* situation S_0. Arbitrary situations can be constructed this way. *Fluents* are functions that represent properties of the world. Their value may be modified by performing actions, so they have situation terms as their last argument. We require that the initial situation be fully axiomatised—every fluent must have a known value in S_0. *Precondition axioms* defining $poss(a, s)$ specify the conditions under which action a may be performed in situation s. *Effect axioms* specify the resulting fluent values after performing an action a in situation s.

A situation-suppressed fluent f represents the partial function $f(s)$ without its situation term. A situation-independent formula does not expect a situation term. A formula is *uniform in s* if it does not mention *poss* or *do*, or quantify over situations, and s is the only situation term that occurs. Essentially it restricts the use of the situation terms to querying the current value of fluents in only that situation.

Note that rather than use effect axioms we adopt Reiter's *successor state axioms* (SSAs)—one per fluent—that provide a solution to the frame problem.[1] SSAs can be automatically generated from effect axioms. This gives the following 'template' for successor state axioms:

$$f(do(a, s)) = y \equiv \gamma^{+}(a, y, s) \lor (f(s) = y \land \neg \exists z \,.\, \gamma^{-}(a, z, s))$$

where γ^{+}, γ^{-} are the positive and negative effects respectively—the conditions that describe when a value becomes true or false. This formula expresses that fluent f has the value y after performing action a in situation s whenever the conditions that make the fluent assume that value hold (γ^{+}) or f already has the value y and nothing occurs to make it false (γ^{-}). Note that we will only consider functional fluents in this paper, so a positive effect for one value must coincide with negative effects for all other values in the domain of that fluent.

The above SSA describes the value of fluent f in situation $do(a, s)$ given its previous value $f(s)$ and the conditions that update its value (γ^{+}) or the absence of effects that make it false (γ^{-}). This last component provides 'inertia'—the

---

[1] A discussion of the frame problem is beyond the scope of this paper, suffice to say that it represents an explosion of axioms required to logically model fluents in dynamic systems that do not change their value as a result of performing an action.

logical assertion that fluents are not affected by unrelated actions. For example the colour of a block does not change as a result of picking it up.

Systems always start in the distinguished initial situation S_0. The value of a fluent f in a later situation s can be determined by regressing[2] back to S_0 and simulating the update axioms on f for each action in s. We are now in a position to define a *basic action theory* which represents a rudimentary type of action theory.

Definition 1. *A finite basic action theory (BAT) is a bounded theory in the situation calculus $\mathcal{B} = \Sigma_{sitcalc} \cup \mathcal{B}_{una} \cup \mathcal{B}_{ss} \cup \mathcal{B}_{ap} \cup \mathcal{B}_{S_0}$, where:*

1. *$\Sigma_{sitcalc}$ are the foundational axioms for the situation calculus that logically describe what a situation looks like.*
2. *$\mathcal{B}_{una}$ are unique names assumptions—logical assertions that distinct names (including fluent, action, and object names) are treated as distinct concepts.*
3. *$\mathcal{B}_{ss}$ is a finite set of successor state axioms of the form above.*
4. *$\mathcal{B}_{ap}$ is a finite set of action precondition axioms of the form $poss(a, s) \equiv \Pi_a(s)$, one for each action. Π_a must be uniform in s.*
5. *$\mathcal{B}_{S_0}$ is a finite set of first-order sentences defining the initial situation S_0.*
6. *There are three finite sets of fluent names, actions, and objects. All ground fluents and actions are constructed from these sets. We shall refer to the sets of* ground *terms as $\mathcal{F}$, $\mathcal{A}$, and $\mathcal{O}$, and assume that each has an arbitrary but fixed ordering.*

For an arbitrary machine M, we say its language $\mathcal{L}(M)$ is some (possibly infinite) set of accepted words $w \in \mathcal{L}(M)$. We will consider a BAT of the situation calculus to be one such machine, as well as more conventional machines like Deterministic Finite-State Automata (DFAs). Both machines depend on a transition mechanism. DFAs use a function δ to map states and symbols to new states. The situation calculus analogue is the function *do* that maps situations and actions to new situations.

Definition 2. *A DFA D is a 5-tuple:*

$$D = \langle Q, \Sigma, q_0, \delta, F \rangle$$

where Q is a finite, non-empty set of states, Σ is the input alphabet. The (total) transition function $\delta : Q \times \Sigma \rightarrow Q$ maps a state and a symbol to a new state. The system starts in state q_0 and transitions according to δ. The set $F \subseteq Q$ identifies the final (accepting) states of Q. If the system reaches one of these states it may *accept—the current string of symbols is a word in the language—or continue running to accept a longer string.*

The transition function of an automaton encapsulates the pre- and post- conditions of a transition. These are separated in the situation calculus as action

[2] See Reiter [5]. We leave the alternative method—progression—for future work.

preconditions (*poss*) and fluent successor state axiom postconditions. It is relevant then to define *executable* situations as a sequence of actions where one action's postconditions are sufficient to guarantee the preconditions of the next:

$$exec(do(a, s)) \equiv exec(s) \wedge poss(a, s) \qquad \text{(exec)}$$

and

$$\mathcal{B} \cup (exec) \models exec(s)$$

This is one of Reiter's definitions of executability. Note that $exec(S_0)$ is considered trivially true, though it is also easily derivable from alternate (equivalent) definitions [5].

Many automata explicitly define a set of final states. In contrast, the situation calculus makes no such distinction. To this end, we introduce an artificial finality condition—a situation s is *final* iff a distinguished formula, $f_{halt}(s)$, is true in that situation. This has applications to planning problems, where the goal is often specified by a formula $goal(s)$ identifying final states/situations. It is necessary in this paper for f_{halt} to be uniform in s. This would restrict goal formulas to being Markovian.

Finally, we define acceptance as:

Definition 3. *A DFA accepts a sequence if the corresponding transitions produce a final state. A BAT accepts a situation (sequence of actions) if that situation is executable and satisfies some theory-specific halt (goal) condition.*

$$w \in \mathcal{L}(DFA) \text{ iff } \hat{\delta}(q_0, w) \in F$$
$$w \in \mathcal{E}(BAT) \text{ iff } BAT \models f_{halt}(\hat{do}(w, S_0)) \wedge exec(\hat{do}(w, S_0))$$

where $\hat{\delta}, \hat{do}$ represent repeated applications of the corresponding transition functions.

Informally, $\mathcal{L}(D)$ is the set of words that transition to a final state in DFA D. $\mathcal{E}(B)$ is the set of legal, 'halting' situations in BAT B—to distinguish it from logically derivable sentences $\mathcal{L}$ in a typical knowledge base.

3 DFA—Literal-Based BAT Equivalence

In this section we prove the formal equivalence between DFAs and the 'literal-based' class of basic action theories. Petrick and Levesque defined literal-based BATs while establishing knowledge equivalence [4]. The functional form was left as an exercise and so we provide the following definition:

Definition 4. *A literal-based BAT (LB-BAT) is a finite BAT of the above form, but with restricted use of the situation term in the positive and negative effects (γ^+, γ^-) of the successor state axioms:*

$$\gamma_F^+(y,a,s),\ \gamma_F^-(y,a,s) \stackrel{def}{=} \bigvee_{i=1}^{k} \pi_i(y,a,s)$$

where each π_i *is:*

$$\pi_i(y,a,s) \stackrel{def}{=} \exists \boldsymbol{z}_i\ .\ a = \beta_i(\boldsymbol{z}_i) \wedge \psi_i(y,\boldsymbol{z}_i,a) \wedge P_1(\boldsymbol{z}_i,s) = c_1 \wedge \ldots \wedge P_l(\boldsymbol{z}_i,s) = c_l$$

where $P_1,\ldots,P_l$ *are fluent literals,* $c_1,\ldots,c_l$ *are constants, the (possibly empty) vector of variables* $\boldsymbol{z}_i$ *must be an argument to the action term* β_i*. To maintain the functional property, we also require that all separate* π*s are mutually inconsistent.*

Informally, SSAs may now only use the situation term to mention a finite conjunction of fluent literals, or to implement inertia.

Fluent values are typically encapsulated in situations, which may grow arbitrarily long. However literal-based BATs have finite domains, so we can define equivalence classes of situations that represent the finite state space of distinct fluent values. This formulation is suitable for an automaton.

Definition 5. *state is a function from* executable *situations in BAT B to ordered tuples of fluent values:*

$$state(s) = \begin{cases} \langle\ \mathcal{B} \models f(s)\ \rangle_{f\in\mathcal{F}} & \text{if } \mathcal{B} \models exec(s) \\ \perp & \text{otherwise} \end{cases}$$

where $\mathcal{F}$ *is the finite (ordered) set of unique ground fluents in B. Note that* non-executable *situations all map to a distinguished failure state* $\perp$.

Definition 6. *The* BAT-state construction *takes a literal-based basic action theory B and produces a finite state machine* $D = \langle Q, q_0, \Sigma, \delta, F_D\rangle$ *where:*

$$\begin{aligned} Q &= \{state(s)\,|\,s \text{ is an executable situation}\} \cup \{\perp\} && \text{(set of states)} \\ q_0 &= state(S_0) && \text{(initial state)} \\ \Sigma &= \text{the set of ground action terms} && \text{(alphabet)} \\ \delta(state(s),a) &= state(do(a,s)) && \text{(transition function } Q\times\Sigma\rightarrow Q) \\ \delta(\perp,a) &= \perp \\ F_D &= \{state(s)\,|\,\mathcal{B} \models f_{halt}(s)\} && \text{(final states} \subseteq Q) \end{aligned}$$

Note also that *state* maps infinitely-many situations (histories) to finitely-many 'states'; the domain of each fluent $dom(f)$ is finite, and both literal-based preconditions and successor-state axioms are finite and uniform. The set of states Q in our DFA is an equivalence class of situations with the same fluent-value bindings.

We illustrate this conversion with an example. Our hero, Maxwell, walks down a corridor full of doors on his way to work. At the end of the corridor is a phone booth in which he dials a passcode. Such a procedure may be represented as a literal-based action theory where we keep walking until we non-deterministically decide that we have reached the phone, at which point we enter and dial. This would typically be augmented with sensing actions, though we prefer to save a discussion of this mechanism for Section 5.

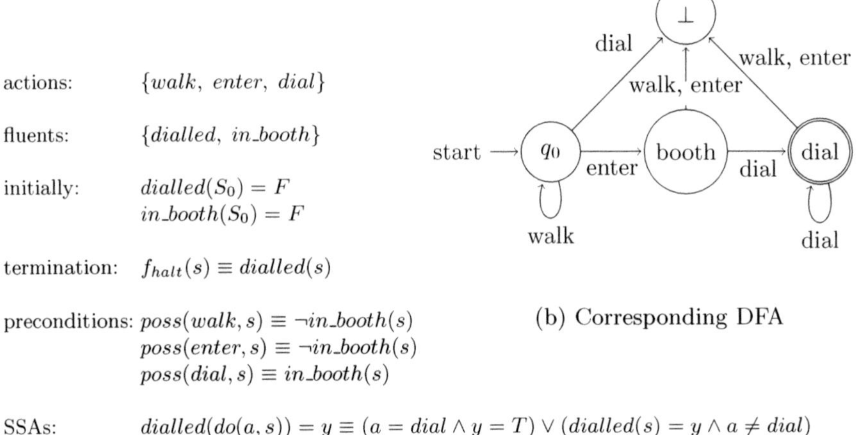

(a) An LB-BAT for 'corridor world'

Fig. 1. Corridor World

We can now apply our construction to this action theory. First, our alphabet is the set of actions $\Sigma = \{walk, enter, dial\}$. Each fluent has a Boolean domain, so our state-space is:

$$Q = \{\langle\rangle, \langle in\_booth\rangle, \langle dialled, in\_booth\rangle, \bot\}$$

For clarity we represent a state as the tuple of true fluents in that state. The initial situation has both fluents false, that is $q_0 = \langle\rangle$. Termination is defined as any states that have *dialled* true:

$$F_D = \{\langle dialled, in\_booth\rangle\}$$

Finally, the update function maps the following states:

$$\delta(\langle\rangle, walk) = \langle\rangle \qquad \delta(\langle in\_booth\rangle, dial) = \langle dialled, in\_booth\rangle$$
$$\delta(\langle\rangle, enter) = \langle in\_booth\rangle \qquad \delta(\langle dialled, in\_booth\rangle, dial) = \langle dialled, in\_booth\rangle$$

All other transitions are illegal (violate *poss* axioms).

Theorem 1. *If B is a literal-based BAT and D is the DFA obtained by applying the BAT-state construction on B, then* $\mathcal{E}(B) = \mathcal{L}(D)$.

We propose a similar construction in the reverse direction:

Definition 7. *The* fluent construction *takes a DFA D and produces a literal-based BAT* $B = \Sigma_{sitcalc} \cup \mathcal{B}_{ss} \cup \mathcal{B}_{ap} \cup \mathcal{B}_{una} \cup \mathcal{B}_{S_0}$ *where:*

1. $flu : Q \rightarrow \mathcal{F}$ *maps states to unique fluent names.*
2. $\Sigma_{sitcalc} \equiv$ *situation calculus foundational axioms.*

3. $\mathcal{B}_{ss} \equiv \bigcup_{f \in flu} SSA_f$; *the effect axioms are the set of SSAs for each fluent name in* flu.
4. $\mathcal{B}_{ap} \equiv \bigcup_{a \in \Sigma_D} poss(a,s)$; *the action preconditions are the set of poss axioms for each action in the DFA alphabet* Σ_D.
5. $poss(a,s) \equiv \bigvee_q flu_q(s) = T$ *where* $\delta(q,a) \neq \bot$; *action* a *is possible in any state where it won't transition to the sink.*
6. $\gamma_f^+(T,\alpha,s) \equiv \gamma_f^-(F,\alpha,s) \equiv \bigvee_q (\alpha = a \wedge flu_q(s) = T)$ *where* $f = flu_{\delta(q,a)}$; *action a makes fluent f true if the current* f' *state* ($q \Rightarrow f'(s)$) *would transition to the f state* ($\delta(q,a) \Rightarrow f(do(a,s))$).
7. $\gamma_f^-(F,\alpha,s) \equiv \gamma_f^+(T,\alpha,s) \equiv$ *dual of* $\gamma_f^+(F,\alpha,s)$.
8. $\mathcal{B}_{S_0} \equiv$ *initial situation. The initial state's fluent is true:* $flu_{q_0}(S_0) = T$. *All other fluents are false:* $f(S_0) = F$. *Also, the initial situation is final iff the initial state is:* $f_{halt}(S_0) \equiv q_0 \in F_D$.
9. $f_{halt}(s) \equiv \bigvee_{q \in F_D} flu_q(s) = T$; *the halt condition holds iff any final state's fluent does.*

And $\mathcal{B}_{una}$ *ensures that the actions* $a \in \Sigma_D$ *for distinct transition labels and the fluents* $f \in flu \cup \{f_{halt}\}$ *for distinct states are all logically unique.*

Note that the preconditions are uniform, the successor state axioms are uniform and literal-based, and $f_{halt} \notin flu$.

Theorem 2. *If D is a DFA and B is the literal-based BAT obtained by applying the fluent construction on D, then* $\mathcal{L}(D) = \mathcal{E}(B)$.

Corollary 1. *DFAs and LB-BATs are equivalent:*

$$\mathcal{L}(DFA) = \mathcal{E}(LB\text{-}BAT).$$

4 Lattice DFA–Context-Free BAT Equivalence

The 'context free' successor state axioms are a weak special case of the literal-based SSAs. This means that the DFA construction in Definition 6 still applies. The DFA that we get will be severely restricted though—a visual indication of the restrictiveness of context-free BATS (CF-BATs). We identify this class as 'Lattice' DFAs because of their high geometric symmetry, and prove their equivalence to CF-BATs.

Definition 8. *A context-free BAT (CF-BAT) is a special case of LB-BAT with additional restrictions on SSAs—the positive and negative effects* (γ^+, γ^-) *can have no situation dependence:*

$$f(do(a,s)) = y \equiv \gamma_f^+(a,y) \vee (f(s) = y \wedge \neg\exists z \,.\, \gamma_f^-(a,z))$$

Definition 9. *A Lattice DFA is a DFA whose states can be partitioned in one or more ways such that all reachable transitions satisfy the following conditions:*

1. *The transitions in a partition wrt an action (an $(action, partition)$ pair) must be 'fixed' or 'inertial':*
   (a) Fixed — all states transition into the same part.
   (b) Inertial — no state can transition between parts.
2. *If every partition is Inertial wrt action a, then a can only label self-loops.*
3. *If every partition is Fixed wrt action a, then a must always transition to the same state.*
4. *There cannot be more partitions than states.*
5. *There cannot be more parts in a partition than the number of actions.*
6. *No two states can appear in the same part across every partition (indistinguishable).*

All other transitions must enter the inescapable sink state $\bot$. Note that Conditions 2 and 3 motivate multiple partitions for most useful Lattice DFAs.

Each partition is a distinct view of the same set of states and represents how states are distinguishable under action transitions—both as a dimension of symmetry in the DFA, and the domain of a fluent in the corresponding CF-BAT.

The following heuristic is useful for categorising $(action, partition)$ pairs:

1. If action a transitions between two distinct parts of partition p, then p must be Fixed wrt a;
2. Otherwise, if there are transitions starting within two separate parts of partition p, then p must be Inertial wrt a; and,
3. Otherwise, it can be either, subject to Conditions 2+3.

Definition 10. *$part_f$ takes a state and returns the index of its part along the fth partition:*

$$part_f : Q \rightarrow \mathbb{N}$$

Note that the conjunction of part indices across all partitions uniquely defines the state, $q \equiv \bigwedge_f part_f(q)$.

The Lattice DFA is heavily restricted and impractical as a modelling tool, but we show that it can represent any context-free BAT. Note first that the BAT-state construction from Definition 6 is sufficient to convert CF-BATs into DFAs since CF-BATS are a special case of LB-BAT. The partitions of the DFA in this case correspond directly to the fluents. If you construct a DFA from a CF-BAT with n fluents, such that states with the same fluent-value all lie on an $(n-1)$-dimensional hyperplane, then you get an n-dimensional lattice. The distinct parts of each partition (the hyperplanes) represent the domain of the corresponding fluent. This geometrical interpretation represents the restrictions that context-free SSAs place on transitions—we *can not* compile out the fluents, so instead the transitions exhibit high dimensional symmetry, and the fluents remain partially represented in the physical construction of the automata.

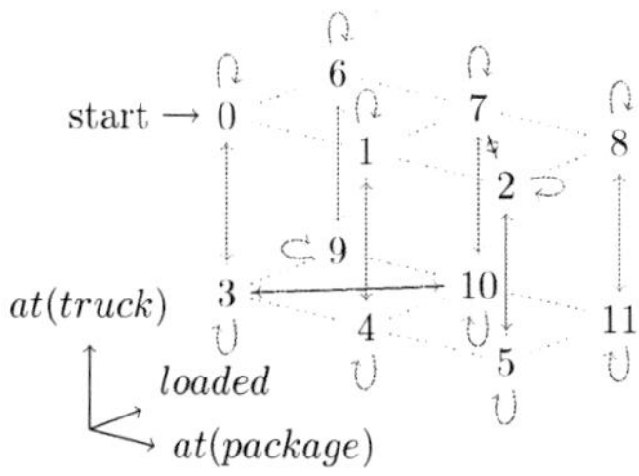

Fig. 2. A 3D Lattice DFA

Table 1. The (*action, partition*) types

		partition		
		loaded	*at(package)*	*at(truck)*
action	$drop_{depot}$	fixed	fixed	either
	pickup	fixed	fixed	inertial
	$drop_{Perth}$	fixed	fixed	either
	$drive_{depot}$	inertial	inertial	fixed
	$drive_{Perth}$	inertial	inertial	fixed

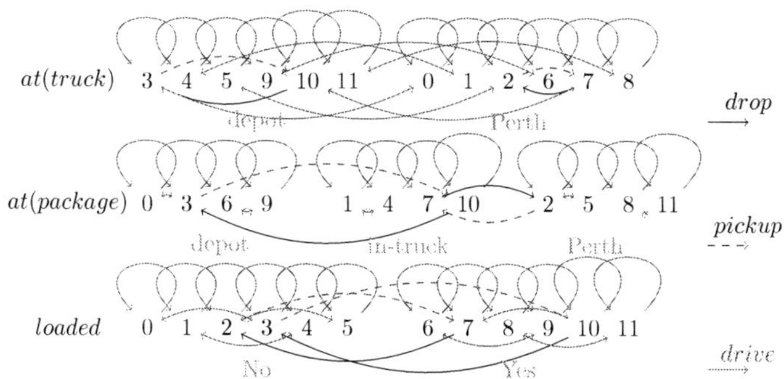

Fig. 3. The partitions of the DFA

We illustrate with a reduced Depot Problem. The part labels in Figure 3 (eg. *Perth* under *at*(*truck*)) indicate that the start state (0) has the truck at Perth and the package unloaded, at the depot. Note that all *drive*-transitions stay in their original part in the *loaded* and *at(package)* partitions and hence are inertial—the *drive* action does not affect these fluents. Conversely, the two *drop* actions may be either inertial or fixed on the *at*(*truck*) partition as the truck's location does not change (inertial), but may equivalently be *set* by the action effects (fixed).

Theorem 3. *If B is a context-free BAT and D is the DFA obtained by applying BAT-state construction on B, then D will be a Lattice DFA and* $\mathcal{L}(D) = \mathcal{E}(B)$.

Definition 11. *The Lattice-BAT construction takes a Lattice DFA D and produces a context-free BAT B such that:*

1. $actions = \Sigma_B$
2. $fluents = \{f_i \,|\, 0 \leq i < |partitions\ of\ D|\}$
3. $poss(a, s) \equiv \bigvee_{q \in Q} \bigwedge_i f_i(s) = part_{f_i}(q) \wedge \delta(q, a) \neq \bot$
4. $f_i(do(a, s)) = y \equiv \gamma^+_{f_i}(a, y) \vee (f_i(s) = y \wedge \neg\exists z \,.\, \gamma^-_{f_i}(a, z))$
   where $\gamma^+_{f_i}(a, y) \equiv \exists q \,.\, \delta(q, a) \neq q \wedge \delta(q, a) \neq \bot \wedge part_{f_i}(\delta(q, a)) = y$
   $\gamma^-_{f_i}(a, y) \equiv$ *dual of* $\gamma^+_{f_i}$ *if it exists*

5. $\mathcal{D}_{S_0} \equiv \{f_i(S_0) = part_{f_i}(q_0) \,|\, 0 \leq i < |partitions\ of\ D|\}$
6. $f_{halt}(s) \equiv \bigvee_{q \in F_D} \bigwedge_{f_i} f_i(S_0) = part_{f_i}(q)$.

Note that the γ^+ construction only exists if the transition has a change of state — a Fixed transition. Inertial transitions are handled by the absence of γ^+ and γ^- axioms for that action. There can be at most one γ^+ and one γ^- for an action because the arguments do not provide greater granularity. This is why each $(action, partition)$ pair must satisfy either the Fixed or Inertial requirements.

Theorem 4. *Every Lattice DFA D can be converted to an equivalent context-free BAT B such that* $\mathcal{L}(D) = \mathcal{E}(B)$.

Corollary 2. *Lattice DFAs and CF-BATs are equivalent:*

$$\mathcal{L}(Lattice\ DFA) = \mathcal{E}(CF\text{-}BAT).$$

5 Conclusion

The situation calculus contains other types of actions beside the 'primitive' actions used above. Sensing actions update fluents directly based on a sensed value, rather than by an effect axiom. Exogenous actions are actions performed externally, but whose effects must be detected so that the internal model can remain consistent with the state of the actual world.

These are technically extra-automata features, however, the situation calculus model is from 'god's eye' or meta view—we assume the logical state mirrors the world it models. In this light exogenous actions are simply regular actions fired by a different hand—a distinction that is irrelevant from an automaton perspective. Similarly, sensing actions can be modeled by non-deterministically selecting a sensing action that returns the correct result. Introducing Lin's indeterminate effects axioms [2] to the situation calculus should facilitate this aspect.

We also intend to investigate the application of these results to planning problems. We believe that analysing special cases like the Lattice DFA will help identify tractable domains.

References

1. Levesque, H.J., Reiter, R., Lespérance, Y., Lin, F., Scherl, R.B.: GOLOG: A logic programming language for dynamic domains. J. Log. Program. 31(1-3), 59–83 (1997)
2. Lin, F.: Embracing causality in specifying the indeterminate effects of actions. In: Proceedings of the Thirteenth National Conference on Artificial Intelligence, AAAI 1996, vol. 1, pp. 670–676. AAAI Press (1996)
3. Lötzsch, M., Bach, J., Burkhard, H.-D., Jüngel, M.: Designing Agent Behavior with the Extensible Agent Behavior Specification Language XABSL. In: Polani, D., Browning, B., Bonarini, A., Yoshida, K. (eds.) RoboCup 2003. LNCS (LNAI), vol. 3020, pp. 114–124. Springer, Heidelberg (2004)
4. Petrick, R.P.A., Levesque, H.J.: Knowledge equivalence in combined action theories. In: KR 2002, pp. 303–314 (2002)
5. Reiter, R.: Knowledge in Action: Logical Foundations for Specifying and Implementing Dynamical Systems. illustrated edn. The MIT Press (September 2001)

Birdsong Acquisition Model by Sexual Selection Focused on Habitat Density

Yohei Fujita, Atsuko Mutoh, and Shohei Kato

Dept. of Computer Science and Engineering, Graduate School of Engineering,
Nagoya Institute of Technology,
Gokiso-cho, Showa-ku, Nagoya 466–8555, Japan
{yfujita,atsuko,shohey}@juno.ics.nitech.ac.jp

Abstract. We describe a simulation model based on an avian ecosystem for determining what causes birdsong evolution. It is already known that songbirds communicate with a "birdsong." This birdsong is used in territorial and courtship behaviors. Some previous researches have suggested that songs related to territorial behaviors should have simple structures while those related to courtship behaviors should have complex ones. We suspect that birdsongs are constantly evolving to achieve a suitable balance between the two behaviors while considering the surrounding environment. We consider avian habitat density to be one of the most important environmental factors influencing birdsong evolution and therefore created different densities in a simulation model. In this paper, we propose a birdsong acquisition model by sexual selection that contains both territorial and courtship behaviors. We conducted simulations with the proposed model and determined that the evolution of birdsongs differs depending on a bird's habitat density.

Keywords: Artificial Life, Sexual Selection, Birdsong, Evolutionary Simulation.

1 Introduction

Songbirds that belong to the passerine order communicate with each other using their voices. Bird vocalizations include both birdsongs and calls; birdsongs are acquired after birth and sung by only the male birds while calls are inherent. Songs are long and complex and are associated with territorial and courtship behaviours, while the shorter calls tend to functions as simple signals or alarms. In this paper, we focus on the songs. It is currently thought that the territorial and courtship behaviours affect sexual selection for the following reasons [4].

- Songbirds' territorial behaviour
  The male birds display territorial behaviour by singing songs. Those that sing short, simple, and stereotyped songs have an advantage in that they can be easily recognized by their neighbors. It is assumed that the effect is one of the results of Dear Enemy Effect that reduces aggressiveness to neighbors for the energy saving [3].

D. Wang and M. Reynolds (Eds.): AI 2011, LNAI 7106, pp. 687–696, 2011.

- Songbirds' courtship behaviour
  The female birds hear the songs of the male birds and use them to search for and select their mates. They tend to prefer longer and more complex songs, which is non-adaptive to survival and caused by the handicap principle [1].

The birds sing different songs according to their habitat — like dialects, even if they are closely related species — and observers expect that the songs evolve to achieve a more suitable balance of the two behaviors depending on the surrounding environment [6]. However, it is difficult to confirm this expectation due to the necessity of long-term experimentation.

Using an artificially designed computational model to simulate real life is one technique for studying systems related to life [2]. The rapid evolution of software agents makes it possible to simulate challenging experiments like evolution observations [8]. There have been previous studies that use the ecological model for birdsongs [5], and birdsongs can be effectively described with a finite-state grammar. Sasahara and Ikegami have suggested a model in which the grammar of the songs sung by male birds and the grammar of the songs preferred by female birds are expressed as automatons, and they also showed that the courtship behaviour requires birdsongs to have complex structure [10]. However, their model did not deal with the territorial behaviour of birds because they were examining the Bengalese finch, which is a domesticated species. Ritchie and Kirby suggested a model in which the hearing function of birds is expressed as filters [9]. They showed that the territorial behaviour causes the hearing function of a bird to prefer a certain type of song. However, their model did not deal with the courtship behaviour of birds.

In this paper, we describe an evolutionary model of birdsongs that draws on both the Sasahara and Ikegami's model and Ritchie and Kirby's model. We express the grammar of the songs as automatons and hearing functions of the birds as filters. The model includes a song pool environment for sharing songs with bird agents in order to design the territorial and courtship behaviour in the avian ecosystem. In addition, we focus on habitat density as one of the key factor of song evolution. We conducted simulations with the proposed model and determined that the evolution of birdsongs differs depending on the habitat density.

2 Conventional Model

In this section, we describe the parts of the other two models we adapted to our own model.

2.1 Sasahara and Ikegami's Model

Recent analysis of birdsongs has shown that the songs consist of a regular order of syllables. Therefore, a song s_x can be defined as

$$s_x = ch_0ch_1..ch_y..ch_{z-1} \ \ (ch_i \in \{\mathbf{a}, \mathbf{b}, \mathbf{c}, \mathbf{d}, \mathbf{e}\}), \tag{1}$$

Table 1. Example of a Filter for Recongnizing Conspecific Songs

	a	b	c	d	e	E
S	0.08	0.15	0.52	0.07	0.08	0.10
a	0.05	0.84	0.00	0.00	0.05	0.06
b	0.05	0.10	0.18	0.45	0.22	0.00
c	0.82	0.09	0.00	0.09	0.00	0.00
d	0.22	0.30	0.00	0.08	0.00	0.40
e	0.12	0.00	0.30	0.30	0.05	0.23

where ch_i is the syllable ("chunk") that is expressed by characters from '**a**' to '**e**,' and z is the number of chunks composing a song s_x.

It is expressed by individual automatons for the grammar of the songs sung by the agents of male birds and the grammar of the songs preferred by the agents of female birds. In other words, the male bird agents sing a song in accordance with the grammar, and the female bird agents prefer the song in accordance with the grammar. The male agents court the female agents by generating their songs with an automaton for song expression, and the female agents select their mates by valuing songs with an automaton for song preference.

As mentioned earlier, Sasahara and Ikegami's model does not deal with territorial behaviour because the bird they modeled, the Bengalese finch, is a domesticated species. Their model also does not deal with song learning which is an important factor for the avian ecosystem. It is therefore necessary to add territorial behaviour and song learning structures to their model to design a general evolution model.

2.2 Ritchie and Kirby's Model

Songbirds that are not born with the ability to sing learn their songs from songs they hear in childhood. They memorize songs they select by a unique hearing function that determines whether the song is conspecific or not. The male birds generate their template of song expression and the female birds generate their template of song preference [7]. Ritchie and Kirby suggested that the hearing function is expressed by filter: conspecific songs are easily memorized through the hearing function, while another species' songs are difficult to memorize because the hearing function blocks them. The filter is a table consisting of the transition probability from chunk to chunk; an example is shown in Table 1. **S** indicates the start of the song and **E** indicates the end. **a**, **b**, **c**, **d**, **e** indicate individual chunks.

A preference $prefer(filter_i, s_x)$, which expresses if a song s_x is conspecific for filter $filter_i$, is defined as

$$prefer(filter_i, s_x) = \frac{\sum_{y=0}^{n} ft_i(t_y)}{n}, \quad (2)$$

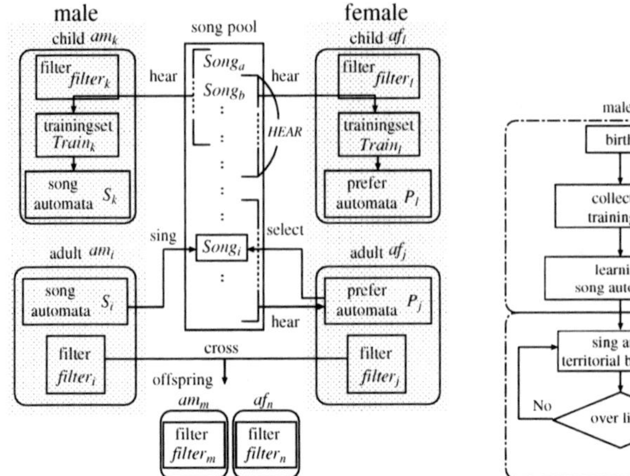

Fig. 1. Configuration diagram of the proposed model

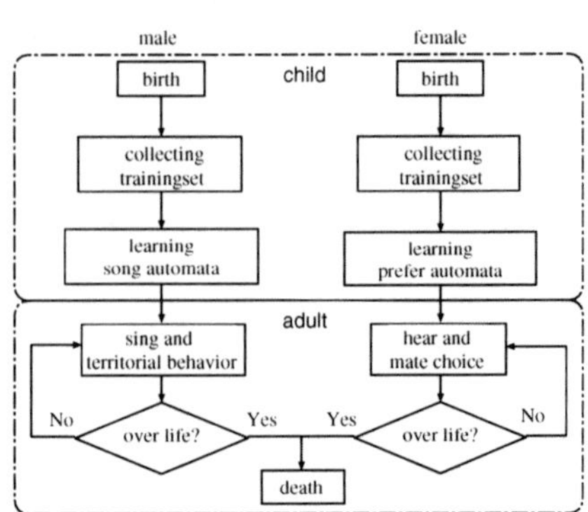

Fig. 2. A flowchart of agent's life cycle

where n is the number of chunk transitions in a song s_x and $ft_i(t_y)$ is the function that refers to the probability t_y of the yth chunk transition of the song s_x in the $filter_i$. A preference $prefer(filter_i, s_x)$ depends on the selection of training songs and the evaluation of territorial behaviour. It is necessary that Ritchie and Kirby's model be adapted to include the function of courtship behaviour for a general evolutionary model.

3 Composition of Computational Model

Our model includes agents as songbirds, song sets as the song repertoire of a songbird, and a song pool as an environment in which to share songs. The configuration diagram of our model — agents, song set, and song pool — is shown in Fig. 1.

A male agent am_k and a female agent af_l in childhood refer to the song set *Song* in the song pool *Song pool*, and each generate either a song expression automaton S_k or song preference automaton P_l from their training song set acquired though their filter. A male agent in adulthood am_i stores its song set $Song_i$, which consists of its songs sung by song expression automaton S_i, into the song pool *Song pool*, while a female agent in adulthood af_j selects its mate by evaluating some song sets *Song* in the song pool *Song pool* by its song preference automaton P_j. Details of each definition are shown below.

3.1 Agent

A male agent am_i and a female agent af_j (i,j is the identifier) are expressed by

$$am_i(filter_i, Train_i, sa_i, risk_i) \tag{3}$$

$$af_j(filter_j, Train_j, pa_j), \tag{4}$$

where $filter_i, filter_j$ is the filter meaning inherent to the hearing function, $Train_i, Train_j$ is the training song set for song learning, sa_i is the song expression automaton for generating songs, pa_j is the song preference automaton for evaluating songs, and $risk_i$ is the expectation value of risk due to territorial behaviour with other agents. The agent's actions — song learning, territorial behaviour, and courtship behaviour — are described in detail later.

3.2 Song Set

A song set $Song_i$ which is the set of songs generated by a male agent am_i, is defined as

$$Song_i = \{s_x | x = 1 \ldots SONG\}, \quad (5)$$

where $SONG$ is the number of songs that a male generates.

3.3 Song Pool

The song pool, which is a multiset of song sets, is updated each simulation step (approximating one year each). All song sets generated by male agents are stored in the song pool and referred to by other agents. The song pool is defined as

$$Song\ pool = \{Song_i | i = 1...AM\}, \quad (6)$$

where AM is the number of male agents.

3.4 Habitat Density

Habitat density is the population density of all birds in a given area. We believe that habitat density depends on the number of birds that are around a particular bird, in other words, the high habitat density makes birds hear many other songs. In our model, we define $HEAR$ as the number of song sets referred to by an agent in one step for the purpose of expressing habitat density.

4 The Life Cycle of an Agent

The life cycle flowchart of each agent is shown in Fig. 2. An agent's life consists of a childhood phase and an adulthood phase. In childhood, each agent learns songs and in adulthood, male agents generate their songs and calculate the risk of territorial behaviour, while female agents evaluate the male agent's song sets and select their mates. When an agent reaches the end of its lifetime, it dies.

4.1 Agent in Childhood

In songbirds ecology, it is known that songbirds make the mold of song after hearing it from some male birds through their inherent hearing function. In our model, an agent collects songs from the song pool through the filter, which is an inherent component, and generates the song expression or song preference automaton from the collected songs. An agent hears $HEAR$ agents' song sets that are randomly selected in the song pool and calculate each of the song's preference through the filter. The training songs we used included five songs that were preferred by the filter.

Next, an agent generates a song expression or song preference automaton from the training songs with the minimum description length (MDL) algorithm This algorithm, which is a compression algorithm for automatons based on the minimum description length principle, is generally used as a model of language acquisition [11]. Any training songs are accepted by the acquired automaton, and an agent thus develops into adulthood after it finishes learning the songs.

4.2 Male Agent in Adulthood

A male agent in adulthood generates a song set that is the set of $SONG$ songs accepted by its song expression automaton. However, it is stochastically difficult to generate songs that have a low preference calculated with the filter. Next, the potential risk of territorial behaviour of each male agent is calculated (NOTE: for real songbirds it is better that they memorize simple and recognizable songs to reduce the risk of territorial behaviour). In our model, the risk is calculated by song complexity and song recognition relations with among other agents. A risk $risk_i$ of male agent am_i is defined as

$$risk_i = \frac{1}{2}\Big(HEAR \cdot complex(Song_i) + \sum_{j=0}^{HEAR} cognit(am_i, am_j)\Big), \tag{7}$$

where $complex(Song_i)$ is the complexity of songs in the song set $Song_i$ and $cognit(am_i, am_j)$ is the song recognition relations between am_i and am_j. It is apparent that the higher the value $HEAR$ expressing habitat density has, the higher value $risk_i$ has. The details of $complex(Song_i)$ are shown as

$$complex(Song_i) = \frac{\sum_{s_x \in Song_i} complex\_s(s_x)}{SONG}, \tag{8}$$

$$complex\_s(s_x) = \frac{1}{2}\left(\frac{cht(s_x)}{L\_MAX} + \frac{ch(s_x)}{CHUNK}\right), \tag{9}$$

where $complex\_s(s_x)$ is the complexity of song s_x in the song set $Song_i$, $cht(s_y)$ and $ch(s_y)$ is the number of chunk transition patterns and the number of chunk types included in song s_y respectively, L_MAX and $CHUNK$ is the maximum number of chunk transitions patterns and the maximum number of chunk types our model allows respectively.

$cognit(am_i, am_j)$ is calculated by

$$cognit(am_i, am_j) = 1 - \frac{1}{2}\Big(ps(filter_i, Song_j) + ps(filter_j, Song_i)\Big), \tag{10}$$

where $ps(filter_i, Song_j)$ is the average of $prefer(filter_i, s_x)$ among s_x in the $Song_j$, defined as

$$ps(filter_i, Song_j) = \frac{\sum_{s_x \in Song_j} prefer(filter_i, s_x)}{SONG}. \tag{11}$$

4.3 Female Agent in Adulthood

A female agent in adulthood evaluates song sets in the song pool by a song preference automaton to select its mate (NOTE: real songbirds, the female birds prefer more complex songs learned in their childhood, and they consider territorial quality [7]). The value of these multiple cues is defined and a female agent hear $HEAR$ potential mates' song sets that are randomly selected in the song pool. The male agent that is given the highest value is selected as the female agent's mate. $value(af_j, am_i)$, which is a female agent af_j's assessment of a male agent am_i, is defined as

$$value(af_j, am_i) = \alpha \cdot sc\_s(pa_j, Song_i) + (1 - \alpha)\ sc\_t(risk_i), \tag{12}$$

where $sc\_s(pa_j, Song_i)$ is the score of the song, $sc\_t(risk_i)$ is the score of the territorial behaviour, and α is the invariable to normalize these scores, resulting in $alpha = 0.9$ in our experiment.

$sc\_s(pa_j, Song_i)$, which consists of the song complexity and acceptability of the song set, is defined as

$$sc\_s(pa_j, Song_i) = complex(Song_i) + \frac{\sum_{s_x \in Song_i} accept(pa_j, s_x)}{SONG}, \tag{13}$$

where $accept(pa_j, s_x)$ is results in an output 1 if the song s_x can be accepted by song preference automaton pa_j. If it cannot, the output is 0.

$sc\_t(risk_i)$, which is the difference between ave_risk (the average of all male agent's risks) and $risk_i$ (a male agent am_i's risk) , is define as

$$sc\_t(risk_i) = ave\_risk - risk_i. \tag{14}$$

A female agent af_j selecting its mate am_i generates the next generation am_n, af_o, whose filters $filter_n, filter_o$ are given to operate both am_i's filter and af_j's filter genetically. Agents of the next generation am_n, af_o are defined as

$$am_n(filter_n, \emptyset, \emptyset, \emptyset), \tag{15}$$

$$af_o(filter_o, \emptyset, \emptyset). \tag{16}$$

The next generations are exposed to the selection pressure by roulette-wheel, in which their fitness depends on the value of their parents' value $value(af_j, am_i)$. The AM male agents and the AF female agents in next generation survive, and the others are remove out.

Table 2. Simulation Condition

Description	Value
Number of songs which bird hears: $HEAR$	5, 20, 40
Number of step in simulation	3000
Number of male agent: AM	100
Number of female agent: AF	100
Number of an agent's song: $SONG$	5
Maximum length of song: L_MAX	10
Kind number of chunk: $CHUNK$	5
Number of step in agent life-time: $LIFE$	2

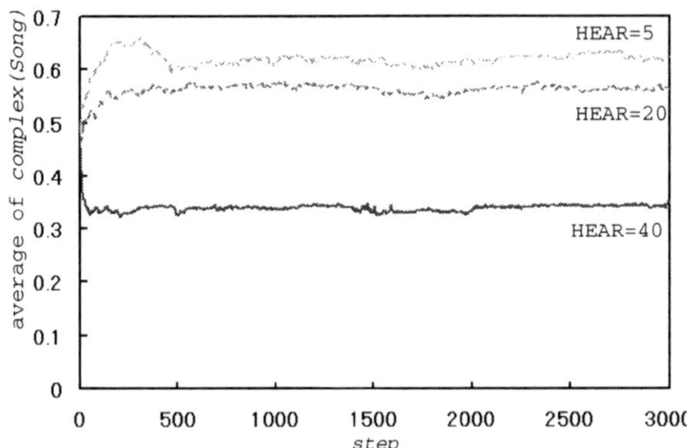

Fig. 3. Average values of $complex(Song)$ in simulations at each $HEAR$

5 Simulation and Discussion

We conducted a simulation using our model under the conditions listed in Table Table 2. Our particular focus was on the influence of avian habitat density on song evolution. As described in section 3.4, avian habitat density is expressed as the number of song sets referred to by an agent in one step $HEAR$. We changed the value of $HEAR$ with 5, 20, and 40 to examine the agent's song complexity $complex(Song)$ acquired by evolution.

5.1 Results

Fig. 3 shows the process of the mean value of $complex(Song)$ in each value of $HEAR$. The value of $complex(Song)$ is lower as the value of $HEAR$ is higher. In other words, simple songs are acquired when the agents are closely spaced, and the complex songs are acquired when the agents are sparse. We conclude that agents change the weight of their behaviour (territory or courtship) depending on the habitat. From equation (7), the force to reduce the value of $risk_i$ arises if the value of $HEAR$ is higher because of the range expansion in the value of $risk_i$.

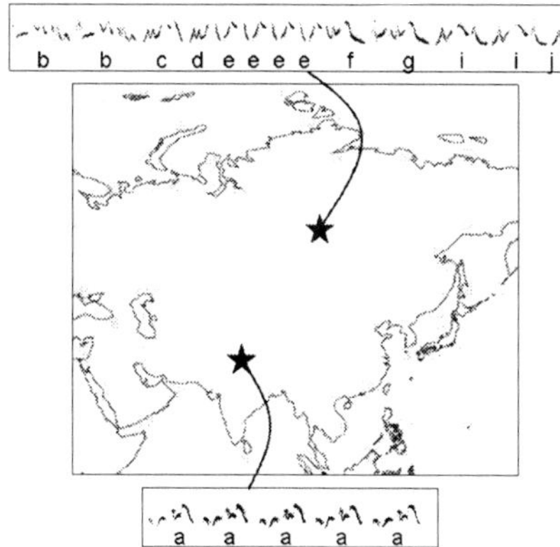

Fig. 4. Example of Greenish Warbler's song

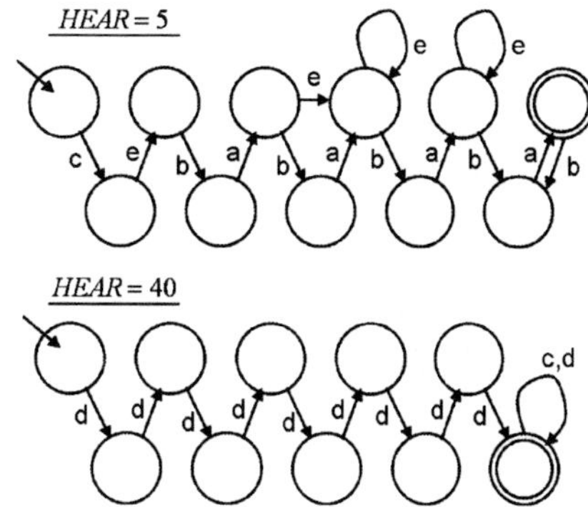

Fig. 5. Example of the agent automaton in the simulation of $HEAR = 5, HEAR = 40$

Table 3. Relation Between Habitat Density and Song

habitat density	priority behavior	song complexity
low	courtship	complex
high	territorial	simple

5.2 Comparison to the Greenish Warbler

The Greenish Warbler (*Phylloscopus trochiloides*) inhabits forests in much of northern and central Asia, and their songs differ depending on the habitat. Fig. 4 shows the example of Greenish Warbler songs in two different habitats [6]. The songs obtained at the mark on the map are expressed by spectrogram and labeled by the characters of each chunk. The Greenish Warbler in north Asia sings a long complex song, while the one in central Asia sings a stereotyped simple song. It has been theorized that the Greenish Warbler in north Asia acquired complex songs because of the priority given to courtship behaviour based on lower habitat density [6]. The relationship between habitat density and song complexity is shown in Table 3.

Fig. 5 shows examples of automaton obtained in the simulation experiments of $HEAR = 5$ and $HEAR = 40$. The automaton in $HEAR = 5$ can output songs consisting of more complex chunk transitions than the automaton in $HEAR = 40$. Compared to the songs of the Greenish Warbler shown in Fig. 4, there are similar relations between song complexity and some difference between song length. We therefore conclude that the song complexity of birds depends on the avian habitat density.

6 Conclusion

In this paper, we examined how a songbird's song evolves due to sexual selection with a focus on habitat density. We conducted an artificial life, multi-agent simulation and determined that the relation between habitat density and song complexity depends on the balance between courtship and territorial behaviours. Bird agents equipped with an inherent hearing function acquired song expression and song preference automatons and then communicated with songs. Influenced the habitat density, which is the number of other birds in the environment, the agents engaged in territorial and courtship behaviour and song learning. When we changed the habitat density, a higher density made the agent's songs simpler and a lower density made them more complex. The simulation was similar to real ecology in term of the relation between habitat density and song complexity. Though it is necessary to mention whether the more data is same with those in real ecology, they are hard to be justified. Our future work will focus on the justification of the proposed model.

References

1. Amots, Z., Avishag, Z.: The Handicap Principle: A Missing Piece of Darwin's Puzzle. Oxford University Press (1997)
2. Arita, T.: A methodology for ensuring constructive approaches based on artificial life model. Journal of the Japanese Society for Artificial Intelligence 24(2), 253–260 (2009) (in Japanese)
3. Catchpole, C.K.: Recognition and territorial deffence. In: Bird Song: Biological Themes and Variations, pp. 139–170. Cambridge University Press (2008)
4. Collins, R.J., Jefferson, D.R.: The evolution of sexual selection and female choice. In: Toward a Practice of Autonomous Systems: Proceedings of the First European Conference on Artificial Life, pp. 327–336. The MIT Press (1992)
5. Hata, M., Mutoh, A., Kato, S., Itoh, H.: An evolutionary model considering acquired traits of bird for generating songs. In: Proc. 69th Annual Convention IPS Japan, pp. 501–502 (2007) (in Japanese)
6. Irwin, D.E.: Song variation in an avian ring species. Evolution 54(3), 998–1010 (2000)
7. Lauay, C., Gerlach, N.M., Adkins-Regan, E., DeVoogd, T.J.: Female zebra finches require early song exposure to prefer high-quality song as adults. Animal Behaviour 68(6), 1249–1255 (2004)
8. Mutoh, A., Kato, S., Itoh, H.: An expression of various patterns of periodic phenomena of fashion by an agent model with desires for conformity and differentiation. Journal of Japan Society for Fuzzy Theory and Intelligent Informatics 21(6), 1035–1043 (2009) (in Japanese)
9. Ritchie, G., Kirby, S.: Selection, domestication, and the emergence of learned communication systems. In: Second International Symposium on the Emergence and Evolution of Linguistic Communication (2005)
10. Sasahara, K., Ikegami, T.: Song grammars as complex sexual displays. General Systems, 1–6 (2004)
11. Teal, T.K., Taylor, C.E.: Effects of compression on language evolution. Artificial Life 6(2), 129–143 (2000)

Speeding Up Bipartite Graph Visualization Method

Takayasu Fushimi[1], Yamato Kubota[2], Kazumi Saito[1,2], Masahiro Kimura[3], Kouzou Ohara[4], and Hiroshi Motoda[5]

[1] Graduate School of Management and Information of Innovation, University of Shizuoka, 52-1 Yada, Suruga-ku, Shizuoka 422-8526, Japan
{j11507,k-saito}@u-shizuoka-ken.ac.jp

[2] School of Management and Information, University of Shizuoka, 52-1 Yada, Suruga-ku, Shizuoka 422-8526, Japan
{b08038,k-saito}@u-shizuoka-ken.ac.jp

[3] Department of Electronics and Informatics, Ryukoku University, Otsu, Shiga 520-2194, Japan
kimura@rins.ryukoku.ac.jp

[4] Department of Integrated Information Technology, Aoyama Gakuin University, Kanagawa 229-8558, Japan
ohara@it.aoyama.ac.jp

[5] Institute of Scientific and Industrial Research, Osaka University, Osaka 567-0047, Japan
motoda@ar.sanken.osaka-u.ac.jp

Abstract. We address the problem of visualizing structure of bipartite graphs such as relations between pairs of objects and their multi-labeled categories. For this task, the existing spherical embedding method, as well as the other standard graph embedding methods, can be used. However, these existing methods either produce poor visualization results or require extremely large computation time to obtain the final results. In order to overcome these shortcomings, we propose a new spherical embedding method based on a power iteration, which additionally performs two operations on the position vectors: double-centering and normalizing operations. Moreover, we theoretically prove that the proposed method always converges. In our experiments using bipartite graphs constructed from the Japanese sites of Yahoo!Movies and Yahoo!Answers, we show that the proposed method works much faster than these existing methods and still the visualization results are comparable to the best available so far.

1 Introduction

Visualization by embedding graphs into a low dimensional Euclidean space plays an important role to intuitively understand the essential structure of graphs (networks). To this end, various graph embedding methods have been proposed in the past that include multi-dimensional scaling [6], spectral embedding [1], spring force embedding [2], cross-entropy embedding [7]. Each method has its own advantages and disadvantages.

In this paper, we address the problem of visualizing structure of bipartite graphs such as relations between pairs of objects and their multi-labeled categories. More specifically, relations of this kind include pairs of movies and their associated genres, pairs

D. Wang and M. Reynolds (Eds.): AI 2011, LNAI 7106, pp. 697–706, 2011.

of persons and their interested genres, pairs of researchers and their coauthoring papers, pairs of words and their appearing documents, and many more. Clearly, we can straightforwardly apply any one of the above-mentioned embedding methods for the visualization. However, we note that these standard methods have an intrinsic limitation because they cannot make much use of the essential structure of bipartite graphs. Indeed, the existing spherical embedding method has been proposed for the purpose of visualizing bipartite graphs [5]. In this method, the position vectors are embedded on two concentric spheres (circles) with different radii. We consider that such a spherical embedding can be a natural representation for bipartite graphs. However, the biggest problem with the existing method is that it often requires an extremely large computation time to obtain the final visualization results.

In this paper, to overcome these shortcomings, we propose a new spherical embedding method based on a power iteration, which adopts two operations to iteratively adjust the positioning vectors: double-centering and normalizing operations. We further show theoretically that the convergence of the proposed algorithm is always guaranteed. In our experiments that use bipartite graphs constructed from the Japanese sites of Yahoo!Movies and Yahoo!Answers, we show that the proposed method works much faster than these existing methods, and yet the visualization results are comparable to the best available so far.

2 Problem Framework

We describe the problem framework of embedding the bipartite graph $G = (V, E)$ into a K-dimensional Euclidean space, where $V = V_A \cup V_B$, $V_A \cap V_B = \emptyset$, and $E \subset V_A \times V_B$. For the sake of technical convenience, we identify each set of the nodes, V_A and V_B, by two different series of positive integers, i.e., $V_A = \{1, \cdots, m, \cdots, M\}$ and $V_B = \{1, \cdots, n, \cdots, N\}$. Here M and N are the numbers of the nodes in V_A and V_B , i.e., $|V_A| = M$ and $|V_B| = N$, respectively. Then, we can define the $M \times N$ adjacency matrix $\mathbf{A} = \{a_{m,n}\}$ by setting $a_{m,n} = 1$ if $(m, n) \in E$; $a_{m,n} = 0$ otherwise. We denote the K-dimensional embedding position vectors by $\mathbf{x}_m$ for the node $m \in V_A$ and $\mathbf{y}_n$ for the node $n \in V_B$. Then we can construct $M \times K$ and $N \times K$ matrices consisting of these position vectors, i.e., $\mathbf{X} = (\mathbf{x}_1, \cdots \mathbf{x}_M)^T$ and $\mathbf{Y} = (\mathbf{y}_1, \cdots \mathbf{y}_N)^T$. Here $\mathbf{X}^T$ stands for the transposition of $\mathbf{X}$.

According to the work on the existing spherical embedding method [5], we explain the framework of spherical embedding of bipartite graph. In Fig. 1, we show an example in a two-dimensional Euclidean space, i.e., unlike the standard visualization scheme shown in Fig. 1a, we consider locating the position vectors on two concentric spheres (circles) as shown in Fig. 1b. We believe that this kind of spherical embedding is natural to represent bipartite graphs, and its usefulness has been reported [5]. Hereafter, we assume that nodes in subset V_A are located on the inner circle θ_A with radius $r_A = 1$, while nodes in V_B are located on the outer circle θ_B with radius $r_B = 2$. Note that $||\mathbf{x}_m|| = 1$, $||\mathbf{y}_n|| = 2$. Then, our aim is to locate the position vectors of the nodes having similar connection patterns closely to each other.

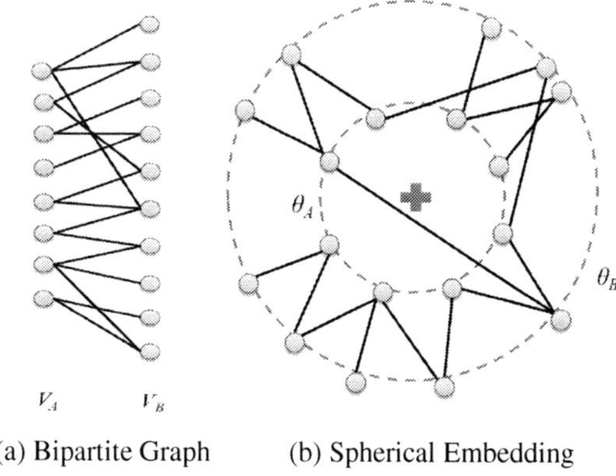

(a) Bipartite Graph (b) Spherical Embedding

Fig. 1. Spherical Embedding for Bipartite Graph

3 Proposed Method

3.1 Proposed Algorithm

The new spherical embedding method is based on a power iteration. It has two operations on the positioning vectors which we call double-centering operation and normalizing operation. In order to describe our algorithm, we need to introduce the centering matrices and normalizing operations. The centering (Young-Householder transformation) matrices are defined as $\mathbf{H}_M = \mathbf{I}_M - \frac{1}{M}\mathbf{1}_M\mathbf{1}_M^T$, $\mathbf{H}_N = \mathbf{I}_N - \frac{1}{N}\mathbf{1}_N\mathbf{1}_N^T$ where $\mathbf{I}_M$ and $\mathbf{I}_N$ stands for $M \times M$ and $N \times N$ identity matrices, respectively, and $\mathbf{1}_M$ and $\mathbf{1}_N$ are M- and N-dimensional vectors whose elements are all one. Clearly, the mean vector of the resulting position vectors becomes $\mathbf{0}$ by the operations $\mathbf{H}_M\mathbf{X}$ and $\mathbf{H}_N\mathbf{Y}$. On the other hand, the normalizing operations are defined as $\mathbf{\Lambda}_M(\mathbf{X}) = r_A\text{diag}(\mathbf{X}\mathbf{X}^T)^{-1/2}\mathbf{X}$, $\mathbf{\Lambda}_N(\mathbf{Y}) = r_B\text{diag}(\mathbf{Y}\mathbf{Y}^T)^{-1/2}\mathbf{Y}$, where diag$(\cdot)$ is an operation to set all the non-diagonal elements to zero, i.e., diag$(\mathbf{X}\mathbf{X}^T)$ is a diagonal matrix whose m-th element is $\mathbf{x}_m^T\mathbf{x}_m$.

Intuitively, the basic procedure of our proposed algorithm is that the position vector $\mathbf{x}_m$ is repeatedly moved to the position calculated by adding the position vectors $\{\mathbf{y}_n\}$ that are connected to $\mathbf{x}_m$. Of course, we need to perform a normalizing operation so as to satisfy the spherical constraints. Below we describe our proposed algorithm.

1. Initialize the matrix $\mathbf{X}$ and $\mathbf{Y}$.
2. Update the matrix $\mathbf{X} \leftarrow \mathbf{\Lambda}_M(\mathbf{H}_M\mathbf{A}\mathbf{H}_N\mathbf{Y})$.
3. Update the matrix $\mathbf{Y} \leftarrow \mathbf{\Lambda}_N(\mathbf{H}_N\mathbf{A}^T\mathbf{H}_M\mathbf{X})$.
4. Terminate if the changes for the position vectors $\mathbf{X}$ and $\mathbf{Y}$ are small.
5. Return to the step 2.

As the basic framework, our proposed algorithm employs a power iteration, just like the HITS algorithm [3], which utilizes $\mathbf{A}$ and $\mathbf{A}^T$, does. However, the main differences are use of the double-centering operations by $\mathbf{H}_M$ and $\mathbf{H}_N$ and the normalizing operations by $\mathbf{\Lambda}_M(\cdot)$ and $\mathbf{\Lambda}_N(\cdot)$. Here note that the double-centering operation is also employed in the standard multidimensional scaling method [6].

Now we briefly mention the computational complexity of our algorithm. Clearly, the main computational complexity of one-iteration comes from the multiplication by the matrix $\mathbf{A}$ (or $\mathbf{A}^T$) which is the most intensive part and is proportional to the number of links in the bipartite graph. Thus, the proposed algorithm is expected to work much faster especially for a sparse bipartite graph, compared with the existing spherical embedding algorithm that require a nonlinear optimization just like a spring force embedding [2] does. In fact, it has been well known that the PageRank algorithm based on a power iteration works very fast for a large and sparse network [4].

3.2 Convergence Proof

We prove the convergence property of the algorithm. To do this, we first introduce the double-centered matrix $\mathbf{B} = \{b_{m,n}\}$ that is calculated from the adjacency matrix $\mathbf{A}$ i.e., $\mathbf{B} = \mathbf{H}_M\mathbf{A}\mathbf{H}_N$. Then, by using the matrix $\mathbf{B}$, we can consider the following objective function with respect to the position vectors $\mathbf{X} = (\mathbf{x}_1, \cdots, \mathbf{x}_M)^T$ and $\mathbf{Y} = (\mathbf{y}_1, \cdots, \mathbf{y}_N)^T$.

$$J(\mathbf{X}, \mathbf{Y}) = \sum_{m=1}^{M}\sum_{n=1}^{N} b_{m,n}\frac{\mathbf{x}_m^T}{r_A}\frac{\mathbf{y}_n}{r_B} + \frac{1}{2}\sum_{m=1}^{M}\lambda_m(r_A^2 - \mathbf{x}_m^T\mathbf{x}_m) + \frac{1}{2}\sum_{n=1}^{N}\mu_n(r_B^2 - \mathbf{y}_n^T\mathbf{y}_n), \quad (1)$$

where $\{\lambda_m \mid m = 1, \cdots, M\}$ and $\{\mu_n \mid n = 1, \cdots, N\}$ correspond to Lagrange multipliers for the spherical constraints, i.e., $\mathbf{x}_m^T\mathbf{x}_m = r_A^2$ and $\mathbf{y}_n^T\mathbf{y}_n = r_B^2$ for $1 \leq m \leq M$ and $1 \leq n \leq N$.

Now we consider maximizing $J(\mathbf{X}, \mathbf{Y})$ defined in Equation (1) by use of a coordinate strategy. Note that maximizing $J(\mathbf{X}, \mathbf{Y})$ pushes the pairs $\mathbf{x}_m$ and $\mathbf{y}_n$ to the same direction if they are connected and pushes them to the opposite direction if they are unconnected, and realizes the intended visualization. We repeat the following two steps: maximizing $J(\mathbf{X}, \mathbf{Y})$ with respect to $\mathbf{X}$ by fixing the matrix $\mathbf{Y}$ first, and maximizing $J(\mathbf{X}, \mathbf{Y})$ with respect to $\mathbf{Y}$ by fixing the matrix $\mathbf{X}$ next. If the maximization of these steps are achieved by the above algorithm's step 2 and 3, respectively, we can guarantee the convergence of our proposed algorithm.

In order to confirm these facts, we consider the following gradient vector of the objective function $J(\mathbf{X}, \mathbf{Y})$ with respect to $\mathbf{x}_m$.

$$\frac{\partial J(\mathbf{X}, \mathbf{Y})}{\partial \mathbf{x}_m} = \frac{1}{r_A r_B}\sum_{n=1}^{N} b_{m,n}\mathbf{y}_n - \lambda_m\mathbf{x}_m. \quad (2)$$

Thus, for a fixed matrix $\mathbf{Y}$, we obtain the optimal position vector $\mathbf{x}_m$ which maximizes the objective function $J(\mathbf{X}, \mathbf{Y})$ as $\mathbf{x}_m = \frac{r_A}{||\tilde{\mathbf{x}}_m||}\tilde{\mathbf{x}}_m$, where $\tilde{\mathbf{x}}_m = \sum_{n=1}^{N} b_{m,n}\mathbf{y}_n$. Here note that the optimal vector $\mathbf{x}_m$ is calculated by using the matrix $\mathbf{Y}$ only. Thus, for $m = 1, \cdots, M$, by using the normalizing operation $\mathbf{\Lambda}_M(\cdot)$ whose diagonal elements become $r_A/||\tilde{\mathbf{x}}_1||, \cdots, r_A/||\tilde{\mathbf{x}}_M||$, we can obtain the solution in the matrix representation, i.e.,

$$\mathbf{X} = \mathbf{\Lambda}_M(\mathbf{B}\mathbf{Y}) = \mathbf{\Lambda}_M(\mathbf{H}_M\mathbf{A}\mathbf{H}_N\mathbf{Y}). \quad (3)$$

1 Science Fiction/Fantasy	● red circle	9 Documentary	▶ olive triangle–right
2 Action/Adventure	■ black square	10 Drama	× lime cross
3 Animation	◆ green diamond	11 Family	+ darkgold plus
4 Comedy	★ blue star	12 Horror	✳ darkcyan asterisk
5 Suspense	✱ maroon hexagon	13 Musical	● magenta circle
6 Teen	▲ orange triangle–up	14 Romance	■ cyan square
7 Western	▼ purple triangle–down	15 Special Effects	◆ yellow diamond
8 War	◀ navy triangle–left	16 Others	★ gray star

Fig. 2. category names in Japanese Yahoo!Movies site

Recall that Equation (3) performs centering the matrix $\mathbf{Y}$ by the matrix $\mathbf{H}_N$, multiplies the adjacency matrix $\mathbf{A}$, performs re-centering the matrix by multiplying the matrix $\mathbf{H}_M$, and normalizes so as to guarantee spherical constraints. By this formula, we can obtain the optimal solution of position vectors $\mathbf{X}$ by fixing the matrix $\mathbf{Y}$.

Similarly, we can also obtain the following optimal solution of position vector $\mathbf{y}_n$ by fixing the matrix $\mathbf{X}$ as $\mathbf{y}_n = \frac{r_B}{\|\tilde{\mathbf{y}}_n\|}\tilde{\mathbf{y}}_n$, where $\tilde{\mathbf{y}}_n = \sum_{m=1}^{M} b_{m,n}\mathbf{x}_m$. Thus, for $n = 1, \cdots, N$, by using the normalizing operation $\mathbf{\Lambda}_N(\cdot)$ whose diagonal elements become $r_B/\|\tilde{\mathbf{y}}_1\|, \cdots, r_B/\|\tilde{\mathbf{y}}_N\|$, we can obtain the solution in the matrix representation, i.e.,

$$\mathbf{Y} = \mathbf{\Lambda}_N(\mathbf{B}^T\mathbf{X}) = \mathbf{\Lambda}_N(\mathbf{H}_N\mathbf{A}^T\mathbf{H}_M\mathbf{X}). \tag{4}$$

Therefore, since the finite objective function $J(\mathbf{X}, \mathbf{Y})$ defined in Equation (1) has the analytical optimal solution under the condition that either $\mathbf{X}$ or $\mathbf{Y}$ is fixed, and is always maximized by performing the step 2 and 3 of the algorithm, we can guarantee that the algorithm always converges.

4 Evaluation by Experiments

4.1 Network Data

We constructed the bipartite graphs from the Japanese sites of Yahoo!Movies and Yahoo!Answers, and experimentally evaluated the proposed method by comparing it with the existing embedding methods in terms of both the efficiency of the algorithms and ease of interpretability of the visualization results.

We regard the movies as nodes in V_B, and their genres as nodes in V_A for the Japanese Yahoo!Movies site [1]. Note that each movie is associated with more than or equal to one genre. In Fig. 2, we show their genre names used in our experiments, and for our visual analyses purpose, we assign an individual marker with a different color to each genre as shown in this figure. In order to evaluate our proposed method by using a set of different bipartite graphs, we classify these movies into 7 groups according to their release dates(1950-59, 1960-69, 1970-79, 1980-89, 1990-99, 2000-04 and 2005-09). Here the number of genres is $|V_A| = 16$ for all the periods, the numbers of movies

[1] `http://movies.yahoo.co.jp/`

$|V_B|$ are 594, 1079, 1314, 1805, 2659, 2948 and 3264, and the numbers of links $|E|$ are 899, 1617, 2071, 2994, 4424, 6057 and 6564 for each period.

We regard the users who answered questions as nodes in V_B, and the genres of these questions as nodes in V_A for the Japanese Yahoo!Answers site [2]. Note that although each question belongs to only one genre, the same user frequently answers several questions belonging to a wide variety of genres. Thus we can obtain bipartite graphs between the pairs of the users and the genres they answered. In our experiments, we utilized a set of data from April, 2004, to October, 2005. Again, in order to evaluate our proposed method by using a set of different bipartite graphs, we classify these questions into 6 groups according to their submission dates(2004-2nd, 3rd, 4th, 2005-1st, 2nd and 3rd). Here the number of genres is $|V_A| = 10$ for all the periods, the numbers of users $|V_B|$ are 11871, 27446, 35907, 39451, 42884 and 46834, and the numbers of links $|E|$ are 30849, 80664, 96926, 95714, 102086 and 112548 for each period.

4.2 Brief Description of Other Visualization Methods Used for Comparison

We first explain the existing spherical embedding method as our primal comparison method, whose problem framework is the same to ours. In this method the following objective function is directly minimized with respect to the position vectors $\mathbf{X} = (\mathbf{x}_1, \cdots, \mathbf{x}_M)^T$ and $\mathbf{Y} = (\mathbf{y}_1, \cdots, \mathbf{y}_N)^T$ under the constraints that $\mathbf{x}_m^T \mathbf{x}_m = r_A^2$ and $\mathbf{y}_n^T \mathbf{y}_n = r_B^2$ for $1 \leq m \leq M$ and $1 \leq n \leq N$. The objective function is defined as $\mathcal{J}(\mathbf{X}, \mathbf{Y}) = \frac{1}{2} \sum_{m=1}^{M} \sum_{n=1}^{N} \left(c_{m,n} r_A r_B - \mathbf{x}_m^T \mathbf{y}_n\right)^2$, where $c_{m,n} = 2a_{m,n} - 1$, i.e., $c_{m,n} = 1$ if $(m, n) \in E$; $c_{m,n} = -1$ otherwise. In order to obtain the solution vectors, this method repeatedly moves each position vector by using the Newton method in a framework of nonlinear optimization, i.e., it repeats the following two steps: First, minimizing $\mathcal{J}(\mathbf{X}, \mathbf{Y})$ for $\mathbf{x}_m$ by fixing $\{\mathbf{x}_1, \cdots \mathbf{x}_M\} \setminus \mathbf{x}_m$ and $\{\mathbf{y}_1, \cdots \mathbf{y}_N\}$, and next minimizing $\mathcal{J}(\mathbf{X}, \mathbf{Y})$ for $\mathbf{y}_n$ by fixing $\{\mathbf{x}_1, \cdots \mathbf{x}_M\}$ and $\{\mathbf{y}_1, \cdots \mathbf{y}_N\} \setminus \mathbf{y}_n$. Thus this method requires an extremely large computation time to obtain the final results.

We have further compared the proposed method with the four well known embedding methods: multi-dimensional scaling [6], spectral embedding [1], spring force embedding [2], and cross-entropy embedding [7]. Here the former two perform a power iteration with respect to either a double-centered distance matrix or a graph Laplacian matrix which is calculated from a given graph, just like our proposed spherical embedding method does, while the latter two repeatedly move each position vector by using the Newton method in a framework of nonlinear optimization, just like the existing spherical embedding method does. Note that these four methods are not designed for embedding bipartite graphs, but as mentioned earlier, we can straightforwardly apply them for our purpose because a bipartite graph is regarded as an instance of general undirected graph.

In what follows in this subsection, we regard a bipartite graph as an undirected graph $G = (V, E)$ to describe the basic ideas of these standard embedding methods, and then consider a framework of embedding it into a K-dimensional Euclidean space. In this framework, we identify the set of the nodes by a positive integer, i.e., $V = \{1, \cdots, l, \cdots, L\}$, $|V| = L$ and $L = M + N$. Then, we can define the $L \times L$ adjacency

[2] http://chiebukuro.yahoo.co.jp/

matrix $\mathbf{A} = \{a_{m,n}\}$ by setting $a_{m,n} = 1$ if $(m, n) \in E$; $a_{m,n} = 0$ otherwise. We denote the K-dimensional embedding position vectors by $\mathbf{x}_m$ for the node $m \in V$, and then construct an $L \times K$ matrix consisting of these position vectors, i.e., $\mathbf{X} = (\mathbf{x}_1, \cdots \mathbf{x}_L)^T$. We also denote the graph distance matrix by $\mathbf{G} = \{g_{m,n}\}$, each element of which is the minimum path length between node m and node n.

Multi-dimensional scaling method [6] first calculates the distance matrix $\mathbf{G}$, and performs the double centering operation ($\mathbf{H}_L = \mathbf{I}_L - \frac{1}{L}\mathbf{1}_L\mathbf{1}_L^T$) to the distance matrix. Mathematically it is formulated as minimizing $\mathcal{M}(\mathbf{X}) = \frac{1}{2}\sum_{k=1}^{K} \mathbf{z}_k^T(\mathbf{H}_L\mathbf{G}\mathbf{H}_L)\mathbf{z}_k$, where $\mathbf{z}_k = (x_{1,k}, \cdots, x_{L,k})^T$, and $\{\mathbf{z}_1, \cdots, \mathbf{z}_K\}$ need to be orthonormal vectors, i.e., $\mathbf{z}_k^T\mathbf{z}_k = 1$ and $\mathbf{z}_k^T\mathbf{z}_{k'} = 0$ if $k \neq k'$. Spectral embedding method [1] tries to directly minimize distances between position vectors of connecting nodes. Mathematically it is formulated as minimizing $\mathcal{S}(\mathbf{X}) = \sum_{k=1}^{K} \mathbf{z}_k^T(\mathbf{D} - \mathbf{A})\mathbf{z}_k$, where $\mathbf{D}$ is a diagonal matrix each element of which is the degree of node (number of links). Note that $(\mathbf{D} - \mathbf{A})$ is referred to as a graph Laplacian matrix. Again, we set $\mathbf{z}_k = (x_{1,k}, \cdots, x_{L,k})^T$, and $\{\mathbf{z}_1, \cdots, \mathbf{z}_K\}$ need to be orthonormal vectors, which excludes the trivial vector expressed as $\mathbf{z} \propto \mathbf{1}_L$. Spring force embedding method [2] assumes that there is a hypothetical spring between each connected node pair and locates nodes such that the distance of each node pair is closest to its minimum path length at equilibrium. Mathematically it is formulated as minimizing $\mathcal{K}(\mathbf{X}) = \sum_{m=1}^{L-1}\sum_{n=m+1}^{L} \alpha_{m,n}(g_{m,n} - \|\mathbf{x}_m - \mathbf{x}_n\|)^2$, where $\alpha_{m,n}$ is a spring constant which is normally set to $1/(2g_{u,v}^2)$. Cross-entropy embedding method [7] first defines a similarity $\rho(\mathbf{x}_m, \mathbf{x}_n)$ between the embedding positions $\mathbf{x}_m$ and $\mathbf{x}_n$ and uses the corresponding element $a_{m,n}$ of the adjacency matrix as a measure of distance between the node pair, and tries to minimize the total cross entropy between these two. Mathematically it is formulated as minimizing $C(\mathbf{X}) = -\sum_{m=1}^{L-1}\sum_{n=m+1}^{L} (a_{m,n}\log\rho(\mathbf{x}_m, \mathbf{x}_n) + (1 - a_{m,n})\log(1 - \rho(\mathbf{x}_m, \mathbf{x}_n)))$. Here, note that we used the function $\rho(\mathbf{x}_m, \mathbf{x}_n) = \exp(-\frac{1}{2}\|\mathbf{x}_m - \mathbf{x}_n\|^2)$ in our experiments.

4.3 Experimental Results

We first evaluated the efficiency of our proposed method in comparison with the existing methods. We show our experimental results in Fig. 3, where Spec, MDS, SF, CE, eSE and pSE stand for the spectral embedding, multi-dimensional scaling, spring force embedding, cross-entropy embedding, existing spherical embedding and proposed spherical embedding methods, respectively (machine used is Intel(R) Xeon(R) CPU X5472 @3.0GHz with 64GB memory). Here Figs. 3a and 3b correspond to the results by using the bipartite graphs constructed from the Yahoo!Movies and Yahoo!Answers sites, respectively. In these figures, we plotted the average processing time (sec.) over 10 trials by changing the initial position vectors, where the horizontal and vertical axes stand for the number of nodes in V_B and the processing times, respectively. Here recall that the number of nodes in V_B is different for each bipartite graph as mentioned above.

As expected, these figures show that our proposed spherical embedding (pSE) method works much faster than all the existing methods we compared. More specifically, the spectral embedding (Spec) method works comparable to our method. This is because these methods perform a power iteration on a sparse adjacency matrix. In fact, the multi-dimensional scaling (MDS) method requires a substantially large computation time because it needs to perform a power iteration on a full distance matrix. All the other methods including the existing spherical embedding (eSE) method, which repeatedly

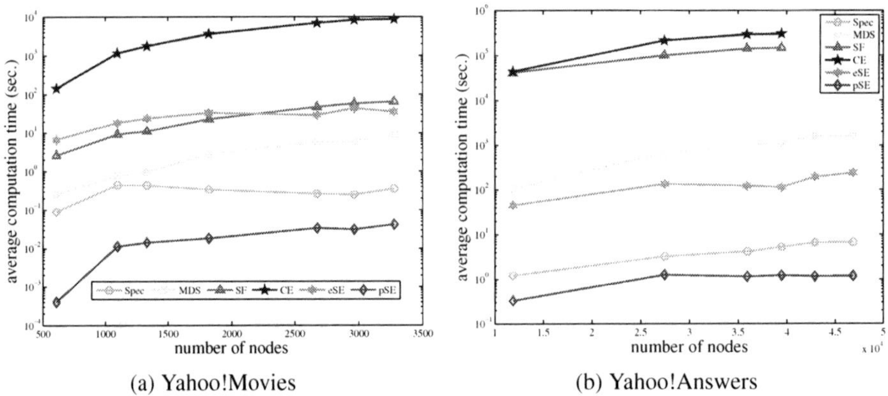

(a) Yahoo!Movies

(b) Yahoo!Answers

Fig. 3. Comparison of processing times

move each position vector by using the Newton method, generally require an extremely large computation time before the final results are obtained. Especially, both the spring force embedding (SF) and cross-entropy embedding (SE) methods require more than three days to obtain the final results even for one trial when the numbers of nodes for the Yahoo!Answers graphs become more than 40,000; thus we omitted these results in Fig. 3b. Here we should emphasize that the scale of the vertical axis of these figures is logarithmic.

Next we evaluated the visualization results of our proposed method in comparison with the existing methods. Due to a space limitation, we only show our experimental results obtained for a bipartite graph constructed from the Japanese Yahoo!Movies sites in Fig. 4. Here recall that the genre information has been shown in Fig. 2. In Fig. 4a, we show the visualization result by our proposed method, which we consider intuitively natural. Actually, we can see that the genre nodes of Action/Adventure (black square) and Suspense (maroon hexagon) are located in near positions at the right-side of the inner circle (θ_A), while at the opposite left-side of this circle, the genre nodes of Teen (orange triangle_up) and Romance (cyan square) are located in near positions. Overall, we can observe that the similar genres are located closely on the inner circle (θ_A).

Now we compare the above results with the five existing methods. The first one is the visualization result by the existing spherical embedding method shown in Fig. 4b. We see that there are several minor differences but we consider this result comparable to the result by our method. However, this one is very slow and inefficient. Our method is much faster. The second one is the visualization result by the multidimensional scaling method shown in Fig. 4c. We can observe some clusters of genres. Although this result might indicate some intrinsic property, we feel that the spherical embedding scheme is a more natural representation of bipartite graphs. The third one is the visualization result by the spectral embedding method shown in Fig. 4d. This one is relatively poor in our own experiments. In fact, the two genres of Drama (lime cross) at the bottom-right and Documentary (Olive triangle_right) at the top-left are too much isolated, although this method works reasonably fast among the existing methods.The fourth and the fifth

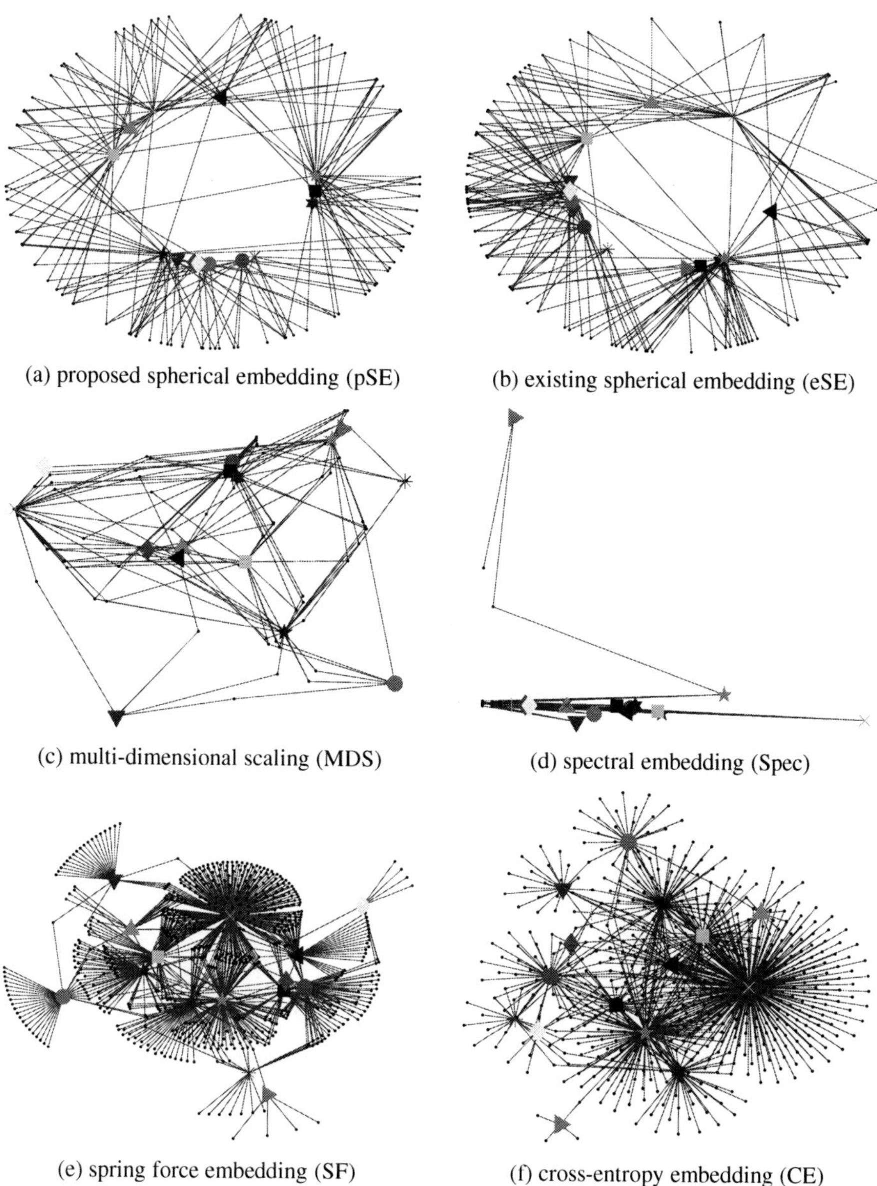

(a) proposed spherical embedding (pSE)

(b) existing spherical embedding (eSE)

(c) multi-dimensional scaling (MDS)

(d) spectral embedding (Spec)

(e) spring force embedding (SF)

(f) cross-entropy embedding (CE)

Fig. 4. experimental results obtained for a bipartite graph constructed from the Japanese Yahoo!Movies sites(1950 - 1959)

ones are the visualization results by the spring force embedding method and the cross-entropy embedding method shown in Figs. 4e and 4f. We can observe a similar tendency between these two, *e.g.*, we can easily see that the genre node of Drama (lime cross) is much isolated in both. The main difference in these methods is that we can observe that some genre nodes are clustered for the spring force embedding method, but there are no such clusters and all the genres are scattered for the cross-entropy embedding method. Overall, although each embedding method might have its own characteristics that are both advantageous and disadvantageous, we believe that our proposed spherical embedding method is most effective for visualizing bipartite graphs in terms of efficiency and interpretability.

Last but not least, we evaluated our proposed method only in the case of two-dimensional embedding for our visualization purpose, but this does not mean that it is limited to two-dimensional embedding. It is quite easy to extend it to the general K-dimension embedding. We plan to evaluate our method as a powerful technique for both dimensional reduction and clustering as a future work.

5 Conclusion

In this paper, we addressed the problem of visualizing structure of bipartite graphs such as relations between pairs of objects and their multi-labeled categories, and proposed a new spherical embedding method that is based on a power iteration. The key features of this method is that it employs two operations on the positioning vectors, one called double-centering operation and the other called normalizing operation. This enables the iterative approach to be equivalent to maximizing an objective function which is guaranteed to converge. Thus, our algorithm is theoretically guaranteed to converge. We applied our method to a set of bipartite graphs with different sizes and connections, and compared the results with five existing visualization methods. The results showed that the proposed method works much faster than all the five existing methods, and the visualization results are intuitively understandable and comparable to the best available so far known. In future, we plan to apply the new method to evaluate its performance and robustness for a wide variety of bipartite graphs.

References

1. Chung, F.R.K.: Spectral Graph Theory. CBMS Regional Conference Series in Mathematics, vol. (92). American Mathematical Society (February 1997)
2. Kamada, T., Kawai, S.: An algorithm for drawing general undirected graphs. Inf. Process. Lett. 31, 7–15 (1989)
3. Kleinberg, J.M.: Authoritative sources in a hyperlinked environment. J. ACM 46, 604–632 (1999)
4. Langville, A.N., Meyer, C.D.: Deeper inside pagerank. Internet Mathematics 1 (2004)
5. Naud, A.P., Usui, S., Ueda, N., Taniguchi, T.: Visualization of documents and concepts in neuroinformatics with the 3d-se viewer. Frontiers in Neuroinformatics 1, Article 7 (2007)
6. Torgerson, W.S.: Theory and methods of scaling. John Wiley & Sons Inc. (1958)
7. Yamada, T., Saito, K., Ueda, N.: Cross-entropy directed embedding of network data. In: Proceedings of the 20th International Conference on Machine Learning (ICML 2003), pp. 832–839 (2003)

Enhancement of Learning Experience Using Skill-Challenge Balancing Approach

Norliza Katuk[1], Ruili Wang[1], and Hokyoung Ryu[2]

[1] School of Engineering and Advanced Technology, Massey University, New Zealand
{n.katuk,r.wang}@massey.ac.nz
[2] Dept. of Industrial Engineering, College of Engineering, Hanyang University, Korea
hryu@hanyang.ac.kr

Abstract. This paper addresses the issue of content sequencing in computer-based learning (CBL). In doing so, it proposes a Skill-Challenge Balancing (SCB) approach as a way to enhance the CBL experience. The approach is based on the Flow Theory, allowing self-adjustment of the given levels of challenges in a given learning tasks so that the learner will consistently be adaptively able to engage in the CBL activity. An empirical study with 70 students suggested that the SCB-based learners were significantly better in their learning experience specifically in their focus of attention and intrinsic interests compared to the learners in the system without SCB. The results also revealed that SCB was fully utilised by the learners to regulate the levels of difficulty of the CBL tasks.

Keywords: Flow theory, learning experience, skill-challenge balancing, computer-based learning.

1 Introduction

Content sequencing is a common topic of research in the area of computer-based learning (CBL). The basic idea of content sequencing is to help learners to find an appropriate learning path which meets certain factors such as their prior knowledge, learning style and preferences [1]. The sequencing technique is mainly achieved using some computational methods and artificial intelligence (AI) techniques such as genetic algorithm [1], particle swam optimisation [2], rule-based [3], and neural network [4]. There is no doubt that the sequencing techniques are robust in organising learning contents; however, little is known about the effectiveness of the complex techniques in the real CBL setting. To be precise, the answer to the question '*do the techniques improve cognitive engagement in performing CBL tasks?*' is still elusive.

To address this issue, we performed an empirical study to understand the effectiveness of the content sequencing approach with regard to the learning experience [5, 6]. We assumed that the content sequencing approach would be able to partially optimise CBL experience, via balancing between the learner's skills or knowledge against the challenges given by the system.

D. Wang and M. Reynolds (Eds.): AI 2011, LNAI 7106, pp. 707–716, 2011.

2 Learning Experience and Flow Theory

In the context of CBL environment, learning experience is an important factor that reflects the acceptance, adoption and future use of the systems [7]. CBL systems must be able to improve learners' performance and give them a personally satisfying experience so that the systems could sustain. A number of studies had already investigated the CBL experience, e.g., [8, 9]. Our approach in this study is to some extent different from the previous studies. This article uses the results of our previous studies [5, 6] in order to develop a pragmatic method to improve the CBL experience.

We adapted Csikszentmihalyi's *Flow Theory* [10, 11] as the basis to define CBL learning experience. It is also an underlying principle for developing the Skill-Challenge Balancing technique as described in Section 3. Basically, the theory suggests the flow condition; a mental state in which a person is totally absorbed in a particular activity. The flow condition gives a person a very rewarding experience and a feeling of enjoyment which is called '*optimal experience*'. Optimal experience is believed to be an important factor to improve human quality of life and achieve happiness. In the context CBL, optimal experience gives learners with enjoyable learning experience that subsequently fosters independent learning.

In spite of flow, both boredom and anxiety are two opposite mental states that could change the quality of learning experience. These three mental states are identified through assessment of one's current levels of skills against the given levels of challenges of an activity. Figure 1 shows four points of the mental states (A_1, A_2, A_3, and A_4) that one may experience when engaging in a learning activity. The *flow* state is achieved when there is a balance between one's skills and the given challenges. The states are represented by points A_1 and A_4 in Figure 1. When a person's levels of skills are not sufficient to satisfy the given levels of challenges, he or she is in the state of *anxiety* (i.e. A_3). If a person has a high level of skills, a low level of challenges given to him or her can cause *boredom* (i.e. A_4). In order to obtain *flow*, a balance between the given levels of challenges and one's skills is required.

Flow Theory emphasises that an equal skills and challenges is the key principle to achieve the optimal experience. For this reason, we exploited the theory to develop the *Skill-Challenge Balancing (SCB)* approach, which is a new method to improve the CBL experience.

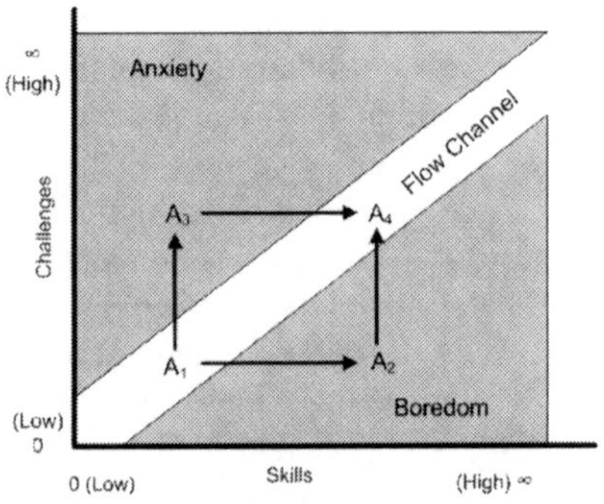

Fig. 1. Changes of mental states based on Flow Theory

3 Skill-Challenge Balancing Approach

The aim of the *Skill-Challenge Balancing* (SCB) approach is to improve interactions between learners and CBL systems so that learners obtain satisfying and engaging CBL experience. The SCB is developed based on one of the flow theory's assumptions that optimal experience could be achieved when the level of the given challenges matches the individual's level of skills. This paper attempts to answer *"how to incorporate the theory in the design of CBL systems"*

There are two approaches to serve this purpose: software-based [12-14] and hardware-based [15-17]. The hardware-based approach uses special devices or sensors for automatic detection of a person's affective states. Although the devices can accurately detect the affective states, they are very expensive and not yet available commercially. In contrast, the software-based approach seems to be more pragmatic as it is much easier, cheaper and feasible to be implemented using the existing computer infrastructure, thereby the underlying principle for our approach.

The main SCB concept is to allow a flexible adjustment of the given level of challenges. In CBL, the levels of challenges are characterised by the increasing level of difficulty of a learning content. In order to keep learners engaged, the given levels of challenges must always equivalent to learners' current level of knowledge. In doing so, the SCB technique allows learners to have self-assessment of their individual level of knowledge. Learners are given chances to evaluate whether the learning unit is too easy or too difficult for them. If a learner finds that the learning unit is too easy, he or she can choose to move forward to a higher level of difficulty of the learning unit. On the other hand, if the learner finds that the learning unit is too difficult, he or she can move backward to the lower level of difficulty of the learning unit.

Our approach introduces "*flow buttons*" in the CBL *user interface* to support the self-assessment capability. The buttons comprise of two types; the "*anxiety button*" comes along with the tutorial questions and the "*boredom button*" appears with the explanation of the concept. The tutorial questions are the tool to measure learner's current knowledge. The decision to move forward to a higher level is depending on the learner's answers in the tutorial session. The correct answers will direct the learner to a higher level of learning. In the case that the wrong answer is given, the learner will be presented with the explanation associated to the question.

The "*boredom button*" accompanies the tutorial questions with the purpose to avoid novice learners from lost in their learning path. As the difficulty level of learning is increasing along with the tutorial questions, the system forces the learners to answer the tutorial in order to move to a higher level so that their current levels of knowledge are accurate. The "*anxiety button*" appears along with the tutorial questions to allow learners to move backward to a lower level of learning. Hence, they will be able to browse the explanation for the question. Figure 2 shows the learning process with the present of the "*flow button*".

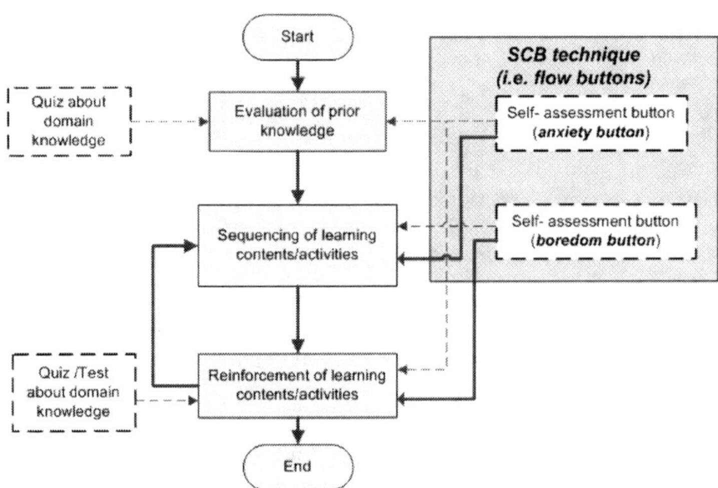

Fig. 2. SCB technique learning process

A prototype has been developed to demonstrate how the SCB technique would work with a realistic learning situation. In doing so, we reused most of the software components of the current version of IT-Tutor system [5, 6] including the user interface layout, the databases, and the functions. The prototype was developed within the .NET platform and set to be accessible through the Internet.

The implementation of "*flow buttons*" has been simplified to avoid confusion among learners. In doing so, more understandable words were used and printed on the buttons. In the case of the "*anxiety button*", the authors use the text "*Click here if you do not know the answer*". For the "*boredom button*", the text "*Click here if you think the section is too easy*" is used. The buttons in the red dotted line in Figure 3 and Figure 4 show the screen shot examples containing the "*anxiety button*" and the "*boredom button*", respectively. The interaction of these buttons with the domain knowledge repository is accomplished by a set of pre-programmed rules using the following algorithm:

```
Present the <tutorial questions>
If <the anxiety button> is pressed then
     Present the associated learning contents
     If <the boredom button > is pressed then
       Test <learner's current knowledge>
       If <learner's current knowledge> is <insufficient> then
            Give feedback to learner
               Present the sequence of learning contents
            Test <learner's current knowledge>
            If <learner's current knowledge> is <sufficient> then
                    Give feedback to learners
                Proceed to the next level of <tutorial questions>
                Test <learner's current knowledge>
                ..................................
                ..................................
```

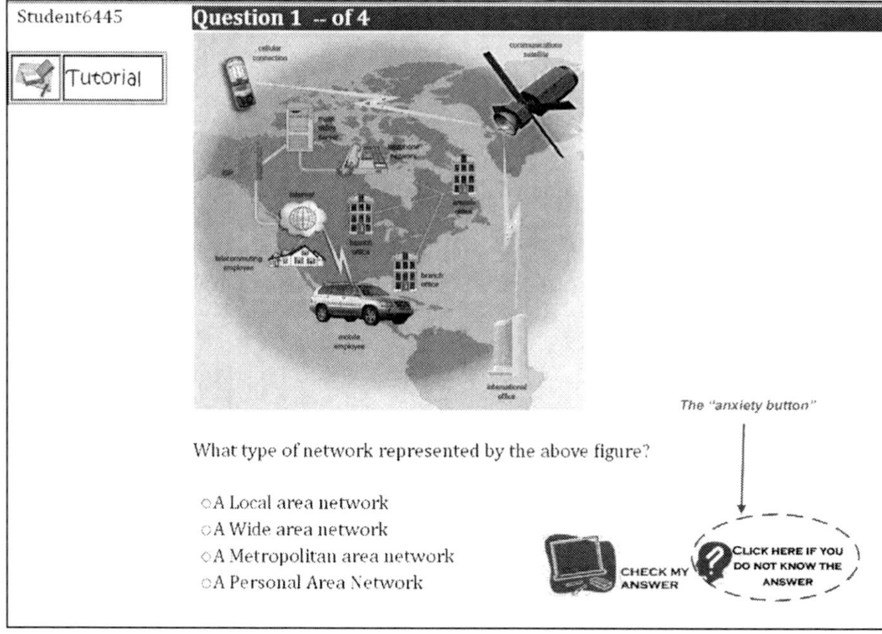

Fig. 3. The "*anxiety button*" in the IT-Tutor interface

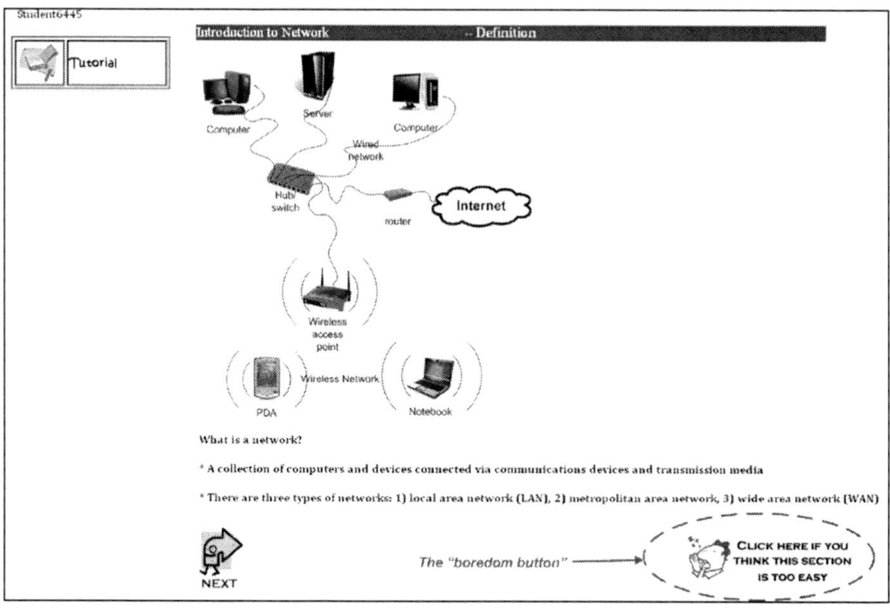

Fig. 4. The "*boredom button*" in the IT-Tutor interface

4 Evaluation of Skill-Challenge Balancing Approach

4.1 Method

Subjects

The subjects were recruited among students from two universities: Massey University (New Zealand) and Northern University of Malaysia through advertisements in the learning management systems of the corresponding universities for some selected courses. Ninety-two students participated on a voluntary basis. However, only seventy of them completed the given tasks. Among them were eighteen males and fifty-two females. 85% of the participants were students of Northern University of Malaysia with 80% of them were undergraduate students. The participants were randomly assigned into one of two groups (i.e. the experimental group and the control group). This experimental study was conducted between March and April 2011.

Apparatus

Two types of materials were used in this study: CBL systems and a set of questionnaire. The CBL systems were comprised of two types: IT-Tutor with SCB and IT-Tutor without SCB (i.e. the older version of the system as reported in [5, 6]). IT-Tutor with SCB was used by the participants of the experimental group, while the control group used IT-Tutor without SCB.

The tutorial session in both types of CBL systems comprised of four questions. As the SCB technique used a couple of "*flow buttons*" that allowed the learners to flexibly move between questions and explanations, the stages of the tutorial in this version of the CBL system was not transparent to the learners. On the other hand, the two stages of tutorial were clearly shown in IT-Tutor without SCB. From the two stages of the tutorial session, Stage 1 was used to evaluate learner's prior knowledge to generate a learning path for the learners, while Stage 2 of the tutorial served as a reinforcement stage.

A learning experience questionnaire was adopted from Park *et al.*[18]. It comprised of four components: demographic information (10 items), learning experience (12 items), and usability (2 items). For the usability questionnaire, it was adopted from Chiu *et al.* [19]. The learners were asked to rate their learning experience and usability questionnaire using 5-point Likert Scale (i.e. 1 represents strongly disagree and 5 represents strongly agree).

Experimental Design

A one-way between-subjects design was used in this study. The independent variable was the two *types of CBL systems* (i.e. IT-Tutor with SCB and IT-Tutor without SCB). The dependent variables were comprised of the *learning experience and usability*. For the case of IT-Tutor with SCB, we analysed the *SCB usage* in order to understand whether or not the "*flow buttons*" were effectively used by the learners.

Procedure

This study was conducted in an unsupervised online mode. All materials were pre-programmed in a form of a web application. The participants were given a URL (an Internet address) to access the materials. Firstly, they were given the research information sheet. As they consented to participate in the research, the system had randomly assigned them into one of two groups of the CBL systems. Then, they were redirected to the corresponding CBL systems. The learners were required to undergo a virtual tutorial session in the corresponding CBL systems and follow the given instructions as they were interacting with the system. As soon as the participants completed the tutorial session, they were given the questionnaire. All participants performed the tasks at their own paces and their own convenience. In order to retain the reliability of the study, the participant will be logged off from the system when they were inactive[1] for five minutes.

4.2 Results and Discussions

The demographic information analysis showed that the average age of the participants was 25.20 years with approximately 85% of them were aged 17 to 30. About 75% of them had more than 3 years experience in using the computer and at least 60% of them had used other CBL systems before. Apart from that, about 64% of the participants classified themselves as beginners, while the rest had learned about the course before. None of the participants classified themselves as experts in the area of the subject of this study (i.e. Computer Networks).

Learning Experience & Usability

The learning experience information was derived from the questionnaire. It was measured in four dimensions: control, attention focus, curiosity, and intrinsic interests. On the other hand, usability measured how useful the corresponding CBL systems in improving the learners' performance and the systems suitability with the learners' learning styles.

A series of *Kolmogorov-Smirnov tests* suggested that the data were not normally distributed. Hence, simpler non-parametric tests were used to analyse the data. The learning experience and usability data were relatively high in their internal consistency, and Cronbach's Alpha coefficient (0.828) confirmed this. The means and mean ranks for each dimension of the learning experience including usability were calculated and presented in Table 1.

Table 1 shows that the experimental group learners (i.e. IT-Tutor with SCB) rated higher in all dimensions of the learning experience and usability compared to that of the counterpart group. For IT-Tutor with SCB, intrinsic interests received the highest ratings (*3.90*), followed by usability (*3.87*), and curiosity (*3.68*). In contrast, attention focused (*3.25*) had received the lowest ratings among learners in this group. For the

[1] Inactive is the situation in which no interaction has occurred (e.g. no clicking buttons, no moving mouse, etc.)

Table 1. Means and mean ranks for the learning experience dimensions and usability

Dimensions of experience	IT-Tutor with SCB (n=35)		IT-Tutor without SCB (n=35)		Significant level
	Mean	Mean rank	Mean	Mean rank	
Control (CO)	3.42	39.07	3.13	31.93	(z=-1.498, p=0.136) n.s.
Attention Focus (AF)	3.25	40.36	2.86	30.64	(z=-2.041, p=0.041) p<0.05
Curiosity (CU)	3.68	37.66	3.52	33.34	(z=-0.902, p=0.371) n.s.
Intrinsic Interests (II)	3.90	40.34	3.58	30.66	(z=-2.020, p=0.043) p<0.05
Average experience	3.56	41.70	3.27	29.30	(z=-2.557, p=0.010) p<0.05
Usability	3.87	39.34	3.60	31.66	(z=-1.613, p=0.108) n.s.

other group (i.e. IT-Tutor without SCB), usability (*3.60*) had received the highest ratings, followed by intrinsic interests (*3.58*). The ratings for attention focus in the control group were also the lowest in the counterpart group.

In order to understand whether or not the SCB technique was effective in improving the learning experience, a series of *Mann-Whitney U* tests had been performed. The test results suggested that attention focus, intrinsic interests, and the overall learning experience for the IT-Tutor with SCB were significantly higher than the opposite group. Although the IT-Tutor with SCB ratings were higher for control, curiosity and usability compared to the counterpart, the differences were not statistically significant. Hence, it can be said that, the SCB technique improved learners' overall learning experience specifically from the context of their attention focus and intrinsic interests.

"*Flow buttons*" usage

The log data analysis showed that 77% (26 out of 35) students from the experimental group used the "*anxiety button*" with 34 hits in Stage 2 and 9 hits in Stage 1. For the case of the "*boredom button*", 34% of the students used this facility with majority accesses came from Stage 2 (i.e. 17 hits). The bar chart in Fig. 5 shows the hits of the buttons in the two stages of the tutorial. From the graph, it clearly shows that the "*anxiety button*" has been used extensively by the learners in comparison to the "*boredom button*". This could be justified by the demographic backgrounds of the participants which comprised of novice and intermediate learners.

The results suggest that the "*anxiety button*" allowed the learners to adjust the difficulty levels of the tutorial by moving backward to the lower one which consequently giving them a better learning experience. On the other hand, the "*boredom button*" helped learners to move to a higher level of learning to prevent them from becoming bored due to the familiar learning content. The analysis on the usage data had suggested that both buttons (i.e. the boredom button and the anxiety button) were needed by learners in order for them to adjust their own learning path flexibly.

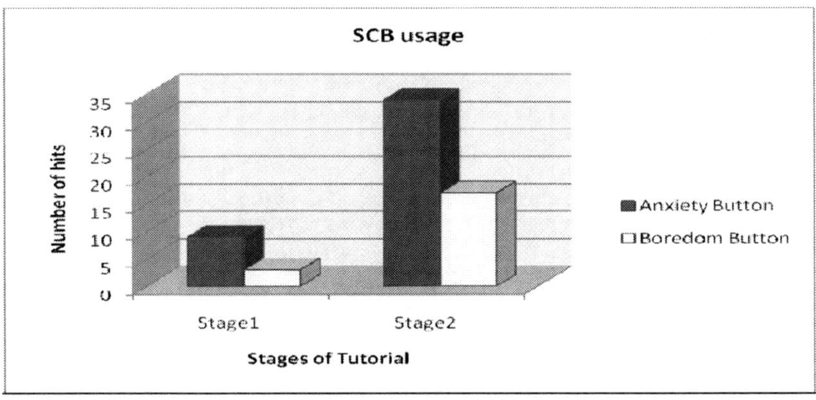

Fig. 5. The "*flow buttons*" usage according to the two stages of tutorial

5 Conclusion and Future Works

We have described in Section 3 and 4 of this paper about the SCB design and evaluation. In general, the SCB approach for sequencing learning content seemed to improve the overall learning experience in comparison to the older version of the content sequencing system. Given that no expert learners were recruited, the effect of the SCB is still not fully discovered. It is our plan in the near future to replicate the research by recruiting expert learners so that the effectiveness of SCB in managing learners with different backgrounds is known. The self-adjustment of levels of challenges seems to be an ideal approach to learners regardless of their prior knowledge in a particular domain. Through this way, it helps learners to engage in the learning tasks constantly which consequently giving them a pleasant learning experience.

References

1. Chen, C.M.: Intelligent Web-Based Learning System with Personalized Learning Path Guidance. Computers & Education 51(2), 787–814 (2008)
2. de Marcos, L., Martinez, J.J., Gutierrez, J.A., Barchino, R., Gutierrez, J.M.: A New Sequencing Method in Web-Based Education. In: IEEE Congress on Evolutionary Computation (CEC 2009), pp. 3219–3225 (2009)
3. Chi, Y.L.: Ontology-Based Curriculum Content Sequencing System with Semantic Rules. Expert Systems with Applications 36(4), 7838–7847 (2009)
4. Idris, N., Yusof, N., Saad, P.: Adaptive Course Sequencing for Personalization of Learning Path Using Neural Network. International Journal of Advanced Soft Computing Applications 1(1), 49–61 (2009)
5. Katuk, N., Ryu, H.: Finding an Optimal Learning Path in Dynamic Curriculum Sequencing: An Account of the Flow Learning Experience. In: 2010 International Conference on Computer Applications & Industrial Electronics (ICCAIE 2010), pp. 227–232. IEEE Computer Society, New York (2010)

6. Katuk, N., Ryu, H.: Does a Longer Usage Mean Flow Experience? An Evaluation of Learning Experience with Curriculum Sequencing Systems (CSS). In: Gupta, G.S., Bailey, D., Demidenko, S., Osseiran, A., Renovell, M. (eds.) Sixth IEEE International Symposium on Electronic Design, Test and Application, pp. 13–18. IEEE Computer Society, New York (2011)
7. Sun, P.C., Tsai, R.J., Finger, G., Chen, Y.Y., Yeh, D.: What Drives a Successful E-Learning? An Empirical Investigation of the Critical Factors Influencing Learner Satisfaction. Computers & Education 50(4), 1183–1202 (2008)
8. Deepwell, F., Malik, S.: On Campus, but out of Class: An Investigation into Students' Experiences of Learning Technologies in Their Self-Directed Study. Research in Learning Technology 16(1), 5–14 (2008)
9. Engelbrecht, E.: Adapting to Changing Expectations: Post-Graduate Students' Experience of an E-Learning Tax Program. Computers & Education 45(2), 217–229 (2005)
10. Csikszentmihalyi, M.: Beyond Boredom and Anxiety. Jossey-Bass Publishers, San Francisco (1975)
11. Csikszentmihalyi, M.: Flow: The Psychology of Optimal Experience. Harper & Row Publishers, New York (1990)
12. Leontidis, M., Halatsis, C., Grigoriadou, M.: Mentoring Affectively the Student to Enhance His Learning. In: Aedo, I., Chen, N.S., Kinshuk, Sampson, D., Zaitseva, L. (eds.) 9th IEEE International Conference on Advanced Learning Technologies (ICALT 2009), pp. 455–459. IEEE Computer Society, New York (2009)
13. Sabine, A.M.: A Learner, Is a Learner, Is a User, Is a Customer: QOS-Based Experience-Aware Adaptation. In: 16th ACM International Conference on Multimedia, pp. 1035–1038. ACM, Vancouver (2008)
14. Ryoo, W., Jung, H., Yoo, M., Hwang, S.: Development of Motivational E-Learning Environment Based on Flow Theory. In: The World Conference on E-Learning in Corporate, Government, Healthcare, and Higher Education 2008, pp. 1208–1211 (2008)
15. Woolf, B., Arroyo, I., Cooper, D., Burleson, W., Muldner, K.: Affective Tutors: Automatic Detection of and Response to Student Emotion. In: Nkambou, R., Bourdeau, J., Mizoguchi, R. (eds.) Advances in Intelligent Tutoring Systems. SCI, vol. 308, pp. 207–227. Springer, Heidelberg (2010)
16. Muldner, K., Burleson, W., VanLehn, K.: "Yes!": Using Tutor and Sensor Data to Predict Moments of Delight During Instructional Activities. In: De Bra, P., Kobsa, A., Chin, D., Muldner, K., Burleson, W., VanLehn, K. (eds.) UMAP 2010. LNCS, vol. 6075, pp. 159–170. Springer, Heidelberg (2010)
17. Kaklauskas, A., Krutinis, M., Seniut, M.: Biometric Mouse Intelligent System for Student's Emotional and Examination Process Analysis. In: Aedo, I., Chen, N.S., Kinshuk, Sampson, D., Zaitseva, L. (eds.) 9th IEEE International Conference on Advanced Learning Technologies (ICALT 2009), pp. 189–193. IEEE Computer Society, New York (2009)
18. Park, J., Parsons, D., Ryu, H.: To Flow and Not to Freeze: Applying Flow Experience to Mobile Learning. IEEE Transactions on Learning Technologies 3(1), 56–67 (2010)
19. Chiu, C.M., Hsu, M.H., Sun, S.Y., Lin, T.C., Sun, P.C.: Usability, Quality, Value and E-Learning Continuance Decisions. Computers & Education 45(4), 399–416 (2005)

A Divide-and-Conquer Tabu Search Approach for Online Test Paper Generation

Minh Luan Nguyen[1], Siu Cheung Hui[1], and Alvis C.M. Fong[2]

[1] Nanyang Technological University, Singapore
{NGUY0093,asschui}@ntu.edu.sg
[2] Auckland University of Technology, New Zealand
acmfong@gmail.com

Abstract. Online Test Paper Generation (Online-TPG) is a promising approach for Web-based testing and intelligent tutoring. It generates a test paper automatically online according to user specification based on multiple assessment criteria, and the generated test paper can then be attempted over the Web by user for self-assessment. Online-TPG is challenging as it is a multi-objective optimization problem on constraint satisfaction that is NP-hard, and it is also required to satisfy the online runtime requirement. The current techniques such as dynamic programming, tabu search, swarm intelligence and biologically inspired algorithms are ineffective for Online-TPG as these techniques generally require long runtime for generating good quality test papers. In this paper, we propose an efficient approach, called DAC-TS, which is based on the principle of constraint-based divide-and-conquer (DAC) and tabu search (TS) for constraint decomposition and multi-objective optimization for Online-TPG. Our empirical performance results have shown that the proposed DAC-TS approach has outperformed other techniques in terms of runtime and paper quality.

Keywords: Online test paper generation, multi-objective optimization, web-based testing, intelligent tutoring system.

1 Introduction

With the rapid growth of E-learning, Web-based testing and intelligent tutoring [2, 5, 13] have become popular for self-assessment and learning in an educational environment. To support Web-based testing and intelligent tutoring, Online Test Paper Generation (Online-TPG) is a promising approach which generates a test paper automatically online according to user specification based on multiple assessment criteria, and the generated test paper can then be attempted over the Web by user. More specifically, Online-TPG aims to find an optimal subset of questions from a question database to form a test paper based on criteria such as total time, topic distribution, difficulty degree, discrimination degree, etc.

Online-TPG is a challenging problem. Firstly, TPG is categorized as a multi-objective optimization problem on constraint satisfaction which is NP-hard [10]. Secondly, the current TPG techniques [6–10, 12, 16] have not taken the online generation requirement into consideration as TPG is traditionally considered as an offline process similar to other multi-objective optimization problems such as timetabling and job-shop scheduling [4].

D. Wang and M. Reynolds (Eds.): AI 2011, LNAI 7106, pp. 717–726, 2011.

These current techniques have optimized an objective function based on multi-criteria constraints and weighting parameters for test paper quality. However, determining appropriate weighting parameters is quite difficult and computationally expensive. And these techniques generally require long runtime for generating good quality test papers.

In this paper, we propose an efficient approach, called DAC-TS, which is based on the principle of constraint-based divide-and-conquer (DAC) and tabu search (TS) for Online-TPG. The rest of the paper is organized as follows. Section 2 reviews the related techniques for automatic test paper generation. Section 3 gives the problem specification. The proposed DAC-TS approach for Online-TPG is presented in Section 4. Section 5 gives the performance evaluation of the proposed approach and its comparison with other TPG techniques. Finally, Section 6 gives the conclusion.

2 Related Work

In [10], tabu search (TS) was proposed to construct test papers by defining an objective function based on multi-criteria constraints and weighting parameters for test paper quality. TS optimized test paper quality by the evaluation of the objective function. In [8], dynamic programming optimized an objective function incrementally based on the recursive optimal relation of the objective function. In [9], a genetic algorithm (GA) was proposed to generate quality test papers by optimizing a fitness ranking function based on the principle of population evolution. In [16], differential evolution (DE) was proposed for test paper generation. DE is similar to the spirit of GA with some modifications on solution representation, fitness ranking function, and the crossover and mutation operations to improve the performance. In [12], an artificial immune system (AIS) was proposed to use the clonal selection principle to deal with the highly similar antibodies for elitist selection in order to maintain the best test papers for different generations.

In addition, swarm intelligence algorithms such as particle swarm optimization and ant colony optimization have also been investigated for TPG. In [6], particle swarm optimization (PSO) was proposed to generate multiple test papers by optimizing a fitness function which is defined based on multi-criteria constraints. In [7], ant colony optimization (ACO) was proposed to generate quality test papers by optimizing an objective function which is based on the simulation of the foraging behavior of real ants.

3 Problem Specification

Let $\mathcal{Q} = \{q_1, q_2, .., q_n\}$ be a dataset consisting of n questions, $\mathcal{C} = \{c_1, c_2, .., c_m\}$ be a set of m topics, and $\mathcal{Y} = \{y_1, y_2, .., y_k\}$ be a set of k question types such as multiple choice questions, fill-in-the-blanks and long questions. Each question $q_i \in \mathcal{Q}$, where $i \in \{1, 2, .., n\}$, has 8 attributes $\mathcal{A} = \{q, o, a, e, t, d, c, y\}$, where q is the question identity, o is the question content, a is the question answer, e is the discrimination degree, t is the question time, d is the difficulty degree, c is the related topic and y is the question type. Table 1 shows a sample Math question dataset.

A *test paper specification* $\mathcal{S} = \langle N, T, D, C, Y \rangle$ is a tuple of 5 attributes which are defined based on the attributes of the selected questions as follows: N is the number of questions, T is the total time, D is the average difficulty degree, $C = \{(c_1, pc_1), ..., (c_M,$

Table 1. An Example of Math Dataset

(a) Question Table

$\mathcal{Q}$_ID	o	a	e	t	d	c	y
q_1	...	...	4	8	1	c_1	y_2
q_2	...	...	7	9	2	c_1	y_2
q_3	...	...	4	6	6	c_2	y_1
q_4	...	...	5	9	9	c_2	y_2
q_5	...	...	4	7	4	c_1	y_1
q_6	...	...	7	4	7	c_2	y_1

(b) Topic Table

$\mathcal{C}$_ID	$name$
c_1	Integration
c_2	Differentiation

(c) Question Type Table

$\mathcal{Y}$_ID	$name$
y_1	Multiple choice
y_2	Fill-in-the-blank

$pc_M)\}$ is the specified proportion for topic distribution and $Y = \{(y_1, py_1),..,((y_K, py_K)\}$ is the specified proportion for question type distribution.

The test paper generation process aims to find a subset of questions from a question dataset $\mathcal{Q} = \{q_1, q_2, .., q_n\}$ to form a test paper P with specification $\mathcal{S}_P$ that maximizes the average discrimination degree and satisfies the test paper specification such that $\mathcal{S}_P = \mathcal{S}$. It is important to note that the test paper generation process occurs over the Web where user expects to generate a test paper within an acceptable response time. Therefore, Online-TPG is as hard as other optimization problems due to its computational NP-hardness, and it is also required to be solved efficiently in runtime.

4 Proposed Approach

In this paper, we propose a constraint-based Divide-And-Conquer Tabu Search (DAC-TS) approach for Online-TPG. As the constraints specified in the test paper specification can be formulated as a standard 0-1 fractional Integer Linear Programming (ILP) problem [10] in the form of linear equality constraints, we can decompose the constraints into two independent subsets, namely *content constraints* and *assessment constraints*, which can then be solved separately and progressively. In the test paper specification $\mathcal{S} = \langle N, T, D, C, Y\rangle$, the content constraints include the constraints on topic distribution C and question type distribution Y, whereas the assessment constraints include the constraints on total time T and average difficulty degree D.

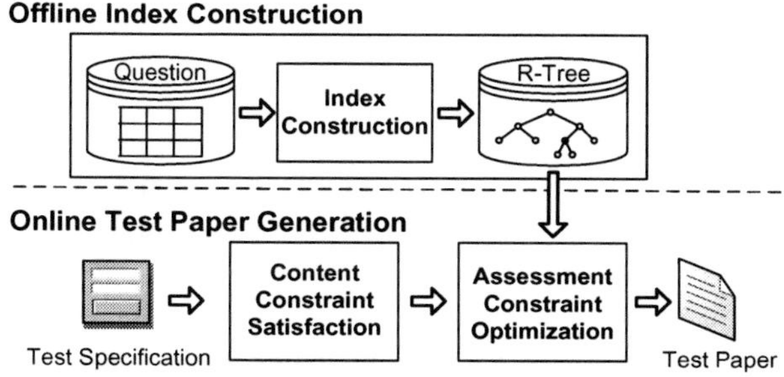

Fig. 1. The Proposed DAC-TS Approach

The proposed DAC-TS approach, as shown in Figure 1, consists of 2 main processes: Offline Index Construction and Online Test Paper Generation. In the Offline Index Construction process, it constructs an effective indexing structure for supporting tabu search to improve the quality of the generated paper. In the Online Test Paper Generation process, it generates a high quality test paper that satisfies the specified content constraints and assessment constraints. As illustrated in Figure 1, it consists of 2 major steps: Content Constraint Satisfaction and Assessment Constraint Optimization.

4.1 R-Tree Index Construction

We propose to use an effective 2-dimensional data structure, called R-Tree, to store questions based on the time and difficulty degree attributes. R-Tree has been widely used for processing queries on 2-dimensional spatial databases. As there is no specified rule on grouping of data into nodes in R-Tree, different versions of R-tree have been proposed [1, 14]. The R-Tree used here is similar to the R-tree version discussed in [1], with some modifications on index construction in order to enhance the efficiency. Some of the modified operations include insertion, subtree selection, overflow handling, and node splitting for index construction. Each leaf node in a R-Tree is a Minimum Bounding Rectangle (MBR) which is the smallest rectangle in the spatial representation that tightly encloses all data points located in the leaf node. Each non-leaf node has child nodes, which contain MBRs at the lower level. Figure 2 illustrates the R-Tree constructed from the Math dataset.

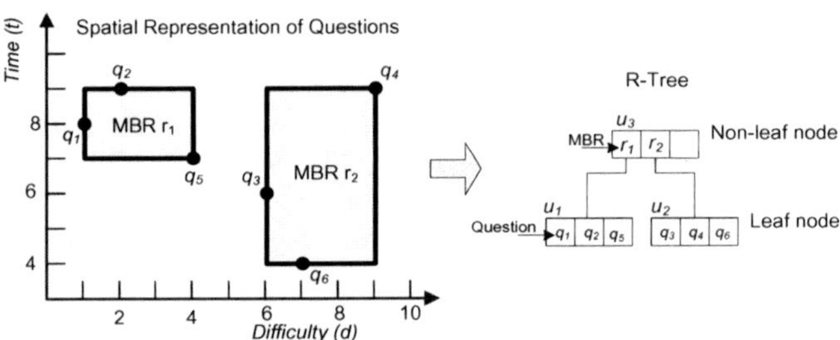

Fig. 2. An Example R-Tree

4.2 Content Constraint Satisfaction

It is quite straightforward to generate an initial test paper that satisfies the content constraints based on the number of questions N. Specifically, the number of questions of each topic c_l is $pc_l * N$, $l = 1..M$. Similarly, the number of questions of each question type y_j is $py_j * N$, $j = 1..K$. There are several ways to assign the N pairs of topic-question type to satisfy the content constraints. Here, we have devised an approach which applies a heuristic to try to achieve the specified total time early. To satisfy the content constraints, the round-robin technique is used for question selection. More specifically, for each topic c_l, $l = 1..M$, we assign questions alternately with various question types y_j, $j = 1..K$, as much as possible according to the number of questions.

Then, for each of the N pairs of topic-question type (c_l, y_j) obtained from the round-robin selection step, we assign a question q from the corresponding topic-question type (c_l, y_j) that has the highest question time to satisfy the total time early.

4.3 Assessment Constraint Optimization

Assessment Constraint Violation indicates the differences between the test paper specification and the generated test paper according to the total time constraint $\triangle T(\mathcal{S}_P, \mathcal{S})$ and the average difficulty degree constraint $\triangle D(\mathcal{S}_P, \mathcal{S})$ as follows:

$$\triangle T(\mathcal{S}_P, \mathcal{S}) = \frac{|T_P - T|}{T} \quad \text{and} \quad \triangle D(\mathcal{S}_P, \mathcal{S}) = \frac{|D_P - D|}{D}$$

A generated test paper P with specification $\mathcal{S}_P = \langle N, T_P, D_P, C_P, Y_P \rangle$ is said to satisfy the assessment constraints in $\mathcal{S}$ if $\triangle T(\mathcal{S}_P, \mathcal{S}) \leq \alpha$ and $\triangle D(\mathcal{S}_P, \mathcal{S}) \leq \beta$, where α and β are two predefined thresholds that indicate the acceptable quality satisfaction on total time and average difficulty degree respectively.

In addition, an objective function is defined for evaluating the quality of test papers based on assessment constraint violations. The quality of a generated test paper P is defined by the following objective function:

$$f(P) = \triangle T(\mathcal{S}_P, \mathcal{S})^2 + \triangle D(\mathcal{S}_P, \mathcal{S})^2$$

In Assessment Constraint Optimization, we conduct tabu search to improve the quality of the test paper by minimizing assessment constraint violations. This optimization process is repeated until the termination conditions are reached.

4.4 Tabu Search

Tabu search [3] is an iterative search method, which aims to find better questions to substitute the existing questions in the test paper in order to minimize assessment constraint violation. To form a new test paper, each question q_k in the original test paper P_0 is substituted by another better question q_m which has the same topic and question type such that assessment constraint violations are minimized. The tabu search comprises a local search with 3 strategies: Memory Usage, Up-hill Movement and Memory Relaxation. The termination conditions for the tabu search are based on the quality satisfaction and the maximum number of iterations in which no better test paper can be found.

In *memory usage*, DAC-TS uses a short-term memory and a long-term memory to avoid visiting a solution repeatedly. The *recency-based short-term memory* is used to prevent the substitution of a specific question in the current test paper for some steps after it has just been substituted. This short-term memory, namely TS, is implemented as follows: when a question q_i is substituted, the position i of that question is put into the short-term tabu list TS with a tenure t_{TS}. After each move, the tenure of the current entries in the TS is decreased by 1 and those entries with zero tenure are dropped from the TS. Whereas the *transitional frequency-based long-term memory* is used to dynamically avoid using over-active questions that have a specific topic-question type in order to help diversification and prevent cycling. To achieve this, a Move Frequency Table (MFT) has been incorporated into the tabu search process to store the move frequency of each topic-question type. This long-term memory, namely TL, is implemented as follows: when a question q_i is substituted, the move frequency of the topic-question type of that question is incremented by 1. If an entry x has been moved more than two times and TL is not full, it will be put into TL. If TL is full and some entries y in TL have a lower move frequency than x, we remove y from TL and add x into TL.

In *up-hill movement*, tabu search can accept a move even if the quality of the next solution is worse than that of the current solution. The reason is to escape the local optimal region and explore other new promising regions in the search space. However, to ensure that the up-hill process will not go too far from the current best solution, we set the following condition: $\frac{f(P)-f_{best}}{f_{best}} \leq r$ where r is a predefined threshold, and $f(P)$ and f_{best} are the values of the objective function of the current test paper P and the current best solution respectively.

Finally, *memory relaxation* is used to relax the tabu lists. If a given number of iterations has elapsed and TL is full since the last best solution was found, or if the current solution is much worse than the last best solution, we empty all entries in both TS and TL. Relaxation of the tabu lists will change the neighborhood of the current solution drastically, which may drive the search into a new promising region and increase the likelihood of finding a better solution.

Pruning Search Space. As the neighborhood region is very large, we need to prune the search space to find a 2-dimensional region W that contains possible questions for substitution. Let $\mathcal{S}_{P_0} = \langle N, T_0, D_0, C_0, Y_0 \rangle$ be the specification of a test paper P_0 generated from a specification $\mathcal{S} = \langle N, T, D, C, Y \rangle$. Let P_1 be the test paper created after substituting a question q_k of P_0 by another question $q_m \in \mathcal{Q}$ with $\mathcal{S}_{P_1} = \langle N, T_1, D_1, C_1, Y_1 \rangle$. The relations of total time and average difficulty degree between P_1 and P_0 can be expressed as follows:

$$T_1 = T_0 + t_m - t_k \quad (1)$$
$$D_1 = D_0 + \frac{d_m}{N} - \frac{d_k}{N} \quad (2)$$

where t_k and t_m are the question time of q_k and q_m respectively, and d_k and d_m are the difficulty degree of q_k and q_m respectively.

Let's consider the total time violation of P_0. If $\triangle T\ (\mathcal{S}_{P_0}, \mathcal{S}) = \frac{|T_0 - T|}{T} \geq \alpha$ and $T_0 \leq T$, where α is the predefined threshold. To improve the total time satisfaction of P_1, q_m should have the question time value of $t_k + (T - T_0)$ such that $\triangle T(\mathcal{S}_{P_1}, \mathcal{S})$ is minimized. Furthermore, as $\triangle T(\mathcal{S}_{P_1}, \mathcal{S}) = \frac{|T_1 - T|}{T} \leq \alpha$, q_m should have the total time t_m in the interval $t_k + (T - T_0) \pm \alpha T$. Therefore, we have $t_m \in [t_k + T - T_0 - \alpha T,$ $t_k + T - T_0 + \alpha T]$. If $\triangle T(\mathcal{S}_{P_0}, \mathcal{S}) = \frac{|T_0 - T|}{T} \geq \alpha$ and $T_0 > T$, we can also derive the same result. Similarly, we can derive the result for the difficulty degree of q_m: $d_m \in [d_k + N(D - D_0) - \beta ND, d_k + N(D - D_0) + \beta ND]$, where D_0, D and β are the average difficulty degree of P_0 and $\mathcal{S}$, and the predefined threshold respectively.

Finding the Best Question for Substitution. Among all the questions located in the 2-dimensional region W, it finds the best question that minimizes the objective function in order to enhance the test paper quality. Consider question q_m as a pair of variables on its question time t and difficulty degree d. The objective function $f(P_1)$ can be considered as a multivariate function $f(t, d)$:

$$f(P_1) = f(t,d) = \Delta T(\mathcal{S}_{P_1}, \mathcal{S})^2 + \Delta D(\mathcal{S}_{P_1}, \mathcal{S})^2 = (\frac{T_1 - T}{T})^2 + (\frac{D_1 - D}{D})^2$$

From Equations (1) and (2), we have:

$$f(t,d) = \frac{(t - T + T_0 - t_k)^2}{T^2} + \frac{(d - ND + ND_0 - d_k)^2}{D^2} = \frac{(t-t^*)^2}{T^2} + \frac{(d-d^*)^2}{D^2}$$
$$\geq \frac{(t-t^*)^2 + (d-d^*)^2}{T^2 + D^2} = \frac{distance^2(q_m, q^*)}{T^2 + D^2}$$

where q^* is a question having question time $t^* = T - T_0 + t_k$ and difficulty degree $d^* = ND - ND_0 + d_k$.

As T and D are predefined constants and q^* is a fixed point in the 2-dimensional space, the good question q_m to replace question q_k in P_0 is the question point that is the nearest neighbor to the point q^* (i.e., the minimum value of the function $f(P_1)$) and located in the region W. To find the good question q_m for substitution efficiently, we perform the Best First Search (BFS) [15] with the R-Tree. BFS recursively visits the nearest question whose region is close to q^*. For efficiency, BFS uses a memory-resident heap $\mathcal{H}$ to manage all the questions in the R-tree that have been accessed. This continues until a question de-heaped from $\mathcal{H}$ is located in W. We note that because there may be more than one good question found as mentioned above, the actual best question should has the maximum discrimination degree among these questions such that the average discrimination degree of the generated test paper is maximized. Algorithm 1 presents the overall Tabu Search algorithm for the assessment constraint optimization.

5 Performance Evaluation

As there is no benchmark datasets available in the research community, we generate 4 large-sized synthetic datasets, namely D_1, D_2, D_3 and D_4 with number of questions of 20000, 30000, 40000 and 50000 respectively for performance evaluation. The values of each attribute in the 4 datasets are generated according to a normal distribution. Table 2 shows the summary of the 4 datasets. In addition, we have designed 12 test specifications with different parameters. The experiments are conducted in the Windows XP environment running on an Intel Core 2 Quad 2.66 GHz CPU with 3.37 GB memory. We evaluate the performance based on the 12 test specifications for each of the following 6 algorithms: GA, PSO, DE, ACO, TS and DAC-TS. We measure the runtime and quality of the generated test papers for each experiment. The 3 parameters of the DAC-TS are set experimentally as follows: $t_{TS} = 30$, $l_{TL} = 200$, $r = 0.6$.

To evaluate the quality of k generated test papers on a dataset $\mathcal{D}$ w.r.t. any arbitrary test paper specification $\mathcal{S}$, we use Mean Discrimination Degree and Mean Constraint Violation. Let $P_1, P_2, ..., P_k$ be the generated test papers on a question dataset $\mathcal{D}$ w.r.t.

Table 2. Test Datasets

	D_1	D_2	D_3	D_4
Number of Questions	20000	30000	40000	50000
Number of Topics	40	50	55	60
Number of Question Types	3	3	3	3

Algorithm 1 . Tabu Search for Assessment Constraint Satisfaction

Input:

$\mathcal{S} = (N, T, D, C, Y)$ - test paper specification; $P_0 = \{q_1, q_2, .., q_N\}$ - initial test paper; t_{TS} - short-term memory tenure; l_{TL} - long-term memory length; r - relaxation ratio; $\mathcal{R}$ - R-Tree index

Output:

P^* - Improved test paper

Process:

$\mathcal{P} \leftarrow \{P_0\}$; $MFT \leftarrow \emptyset$; $TS \leftarrow \emptyset$; $TL \leftarrow \emptyset$; $nbmax = 3l_{TL}$; $nbiter = bestiter = 0$
while P_{best} is not satisfied **and** $(nbiter - bestiter) < nbmax$ **do**
  $nbiter := nbiter + 1$; $optiter := optiter + 1$;
  for each q_i **in** P_0 **do**
    Compute 2-dimensional range W /* pruning search space*/
    $q_m \leftarrow$ **Best_First_Search**$(q_i, W, \mathcal{R})$;
    $P_1 \leftarrow \{P_0 - \{q_i\}\} \cup \{q_m\}$
    if $(q_i \notin TS$ **and** $(c_i, y_i) \notin TL)$ **or** $f(P_1) < f(P_{best})$ **then**
      Inserting new test paper P_1 into $\mathcal{P}$
    end if
  end for
  $P^* \leftarrow \underset{P' \in \mathcal{P}}{\text{argmin}}\ f(P_1)$; $\mathcal{P} \leftarrow \{P^*\}$ /* best move*/
  Update $MFT(c_m, y_m)$, Update $TS(q_i)$, Update $TL(c_m, y_m)$;
  if $f(P^*) < f(P_{best})$ **then**
    $P_{best} = P^*$; $bestiter = nbiter$; $optiter := 0$
  else if $optiter > 2l_{TL}$ **or** $\frac{f(P) - f_{best}}{f_{best}} > r$ **then**
    $optiter := 0$; $TS \leftarrow \emptyset$; $TL \leftarrow \emptyset$ /* memory relaxation*/
  end if
end while
return P^*

different test paper specifications $\mathcal{S}_i$, $i = 1..k$. Let E_{P_i} be the average discrimination degree of P_i. The *Mean Discrimination Degree* $\mathcal{M}_d^{\mathcal{D}}$ is defined as:

$$\mathcal{M}_d^{\mathcal{D}} = \frac{\sum_{i=1}^{k} E_{P_i}}{k}$$

The Mean Constraint Violation consists of two components: Assessment Constraint Violation and Content Constraint Violation. In Content Constraint Violation, Kullback-Leibler (KL) Divergence [11] is used to measure the topic distribution violation $\triangle C(\mathcal{S}_P, \mathcal{S})$ and question type distribution violation $\triangle Y(\mathcal{S}_P, \mathcal{S})$ between the generated test paper specification $\mathcal{S}_P$ and the test paper specification $\mathcal{S}$ as follows:

$$\triangle C(\mathcal{S}_P, \mathcal{S}) = D_{KL}(pc_p || pc) = \sum_{i=1}^{M} pc_p(i) \log \frac{pc_p(i)}{pc(i)}$$
$$\triangle Y(\mathcal{S}_P, \mathcal{S}) = D_{KL}(py_p || py) = \sum_{j=1}^{K} py_p(j) \log \frac{py_p(j)}{py(j)}$$

The Constraint Violation (CV) of a generated test paper P w.r.t. $\mathcal{S}$ is defined as:

$$CV(P, \mathcal{S}) = \frac{\lambda * \triangle T + \lambda * \triangle D + \log \triangle C + \log \triangle Y}{4}$$

where $\lambda = 100$ is a constant used to scale the value to a range between 0-100. The *Mean Constraint Violation* $\mathcal{M}_c^{\mathcal{D}}$ of k generated test papers $P_1, ..., P_k$ on a question dataset $\mathcal{D}$ w.r.t different test paper specifications $\mathcal{S}_i$, $i = 1..k$, is defined as:

$$\mathcal{M}_c^{\mathcal{D}} = \frac{\sum_{i=1}^{k} CV(P_i, \mathcal{S}_i)}{k}$$

Figure 3 gives the runtime performance of the proposed approach in comparison with other techniques on the 4 datasets. The results have shown that DAC-TS outperforms other techniques in runtime. In Figure 3, it also shows that DAC-TS satisfies the runtime requirement as it generally requires less than 2 minutes to complete the paper generation process for various dataset sizes. In addition, the DAC-TS approach is scalable in runtime. Figure 4 shows the quality performance of DAC-TS and other techniques based on Mean Discrimination Degree $\mathcal{M}_d^{\mathcal{D}}$ and Mean Constraint Violation $\mathcal{M}_c^{\mathcal{D}}$ for the

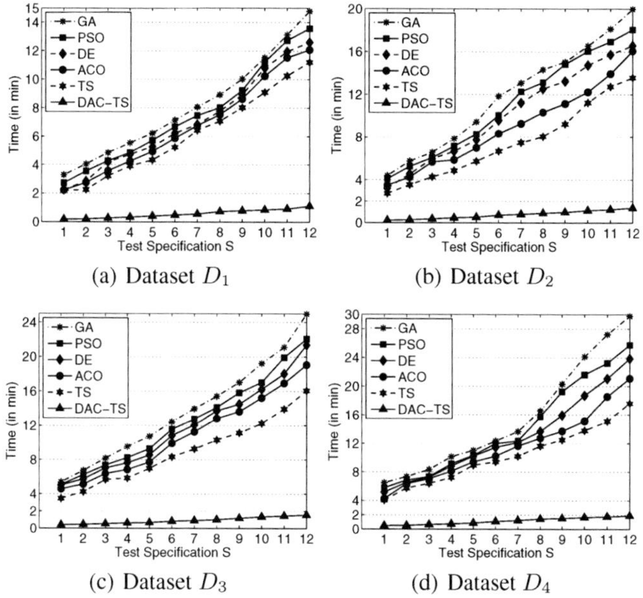

(a) Dataset D_1 (b) Dataset D_2

(c) Dataset D_3 (d) Dataset D_4

Fig. 3. Performance Results Based on Runtime

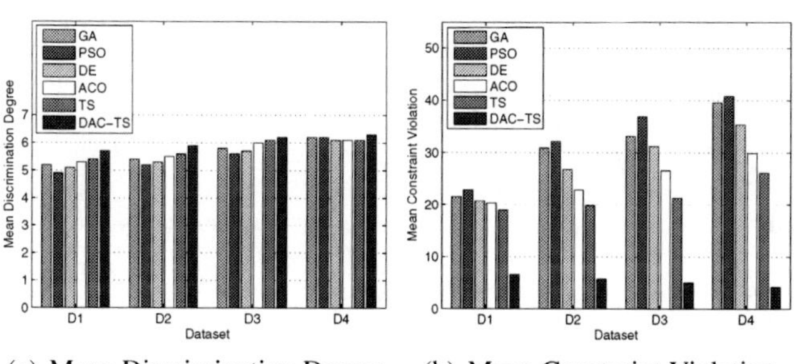

(a) Mean Discrimination Degree (b) Mean Constraint Violation

Fig. 4. Performance Results based on Quality

4 datasets. As can be seen, DAC-TS has consistently outperformed other techniques. As such, DAC-TS is able to generate higher quality test papers than other techniques.

6 Conclusion

In this paper, we have proposed an efficient constraint-based Divide-And-Conquer Tabu Search (DAC-TS) approach for online test paper generation. The performance results have shown that the DAC-TS approach has not only achieved good quality test papers, but also satisfied the online runtime requirement even for large datasets in comparison with other techniques. Thus, the proposed research is particularly useful for Web-based testing and intelligent tutoring in an educational environment. For future work, we would like to combine the DAC-TS with the integer linear programming to further enhance the constraint satisfaction and runtime efficiency of the DAC-TS approach.

References

1. Beckmann, N., Kriegel, H.P., Schneider, R., Seeger, B.: The r*-tree: an efficient and robust access method for points and rectangles. ACM SIGMOD Record 19(2), 322–331 (1990)
2. Conejo, R., Guzmn, E., Milln, E., Trella, M., Prez-De-La-Cruz, J.L., Ros, A.: Siette: a web-based tool for adaptive testing. International Journal of Artificial Intelligence in Education 14(1), 29–61 (2004)
3. Glover, F., Laguna, F.: Tabu Search. Kluwer Academic Publishers (1997)
4. Gonzalez, T.F.: Handbook of Approximation Algorithms and Metaheuristics. Chapman & Hall/Crc Computer & Information Science Series (2007)
5. Guzman, E., Conejo, R.: Improving student performance using self-assessment tests. IEEE Intelligent Systems 22(4), 46–52 (2007)
6. Ho, T.F., Yin, P.Y., Hwang, G.J., Shyu, S.J., Yean, Y.N.: Multi-objective parallel test-sheet composition using enhanced particle swarm optimization. Journal of ETS 12(4), 193–206 (2008)
7. Hu, X.M., Zhang, J., Chung, H.S.H., Liu, O., Xiao, J.: An intelligent testing system embedded with an ant-colony-optimization-based test composition method. IEEE Transactions on Systems, Man, and Cybernetics 39(6), 659–669 (2009)
8. Hwang, G.J.: A test-sheet-generating algorithm for multiple assessment requirements. IEEE Transactions on Education 46(3), 329–337 (2003)
9. Hwang, G.J., Lin, B., Tseng, H.H., Lin, T.L.: On the development of a computer-assisted testing system with genetic test sheet-generating approach. IEEE Transactions on Systems, Man, and Cybernetics 35(4), 590–594 (2005)
10. Hwang, G.J., Yin, P.Y., Yeh, S.H.: A tabu search approach to generating test sheets for multiple assessment criteria. IEEE Transactions on Education 49(1), 88–97 (2006)
11. Kullback, S.: Information theory and statistics. Dover Publisher (1997)
12. Lee, C.-L., Huang, C.-H., Lin, C.-J.: Test-Sheet Composition Using Immune Algorithm for E-Learning Application. In: Okuno, H.G., Ali, M. (eds.) IEA/AIE 2007. LNCS (LNAI), vol. 4570, pp. 823–833. Springer, Heidelberg (2007)
13. Li, Q.: Guest editors' introduction: Emerging internet technologies for e-learning. IEEE Internet Computing 13(4), 11–17 (2009)
14. Manolopoulos, Y., Nanopoulos, A.: R-trees: Theory and Applications. Springer, Heidelberg (2006)
15. Roussopoulos, N., Kelley, S., Vincent, F.: Nearest neighbor queries. In: Proceedings of the ACM SIGMOD, pp. 71–79 (1995)
16. Rui, W.F., Hong, W.W., Ke, P.Q., Chao, Z.F., Liang, J.J.: A novel online test-sheet composition approach for web-based testing. In: Symposium on IT in Medicine & Education, pp. 700–705 (2009)

On the Interactions of Awareness and Certainty

Hans van Ditmarsch[1,2] and Tim French[3]

[1] Department of Logic, University of Sevilla, Spain
hvd@us.es
[2] Institute of Mathematical Sciences Chennai, India
[3] Computer Science and Software Engineering,
The University of Western Australia
tim.french@uwa.edu.au

Abstract. We examine the interactions of knowledge and awareness in dynamic epistemic logic. Implicit knowledge describes the things that an agent could infer from what is known, if the agent were aware of the necessary concepts. Reasoning techniques that are robust to incomplete awareness are important when considering interactions of automated agents in complex dynamic environments, such as the semantic web. Here we revisit Hector Levesque's original motivation of implicit knowledge and consider several contemporary realizations of implicit knowledge. We present a framework to compare different interactions of knowledge and awareness in the context of public announcements, and introduce a new formalism for *tacit knowledge.*

1 Introduction

The term *implicit belief* was introduced by Hector Levesque [13] to describe "not what an agent believes directly, but what the world would be like if what he believed were true."

Implicit knowledge has come to describe the things that an agent knows, but of which the agent is not fully aware [6]. Implicit knowledge has limited significance with respect to static knowledge and awareness, where explicit knowledge suffices for most applications. However, implicit knowledge is essential for the formalization of dynamic epistemic systems, where the knowledge and awareness of agents changes over time. Implicit knowledge may be used to model latent facts that the agent subconsciously recognizes, inferences that the agent can make, but has not yet made, or aspects of the agent's knowledge that the agent does not yet have the vocabulary to express.

Certainty and awareness are two separate and important facets of knowledge. Certainty describes an agent's confidence in a given scenario, and awareness describes an agent's ability to perceive a given scenario. Epistemic logic is the logic of knowledge and is very much focussed on the element of certainty (or uncertainty). It is very well studied [7] and has applications in reasoning about security in information systems. While epistemic logic presumes logically omniscient agents [6], practical agents are not able to hold all relevant facts on their

D. Wang and M. Reynolds (Eds.): AI 2011, LNAI 7106, pp. 727–738, 2011.

mind at once, and do not have the computational resources to instantly make all valid inferences from their knowledge base. This has lead to several models of incomplete reasoning applied to human activities (particularly commercial transactions) [13,14,10].

One approach to address these limitations has been to temper an agent's knowledge with an element of awareness. Thus, we can have an agent who is unaware of a proposition (that is, oblivious to its existence), as opposed to simply uncertain as to its interpretation. This is a simple and elegant approach that allows us to differentiate information that an agent explicitly knows from information that may be implicit in the agent's internal state.

In this paper we will examine dynamic interactions between awareness and knowledge. As with dynamic epistemic logic [5] where an agent may learn new facts about the world, we may consider awareness dynamics where an agent may become aware of new properties in the world. Becoming aware should leave the agent in an enlightened state, and able to accumulate new knowledge (with respect to the properties of which the agent is newly aware) that is compatible with what the agent already knows. The notion of implicit knowledge is essential to capture this compatibility between knowledge states. We will give a uniform framework to consider these different notions and examine their expressivity in the context of dynamic epistemic operations.

2 The Logics of Dynamic Knowledge and Awareness

The properties of knowledge and belief in the presence of full awareness have been extensively studied and are well-understood. They are effectively modelled by normal modal operators that are reflexive, transitive and symmetric (for knowledge) or serial and Euclidean (for belief) [7]. However without full awareness things become more complicated. The first issue to arise is that the logic of explicit knowledge is no longer a normal modal logic. For example, if an agent is not aware of the $\varphi = p \vee \neg p$, then $\mathbf{K}_i\varphi$ is not true even though φ is valid. That is, the necessitation rule fails for explicit knowledge.

2.1 Language

We suppose that we are extending a multi-agent propositional epistemic logic, defined over a set of agents N and and set of atomic propositions, P. We augment multi-agent epistemic logic with a new operator: $\mathbf{A}_i\varphi$, to mean that agent i is aware of all the concepts in φ. We note, some treatments of knowledge and awareness do not have an explicit operator for awareness, preferring to treat it as an abbreviation. The construct $\mathbf{K}_i\varphi$, "agent i knows φ" stands in our case for "agent i explicitly knows φ", and we add an additional operator, $\mathbb{K}_i\varphi$ to mean "Agent i implicitly knows φ". We also use the notation $[\psi]\varphi$ to mean, "After ψ is announced, φ is true". The effect of this public announcement is two-fold.

First, it brings ψ to the attention of all agents. Any agent who was previously not aware of ψ or it's constituent parts, now is. Secondly, it informs the agent that ψ is indeed true, so any world the agent considered where ψ is not true (given the newly found awareness) is discounted.

Definition 1 (Language). *Given are a countably infinite set of propositional variables (facts) P, and a (disjoint) countably infinite set of agents N. The language $\mathcal{L}$ is defined as*

$$\varphi ::= \top \mid p \mid \varphi \wedge \varphi \mid \neg\varphi \mid \mathbf{K}_i\varphi \mid \mathbb{K}_i\varphi \mid \mathbf{A}_i\varphi \mid [\varphi]\varphi$$

where $i \in N$ and $p \in P$. Implication $\rightarrow$, disjunction $\vee$, and equivalence $\leftrightarrow$ are defined by abbreviation.

Finally we will consider a base semantics as a common framework for considering various interactions of knowledge and awareness. The semantics will extend those for multi-agent epistemic (modal) logic with an awareness function that assigns for each agent and each possible world, the set of formulae of which that agent is aware of in that world.

Definition 2 (Epistemic awareness model). *An* epistemic awareness model *for N and P is a tuple $M = (S, R, \mathcal{A}, V)$ that consists of:*

- *a* domain S *of (factual)* states *(or 'worlds');*
- *an* accessibility function $R : N \rightarrow \mathcal{P}(S \times S)$;
- *an* awareness function $\mathcal{A} : N \rightarrow S \rightarrow \mathcal{P}(\mathcal{L})$; *and*
- *a* valuation function $V : P \rightarrow \mathcal{P}(S)$.

For $R(i)$ we write R_i and for $\mathcal{A}(i)$ we write $\mathcal{A}_i$; accessibility function R can be seen as a set of accessibility relations R_i, *and V as a set of* valuations $V(p)$. *A pointed epistemic awareness model (M, s) is an* epistemic awareness state.

The awareness function is defined such that $\mathcal{A}_i(s)$ is the set of propositions that the agent i is aware of in state s, and as such we will require that it is closed under subformulas (so $\varphi \in \mathcal{A}_i(s)$ implies $\psi \in \mathcal{A}_i(s)$ for all subformulas ψ of φ). In [6] various other closure properties are considered for agent awareness. As we will consider dynamic models it is convenient to denote the elements of the models M as $(S^M, R^M, \mathcal{A}^M, V^M)$, and to denote pointed models (M, s) as M_s. The conditions on the accessibility function R^M may be varied to reflect different interpretations of knowledge and belief. Some semantics are defined with respect to general frame conditions (**K**) [13,4], some are defined with respect to belief (**KD45**) [6] and some are defined with respect to knowledge (**S5**) [14,2]. Furthermore, some semantics are only defined in the instance where a frame has a single accessibility relation [13,11]. The semantics presented here are compatible with each of these restrictions.

Definition 3 (Semantics). *Let $M = (S, R, \mathcal{A}, V)$ be given, and suppose $s \in S$.*

$$
\begin{array}{lll}
(M,s) \models \top & & \\
(M,s) \models p & \textit{iff } s \in V(p) & \\
(M,s) \models \varphi \wedge \psi & \textit{iff } (M,s) \models \varphi \textit{ and } (M,s) \models \psi & \\
(M,s) \models \neg\varphi & \textit{iff } (M,s) \not\models \varphi & \\
(M,s) \models \mathbf{A}_i\varphi & \textit{iff } \varphi \in \mathcal{A}_i(s) & \\
(M,s) \models \mathbf{K}_i\varphi & \textit{iff } \varphi \in \mathcal{A}_i(s) \textit{ and } \forall t \in sR_i,\ (M,t) \models \varphi & \\
(M,s) \models [\psi]\varphi & \textit{iff } s \in S^{M^\psi} \Rightarrow (M^\psi, s) \models \varphi & \textit{(see below)}
\end{array}
$$

The semantic definition for public announcements requires us to adjust the model to reflect the agents' newly acquired knowledge. On receiving the public announcement that ψ is true, all agents may disregard any state where ψ is not true. Furthermore, they become aware of the formula ψ, if they were not already [17], [2] . The updated model is M^ψ where

$$
\begin{array}{l}
S^{M^\psi} = \{s \in S^M \mid (M,s) \models \varphi\} \\
R^{M^\psi} = R^M \cap S^{M^\psi} \times S^{M^\psi} \\
\forall i \in N,\ \forall s \in S^{M^\psi},\ \mathcal{A}_i^{M^\psi}(s) = \mathcal{A}_i^M(s) \cup \{\psi' \mid \psi' \subseteq \psi\} \\
\forall p \in P, V^{M^\psi}(p) = V^M(p) \cap S^{M^\psi}.
\end{array}
$$

We will consider several interpretations for implicit knowledge, but in all cases we have that explicit knowledge is simply defined as implicit knowledge plus awareness, so $\mathbf{K}_i\varphi \doteq \mathbb{K}_i\varphi \wedge \mathbf{A}_i\varphi$ (where $\doteq$ is used to define syntactic abbreviations). We define the dual notions of explicit and implicit knowledge as $\mathbf{L}_i\varphi$ (agent i explicitly suspects φ) and $\mathbb{L}_i\varphi$ (agent i implicitly suspects φ). Note that as it is intuitive that explicit suspicion requires that the agent is aware of φ, $\mathbf{L}_i$ is not the exact dual of $\mathbf{K}_i$, but rather defined as implicit suspicion plus awareness. Consequently we may define $\mathbb{K}_i$ and $\mathbf{A}_i$ as the only atomic operators, and use the abbreviations: $\mathbf{K}_i\varphi \doteq \mathbb{K}_i\varphi \wedge \mathbf{A}_i\varphi$, $\mathbb{L}_i\varphi \doteq \neg\mathbb{K}_i\neg\varphi$, and $\mathbf{L}_i\varphi \doteq \mathbb{L}_i\varphi \wedge \mathbf{A}_i\varphi$.

We will give a simple example of how public announcements update the knowledge state of an agent. Suppose that Alice and Bob are ordering a meal at a restaurant. On reading the menu, Bob notes that Chicken and Beef is available, but he does not know which one Alice will order. He is not aware that the soup of the day is pumpkin soup. Alice announces she will have either the chicken or the pumpkin soup. After this exchange Bob knows she will not have the beef, and he becomes aware of the pumpkin soup. This situation in depicted in Figure 1. The three worlds labelled c, b and p correspond to the worlds where Alice orders chicken beef or pumpkin respectively, and the atoms each agent is aware of are marked at each world (we assume that agents are aware of all propositions made of these atoms). The agents' knowledge relations are the transitive, reflexive closure of the relations shown. (So the arrow labelled *Bob* between worlds **c** and **p** indicates that Bob cannot distinguish the world where Alice orders beef from the world where Alice orders chicken.)

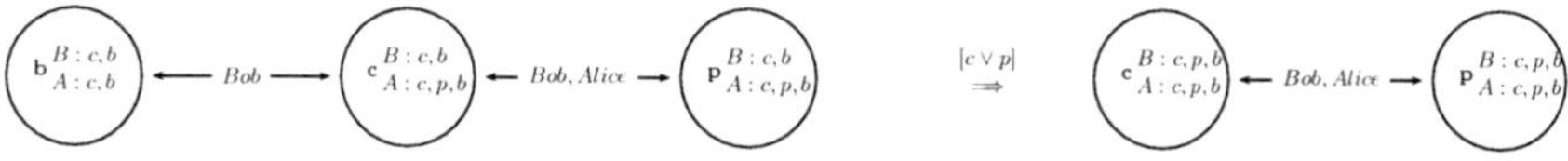

Fig. 1. The scenario before and after Alice makes her announcement

3 Implicit Knowledge

In this section we will examine three different variations of implicit knowledge and relate them to existing logics of knowledge, belief and awareness.

3.1 Strong Implicit Knowledge

Levesque's [13] work introduced the concept implicit belief, and was one of the first papers to separate the concerns of awareness and certainty. His response to the problem of logical omniscience was to separate implicit belief from explicit belief. from [13]: "...a sentence is explicitly believed when it is actively held to be true by an agent and implicitly believed when it follows from what is believed". The semantic formulation for this logic is given through situations rather than possible worlds, and for each formula a situation may support the truth of that formula, the falsity of that formula or neither. These situations were potentially *incoherent* in that a situation may support the truth of both φ and $\neg\varphi$ for some formula φ. This gave a very general approach for agents without logical omniscience. The semantics structures were also kept as general as possible. To define implicit belief, a compatibility relation was required to determine which situations are compatible with the agent's explicit beliefs. An agent implicitly believes a proposition if it is true in all compatible situations.

This approach combines elements of modal logic and situation calculus to reason about the knowledge of imperfect reasoners. A subsequent, and more involved, approach along these lines is given by Cadoli and Schaerf [3] which separate interpretations for situations which are coherent, but incomplete, and those which are complete, but incoherent. See [12] for further discussion and generalizations.

In our setting we present an analogous version of implicit knowledge by refining what is meant by a "compatible situation". Levesque's original version defined compatible to be propositionally consistent with respect to propositions of which the agent was aware. As we are using modal logic we have a well established form of compatibility available, bisimulation, which is known to relate two finite models exactly when they satisfy identical sets of formulas. In [4] the concept of a bisimulation was adjusted to reflect the agent's awareness states.

Awareness bisimulations capture the notion of equivalence in a model, up to an agent's state of awareness. Note that the definition is recursive, so that if agent A considers agent B's explicit knowledge of φ, then both agents must be aware of φ.

Definition 4 (Awareness bisimulation). *Suppose we are given epistemic awareness models* $M = (S, R, \mathcal{A}, V)$ *and* $M' = (S', R', \mathcal{A}', V')$. *For all subformula-closed* $A \subseteq \mathcal{L}$ *we say a relation* $\mathfrak{B}[A]$ *is an* A-awareness bisimulation *iff for all* $(s, s') \in \mathfrak{B}[A]$:

atoms *for all* $p \in A$, $s \in V(p)$ *iff* $s' \in V'(p)$;
aware *for all* $i \in N$, $\mathcal{A}_i(s) \cap A = \mathcal{A}'_i(s') \cap A$;
forth *for all* $i \in N$, *if* $t \in S$ *and* $R_i(s,t)$ *then there is a* $t' \in S'$ *such that* $R'_i(s', t')$ *and* $(t, t') \in \mathfrak{B}[A^i \cap \mathcal{A}_i(s)]$;
back *for all* $i \in N$, *if* $t' \in S'$ *and* $R'_i(s', t')$ *then there is a* $t \in S$ *such that* $R_i(s, t)$ *and* $(t, t') \in \mathfrak{B}[A^i \cap \mathcal{A}'_i(s')]$.

where $A^i = \{\psi \subseteq \varphi \mid \mathbf{K}_i\varphi \in A \text{ or } \mathbb{K}_i\varphi \in A\}$. *An epistemic awareness state* (M', s') *is* A-awareness-bisimilar *to an epistemic awareness state* (M, s) *(written* $(M', s') \leftrightarroweq_A (M, s)$*) iff* $(s, s') \in \mathfrak{B}[A]$.

Note the bisimulation is given modulo a subformula closed set of propositions, so an agent aware of propositions P, would consider structures M_s and $M'_{s'}$ equivalent (as far as he explicitly knows), if they agree on the interpretation of all propositions in P. Awareness bisimulations can be seen to be reflexive, symmetric and transitive, so we have the following proposition.

Proposition 1. *For all* $A \subseteq \mathcal{L}$ *A-awareness bisimulation is an equivalence relation.*

The strong implicit semantics are given below. Note that as all semantics differ only in their interpretation of implicit knowledge we will only distinguish the implicit knowledge operators ($\mathbb{K}^{\mathfrak{I}}$, $knowI^{\mathfrak{L}}$ and $\mathbb{K}^{\mathfrak{T}}$). When the semantics of the operator is clear from its context, we will just use $\mathbb{K}$.

Definition 5 (Strong implicit semantics). *Let* $M = (S, R, \mathcal{A}, V)$ *be given. We define* $(M, s) \models \mathbb{K}_i^{\mathfrak{I}}\varphi$ *if and only if for all* $t \in sR_i$, *for all* $(M', t') \leftrightarroweq_{\mathcal{A}_i(s)} (M, t)$ *we have* $(M', t') \models \varphi$.

The main innovation in these semantics is the treatment of knowledge. An agent knows φ only if in all accessible states φ remains true for every possible interpretation of all concepts that she is unaware of. We achieve this by extending the agent's accessibility relation by composing it with bisimulation modulo those concepts of which the agent is unaware.

The following lemma shows the correspondence between strong implicit knowledge and Levesque's original motivation, and may be proven by induction over the complexity of φ.

Lemma 1. *Suppose that* $\varphi \in A$ *and* $M_s \leftrightarroweq_A N_t$. *Then in the strong implicit semantics,* $M_s \models \varphi$ *iff* $N_t \models \varphi$.

As an agent aware only of the formulas in A will be unable to distinguish any two models that are A-bisimilar it follows that an agent will implicitly know φ if and only if φ is true in every situation that is indistinguishable to the current situation for that agent.

3.2 Latent Knowledge

A simpler approach to handling interactions of knowledge and awareness is presented in Fagin's and Halpern's seminal paper [6]. Their approach is to generalize epistemic logic [7] to also account for the agent's awareness, rather than to invent a wholly new approach. Fagin and Halpern presented a generalization of Levesque's approach, where models are as described in Definition 2 and the semantics are as in Definition 3 as well as:

Definition 6 (Latent semantics). *Let* $M = (S, R, \mathcal{A}, V)$ *be given. Then* $(M, s) \models \mathbb{K}_i^{\mathfrak{L}}\varphi$ *if and only if* $\forall t \in sR_i$, $(M, t) \models \varphi$.

These semantics present an elegantly simple way to address the gap between uncertainty and unawareness. Implicit knowledge is modelled simply as knowledge with or with out awareness. In essence, the logic presents a syntactic restriction for explicit knowledge (so that *knowledge* is only meaningfully considered for terms the agent is aware of) and for all other terms implicit knowledge stands as a placeholder for *knowledge without awareness.*

There were actually three different interpretations of awareness presented in [6], where the approach reported above is the second (the logic of general awareness). The logics presented in [6] are well-defined for multiple agents, nested belief, and do not permit incoherent situations, as Levesque's approach does. However, these practicalities come at a cost. Whereas Levesque's logic of explicit and implicit belief had a clear definition of implicit belief: "a sentence... is implicitly believed when it follows from what is (explicitly) believed"; there is no such such motivation or even definition in the work of Fagin and Halpern.

The elegant simplicity of this approach contrasts the complexity of awareness bisimulations. However, it comes at the expense of having an essential interpretation of implicit knowledge. Indeed the only motivation for implicit belief is given in terms of the abstract semantics: "Implicit belief differs from explicit belief in that for implicit belief we do not take the awareness function into account". Thus we might infer that implicit belief is the belief an agent would have were they fully aware, but there is no understanding given for how this state of parallel awareness might actually manifest itself. For example, $\mathbf{K}_i(\mathbb{K}_j^{\mathfrak{I}}\varphi \wedge \neg\mathbf{K}_j\varphi)$ describes a situation where agent, i knows that j implicitly knows φ, and also that j does not explicitly know φ. This presents a strange situation where other people may know more about what you know than you do. There are various interpretations that could support such a scenario (resource-bounded reasoning, absent-mindedness etc), but there is no general property of knowledge, belief and awareness that matches this notion of implicit knowledge. Rather, it is a convenient semantic device to ensure an agent's explicit knowledge, whilst possibly incomplete, is consistent.

3.3 Tacit Knowledge

Tacit knowledge is a weaker version of latent knowledge which differs in its treatment of information of which an agent is unaware. It occupies a middle ground between latent knowledge (which describes exactly what an agent should know and should not know) and strong implicit knowledge (which assumes an agent knows nothing that can't be derived from explicit knowledge). The term *tacit knowledge* was introduced by Polanyi [15] to describe knowledge that was held by someone, but which was very hard to communicate or make explicit. The example of how to ride a bike is often given. Many people know how to ride a bike, but if they were asked to write down exactly how they are able to, they would struggle to share their knowledge.

Tacit knowledge allows for "unspoken knowledge" things an agent can or should know, but of which the agent is not aware. However it does not extend this to tacit ignorance as latent knowledge does. Thus we may permit an agent who does not have the facility of geometry to tacitly know that two distinct parallel lines will never meet, as this is true in every world the agent considers possible (the agent does not consider non-Euclidean geometries possible). The agent is not aware of concepts such as "parallel" so this is not explicit knowledge. Tacit knowledge captures unspoken truths of which an agent is not aware, but which persist nonetheless in all worlds all worlds the agent considers possible. The mechanism by which this is achieved in the the refinement, which is closely related to the bisimulation.

Definition 7 (Awareness Refinement). *Let epistemic awareness models* $M = (S, R, \mathcal{A}, V)$ *and* $M' = (S', R', \mathcal{A}', V')$ *be given. For all* $A \subseteq \mathcal{L}$ *we say the relation* $\mathfrak{R}[A] \subseteq S \times S'$ *is an* A-awareness refinement *iff for all* $(s, s') \in \mathfrak{R}[A]$ *:*

atoms *for all* $p \in A$, $s \in V(p)$ *iff* $s' \in V'(p)$;
aware *for all* $i \in N$, $\mathcal{A}_i(s) \cap A = \mathcal{A}'_i(s') \cap A$;
back *for all* $i \in N$, *if* $t' \in S'$ *and* $R'_i(s', t')$ *then there is a* $t \in S$ *such that* $R_i(s, t)$ *and* $(t, t') \in \mathfrak{R}[P]$.
forth *for all* $i \in N$, *if* $t \in S$ *and* $R_i(s, t)$ *then there is a* $t' \in S'$ *such that* $R_i(s', t')$ *and* $(t, t') \in \mathfrak{R}[A^i \cap \mathcal{A}_i(s)]$.

Epistemic awareness state (M', s') *is an* A-awareness-refinement *of epistemic awareness state* (M, s) *(written* $(M', s') \preceq_A (M, s)$*) iff* $(s, s') \in \mathfrak{R}[A]$.

Note that as with Awareness Bisimulation (Definition 4 the Awareness Refinement is restricted to range over models of the logic, so models satisfying the **S5** axioms for epistemic logic, and the **KD45** axioms for doxastic logic. It is also interesting to note that the definition of refinement has both the **back** and **forth** relations, but they are not symmetrical. The condition **back** is defined with respect to the full language, so that worlds the agent considers must come from the original model (up to bisimilarity). However **forth** is defined only with respect to the atoms of which the agent is aware, so the agent may discount some worlds of the original model where it does not affect his explicit knowledge. We then describe tacit knowledge as.

Definition 8 (Tacit semantics). *Let* $M = (S, R, \mathcal{A}, V)$ *be given. The semantics are as in Definition 3 and then* $(M, s) \models \mathbb{K}_i^{\mathfrak{T}} \varphi$ *if and only if for all* $t \in sR_i$, *for all* (M', t') *where* $(M, t) \preceq_{\mathcal{A}_i(s)} (M', t')$, *we have* $(M', t') \models \varphi$.

Tacit knowledge is a compromise between rigidity of latent knowledge and the vagueness of strong implicit knowledge. If we consider an agent who is unaware of a proposition, we may not know exactly how an agent may become aware of that proposition, but we may know of some intrinsic relation between that proposition and some other proposition the agent is aware of, which will constrain how an agent may become aware of the proposition. For example an agent observing another agent may know that the second agent tacitly knows summer is hot, even through the second agent is not aware of the concept of summer. As soon as the second agent becomes aware of the concept of summer, the proposition becomes evident. However, if the second agent did not tacitly know summer is hot the first agent still considers it possible that the second agent tacitly knows summer is hot. Although not explicitly stated, tacit knowledge is evident in Heifetz, Meier and Schipper's iterative unawareness [10] which permits a plurality of states of higher awareness for an agent, built upon a set of ground truths, which is the essence of tacit knowledge.

We illustrate the difference between latent, strong implicit and tacit knowledge with the simple example in Figure 2.

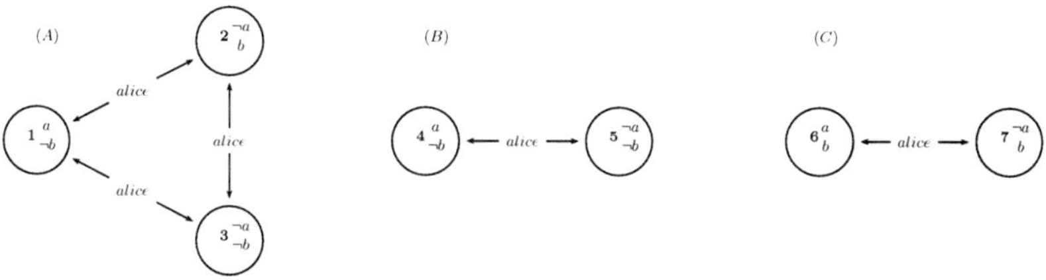

Fig. 2. Suppose that Alice is aware of the atom a, but not of b. Given the model (A) Alice may *latently* consider the worlds 1, 2, or 3 possible. However, she may also *tacitly* consider the worlds 4 and 5 possible in model (B). This is because the worlds 2 and 3 may both be related to the world 5 through a refinement (Alice has no tacit knowledge of b when a is false, but she tacitly knows that if a is true, then b must be false). Finally, Alice may consider the worlds 6 and 7 in model (C) possible in the *strong implicit* semantics as the interpretation of the atom b is completely unconstrained by the model (A). Note that none of the models change the state of Alice's explicit uncertainty about a.

4 Comparative Analysis of Latent, Tacit and Strong Implicit Knowledge

From now on, we will use the terms *strong implicit*, *tacit*, *latent* and *explicit* in the context of the definitions above. Negative awareness introspection is the property that an agent knows when it is not aware of a proposition. We note an interesting distinction between the levels of knowledge that can be seen in the context of negative awareness introspection: latent knowledge does have negative awareness introspection ($\models \neg\mathbf{A}_i\varphi \rightarrow \mathbb{K}_i^{\mathfrak{L}}\neg\mathbf{A}_i\varphi$); strong implicit knowledge does not permit negative awareness introspection ($\models \neg \mathbf{A}_i\varphi \rightarrow \neg \mathbb{K}_i^{\mathfrak{I}} \mathbf{A}_i\varphi$); and for tacit knowledge negative awareness introspection is satisfiable, but not valid (so neither $\models \neg \mathbf{A}_i\varphi \rightarrow \mathbb{K}_i^{\mathfrak{T}} \mathbf{A}_i\varphi$ nor $\models \neg \mathbf{A}_i\varphi \rightarrow \neg \mathbb{K}_i^{\mathfrak{T}} \mathbf{A}_i\varphi$ are true).

We consider the application of the various notions of knowledge, by considering a limited form of awareness change. We show how public announcements allows us to express the potential knowledge of an agent, and how this can be related to the levels of knowledge we have described.

Latent knowledge. When an agent, i, becomes aware of a proposition φ through a public announcement, they acquire a knowledge state of the propositions in φ that is equivalent to its latent knowledge of these propositions with respect to φ. That is: $M, s \models [\varphi]\mathbf{K}_i\psi$ if and only if $M, s \models \mathbb{K}_i^{\mathfrak{L}}(\varphi \rightarrow \psi)$.

Tacit knowledge. When an agent, i, receives a public announcement, φ, they learn φ is true and also become aware of the subformulas of φ. Their knowledge of these subformulas is equivalent to their latent knowledge of φ. However, the agent i would not tacitly know this to be true before the public announcement. Rather they would (tacitly) know that on becoming aware of φ they would acquire a knowledge state that is consistent with their explicit knowledge, and one which preserves their tacit knowledge.

Strong implicit knowledge. When an agent i, receives a public announcement, φ, they learn φ is true, and become aware of the subformulas of φ. Their knowledge of these subformulas is described by the set of worlds the agent considers possible, as with latent knowledge. However, as with tacit knowledge the agent would not know this prior to the announcement. The agent implicitly knows that becoming aware of φ, could lead to any new knowledge that is consistent with their current (explicit) knowledge state.

The following lemma establishes the relative strength of each type of knowledge.

Lemma 2. *Let (M, s) be a pointed awareness model, and φ be a formula. Then:*

$$M, s \models \mathbf{K}_i\varphi \Longrightarrow M, s \models \mathbb{K}_i^{\mathfrak{I}}\varphi$$
$$M, s \models \mathbb{K}_i^{\mathfrak{I}}\varphi \Longrightarrow M, s \models \mathbb{K}_i^{\mathfrak{T}}\varphi$$
$$M, s \models \mathbb{K}_i^{\mathfrak{T}}\varphi \Longrightarrow M, s \models \mathbb{K}_i^{\mathfrak{L}}\varphi$$

Proof. For the first implication it is sufficient to note that if the agent explicitly knows φ, then the agent is aware of φ, and so in the context of φ the awareness bisimulation is simply a bisimulation (see Definition 4). As bisimulations preserve the interpretation of modal formulas it follows that the agent implicitly knows φ.

If the agent implicitly knows φ then in every accessible world $t \in sR^M$, every model N_u that is $\mathcal{A}_i(s)$ bisimilar to M_t, we have $N_u \models \varphi$. Now by Definitions 4 and 7 any model N'_v that is a $\mathcal{A}_i(s)$ refinement of M_t, is also a $\mathcal{A}_i(s)$ bisimulation of M_t, and thus $N'_v \models \varphi$, so $M_s \models \mathbb{K}_i^{\mathfrak{T}}\varphi$. Finally if $M_s \models \mathbb{K}_i^{\mathfrak{T}}\varphi$ it follows that $M_s \models \mathbb{K}_i^{\mathfrak{L}}\varphi$ as the awareness refinement relation is clearly reflexive.

Lemma 3. *For all variations of implicit knowledge, $\mathbb{K}_i\varphi \wedge \mathbf{A}_i\varphi$ is satisfied by a pointed model if and only if $\mathbf{K}_i\varphi$ is.*

Proof This can be seen by noting in the case of both awareness refinement and awareness bisimulations, when the agent is fully aware of a formula φ, with respect to φ both relations are the same as a bisimulation, which preserves the interpretation of modal formulas.

5 Related Approaches

Here we mention some additional work on awareness. Although these contributions do not add to the variety of semantics for interpreting awareness and implicit knowledge, they certainly provide an appropriate context to consider the advantages and disadvantages of each approach. Hill [11] has examined the dynamics of awareness in the single agent case.

Agotnes and Alechina [1] have considered an expressive extension of logics of knowledge and awareness that allow us to quantify over formulas an agent is unaware of. Particularly, this allows an agent to know another agent knows something that the first agent does not. This expressive ability is also considered by Halpern and Rego [9].

Sillari [16], has also examined the question of knowledge of unawareness, and addresses this question in the context of epistemic first order logic. Also, strong implicit knowledge has previously been examined and axiomatized in [4].

A novel approach to distinguishing what an imperfect agent might know as compared to a logically omniscient agent can be found in [18,2]. Here a dynamic epistemic logic is used to model deductive step an agent makes, so we might find an agent who is able to deduce a fact and may indeed be in the process of deducing a fact, but has not yet come to the final realization that the fact is true. We note that [18] gives an axiomatization for such awareness change in the multi-agent case.

In [8] Grossi and Velazquez expand on these ideas using the ideas of multi-valued logic. They differentiate between formulas agents are *aware of* and formulas agents have *access to*, such that access is stronger than awareness and captures the idea that a deduction has occurred that has bought the truth of a formula to the attention of an agent, and thus the agent may proceed to use this formula in future deductions.

Heifetz, Meier and Schipper [10] presented a complete framework for describing the interactive knowledge of multiple agents with differing levels of awareness. By interactive knowledge, we mean one agent may know about the knowledge and awareness state of another agent. This was an extension of Modica's and Rustichini's theory of unforeseen contingencies [14].

6 Conclusion and Future Work

In this paper we have presented a comparative analysis of the way that awareness and knowledge may interact in dynamic settings. This is an important analysis as different situations will require different types of knowledge and awareness change: modelling secure communications may suit latent knowledge; modelling commercial transactions may suit tacit knowledge; and modelling automated reasoning may be better suited to strong implicit knowledge. The semantic model for tacit knowledge presented here is particularly interesting as tacit knowledge is a well known concept of epistemiology, but while strong implicit and latent forms of knowledge have been previously examined in the context of modal logic, tacit has not.

References

1. Agotnes, T., Alechina, N.: Full and relative awareness: a decidable logic for reasoning about knowledge of unawareness. In: Proceedings of TARK XI, pp. 6–14. ACM, New York (2007)
2. van Benthem, J., Velazquez-Quesada, F.: Inference, promotion and the dynamics of awareness. In: Knowledge, Rationality and Action (2009)
3. Cadoli, M., Schaerf, M.: Approximate reasoning and non-omniscient agents. In: TARK (1992)
4. van Ditmarsch, H., French, T.: Becoming Aware of Propositional Variables. In: Banerjee, M., Seth, A. (eds.) Logic and Its Applications. LNCS, vol. 6521, pp. 204–218. Springer, Heidelberg (2011)
5. van Ditmarsch, H., van der Hoek, W., Kooi, B.: Dynamic Epistemic Logic, Synthese Library, vol. 337. Springer, Heidelberg (2007)
6. Fagin, R., Halpern, J.: Belief, awareness, and limited reasoning. Artificial Intelligence 34(1), 39–76 (1988)
7. Fagin, R., Halpern, J., Moses, Y., Vardi, M.: Reasoning about Knowledge. MIT Press, Cambridge (1995)
8. Grossi, D., Velázquez-Quesada, F.R.: *Twelve Angry Men*: A Study on the Fine-Grain of Announcements. In: He, X., Horty, J., Pacuit, E. (eds.) LORI 2009. LNCS, vol. 5834, pp. 147–160. Springer, Heidelberg (2009)
9. Halpern, J., Rego, L.: Reasoning about knowledge of unawareness. Games and Economic Behavior 67(2), 503–525 (2009)
10. Heifetz, A., Meier, M., Schipper, B.: Interactive unawareness. Journal of Economic Theory 130, 78–94 (2006)
11. Hill, B.: Awareness dynamics. Journal of Philosophical Logic 39, 113–137 (2010)
12. Sim, K.M.: Epistemic logic and logical omniscience ii: A unifying framework. International Journal of Intelligent Systems 15 (2000)
13. Levesque, H.: Logic of implicit and explicit belief. In: Proceedings of AAAI 1984, pp. 198–202 (1984)
14. Modica, S., Rustichini, A.: Unawareness and partitional information structures. Games and Economic Behavior 27, 265–298 (1999)
15. Polanyi, M.: The logic of tacit inference. Philosophy 41, 1 (1966)
16. Sillari, G.: Quantified logic of awareness and impossible possible worlds. Review of Symbolic Logic 1(4), 514–529 (2008)
17. Velázquez-Quesada, F.R.: Inference and update. Synthese (Knowledge, Rationality and Action) 169(2), 283–300 (2009)
18. Velazquez-Quesada, F.: Small steps in dynamics of information. Ph.D. thesis, University of Amsterdam (2011), iLLC Dissertation Series DS-2011-02

On the Implementation of a Theory of Perceptual Mapping

Wai Kiang Yeap, M. Zulfikar Hossain, and Thomas Brunner

Centre for Artificial Intelligence Research,
Auckland University of Technology, Auckland, New Zealand
Wai.yeap@aut.ac.nz

Abstract. A recent theory of perceptual mapping argues that humans do not integrate successive views using a mathematical transformation approach to form a perceptual map. Rather, it is formed from integrating views at limiting points in the environment. Each view affords an adequate description of the spatial layout of a local environment and its limiting point is detected via a process of recognizing significant features in it and tracking them across views. This paper discusses the implementation of this theory on a laser-ranging mobile robot. Two algorithms were implemented to produce two different kinds of maps; one which is sparse and fragmented, and the other which is dense and detailed. Both algorithms successfully generated maps that preserve well the layout of the environment. The implementation provides insights into the problem of loop closing, moving in featureless environments, seeing a stable world, and augmenting mapping with commonsense knowledge.

Keywords: Perceptual map, SLAM, cognitive agent, spatial layout.

1 Introduction

In the past 15 years, robotics researchers have made significant advances in solving the simultaneous localization and mapping (SLAM) problem [1, 2]. They argue that the standard state-space approach to SLAM is well understood, and future key challenges lie in developing larger, more persuasive demonstrations which involve mapping a city or structures such as the Barrier Reef or the Mars surface. Although these challenging mapping projects will drive the development of useful robots, this paper explores the more fundamental question of how nature solves the mapping problem of one's own environment. In particular, we ask: do humans (and animals) employ a SLAM-like algorithm to map their environment? If not, why not and what would their algorithm be like?

Intuitively, the answer is no since it is unlikely that humans process a map like that of current robots. Researchers with disparate backgrounds, ranging from psychologists to geographers and urban designers, have been studying this problem and refer to it as a cognitive mapping process and its product, a cognitive map [3-5]. These studies show much of what we remember is an inaccurate (in metric terms), incomplete, and fragmented representation. Yet, as we explore an environment, we,

D. Wang and M. Reynolds (Eds.): AI 2011, LNAI 7106, pp. 739–748, 2011.

like the robots, need to integrate successive views to form a representation of our immediate surrounding, i.e. our perceptual map. This representation is precise enough to point to unseen locations just visited and can be learned early during exposure to a new environment [6, 7], without which we would 'forget' where things are the moment we look away [8]. Many researchers interested in human spatial perception implicitly or explicitly reason that some kind of a SLAM-like algorithm is used to compute a perceptual map [9-12].

Probabilistic solutions to the SLAM problem are a significant achievement, because it shows how an accurate map can be computed and how errors in one's perception can be handled in the integration process to produce a useful map. The latter is important if a SLAM-like solution is used to explain how humans integrate successive views to produce a perceptual map because human perception is an incomplete geometrical description of their environment [8, 13, 14] and illusory [15, 16]. If humans integrate successive views to produce a perceptual map, then a solution at least as powerful as probabilistic SLAM is needed.

Alas, nature appears to have found a different solution. Judging from past AI research in other areas such as stereo vision and bipedal motion, this comes as no surprise. In stereo vision, the matching problem was first thought to be solved more appropriately at the object level rather than the image level [17] and in locomotion, the early machines built were on multiple legs rather than two. In the integration problem, human input is more complex than for robotics. For example, humans have high visual acuity only in the small foveal region of the retina and thus a large part of the input lacks clarity and details. Furthermore, human eyes make rapid movements (known as saccades) to focus on different regions. It was thought that successive retinal-displaced views were integrated to produce the expected richly detailed and stable world representation [18, 19]. However, subsequent tests of this idea discovered that we are often insensitive to changes occurring between saccades and even between views in natural settings [20-23]. This phenomenon, referred to as "change blindness", presents a strong case against the idea of integrating successive views to form a single unified representation.

Thus, there is an apparent paradox in the way in which humans compute a representation of their environment. On the one hand, we need to integrate successive views and a probabilistic SLAM approach would be ideal since it copes with the presence of errors in one's sensing of the environment. On the other hand, our cognitive map bears little resemblance to an accurate metric map and change blindness argue against the use of an integrative approach like SLAM. For instance, Yeap [24] presented a theory of perceptual mapping that does not use the standard mathematical transformation approach to compute a perceptual map. Instead, it integrates successive "local environments" obtained from views at limiting points in the environment. Here, we implemented that theory on a laser-ranging mobile robot. Two algorithms were implemented and tested on two indoor environments: one computes the map in a cognitive manner and the other in a robotic manner. Both algorithms successfully generated maps that well preserve the layout of the environment and provided insights into the nature of perceptual mapping.

2 A Theory of Perceptual Mapping

In developing his theory of perceptual mapping, Yeap [24] made two observations. First, he observed that a view affords us more than a description of the surfaces in front of us. It tells us what and where things are, where we can move to next, what events are unfolding, where there might be dangers, and others [25]. In short, a view is in fact a significant representation of a local environment and it should be made explicit in the map as a description of a local environment rather than as some spatially organized surfaces. Second, he observed that the world we live in is relatively stable. That is, it does not change much when we blink our eyes or take a few steps forward. As such, there is no immediate need to update the view in our perceptual map as we move. For example, consider your first view of a corridor when entering it and assume an exit can be seen at the other end. If you walk down this corridor to the exit, then the description of the corridor space afforded in the first view adequately describes the local environment you were going through. Updating this description to include, for example, a view of a room besides the corridor as you walk past it will enrich the description, but is unnecessary if the room is not entered.

Combining both observations, Yeap [24] suggested that one's perceptual map is computed by integrating views only when one is about to move out of the local environment afforded in an earlier view. This immediately pose two problems: if we do not update our map as we move, how do we know where we are in our map and more importantly, how do we update the map with the next view when the need arises? Yeap argued that humans recognize where they are by recognizing familiar objects in their environment. Again consider the corridor example above and assuming you are half way down the corridor. How do you then know where you are in the map? If you recognize the exit at the end of the corridor to be the same exit in the initial view, then, using triangulation, you can locate your approximate position in the map. Thus, one possible solution is to keep track of objects seen in the remembered view in the current view. These objects are referred to as reference targets. If some could be found, one could triangulate one's position in the map and thus localize oneself. However, at some points, one will not be able to do so and this is when one needs to expand the map to include a new view (albeit, a new local environment). These points are known as limiting points. If the new view to be added is selected at a point just before reaching a limiting point, it could be added to the map using the same method of triangulation.

3 Implementations and Results

We tested Yeap's theory using a Pioneer 3 mobile robot (dimension 40×45 cm) equipped with SICK laser rangefinder (laser beam resolution 0.5 degree, maximum range 30 meter, view angle 180 degree) mapping two different indoor environments as shown in Figure 1. In Figure 1a, the route taken was about 100m long and in Figure 1b, the route was about 170m long. The theory leaves open three implementation issues, namely how and what reference targets are selected, when a new view (a collection of surfaces) should be added to the perceptual map, and how much information in each view is combined with what is in the map.

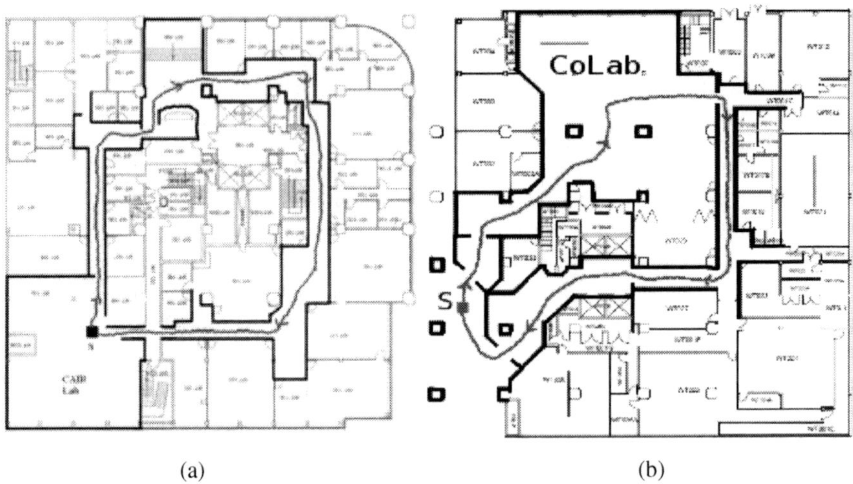

(a) (b)

Fig. 1. Two test environments

Implementing the theory on a laser-ranging mobile robot means that our choice of a reference target is limited to perceived surfaces in view, each being represented as a 2D line segment. Since these lines are recovered from laser points, their shape varies between views depending on the robot vantage point. Consequently, we choose reference targets with a minimum length of 40cm, which has at least an occluding edge. The latter ensures a reference point exists on the surface for relative positioning of new surfaces into the map. A minimum of two such points is needed.

Since the robot cannot recognize these reference targets directly from one view to the next, their recognition has to be "simulated" or done indirectly. For our implementation, we use the transformation method. Transforming between two successive views, the robot can predict where these reference targets will be. They are then "recognized" via the use of some heuristics such as the proximity of two surfaces, the sudden appearance of a new surface in front of another, and others. It is emphasized that the transformation method is used here because of the lack of any recognition ability of our robot. It is not part of the theory.

In terms of when a view is to be added to the perceptual map, we implemented two choices: the first adds a new view at the limiting point and the second adds a view as soon as it becomes available. The former should produce a map more akin to the kind produced in cognitive beings i.e. sparse and fragmented. The latter should produce a map more akin to the kind produced in robots i.e. dense and detailed. We refer to the two algorithms as algorithm #1 and algorithm #2, respectively.

3.1 Algorithm #1

Let PM be the perceptual map, V_0 be one's initial view, and R be some reference targets identified in V_0. Initialize PM with V_0. For each move through the environment, do:

1. Execute move instruction and get a new view, V_n.
2. If it is a turn instruction, use V_n to expand PM and create a new R. Go to step 1.
3. Search for the reference targets in V_n by transforming V_{n-1} to V_n using the mathematical transformation approach. [recognition]
4. If two or more targets are found, remove unseen targets from R. Go to step 1. [tracking]
5. Add V_{n-1} to PM. This is achieved by first locating the robot's position and orientation in PM and then replaces what is in front of the robot in PM with what is seen in V_{n-1}. [expanding]
6. Create a new set of R from V_{n-1}. Go to step 1.

One key feature of this algorithm is that it does not perform much updating of surfaces in the map. What is added as the new local environment is exactly what is seen in the current view. The rationale here is that details are unimportant as long the overall shape of the environment is maintained. An exception is the updating of the length of surfaces that extend from the previous view to the current view. All computations are approximate only. This implementation is thus best suited for cognitive agents where they don't need to remember everything that they have perceived.

Figure 2a shows a perceptual map produced as the robot traversed the path through the environment shown in Figure 1a (in a clockwise direction). We have tested the robot with a total of 7 different start locations and all the maps produced show a good layout of the environment.

3.2 Algorithm #2

Let PM be the perceptual map, and V_0 be one's initial view. Initialize PM with V_0. For each move through the environment, do:

1. Execute move instruction and get a new view, V_n.
2. Find all common surfaces between V_n and V_{n-1} by transforming previous view to the new view using the mathematical transformation approach. Update description of these surfaces in the map if necessary.
3. Identify the best reference target, R, from all found matches in the PM.
4. For all new surfaces in V_n, add them as new to the PM using the reference target R.
5. Remove redundant surfaces in PM – compare all surfaces inside PM with those just added to see if they are viewed as the same surface. If they are, remove the newly added surface.
6. Re-position robot in PM using the reference target, R. Go to step 1.

The key feature of this algorithm is that it tries to find the best reference target to update the map at each view. Such an implementation is suited for robots since they need not forget what they have perceived. Figure 2b shows a perceptual map produced as the robot traversed the path through the environment shown in Figure 1a (and in a clockwise direction). We have tested the robot with a total of three different start locations and all the maps produced show a good layout of the environment.

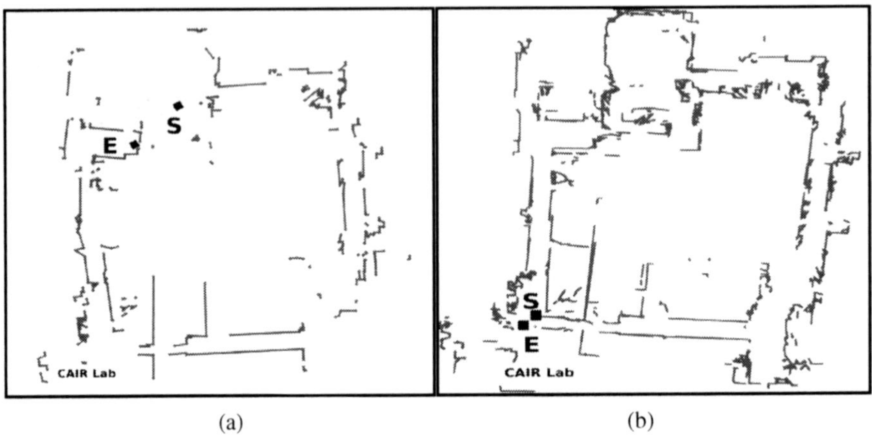

(a) (b)

Fig. 2. Perceptual maps for the environment shown in Figure 1a

4 Discussions

Both algorithms produce maps that capture a good spatial layout of the environment traversed despite no sensing-error corrections. The map produced using the first algorithm is sparser than that produced using the second algorithm. However, the latter shows many surface overlaps. This results from a failure to recognize many surfaces as the same surface and is a consequence of a poor recognition algorithm. It is expected that surface overlaps will be avoided in future recognition algorithms, producing a cleaner, more usable map. Otherwise, one needs a post-processing algorithm to identify useful empty regions for robot planning and navigation in the environment.

When testing the first algorithm, we observed how the use of a single view to describe one's local environment created stable descriptions of local environments. This is in contrast to the map produced using the traditional transformation approach. The latter keeps changing due to error correction. For example, even though one sees a straight corridor from its entrance, one could end up with a curved corridor at the end because of errors. With our approach, one sees and remembers a straight corridor as the local environment that one is entering into and that memory did not change as one move in it. Psychologists refer to this phenomenon as our perception of a stable world.

Given that we produce stable maps, we can address a question which robotics researchers are most interested in: how do we close the loop? This question relates to the problem of incorrectly creating a new part of the map when re-visiting familiar parts of the environment. Robotics researchers are concerned with producing an accurate map. Therefore, they are concerned with developing algorithms to match and fit the incoming map with the existing one so that a coherent map is produced. However, observing how the first algorithm updates its map, we argue there is a useful alternative: re-visiting or not, one deletes old memories of what was there and replaces it with what is in view. This idea is comes from the observation that your current view should provide the best description, in terms of most updated, of what is in the environment, not your memory of it. Figure 3 shows what happen to the perceptual map when you continuously go round the same environment. In the top-left map, the robot (at point E) is about to complete one round of the environment. In the top-right map, it has re-entered a familiar part of the environment looking at it in the same direction when it was first viewed. Note that part of the corridor is now wiped out of memory but nonetheless the robot is aware of what lies ahead. It continues its journey and all along, it maintains an accurate spatial layout of the environment.

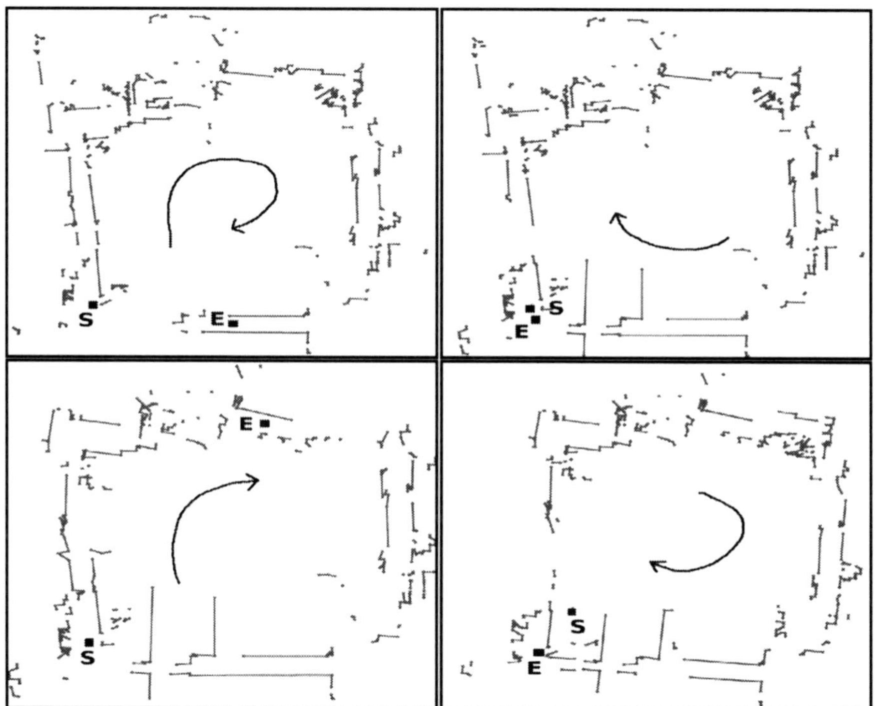

Fig. 3. Continuous mapping of the environment shown in Figure 1a

A plot of the actual physical position of the robot in space (see Figure 4) reveals a physical shift but this has no effect on the map produced. Thus, unlike the transformation method, one can begin a journey at any time without having to worry about synchronizing one's view with the map in the head or remembering one's last position in space.

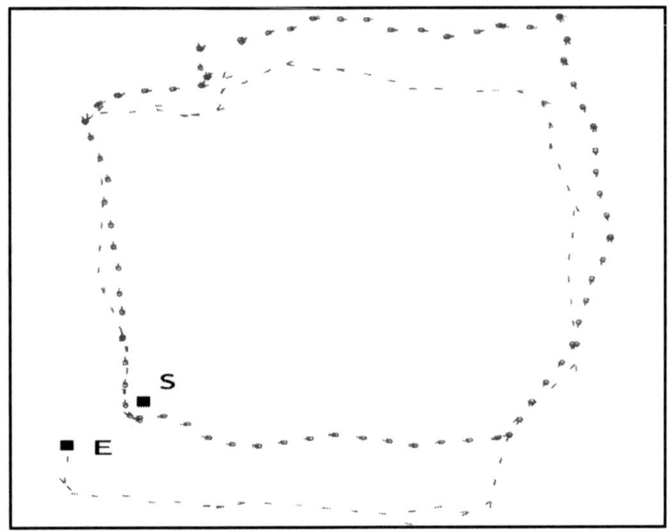

Fig. 4. Robot's absolute path, where the "—" marks the repeated path

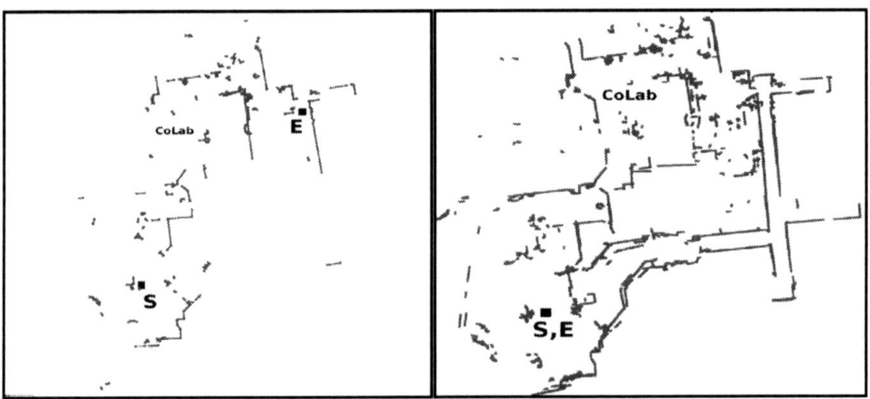

Fig. 5. Perceptual maps produced for the environment shown in Figure 1b

We tested both algorithms using another environment (Figure 1b) and it produced the result shown in Figure 5. Notice that the robot using the first algorithm fails to produce a map (left) for the straight path down a long corridor on the right of the environment. This corridor has no distinguishing feature and consequently the robot could not find a reference target in it. It is interesting to note that humans often are disoriented in featureless environment (such as walking down a spiral corridor) and in such situations, one has to use additional higher reasoning to cope with the situations. The robot using the second algorithm also failed to find a reference target when it first entered the corridor. However, it took the next step forward and managed to find some reference targets. It combined the two steps as one and managed to update its map and eventually successfully produce a map (right).

These tests highlight situations whereby it shows a need to use much of one's higher level or common sense reasoning to deal with the learning a new environment. These situations include the recognition of objects, the use of the perceptual map when returning to a familiar part of the environment, and traversing in featureless environments. The map produced using our algorithm provides an adequate basis for such reasoning.

5 Conclusion

This paper shows the implementation of a new theory of perceptual mapping using a laser-ranging mobile robot. The theory was developed to explain how humans and animals compute their perceptual map. Two algorithms were implemented to produce two different kinds of maps: one which is sparse and fragmented, and the other which is dense and detailed. The results showed that despite no error correction, a useful layout of the environment is easily computed if good reference targets are made available. However, in one of the environments, the lack of good reference targets caused a failure to map the environment. This demonstrates that the algorithm, like humans, is not robust and one could get lost. In situation like this, one needs to exploit other forms of knowledge to resolve the problem.

Future work will focus on developing more robust algorithms based upon this new theory of perceptual mapping for robot mapping, and designing new experiments on human and animal cognitive mapping processes to test the validity of the theory as a theory of human/animal perceptual mapping.

References

1. Bailey, T., Durrant-Whyte, H.: Simultaneous localization and mapping (SLAM): part II. IEEE Robotics & Automation Magazine 13(3), 108–117 (2006)
2. Durrant-Whyte, H., Bailey, T.: Simultaneous localization and mapping: part I. IEEE Robotics & Automation Magazine 13(2), 99–110 (2006)
3. Downs, R.M., Stea, D.: Image and Environment. Cognitive Mapping and Spatial Behavior. Science, Aldine, Chicago (1973)
4. O'Keefe, J., Nadel, L.: The Hippocampus as a Cognitive Map. Behavioral and Brain Sciences 2(4), 487–533 (1979)
5. Tolman, E.C.: Cognitive maps in rats and men. Psychological Review 55(4), 189–208 (1948)
6. Buchner, A., Jansen-Osmann, P.: Is Route Learning More Than Serial Learning? Spatial Cognition & Computation: An Interdisciplinary Journal 8(4), 289–305 (2008)
7. Ishikawa, T., Montello, D.R.: Spatial knowledge acquisition from direct experience in the environment: Individual differences in the development of metric knowledge and the integration of separately learned places. Cognitive Psychology 52(2), 93–129 (2006)

8. Glennerster, A., Hansard, M.E., Fitzgibbon, A.W.: View-Based Approaches to Spatial Representation in Human Vision. In: Cremers, D., Rosenhahn, B., Yuille, A.L., Schmidt, F.R. (eds.) Statistical and Geometrical Approaches to Visual Motion Analysis. LNCS, vol. 5604, pp. 193–208. Springer, Heidelberg (2009)
9. Burgess, N.: Spatial memory: how egocentric and allocentric combine. Trends in Cognitive Sciences 10(12), 551–557 (2006)
10. Mou, W., McNamara, T.P., Valiquette, C.M., Rump, B.: Allocentric and Egocentric Updating of Spatial Memories. Journal of Experimental Psychology: Learning, Memory, and Cognition 30(1), 142–157 (2004)
11. Rump, B., McNamara, T.P.: Updating in Models of Spatial Memory. In: Barkowsky, T., Knauff, M., Ligozat, G., Montello, D.R. (eds.) Spatial Cognition 2007. LNCS (LNAI), vol. 4387, pp. 249–269. Springer, Heidelberg (2007)
12. Wang, R.F., Spelke, E.S.: Updating egocentric representations in human navigation. Cognition 77(3), 215–250 (2000)
13. Bridgeman, B., Hoover, M.: Processing spatial layout by perception and sensorimotor interaction. Psychology Press, London (2008)
14. Fermüller, C., Cheong, L., Aloimonos, Y.: Visual Space Distortion. Biological Cybernetics 77, 323–337 (1997)
15. Hurlbert, A.C.: Visual Perception: Knowing is seeing. Current Biology 4(5), 423–426 (1994)
16. Snowden, R.J.: Visual perception: Here's mud in your mind's eye. Current Biology 9(9), R336–R337 (1999)
17. Frisby, J.P.: Seeing: illusion, brain, and mind. Oxford University Press, Oxford (1980)
18. Davidson, M., Fox, M.-J., Dick, A.: Effect of eye movements on backward masking and perceived location. Attention, Perception, & Psychophysics 14(1), 110–116 (1973)
19. McConkie, G.W., Rayner, K.: Identifying the span of the effective stimulus in reading: Literature review and theories of reading. In: Singer, H., Ruddell, R.B. (eds.) Theoretical Models and Processes of Reading, pp. 137–162. International Reading Association (1976)
20. Intraub, H.: The representation of visual scenes. Trends in Cognitive Sciences 1(6), 217–222 (1997)
21. Irwin, D.E.: Integrating Information Across Saccadic Eye Movements. Current Directions in Psychological Science 5(3), 94–100 (1996)
22. Irwin, D.E., Zelinsky, G.J.: Eye movements and scene perception: memory for things observed. Perception & Psychophysics 64(6), 882–895 (2002)
23. Simons, D.J., Rensink, R.A.: Change blindness: past, present, and future. Trends in Cognitive Sciences 9(1), 16–20 (2005)
24. Yeap, W.K.: A computational theory of humans perceptual mapping. In: 33rd Annual Conference of the Cognitive Science Society (in press, 2011)
25. Gibson, J.J.: The perception of the visual world. Houghton Mifflin, Boston (1950)

Robotic Communications and Surveillance – The DARPA LANdroids Program

Dan R. Corbett[1], Douglas W. Gage[2], and Douglas D. Hackett[3]

[1] Research In Motion, Inc., Waterloo, Ontario, Canada
dcorbett@acm.org
[2] XPM Technologies, Arlington, Virginia, USA
douggage@san.rr.com
[3] Griffin Technologies, Fort Washington, Pennsylvania, USA
dhackett@griffin-technologies.com

Abstract. The principal goal of the LANdroids program (2007-2010) was to validate the concept that mobile tactical radio relay platforms can provide improved communications connectivity in non-line-of-sight communications environments such as urban terrain. The first phase of the program demonstrated that intelligent mobile relays can provide improved system performance in network configuration, optimization, and self-healing, and the second phase added additional capabilities including intruder detection and situational awareness, and included a real-world demonstration to potential users.

Keywords: robotic communications relays, MANET, surveillance, autonomy.

1 Introduction

One of the major challenges facing dismounted warfighters operating in urban environments is the unreliability of radio communications. Buildings and other obstacles absorb, reflect, and diffract radio signals, causing signal loss or attenuation, and creating a generally highly complex signal propagation environment that is difficult to accurately predict. The LANdroids solution to this problem is to deploy small inexpensive robotic relay nodes that position themselves intelligently to relay signals to warfighters moving through these settings to conduct their operations. The significance of the LANdroids program is that several techniques for intelligent robot movement were evaluated, including a kinetic-state machine approach, that implemented a decentralized decision-making process and a "playbook" technique, that modified pre-defined configurations according to the situation and environment.

1.1 The Challenging Urban RF Environment

The LANdroids concept is designed to exploit the fact that a given emitter's signal strength depends sensitively on the position of the receiver. Fig. 1 shows a map of the

D. Wang and M. Reynolds (Eds.): AI 2011, LNAI 7106, pp. 749–758, 2011.

signal strength around a building of an FM radio signal broadcast from the other side of an urban environment. This signal strength map contains instances of both multi-path fading and shadowing. Multi-path fading typically results in large variations within small areas. In this example, a 20dB change in signal strength was measured over a distance of only 10 feet (one wavelength). At higher frequencies (e.g., 2.4 GHz), the shorter wavelength will cause variations to occur on sub-meter scales. LANdroids mobile relay nodes would therefore be able to access locations with greatly increased signal strength by making only small changes in location.

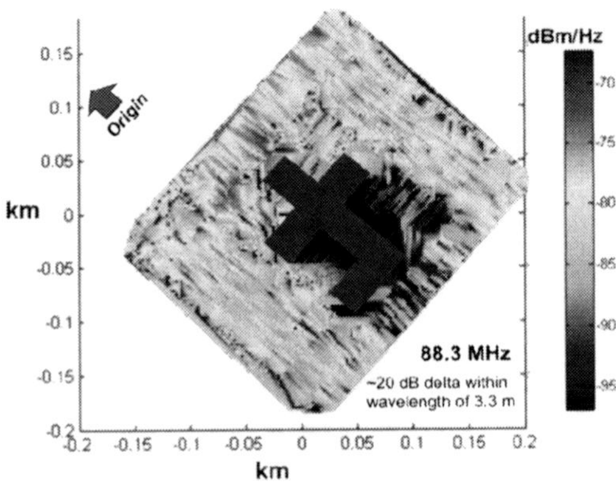

Fig. 1. Signal strength map around a facility in an urban setting

Fig. 1 also shows shadowing, illustrated by the deep blue regions in which the building is blocking the signal. In these regions, the average signal strength is much weaker and the area of poor communications performance can be much larger than is caused by multi-path fading. Shadows are another phenomenon that LANdroids relays can deal with by carefully choosing their locations.

The signal propagation characteristics within an urban environment are not just poor; they are also usually very difficult to predict. Angle of incidence matters, where an emitter is located matters, even the building materials themselves can impact signal strength. We often deal with this in the civilian world through an iterated manual measure, test, and improve cycle – for example, in optimizing cell phone tower placement ("can you hear me now?"). LANdroids, however, must determine where exactly to "sit" based on the situation – not based on pre-programmed maps or other approaches that require detailed knowledge *a priori*. Such knowledge may augment a given approach, but there will always be an element of deciding in real time where a given LANdroid node should locate itself. Another important concept is that dynamical changes in the world can impact what constitutes a good location. Moving a convoy of trucks, removing a structure or building, or putting a new structure or building in place, can degrade RF connectivity, as can changes in RF noise or in the positions of other emitters in the environment. A LANdroid that finds a good position in which to sit at time t_1 may well need to adjust its position at time t_2.

All of these phenomena translate into a need for relay node mobility to achieve basic communications coverage. Once a capability for intelligent autonomous movement is incorporated into the relays, other system-level capabilities, such as self-healing in the face of node loss, can also be implemented.

1.2 The LANdroids System Concept

A LANdroid is a small, inexpensive, smart, mobile radio relay node, comprised of a robotic platform (providing appropriate mobility and sensing capabilities), processor, radio, and power source. The LANdroids system concept is that dismounted warfighters (or police, firefighters, etc.) will each carry a number of these LANdroids and deploy them as they move through a mission operations area. Following deployment, each LANdroid will then move so that as a group they configure themselves to form a mesh network over the area of operations, providing a temporary communication infrastructure that covers the initial warfighters, subsequent warfighters, deployed sensors, UGVs, etc. Any device operating in the area will be able to maintain communications via the LANdroids mesh network.

The advantage of a mesh network approach to communications is resiliency – the network is multi-path, multi-hop, and multiply connected. If any one node should go down or be taken out by an enemy, the packets from other nodes will find other routes to reach the gateway. An additional strength specific to the LANdroids system is that the remaining nodes can in fact move to create new links – the network as a whole can self-heal to cover the area. LANdroids can also exploit movement in order to implement "tethering" or network stretching to keep warfighters or devices covered as they move and expand their area of operations.

Others [1, 2, 3] have investigated the real-time deployment of static relays during a mission, both by humans and by robots; the point of the LANdroids Program was to demonstrate the value of relay mobility following deployment [4].

2 Landroids Software and Hardware

2.1 LANdroids Software Capabilities and Development Efforts

The Control Software developed under LANdroids was required to address capabilities of *Self-Configuration* (self organize to form a mesh network over the coverage region, including detecting neighboring nodes, establishing connections to one or more gateways, and ensuring that the region is fully covered), *Self-Optimization* (continue to make movements 'in the small' to try to find locations with higher signal strength even after a network is formed), *Self-Healing* (detect whenever a gap in the coverage region is created because a LANdroid node is destroyed by an enemy, powers down, or otherwise fails, and self-heal to the best extent possible), *Tethering* (the network itself should adapt and stretch to keep users covered when they move out of the coverage area) and *Intelligent Power Management* (reason about power conservation and make explicit decisions about whether or not to move, and whether it is possible to power down because another LANdroid is covering the same area).

Five Phase I contracts were awarded for the development of Control Software, and two of these five contractors were selected to continue into Phase II: Intelligent Automation, Inc. (IAI), and Lockheed Martin Space Systems Company (LM-SSC). These performers explored a variety of approaches to the issues of short range motion to find local signal strength maxima, long range motion to establish links and optimize network topology, communications between LANdroids to coordinate their movements, and management of the information necessary to support this whole process [5,6].

The IAI approach modeled each robot as a kinetic state machine. A distributed graph-theoretic formulation is adopted for representing the problem and graph metrics were used to determine various state transitions [6]. These transitions happen independently and asynchronously based solely on local information. There is no global view of the network, but the robot nodes are implicitly coordinated by sharing information that gives the robots a common view of the network topology. In this control mechanism, the assumption is that every robot has the same information and view of the overall current state of the network. A robot can therefore make a high-level decision based on this global view and assume that the other robots will come to the same decision. The decision making is therefore decentralized [6].

In the LM-SSC approach, each robot node was given a set of skills, such as corner probe (to map or move away from a hallway intersection), wall-follow, and signal-strength readings. The robots would execute sequences of various skills assembled into set "plays", such as optimize signal strength, spread out, and follow.

2.2 LANdroids Hardware Requirements and Prototype Platforms

Because a dismounted warfighter is obliged to physically carry a lot of gear, including weapons, food, water, and ammunition, any prospective additional load is judged extremely critically. For LANdroids to be accepted in the field, they must provide value commensurate with their weight, which must be minimized. It is obviously critical that LANdroids must be able to support communications for the complete duration of a mission, so batteries must afford maximum mass-energy density. Deployed LANdroids must be expendable – dismounted warfighters must be able to drop and go – to use the communications infrastructure while it is in place, but not have to move back into harm's way to retrieve the robots. The LANdroids program therefore undertook to develop robotic platforms that could have a final production cost of $100 per LANdroid at modest production volumes (e.g., one thousand units), assuming that technology will continue to provide greatly increased performance and reduced cost (and mass) in processing and radios, and, to some degree, in batteries.

A single contract was awarded for the development of the LANdroids robot platform to iRobot, Inc. The LANdroids Phase I prototype robot platform developed by iRobot was the Ember, a small tracked robot with flipper arms extending from the rear wheel hubs (Figure 2). Ember resembles a greatly downsized iRobot PackBot, except that it does not have powered tracks on its flippers. These flippers, however, because they are actively positioned, can be used as levers to greatly enhance the robot's mobility capabilities.

The Ember's external sensors parallel those of the iRobot Create [7,8] robot used as a surrogate platform by the LANdroids software performers in Phase I. They include two infrared cliff sensors located at the bottom-front edge of the robot, one infrared wall sensor on the right side, optical wheel encoders on each track (left and right) and a single camera. Navigation sensors include a 3-axis accelerometer, and a yaw sensor. The power source is an internal 12.8V rechargeable Lithium-Ion battery, which is permanently installed, and can be recharged only in place.

Fig. 2. Ember, the Phase I LANdroid prototype robot developed by iRobot (left), the Phase II LANdroid Development robot (LDR) developed by iRobot – standard top (middle) and sensor development payload bay top (right)

Ember ran OpenEmbedded Linux on a Freescale i.MX-31 processor, which hosted all of the high-level processing, while a separate AVR microcontroller was used for low-level motor control and sensing. The two processors communicate via a serial link at 38.4 kbps.

For Phase II, iRobot iterated its Ember design in the LANdroids Development Robot (LDR), also shown in Figure 2. In addition to resolving a number of mechanical and heat-related problems encountered in the Ember (inevitable in any de novo prototype), the LDR added a microphone and loudspeaker and a set of four cameras to the Ember sensor suite, and included updated processing and radio hardware.

Both the Ember and the LDR were designed to operate autonomously, including software to execute basic low-level behaviors such as self-righting (to automatically use its flippers to right itself when upside-down or on its side), as well as simple forward/reverse and left/right commands. Also included were several behaviors that can be invoked from a simple Operator Control Unit (OCU) via 802.11 communications, including demonstration programs of figure-eight and wall-following, and simple manual driving (forward/reverse and left/right commands).

3 LANdroids Testing

Formal evaluation of both the LANdroids software and hardware was performed at the end of both Phase 1 and Phase II, and, in addition, the LANdroids concept was demonstrated in an operational setting to potential military users at the end of Phase II. The software evaluation effort was executed by CenGen, Inc., the hardware evaluation by Southwest Research Institute (SwRI) [9].

3.1 Phase I Formal Evaluation

The LANdroids Phase I software evaluation was performed by CenGen at their facilities in Columbia, MD, in January 2009. Several types of scenarios focused on the evaluation of each contractor's software for self configuration, self optimization, and self-healing. Details of the Phase I Evaluation process and its specific results are reported in reference [10].

3.2 Phase II Formal Evaluation

The LANdroids Phase 2 software evaluation was performed by CenGen at the Howard Community College in Columbia, MD, in July 2010. Performer team software was loaded onto 15 LANdroids robots distributed among three floors of a large building. Three types of scenarios were established to test the software's capabilities:

- Optimizing – LANdroids moved to establish and maintain an optimal network between four warfighter nodes and a gateway node
- Self-Healing – the LANdroids network attempted to reconnect itself after some LANdroids were intentionally disabled
- Tethering – the LANdroids network accommodated a moving warfighter by either stretching of the LANdroids network (by moving the nodes), or signaling the warfighter to power up and drop new LANdroids nodes

An example Optimizing scenario is shown in Fig. 3. This configuration simulated a fire team placing LANdroids near the doorways of a building. When turned on, the LANdroids (red) autonomously moved into the building to connect the warfighter nodes (blue) to the gateway node (green).

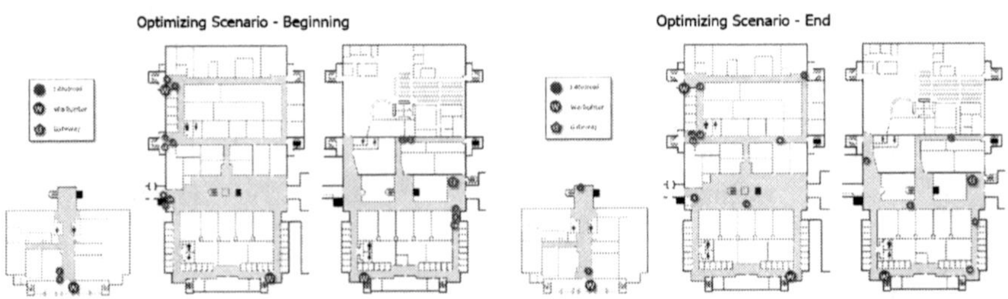

Fig. 3. An example Optimizing scenario – LANdroids moved from the initial positions (in red) to connect all four warfighters (blue) to the gateway (green)

A thorough set of metrics were collected on network performance: Connectedness, Coverage, Goodput Margin, Mission Bits, Coordination Overhead, Convergence Time, Total Energy, and Bits per Energy. Although the number of test runs was of necessity very limited (2-7), the standard deviations of the results were small and therefore the results can be considered statistically meaningful.

Average network coverage for scenarios where the starting configuration had poor connectivity (50-60%) was almost as good as those starting with excellent coverage (60-70%), showing that LANdroids movement optimized the network.

Self-healing was demonstrated to be possible, but the network reconverged slowly. In the Tethering scenario, the LANdroids did not move rapidly enough to stretch the network. Extra nodes were either dropped too soon or not at all.

In summary, the evaluation showed solid performance in optimizing the network. Significant work remains to be done for self-healing and tethering scenarios.

3.3 Phase II Field Demonstration

The LANdroids program culminated in a field demonstration held at the McKenna MOUT site at Fort Benning, Georgia, on September 7-10, 2010. The goal of this event was to showcase LANdroids capabilities for potential military users and partners in a realistic, urban tactical environment.

The scenario was as follows: two fire teams each deployed a set of LANdroids at the door of two separate buildings. The LANdroids distributed themselves throughout each building to establish a communications network. Communications were passed back to the command center through three iRobot PackBot robots. Each PackBot was equipped with a LANdroids-compatible radio.

The fire teams entered and cleared the building, using the LANdroids to maintain communications. Once the building was cleared, the LANdroids were commanded to assume their leave-behind sensor mode, allowing the command center to monitor the building for intruders.

To implement the leave-behind mode, LANdroids were augmented with four sensors: a Hokuyo lidar, a Sperient ultrawideband radar, passive infrared detectors,

Fig. 4. LANdroids deploying in buildings (left), and traversing underground tunnels (right) at the Ft. Benning MOUT site

and a pair of microphones. This package allowed the LANdroids to remain in place and monitor for intruders. Alerts, audio, and video were passed back to the command center through the LANdroids network.

In separate demonstrations, the LANdroids were teleoperated into confined spaces such as tunnels, relaying back video through their linked radios. Fig. 4 shows elements of the LANdroids demonstration.

At Ft. Benning, the LANdroids program demonstrated the following:

- Utility as a stand-alone UGV platform, teleoperated into confined spaces, sensing and serving as radio relays in MOUT features (buildings, tunnels)
- Autonomous dispersal of LANdroids, from the doors of buildings, to establish a communications network in the MOUT area buildings
- A "leave behind" intruder detection sensor capability, with alerts, audio, and videos relayed through the optimized LANdroids network
- A rudimentary LANdroid-based mapping of interiors, based on odometry and radio signal strength

As a research platform, the LANdroids network fell short of demonstrating a full military capability in several ways. First, the LANdroids were outfitted with 802.11 radios, not a military waveform. Second, the LANdroid robots were not sufficiently tall to climb stairs, and so were unable to disperse throughout the upper floors of a multi-story building. Third, the developed platform's mobility, combined with its maximum speed of 0.5 m/s, meant that the LANdroids fall far short of being able to maintain a realistic operational tempo (OPTEMPO).

Nonetheless, the military community expressed interest in several important elements of the LANdroids development. Some of the attendees were attracted to specific capabilities of the LANdroid robots. Specifically, the use of the platforms as tethered nodes to keep up with warfighters, for mapping unknown environments before sending warfighters in, and in the localization and registration of assets. There was also interest in using LANdroids to maintain connectivity of dismounted infantry to the network in the Land Warrior environment.

4 Conclusions: Issues and Next Steps

The LANdroids program has validated the concept that employing mobile tactical radio relay platforms can provide improved communications connectivity in NLOS communications environments such as urban terrain. The LANdroids program showed in Phase I that "intelligent" mobile relays can provide improved system performance in network configuration, optimization, and self-healing.

The distributed graph-theoretic formulation that produced a decentralized-but-global decision making process was shown to be highly successful. The robots using this technique demonstrated that they were able to configure, relay data and map an area. The playbook technique, that demonstrated robots executing sequences of various skills assembled into set "plays", was also validated in a real-world environment.

One explicit challenge addressed by Phase II of the LANdroids program was the need to articulate a specific concept of operations (CONOPS) that could gain acceptance with identified military customers, and to modulate the design of the robotic platform to specifically address this CONOPS. The inclusion of additional platform capabilities through an augmented sensor suite made the LDR a useful tool for surveillance, intruder detection and situational awareness. Moreover, sensors added as payload were exploited to enhance the robots' navigational capabilities. It is possible that a future CONOPS will involve outdoor areas of operation on the scale of hundreds or thousands of meters, rather than the indoor areas on the scale of tens of meters considered so far. Hence, the scale and nature of the navigational problem would definitely be much different, with obvious impacts on software development. Further, greater mobility capabilities (and a consequent possible increase in robot size) would likely be required in this case.

In fact, determining the appropriate scale for LANdroids robotic platforms is a major challenge in several respects. The original LANdroids plan called for a robot the size of a deck of cards, so that each dismounted warfighter could carry and deploy several of them, and such a small platform would also be quite covert. However, Phase II has demonstrated that a robot of this small size cannot currently carry the sensors needed to support localization (much less intruder detection), the batteries required to provide mission-length endurance, or a radio compatible with military use. Moreover, even the current larger LANdroids prototypes are not capable of climbing an 8-inch curb or stair riser.

Another challenge is to identify an appropriate target radio and military network for a future LANdroids system. Military tactical radios currently in use and under development are all far too large for the current LANdroids platform concept, but deployed LANdroids will have to be able to interoperate with the communications gear that the forces actually use. A decision to implement LANdroids as a "black" subnet implemented with COTS radio equipment (such as the 802.11 devices used in Phase I) would have the advantage of eliminating the need to deal with crypto and other security concerns, and would allow the continued direct leveraging of rapidly evolving COTS radio systems, ensuring a higher level of performance by avoiding the lengthy development cycle required to produce a military radio.

One very tangible result of the LANdroids program is that iRobot has developed a new commercially-available robot platform based on the LDR: the iRobot 110 FirstLook [11]. This is a throwable robot, larger than the Ember and LDR, that weighs less than 5 pounds and can be rapidly deployed to acquire situational awareness, to investigate confined spaces, or to perform persistent observation.

Acknowledgements. The authors thank all the participants in the LANdroids Program for their efforts, in Phase I, Phase II and in predecessor seedling projects. We also thank Jonathan Smith, Tom Wagner, Mark McClure and Robbie Mandelbaum for their work in initiating and managing this program.

The views, opinions, and/or findings contained in this article/presentation are those of the author/presenter and should not be interpreted as representing the official views or policies, either expressed or implied, of the Defense Advanced Research Projects Agency or the U.S. Department of Defense.

References

1. Souryal, M.R., Geissbuehler, J., Miller, L.E., Moayeri, N.: Real-time deployment of multihop relays for range extension. In: Proc. ACM MobiSys (June 2007)
2. Nguyen, H.G., Farrington, N., Pezeshkian, N.: Maintaining Communication Link for Tactical Ground Robots. In: AUVSI Unmanned Systems North America 2004, Anaheim, CA, August 3-5 (2004), http://www.spawar.navy.mil/robots/pubs/auvsi04_amcr.pdf
3. Pezeshkian, N., Nguyen, H.G., Burmeister, A.: Unmanned Ground Vehicle Radio Relay Deployment System for Non-line-of-sight Operations. In: Proc. 13th IASTED Int. Conf. on Robotics and Applications, Wuerzburg, Germany, August 29-31 (2007), http://www.spawar.navy.mil/robots/pubs/IASTED_ADCR_2007.pdf
4. BAA 07-46 LANdroids Broad Agency Announcement (BAA) for Information Processing Technology Office (IPTO) Defense Advanced Research Projects Agency (DARPA), June 5 (2007), http://www.darpa.mil/ipto/solicit/baa/BAA-07-46_PIP.pdf
5. Chiu, H.C.-H., Ryu, B., Zhu, H., Szekely, P., Maheswaran, R., Rogers, C., Galstyan, A., Salemi, B., Rubenstein, M., Shen, W.-M.: TENTACLES: Self-Configuring Robotic Radio Networks in Unknown Environments. In: IROS 2009, St Louis MO, October 11-15 (2009)
6. Mayhew, D., Judkins, T., Abeles, P., Manikonda, V.: Agile Robot Teams for Mobile Networking In Urban Environments. In: Proc. AUVSI Unmanned Systems North America (August 2010)
7. iRobot Corporation, iRobot Create Owners Guide (2006), http://www.irobot.com/filelibrary/pdfs/hrd/create/Create%20Manual_Final.pdf
8. iRobot Corporation, iRobot Create Open Interface (2006), http://www.irobot.com/filelibrary/pdfs/hrd/create/Create%20Open%20Interface_v2.pdf
9. http://www.swri.org/4org/d14/ElecPow/SmalRobo.htm
10. McClure, M., Corbett, D.R., Gage, D.W.: The DARPA LANdroids program. In: Unmanned Systems Technology XI, Orlando FL. SPIE Proceedings, vol. 7332 (April 2009)
11. http://www.irobot.com/gi/ground/110_FirstLook

Data Extraction for Search Engine Using Safe Matching

Jer Lang Hong[1], Ee Xion Tan[2], and Fariza Fauzi[2]

[1] School of Computing and IT, Taylor's University
jerlang.hong@taylors.edu.my
[2] School of IT, Monash University
{tan.ee.xion,wan.fariza}@monash.edu

Abstract. Our study shows that algorithms used to check the similarity of data records affect the efficiency of a wrapper. A closer examination indicates that the accuracy of a wrapper can be improved if the DOM Tree and visual properties of data records can be fully utilized. In this paper, we develop algorithms to check the similarity of data records based on the distinct tags and visual cue of the tree structure of data records and the voting algorithm which can detect the similarity of data records of a relevant data region which may contain irrelevant information such as search identifiers to distinguish the potential data regions more correctly and eliminate data region only when necessary. Experimental results show that our wrapper performs better than state of the art wrapper WISH and it is highly effective in data extraction. This wrapper will be useful for meta search engine application, which needs an accurate tool to locate its source of information.

Keywords: Information Extraction, Automatic Wrapper, Search Engines.

1 Introduction

A computer user is able to obtain relevant information from the World Wide Web simply and quickly due to the advent of Information Technology. As the World Wide Web contains a huge amount of data, the extraction of required information has been significantly simplified as the user needs to enter only search queries for the database servers to generate the information needed and deliver directly to the user. The generated information is usually enwrapped in HTML (HyperText Markup Language) pages as data records and it forms the hidden web (or deep web or invisible web). As the generated data records from the deep web is highly dynamic, it is difficult for the current search engines to index these HTML pages. Thus, these web pages are called deep web pages. Data records presented in a web page are usually presented using a predefined template, and these data records normally possess similar structures and patterns to form groups of data called data region. Specific information such as relevant data records from the deep web pages are the main source of information for a meta search engine. However, before data records can be used in a meta search engine, they need to be extracted from the search engine results page and converted to a machine readable form. Automatic wrapper is the tool developed for this purpose and it is used to automate meta search engine to increase the speed and efficiency of these search engines [1], [2].

D. Wang and M. Reynolds (Eds.): AI 2011, LNAI 7106, pp. 759–768, 2011.

A meta search engine normally receives a user's query and disambiguate the query for further processing. The query will then be passed to other search engines after they are classified based on the search query. The search engines will generate search results based on the query and this needs a tool (wrapper) to extract the information before it is sent back to the original server. Meta search engine will then compile the returned information, filter out the irrelevant search results, rank them and display the final results to the user. An accurate and fast tool is required to extract the relevant information from search engine results pages so that the meta search engines can be more efficient in ranking and filtering the search results in a timely manner. Thus, the robustness and accuracy of a wrapper will greatly affect the performance of a meta search engine.

Current wrappers use DOM Tree and Visual Cue to extract relevant data region from search engine results pages. These wrappers use the regularity of the structure and layout of data records for data extraction. Our observations show that 30% of the relevant data regions contain irrelevant information such as search result identifier (*e.g. Search returns 10 records*). Current wrappers (MDR[1], DEPTA[8], WISH[4]) are used to extract data records in a sequential order (first to last data records) and they are not designed to extract the said data regions as these data regions contain a mixture of similar and dissimilar data records. In this paper, we propose a novel and robust wrapper to extract relevant data region from search engine results pages. We use algorithms which are able to check the similarity of data records more accurately. Our approach is to use algorithms to check the similarity of data records using the DOM Tree and visual cue properties of these records which are unique and can be recognized easily. In order not to exclude data regions unnecessarily, we also develop a voting algorithm to distinguish the relevant data region, that is, if 85% of its data records are similar, the data region will be treated as potential relevant data region and retained for further processing. Our wrapper then extracts the relevant data region from the list of available data regions. Finally, we use a filtering technique to remove irrelevant data (search identifiers) from the relevant data region. Our wrapper is called **SafeMatch**ing Wrapper (**SafeMatch**).

Our wrapper is divided into 2 components. We first discuss the data extraction module of our wrapper. Given a web page, our approach is to parse the page and arrange it into a DOM Tree. We then use an Adaptive Search Algorithm to label and detect the correct data region. Our algorithm uses a few filtering stages which are able to group the list of data records available in a web page and filter out irrelevant information such as menu bars. Potential data records are then passed through the similarity check and data record detection filters to further exclude irrelevant data to obtain the correct data region. Tests on datasets obtained from various sources and comparison with other existing wrappers show that our wrapper is highly effective in data extraction.

This paper contains several sections. Section 2 describes the current work that is related to ours. Section 3 gives the implementation details of our wrapper. Section 4 provides the experimental tests conducted on our wrapper and finally Section 5 summarizes our work.

2 Related Work

DEPTA [8] uses a bottom up tree matching algorithm to match tree structures of data records. A tree matching algorithm matches two tree structures and determines how the first tree can be transformed into the second tree. DEPTA's tree matching algorithm determines the maximum matches between two trees by comparing the location and identity of the nodes in the tree structures. DEPTA checks the similarity of two trees using the percentage similarity of the trees.

ViNT [3] extracts content line features from the HTML page, where a content line is a type of text which can be visually bounded by a rectangular box. Content lines are categorized into 8 types, each with their own distinguishing characteristics and features, which are grouped to form content blocks. ViNT parses these content blocks to identify the data records. Essentially, ViNT defines a data record as a content block containing a specific ordering of content lines.

ViPER [5] takes a more "natural" approach by projecting the contents of the HTML page onto a 2-dimensional X/Y co-ordinate plane, effectively simulating how the HTML page may be rendered on a printed hard-copy. This enables ViPER to compute two content graph profiles, one for each X and Y planes, which it uses to detect data regions by locating valleys between the peaks as the separation point between two data records (valleys are usually the space within two data records, separating them apart).

WISH [4] uses frequency measures to match the tree structures of data records. WISH works in a time complexity of O(n) and is able to match tree structures of data records containing iterative and disjunctive data. However, tree matching algorithm of WISH is not able to match data records with dissimilar tree structures.

Recently, ODE wrapper [7] uses ontology technique to extract, align and annotate data from search engine results pages. However, ODE requires training data to generate the domain ontology. ODE is also only able to extract a specific type of data records (single section data records), thus it is not able to extract irregular data records such as multiple sections data records and loosely structured data records.

3 SafeMatch Wrapper

3.1 Overview of SafeMatch

SafeMatch requires that the HTML web page of the search engine result pages are parsed and stored in a DOM Tree. In order to simplify our wrapper, we assume that the page under extraction must contain at least 3 repetitive patterns. This assumption is based on our observations that the majority of search engine result pages contain more than 3 repetitive patterns. These useful repetitive patterns are potential data records. SafeMatch consists of two main components. The first component involves parsing the HTML page and organizing it into Document Object Model (DOM) tree representation. In the second component, SafeMatch extracts data records using visual cue and DOM properties of the data records. In Component 2, SafeMatch goes through four stages for the extraction of data records. The initial stage is to come out with a list of data records. The initial list before the filtering processes is usually large. At each step from Stages 1 to 4 of the filtering processes, SafeMatch reduces the list by removing irrelevant data records in each data region. The underlying implementation varies in each of these stages and will be explained in detail in Section 3.2. At the end of the filtering processes, if successful there will only be one data region with only the correct data records left.

3.2 SafeMatch Extraction Module

Overview

Once the DOM tree is constructed, it is passed through a three filtering stages to filter out irrelevant information and identify the correct data region which also contains the data records. Before the filtering processes can be carried out, we use the Adaptive Search Algorithm to detect and label potential data records in a DOM Tree. Details of these works are presented in the following sections.

Assumptions

We have made several important observations on several unique features inherent to a data record. Based on these observations, we come out with a way to correctly extract data records. The following are the observations made by us:

Observation 1
The size of the data records in a search engine results page is usually large in relation to the size of the whole page.

Observation 2
Data records usually occur three or more times in a given search engine results page.

Observation 3
Data records usually conform to a specific regular expression rule to represent their individual data, hence they have nearly similar tree structure.

We examine carefully these three observations and found that these criteria could be formulated using visual cue and DOM tree structure of data records. Three steps of filtering rules are proposed, each of them considering the above observations.

Adaptive Search

The inclusion of this stage is to detect and label the different groups of potential data records. Groups of data records can be defined as a set of data records having similar parent HTML tag, containing repetitive sequence of HTML tags and are located in the same level of the DOM tree. SafeMatch uses the Adaptive Search extraction technique to determine and label potential tree nodes that represent data records. Subtrees which store data records may be contained in potential tree nodes. The nodes in the same level of a tree are checked to determine their similarity (whether they have the same contents). If none of the nodes can satisfy this criterion, the search will go one level lower and perform the search again on all the lower level nodes. Our method involves the detection of repetitive nodes which may contain data records and the rearrangement of these nodes to form groups of potential records in a list in 2 steps:

1. In a particular tree level, if there are more than 2 nodes and a particular node occurs more than 2 times in this level, SafeMatch will treat it as a potential data record irrespective of the distance between the nodes.

2. These potential data records identified in this tree level are then grouped and stored in a list. The potential data records in this list are identified by the notation $[A_1, A_2, \cdots A_n]$ where A_1 denotes the position of a node in the potential data records where it first appears, A_2 is the position where the same node appears the second time and so on. Fig. **1** shows an example where nodes A, B, C are grouped and stored in list 1.

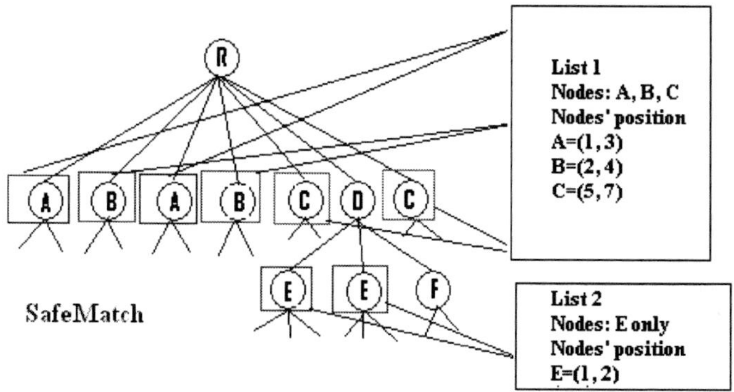

Fig. 1. Potential data records in SafeMatch

Overview of SafeMatch Extraction Rules

After going through the Adaptive Search stage, SafeMatch will have a list of data regions. Our examination shows that data regions fall into one of several groups. We group the first set of potential data regions as menus, this group of data regions determines the layout of HTML pages and is usually large in size and highly dissimilar. The second group is advertisements, regions of this group are highly similar but with simple structures. The third group consists of menu bars, these regions are simple but nearly similar in structure. It is the last group of data records that are relevant to our work, the search engine results output, these regions are highly similar in structure and large in size. We aim to design our wrapper so that it can extract this last group of data regions, while removing the other irrelevant ones. We used filtering stage 1 to remove menus which determine the layout of the HTML page, and filtering stage 3 to remove the remaining irrelevant data (e.g. advertisements). Filtering stage 2 is designed to remove data records which occur less frequently, as observed by author of [8].

Stage 1: Similarity Filter

Proposed Algorithms

In this section, we introduce our algorithms which are able to check the similarity of data records more accurately. Our algorithms include the DOM tree based and visual cue similarity check and a voting algorithm. We derived the DOM tree based matching algorithm based on Observation 3 and our finding that data records share an important characteristic, i.e. the distinct tags of a tree and the total number of distinct tags in each level of the tree are nearly similar to those of the other trees of the group. Thus we are able to formulate a similarity check algorithm which can mimic the behavior of a full tree matching algorithm. Our approach is to carry out the similarity check of two trees by examining the distinct tags and comparing the total number of distinct tags in all levels of the trees. Our algorithm is simple but efficient and it can obtain similar results as those of a tree matching algorithm but it has a reduced time complexity. Our further investigation shows that visual information of data records is also useful in checking the similarity of these data records. This can be achieved by comparing the sizes of the bounding boxes which contain the data records. As noted, not all the data records in the relevant data region are similar, for example, some pages contain relevant data region with search identifiers. It is considered appropriate that if 85% of the data records in a data region are similar, this data region can be treated as relevant and retained for further processing. Otherwise, the data region will be discarded. Our voting algorithm is designed to check whether more than 85% of data records in a data region are similar. It is used together with the DOM Tree and visual cue algorithms to check the similarity of data records. Our similarity check works as follows:

1. The tree structure of data records in a data region are first checked for similarity using the DOM Tree based algorithm. If they pass the test, the voting algorithm will be used to further check the similarity of the data records (i.e. whether more than 85% of the data records are similar)
2. If the trees in the data records are similar, then the trees of Item 1 will be used for visual similarity check and if they pass the test, the voting algorithm will be applied for further test and if they pass the test again, they are considered similar.

The details of our algorithm and its use in detecting similarity of data records and filtering dissimilar data regions are presented in the following subsections.

DOM Tree Based Similarity Check

Our Tree Matching algorithm consists of a two stage screening procedure to check the similarity of a group of trees. Given a number of trees, our algorithm first examines the distinct tags of the first tree and those of the second tree. If almost all the distinct tags occur concurrently in the two trees (overall with say only one element different), then the trees pass the similarity test of the first stage and they are used for the second stage similarity test. In the second stage, we calculate the total number of distinct tags in all the levels of the first tree and that of the second tree. If the first two trees have almost equal number of distinct tags in all levels of the trees (overall with a difference of only one tag), then the two trees are considered similar according to the stage two criterion. The first two trees are similar only if they pass the screening procedures of both stages. If the first two trees are similar, the first tree is retained for further processing and the second tree is then compared with the third tree of the group to check their similarity using Stages 1 and 2 of our screening algorithm. On the other hand, if the first two trees are not similar, our voting algorithm will mark this data region as having one dissimilar data record and the second tree will be compared with the third tree to check their similarity. The screening procedures for both the above cases are repeated until the last tree is used for comparison.

Fig. 2 shows data records presented in a tree form obtained from the DOM Tree of HTML pages. For simplicity, we show only two trees in each figure. We calculate the similarity of the two trees of Fig. 2 using our Tree Matching algorithm. In Fig. 2, the distinct tags are <table, tr, td, div, a, p, b> for both the left and the right trees. The first screening procedure shows that the trees are similar. The total number of distinct tags in all levels is 7 for the left tree and 7 for the right tree respectively (1 <table> tag in level 1, 1 <tr> tag in level 2, 1 <td> tag in level 3, 1 <div> tag in level 4, 1 <a> tag and 1 <p> tag in level 5, 1 <b> tag in level 6 of the trees). Therefore, the left tree is retained for further processing as the two trees are similar. The screening procedures will be repeated using the second tree and third tree and so on until the last tree of the group is used if there are more than 2 trees.

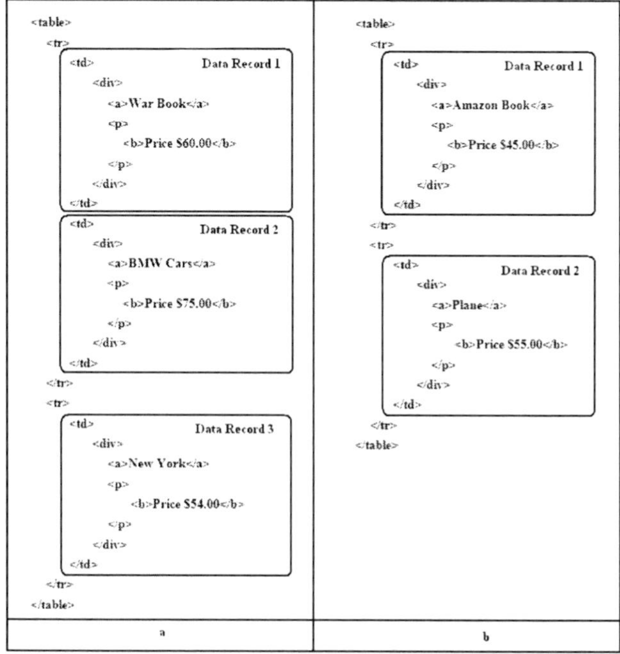

Fig. 2. Two trees with similar structures

Visual Cue Similarity Check

Fig. 3 shows the Lycos search engine results page. As can be seen from Fig. 3, Data Region 1 (solid rectangles in Fig. 3 are considered not similar because they have bounding boxes with different sizes. Data records in Data Region 2, which are represented by the dotted rectangles in Fig. 3 are similar because they are having bounding boxes of similar sizes. The same applies to Data Regions 3, 4 and 5. SafeMatch will determine the bounding box of each data records and the voting algorithm will mark the data region with one dissimilar data record if the data records differ greatly (say more than 50 pixels) in width and height compared to other data records.

Stage 2: Number of Nodes Filter

Stage 2 of the filtering processes involves removing data regions with less than 3 data records. This Stage is carried out based on Observation 2.

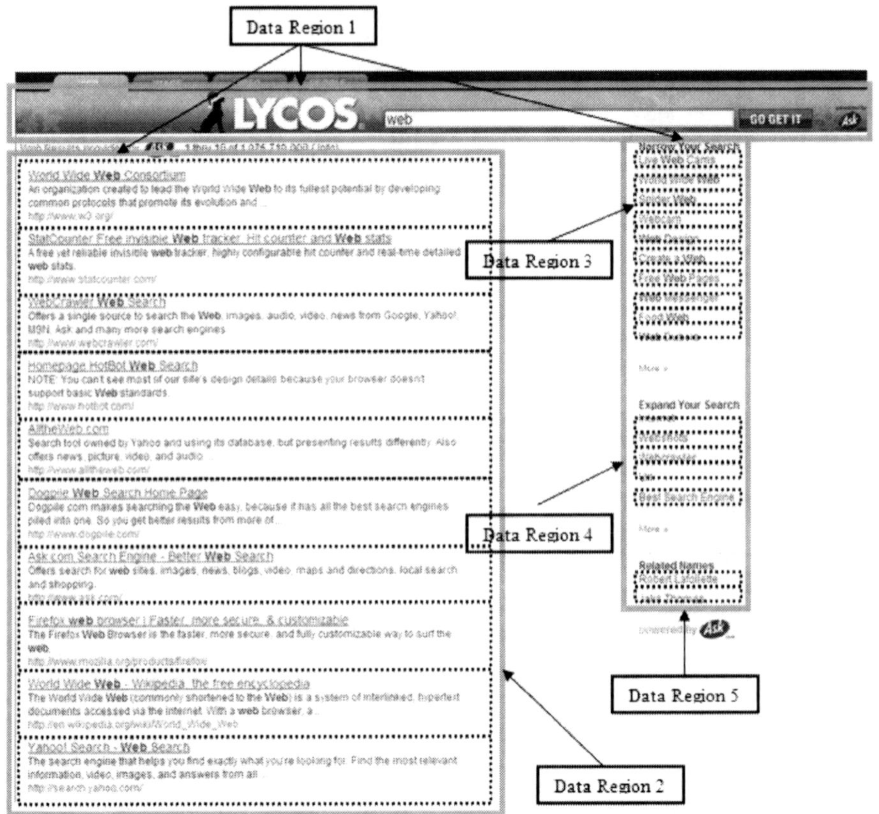

Fig. 3. An example of HTML page containing data regions with similar and dissimilar data records

Stage 3: Largest Scoring Function Filter

After passing through the 3 Stages, SafeMatch will have a list of data regions. It is assumed that these data regions contain data records which are visually and structurally similar as they survived the filtering Stages 1 to 3. Menus which represent the layout of the web page are removed in Stage 2 of the filtering rules. Some of the possible data regions available are the menu bars, advertisements, and data records which are relevant to our work. Each of these data regions is assigned a scoring function. The correct data region is assumed to have the largest score value. Stage 4 is carried out based on Observation 1. We measure the sizes of texts and images of data regions. The sizes of texts and images are measured based on the bounding box of HTML Text and HTML <IMG> tag. Once the area of the bounding box is ascertained, they are summed up to give a final score value. Each of the data regions has its own score value. It is noted that HTML separator nodes such as <BR> also contribute to the space occupied in a data record. SafeMatch will therefore measure the size of the bounding box of these nodes.

The scoring function used by SafeMatch wrapper is given below:
a=Size of Texts in a Data Region
b=Size of IMG tags in a Data Region
c=Size of HTML separator nodes in a Data Region
x=Data region

$$Score(x) = a + b + c \quad (1)$$

Stage 4: Removing Irrelevant Data from the Relevant Data Region

In this stage, we use the same tree matching algorithm mentioned previously to match all the data records in the relevant data region. We also match data records based on their visual boundaries. Data records that matched each other are put into the list of correct data records. Those data records that are not similar are removed from the relevant data region.

4 Experimental Tests

The dataset used in this study is taken from complete planet repositories (www.completeplanet.com). This dataset contains 250 web pages. The distribution of data for the data set varies, ranging from academic sites, general sites to governmental sites. We compare our work with state of the art wrapper, WISH [4] using this dataset. We do not make comparison with other state of the art wrappers such as ViNT [3] and DEPTA [8] as study in [4] shows that WISH performs comparatively better than ViNT and DEPTA. The measures of wrapper's efficiency are based on three factors, the number of actual data records to be extracted, the number of extracted data records from the test cases, and the number of correct data records extracted from test cases. Based on these three values, precision and recall are calculated according to the formula:

$$Recall=Correct/Actual*100 \quad (2)$$
$$Precision=Correct/Extracted*100 \quad (2)$$

SafeMatch takes about 400 milliseconds on average to generate a result for a web page and achieves a high recall and precision rate (Table 1). ViNT [3] (state of the art visual assisted wrapper) takes 1200 milliseconds to generate a result page. This shows that the speed of SafeMatch is better than other existing state of the art system (ViNT [3]) while achieving higher accuracy than WISH [4]. SafeMatch outperforms WISH in terms of recall rate and has precision rate comparable to WISH. Our dataset contains mostly complicated web pages, particularly web pages containing a number

Table 1. Experimental result for SafeMatch and WISH

Term	SafeMatch	WISH [4]
Actual	4139	4139
Extracted	3909	3363
Correct	3758	3288
Recall	90.79%	79.43%
Precision	96.13%	97.76%

of data regions which are similar torelevant data region. As SafeMatch uses visual cue that measures text and image size, it will be able to distinguish more efficiently correct data region from incorrect ones. Unlike WISH, SafeMatch is able to match data records with dissimilar tree structures.

5 Conclusions

In this study, we develop a wrapper (SafeMatch) which is able to extract data records from deep webs more efficiently than existing state of the art wrapper WISH. Unlike existing works, our wrapper is able to extract relevant data region which contains search identifiers. We use our voting algorithm in addition to tree matching algorithm and visual cue to extract the relevant data records and remove irrelevant data from the data region accordingly. We also use text and image size to locate and extract the correct data region containing data records. The exact measurements of text and image and the use of visual information to remove dissimilar data records improve the accuracy of our wrapper in data extraction. The accuracy and speed of our wrapper will be useful in meta search engines application.

References

1. Liu, B., Grossman, R., Zhai, Y.: Mining data records in Web pages. ACM SIGKDD, 601–606 (2003)
2. Miao, G., Tatemura, J., Hsiung, W.-P., Sawires, A., Moser, L.E.: Extracting Data Records from the Web Using Tag Path Clustering. ACM WWW, 981–990 (2009)
3. Zhao, H., Meng, W., Wu, Z., Raghavan, V., Yu, C.: Fully automatic wrapper generation for search engines. ACM WWW, 66–75 (2005)
4. Hong, J.L., Siew, E., Egerton, S.: Information Extraction for Search Engines using Fast Heuristic Techniques. DKE 69(2), 169–196 (2010)
5. Simon, K., Lausen, G.: ViPER: augmenting automatic information extraction with visual perceptions. ACM CIKM, 381–388 (2005)
6. Liu, W., Meng, X., Meng, W.: ViDE: A Vision-based Approach for Deep Web Data Extraction. IEEE TKDE 22(3), 447–460 (2009)
7. Su, W., Wang, J., Lochovsky, F.H.: ODE: Ontology-assisted Data Extraction. ACM TODS 34(12) (2009)
8. Zhai, Y., Liu, B.: Web data extraction based on partial tree alignment. ACM WWW, 76–85 (2005)

Automatic Static Feature Generation for Compiler Optimization Problems

Abid M. Malik

Department of Computer Science
Rice University, Houston, TX
Abid.M.Malik@rice.edu

Abstract. Modern compilers have many optimization passes which help to get a better binary code for a given program. These optimizations are NP-hard. People use different heuristics to get a near optimal solution. These heuristics are designed by a compiler expert after examining sample programs. This is a challenging task. Recently, people have used machine learning techniques instead of heuristics for compiler optimizations. Machine learning techniques have not only eliminated the human efforts but have also out-performed human made huristics. However, the human efforts have now been moved from creating heuristics to selecting good features. Selecting right set of features is important for machine learning techniques since no machine learning tool will work well with poorly choosen features. This paper introduces a noval approach to generate features for machine learning for compiler optimization problems with out any human involvement.

1 Introduction

Modern compilers provide large number of optimization passes to get a better binary code for a target machine. All-most all optimizations are NP-hard [14]. There are no deterministic algorithums for these optimizations to get an optimal solution. People use heuristics to find a near optimal solution for these optimization problems. These heuristics are created by a compiler expert by observing various program applications. It is a challenging task and requires many man hours. If one is able to fine tune a heuristic for a given architecture, when a new processor comes to the market, compiler expert has to repeat the whole process again to fine tune the heuristic for the new architecture. A company's greatest interest is to shorten this process in order to market the new product as soon as possible. Recently, people have used machine learning techniques to reduce this time cycle. The ultimate goal of using machine learning techniques is to learn a heuristic for a new enviorment at the press of a botton. Also, in practice, machine learning techniques have given better performance than their human created counter parts [16].

In a typical machine learning enviornment for a compiler optimization problem, number of programs are transfored into input vectors. For each program, we determine a desired output vector. Input vectors along with output vectors

D. Wang and M. Reynolds (Eds.): AI 2011, LNAI 7106, pp. 769–778, 2011.

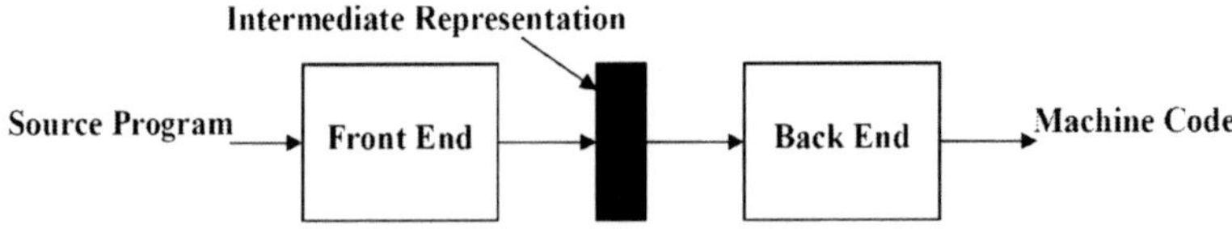

Fig. 1. Typical compilation path in a compiler for modern processors

form a training set. A machine learning tool will then try to construct a model which maps input vectors to output vectors. An input vector is a set of features which captures charateristics of a program. Now, an important question is; what are the best features for a given machine learning approach? A compiler expert has to decide set of features keeping in view the target compiler optimization and architecture. This is a difficult task. Every machine learning tool is bounded by the performance of the input features. Hence, selecting good features is an important research area in the field of applying machine learning techniques for various compiler optimization problems.

A typical compilation path in a modern compiler is shown in Figure 1. A compiler takes a source program written in a high-level language as an input. It performs lexical analysis (scanning), syntax analysis (parsing) and semantic analysis (type checking) in the front-end. It converts a given program into an intermediate representation (IR). IR is a machine and language independent version of the original source code. IR is used to create various data structures like abstract syntex tree (AST), control flow graph (CFG), data dependency graph (DDG) etc. These data structures are taken by the back-end to apply various optimizations. These data structures contain wealth of information about programs. People use these data structures to create features for machine learning techniques. Although machine learning alleviates a compiler expert from the task of building a heuristic, but it put on him another challenging task of reducing the wealth of information to a small set of features. This paper introduces an approach for autmatically creating a feature set with out the involement of a compiler expert by considering a compiler optimization problem as a classification problem.

The rest of the paper is organized as follow. Section 2 gives the related work in this area. Section 3 discusses our approach. Section 4 talks about the experimental set up for the work. Section 5 talks about the results. Section 6 concludes our contributions and talks about our future line of action in this area.

2 Related Work

One of the first researchers to incorporate machine learning into compiler for optimization problems were McGovern et al. [13] who used reinforcement learning for scheduling of straight-line code. Cavazos et al. [5] extended this idea by learning whether or not to apply instruction scheduling. Stephenson et al. [17]

looked at tuning the unroll factor using supervised classification techniques such as K-nearest neighbor (KNN) and support vector machines (SVM).

Subsequent researchers have considered predictive models using machine learning techniques to automatically tune a compiler for an existing micro-architecture. These models use programs features to focus the search of optimization space in promising areas. Agakov et al. [2] use static code features to characterize a given program while Cavazos et al. [4] investigate the use of hardware performance counters. Leather et al. [16] give grammar to select the features to represent a program. Christophe et al. [8] use hardware features for selecting the best compiler options for a given architecture. The work by Ganapathi et al. [10] applies machine learning for compiler optimization problems for multi-core architectures. Malik [11] tries to capture spatial information of DDGs for the machine learning techniques. Yoki et al. [19] give static features for machine learning for tile selection problem.

Recently, MILEPOST-GCC framework has been developed by IBM Haifa to drive the compiler optimization process based on machine learning. The framework gives features which are very comprehensive in terms of capturing all important characteristics of a given program. Interested readers can consult the work by Fursin et al. [9] for complete list of features. In this paper, we compare the performance of our approach against MILEPOST-GCC framework.

3 Our Approach

Previous work in this area needs lot of compiler expert involvement in crafting or selecting the best features for a machine learning technique for a given compiler problem. The main contribution of this work is an approach that does not require this involvement at any stage.

Figure 2(a) is an IR for the C code in Figure 2(b). Due to the space constraints, we are not re-producing the whole IR for the code[1]. We are showing only that part of IR that is good enough to establish our point of view. Figure 2(a) has two types of branch instructions. Branch instruction, **br label register**, takes one argument. Branch instruction, **br i1 register label register label register**, takes three arguments. The branch instruction with one argument is an unconditional branch instruction. The branch instruction with three arguments is a conditional branch instruction. The two branch instructions change the control flow of a program in a different way. The unconditional branch changes the control of a program with-out any testing. The conditional branch first does a test in the first argument and then changes the control of a program to the second or third argument depending upon the out come of the test. These instructions are categorized as one class in the previous work. However, both the branch instructions capture different semantics of a given program. An unconditional branch instruction may represent an endless loop while a conditional branch instruction shows a loop with limited iterations in a program. If one develops a feature set by considering such instructions as one class, it will be hard for a machine

[1] Interested readers can consult the GCC documentation [1] to reproduce it.

```
define i32 @main() nounwind {
entry:
 ........................
 ........................
  br label %for.cond

for.cond:
  ........................
  ........................
  br i1 %cmp, label %for.body,
                    label %for.end28
for.body:
  ..............
  ..............
  br label %for.cond1

for.cond1:
  ..................
  ..................
  br i1 %cmp3, label %for.body4,
                    label %for.end
for.body4:
  ..................
  ..................
  br label %for.inc

for.inc:
  ..................
  ..................
  br label %for.cond1

for.end:
  ...................
  ...................
  br label %for.inc25

for.inc25:
  ..............
 ..............
  br label %for.cond

for.end28:
  ret i32 1
}
```

(a)

```
#include<stdio.h>
#include<stdlib.h>

      int a[10][10];
      int b[10][10];
      int c[10][10];
      int d[10][10];

      int i, j;
void initial(){
      for ( i = 0; i < 10; i++)
      for ( j = 0; j < 10; j++){
           a[i][j]=1;
           b[i][j]=2;
           c[i][j]=3;
           d[i][j]=4;
      }

}
int main ( void ) {


      initial();

      for ( i = 0; i < 10; i++)
      for ( j = 0; j < 10; j++)
           a[i][j]=b[i][j]
            +c[i][j]+d[i][j];
      return 1;
}
```

(b)

Fig. 2. (a) IR code for the C code using the GCC compiler. Most parts of IR have been replaced by the dotted lines due to space constraints (b) C code program.

learning tool to differentiate between semantics of different programs. With this approach, there is a possibility that two programs might have different semantic meanings but have similar feature vectors.

For the approach in this paper, we borrow the idea from the feature selection approaches that are being used in the text classification problem. A text document consists of words which capture it's semantics. In the text classification problem, a text document is represented by a set of words which are considered the most helpful in classifying the document with respect to a given class. Many statistical approaches are used to select the best features [20]. In the approach, we consider IR of a program as a text document. Instead of collapsing certain types of instruction as one instruction in order to reduce the dimensionality of search space, we treat each instruction type at IR level as one feature. A compiler has limited number of instruction types at IR level. In this work, we use the GCC compiler [1] which has 200 intruction types. One can use all 200 intruction types as a feature set for a machine learning tool for any optimization problem. However, this may lead to the curse of dimensionality problem which decreases the performance of a machine learning tool [12].

To select the best features with respect to a compiler optimization problem, we used feature selection methodology adopted by people for the text classification problem. For this, we first defined a compiler optimization problem as a classification problem. In this work, we applied the approach to the best optimization options selection problem. Modern compilers have more than 100 options and selecting the best option for a given program is NP-hard. People use heuristic to find the best options for a given problem. If each optimization option is considered as a class, then the best compiler options selection problem can be defined as a classification problem as follow:

Given a progarm P and compiler optimization class C, determine whether P belongs to C or not.

For simplicity, we assume that each optimization is independent of each other. We use the following criteria from the work [20] to select the best 30 features [2] for our feature vector representation.

3.1 Frequency Thresholding

The number of times an instruction type occurs in the training set. The basic assumptions is that rare instructions are not influential for compiler optimization options. Rare instruction removal reduces the dimensionality of the feature space. This is the simplest technique but not a very good criterion to pick the best features.

3.2 Information Gain(IG)

Information gain is frequently employed as a term-goodness criterion in the field of machine learning. It measures the number of bits of information obtained for category prediction by knowing the presence and absence of a term. Let $\{C_i\}_{i=1}^{m}$ gives the set of compiler options available in a compiler. The information gain of an instruction type t is defined by Equation 1.

$$G(t) = -\sum_{i=1}^{m} P_r(C_i)logP_r(C_i) + P_r(t)P_r(C_i|t)logP_r(C_i|t) + P_r(\bar{t})P_r(C_i|\bar{t})logP_r(C_i|\bar{t}) \quad (1)$$

$P_r(C_i)$ gives the probablity of compiler option C_i being turned ON in the training set. $P_r(t)$ gives the probablity of an instruction type in the training set . $P_r(C_i|t)$ gives the probability of C_i turned ON given an instruction type t. $P_r(C_i|\bar{t})$ gives the probability of C_i turned ON given an instruction type t is absent.

[2] We want to see how robust the approach is as compare to MILEPOST-GCC framework. MILEPOST-GCC gives 55 hand made features.

3.3 χ^2 Statistics (CHI)

CHI measures the lack of independence between two terms. Equation 2 gives CHI for an instruction type t with respect to a compiler option C_i.

$$\chi^2_{avg}(t, Ci) = \frac{N \times (AD - CB)^2}{(A + C) \times (B + D) \times (A + B) \times (C + D)} \tag{2}$$

Where A is the number of times an instruction type t and compiler option C_i being turned ON co-occur for a program. B is the number of times the instruction type t occurs with the compiler option C_i being turned OFF. C is the number of times option C_i is turned ON with out instruction type t. D is the number of times neither instruction type t is present nor option C_i is turned ON. N is the total number of programs in the training set. We calculate the average value of CHI for each instruction type using Equation 3.

$$\chi^2_{avg}(t) = \sum_{i=1}^{m} P_r(C_i)\chi^2(t, C_i) \tag{3}$$

Where m is the total number of compiler options and $P_r(C_i)$ gives the probablity of compiler option C_i being turned ON in the training set.

4 Experimental Setup

The tools, benchmarks, architecture and environment used for the work are briefly described in this section.

4.1 Compiler

The GCC was selected as it is a mature and popular open-source optimizing compiler that supports many languages, has a large community, is competitive with the best commercial compilers, and features a large number of program transformation techniques. The GCC is the only open source compiler that supports more than 30 processor families. For our work, we selected the latest GCC 4.4x version.

4.2 Flags

In the latest version of GCC, there are about 100 flags. It is impossible to validate all combinations of optimization of flags. Most of these flags are considered with the global GCC optimization levels ,i.e., -O1, -O2 and -O3. For our work, we considered the GCC optimization level -O3 and then considered a particular optimization by tunning it ON or OFF through a corresponding flags $-f < optimization-name >$ and $-fno- < optimization-name >$ flags respectively. Certain combination of flags cause the compiler to break or produce incorrect program execution and hence incorrect result. We reduced the probability of such cases by comparing outputs of program with the reference ouputs.

4.3 Plateforms

We used Intel Dual core running at 2.0 GHz with 4.0 GB of main memory and 1 MB of $L2$ cache, runing Ubuntu Linux .

4.4 Benchmarks

We used the MiBench and SPEC2006 benchmark suites. The MiBench consists of six categories of C programs. These categories offer different program characteristics that enable researchers in architecture and compiler to examine their design more effectively. The SPEC2006 benchmark consists of 39 applications both in FORTRAN and C. It is a standard benchmark which is used to verify various compiler optimization techniques. We only considered C applications from the SPEC2006.

4.5 Experiments

First, we identified hot functions in the benchmark applications. We define a hot function as one which is mostly executed in a given program application. We used the **gprof** tool to determine hot function for a given application. The features extracted from a hot function was used to build a feature vector for machine learning. Feature vector of a hot function was used to represent a program instance. Each feature in a feature vector was weighted using a novel approach given by Malik [11].

We build a training set using the SPEC2006 C applications. We used the genetic algorithm (GA) from work [6] to get the best compiler options for each application. We used 1000 evolutions for GA approach. Each run was repeated five times so that the speedups were not cacused by cache priming etc. This was the most time consuming part of our work. In some cases, it took more than a day to find the best compiler options for an application.

For our work, we selected two machine learnng techniques; decision tree (DT) and support vector machines (SVM). We used MILEPOST-GCC work [9] as a reference to compare the quality of our work as it gives the most detailed man made features for machine learning techniques. We used the C4.5 algorithm [7] for DT implementation. We used the default values for various parameters given by the developer. We implemented both the linear and non-linear SVMs. We used the **SVMlight** tool which is freely availabe on the web [18]. Again, we used the default values for various parameters set by the **SVMlight** tool. However, for the non-linear SVMs, we used radial basis function kernel with $\sigma = 1$ and the upper bound parameter (C) of SVMs equal to 10.

DT learning took about a minute to build a model. However, SVM learning took 2 to 5 minutes to build a model for the best compiler flag selection problem.

5 Experimental Results

We compare the performance of the iterative approach using GA with the performance of the two machine learning techniques using MILEPOST-GCC framework

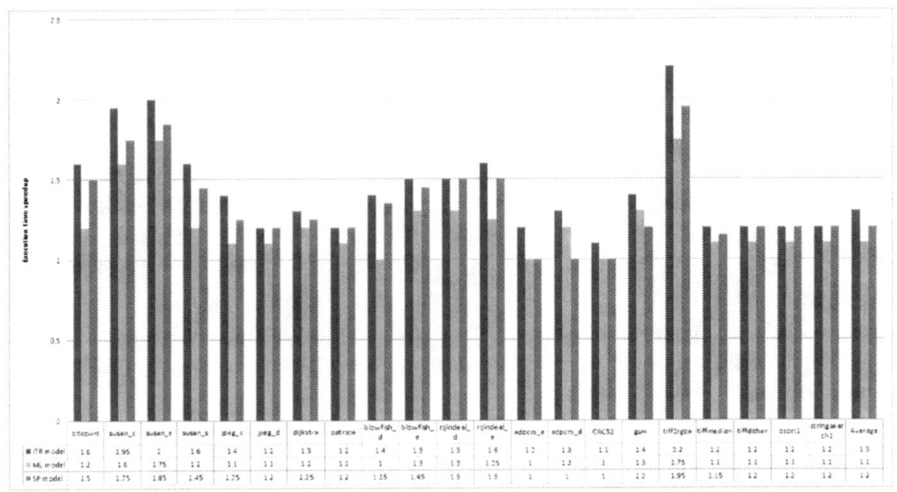

Fig. 3. Performance of the non-linear SVMs learning using χ^2 statistics (CHI) as a selection criterion for our framework

and our approach. Our approach did better on both DT and SVM techniques. However, due to the space constraints, we will discuss the non-linear SVM results.

Figure 3 compares the performance of non-linear SVM using MILEPOST-GCC (ML) and our approah (SP) using CHI criterion againt the iterative model (ITR) using GA. In Figure 3 the horizontal axis gives name of each benchmark application used for testing and its performance using ITR, ML and SP models. The vertical axis gives the speed up over the running time when the same application is compiled with the -O3 GCC optimization level [3]. For the bitcount application using ITR model, we are able to find a binary code which is 1.6 times faster than a binary code when one compiles it with the -O3 GCC optimization level. Figure 3 shows that with CHI selection criterion, our automatic approach outclasses MILEPOST-GCC framework. On average, our approach out performs MILEPOST-GCC framework on average by 6%. The reason of good performace using CHI criterion is its ability to capture the inter-dependency to some extent between different optimization options. Note, CHI criterion calculates the weightage of each instruction type by determining its presence and absence in each compiler option. This information is missing in MILEPOST-GCC framework.

Figure 4 shows the performace of our framework using different selection criteria with the non-linear SVM learning. CHI criterion gives the best performace while the frequeny based selection gives the worst. The information gain performace is reasonable but is not as good as CHI.

[3] -O3 is the highiest optimization level in the GCC compiler.

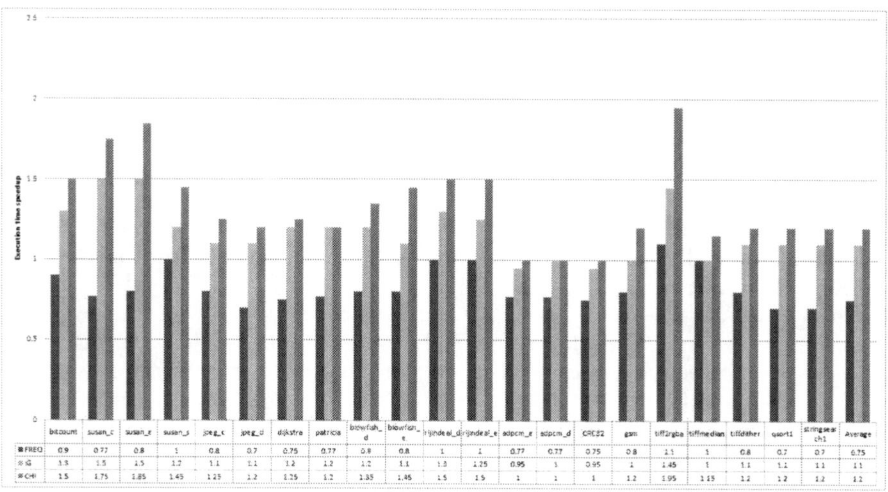

Fig. 4. Performance of the non-linear SVMs learning using χ^2 statistics (CHI), information gain (IG) and frequency (FREQ) as selection criteria for our framework

6 Conclusion and Future Work

Applying machine learning techniques to compilers require experts to generate features. At no point could the expert be sure that they have the best set of features to assist the learner. We presented in this paper a novel technique to generate features automatically with out the assistence of compiler expert. We tested the approach extensively using the two standard benchmarks and compared it against human created features from IBM MILEPOST-GCC framework. The results showed that our framework clearly out-performed IBM MILEPOST-GCC framework on almost all benchmark applications using SVM and DT learning on the best compiler option selection problem.

In future, we plan to use the approach to investigate the automaic selection of better order of optimization passes and fine-grained tuning of transformation parameters for important optimization,e.g., unrolling factor of loop unrolling optimization.

References

1. `http://gcc.gnu.org/`
2. Agakov, F., Bonilla, E., Cavazos, J., Franke, B., Fursin, G., O'Boyle, M., Thomson, J., Toussaint, M., Williams, C.: Using machine learning to focus iterative optimization. In: Proceedings of the International Symposium on Code Generation and Optimization, CGO 2006 (2006)
3. Bodin, F., Kisuki, T., Knijnenburg, P.M.W., O'Boyle, M., Rohou, E.: Iterative compilation in a non-linear optimization space. In: Workshop on Profile Directed Feedback-Compilation, PACT 1998 (1998)

4. Cavazos, J., O'Boyle, M.: Method-specific dynamic compilation using logistic regression. In: Proceedings of the 21st Annual ACM SIGPLAN Conference on Object-Oriented Programming Systems, Languages, and Applications, OOPSLA 2006 (2006)
5. Cavazos, J., Moss, J.: Inducing heuristics to decide whether to schedule. In: Proceedings of the ACM SIGPLAN Conference on Programming Language Design and Implementation, PLDI 2004 (2004)
6. Cooper, K.D., Schielke, P.J., Subramanian, D.: Optimizing for Reduced Code Space using Genetic Algorithms. In: Workshop on Languages, Compilers, and Tools for Embedded Systems, LCTES 1999 (1999)
7. http://cis.temple.edu/~ingargio/cis587/readings/id3-c45.html
8. Dubach, C., Jones, T.M., Bonilla, E.V., Fursin, G., O'Boyle, M.F.: Portable Compiler optimization across embedded programs and micro-architectures using machine learning. In: Proceedings of the 42nd IEEE/ACM International Symposium on Micro-architecture (2009)
9. Fursin, G., Miranda, C., Temam, O., Namolaru, M., Yom-Tov, E., Zaks, A., Mendelson, B., Barnard, P., Ashton, E., Courtois, E., Bodin, F., Bonilla, E., Thomson, J., Leather, H., Williams, C., O'Boyle, M.: MILEPOST GCC: machine learning based research compiler. In: Proceedings of the GCC Developers' Summit, GCC 2008 (2008)
10. Ganapathi, A., Datta, K., Fox, A., Patterson, D.: A case for machine learning to optimize multicore performance. In: Proceedings of the First USENIX Conference on Hot Topics in Parallelism, HotPar 2009 (2009)
11. Malik, A.M.: Spatial Based Feature Generation for Machine Learning Based Optimization Compilation. In: Proceedings of the 9th IEEE International Conference on Machine Learning and Applications, ICMLA 2010 (2010)
12. Mitchell, T.: Machine Learning. McGraw-Hill (1997)
13. McGovern, A., Moss, E.: Scheduling straight-line code using reinforcement learning and rollouts. In: Proceedings of Neural Information Processing Symposium, NIPS 1998 (1998)
14. Muchnick, S.: Compiler Optimization for Modern Compilers. Morgan Kaufmann (1997)
15. Ipek, E., Mckee, S.A.: Efficently exploring architectural design spaces via predictive modeling. In: Proceedings of Architectural Support for Programming Languages and Operating Systems, ASPLOS 2006 (2006)
16. Leather, H., Bonilla, E., O'Boyle, M.: Automatic feature generation for machine learning based optimizing compilation. In: Proceedings of the International Symposium on Code Generation and Optimization, CGO 2009 (2009)
17. Stephenson, M., Amarasinghe, S., Martin, M., O'Relly, U.M.: Meta optimization: Improving compiler heuristics with machine learning. In: Proceedings of the ACM SIGPLAN Conference on Programming Language Design and Implementation, PLDI 2003 (2003)
18. http://svmlight.joachims.org/
19. Yuki, T., Renganarayanan, L., Rajopadhye, S., Anderson, C., Eichenberger, A., O'Brien, K.: Automatic Creation of Tile Size Selection Models. In: Proceedings of the International Symposium on Code Generation and Optimization, CGO 2010 (2010)
20. Yang, Y., Pedersen, J.O.: A Comparative Study on Feature Selection in Text Categorization. In: Proceedings of the Fourteenth International Conference on Machine Learning, ICML 1997 (1997)

Multi-objective Optimisation of Power Restoration in Electricity Distribution Systems

Alexandre Mendes[1] and Natashia Boland[2]

[1] Centre for Intelligent Electricity Networks (CIEN), School of Electrical Engineering and Computer Science, Faculty of Engineering and Built Environment, The University of Newcastle, Callaghan, NSW, 2308, Australia
Alexandre.Mendes@newcastle.edu.au

[2] School of Mathematical and Physical Sciences, Faculty of Science and Information Technology, The University of Newcastle, Callaghan, NSW, 2308, Australia
Natashia.Boland@newcastle.edu.au

Abstract. This paper proposes a new multi-objective approach for the problem of power restoration in (n-1) contingency situations. It builds on a previous, mono-objective approach introduced in Mendes et al. (2010) [14]. Power restoration normally relies on network reconfiguration, and typically involves re-switching and adjustment of tap-changers and capacitor banks. In this work, we focus on re-switching strategies. The quality of the re-switching strategy is measured in terms of voltage deviations, number of consumers still affected after the reconfiguration, number of overloaded branches and number of switches changes. Due to the number of criteria and conflicting objectives, power restoration is a prime candidate for multi-objective optimisation. The method studied is based on a genetic algorithm and was tested using two real-world networks, with up to of 1,645 branches and 158 switches. We present a contingency example for each network and discuss the results obtained. Finally, we discuss the approach's convergence by analysing the evolution of the solutions that compose the Pareto frontier.

Keywords: Multi-objective optimisation, genetic algorithms, power distribution, electricity networks.

1 Introduction

Contingency situations are caused by a single failure (n-1), or multiple failures (n-k) of equipment in the distribution network. They are a relatively common occurrence and electricity companies must quickly implement a contingency plan to re-establish power supply to the consumers affected. In the scientific literature, (n-1) contingency problems can be categorized as either mono- or multi-objective; and the solution methods can be either exact, heuristic, or hybrid. In this work we present a new heuristic based on genetic algorithms to deal with multi-objective version of the (n-1) contingency problem.

D. Wang and M. Reynolds (Eds.): AI 2011, LNAI 7106, pp. 779–788, 2011.

A literature review on the (n-1) contingency problem applied to electricity distribution networks shows several studies dealing with the mono-objective version. From 1987 to 1989, Aoki et al. [1,3,2] have published a series of papers dealing with the electricity distribution restoration problem and the maximization of total load restored. All studies used different types of heuristics. In Dyalinas et al. (1989) [7], a heuristic was implemented to minimize the number of switch operations. In 1992, two works addressed the reconfiguration problem again minimizing the number of switch operations (Kim et al. [13], Fujii et al. [9]). In 2001, Ferreira et al. [8] worked on the minimization of out-of-service load using a genetic algorithm. Finally, Mendes et al. (2010) [14] studied the problem of power supply restoration minimizing number of disconnected buses, cable overloads and switching operations, combined into a mono-objective function.

In terms of multi-objective approaches, there are four works on service restoration. In 1998, Tourne et al. [19] worked on the optimisation of load balance and voltage levels. In the same year, Miu et al. [15] addressed the maximization of load restored, with and without priorities, and number of switch operations, using a local search heuristic. In 2001, Augugliaro et al. [4] addressed load supply and power losses with a fuzzy genetic algorithm. More recently, Garcia and Franca (2008) [10], addressed the multi-objective problem by minimizing the number of affected consumers and the number of switch operations. They used a local search heuristic. For a broad picture of the area, we refer the reader to the two review papers of Perrier et al. (2010a,b) [16,17].

From the perspective of Artificial Intelligence (AI), two works are particularly interesting as they integrate diagnosis and repair of power supply; a more complete scenario than that considered here [18,5]. The main issue with some AI approaches, though, is that restoration is treated mainly as a topological problem – the network must be reconnected and radiality must be maintained. Not much emphasis is given to power quality, e.g. voltage deviations, equipment overloading, etc. The approach presented in this paper tries to address network re-connectivity as well as power quality in a more balanced way.

In this study we consider distribution networks composed of generators, buses, loads, switches and branches. Even though the network has to operate in a radial topology, there is a level of redundancy. That is, power supply can flow through different paths to reach the same customer. This excess connectivity means that if all switches in the network are closed, several loops might be formed, thus violating the radiality requirement. Under normal circumstances, a feasible switching state will have a mix of open and closed switches.

The problem of finding alternative routes for the power distribution is very complex. Alternative routes are determined by re-switching the network, which implies searching through the solution space of switches states. That is a high complexity task; if there are k switches present in the network, the search space has a size 2^k, corresponding to each switch either being *closed* or *open*. Due to the exponential increase in the search space, the presence of a few dozen switches are sufficient to require the use of a heuristic such as genetic algorithms [12] to reach high quality solutions in short computational times.

The multi-objective version of the (n-1) contingency problem addressed in this work considers four criteria: buses voltage deviations, number of disconnected buses, number of overloaded cables and number of switch operations. It uses a genetic algorithm to evolve a population of non-dominated solutions, which compose the Pareto front for the problem. This preliminary study used two real-world networks. The first, *network A*, has with 96 buses, 16 switches and two generators; the second, *network B*, has 1,645 buses, 158 switches and 4 generators. We present computational results for one contingency example for each network; and for the larger network, we show how the Pareto front evolved with the generations.

2 The Network Reconfiguration Problem

The reconfiguration problem addressed in this study can be described as follows. Given an input distribution network in some initial state, consider the loss of a single branch. If any bus in the network becomes disconnected, find a re-switching strategy that will send the power flow back to the affected buses, taking into account operational limits for voltage and load in all sections of the network. The goal in terms of voltages is to minimize the number of buses without power; and for those buses being supplied, minimize the voltage deviation from 1.0 (measured as per-unit). In addition, minimize the number of branches with load above the operational limits. The final solution should also be reached with a minimum number of switch operations.

The genetic algorithm receives as input the physical network and the current state of the switches. Then, a given branch is removed from the network (i.e. all references to it in the network model are removed), simulating an outage. This represents a situation in which the faulty branch has been identified and isolated from the rest of the network. The criteria we use to define high quality reconfigurations are:

- *Topology*: The network has radial topology.
- *Load*: The load on any cable (branch) does not exceed a specific limit. In our tests, that limit was set at 120% of the transmission capacity.
- *Voltage*: The voltage at any bus of the network lies between 0.9 and 1.1 (measured as per-unit). Disconnected buses have zero voltage.
- *Switches*: The reconfigured network state should be reached with a minimum number of switch operations.

2.1 Multi-objective Approach

The multi-objective approach optimizes four quality criteria $[c_1, c_2, c_3, c_4]$:

- c_1: *Buses voltages* $= \sum_{i=1}^{n_{buses}} \left| v_{dev}(i, s) \right|$
- c_2: *Buses disconnected* $= n_{volt_{out}}(s)$
- c_3: *Branches overloaded* $= n_{load_{out}}(s)$
- c_4: *Switch operations* $= n_{switches}(s)$

Where:

- $s \rightarrow$ a solution, i.e. a switching configuration;
- $n_{buses} \rightarrow$ number of buses in the network;
- $v_{dev}(i, s) \rightarrow$ voltage deviation from 1.0 p.u. at bus i in solution s;
- $n_{volt_{out}}(s) \rightarrow$ number of buses without power in s;
- $n_{load_{out}}(s) \rightarrow$ number of overloaded branches in s;
- $n_{switches}(s) \rightarrow$ number of switch changes in s.

In addition to that, a penalty $P(s)$ is added to all four criteria c_i whenever a reconfiguration induces loops in the network. That penalty is calculated as the number of loops in the network multiplied by a large constant, i.e. $P(s) = n_{loops}(s) * M$. That penalization guarantees that after the first radial solution is obtained, only radial solutions will be present in the Pareto frontier. Non-radial solutions will be always dominated by any radial solution. On the other hand, this scheme allows the presence of non-radial solutions in the beginning of the evolutionary process, which is an important feature to improve the initial convergence of the population. For more information about the concept of Pareto frontier and solution dominance, and how they are applied to optimisation, we refer the reader to reference [6].

3 Genetic Algorithm Approach

Genetic algorithms are population-based search methods that use analogies from the Theory of Evolution to find high quality solutions for complex computational problems [6,12]. Normally, GAs start with a population of low quality solutions, usually randomly generated, and then 'evolve' this population via genetic operators, i.e. crossover, mutation and selection, towards better quality individuals, corresponding to solutions with better objective function values. The genetic algorithm used in this study is described next.

3.1 Pseudocode

The first part of the pseudocode (Figure 1) creates an initial random population of solutions, followed by the calculation of the initial Pareto frontier. Then, in the main loop section, solutions are created via crossover and mutation. If a new solution is not dominated by any solution in the current Pareto frontier, it is inserted into the population, triggering a check and removal of any existing solution that became dominated by the new one. The main loop continues until a time limit is reached. Next we will describe the main elements of the genetic algorithm.

3.2 Objective Function and Pareto's Dominance

The objective function of a solution s is represented as a vector of real numbers, with each position associated to one of the four criteria, i.e. $f(s) = [c_1, c_2, c_3, c_4]$. The Pareto's dominance criterion states that solution s' dominates a solution s'' if $c_i(s') \leq c_i(s''); \forall\ c_i, 1 \leq i \leq 4$.

```
Method. MultiObjGeneticAlgorithm
begin
  initializePopulation(pop);
  updateParetoFrontier(pop);
  do % main loop
    parents = selectParents(pop);
    newSolution = generateOffspring(parents);
    newSolution = mutate(mut_rate, newSolution);
    insertNewSolution(pop, newSolution);
    if (inserted) updateParetoFrontier(pop);
  while(cpuTime < limit)
end
```

Fig. 1. Pseudo-code of the genetic algorithm implemented. Initially, a population of random solutions representing switching strategies is created; and then the algorithm enters the main loop. The main loop iteratively creates new solutions which are inserted into the population depending of whether they are dominated or not by another solution already present. If the new solution is not dominated, it is inserted and the Pareto frontier is then updated to remove if any existing solutions that became dominated by the new one. This process is repeated until a time limit is reached.

3.3 Representation and Initialization

The representation and initialization used in this work is the same of Mendes et al. (2010) [14]. Solutions for the problem of switching reconfigurations have a binary representation, with an array of bits of size n representing the states of the n switches. The value '0' indicates open and '1' indicates closed. The initial population is composed of random solutions created by assigning values 0 or 1 to each switch, uniformly at random. The probability is 20% for any given switch to be open, and 80% for it to be closed. This proportion was chosen to match the proportion of open and closed switches in the real networks; and aim at reducing the likelihood of creating solutions with loops or disconnected sections.

3.4 Recombination – Selection, Crossover and Mutation

Parents are selected from solutions in the Pareto frontier. When there is more than one solution in the Pareto frontier, two solutions are randomly selected and a new solution is created via crossover. If there is only one solution in the Pareto frontier, then that solution becomes one of the parents, and the second parent will be a randomly generated solution.

The crossover method implemented is a Uniform Crossover (UX) [11], where the value of each switch state in the child solution is chosen uniformly at random from one of its parents. If both parents have the same state for a specific switch (either 0 or 1), the child will inherit that state. If each parent has a different value, then the value inherited can be 0 or 1, with equal probability.

As the representation is an array of bits, the logical choice for mutation is the bit-swap [11]. If a solution is selected to go through mutation (according to a probability mut_{rate}), a switch is chosen uniformly at random, and its state is swapped, either $0 \rightarrow 1$ or $1 \rightarrow 0$.

3.5 Acceptance Policy

After a new solution is created, because of the multi-objective nature of the problem and Pareto's optimality criterion, two steps have to be followed. First, the algorithm verifies if the new solution is dominated by any other solution already present in the Pareto frontier. If so, the new solution is discarded and the algorithm continues. On the other hand, if the new solution is not dominated, it is inserted into the population. After insertion, the algorithm checks whether the new solution now dominates any solution already present. If so, those dominated solutions are removed from the population and the algorithm continues. This acceptance policy guarantees two things. First, the population will not have duplicated individuals. Second, all individuals in the population belong to the Pareto frontier.

4 Computational Results

In this section we will show two illustrative contingency examples for the test networks. In Figure 2a we show the small network's state (network A) as it is normally operated. The network has two generators ('A' and 'B'), marked as rectangles. Hexagons represent switches. Open switches are indicated in white (30, 50 and 92), whereas yellow indicates closed switches (22, 25, 33, 37, 43, etc). The circles

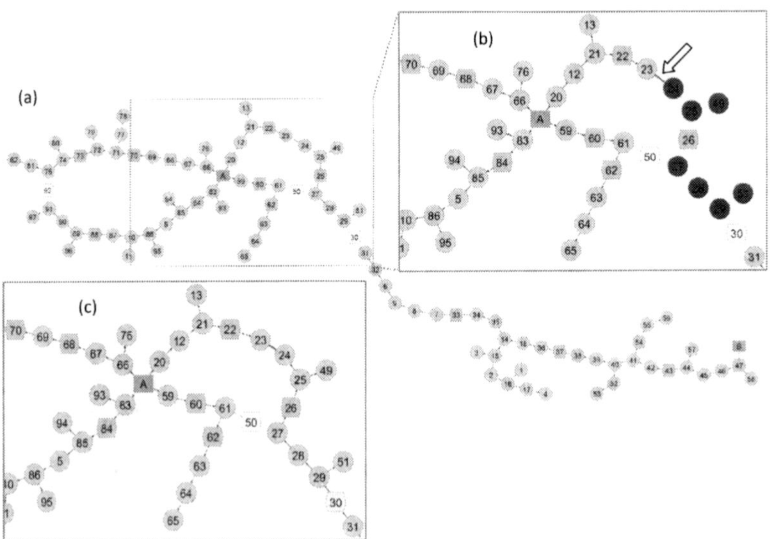

Fig. 2. *Network A*: (a) Diagram of the test network. The two rectangles labeled 'A' and 'B' represent generators. Yellow/white hexagons represent closed/open switches, respectively. Circles indicate ordinary buses. (b) Depiction of a fault in the cable connecting buses 23 and 24. That fault will cut power supply to the buses indicates in red (24, 25, 27, 28, 29, 49 and 51). (c) Solution 2 from Table 1, with 1 switch operation (switch 30 is now closed) and no disconnected buses.

indicate ordinary buses. The network has all buses connected, voltages are within operational limits and no cables are overloaded. In Figure 2b, we introduce a fault in the cable connecting buses 23 and 24. That fault will cut power supply to the buses in red (24, 25, 27, 28, 29, 49 and 51). After running the multi-objective approach, two solutions compose the final Pareto front (see Table 1). The first one represents the 'do-nothing' strategy, with zero switch changes. The second solution is depicted in Figure 2c and requires 1 switch operation. Switch 30 is closed thus redirecting power supply to the affected buses from generator B. That solution has no buses outside operational limits and no overloaded branches.

Table 1 shows the reconfiguration results for networks A and B, after an (n-1) contingency situation happens to each of them. The table shows results for the four quality criteria described before. For each of the two networks, we first show the values of those criteria when the network is operating normally, and after the fault is introduced. For network A, the fault cut power supply to 7 buses. For network B, the fault affected the supply to 22 buses. Notice that network B has three overloaded cables under normal operations and the fault reduced that number to two. Then, under the 'After reconfiguration' labels, we present the solutions that compose the final Pareto frontier. For network A, there are two solutions; one with no switch operations ('do-nothing' strategy) and another with one, corresponding to the solution depicted in Figure 2c. For network B, the final Pareto frontier is composed of four solutions, with zero, one, two and five

Table 1. Reconfiguration results for networks A and B, after an (n-1) contingency situation. For each network, we show the four quality criteria: buses voltages, buses disconnected, branches overloaded and number of switch operations. The table yields the original values for those criteria when the network is operating normally; after the fault is introduced; and for the solutions obtained by the genetic algorithm. For network A, the final Pareto frontier has two solutions; one with no switch changes and another with a single one. For network B there are four solutions, with zero, one, two and five switch changes. Two of those bring power supply back to all buses.

	Network A - 96 branches, 16 switches			
	Criteria			
Configuration	**Buses voltages**	**Buses disconnected**	**Branches overloaded**	**Number of switch operations**
Original (Fig. 2a)	1.33	0	0	–
After outage (Fig. 2b)	8.32	7	0	–
After reconfiguration				
Solution 1	8.32	7	0	0
Solution 2 (Fig. 2c)	1.31	0	0	1

	Network B - 1,645 branches, 158 switches			
	Criteria			
Configuration	**Buses voltages**	**Buses disconnected**	**Branches overloaded**	**Number of switch operations**
Original	29.85	0	3	–
After outage	50.73	22	2	–
After reconfiguration				
Solution 1	50.73	22	2	0
Solution 2	30.40	0	3	1
Solution 3	30.12	0	3	2
Solution 4	50.16	12	2	5

switch operations. Two of those bring power supply back to all buses (solutions 2 and 3). Solution 3, with two switch operations, has a slightly better overall voltage deviation than Solution 2, which has a single switch operation. The solution with five switch operations reduces the number of buses disconnected but does not eliminate them.

To illustrate how the quality of the solutions improve as the genetic algorithm evolves, we present Figure 3. That figure depicts the values for the four quality criteria separately (y-axis), and the number of generations in the x-axis. Values were taken from the test for network B. Each data point represents the value for that criterion, for a solution present in the Pareto frontier, in that particular generation. Notice that in generation 1 the values for the Pareto frontier solutions are all very poor. Voltage deviations are very high; the number of buses disconnected is over 450; and the number of switch operations is also very high. As the genetic search progresses, we see a steady decline in the number of switch operations from over 40 down to less than 10 by generation 50. Voltage deviations and number of disconnected buses decline as well, but at a slower pace, until between generations 40 and 50, when there is a steep drop. The four criteria continue to improve until generation 220, when we obtained the final Pareto frontier configuration.

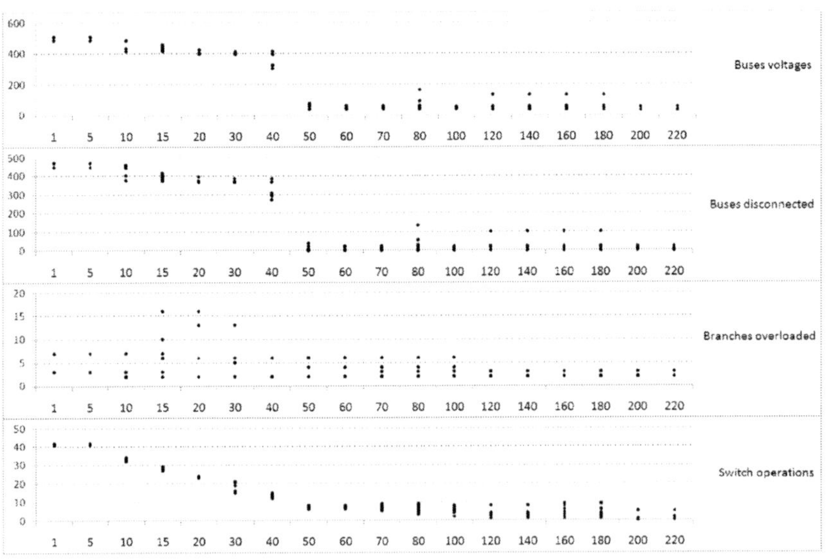

Fig. 3. Quality improvement of the solutions in the Pareto frontier for *network B*. The four quality criteria are depicted separately. The x-axis depicts the number of generations. Each data point represents the value for that criterion, for a solution present in the Pareto frontier, in that particular generation. Notice the steady improvement of the four criteria, as the random solutions from generation 1 evolve into higher quality solutions by generation 220.

The configuration of the genetic algorithm is as follows. In each generation, 20 new individuals are created, and each of them has a 1% change of going through mutation ($mut_{rate} = 0.01$). The number of individuals created in each generation is not critical in our genetic algorithm design, as no intermediate population is created between generations. Individuals are continuously being created/inserted/removed from the population. The value of 20 new individuals per generation was determined because the tests were conducted in a multi-threading environment, with 20 threads executing independent power flow calculations. The CPU time for each of the tests was small. Network A required less than 30 seconds to converge, whereas network B required less than 4 minutes. The two tests shown in this study were run on a Dual Intel Xeon X5650 six-core computer. The software was implemented in Java, version 1.6.0_21-ea 64-bit.

5 Conclusion

This paper introduced a genetic algorithm approach for the multi-objective version of the power restoration problem after an (n-1) contingency situation. Two real-world networks were tested; one with 96 buses and 16 switches, and another with 1,645 buses and 158 switches. The multi-objective approach produced a population of non-dominated solutions in which four criteria are optimized: summation of voltage deviations, number of buses still without supply after the reconfiguration, number of overloaded branches and number of switches operations. A table with the solutions obtained for the two tests is shown. For the large network, we describe the convergence of the genetic algorithm by plotting the evolution of the four criteria in the individuals composing the Pareto frontier, as the generations progress. The ability of the genetic algorithm to consistently improve the solutions is clearly observed. As future research we will add the possibility of tap-changer transformers adjustments as part of the network reconfiguration strategy.

Acknowledgement. The authors wish to thank Ausgrid for providing the network data for this study.

References

1. Aoki, K., Kuwabara, H., Satoh, T., Kanezashi, M.: Outage state optimal load allocation by automatic sectionalizing switches operation in distribution systems. IEEE Transactions on Power Delivery 2, 1177–1185 (1987)
2. Aoki, K., Nara, K., Itoh, M., Satoh, T., Kuwabara, H.: A new algorithm for service restoration in distribution systems. IEEE Transactions on Power Delivery 4, 1832–1839 (1989)
3. Aoki, K., Satoh, T., Itoh, M., Kuwabara, H., Kanezashi, M.: Voltage crop constrained restoration of supply by switch operation in distribution systems. IEEE Transactions on Power Delivery 3, 1267–1274 (1988)
4. Augugliaro, A., Dusonchet, L., Sanseverino, E.R.: Evolving non-dominated solutions in multiobjective service restoration for automated distribution networks. Electric Power Systems Research 59, 185–195 (2001)

5. Bertoli, P., Cimatti, A., Slaney, J., Thibaux, S.: Solving power supply restoration problems with planning via symbolic model-checking. In: Proceedings of the 15th European Conference on Artificial Intelligence, pp. 576–580. IOS Press (2002)
6. Deb, K.: Multi-Objective Optimization Using Evolutionary Algorithms. John Wiley & Sons, USA (2009)
7. Dialynas, E.N., Michos, D.G.: Interactive modeling of supply restoration procedures in distribution system operation. IEEE Transactions on Power Delivery 4, 1847–1854 (1989)
8. Ferreira, L.A.F.M., Grave, S.N.C., Barruncho, L.M.F., Jorge, L.A., Quaresma, E., Carvalho, P.M.S., Martins, J.A., Branco, F.C., Mira, F.: Optimal distribution planning - increasing capacity and improving efficiency and reliability with minimal-cost robust investment. In: Proceedings of the 16th International Conference and Exhibition on Electricity Distribution, pp. 5.21.1–5.21.5. IEE Press (2001)
9. Fujii, Y., Miura, A., Tsukamoto, J., Youssef, M.G., Noguchi, Y.: On-line expert system for power distribution system control. Electrical Power & Energy Systems 14, 45–53 (1992)
10. Garcia, V.J., Franca, P.: Multiobjective service restoration in electric distribution networks using a local search based heuristic. European Journal of Operational Research 189, 649–705 (2008)
11. Goldberg, D.: Genetic Algorithms in Search, Optimization, and Machine Learning. Addison-Wesley Professional, USA (1989)
12. Goldberg, D., Sastry, K.: Genetic Algorithms: The Design of Innovation, 2nd edn. Springer, USA (2010)
13. Kim, H., Ko, Y., Jung, K.-H.: Algorithm of transferring the load of the faulted substation transformer using the best-first search method. IEEE Transactions on Power Delivery 7, 1434–1442 (1992)
14. Mendes, A., Boland, N., Guiney, P., Riveros, C.: (n-1) contingency planning in radial distribution networks using genetic algorithms. In: Proceedings of the IEEE/PES Transmission and Distribution Conference and Exposition: Latin America, pp. 290–297. IEEE Press (2010)
15. Miu, K.N., Chiang, H.-D., Bentao, B., Darling, G.: Fast service restoration for large-scale distribution systems with priority customers and constraints. IEEE Transactions on Power Systems 13, 789–795 (1998)
16. Perrier, N., Agard, B., Baptiste, P., Frayret, J.M., Langevin, A., Pellerin, R., Riopel, D., Trepanier, M.: A survey of models and algorithms for emergency response logistics in electric distribution systems - part I: Reliability planning with fault considerations. Technical Report n. 2010-05 - Interuniversity Research Centre on Enterprise Networks, Logistics and Transportation (CIRRELT), pp. 1–34 (2010)
17. Perrier, N., Agard, B., Baptiste, P., Frayret, J.M., Langevin, A., Pellerin, R., Riopel, D., Trepanier, M.: A survey of models and algorithms for emergency response logistics in electric distribution systems - part II: Contingency planning level. Technical Report n. 2010-06 - Interuniversity Research Centre on Enterprise Networks, Logistics and Transportation (CIRRELT), pp. 1–41 (2010)
18. Thibaux, S., Cordier, M.O., Jehl, O., Krivine, J.P.: Supply restoration in power distribution systems – a case study in integrating model-based diagnosis and repair planning. In: Proceedings of the 12th Conference on Uncertainty in Artificial Intelligence, pp. 525–532. Morgan Kaufmann (1996)
19. Toune, S., Fudo, H., Genji, T., Fukuyama, Y., Nakanishi, Y.: A reactive tabu search for service restoration in electric power distribution systems. In: Proceedings of the IEEE Intl. Conf. on Evolutionary Computation, pp. 763–768. IEEE Press (1998)

Reasoning over OWL/SWRL Ontologies under CWA and UNA for Industrial Applications

Qingmai Wang and Xinghuo Yu

School of Electrical and Computer Engineering, RMIT University, Melbourne, VIC 3001, Australia
qingmai.w@gmail.com, x.yu@rmit.edu.au

Abstract. As expressive schema languages, Web Ontology Language (OWL) and Semantic Web Rule Language (SWRL) have been widely introduced to many industrial applications. Most existing OWL reasoners hold Open World Assumption (OWA) and do not hold Unique Name Assumption (UNA). They lack efficiency when they are applied to industrial models which capture information under Closed World Assumption (CWA) and UNA. To overcome the problem, this paper proposes a novel backward chained ABox reasoner which efficiently reasons through OWL and SWRL under CWA and UNA.

Keywords: Ontology and rules, ABox query, Backward chaining.

1 Introduction

Ontology-based approaches have been widely used in many industrial applications. In those applications, OWL is mostly used to represent concepts and their relationships, and SWRL is usually used as an rule extension of OWL to improve the expressivity. Existing OWL reasoners, such as Pellet [11], Racer [4], and Kaon2 [7], do not make CWA and UNA. This is reasonable because OWL and SWRL originally focus on the Semantic Web (SW) in which the internet can be seen as an unlimited knowledge resource. However, when they are applied to some industrial areas where the information is usually captured under CWA and UNA, some incorrect results may be produced.

For example, OWL and SWRL have been used a lot in computer-aided Product Design and Manufacturing (PDM) to address the semantic interoperability issues [5] [12] [1] [13] [2]. In PDM, the STEP standard [9] captures information of a product under CWA and UNA. Those existing works mainly focus on how to map the STEP-based product information to the ontology while the reasoning issues are rarely mentioned. Most of them simply choose one of the general ontology reasoners for their reasoning tasks without evaluating the reasoner, whereas the reasoner may result some errors due to CWA/OWA and UNA issues, e.g., for an geometrical ontology which captures information of STEP-based product shape, existing reasoners cannot automatically find out any 4-edge face because the reasoners always assume that a face may have countless unknown edges which are not explicitly expressed in the STEP file.

D. Wang and M. Reynolds (Eds.): AI 2011, LNAI 7106, pp. 789–798, 2011.

This paper proposes a novel Backward Chained ABox Reasoner (BCAR) to address the above problem. The objective of BCAR is to provide efficient ABox query reasoning for ontology-based industrial models. Features of BCAR are highlighted below:

1. BCAR partly interprets OWL and SWRL under CWA and UNA (only for ABox) to improve the reasoning for closed world based industrial models.
2. Some existing OWL reasoners support SWRL by translating SWRL to some other formats (Jess, Jena or Seasame) [6] [8]. BCAR directly works on SWRL without any translation, which improves the efficiency.
3. BCAR adopts backward reasoning similar to the Prolog derivation tree. Reordering technique from Optimized Conjunctive Query (OCQ) [10] is integrated to improve the performance.

The rest of the paper is organized as follows: section 2 introduces technical details of BCAR; section 3 discusses the experiments and section 4 concludes the paper.

2 Technical Details

BCAR only focuses on ABox query reasoning for two reasons: 1)In most of the industrial applications, TBox is always decided by the valid and fixed schema of mature information models, e.g. STEP. The TBox reasoning is not necessary in this case. 2)The TBox reasoning under CWA may result inconsistence (ABox can completely determine its TBox under CWA, which may conflict with the source schema). Generally, the objective of BCAR is to help users to find out implicit instances or fillers for some defined classes or properties. For this purpose, BCAR holds following assumptions:

1. Assumption 1: The ontology contains only the explicitly expressed individuals. There is not any unknown individuals.
2. Assumption 2: Only defined classes (or properties) are allowed to have implicit instances (or fillers). Other classes (or properties) only contains instances (or fillers) that are explicitly expressed.
3. Assumption 3: If a class (or property) cannot be proved to contain an instance (or filler), then the class (or property) does not contain the instance (or filler). If a class (or property) cannot be proved to contain any instance (or filler), then the class (or property) contains nothing (negation as failure).
4. Assumption 4: Two individuals are same if and only if their names are same.

In OWL/SWRL ontologies, the most two popular methods to define classes and properties are SWRL rules and OWL Equivalent Class Axiom (ECA). BCAR only considers classes (or properties) which have corresponding ECAs or appear in heads of SWRL rules as defined classes (or properties), and the corresponding ECAs and SWRL rules are their definitions.

In the rest of this section, the concept framework of BCAR is firstly given; how to create a rule base is secondly introduced followed by details of reasoning algorithm; some special cases of reasoning are finally discussed.

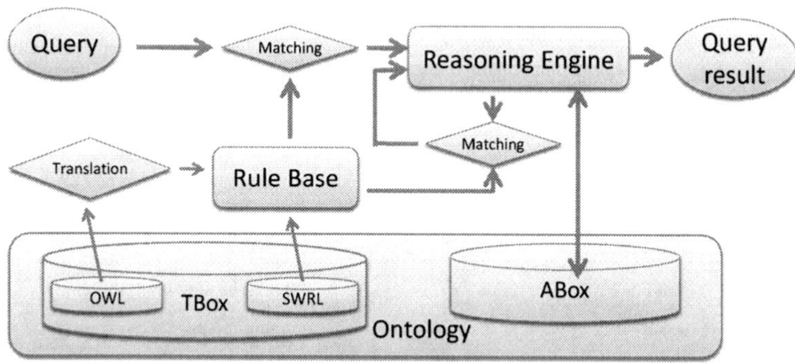

Fig. 1. Cocnept framework of BCAR

2.1 Concept Framework

Fig.1 shows the concept framework. BCAR firstly creates a unified rule base with SWRL rules and OWL ECAs. Once a query is inputted, the rule which matches the query fires. The reasoning engine is then applied to the firing rule to search for solutions for the firing rule. The engine requires information from the ABox and also may fire some other rules if required. After the searching is finished, query result is then generated based on the achieved solutions.

2.2 Building Rule Base

As mentioned above, BCAR only accepts two ways to define classes and properties: SWRL rule and ECA. Since that the backward chained reasoning engine requires a unified rule base, there is a need to translate ECAs to SWRL-like rules. The translation is generally based on the FOL semantics of the OWL constructs, and is described in Tab.1. (A, B, C are classes, P is a property, I is an individual and n is a number).

Normally, ECAs cannot be translated to rules directly since that they represent bidirectional relationships. However, BCAR consider ECAs only as class definition rules, and process ECAs equally with SWRL rules. In this case, interpret ECA as one direction logic, from definition to the class, is reasonable.

2.3 Reasoning Algorithm

In BCAR, retrieving instances (or fillers) of defined classes (or properties) fires their definition rules. The process of query is then transformed to a process of searching the ABox for solutions for the definition rule. Reasoning results are generated based on the solutions. In the following discussion, the solution is firstly defined, the searching algorithm is secondly given, how to generate results based on solutions is then described, the backward chaining in the rule base is finally discussed.

Table 1. Translating ECA to SWRL-like rules

Construct Name	ECA Syntax	Rules
owl: intersectionOf	C = A and B	A(?x) ^B(?x) -> C(?x)
owl: unionOf	C = A or B	A(?x) -> C(?x) B(?x) -> C(?x)
owl: complementOf	C = not A	Not (A(?x))-> C(?x)
owl: someValuesFrom	C = R some A	R(?x,?y)^A(?y) -> C(?x)
owl: allValuesFrom	C = R only A	∀ ?y(R(?x,?y)^A(?y))->C(?x)
owl: hasValue	C = R has I	R(?x, I) -> C(?x)
owl: minCardinality	C = R min *n*	(R >= *n*) (?x) ->C(?x)
owl: maxCardinality	C = R max *n*	(R <= *n*) (?x) -> C(?x)
owl: cardinality	C = R exactly *n*	(R = *n*) (?x) -> C(?x)
Complex ECA	C = A and (R some B)	R(?x,?y)^B(?y) -> H(?x) (create an intermediate class H) A(?x)^H(?x) -> C(?x)

For the purpose of simplicity, we introduce following terms to represent reasoning tasks ($\mathscr{C}$ and $\mathscr{P}$ are defined class and property respectively):

$$\mathscr{C}(?) : \text{query for all the instances of } \mathscr{C}$$
$$\mathscr{C}(? = a) : \text{check whether } a \text{ is an instance of } \mathscr{C}$$
$$\mathscr{P}(?, ?) : \text{query for all the fillers of property } \mathscr{P}$$
$$\mathscr{P}(? = a, ?) : \text{query for the instances which are the property values of } a \text{ for } \mathscr{P}$$
$$\mathscr{P}(?, ? = a) : \text{query for the instances whose property value is } a \text{ for } \mathscr{P}$$
$$\mathscr{P}(? = a, ? = b) : \text{check whether } (a, b) \text{ is a filler of property } \mathscr{P}$$

The example ontology given below is for the following algorithm discussion:

TBox
Atomic Classes:$A; B; C; D$
Defined Class:$E = OP4\ some\ C;$
Object Properties:$OP1(Domain : A, Range : B); OP2(Domain : B, Range : C, D)$
Defined Object Property:$OP4(Domain : A, Range : C)$
rule1(SWRL):$OP1(?x, ?y) \wedge OP2(?y, ?z) \wedge C(?z) \rightarrow OP4(?x, ?z)$
ABox
$A\{A1; A2; A3\}; B\{B1; B2; B3; B4\}; C\{C1; C2; C3; C4\}; D\{D1; D2\}$
$OP1\{(A1, B1); (A1, B2); (A2, B3); (A3, B3); (A3, B4)\}$
$OP2\{(B2, C2); (B2, C3); (B3, D1); (B4, C4)\}$

Solution. In a solution, all the variables in the rule are bound to a value (value can be individuals or datatype values such as integer and string), which makes all the atoms in the rule body hold true. In the example ontology, rule1 has three variables: $?x, ?y, ?z$. With the above ABox, solutions for rule1 are:

$$\{?x \leftarrow A1; ?y \leftarrow B2; ?z \leftarrow C2\}$$
$$\{?x \leftarrow A1; ?y \leftarrow B2; ?z \leftarrow C3\}$$
$$\{?x \leftarrow A3; ?y \leftarrow B4; ?z \leftarrow C4\}$$

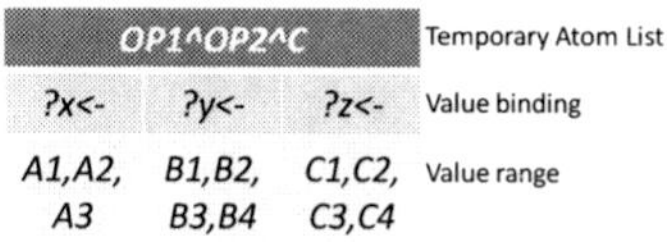

OP1^OP2^C			Temporary Atom List
?x<-	?y<-	?z<-	Value binding
A1,A2, A3	B1,B2, B3,B4	C1,C2, C3,C4	Value range

Fig. 2. The root of searching tree

All the reasoning results of BCAR is generated based on solutions, details of which is discussed later.

Searching for Solutions. Searching for solutions is the key of BCAR. The searching process in BCAR is generally an OCQ process based on the well-known Prolog derivation tree. The idea of atoms reordering [10] is integrated to improve the performance.

Taking the example ontology, assuming the reasoning task is $OP4(?,?)$, rule1 fires. The search is then started with the following steps:

1. **Preprocessing:** Build a Temporary Atom List (TAL) from the rule body. Define a Value Range (VR) for each variable based on the Assumption 2. In the example, from rule 1 it is easy to find out that $?x$ belongs to A, $?y$ belongs to B, and $?z$ belongs to C. Based on Assumption 2, $?x$, $?y$ and $?z$ can only be the explicitly asserted members of A, B and C respectively. Fig.2 shows the TAL and VRs of the example. Other cases are listed below:
   (a) The VR of $?x$ (or $?y$) contains only $A1$ (or $B1$) if the reasoning task is $OP4(? = A1, ?)$ (or $OP4(?, ? = B1)$);
   (b) For constant, BCAR create new variables whose VR only contain the constant's value;
   (c) For variable belonging to a defined class, the VR $= \top$ (Assumption3);
   (d) For datatype variable, the VR $= \infty$
2. **Variable choosing and branching:** The performance of CQ heavily relies on the query ordering. In BCAR, the reasoning always start from the variable which has minimal size of VR and have not been bound to a value, so that the number of branches of the searching tree is minimized. The selected variable is called SV. In the example, $?x$ is SV since it has minimal size of VR among all the variables. BCAR then generate branches for each value in VR of $?x$.
3. **Binding and intersecting:** In each branch, the SV is bound to a value. BCAR then processes the atoms related to the SV in TAL, which may reduce the VR of SV or other variables which are related to SV by an intersecting process. After that, the processed atoms are removed from TAL. In the example, $?x$ is bound to $A1$ in the first branch. BCAR then processes $OP1(?x(= A1), ?y)$ which is the only atom related to $?x$ in TAL. Based on the ABox and Assumption 3, $?y$ can only be either $B1$ or $B2$. BCAR then intersects $\{B1, B2\}$ with $?y$'s original VR $\{B1, B2, B3, B4\}$ to be the new VR of $?y$, as Fig.3 shows:

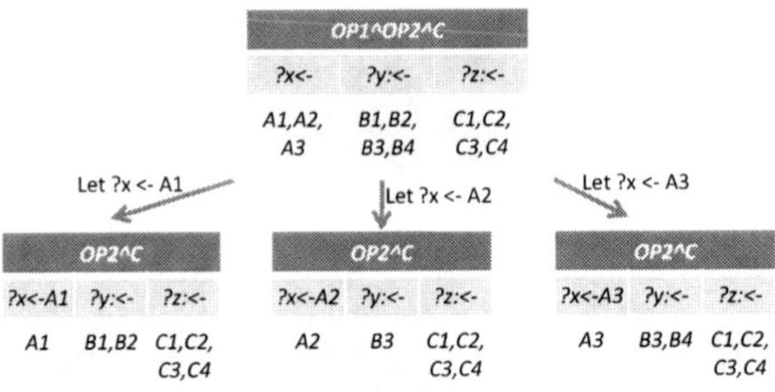

Fig. 3. Generating branches

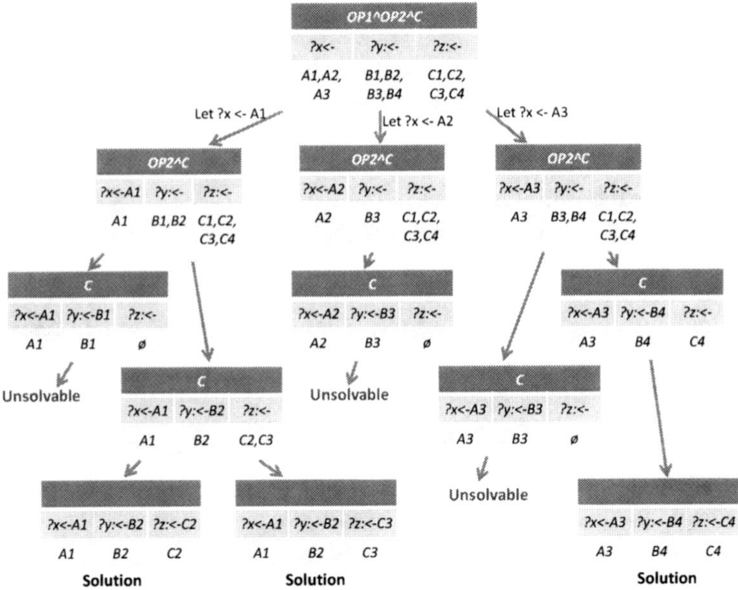

Fig. 4. Searching tree

4. **Termination:** BCAR repeats the above step 2 and step 3 until: 1)VR of any variable turns out to be empty (based on Assumption 1&2, it means this variable is unsolvable in this branch); 2)All the variables have been bound to a value (a solution has been found in this branch); Fig4. shows how the search tree find solutions and how it is terminated.

Generating Reasoning Results. With solutions, BCAR generate results for every reasoning tasks as follows (assuming the head of the definition rule of $\mathscr{C}$ and $\mathscr{P}$ are $\mathscr{C}(?x)$ and $\mathscr{P}(?x, ?y)$ respectively):

1. $\mathscr{C}(?)$: return all the values of $?x$ from all the solutions;
2. $\mathscr{C}(? = a)$: set the initial VR of $?x$ to be $\{a\}$, check wether a solution can be found;

3. $\mathscr{P}(?,?)$: pick up all the values of $?x$ and $?y$ from all the solutions;
4. $\mathscr{P}(?=a,?)(or\mathscr{P}(?,a))$: set the initial VR of $?x$ (or $?y$) to be $\{a\}$, return all the values of $?y$ (or $?x$) from all the solutions;
5. $\mathscr{P}(?=a,?=b)$: set the initial VR of $?x$ and $?y$ to be $\{a\}$ and $\{b\}$ respectively, check wether a solution can be found;

Backward Chaining. In the "Binding and intersecting" step of the searching process, if BCAR need to process an atom corresponding to a defined class or property, another rule fires. For example, considering defined class $E = OP4\ some\ A$, from Tab.1 the translation rule is $OP4(?x,?y) \wedge D(?y) \rightarrow E(?x)$. Assuming the reasoning task (goal) is $E(?=A1)$, when searching for the solutions for the translation rule, the "Binding and intersecting" step will add another reasoning task (new goal) $OP4(?=A1,?)$ and rule 1 consequently fires. This is so-called backward chained reasoning.

2.4 Handling Special Rules and Atoms

In Tab.1, some of the translation rules can be solved normally using the above algorithm (e.g. owl:intersectionOf; owl:unionOf; owl:someValuesFrom), while the others can not. This section generally discusses how BCAR handles some special rules and atoms.

owl:complementOf. $C = notA$ is translated to $Not(A(?x)) \rightarrow C(?x)$. Based on Assumption 3 (negation as failure), BCAR handles the translation rule as follows:

1. $C(?)$: return the individuals which cannot be proved to be instances of A;
2. $C(?=a)$: return false if a is proved to be an instance of A, otherwise return true;

owl:allValuesFrom. $C = R\ only\ A$ is translated to $\forall ?y(R(?x,?y) \wedge A(?y)) \rightarrow C(?x)$. Based on Assumption 2, BCAR handles the translation rule as follows:

1. $C(?)$: return the individuals which are proved to have some property values for R and all these values are proved to be instances to A;
2. $C(?=a)$: return true if a is proved to have some property values for R and all these values are proved to be instances of A, otherwise return false;

owl:minCardinality/maxCardinality/cardinality. $C = min$ or max or $exactly\ a$ is translated to $(R \geq$ or $\leq$ or $= a)(?x) \rightarrow c(?x)$. Based on Assumption 2, BCAR handles the translation rule as follows:

1. $C(?)$: return the individuals which are proved to have more than or less than or exactly a different property values for R;
2. $C(?=a)$: return true if a is proved to have more than or less than or exactly a property values for R, otherwise return false;

Complex ECA. A complex ECA is a combination of basic OWL constructs. As Tab.1 shows, BCAR transform complex ECA to multiple basic rules with self-created intermediate classes between them.

swrl:differentFrom/sameAs. An SWRL rule may contains atoms such as "differentFrom(?x,?y)" and "sameAs(?x,?y)". In this case, BCAR firstly remove these atoms from TAL before searching for solutions. After solutions are found, these comparison atoms are used to validate each solution based on Assumption 4. Only validated solutions are outputted in the end.

swrl:built-in. BCAR only supports built-Ins for comparison in current stage. BCAR process built-In atom similar to the "differentFrom/sameAs" atoms. The only difference is "differentFrom/sameAs" atoms focus on comparing individuals while built-In atoms focus on comparing datatype values.

3 Experiments

In this section, two experiments has been done to test the performance of searching and the effectiveness of BCAR in industrial applications.

The first experiment tests BCAR with LUBM[1,0] [3] which contains 103074 triples, and compares the performance with popular reasoner Pellet and RacerPro. The 14 standard quires are transferred to rules for BCAR. An example of transfer is shown below:

$$ub : GraduateStudent(?x) \wedge ub : takesCourse(?x, http : \\ //www.Department0.University0.edu/GraduateCourse0) \rightarrow Query01(?x)$$

As mentioned before, in BCAR, generating results for $Query01(?)$ is the process of searching for solutions for rule body. And searching for solutions is essentially a conjunctive query of body atoms. In this way, the 14 standard quires of LUBM are used to test the searching performance of BCAR.

Table 2. Experimental result (Unit: ms)

	Q1	Q2	Q3	Q4	Q5	Q6	Q7	Q8	Q9	Q10	Q11	Q12	Q13	Q14
BCAR	<1	340	4	<1	112	670	12	966	380	1	N/A	6	7	660
Pellet	78	536	2	8	34	97	11	1102	867	3	35	38	2	88
RacerPro	308	896	24	201	66	899	1033	1297	1156	92	43	881	306	630

The experimental result is shown in Tab.2. Currently BCAR doesn't support transitive roles, so no result can be found for Query 11. Generally BCAR shows a decent performance comparing with other two reasoners. Especially for quires which require only small part of ABox information, BCAR performs very

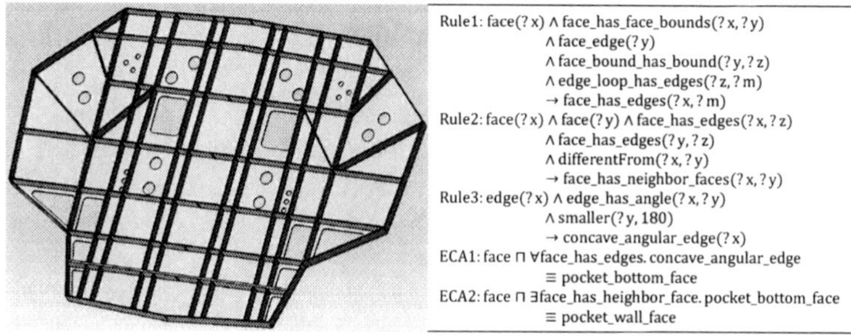

Fig. 5. Input model and rules

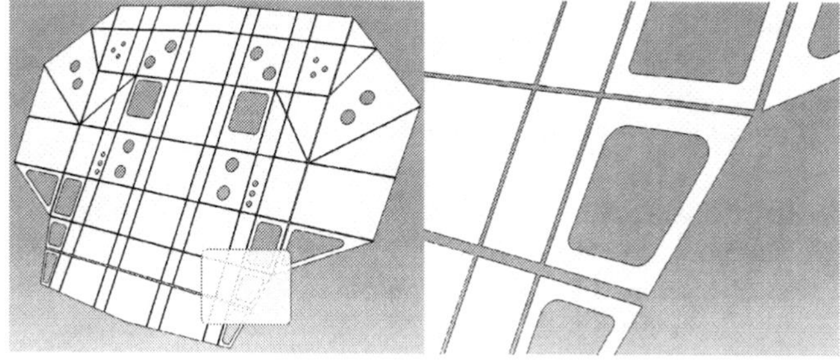

Fig. 6. Visualization of query result

good since that backward reasoning guarantees only necessary information is processed.

Another experiment tests the effectiveness of applying BCAR to an industrial model as shown in the left of Fig.5. The model comes from a STEP file which captures the geometric information of a product. After transferring from STEP to OWL, the ontology contains classes such as faces, edges and points. Instances of the STEP file are mapped to OWL individuals. Several SWRL rules and OWL ECAs (shown in right of Fig.5) are added to the ontology to define a class *pocket_bottom_face*, and BCAR is applied to find out all the members of *pocket_bottom_face*. The visualization of query result is shown in Fig.6.

Other reasoners are not applicable in this case for the UNA and OWA/CWA issues. For example, without making UNA the atom $differentFrom(?x, ?y)$ in Rule2 always returns false because the STEP file doesn't explicitly mention that an instance is different from another instance. Similarly, the universal quantification in ECA1 always returns false, because under OWA the reasoners always believe there must exist some unknown edges.

4 Conclusion

This paper has proposed a novel Backward Chained ABox reasoner which provides efficient ABox query reasoning under CWA and UNA for industrial applications. The reasoner firstly translated OWL ECAs to SWRL-like rules, and then backward reason through the unified rule base of the ontology to retrieval instances for query tasks.

Experiments shows that the BCAR can effectively execute queries for industrial models under CWA and UNA, and also demonstrates that it has a decent performance. Future works will focus on improving the reasoner to support more OWL constructs and evaluating the reasoner using more practical industrial cases.

References

1. Abdul-Ghafour, S., Ghodous, P., Shariat, B., Perna, E.: Towards an intelligent cad models sharing based on semantic web technologies. In: Curran, R., Chou, S.-Y., Trappey, A. (eds.) Collaborative Product and Service Life Cycle Management for a Sustainable World. Advanced Concurrent Engineering, pp. 195–203. Springer, Heidelberg (2008)
2. Andersen, O.A., Vasilakis, G.: Building an ontology of cad model information. In: Hasle, G., Lie, K.-A., Quak, E. (eds.) Geometric Modelling, Numerical Simulation, and Optimization, pp. 11–40. Springer, Heidelberg (2007)
3. Guo, Y., Pan, Z., Heflin, J.: Lubm: A benchmark for owl knolwedge base systems. Journal of Web Semantics 3, 158–182 (2005)
4. Haarslev, V., Moller, R.: Racer: An owl reasoning agent for the semantic web. In: Proceedings of the International Workshop on Applications, Products and Services of Web-based Support Systems, pp. 91–95 (2003)
5. Kim, K.Y., Manley, D.G., Yang, H.: Ontology-based assembly design and information sharing for collaborative product development. Computer-Aided Design 38, 1233–1250 (2006)
6. Mei, J., Paslaru, E.B.: Reasoning paradigms for swrl-enabled ontologies (2005), `http://www.inf.fu-berlin.de/inst/pubs/tr-b-04`
7. Motik, B.: Reasoning in Description Logics using Resolution and Deductive Databases. Ph.D. thesis, University of KarLsruhe, Karlsruhe, Germany (January 2006)
8. O'Connor, M.F., Knublauch, H., Tu, S., Grosof, B.N., Dean, M., Grosso, W., Musen, M.A.: Supporting Rule System Interoperability on the Semantic Web with SWRL. In: Gil, Y., Motta, E., Benjamins, V.R., Musen, M.A. (eds.) ISWC 2005. LNCS, vol. 3729, pp. 974–986. Springer, Heidelberg (2005)
9. Owen, J.: STEP: An Introduction. Information Geometers (1997)
10. Sirin, E., Parsia, B.: Optimizations for answering conjunctive abox queries: First results. In: Proc. of the Int. Description Logics Workshop, DL (2006)
11. Sirin, E., Parsia, B., Grau, B.C., Kalyanpur, A., Katz, Y.: Pellet: A practical owl-dl reasoner. Web Semantics: Science, Services and Agents on the World Wide Web 55, 51–53 (2007)
12. Yang, D., Dong, M., Miao, R.: Development of a product configuration system with an ontology-based approach. Computer-Aided Design 40, 863–878 (2008)
13. Zhao, W., Liu, J.K.: Owl/swrl representation methodology for express-driven product information model: Part 1. implementation methodology. Computers in Industry 59, 580–589 (2008)

Tracking the Preferences of Users Using Weak Estimators

Anis Yazidi[1], Ole-Christoffer Granmo[1], and B. John Oommen[2,*]

[1] Dept. of ICT, University of Agder, Grimstad, Norway
[2] School of Computer Science, Carleton University, Ottawa, Canada

Abstract. Since a social network, by definition, is so diverse, the problem of estimating the preferences of its users is becoming increasingly essential for personalized applications which range from service recommender systems to the targeted advertising of services. However, unlike traditional estimation problems where the underlying target distribution is stationary, estimating a user's interests, typically, involves non-stationary distributions. The consequent time varying nature of the distribution to be tracked imposes stringent constraints on the "*unlearning*" capabilities of the estimator used. Therefore, resorting to strong estimators that converge with probability 1 is inefficient since they rely on the assumption that the distribution of the user's preferences is stationary. In this vein, we propose to use a family of stochastic-learning based *Weak* estimators for learning and tracking user's time varying interests. Experimental results demonstrate that our proposed paradigm outperforms some of the traditional legacy approaches that represent the state-of-the-art.

Keywords: Weak estimators, User's Profiling, Time Varying Preferences.

1 Introduction

Utilizing the power of the Internet to affect marketing, business and politics *via* strategies applicable for social networking, is becoming increasingly important, especially in a user-driven universe. Over the last few years, the issue of maintaining users' profiles has become more crucial for designing and streamlining personalized applications ranging from service recommender systems to the advertising of targeted services. Mastering and optimally utilizing the knowledge about a user's interests has led to promising applications in filtering and recommending documents [2], multimedia [4] and TV programs [14], based on their respective contents.

---

* Chancellor's Professor; Fellow : IEEE and Fellow : IAPR. The Author also holds an Adjunct Professorship with the Dept. of ICT, University of Agder, Norway. The first author gratefully acknowledges the financial support of the *Ericsson Research*, Aachen, Germany, and the third author is grateful for the partial support provided by NSERC, the Natural Sciences and Engineering Research Council of Canada.

D. Wang and M. Reynolds (Eds.): AI 2011, LNAI 7106, pp. 799–808, 2011.

Usually, constructing a user's profile involves applying estimation techniques to leverage the knowledge about his interests, which, in turn, is gleaned from the history of the services that he utilizes [4,5]. A number of previous studies [8] have shown that a user's interests are not constant over time, and consequently, paradigms which are to be promising, should take into account the drift of these interests. The time varying nature of the distribution of the user's interests renders the problem of estimating them both difficult and non-trivial.

Recently, Oommen and Rueda [16] have proposed a strategy by which the parameters of a binomial/multinomial distribution can be estimated when the underlying distribution is non-stationary. The method is referred to as Stochastic Learning Weak Estimation (SLWE), and is based on the principles of stochastic Learning Automata (LA) [13,20]. The SLWE has found successful applications in many real-life problems that involve estimating distributions in non-stationary environments such as in adaptive encoding [17], route selection in mobile ad-hoc networks [15], and topic detection and tracking in multilingual online discussions [19]. Motivated by these successful applications of the SLWE in various areas, in the course of this study, we consider employing the SLWE for solving the intriguing problem of tracking user's interests. The objective of the paper is to present a personalized *Learning Preferences Manager*, a *modus operandus* for capturing user's preferences. The latter will be able to cope with changes brought about by variations in the distribution of the user's interests, which will be where the SLWE plays a prominent part.

2 State of the Art

The core function of a personalized *Learning Preferences Manager* is to update the user's profile in a dynamic and incremental way. This is done so that the "Manager" can closely follow the real-time evolution of the user's interests. In fact, any user's interests are not constant over time, and therefore it is imperative that the system takes the profile's drift into account. In this sense, whenever one attempts to represent the user's *current* interests, the most recent observations are more reliable than older ones. From a more general perspective, the task of learning the drifts in the user's interests corresponds to the problem of learning evolving concepts [21]. There are several studies that have dealt with the task of learning a user's interests. These include the use of a sliding window [11], aging examples [9], and a Gradual Forgetting (GF) function [6,7,8] etc. However, of all these, a sliding window approach is the most popular one. It consists of learning the description of the user's interests from the most recent observations, and thereafter, of discarding the observations that fall outside the window. A substantial shortcoming of the sliding window approach is the choice of the window size. In [11], the authors adopted a fixed-size time window in order to learn a user's scheduling preferences. They empirically determined that a window size of 180 was a proper choice for their particular scheduling application. The GF, on the other hand, relies on assigning weights to the observations that decrease over time. Hence, the influence of older (more "stale") observations on the running estimates, decreases with time. The authors of [8] suggested a linearly-decreasing function, $w = f(t)$, for decaying the relative weights of the GF as follows:

$$w_i = \frac{-2k}{n-1}(i-1) + 1 + k, \tag{1}$$

where i denotes a counter of observations starting from the most recent one, n is the number of observations, $k \in [0,1]$ is a parameter that represents the percentage by which the weight of any subsequent observation is decreased, and consequently the percentage by which the weight of the most recent one, in comparison to the average, is increased. Thus k is a parameter that controls the slope of the forgetting function. In order to achieve a synergy between both the two approaches, namely GF and sliding window, Koychev in [8], proposed to apply the GF *within each sliding window*. Thus, in this case, the parameter n (i.e., the length of the observation sequence) in equation (1) was set to be equal to L, where L denotes the length of the window. Apart from the sliding window and GF schemes, other approaches, which also deal with *change detection*, have also emerged. In general, there are two major competitive sequential change-point detection algorithms: Page's cumulative sum (CUSUM) [1] detection procedure and the Shiryaev–Roberts–Pollak detection procedure. In [18], Shiryayev used a Bayesian approach to detect changes in the parameters distribution, where the change points were assumed to obey a geometric distribution. CUMSUM is motivated by a maximum likelihood ratio test for the hypotheses that a change occurred. Both approaches utilize the log-likelihood ratio for the hypotheses that the change occurred at the point, and that there is no change. Inherent limitations of CUMSUM and the Shiryaev–Roberts–Pollak approaches for on-line implementation are the demanding computational and memory requirements. In contrast to the CUMSU and the Shiryaev–Roberts–Pollak, the SLWE avoids the intensive computations of ratios, and do not invoke hypothesis testing. A particularly interesting recent study for learning user's interests in ambient media services (and in, consequently, locating relevant services) was reported in [5]. Hossain *et al* devised the so-called Ambient Media Score Update method, which we shall refer to as SU for the rest of the paper. The SU method was used to learn a user's changing interests [4,5] by recording the so-called "scores", which represented his/her affinity of interests. In order to follow closely the evolution of the scores, the authors of [5] refined their proposed updating method defined earlier in [4] and updated the scores of the services at every time instant whenever the service was used. This was done instead of performing updates in a batch mode [4].

3 SLWE-Based Solution to Adaptation to User's Interests Drift

In this section, we devise a *Learning Preferences Manager* which takes advantage of the SLWE updating scheme [16], so as to accurately estimate the user's interest affinity in non-stationary environments. First, we will present our adapted model, as it pertains to the presentation of the user's profile. Thereafter, we introduce two profile update methodologies based on whether the data items attached to an attribute are disjunctive or conjunctive.

3.1 Profile Representation

An essential element of the Learning Preference Manager is the *Profile Representation*. For instance, a possible representation model for a user's interests can be in terms of the topic hierarchies [3,10]. We adopt the Profile Representation Model advocated by Hossain and his co-authors in [4,5]. It is important to remark that in these publications, the latter Profile Representation Model was mainly devised for representing the user's preferences in content media. Nevertheless, the model can be easily applied to encompass a wider set of interests. It should also be noted that the model reported in [4,5] is similar to that of [23] in the sense that it is based on <feature, weight> pairs, except that in [4,5], the authors have invoked a normalized score for the data items. We shall first briefly present the Profile Representation Model reported in [4,5]. The user's affinity of interests in a service type, such as movies, or restaurants, is represented by a set of *attributes*. For example, for a repository of services of type movie, the set of possible attributes could be {movie genre, director name, etc.}. An attribute, in turn, possesses a set of *data items*. For example, if the movie attribute "genre" has two data items, namely "action" and "comedy", a vector associated with the attribute (comedy affininity=0.7, action affininity=0.3) reflects that the user likes comedy movies more than action movies, with a relative weighting of 0.7 to 0.3. The update of the weights of the data items for a particular attribute is done in an incremental manner.

3.2 Profile Updating Method

In the quest to learn the user's dynamic profile, the *Learning Preferences Manager* is guided by so-called *Relevance Feedback* (RF) [12]. In this paper, we rely on the *Service Usage History* (analogous to the history maintained by the authors of [4,5]) as the main source of the RF. In fact, a common approach towards constructing a user's profile is through non-intrusively monitoring the history of the usage of his services. A *Service Usage History* (also known as the *Interaction History*), contains the history of the services used by the user over time. For example, when the user has used a certain service at a certain time instant, the *Learning Preferences Manager* refines and revises the user's profile based on the current instance of the usage history, which, in turn, is automatically and unobtrusively observed in the background. To obtain an index to measure this, the sum of the scores of a data item for a given attribute is made to be equal to unity. To now quantify this, we have opted to use the SLWE [16], so as to update the score of the data item based on the usage history. Whenever a user selects a service, the metadata describing the service is used to update the score of the data item. Thus, for example, if a user currently views an "action" movie, the scheme would increase the weight associated with the data item "action". Apart from the updating mechanism, our strategy can also be seen to be philosophically related to the approach presented in [4,5] in which the authors utilized the history to update the affinity of the user's interests. We believe that this will facilitate the ease of the retrieval of personalized information, and help alleviate the user's cognitive load, i.e., that which is needed to locate relevant information. At this juncture, we distinguish two classes of data items that, in turn, require two different forms of update mechanisms. In fact, the data items related to a given attribute could be either semantically **disjunctive** or semantically **conjunctive**. We illustrate what we mean by the latter concepts by alluding to two simple examples.

Profile Update for Disjunctive Data Items. Data items of a particular attribute are said to be *disjunctive* if every service usage history can only be instantiated with the exclusive realization of one of the data items at a time. To illustrate the idea in simpler terms, consider the example of learning a user's preferences when it concerns a type of services such as restaurants. In this case, we can consider the attribute genre of the restaurant, with the data items being, for example, Chinese, Italian, Indian, French etc. The latter data items correspond to a possible semantic taxonomy of restaurants according to their genre. Whenever a user interacts with a service of type restaurant, a *Service Usage History* instance is submitted to the *Learning Preference Manger* where the restaurant is described by a single exclusive attribute, such as Italian. Consequently, the weight of the latter data item can be incremented while the weight of the remaining data items of the same attribute can be decremented. Therefore, a multinomial SLWE is a viable option for estimating the evolving weights of the data items. Proceeding to make inferences from these weak estimators becomes then a suitable choice for managing the time-varying preferences. It is crucial for the reader to observe that the SU approach presented in [4,5] deals only with this specific case, i.e., of disjunctive data items.

Profile Update for Conjunctive Data Items. Data items of a particular attribute are said to be *conjunctive* whenever every service usage history can be instantiated with one *or more* data items at a time. To illustrate this, consider the example of the service usage history corresponding to the services for movies. The attribute movie genre is associated with the data item set $S_{genre} = \{action, romantic, comedy, horror\}$. The latter data items are conjunctive (not disjunctive) in the sense that a movie's genre can be described with more than a single data item at a time. For instance, a movie genre could be "romantic" and "action packed" at the same time. Suppose that the user watches a movie that belongs to the genres *action* and *romance* at a given time instant 'n'. In this case, the weights of both the data items *action* and *romance* can be increased at time '$n+1$'. In this case, a multinomial SLWE will not be able to update the different weights of the data items because it is not designed to increase the weights of more than a component at a time. Thus, a different methodology for updating the weights of the data items is needed, where more than a single data item's weight can be incremented at a time. To solve the problem, we propose to attach a binomial SLWE to each data items instead of having a multinonmial probability vector for each attribute, as in the case of disjunctive data items. In other words, a binomial probability vector will be attached to each of data items in S_{genre}. For the sake of clarity, we consider the above-mentioned example and describe the update at the subsequent instant '$n+1$' of each binomial probability vector as:

$$p_{action}(n+1) \leftarrow 1 - \lambda(1 - p_{action}(n)) \tag{2}$$

$$p_{romantic}(n+1) \leftarrow 1 - \lambda(1 - p_{romantic}(n)) \tag{3}$$

$$p_{comedy}(n+1) \leftarrow \lambda p_{comedy}(n) \tag{4}$$

$$p_{horror}(n+1) \leftarrow \lambda p_{horror}(n) \tag{5}$$

Once these binomial-based computations have been achieved, we then resort to an additional computation in order to normalize the weights of each data items. The normalization is, quite simply, given by: For $k \in S_{genre} = \{action, romantic, comedy, horror\}$

$$W_k(n+1) = \frac{p_k(n+1)}{\sum_{j \in S} p_j(n+1)} \tag{6}$$

Consequently W_k tracks, with a SLWE-philosophy, the ratio of the number of times the particular data items ($k \in S_{genre} = \{action, romantic, comedy, horror\}$) of the particular attribute (movie's genre) appears in the service usage within a given number of usage records, to the total number of occurrences of the data items of S_{genre}. In order to model this in a "tangible" (or realistic) way, we suppose that the occurrence of each data item in the usage history is controlled by a binomial distribution. We further suppose, that the occurrence of the data items is independent of each other. Let s_k be the binomial parameter that describes the occurrence of data item k in the usage history, where $k \in S_{genre}$. With these assumptions, based on the results of the previous subsection, we easily derive the asymptotic weight:

$$\mathrm{E}\left[W_k(\infty)\right] = \frac{s_k}{\sum_{j \in S} s_j}. \tag{7}$$

It is worth noting that whenever the data items corresponding to a given attribute are disjunctive, it is computationally more efficient (although only marginally) to employ a multinomial SLWE – instead of a set of binomial SLWEs.

Modelling changes in the Interests. We suppose that at every time instance 'n', the *Learning Preferences Manager* is fed by a service usage instance. We further assume that the distribution of the user's interests, relative to a given attribute, undergoes an abrupt change at a random time instance with an unknown probability p. In the case of disjunctive data items, we assume that the parameters of the multinomial distribution change to yield a new distribution.

4 Experimental Results

To verify our computational model and our proposed solution, we have performed extensive simulations. However, in the interest of space and brevity, we report here only a subset of these results. Due to space limitations, experimental results concerning the case of disjunctive data items are omitted in this paper and are found in [22]. We emphasize though that these results are both representative and typical. The obtained experimental results are conclusive, and demonstrate that our SLWE-based update schemes, when applied to tracking users' interests, outperforms the GF approach, the sliding window, and the SU. In order to model the changes in the interests' distribution, we assume that at any given time instant, the distribution of the user's preferences changes with probability 0.02. This implies that on *average*, a change occurs every 50 time instants. The reader should observe that our experimental results are based on synthetic data due to fact that it is difficult (if not *impossible*) to obtain real-life data that describe user's preferences. Indeed, no existing organization will disclose or share such data because of the implied privacy and security considerations. However, we believe that the model which we have used to "artificially" indicate the changes in the user's interest distributions is strong enough to mimic real-life settings. To study the case of disjunctive data items, we assume that we are dealing with estimating the evolving user's interests' weights of data items of this type, namely, those which are associated with a given attribute. In the interest of completeness, we will present separate experimental results for the multinomial case. We considered a multinomial random variable, X, which can take any of four different values, namely '1', '2', '3' or '4', whose characterizing parameters changed (randomly) at random time instants. We ran the estimators for 400 steps, repeated this $1,000$ times, and then computed the corresponding ensemble averages. For each experiment, we computed $||P - S||$, the *Euclidean distance* between P and S, which we reckoned as a measure of how good our estimate, P, was of S. The plots of the latter distance obtained from the SLWE, the GF and the SU are depicted in Figures 1(a), 1(b), 1(c) and 1(d), where the values of λ were 0.908, 0.903, 0.952 and 0.948, and the sizes of the windows were 35, 44, 63 and 76 respectively. The values for λ and the window size were obtained randomly from a uniform distribution in $[0.9, 0.99]$ and $[20, 80]$ respectively. From these figures, we observe that the GF, the SU and the SLWE converge to zero relatively quickly prior to the first instant when the distribution changes. However, this behavior is not present for subsequent (successive) distribution "switches". Rather, we notice that the GF is capable of tracking the changes of the parameters when the size of the window is small, or at least smaller than the intervals of constant probabilities. It is, however, not able to track the changes properly when the window size is relatively large. Since neither the magnitude nor the instants of the changes is known *a priori*, this scenario demonstrates the weakness of the GF, and its dependence on the knowledge of the input parameters. Again, such observations are typical. In Table 1, we report the error rates associated with the experiments plotted in Figures 1(a), 1(b), 1(c) and 1(d). We also include the error rates for the MLE augmented with a sliding window in Table 1. Clearly, one observes that the SLWE exhibits a lower error rate than the GF, the SU and the MLE.

Table 1. The effects of varying the window size and the updating parameter on the error rates for the various schemes investigated for disjunctive data items

Figure No.	Error rate: SLWE	Error rate: GF	Error rate: SW	Error rate: SU
Figure 1(a)	0.0612	0.0724	0.0836	0.4606
Figure 1(b)	0.0665	0.1006	0.1152	0.4037
Figure 1(c)	0.1601	0.1893	0.2074	0.4175
Figure 1(d)	0.0507	0.0567	0.0672	0.4165

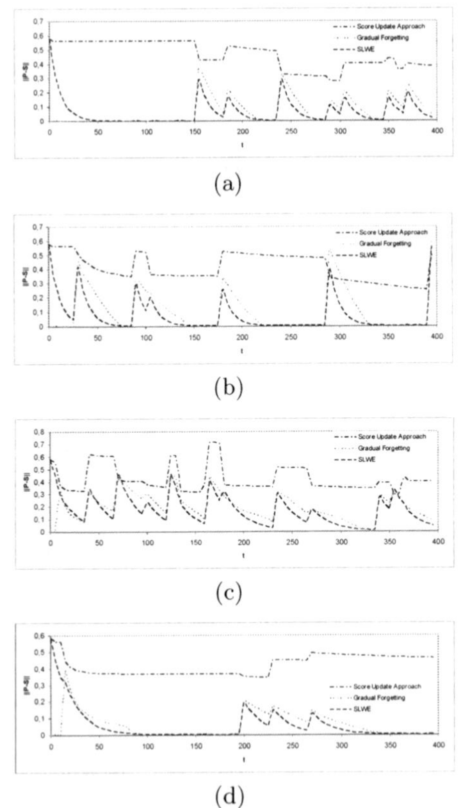

(a) (b) (c) (d)

Fig. 1. Plot of the Euclidean norm $||P - S||$ (the Euclidean distance between P and S) for disjunctive data items, for the SLWE, the GF and the SU, where (a) $\lambda = 0.908$ and w = 35, (b) $\lambda = 0.903$ and w = 44, (c) $\lambda = 0.952$ and w = 63 and (d) $\lambda = 0.948$ and w = 76

5 Conclusions

In this paper we have studied the complex problem of having a social network adapt with the preferences of its users. The premise for this study is that the diversity of a social network cannot be accurately modeled by a *static* set of preferences. Thus, the problem of "estimating" the preferences of its users is becoming increasingly essential for personalized applications which range from service recommender systems to the targeted advertising of services. This being the case, one observes that a traditional estimation strategy, (for estimating the user's interests) which works when the underlying target distribution is stationary, is unsuitable for dynamic non-stationary environments. We have therefore argued that resorting to strong estimators that converge with probability 1 is inefficient since they rely on the assumption that the distribution of the user's preferences is stationary. Consequently, we have proposed the use of a family of stochastic-learning based *weak* estimators for learning and tracking the user's time varying interests. To solve the problem, we have approached the problem by modeling the user's interests using the concept of data items. Thereafter, we have devised two cohesive models for updating the score of the data items in the user's profile depending on whether the data items associated with a given attribute are disjunctive or conjunctive. Simulations results based on synthetic data demonstrates the superiority of our proposed weak estimator-based update methods when compared to the state-of-the-art methods involving "Gradual Forgetting", the Ambient Media Score Update method (SU), and the Maximum Likelihood Estimation (MLE) scheme augmented with a sliding window. The problem of utilizing of the learned profiles in order to perform efficient matchmaking between available services and the user's profile is a potential avenue for future research, for which we do, indeed, have some very promising initial results.

References

1. Basseville, M., Nikiforov, I.V.: Detection of abrupt changes: theory and application. Prentice-Hall, Inc. (1993)
2. Godoy, D., Amandi, A.: User profiling in personal information agents: a survey. Knowl. Eng. Rev. 20, 329–361 (2005)
3. Godoy, D., Schiaffino, S., Amandi, A.: Interface agents personalizing web-based tasks. Cognitive Systems Research, Special Issue on Intelligent Agents and Data Mining for Cognitive Systems 5(3), 207–222 (2004)
4. Hossain, M.A., Atrey, P.K., El Saddik, A.: Gain-based selection of ambient media services in pervasive environments. Mob. Netw. Appl. 13(6), 599–613 (2008)
5. Hossain, M.A., Parra, J., Atrey, P.K., El Saddik, A.: A framework for human-centered provisioning of ambient media services. Multimedia Tools and Applications 44, 407–431 (2009)
6. Koychev, I.: Gradual forgetting for adaptation to concept drift. In: Proceedings of ECAI 2000 Workshop Current Issues in Spatio-Temporal Reasoning, pp. 101–106 (2000)

7. Koychev, I., Lothian, R.: Tracking drifting concepts by time window optimisation. In: Bramer, M., Coenen, F., Allen, T. (eds.) Research and Development in Intelligent Systems XXII, pp. 46–59. Springer, London (2006)
8. Koychev, I., Schwab, I.: Adaptation to drifting user's interests. In: Proceedings of ECML 2000 Workshop: Machine Learning in New Information Age, pp. 39–46 (2000)
9. Maloof, M.A., Michalski, R.S.: Selecting examples for partial memory learning. Machine Learning 41, 27–52 (2000)
10. Middleton, S.E., Shadbolt, N.R., De Roure, D.C.: Ontological user profiling in recommender systems. ACM Trans. Inf. Syst. 22(1), 54–88 (2004)
11. Mitchell, T.M., Caruana, R., Freitag, D., McDermott, J., Zabowski, D.: Experience with a learning personal assistant. Commun. ACM 37(7), 80–91 (1994)
12. Montaner, M., Lpez, B., de la Rosa, J.L.: A taxonomy of recommender agents on the internet. Artificial Intelligence Review 19, 285–330 (2003)
13. Narendra, K.S., Thathachar, M.A.L.: Learning Automata: An Introduction. Prentice Hall (1989)
14. Naudet, Y., Aghasaryanb, A., Mignon, S., Toms, Y., Senot, C.: Ontology-Based Profiling and Recommendations for Mobile TV. In: Wallace, M., Anagnostopoulos, I.E., Mylonas, P., Bielikova, M. (eds.) Semantics in Adaptive and Personalized Services. SCI, vol. 279, pp. 23–48. Springer, Heidelberg (2010)
15. Oommen, B.J., Misra, S.: Fault-tolerant routing in adversarial mobile ad hoc networks: an efficient route estimation scheme for non-stationary environments. Telecommunication Systems 44, 159–169 (2010), 10.1007/s11235-009-9215-4
16. Oommen, B.J., Rueda, L.: Stochastic learning-based weak estimation of multinomial random variables and its applications to pattern recognition in non-stationary environments. Pattern Recogn. 39(3), 328–341 (2006)
17. Rueda, L., Oommen, B.J.: Stochastic automata-based estimators for adaptively compressing files with nonstationary distributions. IEEE Transactions on Systems, Man, and Cybernetics, Part B: Cybernetics 36(5), 1196–1200 (2006)
18. Shiryayev, A.N.: Optimal Stopping Rules. Springer, Heidelberg (1978)
19. Stensby, A., Oommen, B.J., Granmo, O.-C.: Language Detection and Tracking in Multilingual Documents Using Weak Estimators. In: Hancock, E.R., Wilson, R.C., Windeatt, T., Ulusoy, I., Escolano, F. (eds.) SSPR&SPR 2010. LNCS, vol. 6218, pp. 600–609. Springer, Heidelberg (2010)
20. Thathachar, M.A.L., Sastry, P.S.: Networks of Learning Automata: Techniques for Online Stochastic Optimization. Kluwer Academic Publishers (2004)
21. Widmer, G.: Tracking context changes through meta-learning. Mach. Learn. 27(3), 259–286 (1997)
22. Yazidi, A., Granmo, O.C., Oommen, B.J.: An adaptive approach to learning the preferences of users in a social network using weak estimators. Unabridged version of this paper (submitted for publication)
23. Yu, Z., Zhou, X., Zhang, D., Chin, C.Y., Wang, X., Men, J.: Supporting context-aware media recommendations for smart phones. IEEE Pervasive Computing 5, 68–75 (2006)

Dynamic Auction for Efficient Competitive Equilibrium under Price Rigidities*

Junwu Zhu[1,2] and Dongmo Zhang[1]

[1] Intelligent Systems Laboratory, School of Computing and Mathematics, University of Western Sydney, Australia
[2] School of Information Engineering, Yangzhou University, China

Abstract. In an auction market where the price of each selling item is restricted to an admissible interval (price rigidities), a Walrasian equilibrium usually fails to exist. Dreze (1975) introduced a variant concept of Walrasian equilibrium based on rationing systems, named constrained Walrasian equilibrium, for modelling an economy with price rigidities. Talman and Yang (2008) further refined the concept and proposed a dynamic auction procedure that converges to a constrained Walrasian equilibrium. However, a constrained Walrasian equilibrium does not guarantee market efficiency. In other words, a constrained Walrasian equilibrium allocation does not necessarily lead to the best market value. In this paper, we introduce a concept of competitive equilibrium by weakening the concept of constrained Walrasian equilibrium and devise an dynamic auction procedure that generates an efficient competitive equilibrium.

1 Introduction

Auctions have been widely used for discovering market-clearing prices and efficient allocations [1]. However, in many market situations, the price of an item cannot be fully determined by its market. There are certain exogenous reasons that could cause the price of a selling item not completely flexible. For instance, price ceilings and floors in stock markets to prevent breakdown; price controls to reduce inflation or deflation; and imposing upper prices to protect low-income buyers [2,3,4]. Such a phenomenon is normally referred to as *price rigidities* in economics.

In a market with price rigidities, certain rationing mechanism is normally needed to facilitate the distribution of commodities among agents in additional to the price leverage. Dreze (1975) introduced a variant concept of Walrasian equilibrium based on rationing, named constrained Walrasian equilibrium, for economies with price rigidities [2]. Talman and Yang (2008) further refined the concept and proposed a dynamic auction procedure that produces a constrained Walrasian equilibrium outcome in a finite number of steps [4]. However, as we will show in this paper, a constrained equilibrium under Talman and Yang's definition does not guarantee market efficiency.

At first glance, it seems impossible for a dynamic auction procedure to achieve market efficiency in a market with price rigidities because the market value of an item over

* This work was partly supported by the Australian Research Council through project LP0883646.

D. Wang and M. Reynolds (Eds.): AI 2011, LNAI 7106, pp. 809–818, 2011.

its price upper limit can never be discovered by any auction procedure. However, if we count bidders' contributions to the market value within the constraint price intervals, market efficiency will be achievable.

Similar to Talman and Yang (2008), we study the market situations where the following conditions hold:

- The commodities to be sold are heterogeneous and indivisible, such as cars and houses;
- Each buyer can buy at most one item at each auction;
- The price restriction on each item is represented by an interval, the lower bound and the upper bound, which is given to the auctioneer as a reservation at the beginning of an auction procedure.

The rest of the paper is organised in the following. Section 2 sets up the model of the underlying markets. Section 3 presents our dynamic auction procedure and prove that the procedure can find an efficient competitive equilibrium in a finite number of steps. Section 4 gives an example of how the procedure runs. Finally we conclude the work with brief remarks on the related work.

2 The Market Model

Consider a market situation where a seller wishes to sell a finite set of items to a finite number of buyers. Each item is indivisible and the items are heterogeneous. Each buyer has a private value over each item. Formally, let X be the set of items on offer, N the set of buyers, and v^i the value function of buyer i ($i \in N$). We assume that the seller values each item in X at zero. We also assume that among the items in X, there is a specific item, called the dummy item, which value is zero to each buyer and the seller. For sake of simplicity, we let $N = \{1, 2, \cdots, n\}$ and $X = \{0, 1, \cdots, m\}$, where item 0 represents the dummy item.

We assume that each buyer i has an integer value function, i.e., $v^i : X \to \mathbb{Z}^+$, which assigns each item $j \in X$ an integer $v^i(j)$ (in the unit of money) with $v^i(0) = 0$.

A *price vector* $\mathbf{p}$ is a function $\mathbf{p} : X \to \mathbb{Z}^+$ that assigns a non-negative integer to each item in X. For each $j \in X$, we write p_j, instead of $\mathbf{p}(j)$, to indicate the price of item j under the price vector $\mathbf{p}$.

As we have mentioned in the previous section, we will consider in this paper the problem of price discovery under price rigidities. We assume that the price of each item $j \in X$ is restricted to an interval $[\underline{p}_j, \overline{p}_j]$, where $\underline{p}_j$ and $\overline{p}_j$ are integers and $0 \leq \underline{p}_j \leq \overline{p}_j < +\infty$. Specifically, we assume that $\underline{p}_0 = \overline{p}_0 = 0$, which means that the price of the dummy item can only be zero. $\overline{p}_j = +\infty$ means that there is no upper bound limit of price to item j. We say that a price vector $\mathbf{p}$ is *admissible* if $\underline{p}_j \leq p_j \leq \overline{p}_j$ for all $j \in X$.

Traditionally, the following defines the *demand correspondence* of bidder i at price vector $\mathbf{p}$:

$$D^i(\mathbf{p}) = \{j \in X \mid v^i(j) - p_j \geq v^i(k) - p_k, \forall k \in X\}. \tag{1}$$

The following defines all the items that the bidder i would demand thus we call it the *demand set* of bidder i at $\mathbf{p}$:

$$M^i(\mathbf{p}) = \{j \in X \setminus \{0\} \mid v^i(j) \geq p_j\} \tag{2}$$

In a competitive market, it is possible that one item is demanded by more than one bidders. The following notation represents all the items that are demanded by more than one buyers at price vector **p**, called *over-demanded set*:

$$O(\mathbf{p}) = \{j \in X : \exists i, i' \in N(i \neq i' \,\&\, j \in M^i(\mathbf{p}) \cap M^{i'}(\mathbf{p}))\} \tag{3}$$

Following the assumption of Talman and Yang [4], we assume in this paper that each buyer can only receive one item and each item, except the dummy item, can only be allocated to one buyer. Based on the assumption, an *allocation* of X can be represented as a function $\pi : N \to X$ that satisfies the following condition:

- If $\pi(i) = \pi(i')$ and $i \neq i'$, then $\pi(i) = \pi(i') = 0$.

Traditionally, an allocation π^* being efficient means that it gives the best market value, i.e., for any allocation π of X in N,

$$\sum_{i \in N} v^i(\pi^*(i)) \geq \sum_{i \in N} v^i(\pi(i))$$

However, such a traditional definition of efficiency is not applicable to the markets with price rigidities because the market value of an item over its price upper limit can never be discovered by any auction procedure. For this reason, we redefine the concept of market efficiency as follows.

Definition 1. *Let π be an allocation π of X, the market value of π at price vector* **p** *under price rigidities is $\sum_{i \in N}(min(v^i(\pi(i)), \bar{p}_{\pi(i)}) - \underline{p}_{\pi(i)})$, where v^i is the value function of bidder i.*

Note that the market rule means the totally value the market generates. No value can be generated under the lower bound.

Definition 2. *An allocation π^* of X is* efficient *if, for any allocation π of X,*

$$\sum_{i \in N} (\min\{v^i(\pi^*(i)), \bar{p}_{\pi^*(i)}\} - \underline{p}_{\pi^*(i)}) \geq \sum_{i \in N} (\min\{v^i(\pi(i)), \bar{p}_{\pi(i)}\} - \underline{p}_{\pi(i)}) \tag{4}$$

In economics, a rationing system describes a set of market rules. Formally, a *rationing system* $R = (R^i_j)_{i \in N, j \in X}$ is a $|N| \times |X|$ matrix, which element has a value either 1 or 0. For each $i \in N$ and $j \in X$, $R^i_j = 1$ means that buyer i has right to buy item j while $R^i_j = 0$ indicates that buyer i is prohibited to buy item j. With a rationing system R, the demand correspondence can be re-defined as follows:

$$D^i(\mathbf{p}, R) = \{j \in X \mid R^i_j = 1 \text{ and } \min\{v^i(j), \bar{p}_j\} - p_j \geq \max\{\min\{v^i(h), \bar{p}_h\} - p_h \mid R^i_h = 1\}\} \tag{5}$$

Based on a rationing system, Talman and Yang [4] gave the following variation of Walrasian equilibrium:

Definition 3. [4] *A tuple $(\mathbf{p}^*, \pi^*, R^*)$ is a constrained Walrasian equilibrium if*

1. *π^* is an allocation, $\mathbf{p}^*$ is an admissible price vector, and R^* is a rationing system;*
2. *$\pi^*(i) \in D_i(\mathbf{p}^*, R^*)$ for all $i \in N$;*

3. $p_j^* = \underline{p}_j$, *if* $\pi^*(i) \neq j$ *for all* $i \in N$;
4. $p_j^* = \bar{p}_j$ *and* $\pi(h) = j$ *for some* $h \in N$ *if* $R_j^{*i} = 0$ *for some* $i \in N$;
5. $j \in D^i(\mathbf{p}^*, R_{-j}^{*i})$ *if* $R_j^{*i} = 0$.

where R_{-j}^{*i} *denote that* R_j^i *is being ignored and bidder* i *is allowed to demand item* j *(see [4]).*

Talman and Yang devised a dynamic auction procedure that can generate a constrained Walrasian equilibrium [4].

Theorem 1. [4] *There exists at least one constrained Walrasian equilibrium in the model under price rigidities.*

Unfortunately, the allocation of a constrained Walrasian equilibrium is not necessarily efficient.

Example 1. Suppose that $N = \{1,2,3\}$ and $X = \{0,1,2,3\}$. The lower and upper bound of prices are $\underline{\mathbf{p}} = \{0,0,0,0,\}$ and $\overline{\mathbf{p}} = \{0,10,10,30\}$. Bidders' values are given as follows:

	item 0	item 1	item 2	item 3
Bider 1	0	6	7	38
Bider 2	0	8	6	40
Bider 3	0	0	0	28

There are two constrained Walrasian equilibria. The price vector of both equilibria is $\mathbf{p}^* = (0,0,30)$. The allocation and rationing system of the first equilibrium is $\pi^* = (2,3,1)$ and $R^* = ((1,1,0),(1,1,1),(1,1,0))$. The equilibrium gives a market value $7+30+0 = 37$. The other equilibrium is $\mathbf{p}'^* = (0,0,30)$, $\pi'^* = (3,1,2)$ and $R'^* = ((1,1,1),(1,1,0),(1,1,0))$. The market value of this equilibrium is $30+8+0 = 38$. However, if the allocation is $(2,1,3)$, which is not a constrained Walrasian equilibrium allocation, the market value can be $7+8+28 = 44$.

The above example shows that an efficient allocation may not be a constrained Walrasian equilibrium allocation. Therefore, if our target is to get an efficient allocation, we have to weaken the concept of constrained Walrasian equilibrium. The following definition of competitive equilibrium is actual a weak version Talman and Yang's concept of constrained Walrasian equilibrium:

Definition 4. A competitive equilibrium with rationing *is a triple* $(\mathbf{p}^*, \pi^*, R^*)$, *where* $\mathbf{p}^*$ *is an admissible price vector,* π^* *is an allocation and* R^* *is a rationing scheme at* $\mathbf{p}^*$ *such that*

1. $\pi^*(i) \in D^i(\mathbf{p}^*, R^*)$ *for all* $i \in N$.
2. $p_j^* = \underline{p}_j$, *if* $\pi^*(i) \neq j$ *for all* $i \in N$;
3. $\min(v^i(j), \bar{p}_j) \geq \min(v^{i'}(j), \bar{p}_j)$ *and* $\pi^*(i') = j$ *if* $R_j^{*i} = 0$ *for some* $i \in N$.

A competitive equilibrium with rationing $(\mathbf{p}^*, R^*, \pi^*)$ *is efficient if the allocation* π^* *is efficient.*

The first two conditions are exactly the same as Talman and Yang's definition. The third reads a bid strange: only the bidders who has the highest value on the item get rationed. In fact, rationing is a privilege of the auctioneer to govern a market and only those bidders who have high valuation on an item need to be rationed (because they are more likely to get the goods). For instance, a government could ban high-income people applying for public housing.

Lemma 1. *Each constrained Walrasian equilibrium is a competitive equilibrium with rationing.*

Proof. Assume that $(\mathbf{p}^*, R^*, \pi^*)$ is a constrained Walrasian equilibrium. Then the first two conditions for a competitive equilibrium with a rationing system is satisfied. Now we prove that the third condition holds. Let $R_j^{*i} = 0$ for some $i \in N$. According to Condition 4 for a constrained Walrasian equilibrium, we have $p_j^* = \bar{p}_j$ and $\pi(h) = j$ for some $h \neq i$. It turns out that $min(v^i(j), \bar{p}_j) = \bar{p}_j$ and $min(v^h(j), \bar{p}_j) = \bar{p}_j$. Consequently, $min(v^i(j), \bar{p}_j) = min(v^h(j), \bar{p}_j)$, as desired. □

As we have shown in above example, an efficient competitive equilibrium with rationing is not necessarily a constrained Walrasian equilibrium.

3 Dynamic Auction Procedure under Price Rigidities

In this section, we will introduce a dynamic auction procedure that can generate an efficient competitive equilibrium with rationing.

Given a set $N = \{1, 2, \cdots, n\}$ of bidders and a set $X = \{0, 1, 2, \cdots, m\}$ of items on offer, where 0 is a dummy item which can be allocated to more than one bidders. $\bar{p}$ and $\underline{p}$ are the upper price bound and the lower price bound respectively. The dynamic auction procedure consists of the following steps:

Step 1. Set the initial price vector $\mathbf{p} := \underline{\mathbf{p}}$ and the initial rationing scheme $R_j^i = 1$ for all $i \in N, j \in X$. Let $S = (S_{i,j})_{i \in N, j \in X}$ be a $n \times m$ matrix initiated as follows:

$$S = \begin{pmatrix} 0 & -\infty & -\infty & -\infty & -\infty \\ 0 & -\infty & -\infty & -\infty & -\infty \\ 0 & -\infty & -\infty & -\infty & -\infty \\ 0 & -\infty & -\infty & -\infty & -\infty \\ 0 & -\infty & -\infty & -\infty & -\infty \end{pmatrix} \tag{6}$$

In this matrix, the rows represent the bidders and the column for items. The elements are initiated by $S_{i,j} := -\infty$ for all $i \in N, j \in X$ except zero for the dummy item.

Step 2. Auctioneer announces the price vector $\mathbf{p}$ and invites all the buyers to submit their demand set $M^i(\mathbf{p})$. For all $j \in M^i(\mathbf{p})$, $S_{i,j} := p_j$.

Step 3. Calculate over-demanded set $O(\mathbf{p})$. If $O(\mathbf{p}) \neq \emptyset$ and $p_j < \bar{p}_j$ for all $j \in O(\mathbf{p})$, then go to Step 4. Otherwise go to Step 5.

Step 4. For all $j \in O(\mathbf{p})$ such that $p_j < \bar{p}_j$, let $p_j := p_j + 1$. Go back to Step 2.

Step 5. Construct a weighted bipartite graph $G = (N \cup X, E, W)$, where
- $E \subseteq N \times X$ such that $e_{i,j} \in E$ iff $S_{i,j} \neq -\infty$ for all $i \in N, j \in X$
- $W : E \to \mathbb{Z}$ such that $W(e_{i,j}) = S_{i,j} - \underline{p}_j$ for each $e_{i,j} \in E$.

Step 6. Let $\Omega \subseteq E$ be a maximum weighted bipartite matching in G[1].

Step 7. For each $i \in N$, if $e_{i,j} \in \Omega$, let $\pi^*(i) = j$ and $p_j^* = S_{i,j}$; Meanwhile, for each $k \in N$ such that $k \neq i$ and $S_{k,\pi^*(i)} \geq S_{i,\pi^*(i)}$, let $R_{\pi^*(i)}^{*k} := 0$.

Since each item has a finite price upper bound, the above dynamic auction procedure terminates in finite number of steps. Let $(\mathbf{p}^*, R^*, \pi^*)$ be the outcome of the procedure when it terminates.

Lemma 2. $\pi^*(i) \in D^i(\mathbf{p}^*, R^*)$.

Proof. Let $\pi^*(i) = j$. Firstly, $S_{i,j} \neq -\infty$ because no edge links between i and j in the associated bipartite graph. For all $k \in N$, if $k \neq i$ and $S_{k,j} \geq S_{i,j}$, then $R_j^{*k} = 0$. Assume that there is a $h \in X$ such that $\min\{v^i(h), \bar{p}_h\} - p_h^* > \min\{v^i(j), \bar{p}_j\} - p_j^*$ and $R_h^i = 1$. If $p_h^* < \bar{p}_h$, it means that only i bids for h at price p_h. We change the matching from $\pi^*(i) = j$ to $\pi^*(i) = h$ and keep the other allocation unchanged. We can then increase the weight of the matching, which contradicts the fact that π^* is a maximum weighted matching. If $p_h^* = \bar{p}_h$, then we have $v^i(j) < p_j^*$. It implies that $S_{i,j} < p_j^*$. By the construction of the rationing system, $R_j^{*k} = 0$ for all $k \neq i$ and $S_{k,j} >= S_{i,j}$. In other words, $v^i(j) - p_j^* \geq \max\{v^i(h) - p_h^* \mid R_h^i = 1\}$. Therefore we have $\pi^*(i) \in D^i(p^*, R^*)$. □

Theorem 2. *(p^*, R^*, π^*) is an efficient competitive equilibrium.*

Proof. Lemma 2 has shown that the dynamic auction mentioned above can yield a competitive equilibrium (p^*, R^*, π^*). We now prove that π^* is an efficient allocation.

π^* is the maximum weighted matching of the weighted graph $G = (N \cup X, E, W)$ as defined in the auction procedure. Assume that there is $\pi^{'}$ is efficient allocation, which obviously satisfies the following inequality:

$$\sum_{i \in N}(min\{v^i(\pi^{'}(i)), \bar{p}_{\pi'(i)}\} - \underline{p}_{\pi'(i)}) > \sum_{i \in N}(min\{v^i(\pi^*(i)), \bar{p}_{\pi^*(i)}\} - \underline{p}_{\pi(i)}) \quad (7)$$

Note that the allocation π' also determines a matching in the weighted bipartite graph unless there is an i such that $S_{i,\pi'(i)} = -\infty$. In this case, $v_i(\pi'(i)) < \underline{p}_{\pi(i)}$. Now we define a new allocation π'' such that $\pi''(i) = 0$ and $\pi''(j) = \pi'(j)$ for all $j \neq i$. It turns out that π'' can implement more market value than π', which contradicts to the assumption. On the other hand, π' cannot be a maximum weighted matching of G because otherwise π will not be a maximum weighted matching of G. Therefore $(\mathbf{p}^*, R^*, \pi^*)$ is an efficient competitive equilibrium. □

[1] We omit the algorithm for finding a maximum weighted matching in a bipartite graph. In fact, any maximum weighted bipartite matching algorithm is applicable. The reader is referred to the algorithm in [5].

4 Calculation and Comparison

To compare our auction procedure with Talman and Yang's, we use the same example that has been used in [4] to demonstrate how to calculate an efficient competitive equilibrium with rationing by using the dynamic auction procedure introduced in the previous section.

Example 2. Suppose that there are five bidders $N = \{a, b, c, d, e\}$ and five items $X = \{0, 1, 2, 3, 4\}$ in a market, where 0 is a dummy item and the others are real items. The lower and upper price vectors are $\underline{\mathbf{p}} = (0, 5, 4, 1, 5)$ and $\bar{\mathbf{p}} = (0, 6, 6, 4, 7)$, respectively. Bidders' values are given by the following table.

Item	dummy	1	2	3	4
Bidder a	0	4	3	5	7
Bidder b	0	7	6	8	3
Bidder c	0	5	5	7	7
Bidder d	0	9	4	3	2
Bidder e	0	6	2	4	10

Initially, we set $\mathbf{p} = \underline{\mathbf{p}}$ and S as follows.

$$S = \begin{pmatrix} 0 & -\infty & -\infty & -\infty & -\infty \\ 0 & -\infty & -\infty & -\infty & -\infty \\ 0 & -\infty & -\infty & -\infty & -\infty \\ 0 & -\infty & -\infty & -\infty & -\infty \\ 0 & -\infty & -\infty & -\infty & -\infty \end{pmatrix} \tag{8}$$

After $\mathbf{p}$ is announced to all bidders by the auctioneer, they submit their $M^i(\mathbf{p})$ respectively:

$M^a(\mathbf{p})=\{3, 4\}$
$M^b(\mathbf{p})=\{1, 2, 3\}$
$M^c(\mathbf{p})=\{2, 3, 4\}$
$M^d(\mathbf{p})=\{1, 3\}$
$M^e(\mathbf{p})=\{1, 3, 4\}$.

For each $j \in M^i(\mathbf{p})$, let $S_{i,j} := p_j$, then the matrix S becomes:

$$S = \begin{pmatrix} 0 & -\infty & -\infty & 1 & 5 \\ 0 & 5 & 4 & 1 & -\infty \\ 0 & 5 & 4 & 1 & 5 \\ 0 & 5 & 4 & 1 & -\infty \\ 0 & 5 & -\infty & 1 & 5 \end{pmatrix} \tag{9}$$

Now $O(\mathbf{p}) = \{1, 2, 3, 4\}$, i.e., all the items except the dummy item are over-demanded. $p_1 < \bar{p}_1$, $p_2 < \bar{p}_2$, $p_3 < \bar{p}_3$ and $p_4 < \bar{p}_4$. We then let $p_1 := p_1 + 1$, $p_2 := p_2 + 1$,

$p_3 := p_3 + 1$, $p_4 := p_4 + 1$, and the price vector is adjusted to $\mathbf{p} = (0, 6, 5, 2, 6)$. The auctioneer announces the new price vector and asks all bidders resubmit their demand sets, which are$M^a(\mathbf{p}) = \{3, 4\}$, $M^b(\mathbf{p}) = \{1, 2, 3\}$, $M^c(\mathbf{p}) = \{3, 4\}$, $M^d(\mathbf{p}) = \{1, 3\}$, and $M^e(\mathbf{p}) = \{3, 4\}$.

Use the demand sets and the new price to update matrix S as follows:

$$S = \begin{pmatrix} 0 & -\infty & -\infty & 2 & 6 \\ 0 & 6 & 5 & 2 & -\infty \\ 0 & 5 & 5 & 2 & 6 \\ 0 & 6 & 4 & 2 & -\infty \\ 0 & 6 & -\infty & 2 & 6 \end{pmatrix} \tag{10}$$

In this case, $O(\mathbf{p}) = \{1, 3, 4\}$, $p_1 = \bar{p}_1$, $p_2 < \bar{p}_2$, $p_3 < \bar{p}_3$ and $p_4 < \bar{p}_4$. Let $p_2 := p_2 + 1$, $p_3 := p_3 + 1$ and $p_4 := p_4 + 1$. The auctioneer announces the new price vector $\mathbf{p} = (0, 6, 6, 3, 7)$ and requests the bidders to report their new demands. Assume the new demand sets are:

$M^a(\mathbf{p}) = \{3\}$,
$M^b(\mathbf{p}) = \{1, 2, 3\}$,
$M^c(\mathbf{p}) = \{3\}$,
$M^d(\mathbf{p}) = \{1\}$ and
$M^e(\mathbf{p}) = \{4\}$.

Then the matrix S becomes

$$S = \begin{pmatrix} 0 & -\infty & -\infty & 3 & 7 \\ 0 & 6 & 6 & 3 & -\infty \\ 0 & 5 & 5 & 3 & 7 \\ 0 & 6 & 4 & 3 & -\infty \\ 0 & 6 & -\infty & 3 & 7 \end{pmatrix} \tag{11}$$

Increase one unit of the price of item 3 because $O(\mathbf{p}) = \{1, 3\}$ and $p_3 < \bar{p}_3$. After the auctioneer announces $\mathbf{p} = (0, 6, 6, 4, 7)$, the bidders report their demands again: $M^a(\mathbf{p}) = \{3\}$, $M^b(\mathbf{p}) = \{1, 2, 3\}$, $M^c(\mathbf{p}) = \{3\}$, $M^d(\mathbf{p}) = \{1\}$ and $M^e(\mathbf{p}) = \{4\}$. The matrix S can be rebuilt according to the current demands and price vector $\mathbf{p}$.

$$S = \begin{pmatrix} 0 & -\infty & -\infty & 4 & 7 \\ 0 & 6 & 6 & 4 & -\infty \\ 0 & 5 & 5 & 4 & 7 \\ 0 & 6 & 4 & 3 & -\infty \\ 0 & 6 & -\infty & 4 & 7 \end{pmatrix} \tag{12}$$

At this time, we find that only item 1 and item 3 are over-demanded, but their prices have both reached their upper bound. The auction procedure stops because $p_j = \bar{p}_j$ for all $j \in O(\mathbf{p})$.

According to matrix S, we build a weighted graph $G = (N \cup X, E, W)$, where $S_{i,j} \in E$ iff $S_{i,j} \neq -\infty$ and $W(e_{i,j}) = S_{i,j}$ for each $i \in N, j \in X$. It is easy to verify that $\Omega = \{(a, 3), (b, 2), (c, 0), (d, 1), (e, 4)\}$ is a maximum weight matching, which determines an allocation $\pi^* = (3, 2, 0, 1, 4)$. The following picture shows the weighted graph. Bold lines represent the maximum weight matching.

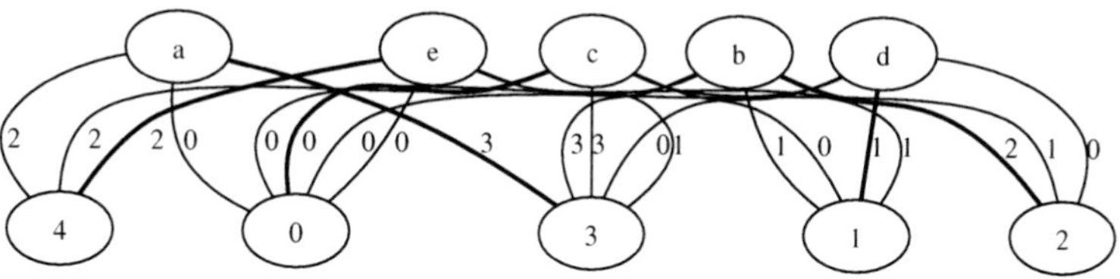

The rationing system is then $R^* = (R^{*a}, R^{*b}, R^{*c}, R^{*d}, R^{*e})$ where
$R^{*a} = (1, 1, 1, 1, 0)$
$R^{*b} = (1, 0, 1, 0, 1)$
$R^{*c} = (1, 1, 1, 0, 0)$
$R^{*d} = (1, 1, 1, 1, 1)$
$R^{*e} = (1, 0, 1, 0, 1)$.
The equilibrium price is $P^* = (0, 6, 6, 4, 7)$. The equilibrium implements a market value at 8.

We would like to remark that the constrained Walrasian equilibrium allocation under Talman and Yang's definition is $(0, 3, 2, 1, 4)$ at price $(0, 5, 4, 2, 5)$. The associated market value is 7, which is lower than the efficient competitive equilibrium.

5 Conclusion and Related Work

In this paper, we have introduced a concept of competitive equilibrium by weakening the concept of constrained Walrasian equilibrium. We have devised an dynamic auction procedure and prove that it can generates an efficient competitive equilibrium for any economy for selling in dividable items with price rigidities.

For the purpose of controlling price macroscopically, preventing speculation or protecting the profits of low-incoming buyers, price rigidity is widely adopted to restrict the price of each item to an interval. The phenomenon of price rigidity, i.e., the persistence of price at which supply and demand are not equal, is frequently observed, and plays an important role in some macro-economic models [2]. After investigating the ability of nominal price rigidity, a dynamic general equilibrium model is constructed by [6] with the introduction of monopolistic competition and nominal price rigidity in a standard real business cycle model, allowing for an endogenous money supply rule. From the aspect of banking industry, the price rigidity is significantly greater in markets characterised by higher levels of concentration [7].

Ausubel proposed a dynamic auction procedure for auctioning multiple heterogeneous commodities, and this auction yields a Walrasian equilibrium price and an efficient allocation without considering price rigidities [8]. The Vickrey and Groves-Clarke auctions can be generalised to attain efficiency when there are common values, if each buyers' information can be represented as a one-dimensional signal. Also, when a buyer's information is multidimensional, no auction is generally efficient [9].

Subsequently, Talman and Yang proposed a dynamic auction for differentiated items under price rigidities and by which yielded a constrained Walrasian equilibrium in finite steps [4]. As can be seen from the procedure of dynamic auction, a group of constrained

Walrasian equilibria taking the form (p^*, π^*, R^*) can be generated. It is obviously that each of their social efficiencies at certain price vector $\mathbf{p}$ can be computed, but not all of them have the same efficiency. So, these constrained Walrasian equilibria are not efficient.

Motivated by the difficulty to achieve the social efficiency under price rigidities, the dynamic auction, suggested by this paper, invite bidders to present their demand set for all items so as to promote the possibility to be allotted an item and drive price ascending under over demands. The efficient competitive equilibrium can be found by the dynamic auction procedure in a finite number of step. Also, this dynamic auction procedure is useful to discover the social revenue of auctioneer. For the further research, we will devote ourself to analysis, present and value the relations among different items, such as substitute relation and complement relation, because these relations effect bidders' strength of demands and the distance of price ascending.

References

1. Sun, N., Yang, Z.: A double-track adjustment process for discrete markets with substitutes and complements. Econometrica 77(3), 933–952 (2009)
2. Dreze, J.H.: Existence of an exchange equilibrium under price rigidities. International Economic Review 16(2), 301–320 (1975)
3. Herings, J.J., Talman, D., Yang, Z.: The computation of a continuum of constrained equilibria. Mathematics of Operations Research 21(3) (1996)
4. Talman, D., Yang, Z.: A dynamic auction for differentiated items under price rigidities. Economics Letters 99(2), 278–281 (2008)
5. Schrijver, A.: Combinatorial Optimization: Polyhedra and Efficiency, vol. A,B,C. Springer, Heidelberg (2004)
6. Yun, T.: Nominal price rigidity, money supply endogeneity, and business cycles. Journal of Monetary Economics 37(2-3), 345–370 (1996)
7. Hannan, T.H., Berger, A.N.: The rigidity of prices: Evidence from the banking industry. American Economic Review 81(4), 938–945 (1991)
8. Ausubel, L.M.: An efficient dynamic auction for heterogeneous commodities. The American Economic Review 96(3), 602–629 (2006)
9. Dasgupta, P., Maskin, E.: Efficient auctions. The Quarterly Journal of Economics 115(2), 341–388 (2000)

Author Index